AF893546

Birkhäuser

Progress in Mathematics

Volume 362

Series Editors

Antoine Chambert-Loir, Université Paris-Diderot, Paris, France

Jiang-Hua Lu, The University of Hong Kong, Hong Kong SAR, Hong Kong

Michael Ruzhansky, Ghent University, Belgium and
Queen Mary University of London, London, UK

Yuri Tschinkel, Courant Institute of Mathematical Sciences, New York, USA

Progress in Mathematics is a series of books intended for professional mathematicians and scientists, encompassing all areas of pure mathematics. This distinguished series, which began in 1979, includes research level monographs, polished notes arising from seminars or lecture series, graduate level textbooks, and proceedings of focused and refereed conferences. It is designed as a vehicle for reporting ongoing research as well as expositions of particular subject areas.

Marius Mitrea · Pedro Takemura

Singular Integrals, Herz-Type Function Spaces, and Boundary Problems

Birkhäuser

Marius Mitrea
Department of Mathematics
Baylor University
Waco, TX, USA

Pedro Takemura
Department of Mathematics
Baylor University
Waco, TX, USA

ISSN 0743-1643 ISSN 2296-505X (electronic)
Progress in Mathematics
ISBN 978-3-032-12515-6 ISBN 978-3-032-12516-3 (eBook)
https://doi.org/10.1007/978-3-032-12516-3

Mathematics Subject Classification: 42B20, 42B25, 35J25, 35J46, 35J47

© The Editor(s) (if applicable) and The Author(s), under exclusive license to Springer Nature Switzerland AG 2026

This work is subject to copyright. All rights are solely and exclusively licensed by the Publisher, whether the whole or part of the material is concerned, specifically the rights of translation, reprinting, reuse of illustrations, recitation, broadcasting, reproduction on microfilms or in any other physical way, and transmission or information storage and retrieval, electronic adaptation, computer software, or by similar or dissimilar methodology now known or hereafter developed.
The use of general descriptive names, registered names, trademarks, service marks, etc. in this publication does not imply, even in the absence of a specific statement, that such names are exempt from the relevant protective laws and regulations and therefore free for general use.
The publisher, the authors and the editors are safe to assume that the advice and information in this book are believed to be true and accurate at the date of publication. Neither the publisher nor the authors or the editors give a warranty, expressed or implied, with respect to the material contained herein or for any errors or omissions that may have been made. The publisher remains neutral with regard to jurisdictional claims in published maps and institutional affiliations.

This book is published under the imprint Birkhäuser, www.birkhauser-science.com by the registered company Springer Nature Switzerland AG
The registered company address is: Gewerbestrasse 11, 6330 Cham, Switzerland

If disposing of this product, please recycle the paper.

Preface

This monograph stands at the crossroads of several branches of mathematics, including Harmonic Analysis, Functional Analysis, Partial Differential Equations, and Geometric Measure Theory. One of its principal aims is to establish a harmonic-analysis-oriented functional analytic framework for a diverse range of Herz-type spaces within the general geometric setting of Ahlfors regular sets. This portion of the theory encompasses embedding and duality results, boundedness of the Hardy–Littlewood maximal operator, interpolation and extrapolation techniques, atomic and molecular decompositions, and connections with Generalized Banach Function Spaces (aka Köthe theory). Subsequently, a comprehensive Calderón–Zygmund theory on uniformly rectifiable sets, tailored for singular integral operators (SIO's) that act on, and between, various Herz-type spaces is systematically developed. This begins with convolution-type singular integrals and then proceeds to more specialized operators, such as generalized double layers. The latter class incorporates the geometry-sensitive chord-dot-normal type SIO's, which play a pivotal role in the present analysis.

A central theme of this work is refitting the boundary layer potential method as to turn it into a powerful and effective tool for solving a broad class of boundary value problems with data prescribed in Herz-type spaces. This highlights the effectiveness of single and double layer potentials in addressing classical problems such as the Dirichlet, Neumann, Regularity, and Transmission boundary value problems. To unify these developments, the monograph proposes a general functional analytic blueprint capable of generating a wide array of Herz-flavored function spaces. Such considerations are compatible with the Calderón–Zygmund machinery, facilitating the formulation and resolution of the aforementioned boundary value problems.

The most sensitive portion of our analysis is carried out within the context of δ-Ahlfors regular (δ-AR) domains $\Omega \subseteq \mathbb{R}^n$, where the parameter $\delta > 0$ regulates the size of the BMO seminorm of the (geometric measure theoretic) outward unit normal vector ν to the set Ω. The staple assumption is that $\delta \in (0, 1)$ is sufficiently small relative to other (large geometry) characteristics, a feature making this a subclass of the category of two-sided non-tangentially accessible (NTA) domains with

Ahlfors regular boundaries. Such a choice aligns the present work with the research program undertaken by D. Mitrea, I. Mitrea, and M. Mitrea in the series "*Geometric Harmonic Analysis*", volumes I-V, 2022-2023, which seeks to develop real-variable and geometric-measure-theoretic tools for solving boundary value problems for elliptic systems with complex coefficients via layer potentials in broad analytic and geometric environments. In the larger picture, this study builds on the foundational contributions of E. Fabes, M. Jodeit, and N. Rivière, whose seminal 1978 paper expanded the use of layer potentials to boundary value problems in C^1 domains. That influential work has inspired an expansive and diverse body of research, within which the current work is also situated.

Acknowledgments: The authors acknowledge partial support from the Simons Foundation through grant # 637481.

Waco, TX, *Marius Mitrea*
Waco, TX, *Pedro Takemura*
July 2025

Contents

Chapter 1
Introduction

1.1 Motivation and Goals

The present monograph lies at the crossroads between several branches of mathematics, including Harmonic Analysis, Functional Analysis, Partial Differential Equations, and Geometric Measure Theory. The primary objectives are:

⋄ develop a functional analytic framework (Harmonic Analysis oriented) for a diversity of Herz-type spaces in the general geometric setting of Ahlfors regular sets. This includes embeddings and duality results, boundedness of the Hardy-Littlewood maximal operator, interpolation and extrapolation results, atomic and molecular decompositions, connections with Banach Function Spaces (a.k.a. Köthe theory), among other features.

⋄ develop a comprehensive Calderón-Zygmund theory on uniformly rectifiable sets for singular integral operators (SIO's) acting on, and from, a multitude of Herz-type spaces. This starts by treating convolution-type singular integrals, then the more specialized class of generalized double layers, hence paving the road for considering the category of singular integral operators of chord-dot-normal type, which are the SIO's most sensitive to geometry (and are thus most relevant for our work).

⋄ make the method of boundary layer potentials an effective tool for tackling a large family of boundary value problems with data prescribed in Herz-type spaces. This substantiates the efficacy of the single and double layers in the treatment of the Dirichlet, Neumann, Regularity, and Transmission boundary value problems.

⋄ unify the mosaic of Herz-type spaces by identifying the most inclusive functional analytic blueprint capable of generating an abundance of Herz-flavored function spaces, which work in conjunction with the Calderón-Zygmund theory mentioned above, and can be used in the formulation of the aforementioned boundary value problems.

© The Author(s), under exclusive license to Springer Nature Switzerland AG 2026
M. Mitrea and P. Takemura, *Singular Integrals, Herz-Type Function Spaces, and Boundary Problems*, Progress in Mathematics 362,
https://doi.org/10.1007/978-3-032-12516-3_1

This aligns our work with the comprehensive program undertaken by D. Mitrea, I. Mitrea, and M. Mitrea recently in [GHA-I]-[GHA-V]. One of the main purposes of the previously mentioned references was

to develop tools in Real and Harmonic Analysis, of geometric measure theoretic flavor, capable of treating a broad spectrum of boundary value problems, via layer potentials method, formulated in a rather general geometric and analytic settings. (1.1)

Our work also falls within the scope of the program pioneered by E. Fabes, M. Jodeit, and N. Rivière in 1978 [FaJoRi78] regarding the use of layer potentials to solve boundary problems in $\mathscr{C}^1$ domains. This influential article has motivated a large and diverse body of work pertaining to boundary layer potentials in $\mathscr{C}^1$ domains and Lipschitz domains (a topic we elaborate more later on).

To describe the method of boundary layer potentials we need some notation. Fix $n \in \mathbb{N}$ with $n \geq 2$, along with some $M \in \mathbb{N}$. Consider a second-order, homogeneous, constant complex coefficient, $M \times M$ system L in $\mathbb{R}^n$, say

$$L := A_{jk}\partial_j\partial_k, \tag{1.2}$$

for some matrices $A_{jk} = \big(a_{jk}^{\alpha\beta}\big)_{1\leq\alpha,\beta\leq n} \in \mathbb{C}^{M\times M}$, with $j, k \in \{1, \dots, n\}$. Here and elsewhere we adopt the summation convention over repeated indices. Frequently, we shall assume that L is weakly elliptic, a condition requiring that its characteristic matrix

$$L(\xi) := -\xi_j\xi_k A_{jk} \text{ is invertible for each } \xi = (\xi_1, \dots, \xi_n) \in \mathbb{R}^n \setminus \{0\}. \tag{1.3}$$

The block-matrix $A := \big(A_{jk}\big)_{1\leq j,k\leq n} \in \big(\mathbb{C}^{M\times M}\big)^{n\times n}$ is going to be referred to as a coefficient tensor representing L, and we denote by $\mathfrak{A}_L$ the "library" of all such coefficient tensors. Note that $\mathfrak{A}_L$ has infinite cardinality, since along with $A = \big(A_{jk}\big)_{1\leq j,k\leq n} \in \mathfrak{A}_L$, this contains any $\widetilde{A} := \big(A_{jk} + B_{jk}\big)_{1\leq j,k\leq n}$, where each $B_{jk} \in \mathbb{C}^{M\times M}$ is such that $B_{kj} = -B_{jk}$. To summarize, $\#\mathfrak{A}_L = \infty$ and $L = A_{jk}\partial_j\partial_k$ for each $A = \big(A_{jk}\big)_{1\leq j,k\leq n} \in \mathfrak{A}_L$.

The relevance of the weakly ellipticity condition in L is quickly apparent. Most prominently, work in [Mi18, Theorem 11.1, p. 393] ensures that a given system L as above possesses a canonical fundamental solution $E_L = (E_{\alpha\beta})_{1\leq\alpha,\beta\leq M}$ which belongs to $\big[\mathscr{C}^\infty(\mathbb{R}^n \setminus \{0\})\big]^M$ and whose first-order derivatives are odd and positive homogeneous of degree $1 - n$. It should be stressed here that such a matrix-valued function E_L is independent of the choice of the coefficient tensor A used to represent L. The availability of this fundamental solution plays a key role in the development of the method of layer potentials. To be specific, consider a uniformly rectifiable domain $\Omega \subseteq \mathbb{R}^n$ (in the sense of [HoMiTa10, Definition 3.7, p. 2631]; cf. Definition 2.4), and set $\sigma := \mathcal{H}^{n-1}\lfloor\partial\Omega$ for the "surface" measure on $\partial\Omega$, where $\mathcal{H}^{n-1}$ is the $(n-1)$-dimensional Hausdorff measure in $\mathbb{R}^n$. Also, we shall consistently use the symbol $\nu = (\nu_1, \dots, \nu_n)$ to denote the geometric measure theoretic outward unit normal to Ω. Recall from [GHA-IV, (1.3.18), p. 27, and (1.3.68), p. 38] the boundary-to-domain and boundary-to-boundary double layer potentials associated with any given coefficient tensor $A \in \mathfrak{A}_L$. These act on each vector-valued function

$$g \in \Big[L^1\Big(\partial\Omega, \frac{\sigma(x)}{1+|x|^{n-1}}\Big)\Big]^M \tag{1.4}$$

as

$$\big(\mathcal{D}_A g\big)(x) := -\int_{\partial\Omega} \nu_k(y)(\partial_j E_L)(x-y)A_{jk}g(y)\,d\sigma(y) \tag{1.5}$$

for every $x \in \Omega$, and respectively as

$$\big(K_A g\big)(x) := -\lim_{\varepsilon\to 0^+} \int\limits_{\substack{y\in\partial\Omega\\ |x-y|>\varepsilon}} \nu_k(y)(\partial_j E_L)(x-y)A_{jk}g(y)\,d\sigma(y) \tag{1.6}$$

at σ-a.e. $x \in \partial\Omega$. The significance of the weighted L^1 space in (1.4) is that this constitutes the largest functional analytic environment where these integral operators are meaningfully defined. Note the disparity between the roles played by the indices $j, k \in \{1, \dots, n\}$ both in (1.5) and (1.6). Indeed, j labels ∂_j, while k labels ν_k. As such, additively perturbing a coefficient tensor $A = (A_{jk})_{1\le j,k\le n} \in \mathfrak{A}_L$ with some block-matrix $B = (B_{jk})_{1\le j,k\le n} \in \big(\mathbb{C}^{M\times M}\big)^{n\times n}$ which is antisymmetric in j and k, produces different double layers. Schematically,

a given weakly elliptic system L	$\Longrightarrow$	infinitely many writings of the system L as in (1.2)	$\Longrightarrow$	infinitely many double layers.	(1.7)

The significance of the operators in (1.5)–(1.6) within the realm of boundary value problems in this general geometric setting is apparent from the work of D. Mitrea, I. Mitrea, and M. Mitrea in [GHA-V]. In the aforementioned reference, a plethora of fundamental properties is established, including the existence of principal-value singular integrals, nontangential maximal function estimates, jump-formulas, and boundedness results. One particular feature we wish to highlight is that the (boundary-to-domain) double layer potential in (1.5) serves as a tool for generating a multitude of smooth null solutions to the system L in Ω. Specifically, for any g as in (1.4), the function $\mathcal{D}_A g$ satisfies

$$\mathcal{D}_A g \in \big[\mathscr{C}^\infty(\Omega)\big]^M \quad\text{and}\quad L(\mathcal{D}_A g) = 0 \ \text{ in } \Omega. \tag{1.8}$$

Moreover, the operators in (1.5)-(1.6) are intimately related via the jump-formula

$$\big(\mathcal{D}_A g\big)\Big|_{\partial\Omega}^{\kappa-\mathrm{n.t.}} = (\tfrac{1}{2}I + K_A)g \quad \text{pointwise } \sigma\text{-a.e. on } \partial\Omega, \tag{1.9}$$

where $(\cdot)\big|_{\partial\Omega}^{\kappa-\mathrm{n.t.}}$ denotes the nontangential boundary trace operator (as described in (2.9)), and $\kappa \in (0, \infty)$ is a fixed aperture parameter. Properties (1.8)-(1.9) are particularly relevant in the treatment of boundary value problems, such as the Dirichlet Problem for L in a uniformly rectifiable domain $\Omega \subseteq \mathbb{R}^n$ with boundary data in Lebesgue spaces $L^p(\partial\Omega, \sigma)$ with $p \in (1, \infty)$, formulated as:

$$\begin{cases} u \in [\mathscr{C}^\infty(\Omega)]^M, \quad Lu = 0 \text{ in } \Omega, \\ \mathcal{N}_\kappa u \in L^p(\partial\Omega, \sigma), \\ u\big|_{\partial\Omega}^{\kappa-\text{n.t.}} = f \in [L^p(\partial\Omega, \sigma)]^M. \end{cases} \tag{1.10}$$

Here, $\mathcal{N}_\kappa$ denotes the nontangential maximal operator (cf. (2.6)) on $\partial\Omega$. Essentially, employing layer potentials to solve (1.10) starts by considering a candidate for a solution defined as $u := \mathcal{D}_A g$ in Ω. Due to the jump relation (1.9), solving (1.10) is therefore reduced to the task of solving the following boundary integral equation for a density g to be yet determined:

$$(\tfrac{1}{2}I + K_A)g = f. \tag{1.11}$$

This brings into focus the issue of whether the operator $\frac{1}{2}I + K_A$ is invertible in the space $[L^p(\partial\Omega, \sigma)]^M$. The approach taken in [GHA-V], as well as in [MMMi³], for proving the existence of solutions for (1.10) hinges on the ability to prove that the operator norm $\|K_A\|_{[L^p(\partial\Omega,\sigma)]^M \to [L^p(\partial\Omega,\sigma)]^M}$ is sufficiently small as a means of ensuring the invertibility of the operator

$$\tfrac{1}{2}I + K_A : \big[L^p(\partial\Omega, \sigma)\big]^M \to \big[L^p(\partial\Omega, \sigma)\big]^M. \tag{1.12}$$

The difficulty encountered in proving this stems from the fact that this can only happen for certain specialized choices of coefficient tensors A. In this regard, a striking result obtained in [GHA-V] provides purely algebraic conditions on the coefficient tensors A that make the operator norm of the corresponding double layer potential K_A sensitive to the geometry of the underlying domain. More concretely, following [GHA-V, Definition 1.2.2, p. 17] and [GHA-V, (1.2.13) p. 13], one calls $A = (A_{jk})_{1 \leq j,k \leq n} \in \mathfrak{A}_L$ a distinguished coefficient tensor for L if

$$\xi_k (\partial_j E_L)(x) A_{jk} = 0 \in \mathbb{C}^{M \times M} \quad \text{for all } x, \xi \in S^{n-1} \text{ with } x \cdot \xi = 0, \tag{1.13}$$

where E_L is the canonical fundamental solution for L, as discussed earlier. With this piece of terminology, the aforementioned result asserts that if A is a distinguished coefficient tensor for L, then there exists some constant $C \in (0, \infty)$ with the property that

$$\begin{aligned} &\|K_A\|_{[L^p(\partial\Omega,\sigma)]^M \to [L^p(\partial\Omega,\sigma)]^M} \\ &\qquad \leq C \cdot \|\nu\|_{[\mathrm{BMO}(\partial\Omega,\sigma)]^n} \cdot \ln\big(e/\|\nu\|_{[\mathrm{BMO}(\partial\Omega,\sigma)]^n}\big), \end{aligned} \tag{1.14}$$

where $\mathrm{BMO}(\partial\Omega, \sigma)$ stands for the John-Nirenberg space of functions of bounded mean oscillations on $\partial\Omega$ (cf. (2.33)). For a stronger result of this flavor see [MMMi³, Theorem 4.7, p. 303]. As a consequence of this, if $\|\nu\|_{[\mathrm{BMO}(\partial\Omega,\sigma)]^n}$ is sufficiently small to begin with, then the right-hand side of (1.14) becomes small, and the invertibility result mentioned in relation to (1.12) may be obtained via a Neumann series. In turn, this permits one to produce a solution for (1.10) using the template

$$u := \mathcal{D}_A\big[(\tfrac{1}{2}I + K_A)^{-1}f\big] \quad \text{in } \Omega. \tag{1.15}$$

The table below offers some comparison of work done [GHA-V], [MMMi³], [FaJoNe76] and [DaKe87] in relation to boundary value problems:

In...	Geometry of domains	Boundary data in...
[FaJoRi78]	Ω is bounded $\mathscr{C}^1$ domain ($\Rightarrow$ dist$(\nu, \text{VMO}) = 0$)	Lebesgue spaces L^p for any $p \in (1, \infty)$
[DaKe87]	Ω is either upper-graph Lipschitz domain ($\Rightarrow \|\nu\|_{\text{BMO}} \ll 1$, if Lip. constant is small) or bounded Lipschitz domain ($\Rightarrow$ dist$(\nu, \text{VMO}) \ll 1$, if Lip. constant is small)	Lebesgue spaces L^p with necessary restrictions on p
[MMMi³]	Ω is AR domain, $\partial\Omega$ is unbounded with $\|\nu\|_{\text{BMO}} \ll 1$	many function spaces
[GHA-V]	Ω is AR domain, with either $\partial\Omega$ unbounded and $\|\nu\|_{\text{BMO}} \ll 1$, or $\partial\Omega$ compact and dist$(\nu, \text{VMO}) \ll 1$	many function spaces, such as GBFS

As illustrated in the previous paragraph, estimates of similar flavor to (1.14) are powerful tools for establishing existence results for boundary problems in other function spaces. Remarkably, the largest class of SIO's for which a geometric-analytic phenomenon as (1.14) occurs has been identified by D. Mitrea, I. Mitrea, and M. Mitrea in [GHA-V]. There, the authors have introduced and systematically studied, the class of singular integral operators of *chord-dot-normal type*, i.e., singular integral operators T on $\partial\Omega$ whose integral kernel is of the format $\langle x-y, \nu(y)\rangle k(x-y)$ with $x, y \in \partial\Omega$, for some sufficiently smooth integral kernel k in $\mathbb{R}^n \setminus \{0\}$ that is even and positive homogeneous of degree $-n$. The significance of such a class of integral operators is highlighted by the fact that they are precisely the ones for which geometrically sensitive estimates, such as (1.14), hold (see [GHA-V, Chapter 5]). It turns out that the class of chord-dot-normal SIO's interfaces tightly with the concept of distinguished coefficient tensor. Denoting by $\mathfrak{A}_L^{\text{dis}}$ the totality of all such distinguished coefficient tensors for a given weakly elliptic system L, the culmination of work in [GHA-V] is that

$$A \in \mathfrak{A}_L^{\text{dis}} \quad \Longleftrightarrow \quad \begin{array}{l}\text{the double layer } K_A \text{ associated with } A \text{ (as in (1.6))}\\ \text{is of chord-dot-normal type.}\end{array} \tag{1.16}$$

This specialized class of coefficient tensors play a decisive role when obtaining estimates of the sort described in (1.14). Conceptually,

$$\begin{matrix}\text{a weakly elliptic} \\ \text{system } L \text{ with } \mathfrak{A}_L^{\text{dis}} \neq \varnothing\end{matrix} \Longrightarrow \begin{matrix}\text{double layers of} \\ \text{chord-dot-normal type}\end{matrix} \Longrightarrow \begin{matrix}\text{finer operator} \\ \text{norm estimates.}\end{matrix} \tag{1.17}$$

We elaborate on the finer operator norm estimates alluded to above in Theorem 8.1, Theorem 15.1, Theorem 15.6, Theorem 15.8, Theorem 18.4, and Theorem 22.1, which are key ingredients for the present work.

Given how much rides on the existence of distinguished coefficient tensors, it is then natural to ask what systems possess them. In this regard, the authors in [MMMi^3] provide a sizable list of weakly elliptic operators that have a distinguished coefficient tensor. In particular, it was shown in [MMMi^3, Example 3.3, pp. 203–206] that all scalar second-order, homogeneous, constant complex coefficient, weakly elliptic operators $L = \operatorname{div}(A\nabla)$ in $\mathbb{R}^n$, with $n \geq 3$ and $A \in \mathbb{C}^{n\times n}$, have precisely one distinguished coefficient tensor, explicitly given by $(A + A^\top)/2$. A prominent example is, of course, the Laplacian $\Delta = \sum_{j=1}^n \partial_j^2$ in $\mathbb{R}^n$, in which case its unique distinguished coefficient tensor is the $n \times n$ identity matrix. Other examples include the "complex" Lamé system of elasticity in $\mathbb{R}^n$, for $n \geq 2$, defined as

$$L_{\mu,\lambda} := \mu\Delta + (\lambda + \mu)\nabla \operatorname{div}, \tag{1.18}$$

where the Lamé moduli λ and μ are allowed to be complex numbers. Work done in [GHA-V, Remark 1.6.10, p. 60] and [MMMi^3, Example 3.4, pp. 208–210] ensures that $L_{\mu,\lambda}$ has a distinguished coefficient tensor whenever the Lamé moduli satisfy

$$\mu \neq 0, \quad 2\mu + \lambda \neq 0, \quad \text{and} \quad 3\mu + \lambda \neq 0. \tag{1.19}$$

For a more thorough discussion on this topic, see [GHA-V, Chapter 1] and [MMMi^3, § 3.3].

Significantly, the layer potentials approach used for solving boundary problems in UR domains continues to be effective in various functional analytic contexts, other than the standard scale of Lebesgue spaces, and for other types of boundary conditions (in addition to Dirichlet). One particular case we wish to highlight is that of the Neumann Problem. Not only is this meaningfully formulated in Lebesgue spaces L^p with $p > 1$, but now this has the added benefit of allowing one to prescribe boundary data in the scale of Hardy spaces H^p, which serves as a suitable substitute for L^p in the regime $0 < p \leq 1$. This is natural and important, especially from the standpoint of Harmonic Analysis and PDE's. An in-depth discussion of the theory of (real) Hardy spaces in the general geometric setting of Ahlfors regular sets can be found in [AlMi15]. There, a long-standing open question has been solved, regarding the nature of the range of integrability exponents p for which a natural scale of Hardy spaces H^p may be considered in an environment (X, ρ, μ) that is a space of homogeneous type (in the sense of R.R. Coifman and G. Weiss; see [CoWe71] and [CoWe77]). Here, we will primarily be concerned with a closed unbounded Ahlfors regular set $\Sigma \subseteq \mathbb{R}^n$ as a background geometric setting. This becomes a space of homogeneous type when equipped with the Euclidean distance and the (doubling) measure $\sigma := \mathcal{H}^{n-1}\lfloor\Sigma$. Recall that for each $p \in (0, \infty)$, and each parameter $\gamma \in (0,1)$, the Hardy space $H^p(\Sigma,\sigma)$ on Σ is defined as

$$H^p(\Sigma,\sigma) := \big\{ f \in \big(\mathrm{Lip}_c(\Sigma)\big)' : f^{\#}_{\gamma} \in L^p(\Sigma,\sigma) \big\}. \tag{1.20}$$

Here $\big(\mathrm{Lip}_c(\Sigma)\big)'$ denotes the dual of the space $\mathrm{Lip}_c(\Sigma)$ of compactly supported Lipschitz functions on Σ, and $f^{\#}_{\gamma}$ is the Fefferman-Stein grand maximal function of a distribution $f \in \big(\mathrm{Lip}_c(\Sigma)\big)'$, given by

$$f^{\#}_{\gamma}(x) := \sup_{\psi \in \mathcal{B}^{\gamma}_{x}(\Sigma)} |\langle f, \psi \rangle| \quad \text{for each } x \in \Sigma, \tag{1.21}$$

where $\mathcal{B}^{\gamma}_{x}(\Sigma) \subseteq \mathrm{Lip}_c(\Sigma)$ is a collection of "bump" functions centered at x suitably normalized in the homogeneous Hölder space $\dot{\mathscr{C}}^{\gamma}(\Sigma)$ (see Definition 4.1). It has been shown in [AlMi15] that the natural range of p's for which one has a rich theory of Hardy spaces in this geometric setting is

$$\tfrac{n-1}{n} < p \leq 1, \tag{1.22}$$

while in the regime $p \in (1,\infty)$, the Hardy space $H^p(\Sigma,\sigma)$ actually coincides with the Lebesgue space $L^p(\Sigma,\sigma)$.

The above considerations are particularly pertinent to the case when $\Sigma := \partial\Omega$, where $\Omega \subseteq \mathbb{R}^n$ is a UR domain. In such a setting, given a weakly elliptic system L as in (1.2) written with some choice of a coefficient tensor $A \in \mathfrak{A}_L$, and fixed $\kappa \in (0,\infty)$, the H^p-Neumann Problem is formulated as

$$\begin{cases} u \in [\mathscr{C}^{\infty}(\Omega)]^M, \quad Lu = 0 \ \text{ in } \Omega, \\ \mathcal{N}_{\kappa}(\nabla u) \in L^p(\partial\Omega,\sigma), \\ \partial^A_{\nu} u = f \in [H^p(\partial\Omega,\sigma)]^M, \end{cases} \tag{1.23}$$

where p is as in (1.22), and $\partial^A_{\nu} u$ denotes the conormal derivative of u associated with A. The latter is defined in a nontangential pointwise sense as in (7.55) when $p = 1$, and is replaced by the weak conormal derivative $\dot{\partial}^A_{\nu} u$ from [GHA-II, Definition 10.2.18, p. 658] when $\frac{n}{n-1} < p < 1$.

In the present work, we are concerned with versions of (1.23) in which the scale of Hardy spaces is replaced by various Herz-type Hardy spaces. We adapt the approach devised in [GHA-IV, GHA-V] where (1.23) has been successfully addressed for the full scale of Hardy spaces H^p, with p as in (1.22). As such, it is instructive to describe in greater detail the main steps taken in [GHA-IV, GHA-V] in the treatment of (1.23) for $p = 1$. The essence of the boundary layer method in the context of the Neumann Problem (1.23) with $p = 1$ is to look for a candidate for a solution u in the format of a modified single layer potential:

$$u(x) := \mathscr{S}_{\mathrm{mod}} g(x) := \int_{\partial\Omega} \big\{ E_L(x-y) - E_L(y) \cdot \mathbb{1}_{\mathbb{R}^n \setminus \overline{B(0,1)}}(y) \big\} g(y)\, d\sigma(y) \tag{1.24}$$

for each $x \in \Omega$, and for some yet-to-be-determined function $g \in [H^1(\partial\Omega, \sigma)]^M$. The reason this is a good candidate for a solution of (1.23) is that the single layer operator $\mathscr{S}_{\rm mod}$ satisfies

$$\mathscr{S}_{\rm mod}g \in \big[\mathscr{C}^\infty(\Omega)\big]^M \quad \text{and} \quad L(\mathscr{S}_{\rm mod}g) = 0 \ \text{ in } \Omega, \tag{1.25}$$

for each g as in (1.4). In other words, $\mathscr{S}_{\rm mod}$ serves as a mechanism to manufacture infinitely many smooth null solutions to the system L. Moreover, the gradient of $\mathscr{S}_{\rm mod}g$ may be computed via

$$\nabla(\mathscr{S}_{\rm mod}g)(x) = \int_{\partial\Omega} (\nabla E_L)(x-y)g(y)\,d\sigma(y) \ \text{ for each } x \in \Omega. \tag{1.26}$$

It turns out that the matrix-valued function ∇E_L has the correct parity and homogeneity. Specifically, its entries are odd and positive homogeneous of degree $1-n$. This allows one to employ the Calderón-Zygmund theory in UR domains from [GHA-IV] to justify the size condition formulated in the second line of (1.23). Crucially, $\mathscr{S}_{\rm mod}$ satisfies the following jump-relation:

$$\partial_\nu^A(\mathscr{S}_{\rm mod}g) = (-\tfrac{1}{2}I + K^\#_{A^\top})g \quad \text{pointwise } \sigma\text{-a.e. on } \partial\Omega, \tag{1.27}$$

for each g as in (1.4), where $K^\#_{A^\top}$ is the "formal" transpose double layer potential in (1.6) associated with the coefficient tensor $A^\top$ (cf. (7.24)). Granted these, the boundary value problem (1.23) with $p = 1$ is then reduced to solving the following boundary integral equation:

$$(-\tfrac{1}{2}I + K^\#_{A^\top})g = f, \tag{1.28}$$

where g is an unknown function, and f is the boundary datum arbitrarily specified in $\big[H^1(\partial\Omega, \sigma)\big]^M$. Then the crux of the matter is to show that the operator

$$-\tfrac{1}{2}I + K^\#_{A^\top} : \big[H^1(\partial\Omega, \sigma)\big]^M \to \big[H^1(\partial\Omega, \sigma)\big]^M \quad \text{is invertible.} \tag{1.29}$$

The strategy for proving said invertibility result is to obtain an operator norm estimate such as (1.14) within the context of $H^1(\partial\Omega, \sigma)$, i.e.,

$$\begin{aligned} &\|K^\#_{A^\top}\|_{[H^1(\partial\Omega,\sigma)]^M \to [H^1(\partial\Omega,\sigma)]^M} \\ &\qquad \leq C \cdot \|\nu\|_{[\mathrm{BMO}(\partial\Omega,\sigma)]^n} \cdot \ln\big(e/\|\nu\|_{[\mathrm{BMO}(\partial\Omega,\sigma)]^n}\big). \end{aligned} \tag{1.30}$$

This strategy is successful when $A^\top$ is a distinguished coefficient tensor for $L^\top$, i.e., the transpose of the system L. Having $\|\nu\|_{[\mathrm{BMO}(\partial\Omega,\sigma)]^n}$ sufficiently small then allows one to obtain the invertibility (1.29) via a Neumann series. This ultimately permits one to obtain a solution of (1.23) as

$$u := \mathscr{S}_{\rm mod}\big[(-\tfrac{1}{2}I + K^\#_{A^\top})^{-1}f\big] \quad \text{in } \Omega. \tag{1.31}$$

In this line of work, (1.30) is one of the key estimates as far as the existence of a solution for the Neumann Problem (1.23) is concerned.

Having discussed the interplay between geometry, layer potentials, and boundary problems, we now shift our focus to function spaces. As already apparent from the exposition so far, the specification of a functional analytic setting (through the choice of particular function spaces for boundary data and for measuring the size/smoothness of the solution) fundamentally affects the format of the boundary value problem being considered. The question of determining appropriate function spaces for which the method of layer potentials can be successfully implemented has been around for a long time. Famously, it has been raised by A.P. Calderón in [Ca78, p. 95]. The present work focuses on scales of Herz-type spaces and their associated Hardy spaces.

In essence, the (quasi)norm for a Herz-type space is defined in the spirit of the Littlewood-Paley theory, in which one localizes functions on concentric dyadic annuli, measures their size using function spaces (e.g., Lebesgue spaces), and assembles these numerical pieces all together using a "weighted" sequence quasinorm (e.g., weighted ℓ^q spaces). This is in line with one of the most fundamental principles in Harmonic Analysis (going back to J. Fourier himself):

breaking down a mathematical object into simpler pieces, examining these individually, and then recombining them to derive global information about the original object. (1.32)

The exposition in the present monograph follows the natural progression of the theory of Herz spaces and examines how this dynamic is mirrored in the realm of boundary value problems. We begin by studying the most primordial version of Herz spaces, called Beurling Algebras A^p, move on to what we refer to as the first-generation Herz spaces $\mathrm{A}_{p,q}$, and subsequently discuss the second-generation Herz spaces $\mathrm{K}^{\alpha}_{p,q}$. Finally, we introduce a new class of Herz-type spaces which we name Composite Herz spaces. Structuring the monograph like so has several advantages. For one thing, this offers the reader direct access to a particular subgenre of the general theory, without the effort of having to deal with unnecessary jargon. For another, this allows us to build up the theory in stages, without having to face all difficulties at once. In view of the fact that the proofs are (necessarily) intricate, this has the benefit of permitting us to emphasize only the novel difficulties to overcome at each stage as we move up the ladder.

1.2 Beurling Algebras and their Associated Hardy Spaces

The theory of Herz spaces has elicited renewed interest in recent years, and its most basic version is referred to as the Beurling Algebras $\mathrm{A}^p(\mathbb{R}^n)$, for $p \in (1,\infty)$. The space $\mathrm{A}^p(\mathbb{R}^n)$, and its dual $\mathrm{B}^q(\mathbb{R}^n)$, where $p, q \in (1,\infty)$ satisfy $1/p + 1/q = 1$, have been first introduced by A. Beurling in 1964 (cf. [Be64]) to study certain convolution algebras of functions and mapping properties of the Fourier transform. Beurling's work shows that the Lebesgue spaces $L^1(\mathbb{R}^n)$ and $L^\infty(\mathbb{R}^n)$ arises as the limiting case of $\mathrm{A}^p(\mathbb{R}^n)$ when $p \to 1^+$ and $\mathrm{B}^q(\mathbb{R}^n)$ when $q \to \infty$, respectively. In this regard, the A^p, B^q theory may be thought of as a parallel theory to L^1, L^∞. In fact, Beurling asserts (see [Be64, p. 11]) that

$$\text{“it is to be expected that for all } p \text{ the properties of } \mathrm{A}^p \text{ are very closely related to those of } L^1\text{”.} \tag{1.33}$$

Hence, heuristically, one may therefore think of A^p as an L^p-flavored L^1 space. Later in 1987, H. G. Feichtinger in [Fe87] has obtained an equivalent description of the spaces $\mathrm{A}^p(\mathbb{R}^n)$ (respectively $\mathrm{B}^p(\mathbb{R}^n)$) in terms of a dyadic weighted ℓ^1-norm (respectively ℓ^∞-norm) of numerical sequences of L^p-norms localized on dyadic annuli. Specifically, Feichtinger showed that given a Lebesgue measurable function f in $\mathbb{R}^n$, for $p \in [1, \infty)$ one has

$$f \in \mathrm{A}^p(\mathbb{R}^n) \quad \Longleftrightarrow \quad \|f\|_{\mathrm{A}^p(\mathbb{R}^n)} := \sum_{k=0}^{\infty} 2^{kn(1-1/p)} \|f \cdot \mathbb{1}_{C_k}\|_{L^p(\mathbb{R}^n)} < \infty, \tag{1.34}$$

while for $q \in (1, \infty]$ one has

$$f \in \mathrm{B}^q(\mathbb{R}^n) \quad \Longleftrightarrow \quad \|f\|_{\mathrm{B}^q(\mathbb{R}^n)} := \sup_{k \in \mathbb{N}_0} \left\{2^{-kn/q} \|f \cdot \mathbb{1}_{C_k}\|_{L^q(\mathbb{R}^n)}\right\} < \infty. \tag{1.35}$$

Incidentally, these descriptions offer a transparent way of identifying L^1 and L^∞ as endpoint cases for the scales $\{\mathrm{A}^p\}_{1 \le p < \infty}$ and $\{\mathrm{B}^q\}_{1 < q \le \infty}$, i.e.,

$$\mathrm{A}^1(\mathbb{R}^n) = L^1(\mathbb{R}^n) \quad \text{and} \quad \mathrm{B}^\infty(\mathbb{R}^n) = L^\infty(\mathbb{R}^n). \tag{1.36}$$

However, the most relevant aspect of describing A^p and B^q scales as in (1.35)-(1.36) is that this opens the door for considering such scales in geometric settings that are much more general than the whole Euclidean space. Here we shall take advantage of this feature (for more on this, see below).

Connections between Beurling Algebras and Hardy spaces are particularly noteworthy. Recall that the latter are spaces of distributions for which a size condition on the grand maximal function is imposed through its membership to some “base” space (classically, L^p with $0 < p \le 1$). Given the affinity between Beurling Algebras and Lebesgue spaces, A^p serves well as a base space for a new genre of Hardy spaces. Specifically, choosing the base space to be A^p gives rise to Beurling-Hardy spaces HA^p which have been studied on the real line by Y.Z. Chen and K.-S. Lau in 1989 in their early work [ChLa89]. These authors introduced the Beurling-Hardy spaces $\mathrm{HA}^p(\mathbb{R})$, established an atomic decomposition when $1 < p \le 2$, and identified their duals as the space of bounded central p'-mean oscillations functions $\mathrm{BCMO}_{p'}(\mathbb{R})$. In 1989, this theory has been further extended to the whole n-dimensional Euclidean ambient by J. Garcia-Cuerva [Ga89], where atomic decompositions, several maximal characterizations, duality, and harmonic extensions results, have been proved for $\mathrm{HA}^p(\mathbb{R}^n)$ in the entire range $1 < p < \infty$. In particular, Garcia-Cuerva has shown that $\mathrm{HA}^p(\mathbb{R}^n)$ may be characterized as the space of all (locally integrable) functions in $\mathbb{R}^n$ whose grand-maximal function belongs to the Beurling Algebra $\mathrm{A}^p(\mathbb{R}^n)$. Alternatively, the membership of a function f to $\mathrm{HA}^p(\mathbb{R}^n)$ is equivalent to having a (central) atomic decomposition in a quantitative fashion:

$$f = \sum_{k=0}^{\infty} \lambda_k a_k \quad \text{in } L^1(\mathbb{R}^n), \tag{1.37}$$

where $\{\lambda_k\} \in \ell^1(\mathbb{N}_0)$ and $\{a_k\}_{k\in\mathbb{N}_0}$ is a sequence of central $(1, p)$-atoms. The nature of such central atoms resembles those used in the atomic description of the classical Hardy space $H^1(\mathbb{R}^n)$, except that now these are supported in "large" balls centered at the origin. This mirrors the centrality of the localizations over annuli used to define $\mathrm{A}^p(\mathbb{R}^n)$ (cf. (1.34)), and it suggests some similarities between $\mathrm{HA}^p(\mathbb{R}^n)$ and $H^1(\mathbb{R}^n)$, although these brands of Hardy spaces are vastly different.

In the first portion of this monograph (comprising Chapters 3–9), we are concerned with studying Beurling-Hardy spaces, and their dual spaces, on unbounded closed Ahlfors regular subsets of $\mathbb{R}^n$. This takes advantage of the flexible nature of the equivalent norm in (1.34). To be specific, consider $\Sigma \subseteq \mathbb{R}^n$ an unbounded closed Ahlfors regular set, and abbreviate $\sigma := \mathcal{H}^{n-1}\lfloor\Sigma$. Fix a reference point $x_0 \in \Sigma$, and a dilation factor $2_* > 1$ (depending on the geometry of Σ). Consider the family of "surface" annuli $\{C_k\}_{k\in\mathbb{N}_0}$ on Σ given by

$$\begin{aligned} C_0 &:= B(x_0, 2_*) \cap \Sigma, \\ C_k &:= \big(B(x_0, 2_*^{k+1})\setminus B(x_0, 2_*^k)\big) \cap \Sigma, \quad \text{for each } k \in \mathbb{N}. \end{aligned} \tag{1.38}$$

In this scenario, we define the *Beurling space* $\mathrm{A}^p(\Sigma, \sigma)$ as the collection of all σ-measurable functions f on Σ satisfying

$$\|f\|_{\mathrm{A}^p(\Sigma,\sigma)} := \sum_{k=0}^{\infty} 2_*^{k(n-1)(1-1/p)} \|f \cdot \mathbb{1}_{C_k}\|_{L^p(\Sigma,\sigma)} < \infty. \tag{1.39}$$

The corresponding *Beurling-Hardy space* on Σ is then defined following the same recipe (1.20) as in the classical setting of Lebesgue spaces, i.e., by demanding the membership of the grand maximal function (1.21) to $\mathrm{A}^p(\Sigma, \sigma)$. More precisely, we introduce the Beurling-Hardy space on Σ as

$$\mathrm{HA}^p(\Sigma, \sigma) := \big\{f \in \big(\mathrm{Lip}_c(\Sigma)\big)' : f_\gamma^\# \in \mathrm{A}^p(\Sigma, \sigma)\big\}. \tag{1.40}$$

Even though this is defined as a space of distributions, the richness of $\mathrm{A}^p(\Sigma, \sigma)$ as a functional analytic setting forces the elements of $\mathrm{HA}^p(\Sigma, \sigma)$ to be of function type (see (4.10)). Furthermore, in this general geometric framework, we are able to develop a versatile atomic and molecular theory for $\mathrm{HA}^p(\Sigma, \sigma)$, thus laying the groundwork for studying a variety of singular integral operators acting on these spaces, which are particularly relevant from the point of view of Harmonic Analysis, and Partial Differential Equations.

Inspired by the work of B. Dahlberg and C. Kenig [DaKe87] regarding the Neumann Problem on the classical Hardy space H^1 in Lipschitz domains, and bearing in mind the earlier discussion concerning the efficacy of the layer potentials approach as a means of treating boundary problems, it is natural to formulate the Neumann Problem on Beurling-Hardy spaces HA^p and attempt solving it using boundary layer methods. This is one of our main goals in the first portion of the present monograph, that is *to successfully implement layer potential methods in establishing well-posedness results for boundary problems with data prescribed in Beurling-Hardy spaces.*

To expand on this matter, consider a uniformly rectifiable (UR for short) domain $\Omega \subseteq \mathbb{R}^n$ with surface measure $\sigma := \mathcal{H}^{n-1}\lfloor\partial\Omega$. Fix an integrability exponent $p \in (1, \infty)$ and an aperture parameter $\kappa \in (0, \infty)$. Consider a second-order, homogeneous, constant complex coefficient, weakly elliptic, $M \times M$ system L in $\mathbb{R}^n$ (as in (1.2)), and formulate the Neumann Problem for L in Ω with boundary data in HA^p as

$$\begin{cases} u \in [\mathscr{C}^\infty(\Omega)]^M, \quad Lu = 0 \ \text{ in } \Omega, \\ \mathcal{N}_\kappa(\nabla u) \in \mathrm{A}^p(\partial\Omega, \sigma), \\ \partial_\nu^A u = f \in [\mathrm{HA}^p(\partial\Omega, \sigma)]^M. \end{cases} \tag{1.41}$$

The earlier discussion on the single layer method (see (1.24), (1.25), (1.26), and (1.27)) brings into focus the issue of whether the operator $-\frac{1}{2}I + K^\#_{A^\top}$ is invertible on the Beurling-Hardy space $[\mathrm{HA}^p(\partial\Omega, \sigma)]^M$. Motivated by estimate (1.30) proved in [GHA-V], we succeed in showing that, whenever $A^\top$ is a distinguished coefficient tensor for $L^\top$, there exists some constant $C \in (0, \infty)$ for which

$$\begin{aligned} &\|K^\#_{A^\top}\|_{[\mathrm{HA}^p(\partial\Omega,\sigma)]^M \to [\mathrm{HA}^p(\partial\Omega,\sigma)]^M} \\ &\qquad \le C \cdot \|\nu\|_{[\mathrm{BMO}(\partial\Omega,\sigma)]^n} \cdot \ln\big(e/\|\nu\|_{[\mathrm{BMO}(\partial\Omega,\sigma)]^n}\big). \end{aligned} \tag{1.42}$$

For a more precise statement, see Theorem 8.1. This is one of the main results of the current monograph, which opens the door for solving the boundary problem formulated in (1.41), by allowing us to invert $-\frac{1}{2}I + K^\#_{A^\top}$ via a Neumann series (which converges if $\|\nu\|_{[\mathrm{BMO}(\partial\Omega,\sigma)]^n}$ is sufficiently small). Such invertibility thus permits us to write a solution of (1.41) in the form

$$u := \mathscr{S}_{\mathrm{mod}}\big[(-\tfrac{1}{2}I + K^\#_{A^\top})^{-1} f\big] \quad \text{in } \Omega. \tag{1.43}$$

Note that (1.42) is not a mere boundedness statement for the operator $K^\#_{A^\top}$ on $[\mathrm{HA}^p(\partial\Omega, \sigma)]^M$ involving a large geometric constant in the right-hand side. Indeed, the right-hand side in (1.42) is sensitive to the "flatness" (or rather oscillatory nature) of $\partial\Omega$ through the presence of the BMO-seminorm of the outward unit normal ν to Ω. To obtain it, we first develop a comprehensive atomic and molecular theory for Beurling-Hardy spaces HA^p on Ahlfors regular sets (see Section 4.2, and Section 4.3), and follow the classical strategy of showing that the action of $K^{\#}_{A^\top}$ on each atom is a fixed multiple of a suitable molecule. The novel challenge here is to bound all pieces involved using constants which depend on geometry according to the template in the right-hand side of (1.42). The presence of just one constant deviating from this blue-print, which in principle still keeps us on track for showing the boundedness of $K^{\#}_{A^\top}$ on $[\mathrm{HA}^p(\partial\Omega,\sigma)]^M$, would derail our efforts to obtain the sort of operator norm estimate stated in (1.42). In this endeavor, the algebraic chord-dot-normal structure of $K^{\#}_{A^\top}$ (ensured by $A^\top$ being a distinguished coefficient tensor for $L^\top$) plays a crucial role, and this is indeed the difference maker. For details, see Theorem 6.4 and Theorem 8.1.

A key observation we wish to make in relation to (1.41) is that the membership of the boundary datum to the Beurling-Hardy space HA^p is a necessary condition, as far as solvability of (1.41) is concerned. More specifically, we succeed in proving a Fatou-type result in the context of Beurling-Hardy spaces, described as follows (see Theorem 9.3). Continue to assume $\Omega \subseteq \mathbb{R}^n$ is a UR domain and set $\sigma := \mathcal{H}^{n-1}\lfloor\partial\Omega$. Then whenever $p \in (1,\infty)$ and u is a function satisfying

$$\begin{cases} u \in [\mathscr{C}^\infty(\Omega)]^M, \quad Lu = 0 \text{ in } \Omega, \\ \mathcal{N}_\kappa(\nabla u) \in \mathrm{A}^p(\partial\Omega,\sigma), \end{cases} \tag{1.44}$$

it follows that the conormal derivative $\partial^A_\nu u$ exists σ-a.e. on $\partial\Omega$ and, as a function, it belongs to $[\mathrm{HA}^p(\partial\Omega,\sigma)]^M$. Hence, the boundary condition in (1.41) is meaningfully formulated. In the end-point case $p = 1$, when $L = \Delta$, and Ω is a Lipschitz domain this corresponds to the Fatou-type result in the classical Hardy space $H^1(\partial\Omega,\sigma)$ established by B. Dahlberg and C. Kenig in [DaKe87, Lemma 2.10, p. 446].

In many ways, the scale of Beurling-Hardy spaces HA^p, with $p \in (1,\infty)$, exhibits similar properties to that of the standard Hardy space H^1. One particular feature we wish to highlight is the fact that the dual of the Beurling Hardy space is the space of bounded central mean oscillations functions. More precisely, consider a closed unbounded Ahlfors regular set $\Sigma \subseteq \mathbb{R}^n$ and abbreviate $\sigma := \mathcal{H}^{n-1}\lfloor\Sigma$. In this setting, whenever $p, p' \in (1,\infty)$ are Hölder conjugate exponents, we have

$$\big(\mathrm{HA}^p(\Sigma,\sigma)\big)^* = \widetilde{\mathrm{BCMO}}_{p'}(\Sigma,\sigma) := \mathrm{BCMO}_{p'}(\Sigma,\sigma)/\mathbb{C}, \tag{1.45}$$

where $\mathrm{BCMO}_{p'}(\Sigma,\sigma)$ is the *space of bounded central p'-mean oscillation functions* (cf. Definition 5.1), which is discussed at length in Chapter 5. The aforementioned duality should be understood in the sense described in Theorem 5.1. This is reminiscent of the classical Fefferman H^1 – BMO duality result. There are, however, stark differences between the BCMO scale and the classical John-Nirenberg space BMO, most notably the absence of a John-Nirenberg inequality in the former setting.

One second major achievement in the first portion of the current work is that we also develop tools to deal with boundary value problems in the context of functions of bounded central mean oscillations. We not only develop a functional analytic framework for this brand of spaces in the setting of unbounded Ahlfors regular set, but we also build up Calderón-Zygmund theory, paving the way for employing boundary layer potentials in the treatment of boundary value problems formulated in terms of BCMO spaces.

Significantly, the body of results and techniques developed here allows us to solve the Dirichlet Problem for weakly elliptic systems on BCMO spaces. To expand on this, let $\Omega \subseteq \mathbb{R}^n$ be a UR domain, with $\partial\Omega$ unbounded, and set $\sigma := \mathcal{H}^{n-1}\lfloor\partial\Omega$. Fix a second-order, homogeneous, constant complex coefficient, weakly elliptic, $M \times M$ system L in $\mathbb{R}^n$ (as in (1.2)), some aperture parameter $\kappa \in (0, \infty)$, and an exponent $p \in (1, \infty)$. The Dirichlet Problem for L in Ω with boundary data in BCMO_p is formulated as

$$\begin{cases} u \in [\mathscr{C}^\infty(\Omega)]^M, \quad Lu = 0 \ \text{ in } \Omega, \\ |\nabla u|^p \operatorname{dist}(\cdot, \partial\Omega)^{p-1} \, d\mathcal{L}^n \ \text{ is a central Carleson measure in } \Omega, \\ u\big|_{\partial\Omega}^{\kappa-\text{n.t.}} = f \in [\mathrm{BCMO}_p(\partial\Omega, \sigma)]^M, \end{cases} \tag{1.46}$$

subject to the central Carleson measure estimate

$$\sup_{R \geq 2_*} \left\{ R^{1-n} \int_{B(x_0,R)\cap\Omega} |\nabla u|^p \operatorname{dist}(\cdot, \partial\Omega)^{p-1} \, d\mathcal{L}^n \right\}^{1/p} \leq C\|f\|_{[\mathrm{BCMO}_p(\partial\Omega,\sigma)]^M}. \tag{1.47}$$

The notion of *central Carleson measures* (cf. Definition 6.1) is closely related to BCMO spaces. Indeed, we prove in Theorem 6.8 that (modified) generalized double layer potentials acting on functions of bounded central mean oscillations are densities for central Carleson measures, with naturally accompanying estimates of the sort described in (1.47).

Significantly, layer potential methods continue to be effective in this context, and we are able to demonstrate that the boundary problem in (1.46) is solvable whenever L has a distinguished coefficient tensor (see Theorem 9.5). At the core of our approach for solving (1.46) is a duality argument, which, starting with (1.42), yields geometrically sensitive operator norm estimates for modified double layer potentials $K_{A,\mathrm{mod}}$ (see (7.29)) associated with a distinguished coefficient tensor A for L on $\mathrm{BCMO}_p(\partial\Omega, \sigma)$ of the sort

$$\begin{aligned} &\|[K_{A,\mathrm{mod}}]\|_{[\widetilde{\mathrm{BCMO}}_p(\partial\Omega,\sigma)]^M \to [\widetilde{\mathrm{BCMO}}_p(\partial\Omega,\sigma)]^M} \\ &\qquad \leq C \cdot \|\nu\|_{[\mathrm{BMO}(\partial\Omega,\sigma)]^n} \cdot \ln\left(e/\|\nu\|_{[\mathrm{BMO}(\partial\Omega,\sigma)]^n}\right), \end{aligned} \tag{1.48}$$

for some constant $C \in (0, \infty)$ depending only on the "large" geometric features of Ω. Having $\|\nu\|_{[\mathrm{BMO}(\partial\Omega,\sigma)]^n}$ small enough then allows us to invert $\frac{1}{2}I + [K_{A,\mathrm{mod}}]$ on $[\widetilde{\mathrm{BCMO}}_p(\partial\Omega, \sigma)]^M$ (see Theorem 8.3), which ultimately produces a solution of

(1.46) in the format of a modified boundary-to-domain double layer potential

$$\begin{aligned} & u := \mathcal{D}_{A,\mathrm{mod}} g + c \quad \text{in } \Omega \\ & \text{for some } g \in [\mathrm{BCMO}_p(\partial\Omega, \sigma)]^M \text{ and } c \in \mathbb{C}^M \\ & \text{such that } (\tfrac{1}{2} I + K_{A,\mathrm{mod}}) g = f - c \end{aligned} \tag{1.49}$$

See Theorem 9.5 for the solvability of the Dirichlet Problem formulated in (1.46).

The boundary problem formulated in (1.46) is of same nature as the Dirichlet Problem in the (classical) John-Nirenberg space $\mathrm{BMO}(\partial\Omega, \sigma)$, namely

$$\begin{cases} u \in [\mathscr{C}^\infty(\Omega)]^M, \quad Lu = 0 \ \text{ in } \Omega, \\ |\nabla u|^p \operatorname{dist}(\cdot, \partial\Omega)^{p-1} \, d\mathcal{L}^n \ \text{ is a Carleson measure in } \Omega, \forall \, p \in (1, \infty) \\ u\big|_{\partial\Omega}^{\kappa-\mathrm{n.t.}} = f \in [\mathrm{BMO}(\partial\Omega, \sigma)]^M. \end{cases} \tag{1.50}$$

This boundary problem has been solved by D. Mitrea, I. Mitrea, and M. Mitrea [GHA-V] in the context of δ-AR domains. When Ω is a Lipschitz domain or a $\mathscr{C}^1$ domain, and $L := \Delta$ (the Laplacian), this problem has been treated by E. Fabes and U. Neri [FaNe80] in 1980, and E. Fabes and C. Kenig [FaKe81] in 1981.

Note that the flexibility in the choice of the exponent p in the formulation of (1.50) reflects the fact that, thanks to the John-Nirenberg inequality, $\mathrm{BMO}(\partial\Omega, \sigma)$ coincides, for all $p \in (1, \infty)$, with $\mathrm{BMO}_p(\partial\Omega, \sigma)$ (where the seminorm is defined in terms of p-mean oscillations). This stands in stark contrast to the nature of $\mathrm{BCMO}_p(\partial\Omega, \sigma)$ which strongly depends on the exponent p, as showed in [ChLa89] and [AlLaGu00]. This phenomenon explains why the exponent p from the central Carleson measure condition in (1.46) is tied up with the exponent p from $\mathrm{BCMO}_p(\partial\Omega, \sigma)$, the space of boundary data in (1.50).

The notion of central Carleson measures was first introduced by Y. Z. Chen and K.-S. Lau [ChLa89] in the upper-half plane, and they use it to characterize $\mathrm{BCMO}_2(\mathbb{R})$. This characterization was further extended to the n-dimensional Euclidean space by S. Lu and D. Yang [LuYa95-02] in 1995, where they also studied mapping properties of Littlewood-Paley operators on $\mathrm{BCMO}_p(\mathbb{R}^n)$.

1.3 Herz Spaces of First Generation

While Beurling algebras play a significant role in the study of Hardy spaces, they are not stable under the action of Calderón-Zygmund singular integral operators, or the Hardy-Littlewood maximal operator. One reason for this is the fact that Beurling algebras behave as the Lebesgue space L^1 (see (1.33)). The remedy is to consider the scale which has Beurling algebras on its fringes, something we shall refer to as Herz spaces of first generation.

Specifically, consider a closed unbounded Ahlfors regular set Σ and abbreviate $\sigma := \mathcal{H}^{n-1}\lfloor\Sigma$. Fix two exponents $p \in (1,\infty)$ and $q \in (0,\infty)$, and let $\{C_k\}_{k\in\mathbb{N}_0}$ be as in (1.38). Introduce the *first-generation Herz space* $\mathrm{A}_{p,q}(\Sigma,\sigma)$ as the collection of all σ-measurable functions f on Σ satisfying

$$\|f\|_{\mathrm{A}_{p,q}(\Sigma,\sigma)} := \left[\sum_{k=0}^{\infty}\left(2^{k(n-1)(1/q-1/p)}\|f\cdot\mathbb{1}_{C_k}\|_{L^p(\Sigma,\sigma)}\right)^q\right]^{1/q} < \infty. \tag{1.51}$$

Naturally, one can associate to $\mathrm{A}_{p,q}(\Sigma,\sigma)$ the *first-generation Herz-type Hardy space* defined as

$$\mathrm{HA}_{p,q}(\Sigma,\sigma) := \left\{f \in \left(\mathrm{Lip}_c(\Sigma)\right)' : f^{\#}_{\gamma} \in \mathrm{A}_{p,q}(\Sigma,\sigma)\right\}. \tag{1.52}$$

There are several reasons to bring in these scales of spaces. For one, the above definitions generalize those introduced by J. Garcia-Cuerva and M.-J.L. Herrero [GaHe94] in 1994, and independently by S. Lu and D. Yang [LuYa] in 1995, in the Euclidean ambient. For another, these are natural generalizations of the corresponding Beurling spaces, defined via (1.39)-(1.40). Precisely, it is apparent from (1.51) that when $q = 1$ the Herz space $\mathrm{A}_{p,1}(\Sigma,\sigma)$ becomes the Beurling space $\mathrm{A}^p(\Sigma,\sigma)$, while $\mathrm{HA}_{p,1}(\Sigma,\sigma)$ coincides with the Beurling-Hardy space $\mathrm{HA}_p(\Sigma,\sigma)$. It is worth noting that $\mathrm{A}_{p,1}(\Sigma,\sigma)$ generalizes the Lebesgue scale, since $\mathrm{A}_{q,q}(\Sigma,\sigma) = L^q(\Sigma,\sigma)$ whenever $p = q \in (1,\infty)$. This is indicative of the fact that the scale of Herz spaces $\mathrm{A}_{p,q}$ is closely related to Lebesgue spaces L^q.

Having introduced $\mathrm{A}_{p,q}(\Sigma,\sigma)$ and $\mathrm{HA}_{p,q}(\Sigma,\sigma)$ in (1.51)-(1.52), the challenge in the second portion of the present monograph (comprising Chapters 10–16), is *to develop a Calderón-Zygmund theory for singular integral operators acting on, and from, Herz spaces (of first generation) and their related Hardy spaces, and implement layer potentials to deal with boundary value problems in this context.* As a preamble, we need to first anchor more firmly the scale of first-generation Herz spaces $\mathrm{A}_{p,q}$ into the realms of Functional Analysis and Harmonic Analysis. In this vein, three basic aspects are:

- establishing that for all $p, q \in (1,\infty)$ the space $\mathrm{A}_{p,q}(\Sigma,\sigma)$ is a Generalized Banach Function Space (cf. Proposition 10.11);
- identifying the Köthe dual of $\mathrm{A}_{p,q}(\Sigma,\sigma)$, for all $p, q \in (1,\infty)$, as $\mathrm{A}_{p',q'}(\Sigma,\sigma)$, where $p', q' \in (1,\infty)$ are the Hölder conjugate exponent of p, q, respectively (cf. Proposition 10.12);
- obtaining the boundedness of the Hardy-Littlewood maximal operator on the space $\mathrm{A}_{p,q}(\Sigma,\sigma)$ for each $p, q \in (1,\infty)$ (cf. Theorem 10.1).

These key features of first-generation Herz spaces permit us to obtain a powerful extrapolation result described in Theorem 10.3. Loosely speaking, said extrapolation result converts estimates on Muckenhoupt weighted Lebesgue spaces into estimates involving the Herz spaces $A_{p,q}$.

Recently, a comprehensive Calderón-Zygmund theory for singular integral operators in UR sets on Muckenhoupt weighted Lebesgue spaces in the setting of Ahlfors regular sets, and their application to the formulation of boundary problems has been established in [MMMi[3]]. This development, combined with our extrapolation result, paves the way for developing a robust Calderón-Zygmund theory for singular integral operators on $A_{p,q}$ spaces. Concretely, in Section 13.1, we study convolution-type singular integral operators, commutators, chord-dot-normal SIO's, and layer potentials in the context of first-generation Herz spaces. One of the main results we wish to highlight here is in relation to geometrically sensitive operator norm estimates for double layers. Specifically, assume Ω is a UR domain, abbreviate $\sigma := \mathcal{H}^{n-1}\lfloor\partial\Omega$, and denote by ν its geometric measure theoretic outward unit normal. Consider a second-order, homogeneous, constant complex coefficient, weakly elliptic, $M \times M$ system L in $\mathbb{R}^n$ (as in (1.2)), and assume that A is a distinguished coefficient tensor for L. Then, given two exponents $p, q \in (1, \infty)$, there exists a constant $C \in (0, \infty)$, depending on the "large" geometric characters of Ω, with the property that

$$\begin{aligned}\|K_A\|_{[A_{p,q}(\partial\Omega,\sigma)]^M \to [A_{p,q}(\partial\Omega,\sigma)]^M} \\ \leq C \cdot \|\nu\|_{[\mathrm{BMO}(\partial\Omega,\sigma)]^n} \cdot \ln\big(e/\|\nu\|_{[\mathrm{BMO}(\partial\Omega,\sigma)]^n}\big).\end{aligned} \tag{1.53}$$

This estimate highlights the deep connection between the geometry of the domain Ω and double layers in the functional analytic setting of Herz spaces $A_{p,q}(\partial\Omega, \sigma)$. Requiring that $\|\nu\|_{[\mathrm{BMO}(\partial\Omega,\sigma)]^n}$ is sufficiently small, then guarantees that the operator $\frac{1}{2}I + K_A$ is invertible on $[A_{p,q}(\partial\Omega, \sigma)]^M$, via a Neumann series. This invertibility is a pivotal tool for establishing existence of solutions to the Dirichlet Problem for L in Ω with boundary data in $A_{p,q}$, formulated as:

$$\begin{cases} u \in [\mathscr{C}^\infty(\Omega)]^M, \quad Lu = 0 \ \text{ in } \Omega, \\ \mathcal{N}_\kappa u \in A_{p,q}(\partial\Omega, \sigma), \\ u\big|_{\partial\Omega}^{\kappa-\mathrm{n.t.}} = f \in [A_{p,q}(\partial\Omega, \sigma)]^M. \end{cases} \tag{1.54}$$

A well-posedness result for this Dirichlet Problem is presented in Theorem 16.2. In addition, we succeed in establishing well-posedness results for the Neumann Problem, Regularity Problem, and Transmission Problem with boundary data arbitrarily prescribed in $A_{p,q}$ (see Theorems 16.3, 16.4, 16.5, and 16.6). These well-posedness results complement the work done on Lebesgue spaces in [MMMi[3]]. The figure below puts this in perspective, by illustrating how the Lebesgue scale fits within the framework of first-generation Herz spaces. As legend for the map below, we have the identification of a point having coordinates $(1/p, 1/q)$ with the Herz space $A_{p,q}$ if $q > 1$, and the Herz-based Hardy space $HA_{p,q}$ if $q \leq 1$:

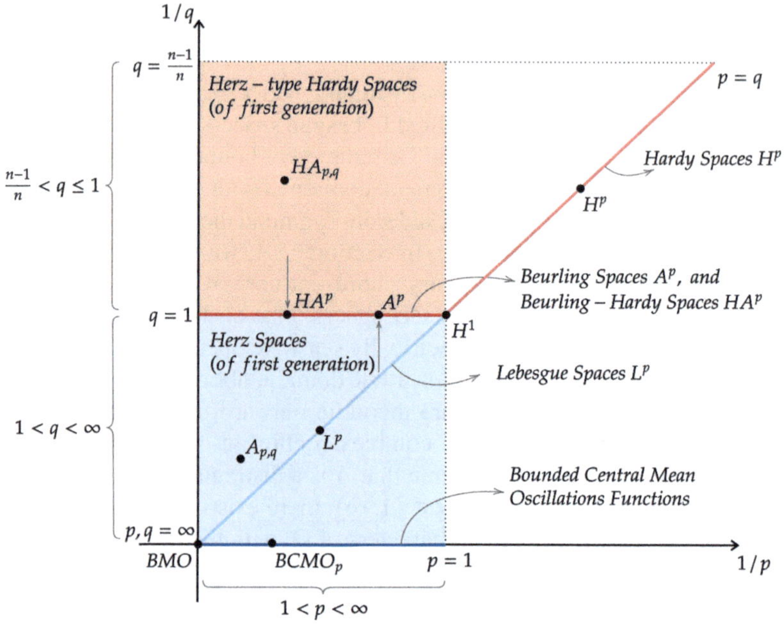

Fig. 1.1: Contrast between the Herz-type scales $A_{p,q}$ and $HA_{p,q}$, the Lebesgue scale L^p, and (classical) Hardy spaces H^p. Note the bordering scales: $BCMO_p$, BMO, and H^1.

It turns out that the range of p's and q's for which there exists a satisfactory theory of Herz-type Hardy spaces (of first generation) is intimately related to that of the theory of (ordinary) Hardy spaces in the general geometric setting of unbounded Ahlfors regular sets. Specifically, we show in Theorem 11.2 that given $p \in (1, \infty)$ and $q \in \big(\frac{n-1}{n}, 1\big]$, the space $\mathrm{HA}_{p,q}(\Sigma, \sigma)$ has an atomic characterization described as follows. A distribution $f \in \big(\mathrm{Lip}_c(\Sigma)\big)'$ belongs to $\mathrm{HA}_{p,q}(\Sigma, \sigma)$ if and only if

$$f = \sum_{k=0}^{\infty} \lambda_k a_k \quad \text{in } L^1(\Sigma, \sigma), \tag{1.55}$$

where $\{\lambda_k\} \in \ell^q(\mathbb{N}_0)$ and $\{a_k\}_{k \in \mathbb{N}_0}$ is a sequence of central (q, p)-atoms. Notably, the range $q \in \big(\frac{n-1}{n}, 1\big]$ matches the range for which a rich theory of $H^q(\Sigma, \sigma)$ on Ahlfors regular sets can be developed (see [AlMi15]).

The atomic (and molecular) theory for $\mathrm{HA}_{p,q}(\Sigma, \sigma)$ developed in Sections 11.2 and 11.3 is a crucial step, as far as boundedness and invertibility results of layer potentials on these spaces are concerned. To highlight one of our main results on this topic, consider Ω a uniformly rectifiable domain, with surface measure $\sigma := \mathcal{H}^{n-1} \lfloor \partial\Omega$. Given a weakly elliptic system L written with a coefficient tensor A as (1.2), assume that $A^\top$ is a distinguished coefficient tensor for $L^\top$. Then, given

$p \in (1,\infty)$ and $q \in \big(\frac{n-1}{n},1\big]$, there exists a constant $C \in (0,\infty)$ depending on the geometric character of Ω with the property that

$$\begin{aligned}&\|K^{\#}_{A^\top}\|_{[\mathrm{HA}_{p,q}(\partial\Omega,\sigma)]^M\to[\mathrm{HA}_{p,q}(\partial\Omega,\sigma)]^M}\\&\qquad\le C\cdot\|\nu\|_{[\mathrm{BMO}(\partial\Omega,\sigma)]^n}\cdot\ln\big(e/\|\nu\|_{[\mathrm{BMO}(\partial\Omega,\sigma)]^n}\big).\end{aligned} \tag{1.56}$$

See Theorem 15.6 and Theorem 13.7. When $q=1$, we recover our earlier result on Beurling-Hardy spaces (see Theorem 8.1). From the point of view of boundary value problems, this opens the door for employing layer potential techniques to deal with the Neumann Problem for L in Ω with boundary data in $\mathrm{HA}_{p,q}$ formulated as

$$\begin{cases}u\in[\mathscr{C}^\infty(\Omega)]^M, \quad Lu=0 \text{ in } \Omega,\\ \mathcal{N}_\kappa(\nabla u)\in \mathrm{A}_{p,q}(\partial\Omega,\sigma),\\ \partial^A_\nu u=f\in[\mathrm{HA}_{p,q}(\partial\Omega,\sigma)]^M.\end{cases} \tag{1.57}$$

Specifically, the operator $-\frac{1}{2}I+K^{\#}_{A^\top}$ is invertible, via a Neumann series, on the Herz-type Hardy space $[\mathrm{HA}_{p,q}(\partial\Omega,\sigma)]^M$ if $\|\nu\|_{[\mathrm{BMO}(\partial\Omega,\sigma)]^n}$ is sufficiently small, the Neumann Problem (1.57) is well-posed (cf. Theorem 16.8), and its unique solution (modulo constants) is given in terms of a single layer potential:

$$u:=\mathscr{S}_{\mathrm{mod}}\big[(-\tfrac{1}{2}I+K^{\#}_{A^\top})^{-1}f\big] \quad \text{in } \Omega. \tag{1.58}$$

A natural Fatou-type result holds in relation to (1.57). Namely, for any function u satisfying

$$\begin{cases}u\in[\mathscr{C}^\infty(\Omega)]^M, \quad Lu=0 \text{ in } \Omega,\\ \mathcal{N}_\kappa(\nabla u)\in \mathrm{A}_{p,q}(\partial\Omega,\sigma),\end{cases} \tag{1.59}$$

it follows that the conormal derivative $\partial^A_\nu u$ is well defined σ-a.e. on $\partial\Omega$ and belongs to $[\mathrm{HA}_{p,q}(\partial\Omega,\sigma)]^M$ (see Theorem 16.7). This proves the optimality of the Herz-based Hardy space in the formulation of the Neumann Problem (1.57).

1.4 Herz Spaces of Second Generation

Classically, the Lebesgue scale is further enhanced through the consideration of Lorentz spaces, which, e.g., arise as intermediate spaces for the real interpolation method implemented in the context of Lebesgue spaces. Likewise, the fact that the $\mathrm{A}_{p,q}$ family is not stable under the real method of interpolation offers us the chance to further expand it, by considering

$$\mathrm{K}^{\alpha}_{p,q}:=\big(\mathrm{A}_{p,q_0},\ \mathrm{A}_{p,q_1}\big)_{\theta,q}. \tag{1.60}$$

In view of the Reiteration Theorem, $\mathrm{K}^{\alpha}_{p,q}$ thus becomes the smallest enhancement of the Herz scale $\mathrm{A}_{p,q}$ which is stable under real interpolation. This should be regarded as a "motivational" definition of $\mathrm{K}^{\alpha}_{p,q}$, which we call the scale of *second-generation Herz spaces*.

An intrinsic definition of this scale, more akin to A^p in (1.39), and $\mathrm{A}_{p,q}$ in (1.51), goes as follows. Consider a closed unbounded Ahlfors regular set $\Sigma \subseteq \mathbb{R}^n$ and abbreviate $\sigma := \mathcal{H}^{n-1}\lfloor\Sigma$. Fix two exponents $p \in (1, \infty)$, and $q \in (0, \infty]$, along with an index $\alpha \in \mathbb{R}$. With the family of annuli $\{C_k\}_{k\in\mathbb{N}_0}$ defined in relation to some fixed reference point $x_0 \in \Sigma$ as in (1.38), the inhomogeneous Herz space $\mathrm{K}^{\alpha}_{p,q}(\Sigma, \sigma)$ is defined as the space of all σ-measurable functions f on Σ for which

$$\|f\|_{\mathrm{K}^{\alpha}_{p,q}(\Sigma,\sigma)} := \left[\sum_{k=0}^{\infty} \left(2_*^{k\alpha}\|f \cdot \mathbb{1}_{C_k}\|_{L^p(\Sigma,\sigma)}\right)^q\right]^{1/q} < \infty. \tag{1.61}$$

There is another closely related brand of Herz space in which we not only zoom away from the reference point x_0, but also zoom in towards it. Specifically, the homogeneous Herz space $\dot{\mathrm{K}}^{\alpha}_{p,q}(\Sigma, \sigma)$ is defined as the collection of all σ-measurable functions f on Σ such that

$$\|f\|_{\dot{\mathrm{K}}^{\alpha}_{p,q}(\Sigma,\sigma)} := \left[\sum_{k=-\infty}^{\infty} \left(2_*^{k\alpha}\|f \cdot \mathbb{1}_{C_k}\|_{L^p(\Sigma,\sigma)}\right)^q\right]^{1/q} < \infty, \tag{1.62}$$

where now the family of annuli $\{C_k\}_{k\in\mathbb{Z}}$ under consideration is defined as

$$C_k := \big(B(x_0, 2_*^{k+1})\backslash B(x_0, 2_*^k)\big) \cap \Sigma, \quad \text{for each } k \in \mathbb{Z}. \tag{1.63}$$

In contrast with (1.39) and (1.51), the index α in (1.61) adds an extra degree of flexibility to the family of Herz spaces $\mathrm{K}^{\alpha}_{p,q}$. It is visible that this scale reduces precisely to the $\mathrm{A}_{p,q}$ spaces for the choice $\alpha := (n-1)(1/q - 1/p)$. When $p = q$, and $\alpha = 0$, it is evident from (1.61) and (1.62) that $\mathrm{K}^{0}_{p,p}$ and $\dot{\mathrm{K}}^{0}_{p,p}$ coincide with the Lebesgue space L^p. As such, the second-generation Herz spaces $\mathrm{K}^{\alpha}_{p,q}$ and $\dot{\mathrm{K}}^{\alpha}_{p,q}$ are natural generalizations of L^p spaces.

With these definitions in place, in the third portion of the present monograph (comprising Chapters 17–22) we *develop a functional analytic framework for second-generation Herz spaces in the setting of unbounded Ahlfors regular sets as well as a potent Calderón-Zygmund theory for singular integrals, with the ultimate goal of treating boundary value problems formulated in terms of* $\mathrm{K}^{\alpha}_{p,q}$ *and* $\dot{\mathrm{K}}^{\alpha}_{p,q}$ *via layer potentials.* The novel aspect here is that, while extrapolation results played an important role in the $\mathrm{A}_{p,q}$ theory, our approach now relies on the real interpolation method (a point of view consistent with (1.60)).

To further elaborate on this, retain the same geometric setting as before. In Theorem 17.3 we prove that second-generation Herz spaces are stable under the real method of interpolation. A precise formulation is as follows. Select exponents

$p \in (0, \infty)$ and $q, q_0, q_1 \in (0, \infty]$, along with indices $\alpha_0, \alpha_1 \in \mathbb{R}$ with $\alpha_0 \neq \alpha_1$. Then whenever $\theta \in (0, 1)$ and $\alpha := (1 - \theta)\alpha_0 + \theta\alpha_1$ it follows that

$$\big(\dot{\mathrm{K}}^{\alpha_0}_{p,q_0}(\Sigma, \sigma), \, \dot{\mathrm{K}}^{\alpha_1}_{p,q_1}(\Sigma, \sigma)\big)_{\theta,q} = \dot{\mathrm{K}}^{\alpha}_{p,q}(\Sigma, \sigma). \tag{1.64}$$

A similar result holds for inhomogeneous Herz spaces $\mathrm{K}^{\alpha}_{p,q}$. A remarkable consequence of this interpolation result is the characterization of $\dot{\mathrm{K}}^{\alpha}_{p,q}$ as the intermediate space between weighted Lebesgue spaces. Specifically for the homogeneous case, we show that if $p \in (1, \infty)$, $q \in (0, \infty]$, and $\alpha \in \mathbb{R}$, then whenever $a, b \in \mathbb{R}$ satisfy $a < \alpha p < b$ there exists $\theta \in (0, 1)$ for which

$$\dot{\mathrm{K}}^{\alpha}_{p,q}(\Sigma, \sigma) = \Big(L^p\big(\Sigma, |x - x_0|^a \sigma(x)\big), \, L^p\big(\Sigma, |x - x_0|^b \sigma(x)\big)\Big)_{\theta,q}. \tag{1.65}$$

This identification is relevant since it provides a link to the extensive body of work on weighted Lebesgue spaces, particularly Lebesgue spaces with power weights (see, e.g., [St57], [GaRu85]). A known result on this topic (see, for instance, [GeGoKoKr97, DyIhLeTuVa19, MMMi³]) establishes that the function

$$\begin{aligned} &|\cdot - x_0|^a \text{ belongs to the Muckenhoupt class } A_p(\Sigma, \sigma) \\ &\text{if and only if } -(n-1) < a < (n-1)(p-1). \end{aligned} \tag{1.66}$$

Since $A_p(\Sigma, \sigma)$ is the largest class of weights for which the Hardy-Littlewood maximal operator is bounded on the corresponding weighted Lebesgue spaces, (1.65)-(1.66) are indicative of the fact that the largest range of α's for which a satisfactory theory for $\dot{\mathrm{K}}^{\alpha}_{p,q}(\Sigma, \sigma)$ can be developed is

$$\frac{-(n-1)}{p} < \alpha < \frac{(n-1)(p-1)}{p}. \tag{1.67}$$

Indeed, in this regime we can choose $a, b \in \mathbb{R}$ satisfying

$$-(n-1) < a < \alpha p < b < (n-1)(p-1), \tag{1.68}$$

a scenario that ensures that $|\cdot - x_0|^a$ and $|\cdot - x_0|^b$ belongs to the Muckenhoupt class $A_p(\Sigma, \sigma)$. As such, (1.65) guarantees that second-generation Herz spaces can be identified as intermediate spaces between Muckenhoupt weighted Lebesgue spaces. This highlights the importance of Muckenhoupt weights in the theory of second-generation Herz spaces.

The characterization described in (1.65) serves as the starting point for developing a Calderón-Zygmund theory for singular integral operators on second-generation Herz spaces. By combining this characterization with boundedness results for singular integral operators in UR sets acting on Muckenhoupt weighted Lebesgue spaces from [MMMi³], we are able to establish similar boundedness results in the context of $\dot{\mathrm{K}}^{\alpha}_{p,q}$ and $\mathrm{K}^{\alpha}_{p,q}$. See Chapter 18 for a thorough discussion on this topic, involving

SIO's of convolution-type, chord-dot-normal type, and of boundary layer potentials type.

One specific result we wish to highlight concerns operator norm estimates for chord-dot-normal singular integral operators which are sensitive to the "flatness" of the boundary of the underlying domain. Let us give a more precise description of such a result. Suppose $\Omega \subseteq \mathbb{R}^n$ is a UR domain, with $\partial\Omega$ unbounded, and surface measure $\sigma := \mathcal{H}^{n-1}\lfloor\partial\Omega$. Consider T a singular integral operator of chord-dot-normal type (cf. (6.103)), associated with some kernel k that is smooth in $\mathbb{R}^n \setminus \{0\}$, even, and positive homogeneous of degree $-n$. Select indices $p \in (1, \infty)$, $q \in (0, \infty]$, and $\alpha \in \mathbb{R}$ satisfying (1.67). Then there exists a constant $C \in (0, \infty)$ depending only on the "large" geometric features of Ω for which

$$\begin{aligned} &\|T\|_{\dot{\mathrm{K}}^{\alpha}_{p,q}(\partial\Omega,\sigma)\to\dot{\mathrm{K}}^{\alpha}_{p,q}(\partial\Omega,\sigma)} \\ &\qquad \leq C \cdot \|\nu\|_{[\mathrm{BMO}(\partial\Omega,\sigma)]^n} \cdot \ln\big(e/\|\nu\|_{[\mathrm{BMO}(\partial\Omega,\sigma)]^n}\big). \end{aligned} \tag{1.69}$$

See Theorem 18.2. A similar result is obtained for inhomogeneous Herz spaces $\mathrm{K}^{\alpha}_{p,q}$. The geometric estimate (1.69) is one of our main results in the Calderón-Zygmund theory on $\dot{\mathrm{K}}^{\alpha}_{p,q}$ spaces, and prefigures the role that layer potentials will eventually play in the treatment of boundary value problems in second-generation Herz spaces.

To shed some light on this, retain the same geometric background assumptions as above. Fix a second-order, homogeneous, constant complex coefficient, weakly elliptic $M \times M$ system L in $\mathbb{R}^n$ as in (1.2) written for a choice of coefficient tensor $A \in \mathfrak{A}_L$. Select indices $p \in (1, \infty)$, $q \in (0, \infty]$, and $\alpha \in \mathbb{R}$ satisfying (1.67), and fix an aperture parameter $\kappa \in (0, \infty)$. We then formulate the Dirichlet Problem for L in Ω with boundary data in $\dot{\mathrm{K}}^{\alpha}_{p,q}$ as follows:

$$\begin{cases} u \in [\mathscr{C}^\infty(\Omega)]^M, \quad Lu = 0 \ \text{ in } \Omega, \\ \mathcal{N}_\kappa u \in \dot{\mathrm{K}}^{\alpha}_{p,q}(\partial\Omega,\sigma), \\ u\big|^{\kappa-\mathrm{n.t.}}_{\partial\Omega} = f \in [\dot{\mathrm{K}}^{\alpha}_{p,q}(\partial\Omega,\sigma)]^M. \end{cases} \tag{1.70}$$

The key aspect of A being a distinguished coefficient tensor for L is that the entries of the boundary-to-boundary double layer potential K_A become singular integral operators of chord-dot-normal type. In this scenario, (1.69) applies and yields

$$\begin{aligned} &\|K_A\|_{[\dot{\mathrm{K}}^{\alpha}_{p,q}(\partial\Omega,\sigma)]^M\to[\dot{\mathrm{K}}^{\alpha}_{p,q}(\partial\Omega,\sigma)]^M} \\ &\qquad \leq C \cdot \|\nu\|_{[\mathrm{BMO}(\partial\Omega,\sigma)]^n} \cdot \ln\big(e/\|\nu\|_{[\mathrm{BMO}(\partial\Omega,\sigma)]^n}\big). \end{aligned} \tag{1.71}$$

The upshot of this is that having $\|\nu\|_{[\mathrm{BMO}(\partial\Omega,\sigma)]^n}$ sufficiently small then implies that the operator $\frac{1}{2}I + K_A$ is invertible on $[\dot{\mathrm{K}}^{\alpha}_{p,q}(\partial\Omega,\sigma)]^M$, via a Neumann series. Ultimately, this permits us to obtain a solution of (1.70) in the format

$$u := \mathcal{D}_A\big[\big(\tfrac{1}{2}I + K_A\big)^{-1} f\big] \quad \text{in } \Omega, \tag{1.72}$$

where $\mathcal{D}_A$ is the boundary-to-domain double layer as in (1.5). In fact, we establish a genuine well-posedness result for (1.70) in Theorem 19.1. Our approach is robust and continues to be effective in the case of Neumann, Regularity, and Transmission Problems with boundary data in $\dot{\mathrm{K}}^{\alpha}_{p,q}$ (see Theorem 19.2, Theorem 19.4, and Theorem 19.5). Similar results are also obtained on inhomogeneous Herz spaces $\mathrm{K}^{\alpha}_{p,q}$.

In Section 17.6 we present a more flexible version of (1.62) in which the family of annuli $\{C_k\}_{k\in\mathbb{Z}}$ is replaced by a more general decomposition of Σ into dyadic "tubular" neighborhoods $\{C^E_k\}_{k\in\mathbb{Z}}$ associated with some d-dimensional subset E of Σ with $d \in [0, n-1)$. This gives rise to a new class of Herz-type spaces, which we refer to as geometric Herz spaces. Also, in Section 17.7 we introduce the scale $\dot{\mathrm{K}}^{\alpha}_{\mathbb{X},q}$ where the quasinorm is constructed according to the blueprint described in (1.62) in which L^p is replaced by a generic Generalized Banach Function Space $\mathbb{X}$.

In anticipation to discussing the next stage in the development of our theory in Section 1.6 we present a brief historical overview of Herz spaces.

1.5 A Brief Historical Overview of Herz Spaces

The standard scale of $\mathrm{K}^{\alpha}_{p,q}$ spaces (in the whole Euclidean setting) has been first introduced by C. Herz [He68] in 1969 as a means to study mapping properties of the Fourier transform. In 1974, T. M. Flett [Fl74] improved some of Herz's results by employing the norm characterizations of $\mathrm{K}^{\alpha}_{p,q}$ obtained by R. Johnson [Jo74] earlier that year. Subsequently, in 1978, R. Johnson [Jo78] showed that certain multipliers on ordinary Hardy spaces H^p can be described in terms $\mathrm{K}^{\alpha}_{p,q}$ spaces, for specific indices p, q, and α, thus establishing a link between Herz spaces $\mathrm{K}^{\alpha}_{p,q}$ and classical Hardy spaces. The work of A. Baernstein II and E. T. Sawyer [BaSa85] in 1985, gave further impetus and accelerated the development of Herz spaces. Among other things, they introduced the homogeneous version of Herz spaces $\dot{\mathrm{K}}^{\alpha}_{p,q}$ and used them to obtain sharp embedding results into H^p spaces, improving earlier results by M. Taibleson and G. Weiss [TaWe80].

Motivated by the work of Y.Z. Chen and K-S. Lau [ChLa89] (on Beurling-Hardy spaces HA^p on the real line), and J. Garcia-Cuerva [Ga89] (on HA^p in $\mathbb{R}^n$ with $n \geq 2$), the authors J. Garcia-Cuerva and M.J. Herrero [GaHe94] in 1995, and independently S. Lu and D. Yang [LuYa], initiated the study of Herz-type Hardy spaces $\mathrm{HK}^{(n-1)(1/q-1/p)}_{p,q}$. In particular, Garcia-Cuerva and Herrero contributed several real variable characterizations for $\mathrm{HK}^{(n-1)(1/q-1/p)}_{p,q}$, including an atomic decomposition, while Lu and Yang provided a molecular theory, and boundedness results for certain singular integral operators on $\mathrm{HK}^{(n-1)(1/q-1/p)}_{p,q}$. Such an atomic and molecular theory solidified connections with Harmonic Analysis. Subsequently, this has stimulated further developments in a number of directions, mostly from the perspective of Harmonic Analysis. Here we shall mention some of them.

In a series of papers, S. Lu and D. Yang introduced and studied the weighted version of Herz spaces $\dot{K}^{\alpha}_{p,q}(\omega_1, \omega_2)$, and weighted Herz-type Hardy spaces $H\dot{K}^{\alpha}_{p,q}(\omega_1, \omega_2)$, for some A_1-weights ω_1, ω_2 in $\mathbb{R}^n$. See [LuYa95], [LuYa95-01], and [LuYa97]. A local version of Herz-type Hardy space $h\dot{K}^{\alpha}_{p,q}(\omega_1, \omega_2)$ was later introduced by D. Fan and D. Yang [FaYa97], in the spirit of the local Hardy spaces h^p of D. Goldberg [Go79]. Additionally, E. Hernández, G. Weiss, and D. Yang [HeWeYa96] established a ϕ-transform characterization of $H\dot{K}^{\alpha}_{p,q}(\omega_1, \omega_2)$, and a wavelet characterization of $H\dot{K}^{\alpha}_{p,q}(\mathbb{R}^n)$. E. Hernández and D. Yang in [HeYa98] and [HeYa99] also characterized the intermediate spaces, via the complex method of interpolation, for the scales of (unweighted) Herz spaces, and Herz-type Hardy spaces.

Additional contributions, focusing on boundedness properties on Herz spaces $\dot{K}^{\alpha}_{p,q}$ and Herz-type Hardy spaces and $H\dot{K}^{\alpha}_{p,q}$ for certain classes of sublinear operators, fractional-type sublinear operators, bilinear operators, vector-valued operators, and commutator, have been made in instance, see [LuYaYa00], [Li00], [GrLiYa98], [LiYa96], [LuYaYa00], and [TaYa00]; see also the references therein.

In 2005, S. Lu and L. Xu [LuXu05] introduced the scale of Morrey-Herz spaces $M\dot{K}^{\alpha,\lambda}_{p,q}$ and proved boundedness results for some sublinear operators. Additional boundedness results on this scale of spaces appeared in [Wu09] and [TaShZh09]. In 2009, M. Izuki [Iz09-I] introduced Herz spaces with variable exponents $\dot{K}^{\alpha}_{p(\cdot),q}$ and proved boundedness results for sublinear operators on this scale in [Iz10-I]. Other types of Herz spaces with variable exponents, where α and q are allowed to be variable, have been considered by A. Almeida and D. Drihem [AlDr12], and S. Samko [Sa13]. Izuki also introduced the Herz-Morrey spaces with variable exponents $M\dot{K}^{\alpha,\lambda}_{p(\cdot),q}$ in [Iz09-II] (see also [Iz10-II]) and studied certain sublinear operators on them. Weighted versions of Herz spaces with variable exponents have been studied by M. Izuki and T. Noi in [IzNo16] and [IzNo17]. Subsequently, L. Wang, and L. Shu defined the weighted Herz-Morrey spaces with variable exponent $M\dot{K}^{\alpha,\lambda}_{p(\cdot),q}(\omega)$ in [WaSh18], which were further studied by S. Wang and J. Xu [WaXu21].

As part of the development of Herz-type scales, H. Nafis, H. Rafeiro, and M.A. Zaighum [NaRaZa20] introduced the grand variable Herz spaces, denoted by $\dot{K}^{\alpha,\theta}_{p(\cdot),q)}$. Subsequent work done on this brand of spaces includes [NaRaZa21] and [NaRaZa22]. Around the same time, H. Rafeiro and S. Samko [RaSa20] also defined the local and global generalized Herz spaces, denoted by $\mathbf{H}^{p,q}_{\omega}$ and $\mathbf{H}^{p,q}_{\omega,0}$, respectively. The variable exponent version of these scales has been considered by Rafeiro and Samko in [RaSa22]. A real-variable theory of Hardy spaces associated with local and global generalized Herz spaces has been established by Y. Li, D. Yang, and L. Huang in the monograph [LiYaHu23]. The scale of single-weighted grand Herz spaces with variable exponents $\dot{K}^{\alpha,\theta}_{p(\cdot),q)}(\omega)$ have been introduced in [SuSuMeAzAlMl22]. In addition, the doubly-weighted version of grand Herz spaces (with constant exponents) has been considered by X. Yu and Z. Liu [YuLi22]. The grand variable Herz-Morrey spaces have been introduced in [SuAzSuMeMl22], and the grand weighted Herz-Morrey spaces with variable exponents have been defined in [SuAzSuMaMlSo22].

1.6 A New Class: Composite Herz Spaces

The fourth and final portion of the present monograph (comprising Chapters 23–25) *focuses on identifying the most inclusive functional analytic framework, referred to as Composite Herz spaces, that generalizes and unifies the large family of Herz-type spaces, with the ultimate goal of using them in connection with Calderón-Zygmund theory and boundary value problems.*

While there exists an abundance of Herz-type scales, all share a fundamental structure. Specifically, the quasinorm used to define them involves

(i) a decomposition of the underlying set into a family of dyadic annuli,

(ii) an appropriate function space (e.g., Lebesgue spaces $L^p(\mathbb{R}^n)$ in the case of (1.62)), as a means of measuring the size of a given function localized on said annuli,

(iii) a "weighted" sequence space (e.g., the weighted ℓ^q appearing in (1.62)) used to assemble the above pieces all together.

The rather large variety of scales of Herz-type spaces (reviewed in Section 1.5) then raises the question of whether there exists a functional analytic framework that unifies and generalizes them. To address this question, we bring in the notion of Generalized Banach Function Space (GBFS) from [GHA-II] to do the job in item (ii) above, and rely on the concept of Generalized Quasi-Banach Sequence Space (GQBSS) to do the job in item (iii) above; see Definition 10.2 and Definition 23.1.

To delve deeper into this, let $\Sigma \subseteq \mathbb{R}^n$ be a closed unbounded Ahlfors regular set, with surface measure $\sigma := \mathcal{H}^{n-1}\lfloor\Sigma$. Consider a Generalized Banach Function Space $(\mathbb{X}, \|\cdot\|_{\mathbb{X}})$ over (Σ, σ), and $(\mathbb{Y}, \|\cdot\|_{\mathbb{Y}})$ a Generalized Quasi-Banach Sequence Space on $\mathbb{Z}$. With $\{C_k\}_{k\in\mathbb{Z}}$ the family of annuli as in (1.63), for each σ-measurable function f on Σ we introduce the "quasinorm"

$$\|f\|_{\dot{\mathbb{K}}(\Sigma;\mathbb{X};\mathbb{Y})} := \Big\|\big\{\|f \cdot \mathbb{1}_{C_k}\|_{\mathbb{X}}\big\}_{k\in\mathbb{Z}}\Big\|_{\mathbb{Y}} \in [0, \infty]. \tag{1.73}$$

We then introduce the *Composite Herz space* on Σ with base spaces $\mathbb{X}$ and $\mathbb{Y}$, denoted by $\dot{\mathbb{K}}(\Sigma; \mathbb{X}; \mathbb{Y})$, as the collection of all σ-measurable functions f defined on Σ for which $\|f\|_{\dot{\mathbb{K}}(\Sigma;\mathbb{X};\mathbb{Y})} < \infty$. Many of the Herz-type spaces existing in the literature are equipped with a quasinorm fitting in the template displayed in (1.73).

To offer a concrete example in the Euclidean setting $\Sigma := \mathbb{R}^{n-1}$, take the Lebesgue space $\mathbb{X} := L^p(\mathbb{R}^{n-1})$ for $p \in (1, \infty)$ and consider for $q \in (0, \infty]$ and $\alpha \in \mathbb{R}$, the weighted sequence space $\mathbb{Y} := \ell^q_\alpha(\mathbb{Z})$, that is, the collection of sequences $\{a_k\}_{k\in\mathbb{Z}} \subseteq \mathbb{R}$ for which

$$\|\{a_k\}_{k\in\mathbb{Z}}\|_{\ell^q_\alpha(\mathbb{Z})} := \|\{2^{k\alpha} a_k\}_{k\in\mathbb{Z}}\|_{\ell^q(\mathbb{Z})} < \infty. \tag{1.74}$$

Then (1.73) reads as

$$\|f\|_{\dot{\mathrm{K}}(\mathbb{R}^{n-1};L^p(\mathbb{R}^{n-1});\ell^q_\alpha(\mathbb{Z}))} = \begin{cases} \left[\displaystyle\sum_{k=-\infty}^{\infty} \big(2^{k\alpha}\|f\cdot \mathbb{1}_{C_k}\|_{L^p(\mathbb{R}^{n-1})}\big)^q\right]^{1/q} & \text{if } q<\infty, \\ \displaystyle\sup_{k\in\mathbb{Z}} \big\{2^{k\alpha}\|f\cdot \mathbb{1}_{C_k}\|_{L^p(\mathbb{R}^{n-1})}\big\} & \text{if } q=\infty. \end{cases} \tag{1.75}$$

In such a case, the Composite Herz space in $\mathbb{R}^{n-1}$ with base spaces $L^p(\mathbb{R}^{n-1})$ and $\ell^q_\alpha(\mathbb{Z})$ coincides the standard Herz space:

$$\dot{\mathrm{K}}\big(\mathbb{R}^{n-1}; L^p(\mathbb{R}^{n-1}); \ell^q_\alpha(\mathbb{Z})\big) = \dot{\mathrm{K}}^\alpha_{p,q}(\mathbb{R}^{n-1}). \tag{1.76}$$

As mentioned earlier, this scale of spaces has been extensively studied (see for instance [LuYa95], [LuYa95-01], [LiYa96], [LuYa97], [HeYa98], [HeYa99], [LuYaYa00], [Li00], [GrLiYa98], [LuYaYa00], [TaYa00], and the references therein). It turns out that in the regime

$$1<p<\infty, \quad 0<q\leq\infty \quad \text{and} \quad -\tfrac{n-1}{p} < \alpha < \tfrac{n-1}{p'}, \tag{1.77}$$

where $p' \in (1,\infty)$ is the Hölder conjugate exponent of p, there are many operators bounded on $\dot{\mathrm{K}}^\alpha_{p,q}(\mathbb{R}^{n-1})$. For instance, this is the case for the singular integral operators for which we have developed a Calderón-Zygmund theory on $\dot{\mathrm{K}}^\alpha_{p,q}$ (see Chapter 18). This also happens for any sublinear operator T satisfying the pointwise size condition

$$|Tf(x)| \lesssim \int_{\mathbb{R}^{n-1}} \frac{|f(y)|}{|x-y|^{n-1}}\,dy \quad \text{for all } f\in L^1_{\mathrm{comp}}(\mathbb{R}^{n-1}) \text{ and } x\notin \operatorname{supp} f, \tag{1.78}$$

which is also bounded on $L^p(\mathbb{R}^{n-1})$ (see[1] [LiYa96]). See also Theorem 23.3 (and (1.83) below) for a version of this result in the context of Composite Herz spaces. Bearing in mind that $\dot{\mathrm{K}}^\alpha_{p,q}(\mathbb{R}^{n-1})$ is a Composite Herz space $\dot{\mathrm{K}}(\mathbb{R}^{n-1};\mathbb{X};\mathbb{Y})$ when $\mathbb{X} := L^p(\mathbb{R}^{n-1})$ and $\mathbb{Y} := \ell^q_\alpha(\mathbb{Z})$ (cf. (1.76)), this brings into focus the latter pair of spaces, and schematically we have:

[1] this builds on the work [HeYa99] and [LuYa95]. The proof in [LiYa96] has a gap, in that the sublinearity of T is used to commute T with an infinite series.

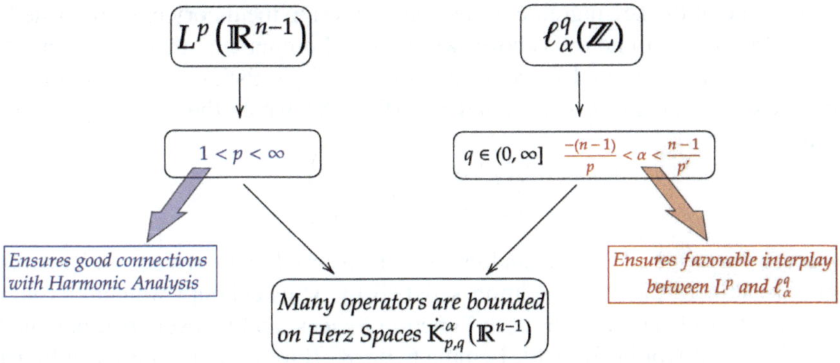

Fig. 1.2: Key ingredients for standard Herz spaces

From this perspective, a natural question arises: what are the conditions on the base spaces $\mathbb{X}$ and $\mathbb{Y}$ that yield a satisfactory theory for the corresponding Composite Herz space $\dot{\mathrm{K}}(\Sigma; \mathbb{X}; \mathbb{Y})$? To gain some insight into this question, we propose to regard the condition $1 < p < \infty$ as the manifestation of the fact that L^p is a "good" functional analytic environment for doing Harmonic Analysis. Specifically in this regime, we have that the Hardy-Littlewood maximal operator $\mathcal{M}_{\mathbb{R}^{n-1}}$ is bounded both on $\mathbb{X} := L^p(\mathbb{R}^{n-1})$ and on $\mathbb{X}' := L^{p'}(\mathbb{R}^{n-1})$, where $1/p + 1/p' = 1$. Also, the role of the conditions

$$0 < q \leq \infty \quad \text{and} \quad -\tfrac{n-1}{p} < \alpha < \tfrac{n-1}{p'} \tag{1.79}$$

is to ensure a "favorable" interplay between the function space $\mathbb{X} := L^p(\mathbb{R}^{n-1})$ and the sequence space $\mathbb{Y} := \ell^q_\alpha(\mathbb{Z})$. The manner in which the aforementioned favorable interplay should be interpreted for two generic spaces $\mathbb{X}$ and $\mathbb{Y}$ is a subtle issue. Elucidating this led us to introduce two key concepts:

(i) the A_∞ index associated with a Generalized Banach Function Space; and

(ii) the compliancy of (weighted) sequence quasinorms.

To elaborate on these aspects, assume $\Sigma \subseteq \mathbb{R}^n$ is a closed unbounded Ahlfors regular set, and abbreviate $\sigma := \mathcal{H}^{n-1}\lfloor\Sigma$. Let $\mathbb{X}$ be a Generalized Banach Function Space on (Σ, σ). It turns out that many Harmonic Analysis flavored results, e.g., extrapolation and boundedness of SIO's (see [GHA-II, GHA-III]), can be obtained on $\mathbb{X}$ provided

$$\begin{gathered} \mathcal{M}_\Sigma : \mathbb{X} \to \mathbb{X} \quad \text{and} \quad \mathcal{M}_\Sigma : \mathbb{X}' \to \mathbb{X}' \\ \text{are well-defined and bounded operators,} \end{gathered} \tag{1.80}$$

where $\mathcal{M}_\Sigma$ is the Hardy-Littlewood maximal operator on Σ. As such, this assumption establishes good connections between $\mathbb{X}$ and Harmonic Analysis. Also, we find that

(1.80) is one of the key ingredients needed to develop the theory of Composite Herz spaces. In relation to item (i) above, we show in Lemma 23.1 that whenever $\mathcal{M}_\Sigma$ is bounded on the associated space $\mathbb{X}'$ the norm $\|\cdot\|_{\mathbb{X}}$ satisfies an A_∞ condition, i.e., there exist constants $C \in (0, \infty)$ and $\tau \in (0, 1)$ for which the following inequality holds:

$$\frac{\|\mathbb{1}_E\|_{\mathbb{X}}}{\|\mathbb{1}_\Delta\|_{\mathbb{X}}} \leq C\left(\frac{\sigma(E)}{\sigma(\Delta)}\right)^\tau, \tag{1.81}$$

for every surface ball $\Delta \subseteq \Sigma$ and every σ-measurable subset $E \subseteq \Delta$. This result, which is akin to the A_∞ condition satisfied by Muckenhoupt weights, is new in the context of Generalized Banach Function Spaces, and recovers that obtained by M. Izuki and T. Noi in [IzNo17, Lemma 6, pp. 8–9] for (classical) Banach Function Spaces (see also [CrHeMa16, Lemma 3.3, p. 363]). This motivates us to introduce the A_∞ *index* of $\mathbb{X}$ by setting

$$\Upsilon_{\mathbb{X}} := \sup\big\{\tau > 0 : \ (1.81) \text{ holds for some } C \in (0, \infty)\big\} \in (0, \infty] \tag{1.82}$$

as a means of quantifying the best exponent τ that can be used in (1.81). In particular, (1.81) holds whenever $\tau \preceq \Upsilon_{\mathbb{X}}$, where the symbol $\preceq$ stands for "$\leq$" if $\Upsilon_{\mathbb{X}} < \infty$, and "$<$" if $\Upsilon_{\mathbb{X}} = \infty$.

Regarding the second item (ii) listed earlier, consider a Generalized Quasi-Banach Sequence Space $\mathbb{Y}$ over $\mathbb{Z}$. The (new) notion of compliancy of $\|\cdot\|_{\mathbb{Y}}$ (see Definition 23.5) can be understood as a boundedness property on $\mathbb{Y}$ under forward and backward shifts, as well as for numerical sequences consisting of tails of weighted series (see (C3)-(C4) in Definition 23.5). Significantly, in Section 23.4 we take up the task of identifying the weighted sequence spaces used to define Herz-type spaces, and show that they satisfy our compliancy condition. Having elaborated on the notions of A_∞ index of a Generalized Banach Function Space $\mathbb{X}$ and the compliancy property of a Generalized Quasi-Banach Sequence Space $\mathbb{Y}$, we shall interpret the favorable interplay between $\mathbb{X}$ and $\mathbb{Y}$ as the demand that $\mathbb{Y}$ is $\mathbb{X}$-compliant (cf. Definition 23.7). Schematically, the following abstraction of the diagram in Figure 1.2 now emerges:

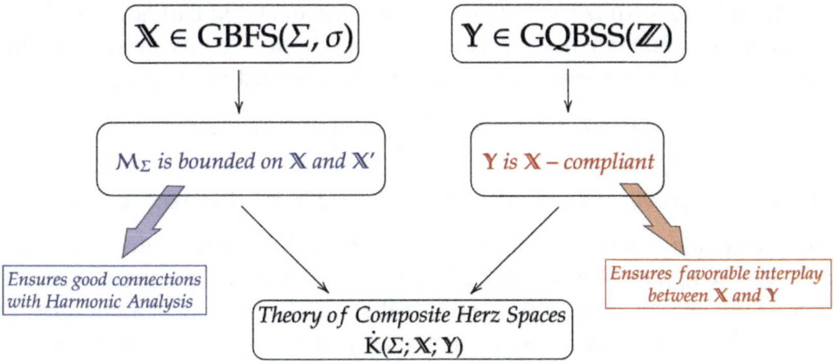

Fig. 1.3: Key ingredients for Composite Herz spaces

This diagram displays the key requirements on $\mathbb{X}$ and $\mathbb{Y}$ ensuring that a rich and robust theory for the corresponding Composite Herz space $\dot{\mathrm{K}}(\Sigma;\mathbb{X};\mathbb{Y})$ can be developed. For instance, in Theorem 23.1 we show that if $\mathbb{X}$ is a Generalized Banach Function Space on (Σ,σ) such that the Hardy-Littlewood maximal operator $\mathcal{M}_\Sigma$ in bounded on $\mathbb{X}$ and $\mathbb{X}'$, and if $\mathbb{Y}$ is a Generalized Quasi-Banach Sequence Space on $\mathbb{Z}$ which is $\mathbb{X}$-compliant, then $\mathcal{M}_\Sigma$ is also bounded on $\dot{\mathrm{K}}(\Sigma;\mathbb{X};\mathbb{Y})$. We also prove a more general result in Theorem 23.3, which is described as follows. Assume $T : L^1(\Sigma, \frac{\sigma(x)}{1+|x|^{n-1}}) \longrightarrow \mathscr{M}(\Sigma,\sigma)$ is a sublinear operator satisfying the size condition (1.78), and such that $T : \mathbb{X} \to \mathbb{X}$ is bounded. Then

$$\begin{gathered} T : \dot{\mathrm{K}}(\Sigma;\mathbb{X};\mathbb{Y}) \longrightarrow \dot{\mathrm{K}}(\Sigma;\mathbb{X};\mathbb{Y}) \\ \text{is a well-defined and bounded operator.} \end{gathered} \tag{1.83}$$

This boundedness result opens the door for developing a Calderón-Zygmund theory for layer potentials on Composite Herz spaces (cf. Section 23.8). The novel aspect of this is that said result allows us to simultaneously treat a large variety of Herz-type spaces in a unified fashion. Moreover, this is new even when the underlying ambient is the Euclidean space.

To illustrate the scope of the boundedness result advertised in (1.83), let $\Sigma := \mathbb{R}^{n-1}$ and consider $\mathbb{X} := L^p(\mathbb{R}^{n-1})$ for some exponent $p \in (1,\infty)$. Then $L^p(\mathbb{R}^{n-1})$ is a Generalized Banach Function Space on $\mathbb{R}^{n-1}$, and its associated space is $L^{p'}(\mathbb{R}^{n-1})$, where $p' \in (1,\infty)$ is the Hölder conjugate exponent of p. Also, it is well known that the Hardy-Littlewood maximal operator in $\mathbb{R}^{n-1}$ is bounded both on $L^p(\mathbb{R}^{n-1})$ and $L^{p'}(\mathbb{R}^{n-1})$. Next, given $q \in (0,\infty]$ and $\alpha \in \mathbb{R}$, take $\mathbb{Y} := \ell^q_\alpha(\mathbb{Z})$. It turns out that $\ell^q_\alpha(\mathbb{Z})$ is a Generalized Quasi-Banach Sequence Space on $\mathbb{Z}$ (see Example 23.1), and

$$\ell^q_\alpha(\mathbb{Z}) \text{ is } L^p(\mathbb{R}^{n-1})\text{-compliant} \iff -\tfrac{n-1}{p} < \alpha < \tfrac{n-1}{p'}. \tag{1.84}$$

Notice that this range of α's coincides with that encountered earlier in (1.77). Then our result ensures that whenever T is a sublinear operator satisfying (1.78) which is bounded on $L^p(\mathbb{R}^{n-1})$, then T is bounded on the standard Herz space

$$\dot{\mathrm{K}}^{\alpha}_{p,q}(\mathbb{R}^{n-1}) = \dot{\mathrm{K}}\big(\mathbb{R}^{n-1}, L^p(\mathbb{R}^{n-1}), \ell^q_\alpha(\mathbb{Z})\big). \tag{1.85}$$

This recovers[2] a previous result obtained by Li and Yang [LiYa96] in 1996.

Other examples of Generalized Banach Function Spaces that can be utilized in our template of Composite Herz spaces include Lebesgue spaces with variable exponents. Specifically, consider the variable exponent Lebesgue space $\mathbb{X} := L^{p(\cdot)}(\mathbb{R}^{n-1})$ for an exponent function $p(\cdot) : \mathbb{R}^{n-1} \to (1, \infty)$. When $p(\cdot)$ stays away from 1 and ∞, and also satisfies a certain log-Hölder continuity condition (cf. (23.255)), the Hardy-Littlewood maximal operator is bounded both on $L^{p(\cdot)}(\mathbb{R}^{n-1})$ and on its associated space $L^{p'(\cdot)}(\mathbb{R}^{n-1})$, where $p'(\cdot) := \frac{p(\cdot)}{p(\cdot)-1}$ in $\mathbb{R}^{n-1}$ is the Hölder conjugate exponent function of $p(\cdot)$. Then there exist $\tau_1, \tau_2 \in (0, 1)$ for which (1.81) holds for $L^{p(\cdot)}(\mathbb{R}^{n-1})$ and $L^{p'(\cdot)}(\mathbb{R}^{n-1})$, respectively. It turns out that $\ell^q_\alpha(\mathbb{Z})$ is $L^{p(\cdot)}(\mathbb{R}^{n-1})$-compliant whenever $q \in (0, \infty]$, and $\alpha \in \mathbb{R}$ satisfies

$$-(n-1)\tau_1 < \alpha < (n-1)\tau_2. \tag{1.86}$$

In this context, if T is a sublinear operator satisfying (1.78) which is bounded on $L^{p(\cdot)}(\mathbb{R}^{n-1})$, then our boundedness result ensures that T is bounded on the variable exponent Herz space

$$\dot{\mathrm{K}}^{\alpha}_{p(\cdot),q}(\mathbb{R}^{n-1}) = \dot{\mathrm{K}}\big(\mathbb{R}^{n-1}; L^{p(\cdot)}(\mathbb{R}^{n-1}); \ell^q_\alpha(\mathbb{Z})\big). \tag{1.87}$$

This in particular recovers (and corrects) a result obtained by M. Izuki in [Iz10-I].

In Section 23.4 and Section 23.5 we provide a wealth of examples of Generalized Banach Function Spaces and Generalized Quasi-Banach Sequence Spaces that can be used in our theory of Composite Herz spaces. We thus recover many existing Herz-type spaces in the literature. Some of the specific choices of the base spaces $\mathbb{X}$ and $\mathbb{Y}$ producing Composite Herz spaces that appeared earlier in the literature are presented in the table below.

[2] and corrects

$\mathbb{X} \in$ GBFS	$\mathbb{Y} \in$ GQBSS	Resulting Herz Space	Introduced in
$L^p(\mathbb{R}^n)$	$\ell^q_\alpha(\mathbb{Z})$	$\dot{\mathrm{K}}^\alpha_{p,q}(\mathbb{R}^n)$	[He68], [Fe87], [Fl74]
$L^{p(\cdot)}(\mathbb{R}^n)$	$\ell^q_\alpha(\mathbb{Z})$	$\dot{\mathrm{K}}^\alpha_{p(\cdot),q}(\mathbb{R}^n)$	[Iz09-I]
$L^{p,r}(\mathbb{R}^n)$	$\ell^q_\alpha(\mathbb{Z})$	$\dot{\mathrm{K}}^{\alpha,r}_{p,q}(\mathbb{R}^n)$	[WeYa22], [BhKoKo23]
$L^p(\mathbb{R}^n, w)$	$\ell^q_\alpha(\mathbb{Z}; w_1)$	$\dot{\mathrm{K}}^\alpha_{p,q}(\mathbb{R}^n; w; w_1)$	[LuYa95]
$L^{p(\cdot)}(\mathbb{R}^n)$	$\ell^{q),\theta}_\alpha(\mathbb{Z})$	$\dot{\mathrm{K}}^{\alpha,\theta}_{p(\cdot),q)}(\mathbb{R}^n)$	[NaRaZa20]
$L^p(\mathbb{R}^n)$	$\ell^q_{\omega(\cdot)}(\mathbb{Z})$	$\dot{\mathrm{K}}^{\omega(\cdot)}_{p,q}(\mathbb{R}^n)$	[RaSa20]
$L^p(\mathbb{R}^n)$	$\ell^{\infty,q}_{\alpha,\lambda}(\mathbb{Z})$	$\mathrm{M}\dot{\mathrm{K}}^{\alpha,\lambda}_{p,q}(\mathbb{R}^n)$	[LuXu05]
$L^{p(\cdot)}(\mathbb{R}^n)$	$\ell^{\infty,q}_{\alpha,\lambda}(\mathbb{Z})$	$\mathrm{M}\dot{\mathrm{K}}^{\alpha,\lambda}_{p(\cdot),q}(\mathbb{R}^n)$	[Iz09-II]
$\mathbb{X}$ such that $\mathcal{M}$ is bounded on $\mathbb{X}$ and $\mathbb{X}'$	$\mathbb{Y}$ is $\mathbb{X}$-compliant	$\dot{\mathrm{K}}(\mathbb{R}^n; \mathbb{X}; \mathbb{Y})$	[Mitrea–Takemura] present work

Table 1.1: Choices for GBFS and GQBSS yielding Herz-type space in the literature.

The novel aspect in relation to these existing Herz-type scales is that for the first time they are going to be used in connection with boundary value problems (more on this later).

While Table 1.1 above emphasizes the resulting Herz-type space (on the third column) and where they have first appeared and/or studied, Table 1.2 below offers the concrete ranges of indices for which the Hardy-Littlewood maximal operator $\mathcal{M}_{\mathbb{R}^n}$ is bounded on $\mathbb{X}$ and where the compliancy of $\mathbb{Y}$ with respect to $\mathbb{X}$ holds.

$\mathbb{X} \in$ GBFS	$\mathbb{Y} \in$ GQBSS	Regime where $\mathcal{M}_{\mathbb{R}^n}$ is bounded on $\mathbb{X}$ and $\mathbb{X}'$ and where $\mathbb{Y}$ is $\mathbb{X}$-compliant
$L^p(\mathbb{R}^n)$	$\ell^q_\alpha(\mathbb{Z})$	$p \in (1,\infty),\ \ q \in (0,\infty]$, and $-\frac{n}{p} < \alpha < \frac{n}{p'}$
$L^{p(\cdot)}(\mathbb{R}^n)$	$\ell^q_\alpha(\mathbb{Z})$	$p(\cdot) : \mathbb{R}^n \to (1,\infty),\ \ q \in (0,\infty]$, and $-n\tau_{p(\cdot)} < \alpha < n\tau_{p'(\cdot)}$ with $\tau_{p(\cdot)} \precsim \Upsilon_{L^{p(\cdot)}}$ and $\tau_{p'(\cdot)} \precsim \Upsilon_{L^{p'(\cdot)}}$
$L^{p,r}(\mathbb{R}^n)$	$\ell^q_\alpha(\mathbb{Z})$	$p, r \in (1,\infty),\ \ q \in (0,\infty]$, and $-\frac{n}{p} < \alpha < \frac{n}{p'}$
$L^p(\mathbb{R}^n, w)$ $w \in A_p(\mathbb{R}^n)$	$\ell^q_\alpha(\mathbb{Z}; w_1)$ $w_1 \in A_{p_1}(\mathbb{R}^n)$	$p, p_1 \in (1,\infty),\ \ q \in (0,\infty]$, and $-\frac{n\tau}{p_1 p} < \alpha < \frac{n\tau'}{p_1 p'}$ with $\frac{\tau}{p} \precsim \Upsilon_{L^p_w}$ and $\frac{\tau'}{p'} \precsim \Upsilon_{L^{p'}_{w'}}$
$L^{p(\cdot)}(\mathbb{R}^n)$	$\ell^{q),\theta}_\alpha(\mathbb{Z})$	$p(\cdot) : \mathbb{R}^n \to (1,\infty),\ \ q \in (1,\infty)$, $\theta \in (0,\infty)$, and $-n\tau_{p(\cdot)} < \alpha < n\tau_{p'(\cdot)}$ with $\tau_{p(\cdot)} \precsim \Upsilon_{L^{p(\cdot)}}$ and $\tau_{p'(\cdot)} \precsim \Upsilon_{L^{p'(\cdot)}}$
$L^p(\mathbb{R}^n)$	$\ell^q_{\omega(\cdot)}(\mathbb{Z})$	$p \in (1,\infty), q \in (0,\infty]$, $-\frac{n}{p} < m_0(\omega) \leq M_0(\omega) < \frac{n}{p'}$ and $-\frac{n}{p} < m_\infty(\omega) \leq M_\infty(\omega) < \frac{n}{p'}$
$L^p(\mathbb{R}^n)$	$\ell^{\infty,q}_{\alpha,\lambda}(\mathbb{Z})$	$p \in (1,\infty), q \in (0,\infty], \lambda \in (0,\infty)$, and $-\frac{n}{p} + \lambda < \alpha < \frac{n}{p'} + \lambda$
$L^{p(\cdot)}(\mathbb{R}^n)$	$\ell^{\infty,q}_{\alpha,\lambda}(\mathbb{Z})$	$p(\cdot) : \mathbb{R}^n \to (1,\infty), q \in (0,\infty], \lambda \in (0,\infty)$ $-n\tau_{p(\cdot)} + \lambda < \alpha < n\tau_{p'(\cdot)} + \lambda$ with $\tau_{p(\cdot)} \precsim \Upsilon_{L^{p(\cdot)}}$ and $\tau_{p'(\cdot)} \precsim \Upsilon_{L^{p'(\cdot)}}$

Table 1.2: Relationship between GBFS and GQBSS through compliancy

These examples (and others) are discussed in detail in Section 23.9. Here we only wish to mention that the third column in Table 1.2 provides a clear and transparent relationship between the indices of the function spaces and sequence spaces under consideration. See also Corollaries 23.5, 23.6, 23.7, 23.8, 23.9.

Having identified and studied the key concepts that contribute to a robust theory of Composite Herz spaces, the next order of business is to determine the suitability of this scale of spaces in connection with boundary value problems. It turns out that Generalized Banach Function Spaces on which the Hardy-Littlewood maximal operator happens to be bounded are natural function analytic settings in which boundary value problems can be formulated and solved [GHA-V]. It is therefore imperative to determine whether Composite Herz spaces themselves are Generalized Banach Function Spaces.

Strictly speaking, Composite Herz spaces $\dot{\mathrm{K}}(\Sigma;\mathbb{X};\mathbb{Y})$ do not fall under the umbrella of GBFS simply because $\|\cdot\|_{\dot{\mathrm{K}}(\Sigma;\mathbb{X};\mathbb{Y})}$ may be a quasinorm, a property that is inherited from the sequence quasinorm $\|\cdot\|_{\mathbb{Y}}$. However, when both base spaces are equipped with genuine norms, the corresponding Composite Herz space does become a Generalized Banach Function Space. In precise terms, given a closed unbounded Ahlfors regular set Σ, Proposition 23.4 shows that if $\mathbb{X}$ is a Generalized Banach Function Space on (Σ,σ), and $\mathbb{Y}$ is a Generalized Banach Sequence Space on $\mathbb{Z}$, then the corresponding Composite Herz space $\dot{\mathrm{K}}(\Sigma;\mathbb{X};\mathbb{Y})$ is a Generalized Banach Function Space on (Σ,σ).

Once the GBFS nature of Composite Herz spaces has been elucidated, the next step is understanding their Köthe duals. In this regard, we succeed in identifying the associated space of a Composite Herz space with base spaces $\mathbb{X}$ and $\mathbb{Y}$ as being the Composite Herz space with base spaces $\mathbb{X}'$ and $\mathbb{Y}'$, i.e.,

$$\big(\dot{\mathrm{K}}(\Sigma;\mathbb{X};\mathbb{Y})\big)' = \dot{\mathrm{K}}(\Sigma;\mathbb{X}';\mathbb{Y}'). \tag{1.88}$$

See Proposition 23.5. This identification, together with the GBFS nature of Composite Herz spaces, and the boundedness of the Hardy-Littlewood maximal operator pave the way for treating boundary value problems for weakly elliptic systems with boundary data in Composite Herz spaces.

Let us briefly elaborate on the Dirichlet Problem. Assume $\Omega \subseteq \mathbb{R}^n$ is a UR domain with surface measure $\sigma := \mathcal{H}^{n-1}\lfloor\partial\Omega$. Suppose $\mathbb{X}$ is a Generalized Banach Function Space on $\partial\Omega$ such that the Hardy-Littlewood maximal operator $\mathcal{M}_{\partial\Omega}$ on $\partial\Omega$ satisfies

$$\begin{gathered} \mathcal{M}_{\partial\Omega} : \mathbb{X} \to \mathbb{X} \quad \text{and} \quad \mathcal{M}_{\partial\Omega} : \mathbb{X}' \to \mathbb{X}' \\ \text{are well defined and bounded.} \end{gathered} \tag{1.89}$$

In addition, assume $\mathbb{Y}$ is a Generalized Banach Sequence Space on $\mathbb{Z}$ which is $\mathbb{X}$-compliant. Consider a second-order, homogeneous, constant complex coefficient, weakly elliptic, $M \times M$ system L in $\mathbb{R}^n$. Given an aperture parameter $\kappa \in (0,\infty)$, formulate the following Dirichlet Problem:

$$\begin{cases} u \in [\mathscr{C}^\infty(\Omega)]^M, \quad Lu = 0 \ \text{ in } \Omega, \\ \mathcal{N}_\kappa u \in \dot{\mathrm{K}}(\partial\Omega;\mathbb{X};\mathbb{Y}), \\ u\big|_{\partial\Omega}^{\kappa-\text{n.t.}} = f \in [\dot{\mathrm{K}}(\partial\Omega;\mathbb{X};\mathbb{Y})]^M. \end{cases} \tag{1.90}$$

Then, assuming both L and $L^\top$ possess distinguished coefficient tensors, this problem is well posed whenever $\|\nu\|_{[\mathrm{BMO}(\partial\Omega,\sigma)]^n}$ is sufficiently small (see Theorem 23.11). Specifically, it turns out that if A is a distinguished coefficient tensor for L, then there exists some constant $C \in (0,\infty)$ depending on the "large" geometric features of Ω for which

$$\|K_A\|_{[\dot{\mathbb{K}}(\partial\Omega;\mathbb{X};\mathbb{Y})]^M\to[\dot{\mathbb{K}}(\partial\Omega;\mathbb{X};\mathbb{Y})]^M}$$
$$\leq C\cdot\|\nu\|_{[\mathrm{BMO}(\partial\Omega,\sigma)]^n}\cdot\ln\big(e/\|\nu\|_{[\mathrm{BMO}(\partial\Omega,\sigma)]^n}\big), \tag{1.91}$$

where K_A is the boundary-to-boundary double layer potential associated with A (cf. (1.6)). Having $\|\nu\|_{[\mathrm{BMO}(\partial\Omega,\sigma)]^n}$ small enough then implies $\frac{1}{2}I + K_A$ is invertible on $[\dot{\mathbb{K}}(\partial\Omega;\mathbb{X};\mathbb{Y})]^M$ via a Neumann series. In this case, the following integral representation formula to the (unique) solution of (1.90) holds:

$$u := \mathcal{D}_A\big[(\tfrac{1}{2}I + K_A)^{-1}f\big] \quad \text{in } \Omega, \tag{1.92}$$

where $\mathcal{D}_A$ is the boundary-to-domain double layer (as in (1.5)).

Remarkably, the diverse array of choices for $\mathbb{X}$ and $\mathbb{Y}$ (as detailed in Table 1.1; see also Sections 23.4 and 23.5) makes the scale of Composite Herz spaces a fruitful and inclusive functional analytic framework yielding to well-posed Dirichlet Problems (and other boundary value problems; cf. Section 23.13). Importantly, through the lens of Composite Herz spaces, a wide variety of existing Herz-type spaces in the literature can now be used for the first time in the context of boundary value problems. In closing, we note that such well-posedness results are new even when the underlying domain Ω is the upper-half space $\mathbb{R}^n_+$ and L is the Laplacian Δ.

We close this section with a discussion pertaining to the geometry of the underlying domains in this work. As described earlier, estimates of similar flavor to (1.14) and (1.30) become powerful tools to prove the existence of solutions for boundary problems. The smallness of the geometric entity in the right-hand side of said estimates has led the authors of [GHA-I]-[GHA-V] to introduce the class of δ-AR domains. More precisely, given $\delta \in (0,1)$, an open nonempty proper subset Ω of $\mathbb{R}^n$ is called a *δ-flat Ahlfors regular domain* (or simply a δ-AR *domain*) provided $\partial\Omega$ is an Ahlfors regular set for which $\mathcal{H}^{n-1}(\partial_*\Omega \setminus \partial\Omega) = 0$, and the geometric measure theoretic outward unit normal ν to Ω satisfies

$$\|\nu\|_{[\mathrm{BMO}(\partial\Omega,\sigma)]^n} < \delta, \tag{1.93}$$

where $\sigma := \mathcal{H}^{n-1}\lfloor\partial\Omega$. Significantly, demanding that δ is sufficiently small has topological and metric implications for the underlying domain. Specifically, this forces Ω to be a two-sided NTA domain, which is a connected unbounded open set, with a connected unbounded boundary, and unbounded connected complement. In particular, this ensures that Ω is a uniformly rectifiable domain (UR domain for short) with an unbounded topological boundary. Heuristically speaking,

> *the class of δ-AR domains is the sharp version, from a geometric measure theoretic point of view, of the class of upper-graph Lipschitz domains with small Lipschitz constants.* (1.94)

Observe that the upper-half space $\mathbb{R}^n_+ := \mathbb{R}^{n-1} \times (0,\infty) \subseteq \mathbb{R}^n$ is a δ-AR domain for any $\delta \in (0,1)$, since its geometric measure theoretic outward unit normal is the is constant vector $\nu = e_n := (0,\dots,0,1)$ on the boundary $\partial\mathbb{R}^n_+ \equiv \mathbb{R}^{n-1}$, and in particular $\|\nu\|_{[\mathrm{BMO}(\partial\Omega,\sigma)]^n} = 0$. The authors of [MMMi³] provide a long list of examples of δ-AR domains. To mention some, it has been shown in [MMMi³, Example 2.1, p. 86] that if $\delta \in (0, 1)$, then the infinite sector with aperture angle $\theta \in (0, 2\pi)$ given by $\Omega_\theta := \{x \in \mathbb{R}^n \setminus \{0\} : x_n/|x| > \cos(\theta/2)\}$ is a δ-AR domain if and only if $\delta > |\cos(\theta/2)|$, that is, θ is sufficiently close to π.

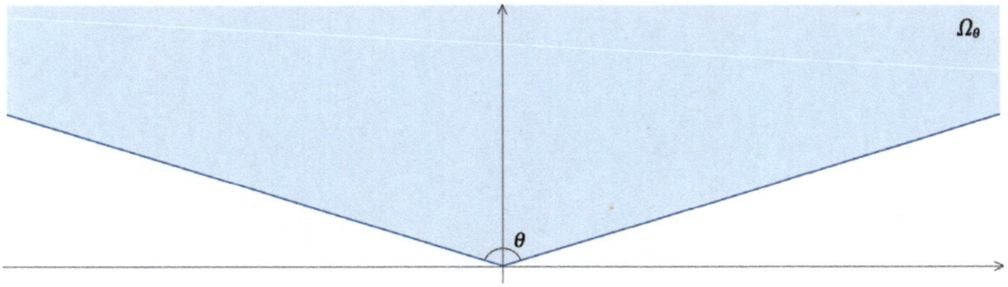

Fig. 1.4: An infinite sector Ω_θ of aperture angle θ. This is a δ-AR domain if θ is close enough to π.

Moreover, the class of upper-graph Lipschitz domains with small Lipschitz constant falls within the scope of δ-AR domains. Specifically, given $\delta \in (0, 1)$ and a Lipschitz function $\varphi : \mathbb{R}^{n-1} \to \mathbb{R}$ with $\|\nabla\varphi\|_{L^\infty(\mathbb{R}^n)} < 2^{-3/2}\delta$, the region above the graph of φ defined as $\Omega := \{(x', t) \in \mathbb{R}^{n-1} \times \mathbb{R} : t > \varphi(x')\}$ is a δ-AR domain (cf. [MMMi³, Example 2.3, p. 87]). The quantitative flatness encapsulated in (1.93) allows such domains to have "high peaks" and "deep valleys" as illustrated in the figure below:

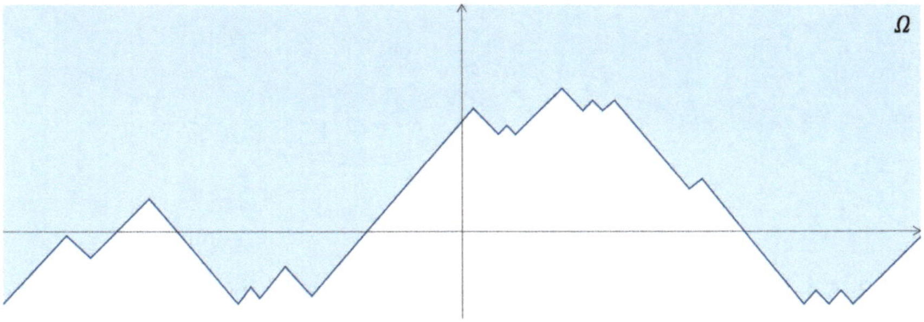

Fig. 1.5: An example of an upper-graph Lipschitz domain with a piecewise linear boundary. This is a δ-AR domain if the slopes of these lines are sufficiently small.

Remarkably, the class of δ-AR domains goes beyond the class of domains of upper-graph type. For instance, an open set Ω whose boundary $\partial\Omega$ coincides with a chord-arc curve with constant $\chi > 0$ is a δ-AR domain, provided $\delta > 2\sqrt{\chi(2+\chi)}$ (cf. [MMMi^3, Example 2.7, p. 92]). This class contains domains which fail to be of upper-graph type since they are allowed to have spiral points on their boundary (see figure below).

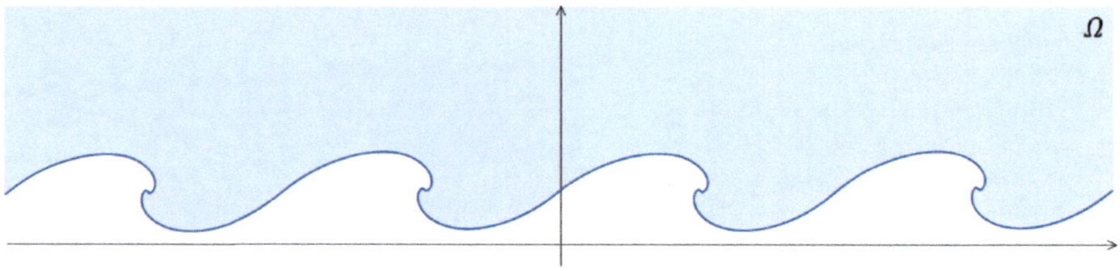

Fig. 1.6: A δ-AR domain which is not of upper-graph type.

That δ-AR domains are not necessarily upper-graph Lipschitz domains may also be seen by considering the class of BMO_1 domains (cf. [MMMi^3, Example 2.5, p. 89]). Specifically, if $\phi \in L^1_{\mathrm{loc}}(\mathbb{R}^{n-1}, \mathcal{L}^{n-1})$ has the property that $\nabla\phi$ belongs to $[\mathrm{BMO}(\mathbb{R}^{n-1}, \mathcal{L}^{n-1})]^{n-1}$ and $\|\nabla\phi\|_{[\mathrm{BMO}(\mathbb{R}^{n-1}, \mathcal{L}^{n-1})]^{n-1}} < \min\{1, \delta/C_n\}$ for some dimensional constant $C_n \in (0,\infty)$, the region $\Omega := \{(x',t) \in \mathbb{R}^{n-1} \times \mathbb{R} : t > \phi(x')\}$ above the graph of ϕ is a δ-AR domain. A concrete example of a function ϕ in BMO_1 that is not Lipschitz has been provided in [MMMi^3, Example 2.5, p. 89] and its corresponding epigraph is illustrated below.

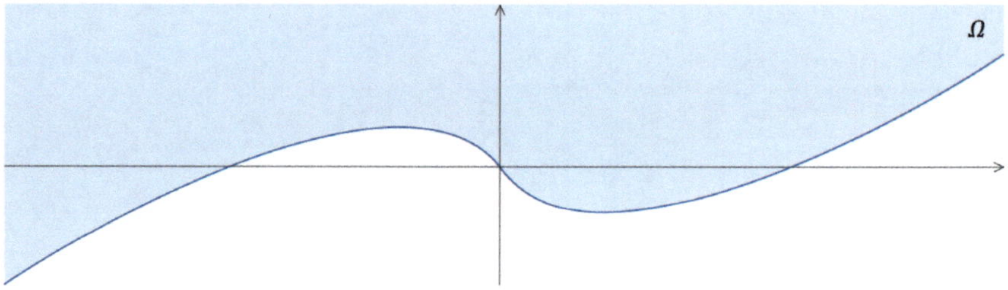

Fig. 1.7: A δ-AR domain which is the upper-graph of a non-Lipschitz BMO_1 function.

1.7 A Broader Historical Perspective

The types of boundary value problems formulated in (1.10) and (1.23) have a long and rich history, and the goal here is to place this topic into a broader context. There is an extensive and distinguished tradition when it comes to the use of layer potentials in connection with boundary value problems, going back to the pioneering work of C.F. Gauss, H. Poicaré, D. Hilbert, C. Neumann, E.I. Fredholm, and T. Carleman. The smoothness of the underlying domain and the nature of the function spaces involved in the formulation of the problem have been issues of major importance from the very outset of the theory. For example, H. Lebesgue has shown that the classical Dirichlet boundary value problem for the Laplacian

$$\begin{cases} u \in \mathscr{C}^\infty(\Omega) \cap \mathscr{C}^0(\overline{\Omega}), & \Delta u = 0 \ \text{ in } \Omega, \\ u\big|_{\partial\Omega} = f \in \mathscr{C}^0(\partial\Omega), \end{cases} \tag{1.95}$$

fails to be solvable in domains with cusps[3]. In their influential 1952 monograph on functional analysis, republished in [RiNa52], F. Riesz and B. Sz.-Nagy examine the Dirichlet and Neumann problems in potential theory in Section 81, addressing their resolution via the Fredholm method. On page 224, they mention that although this method is elegant, its applicability is constrained by stringent conditions on the domain's boundary. They attribute these limitations to the method itself, and argue that the use of boundary layer potentials requires rather regular curves or surfaces. In the planar case, Radon broadened the scope of Fredholm theory to what Riesz and Sz.-Nagy regarded at the time as its natural boundary: domains enclosed by curves of bounded rotation. Towards the conclusion of Section 91, on page 226, these authors remark that, in the spatial case, no analogue to curves with bounded rotation has yet been identified. This sentiment largely defined scholarly consensus for nearly four decades following Radon's pioneering contributions. The 1960's marked a shift, ushering in several significant developments. Researchers such as Yu.D. Burago, V.G. Maz'ya, V.D. Sapozhnikova ([BMS62], [BMS66]) and J. Král ([Kr64], [Kr65]) proposed refinements to Radon's framework. Their approach expanded the allowable class of curves and extended the theory into higher-dimensional settings by substituting the condition of "bounded rotation" with suitable demands on the variation of the angle at which the subsets of the curve are seen from an arbitrary point on it. In particular, the 1969 book of Yu.D. Burago and V.C. Maz'ya [BuMa69] and the 1980 book [Kr80] by J. Král are some of the early monographs fully devoted to potential theory in rough geometric settings.

A major breakthrough came in 1977, when A.P. Calderón [Ca77] succeeded in proving the boundedness of the Cauchy integral operator on Lipschitz curves with small constants. In 1978, E. Fabes, M. Jodeit Jr., and N.M. Rivière [FaJoRi78] succeeded in employing Calderón's result to prove the well-posedness of the Dirichlet

[3] Lebesgue's 1912 [Le13] counterexample involved what is now commonly referred to as Lebesgue spine domains.

Problem for the Laplacian in bounded $\mathscr{C}^1$ domains with data in Lebesgue spaces L^p, for arbitrary $p \in (1, \infty)$, using layer potential techniques. They also tackled the Neumann Problem and the so-called Regularity Problem. In a nutshell, the idea is as follows. Given a bounded $\mathscr{C}^1$ domain in the n-dimensional Euclidean space and a boundary point $P \in \partial\Omega$, choose a system of coordinates with P as the origin and such that the tangent plane to $\partial\Omega$ at P is $\mathbb{R}^{n-1} \times \{0\} \equiv \mathbb{R}^{n-1}$. Then, near $P \equiv 0$, the boundary of Ω may be described as the graph of a $\mathscr{C}^1$ function $\varphi : \mathbb{R}^{n-1} \to \mathbb{R}$ with $\varphi(0) = 0$ and $\nabla\varphi(0) = 0$. Since $\nabla\varphi$ is continuous, it follows that for every $\varepsilon \in (0, \infty)$, we can arrange things so that $|\nabla\varphi(x')| < \varepsilon$ for x' near $0' \in \mathbb{R}^{n-1}$. Thus, $\partial\Omega$ can be identified near $P \equiv 0$ with the upper-graph of a Lipschitz function with small Lipschitz constant, a scenario in which Calderón's result applies.

The smallness restriction on Calderón's result was one impediment to employing layer potential techniques on general Lipschitz domains (since in the above reasoning one now lacks the continuity of $\nabla\varphi$). A crucial step to overcome this was made by R. Coifman, A. McIntosh, and Y. Meyer [CoMcMe82] in 1982, when they established L^2-boundedness results for the Cauchy integral on Lipschitz curves with arbitrarily large Lipschitz constants, thus removing said restriction. We shall return to this momentarily, but for now, let us briefly elaborate B. Dahlberg's work on the Dirichlet Problems in $\mathscr{C}^1$ and Lipschitz domains based on harmonic measures techniques.

The starting point is the Poisson integral representation formula in a domain $\Omega \subseteq \mathbb{R}^n$ with outward unit normal ν and surface measure σ valid for functions $u \in \mathscr{C}^\infty(\Omega) \cap \mathscr{C}^0(\overline{\Omega})$ such that $\Delta u = 0$ in Ω:

$$u(x_0) = \int_{\partial\Omega} (\partial_\nu G^{x_0})(y)\big(u\big|_{\partial\Omega}\big)(y)\, d\sigma(y) \tag{1.96}$$

where G^{x_0} is the Green function for Δ with pole at $x_0 \in \Omega$. A sharp version of (1.96) is given in [GHA-III, Theorem 4.4, pp. 799–801]. In particular, if one prescribes $u\big|_{\partial\Omega} = f$ (as one does in the Dirichlet Problem), then

$$u(x_0) = \int_{\partial\Omega} (\partial_\nu G^{x_0})(y) f(y)\, d\sigma(y). \tag{1.97}$$

This should be compared with the harmonic measure representation of u, namely

$$u(x_0) = \int_{\partial\Omega} f(y)\, d\omega^{x_0}(y) \tag{1.98}$$

where ω^{x_0} is the harmonic measure for Ω with pole at $x_0 \in \Omega$. If Ω is a $\mathscr{C}^\infty$ smooth domain then $G^{x_0} \in \mathscr{C}^\infty(\overline{\Omega} \setminus \{x_0\})$ and (1.97)-(1.98) imply that ω^{x_0} and σ are mutually absolutely continuous and $\frac{d\omega^{x_0}}{d\sigma} = \partial_\nu G^{x_0}$. Moreover, the Hopf-Oleinik boundary point principle shows $0 < c \le \partial_\nu G^{x_0} \le C < \infty$ for some constants $c, C \in (0, \infty)$. Thus, ω^{x_0} and σ are morally the same. This relationship between the harmonic measure and the surface measure actually survives for domains with as little smoothness as $\mathscr{C}^{1,\alpha}$ for $\alpha > 0$ (see the discussion in [Ke87, p. 142]), though matters become significantly more subtle for $\mathscr{C}^1$ and Lipschitz domains.

In 1977, B. Dahlberg [Da77] showed that on a $\mathscr{C}^1$ (or even a Lipschitz) domain ω^{x_0} and σ remain absolutely continuous with respect to one another, in a quantitative scale-invariant way. Namely, the Radon-Nikodym derivative

$$\frac{d\omega^{x_0}}{d\sigma} = \partial_\nu G^{x_0} \tag{1.99}$$

satisfies a reverse Hölder condition RH_2. The latter is the main ingredient in Dahlberg's subsequent paper [Da79] where it was shown that having $\frac{d\omega^{x_0}}{d\sigma} \in RH_2$ implies the well-posedness of the L^2-Dirichlet Problem for the Laplacian in Lipschitz domains (cf. [Ke94, Theorem 1.7.3, p. 29]). By the self-improving property of reverse Hölder classes and interpolation with the L^∞-Dirichlet Problem (which is essentially the Maximum Principle), one then obtains the well-posedness of the L^p-Dirichlet Problem for $2-\varepsilon < p < \infty$, for some small $\varepsilon = \varepsilon(\Omega) \in (0,\infty)$. Here, Dahlberg also observed that if Ω is of class $\mathscr{C}^1$, one can allow the full range $1 < p < \infty$. Note that this result is in agreement with the work of Fabes-Jodeit-Rivière [FaJoRi78], albeit via a different method.

In 1980, D. Jerison and C. Kenig [JeKe80] used a novel integral formula involving the harmonic measure to recover Dahlberg's result. Subsequently, in 1981, D. Jerison and C. Kenig [JeKe81] employed Rellich identities to solve the Neumann Problem in L^2 and the Regularity Problem in L^2_1 in the setting of Lipschitz domains. Rellich identities have been used before, notably by J. Nečas in his pioneering monograph 1967 [Ne67] and also by L. Payne and H. Weinberger [PaWe54] in 1954. This being said, the use of Rellich identities in the work of Jerison and Kenig (who actually re-discovered them) marks a milestone in the development of potential theory in Lipschitz domains. For one thing, while effective for the Dirichlet Problem for the Laplacian, methods based on harmonic measures do not apply (directly) to other types of boundary conditions like Neumann, or Regularity, nor do they apply for PDE's with complex coefficients, or systems of PDE's. Thus [JeKe81] constitutes genuine progress. For another thing, the notion of Rellich identities is robust and adaptable. Over time, it will be successfully implemented in connection with boundary integral methods for elliptic PDE's in Lipschitz domains for a remarkable array of problems.

First, it was G. Verchota [Ve84] who, in 1984, figured out how to use the Rellich identities employed earlier by Jerison and Kenig in order to invert $\frac{1}{2}I + K_\Delta$ in $L^2(\partial\Omega,\sigma)$ for an arbitrary Lipschitz domain $\Omega \subseteq \mathbb{R}^n$. Here, the Coifman-McIntosh-Meyer result regarding the L^2-boundedness of the Cauchy operator on arbitrary Lipschitz curves also played a crucial role. As a by-product, this also yields the integral representation formula

$$u = \mathcal{D}_\Delta\big[(\tfrac{1}{2}I + K_\Delta)^{-1} f\big] \tag{1.100}$$

for the solution of the L^2-Dirichlet Problem for the Laplacian. In addition, [Ve84] contains the invertibility of the single layer $\mathscr{S}_\Delta : L^p(\partial\Omega,\sigma) \to L^p_1(\partial\Omega,\sigma)$ for

$1 < p \leq 2$, which is a key ingredient in the solvability of the L_1^p-Regularity Problem for the Laplacian in the Lipschitz domain $\Omega \subseteq \mathbb{R}^n$.

A major boost came from the seminal paper [DaKe87] written by B. Dahlberg and C. Kenig in 1987, where several novel techniques and procedures were first introduced. One vastly influential idea is to employ Hardy spaces as endpoints for the L^p-Neumann Problem and L_1^p-Regularity Problem corresponding to $p = 1$, and to utilize atomic and molecular decompositions in the latter settings. This further cemented connections with Harmonic Analysis. Another pioneering idea was to flatten an upper-graph Lipschitz domain at the expense of turning the Laplacian Δ into an operator of divergence form $L = \operatorname{div} A(x)\nabla$ with only bounded measurable coefficients, which opened the door for using the De Giorgi-Nash-Moser theory in this context. The picture that emerged from [DaKe87] is that, in Lipschitz domains, the L^p-Dirichlet Problem is well posed whenever $2 - \varepsilon < p < \infty$, and that the L_1^p-Regularity Problem and the L^p-Neumann Problem is well posed for $1 < p < 2 + \varepsilon$, for some $\varepsilon \in (0, 1)$ small depending on the Lipschitz character of Ω. There are accompanying integral representation formulas, and results addressing the endpoint cases: the Neumann Problem on the Hardy space H^1, and the Dirichlet Problems on the John-Nirenberg Space BMO.

Some of the aforementioned endpoint cases have been considered earlier. For example, in 1980, E. Fabes and U. Neri [FaNe80] used harmonic measure techniques and Poisson integral representation formulas to characterize harmonic functions in Lipschitz domains whose boundary values belong to the John-Nirenberg space BMO. In 1981, E. Fabes and C. Kenig [FaKe81] considered the endpoint cases $p = 1$ and $p = \infty$ by solving the Neumann Problem for the Laplacian in the Hardy space H^1, and the Dirichlet Problem for the Laplacian in the John-Nirenberg space BMO, in the context of $\mathscr{C}^1$ domains (thus prefiguring work in [DaKe87]).

In 1995, R. Brown [Bro95] extracted more out of the approach developed by Dahlberg and Kenig to accommodate the Neumann Problem on star-shaped Lipschitz domains, with data in Hardy spaces H^p for $1 - \varepsilon < p \leq 1$ for some small $\varepsilon \in (0, 1)$, and in dimension $n \geq 3$. For $n = 2$, the well-posedness of the Neumann Problem in H^p, with $\frac{2}{3} - \varepsilon < p \leq 1$ and some $\varepsilon \in (0, \frac{1}{6}]$, is due to D. Mitrea [Mi02]. In 1998, N. Kalton and M. Mitrea [KaMi98] used complex interpolation methods to solve the Neumann Problem for the Laplacian and the Lamé system of elastostatics in Lipschitz domains and boundary data in H^p for $1 - \varepsilon < p \leq 1$, with $\varepsilon \in (0, 1)$.

Since then, extensive work has been devoted to studying boundary value problems in Lipschitz domains, leading to numerous significant developments. B. Dahlberg, C. Kenig, and G. Verchota treated the L^2-Dirichlet Problem for the bi-Laplacian Δ^2 in Lipschitz domains in [DaKeVe86], and the Dirichlet, Neumann, and Regularity Problems for the Lamé system in Lipschitz domains in [DaKeVe88]. In the context of Lipschitz cylinders, R. Brown dealt with initial boundary value problems for the heat operator in [Bro89], while the parabolic version of the Lamé and Stokes systems have been considered by Z. Shen in [Sh91]. E. Fabes, C. Kenig, and G. Verchota initiated in [FaKeVe88] the study of the Stokes systems in Lipschitz domains by treating the Dirichlet, Regularity, and Neumann Problems with L^2-data. The aforementioned boundary value problems with L^p-data, as well as the Transmission Problem for the Stokes systems in Lipschitz domains, have been treated by M. Mitrea and M. Wright in [MiWr12]. Optimal L^p-solvability results for the Transmission Problems for the Laplacian in arbitrary Lipschitz domains have been obtained by L. Escauriaza and M. Mitrea in [EsMi04] via techniques based on Rellich identities and De Giorgi-Nash-Moser theory.

Continuing work initiated by A.P. Calderón in his 1954 paper [Ca54], boundary value problems for the Maxwell system in arbitrary Lipschitz domains in $\mathbb{R}^3$ have been solved by M. Mitrea in [Mi95]. Subsequently, B. Jawerth and M. Mitrea in [JaMi95] extended this to higher dimensions for the version of the Maxwell system in the language of differential forms in arbitrary Lipschitz domains. Optimal L^p-solvability results for the Maxwell system in arbitrary Lipschitz domains in $\mathbb{R}^3$ have been obtained by D. Mitrea, M. Mitrea, and J. Pipher in [MiMiPi97].

The line of work started in [DaKeVe86] in relation to the bi-Laplacian was continued by J. Pipher and G. Verchota in a string of papers [PiVe91], [PiVe92], [PiVe93], [PiVe95-I], and [PiVe95-II]. Z. Shen further contributed in this line of work in [Sh06-I].

In the case of polygonal domains, the theory acquires an even more nuanced format due to the special nature of the singularities. Results on this flavor based on Mellin transform techniques have been produced by E. Fabes, M. Jodeit, and J. Lewis [FaJoLe77] in 1977, J. Lewis [Le90], I. Mitrea [Mi99]-[Mi00], and I. Mitrea and W. Tucker [MiTu03].

Mixed boundary value problems in Lipschitz domains have been considered for the Laplacian by R. Brown in [Bro94], R. Brown and I. Mitrea in [BrMi09] for the Lamé system, and R. Brown, I. Mitrea, M. Mitrea, and M. Wright for the Stokes systems in [BrMiMiWr10]. The aforementioned works deal with the class of "creased" Lipschitz domains, and a more inclusive geometric setting has been considered by R. Brown, K. Ott, and M. Taylor [BrOtTa13], R. Brown and J. Sykes [BrSy01].

A new direction emerged from the work of D. Jerison and C. Kenig in 1995. Specifically, in [JeKe95] they obtained optimal well-posedness results for the Poisson Problem with (a homogeneous) Dirichlet boundary condition for the Laplacian, formulated in Sobolev-Besov spaces in arbitrary Lipschitz domains. Their approach is a blend of harmonic measure techniques, endpoint estimates, and interpolation. A different approach was devised by E. Fabes, O. Mendez, and M. Mitrea [FaMeMi98]. This relies on boundary layer potentials and can handle Neumann boundary conditions (thus answering the open question formulated on Problem 3.2.21 in [Ke94]). The two-dimensional case has been treated by D. Mitrea in [Mi02]. The Poisson Problem with mixed boundary conditions has been solved by I. Mitrea and M. Mitrea [MiMi07] in the class of "creased" domains. Other related results pertaining to elliptic PDE's on Sobolev-Besov spaces in non-smooth domains may be found in [MeMi00], [MiMi02], [MiMi03], [Mi04], [MaRo10], [BaMa16-I], [KoMaRo97].

A novel direction was mixing up potential theory in Lipschitz domains with differential geometry. Originating in the work of M. Mitrea and M. Taylor, starting in the late 1990's, this program has produced [MiTa1]-[MiTa7]. D. Mitrea, M. Mitrea, and M. Taylor further extended the scope of this endeavor by devising suitable Rellich identities for the Hodge-Laplacian acting on differential forms, which allow them to solve boundary value problems for this operator with absolute and relative boundary conditions in Lipschitz subdomains of Riemannian manifolds in [MiMiTa01].

There is also a large body of work on the topic of layer potential and boundary value problems in Lipschitz domains for higher-order systems including [CoGo83], [MiMi13], [BrHu01], [BaMa13], [Sh06-II]. Informative surveys and additional references may be found in [BaMa14] and [BaMa16-II].

As mentioned earlier, the work of Coifman, McIntosh, and Meyer concerning the boundedness of the Cauchy integral operator on arbitrary Lipschitz curves plays a crucial role in the study of boundary problems in rough domains. One important aspect of this theory is correlating the geometry of subsets of $\mathbb{R}^n$ with the analysis of functions and (integral) operators on such sets. In 1984, G. David [Da84] took a decisive step in this direction by showing that for a locally rectifiable curve Σ in the complex plane, the Cauchy integral operator is bounded on $L^2(\Sigma, \sigma)$ if and only if Σ is upper Ahlfors regular (see [Da84, Théorème 2, p. 173]). David's work foreshadowed the crucial role Geometric Measure Theory would eventually play in understanding the boundedness of singular integral operators. Indeed, in 1991, G. David and S. Semmes [DaSe91] introduced the concept of uniformly rectifiability (amounting to Ahlfors regularity and Big Pieces of Lipschitz Images) and showed that this fully characterizes the boundedness of singular integral operators of Calderón-Zygmund type. More precisely, David and Semmes proved that (see [DaSe91, Theorem, pp. 10–14]) given a closed set $\Sigma \subseteq \mathbb{R}^n$ which is Ahlfors regular then

> Σ is a UR set if and only if the truncated singular integral operator $T_{k,\varepsilon} f(x) := \int_{y \in \Sigma \setminus \overline{B(x,\varepsilon)}} k(x-y) f(y)\, d\mathcal{H}^{n-1}(y)$, for $x \in \Sigma$, is bounded on $L^2(\Sigma, \mathcal{H}^{n-1})$ with norm majorized by a constant independent of $\varepsilon \in (0, \infty)$, whenever the integral kernel $k \in \mathscr{C}^\infty(\mathbb{R}^n \setminus \{0\})$ is odd and satisfies $\sup_{x \in \mathbb{R}^n \setminus \{0\}} |x|^{n-1+\ell} |(\nabla^\ell k)(x)| < +\infty$ for each $\ell \in \mathbb{N}_0$. (1.101)

In relation to this, F. Nazarov, X. Tolsa and A. Volberg have proved in [NaToVo14] that

> under the background assumption of Ahlfors regularity on the closed set $\Sigma \subseteq \mathbb{R}^n$, operator norm bounds on $L^2(\Sigma, \mathcal{H}^{n-1} \lfloor \Sigma)$ for the truncated Riesz transforms on Σ (corresponding to the kernels $k_j(x) := x_j/|x|^n$ for $1 \leq j \leq n$) which are in fact uniform with respect to the truncation parameter imply Σ is actually a UR set. (1.102)

This foundational body of work was subsequently used in connection with boundary value problems, for the first time, by S. Hofmann, M. Mitrea, and M. Taylor in [HoMiTa10]. They have articulated their main objective as follows:

$$\text{“\textit{to find the optimal geometric measure theoretic context in which Fredholm theory can be successfully implemented, along the lines of its original development, for solving boundary value problems with } L^p \text{ \textit{data via the method of layer potentials [in domains with compact} UR \textit{boundaries].}”} \tag{1.103}$$

In their work, they obtained L^p-solvability results for boundary value problems (with Dirichlet, Neumann, Regularity, and Transmission boundary conditions) on a subclass of UR domains with compact boundaries, which they called SKT domains, for second-order elliptic operators including the Laplacian, the Lamé system, and the Stokes system. Their approach forged links between Harmonic Analysis, Geometric Measure Theory, and PDE's, which opened the door for further developments.

In this direction, D. Mitrea, I. Mitrea, M. Mitrea, and M. Taylor [MiMiMiTa16] studied boundary value problems with L^p-data for the Hodge-Laplacian on differential forms on regular SKT subdomains of Riemannian manifolds. G. Hoepfner, P. Liboni, D. Mitrea, I. Mitrea, and M. Mitrea [HoLiMiMiMi21] developed a Calderón-Zygmund theory on uniformly rectifiable domains for multilayer potentials as a means to studying boundary value problems for higher-order elliptic systems on bounded UR domains with data in L^p.

The techniques employed by Hofmann, Mitrea, and Taylor [HoMiTa10] rely on the compactness of layer potentials on Lebesgue spaces $L^p(\partial\Omega, \sigma)$ with $p \in (1, \infty)$ in the case when Ω is a UR domain with compact boundary whose geometric measure theoretic outward unit normal ν belongs to the Sarason space $\mathrm{VMO}(\partial\Omega, \sigma)$. This work cannot be directly applied to the case when $\partial\Omega$ is unbounded, as in this case compactness may fail. Significant progress in this regard and other related fronts has been made recently by J.J. Marín, J.M. Martell, D. Mitrea, I. Mitrea, and M. Mitrea [MMMi^3], and D. Mitrea, I. Mitrea, and M. Mitrea [GHA-I]-[GHA-V]. In the aforementioned references, the authors build a comprehensive Calderón-Zygmund theory for singular integral operators of layer potential type on uniformly rectifiable sets and use such a theory to treat a wide range of boundary value problems for weakly elliptic systems (e.g., Dirichlet, Neumann, Regularity, and Transmission Problems) in uniformly rectifiable domains with unbounded boundary, and boundary data taken from a variety of function spaces. At the core of their approach, one finds estimates for integral operators that are sensitive to "flatness" (or rather the oscillatory nature) of $\partial\Omega$, yielding invertibility results of layer potentials on many function spaces (e.g., Lebesgue spaces, Hardy spaces, Morrey spaces, among others). Following this line of work, M. Laurel and M. Mitrea [LaMi24] established well-posedness results, via layer potential methods, for boundary value problems for weakly elliptic systems on UR domains with data in Muckenhoupt weighted Morrey and Block spaces.

Other recent developments pertaining to boundary value problems and Geometric Measure Theory include work by M. Mourgoglou and X. Tolsa [MoTo24-I]-[MoTo24-II], J. Feneuil, S. Mayboroda, and Z. Zhao [FeMaZh21], and J. Azzam, S. Hofmann, J.M. Martell, S. Mayboroda, M. Mourgoglou, X. Tolsa, and A. Volberg [AzHoEtAl16].

Chapter 2
Preliminary Matters

We begin by reviewing some basic notation and conventions used throughout the monograph.

2.1 Fixing Notation

With $\mathbb{N}$ denoting the set of natural numbers, let $\mathbb{N}_0 := \mathbb{N} \cup \{0\}$. For $n \in \mathbb{N}$ with $n \geq 2$, denote by $\mathcal{L}^n$ the n-dimensional Lebesgue measure in $\mathbb{R}^n$. Also, $\mathcal{H}^s$ stands for the s-dimensional Hausdorff measure in $\mathbb{R}^n$.

For an arbitrary set E, use the symbol $\mathbb{1}_E$ to denote the characteristic function of E, i.e., $\mathbb{1}_E(x) := 1$ if $x \in E$ and $\mathbb{1}_E(x) := 0$ if $x \notin E$. Also, δ_{jk} is the Kronecker delta, that is, $\delta_{jk} := 1$ if $j = k$ and $\delta_{jk} := 0$ if $j \neq k$. For each $x \in \mathbb{R}^n$ and $r \in (0, \infty)$ abbreviate $B(x, r) := \{y \in \mathbb{R}^n : |x - y| < r\}$. The dot product of two vectors $u, v \in \mathbb{R}^n$ is denoted by $u \cdot v = \langle u, v \rangle$. Moreover, the set $\mathbb{R}^n_+ := \mathbb{R} \times (0, \infty) \subseteq \mathbb{R}^n$ denotes the upper half-space in $\mathbb{R}^n$. For any two points $x, y \in \mathbb{R}^n$, we let $[x, y]$ denote the line segment between x and y, and $\operatorname{dist}(x, E) := \inf\{|x - y| : y \in E\}$ is the distance from x to a nonempty set $E \subseteq \mathbb{R}^n$.

Consider a generic measure space (X, μ). For each $p \in (0, \infty)$ we denote by $L^p(X, \mu)$ the collection of all μ-measurable functions f on X for which

$$\|f\|_{L^p(X,\mu)} := \Big(\int_X |f|^p \, d\mu\Big)^{1/p} < \infty. \tag{2.1}$$

Corresponding to $p = \infty$, denote by $L^\infty(X, \mu)$ the collection of all μ-measurable functions f on X that are essentially bounded, i.e.,

$$\|f\|_{L^\infty(X,\mu)} := \operatorname*{ess\,sup}_X |f| < \infty. \tag{2.2}$$

© The Author(s), under exclusive license to Springer Nature Switzerland AG 2026
M. Mitrea and P. Takemura, *Singular Integrals, Herz-Type Function Spaces, and Boundary Problems*, Progress in Mathematics 362,
https://doi.org/10.1007/978-3-032-12516-3_2

Given an μ-measurable set $E \subseteq X$ with $0 < \mu(E) < \infty$, for each absolutely integrable function f on E, denote by $\fint_E f\,d\mu := \mu(E)^{-1} \int_E f\,d\mu$ the integral average of f over E. We may also adopt the convention:

$$c \cdot \infty := \begin{cases} 0 & \text{if } c = 0, \\ \infty & \text{if } 0 < c < \infty. \end{cases} \tag{2.3}$$

Next, given a topological space X equipped with a Borel measure μ, for each $p \in (0, \infty]$ we denote by $L^p_{\rm comp}(X, \mu)$ the collection of all compactly supported functions f from $L^p(X, \mu)$. Also, $L^p_{\rm loc}(X, \mu)$ stands for the space of functions that are p-th power integrable over compact sets of X.

Given an open set $\Omega \subseteq \mathbb{R}^n$, denote by $\mathscr{C}^0(\Omega)$ the collection of all continuous functions in Ω. For each $k \in \mathbb{N}$, we let $\mathscr{C}^k(\Omega)$ be the class of all continuously differentiable functions defined in Ω of order $\leq k$. The space of infinitely differentiable functions in Ω is denoted by $\mathscr{C}^\infty(\Omega)$, and $\mathscr{C}^\infty_0(\Omega)$ stands for the class of all compactly supported functions from $\mathscr{C}^\infty(\Omega)$. Define the Fourier transform of a given complex-valued Schwartz function ϕ in $\mathbb{R}^n$ as

$$\widehat{\phi}(\xi) := \int_{\mathbb{R}^n} e^{-i\langle x, \xi\rangle} \phi(x)\,dx \quad \text{for each } \xi \in \mathbb{R}^n. \tag{2.4}$$

The action of the Fourier transform can be extended to tempered distributions via duality, in the usual fashion. Denote by $\mathcal{D}'(\Omega)$ the space of distributions in Ω and, for each $p \in [1, \infty]$ and $k \in \mathbb{N}$, define the local L^p-based Sobolev space of order k in Ω as $W^{k,p}_{\rm loc}(\Omega) := \big\{u \in \mathcal{D}'(\Omega) : \partial^\alpha u \in L^p_{\rm loc}(\Omega, \mathcal{L}^n),\ |\alpha| \leq k\big\}$.

Throughout, whenever A and B are two positive quantities we write $A \lesssim B$ if there exists some constant $C \in (0, \infty)$ for which $A \leq CB$, while $A \approx B$ means that there exist constants $C, c \in (0, \infty)$ such that $cB \leq A \leq CB$.

2.2 Geometric Measure Theory

Consider an arbitrary nonempty open subset Ω of $\mathbb{R}^n$ along with an aperture parameter $\kappa \in (0, \infty)$. For each $x \in \partial\Omega$, define the `nontangential approach region` (with vertex at x) as

$$\Gamma_\kappa(x) := \big\{y \in \Omega : |y - x| < (1 + \kappa)\,\mathrm{dist}\,(y, \partial\Omega)\big\}. \tag{2.5}$$

Use this region to define the `nontangential maximal operator` $\mathcal{N}_\kappa$, whose action on each $\mathcal{L}^n$-measurable function $u : \Omega \to \mathbb{C}$ is given by

$$\big(\mathcal{N}_\kappa u\big)(x) := \|u\|_{L^\infty(\Gamma_\kappa(x), \mathcal{L}^n)} \quad \text{for each } x \in \partial\Omega. \tag{2.6}$$

We agree to use the convention $(\mathcal{N}_\kappa u)(x) := 0$ whenever $x \in \partial\Omega$ is such that $\Gamma_\kappa(x) = \varnothing$. In general, whenever $E \subseteq \Omega$ is an $\mathcal{L}^n$-measurable set, and u is an $\mathcal{L}^n$-measurable function in Ω, we denote by $\mathcal{N}_\kappa^E$ the nontangential maximal operator restricted to E, whose action on u is defined according to

$$(\mathcal{N}_\kappa^E u)(x) := \|u\|_{L^\infty(\Gamma_\kappa(x)\cap E,\mathcal{L}^n)} \quad \text{for each } x \in \partial\Omega. \tag{2.7}$$

In particular, $\mathcal{N}_\kappa^E u = \mathcal{N}_\kappa(u \cdot \mathbb{1}_E)$. Moreover, we denote by $\mathcal{N}_\kappa^\delta$ the nontangential maximal operator corresponding to $E := \{x \in \Omega : \operatorname{dist}(x, \partial\Omega) < \delta\}$ for some $\delta \in (0, \infty)$, i.e.,

$$\mathcal{N}_\kappa^\delta u := \mathcal{N}_\kappa^{O_\delta} u = \mathcal{N}_\kappa\big(u \cdot \mathbb{1}_{O_\delta}\big) \quad \text{where } O_\delta := \{x \in \Omega : \operatorname{dist}(x, \partial\Omega) < \delta\}. \tag{2.8}$$

Work done in [GHA-I, Proposition 8.2.3, p. 684] ensures that if u is an $\mathcal{L}^n$-measurable function defined in an open set $\Omega \subseteq \mathbb{R}^n$, with $\sigma := \mathcal{H}^{n-1}\lfloor\partial\Omega$, then for any aperture parameter $\kappa \in (0, \infty)$, the function $\mathcal{N}_\kappa u : \partial\Omega \to [0, \infty]$ is lower-semicontinuous, hence σ-measurable.

Moving on, consider $x \in \partial\Omega$ such that $x \in \overline{\Gamma_\kappa(x)}$, i.e., x belongs to the closure of the nontangential approach region $\Gamma_\kappa(x)$. Suppose that $u : \Omega \to \mathbb{C}^M$ is a vector-valued $\mathcal{L}^n$-measurable function in Ω. The nontangential limit of u at x exists, and its value is the vector $a \in \mathbb{C}^M$, provided

$$\begin{aligned} &\text{for every } \varepsilon > 0 \text{ there exists some } r > 0 \text{ such that } |u(y) - a| < \varepsilon \\ &\text{for } \mathcal{L}^n\text{-a.e. point } y \in \Gamma_\kappa(x) \cap B(x, r). \end{aligned} \tag{2.9}$$

In this case, we denote by $\big(u\big|_{\partial\Omega}^{\kappa-\text{n.t.}}\big)(x)$ the nontangential limit of u at x from within $\Gamma_\kappa(x)$, whenever it exists. It is visible from definitions that whenever the latter exists we have

$$\Big|\big(u\big|_{\partial\Omega}^{\kappa-\text{n.t.}}\big)(x)\Big| \le \big(\mathcal{N}_\kappa^\delta u\big)(x) \le \big(\mathcal{N}_\kappa u\big)(x) \quad \text{for all } \delta > 0. \tag{2.10}$$

Next, assume that $\Omega \subseteq \mathbb{R}$ is an $\mathcal{L}^n$-measurable set. The geometric measure theoretic boundary of Ω is defined as

$$\begin{aligned} \partial_*\Omega := \Big\{ x \in \partial\Omega : \limsup_{r\to 0^+} \frac{\mathcal{L}^n\big(\Omega \cap B(x,r)\big)}{r^n} > 0, \\ \text{and} \quad \limsup_{r\to 0^+} \frac{\mathcal{L}^n\big(B(x,r)\setminus\Omega\big)}{r^n} > 0 \Big\}. \end{aligned} \tag{2.11}$$

Recall that Ω is said to be a set of locally finite perimeter provided

$$\mathcal{H}^{n-1}\big(\partial_*\Omega \cap K\big) < +\infty \quad \text{for each compact } K \subseteq \mathbb{R}^n. \tag{2.12}$$

Equivalently, Ω is of locally finite perimeter if

$$\mu_\Omega := \nabla \mathbb{1}_\Omega \text{ is a } \mathbb{R}^n\text{-valued Borel measure in } \mathbb{R}^n \text{ of} \\ \text{locally finite total variation,} \tag{2.13}$$

where the gradient is taken in the sense of distributions. In this context, work done by De Giorgi-Federer ensures there exists some $\mathbb{R}^n$-valued function

$$\nu \in \left[L^\infty(\partial_*\Omega, \mathcal{H}^{n-1})\right]^n \text{ satisfying} \\ |\nu(x)| = 1 \text{ at } \mathcal{H}^{n-1}\text{-a.e. point } x \in \partial_*\Omega, \tag{2.14}$$

and with the property that

$$\mu_\Omega = \nabla \mathbb{1}_\Omega = -\nu |\nabla \mathbb{1}_\Omega|, \tag{2.15}$$

where $|\nabla \mathbb{1}_\Omega|$ is the total variation measure of $\nabla \mathbb{1}_\Omega$, i.e., $|\nabla \mathbb{1}_\Omega| = \mathcal{H}^{n-1} \lfloor \partial_*\Omega$. We refer to this function ν as the `geometric measure theoretic outward unit normal` to Ω .

Definition 2.1 A set $\Sigma \subseteq \mathbb{R}^n$ is called `Ahlfors regular` (or `Ahlfors-David regular`) provided Σ is a closed set in $\mathbb{R}^n$, and there exist constants $C_\Sigma, c_\Sigma \in (0, \infty)$ with the property that for each $x \in \Sigma$, and each $R \in \big(0, 2\,\mathrm{diam}(\Sigma)\big)$ one has

$$c_\Sigma R^{n-1} \le \mathcal{H}^{n-1}\big(B(x,R) \cap \Sigma\big) \le C_\Sigma R^{n-1}, \tag{2.16}$$

where $\mathcal{H}^{n-1}$ stands for the $(n-1)$-dimensional Hausdorff measure in $\mathbb{R}^n$. The constants c_Σ and C_Σ are called the lower and upper Ahlfors regularity constant of Σ, respectively.

Given a closed set $\Sigma \subseteq \mathbb{R}^n$, a location $x \in \Sigma$, and a radius $r \in \big(0, 2\,\mathrm{diam}(\Sigma)\big)$, we agree to abbreviate $\Delta(x, r) := \Sigma \cap B(x, r)$, the `surface ball` on Σ with center x and radius r. In essence, a closed set $\Sigma \subseteq \mathbb{R}^n$ being Ahlfors regular means that Σ behaves like a $(n-1)$-dimensional surface in $\mathbb{R}^n$ in a quantitative, scale-invariant fashion, with respect to $\mathcal{H}^{n-1}$.

The notion of Ahlfors regular domain introduced below is due to S. Hofmann, M. Mitrea, and M. Taylor (see [HoMiTa10]).

Definition 2.2 Consider an open, nonempty, and proper subset Ω of $\mathbb{R}^n$. Call Ω an `Ahlfors regular domain` (`AR domain` for short) provided $\partial\Omega$ is an Ahlfors regular set (in the sense of Definition 2.1) and $\mathcal{H}^{n-1}(\partial\Omega \setminus \partial_*\Omega) = 0$.

In light of the discussion pertaining to (2.11)-(2.15), the demand that the geometric measure theoretic boundary $\partial_*\Omega$ of Ω has full measure (with respect to $\mathcal{H}^{n-1}$) in the topological boundary $\partial\Omega$ in Definition 2.2 ensures that the geometric measure theoretic outward unit normal (cf. (2.14)) to an Ahlfors regular domain Ω is well defined $\mathcal{H}^{n-1}$-a.e. on $\partial\Omega$.

We next recall the notion of uniformly rectifiability, first introduced by G. David and S. Semmes in [DaSe93].

Definition 2.3 A closed set $\Sigma \subseteq \mathbb{R}^{n-1}$ is said to be a `uniformly rectifiable set` (or simply `UR set`) provided Σ is an Ahlfors regular set and it has "big pieces of Lipschitz images" in a uniform, quantitative, and scale-invariant fashion, i.e., there exist two constants $\varepsilon, M \in (0, \infty)$ such that for each $x \in \Sigma$ and each $R \in \big(0, 2\,\mathrm{diam}(\Sigma)\big)$ one can find a Lipschitz function $\varphi : B_R^{n-1} \to \mathbb{R}^n$, with Lipschitz constant $\leq M$, satisfying

$$\mathcal{H}^{n-1}\big(\Sigma \cap B(x, R) \cap \varphi(B_R^{n-1})\big) \geq \varepsilon \cdot R^{n-1} \tag{2.17}$$

where B_R^{n-1} is a ball in $\mathbb{R}^{n-1}$ with radius R.

One of the main achievements in [DaSe91] is identifying uniformly rectifiable sets as the optimal setting where one can develop a robust Calderón-Zygmund theory for singular integral operators.

The notion of a uniformly rectifiable domain, due to S. Hofmann, M. Mitrea, and M. Taylor (see [HoMiTa10]) plays a fundamental role in this work.

Definition 2.4 An open, nonempty, proper subset Ω of $\mathbb{R}^n$ is called a `uniformly rectifiable domain` (`UR domain` for short) if its topological boundary $\partial\Omega$ is a UR set (in the sense of Definition 2.3) and $\mathcal{H}^{n-1}(\partial\Omega \setminus \partial_*\Omega) = 0$.

The demand $\mathcal{H}^{n-1}(\partial\Omega \setminus \partial_*\Omega) = 0$ on Definition 2.4 ensures that the geometric measure theoretic outward unit normal to Ω is meaningfully defined $\mathcal{H}^{n-1}$-a.e. on $\partial\Omega$. In addition, observe that every UR domain is also an Ahlfors regular domain.

The following definition may be regarded as a quantitative openness condition.

Definition 2.5 Fix two numbers $R \in (0, \infty]$ and $c \in (0, 1)$, and consider a nonempty proper subset Ω of $\mathbb{R}^n$. The subset Ω is said to satisfy the (R, c)-`corkscrew condition` (or, simply, a `corkscrew condition` if the particular values of R, c are not relevant) if for each location $x \in \partial\Omega$ and each scale $r \in (0, R)$ there exists some $z \in \Omega$ (called a corkscrew point relative to x and r) for which $B(z, c\,r) \subseteq B(x, r) \cap \Omega$.

Moreover, the subset Ω is said to satisfy the (R, c)-`two-sided corkscrew condition` provided both Ω and $\mathbb{R}^n \setminus \Omega$ satisfy the (R, c)-corkscrew condition (with the same convention regarding the omission of R, c).

It is evident from definitions that whenever Ω satisfies a two-sided corkscrew condition, the geometric measure theoretic boundary of Ω coincides with its topological boundary, i.e., $\partial_*\Omega = \partial\Omega$. Work in [DaJe90] implies that, in a quantitative fashion,

> whenever $\Omega \subseteq \mathbb{R}^n$ is a nonempty proper open subset that satisfies a two-sided corkscrew condition and whose boundary is an Ahlfors regular set, then Ω is actually a UR domain. (2.18)

The next definition may be regarded as a quantitative notion of local connectivity for an open set.

Definition 2.6 Pick $R \in (0, \infty]$ and $N \in \mathbb{N}$. An open set $\Omega \subseteq \mathbb{R}^n$ is said to satisfy the (R, N)-`Harnack chain condition` (or, simply, a `Harnack chain condition` if the values of R, N are not important) provided

$$\begin{aligned} &\text{whenever } \varepsilon > 0,\ k \in \mathbb{N},\ z \in \partial\Omega, \text{ and } x, y \in \Omega \text{ with} \\ &\max\big\{|x-z|, |y-z|\big\} < R/4 \text{ as well as } |x-y| \leq 2^k\varepsilon \\ &\text{and } \min\big\{\operatorname{dist}(x, \partial\Omega), \operatorname{dist}(y, \partial\Omega)\big\} \geq \varepsilon, \end{aligned} \tag{2.19}$$

one may find a chain of balls $B_1, B_2, \dots, B_K$ with $K \leq Nk$, such that $x \in B_1$, $y \in B_K$, $B_i \cap B_{i+1} \neq \varnothing$ for each $i \in \{1, \dots, K-1\}$, and

$$N^{-1} \cdot \operatorname{diam}(B_i) \leq \operatorname{dist}(B_i, \partial\Omega) \leq N \cdot \operatorname{diam}(B_i), \tag{2.20}$$

$$\operatorname{diam}(B_i) \geq N^{-1} \cdot \min\big\{\operatorname{dist}(x, B_i), \operatorname{dist}(y, B_i)\big\}, \tag{2.21}$$

for each $i \in \{1, \dots, K\}$.

As observed in [MMMi^3, (2.52), p. 38], whenever Ω is an open set satisfying an (∞, N)-Harnack chain condition (for some $N \in \mathbb{N}$), it follows that Ω is pathwise connected (hence also connected) in a quantitative fashion.

The relevance of the corkscrew condition and the Harnack chain condition is that these are used to define the class of NTA domains, first introduced in [JeKe82].

Definition 2.7 Fix two values $R \in (0, \infty]$ and $N \in \mathbb{N}$. An open, nonempty, proper subset Ω of $\mathbb{R}^n$ is called an (R, N)-`nontangentially accessible domain` (or simply an NTA `domain` if the particular values of R, N are irrelevant) if Ω satisfies both the (R, N^{-1})-two-sided corkscrew condition and the (R, N)-Harnack chain condition.

Call Ω an (R, N)-`two-sided nontangentially accessible domain` (or, simply, a `two-sided NTA domain` if the particular values of R, N are unimportant) provided Ω is an open, nonempty, proper subset of $\mathbb{R}^n$ such that both Ω and $\mathbb{R}^n \setminus \overline{\Omega}$ are (R, N)-nontangentially accessible domains.

A set $\Omega \subseteq \mathbb{R}^n$ is said to be an (R, N)-`one-sided` NTA `domain` provided Ω satisfies the (R, N)-Harnack chain condition and the (R, N^{-1})-corkscrew condition (once again, with the convention that the parameters R, N are dropped if theirs values are not relevant).

Finally, it is agreed that, in all cases, one takes $R = \infty$ if $\partial\Omega$ is unbounded.

Next, we record a quantitative notion of local path connectivity, first introduced by S. Hofmann, M. Mitrea, and M. Taylor in [HoMiTa10].

Definition 2.8 An open, nonempty, proper subset Ω of $\mathbb{R}^n$ is said to satisfy a `local John condition` if there exist two parameters, $\theta \in (0, 1)$ and $R \in (0, \infty]$ (with the requirement that $R = \infty$ if $\partial\Omega$ is unbounded) such that for every location $x \in \partial\Omega$

and every scale $r \in (0, R)$ one may find $x_r \in B(x,r) \cap \Omega$ with the property that $B(x_r, \theta r) \subseteq \Omega$ and such that for each $y \in B(x,r) \cap \partial\Omega$ there exists a rectifiable path $\gamma_y : [0,1] \to \overline{\Omega}$ whose length is $\leq \theta^{-1} r$ and satisfies

$$\gamma_y(0) = y, \quad \gamma_y(1) = x_r, \quad \operatorname{dist}\big(\gamma_y(t), \partial\Omega\big) > \theta|\gamma_y(t) - y| \text{ for all } t \in (0,1]. \tag{2.22}$$

Finally, a nonempty open set $\Omega \subseteq \mathbb{R}^n$ which is not dense in $\mathbb{R}^n$ is said to satisfy a `two-sided local John condition` if both Ω and $\mathbb{R}^n \setminus \overline{\Omega}$ satisfy a local John condition.

From the definitions, it is visible that, quantitatively,

> any set satisfying a local John condition (respectively, a two-sided local John condition) also satisfies a corkscrew condition (respectively, a two-sided corkscrew condition), (2.23)

In addition, given $R \in (0, \infty]$ and $N \in \mathbb{N}$, work done in [HoMiTa10, Lemma 3.13, p. 2634] ensures that

> any (R, N)-nontangentially accessible domain satisfies a local John condition, and any (R, N)-two-sided nontangentially accessible domain satisfies a two-sided local John condition. (2.24)

2.3 Some Tools from Harmonic Analysis

In this section, we collect some definitions and results from Harmonic Analysis. First, we discuss the Hardy-Littlewood maximal operator on Ahlfors regular sets and some of its properties. We then recall the classical John-Nirenberg space, and the class of δ-flat AR domains (in the sense of [MMMi^3]). Lastly, we record some basic properties of Muckenhoupt weights on Ahlfors regular sets, which are going to play an important role in subsequent chapters.

2.3.0.1 The Hardy-Littlewood Maximal Operator

Consider an Ahlfors regular set $\Sigma \subseteq \mathbb{R}^n$ and abbreviate $\sigma := \mathcal{H}^{n-1}\lfloor\Sigma$. Also, fix a σ-measurable function $f : \Sigma \to \mathbb{R}$. The action of the `(non-centered) Hardy-Littlewood maximal operator` M_Σ on f is defined as

$$(M_\Sigma f)(x) := \sup_{\Delta \ni x} \fint_\Delta |f|\, d\sigma \quad \text{at each } x \in \Sigma, \tag{2.25}$$

where the supremum is taken over all surface balls $\Delta \subseteq \Sigma$ containing x. Moreover, the (centered) Hardy-Littlewood maximal operator M_Σ acts on f via

$$(\mathcal{M}_\Sigma f)(x) := \sup_{R>0} \fint_{\Delta(x,R)} |f|\, d\sigma \quad \text{at each } x \in \Sigma. \tag{2.26}$$

A standard argument shows that

$$(M_\Sigma f)(x) \approx (\mathcal{M}_\Sigma f)(x) \quad \text{uniformly in } x \in \Sigma \tag{2.27}$$

with proportionality constants depending only on the dimension n and the Ahlfors regularity character of Σ.

For a fixed exponent $s \in (0, \infty)$, define the action of the L^s-based Hardy-Littlewood maximal operator $\mathcal{M}_{\Sigma,s}$ on a σ-measurable function f on Σ as

$$(\mathcal{M}_{\Sigma,s} f)(x) := \sup_{R>0} \Big(\fint_{\Delta(x,R)} |f|^s\, d\sigma \Big)^{1/s} \quad \text{at each } x \in \Sigma. \tag{2.28}$$

It is clear that corresponding $s = 1$, we have $\mathcal{M}_{\Sigma,1} = \mathcal{M}_\Sigma$.

It is useful to recall a result implied by [GHA-I, Proposition 7.6.8, p. 632] regarding pointwise estimates for the L^s-based Hardy-Littlewood maximal operator.

Proposition 2.1 *Consider $\Sigma \subseteq \mathbb{R}^n$ a closed Ahlfors regular set, and abbreviate $\sigma := \mathcal{H}^{n-1}\lfloor\Sigma$. Also, fix an exponent $s \in (0, \infty)$. Then*

$$\begin{gathered} \big(\mathcal{M}_{\Sigma,s}(\mathbb{1}_{\Delta(x_0,R)})\big)(x) \approx \frac{R^{(n-1)/s}}{(R + |x - x_0|)^{(n-1)/s}} \\ \textit{uniformly for } x_0, x \in \Sigma, \textit{ and } R \in (0, \infty). \end{gathered} \tag{2.29}$$

In particular, corresponding to $s := 1$ one has

$$\begin{gathered} \big(\mathcal{M}_\Sigma(\mathbb{1}_{\Delta(x_0,R)})\big)(x) \approx \frac{R^{n-1}}{(R + |x - x_0|)^{n-1}} \\ \textit{uniformly for } x_0, x \in \Sigma, \textit{ and } R \in (0, \infty). \end{gathered} \tag{2.30}$$

Moreover, corresponding to $R := 1$ one has

$$\begin{gathered} \big(\mathcal{M}_\Sigma(\mathbb{1}_{\Delta(x_0,1)})\big)(x) \approx \frac{1}{1 + |x - x_0|^{n-1}} \\ \textit{uniformly for } x_0, x \in \Sigma. \end{gathered} \tag{2.31}$$

Another valuable tool is the Fefferman-Stein Maximal Inequality, which is stated below in the context of Ahlfors regular sets. A more general version of this in the setting of spaces of homogeneous type may be found in [GHA-I, Proposition 7.6.7, p. 628].

Proposition 2.2 *Consider $\Sigma \subseteq \mathbb{R}^n$ a closed Ahlfors regular set, and abbreviate $\sigma := \mathcal{H}^{n-1}\lfloor\Sigma$. Then for every $s \in (0,1)$ there exists a constant $C_s \in (0,\infty)$ that depends only on n, s, and the Ahlfors regularity character of Σ with the property that for all σ-measurable functions $f, g : \Sigma \to [0,\infty]$ one has*

$$\int_\Sigma (\mathcal{M}_{\Sigma,s} f) g \, d\sigma \le C_s \int_\Sigma f (\mathcal{M}_\Sigma g) \, d\sigma. \tag{2.32}$$

2.3.0.2 The John-Nirenberg Space on Ahlfors Regular Sets

We begin by recalling the John-Nirenberg space of functions of bounded mean oscillations. Let $\Sigma \subseteq \mathbb{R}^n$ be a closed Ahlfors regular set, and denote $\sigma := \mathcal{H}^{n-1}\lfloor\Sigma$. For each $f \in L^1_{\text{loc}}(\Sigma, \sigma)$ define

$$\|f\|_{\text{BMO}(\Sigma,\sigma)} := \sup_{\Delta \subseteq \Sigma} ⨍_\Delta |f - f_\Delta| \, d\sigma \in [0,\infty] \tag{2.33}$$

where the supremum is taken over all surface balls $\Delta \subseteq \Sigma$, and $f_\Delta := ⨍_\Delta f \, d\sigma$. The `John-Nirenberg space` on Σ is defined as

$$\text{BMO}(\Sigma, \sigma) := \big\{ f \in L^1_{\text{loc}}(\Sigma, \sigma) : \|f\|_{\text{BMO}(\Sigma,\sigma)} < \infty \big\}. \tag{2.34}$$

A remarkable consequence of the John-Nirenberg inequality is that size in the space $\text{BMO}(\Sigma, \sigma)$ can be characterized in terms of L^p-averages. More precisely, given an integrability exponent $p \in [1,\infty)$, and a function $f \in L^1_{\text{loc}}(\Sigma, \sigma)$ introduce

$$\|f\|_{\text{BMO}_p(\Sigma,\sigma)} := \sup_{\Delta \subseteq \Sigma} \left(⨍_\Delta |f - f_\Delta|^p \, d\sigma \right)^{1/p} \tag{2.35}$$

where the supremum is taken over all surface balls $\Delta \subseteq \Sigma$. Then for each integrability exponent $p \in [1,\infty)$ one has

$$\|f\|_{\text{BMO}(\Sigma,\sigma)} \approx \|f\|_{\text{BMO}_p(\Sigma,\sigma)} \quad \text{for all } f \in L^1_{\text{loc}}(\Sigma, \sigma), \tag{2.36}$$

with implicit proportionality constants depending only on n, p, and the Ahlfors regularity constants of Σ. Hence, a function $f \in L^1_{\text{loc}}(\Sigma, \sigma)$ belongs to $\text{BMO}(\Sigma, \sigma)$ if and only if $\|f\|_{\text{BMO}_p(\Sigma,\sigma)} < \infty$. In particular, the membership of f to $\text{BMO}(\Sigma, \sigma)$ is independent of the integrability exponent p used to define the integral averages in (2.35).

With this definition in place, we next introduce the class of δ-AR domains.

Definition 2.9 Fix a parameter $\delta > 0$. A nonempty, proper subset Ω of $\mathbb{R}^n$ is called a δ-`flat Ahlfors regular domain` (or δ-`flat AR domain`, or simply δ-`AR domain`) provided Ω is an Ahlfors regular domain (in the sense of Definition 2.2) whose geometric measure theoretic outward unit normal ν satisfies (with $\sigma := \mathcal{H}^{n-1}\lfloor\partial\Omega$)

$$\|\nu\|_{[\mathrm{BMO}(\partial\Omega,\sigma)]^n} < \delta. \tag{2.37}$$

It has been shown in [MMMi³, Lemma 2.8, p. 52] that, in the class of Ahlfors regular domains, one always has $\|\nu\|_{[\mathrm{BMO}(\partial\Omega,\sigma)]^n} \leq 1$. As such, when $\delta > 1$, condition (2.37) becomes redundant. For this reason, we will be mainly interested when δ is sufficiently small, say $\delta \in (0, 1)$.

2.3.0.3 Muckenhoupt Weights on Ahlfors Regular Sets

Assume that $\Sigma \subseteq \mathbb{R}^n$ is a closed Ahlfors regular set, and abbreviate $\sigma := \mathcal{H}^{n-1}\lfloor\Sigma$. Let w be a `weight` on Σ, i.e., a σ-measurable function satisfying $0 < w(x) < \infty$ for σ-a.e. point $x \in \Sigma$. We agree to also use the symbol w for the weighted measure $w\,\sigma$, i.e.,

$$w(E) := \int_E w\,d\sigma \quad \text{for each } \sigma\text{-measurable set } E \subseteq \Sigma. \tag{2.38}$$

Given an integrability exponent $p \in (1, \infty)$, we say that the weight w belongs to the `Muckenhoupt class` $A_p(\Sigma, \sigma)$ if

$$[w]_{A_p} := \sup_{\Delta\subseteq\Sigma}\left(\fint_\Delta w(x)\,d\sigma(x)\right)\left(\fint_\Delta w(x)^{1-p'}\,d\sigma(x)\right)^{p-1} < \infty, \tag{2.39}$$

where$p' \in (1, \infty)$ is the Hölder conjugate exponent of p (i.e., $1/p + 1/p' = 1$) and the supremum runs over all surface balls Δ on Σ.

Corresponding to the end-point case $p = 1$, we say that $w \in A_1(\Sigma, \sigma)$ if

$$[w]_{A_1} := \sup_{\Delta\subseteq\Sigma}\left(\operatorname*{ess\,inf}_{x\in\Delta} w(x)\right)^{-1}\left(\fint_\Delta w\,d\sigma\right) < \infty. \tag{2.40}$$

In relation to this, observe that

a weight w on Σ belongs to $A_1(\Sigma, \sigma)$ if and only if there exists some constant $C \in (0, \infty)$ with the property that for every surface ball $\Delta \subseteq \Sigma$ one has $\fint_\Delta w\,d\sigma \leq Cw(x)$ at σ-a.e. $x \in \Delta$, and the best constant is actually $[w]_{A_1}$. (2.41)

Moreover, with $\mathcal{M}_\Sigma$ denoting the Hardy-Littlewood maximal operator on Σ, the following equivalence also holds:

a weight w on Σ belongs to $A_1(\Sigma, \sigma)$ if and only if there exists a constant $C \in (0, \infty)$ with the property that $\mathcal{M}_\Sigma w(x) \leq Cw(x)$ at σ-a.e. point $x \in \Sigma$, and the best constant is actually $[w]_{A_1}$.
(2.42)

Corresponding to $p = \infty$, define the class

$$A_\infty(\Sigma, \sigma) := \bigcup_{p \in [1,\infty)} A_p(\Sigma, \sigma). \tag{2.43}$$

Whenever $p \in [1, \infty)$, refer to $[w]_{A_p}$ as the A_p Muckenhoupt characteristic of w.

Consider the Muckenhoupt weighted Lebesgue space $L^p(\Sigma, w) := L^p(\Sigma, w\sigma)$ with $p \in (1, \infty)$ and $w \in A_p(\Sigma, \sigma)$. A classical result is that given a weight w on Σ along with an integrability exponent $p \in (1, \infty)$, the Hardy-Littlewood maximal operator

$$\mathcal{M}_\Sigma \text{ is bounded on } L^p(\Sigma, w) \text{ if and only if } w \in A_p(\Sigma, \sigma). \tag{2.44}$$

In this case, there exists some constant $C = C(\Sigma, n, p) \in (0, \infty)$ with the property that

$$\|\mathcal{M}_\Sigma f\|_{L^p(\Sigma,w)} \leq C[w]_{A_p}^{1/(p-1)} \|f\|_{L^p(\Sigma,w)} \text{ for each } f \in L^p(\Sigma, w) \tag{2.45}$$

(see, e.g., [HyKa12, Proposition 7.13]). Corresponding to the end-point case $p = 1$, the operator $\mathcal{M}_\Sigma$ satisfies the weak-type $(1, 1)$ inequality

$$\begin{aligned} &\sup\nolimits_{0<\lambda<\infty} \lambda \cdot w\big(\{x \in \Sigma : \mathcal{M}_\Sigma f(x) > \lambda\}\big) \leq C\|f\|_{L^1(\Sigma,w)} \\ &\text{for all } f \in L^1(\Sigma, w), \text{ with } C \in (0, \infty) \text{ independent of } f, \end{aligned} \tag{2.46}$$

if and only if $w \in A_1(\Sigma, \sigma)$.

The following proposition summarizes various properties of Muckenhoupt weights on Ahlfors regular sets.

Proposition 2.3 *Consider a closed Ahlfors regular set $\Sigma \subseteq \mathbb{R}^n$, and abbreviate $\sigma := \mathcal{H}^{n-1}\lfloor\Sigma$. Then the following properties hold.*

(1) [Openness/Self-Improving] If $w \in A_p(\Sigma, \sigma)$ with $p \in (1, \infty)$ then there exist some $\tau \in (1, \infty)$ and some $\varepsilon \in (0, p - 1)$ (both of which depend only on p, $[w]_{A_p}$, n, and the Ahlfors regularity constant of Σ) such that

$$w^\tau \in A_p(\Sigma, \sigma) \ \text{ and } \ w \in A_{p-\varepsilon}(\Sigma, \sigma). \tag{2.47}$$

In addition, both $\left[w^\tau\right]_{A_p}$ and $[w]_{A_{p-\varepsilon}}$ are controlled in terms of p, $[w]_{A_p}$, n, and the Ahlfors regularity constant of Σ. In fact, matters may be arranged so that, in a quantitative fashion,

$$w^\theta \in A_q(\Sigma, \sigma) \ \text{ for each } \ \theta \in (\tau^{-1}, \tau) \ \text{ and } \ q \in (p - \varepsilon, \infty). \tag{2.48}$$

(2) [Monotonicity] If $1 \leq p \leq q \leq \infty$ then $A_p(\Sigma, \sigma) \subseteq A_q(\Sigma, \sigma)$ and if $q < \infty$ then $[w]_{A_q} \leq [w]_{A_p}$ for each $w \in A_p(\Sigma, \sigma)$.

(3) [*Dual Weights*] *Given any* $w \in A_p(\Sigma, \sigma)$ *with* $p \in (1, \infty)$, *it follows that* $w^{1-p'}$ *belongs to* $A_{p'}(\Sigma, \sigma)$ *and* $[w^{1-p'}]_{A_{p'}} = [w]_{A_p}^{1/(p-1)}$, *where* $p' \in (1, \infty)$ *is the Hölder conjugate exponent of* p.

(4) [*Products/Factorization*] *If* $w_1, w_2 \in A_1(\Sigma, \sigma)$ *then for every* $p \in (1, \infty)$ *one has* $w_1 \cdot w_2^{1-p} \in A_p(\Sigma, \sigma)$ *and* $[w_1 \cdot w_2^{1-p}]_{A_p} \leq [w_1]_{A_1} \cdot [w_2]_{A_1}^{p-1}$. *Also, given any two weights* $w_1, w_2 \in A_p(\Sigma, \sigma)$ *with* $p \in (1, \infty)$ *along with some* $\alpha \in [0, 1]$, *it follows that* $w_1^{\alpha} \cdot w_2^{1-\alpha} \in A_p(\Sigma, \sigma)$ *and* $[w_1^{\alpha} \cdot w_2^{1-\alpha}]_{A_p} \leq [w_1]_{A_p}^{\alpha} \cdot [w_2]_{A_p}^{1-\alpha}$.

(5) [*Doubling*] *If* $w \in A_p(\Sigma, \sigma)$ *with* $p \in (1, \infty)$ *then for every surface ball* $\Delta \subseteq \Sigma$ *and every* σ*-measurable set* $E \subseteq \Delta$ *one has*

$$\left(\frac{\sigma(E)}{\sigma(\Delta)}\right)^p \leq [w]_{A_p} \cdot \frac{w(E)}{w(\Delta)} \tag{2.49}$$

In particular, the measure w *is doubling, that is, there exists some* $C \in (0, \infty)$ *which depends only on* p, n, *and the Ahlfors regularity constant of* Σ, *such that* $w(2\Delta) \leq C[w]_{A_p} \cdot w(\Delta)$ *for every surface ball* $\Delta \subseteq \Sigma$. *More generally, with the constant* $C \in (0, \infty)$ *of the same nature as above, for each* $\lambda \in (1, \infty)$ *and each surface ball* $\Delta \subseteq \Sigma$ *one has the inequality* $w(\lambda\Delta) \leq C[w]_{A_p} \cdot \lambda^{p(n-1)} \cdot w(\Delta)$, *where* $\lambda\Delta$ *denotes the concentric dilate of* Δ *by a factor of* λ.

(6) [*Reverse Hölder Inequalities*] *For every* $w \in A_\infty(\Sigma, \sigma)$ *there exist some exponent* $q \in (1, \infty)$ *and some constant* $C \in (0, \infty)$ *(which both depend only on* p, $[w]_{A_p}$, n, *and the Ahlfors regularity constant of* Σ, *for some* $p \in (1, \infty)$ *for which* $w \in A_p(\Sigma, \sigma)$*) such that*

$$\left(⨍_\Delta w^q \, d\sigma\right)^{1/q} \leq C ⨍_\Delta w \, d\sigma, \tag{2.50}$$

for every surface ball $\Delta \subseteq \Sigma$. *This has several remarkable consequences. First, there exist some power* $\tau > 0$ *and some constant* $C \in (0, \infty)$ *(in fact,* C *is the same as in* (2.50) *and* $\tau = 1/q'$ *where* q' *is the Hölder conjugate of the exponent* q *from* (2.50)*) such that*

$$\frac{w(E)}{w(\Delta)} \leq C\left(\frac{\sigma(E)}{\sigma(\Delta)}\right)^\tau \tag{2.51}$$

for every surface ball $\Delta \subseteq \Sigma$ *and every* σ*-measurable set* $E \subseteq \Delta$.

(7) [*Additivity*] *Select* $p \in [1, \infty)$. *Then for every* $w_0, w_1 \in A_p(\Sigma, \sigma)$, *it follows that* $w_0 + w_1 \in A_p(\Sigma, \sigma)$ *and* $[w_0 + w_1]_{A_p} \leq [w_0]_{A_p} + [w_1]_{A_p}$. *In particular, for every* $w \in A_p(\Sigma, \sigma)$ *it follows that* $1 + w \in A_p(\Sigma, \sigma)$. *Moreover, if* $p_0, p_1 \in [1, \infty)$, *and* $w_j \in A_{p_j}(\Sigma, \sigma)$, *with* $j \in \{0, 1\}$, *it follows that* $w_0 + w_1 \in A_{\max\{p_0, p_1\}}(\Sigma, \sigma)$.

(8) [*Homogeneity*] *Given an exponent* $p \in [1, \infty)$, *a weight* $w \in A_p(\Sigma, \sigma)$, *and a constant* $\lambda \in (0, \infty)$, *it follows that* $\lambda w \in A_p(\Sigma, \sigma)$ *and* $[\lambda w]_{A_p} = [w]_{A_p}$.

(9) [*Cone Property*] *Pick* $p \in [1, \infty)$. *Then the Muckenhoupt class* $A_p(\Sigma, \sigma)$ *is a cone. Specifically,* $w_0 + w_1 \in A_p(\Sigma, \sigma)$ *for every* $w_0, w_1 \in A_p(\Sigma, \sigma)$, *and* $\lambda w \in A_p(\Sigma, \sigma)$ *for every* $w \in A_p(\Sigma, \sigma)$ *and* $\lambda \in (0, \infty)$.

(10) [*Minimum Property*] *Whenever* $p \in (1, \infty)$, *and* $w_0, w_1 \in A_p(\Sigma, \sigma)$ *it follows that* $\min\{w_0, w_1\} \in A_p(\Sigma, \sigma)$. *In particular,* $\min\{w, N\} \in A_p(\Sigma, \sigma)$ *for all* $w \in A_p(\Sigma, \sigma)$ *and* $N \in (0, \infty)$.

(11) [A_1 *Weights*] *A weight* w *on* Σ *belongs to* $A_1(\Sigma, \sigma)$ *if and only if there exists a constant* $C \in (0, \infty)$ *with the property that* $w(x) \le \mathcal{M}_\Sigma w(x) \le C w(x)$ *at* σ*-a.e. point* $x \in \Sigma$.

(12) [*Pointwise Convergence of* A_1 *Weights*] *Consider a sequence* $\{w_j\}_{j\in\mathbb{N}}$ *contained in* $A_1(\Sigma, \sigma)$. *Suppose that* $\sup_{j\in\mathbb{N}} [w_j]_{A_1} < \infty$ *and* $w_j \to w$ *pointwise* σ*-a.e. on* Σ *as* $j \to \infty$ *for some weight* w *on* Σ. *Then* $w \in A_1(\Sigma, \sigma)$ *quantitatively, i.e.,* $[w]_{A_1} \le \sup_{j\in\mathbb{N}} [w_j]_{A_1}$.

Proof The properties listed on items *(1)-(6)* can be found on [GHA-I, Lemma 7.7.1, pp. 641–643]. Here we present the proof of the remaining properties. We begin by verifying the claims made in *(7)*. Select two weights $w_0, w_1 \in A_p(\Sigma, \sigma)$, and assume first $1 < p < \infty$. Since $w_0 + w_1 \ge w_j$ for each $j \in \{0, 1\}$, and $1 - p < 0$, we see that

$$(w_0 + w_1)^{1/(1-p)} \le w_j^{1/(1-p)} \quad \text{on } \Sigma \text{ for } j \in \{0, 1\}. \tag{2.52}$$

Hence for every surface ball $\Delta \subseteq \Sigma$ we have

$$\begin{aligned} &\Big(⨍_\Delta (w_0 + w_1)^{1/(1-p)}\, d\sigma\Big)^{p-1} \\ &\qquad \le \min\Big\{\Big(⨍_\Delta w_0^{1/(1-p)}\, d\sigma\Big)^{p-1}, \Big(⨍_\Delta w_1^{1/(1-p)}\, d\sigma\Big)^{p-1}\Big\}. \end{aligned} \tag{2.53}$$

This further implies

$$\begin{aligned} &\Big(⨍_\Delta (w_0 + w_1)\, d\sigma\Big)\Big(⨍_\Delta (w_0 + w_1)^{1/(1-p)}\, d\sigma\Big)^{p-1} \\ &\le \Big(⨍_\Delta w_0\, d\sigma\Big)\Big(⨍_\Delta w_0^{1/(1-p)}\, d\sigma\Big)^{p-1} + \Big(⨍_\Delta w_1\, d\sigma\Big)\Big(⨍_\Delta w_1^{1/(1-p)}\, d\sigma\Big)^{p-1} \\ &\le [w_0]_{A_p} + [w_1]_{A_p}. \end{aligned} \tag{2.54}$$

After taking the supremum over all surface balls $\Delta \subseteq \Sigma$ in the right-hand of (2.54) we arrive at

$$[w_0 + w_1]_{A_p} \le [w_0]_{A_p} + [w_1]_{A_p} < \infty \quad \Longrightarrow \quad w_0 + w_1 \in A_p(\Sigma, \sigma), \tag{2.55}$$

as desired. Corresponding to the case $p = 1$, use (2.42) to ensure that

$$\mathcal{M}_\Sigma w_j \leq c_j w_j \quad \text{on } \Sigma \tag{2.56}$$

for each $j \in \{0, 1\}$, and for some constants $c_0, c_1 \in (0, \infty)$. This and the sub-linearity of $\mathcal{M}_\Sigma$ give

$$\mathcal{M}_\Sigma(w_0 + w_1) \leq \mathcal{M}_\Sigma w_0 + \mathcal{M}_\Sigma w_1 \leq \max\{c_0, c_1\}(w_0 + w_1). \tag{2.57}$$

As such, (2.42) yields $w_0 + w_1 \in A_1(\Sigma, \sigma)$ and

$$[w_0 + w_1]_{A_1} \leq [w_0]_{A_1} + [w_1]_{A_1}. \tag{2.58}$$

This justifies the first claim made in *(7)*. As a consequence of (2.55) and (2.58), and bearing in mind that constant weights belong to $A_p(\Sigma, \sigma)$, it is evident that the weight $1 + w \in A_p(\Sigma, \sigma)$ for every $w \in A_p(\Sigma, \sigma)$, and every $p \in [1, \infty)$.

Going further, pick two exponents $p_0, p_1 \in [1, \infty)$, along with two arbitrary weights $w_0 \in A_{p_0}(\Sigma, \sigma)$ and $w_1 \in A_{p_1}(\Sigma, \sigma)$. Then the monotonicity property described in *(2)* yields

$$A_{p_0}(\Sigma, \sigma) \subseteq A_{\max\{p_0, p_1\}}(\Sigma, \sigma) \quad \text{and} \quad A_{p_1}(\Sigma, \sigma) \subseteq A_{\max\{p_0, p_1\}}(\Sigma, \sigma). \tag{2.59}$$

From what we have proved so far, we find that

$$w_0 + w_1 \in A_{\max\{p_0, p_1\}}(\Sigma, \sigma). \tag{2.60}$$

This justifies the second claim made in *(7)*.

The homogeneity property in item *(8)* is visible from (2.39)-(2.40), and the cone property in item *(9)* is a direct consequence of *(7)* and *(8)*.

To justify the claim made in *(10)*, select two weights $w_0, w_1 \in A_p(\Sigma, \sigma)$. Consider first the case $1 < p < \infty$. Given a surface ball $\Delta \subseteq \Sigma$, compute

$$\begin{aligned}
&\Big(⨍_\Delta \min\{w_0, w_1\}\, d\sigma\Big)\Big(⨍_\Delta \big[\min\{w_0, w_1\}\big]^{1/(1-p)}\, d\sigma\Big)^{p-1} \\
&\quad \leq \Big(⨍_\Delta (w_0 + w_1)\, d\sigma\Big)\Big(⨍_\Delta \min\big\{w_0^{1/(1-p)}, w_1^{1/(1-p)}\big\}\, d\sigma\Big)^{p-1} \\
&\quad \leq \Big(⨍_\Delta w_0\, d\sigma\Big)\Big(⨍_\Delta w_0^{1/(1-p)}\, d\sigma\Big)^{p-1} + \Big(⨍_\Delta w_1\, d\sigma\Big)\Big(⨍_\Delta w_1^{1/(1-p)}\, d\sigma\Big)^{p-1} \\
&\quad \leq [w_0]_{A_p} + [w_1]_{A_p}.
\end{aligned} \tag{2.61}$$

Since this holds for every surface ball $\Delta \subseteq \Sigma$, we find that

$$\begin{aligned}
&\big[\min\{w_0, w_1\}\big]_{A_p} \leq [w_0]_{A_p} + [w_1]_{A_p} < \infty \\
&\quad \Longrightarrow \quad \min\{w_0, w_1\} \in A_p(\Sigma, \sigma).
\end{aligned} \tag{2.62}$$

Now consider the case $p = 1$. Observe that, given a surface ball $\Delta \subseteq \Sigma$ we have

$$\begin{aligned} \fint_{\Delta} \min\{w_0, w_1\}\, d\sigma &\leq \Big(\fint_{\Delta} w_0\, d\sigma\Big) \leq [w_0]_{A_1} w_0 \\ &\leq \big([w_0]_{A_1} + [w_1]_{A_1}\big) w_0 \end{aligned} \tag{2.63}$$

at σ-a.e. point on Δ, and similarly

$$\fint_{\Delta} \min\{w_0, w_1\}\, d\sigma \leq \big([w_0]_{A_1} + [w_1]_{A_1}\big) w_1 \tag{2.64}$$

at σ-a.e. point on Δ. Hence,

$$\fint_{\Delta} \min\{w_0, w_1\}\, d\sigma \leq \big([w_0]_{A_1} + [w_1]_{A_1}\big) \min\{w_0, w_1\} \quad \text{pointwise } \sigma\text{-a.e. on } \Delta. \tag{2.65}$$

Because $\Delta \subseteq \Sigma$ was taken arbitrary, (2.41) ensures that $\min\{w_0, w_1\} \in A_1(\Sigma, \sigma)$ with $\big[\min\{w_0, w_1\}\big]_{A_1} \leq [w_0]_{A_1} + [w_1]_{A_1}$. The last property in item *(10)* follows from what we have proved so far with $w_0 := w$ and $w_1 := N$.

Notice that the claim made in *(11)* is a consequence of (2.42) and the Lebesgue Differentiation Theorem, which implies that for every $w \in A_p(\Sigma, \sigma) \subseteq L^1_{\mathrm{loc}}(\Sigma, \sigma)$ we have

$$w(x) = \lim_{r \to 0^+} \fint_{\Delta(x,r)} w\, d\sigma \leq \mathcal{M}_\Sigma(x) \quad \text{at } \sigma\text{-a.e. } x \in \Sigma. \tag{2.66}$$

Lastly, to justify the claim made in item *(12)* fix $x_0 \in \Sigma$ and $R \in (0, \infty)$. Then Fatou's Lemma ensures that

$$\fint_{\Delta(x_0,R)} w\, d\sigma = \fint_{\Delta(x_0,R)} \lim_{j \to \infty} w_j\, d\sigma \leq \liminf_{j \to \infty} \fint_{\Delta(x_0,R)} w_j\, d\sigma. \tag{2.67}$$

Based on (2.42), for every $j \in \mathbb{N}$ we have

$$\fint_{\Delta(x_0,R)} w_j\, d\sigma \leq (\mathcal{M}_\Sigma w_j)(x_0) \leq [w_j]_{A_1} \cdot w_j(x_0) \leq \big(\sup_{k \in \mathbb{N}} [w_k]_{A_1}\big) \cdot w_j(x_0). \tag{2.68}$$

This, (2.67), and the fact that $w_j \to w$ as $j \to \infty$ pointwise σ-a.e. on Σ yields

$$(\mathcal{M}_\Sigma w)(x_0) \leq \big(\sup_{k \in \mathbb{N}} [w_k]_{A_1}\big) \cdot w(x_0), \tag{2.69}$$

which holds for σ-a.e. $x_0 \in \Sigma$. Hence, $w \in A_1(\Sigma, \sigma)$ and $[w]_{A_1} \leq \sup_{j \in \mathbb{N}} [w_j]_{A_1}$ (cf. (2.42)). This concludes the proof of Proposition 2.3. □

Example 2.1 Let $\Sigma \subseteq \mathbb{R}^n$ be a closed Ahlfors regular set, and set $\sigma := \mathcal{H}^{n-1} \lfloor \Sigma$. Fix some $p \in (1, \infty)$ along with an arbitrary point $x_0 \in \Sigma$ and a power $a \in \mathbb{R}$. Then the function

$$
\begin{aligned}
&|\cdot - x_0|^a \text{ belongs to the Muckenhoupt class } A_p(\Sigma, \sigma) \\
&\quad \text{if and only if } -(n-1) < a < (p-1)(n-1).
\end{aligned}
\tag{2.70}
$$

Also, whenever this happens, it follows that the A_p characteristic of $|\cdot - x_0|^a$ depends only on the Ahlfors regularity constant of Σ, p, and a. In the classical Euclidean setting, (2.70) is well known. For a more general geometric setting see, e.g., [GeGoKoKr97, Proposition 1.5.9, p. 42].

In particular, if $-(n-1) < a < (p-1)(n-1)$ one may also conclude that

$$
\begin{aligned}
&\text{the functions } \ 1 + |\cdot - x_0|^a \ \text{ and } \ (1 + |\cdot - x_0|)^a \\
&\quad \text{belong to the Muckenhoupt class } A_p(\Sigma, \sigma).
\end{aligned}
\tag{2.71}
$$

Indeed, the membership of $1 + |\cdot - x_0|^a$ to $A_p(\Sigma, \sigma)$ is a consequence of (2.70) and item *(7)* in Proposition 2.3. Regarding the function $(1 + |\cdot - x_0|)^a$, consider the case $0 \le a < (n-1)(p-1)$. Since $1 + |\cdot - x_0|^a \in A_p(\Sigma, \sigma)$ and

$$
(1 + |x - x_0|)^a \approx 1 + |x - x_0|^a \quad \text{uniformly in } x \in \Sigma, \tag{2.72}
$$

it follows that $(1 + |\cdot - x_0|)^a \in A_p(\Sigma, \sigma)$. Assume next $-(n-1) < a < 0$, and set $b := a(1 - p')$, where $p' := \frac{p}{p-1} \in (1, \infty)$ is the Hölder conjugate exponent of p. Evidently, $0 < b < (n-1)(p'-1)$. Hence, from what we have proved so far,

$$
(1 + |\cdot - x_0|)^b \in A_{p'}(\Sigma, \sigma). \tag{2.73}
$$

As such, thanks to the duality property from item *(5)* in Proposition 2.3, we obtain

$$
(1 + |\cdot - x_0|)^a = \left[(1 + |\cdot - x_0|)^b\right]^{1/(1-p')} \in A_p(\Sigma, \sigma). \tag{2.74}
$$

This establishes the claim (2.71).

In a more general geometric setting, consider the following result implied by work in [DyIhLeTuVa19].

Proposition 2.4 *Assume $\Sigma \subseteq \mathbb{R}^n$ is a closed set which is Ahlfors regular, and abbreviate $\sigma := \mathcal{H}^{n-1} \lfloor \Sigma$. Consider a d-*`Ahlfors regular set` *$E \subseteq \Sigma$ for some $d \in [0, n-1)$, i.e., E is a closed subset of Σ, and has the property that*

$$
\mathcal{H}^d\big(B(x,R) \cap E\big) \approx R^d \quad \textit{uniformly for } x \in E \textit{ and } R \in \big(0, 2\mathrm{diam}(E)\big), \tag{2.75}
$$

where $\mathcal{H}^d$ denotes the d-dimensional Hausdorff measure in $\mathbb{R}^n$, and $B(x,R)$ denotes the n-dimensional ball in $\mathbb{R}^n$ with center x and radius R.

Then for each exponent $p \in (1, \infty)$ and each power $a \in \mathbb{R}$ satisfying

$$
-(n-1-d) < a < (p-1)(n-1-d), \tag{2.76}
$$

it follows that the function $\operatorname{dist}(\cdot, E)^a$ *is a Muckenhoupt weight in the class* $A_p(\Sigma, \sigma)$, *and its* A_p *characteristic depends only on the Ahlfors regularity constants of* Σ, *the proportionality constants in* (2.75), d, p, *and* a.

Remark 2.1 In the context of Proposition 2.4, if $p \in (1, \infty)$ and $a \in \mathbb{R}$ satisfy (2.76) then

$$\begin{gathered}\text{the functions} \quad 1 + \operatorname{dist}(\cdot, E)^a \quad \text{and} \quad \big(1 + \operatorname{dist}(\cdot, E)\big)^a \\ \text{belong to the Muckenhoupt class } A_p(\Sigma, \sigma).\end{gathered} \tag{2.77}$$

To justify this, first use Proposition 2.4 and item *(7)* in Proposition 2.3 to obtain $1 + \operatorname{dist}(\cdot, E)^a \in A_p(\Sigma, \sigma)$. In the regime $0 \leq a < (p-1)(n-1-d)$ we have

$$\big(1 + \operatorname{dist}(x, E)\big)^a \approx 1 + \operatorname{dist}(x, E)^a \quad \text{uniformly in } x \in \Sigma. \tag{2.78}$$

Hence, since $1 + \operatorname{dist}(\cdot, E)^a \in A_p(\Sigma, \sigma)$, we have $\big(1 + \operatorname{dist}(\cdot, E)\big)^a \in A_p(\Sigma, \sigma)$. In the case when $-(n-1-d) < a < 0$, let $b := a(1-p')$, where $p' := \frac{p}{p-1} \in (1, \infty)$ is the Hölder conjugate exponent of p. With this piece of notation, it is visible that $0 < b < (n-1-d)(p'-1)$. From the proof so far, we may conclude that the function $\big(1 + \operatorname{dist}(\cdot, E)\big)^b$ belongs to $A_{p'}(\Sigma, \sigma)$. Via the duality property described in item *(5)* in Proposition 2.3, this entails

$$\big(1 + \operatorname{dist}(\cdot, E)\big)^a = \Big[\big(1 + \operatorname{dist}(\cdot, E)\big)^b\Big]^{1/(1-p')} \in A_p(\Sigma, \sigma). \tag{2.79}$$

This concludes the proof of the memberships claimed in (2.77).

Let us also recall the following embedding result for Muckenhoupt weighted Lebesgue spaces (cf. [GHA-I, Lemma 7.7.13, p. 659]).

Proposition 2.5 *Suppose* $\Sigma \subseteq \mathbb{R}^n$, *where* $n \in \mathbb{N}$ *with* $n \geq 2$, *is a closed set which is Ahlfors regular, and abbreviate* $\sigma := \mathcal{H}^{n-1}\lfloor\Sigma$. *In this setting, fix some* $p \in (1, \infty)$ *along with a Muckenhoupt weight* $w \in A_p(\Sigma, \sigma)$. *Then*

$$\int_\Sigma \frac{w(x)}{(1 + |x|^{n-1})^p}\, d\sigma(x) < \infty. \tag{2.80}$$

Also, there exists $\varepsilon \in (0, 1)$ *such that*

$$L^p(\Sigma, w) \hookrightarrow L^1\Big(\Sigma, \frac{\sigma(x)}{1 + |x|^{n-1-\varepsilon}}\Big), \tag{2.81}$$

and there exists an exponent $p_o \in (1, p]$ *with the property that*

$$\begin{gathered} L^p(\Sigma, w) \hookrightarrow L^q\Big(\Sigma, \frac{\sigma(x)}{1 + |x|^{n-1}}\Big) \\ \textit{continuously for each fixed } q \in (0, p_o). \end{gathered} \tag{2.82}$$

As a consequence,

$$L^p(\Sigma, w) \hookrightarrow L^1\Big(\Sigma, \frac{\sigma(x)}{1+|x|^{n-1}}\Big) \text{ continuously, and} \tag{2.83}$$

$$L^p(\Sigma, w) \subseteq L^p_{\text{loc}}(\Sigma, w) \subseteq \bigcup_{1<q<p} L^q_{\text{loc}}(\Sigma, \sigma) \subseteq L^1_{\text{loc}}(\Sigma, \sigma). \tag{2.84}$$

Chapter 3
Beurling Algebras

The goal here is to develop a theory for the scale of Beurling spaces A^p, and the spaces B^q, in the context of unbounded Ahlfors regular sets (as a replacement for the Euclidean ambient). In Section 3.1 we establish some functional analytic properties of these spaces, while in Section 3.2 we study the duality between A^p and B^q spaces. In Section 3.3 we explore the action of the Hardy-Littlewood maximal operator, and its L^s-based version, on these scales of spaces, and prove some boundedness results. This provides the groundwork for studying a certain class of Hardy spaces associated with A^p, which will be discussed at length in Chapter 4.

3.1 Definitions and Basic Properties

In this section, we establish various properties of Beurling spaces A^p, and other closely related spaces. This includes embedding results, completeness, norm characterizations, an analog of Lebesgue's Dominated Convergence Theorem on A^p, among other things. We begin with the following definition.

Definition 3.1 Consider $\Sigma \subseteq \mathbb{R}^n$ an unbounded closed Ahlfors regular set, and abbreviate $\sigma := \mathcal{H}^{n-1}\lfloor\Sigma$. Denote by $c_\Sigma, C_\Sigma \in (0,\infty)$ the lower and upper Ahlfors regularity constants of Σ, respectively. Also, pick a reference point $x_0 \in \Sigma$, and a dilation factor $2_* > (C_\Sigma/c_\Sigma)^{1/(n-1)}$. Introduce the following sets:

$$\begin{aligned} C_0 &:= \Delta(x_0, 2_*), \\ C_k &:= \Delta(x_0, 2_*^{k+1}) \setminus \Delta(x_0, 2_*^k), \quad \text{for each } k \in \mathbb{N}. \end{aligned} \tag{3.1}$$

Select an integrability exponent $p \in (0,\infty)$. For each σ-measurable function $f : \Sigma \to \mathbb{R}$, denote

© The Author(s), under exclusive license to Springer Nature Switzerland AG 2026
M. Mitrea and P. Takemura, *Singular Integrals, Herz-Type Function Spaces, and Boundary Problems*, Progress in Mathematics 362,
https://doi.org/10.1007/978-3-032-12516-3_3

$$\|f\|_{\mathrm{A}^p(\Sigma,\sigma)} := \sum_{k=0}^{\infty} 2_*^{k(n-1)(1-1/p)} \|f \cdot \mathbb{1}_{C_k}\|_{L^p(\Sigma,\sigma)} \in [0, \infty]. \tag{3.2}$$

The `Beurling space` $\mathrm{A}^p(\Sigma, \sigma)$ is defined according to

$$\mathrm{A}^p(\Sigma, \sigma) := \{f : \Sigma \to \mathbb{R} : f \text{ is } \sigma\text{-measurable and } \|f\|_{\mathrm{A}^p(\Sigma,\sigma)} < \infty\}. \tag{3.3}$$

In addition, for each σ-measurable function $f : \Sigma \to \mathbb{R}$ introduce

$$\|f\|_{\mathrm{B}^p(\Sigma,\sigma)} := \sup_{k\in\mathbb{N}_0} \left(2_*^{-k(n-1)/p} \|f \cdot \mathbb{1}_{C_k}\|_{L^p(\Sigma,\sigma)}\right) \in [0, \infty]. \tag{3.4}$$

Define the space of functions $\mathrm{B}^p(\Sigma, \sigma)$ as

$$\mathrm{B}^p(\Sigma, \sigma) := \{f : \Sigma \to \mathbb{R} : f \text{ is } \sigma\text{-measurable and } \|f\|_{\mathrm{B}^p(\Sigma,\sigma)} < \infty\}. \tag{3.5}$$

Some comments regarding the nature of the spaces A^p, B^p are in order.

Remark 3.1 It is evident from (3.2) that corresponding to $p = 1$ the space $\mathrm{A}^1(\Sigma, \sigma)$ coincides the Lebesgue space $L^1(\Sigma, \sigma)$.

Remark 3.2 It is visible that $\big(\mathrm{A}^p(\Sigma, \sigma), \|\cdot\|_{\mathrm{A}^p(\Sigma,\sigma)}\big)$ and $\big(\mathrm{B}^p(\Sigma, \sigma), \|\cdot\|_{\mathrm{B}^p(\Sigma,\sigma)}\big)$ introduced above are well-defined normed vector spaces.

Remark 3.3 In the same context of Definition 3.1, notice that the spaces $\mathrm{A}^p(\Sigma, \sigma)$ and $\mathrm{B}^p(\Sigma, \sigma)$ are lattices. Precisely,

$$\begin{array}{l} \text{if } f, g \text{ are } \sigma\text{-measurable functions on } \Sigma \text{ such that } |f| \leq |g| \text{ at} \\ \sigma\text{-a.e. point on } \Sigma, \text{ and } g \in \mathrm{A}^p(\Sigma, \sigma) \text{ then } f \in \mathrm{A}^p(\Sigma, \sigma) \text{ and} \\ \|f\|_{\mathrm{A}^p(\Sigma,\sigma)} \leq \|g\|_{\mathrm{A}^p(\Sigma,\sigma)} \end{array} \tag{3.6}$$

and similarly for $\mathrm{B}^p(\Sigma, \sigma)$.

Remark 3.4 Perhaps a more precise piece of notation for the A^p and B^p scale of spaces would be $\mathrm{A}^p_{x_0,2_*}$ and $\mathrm{B}^p_{x_0,2_*}$ as to emphasize the dependence on the reference point x_0 and the dilation factor 2_*. However, we will show that these spaces are actually independent of said entities (cf. Proposition 3.2 and Remark 3.10). For this reason, our current notation does not reflect the dependence on x_0 and 2_*.

Remark 3.5 We now make an observation regarding the nature of the dilation factor 2_* appearing in Definition 3.1. Under the same background assumptions of Definition 3.1, we claim that the condition

$$2_* > (C_\Sigma/c_\Sigma)^{1/(n-1)} \tag{3.7}$$

is sufficient to guarantee $\sigma(C_k) > 0$ for every $k \in \mathbb{N}$. Indeed, since Σ is Ahlfors regular, for each $k \in \mathbb{N}$ we have

$$
\begin{aligned}
\sigma(C_k) &= \sigma\big(\Delta(x_0, 2_*^{k+1})\backslash\Delta(x_0, 2_*^k)\big) \\
&= \sigma\big(\Delta(x_0, 2_*^{k+1})\big) - \sigma\big(\Delta(x_0, 2_*^k)\big) \\
&\geq c_\Sigma 2_*^{(k+1)(n-1)} - C_\Sigma 2_*^{k(n-1)} \\
&= 2_*^{k(n-1)}(c_\Sigma 2_*^{n-1} - C_\Sigma).
\end{aligned} \tag{3.8}
$$

This and (3.7) implies $\sigma(C_k) > 0$ for every $k \in \mathbb{N}$, as desired. Corresponding to case $k = 0$ we have $\sigma(C_0) \approx 2_*^{n-1} > 0$, which follows from the Ahlfors regularity of Σ.

Here we provide some concrete examples of A^p functions.

Example 3.1 Assume $\Sigma \subseteq \mathbb{R}^n$ is a closed unbounded Ahlfors regular set, and abbreviate $\sigma := \mathcal{H}^{n-1}\lfloor\Sigma$. With $c_\Sigma, C_\Sigma \in (0, \infty)$ denoting the lower and upper Ahlfors regularity constant of Σ, respectively, respectively, consider a dilation factor $2_* > (C_\Sigma/c_\Sigma)^{1/(n-1)}$. Also, fix a reference point $x_0 \in \Sigma$. Having fixed $R \in (0, \infty)$ and $s \in (0, 1)$ define the function $\phi_{R,s} : \Sigma \longrightarrow \mathbb{R}$ by

$$
\phi_{R,s}(x) := \frac{1}{(R + |x - x_0|)^{(n-1)/s}} \quad \text{for all } x \in \Sigma. \tag{3.9}
$$

Given an integrability exponent $p \in (1, \infty)$, we claim that $\phi_{R,s} \in \mathrm{A}^p(\Sigma, \sigma)$. Indeed, since Σ is Ahlfors regular for each $k \in \mathbb{N}$ we have

$$
\begin{aligned}
\|\phi_{R,s} \cdot \mathbb{1}_{C_k}\|_{L^p(\Sigma,\sigma)} &= \Big(\int_{C_k} \frac{d\sigma(x)}{(R + |x - x_0|)^{p(n-1)/s}}\Big)^{1/p} \\
&\leq \Big(\int_{C_k} \frac{d\sigma(x)}{|x - x_0|^{p(n-1)/s}}\Big)^{1/p} \\
&\leq \frac{\sigma(C_k)^{1/p}}{2_*^{k(n-1)/s}} \leq C\frac{2_*^{k(n-1)/p}}{2_*^{k(n-1)/s}},
\end{aligned} \tag{3.10}
$$

for some constant $C = C(\Sigma, p, n) \in (0, \infty)$. Corresponding to the case $k = 0$, use the Ahlfors regularity of Σ to estimate

$$
\begin{aligned}
\|\phi_{R,s} \cdot \mathbb{1}_{C_0}\|_{L^p(\Sigma,\sigma)} &= \Big(\int_{\Delta(x_0,2_*)} \frac{d\sigma(x)}{(R + |x - x_0|)^{p(n-1)/s}}\Big)^{1/p} \\
&\leq \frac{\sigma\big(\Delta(x_0, 2_*)\big)^{1/p}}{R^{(n-1)/s}} \leq C\frac{2_*^{(n-1)/p}}{R^{(n-1)/s}},
\end{aligned} \tag{3.11}
$$

for some constant $C = C(\Sigma, p, n) \in (0, \infty)$. Thanks to estimates (3.10)-(3.11), and keeping in mind that $s \in (0, 1)$ we see that

$$
\begin{aligned}
\|\phi_{R,s}\|_{A^p(\Sigma,\sigma)} &= \sum_{k=0}^{\infty} 2_*^{k(n-1)/p'} \|\phi_{R,s} \cdot \mathbb{1}_{C_k}\|_{L^p(\Sigma,\sigma)} \\
&= \|\phi_{R,s} \cdot \mathbb{1}_{C_0}\|_{L^p(\Sigma,\sigma)} + \sum_{k=1}^{\infty} 2_*^{k(n-1)/p'} \|\phi_{R,s} \cdot \mathbb{1}_{C_k}\|_{L^p(\Sigma,\sigma)} \\
&\leq C \frac{2_*^{(n-1)/p}}{R^{(n-1)/s}} + C \sum_{k=1}^{\infty} 2_*^{k(n-1)/p'} \frac{2_*^{k(n-1)/p}}{2_*^{k(n-1)/s}} \\
&= C \frac{2_*^{(n-1)/p}}{R^{(n-1)/s}} + C \sum_{k=1}^{\infty} \frac{2_*^{k(n-1)}}{2_*^{k(n-1)/s}} \\
&= C \frac{2_*^{(n-1)/p}}{R^{(n-1)/s}} + C \sum_{k=1}^{\infty} 2_*^{k(n-1)(1-1/s)} < \infty.
\end{aligned}
\tag{3.12}
$$

This shows that $\phi_{R,s} \in A^p(\Sigma, \sigma)$ as desired.

It is also of interest to provide examples of functions that do not belong to Beurling spaces.

Example 3.2 Let $\Sigma \subseteq \mathbb{R}^n$ be a closed unbounded Ahlfors regular set, and abbreviate $\sigma := \mathcal{H}^{n-1}\lfloor\Sigma$. Having selected a reference point $x_0 \in \Sigma$ and a radius $R \in (0, \infty)$, define the function $\psi_R : \Sigma \to \mathbb{R}$ by

$$
\psi_R(x) := \frac{1}{(R + |x - x_0|)^{n-1}} \quad \text{for all } x \in \Sigma. \tag{3.13}
$$

We claim that whenever $p \in (1, \infty)$ one has $\psi_R \notin A^p(\Sigma, \sigma)$. To justify this claim, consider $2_* \in (1, \infty)$ as in Definition 3.1, and select $\ell \in \mathbb{N}_0$ large enough so that $R \leq 2_*^{\ell+1}$. Keeping in mind (3.1), and using the Ahlfors regularity of Σ, for each $k \geq \ell$ we estimate

$$
\begin{aligned}
\|\psi_R \cdot \mathbb{1}_{C_k}\|_{L^p(\Sigma,\sigma)} &= \Big(\int_{C_k} \frac{d\sigma(x)}{(R + |x - x_0|)^{(n-1)p}} \Big)^{1/p} \geq \frac{\sigma(C_k)^{1/p}}{(2_*^{\ell+1} + 2_*^{k+1})^{n-1}} \\
&\geq \frac{C 2_*^{k(n-1)/p}}{(2_*^{k+1} + 2_*^{k+1})^{n-1}} = \frac{C}{2_*^{k(n-1)(1-1/p)}}
\end{aligned}
\tag{3.14}
$$

for some constant $C = C(\Sigma, n, p) \in (0, \infty)$. This therefore implies

$$
C \leq 2_*^{k(n-1)(1-1/p)} \|\psi_R \cdot \mathbb{1}_{C_k}\|_{L^p(\Sigma,\sigma)} \tag{3.15}
$$

for every $k \geq \ell$. This and (3.2) readily give

$$
\|\psi_R\|_{A^p(\Sigma,\sigma)} = \sum_{k=0}^{\infty} 2_*^{k(n-1)/p'} \|\psi_R \cdot \mathbb{1}_{C_k}\|_{L^p(\Sigma,\sigma)} = \infty. \tag{3.16}
$$

Hence $\psi_R \notin \mathrm{A}^p(\Sigma, \sigma)$, as desired.

The next result provides a characterization of the B^p-norm in terms of L^p-averages over balls centered at a fixed reference point.

Proposition 3.1 *Let $\Sigma \subseteq \mathbb{R}^n$ be a closed unbounded Ahlfors regular set, and abbreviate $\sigma := \mathcal{H}^{n-1}\lfloor\Sigma$. Fix a reference point $x_0 \in \Sigma$ and select a dilation factor $2_* > \big(C_\Sigma/c_\Sigma\big)^{1/(n-1)}$, where $c_\Sigma, C_\Sigma \in (0,\infty)$ are the lower and upper Ahlfors regularity constant of Σ, respectively. Finally, pick an integrability exponent $p \in (1,\infty)$. Then for every σ-measurable function $f : \Sigma \to \mathbb{R}$ one has*

$$\|f\|_{\mathrm{B}^p(\Sigma,\sigma)} \approx \sup_{R \geq 2_*} \Big(\fint_{\Delta(x_0,R)} |f|^p \, d\sigma \Big)^{1/p} \tag{3.17}$$

with implicit proportionality constants depending only on n, p, and the Ahlfors regularity character of Σ.

Proof To begin, fix a finite radius $R \geq 2_*$, and select the unique integer $j \in \mathbb{N}$ for which $2_*^j \leq R < 2_*^{j+1}$. Then from this, the Ahlfors regularity of Σ, and the fact that $\Delta(x_0, R) \subseteq \Delta(x_0, 2_*^{j+1}) = \cup_{k=0}^{j} C_k$, estimate

$$\begin{aligned}
\Big(\fint_{\Delta(x_0,R)} |f|^p \, d\sigma \Big)^{1/p} &\leq \frac{C}{R^{(n-1)/p}} \sum_{k=0}^{j} \Big(\int_{C_k} |f|^p \, d\sigma \Big)^{1/p} \\
&\leq \frac{C}{2_*^{j(n-1)/p}} \sum_{k=0}^{j} 2_*^{k(n-1)/p} \big(2_*^{-k(n-1)/p} \|f \cdot \mathbb{1}_{C_k}\|_{L^p(\Sigma,\sigma)}\big) \\
&\leq C \Big(\sum_{k=0}^{j} 2_*^{(k-j)(n-1)/p} \Big) \|f\|_{\mathrm{B}^p(\Sigma,\sigma)} \\
&\leq C \Big(\sum_{\ell=0}^{\infty} 2_*^{-\ell(n-1)/p} \Big) \|f\|_{\mathrm{B}^p(\Sigma,\sigma)} \\
&= C \|f\|_{\mathrm{B}^p(\Sigma,\sigma)},
\end{aligned} \tag{3.18}$$

for some constant $C = C(\Sigma, n, p) \in (0,\infty)$. After taking the supremum over all radii $R \in [2_*, \infty)$, this yields

$$\sup_{R \geq 2_*} \Big(\fint_{\Delta(x_0,R)} |f|^p \, d\sigma \Big)^{1/p} \leq C \|f\|_{\mathrm{B}^p(\Sigma,\sigma)}, \tag{3.19}$$

which, since σ is doubling, readily implies the right-pointing inequality in (3.17).

In the opposite direction, keeping in mind that $C_k \subseteq \Delta(x_0, 2_*^{k+1})$ and using the Ahlfors regularity of Σ, for each $k \in \mathbb{N}_0$ we bound

$$
\begin{aligned}
\|f \cdot \mathbb{1}_{C_k}\|_{L^p(\Sigma,\sigma)} &= \Big(\int_{C_k} |f|^p \, d\sigma\Big)^{1/p} \leq \Big(\int_{\Delta(x_0, 2_*^{k+1})} |f|^p \, d\sigma\Big)^{1/p} \\
&\leq C 2_*^{k(n-1)/p} \Big(\fint_{\Delta(x_0, 2_*^{k+1})} |f|^p \, d\sigma\Big)^{1/p} \\
&\leq C 2_*^{k(n-1)/p} \sup_{R \geq 2_*} \Big(\fint_{\Delta(x_0,R)} |f|^p \, d\sigma\Big)^{1/p}, \qquad (3.20)
\end{aligned}
$$

for some constant $C = C(\Sigma, n, p) \in (0, \infty)$. As such, this gives

$$
\begin{aligned}
\|f\|_{\mathrm{B}^p(\Sigma,\sigma)} &= \sup_{k \in \mathbb{N}_0} \big(2_*^{-k(n-1)/p} \|f\|_{L^p(C_k,\sigma)}\big) \\
&\leq C \cdot \sup_{R \geq 2_*} \Big(\fint_{\Delta(x_0,R)} |f|^p \, d\sigma\Big)^{1/p}, \qquad (3.21)
\end{aligned}
$$

from which the left-pointing inequality in (3.17) follows again in view of the doubling character of Σ. □

As apparent from Definition 3.1, the spaces A^p and B^p depend a priori on the reference point $x_0 \in \Sigma$ and dilation factor $2_* \in (0, \infty)$. The purpose of the result recorded below is to show the independence of B^p on the parameters x_0 and 2_*. A similar result for A^p will be discussed in Section 3.2.

Proposition 3.2 *Let $\Sigma \subseteq \mathbb{R}^n$ be a closed unbounded Ahlfors regular set, and abbreviate $\sigma := \mathcal{H}^{n-1} \lfloor \Sigma$. Fix two reference points $x_0, x_1 \in \Sigma$ along two numbers $2_*, \widetilde{2}_* \in (0, \infty)$, and select an integrability exponent $p \in (1, \infty)$.*

Then whenever $f : \Sigma \to \mathbb{R}$ is a σ-measurable function one has

$$
\sup_{R \geq 2_*} \Big(\fint_{\Delta(x_0,R)} |f|^p \, d\sigma\Big)^{1/p} \approx \sup_{R \geq \widetilde{2}_*} \Big(\fint_{\Delta(x_1,R)} |f|^p \, d\sigma\Big)^{1/p} \qquad (3.22)
$$

with implicit proportionality constants depending only on n, p, 2_*, $\widetilde{2}_*$, x_0, x_1, and the Ahlfors regularity character of Σ. Consequently, the definition of $\mathrm{B}^p(\Sigma, \sigma)$ is independent of the choice of the reference point $x_0 \in \Sigma$ and the dilation factor $2_* > (C_\Sigma/c_\Sigma)^{1/(n-1)}$, where $c_\Sigma, C_\Sigma \in (0, \infty)$ denote the lower and upper Ahlfors regularity constants of Σ, respectively.

Proof To get started, pick a radius a finite radius $R \geq 2_*$. Choose $\lambda \in (0, \infty)$ satisfying

$$
\lambda > \max \Big\{2, \frac{2}{2_*}|x_0 - x_1|, \frac{\widetilde{2}_*}{2_*}\Big\} \qquad (3.23)
$$

and define

$$
\widetilde{R} := \lambda R \in (0, \infty). \qquad (3.24)
$$

Then whenever $x \in \Delta(x_0, R)$ we have

$$|x - x_1| \le |x - x_0| + |x_0 - x_1| \le R + \frac{\lambda}{2} 2_* \le \frac{\lambda}{2} R + \frac{\lambda}{2} R = \lambda R = \widetilde{R}. \tag{3.25}$$

Hence, $\Delta(x_0, R) \subseteq \Delta(x_1, \widetilde{R})$. It is visible from (3.23) that $\widetilde{R} \ge \widetilde{2}_*$. Thanks to this, (3.24), and the Ahlfors regularity of Σ, estimate

$$\begin{aligned} \Big(⨍_{\Delta(x_0,R)} |f|^p \, d\sigma \Big)^{1/p} &\le C \Big(\frac{1}{R^{n-1}} \int_{\Delta(x_1,\widetilde{R})} |f|^p \, d\sigma \Big)^{1/p} \\ &= C \Big(\frac{\lambda^{n-1}}{\widetilde{R}^{n-1}} \int_{\Delta(x_1,\widetilde{R})} |f|^p \, d\sigma \Big)^{1/p} \\ &\le C \Big(\frac{1}{\sigma(\Delta(x_1, \widetilde{R}))} \int_{\Delta(x_1,\widetilde{R})} |f|^p \, d\sigma \Big)^{1/p} \\ &\le C \cdot \sup_{R \ge \widetilde{2}_*} \Big(⨍_{\Delta(x_1,R)} |f|^p \, d\sigma \Big)^{1/p}, \end{aligned} \tag{3.26}$$

for some constant $C \in (0, \infty)$ depending only on $n, p, x_0, x_1, 2_*, \widetilde{2}_*$, and the Ahlfors regularity character of Σ. Since $R \ge 2_*$ was taken arbitrary to begin with, this yields

$$\sup_{R \ge 2_*} \Big(⨍_{\Delta(x_0,R)} |f|^p \, d\sigma \Big)^{1/p} \le C \cdot \sup_{R \ge \widetilde{2}_*} \Big(⨍_{\Delta(x_1,R)} |f|^p \, d\sigma \Big)^{1/p}. \tag{3.27}$$

The opposite inequality is obtained via the same exact argument by switching the roles of 2_* and $\widetilde{2}_*$. This establishes (3.22). As a consequence (3.22) and Proposition 3.1, the definition of $\mathrm{B}^p(\Sigma, \sigma)$ is independent of x_0 and 2_*. This concludes the proof of Proposition 3.2. □

In the following result, we collect some useful embeddings for the Beurling spaces A^p.

Proposition 3.3 *Suppose $\Sigma \subseteq \mathbb{R}^n$ is a closed unbounded Ahlfors regular set, and abbreviate $\sigma := \mathcal{H}^{n-1} \lfloor \Sigma$. Fix an integrability exponent $p \in (0, \infty)$. Then whenever $p < q < \infty$ one has*

$$\mathrm{A}^q(\Sigma, \sigma) \hookrightarrow \mathrm{A}^p(\Sigma, \sigma) \quad \textit{continuously}, \tag{3.28}$$

i.e., the scale of Beurling spaces $\mathrm{A}^p(\Sigma, \sigma)$ is (topologically) nested with respect to p. Also, if $1 \le p < \infty$, the following embeddings are continuous:

$$\mathrm{A}^p(\Sigma, \sigma) \hookrightarrow L^1(\Sigma, \sigma) \quad \textit{and} \quad \mathrm{A}^p(\Sigma, \sigma) \hookrightarrow L^p(\Sigma, \sigma). \tag{3.29}$$

Consequently, whenever $1 \le r \le p$

$$\mathrm{A}^p(\Sigma, \sigma) \hookrightarrow L^r(\Sigma, \sigma) \quad \textit{continuously}. \tag{3.30}$$

Moreover, whenever $1 \le p < \infty$, *one also has*

$$\mathrm{A}^p(\Sigma, \sigma) \hookrightarrow L^1\Big(\Sigma, \frac{\sigma(x)}{1+|x|^{n-1}}\Big) \quad \textit{continuously,} \tag{3.31}$$

and the following inclusions hold

$$\bigcup_{p<q<\infty} \mathrm{A}^q(\Sigma, \sigma) \subseteq \mathrm{A}^p(\Sigma, \sigma) \subseteq L^p_{\mathrm{loc}}(\Sigma, \sigma) \subseteq L^1_{\mathrm{loc}}(\Sigma, \sigma). \tag{3.32}$$

Proof Throughout, let $(\cdot)'$ indicate Hölder conjugation of exponents. To justify the claim made in (3.28), fix $q \in (p, \infty)$, and pick $f \in \mathrm{A}^q(\Sigma, \sigma)$. Denoting by $r := (q/p)' \in (1, \infty)$, we have in particular $1/p - 1/q = 1/(rp)$. Keeping this in mind, using Hölder's inequality, and the Ahlfors regularity of Σ, for each $k \in \mathbb{N}_0$ we estimate

$$\begin{aligned}
\|f \cdot \mathbb{1}_{C_k}\|_{L^p(\Sigma,\sigma)} &= \Big(\int_{C_k} |f|^p \, d\sigma\Big)^{1/p} \\
&\le \Big[\sigma(C_k)^{1/r}\Big(\int_{C_k} |f|^q \, d\sigma\Big)^{p/q}\Big]^{1/p} \\
&= \sigma(C_k)^{1/p-1/q}\|f \cdot \mathbb{1}_{C_k}\|_{L^q(\Sigma,\sigma)} \\
&\le C 2_*^{k(n-1)(1/p-1/q)}\|f \cdot \mathbb{1}_{C_k}\|_{L^q(\Sigma,\sigma)},
\end{aligned} \tag{3.33}$$

for some constant $C \in (0, \infty)$ depending only on p, q and the Ahlfors regularity of Σ, where $p', q' \in (1, \infty)$ are the Hölder conjugate exponents of p and q, respectively. This permits us to write

$$\begin{aligned}
\|f\|_{\mathrm{A}^p(\Sigma,\sigma)} &= \sum_{k=0}^{\infty} 2_*^{k(n-1)(1-1/p)}\|f \cdot \mathbb{1}_{C_k}\|_{L^p(\Sigma,\sigma)} \\
&\le C \sum_{k=0}^{\infty} 2_*^{k(n-1)(1-1/q)}\|f \cdot \mathbb{1}_{C_k}\|_{L^q(\Sigma,\sigma)} = \|f\|_{\mathrm{A}^q(\Sigma,\sigma)} < \infty.
\end{aligned} \tag{3.34}$$

Hence, the embedding (3.28) is established.

Recall from Remark 3.1 that $A^1(\Sigma, \sigma) = L^1(\Sigma, \sigma)$. So assume that $p \in (1, \infty)$, and select an arbitrary function $f \in \mathrm{A}^p(\Sigma, \sigma)$. Observe that by Hölder's inequality and the Ahlfors regularity of Σ we have

$$\begin{aligned}
\|f\|_{L^1(\Sigma,\sigma)} &= \sum_{k=0}^{\infty} \int_{C_k} |f| \, d\sigma \le \sum_{k=0}^{\infty} \sigma(C_k)^{1/p'}\Big(\int_{C_k} |f|^p \, d\sigma\Big)^{1/p} \\
&\le C \sum_{k=0}^{\infty} 2_*^{k(n-1)/p'}\|f \cdot \mathbb{1}_{C_k}\|_{L^p(\Sigma,\sigma)} = C\|f\|_{\mathrm{A}^p(\Sigma,\sigma)}.
\end{aligned} \tag{3.35}$$

Furthermore, the simple fact that $2_* > 1$ yields

$$\begin{aligned}\|f\|_{L^p(\Sigma,\sigma)} &\leq \sum_{k=0}^{\infty} \|f \cdot \mathbb{1}_{C_k}\|_{L^p(\Sigma,\sigma)} \\ &\leq \sum_{k=0}^{\infty} 2_*^{k(n-1)/p'} \|f \cdot \mathbb{1}_{C_k}\|_{L^p(\Sigma,\sigma)} = \|f\|_{\mathrm{A}^p(\Sigma,\sigma)}.\end{aligned} \tag{3.36}$$

This proves the embeddings claimed in (3.29). Via real interpolation, (3.30) becomes a consequence of (3.29).

Going further, the continuous embedding claimed in (3.31) follows from the first embedding in (3.29) along with the fact that $L^1(\Sigma,\sigma) \hookrightarrow L^1\big(\Sigma, \frac{\sigma(x)}{1+|x|^{n-1}}\big)$ continuously. Lastly, the inclusions in (3.32) are implied by (3.28)-(3.29). □

Remark 3.6 The Beurling space $\mathrm{A}^p(\Sigma,\sigma)$ is strictly smaller than $L^p(\Sigma,\sigma)$. Indeed, while $\mathrm{A}^p(\Sigma,\sigma) \hookrightarrow L^p(\Sigma,\sigma)$ continuously (see (3.29)), the function provided in Example 3.2 (see (3.13)) belongs to $L^p(\Sigma,\sigma)$, but does not belong to $\mathrm{A}^p(\Sigma,\sigma)$.

Concerning embeddings for the B^p spaces, we have the following result:

Proposition 3.4 *Assume $\Sigma \subseteq \mathbb{R}^n$ is a closed Ahlfors regular set that is unbounded, and abbreviate $\sigma := \mathcal{H}^{n-1}\lfloor\Sigma$. Select an integrability exponent $p \in (0,\infty)$. Then whenever $p < q < \infty$ one has*

$$\mathrm{B}^q(\Sigma,\sigma) \hookrightarrow \mathrm{B}^p(\Sigma,\sigma) \ \ \textit{continuously}, \tag{3.37}$$

i.e., the scale of B^p spaces is (topologically) nested with respect to p. Also, the following embeddings are continuous:

$$L^\infty(\Sigma,\sigma) \hookrightarrow \mathrm{B}^p(\Sigma,\sigma) \ \ \textit{and} \ \ L^p(\Sigma,\sigma) \hookrightarrow \mathrm{B}^p(\Sigma,\sigma). \tag{3.38}$$

Moreover, whenever $1 \leq p < \infty$ one has

$$\begin{gathered}\mathrm{B}^p(\Sigma,\sigma) \hookrightarrow L^1\Big(\Sigma, \frac{\sigma(x)}{1+|x|^{n-1+\varepsilon}}\Big) \\ \textit{continuously for every } \varepsilon \in (0,\infty),\end{gathered} \tag{3.39}$$

and the following inclusions hold

$$\bigcup_{p<q<\infty} \mathrm{B}^q(\Sigma,\sigma) \subseteq \mathrm{B}^p(\Sigma,\sigma) \subseteq L^p_{\mathrm{loc}}(\Sigma,\sigma) \subseteq L^1_{\mathrm{loc}}(\Sigma,\sigma). \tag{3.40}$$

Proof To justify the monotonicity property claimed in (3.37), we recycle (3.33) to derive

$$\|f \cdot \mathbb{1}_{C_k}\|_{L^p(\Sigma,\sigma)} \leq C 2_*^{k(n-1)(1/p-1/q)} \|f \cdot \mathbb{1}_{C_k}\|_{L^q(\Sigma,\sigma)}, \tag{3.41}$$

for every $k \in \mathbb{N}_0$, for any σ-measurable function f defined on Σ, and for some constant $C \in (0, \infty)$ depending only on n, p, q, and the Ahlfors regularity character of Σ. This and (3.4) immediately imply (3.37).

Moving on to the claims made in (3.38), fix a σ-measurable function f on Σ. Since Σ is Ahlfors regular, we have

$$\begin{aligned} \|f\|_{\mathrm{B}^p(\Sigma,\sigma)} &= \sup_{k\in\mathbb{N}_0} 2_*^{-k(n-1)/p}\|f\cdot\mathbb{1}_{C_k}\|_{L^p(\Sigma,\sigma)} \\ &\leq \Big(\sup_{k\in\mathbb{N}_0} 2_*^{-k(n-1)/p}\sigma(C_k)^{1/p}\Big)\|f\|_{L^\infty(\Sigma,\sigma)} \\ &\leq C\Big(\sup_{k\in\mathbb{N}_0} 2_*^{-k(n-1)/p}2_*^{k(n-1)/p}\Big)\|f\|_{L^\infty(\Sigma,\sigma)} = C\|f\|_{L^\infty(\Sigma,\sigma)}, \end{aligned} \tag{3.42}$$

for some constant $C \in (0, \infty)$ depending only on n, p, and the Ahlfors regularity of Σ. Bearing in mind that $2_* > 1$, we also have

$$\begin{aligned} \|f\|_{\mathrm{B}^p(\Sigma,\sigma)} &= \sup_{k\in\mathbb{N}_0} 2_*^{-k(n-1)/p}\|f\cdot\mathbb{1}_{C_k}\|_{L^p(\Sigma,\sigma)} \\ &\leq \Big(\sup_{k\in\mathbb{N}_0} 2_*^{-k(n-1)/p}\Big)\|f\|_{L^p(\Sigma,\sigma)} = \|f\|_{L^p(\Sigma,\sigma)}. \end{aligned} \tag{3.43}$$

Hence, the embeddings in (3.38) follow from (3.42)-(3.43).

Let us now verify the claim made in (3.39). To this goal, select a reference point $x_0 \in \Sigma$, and a dilation factor 2_* as in Definition 3.1. Pick $\varepsilon \in (0, \infty)$ and a function $f \in \mathrm{B}^p(\Sigma, \sigma)$. Using the fact that

$$1 + |x|^{n-1+\varepsilon} \approx 1 + |x - x_0|^{n-1+\varepsilon} \text{ uniformly in } x \in \Sigma, \tag{3.44}$$

employing Hölder's inequality, the Ahlfors regularity of Σ, and Proposition 3.1, we may estimate

$$
\begin{aligned}
&\int_\Sigma \frac{|f(x)|}{1+|x|^{n-1+\varepsilon}}\,d\sigma(x)\\
&\le \int_{\Delta(x_0,2_*)} \frac{|f(x)|}{1+|x|^{n-1}}\,d\sigma(x) + C\sum_{k=1}^{\infty}\int_{C_k}\frac{|f(x)|}{1+|x-x_0|^{n-1+\varepsilon}}\,d\sigma(x)\\
&\le \int_{\Delta(x_0,2_*)}|f|\,d\sigma + C\sum_{k=1}^{\infty}\frac{1}{2_*^{k(n-1+\varepsilon)}}\int_{\Delta(x_0,2_*^{k+1})}|f|\,d\sigma\\
&\le \Big(\int_{\Delta(x_0,2_*)}|f|^p\,d\sigma\Big)^{1/p}\sigma\big(\Delta(x_0,2_*)\big)^{1/p'}\\
&\quad + \sum_{k=1}^{\infty}\frac{1}{2_*^{k(n-1+\varepsilon)}}\Big(\int_{\Delta(x_0,2_*^{k+1})}|f|^p\,d\sigma\Big)^{1/p}\sigma\big(\Delta(x_0,2_*^{k+1})\big)^{1/p'}\\
&\le \sigma\big(\Delta(x_0,2_*)\big)\Big(⨍_{\Delta(x_0,2_*)}|f|^p\,d\sigma\Big)^{1/p}\\
&\quad + \sum_{k=1}^{\infty}\frac{\sigma\big(\Delta(x_0,2_*^{k+1})\big)}{2_*^{k(n-1+\varepsilon)}}\Big(⨍_{\Delta(x_0,2_*^{k+1})}|f|^p\,d\sigma\Big)^{1/p}\\
&\le C\Big(\sum_{k=0}^{\infty}\frac{2_*^{k(n-1)}}{2_*^{k(n-1+\varepsilon)}}\Big)\sup_{R\ge 2_*}\Big(⨍_{\Delta(x_0,R)}|f|^p\,d\sigma\Big)^{1/p}\\
&\le C\Big(\sum_{k=0}^{\infty}\frac{1}{2_*^{k\varepsilon}}\Big)\|f\|_{\mathrm{B}^p(\Sigma,\sigma)}
\end{aligned}
\tag{3.45}
$$

for some constant $C \in (0,\infty)$ depending only on n, p, ε, and the Ahlfors regularity of Σ. Since $2_* > 1$, the series in (3.45) converges. This finishes the justification of (3.39).

Finally, the first inclusion in (3.40) is a direct consequence of (3.37), so we only need to verify the second inclusion. To this end, consider an arbitrary compact set $K \subseteq \Sigma$, and let N be sufficiently large so that $K \subseteq \Delta(x_0, 2_*^N) = \cup_{k=0}^N C_k$. Then, for any $f \in \mathrm{B}^p(\Sigma,\sigma)$, we have

$$
\begin{aligned}
\int_K |f|^p\,d\sigma &\le \sum_{k=0}^{N} 2_*^{k(n-1)/p}\big(2_*^{-k(n-1)/p}\|f\cdot \mathbb{1}_{C_k}\|_{L^p(\Sigma,\sigma)}\big)\\
&\le \Big(\sum_{k=0}^{N} 2_*^{k(n-1)/p}\Big)\|f\|_{\mathrm{B}^p(\Sigma,\sigma)} < \infty.
\end{aligned}
\tag{3.46}
$$

Hence, $f \in L^p_{\mathrm{loc}}(\Sigma,\sigma) \subseteq L^1_{\mathrm{loc}}(\Sigma,\sigma)$. □

The space B_0^p in the Euclidean setting is defined as the closure of $\mathscr{C}_0^\infty(\mathbb{R}^n)$ in B^p (see, e.g., [Ga89]). In the setting of unbounded Ahlfors regular sets, the correct substitute for the space of smooth functions with compact support is the space of compactly supported Lipschitz functions. This is made precise in the following definition:

Definition 3.2 Consider $\Sigma \subseteq \mathbb{R}^n$ a closed Ahlfors regular set that is unbounded, and abbreviate $\sigma := \mathcal{H}^{n-1}\lfloor\Sigma$. Fix an integrability exponent $p \in (0,\infty)$. Define $\mathrm{B}^p_0(\Sigma,\sigma)$ as the closure of the space $\mathrm{Lip}_c(\Sigma)$ in the space $\mathrm{B}^p(\Sigma,\sigma)$, i.e.,

$$\mathrm{B}^p_0(\Sigma,\sigma) := \overline{\mathrm{Lip}_c(\Sigma)}^{\|\cdot\|_{\mathrm{B}^p(\Sigma,\sigma)}}. \tag{3.47}$$

In other words, $\mathrm{B}^p_0(\Sigma,\sigma)$ is the collection of all functions $f \in \mathrm{B}^p(\Sigma,\sigma)$ for which there exists a sequence of functions $\{f_j\}_{j\in\mathbb{N}} \subseteq \mathrm{Lip}_c(\Sigma)$ with the property that

$$f_j \to f \text{ in } \mathrm{B}^p(\Sigma,\sigma) \text{ as } j \to \infty. \tag{3.48}$$

Equip the space $\mathrm{B}^p_0(\Sigma,\sigma)$ with the norm inherited from $\mathrm{B}^p(\Sigma,\sigma)$.

Remark 3.7 The above definition is meaningful since for every $p \in [1,\infty]$ we have $\mathrm{Lip}_c(\Sigma) \subseteq L^p(\Sigma,\sigma) \subseteq \mathrm{B}^p(\Sigma,\sigma)$, where the second inclusion follows from Proposition 3.4. Moreover, by design $\mathrm{B}^p_0(\Sigma,\sigma)$ is a closed subspace of $\mathrm{B}^p(\Sigma,\sigma)$ in which $\mathrm{Lip}_c(\Sigma)$ is a dense subspace.

The next result provides several characterizations of B^p_0. As a preamble, the reader is reminded that L^p_{comp} is the subspace of L^p consisting of functions with compact support.

Proposition 3.5 *Assume $\Sigma \subseteq \mathbb{R}^n$ is a closed Ahlfors regular set that is unbounded, and abbreviate $\sigma := \mathcal{H}^{n-1}\lfloor\Sigma$. Select a reference point $x_0 \in \Sigma$, and a finite dilation factor $2_* > (C_\Sigma/c_\Sigma)^{1/n-1}$, where C_Σ and c_Σ denote the upper and lower Ahlfors regularity constant of Σ, respectively. Also, recall the family of annuli $\{C_k\}_{k\in\mathbb{N}_0}$ from* (3.1). *Fix an integrability exponent $p \in (1,\infty)$ along with an arbitrary function $f \in \mathrm{B}^p(\Sigma,\sigma)$. Then the following statements are equivalent.*

(a) The following holds:

$$2_*^{-k(n-1)/p}\|f \cdot \mathbb{1}_{C_k}\|_{L^p(\Sigma,\sigma)} \to 0 \text{ as } k \to \infty. \tag{3.49}$$

(b) The following holds:

$$\Big(⨍_{\Delta(x_0,R)} |f|^p \, d\sigma\Big)^{1/p} \to 0 \text{ as } R \to \infty. \tag{3.50}$$

(c) There exists $\{\phi_j\}_{j\in\mathbb{N}} \subseteq L^\infty_{\mathrm{comp}}(\Sigma,\sigma)$ satisfying

$$\phi_j \to f \text{ in } \mathrm{B}^p(\Sigma,\sigma) \text{ as } j \to \infty. \tag{3.51}$$

(d) There exists $\{\psi_j\}_{j\in\mathbb{N}} \subseteq \mathrm{Lip}_c(\Sigma)$ *satisfying*

$$\psi_j \to f \ \text{ in } \mathrm{B}^p(\Sigma,\sigma) \text{ as } j \to \infty. \tag{3.52}$$

(e) There exists $\{\eta_j\}_{j\in\mathbb{N}} \subseteq L^p(\Sigma,\sigma)$ *satisfying*

$$\eta_j \to f \ \text{ in } \mathrm{B}^p(\Sigma,\sigma) \text{ as } j \to \infty. \tag{3.53}$$

(f) There exists $\{\xi_j\}_{j\in\mathbb{N}} \subseteq L^p_{\mathrm{comp}}(\Sigma,\sigma)$ *satisfying*

$$\xi_j \to f \ \text{ in } \mathrm{B}^p(\Sigma,\sigma) \text{ as } j \to \infty. \tag{3.54}$$

(g) There exists $\{\zeta_j\}_{j\in\mathbb{N}} \subseteq L^r(\Sigma,\sigma)$*, for some* $r \in [p,\infty)$*, satisfying*

$$\zeta_j \to f \ \text{ in } \mathrm{B}^p(\Sigma,\sigma) \text{ as } j \to \infty. \tag{3.55}$$

(h) There exists $\{\theta_j\}_{j\in\mathbb{N}} \subseteq L^q_{\mathrm{comp}}(\Sigma,\sigma)$*, for some* $q \in [p,\infty]$*, satisfying*

$$\theta_j \to f \ \text{ in } \mathrm{B}^p(\Sigma,\sigma) \text{ as } j \to \infty. \tag{3.56}$$

Proof We begin by justifying the implication *(a)⇒(f)*. To this end, for each $j \in \mathbb{N}_0$ abbreviate

$$\xi_j := f \cdot \mathbb{1}_{\Delta(x_0, 2_*^{j+1})} \quad \text{on } \Sigma. \tag{3.57}$$

Since $f \in \mathrm{B}^p(\Sigma,\sigma) \subseteq L^p_{\mathrm{loc}}(\Sigma,\sigma)$ (cf. (3.40)), it is clear that $\xi_j \in L^p_{\mathrm{comp}}(\Sigma,\sigma)$ for each $j \in \mathbb{N}_0$. Recall (3.1) and notice that for each $j \in \mathbb{N}_0$ we have

$$\xi_j \cdot \mathbb{1}_{C_k} = \begin{cases} f \cdot \mathbb{1}_{C_k}, & \text{if } k \in \{0,1,\dots,j\}, \\ 0, & \text{if } k \in \mathbb{N}_0 \setminus \{0,1,\dots,j\}. \end{cases} \tag{3.58}$$

Fix $\varepsilon \in (0,\infty)$ arbitrary. In view of (3.49), select $k_0 \in \mathbb{N}$ such that

$$\sup_{k \geq k_0} \big(2_*^{-k(n-1)/p} \|f \cdot \mathbb{1}_{C_k}\|_{L^p(\Sigma,\sigma)}\big) < \varepsilon. \tag{3.59}$$

Thanks to this and (3.58), whenever $j \in \mathbb{N}$ and $j \geq k_0$ we have

$$\begin{aligned} \|f - \xi_j\|_{\mathrm{B}^p(\Sigma,\sigma)} &= \sup_{k\in\mathbb{N}_0} \big(2_*^{-k(n-1)/p} \|(f-\xi_j)\mathbb{1}_{C_k}\|_{L^p(\Sigma,\sigma)}\big) \\ &= \sup_{k>j} \big(2_*^{-k(n-1)/p} \|f \cdot \mathbb{1}_{C_k}\|_{L^p(\Sigma,\sigma)}\big) \\ &\leq \sup_{k\geq k_0} \big(2_*^{-k(n-1)/p} \|f \cdot \mathbb{1}_{C_k}\|_{L^p(\Sigma,\sigma)}\big) < \varepsilon. \end{aligned} \tag{3.60}$$

This proves (3.54), which establishes *(a)*⇒*(f)*. The implication *(f)*⇒*(e)* follows from the fact that $L^p_{\rm comp}(\Sigma,\sigma)\subseteq L^p(\Sigma,\sigma)$ (by simply choosing $\eta_j:=\xi_j$).

We next verify *(e)*⇒*(d)*. Consider $\{\eta_j\}_{j\in\mathbb{N}}\subseteq L^p(\Sigma,\sigma)$ such that (3.53) holds. Since $\mathrm{Lip}_c(\Sigma)$ is dense in $L^p(\Sigma,\sigma)$ (cf. [GHA-I, Proposition 3.7.1, p. 279]), for each $j\in\mathbb{N}$ choose $\psi_j\in\mathrm{Lip}_c(\Sigma)$ satisfying

$$\|\eta_j-\psi_j\|_{L^p(\Sigma,\sigma)}<\frac{1}{j}. \tag{3.61}$$

Bering in mind that $L^p(\Sigma,\sigma)\hookrightarrow \mathrm{B}^p(\Sigma,\sigma)$ continuously (cf. Proposition 3.4; see also (3.43)), this yields

$$\|\eta_j-\psi_j\|_{\mathrm{B}^p(\Sigma,\sigma)}<\frac{1}{j}\quad\text{for all } j\in\mathbb{N}. \tag{3.62}$$

In view of (3.53), given $\varepsilon\in(0,\infty)$, it is possible to select $j_0\in\mathbb{N}$ for which

$$\|f-\eta_j\|_{\mathrm{B}^p(\Sigma,\sigma)}<\varepsilon\quad\text{whenever } j\geq j_0. \tag{3.63}$$

Let $j_0\in\mathbb{N}$ be large enough so that $\frac{1}{j}<\varepsilon$ for all $j\geq j_0$. From this, (3.62)-(3.63), we obtain

$$\|f-\psi_j\|_{\mathrm{B}^p(\Sigma,\sigma)}\leq\|f-\eta_j\|_{\mathrm{B}^p(\Sigma,\sigma)}+\|\eta_j-\psi_j\|_{\mathrm{B}^p(\Sigma,\sigma)}<2\varepsilon, \tag{3.64}$$

for every $j\in\mathbb{N}$ with $j\geq j_0$. This establishes *(e)*⇒*(d)*.

The implication *(d)*⇒*(c)* follows by simply choosing $\phi_j:=\psi_j$, keeping in mind that $\mathrm{Lip}_c(\Sigma)\subseteq L^\infty_{\rm comp}(\Sigma,\sigma)$. To proceed, consider the implication *(c)*⇒*(b)*. Pick $\varepsilon\in(0,\infty)$, and select $j_0\in\mathbb{N}$ for which

$$\|f-\phi_{j_0}\|_{\mathrm{B}^p(\Sigma,\sigma)}<\varepsilon. \tag{3.65}$$

Recall that ϕ_{j_0} is a compactly supported function. As such, we have ϕ_{j_0} is supported in $\Delta(x_0,r)$ for some large radius $r\in(1,\infty)$. Denote by $M:=\sup_{j\in\mathbb{N}}\|\phi_j\|_{\mathrm{B}^p(\Sigma,\sigma)}$, which is a positive and finite constant since $\{\phi_j\}_{j\in\mathbb{N}}$ is a convergent sequence in $\mathrm{B}^p(\Sigma,\sigma)$ (see (3.51)). With this piece of notation, using the Ahlfors regularity of Σ, Proposition 3.1, and (3.65), for every $R\geq\max\{2_*,\varepsilon^{-p/(n-1)}r\}$ we estimate

$$
\begin{aligned}
&\Big(⨍_{\Delta(x_0,R)} |f|^p\,d\sigma\Big)^{1/p} \\
&\le \Big(⨍_{\Delta(x_0,R)} |f-\phi_{j_0}|^p\,d\sigma\Big)^{1/p} + \Big(⨍_{\Delta(x_0,R)} |\phi_{j_0}|^p\,d\sigma\Big)^{1/p} \\
&\le \sup_{R\ge 2_*}\Big(⨍_{\Delta(x_0,R)} |f-\phi_{j_0}|^p\,d\sigma\Big)^{1/p} \\
&\quad + C\frac{\sigma\big(\Delta(x_0,r)\big)^{1/p}}{\sigma\big(\Delta(x_0,R)\big)^{1/p}}\Big(⨍_{\Delta(x_0,r)} |\phi_{j_0}|^p\,d\sigma\Big)^{1/p} \\
&\le C\|f-\phi_{j_0}\|_{\mathrm{B}^p(\Sigma,\sigma)} + \frac{Cr^{(n-1)/p}}{R^{(n-1)/p}}\Big(⨍_{\Delta(x_0,r)} |\phi_{j_0}|^p\,d\sigma\Big)^{1/p} \\
&\le C\varepsilon + \frac{Cr^{(n-1)/p}}{R^{(n-1)/p}}\|\phi_{j_0}\|_{B^p(\Sigma,\sigma)} \\
&\le C(1+M)\varepsilon
\end{aligned}
\tag{3.66}
$$

for some constant $C \in (0,\infty)$ that depends only on n, p, and the Ahlfors regularity of Σ. This shows (3.50), thus justifying *(c)⇒(b)*.

To prove *(b)⇒(a)*, we recycle the computations made in (3.20) to derive

$$
\begin{aligned}
&2_*^{-k(n-1)/p}\|f\cdot \mathbb{1}_{C_k}\|_{L^p(\Sigma,\sigma)} \\
&\qquad \le C\Big(⨍_{\Delta(x_0,2_*^{k+1})} |f|^p\,d\sigma\Big)^{1/p} \quad \text{for all } k\in\mathbb{N}_0.
\end{aligned}
\tag{3.67}
$$

Letting $k\to\infty$ in (3.67) and making use of (3.50) we arrive at (3.49).

Next, note that the implication *(c)⇒(h)* is obtained by taking $q := \infty$ and $\theta_j := \phi_j$. The fact that *(h)⇒(g)* may be justified as follows. In the case when $q \neq \infty$, choose $r := q$ and $\zeta_j := \theta_j \in L^q_{\mathrm{comp}}(\Sigma,\sigma) \subseteq L^q(\Sigma,\sigma)$ for each $j\in\mathbb{N}$. When $q=\infty$, pick $r := p$ and $\zeta_j := \theta_j \in L^\infty_{\mathrm{comp}}(\Sigma,\sigma)\subseteq L^p(\Sigma,\sigma)$.

Lastly, to see that the implication *(g)⇒(b)* holds, fix $\varepsilon\in(0,\infty)$ and select $j_0\in\mathbb{N}$ so that

$$
\|f-\zeta_{j_0}\|_{\mathrm{B}^p(\Sigma,\sigma)} < \varepsilon. \tag{3.68}
$$

Using the Ahlfors regularity of Σ, Hölder's inequality, Proposition 3.1, and (3.68), for every $R \ge \max\Big\{2_*, \varepsilon^{-r/(n-1)}(1+\|\zeta_{j_0}\|_{L^r(\Sigma,\sigma)})^{r/(n-1)}\Big\}$ we have

$$
\begin{aligned}
&\Big(\fint_{\Delta(x_0,R)} |f|^p\,d\sigma\Big)^{1/p} \\
&\leq \Big(\fint_{\Delta(x_0,R)} |f-\zeta_{j_0}|^p\,d\sigma\Big)^{1/p} + \Big(\fint_{\Delta(x_0,R)} |\zeta_{j_0}|^p\,d\sigma\Big)^{1/p} \\
&\leq \sup_{R\geq 2_*} \Big(\fint_{\Delta(x_0,R)} |f-\zeta_{j_0}|^p\,d\sigma\Big)^{1/p} + \Big(\fint_{\Delta(x_0,R)} |\zeta_{j_0}|^r\,d\sigma\Big)^{1/r} \\
&\leq C\|f-\zeta_{j_0}\|_{\mathrm{B}^p(\Sigma,\sigma)} + \frac{C}{R^{(n-1)/r}}\|\zeta_{j_0}\|_{L^r(\Sigma,\sigma)} \\
&\leq C\varepsilon + C\frac{\|\zeta_{j_0}\|_{L^r(\Sigma,\sigma)}}{1+\|\zeta_{j_0}\|_{L^r(\Sigma,\sigma)}}\varepsilon \\
&\leq C\varepsilon, \qquad (3.69)
\end{aligned}
$$

for some constant $C \in (0,\infty)$ depending only n, p, r, and the Ahlfors regularity of Σ. By the arbitrariness of ε, this proves (3.50), which justifies *(g)⇒(b)*. This concludes the proof of Proposition 3.5. □

Remark 3.8 In the context of Proposition 3.5, we see that item *(d)* is what describes the membership to $\mathrm{B}^p_0(\Sigma,\sigma)$. Hence, Proposition 3.5 ensures that given a function $f \in \mathrm{B}^p(\Sigma,\sigma)$ we have $f \in \mathrm{B}^p_0(\Sigma,\sigma)$ if and only if one of the items *(a)–(h)* holds.

Below, we establish a Hölder-type inequality in the context of A^p and B^p spaces.

Proposition 3.6 *Assume $\Sigma \subseteq \mathbb{R}^n$ is a closed unbounded Ahlfors regular set, and abbreviate $\sigma := \mathcal{H}^{n-1}\lfloor\Sigma$. Pick an integrability exponent $p \in (1,\infty)$, and denote by $p' \in (1,\infty)$ its Hölder conjugate exponent, i.e., $1/p + 1/p' = 1$. Then for all σ-measurable functions $f, g : \Sigma \to \mathbb{R}$ one has*

$$
\int_\Sigma |fg|\,d\sigma \leq \|f\|_{\mathrm{A}^p(\Sigma,\sigma)}\|g\|_{B^{p'}(\Sigma,\sigma)}. \qquad (3.70)
$$

In particular, if $f \in \mathrm{A}^p(\Sigma,\sigma)$ and $g \in \mathrm{B}^{p'}(\Sigma,\sigma)$, it follows that $fg \in L^1(\Sigma,\sigma)$.

Proof Fix σ-measurable functions f and g defined on Σ. By using Hölder's inequality, we may estimate

$$
\begin{aligned}
&\int_{\Sigma} |fg|\, d\sigma \\
&= \sum_{k=0}^{\infty} \int_{C_k} |fg|\, d\sigma \leq \sum_{k=0}^{\infty} \|f \cdot \mathbb{1}_{C_k}\|_{L^p(\Sigma,\sigma)} \|g \cdot \mathbb{1}_{C_k}\|_{L^{p'}(\Sigma,\sigma)} \\
&= \sum_{k=0}^{\infty} \Big\{ 2_*^{k(n-1)(1-1/p)} \|f\|_{L^p(C_k,\sigma)} 2_*^{-k(n-1)/p'} \|g\|_{L^{p'}(C_k,\sigma)} \Big\} \\
&\leq \Big(\sum_{k=0}^{\infty} 2_*^{k(n-1)(1-1/p)} \|f\|_{L^p(C_k,\sigma)} \Big) \sup_{k \in \mathbb{N}_0} \Big(2_*^{-k(n-1)/p'} \|g\|_{L^{p'}(C_k,\sigma)} \Big) \\
&= \|f\|_{\mathrm{A}^p(\Sigma,\sigma)} \|g\|_{B^{p'}(\Sigma,\sigma)},
\end{aligned}
\tag{3.71}
$$

which justifies (3.70). □

It is of interest to establish a version of the classical Lebesgue's Dominated Convergence Theorem in the context of Beurling spaces.

Proposition 3.7 *Consider a closed Ahlfors regular set $\Sigma \subseteq \mathbb{R}^n$ that is unbounded, and abbreviate $\sigma := \mathcal{H}^{n-1} \lfloor \Sigma$. Let $\{f_j\}_{j \in \mathbb{N}}$ be a sequence of σ-measurable functions on Σ such that for some σ-measurable function $f : \Sigma \to \mathbb{R}$ one has*

$$
f_j \to f \ \ \text{at } \sigma\text{-a.e. point on } \Sigma \text{ as } j \to \infty. \tag{3.72}
$$

Having fixed $p \in (0, \infty)$, assume there exists $g \in \mathrm{A}^p(\Sigma, \sigma)$ for which

$$
|f_j(x)| \leq g(x) \ \ \text{at } \sigma\text{-a.e. } x \in \Sigma \text{ and for all } j \in \mathbb{N}. \tag{3.73}
$$

Then

$$
\|f_j - f\|_{\mathrm{A}^p(\Sigma,\sigma)} \to 0 \ \ \text{as } j \to \infty. \tag{3.74}
$$

Proof To get started, (3.72)-(3.73) ensure that for each $k \in \mathbb{N}_0$ fixed we have

$$
\begin{aligned}
&f_j \to f \ \ \text{at } \sigma\text{-a.e. point on } C_k \text{ as } j \to \infty, \text{ and} \\
&|f_j(x)| \leq g(x) \ \ \text{at } \sigma\text{-a.e. } x \in C_k \text{ and for all } j \in \mathbb{N}.
\end{aligned}
\tag{3.75}
$$

Since $g \in \mathrm{A}^p(\Sigma, \sigma)$, it is clear from (3.2) that

$$
g \cdot \mathbb{1}_{C_k} \in L^p(\Sigma, \sigma). \tag{3.76}
$$

This, (3.75), and Lebesgue's Dominated Convergence Theorem imply

$$
\|(f_j - f) \cdot \mathbb{1}_{C_k}\|_{L^p(\Sigma,\sigma)} \to 0 \ \ \text{as } j \to \infty, \text{ for each } k \in \mathbb{N}_0. \tag{3.77}
$$

Hence, having fixed $\varepsilon > 0$ and $k \in \mathbb{N}_0$, there exists $J_k \in \mathbb{N}$ such that

$$
\|(f_j - f) \cdot \mathbb{1}_{C_k}\|_{L^p(\Sigma,\sigma)} < \frac{\varepsilon}{2_*^{k(n-1)(1-1/p)} 2^{k+1}} \quad \text{for all } j \geq J_k, \tag{3.78}
$$

where $p' \in (1, \infty)$ denotes the Hölder conjugate exponent of p. Moreover, (3.72)-(3.73) imply

$$|f(x)| \leq g(x) \quad \text{at } \sigma\text{-a.e. } x \in \Sigma. \tag{3.79}$$

Next, since $g \in \mathrm{A}^p(\Sigma, \sigma)$ there exists some $N \in \mathbb{N}$ sufficiently large for which

$$\sum_{k=N+1}^{\infty} 2_*^{k(n-1)(1-1/p)} \|g \cdot \mathbb{1}_{C_k}\|_{L^p(\Sigma,\sigma)} < \varepsilon. \tag{3.80}$$

Combining (3.78)-(3.80), we deduce that if $j \in \mathbb{N}$ and $j \geq \max\{J_1, \ldots, J_N\}$ then

$$\begin{aligned}
&\|f - f_j\|_{\mathrm{A}^p(\Sigma,\sigma)} \\
&= \sum_{k=0}^{N} 2_*^{k(n-1)(1-1/p)} \|(f_j - f) \cdot \mathbb{1}_{C_k}\|_{L^p(\Sigma,\sigma)} \\
&\quad + \sum_{k=N+1}^{\infty} 2_*^{k(n-1)(1-1/p)} \|(f_j - f) \cdot \mathbb{1}_{C_k}\|_{L^p(\Sigma,\sigma)} \\
&\leq \sum_{k=0}^{N} \frac{\varepsilon}{2^{k+1}} + \sum_{k=N+1}^{\infty} 2_*^{k(n-1)(1-1/p)} \big(\|f_j \cdot \mathbb{1}_{C_k}\|_{L^p(\Sigma,\sigma)} + \|f \cdot \mathbb{1}_{C_k}\|_{L^p(\Sigma,\sigma)}\big) \\
&\leq \sum_{k=0}^{\infty} \frac{\varepsilon}{2^{k+1}} + 2 \sum_{k=N+1}^{\infty} 2_*^{k(n-1)(1-1/p)} \|g \cdot \mathbb{1}_{C_k}\|_{L^p(\Sigma,\sigma)} \leq 3\varepsilon.
\end{aligned} \tag{3.81}$$

Since $\varepsilon \in (0, \infty)$ was taken arbitrary to begin with, this ultimately proves (3.74). □

We now present a version of Lebesgue's Monotone Convergence Theorem in the setting of Beurling spaces.

Proposition 3.8 *Let $\Sigma \subseteq \mathbb{R}^n$ be a closed Ahlfors regular set that is unbounded, and abbreviate $\sigma := \mathcal{H}^{n-1} \lfloor \Sigma$. Assume $\{f_j\}_{j \in \mathbb{N}}$ is a sequence of non-negative σ-measurable functions on Σ such that $f_j \nearrow f$ pointwise σ-a.e. on Σ as $j \to \infty$, for some non-negative σ-measurable function f defined on Σ. Then whenever $p \in (0, \infty)$ one has*

$$\|f_j\|_{\mathrm{A}^p(\Sigma,\sigma)} \nearrow \|f\|_{\mathrm{A}^p(\Sigma,\sigma)} \quad \text{as } j \to \infty. \tag{3.82}$$

Proof To get started, notice the lattice property of $\mathrm{A}^p(\Sigma, \sigma)$ (cf. (3.6)) implies

$$\|f_j\|_{\mathrm{A}^p(\Sigma,\sigma)} \leq \|f_{j+1}\|_{\mathrm{A}^p(\Sigma,\sigma)} \leq \|f\|_{\mathrm{A}^p(\Sigma,\sigma)} \quad \text{for all } j \in \mathbb{N}. \tag{3.83}$$

Hence,

$$\limsup_{j\to\infty} \|f_j\|_{\mathrm{A}^p(\Sigma,\sigma)} \leq \|f\|_{\mathrm{A}^p(\Sigma,\sigma)}. \tag{3.84}$$

In the opposite direction, fix $N \in \mathbb{N}$ and use the classical Lebesgue's Monotone Convergence Theorem to obtain

$$
\begin{aligned}
&\sum_{k=0}^{N} 2_*^{k(n-1)(1-1/p)} \|f \cdot \mathbb{1}_{C_k}\|_{L^p(\Sigma,\sigma)} \\
&\quad = \sum_{k=0}^{N} 2_*^{k(n-1)(1-1/p)} \liminf_{j\to\infty} \|f_j \cdot \mathbb{1}_{C_k}\|_{L^p(\Sigma,\sigma)} \\
&\quad = \liminf_{j\to\infty} \sum_{k=0}^{N} 2_*^{k(n-1)(1-1/p)} \|f_j \cdot \mathbb{1}_{C_k}\|_{L^p(\Sigma,\sigma)} \\
&\quad \leq \liminf_{j\to\infty} \|f_j\|_{\mathrm{A}^p(\Sigma,\sigma)}.
\end{aligned}
\tag{3.85}
$$

After letting $N \to \infty$, we obtain

$$
\|f\|_{\mathrm{A}^p(\Sigma,\sigma)} \leq \liminf_{j\to\infty} \|f_j\|_{\mathrm{A}^p(\Sigma,\sigma)}. \tag{3.86}
$$

Hence, (3.82) follows from (3.83), (3.84), and (3.86). □

In the following result we establish the completeness of the spaces A^p and B^p.

Proposition 3.9 *Consider a closed unbounded Ahlfors regular set $\Sigma \subseteq \mathbb{R}^n$, and abbreviate $\sigma := \mathcal{H}^{n-1}\lfloor\Sigma$. Then for every exponent $p \in (0,\infty)$ the spaces $\big(\mathrm{A}^p(\Sigma,\sigma), \|\cdot\|_{\mathrm{A}^p(\Sigma,\sigma)}\big)$ and $\big(\mathrm{B}^p(\Sigma,\sigma), \|\cdot\|_{\mathrm{B}^p(\Sigma,\sigma)}\big)$ are complete. In particular, $\mathrm{A}^p(\Sigma,\sigma)$ and $\mathrm{B}^p(\Sigma,\sigma)$ are genuine Banach spaces in the regime $p \in [1,\infty)$.*

Proof We first justify the completeness of $\mathrm{A}^p(\Sigma,\sigma)$. Consider $\{f_j\}_{j\in\mathbb{N}}$ a Cauchy sequence in $\mathrm{A}^p(\Sigma,\sigma)$. Having fixed $\varepsilon \in (0,\infty)$ arbitrary, select $j_0 \in \mathbb{N}$ such that whenever $j, \ell \in \mathbb{N}$ and $j, \ell \geq j_0$ we have

$$
\sum_{k=0}^{\infty} 2_*^{k(n-1)(1-1/p)} \|(f_j - f_\ell) \cdot \mathbb{1}_{C_k}\|_{L^p(\Sigma,\sigma)} = \|f_j - f_\ell\|_{\mathrm{A}^p(\Sigma,\sigma)} < \varepsilon. \tag{3.87}
$$

In particular, since

$$
\Big\|f_j\big|_{C_k} - f_\ell\big|_{C_k}\Big\|_{L^p(C_k,\sigma)} = \|(f_j - f_\ell) \cdot \mathbb{1}_{C_k}\|_{L^p(\Sigma,\sigma)} \leq \|f_j - f_\ell\|_{\mathrm{A}^p(\Sigma,\sigma)} \tag{3.88}
$$

for all $j, \ell \in \mathbb{N}$ and $k \in \mathbb{N}_0$, this implies that for any $k \in \mathbb{N}_0$ the sequence $\big\{f_j\big|_{C_k}\big\}_{j\in\mathbb{N}}$ is Cauchy in the quasi-Banach space $L^p(C_k,\sigma)$. Hence, for each $k \in \mathbb{N}$ there exists some function $f^{(k)} \in L^p(C_k,\sigma)$ for which

$$
f_j\big|_{C_k} \to f^{(k)} \text{ in } L^p(C_k,\sigma) \text{ as } j \to \infty. \tag{3.89}
$$

Denoting by $\widetilde{f}^{(k)}$ the extension by zero of $f^{(k)}$ to the entire Σ, define $f : \Sigma \to \mathbb{R}$ by setting

$$f := \sum_{k=0}^{\infty} \widetilde{f}^{(k)}. \tag{3.90}$$

Then f is a well-defined σ-measurable function on Σ. Granted this, if we fix $N \in \mathbb{N}$ and $j \geq j_0$ then (3.87) implies

$$\sum_{k=0}^{N} 2_*^{k(n-1)(1-1/p)} \big\| f_j\big|_{C_k} - f_\ell\big|_{C_k} \big\|_{L^p(C_k,\sigma)} < \varepsilon \quad \text{for every } \ell \geq j_0. \tag{3.91}$$

Letting $\ell \to \infty$, bearing in mind (3.89) and (3.90), we obtain

$$\begin{aligned} &\sum_{k=0}^{N} 2_*^{k(n-1)(1-1/p)} \big\| (f_j - f) \cdot \mathbb{1}_{C_k} \big\|_{L^p(\Sigma,\sigma)} \\ &\quad = \sum_{k=0}^{N} 2_*^{k(n-1)(1-1/p)} \big\| f_j\big|_{C_k} - f^{(k)} \big\|_{L^p(C_k,\sigma)} \leq \varepsilon \end{aligned} \tag{3.92}$$

for all $j \geq j_0$ and $N \in \mathbb{N}_0$. Since this holds for every $N \in \mathbb{N}_0$, it follows that for every $j \geq j_0$ we have

$$\|f_j - f\|_{\mathrm{A}^p(\Sigma,\sigma)} = \sum_{k=0}^{\infty} 2_*^{k(n-1)(1-1/p)} \big\| (f_j - f) \cdot \mathbb{1}_{C_k} \big\|_{L^p(\Sigma,\sigma)} \leq \varepsilon. \tag{3.93}$$

Consequently, $f = (f - f_{j_0}) + f_{j_0} \in \mathrm{A}^p(\Sigma,\sigma)$ and $f_j \to f$ in $\mathrm{A}^p(\Sigma,\sigma)$ as $j \to \infty$. This establishes the completeness of $\mathrm{A}^p(\Sigma,\sigma)$.

Moving on, we prove that $\mathrm{B}^p(\Sigma,\sigma)$ is complete. Fix $\{f_j\}_{j\in\mathbb{N}} \subseteq \mathrm{B}^p(\Sigma,\sigma)$ a Cauchy sequence. Pick $\varepsilon \in (0,\infty)$ arbitrary and let $j_0 \in \mathbb{N}$ be such that

$$\sup_{k\in\mathbb{N}_0} 2_*^{-k(n-1)/p} \|(f_j - f_\ell) \cdot \mathbb{1}_{C_k}\|_{L^p(\Sigma,\sigma)} = \|f_j - f_\ell\|_{\mathrm{B}^p(\Sigma,\sigma)} < \varepsilon \tag{3.94}$$

for all $j, \ell \in \mathbb{N}$ with $j, \ell \geq j_0$. For each $k \in \mathbb{N}_0$ fixed, this entails

$$\begin{aligned} &2_*^{-k(n-1)/p} \big\| f_j\big|_{C_k} - f_\ell\big|_{C_k} \big\|_{L^p(C_k,\sigma)} \\ &\qquad = 2_*^{-k(n-1)/p} \|(f_j - f_\ell) \cdot \mathbb{1}_{C_k}\|_{L^p(\Sigma,\sigma)} < \varepsilon \end{aligned} \tag{3.95}$$

for all $j, \ell \in \mathbb{N}$ with $j, \ell \geq j_0$. Therefore $\big\{f_j\big|_{C_k}\big\}_{j\in\mathbb{N}}$ is a Cauchy sequence in the quasi-Banach space $L^p(C_k,\sigma)$. Hence, for each $k \in \mathbb{N}_0$, there exists a function $f^{(k)} \in L^p(C_k,\sigma)$ such that $f_j\big|_{C_k} \to f^{(k)}$ in $L^p(C_k,\sigma)$ as $j \to \infty$. Hence, retaining the same piece of notation as in (3.90), and letting $\ell \to \infty$ in (3.95) we find that

$$2_*^{-k(n-1)/p} \|(f_j - f) \cdot \mathbb{1}_{C_k}\|_{L^p(\Sigma,\sigma)} = 2_*^{-k(n-1)/p} \big\| f_j\big|_{C_k} - f^{(k)} \big\|_{L^p(C_k,\sigma)} \leq \varepsilon \tag{3.96}$$

for all $j \in \mathbb{N}$ with $j \geq j_0$, and for all $k \in \mathbb{N}_0$. This implies

$$\sup_{k\in\mathbb{N}_0} 2_*^{-k(n-1)/p}\|(f_j - f)\cdot \mathbb{1}_{C_k}\|_{L^p(\Sigma,\sigma)} = \|f_j - f\|_{\mathrm{B}^p(\Sigma,\sigma)} \le \varepsilon \tag{3.97}$$

for all $j \in \mathbb{N}$ with $j \ge j_0$. As such, $f = f_{j_0} - (f_{j_0} - f) \in \mathrm{B}^p(\Sigma, \sigma)$ and $f_j \to f$ in $\mathrm{B}^p(\Sigma, \sigma)$ as $j \to \infty$. This justifies the completeness of $\mathrm{B}^p(\Sigma, \sigma)$. □

Remark 3.9 Upon recalling that $\mathrm{B}^p_0(\Sigma, \sigma)$ is a closed subspace of $\mathrm{B}^p(\Sigma, \sigma)$ (cf. Remark 3.7), Proposition 3.9 implies that $\big(\mathrm{B}^p_0(\Sigma, \sigma), \|\cdot\|_{\mathrm{B}^p(\Sigma,\sigma)}\big)$ is complete for every $p \in (0, \infty)$, hence Banach in the regime $p \in [1, \infty)$.

3.2 Duality in Beurling Spaces

As part of the development of the functional analytic framework of Beurling spaces in the setting of unbounded Ahlfors regular sets, in this section we prove a duality result between the spaces A^p and B^q. More specifically, we have the following proposition.

Proposition 3.10 *Consider a closed Ahlfors regular set $\Sigma \subseteq \mathbb{R}^n$ which is unbounded, and abbreviate $\sigma := \mathcal{H}^{n-1}\lfloor\Sigma$. Select an integrability exponent $p \in (1, \infty)$, and denote by $p' \in (1, \infty)$ its Hölder conjugate exponent. Then the dual of $\mathrm{A}^p(\Sigma, \sigma)$ is $\mathrm{B}^{p'}(\Sigma, \sigma)$, i.e.,*

$$\big(\mathrm{A}^p(\Sigma, \sigma)\big)^* = \mathrm{B}^{p'}(\Sigma, \sigma). \tag{3.98}$$

More precisely, the mapping

$$\mathrm{B}^{p'}(\Sigma, \sigma) \ni f \longmapsto \Phi f := \Lambda_f \in \big(\mathrm{A}^p(\Sigma, \sigma)\big)^* \tag{3.99}$$

where

$$\Lambda_f(g) := \int_\Sigma fg\, d\sigma \quad \textit{for all } g \in \mathrm{A}^p(\Sigma, \sigma) \tag{3.100}$$

is a well-defined, linear, isometric isomorphism.

Proof Well-definiteness of Φ. Fix $f \in \mathrm{B}^{p'}(\Sigma, \sigma)$, and consider the mapping Λ_f on $\mathrm{A}^p(\Sigma, \sigma)$ as in (3.100). Thanks to Proposition 3.6, the integral in (3.100) is absolutely convergent. Hence Λ_f is well defined and linear. Moreover, Λ_f continuous since (3.70) implies

$$|\Lambda_f(g)| \le \int_\Sigma |fg|\, d\sigma \le \|f\|_{\mathrm{B}^{p'}(\Sigma,\sigma)}\|g\|_{\mathrm{A}^p(\Sigma,\sigma)}, \tag{3.101}$$

for each $g \in \mathrm{A}^p(\Sigma, \sigma)$. This shows that the mapping

$$\begin{gathered}\mathrm{A}^p(\Sigma, \sigma) \ni g \longmapsto \Lambda_f(g) = \int_\Sigma fg\, d\sigma \quad \text{is a bounded}\\ \text{linear functional with norm } \|\Lambda_f\|_{(\mathrm{A}^p(\Sigma,\sigma))^*} \le \|f\|_{\mathrm{B}^{p'}(\Sigma,\sigma)}.\end{gathered} \tag{3.102}$$

Surjectivity of Φ. Having fixed $j \in \mathbb{N}_0$, denote by E_j the operator that assign each σ-measurable function $g : C_j \to \mathbb{R}$ to its extension by zero on the entire Σ, i.e.,

$$(E_j g)(x) := \begin{cases} g(x) & \text{if } x \in C_j, \\ 0 & \text{if } x \notin C_j, \end{cases} \qquad \text{for } \sigma\text{a.e. } x \in \Sigma. \tag{3.103}$$

Observe that for each $g \in L^p(C_j, \sigma)$ we have

$$\|E_j g\|_{A^p(\Sigma,\sigma)} = \sum_{k=0}^{\infty} 2_*^{k(n-1)/p'} \|(E_j g) \cdot \mathbb{1}_{C_k}\|_{L^p(\Sigma,\sigma)} = 2_*^{j(n-1)/p'} \|g\|_{L^p(C_j,\sigma)}. \tag{3.104}$$

Hence, the mapping

$$\begin{gathered} E_j : L^p(C_j, \sigma) \to A^p(\Sigma, \sigma) \ \text{ is a bounded linear} \\ \text{operator with norm } \ \|E_j\|_{L^p(C_j,\sigma) \to A^p(\Sigma,\sigma)} = 2_*^{j(n-1)/p'}. \end{gathered} \tag{3.105}$$

In order to prove that Φ (cf. (3.99)) is surjective, fix a continuous linear functional $\Lambda \in \big(A^p(\Sigma, \sigma)\big)^*$. From (3.105) we see that

$$\begin{gathered} \Lambda \circ E_j : L^p(C_j, \sigma) \longrightarrow \mathbb{R} \ \text{ is a well-defined, continuous, linear} \\ \text{functional with norm } \ \|\Lambda \circ E_j\|_{(L^p(C_k,\sigma))^*} \le 2_*^{j(n-1)/p'} \|\Lambda\|_{(A^p(\Sigma,\sigma))^*}. \end{gathered} \tag{3.106}$$

By the Riesz Representation Theorem, there exists some function $f_j \in L^{p'}(C_j, \sigma)$ for which

$$\begin{gathered} (\Lambda \circ E_j)(g) = \int_{C_j} f_j g \, d\sigma \ \text{ for all } g \in L^p(C_k, \sigma), \\ \text{and } \ \|\Lambda \circ E_j\|_{(L^p(C_j,\sigma))^*} = \|f_j\|_{L^{p'}(C_j,\sigma)}. \end{gathered} \tag{3.107}$$

Since for each $j \in \mathbb{N}_0$ the function $E_j f_j$ is supported in C_j, and $\{C_j\}_{j \in \mathbb{N}_0}$ is a disjoint family of sets on Σ, the function $f : \Sigma \to \mathbb{R}$ given by

$$f := \sum_{j=0}^{\infty} E_j f_j \tag{3.108}$$

is well defined and σ-measurable on Σ. We claim that f belongs to $B^{p'}(\Sigma, \sigma)$. To justify this, select $k \in \mathbb{N}_0$ and use (3.103), (3.106), (3.107) to compute

$$\begin{aligned} \|f \cdot \mathbb{1}_{C_k}\|_{L^{p'}(\Sigma,\sigma)} &= \Big\| \sum_{j=0}^{\infty} (E_j f_j) \cdot \mathbb{1}_{C_k} \Big\|_{L^{p'}(\Sigma,\sigma)} = \|(E_k f_k) \cdot \mathbb{1}_{C_k}\|_{L^{p'}(\Sigma,\sigma)} \\ &= \|f_k\|_{L^{p'}(C_k,\sigma)} = \|\Lambda \circ E_k\|_{(L^p(C_k,\sigma))^*} \\ &\le 2_*^{k(n-1)/p'} \|\Lambda\|_{(A^p(\Sigma,\sigma))^*}. \end{aligned} \tag{3.109}$$

Granted this, estimate

$$\begin{aligned}
\|f\|_{B^{p'}(\Sigma,\sigma)} &= \sup_{k\in\mathbb{N}_0} \big(2_*^{-k(n-1)/p'}\|f\cdot \mathbb{1}_{C_k}\|_{L^{p'}(\Sigma,\sigma)}\big) \\
&\leq \sup_{j\in\mathbb{N}_0} \big(2_*^{-k(n-1)/p'}2_*^{k(n-1)/p'}\|\Lambda\|_{(A^p(\Sigma,\sigma))^*}\big) \\
&= \|\Lambda\|_{(A^p(\Sigma,\sigma))^*} < \infty,
\end{aligned} \tag{3.110}$$

which yields the desired membership $f \in B^{p'}(\Sigma,\sigma)$.

Moving forward, we claim the linear functional Λ coincides with the functional Λ_f associated with f given in (3.108). To prove this, pick $g \in A^p(\Sigma,\sigma)$ and decompose it as

$$g = \sum_{k=0}^{\infty} g\cdot \mathbb{1}_{C_k} \quad \text{pointwise on } \Sigma. \tag{3.111}$$

In light of Proposition 3.8, the series (3.111) also converges in $A^p(\Sigma,\sigma)$. From this, (3.107), and Lebesgue's Dominated Convergence Theorem, we find that

$$\begin{aligned}
\Lambda(g) &= \sum_{j=0}^{\infty}\Lambda(g\cdot \mathbb{1}_{C_j}) = \sum_{j=0}^{\infty}(\Lambda\circ E_j)\big(g\big|_{C_j}\big) \\
&= \sum_{j=0}^{\infty}\int_{C_j} f_j\cdot\big(g\big|_{C_j}\big)\,d\sigma = \sum_{j=0}^{\infty}\int_{\Sigma}(E_jf_j)\cdot g\,d\sigma \\
&= \lim_{N\to\infty}\int_{\Sigma}\Big(\sum_{j=0}^{N}E_jf_j\Big)\cdot g\,d\sigma = \int_{\Sigma} fg\,d\sigma = \Lambda_f(g).
\end{aligned} \tag{3.112}$$

This shows $\Lambda = \Lambda_f$, and therefore Φ is a surjective mapping. Moreover, this, (3.102), and (3.110) entail

$$\|f\|_{B^{p'}(\Sigma,\sigma)} \leq \|\Lambda\|_{(A^p(\Sigma,\sigma))^*} = \|\Lambda_f\|_{(A^p(\Sigma,\sigma))^*} \leq \|f\|_{B^{p'}(\Sigma,\sigma)}, \tag{3.113}$$

which further implies

$$\|\Lambda_f\|_{(A^p(\Sigma,\sigma))^*} = \|f\|_{B^{p'}(\Sigma,\sigma)}. \tag{3.114}$$

<u>`Injectivity of` Φ.</u> Pick an arbitrary function $f \in B^{p'}(\Sigma,\sigma)$ and assume that $\Lambda_f = 0$ as functionals in $\big(A^p(\Sigma,\sigma)\big)^*$, i.e.,

$$\Lambda_f(g) = \int_{\Sigma} fg\,d\sigma = 0 \quad \text{for all } g \in A^p(\Sigma,\sigma). \tag{3.115}$$

Notice that for every $g \in L^p(\Sigma,\sigma)$, and $k \in \mathbb{N}_0$, we have $g\cdot\mathbb{1}_{C_k} \in A^p(\Sigma,\sigma)$. Hence,

$$\int_{\Sigma}(f\cdot\mathbb{1}_{C_k})g\,d\sigma=\int_{\Sigma}f(g\cdot\mathbb{1}_{C_k})\,d\sigma=0\quad\text{for all } g\in L^p(\Sigma,\sigma). \tag{3.116}$$

By Riesz Representation Theorem we may then conclude that $f=0$ at σ-a.e. point on C_k, for all $k\in\mathbb{N}_0$, which further entails that f vanishes σ-a.e. on Σ. This shows that Φ (cf. (3.99)) is an injective map. Moreover, this ensures that for every linear functional $\Lambda\in\big(\mathrm{A}^p(\Sigma,\sigma)\big)^*$ there exists a unique function $f\in\mathrm{B}^{p'}(\Sigma,\sigma)$ for which $\Lambda=\Lambda_f$. In particular, this and (3.114) imply that

$$\|\Phi f\|_{(\mathrm{A}^p(\Sigma,\sigma))^*}=\|f\|_{\mathrm{B}^{p'}(\Sigma,\sigma)}\quad\text{for all } f\in\mathrm{B}^{p'}(\Sigma,\sigma). \tag{3.117}$$

As such, Φ is an isometry. This concludes the proof of the fact that Φ is a well-defined, linear, continuous isometric isomorphism, which ultimately proves Proposition 3.10. □

We now identify the dual of the space B^p_0 as the Beurling space $\mathrm{A}^{p'}$. This is made precise in the following result.

Proposition 3.11 *Consider a closed Ahlfors regular set $\Sigma\subseteq\mathbb{R}^n$ which is unbounded, and abbreviate $\sigma:=\mathcal{H}^{n-1}\lfloor\Sigma$. Select an integrability exponent $p\in(1,\infty)$, and denote by $p'\in(1,\infty)$ its Hölder conjugate exponent. Then the dual of $\mathrm{B}^p_0(\Sigma,\sigma)$ is $\mathrm{A}^{p'}(\Sigma,\sigma)$, i.e.,*

$$\big(\mathrm{B}^p_0(\Sigma,\sigma)\big)^*=\mathrm{A}^{p'}(\Sigma,\sigma). \tag{3.118}$$

More precisely, the mapping

$$\mathrm{A}^{p'}(\Sigma,\sigma)\ni g\longmapsto\Psi g:=\Lambda_g\in\big(\mathrm{B}^p_0(\Sigma,\sigma)\big)^* \tag{3.119}$$

where

$$\Lambda_g(f):=\int_{\Sigma}fg\,d\sigma\quad\text{for all } f\in\mathrm{B}^p_0(\Sigma,\sigma) \tag{3.120}$$

is a well-defined, linear, isometric isomorphism.

Proof Well-definiteness of Ψ. Fix a function $g\in\mathrm{A}^{p'}(\Sigma,\sigma)$. Using Proposition 3.6 (much as in (3.100) and (3.101)) we deduce that the functional

$$\begin{gathered}\mathrm{B}^p_0(\Sigma,\sigma)\ni f\longmapsto\Lambda_g(f):=\int_{\Sigma}fg\,d\sigma\quad\text{is well defined,}\\ \text{linear, and bounded with norm }\|\Lambda_g\|_{(\mathrm{B}^p_0(\Sigma,\sigma))^*}\le\|g\|_{\mathrm{A}^{p'}(\Sigma,\sigma)}.\end{gathered} \tag{3.121}$$

Hence, the mapping Ψ given in (3.119) is well defined and linear.

Surjectivity of Ψ. Dealing with the surjectivity of Ψ requires some preparation. Consider the sequence of Banach spaces $\mathscr{A}_\Sigma:=\{L^p(C_k,\sigma)\}_{k\in\mathbb{N}_0}$, and its dual sequence given by $\mathscr{A}^*_\Sigma:=\{L^{p'}(C_k,\sigma)\}_{k\in\mathbb{N}_0}$. Introduce the following sequence spaces:

$$\ell^1(\mathscr{A}_\Sigma^*) := \Big\{F = \{f_k\}_{k\in\mathbb{N}_0} \in \prod_{k\in\mathbb{N}_0} L^{p'}(C_k,\sigma) : \\ \|F\|_{\ell^1(\mathscr{A}_\Sigma^*)} := \sum_{k=0}^{\infty} \|f_k\|_{L^{p'}(C_k,\sigma)} < \infty\Big\}, \tag{3.122}$$

$$\ell^\infty(\mathscr{A}_\Sigma) := \Big\{F = \{f_k\}_{k\in\mathbb{N}_0} \in \prod_{k\in\mathbb{N}_0} L^{p}(C_k,\sigma) : \\ \|F\|_{\ell^\infty(\mathscr{A}_\Sigma^*)} := \sup_{k\in\mathbb{N}_0} \|f_k\|_{L^{p}(C_k,\sigma)} < \infty\Big\}, \tag{3.123}$$

$$c_0(\mathscr{A}_\Sigma) := \Big\{F = \{f_k\}_{k\in\mathbb{N}_0} \in \ell^\infty(\mathscr{A}_\Sigma) : \lim_{k\to\infty} \|f_k\|_{L^p(C_k,\sigma)} = 0\Big\}. \tag{3.124}$$

Then these are Banach spaces (see [Tr95, §1.18.1]), and in particular $c_0(\mathscr{A}_\Sigma)$ is a closed subspace of $\ell^\infty(\mathscr{A}_\Sigma)$. Moreover, from [Tr95, §1.11.] the following duality relation holds:

$$\big(c_0(\mathscr{A}_\Sigma)\big)^* = \ell^1(\mathscr{A}_\Sigma^*), \tag{3.125}$$

in the sense that the mapping

$$\ell^1(\mathscr{A}_\Sigma^*) \ni G \longmapsto {}_{c_0(\mathscr{A}_\Sigma)}(\cdot, G)_{\ell^1(\mathscr{A}_\Sigma^*)} \in \big(c_0(\mathscr{A}_\Sigma)\big)^* \tag{3.126}$$

is a well-defined isomorphism, where

$${}_{c_0(\mathscr{A}_\Sigma)}(F, G)_{\ell^1(\mathscr{A}_\Sigma^*)} := \sum_{k=0}^{\infty} \int_{C_k} f_k g_k \, d\sigma, \tag{3.127}$$

for every $F = \{f_k\}_{k\in\mathbb{N}_0} \in c_0(\mathscr{A}_\Sigma)$ and $G = \{g_k\}_{k\in\mathbb{N}_0} \in \ell^1(\mathscr{A}_\Sigma^*)$.

Next, for each σ-measurable function f defined on Σ introduce

$$\Theta(f) := \{2_*^{-k(n-1)/p} f\big|_{C_k}\}_{k\in\mathbb{N}_0}. \tag{3.128}$$

Notice that whenever $f \in \mathrm{B}^p(\Sigma,\sigma)$ we have

$$\|\Theta(f)\|_{\ell^\infty(\mathscr{A}_\Sigma^*)} = \sup_{k\in\mathbb{N}_0} 2_*^{-k(n-1)/p} \|f|_{C_k}\|_{\mathrm{B}^p(C_k,\sigma)} = \|f\|_{\mathrm{B}^p(\Sigma,\sigma)}. \tag{3.129}$$

As such, this entails

$$\Theta : \mathrm{B}^p(\Sigma,\sigma) \longrightarrow \ell^\infty(\mathscr{A}_\Sigma) \text{ is a well-defined, linear, and isometric operator.} \tag{3.130}$$

As a consequence, Θ is bounded. Bearing in mind Proposition 3.5 item *(a)*, (3.124) and (3.128) we can see that

$$\Theta\big(\mathrm{B}_0^p(\Sigma,\sigma)\big) \subseteq c_0(\mathscr{A}_\Sigma). \tag{3.131}$$

Furthermore, (3.130) readily gives

$$\begin{aligned} &\Theta : \mathrm{B}_0^p(\Sigma,\sigma) \longrightarrow \Theta\big(\mathrm{B}_0^p(\Sigma,\sigma)\big) \\ &\text{is an isometric isomorphism.} \end{aligned} \tag{3.132}$$

We claim that $\Theta\big(\mathrm{B}_0^p(\Sigma,\sigma)\big)$ is a closed subspace of $c_0(\mathscr{A}_\Sigma)$. To establish this, assume $\{f_j\}_{j\in\mathbb{N}} \subseteq \mathrm{B}_0^p(\Sigma,\sigma)$ is a sequence for which there exists some $F \in c_0(\mathscr{A}_\Sigma)$ satisfying

$$\lim_{j\to\infty} \Theta(f_j) = F \ \text{ in } \ell^\infty(\mathscr{A}_\Sigma). \tag{3.133}$$

In particular, $\{\Theta(f_j)\}_{j\in\mathbb{N}}$ is a Cauchy sequence in $\ell^\infty(\mathscr{A}_\Sigma)$. Hence, being Θ is a linear isometry,

$$\begin{aligned} \|f_j - f_\ell\|_{\mathrm{B}^p(\Sigma,\sigma)} &= \|\Theta(f_j - f_\ell)\|_{\ell^\infty(\mathscr{A}_\Sigma^*)} \\ &= \|\Theta(f_j) - \Theta(f_\ell)\|_{\ell^\infty(\mathscr{A}_\Sigma^*)} \quad \text{for all } j,\ell \in \mathbb{N}, \end{aligned} \tag{3.134}$$

which in turn implies that $\{f_j\}_{j\in\mathbb{N}}$ is a Cauchy sequence in $\mathrm{B}_0^p(\Sigma,\sigma)$. From the completeness of $\mathrm{B}_0^p(\Sigma,\sigma)$ (cf. Remark 3.9), there exists $f \in \mathrm{B}_0^p(\Sigma,\sigma)$ such that

$$\lim_{j\to\infty} f_j = f \ \text{ in the norm of } \mathrm{B}^p(\Sigma,\sigma). \tag{3.135}$$

By continuity of Θ (see (3.130)), we obtain

$$\lim_{j\to\infty} \Theta(f_j) = \Theta(f) \ \text{ in } \ell^\infty(\mathscr{A}_\Sigma). \tag{3.136}$$

In concert with (3.133), this implies $\Theta(f) = F$, and therefore $F \in \Theta\big(\mathrm{B}_0^p(\Sigma,\sigma)\big)$. Hence, $\Theta\big(\mathrm{B}_0^p(\Sigma,\sigma)\big)$ is a closed subspace of the Banach $c_0(\mathscr{A}_\Sigma)$. Therefore, $\Theta\big(\mathrm{B}_0^p(\Sigma,\sigma)\big)$ is a Banach space itself. Thanks to this and Remark 3.9, the mapping

$$\begin{aligned} &\Theta : \mathrm{B}_0^p(\Sigma,\sigma) \to \Theta\big(\mathrm{B}_0^p(\Sigma,\sigma)\big) \ \text{ is an isometric} \\ &\text{isomorphism between Banach spaces.} \end{aligned} \tag{3.137}$$

By the Open Mapping Theorem, its inverse

$$\begin{aligned} &\Psi := \Theta^{-1} : \Theta\big(\mathrm{B}_0^p(\Sigma,\sigma)\big) \to \mathrm{B}_0^p(\Sigma,\sigma) \\ &\text{is also a bounded linear operator.} \end{aligned} \tag{3.138}$$

We now move on to the task of proving that Ψ (cf. (3.119)) is surjective. To this end, fix an arbitrary linear functional $\Lambda \in \big(\mathrm{B}_0^p(\Sigma,\sigma)\big)^*$. Observe that the mapping

$$\begin{aligned} &\Lambda \circ \Psi : \Theta\big(\mathrm{B}_0^p(\Sigma,\sigma)\big) \to \mathbb{C} \\ &\text{is a continuous linear functional.} \end{aligned} \tag{3.139}$$

The Hahn-Banach Theorem then entails that $\Lambda \circ \Psi$ extends to a continuous linear functional $\widetilde{\Lambda} \in \big(c_0(\mathscr{A}_\Sigma)\big)^*$ that satisfies

$$\begin{aligned} \widetilde{\Lambda}\big|_{\Theta(\mathrm{B}^p_0(\Sigma,\sigma))} &= \Lambda \circ \Psi \ \text{ in } \Theta(\mathrm{B}^p_0(\Sigma,\sigma)), \\ \text{and } \ \|\widetilde{\Lambda}\|_{(c_0(\mathscr{A}_\Sigma))^*} &= \|\Lambda \circ \Psi\|_{[\Theta(\mathrm{B}^p_0(\Sigma,\sigma))]^*}. \end{aligned} \tag{3.140}$$

Therefore, bearing in mind (3.125)-(3.127), there exists $G = \{g_k\}_{k\in\mathbb{N}_0} \in \ell^1(\mathscr{A}^*_\Sigma)$ such that

$$\begin{aligned} \widetilde{\Lambda}(F) &= \sum_{k=0}^{\infty} \int_{C_k} f_k g_k \, d\sigma \quad \text{for every } \ F = \{f_k\}_{k\in\mathbb{N}_0} \in c_0(\mathscr{A}_\Sigma) \\ \text{and } \ \|\widetilde{\Lambda}\|_{(c_0(\mathscr{A}_\Sigma))^*} &= \|G\|_{\ell^1(\mathscr{A}^*_\Sigma)} = \sum_{k=0}^{\infty} \|g_k\|_{L^p(C_k,\sigma)}. \end{aligned} \tag{3.141}$$

Using this representation of $\widetilde{\Lambda}$, along with (3.138) and (3.140), for each $f \in \mathrm{B}^p_0(\Sigma,\sigma)$ we may derive the following formula for $\Lambda(f)$:

$$\begin{aligned} \Lambda(f) &= (\Lambda \circ \Psi)\big(\Theta(f)\big) = \widetilde{\Lambda}\big(\Theta(f)\big) = \widetilde{\Lambda}\big(\{2_*^{-k(n-1)/p} f|_{C_k}\}_{k\in\mathbb{N}_0}\big) \\ &= \sum_{k=0}^{\infty} \int_{C_k} 2_*^{-k(n-1)/p} (f|_{C_k}) g_k \, d\sigma. \end{aligned} \tag{3.142}$$

Concerning the operator norm of Λ, use (3.137), (3.138), (3.140), and (3.141) to compute

$$\begin{aligned} \|\Lambda\|_{(\mathrm{B}^p_0(\Sigma,\sigma))^*} &= \sup_{\substack{f\in \mathrm{B}^p_0(\Sigma,\sigma) \\ \|f\|_{\mathrm{B}^p(\Sigma,\sigma)}\le 1}} |\Lambda(f)| = \sup_{\substack{f\in \mathrm{B}^p_0(\Sigma,\sigma) \\ \|f\|_{\mathrm{B}^p(\Sigma,\sigma)}\le 1}} \big|(\Lambda \circ \Psi)\big(\Theta(f)\big)\big| \\ &= \sup_{\substack{g\in \Theta(\mathrm{B}^p_0(\Sigma,\sigma)) \\ \|g\|_{\ell^\infty(\mathscr{A}_\Sigma)}\le 1}} |(\Lambda \circ \Psi)(g)| = \|\Lambda \circ \Psi\|_{[\Theta(\mathrm{B}^p_0(\Sigma,\sigma))]^*} \\ &= \|\widetilde{\Lambda}\|_{(c_0(\mathscr{A}_\Sigma))^*} = \|G\|_{\ell^1(\mathscr{A}^*_\Sigma)}. \end{aligned} \tag{3.143}$$

Now, for each $k \in \mathbb{N}_0$ let $\widetilde{g}_k$ be the extension by zero of g_k on Σ, and define the function $g : \Sigma \to \mathbb{R}$ as

$$g := \sum_{k=0}^{\infty} 2_*^{-k(n-1)/p} \widetilde{g}_k. \tag{3.144}$$

Since $\widetilde{g}_k$ is supported in C_k for each $k \in \mathbb{N}_0$, the function g is well defined and σ-measurable on Σ and satisfies

$$g \cdot \mathbb{1}_{C_j} = \sum_{k=0}^{\infty} 2_*^{-k(n-1)/p} (\widetilde{g}_k \cdot \mathbb{1}_{C_j}) = 2_*^{-j(n-1)/p} \cdot \widetilde{g}_j \quad \text{for every } j \in \mathbb{N}_0. \tag{3.145}$$

Granted this, the second line of (3.141), and (3.143), we next compute

$$\begin{aligned}\|g\|_{\mathrm{A}^{p'}(\Sigma,\sigma)} &= \sum_{j=0}^{\infty} 2_*^{j(n-1)/p}\|g\cdot \mathbb{1}_{C_j}\|_{L^{p'}(\Sigma,\sigma)} \\ &= \sum_{j=0}^{\infty} 2_*^{j(n-1)/p}\|2_*^{-j(n-1)/p}\cdot \widetilde{g}_j\|_{L^{p'}(\Sigma,\sigma)} \\ &= \sum_{k=0}^{\infty} \|g_j\|_{L^{p'}(C_j,\sigma)} = \|G\|_{\ell^1(\mathscr{A}_\Sigma^*)} = \|\Lambda\|_{(\mathrm{B}_0^p(\Sigma,\sigma))^*} < \infty. \end{aligned} \tag{3.146}$$

This ultimately establishes the membership $g \in \mathrm{A}^{p'}(\Sigma, \sigma)$.

Lastly, we claim that Λ coincides with the functional Λ_g associated with g defined in (3.144). To see this, pick $f \in \mathrm{B}_0^p(\Sigma, \sigma)$. On account of (3.142), and (3.144) we have

$$\begin{aligned}\Lambda(f) &= \sum_{k=0}^{\infty}\int_{C_k} 2_*^{-k(n-1)/p}(f|_{C_k})g_k\,d\sigma = \sum_{k=0}^{\infty}\int_{\Sigma} 2_*^{-k(n-1)/p} f\widetilde{g}_k\,d\sigma \\ &= \lim_{N\to\infty}\int_{\Sigma} f\Big(\sum_{k=0}^{N} 2_*^{-k(n-1)/p}\widetilde{g}_k\Big)\,d\sigma \\ &= \int_{\Sigma} f\Big(\sum_{k=0}^{\infty} 2_*^{-k(n-1)/p}\widetilde{g}_k\Big)\,d\sigma = \int_{\Sigma} fg\,d\sigma = \Lambda_g(f), \end{aligned} \tag{3.147}$$

where in the fourth equality we have used Lebesgue's Dominated Convergence Theorem, which is applicable thanks to Proposition 3.70. Hence, $\Lambda = \Lambda_g$ as functionals. This shows that the mapping Ψ is surjective, as desired. In particular, this and (3.146) imply

$$\|\Lambda_g\|_{(\mathrm{B}_0^p(\Sigma,\sigma))^*} = \|\Lambda\|_{(\mathrm{B}_0^p(\Sigma,\sigma))^*} = \|g\|_{\mathrm{A}^{p'}(\Sigma,\sigma)}. \tag{3.148}$$

<u>`Injectivity of Ψ.`</u> To show that Ψ is injective, select a function $g \in \mathrm{A}^{p'}(\Sigma, \sigma)$ and suppose $\Lambda_g = 0$ as a functional in $\big(\mathrm{B}_0^p(\Sigma, \sigma)\big)^*$, i.e.,

$$\int_{\Sigma} fg\,d\sigma = 0 \quad \text{for all } f \in \mathrm{B}_0^p(\Sigma, \sigma). \tag{3.149}$$

Notice that every function $f \in L^p(\Sigma, \sigma)$ also belongs to $\mathrm{B}^p(\Sigma, \sigma)$ (cf. Proposition 3.4) and satisfies item *(a)* in Proposition 3.5. Based on this and Remark 3.8 we see that $L^p(\Sigma, \sigma) \subseteq \mathrm{B}_0^p(\Sigma, \sigma)$. As such (3.149) holds for every $f \in L^p(\Sigma, \sigma)$, i.e.,

$$\int_{\Sigma} fg\,d\sigma = 0 \quad \text{for all } f \in L^p(\Sigma, \sigma). \tag{3.150}$$

Thanks to the Riesz Representation Theorem we find that $g = 0$ pointwise σ-a.e. on Σ. This proves that Ψ is injective. Furthermore, this guarantees that every linear functional $\Lambda \in \big(\mathrm{B}_0^p(\Sigma,\sigma)\big)^*$ is uniquely determined by a function $g \in \mathrm{A}^{p'}(\Sigma,\sigma)$ for which $\Lambda = \Lambda_g$. With this and (3.148) in hand, we obtain that

$$\|\Psi g\|_{(\mathrm{B}_0^p(\Sigma,\sigma))^*} = \|g\|_{\mathrm{A}^{p'}(\Sigma,\sigma)} \quad \text{for all } g \in \mathrm{A}^{p'}(\Sigma,\sigma). \tag{3.151}$$

This shows that Ψ is an isometry, which, together with what we have established so far, concludes the proof of Proposition 3.11. □

The duality result established above permits us to obtain the following norm characterization of A^p.

Corollary 3.1 *Consider a closed unbounded Ahlfors regular set $\Sigma \subseteq \mathbb{R}^n$, and abbreviate $\sigma := \mathcal{H}^{n-1}\lfloor\Sigma$. Pick $p \in (1,\infty)$, and denote by $p' \in (1,\infty)$ its Hölder conjugate exponent. Then for every $f \in \mathrm{A}^p(\Sigma,\sigma)$ one has*

$$\|f\|_{\mathrm{A}^p(\Sigma,\sigma)} = \sup\Big\{\Big|\int_\Sigma fg\,d\sigma\Big| : g \in \mathrm{B}^{p'}(\Sigma,\sigma) \textit{ with } \|g\|_{\mathrm{B}^{p'}(\Sigma,\sigma)} \le 1\Big\}. \tag{3.152}$$

Proof Based on Proposition 3.11, whenever $f \in \mathrm{A}^p(\Sigma,\sigma)$ we have

$$\begin{aligned} \|f\|_{\mathrm{A}^p(\Sigma,\sigma)} &= \|\Lambda_f\|_{(\mathrm{B}_0^{p'}(\Sigma,\sigma))^*} \\ &= \sup\Big\{\Big|\int_\Sigma fg\,d\sigma\Big| : g \in \mathrm{B}_0^{p'}(\Sigma,\sigma) \text{ with } \|g\|_{\mathrm{B}^{p'}(\Sigma,\sigma)} \le 1\Big\} \\ &\le \sup\Big\{\Big|\int_\Sigma fg\,d\sigma\Big| : g \in \mathrm{B}^{p'}(\Sigma,\sigma) \text{ with } \|g\|_{\mathrm{B}^{p'}(\Sigma,\sigma)} \le 1\Big\}. \end{aligned} \tag{3.153}$$

The opposite inequality follows from the Hölder-type inequality (3.70). □

Remark 3.10 As seen from Definition 3.1, the space $\mathrm{A}^p(\Sigma,\sigma)$ depends implicitly on the reference point $x_0 \in \Sigma$, and on the dilation factor $2_* > (C_\Sigma/c_\Sigma)^{1/(n-1)}$. Upon recalling Proposition 3.2, the right-hand side of (3.152) is independent of said parameters. As such, the membership to $\mathrm{A}^p(\Sigma,\sigma)$ is independent of the choice of x_0 and 2_*.

3.3 The Hardy-Littlewood Maximal Operator on Beurling Spaces

In this section, we analyze the action of the Hardy-Littlewood maximal operator, and its L^s-styled versions, on A^p, B^p, and B_0^p spaces. Our first result here concerns the boundedness of the Hardy-Littlewood maximal operator on B^p spaces.

Theorem 3.1 *Assume $\Sigma \subseteq \mathbb{R}^n$ is a closed unbounded Ahlfors regular set, and abbreviate $\sigma := \mathcal{H}^{n-1}\lfloor\Sigma$. Then for every integrability exponent $p \in (1,\infty)$, the Hardy-Littlewood maximal operator*

$$\begin{aligned} &\mathcal{M}_\Sigma : \mathrm{B}^p(\Sigma,\sigma) \to \mathrm{B}^p(\Sigma,\sigma) \\ &\textit{is well defined and bounded.} \end{aligned} \tag{3.154}$$

Proof Consider a reference point $x_0 \in \Sigma$, and a dilation factor $2_* > (C_\Sigma/c_\Sigma)^{1/(n-1)}$ as in Definition 3.1. Pick an arbitrary function $f \in \mathrm{B}^p(\Sigma,\sigma)$, and select a finite number $R \geq 2_*$. For each $x \in \Sigma$ write

$$\begin{aligned} (\mathcal{M}_\Sigma f)(x) &\leq \sup_{0<r<3R} ⨍_{\Delta(x,r)} |f|\, d\sigma + \sup_{r\geq 3R} ⨍_{\Delta(x,r)} |f|\, d\sigma \\ &=: (M_\Sigma^{(1)} f)(x) + (M_\Sigma^{(2)} f)(x). \end{aligned} \tag{3.155}$$

With this piece of notation, it is visible that

$$\begin{aligned} \Big(⨍_{\Delta(x_0,R)} (\mathcal{M}_\Sigma f)^p\, d\sigma \Big)^{1/p} &\leq \Big(⨍_{\Delta(x_0,R)} (M_\Sigma^{(1)} f)^p\, d\sigma \Big)^{1/p} \\ &\quad + \Big(⨍_{\Delta(x_0,R)} (M_\Sigma^{(2)} f)^p\, d\sigma \Big)^{1/p}. \end{aligned} \tag{3.156}$$

Let us analyze the second term in the right-hand side of (3.156). To this end, fix a finite radius $r \geq 3R$ and an arbitrary $x \in \Delta(x_0, R)$. Then whenever $z \in \Delta(x, r)$ we have

$$|z - x_0| \leq |z - x| + |x - x_0| < r + R \leq \frac{4}{3} r. \tag{3.157}$$

Hence, $\Delta(x, r) \subseteq \Delta(x_0, \frac{4}{3} r)$. Use this, the Ahlfors regularity of Σ, Hölder's inequality, and Proposition 3.1 to estimate

$$⨍_{\Delta(x,r)} |f|\, d\sigma \leq C \Big(⨍_{\Delta(x_0,4r/3)} |f|^p\, d\sigma \Big)^{1/p} \leq C \|f\|_{\mathrm{B}^p(\Sigma,\sigma)}, \tag{3.158}$$

for some constant $C \in (0, \infty)$ depending only on n, p, and the Ahlfors regularity character of Σ. Taking the supremum over all radii $r \geq 3R$ and recalling (3.155) we obtain

$$(M_\Sigma^{(2)} f)(x) \leq C \|f\|_{\mathrm{B}^p(\Sigma,\sigma)} \text{ for every } x \in \Delta(x_0, R). \tag{3.159}$$

This further implies

$$\Big(⨍_{\Delta(x_0,R)} (M_\Sigma^{(2)} f)^p\, d\sigma \Big)^{1/p} \leq C \|f\|_{\mathrm{B}^p(\Sigma,\sigma)}, \tag{3.160}$$

for some constant $C = C(n, p, \Sigma) \in (0, \infty)$.

To proceed, we turn our attention to the first term in the right-hand side of (3.156). Pick a radius $r \in (0, 3R)$ and a point $x \in \Delta(x_0, R)$. Then for each $z \in \Delta(x, r)$ compute

$$|z - x_0| \leq |z - x| + |x - x_0| < r + R < 4R, \tag{3.161}$$

and conclude that $\Delta(x,r) \subseteq \Delta(x_0, 4R)$. Thanks to this, for each $x \in \Delta(x_0, R)$ we have

$$(M_\Sigma^{(1)} f)(x) = \sup_{0<r<3R} \fint_{\Delta(x,r)} |f \cdot \mathbb{1}_{\Delta(x_0,4R)}|\, d\sigma \leq \big(\mathcal{M}_\Sigma(f \cdot \mathbb{1}_{\Delta(x_0,4R)})\big)(x). \tag{3.162}$$

In view of this, the boundedness of $\mathcal{M}_\Sigma$ on $L^p(\Sigma,\sigma)$ together with the doubling property of σ, and Proposition 3.1 we may now estimate

$$\begin{aligned}
\Big(\fint_{\Delta(x_0,R)} (M_\Sigma^{(1)} f)^p\, d\sigma\Big)^{1/p} &\leq \Big(\fint_{\Delta(x_0,R)} \big(\mathcal{M}_\Sigma(f \cdot \mathbb{1}_{\Delta(x_0,4R)})\big)^p\, d\sigma\Big)^{1/p} \\
&\leq \frac{1}{\sigma\big(\Delta(x_0,R)\big)^{1/p}} \|\mathcal{M}_\Sigma(f \cdot \mathbb{1}_{\Delta(x_0,4R)})\|_{L^p(\Sigma,\sigma)} \\
&\leq \frac{C}{\sigma\big(\Delta(x_0,4R)\big)^{1/p}} \|f \cdot \mathbb{1}_{\Delta(x_0,4R)}\|_{L^p(\Sigma,\sigma)} \\
&= C\Big(\fint_{\Delta(x_0,4R)} |f|^p\, d\sigma\Big)^{1/p} \\
&\leq C\|f\|_{\mathrm{B}^p(\Sigma,\sigma)},
\end{aligned} \tag{3.163}$$

for some constant $C = C(n, p, \Sigma) \in (0, \infty)$. Combining estimates (3.156), (3.159), and (3.163) then yields

$$\Big(\fint_{\Delta(x_0,R)} (\mathcal{M}_\Sigma f)^p\, d\sigma\Big)^{1/p} \leq C\|f\|_{\mathrm{B}^p(\Sigma,\sigma)} \tag{3.164}$$

for some constant $C = C(n, p, \Sigma) \in (0, \infty)$. Taking the supremum over $R \geq 2_*$ and using Proposition 3.1 we finally arrive at

$$\|\mathcal{M}_\Sigma f\|_{\mathrm{B}^p(\Sigma,\sigma)} \leq C\|f\|_{\mathrm{B}^p(\Sigma,\sigma)}. \tag{3.165}$$

This implies (3.154), which in turn concludes the proof of Theorem 3.1. □

As an immediate consequence of Theorem 3.1 and the density of L^p into B_0^p we see that the space $\mathrm{B}_0^p(\Sigma,\sigma)$ is stable under the action of the Hardy-Littlewood maximal operator.

Theorem 3.2 *Consider a closed unbounded Ahlfors regular set $\Sigma \subseteq \mathbb{R}^n$, and abbreviate $\sigma := \mathcal{H}^{n-1}\lfloor\Sigma$. Then for every exponent $p \in (1,\infty)$ the Hardy-Littlewood maximal operator*

$$\mathcal{M}_\Sigma : \mathrm{B}_0^p(\Sigma,\sigma) \to \mathrm{B}_0^p(\Sigma,\sigma) \tag{3.166}$$

is well defined and bounded.

Proof Notice that Proposition 3.5 and Remark 3.8 imply that

$$\overline{L^p(\Sigma,\sigma)}^{\|\cdot\|_{\mathrm{B}^p(\Sigma,\sigma)}} = \mathrm{B}^p_0(\Sigma,\sigma). \tag{3.167}$$

As such, since $\mathcal{M}_\Sigma$ is bounded on $L^p(\Sigma,\sigma)$, standard functional analysis gives the desired result. □

Let us provide an auxiliary result regarding pointwise convergence of the L^s-based Hardy-Littlewood maximal operator acting on sequences of functions.

Lemma 3.1 *Consider a closed unbounded Ahlfors regular set $\Sigma \subseteq \mathbb{R}^n$, and abbreviate $\sigma := \mathcal{H}^{n-1}\lfloor\Sigma$. Assume $\{\phi_N\}_{N\in\mathbb{N}}$ is a sequence of non-negative σ-measurable functions on Σ for which*

$$\phi_N \nearrow \phi \quad \textit{pointwise } \sigma\textit{-a.e. on } \Sigma \textit{ as } N \to \infty \tag{3.168}$$

for some σ-measurable function ϕ on Σ. Then for each $s \in (0,\infty)$ one has

$$\mathcal{M}_{\Sigma,s}\phi_N \nearrow \mathcal{M}_{\Sigma,s}\phi \quad \textit{pointwise on } \sigma\textit{-a.e. on } \Sigma \textit{ as } N \to \infty, \tag{3.169}$$

Proof To get started, notice that $\{\phi_N\}_{N\in\mathbb{N}}$ being a non-decreasing sequence which is pointwise controlled by ϕ (see (3.168)) implies

$$(\mathcal{M}_{\Sigma,s}\phi_N)(x) \le (\mathcal{M}_{\Sigma,s}\phi_{N+1})(x) \le (\mathcal{M}_{\Sigma,s}\phi)(x) \quad \text{for all } x \in \Sigma. \tag{3.170}$$

Hence, for every $x \in \Sigma$ we have

$$0 \le \limsup_{N\to\infty}(\mathcal{M}_{\Sigma,s}\phi_N)(x) \le (\mathcal{M}_{\Sigma,s}\phi)(x). \tag{3.171}$$

Next, pick $x \in \Sigma$. In case $(\mathcal{M}_{\Sigma,s}\phi)(x) = 0$, notice that the condition (3.171) forces $\lim_{N\to\infty}(\mathcal{M}_{\Sigma,s}\phi_N)(x) = (\mathcal{M}_{\Sigma,s}\phi)(x) = 0$. For this reason, assume $\mathcal{M}_{\Sigma,s}\phi(x) > 0$ and choose $\lambda \in (0,\infty)$ for which

$$0 < \lambda < (\mathcal{M}_{\Sigma,s}\phi)(x). \tag{3.172}$$

Then, there exists a radius $R \in (0,\infty)$ with the property that

$$\lambda < \Big(⨍_{\Delta(x,R)} |\phi|^s\,d\sigma\Big)^{1/s} = \lim_{N\to\infty}\Big(⨍_{\Delta(x,R)} |\phi_N|^s\,d\sigma\Big)^{1/s}, \tag{3.173}$$

where the equality follows from Lebesgue's Monotone Convergence Theorem. As such, for some $N_\lambda \in \mathbb{N}$ it follows that

$$\lambda < \Big(⨍_{\Delta(x,R)} |\phi_N|^s\,d\sigma\Big)^{1/s} \le (\mathcal{M}_{\Sigma,s}\phi_N)(x), \quad \text{for all } N \ge N_\lambda. \tag{3.174}$$

which in turn implies

$$\lambda \le \liminf_{N\to\infty}(\mathcal{M}_{\Sigma,s}\phi_N)(x). \tag{3.175}$$

Since this hols for every λ satisfying (3.172), we can let $\lambda \nearrow (\mathcal{M}_{\Sigma,s}\phi)(x)$ and then arrive at

$$(\mathcal{M}_{\Sigma,s}\phi)(x) \leq \liminf_{N\to\infty}(\mathcal{M}_{\Sigma,s}\phi_N)(x). \tag{3.176}$$

Putting it all together, (3.169) is a consequence of (3.170), (3.171), (3.176). This establishes Lemma 3.1. □

We are now in a position to state and prove the boundedness of the L^s-based Hardy-Littlewood maximal operator on A^p.

Theorem 3.3 *Assume $\Sigma \subseteq \mathbb{R}^n$ is a closed unbounded Ahlfors regular set, and abbreviate $\sigma := \mathcal{H}^{n-1}\lfloor\Sigma$. Then for every integrability exponent $p \in [1,\infty)$, every $s \in (0,1)$, the L^s-based Hardy-Littlewood maximal operator*

$$\begin{array}{c}\mathcal{M}_{\Sigma,s} : \mathrm{A}^p(\Sigma,\sigma) \longrightarrow \mathrm{A}^p(\Sigma,\sigma) \\ \textit{is well defined and bounded.}\end{array} \tag{3.177}$$

Moreover the range $s \in (0,1)$ is sharp, in the sense that $\mathcal{M}_{\Sigma,s}$ fails to be bounded on $\mathrm{A}^p(\Sigma,\sigma)$ when $s \geq 1$.

Proof The validity of (3.177) for the case when $p = 1$ follows from the fact that $\mathrm{A}^1(\Sigma,\sigma) = L^1(\Sigma,\sigma)$ together with the boundedness of $\mathcal{M}_{\Sigma,s}$ on $L^1(\Sigma,\sigma)$ for all $s \in (0,1)$ (cf. [GHA-I, Corollary 7.6.2, pp. 622–623]). Assume $1 < p < \infty$. Select an arbitrary function $f \in \mathrm{A}^p(\Sigma,\sigma)$ and without loss of generality assume $f \geq 0$ on Σ, otherwise replace f by $|f|$. Having fixed a reference point $x_0 \in \Sigma$, for each $N \in \mathbb{N}$ and $x \in \Sigma$ define

$$f_N(x) := \begin{cases} f(x), & \text{if } 0 \leq f(x) \leq N \text{ and } |x - x_0| \leq N, \\ 0, & \text{if } f(x) > N \text{ or } |x - x_0| > N. \end{cases} \tag{3.178}$$

Thus by design, the sequence of functions $\{f_N\}_{N\in\mathbb{N}}$ satisfies

$$\begin{array}{c} f_N \nearrow f \ \text{ at every point on } \Sigma \text{ as } N \to \infty, \text{ and} \\ f_N(x) \leq N \cdot \mathbb{1}_{\Delta(x_0,N)}(x) \ \text{ for all } \ x \in \Sigma \ \text{ and } \ N \in \mathbb{N}. \end{array} \tag{3.179}$$

The property in the second line of (3.179) then entails

$$(\mathcal{M}_{\Sigma,s} f_N)(x) \leq N\big(\mathcal{M}_{\Sigma,s}(\mathbb{1}_{\Delta(x_0,N)})\big)(x) \quad \text{for each } x \in \Sigma, \text{ and } N \in \mathbb{N}. \tag{3.180}$$

In light of Proposition 2.1, and Example 3.1, the function in the right-hand side of (3.180) belongs to $\mathrm{A}^p(\Sigma,\sigma)$. This and the lattice property of $\mathrm{A}^p(\Sigma,\sigma)$ (cf. Remark 3.3) give the membership

$$\mathcal{M}_{\Sigma,s} f_N \in \mathrm{A}^p(\Sigma,\sigma) \ \text{ for every } N \in \mathbb{N}. \tag{3.181}$$

Based on Corollary 3.1, the Fefferman-Stein maximal inequality (cf. Proposition 2.2), Proposition 3.6, the boundedness of $\mathcal{M}_\Sigma$ on $\mathrm{B}^{p'}(\Sigma,\sigma)$ (cf. Theorem 3.1), and

(3.179), for each $N \in \mathbb{N}$ we may estimate

$$\begin{aligned}
\|\mathcal{M}_{\Sigma,s} f_N\|_{\mathrm{A}^p(\Sigma,\sigma)} &= \sup_{\substack{g\in \mathrm{B}^{p'}(\Sigma,\sigma)\\ \|g\|_{\mathrm{B}^{p'}(\Sigma,\sigma)}\leq 1}} \left|\int_\Sigma (\mathcal{M}_{\Sigma,s} f_N) g \, d\sigma\right| \\
&\leq \sup_{\substack{g\in \mathrm{B}^{p'}(\Sigma,\sigma)\\ \|g\|_{\mathrm{B}^{p'}(\Sigma,\sigma)}\leq 1}} \int_\Sigma (\mathcal{M}_{\Sigma,s} f_N)|g| \, d\sigma \\
&\leq C\cdot \sup_{\substack{g\in \mathrm{B}^{p'}(\Sigma,\sigma)\\ \|g\|_{\mathrm{B}^{p'}(\Sigma,\sigma)}\leq 1}} \int_\Sigma f_N (\mathcal{M}_\Sigma g) \, d\sigma \\
&\leq C\cdot \sup_{\substack{g\in \mathrm{B}^{p'}(\Sigma,\sigma)\\ \|g\|_{\mathrm{B}^{p'}(\Sigma,\sigma)}\leq 1}} \|f_N\|_{\mathrm{A}^p(\Sigma,\sigma)} \|\mathcal{M}_\Sigma g\|_{\mathrm{B}^{p'}(\Sigma,\sigma)} \\
&\leq C\cdot \|\mathcal{M}_\Sigma\|_{\mathrm{B}^{p'}(\Sigma,\sigma)\to \mathrm{B}^{p'}(\Sigma,\sigma)} \|f\|_{\mathrm{A}^p(\Sigma,\sigma)}. \qquad (3.182)
\end{aligned}$$

Being mindful of the property in the first line of (3.179), Lemma 3.1 ensures that

$$\mathcal{M}_{\Sigma,s} f_N \nearrow \mathcal{M}_{\Sigma,s} f \ \text{ pointwise on } \Sigma \text{ as } N \to \infty. \qquad (3.183)$$

Thanks to (3.182) and (3.183), we may invoke Proposition 3.82 to guarantee that

$$\|\mathcal{M}_{\Sigma,s} f\|_{\mathrm{A}^p(\Sigma,\sigma)} \leq \|\mathcal{M}_\Sigma\|_{\mathrm{B}^{p'}(\Sigma,\sigma)\to \mathrm{B}^{p'}(\Sigma,\sigma)} \|f\|_{\mathrm{A}^p(\Sigma,\sigma)}. \qquad (3.184)$$

This ultimately proves (3.177). To show that the range $s \in (0, 1)$ is sharp, first observe that for every $s \in (1, \infty)$ Hölder's inequality yields

$$\mathcal{M}_\Sigma(\mathbb{1}_{\Delta(x_0,1)}) = \mathcal{M}_{\Sigma,1}(\mathbb{1}_{\Delta(x_0,1)}) \leq \mathcal{M}_{\Sigma,s}(\mathbb{1}_{\Delta(x_0,1)}) \ \text{ everywhere on } \Sigma. \qquad (3.185)$$

It then suffices to verify that $\mathcal{M}_\Sigma$ is not bounded $\mathrm{A}^p(\Sigma, \sigma)$. To see that this is the case, use Proposition 2.1 and Example 3.2 to obtain that $\mathcal{M}_\Sigma(\mathbb{1}_{\Delta(x_0,1)}) \notin \mathrm{A}^p(\Sigma, \sigma)$. The concludes the proof of Theorem 3.3 □

Chapter 4
Beurling-Hardy Spaces

The classical (Lebesgue-based) Hardy space H^p is defined as the space of distributions satisfying a certain size condition by demanding the membership of the grand maximal function to the Lebesgue space L^p. Following this template, we introduce the scale of Beurling-Hardy spaces HA^p on Ahlfors regular sets by requiring that the grand maximal function belongs to Beurling spaces A^p. In this chapter, we take up the task of developing a functional analytic theory for HA^p in which atomic and molecular decompositions play a crucial role (see Section 4.2 and Section 4.3).

4.1 Definitions and Basic Results

In this section we discuss some fundamental properties of HA^p, such as completeness and embeddings. In order to define the Beurling-Hardy spaces, we need additional terminology. Consider $\Sigma \subseteq \mathbb{R}^n$ an arbitrary closed set. Denote by $\mathrm{Lip}_c(\Sigma)$ class of Lipschitz functions with compact support, and equip this space with the seminorm

$$\|\psi\|_{\mathrm{Lip}(\Sigma)} := \sup_{\substack{x,y\in\Sigma\\ x\neq y}} \frac{|\psi(x)-\psi(y)|}{|x-y|} \quad \text{for all } \psi \in \mathrm{Lip}_c(\Sigma). \tag{4.1}$$

Also let $\big(\mathrm{Lip}_c(\Sigma)\big)'$ be the algebraic dual of $\mathrm{Lip}_c(\Sigma)$, i.e., the space of linear functionals defined on $\mathrm{Lip}_c(\Sigma)$. We often refer to elements from $\big(\mathrm{Lip}_c(\Sigma)\big)'$ as distributions.

Recall the Fefferman-Stein grand maximal function in the general geometric setting of Ahlfors regular sets.

Definition 4.1 Assume $\Sigma \subseteq \mathbb{R}^n$ is a closed set which is Ahlfors regular, and abbreviate $\sigma := \mathcal{H}^{n-1}\lfloor\Sigma$. For each $x \in \Sigma$ and each parameter $\gamma \in (0,1)$, denote $\mathcal{B}_x^\gamma(\Sigma)$ the collection of all functions $\psi \in \mathrm{Lip}_c(\Sigma)$ for which there exists a radius $r \in \big(0, 2\mathrm{diam}(\Sigma)\big)$ satisfying:

© The Author(s), under exclusive license to Springer Nature Switzerland AG 2026
M. Mitrea and P. Takemura, *Singular Integrals, Herz-Type Function Spaces, and Boundary Problems*, Progress in Mathematics 362,
https://doi.org/10.1007/978-3-032-12516-3_4

(a) $\operatorname{supp}\psi \subseteq \Delta(x,r)$,

(b) $\sup_{\Sigma} |\psi| \leq \sigma\big(\Delta(x,r)\big)^{-1}$,

(c) $\displaystyle \|\psi\|_{\dot{\mathscr{C}}^{\gamma}(\Sigma)} := \sup_{\substack{z_1,z_2\in\Sigma\\ z_1\neq z_2}} \frac{|\psi(z_1)-\psi(z_2)|}{|z_1-z_2|^{\gamma}} \leq r^{-\gamma}\sigma\big(\Delta(x,r)\big)^{-1}$.

Given a distribution $f \in \big(\mathrm{Lip}_c(\Sigma)\big)'$, introduce the `Fefferman-Stein grand maximal function` of f as being the function $f^{\#}_{\gamma} : \Sigma \to [0,\infty]$ defined at each $x \in \Sigma$ according to

$$f^{\#}_{\gamma}(x) := \sup_{\psi\in\mathcal{B}^{\gamma}_{x}(\Sigma)} |\langle f,\psi\rangle|. \tag{4.2}$$

Here, $\langle\cdot,\cdot\rangle$ denotes the natural duality pairing between $\big(\mathrm{Lip}_c(\Sigma)\big)'$ and $\mathrm{Lip}_c(\Sigma)$.

The result below summarizes some fundamental properties of the grand maximal function. See [GHA-II, §4.1] and references therein for its proof.

Proposition 4.1 *Consider an unbounded closed set $\Sigma \subseteq \mathbb{R}^n$ which is Ahlfors regular, and abbreviate $\sigma := \mathcal{H}^{n-1}\lfloor\Sigma$. Also, fix a parameter $\gamma \in (0,1)$. Then the following statements hold true:*

(a) There exists a geometric constant $C \in (0,\infty)$ with the property that for every $f \in L^1_{\mathrm{loc}}(\Sigma,\sigma)$, which in particular puts f in $\big(\mathrm{Lip}_c(\Sigma)\big)'$, one has

$$|f| \leq C f^{\#}_{\gamma} \quad \text{at } \sigma\text{-a.e. point on } \Sigma. \tag{4.3}$$

(b) For each distribution $f \in \big(\mathrm{Lip}_c(\Sigma)\big)'$ the function $f^{\#}_{\gamma}$ is lower-semicontinuous. Consequently, it is Borel measurable, hence σ-measurable.

(c) For every distribution $f \in \big(\mathrm{Lip}_c(\Sigma)\big)'$ and every exponent $p \in [1,\infty]$ the following implications hold:

$$f^{\#}_{\gamma} \in L^p(\Sigma,\sigma) \implies f \in L^p(\Sigma,\sigma), \tag{4.4}$$

$$f^{\#}_{\gamma} \in L^p_{\mathrm{loc}}(\Sigma,\sigma) \implies f \in L^p_{\mathrm{loc}}(\Sigma,\sigma). \tag{4.5}$$

(d) There exists some constant $C = C(\Sigma,n,\gamma) \in (0,\infty)$ such that for every distribution $f \in \big(\mathrm{Lip}_c(\Sigma)\big)'$ and $\psi \in \mathrm{Lip}_c(\Sigma)$ one has

$$|\langle f,\psi\rangle| \leq C R^{n-1+\gamma} \|\psi\|^{\gamma}_{\mathrm{Lip}(\Sigma)} \big(\sup_{\Sigma}|\psi|\big)^{1-\gamma} f^{\#}_{\gamma}(x), \tag{4.6}$$

whenever $x \in \Sigma$ and $R \in (0,\infty)$ are such that $\operatorname{supp}\psi \subseteq \Delta(x,R)$.

(e) For any function $f \in L^1_{\mathrm{loc}}(\Sigma,\sigma)$ one has

$$f_\gamma^\#(x) \le \mathcal{M}_\Sigma f(x) \quad \text{at } \sigma\text{-a.e. } x \in \Sigma, \tag{4.7}$$

where $\mathcal{M}_\Sigma$ denotes the Hardy-Littlewood maximal operator.

We are now ready to introduce the Beurling-Hardy spaces. The following definition is of basic importance for this work.

Definition 4.2 Consider a closed unbounded Ahlfors regular set $\Sigma \subseteq \mathbb{R}^n$, and abbreviate $\sigma := \mathcal{H}^{n-1}\lfloor\Sigma$. Having fixed a background parameter $\gamma \in (0,1)$, for each $p \in [1,\infty)$ define the `Beurling-Hardy space` as

$$\mathrm{HA}^p(\Sigma,\sigma) := \big\{ f \in \big(\mathrm{Lip}_c(\Sigma)\big)' : f_\gamma^\# \in \mathrm{A}^p(\Sigma,\sigma) \big\}, \tag{4.8}$$

and equip this space with the norm

$$\|f\|_{\mathrm{HA}^p(\Sigma,\sigma)} := \|f_\gamma^\#\|_{\mathrm{A}^p(\Sigma,\sigma)} \quad \text{for all } f \in \mathrm{HA}^p(\Sigma,\sigma). \tag{4.9}$$

Remark 4.1 Upon recalling that corresponding to the endpoint case $p = 1$ we have $\mathrm{A}^1(\Sigma,\sigma) = L^1(\Sigma,\sigma)$ (see Remark 3.1), it is visible from (4.8) that the Beurling-Hardy space $\mathrm{HA}^1(\Sigma,\sigma)$ coincides with the standard Hardy space $H^1(\Sigma,\sigma)$ (as defined in, e.g., [GHA-II, Definition 4.2.1, p. 114]).

A combination of (3.30), (4.4), and (4.8) permits us to characterize $\mathrm{HA}^p(\Sigma,\sigma)$ as

$$\mathrm{HA}^p(\Sigma,\sigma) = \big\{ f \in L^r(\Sigma,\sigma) : f_\gamma^\# \in \mathrm{A}^p(\Sigma,\sigma) \big\} \quad \text{for each } r \in [1,p]. \tag{4.10}$$

In particular, elements from HA^p are genuine functions, as opposed to the classical Hardy spaces H^p for $0 < p < 1$.

The following result provides embeddings for Beurling-Hardy spaces.

Proposition 4.2 *Consider a closed unbounded Ahlfors regular set $\Sigma \subseteq \mathbb{R}^n$, and abbreviate $\sigma := \mathcal{H}^{n-1}\lfloor\Sigma$. Also fix an integrability exponent $p \in [1,\infty)$. Then whenever $p < q < \infty$ it follows that*

$$\mathrm{HA}^q(\Sigma,\sigma) \hookrightarrow \mathrm{HA}^p(\Sigma,\sigma) \quad \textit{continuously}, \tag{4.11}$$

i.e., the scale of Beurling-Hardy spaces $\mathrm{HA}^p(\Sigma,\sigma)$ is (topologically) nested with respect to p. Corresponding to the end-point case $p = 1$, it follows that

$$\mathrm{HA}^q(\Sigma,\sigma) \hookrightarrow H^1(\Sigma,\sigma) \quad \textit{continuously for all } q \in (1,\infty), \tag{4.12}$$

where $H^1(\Sigma,\sigma)$ denotes the (classical) Hardy space (cf. [GHA-II, Definition 4.2.1, p. 114]). Moreover,

$$\mathrm{HA}^p(\Sigma,\sigma) \hookrightarrow \mathrm{A}^p(\Sigma,\sigma) \quad \textit{continuously}. \tag{4.13}$$

As a consequence, whenever $1 \le r \le p$, one also has the continuous embeddings:

$$\mathrm{HA}^p(\Sigma, \sigma) \hookrightarrow L^r(\Sigma, \sigma) \tag{4.14}$$

and

$$\mathrm{HA}^p(\Sigma, \sigma) \hookrightarrow L^1\Big(\Sigma, \frac{\sigma(x)}{1 + |x|^{n-1}}\Big). \tag{4.15}$$

Lastly, the following inclusions hold:

$$\bigcup_{p<q<\infty} \mathrm{HA}^q(\Sigma, \sigma) \subseteq \mathrm{HA}^p(\Sigma, \sigma) \subseteq L^p_{\mathrm{loc}}(\Sigma, \sigma) \subseteq L^1_{\mathrm{loc}}(\Sigma, \sigma). \tag{4.16}$$

Proof All claims follow from (4.10), item *(a)* in Proposition 4.1, Proposition 3.3, and Remark 3.3. □

Remark 4.2 Since the scale HA^p is monotonically with respect to $p \in (1, \infty)$ and has H^1 as its endpoint (corresponding to $p = 1$), the Beurling-Hardy spaces may be regarded as an extension of H^1.

Remark 4.3 As observed in Remark 3.6, the Beurling space $\mathrm{A}^p(\Sigma, \sigma)$ sits properly inside $L^p(\Sigma, \sigma)$. This and (4.13) then imply that $\mathrm{HA}^p(\Sigma, \sigma)$ is strictly smaller than $L^p(\Sigma, \sigma)$.

In the following lemma we prove a weak-$*$ convergence result.

Lemma 4.1 *Consider a closed set $\Sigma \subseteq \mathbb{R}^n$ which is Ahlfors regular, and abbreviate $\sigma := \mathcal{H}^{n-1}\lfloor\Sigma$. Pick a sequence $\{f_j\}_{j\in\mathbb{N}} \subseteq L^1_{\mathrm{loc}}(\Sigma, \sigma)$, and assume there exists $f \in L^1_{\mathrm{loc}}(\Sigma, \sigma)$ for which*

$$f_j \to f \ \text{ in } L^1_{\mathrm{loc}}(\Sigma, \sigma) \text{ as } j \to \infty. \tag{4.17}$$

Then for every $\psi \in \mathrm{Lip}_c(\Sigma)$ one has

$$\langle f_j, \psi\rangle \to \langle f, \psi\rangle \ \text{ as } j \to \infty. \tag{4.18}$$

In particular, (4.18) *is also true whenever the convergence* (4.17) *holds in $L^p(\Sigma, \sigma)$ for some $p \in [1, \infty]$.*

Proof Fix a function $\psi \in \mathrm{Lip}_c(\Sigma)$. In particular, ψ is bounded and $\operatorname{supp}\psi \subseteq \Delta$ for some surface ball $\Delta \subseteq \Sigma$. As such, for each $j \in \mathbb{N}$ we have

$$|\langle f_j, \psi\rangle - \langle f, \psi\rangle| = \Big|\int_\Delta (f_j - f)\psi \, d\sigma\Big| \le \|\psi\|_{L^\infty(\Sigma,\sigma)} \int_\Delta |f_j - f| \, d\sigma. \tag{4.19}$$

Letting $j \to \infty$, then (4.17) implies that the right-hand side of (4.19) tends to zero. This then proves the convergence (4.18).

The very last claim made in the statement of Lemma 4.1 is a consequence of the fact that $L^p(\Sigma,\sigma) \subseteq L^1_{\rm loc}(\Sigma,\sigma)$ for every $p \in [1,\infty]$. □

Here is a result concerning the completeness of Beurling-Hardy spaces.

Proposition 4.3 *Let $\Sigma \subseteq \mathbb{R}^n$ be a closed unbounded Ahlfors regular set, and abbreviate $\sigma := \mathcal{H}^{n-1}\lfloor\Sigma$. For each $p \in (1,\infty)$, the Beurling-Hardy space* $\mathrm{HA}^p(\Sigma,\sigma)$ *is a Banach space when equipped with the norm $\|\cdot\|_{\mathrm{HA}^p(\Sigma,\sigma)}$.*

Proof Consider $\{f_j\}_{j\in\mathbb{N}}$ a Cauchy sequence in $\mathrm{HA}^p(\Sigma,\sigma)$. Having fixed an arbitrary $\varepsilon \in (0,\infty)$, choose $N \in \mathbb{N}$ such that

$$\|f_j - f_\ell\|_{\mathrm{HA}^p(\Sigma,\sigma)} < \varepsilon \quad \text{whenever } j,\ell \geq N. \tag{4.20}$$

Thanks to (4.13), the sequence $\{f_j\}_{j\in\mathbb{N}}$ is also Cauchy in $\mathrm{A}^p(\Sigma,\sigma)$. From the completeness of $\mathrm{A}^p(\Sigma,\sigma)$ (see Proposition 3.9), there exists some $f \in \mathrm{A}^p(\Sigma,\sigma)$ such that

$$f_j \longrightarrow f \quad \text{in } \mathrm{A}^p(\Sigma,\sigma) \text{ as } j \to \infty. \tag{4.21}$$

Since $\{f_j\}_{j\in\mathbb{N}}$ is a Cauchy sequence in $\mathrm{HA}^p(\Sigma,\sigma)$, we may inductively construct a subsequence $\{f_{j_m}\}_{m\in\mathbb{N}}$ satisfying

$$\|f_{j_m} - f_{j_{m+1}}\|_{\mathrm{HA}^p(\Sigma,\sigma)} < \frac{1}{2^m} \quad \text{for all } m \in \mathbb{N}. \tag{4.22}$$

Next, fix $k,\ell \in \mathbb{N}$ such that

$$k \geq N, \quad j_\ell \geq N \quad \text{and} \quad \frac{1}{2^\ell} < \varepsilon. \tag{4.23}$$

Since $\{f_{j_m}\}_{m\in\mathbb{N}}$ also converges to f in $\mathrm{A}^p(\Sigma,\sigma)$, we may write

$$f - f_k = f_{j_\ell} - f_k + \sum_{m=\ell}^{\infty}(f_{j_{m+1}} - f_{j_m}) \;\text{ in } \mathrm{A}^p(\Sigma,\sigma). \tag{4.24}$$

From Proposition 3.3, this convergence holds in $L^1_{\rm loc}(\Sigma,\sigma)$. Hence, Lemma 4.1 ensures that for every $\psi \in \mathrm{Lip}_c(\Sigma)$ we have

$$\langle f - f_k, \psi\rangle = \langle f_{j_\ell} - f_k, \psi\rangle + \sum_{m=\ell}^{\infty}\langle f_{j_{m+1}} - f_{j_m}, \psi\rangle. \tag{4.25}$$

Upon recalling (4.2), we may immediately derive from (4.25) that

$$(f - f_k)^{\#}_{\gamma}(x) \leq (f_{j_\ell} - f_k)^{\#}_{\gamma}(x) + \sum_{m=\ell}^{\infty}(f_{j_{m+1}} - f_{j_m})^{\#}_{\gamma}(x) \quad \text{for all } x \in \Sigma. \tag{4.26}$$

This, (4.20), (4.22), and (4.23) then imply

$$\|f - f_k\|_{\mathrm{HA}^p(\Sigma,\sigma)} \le \|f_{j_\ell} - f_k\|_{\mathrm{HA}^p(\Sigma,\sigma)} + \sum_{m=\ell}^{\infty} \|f_{j_{m+1}} - f_{j_m}\|_{\mathrm{HA}^p(\Sigma,\sigma)}$$

$$< \varepsilon + \frac{1}{2^\ell} \sum_{m=\ell}^{\infty} \frac{1}{2^{m-\ell}} \le 3\varepsilon. \tag{4.27}$$

Therefore $f = (f - f_k) + f_k \in \mathrm{HA}^p(\Sigma, \sigma)$ and $f_k \to f$ in $\mathrm{HA}^p(\Sigma, \sigma)$, as $k \to \infty$. This establishes Proposition 4.3. □

4.2 Atomic Theory in Beurling-Hardy Spaces

The goal here is to establish an atomic decomposition for functions belonging to Beurling-Hardy spaces HA^p. This involves the notion of central atoms on unbounded Ahlfors regular sets from Definition 4.3 below. Our main result in the section is Theorem 4.2, which shows that said central atoms constitute the correct building blocks for Beurling-Hardy spaces.

Definition 4.3 Let $\Sigma \subseteq \mathbb{R}^n$ be a closed unbounded Ahlfors regular subset and abbreviate $\sigma := \mathcal{H}^{n-1}\lfloor\Sigma$. Denote by $C_\Sigma, c_\Sigma \in (0, \infty)$ the upper and lower Ahlfors regularity constant of Σ, respectively. Consider a reference point $x_0 \in \Sigma$ along with a dilation factor $2_* > (C_\Sigma/c_\Sigma)^{1/(n-1)}$. Fix an integrability exponent $p \in (1, \infty)$ with Hölder conjugate exponent $p' \in (1, \infty)$. A σ-measurable function $a : \Sigma \to \mathbb{R}$ is called an x_0-`central` $(1, p)$-`atom` provided there exists some finite radius $R \ge 2_*$ such that the following three properties are satisfied:

(1) Localization Condition:

$$\operatorname{supp} a \subseteq \Delta(x_0, R). \tag{4.28}$$

(2) Normalization Condition:

$$\|a\|_{L^p(\Sigma,\sigma)} \le R^{-(n-1)/p'}. \tag{4.29}$$

(3) Vanishing Moment Condition:

$$\int_\Sigma a \, d\sigma = 0. \tag{4.30}$$

Remark 4.4 It is visible that conditions (4.28)-(4.29) and Hölder's inequality guarantee that central atoms are absolutely integrable functions. As such, the vanishing moment condition in (4.30) is meaningfully formulated.

The following result ensures that central atoms (in the sense of Definition 4.3) are uniformly bounded in Beurling-Hardy spaces.

Theorem 4.1 *Assume that* $\Sigma \subseteq \mathbb{R}^n$ *is a closed unbounded Ahlfors regular set, and abbreviate* $\sigma := \mathcal{H}^{n-1}\lfloor\Sigma$. *Fix a reference point* $x_0 \in \Sigma$, *an integrability exponent* $p \in (1,\infty)$ *and a parameter* $\gamma \in (0,1)$. *Then every* x_0*-central* $(1,p)$*-atom belongs to* $\mathrm{HA}^p(\Sigma,\sigma)$ *in a uniform fashion. Specifically, there exists some constant* $C \in (0,\infty)$ *depending only on* n, p, γ, *and the Ahlfors regularity of* Σ, *such that for every* x_0*-central* $(1,p)$*-atom* $a : \Sigma \to \mathbb{R}$ *one has*

$$\|a\|_{\mathrm{HA}^p(\Sigma,\sigma)} = \|a^\#_\gamma\|_{\mathrm{A}^p(\Sigma,\sigma)} \leq C. \tag{4.31}$$

Proof Recall 2_* from in Definition 4.3. Select an x_0-central $(1,p)$-atom a on Σ, and a finite radius $R \in [2_*,\infty)$ associated with a as in Definition 4.3. Select the unique integer $\ell \in \mathbb{N}$ with the property that

$$2_*^\ell \leq R < 2_*^{\ell+1}. \tag{4.32}$$

Fix $k \in \mathbb{N}_0$. Pick $x \in C_k = \Delta(x_0, 2_*^{k+1})\backslash\Delta(x_0, 2_*^k)$, along with a function $\psi \in \mathscr{B}_x^\gamma(\Sigma)$ associated with a radius $r \in (0,\infty)$ as in Definition 4.1. In particular, $\psi \in \mathrm{Lip}_c(\Sigma)$ and

$$\sup_\Sigma |\psi| \leq \sigma\big(\Delta(x,r)\big)^{-1}, \tag{4.33}$$

$$\operatorname{supp}\psi \subseteq \Delta(x,r), \tag{4.34}$$

$$\|\psi\|_{\dot{\mathscr{C}}^\gamma(\Sigma)} \leq r^{-\gamma}\sigma\big(\Delta(x,r)\big)^{-1}. \tag{4.35}$$

Our goal is to obtain an estimate independent of a for the pairing $\langle a,\psi\rangle$, which in this case is the integral of a against ψ, i.e.,

$$\langle a,\psi\rangle = \int_\Sigma a\psi\, d\sigma. \tag{4.36}$$

Notice that if $\Delta(x,r) \cap \Delta(x_0,R) = \varnothing$, then this integral is automatically zero. Henceforth, assume $\Delta(x,r) \cap \Delta(x_0,R) \neq \varnothing$.

Suppose first that k is an integer such that $k \geq \ell + 2$. Then, keeping in mind (4.32), whenever $y \in \Delta(x,r) \cap \Delta(x_0,R)$, we have

$$r > |x-y| \geq |x-x_0| - |y-x_0| \geq 2_*^k - 2_*^{\ell+1} \geq c2_*^k \tag{4.37}$$

where $c := 1 - 1/2_* > 0$. Use this, properties (2)-(3) in Definition 4.3, (4.32), (4.35), the Ahlfors regularity of Σ, and Hölder's inequality, to estimate

$$
\begin{aligned}
|\langle a,\psi\rangle| &= \Big|\int_\Sigma a(y)\psi(y)\,d\sigma(y)\Big| = \Big|\int_\Sigma a(y)\big(\psi(y)-\psi(x_0)\big)\,d\sigma(y)\Big| \\
&\leq \int_{\Delta(x_0,R)} |a(y)||\psi(y)-\psi(x_0)|\,d\sigma(y) \\
&\leq R^\gamma \int_{\Delta(x_0,R)} |a(y)|\frac{|\psi(y)-\psi(x_0)|}{|y-x_0|^\gamma}\,d\sigma(y) \\
&\leq C2_*^{\ell\gamma}\|\psi\|_{\dot{\mathscr{C}}^\gamma(\Sigma)}\int_{\Delta(x_0,R)}|a(y)|\,d\sigma(y) \\
&\leq \frac{C\cdot 2_*^{\ell\gamma}}{r^\gamma\sigma\big(\Delta(x,r)\big)}\sigma\big(\Delta(x_0,R)\big)^{1/p'}\|a\|_{L^p(\Sigma,\sigma)} \\
&\leq C\frac{2_*^{(\ell-k)\gamma}}{2_*^{k(n-1)}},
\end{aligned}
\tag{4.38}
$$

for some constant $C\in(0,\infty)$ depending only on n, γ, p, and the Ahlfors regularity of Σ. Taking the supremum over all $\psi\in\mathscr{B}_x^\gamma(\Sigma)$ in (4.38) we see that

$$
a_\gamma^\#(x) = \sup_{\psi\in\mathscr{B}_x^\gamma(\Sigma)}|\langle a,\psi\rangle| \leq C\frac{2_*^{(\ell-k)\gamma}}{2_*^{k(n-1)}}, \quad \text{for all } x\in C_k. \tag{4.39}
$$

Bearing this and the Ahlfors regularity of Σ in mind, write

$$
\begin{aligned}
\|a_\gamma^\#\cdot\mathbb{1}_{C_k}\|_{L^p(\Sigma,\sigma)} &= \Big(\int_{C_k}|a_\gamma^\#|^p\,d\sigma\Big)^{1/p} \leq C\frac{2_*^{(\ell-k)\gamma}}{2_*^{k(n-1)}}\sigma(C_k)^{1/p} \\
&\leq C\frac{2_*^{(\ell-k)\gamma}}{2_*^{k(n-1)}}2_*^{k(n-1)/p} = C\frac{2_*^{(\ell-k)\gamma}}{2_*^{k(n-1)/p'}},
\end{aligned}
\tag{4.40}
$$

for all $k\in\mathbb{N}$ with $k\geq\ell+2$, and for some constant $C\in(0,\infty)$ depending on n, γ, p, and the Ahlfors regularity of Σ. As such, we have

$$
\sum_{k=\ell+2}^{\infty} 2_*^{k(n-1)/p'}\|a_\gamma^\#\cdot\mathbb{1}_{C_k}\|_{L^p(\Sigma,\sigma)} \leq C\sum_{k=\ell+2}^{\infty}2_*^{(\ell-k)\gamma} = C\sum_{j=2}^{\infty}\frac{1}{2_*^{j\gamma}} < \infty. \tag{4.41}
$$

Pressing on, consider now the case $0\leq k\leq\ell+1$. Since a is itself a locally integrable function on Σ, from (4.7) we see that

$$
a_\gamma^\# \leq \mathcal{M}_\Sigma a \quad \text{pointwise everywhere on } \Sigma. \tag{4.42}
$$

Thanks to this, (4.32), the normalization property (4.29) of a, the boundedness of $\mathcal{M}_\Sigma$ on $L^p(\Sigma,\sigma)$, we may now estimate

$$
\begin{aligned}
\sum_{k=0}^{\ell+1} 2_*^{k(n-1)/p'} \|a_\gamma^\# \cdot \mathbb{1}_{C_k}\|_{L^p(\Sigma,\sigma)} &\leq \|\mathcal{M}_\Sigma a\|_{L^p(\Sigma,\sigma)} \sum_{k=0}^{\ell+1} 2_*^{k(n-1)/p'} \\
&\leq C\|a\|_{L^p(\Sigma,\sigma)} \sum_{k=0}^{\ell+1} 2_*^{k(n-1)/p'} \\
&\leq C R^{-(n-1)/p'} \sum_{k=0}^{\ell+1} 2_*^{k(n-1)/p'} \\
&\leq C \sum_{k=0}^{\ell+1} 2_*^{(k-\ell)(n-1)/p'} \\
&\leq C \sum_{j=0}^{\infty} 2_*^{-j(n-1)/p'} < \infty, \qquad (4.43)
\end{aligned}
$$

for some constant $C \in (0, \infty)$ depending only on n, p, and the Ahlfors regularity of Σ.

Finally, based on estimates (4.41) and (4.43) we conclude that there exists some $C \in (0, \infty)$ depending solely on n, p, γ, and the Ahlfors regularity of Σ with the property that

$$
\begin{aligned}
\|a\|_{\mathrm{HA}^p(\Sigma,\sigma)} &= \sum_{k=0}^{\ell+1} 2_*^{k(n-1)/p'} \|a_\gamma^\# \cdot \mathbb{1}_{C_k}\|_{L^p(\Sigma,\sigma)} \\
&\quad + \sum_{k=\ell+2}^{\infty} 2_*^{k(n-1)/p'} \|a_\gamma^\# \cdot \mathbb{1}_{C_k}\|_{L^p(\Sigma,\sigma)} \leq C,
\end{aligned} \qquad (4.44)
$$

as wanted. This concludes the proof of Theorem 4.1. □

We now present a crucial result in the development of the present theory, namely the atomic decompositions in Beurling-Hardy spaces.

Theorem 4.2 *Let $\Sigma \subseteq \mathbb{R}^n$ be a closed unbounded Ahlfors regular set, and abbreviate $\sigma := \mathcal{H}^{n-1}\lfloor\Sigma$. Consider a reference point $x_0 \in \Sigma$, and an integrability exponent $p \in (1, \infty)$. Then, in this setting, the following statements are true.*

(a) For every $f \in \mathrm{HA}^p(\Sigma, \sigma)$ there exist a sequence of x_0-central $(1, p)$-atoms $\{a_k\}_{k\in\mathbb{N}_0}$ and a numerical sequence $\{\lambda_k\}_{k\in\mathbb{N}_0} \subseteq \mathbb{R}$ satisfying

$$
f = \sum_{k=0}^{\infty} \lambda_k a_k \qquad (4.45)
$$

with simultaneous convergence to f *in* $L^1(\Sigma,\sigma)$, *in* $L^p(\Sigma,\sigma)$, *and in* $\mathrm{HA}^p(\Sigma,\sigma)$. *Moreover, there exists a constant* $C \in (0,\infty)$ *depending only on n, p, and the Ahlfors regularity character of* Σ *with the property that*

$$\sum_{k=0}^{\infty} |\lambda_k| \leq C\|f\|_{\mathrm{HA}^p(\Sigma,\sigma)}. \tag{4.46}$$

(b) Conversely, assume $f : \Sigma \longrightarrow \mathbb{R}$ *is a* σ*-measurable function for which there exist some sequence of* x_0*-central* $(1, p)$*-atoms* $\{a_k\}_{k\in\mathbb{N}_0}$ *and some numerical sequence* $\{\lambda_k\}_{k\in\mathbb{N}_0} \in \ell^1(\mathbb{N}_0)$ *such that*

$$f = \sum_{k=0}^{\infty} \lambda_k a_k \quad \textit{pointwise } \sigma\textit{-a.e. on } \Sigma. \tag{4.47}$$

Then $f \in \mathrm{HA}^p(\Sigma,\sigma)$ *quantitatively, i.e., there exists some constant* $C \in (0,\infty)$, *depending only on n, p, and the Ahlfors regularity of* Σ *for which*

$$\|f\|_{\mathrm{HA}^p(\Sigma,\sigma)} \leq C \sum_{k=0}^{\infty} |\lambda_k|. \tag{4.48}$$

(c) As a consequence of items (a) and (b), for every $f \in \mathrm{HA}^p(\Sigma,\sigma)$ *it follows that*

$$\|f\|_{\mathrm{HA}^p(\Sigma,\sigma)} \approx \inf\Big\{\sum_{k=0}^{\infty} |\lambda_k| \,:\, f = \sum_{k=0}^{\infty} \lambda_k a_k \textit{ in } L^1(\Sigma,\sigma)\Big\}, \tag{4.49}$$

where the infimum is taken over all atomic representations of f, *and with implicit proportionality constants independent of* f.

Proof As a preamble to the justification of the claim made in item *(a)*, we first construct a smooth partition of unity. Fix a dilation factor $2_* \in (1,\infty)$ as in Definition 4.3, and a number $\varepsilon \in (0,\infty)$ such that

$$0 < \varepsilon < \min\Big\{\frac{2_* - 1}{1 + 2_*^2}, \frac{1}{2_*^2}\Big\}. \tag{4.50}$$

Also, pick two functions $\widetilde{\psi}, \psi : \mathbb{R}^n \longrightarrow \mathbb{R}$ satisfying

$$\begin{gathered} \widetilde{\psi} \in \mathscr{C}_c^\infty(\mathbb{R}^n), \quad 0 \leq \widetilde{\psi} \leq 1, \quad \operatorname{supp}\widetilde{\psi} \subseteq B(0, 2_* + 2_*\varepsilon), \\ \text{and } \widetilde{\psi}(x) = 1 \text{ for all } x \in \mathbb{R}^n \text{ with } |x| \leq 2_*, \end{gathered} \tag{4.51}$$

and

$$\psi \in \mathscr{C}_c^\infty(\mathbb{R}^n),\ 0 \le \psi \le 1,\ \operatorname{supp}\psi \subseteq \{x \in \mathbb{R}^n : 1 - 2_*\varepsilon \le |x| \le 2_* + 2_*\varepsilon\},$$
$$\text{and } \psi(x) = 1 \text{ for all } x \in \mathbb{R}^n \text{ with } 1 \le |x| \le 2_*. \tag{4.52}$$

Notice that our choice of ε in (4.50), implies $1 - 2_*\varepsilon > 0$. Hence the annulus in (4.52) in which ψ is supported is well defined. For each $x \in \mathbb{R}^n$ introduce

$$\begin{aligned} \psi_0(x) &:= \widetilde{\psi}(x - x_0) \\ \psi_k(x) &:= \psi\big(2_*^{-k}(x - x_0)\big) \quad \text{for all } k \in \mathbb{N}. \end{aligned} \tag{4.53}$$

Thanks to (4.51)-(4.52), we see that $\{\psi_k\}_{k\in\mathbb{N}_0} \subseteq \mathscr{C}_c^\infty(\mathbb{R}^n)$. Moreover,

$$\begin{aligned} 0 \le \psi_0 \le 1,\quad \operatorname{supp}\psi_0 \subseteq B(x_0, 2_* + 2_*\varepsilon), \\ \text{and } \psi_0(x) = 1 \text{ whenever } x \in \mathbb{R}^n \text{ and } |x - x_0| \le 2_*, \end{aligned} \tag{4.54}$$

and every $k \in \mathbb{N}$ we have

$$0 \le \psi_k \le 1,\ \operatorname{supp}\psi_k \subseteq \{x \in \mathbb{R}^n : 2_*^k - 2_*^{k+1}\varepsilon \le |x - x_0| \le 2_*^{k+1} + 2_*^{k+1}\varepsilon\},$$
$$\text{and } \psi_k(x) = 1 \text{ whenever } x \in \mathbb{R}^n \text{ satisfies } 2_*^k \le |x - x_0| \le 2_*^{k+1}. \tag{4.55}$$

Next, consider the following sets:

$$\begin{aligned} A_0 &:= B(x_0, 2_* + 2_*\varepsilon), \\ A_k &:= \{x \in \mathbb{R}^n : 2_*^k - 2_*^{k+1}\varepsilon \le |x - x_0| \le 2_*^{k+1} + 2_*^{k+1}\varepsilon\} \quad \text{for each } k \in \mathbb{N}. \end{aligned} \tag{4.56}$$

It is visible that

$$A_k \subseteq B(x_0, 2_*^{k+2}) \quad \text{for every } k \in \mathbb{N}_0 \tag{4.57}$$

since the choice of ε in (4.50) implies that $2_*^{k+1} + 2_*^{k+1}\varepsilon \le 2_*^{k+2}$ for every $k \in \mathbb{N}_0$.

<u>`Claim 1.`</u> *For every $k \in \mathbb{N}_0$ we have $A_k \cap A_{k+1} \neq \varnothing$ and $A_k \cap A_{k+2} = \varnothing$.*

Consider first the case $k = 0$. Since $0 < 2_* - 2_*^2\varepsilon < 2_* + 2_*\varepsilon$, it follows that

$$A_0 \cap A_1 \neq \varnothing. \tag{4.58}$$

On the other hand, our choice of ε in (4.50) permits us to write

$$\begin{aligned} \varepsilon < \frac{2_* - 1}{1 + 2_*^2} \quad &\Longrightarrow \quad 1 + \varepsilon < 2_* - 2_*^2\varepsilon \\ &\Longrightarrow \quad 2_* + 2_*\varepsilon < 2_*^2 - 2_*^3\varepsilon, \end{aligned} \tag{4.59}$$

which in view of (4.56) entails

$$A_0 \cap A_2 = \varnothing. \tag{4.60}$$

Now, consider an arbitrary $k \in \mathbb{N}$. Because $2_*^{k+1} - 2_*^{k+2}\varepsilon < 2_*^{k+1} + 2_*^{k+1}\varepsilon$, a glance at (4.56) reveals that

$$A_k \cap A_{k+1} \neq \varnothing. \tag{4.61}$$

Moreover, (4.59) implies

$$2_*^{k+1} + 2_*^{k+1}\varepsilon < 2_*^{k+2} - 2_*^{k+3}\varepsilon, \tag{4.62}$$

which together with (4.56) forces

$$A_k \cap A_{k+2} = \varnothing. \tag{4.63}$$

Hence, Claim 1 is established.

Claim 2. *For each $k, j \in \mathbb{N}_0$ with $j \geq k+2$ it follows that*

$$A_k \cap A_j = \varnothing. \tag{4.64}$$

To see this, notice that (4.59) implies that for every $k, j \in \mathbb{N}_0$ with $j \geq k+2$ we have

$$2_*^{k+2} + \varepsilon 2_*^{k+2} \leq 2_*^j + \varepsilon 2_*^j < 2_*^{j+1} - 2_*^{j+2}\varepsilon, \tag{4.65}$$

which, together with (4.56), gives (4.64).

Claim 3. *The sum $\sum_{k=0}^{\infty} \psi_k$ is locally finite and is a well-defined smooth function with the property that*

$$1 \leq \sum_{k=0}^{\infty} \psi_k \leq 2 \ \text{ in } \mathbb{R}^n. \tag{4.66}$$

To prove this claim, first observe that $\mathbb{R}^n = \cup_{k=0}^{\infty} A_k$. Then having fixed $x \in \mathbb{R}^n$ there exists $k \in \mathbb{N}_0$ such that $x \in A_k$. In view of Claim 2, $x \notin A_j$ for any $j \in \mathbb{N}$ with $j \geq k+2$. We have the following scenarios

$$\begin{aligned} x \in A_0 \cap A_1, \ \text{ or } \ x \in A_0 \backslash A_1 \ \text{ in case } k = 0, \\ \text{or } \ x \in A_k \cap A_{k+1}, \ \text{ or } \ x \in A_{k-1} \cap A_k, \\ \text{or } \ x \in A_k \backslash (A_{k-1} \cup A_{k+1}) \ \text{ in case } k \neq 0. \end{aligned} \tag{4.67}$$

Case 1: Assume $k = 0$ and $x \in A_0 \cap A_1$. Then from (4.54)-(4.56) we have

$$\sum_{j=0}^{\infty} \psi_j(x) = \psi_0(x) + \psi_1(x) \leq 2. \tag{4.68}$$

Moreover, since $x \in A_0 \cap A_1$ it follows that

$$2_* - 2_*^2\varepsilon \leq |x - x_0| < 2_* + 2_*\varepsilon \leq 2_*^2. \tag{4.69}$$

Clearly, $|x - x_0| < 2_*$ or $|x - x_0| \geq 2_*$. The fact that $\psi_0(x) + \psi_1(x) \geq 1$ can be seen from this, (4.54)-(4.56), and the implications

$$\begin{aligned} |x - x_0| < 2_* \quad &\Longrightarrow \quad \psi_0(x) + \psi_1(x) = 1 + \psi_1(x) \geq 1 \quad \text{and} \\ 2_* \leq |x - x_0| \leq 2_*^2 \quad &\Longrightarrow \quad \psi_0(x) + \psi_1(x) = \psi_0(x) + 1 \geq 1. \end{aligned} \tag{4.70}$$

This and (4.68) justifies (4.66) at the point x in Case 1.

Case 2: Assume $k = 0$ and $x \in A_0 \backslash A_1$. Then thanks to (4.54)-(4.56) we have that $|x - x_0| < 2_* - 2_*^2 \varepsilon < 2_*$, which further implies

$$\sum_{j=0}^{\infty} \psi_j(x) = \psi_0(x) = 1 \leq 2. \tag{4.71}$$

This validates (4.66) at the point x in Case 2.

Case 3: Assume $k \neq 0$ and $x \in A_k \cap A_{k+1}$. Then $x \notin A_j$ for each $j \in \mathbb{N}_0 \backslash \{k, k+1\}$. This and (4.54)-(4.56) give

$$\sum_{j=0}^{\infty} \psi_j(x) = \psi_k(x) + \psi_{k+1}(x) \leq 2. \tag{4.72}$$

Because $x \in A_k \cap A_{k+1}$, the choice of ε in (4.50) allows us to estimate

$$2_*^k \leq 2_*^{k+1} - 2_*^{k+2} \varepsilon \leq |x - x_0| \leq 2_*^{k+1} + 2_*^{k+1} \varepsilon \leq 2_*^{k+2}. \tag{4.73}$$

Obviously $|x - x_0| \leq 2_*^{k+1}$, or $|x - x_0| \geq 2_*^{k+1}$. If $|x - x_0| \leq 2_*^{k+1}$, then

$$2_*^k \leq |x - x_0| \leq 2_*^{k+1}. \tag{4.74}$$

In view of (4.55), we obtain $\psi_k(x) = 1$. Together with (4.72), this permits us to write

$$\sum_{j=0}^{\infty} \psi_j(x) = \psi_k(x) + \psi_{k+1}(x) = 1 + \psi_{k+1}(x) \geq 1. \tag{4.75}$$

A similar argument holds in the case when $|x - x_0| \geq 2_*^{k+1}$. All together, the veracity of (4.66) at the point x in Case 3 has been established.

Case 4: Assume $k \neq 0$ and $x \in A_{k-1} \cap A_k$. We can employ the same reasoning as in Case 3 with A_k replaced by A_{k-1}, and A_{k+1} replaced by A_k, to obtain the desired conclusion.

Case 5: Assume $k \neq 0$ and $x \in A_k \backslash (A_{k-1} \cup A_{k+1})$. Then in light of (4.55) and (4.64) we have $\psi_j(x) = 0$ for all $j \in \mathbb{N}_0$ with $j \neq k$, and

$$2_*^k < 2_*^k + 2_*^k \varepsilon < |x - x_0| < 2_*^{k+1} - 2_*^{k+2} \varepsilon < 2_*^{k+1}. \tag{4.76}$$

From this and (4.55), it follows that

$$\sum_{j=0}^{\infty}\psi_j(x)=\psi_k(x)=1\leq 2. \tag{4.77}$$

At this stage, we conclude that $\sum_{k=0}^{\infty}\psi_k$ is a locally finite sum and that (4.66) holds. Since each ψ_k is smooth, and differentiability is a local property, we can differentiate this series term by term infinitely many times. As such, $\sum_{k=0}^{\infty}\psi_k$ is a well-defined smooth function in $\mathbb{R}^n$. This establishes Claim 3.

To proceed, for each $k\in\mathbb{N}_0$ define the function $\phi_k:\mathbb{R}^n\to\mathbb{R}$ by setting

$$\phi_k:=\frac{\psi_k}{\sum_{j=0}^{\infty}\psi_j}. \tag{4.78}$$

Thanks to Claim 3, (4.72), and (4.54)-(4.56), we have

$$\begin{aligned}&\phi_k\in\mathscr{C}_c^{\infty}(\mathbb{R}^n),\quad 0\leq\phi_k\leq 1,\quad\text{and}\\&\operatorname{supp}\phi_k\subseteq A_k\quad\text{for every } k\in\mathbb{N}_0.\end{aligned} \tag{4.79}$$

Moreover, the sum $\sum_{k=0}^{\infty}\phi_k$ is locally finite and in fact

$$\sum_{k=0}^{\infty}\phi_k=1\ \text{ at every point in }\mathbb{R}^n. \tag{4.80}$$

Upon recalling (4.56), next introduce

$$\begin{aligned}&\widetilde{A}_0:=A_0\cap\Sigma=\Delta(x_0,2_*+2_*\varepsilon)\\&\widetilde{A}_k:=A_k\cap\Sigma=\Delta(x_0,2_*^{k+1}+2_*^{k+1}\varepsilon)\backslash\Delta(x_0,2_*^{k}-2_*^{k+1}\varepsilon)\ \text{ for all } k\in\mathbb{N}.\end{aligned} \tag{4.81}$$

From the proof so far (see Claim 1 and Claim 2), we see that $\{\widetilde{A}_k\}_{k\in\mathbb{N}_0}$ is a family of σ-measurable sets on Σ satisfying the following properties:

$$\begin{aligned}&\Sigma=\textstyle\bigcup_{k\in\mathbb{N}_0}\widetilde{A}_k,\quad\widetilde{A}_k\subseteq\Delta(x_0,2_*^{k+2})\ \text{ for all } k\in\mathbb{N}_0,\\&\text{and }\ \widetilde{A}_k\cap\widetilde{A}_j=\varnothing\ \text{ for all } j,k\in\mathbb{N}_0 \text{ with } j\geq k+2.\end{aligned} \tag{4.82}$$

In particular, with $\{C_k\}_{k\in\mathbb{N}_0}$ as in (3.1) we have

$$\widetilde{A}_0\subseteq C_0\cup C_1,\quad\text{and}\quad\widetilde{A}_k\subseteq C_{k-1}\cup C_k\cup C_{k+1}\quad\text{for all } k\in\mathbb{N}. \tag{4.83}$$

For each $k\in\mathbb{N}_0$ abbreviate

$$\widetilde{\mathbb{1}}_k:=\sigma(\widetilde{A}_k)^{-1}\mathbb{1}_{\widetilde{A}_k}. \tag{4.84}$$

It is visible from (4.81) and (3.1) that

$$C_k\subseteq\widetilde{A}_k\quad\text{for all } k\in\mathbb{N}_0. \tag{4.85}$$

Since $\sigma(C_k) > 0$ (cf. Remark 3.5), we have $\sigma(\widetilde{A}_k) > 0$ for all $k \in \mathbb{N}_0$. As such, for each $k \in \mathbb{N}_0$ the function $\widetilde{\mathbb{1}}_k$ is well defined and not identically zero on Σ.

Since $\operatorname{supp}\psi_k \subseteq A_k$ for all $k \in \mathbb{N}_0$ (cf. (4.54)-(4.56)), then based on (4.81) we have $\operatorname{supp}\big(\psi_k\big|_\Sigma\big) \subseteq \widetilde{A}_k$ for all $k \in \mathbb{N}_0$. This and (4.78) imply

$$\operatorname{supp}\big(\phi_k\big|_\Sigma\big) \subseteq \widetilde{A}_k \quad \text{for all } k \in \mathbb{N}_0, \tag{4.86}$$

which in concert with (4.79) and (4.83) entails

$$0 \le \phi_0\big|_\Sigma \le \mathbb{1}_{C_0} + \mathbb{1}_{C_1} \quad \text{and} \quad 0 \le \phi_k\big|_\Sigma \le \mathbb{1}_{C_{k-1}} + \mathbb{1}_{C_k} + \mathbb{1}_{C_{k+1}} \quad \text{for all } k \in \mathbb{N}. \tag{4.87}$$

Pressing on, fix an arbitrary function $f \in \mathrm{HA}^p(\Sigma, \sigma)$. In light of (4.80), we have

$$f = \sum_{k=0}^{\infty} f\big(\phi_k\big|_\Sigma\big) \quad \text{pointwise } \sigma\text{-a.e. on } \Sigma. \tag{4.88}$$

Moreover, we claim that the above convergence holds in $L^1(\Sigma, \sigma)$ and in $L^p(\Sigma, \sigma)$. Indeed, (4.80) implies that, on Σ,

$$\Big|\sum_{k=0}^{N} f\big(\phi_k\big|_\Sigma\big)\Big| = |f| \sum_{k=0}^{N} \big(\phi_k\big|_\Sigma\big) \le |f| \quad \text{for every } N \in \mathbb{N}. \tag{4.89}$$

Since $f \in L^1(\Sigma, \sigma) \cap L^p(\Sigma, \sigma)$ (cf. Proposition 4.2), Lebesgue's Dominated Convergence Theorem gives

$$f = \sum_{k=0}^{\infty} f\big(\phi_k\big|_\Sigma\big) \quad \text{in } L^1(\Sigma, \sigma) \text{ and in } L^p(\Sigma, \sigma). \tag{4.90}$$

Next, express f as

$$\begin{aligned} f &= \lim_{N\to\infty} \sum_{k=0}^{N} f\big(\phi_k\big|_\Sigma\big) \\ &= \lim_{N\to\infty} \Big\{ \sum_{k=0}^{N} \Big[f\big(\phi_k\big|_\Sigma\big) - \Big(\int_\Sigma f\phi_k \, d\sigma\Big)\widetilde{\mathbb{1}}_k \Big] + \sum_{k=0}^{N} \Big(\int_\Sigma f\phi_k \, d\sigma\Big)\widetilde{\mathbb{1}}_k \Big\}. \end{aligned} \tag{4.91}$$

For each $k \in \mathbb{N}_0$ define the function $g_k : \Sigma \to \mathbb{R}$ by setting

$$g_k := f\big(\phi_k\big|_\Sigma\big) - \Big(\int_\Sigma f\phi_k \, d\sigma\Big)\widetilde{\mathbb{1}}_k. \tag{4.92}$$

Upon recalling (4.79), and (4.81)-(4.84) it is visible that

$$
\begin{aligned}
&\operatorname{supp} g_k \subseteq \Delta(x_0, 2_*^{k+2}) \ \text{ and } \ \int_\Sigma g_k \, d\sigma = 0 \ \text{ for all } k \in \mathbb{N}_0, \\
&\text{and } \ \sum_{k=0}^{\infty} g_k \ \text{ is pointwise convergent } \sigma\text{-a.e. on } \Sigma.
\end{aligned}
\tag{4.93}
$$

Bearing in mind (4.84), (4.86), and using Hölder's inequality, for each $k \in \mathbb{N}_0$ we estimate

$$
\begin{aligned}
\|g_k\|_{L^p(\Sigma,\sigma)} &\leq \|f\phi_k\|_{L^p(\Sigma,\sigma)} + \Big(\int_\Sigma |f\phi_k| \, d\sigma\Big) \|\widetilde{\mathbb{1}}_k\|_{L^p(\Sigma,\sigma)} \\
&= \|f\phi_k\|_{L^p(\Sigma,\sigma)} + \Big(\int_{\widetilde{A}_k} |f\phi_k| \, d\sigma\Big) \sigma(\widetilde{A}_k)^{-1/p'} \\
&\leq \|f\phi_k\|_{L^p(\Sigma,\sigma)} + \|f\phi_k\|_{L^p(\Sigma,\sigma)} \sigma(\widetilde{A}_k)^{1/p'} \sigma(\widetilde{A}_k)^{-1/p'} \\
&= 2\|f\phi_k\|_{L^p(\Sigma,\sigma)}.
\end{aligned}
\tag{4.94}
$$

In concert with (4.87), we then obtain

$$
\begin{aligned}
&\|g_0\|_{L^p(\Sigma,\sigma)} \leq 2\big(\|f \cdot \mathbb{1}_{C_0}\|_{L^p(\Sigma,\sigma)} + \|f \cdot \mathbb{1}_{C_1}\|_{L^p(\Sigma,\sigma)}\big) \ \text{ and} \\
&\|g_k\|_{L^p(\Sigma,\sigma)} \leq 2\big(\|f \cdot \mathbb{1}_{C_{k-1}}\|_{L^p(\Sigma,\sigma)} + \|f \cdot \mathbb{1}_{C_k}\|_{L^p(\Sigma,\sigma)} \\
&\qquad\qquad\qquad\qquad + \|f \cdot \mathbb{1}_{C_{k+1}}\|_{L^p(\Sigma,\sigma)}\big) \ \text{if } k \in \mathbb{N}.
\end{aligned}
\tag{4.95}
$$

Next, introduce the set of indices

$$
\mathcal{J}_1 := \big\{k \in \mathbb{N}_0 : \|g_k\|_{L^p(\Sigma,\sigma)} \neq 0\big\}.
\tag{4.96}
$$

Define the function $a_0 : \Sigma \to \mathbb{R}$ as

$$
a_0 := \begin{cases} \Big[2(2_*^2)^{(n-1)/p'} \sum_{j=0}^{1} \|f \cdot \mathbb{1}_{C_j}\|_{L^p(\Sigma,\sigma)}\Big]^{-1} g_0 & \text{if } 0 \in \mathcal{J}_1, \\ 0 & \text{if } 0 \notin \mathcal{J}_1, \end{cases}
\tag{4.97}
$$

and for each $k \in \mathbb{N}$ set $a_k : \Sigma \to \mathbb{R}$ by

$$
a_k := \begin{cases} \Big[2(2_*^{k+2})^{(n-1)/p'} \sum_{\ell=k-1}^{k+1} \|f \cdot \mathbb{1}_{C_\ell}\|_{L^p(\Sigma,\sigma)}\Big]^{-1} g_k & \text{if } k \in \mathcal{J}_1, \\ 0 & \text{if } k \notin \mathcal{J}_1. \end{cases}
\tag{4.98}
$$

For each $k \in \mathbb{N}_0$ the function a_k is well defined and σ-measurable on Σ. Moreover, in light of (4.93), (4.95), (4.97)-(4.98), for each $k \in \mathbb{N}_0$ the function a_k satisfies

$$
\operatorname{supp} a_k \subseteq \Delta(x_0, 2_*^{k+2}), \ \ \|a_k\|_{L^p(\Sigma,\sigma)} \leq (2_*^{k+2})^{-(n-1)/p'}, \ \ \int_\Sigma a_k \, d\sigma = 0.
\tag{4.99}
$$

These properties ensure that a_k is an x_0-central $(1, p)$-atom for every $k \in \mathbb{N}_0$.

We now introduce the following numbers:

$$\lambda_0 := \begin{cases} 2(2_*^2)^{(n-1)/p'} \sum_{\ell=0}^{1} \|f \cdot \mathbb{1}_{C_\ell}\|_{L^p(\Sigma,\sigma)} & \text{if } 0 \in \mathcal{J}_1, \\ 0 & \text{if } 0 \notin \mathcal{J}_1, \end{cases} \tag{4.100}$$

and for each $k \in \mathbb{N}$ set

$$\lambda_k := \begin{cases} 2(2_*^{k+2})^{(n-1)/p'} \sum_{\ell=k-1}^{k+1} \|f \cdot \mathbb{1}_{C_\ell}\|_{L^p(\Sigma,\sigma)} & \text{if } k \in \mathcal{J}_1, \\ 0 & \text{if } k \notin \mathcal{J}_1. \end{cases} \tag{4.101}$$

Notice that these coefficients are indeed finite. Also, observe that for each $k \in \mathbb{N}_0$, we have

$$\|g_k\|_{L^p(\Sigma,\sigma)}, \quad a_k, \quad \text{and} \quad \lambda_k \quad \text{are all zero if and only if} \quad k \notin \mathcal{J}_1. \tag{4.102}$$

Next, use (4.100)-(4.101) and (4.13) to estimate

$$\begin{aligned} \sum_{k=0}^{\infty} |\lambda_k| &\leq C \sum_{\ell=0}^{1} \|f \cdot \mathbb{1}_{C_\ell}\|_{L^p(\Sigma,\sigma)} + C \sum_{k=1}^{\infty} \sum_{\ell=k-1}^{k+1} 2_*^{k(n-1)/p'} \|f \cdot \mathbb{1}_{C_\ell}\|_{L^p(\Sigma,\sigma)} \\ &\leq C \sum_{j=0}^{\infty} 2_*^{j(n-1)/p'} \|f \cdot \mathbb{1}_{C_j}\|_{L^p(\Sigma,\sigma)} \\ &= C\|f\|_{\mathrm{A}^p(\Sigma,\sigma)} \leq C\|f\|_{\mathrm{HA}^p(\Sigma,\sigma)} < \infty \end{aligned} \tag{4.103}$$

for some constant $C \in (0, \infty)$ depending only on n, p, and the Ahlfors regularity character of Σ. Hence, $\{\lambda_k\}_{k\in\mathbb{N}_0} \in \ell^1(\mathbb{N}_0)$.

Having the sequences $\{a_k\}_{k\in\mathbb{N}_0}$ and $\{\lambda_k\}_{k\in\mathbb{N}_0}$ in hand, use the second line in (4.93), (4.97)-(4.98), and (4.100)-(4.101) to write

$$\sum_{k=0}^{\infty} g_k = \sum_{k=0}^{\infty} \lambda_k a_k \quad \text{pointwise } \sigma\text{-a.e. on } \Sigma. \tag{4.104}$$

Pressing forward, we now focus on the second partial sum inside the last limit in (4.91). For this purpose, recall that whenever $\{s_k\}_{k\in\mathbb{N}_0}$ and $\{t_k\}_{k\mathbb{N}_0}$ are two numerical sequences, and $N \in \mathbb{N}$, the following summation by parts formula[1] holds:

$$\sum_{k=0}^{N} s_k(t_k - t_{k+1}) = (s_0 t_0 - s_N t_{N+1}) + \sum_{k=1}^{N} t_k(s_k - s_{k-1}). \tag{4.105}$$

Having fixed $x \in \Sigma$, choose $t_k := \widetilde{\mathbb{1}}_k(x)$ and $s_k := \sum_{j=0}^{k} \int_\Sigma f \phi_j \, d\sigma$ for each $k \in \mathbb{N}_0$. Given $N \in \mathbb{N}$, use (4.105) to compute

[1] also know as Abel's summation formula

$$
\begin{aligned}
&\sum_{k=0}^{N}\Big(\sum_{j=0}^{k}\int_{\Sigma} f\phi_j\,d\sigma\Big)\big(\widetilde{\mathbb{1}}_k(x)-\widetilde{\mathbb{1}}_{k+1}(x)\big)\\
&\quad=\Big(\int_{\Sigma} f\phi_0\,d\sigma\Big)\widetilde{\mathbb{1}}_0(x)-\Big(\sum_{j=0}^{N}\int_{\Sigma} f\phi_j\,d\sigma\Big)\widetilde{\mathbb{1}}_{N+1}(x)\\
&\qquad+\sum_{k=1}^{N}\widetilde{\mathbb{1}}_k(x)\Big(\sum_{j=0}^{k}\int_{\Sigma} f\phi_j\,d\sigma-\sum_{j=0}^{k-1}\int_{\Sigma} f\phi_j\,d\sigma\Big)\\
&\quad=\Big(\int_{\Sigma} f\phi_0\,d\sigma\Big)\widetilde{\mathbb{1}}_0(x)-\Big(\sum_{j=0}^{N}\int_{\Sigma} f\phi_j\,d\sigma\Big)\widetilde{\mathbb{1}}_{N+1}(x)\\
&\qquad+\sum_{k=1}^{N}\widetilde{\mathbb{1}}_k(x)\Big(\int_{\Sigma} f\phi_k\,d\sigma\Big)\\
&\quad=\sum_{k=0}^{N}\Big(\int_{\Sigma} f\phi_k\,d\sigma\Big)\widetilde{\mathbb{1}}_k(x)-\Big(\sum_{j=0}^{N}\int_{\Sigma} f\phi_j\,d\sigma\Big)\widetilde{\mathbb{1}}_{N+1}(x).
\end{aligned}
\tag{4.106}
$$

For each $N\in\mathbb{N}_0$, define the function $\Phi_N:\mathbb{R}^n\to\mathbb{R}$ as

$$
\Phi_N:=\sum_{k=0}^{N}\phi_k. \tag{4.107}
$$

With this piece of notation, (4.106) yields

$$
\begin{aligned}
&\sum_{k=0}^{N}\Big(\int_{\Sigma} f\phi_k\,d\sigma\Big)\widetilde{\mathbb{1}}_k\\
&\quad=\widetilde{\mathbb{1}}_{N+1}\Big(\sum_{j=0}^{N}\int_{\Sigma} f\phi_j\,d\sigma\Big)+\sum_{k=0}^{N}\Big(\sum_{j=0}^{k}\int_{\Sigma} f\phi_j\,d\sigma\Big)\big(\widetilde{\mathbb{1}}_k-\widetilde{\mathbb{1}}_{k+1}\big)\\
&\quad=\widetilde{\mathbb{1}}_{N+1}\int_{\Sigma} f\Phi_N\,d\sigma+\sum_{k=0}^{N}\Big(\int_{\Sigma} f\Phi_k\,d\sigma\Big)\big(\widetilde{\mathbb{1}}_k-\widetilde{\mathbb{1}}_{k+1}\big)
\end{aligned}
\tag{4.108}
$$

pointwise everywhere on Σ.

Regarding the first term in the right-hand side of (4.108), we claim that

$$
\widetilde{\mathbb{1}}_{N+1}\int_{\Sigma} f\Phi_N\,d\sigma\to 0\quad\text{pointwise everywhere on }\Sigma\text{ as }N\to\infty. \tag{4.109}
$$

To see this, first notice that (4.80) and (4.107) imply that $0\le\Phi_N\le 1$ for every $N\in\mathbb{N}_0$. Using this, (4.84), (4.85), the Ahlfors regularity of Σ, the choice of 2_*, and the fact that $f\in L^1(\Sigma,\sigma)$ (cf. (4.14) with $r:=1$), we may bound

$$\begin{aligned}\Big|\widetilde{\mathbb{1}}_{N+1}\int_\Sigma f\Phi_N\,d\sigma\Big| &\le \sigma(\widetilde{A}_{N+1})^{-1}\|f\|_{L^1(\Sigma,\sigma)}\\ &\le \sigma(C_{N+1})^{-1}\|f\|_{L^1(\Sigma,\sigma)}\\ &\le C2_*^{-N(n-1)}\|f\|_{L^1(\Sigma,\sigma)}\longrightarrow 0 \quad\text{as } N\to\infty.\end{aligned} \tag{4.110}$$

Next, we make the claim that for each $x\in\Sigma$ the series

$$\sum_{k=0}^{\infty}\Big(\int_\Sigma f\phi_k\,d\sigma\Big)\widetilde{\mathbb{1}}_k(x)\ \text{ is absolutely convergent.} \tag{4.111}$$

To justify this, fix $x\in\Sigma$, and use the membership of f to $L^1(\Sigma,\sigma)$ (cf. (4.14) with $r:=1$), (4.79), (4.84), (4.85), the choice of 2_*, the Ahlfors regularity of Σ, to estimate

$$\begin{aligned}\sum_{k=0}^{\infty}\Big|\int_\Sigma f\phi_k\,d\sigma\Big|\widetilde{\mathbb{1}}_k(x) &\le \sum_{k=0}^{\infty}\Big(\int_\Sigma |f||\phi_k|\,d\sigma\Big)\widetilde{\mathbb{1}}_k(x)\\ &\le \|f\|_{L^1(\Sigma,\sigma)}\sum_{k=0}^{\infty}\sigma(\widetilde{A}_k)^{-1}\le C\|f\|_{L^1(\Sigma,\sigma)}\sum_{k=0}^{\infty}\sigma(C_k)^{-1}\\ &\le C\|f\|_{L^1(\Sigma,\sigma)}\sum_{k=0}^{\infty}2_*^{-k(n-1)}<\infty.\end{aligned} \tag{4.112}$$

Thanks to this and (4.109), if we let $N\to\infty$ in (4.108) we find that

$$\sum_{k=0}^{\infty}\Big(\int_\Sigma f\phi_k\,d\sigma\Big)\widetilde{\mathbb{1}}_k=\sum_{k=0}^{\infty}\Big(\int_\Sigma f\Phi_k\,d\sigma\Big)(\widetilde{\mathbb{1}}_k-\widetilde{\mathbb{1}}_{k+1}) \tag{4.113}$$

pointwise everywhere on Σ.

Moving on, for each $k\in\mathbb{N}_0$ define the function $h_k:\Sigma\to\mathbb{R}$ by setting

$$h_k:=\Big(\int_\Sigma f\Phi_k\,d\sigma\Big)(\widetilde{\mathbb{1}}_k-\widetilde{\mathbb{1}}_{k+1}), \tag{4.114}$$

which is a well-defined, σ-measurable function. In view of (4.113), we have

$$\sum_{k=0}^{\infty}h_k\quad\text{converges pointwise everywhere on } \Sigma. \tag{4.115}$$

Taking into account that $\widetilde{A}_k\cup\widetilde{A}_{k+1}\subseteq\Delta(x_0,2_*^{k+3})$ for all $k\in\mathbb{N}_0$ (cf. (4.82)), as well as (4.84) and (4.114), we see that the function h_k satisfies

$$\operatorname{supp}h_k\subseteq\Delta(x_0,2_*^{k+3})\quad\text{and}\quad\int_\Sigma h_k\,d\sigma=0\quad\text{for each } k\in\mathbb{N}_0. \tag{4.116}$$

Let us focus on the function Φ_k defined in (4.107). Notice that (4.57), (4.79), and (4.107) entail

$$\Phi_k \in \mathscr{C}_c^\infty(\mathbb{R}^n) \text{ and } \operatorname{supp}\Phi_k \subseteq \bigcup_{j=0}^{k} A_j \subseteq B(x_0, 2_*^{k+2}) \text{ for each } k \in \mathbb{N}_0. \quad (4.117)$$

Recalling that $\varepsilon < \frac{2_*-1}{1+2_*^2}$ (cf. (4.50)), for each $j, k \in \mathbb{N}_0$ with $j \geq k+1$ we have

$$2_*^k \leq 2_*^{j-1} < 2_*^{j-1} + \varepsilon 2_*^{j-1} < 2_*^j - 2_*^{j+1}\varepsilon. \quad (4.118)$$

Bearing this and (4.56) mind, it follows that $B(x_0, 2_*^k) \cap A_j = \varnothing$ for every $j, k \in \mathbb{N}_0$ with $j \geq k+1$. As such, from (4.55) we see that if $j, k \in \mathbb{N}_0$ with $j \geq k+1$ then $\psi_j = 0$ in $B(x_0, 2_*^k)$ and therefore

$$1 = \sum_{j=0}^{\infty} \phi_j(x) = \sum_{j=0}^{k} \phi_j(x) = \Phi_k(x) \text{ for every } x \in B(x_0, 2_*^k). \quad (4.119)$$

Hence $\nabla\Phi_k = 0$ in $B(x_0, 2_*^k)$ for all $k \in \mathbb{N}_0$. This and the fact that each Φ_k is supported in $B(x_0, 2_*^{k+2})$ (cf. (4.117)) imply

$$\operatorname{supp}(\nabla\Phi_k) \subseteq B(x_0, 2_*^{k+2}) \backslash B(x_0, 2_*^k) \quad \text{for each } k \in \mathbb{N}_0. \quad (4.120)$$

As such, for each $k \in \mathbb{N}_0$ we can write

$$\nabla\Phi_k = (\nabla\Phi_k)\mathbb{1}_{B(x_0,2_*^{k+2})\backslash B(x_0,2_*^k)} = \sum_{j=0}^{k} (\nabla\phi_j)\mathbb{1}_{B(x_0,2_*^{k+2})\backslash B(x_0,2_*^k)} \text{ in } \mathbb{R}^n. \quad (4.121)$$

Corresponding to $k = 0$, this reads

$$\nabla\Phi_0 = (\nabla\phi_0)\mathbb{1}_{B(x_0,2_*^2)\backslash B(x_0,1)} \text{ in } \mathbb{R}^n, \quad (4.122)$$

while corresponding to $k = 1$, we have

$$\nabla\Phi_1 = (\nabla\phi_0 + \nabla\phi_1)\mathbb{1}_{B(x_0,2_*^3)\backslash B(x_0,2_*)} \text{ in } \mathbb{R}^n. \quad (4.123)$$

Consider now $j, k \in \mathbb{N}_0$ and assume $j \leq k-2$. With this in mind, the choice of ε in (4.50) ensures that $2_*^{j+1} + 2_*^{j+1}\varepsilon \leq 2_*^{k-1} + 2_*^{k-1}\varepsilon < 2_*^k$. Recalling (4.56), the latter implies $A_j \cap \big(B(x_0, 2_*^{k+2})\backslash B(x_0, 2_*^k)\big) = \varnothing$. Since $\operatorname{supp}(\nabla\phi_j) \subseteq \operatorname{supp}\phi_j \subseteq A_j$ (cf. (4.79)), it follows that $\nabla\phi_j(x) = 0$ for every $x \in B(x_0, 2_*^{k+2})\backslash B(x_0, 2_*^k)$. Formula (4.121) then reduces to

$$\nabla\Phi_k = (\nabla\phi_{k-1} + \nabla\phi_k)\mathbb{1}_{B(x_0,2_*^{k+2})\backslash B(x_0,2_*^k)} \text{ in } \mathbb{R}^n. \quad (4.124)$$

To proceed, we claim that there exists some $C \in (0, \infty)$ such that

$$\sup_{\mathbb{R}^n} |\nabla \Phi_k| \le C 2_*^{-k} \quad \text{for all } k \in \mathbb{N}_0. \tag{4.125}$$

To verify this, select $j \in \mathbb{N}$. Being mindful of (4.78), for each $x \in \mathbb{R}^n$ compute

$$\begin{aligned}
\nabla \phi_j(x) &= (\nabla \psi_j)(x)\Big(\sum_{\ell=0}^{\infty} \psi_\ell(x)\Big)^{-1} - \psi_j(x)\Big(\sum_{\ell=0}^{\infty}(\nabla \psi_\ell)(x)\Big)\Big(\sum_{\ell=0}^{\infty} \psi_\ell(x)\Big)^{-2} \\
&= 2_*^{-j}(\nabla\psi)(2_*^{-j}(x-x_0))\Big(\sum_{\ell=0}^{\infty} \psi_\ell(x)\Big)^{-1} \\
&\quad - \psi_j(x)\mathbb{1}_{A_j}(x)\Big(\sum_{\ell=0}^{\infty} 2_*^{-\ell}(\nabla\psi)(2_*^{-\ell}(x-x_0))\Big)\Big(\sum_{\ell=0}^{\infty} \psi_\ell(x)\Big)^{-2} \\
&= 2_*^{-j}(\nabla\psi)(2_*^{-j}(x-x_0))\Big(\sum_{\ell=0}^{\infty} \psi_\ell(x)\Big)^{-1} \\
&\quad - \psi_j(x)\mathbb{1}_{A_j}(x)\Big(\sum_{\ell=j-1}^{j+1} 2_*^{-\ell}(\nabla\psi)(2_*^{-\ell}(x-x_0))\Big)\Big(\sum_{\ell=0}^{\infty} \psi_\ell(x)\Big)^{-2}.
\end{aligned} \tag{4.126}$$

The first equality follows from the product rule, and Claim 3 (used to differentiate the series term by term). The second equality follows from (4.53) and the Chain Rule. The third equality follows from the fact that ψ_j is supported in A_j (cf. (4.54)-(4.56)), which intersects at most the supports of $\nabla\psi_{j-1}$, $\nabla\psi_j$, and $\nabla\psi_{j+1}$.

In light of (4.66), (4.72), and using the fact that both ψ and $\nabla\psi$ are bounded (cf. (4.51)-(4.52)), for each $x \in \mathbb{R}^n$ we may now estimate

$$|\nabla \phi_j(x)| \le 2_*^{-j} \cdot \sup_{\mathbb{R}^n} |\nabla\psi| + \sum_{\ell=j-1}^{j+1} 2_*^{-\ell} \cdot \sup_{\mathbb{R}^n} |\nabla\psi| \le C 2_*^{-j} \tag{4.127}$$

for some constant $C \in (0,\infty)$, depending on ψ and 2_*. This together with (4.122)-(4.124) then give (4.125).

Bearing in mind that for each $k \in \mathbb{N}_0$ the function Φ_k is smooth and compactly supported (cf. (4.117)), use the Mean Value Theorem and (4.125) to estimate

$$\|\Phi_k\|_{\mathrm{Lip}(\Sigma)} = \sup_{\substack{x,y\in\Sigma \\ x\neq y}} \frac{|\Phi_k(x) - \Phi_k(y)|}{|x-y|} \le \sup_{\mathbb{R}^n} |\nabla \Phi_k| = C 2_*^{-k} \tag{4.128}$$

for each $k \in \mathbb{N}_0$. In addition, since $0 \le \Phi_k \le 1$ in $\mathbb{R}^n$ (see (4.80) and (4.107)), it is clear that

$$\sup_{\Sigma} |\Phi_k| \le 1 \quad \text{for every } k \in \mathbb{N}_0. \tag{4.129}$$

To proceed, fix an arbitrary point $x \in \Delta(x_0, 2_*^{k+2})$. Simple geometrical considerations show that $\Delta(x_0, 2_*^{k+2}) \subseteq \Delta(x, 2 \cdot 2_*^{k+2})$. Hence, recalling the support property of Φ_k in (4.117), we see that $\operatorname{supp}(\Phi_k\big|_\Sigma) \subseteq \Delta(x, 2 \cdot 2_*^{k+2})$. Thanks to item *(d)* in Proposition 4.1, (4.128), and (4.129), having fixed $\gamma \in (0,1)$ there exists some constant $C \in (0,\infty)$ depending only on n, γ, ψ, and the Ahlfors regularity character of Σ such that for every $x \in \Delta(x_0, 2_*^{k+2})$ we have

$$\begin{aligned}\Big|\int_\Sigma f\Phi_k\,d\sigma\Big| = |\langle f, \Phi_k|_\Sigma\rangle| &\leq C 2_*^{(k+2)(n-1+\gamma)} \|\Phi_k\|_{\mathrm{Lip}(\Sigma)}^\gamma \big(\sup_\Sigma |\Phi_k|\big)^{1-\gamma} \cdot f_\gamma^\#(x)\\ &\leq C 2_*^{k(n-1)+k\gamma} 2_*^{-k\gamma} f_\gamma^\#(x)\\ &= C 2_*^{k(n-1)} f_\gamma^\#(x).\end{aligned} \tag{4.130}$$

From this, and recalling the definition of h_k in (4.114) we have

$$|h_k(x)| \leq C 2_*^{k(n-1)} f_\gamma^\#(x) |\widetilde{\mathbb{1}}_k(x) - \widetilde{\mathbb{1}}_{k+1}(x)| \quad \text{for every } x \in \Sigma \text{ and } k \in \mathbb{N}_0. \tag{4.131}$$

Thanks to (4.83), (4.84), (4.85), the Ahlfors regularity of Σ, and the choice of 2_*, for each $k \in \mathbb{N}$ we may estimate

$$\begin{aligned}|\widetilde{\mathbb{1}}_k - \widetilde{\mathbb{1}}_{k+1}| &\leq \widetilde{\mathbb{1}}_k + \widetilde{\mathbb{1}}_{k+1}\\ &= \frac{1}{\sigma(\widetilde{A}_k)} \mathbb{1}_{\widetilde{A}_k} + \frac{1}{\sigma(\widetilde{A}_{k+1})} \mathbb{1}_{\widetilde{A}_{k+1}}\\ &\leq \frac{1}{\sigma(C_k)} (\mathbb{1}_{C_{k-1}} + \mathbb{1}_{C_k} + \mathbb{1}_{C_{k+1}}) + \frac{1}{\sigma(C_{k+1})} (\mathbb{1}_{C_k} + \mathbb{1}_{C_{k+1}} + \mathbb{1}_{C_{k+2}})\\ &\leq C 2_*^{-k(n-1)} (\mathbb{1}_{C_{k-1}} + \mathbb{1}_{C_k} + \mathbb{1}_{C_{k+1}} + \mathbb{1}_{C_{k+2}}),\end{aligned} \tag{4.132}$$

for some constant $C \in (0,\infty)$ depending only on n, and the Ahlfors regularity of Σ. Corresponding to $k = 0$, a similar computation yields

$$|\widetilde{\mathbb{1}}_{C_0} - \widetilde{\mathbb{1}}_{C_1}| \leq C(\mathbb{1}_{C_0} + \mathbb{1}_{C_1} + \mathbb{1}_{C_2}). \tag{4.133}$$

After combining (4.131), (4.132), and (4.133), we arrive at

$$\begin{aligned}&|h_0| \leq \widetilde{C} f_\gamma^\# \cdot (\mathbb{1}_{C_0} + \mathbb{1}_{C_1} + \mathbb{1}_{C_2}) \quad \text{and}\\ &|h_k| \leq \widetilde{C} f_\gamma^\# \cdot (\mathbb{1}_{C_{k-1}} + \mathbb{1}_{C_k} + \mathbb{1}_{C_{k+1}} + \mathbb{1}_{C_{k+2}}) \quad \text{for all } k \in \mathbb{N},\end{aligned} \tag{4.134}$$

for some constant $\widetilde{C} \in (0,\infty)$ depending only on n, ψ, and the Ahlfors regularity character of Σ. Hence,

$$\begin{aligned}\|h_0\|_{L^p(\Sigma,\sigma)} &\leq \widetilde{C}\sum_{j=0}^{2}\|f_\gamma^\# \cdot \mathbb{1}_{C_j}\|_{L^p(\Sigma,\sigma)} \quad \text{and} \\ \|h_k\|_{L^p(\Sigma,\sigma)} &\leq \widetilde{C}\sum_{j=k-1}^{k+2}\|f_\gamma^\# \cdot \mathbb{1}_{C_j}\|_{L^p(\Sigma,\sigma)} \quad \text{for all } k \in \mathbb{N}.\end{aligned} \tag{4.135}$$

Next, introduce the set of indices

$$\mathcal{J}_2 := \{k \in \mathbb{N}_0 : \|h_k\|_{L^p(\Sigma,\sigma)} \neq 0\}. \tag{4.136}$$

Define the function $b_0 : \Sigma \to \mathbb{R}$ as

$$b_0 := \begin{cases} \Big[\widetilde{C}(2_*^3)^{(n-1)/p'}\sum_{j=0}^{2}\|f_\gamma^\# \cdot \mathbb{1}_{C_j}\|_{L^p(\Sigma,\sigma)}\Big]^{-1} h_0 & \text{if } 0 \in \mathcal{J}_2, \\ 0 & \text{if } 0 \notin \mathcal{J}_2, \end{cases} \tag{4.137}$$

and for each $k \in \mathbb{N}$, define the function $b_k : \Sigma \to \mathbb{R}$ as

$$b_k := \begin{cases} \Big[\widetilde{C}(2_*^{k+3})^{(n-1)/p'}\sum_{j=k-1}^{k+2}\|f_\gamma^\# \cdot \mathbb{1}_{C_j}\|_{L^p(\Sigma,\sigma)}\Big]^{-1} h_k & \text{if } k \in \mathcal{J}_2, \\ 0 & \text{if } k \notin \mathcal{J}_2. \end{cases} \tag{4.138}$$

For each $k \in \mathbb{N}_0$, the function b_k is well defined and σ-measurable on Σ. In addition, from (4.137), (4.138), (4.116), and (4.135), for each $k \in \mathbb{N}_0$ the function b_k satisfies the following properties:

$$\operatorname{supp} b_k \subseteq \Delta(x_0, 2_*^{k+3}), \quad \|b_k\|_{L^p(\Sigma,\sigma)} \leq (2_*^{k+3})^{-(n-1)/p'}, \quad \int_\Sigma b_k\, d\sigma = 0. \tag{4.139}$$

Hence, b_k is an x_0-central $(1, p)$-atom for every $k \in \mathbb{N}_0$.

To proceed, define the following finite numbers:

$$\gamma_0 := \begin{cases} \widetilde{C}(2_*^3)^{(n-1)/p'}\sum_{\ell=0}^{2}\|f_\gamma^\# \cdot \mathbb{1}_{C_\ell}\|_{L^p(\Sigma,\sigma)} & \text{if } 0 \in \mathcal{J}_2, \\ 0 & \text{if } 0 \notin \mathcal{J}_2, \end{cases} \tag{4.140}$$

and for each $k \in \mathbb{N}$ set

$$\gamma_k := \begin{cases} \widetilde{C}(2_*^{k+3})^{(n-1)/p'}\sum_{\ell=k-1}^{k+2}\|f_\gamma^\# \cdot \mathbb{1}_{C_\ell}\|_{L^p(\Sigma,\sigma)} & \text{if } k \in \mathcal{J}_2, \\ 0 & \text{if } k \notin \mathcal{J}_2. \end{cases} \tag{4.141}$$

Then given an $k \in \mathbb{N}_0$,

$$\|h_k\|_{L^p(\Sigma,\sigma)}, \quad b_k, \quad \text{and } \gamma_k \text{ are all zero if and only if } k \notin \mathcal{J}_2. \tag{4.142}$$

We claim that the sequence $\{\gamma_k\}_{k\in\mathbb{N}_0}$ belongs to $\ell^1(\mathbb{N}_0)$. This can be seen from (4.9), (4.140)-(4.141), and

$$
\begin{aligned}
\sum_{k=0}^{\infty}|\gamma_k| &\leq C\sum_{\ell=0}^{2}\|f_\gamma^{\#}\cdot\mathbb{1}_{C_\ell}\|_{L^p(\Sigma,\sigma)} \\
&\quad + C\sum_{k=1}^{\infty}\sum_{\ell=k-1}^{k+2}2_*^{(k+3)(n-1)1/p'}\|f_\gamma^{\#}\cdot\mathbb{1}_{C_\ell}\|_{L^p(\Sigma,\sigma)} \\
&\leq C\sum_{k=0}^{\infty}2_*^{k(n-1)/p'}\|f_\gamma^{\#}\cdot\mathbb{1}_{C_k}\|_{L^p(\Sigma,\sigma)} \\
&= C\|f_\gamma^{\#}\|_{\mathrm{A}^p(\Sigma,\sigma)} = C\|f\|_{\mathrm{HA}^p(\Sigma,\sigma)} < \infty,
\end{aligned}
\tag{4.143}
$$

for some $C\in(0,\infty)$ depending only on n, p, ψ, γ, the Ahlfors regularity character of Σ. Having obtained the sequence of x_0-central $(1,p)$-atoms $\{b_k\}_{k\in\mathbb{N}_0}$ and the numerical sequence $\{\gamma_k\}_{k\in\mathbb{N}_0}\in\ell^1(\mathbb{N}_0)$, and bearing in mind (4.137), (4.138), (4.140), and (4.141) we may recast (4.115) as

$$
\sum_{k=0}^{\infty}h_k=\sum_{k=0}^{\infty}\gamma_k b_k \quad\text{pointwise everywhere on } \Sigma. \tag{4.144}
$$

Combining this with (4.104), (4.91), and (4.92) we finally obtain the following atomic decomposition

$$
f=\sum_{k=0}^{\infty}\lambda_k a_k+\sum_{k=0}^{\infty}\gamma_k b_k \tag{4.145}
$$

with convergence to f both in $L^1(\Sigma,\sigma)$ and $L^p(\Sigma,\sigma)$, which is guaranteed by (4.90). This can be further rewritten as

$$
f=\sum_{k=0}^{\infty}\widetilde{\lambda}_k\widetilde{a}_k \quad\text{in } L^1(\Sigma,\sigma) \text{ and } L^p(\Sigma,\sigma), \tag{4.146}
$$

where $\{\widetilde{a}_k\}_{k\in\mathbb{N}_0}$ is the sequence of x_0-central $(1,p)$-atoms defined as

$$
\widetilde{a}_{2k}:=a_k \quad\text{and}\quad \widetilde{a}_{2k+1}:=b_k \quad\text{for each } k\in\mathbb{N}_0, \tag{4.147}
$$

and $\{\widetilde{\lambda}_k\}_{k\in\mathbb{N}_0}$ is the numerical sequence defined according to

$$
\widetilde{\lambda}_{2k}:=\lambda_k \quad\text{and}\quad \widetilde{\lambda}_{2k+1}:=\gamma_k \quad\text{for each } k\in\mathbb{N}_0. \tag{4.148}
$$

In particular, (4.103) and (4.143) imply that $\{\widetilde{\lambda}_k\}_{k\in\mathbb{N}_0}\in\ell^1(\mathbb{N}_0)$ with norm

$$
\sum_{k=0}^{\infty}|\widetilde{\lambda}_k|\leq C\|f\|_{\mathrm{HA}^p(\Sigma,\sigma)}, \tag{4.149}
$$

for some constant $C\in(0,\infty)$ depending only on n, p, γ, and the Ahlfors regularity of Σ (in particular, independent of f).

Moving on, we claim that the series in (4.146) also converges to f in $\mathrm{HA}^p(\Sigma,\sigma)$. To verify this, fix $N, M \in \mathbb{N}_0$ with $M \geq N$. Thanks to Theorem 4.1, there exists some uniform constant $C \in (0,\infty)$ depending only on n, p, γ, and the Ahlfors regularity constants of Σ such that

$$\begin{aligned}\Big\|\sum_{k=0}^{M}\widetilde{\lambda}_k\widetilde{a}_k-\sum_{k=0}^{N}\widetilde{\lambda}_k\widetilde{a}_k\Big\|_{\mathrm{HA}^p(\Sigma,\sigma)}&=\Big\|\sum_{k=N+1}^{M}\widetilde{\lambda}_j\widetilde{a}_k\Big\|_{\mathrm{HA}^p(\Sigma,\sigma)}\\&\leq\sum_{k=N+1}^{M}|\widetilde{\lambda}_k|\|\widetilde{a}_k\|_{\mathrm{HA}^p(\Sigma,\sigma)}\\&\leq C\sum_{k=N+1}^{\infty}|\widetilde{\lambda}_k|.\end{aligned}\tag{4.150}$$

Since $\{\widetilde{\lambda}_k\}_{k\in\mathbb{N}_0}\in\ell^1(\mathbb{N}_0)$, if we let $N\to\infty$ in the right-hand side of (4.150) then $\sum_{k=N+1}^{\infty}|\widetilde{\lambda}_k|\to 0$. Hence $\big\{\sum_{k=0}^{N}\widetilde{\lambda}_k\widetilde{a}_k\big\}_{N\in\mathbb{N}_0}$ is a Cauchy sequence in $\mathrm{HA}^p(\Sigma,\sigma)$. From the completeness of $\mathrm{HA}^p(\Sigma,\sigma)$ (cf. Proposition 4.3), there exists some function $\widetilde{f}\in\mathrm{HA}^p(\Sigma,\sigma)$ satisfying

$$\widetilde{f}=\sum_{k=0}^{\infty}\widetilde{\lambda}_k\widetilde{a}_k \text{ with convergence in } \mathrm{HA}^p(\Sigma,\sigma).\tag{4.151}$$

In light of the continuous embedding $\mathrm{HA}^p(\Sigma,\sigma)\hookrightarrow L^1(\Sigma,\sigma)$ (cf. Proposition 4.2), this convergence also holds in $L^1(\Sigma,\sigma)$. Consequently, (4.146) and (4.151) imply $\widetilde{f}=f$ at σ-a.e. point on Σ. Hence, the series in (4.146) converges to f in $\mathrm{HA}^p(\Sigma,\sigma)$.

Consider next the claim made in item *(b)*. Fix a σ-measurable function $f:\Sigma\to\mathbb{R}$ such that

$$f=\sum_{j=0}^{\infty}\lambda_j a_j \text{ pointwise } \sigma\text{-a.e. on } \Sigma\tag{4.152}$$

for some sequence $\{a_k\}_{k\in\mathbb{N}_0}$ of x_0-central $(1,p)$-atoms and some numerical sequence $\{\lambda_k\}_{k\in\mathbb{N}_0}\in\ell^1(\mathbb{N}_0)$. A similar reasoning as in (4.150) guarantees that $\big\{\sum_{j=0}^{N}\lambda_j a_j\big\}_{N\in\mathbb{N}_0}$ is a Cauchy sequence in $\mathrm{HA}^p(\Sigma,\sigma)$. In view of the completeness of $\mathrm{HA}^p(\Sigma,\sigma)$ (cf. Proposition 4.3), this sequence of partial sums converges to some function $\widetilde{f}$ in $\mathrm{HA}^p(\Sigma,\sigma)$. Since $\mathrm{HA}^p(\Sigma,\sigma)\hookrightarrow L^1(\Sigma,\sigma)$ continuously, $\sum_{j=0}^{N}\lambda_j a_j\to\widetilde{f}$ in $L^1(\Sigma,\sigma)$ as $N\to\infty$. Restricting to a subsequence if necessary, $\sum_{j=0}^{N}\lambda_j a_j\to\widetilde{f}$ pointwise σ-a.e. on Σ as $N\to\infty$, which in concert with (4.152) gives $\widetilde{f}=f$ at σ-a.e. point on Σ. Hence, $f\in\mathrm{HA}^p(\Sigma,\sigma)$ and for some constant $C\in(0,\infty)$ independent of f we have

$$\|f\|_{\mathrm{HA}^p(\Sigma,\sigma)}=\Big\|\sum_{j=0}^{\infty}\lambda_j a_j\Big\|_{\mathrm{HA}^p(\Sigma,\sigma)}\leq\sum_{j=0}^{\infty}|\lambda_j|\|a_j\|_{\mathrm{HA}^p(\Sigma,\sigma)}\leq C\sum_{j=0}^{\infty}|\lambda_j|,\tag{4.153}$$

where the last inequality relies on Theorem 4.1. This concludes the treatment of item *(b)*. The claim made in item *(c)* follows directly from *(a)* and *(b)*, thus completing the proof of Theorem 4.2. □

An immediate consequence of Theorem 4.2 is the fact that functions from HA^p satisfy a vanishing moment condition.

Corollary 4.1 *Consider a closed unbounded Ahlfors regular set $\Sigma \subseteq \mathbb{R}^n$, and abbreviate $\sigma := \mathcal{H}^{n-1}\lfloor\Sigma$. Then for any $p \in (1, \infty)$ and $f \in \mathrm{HA}^p(\Sigma, \sigma)$ one has $f \in L^1(\Sigma, \sigma)$ and*

$$\int_\Sigma f \, d\sigma = 0. \tag{4.154}$$

Proof Fix a function $f \in \mathrm{HA}^p(\Sigma, \sigma)$. In particular, $f \in L^1(\Sigma, \sigma)$ by Proposition 4.2 (cf. (4.14) with $r := 1$). Next, consider an atomic decomposition for f as in (4.45). Since convergence in (4.45) occurs in the sense of $L^1(\Sigma, \sigma)$, we may write

$$\int_\Sigma f \, d\sigma = \sum_{j=0}^{\infty} \lambda_j \int_\Sigma a_j \, d\sigma = 0 \tag{4.155}$$

with the last equality based on the vanishing moment of central atoms (cf. Definition 4.3 item *(c)*). □

Another useful consequence of the atomic decomposition of Beurling-Hardy spaces is a characterization of the subspace

$$\mathrm{HA}^p_{\mathrm{comp}}(\Sigma, \sigma) := \{f \in \mathrm{HA}^p(\Sigma, \sigma) : f \text{ is compactly supported}\}. \tag{4.156}$$

More precisely, we have the following result.

Corollary 4.2 *Consider a closed unbounded Ahlfors regular set $\Sigma \subseteq \mathbb{R}^n$, and abbreviate $\sigma := \mathcal{H}^{n-1}\lfloor\Sigma$. Select an integrability exponent $p \in (1, \infty)$ and a reference point $x_0 \in \Sigma$. Having fixed a σ-measurable function $f : \Sigma \to \mathbb{R}$, the following conditions are equivalent:*

(a) f belongs to $\mathrm{HA}^p_{\mathrm{comp}}(\Sigma, \sigma)$;

(b) f belongs to $L^p_{\mathrm{comp}}(\Sigma, \sigma)$ and has integral zero;

(c) f is a finite linear combination of x_0-central $(1, p)$-atoms;

(d) f is a multiple of an x_0-central $(1, p)$-atom.

Proof To see that the implication *(a)* ⇒ *(b)* holds, use the fact

$$\mathrm{HA}^p_{\mathrm{comp}}(\Sigma, \sigma) \subseteq \mathrm{HA}^p(\Sigma, \sigma) \subseteq L^p(\Sigma, \sigma), \tag{4.157}$$

(cf. Proposition 4.2) along with (4.156) and Corollary 4.1. To justify the implication *(b)* $\Rightarrow$ *(c)* assume $f \in L^p_{\rm comp}(\Sigma, \sigma)$ and $\int_\Sigma f\, d\sigma = 0$. Suppose $\|f\|_{L^p(\Sigma,\sigma)} \neq 0$, otherwise the desired conclusion is trivial. Pick a finite radius $R \geq 2_*$ such that $\operatorname{supp} f \subseteq \Delta(x_0, R)$. Then the function $a : \Sigma \to \mathbb{R}$ defined as

$$a := \lambda^{-1} f \quad \text{with} \quad \lambda := R^{(n-1)/p'} \|f\|_{L^p(\Sigma,\sigma)} \in (0, \infty) \tag{4.158}$$

satisfies

$$\operatorname{supp} a \subseteq \Delta(x_0, R), \quad \|a\|_{L^p(\Sigma,\sigma)} = R^{-(n-1)/p'}, \quad \int_\Sigma a\, d\sigma = \lambda \int_\Sigma f\, d\sigma = 0. \tag{4.159}$$

Hence, a is an x_0-central $(1, p)$-atom. Thanks to Theorem 4.1 we obtain that a belongs to $\mathrm{HA}^p(\Sigma, \sigma)$, which further implies $f = \lambda a \in \mathrm{HA}^p_{\rm comp}(\Sigma, \sigma)$, as wanted. This establishes *(b)* $\Rightarrow$ *(c)*. The implication *(c)* $\Rightarrow$ *(d)* is obvious. Finally, the implication *(d)* $\Rightarrow$ *(a)* is a direct consequence of Theorem 4.1. This concludes the proof of Corollary 4.2. □

Remark 4.5 In summary, the previous corollary establishes that

$$\begin{aligned} \mathrm{HA}^p_{\rm comp}(\Sigma, \sigma) &= \Big\{ f \in L^p_{\rm comp}(\Sigma, \sigma) : \int_\Sigma f\, d\sigma = 0 \Big\} \\ &= \Big\{ \text{multiples of } x_0\text{-central } (1, p)\text{-atoms} \Big\}. \end{aligned} \tag{4.160}$$

In addition, in view Theorem 4.2, it follows that $\mathrm{HA}^p_{\rm comp}(\Sigma, \sigma)$ is a dense subspace of $\mathrm{HA}^p(\Sigma, \sigma)$.

The description obtained in (4.160) is merely a coincidence of sets, with no quantitative aspect attached. As such, it is useful to provide an atomic decomposition for $\mathrm{HA}^p_{\rm comp}$ with the key feature that said decomposition involves only finitely many terms and is quasi-optimal.

Theorem 4.3 *Assume $\Sigma \subseteq \mathbb{R}^n$ is a closed unbounded Ahlfors regular set, and abbreviate $\sigma := \mathcal{H}^{n-1}\lfloor\Sigma$. Fix an integrability exponent $p \in (1, \infty)$ and a reference point $x_0 \in \Sigma$. Then there exists a constant $C \in (0, \infty)$ depending only on n, p, and the Ahlfors regularity of Σ such that for every $f \in \mathrm{HA}^p_{\rm comp}(\Sigma, \sigma)$ there exist a finite sequence of x_0-central $(1, p)$-atoms $\{a_k\}_{0 \leq k \leq N}$, and a finite numerical sequence $\{\lambda_k\}_{0 \leq k \leq N}$, for some $N \in \mathbb{N}$, with the property that*

$$f = \sum_{k=0}^{N} \lambda_k a_k \quad \text{pointwise } \sigma\text{-a.e. on } \Sigma \text{ and } \quad \sum_{k=0}^{N} |\lambda_k| \leq C \|f\|_{\mathrm{HA}^p(\Sigma,\sigma)}. \tag{4.161}$$

Remark 4.6 At a glance, Theorem 4.3 may seem redundant, since we already have the atomic decomposition for HA^p from Theorem 4.2, and the latter space contains $\mathrm{HA}^p_{\rm comp}$. This being said the main purpose of the current theorem is to obtain a quantitative description of $\mathrm{HA}^p_{\mathrm{comp}}$ involving only finite linear combinations of central atoms.

Let us now prove Theorem 4.3.

Proof of Theorem 4.3 We largely follow the argument presented in the proof of item *(a)* of Theorem 4.2. The strategy is to reconsider the approach developed there, now taking into account that we are dealing with an HA^p function with compact support. In light of (4.45), all we have to do is to ensure that in the current scenario, the atomic decomposition involves only finitely many x_0-central $(1, p)$-atoms and coefficients. With this goal in mind, we only have to pay attention to those aspects of the proof that directly affect the nature of the atoms and coefficients.

To get started, fix a function $f \in \mathrm{HA}^p_{\mathrm{comp}}(\Sigma, \sigma)$ and recall the partition of unity $\{\phi_k\}_{k\in\mathbb{N}_0}$ from (4.78)-(4.80). Since f is compactly supported and each ϕ_k is supported in the annulus A_k for all $k \in \mathbb{N}_0$ (see (4.79)), from the definition of g_k (see (4.92)) we see that, presently,

$$\text{only finitely many functions } g_k \text{ are non-zero.} \tag{4.162}$$

As a consequence, the set of indices $\mathcal{J}_1$ (introduced in (4.96)) is finite, i.e.,

$$\#\mathcal{J}_1 < \infty. \tag{4.163}$$

This plus the definitions of $\{a_k\}_{k\in\mathbb{N}_0}$ and $\{\lambda_k\}_{k\in\mathbb{N}_0}$ (see (4.97), (4.98), (4.100), (4.101)), imply that

$$\begin{array}{l}\text{at most finitely many terms of } \{a_k\}_{k\in\mathbb{N}_0} \text{ and} \\ \{\lambda_k\}_{k\in\mathbb{N}_0} \text{ are non-zero.}\end{array} \tag{4.164}$$

Next, observe that the function Φ_k (defined in (4.107)) has the property that

$$\Phi_k \equiv 1 \text{ on the support of } f \text{ for } k \in \mathbb{N}_0 \text{ sufficiently large.} \tag{4.165}$$

This follows from the fact that f is compactly supported, and Φ_k was constructed in such a way that (cf. (4.117) and (4.119))

$$\operatorname{supp}\Phi_k \subseteq \Delta(x_0, R^{k+2}) \quad \text{and} \quad \Phi_k \equiv 1 \ \text{ on } \Delta(x_0, 2^k_*). \tag{4.166}$$

Bearing in mind the definition of h_k (see (4.114)), using (4.165), and Corollary 4.1,

$$\text{only finitely many functions } h_k \text{ are non-zero.} \tag{4.167}$$

Consequently, the set of indices $\mathcal{J}_2$ (see (4.136)) is finite, i.e.,

$$\#\mathcal{J}_2 < \infty. \tag{4.168}$$

In turn, from this, and the definitions of $\{b_k\}_{k\in\mathbb{N}_0}$ and $\{\gamma_k\}_{k\in\mathbb{N}_0}$ (see (4.137), (4.138), (4.140), and (4.141)), we have that

$$\begin{aligned}&\text{at most finitely many terms of } \{b_k\}_{k\in\mathbb{N}_0}\\ &\text{and } \{\gamma_k\}_{k\in\mathbb{N}_0} \text{ are non-zero.}\end{aligned} \tag{4.169}$$

Finally, (4.164) and (4.169) imply that the sum (4.146) (after making the adjustments as in (4.147)) involves finitely many non-zero terms, i.e.,

$$f = \sum_{k=0}^{N} \widetilde{\lambda}_k \widetilde{a}_k \tag{4.170}$$

for some $N \in \mathbb{N}_0$ sufficiently large, that depends on the support of f. Moreover, estimate (4.149) continues to hold, i.e.,

$$\sum_{k=0}^{N} |\widetilde{\lambda}_k| \leq C\|f\|_{\mathrm{HA}^p(\Sigma,\sigma)}. \tag{4.171}$$

for some constant $C \in (0, \infty)$ depending exclusively on p, n, and the Ahlfors regularity of Σ. In particular, said constant is independent of f. This concludes the proof of Theorem 4.3. □

The result recorded below establishes the separability of Beurling-Hardy spaces.

Proposition 4.4 *Let $\Sigma \subseteq \mathbb{R}^n$ be a closed unbounded Ahlfors regular set, and abbreviate $\sigma := \mathcal{H}^{n-1}\lfloor\Sigma$. Then for any integrability exponent $p \in (1, \infty)$, the Beurling-Hardy space* $\mathrm{HA}^p(\Sigma, \sigma)$ *is separable.*

Proof Fix $p \in (1, \infty)$ and consider a reference point $x_0 \in \Sigma$, along with a dilation factor $2_* > (C_\Sigma/c_\Sigma)^{1/(n-1)}$ as in Definition 3.1. For each $j \in \mathbb{N}$ with $j > 2_*$ abbreviate

$$\Delta_j := B(x_0, j) \cap \Sigma. \tag{4.172}$$

Since the space $L^p(\Delta_j, \sigma)$ is separable [MiMiMiMo13, Theorem 5.6, p. 300], it follows that for each $j \in \mathbb{N}$ with $j > 2_*$ the subspace

$$L^p_0(\Delta_j, \sigma) := \left\{ f \in L^p(\Delta_j, \sigma) : \int_\Sigma f \, d\sigma = 0 \right\} \tag{4.173}$$

is also separable. Hence, for each $j \in \mathbb{N}$ with $j > 2_*$ there exists

$$\mathcal{F}_j \text{ a countable and dense subset of } L^p_0(\Delta_j, \sigma). \tag{4.174}$$

To proceed, extend each function from $\mathcal{F}_j$ by zero to the entire Σ, and denote by $\widetilde{\mathcal{F}}_j$ the collection of all such extensions. Notice that for each $j \in \mathbb{N}$ satisfying $j > 2_*$ every function of $\widetilde{\mathcal{F}}_j$ is a multiple of an x_0-central $(1, p)$-atom. Indeed, having fixed a function $f \in \widetilde{\mathcal{F}}_j$ define $a : \Sigma \to \mathbb{R}$ by

$$a := \frac{j^{(n-1)(1/p-1)}}{1 + \|f\|_{L^p(\Sigma,\sigma)}} f. \tag{4.175}$$

Keeping in mind that $\operatorname{supp} f \subseteq \Delta_j$, and $\int_\Sigma f\, d\sigma = 0$, we see that

$$\operatorname{supp} a \subseteq \Delta_j = \Delta(x_0, j), \quad \|a\|_{L^p(\Sigma,\sigma)} \leq j^{(n-1)(1/p-1)}, \quad \int_\Sigma a\, d\sigma = 0. \tag{4.176}$$

This ensures that a is an x_0-central $(1, p)$-atom. As such, we have $a \in \mathrm{HA}^p(\Sigma, \sigma)$ (cf. Theorem 4.1). Since f is a constant multiple of a (cf. (4.175)), this implies $f \in \mathrm{HA}^p(\Sigma, \sigma)$. Hence, we obtain

$$\widetilde{\mathcal{F}}_j \subseteq \mathrm{HA}^p(\Sigma, \sigma) \quad \text{for all } j \in \mathbb{N} \text{ with } j > 2_*. \tag{4.177}$$

Next, denote by $\mathcal{F}$ the collection of all finite sums of functions in $\cup_{j \in \mathbb{N}} \widetilde{\mathcal{F}}_j$. Since this is a countable union of countable sets, $\mathcal{F}$ is a countable subset of $\mathrm{HA}^p(\Sigma, \sigma)$ which is closed under addition. We claim that

$$\mathcal{F} \text{ is dense in } \mathrm{HA}^p(\Sigma, \sigma). \tag{4.178}$$

To prove this claim, it suffices to show the following:

for every x_0-central $(1, p)$-atom a on Σ, for every scalar $\lambda \in \mathbb{R}$, and $\varepsilon \in (0, \infty)$, there exists $f \in \mathcal{F}$ such that $\|\lambda a - f\|_{\mathrm{HA}^p(\Sigma,\sigma)} < \varepsilon$. (4.179)

As to not disrupt the flow of the presentation, assume momentarily that (4.179) holds. Given $g \in \mathrm{HA}^p(\Sigma, \sigma)$, Theorem 4.2 guarantees the existence of some sequence $\{\lambda_j\}_{j \in \mathbb{N}_0} \in \ell^1(\mathbb{N}_0)$, and some sequence of x_0-central $(1, p)$-atoms $\{a_j\}_{j \in \mathbb{N}_0}$ such that

$$g = \lim_{N \to \infty} g_N \quad \text{in } \mathrm{HA}^p(\Sigma, \sigma), \tag{4.180}$$

where $g_N := \sum_{j=0}^N \lambda_j a_j$ for each $N \in \mathbb{N}$. From (4.179), for each $\varepsilon \in (0, \infty)$ and $j \in \mathbb{N}_0$, we can find a function $f_j \in \mathcal{F}$ satisfying

$$\|\lambda_j a_j - f_j\|_{\mathrm{HA}^p(\Sigma,\sigma)} < \frac{\varepsilon}{2^j}. \tag{4.181}$$

For each $N \in \mathbb{N}$ abbreviate

$$\widetilde{f}_N := \sum_{j=0}^N f_j \in \mathcal{F}. \tag{4.182}$$

Bearing in mind (4.180) and (4.181), for $N \in \mathbb{N}$ sufficiently large we can estimate

$$\begin{aligned} \|g - \widetilde{f}_N\|_{\mathrm{HA}^p(\Sigma,\sigma)} &\leq \|g - g_N\|_{\mathrm{HA}^p(\Sigma,\sigma)} + \Big\| \sum_{j=0}^N (\lambda_j a_j - f_j) \Big\|_{\mathrm{HA}^p(\Sigma,\sigma)} \\ &\leq \varepsilon + \sum_{j=0}^N \|\lambda_j a_j - f_j\|_{\mathrm{HA}^p(\Sigma,\sigma)} \leq \varepsilon + \Big(\sum_{j=0}^\infty \frac{1}{2^j} \Big) \varepsilon = 3\varepsilon. \end{aligned} \tag{4.183}$$

The arbitrariness of ε then establishes the density of $\mathcal{F}$ in $\mathrm{HA}^p(\Sigma,\sigma)$.

At this stage, there remains to prove (4.179). Fix a, λ, and ε as in (4.179). Then there exists $j_0 \in \mathbb{N}$ with $j_0 > 2_*$ with the property that

$$\operatorname{supp} a \subseteq \Delta_{j_0}, \quad \text{and} \quad (\lambda a)\big|_{\Delta_{j_0}} \in L^p(\Delta_{j_0},\sigma). \tag{4.184}$$

By the density of $\mathcal{F}_j$ in $L^p_0(\Delta_{j_0},\sigma)$ (see (4.174)), we can pick a function $f \in \mathcal{F}_{j_0}$ satisfying

$$\|(\lambda a)\big|_{\Delta_{j_0}} - f\|_{L^p(\Delta_{j_0},\sigma)} \leq \varepsilon \cdot j_0^{(n-1)(1/p-1)}. \tag{4.185}$$

Denoting by $\widetilde{f}$ the extension of f by zero to Σ we see that

$$\widetilde{f} \in \widetilde{\mathcal{F}}_{j_0} \subseteq \mathcal{F}. \tag{4.186}$$

Moreover, bearing in mind (4.184) and (4.185), if we set $g := \lambda a - \widetilde{f}$, then

$$\operatorname{supp} g \subseteq \Delta_{j_0}, \quad \int_{\Delta_{j_0}} g\, d\sigma = 0, \quad \text{and} \quad \|g\|_{L^p(\Sigma,\sigma)} < \varepsilon \cdot j_0^{(n-1)(1/p-1)}. \tag{4.187}$$

From this, we see that $\varepsilon^{-1} \cdot g$ is an x_0-central $(1,p)$-atom on Σ. According to Theorem 4.1, there exists a constant $C \in (0,\infty)$ that bounds the norm of every atom in $\mathrm{HA}^p(\Sigma,\sigma)$. In particular, $\|\varepsilon^{-1} \cdot g\|_{\mathrm{HA}^p(\Sigma,\sigma)} \leq C$. This immediately implies

$$\|\lambda a - \widetilde{f}\|_{\mathrm{HA}^p(\Sigma,\sigma)} = \|g\|_{\mathrm{HA}^p(\Sigma,\sigma)} < C \cdot \varepsilon. \tag{4.188}$$

After re-adjusting ε, this completes the proof of (4.179), which in turn finishes the proof of Proposition 4.4 □

4.3 Molecules for Beurling-Hardy Spaces

This section is devoted to developing a molecular theory for Beurling-Hardy spaces, which plays a crucial role in the current work. Central molecules may be regarded as a weakened version of central $(1,p)$-atoms, where now the localization condition is replaced by a decay condition. This is made precise in the definition below.

Definition 4.4 Assume $\Sigma \subseteq \mathbb{R}^n$ is a closed unbounded Ahlfors regular set, and abbreviate $\sigma := \mathcal{H}^{n-1}\lfloor\Sigma$. Denote by $c_\Sigma, C_\Sigma \in (0,\infty)$ the lower and upper Ahlfors regularity constant of Σ, respectively. Fix a reference point $x_0 \in \Sigma$ and a dilation factor $2_* > (C_\Sigma/c_\Sigma)^{1/(n-1)}$. Consider an integrability exponent $p \in (1,\infty)$ along with a parameter $\theta \in \big((n-1)(p-1),\infty\big)$. A σ-measurable function $M : \Sigma \to \mathbb{R}$ is called an x_0-`central` $(1,p,\theta)$-`molecule` provided that there exists a radius $R \in [2_*,\infty)$ such that the following three properties hold:

(1) Normalization Condition:

$$\|M\|_{L^p(\Sigma,\sigma)} \leq R^{-(n-1)(1-1/p)}. \tag{4.189}$$

(2) Decay Condition:

$$\Big(\int_\Sigma |M(x)|^p|x-x_0|^\theta\,d\sigma(x)\Big)^{1/p} \leq R^{-(n-1)(1-1/p)+\theta/p}. \tag{4.190}$$

(3) Vanishing Moment Condition:

$$\int_\Sigma M\,d\sigma = 0. \tag{4.191}$$

Remark 4.7 In the context of Definition 4.4, we claim that the normalization condition (4.189) and decay condition (4.190) imply that $M \in L^1(\Sigma,\sigma)$. To justify this claim, observe that Hölder's inequality and (4.189) imply

$$\int_{\Delta(x_0,R)} |M|\,d\sigma \leq \|M\|_{L^p(\Sigma,\sigma)}\sigma\big(\Delta(x_0,R)\big)^{1/p'} < \infty, \tag{4.192}$$

where $p' \in (1,\infty)$ denotes the Hölder conjugate exponent of p. Moreover, use (4.190), the Ahlfors regularity of Σ, and Hölder's inequality to estimate

$$\begin{aligned}
\int_{\Sigma\setminus\Delta(x_0,R)} |M|\,d\sigma &= \sum_{j=0}^\infty \int_{2^jR\leq|x-x_0|<2^{j+1}R} |M|\,d\sigma\\
&\leq \sum_{j=0}^\infty \Big(\int_{2^jR\leq|x-x_0|<2^{j+1}R} |M|^p\,d\sigma\Big)^{1/p}\sigma\big(\Delta(x_0,2^{j+1}R)\setminus\Delta(x_0,2^jR)\big)^{1/p'}\\
&\leq C\sum_{j=0}^\infty \Big(\int_{2^jR\leq|x-x_0|<2^{j+1}R} |M(x)|^p\frac{|x-x_0|^\theta}{(2^jR)^\theta}\,d\sigma(x)\Big)^{1/p}(2^{j+1}R)^{(n-1)/p'}\\
&\leq CR^{(n-1)/p'-\theta/p}\Big(\int_\Sigma |M(x)|^p|x-x_0|^\theta\,d\sigma(x)\Big)^{1/p}\sum_{j=0}^\infty 2^{j[(n-1)/p'-\theta/p]}\\
&\leq C\sum_{j=0}^\infty 2^{j[(n-1)/p'-\theta/p]}.
\end{aligned} \tag{4.193}$$

This series is convergent, since

$$\theta > (n-1)(p-1) \Longleftrightarrow \frac{n-1}{p'} - \frac{\theta}{p} < 0. \tag{4.194}$$

All together, estimates (4.192) and (4.193) readily imply that $M \in L^1(\Sigma, \sigma)$, as desired. In particular, this ensures that the integral in (4.191) is meaningful.

Remark 4.8 It is of relevance to observe that

$$\text{every } x_0\text{-central } (1,p)\text{-atom is an } x_0\text{-central } (1,p,\theta)\text{-molecule.} \tag{4.195}$$

This follows from the fact that if $a : \Sigma \to \mathbb{R}$ is an x_0-central $(1,p)$-atom, and $\Delta(x_0, R)$ is a surface ball associated with a for some finite $R \geq 2_*$ (in the sense of Definition 4.3), then a satisfies automatically items *(1)* and *(3)* in Definition 4.4. To see that a also satisfies the decay condition in (4.190), use (4.29) to bound

$$\begin{aligned}\Big(\int_\Sigma |a(x)|^p |x-x_0|^\theta \, d\sigma(x)\Big)^{1/p} &= \Big(\int_{\Delta(x_0,R)} |a(x)|^p |x-x_0|^\theta \, d\sigma(x)\Big)^{1/p} \\ &\leq R^{\theta/p} \|a\|_{L^p(\Sigma,\sigma)} \leq R^{-(n-1)/p'+\theta/p}.\end{aligned} \tag{4.196}$$

This establishes (4.195).

It is of interest to have another version of central molecules. The following notion of a central molecule is inspired by [GHA-II, Definition 4.5.1, pp. 174–175].

Definition 4.5 Suppose $\Sigma \subseteq \mathbb{R}^n$ is a closed unbounded Ahlfors regular set, and abbreviate $\sigma := \mathcal{H}^{n-1}\lfloor\Sigma$. Fix a reference point $x_0 \in \Sigma$, along with a dilation factor $2_* > (C_\Sigma/c_\Sigma)^{1/(n-1)}$, where $c_\Sigma, C_\Sigma \in (0,\infty)$ denote the lower and upper Ahlfors regularity constant of Σ, respectively. Pick an exponent $p \in (1,\infty)$ and a parameter $\varepsilon \in (0,\infty)$. A σ-measurable function $M : \Sigma \to \mathbb{R}$ is called an x_0-`central` $(1,p)_\varepsilon$-`molecule` provided there exists a radius $R \in [2_*, \infty)$ with the property that for each $k \in \mathbb{N}_0$ one has

$$\Big(\int_{A_k(x_0,R)} |M|^p \, d\sigma\Big)^{1/p} \leq 2^{k(n-1)(1/p-1-\varepsilon)} \cdot R^{(n-1)(1/p-1)}, \tag{4.197}$$

where

$$A_k(x_0,R) := \begin{cases} \Delta(x_0,R) & \text{if } k=0, \\ \Delta(x_0, 2^k R)\setminus\Delta(x_0, 2^{k-1}R) & \text{if } k \geq 1, \end{cases} \tag{4.198}$$

and

$$\int_\Sigma M \, d\sigma = 0. \tag{4.199}$$

Remark 4.9 In the context of the above definition, we claim that condition (4.197) implies that $M \in L^1(\Sigma,\sigma)$. To see this, use (4.197)-(4.198), Hölder's inequality, and the Ahlfors regularity of Σ to estimate

$$
\begin{aligned}
\int_{\Sigma} |M|\, d\sigma &= \sum_{j=0}^{\infty} \int_{A_k(x_0,R)} |M|\, d\sigma \\
&\leq \sum_{j=0}^{\infty} \Big(\int_{A_k(x_0,R)} |M|^p \, d\sigma \Big)^{1/p} \sigma\big(A_k(x_0,R)\big)^{1/p'} \\
&\leq C \sum_{j=0}^{\infty} 2^{k(n-1)(1/p-1-\varepsilon)} \cdot R^{(n-1)(1/p-1)} \cdot (2^k R)^{(n-1)/p'} \\
&= C \sum_{j=0}^{\infty} 2^{-k(n-1)\varepsilon} < \infty.
\end{aligned} \tag{4.200}
$$

As such, the integral in (4.199) is meaningfully defined.

It turns out that both types of central molecules introduced in Definition 4.5 and Definition 4.4 are compatible, up to multiplicative constants. More precisely, we have the following result.

Proposition 4.5 *Suppose $\Sigma \subseteq \mathbb{R}^n$ is a closed unbounded Ahlfors regular set, and abbreviate $\sigma := \mathcal{H}^{n-1}\lfloor\Sigma$. Fix a reference point $x_0 \in \Sigma$ and select an integrability exponent $p \in (1,\infty)$. Then the following statements are true:*

(a) For every $\varepsilon \in (0,\infty)$ there exists some $\theta = \theta(\varepsilon) \in \big((n-1)(p-1),\infty\big)$, and a constant $C_\varepsilon \in (0,\infty)$ depending on n, p, and ε such that every x_0-central $(1,p)_\varepsilon$-molecule (in the sense of Definition 4.5) is a C_ε-multiple of an x_0-central $(1,p,\theta)$-molecule (in the sense of Definition 4.4).

(b) Conversely, for every $\theta \in \big((n-1)(p-1),\infty\big)$ there exists some $\varepsilon = \varepsilon(\theta) \in (0,\infty)$ and a constant $C_\theta \in (0,\infty)$ depending on n, p, and θ such that every x_0-central $(1,p,\theta)$-molecule (in the sense of Definition 4.4) is a C_θ-multiple of an x_0-central $(1,p)_\varepsilon$-molecule (in the sense of Definition 4.5).

Proof To justify the claim made in item *(a)*, fix an arbitrary $\varepsilon \in (0,\infty)$ and let M be an x_0-central $(1,p)_\varepsilon$-molecule on Σ. Then properties (4.197) and (4.199) in Definition 4.5 hold for some $R \in [2_*,\infty)$. We have to prove that M satisfies the normalization and decay conditions from Definition 4.4. Since

$$
(n-1)(p-1) < (n-1)\Big(\varepsilon + 1 - \frac{1}{p}\Big)p \tag{4.201}
$$

we can pick $\theta \in (0,\infty)$ such that

$$
(n-1)(p-1) < \theta < (n-1)\Big(\varepsilon + 1 - \frac{1}{p}\Big)p. \tag{4.202}
$$

Bearing this in mind, along with (4.198) and (4.197), we estimate

$$
\begin{aligned}
\Big(\int_\Sigma |M(x)|^p|x-x_0|^\theta\,d\sigma(x)\Big)^{1/p} &\le \sum_{k=0}^{\infty}\Big(\int_{A_k(x_0,R)} |M(x)|^p|x-x_0|^\theta\,d\sigma(x)\Big)^{1/p}\\
&\le \sum_{k=0}^{\infty}(2^kR)^{\theta/p}\Big(\int_{A_k(x_0,R)} |M|^p\,d\sigma\Big)^{1/p}\\
&\le \sum_{k=0}^{\infty}(2^kR)^{\theta/p}\cdot 2^{k(n-1)(1/p-1-\varepsilon)}\cdot R^{(n-1)(1/p-1)}\\
&\le R^{\theta/p+(n-1)(1/p-1)}\sum_{k=0}^{\infty}2^{k[\theta/p-(n-1)(\varepsilon+1-1/p)]}<\infty.
\end{aligned}
\tag{4.203}
$$

Next, in view of (4.197) and (4.198) we have

$$
\begin{aligned}
\|M\|_{L^p(\Sigma,\sigma)} = \Big(\int_\Sigma |M|^p\,d\sigma\Big)^{1/p} &\le \sum_{k=0}^{\infty}\Big(\int_{A_k(x_0,R)} |M|^p\,d\sigma\Big)^{1/p}\\
&\le R^{(n-1)(1/p-1)}\cdot\sum_{k=0}^{\infty}2^{k(n-1)(1/p-1-\varepsilon)},
\end{aligned}
\tag{4.204}
$$

and this is a convergent series since $1/p-1-\varepsilon<0$. If we define

$$
C_\varepsilon := \sum_{k=0}^{\infty}2^{k[\theta/p-(n-1)(\varepsilon+1-1/p)]}+\sum_{k=0}^{\infty}2^{k(n-1)(1/p-1-\varepsilon)}\in(0,\infty), \tag{4.205}
$$

then (4.203) and (4.204) imply that the function $C_\varepsilon^{-1}\cdot M$ satisfies

$$
\begin{aligned}
\Big(\int_\Sigma |C_\varepsilon^{-1}\cdot M(x)|^p|x-x_0|^\theta\,d\sigma(x)\Big)^{1/p} &\le R^{\theta/p+(n-1)(1/p-1)}\\
\text{and}\quad \|C_\varepsilon^{-1}\cdot M\|_{L^p(\Sigma,\sigma)} &\le R^{(n-1)(1/p-1)}.
\end{aligned}
\tag{4.206}
$$

Together with the vanishing moment property of M, this ensures that $C_\varepsilon^{-1}\cdot M$ is an x_0-central $(1,p,\theta)$-molecule, for θ as in (4.202). In other words M is a C_ε-multiple of an x_0-central $(1,p,\theta)$-molecule. This establishes the claim made in item *(a)*.

Consider next the claim made in item *(b)*. Fix some $\theta\in\big((n-1)(p-1),\infty\big)$ and assume M is an x_0-central $(1,p,\theta)$-molecule on Σ. Let $R\in[2_*,\infty)$ be a radius associated with M as in Definition 4.4. Also, select

$$
\varepsilon := \frac{\theta}{p(n-1)}+\frac{1}{p}-1\in(0,\infty). \tag{4.207}
$$

Observe that M satisfies (4.197) for $k=0$, since the normalization condition (4.189) yields

$$\Big(\int_{A_0(x_0,R)}|M|^p\,d\sigma\Big)^{1/p}\leq\|M\|_{L^p(\Sigma,\sigma)}\leq R^{(n-1)(1/p-1)}. \tag{4.208}$$

Corresponding to $k\in\mathbb{N}$, use (4.207), which implies $-\frac{\theta}{p}=(n-1)(\frac{1}{p}-1-\varepsilon)$, along with (4.190) and (4.198) to estimate

$$\begin{aligned}\Big(\int_{A_k(x_0,R)}|M|^p\,d\sigma\Big)^{1/p}&\leq(2^{k-1}R)^{-\theta/p}\Big(\int_{A_k(x_0,R)}|M(x)|^p|x-x_0|^\theta\,d\sigma\Big)^{1/p}\\&\leq(2^{k-1}R)^{-\theta/p}\Big(\int_\Sigma|M(x)|^p|x-x_0|^\theta\,d\sigma\Big)^{1/p}\\&\leq(2^{k-1}R)^{-\theta/p}R^{(n-1)(1/p-1)+\theta/p}\\&=2^{\theta/p}\cdot2^{k(n-1)(1/p-1-\varepsilon)}\cdot R^{(n-1)(1/p-1)}.\end{aligned} \tag{4.209}$$

If we set $C_\theta:=2^{\theta/p}\in(0,\infty)$ then (4.208)-(4.209) ensures that $C_\theta^{-1}\cdot M$ satisfies (4.197). This together with the vanishing moment condition of M ensure that $C_\theta^{-1}\cdot M$ is an x_0-central $(1,p)_\varepsilon$-molecule. Hence M is a C_θ-multiple of an x_0-central $(1,p)_\varepsilon$-molecule. This justifies the claim made in *(b)*. □

The result recorded below establishes that there exists some constant that bounds the HA^p norm of every central molecule.

Theorem 4.4 *Assume $\Sigma\subseteq\mathbb{R}^n$ is a closed unbounded Ahlfors regular set, and abbreviate $\sigma:=\mathcal{H}^{n-1}\lfloor\Sigma$. Select an integrability exponent $p\in(1,\infty)$ along with a parameter $\theta\in\big((n-1)(p-1),\infty\big)$. Also, fix a reference point $x_0\in\Sigma$. Then there exists some constant $C\in(0,\infty)$ depending only on n, p, θ, and the Ahlfors regularity character of Σ with the property that for every x_0-central $(1,p,\theta)$-molecule M on Σ (in the sense of Definition 4.4) one has*

$$\|M\|_{\mathrm{HA}^p(\Sigma,\sigma)}\leq C. \tag{4.210}$$

Moreover, the same type of conclusions hold for x_0-central $(1,p)_\varepsilon$-molecules (in the sense of Definition 4.5) for all $\varepsilon\in(0,\infty)$.

Proof Select an x_0-central $(1,p,\theta)$-molecule M on Σ. Pick a radius $R\geq2_*$ associated with M as in Definition 4.4, where $2_*>(C_\Sigma/c_\Sigma)^{1/(n-1)}$ and $c_\Sigma,C_\Sigma\in(0,\infty)$ denote the lower and upper Ahlfors regularity constant of Σ, respectively. For each $k\in\mathbb{N}_0$ define

$$\Delta_{-1}:=\varnothing \tag{4.211}$$

$$\Delta_k:=\Delta(x_0,2_*^kR) \tag{4.212}$$

$$A_k:=\Delta_k\backslash\Delta_{k-1} \tag{4.213}$$

$$M_{A_k}:=⨍_{A_k}M\,d\sigma. \tag{4.214}$$

Thanks to the absolute integrability of M on Σ (cf. Remark 4.7), M_{A_k} is meaningfully defined for all $k \in \mathbb{N}_0$. Notice that $\sigma(A_k) > 0$ for all $k \in \mathbb{N}_0$, since the choice of 2_* and the Ahlfors regularity of Σ imply

$$\begin{aligned}\sigma(A_k) &= \sigma(\Delta_k) - \sigma(\Delta_{k-1})\\ &\geq c_\Sigma(2_*^k R)^{n-1} - C_\Sigma(2_*^{k-1}R)^{n-1}\\ &= (c_\Sigma 2_*^{n-1} - C_\Sigma)(2_*^{k-1}R)^{n-1} > 0.\end{aligned} \tag{4.215}$$

This, (4.212), (4.213), and the Ahlfors regularity of Σ give

$$\sigma(A_k) \approx (2_*^k R)^{n-1} \quad \text{uniformly for } k \in \mathbb{N}_0. \tag{4.216}$$

To proceed, fix $N \in \mathbb{N}$. Using the summation by parts formula from (4.105) for the choices $s_k := \frac{\mathbb{1}_{A_k}}{\sigma(A_k)}$ and $t_k := \int_{\Sigma\setminus\Delta_{k-1}} M\,d\sigma$ with $k \in \mathbb{N}_0$, along with (4.211) and the vanishing moment of M (cf. (4.191)), we compute

$$\begin{aligned}\sum_{k=0}^{N} M_{A_k}\cdot\mathbb{1}_{A_k} &= \sum_{k=0}^{N}\frac{\mathbb{1}_{A_k}}{\sigma(A_k)}\int_{\Delta_k\setminus\Delta_{k-1}} M\,d\sigma\\ &= \sum_{k=0}^{N}\frac{\mathbb{1}_{A_k}}{\sigma(A_k)}\Big(\int_{\Sigma\setminus\Delta_{k-1}} M\,d\sigma - \int_{\Sigma\setminus\Delta_k} M\,d\sigma\Big)\\ &= \frac{\mathbb{1}_{A_0}}{\sigma(A_0)}\int_{\Sigma\setminus\Delta_{-1}} M\,d\sigma - \frac{\mathbb{1}_{A_N}}{\sigma(A_N)}\int_{\Sigma\setminus\Delta_N} M\,d\sigma\\ &\quad + \sum_{k=1}^{N}\Big(\int_{\Sigma\setminus\Delta_{k-1}} M\,d\sigma\Big)\Big(\frac{\mathbb{1}_{A_k}}{\sigma(A_k)} - \frac{\mathbb{1}_{A_{k-1}}}{\sigma(A_{k-1})}\Big)\\ &= -\frac{\mathbb{1}_{A_N}}{\sigma(A_N)}\int_{\Sigma\setminus\Delta_N} M\,d\sigma + \sum_{k=0}^{N-1}\Big(\int_{\Sigma\setminus\Delta_k} M\,d\sigma\Big)\Big(\frac{\mathbb{1}_{A_{k+1}}}{\sigma(A_{k+1})} - \frac{\mathbb{1}_{A_k}}{\sigma(A_k)}\Big).\end{aligned} \tag{4.217}$$

It is visible from (4.211)-(4.213) that $\sqcup_{k=0}^N A_k = \Delta_N$ for all $N \in \mathbb{N}_0$. Hence, for every $N \in \mathbb{N}_0$ we may write

$$\begin{aligned}\sum_{k=0}^{N}(M - M_{A_k})\mathbb{1}_{A_k} &= M\sum_{k=0}^{N}\mathbb{1}_{A_k} - \sum_{k=0}^{N} M_{A_k}\cdot\mathbb{1}_{A_k}\\ &= M\cdot\mathbb{1}_{\Delta_N} - \sum_{k=0}^{N} M_{A_k}\cdot\mathbb{1}_{A_k}.\end{aligned} \tag{4.218}$$

For each $k \in \mathbb{N}_0$ define the functions $\alpha_k, \beta_k : \Sigma \to \mathbb{R}$ by setting

$$\alpha_k := (M - M_{A_k})\mathbb{1}_{A_k} \quad \text{and} \tag{4.219}$$

$$\beta_k := \Big(\frac{\mathbb{1}_{A_{k+1}}}{\sigma(A_{k+1})} - \frac{\mathbb{1}_{A_k}}{\sigma(A_k)}\Big)\int_{\Sigma\setminus\Delta_k} M\,d\sigma. \tag{4.220}$$

In view of (4.217), (4.218), (4.219), and (4.220), for each $N \in \mathbb{N}$ we may express $M \cdot \mathbb{1}_{\Delta_N}$ as

$$M \cdot \mathbb{1}_{\Delta_N} = \sum_{k=0}^{N} \alpha_k + \sum_{k=0}^{N} \beta_k - \frac{\mathbb{1}_{A_N}}{\sigma(A_N)}\int_{\Sigma\setminus\Delta_N} M\,d\sigma. \tag{4.221}$$

To further analyze each term in the right-hand side of (4.221), fix some $k \in \mathbb{N}_0$. Being mindful of (4.212)-(4.214) and (4.219), the function α_k satisfies

$$\operatorname{supp}\alpha_k \subseteq \Delta_k = \Delta(x_0, 2_*^k R) \tag{4.222}$$

and

$$\int_\Sigma \alpha_k\,d\sigma = \int_{A_k}(M - M_{A_k})\,d\sigma = \int_{A_k} M\,d\sigma - \sigma(A_k)M_{A_k} = 0. \tag{4.223}$$

Moreover, use Hölder's inequality to estimate

$$\begin{aligned}
\|\alpha_k\|_{L^p(\Sigma,\sigma)} &= \Big(\int_{A_k} |M - M_{A_k}|^p\,d\sigma\Big)^{1/p} \\
&\le \Big(\int_{A_k} |M|^p\,d\sigma\Big)^{1/p} + \sigma(A_k)^{1/p}|M_{A_k}| \\
&\le \Big(\int_{A_k} |M|^p\,d\sigma\Big)^{1/p} + \sigma(A_k)^{1/p} \fint_{A_k} |M|\,d\sigma \\
&\le \Big(\int_{A_k} |M|^p\,d\sigma\Big)^{1/p} + \sigma(A_k)^{1/p}\Big(\fint_{A_k} |M|^p\,d\sigma\Big)^{1/p} \\
&= 2\Big(\int_{A_k} |M|^p\,d\sigma\Big)^{1/p}.
\end{aligned} \tag{4.224}$$

This and the normalization condition of M (see (4.189)) imply

$$\|\alpha_0\|_{L^p(\Sigma,\sigma)} \le 2\Big(\int_{A_0} |M|^p\,d\sigma\Big)^{1/p} \le 2\|M\|_{L^p(\Sigma,\sigma)} \le 2R^{-(n-1)/p'}, \tag{4.225}$$

where $p' \in (1, \infty)$ is the Hölder conjugate exponent of p. Corresponding to $k \ge 1$, use (4.224) and the decay condition of M (see (4.190)), while keeping in mind (4.213), to obtain

$$
\begin{aligned}
\|\alpha_k\|_{L^p(\Sigma,\sigma)} &\le 2\Big(\int_{A_k} |M|^p \, d\sigma\Big)^{1/p} \\
&\le 2\Big(\int_{A_k} |M(x)|^p |x-x_0|^\theta (2_*^{k-1}R)^{-\theta} \, d\sigma(x)\Big)^{1/p} \\
&\le 2(2_*^{k-1}R)^{-\theta/p}\Big(\int_\Sigma |M(x)|^p |x-x_0|^\theta \, d\sigma(x)\Big)^{1/p} \\
&\le 2(2_*^{k-1}R)^{-\theta/p} R^{-(n-1)/p'+\theta/p} \\
&= C 2_*^{-k\theta/p} R^{-(n-1)/p'}
\end{aligned}
\tag{4.226}
$$

for some $C \in (0,\infty)$ depending only on n, p, and θ. This and (4.225) imply that there exists some $C' \in (0,\infty)$ depending only on n, p, and θ such that

$$
\|\alpha_k\|_{L^p(\Sigma,\sigma)} \le C' 2_*^{-k\theta/p} R^{-(n-1)/p'} \quad \text{for all } k \in \mathbb{N}_0. \tag{4.227}
$$

For each $k \in \mathbb{N}_0$ define the function $a_k : \Sigma \to \mathbb{R}$ according to

$$
a_k := \lambda_k^{-1}\alpha_k \quad \text{with } \lambda_k := C' 2_*^{-k[\theta/p-(n-1)/p']} \in (0,\infty). \tag{4.228}
$$

Based on (4.219), (4.222), (4.223), (4.224), (4.225), and (4.226), for each $k \in \mathbb{N}_0$ we have that a_k is σ-measurable on Σ and satisfies

$$
\operatorname{supp} a_k \subseteq \Delta(x_0, 2_*^k R), \quad \|a_k\|_{L^p(\Sigma,\sigma)} \le (2_*^k R)^{-(n-1)/p'}, \quad \int_\Sigma a_k \, d\sigma = 0. \tag{4.229}
$$

Hence, a_k is an x_0-central $(1,p)$-atom on Σ for all $k \in \mathbb{N}_0$.

Observe that the sequence $\{\lambda_k\}_{k\in\mathbb{N}_0}$ belongs to $\ell^1(\mathbb{N}_0)$, since (4.228) and (4.194) yield

$$
\sum_{k=0}^\infty |\lambda_k| = C' \sum_{k=0}^\infty 2_*^{-k[\theta/p-(n-1)/p']} < \infty. \tag{4.230}
$$

Moving on, fix an arbitrary $k \in \mathbb{N}_0$. Upon recalling (4.212)-(4.213), the function β_k defined in (4.220) satisfies

$$
\operatorname{supp} \beta_k \subseteq \Delta_{k+1} = \Delta(x_0, 2_*^{k+1} R) \tag{4.231}
$$

and

$$
\int_\Sigma \beta_k \, d\sigma = \Big(\int_\Sigma \frac{\mathbb{1}_{A_{k+1}}}{\sigma(A_{k+1})} \, d\sigma - \int_\Sigma \frac{\mathbb{1}_{A_k}}{\sigma(A_k)} \, d\sigma\Big)\Big(\int_{\Sigma\setminus\Delta_k} M \, d\sigma\Big) = 0. \tag{4.232}
$$

Using the Ahlfors regularity of Σ, Hölder's inequality, the decay condition of M (cf. (4.190)), and (4.216) we may estimate

$$
\begin{aligned}
&\|\beta_k\|_{L^p(\Sigma,\sigma)}\\
&\leq \Big(\int_\Sigma \Big|\frac{\mathbb{1}_{A_{k+1}}}{\sigma(A_{k+1})}-\frac{\mathbb{1}_{A_k}}{\sigma(A_k)}\Big|^p\,d\sigma\Big)^{1/p}\int_{\Sigma\setminus\Delta_k}|M|\,d\sigma\\
&\leq \Big(\frac{\|\mathbb{1}_{A_{k+1}}\|_{L^p(\Sigma,\sigma)}}{\sigma(A_{k+1})}+\frac{\|\mathbb{1}_{A_k}\|_{L^p(\Sigma,\sigma)}}{\sigma(A_k)}\Big)\int_{\Sigma\setminus\Delta_k}|M|\,d\sigma\\
&= \big(\sigma(A_{k+1})^{1/p-1}+\sigma(A_k)^{1/p-1}\big)\int_{\Sigma\setminus\Delta_k}|M|\,d\sigma\\
&\leq C(2_*^{k+1}R)^{(n-1)(1/p-1)}\int_{\Sigma\setminus\Delta_k}|M(x)||x-x_0|^{\theta/p}|x-x_0|^{-\theta/p}\,d\sigma(x)\\
&\leq C(2_*^{k+1}R)^{-(n-1)/p'}\\
&\quad\times\Big(\int_{\Sigma\setminus\Delta_k}|M(x)|^p|x-x_0|^\theta\,d\sigma(x)\Big)^{1/p}\Big(\int_{\Sigma\setminus\Delta_k}|x-x_0|^{-\theta p'/p}\,d\sigma(x)\Big)^{1/p'}\\
&\leq C(2_*^{k+1}R)^{-(n-1)/p'}\\
&\quad\times\Big(\int_{\Sigma}|M(x)|^p|x-x_0|^\theta\,d\sigma(x)\Big)^{1/p}\Big(\int_{\Sigma\setminus\Delta_k}|x-x_0|^{-\theta p'/p}\,d\sigma(x)\Big)^{1/p'}\\
&\leq C(2_*^{k+1}R)^{-(n-1)/p'}R^{-(n-1)/p'+\theta/p}\Big(\int_{\Sigma\setminus\Delta_k}|x-x_0|^{-\theta p'/p}\,d\sigma(x)\Big)^{1/p'},
\end{aligned}
\tag{4.233}
$$

for some constant $C\in(0,\infty)$ depending only on n, p, θ, and the Ahlfors regularity constants of Σ. Regarding the integral in the right-most side of (4.233), use the Ahlfors regularity of Σ and (4.194) to estimate

$$
\begin{aligned}
&\Big(\int_{\Sigma\setminus\Delta_k}|x-x_0|^{-\theta p'/p}\,d\sigma\Big)^{1/p'}\\
&\qquad\leq\sum_{\ell=0}^{\infty}\Big(\int_{\Delta(x_0,2^{\ell+1}2_*^kR)\setminus\Delta(x_0,2^\ell 2_*^kR)}|x-x_0|^{-\theta p'/p}\,d\sigma\Big)^{1/p'}\\
&\qquad\leq\sum_{\ell=0}^{\infty}(2^\ell 2_*^kR)^{-\theta/p}\sigma\big(\Delta(x_0,2^{\ell+1}2_*^kR)\setminus\Delta(x_0,2^\ell 2_*^kR)\big)^{1/p'}\\
&\qquad\leq C(2_*^kR)^{-\theta/p}\sum_{\ell=0}^{\infty}(2^\ell)^{-\theta/p}(2^{\ell+1}2_*^kR)^{(n-1)/p'}\\
&\qquad\leq C(2_*^{k+1}R)^{(n-1)/p'-\theta/p}\sum_{\ell=0}^{\infty}2^{\ell[(n-1)/p'-\theta/p]}<\infty,
\end{aligned}
\tag{4.234}
$$

for some $C\in(0,\infty)$ depending only on n, p, θ, and the Ahlfors regularity constants of Σ. A combination of this and (4.233) guarantees that there exists some $C''\in(0,\infty)$

depending only on n, p, θ, and the Ahlfors regularity constants of Σ for which

$$\begin{aligned}\|\beta_k\|_{L^p(\Sigma,\sigma)} &\leq C''(2_*^{k+1}R)^{-(n-1)/p'}R^{-(n-1)/p'+\theta/p}(2_*^{k+1}R)^{(n-1)/p'-\theta/p}\\ &= C''2_*^{-(k+1)\theta/p}R^{-(n-1)/p'}.\end{aligned} \tag{4.235}$$

Pressing forward, for each $k \in \mathbb{N}_0$ define the function $b_k : \Sigma \to \mathbb{R}$ by

$$b_k := \widetilde{\lambda}_k^{-1}\beta_k, \quad \text{where } \widetilde{\lambda}_k := C''2_*^{(k+1)(n-1)/p'}2_*^{-(k+1)\theta/p}. \tag{4.236}$$

Then for all $k \in \mathbb{N}_0$ the function b_k is σ-measurable on Σ, and in light of (4.231), (4.232), and (4.235) we have

$$\operatorname{supp} b_k \subseteq \Delta(x_0, 2_*^{k+1}R), \quad \|b_k\|_{L^p(\Sigma,\sigma)} \leq (2_*^{k+1}R)^{-(n-1)/p'}, \quad \int_\Sigma b_k\,d\sigma = 0, \tag{4.237}$$

for every $k \in \mathbb{N}_0$. Hence b_k is an x_0-central $(1, p)$-atom for every $k \in \mathbb{N}_0$.

Bearing in mind (4.194) along with the definition of $\{\widetilde{\lambda}_k\}_{k\in\mathbb{N}_0}$ in (4.236), we see that

$$\sum_{k=0}^{\infty}|\widetilde{\lambda}_k| = C''\sum_{k=0}^{\infty}2_*^{(k+1)[(n-1)/p'-\theta/p]} < \infty. \tag{4.238}$$

As such, $\{\widetilde{\lambda}_k\}_{k\in\mathbb{N}_0} \in \ell^1(\mathbb{N}_0)$.

At this stage of the proof, recall (4.228) and (4.236) to recast (4.221) as

$$M \cdot \mathbb{1}_{\Delta_N} = \sum_{k=0}^{N}\lambda_k a_k + \sum_{k=0}^{N}\widetilde{\lambda}_k b_k - \frac{\mathbb{1}_{A_N}}{\sigma(A_N)}\int_{\Sigma\setminus\Delta_N} M\,d\sigma \tag{4.239}$$

for each $N \in \mathbb{N}$. Thanks to (4.216), for each $N \in \mathbb{N}$ we have

$$\left|\frac{\mathbb{1}_{A_N}}{\sigma(A_N)}\int_{\Sigma\setminus\Delta_N} M\,d\sigma\right| \leq \frac{1}{\sigma(A_N)}\int_\Sigma |M|\,d\sigma \leq \frac{C\|M\|_{L^1(\Sigma,\sigma)}}{(2_*^N R)^{n-1}}, \tag{4.240}$$

for some constant $C \in (0, \infty)$ depending only on the Ahlfors regularity of Σ. From this we see that

$$\frac{\mathbb{1}_{A_N}}{\sigma(A_N)}\int_{\Sigma\setminus\Delta_N} M\,d\sigma \to 0 \quad \text{pointwise everywhere on } \Sigma \text{ as } N \to \infty. \tag{4.241}$$

Granted this, letting $N \to \infty$ in (4.239) implies

$$M = \sum_{k=0}^{\infty}\lambda_k a_k + \sum_{k=0}^{\infty}\widetilde{\lambda}_k b_k \quad \text{pointwise } \sigma\text{-a.e. on } \Sigma. \tag{4.242}$$

From (4.228), (4.230), (4.237), (4.238), and item *(b)* in Theorem 4.2, we finally obtain that $M \in \mathrm{HA}^p(\Sigma, \sigma)$ and

$$\|M\|_{\mathrm{HA}^p(\Sigma,\sigma)} \leq C\sum_{k=0}^{\infty}|\lambda_k| + C\sum_{k=0}^{\infty}|\widetilde{\lambda}_k|, \tag{4.243}$$

for some constant $C \in (0,\infty)$ independent of M. Recall from (4.230) and (4.238) that

$$\sum_{k=0}^{\infty}|\lambda_k| + \sum_{k=0}^{\infty}|\widetilde{\lambda}_k| = C'\sum_{k=0}^{\infty}2_*^{-k(\theta/p-(n-1)/p')} + C''\sum_{k=0}^{\infty}2_*^{(k+1)[(n-1)/p'-\theta/p]}. \tag{4.244}$$

Since the constants $C', C'' \in (0,\infty)$ appearing in (4.227) and (4.235) depend only on n, p, and θ, the quantity in the right-hand side of (4.243) is independent of M, and in fact, it involves only n, p, θ, and the Ahlfors regularity character of Σ. As such, this establishes (4.210).

The very last claim regarding the validity of (4.210) when M is an arbitrary x_0-central $(1,p)_\varepsilon$-molecule, for $\varepsilon \in (0,\infty)$, follows from what we have proved so far, and the fact that whenever $\varepsilon \in (0,\infty)$ each x_0-central $(1,p)_\varepsilon$-molecule is a fixed constant multiple of an x_0-central $(1,p,\theta)$-molecule for some $\theta \in \big((n-1)(p-1),\infty\big)$ (cf. Proposition 4.5). This concludes the proof of Theorem 4.4. □

The following result is the molecular version of the atomic characterization of HA^p from Theorem 4.2.

Theorem 4.5 *Suppose $\Sigma \subseteq \mathbb{R}^n$ is a closed unbounded Ahlfors regular set, and abbreviate $\sigma := \mathcal{H}^{n-1}\lfloor\Sigma$. Select an integrability exponent $p \in (1,\infty)$ along with a parameter $\theta \in \big((n-1)(p-1),\infty\big)$. Also, fix a reference point $x_0 \in \Sigma$. In this setting, the following statements are true:*

(a) For every $f \in \mathrm{HA}^p(\Sigma,\sigma)$ there exists a sequence of x_0-central $(1,p,\theta)$-molecules $\{M_k\}_{k\in\mathbb{N}_0}$ along with a numerical sequence $\{\lambda_k\}_{k\in\mathbb{N}_0} \in \ell^1(\mathbb{N}_0)$ satisfying

$$f = \sum_{k=0}^{\infty}\lambda_k M_k \tag{4.245}$$

with simultaneous convergence to f in $L^1(\Sigma,\sigma)$, in $L^p(\Sigma,\sigma)$, and in $\mathrm{HA}^p(\Sigma,\sigma)$. In addition, there exists a constant $C \in (0,\infty)$ depending only on n, p, θ, and the Ahlfors regularity character of Σ with the property that

$$\sum_{k=0}^{\infty}|\lambda_k| \leq C\|f\|_{\mathrm{HA}^p(\Sigma,\sigma)}. \tag{4.246}$$

(b) Conversely, assume $f : \Sigma \to \mathbb{R}$ is σ-measurable function for which there exist some sequence of x_0-central $(1,p,\theta)$-molecules $\{M_k\}_{k\in\mathbb{N}_0}$ and some sequence $\{\lambda_k\}_{k\in\mathbb{N}_0} \in \ell^1(\mathbb{N}_0)$ such that

$$f = \sum_{k=0}^{\infty} \lambda_k M_k \quad \textit{pointwise } \sigma\textit{-a.e. on } \Sigma. \tag{4.247}$$

Then $f \in \mathrm{HA}^p(\Sigma, \sigma)$ *quantitatively, i.e., there exists some constant* $C \in (0, \infty)$ *depending solely on* n*,* p*,* θ*, and the Ahlfors regularity character of* Σ *with the property that*

$$\|f\|_{\mathrm{HA}^p(\Sigma,\sigma)} \le C \sum_{k=0}^{\infty} |\lambda_k|. \tag{4.248}$$

(c) As a consequence of items (a) and (b), for every $f \in \mathrm{HA}^p(\Sigma, \sigma)$ *it follows that*

$$\|f\|_{\mathrm{HA}^p(\Sigma,\sigma)} \approx \inf \Big\{ \sum_{k=0}^{\infty} |\lambda_k| \, : \, f = \sum_{k=0}^{\infty} \lambda_k M_k \textit{ in } L^1(\Sigma, \sigma) \Big\} \tag{4.249}$$

where the infimum is taken over all molecular representations of f*, and with proportionality constants depending only on* n*,* p*,* θ*, and the Ahlfors regularity constants of* Σ*.*

Lastly, the same type of conclusions are valid for x_0*-central* $(1, p)_\varepsilon$*-molecules for any* $\varepsilon \in (0, \infty)$ *(in the sense of Definition 4.5) in place of* x_0*-central* $(1, p, \theta)$*-molecules.*

Proof Observe that the claim made in item *(a)* is a direct consequence of Theorem 4.2 and Remark 4.8. To establish the claim made in item *(b)*, employ the same type of reasoning used in the proof of item *(b)* in Theorem 4.2, now bearing in mind Theorem 4.4. The claim made in *(c)* readily follows from items *(a)* and *(b)*. Finally, the validity of items *(a)*-*(c)* written in terms of x_0-central $(1, p)_\varepsilon$-molecules (in the sense of Definition 4.5) follows from what we have proved so far, together with Proposition 4.5. □

Chapter 5
Functions of Bounded Central Mean Oscillations

The goal here is to study the space of functions of bounded central p-mean oscillations BCMO_p and compare it to the classical John-Nirenberg space BMO. In Section 5.1 we establish some basic functional analytic results concerning BCMO_p spaces on unbounded Ahlfors regular sets, including embeddings and norm characterizations. In Section 5.2 we prove the main result of this chapter which identifies the dual of the Beurling-Hardy space as the spaces of bounded central mean oscillations (see Theorem 5.1). This may be regarded as a version of the celebrated Fefferman-Stein duality result between the Hardy space H^1 and the John-Nirenberg space BMO.

5.1 Definitions and Basic Results

We begin by defining the space of bounded central p-mean oscillation functions in the general geometric setting of unbounded Ahlfors regular sets.

Definition 5.1 Assume $\Sigma \subseteq \mathbb{R}^n$ is a closed unbounded Ahlfors regular set, and abbreviate $\sigma := \mathcal{H}^{n-1}\lfloor\Sigma$. Denote by $c_\Sigma, C_\Sigma \in (0,\infty)$ the lower and upper Ahlfors regularity constant of Σ, respectively, and select a dilation factor $2_* > (C_\Sigma/c_\Sigma)^{1/(n-1)}$. Fix a reference point $x_0 \in \Sigma$, and pick an integrability exponent $p \in (1,\infty)$. For each function $f \in L^1_{\mathrm{loc}}(\Sigma,\sigma)$ define

$$\|f\|_{\mathrm{BCMO}_p(\Sigma,\sigma)} := \sup_{R\geq 2_*}\Big(\fint_{\Delta(x_0,R)} |f - f_{\Delta(x_0,R)}|^p\,d\sigma\Big)^{1/p} \in [0,\infty] \tag{5.1}$$

where $f_{\Delta(x_0,R)} = \fint_{\Delta(x_0,R)} f\,d\sigma$. Introduce the `space of functions of bounded central` p`-mean oscillations` on Σ as

$$\mathrm{BCMO}_p(\Sigma,\sigma) := \big\{f \in L^1_{\mathrm{loc}}(\Sigma,\sigma) : \|f\|_{\mathrm{BCMO}_p(\Sigma,\sigma)} < \infty\big\}. \tag{5.2}$$

© The Author(s), under exclusive license to Springer Nature Switzerland AG 2026
M. Mitrea and P. Takemura, *Singular Integrals, Herz-Type Function Spaces, and Boundary Problems*, Progress in Mathematics 362,
https://doi.org/10.1007/978-3-032-12516-3_5

It is evident from (5.1) that $\|\cdot\|_{\mathrm{BCMO}_p(\Sigma,\sigma)}$ is a seminorm, since whenever $f \in L^1_{\mathrm{loc}}(\Sigma, \sigma)$ we have

$$\|f\|_{\mathrm{BCMO}_p(\Sigma,\sigma)} = 0 \quad \text{if and only if } f \text{ is a constant function on } \Sigma. \tag{5.3}$$

For this reason, it is natural to mod out constants from $\mathrm{BCMO}_p(\Sigma, \sigma)$. Specifically, for any two functions $f, g \in L^1_{\mathrm{loc}}(\Sigma, \sigma)$ consider the equivalence relation $\sim$ defined by

$$f \sim g \iff f - g \text{ is a constant.} \tag{5.4}$$

Denote by $[f]$ the equivalence class modulo constants of a given $f \in L^1_{\mathrm{loc}}(\Sigma, \sigma)$. Introduce

$$\widetilde{\mathrm{BCMO}}_p(\Sigma, \sigma) := \mathrm{BCMO}_p(\Sigma, \sigma)/\!\sim\, = \big\{[f] \,:\, f \in \mathrm{BCMO}_p(\Sigma, \sigma)\big\}, \tag{5.5}$$

and equip this space with

$$\big\|[f]\big\|_{\widetilde{\mathrm{BCMO}}_p(\Sigma,\sigma)} := \|f\|_{\mathrm{BCMO}_p(\Sigma,\sigma)} \quad \text{for each } f \in \mathrm{BCMO}_p(\Sigma, \sigma). \tag{5.6}$$

Then $\big(\widetilde{\mathrm{BCMO}}_p(\Sigma, \sigma), \|\cdot\|_{\widetilde{\mathrm{BCMO}}_p(\Sigma,\sigma)}\big)$ becomes a genuine normed space.

The following result provides a natural characterization of the BCMO_p seminorm.

Proposition 5.1 *Consider a closed unbounded Ahlfors regular set $\Sigma \subsetneq \mathbb{R}^n$, and abbreviate $\sigma := \mathcal{H}^{n-1}\lfloor\Sigma$. Fix a reference point $x_0 \in \Sigma$, and a dilation factor 2_* as in Definition 5.1. Pick an integrability exponent $p \in (1, \infty)$. Then for every function $f \in L^1_{\mathrm{loc}}(\Sigma, \sigma)$ one has*

$$\frac{1}{2}\|f\|_{\mathrm{BCMO}_p(\Sigma,\sigma)} \leq \sup_{R \geq 2_*} \inf_{c \in \mathbb{C}} \Big(⨍_{\Delta(x_0,R)} |f - c|^p \, d\sigma \Big)^{1/p} \leq \|f\|_{\mathrm{BCMO}_p(\Sigma,\sigma)}. \tag{5.7}$$

Consequently, a function $f \in L^1_{\mathrm{loc}}(\Sigma, \sigma)$ belongs to $\mathrm{BCMO}_p(\Sigma, \sigma)$ if and only if

$$\sup_{R \geq 2_*} \inf_{c \in \mathbb{C}} \Big(⨍_{\Delta(x_0,R)} |f - c|^p \, d\sigma \Big)^{1/p} < \infty. \tag{5.8}$$

Proof Fix some function $f \in L^1_{\mathrm{loc}}(\Sigma, \sigma)$. The second inequality in (5.7) follows from the fact that for each $R \geq 2_*$ we have

$$\begin{aligned} \inf_{c \in \mathbb{C}} \Big(⨍_{\Delta(x_0,R)} |f - c|^p \, d\sigma \Big)^{1/p} &\leq \Big(⨍_{\Delta(x_0,R)} |f - f_{\Delta(x_0,R)}|^p \, d\sigma \Big)^{1/p} \\ &\leq \|f\|_{\mathrm{BCMO}_p(\Sigma,\sigma)}. \end{aligned} \tag{5.9}$$

To justify the first inequality in (5.7), pick an arbitrary radius $R \geq 2_*$ and a number $c \in \mathbb{C}$. For every $x \in \Delta(x_0, R)$, estimate

$$
\begin{aligned}
|f(x) - f_{\Delta(x_0,R)}| &\le |f(x) - c| + |f_{\Delta(x_0,R)} - c| \\
&= |f(x) - c| + \Big| \fint_{\Delta(x_0,R)} (f - c)\, d\sigma \Big| \\
&\le |f(x) - c| + \fint_{\Delta(x_0,R)} |f - c|\, d\sigma \\
&\le |f(x) - c| + \Big(\fint_{\Delta(x_0,R)} |f - c|^p\, d\sigma \Big)^{1/p}
\end{aligned} \tag{5.10}
$$

where the last inequality follows from Hölder's inequality. Hence,

$$
\Big(\fint_{\Delta(x_0,R)} |f - f_{\Delta(x_0,R)}|^p\, d\sigma \Big)^{1/p} \le 2 \Big(\fint_{\Delta(x_0,R)} |f - c|^p\, d\sigma \Big)^{1/p}. \tag{5.11}
$$

By the arbitrariness of $R \ge 2_*$ and $c \in \mathbb{C}$, we see that

$$
\|f\|_{\mathrm{BCMO}_p(\Sigma,\sigma)} \le 2 \sup_{R \ge 2_*} \inf_{c \in \mathbb{C}} \Big(\fint_{\Delta(x_0,R)} |f - c|^p\, d\sigma \Big)^{1/p}. \tag{5.12}
$$

This establishes the first inequality in (5.7). □

The following lemma plays a basic role in establishing further properties of BCMO_p functions.

Lemma 5.1 *Let $\Sigma \subseteq \mathbb{R}^n$ be a closed Ahlfors regular, and abbreviate $\sigma := \mathcal{H}^{n-1} \lfloor \Sigma$. Then for every function $f \in L^1_{\mathrm{loc}}(\Sigma, \sigma)$ and every surface ball $\Delta \subseteq \Sigma$ one has*

$$
\Big(\fint_\Delta \Big| |f| - \fint_\Delta |f|\, d\sigma \Big|^p d\sigma \Big)^{1/p} \le 2 \Big(\fint_\Delta \Big| f - \fint_\Delta f\, d\sigma \Big|^p d\sigma \Big)^{1/p}. \tag{5.13}
$$

Proof Fix a function $f \in L^1_{\mathrm{loc}}(\Sigma, \sigma)$ and every surface ball $\Delta \subseteq \Sigma$. Then for every $x \in \Sigma$ we have

$$
\begin{aligned}
\Big| |f(x)| - \fint_\Delta |f(y)|\, d\sigma(y) \Big| &= \Big| \fint_\Delta (|f(x)| - |f(y)|)\, d\sigma(y) \Big| \\
&\le \fint_\Delta \big| |f(x)| - |f(y)| \big|\, d\sigma(y) \\
&\le \fint_\Delta |f(x) - f(y)|\, d\sigma(y) \\
&\le \Big| f(x) - \fint_\Delta f\, d\sigma \Big| + \fint_\Delta \Big| f(y) - \fint_\Delta f\, d\sigma \Big|\, d\sigma(y).
\end{aligned} \tag{5.14}
$$

This and Hölder's inequality imply

$$
\begin{aligned}
&\Big(\fint_\Delta \Big||f(x)| - \fint_\Delta |f(y)|\,d\sigma(y)\Big|^p\,d\sigma(x)\Big)^{1/p}\\
&\quad\le \Big(\fint_\Delta \Big|f(x) - \fint_\Delta f\,d\sigma\Big|^p\,d\sigma(x)\Big)^{1/p} + \fint_\Delta \Big|f(y) - \fint_\Delta f\,d\sigma\Big|\,d\sigma(y)\\
&\quad\le \Big(\fint_\Delta \Big|f(x) - \fint_\Delta f\,d\sigma\Big|^p\,d\sigma(x)\Big)^{1/p} + \Big(\fint_\Delta \Big|f(y) - \fint_\Delta f\,d\sigma\Big|^p\,d\sigma(y)\Big)^{1/p}\\
&\quad= 2\Big(\fint_\Delta \Big|f - \fint_\Delta f\,d\sigma\Big|^p\,d\sigma\Big)^{1/p}.
\end{aligned}
\tag{5.15}
$$

This establishes (5.13). □

It is useful to record the following basic properties of the BCMO_p-seminorm.

Proposition 5.2 *Let $\Sigma \subseteq \mathbb{R}^n$ be a closed unbounded Ahlfors regular set, and abbreviate $\sigma := \mathcal{H}^{n-1}\lfloor\Sigma$. Fix an integrability exponent $p \in (1,\infty)$. Then for every $f, g \in L^1_{\mathrm{loc}}(\Sigma,\sigma)$ it follows that*

$$
\big\||f|\big\|_{\mathrm{BCMO}_p(\Sigma,\sigma)} \le 2\|f\|_{\mathrm{BCMO}_p(\Sigma,\sigma)}, \tag{5.16}
$$

$$
\big\|\max\{f,g\}\big\|_{\mathrm{BCMO}_p(\Sigma,\sigma)} \le \frac{3}{2}\Big(\|f\|_{\mathrm{BCMO}_p(\Sigma,\sigma)} + \|g\|_{\mathrm{BCMO}_p(\Sigma,\sigma)}\Big), \tag{5.17}
$$

$$
\big\|\min\{f,g\}\big\|_{\mathrm{BCMO}_p(\Sigma,\sigma)} \le \frac{3}{2}\Big(\|f\|_{\mathrm{BCMO}_p(\Sigma,\sigma)} + \|g\|_{\mathrm{BCMO}_p(\Sigma,\sigma)}\Big). \tag{5.18}
$$

In particular, whenever $f, g \in \mathrm{BCMO}_p(\Sigma,\sigma)$ one has

$$
|f|,\ \max\{f,g\},\ \min\{f,g\} \in \mathrm{BCMO}_p(\Sigma,\sigma). \tag{5.19}
$$

Proof Consider a reference point $x_0 \in \Sigma$, and a dilation factor 2_* as in Definition 5.1. Fix two functions $f, g \in L^1_{\mathrm{loc}}(\Sigma,\sigma)$. To see that (5.16) holds, recall (5.1) and use Lemma 5.1 to estimate

$$
\begin{aligned}
\big\||f|\big\|_{\mathrm{BCMO}_p(\Sigma,\sigma)} &= \sup_{R\ge 2_*}\Big(\fint_{\Delta(x_0,R)} \Big||f| - \fint_{\Delta(x_0,R)} |f|\,d\sigma\Big|^p\,d\sigma\Big)^{1/p}\\
&\le 2\sup_{R\ge 2_*}\Big(\fint_{\Delta(x_0,R)} \Big|f - \fint_{\Delta(x_0,R)} f\,d\sigma\Big|^p\,d\sigma\Big)^{1/p}\\
&= 2\|f\|_{\mathrm{BCMO}_p(\Sigma,\sigma)}.
\end{aligned}
\tag{5.20}
$$

To justify (5.17) and (5.18) recall that $\max\{f,g\} = \frac{1}{2}(f + g + |f-g|)$ and also that $\min\{f,g\} = \frac{1}{2}(f + g - |f-g|)$. Thanks to these properties and (5.20) we have

$$
\begin{aligned}
&\big\|\max\{f,g\}\big\|_{\mathrm{BCMO}_p(\Sigma,\sigma)}\\
&\quad\le \frac{1}{2}\big(\|f\|_{\mathrm{BCMO}_p(\Sigma,\sigma)} + \|g\|_{\mathrm{BCMO}_p(\Sigma,\sigma)} + \big\||f-g|\big\|_{\mathrm{BCMO}_p(\Sigma,\sigma)}\big)\\
&\quad\le \frac{1}{2}\big(\|f\|_{\mathrm{BCMO}_p(\Sigma,\sigma)} + \|g\|_{\mathrm{BCMO}_p(\Sigma,\sigma)} + 2\|f-g\|_{\mathrm{BCMO}_p(\Sigma,\sigma)}\big)\\
&\quad\le \frac{3}{2}\big(\|f\|_{\mathrm{BCMO}_p(\Sigma,\sigma)} + \|g\|_{\mathrm{BCMO}_p(\Sigma,\sigma)}\big),
\end{aligned}
\tag{5.21}
$$

and similarly

$$
\big\|\min\{f,g\}\big\|_{\mathrm{BCMO}_p(\Sigma,\sigma)} \le \frac{3}{2}\big(\|f\|_{\mathrm{BCMO}_p(\Sigma,\sigma)} + \|g\|_{\mathrm{BCMO}_p(\Sigma,\sigma)}\big). \tag{5.22}
$$

This concludes the proof of Proposition 5.2. □

The purpose of the following proposition is to establish that the membership to BCMO_p is independent of the reference point x_0 and the dilation factor 2_* appearing in the definition of the BCMO_p-seminorm (cf. (5.1)).

Proposition 5.3 *Assume $\Sigma \subseteq \mathbb{R}^n$ is a closed unbounded Ahlfors regular set, and abbreviate $\sigma := \mathcal{H}^{n-1}\lfloor\Sigma$. Fix two reference points $x_0, x_1 \in \Sigma$, along with two numbers $2_*, 2_\diamond \in (1,\infty)$. Finally, pick an integrability exponent $p \in (1,\infty)$. Then for every function $f \in L^1_{\mathrm{loc}}(\Sigma,\sigma)$ one has*

$$
\sup_{R\ge 2_*}\Big(⨍_{\Delta(x_0,R)} |f - f_{\Delta(x_0,R)}|^p\,d\sigma\Big)^{1/p} \approx \sup_{R\ge 2_\diamond}\Big(⨍_{\Delta(x_1,R)} |f - f_{\Delta(x_1,R)}|^p\,d\sigma\Big)^{1/p}, \tag{5.23}
$$

with implicit proportionality constants independent of the function f. As a consequence, the BCMO_p-seminorm in (5.1) is independent of the choices of x_0 and 2_*.

Proof Fix a finite radius $R \ge 2_*$ and select a dilation factor $\lambda \in (0,\infty)$ such that

$$
\lambda > \max\Big\{2, \frac{2}{2_*}|x_0 - x_1|, \frac{2_\diamond}{2_*}\Big\}. \tag{5.24}
$$

Define

$$
\widetilde{R} := \lambda R \in (0,\infty). \tag{5.25}
$$

Clearly $\widetilde{R} \ge 2_\diamond$. Notice that whenever $x \in \Delta(x_0,R)$ we have

$$
|x - x_1| \le |x - x_0| + |x_0 - x_1| < R + \frac{\lambda}{2}2_* \le \frac{\lambda}{2}R + \frac{\lambda}{2}R = \widetilde{R}, \tag{5.26}
$$

which implies

$$
\Delta(x_0,R) \subseteq \Delta(x_1,\widetilde{R}). \tag{5.27}
$$

Next, estimate

$$\begin{aligned}
&\Big(\fint_{\Delta(x_0,R)}|f-f_{\Delta(x_0,R)}|^p\,d\sigma\Big)^{1/p}\\
&\quad\leq\Big(\fint_{\Delta(x_0,R)}|f-f_{\Delta(x_1,\widetilde{R})}|^p\,d\sigma\Big)^{1/p}+|f_{\Delta(x_0,R)}-f_{\Delta(x_1,\widetilde{R})}|.
\end{aligned}\tag{5.28}$$

To deal with the first term in the right-hand side of (5.28), use (5.25), (5.27), and the Ahlfors regularity of Σ to obtain

$$\begin{aligned}
&\Big(\fint_{\Delta(x_0,R)}|f-f_{\Delta(x_1,\widetilde{R})}|^p\,d\sigma\Big)^{1/p}\\
&\quad\leq\frac{\sigma\big(\Delta(x_1,\widetilde{R})\big)^{1/p}}{\sigma\big(\Delta(x_0,R)\big)^{1/p}}\Big(\fint_{\Delta(x_1,\widetilde{R})}|f-f_{\Delta(x_1,\widetilde{R})}|^p\,d\sigma\Big)^{1/p}\\
&\quad\leq C\frac{\widetilde{R}^{(n-1)/p}}{R^{(n-1)/p}}\Big(\fint_{\Delta(x_1,\widetilde{R})}|f-f_{\Delta(x_1,\widetilde{R})}|^p\,d\sigma\Big)^{1/p}\\
&\quad=C\lambda^{(n-1)/p}\Big(\fint_{\Delta(x_1,\widetilde{R})}|f-f_{\Delta(x_1,\widetilde{R})}|^p\,d\sigma\Big)^{1/p}\\
&\quad\leq C\lambda^{(n-1)/p}\sup_{r\geq 2_\diamond}\Big(\fint_{\Delta(x_1,r)}|f-f_{\Delta(x_1,r)}|^p\,d\sigma\Big)^{1/p}
\end{aligned}\tag{5.29}$$

for some $C\in(0,\infty)$ depending only on n and the Ahlfors regularity constants of Σ. To estimate the second term in the right-hand side of (5.28) use (5.25), (5.27), the Ahlfors regularity of Σ, and Hölder's inequality to obtain

$$\begin{aligned}
|f_{\Delta(x_0,R)}-f_{\Delta(x_1,\widetilde{R})}|&=\Big|\fint_{\Delta(x_0,R)}(f-f_{\Delta(x_1,\widetilde{R})})\,d\sigma\Big|\\
&\leq\frac{\sigma\big(\Delta(x_1,\widetilde{R})\big)}{\sigma\big(\Delta(x_0,R)\big)}\fint_{\Delta(x_1,\widetilde{R})}|f-f_{\Delta(x_1,\widetilde{R})}|\,d\sigma\\
&\leq C\frac{\widetilde{R}^{n-1}}{R^{n-1}}\fint_{\Delta(x_1,\widetilde{R})}|f-f_{\Delta(x_1,\widetilde{R})}|\,d\sigma\\
&\leq C\lambda^{n-1}\Big(\fint_{\Delta(x_1,\widetilde{R})}|f-f_{\Delta(x_1,\widetilde{R})}|^p\,d\sigma\Big)^{1/p}\\
&\leq C\lambda^{(n-1)}\sup_{r\geq 2_\diamond}\Big(\fint_{\Delta(x_1,r)}|f-f_{\Delta(x_1,r)}|^p\,d\sigma\Big)^{1/p},
\end{aligned}\tag{5.30}$$

for some $C\in(0,\infty)$ depending only on n and the Ahlfors regularity constants of Σ.

Collectively, (5.28), (5.29), and (5.30) ensure the existence of some $C\in(0,\infty)$ depending only on n, p, x_0, x_1, 2_*, $2_\diamond$, and the Ahlfors regularity constants of Σ with the property that

$$\sup_{R\geq 2_*}\Big(⨍_{\Delta(x_0,R)}|f-f_{\Delta(x_0,R)}|^p\,d\sigma\Big)^{1/p}$$
$$\leq C\cdot\sup_{r\geq 2_\diamond}\Big(⨍_{\Delta(x_1,r)}|f-f_{\Delta(x_1,r)}|^p\,d\sigma\Big)^{1/p}, \tag{5.31}$$

The reverse inequality can be obtained via the same reasoning as before, by changing the roles of 2_* with $2_\diamond$, and x_0 with x_1. This completes the proof of (5.23). □

The proposition recorded below gives embedding results for the space of bounded central p-mean oscillations functions BCMO_p.

Proposition 5.4 *Assume $\Sigma\subseteq\mathbb{R}^n$ is a closed unbounded Ahlfors regular set, and abbreviate $\sigma:=\mathcal{H}^{n-1}\lfloor\Sigma$. Fix an integrability exponent $p\in(1,\infty)$. Then whenever $q\in(1,\infty)$ and $q>p$ it follows that*

$$\mathrm{BCMO}_q(\Sigma,\sigma)\hookrightarrow\mathrm{BCMO}_p(\Sigma,\sigma)\quad\textit{continuously,} \tag{5.32}$$

i.e., the scale of spaces of functions of bounded central p-mean oscillations $\mathrm{BCMO}_p(\Sigma,\sigma)$ is (topologically) nested with respect to p.

Moreover, upon recalling the space $\mathrm{B}^p(\Sigma,\sigma)$ (cf. (3.5)), one has

$$\mathrm{B}^p(\Sigma,\sigma)\hookrightarrow\mathrm{BCMO}_p(\Sigma,\sigma)\quad\textit{continuously.} \tag{5.33}$$

Regarding the classical John-Nirenberg space (cf. (2.34)), one has

$$\mathrm{BMO}(\Sigma,\sigma)\hookrightarrow\mathrm{BCMO}_p(\Sigma,\sigma)\quad\textit{continuously.} \tag{5.34}$$

In particular,

$$L^\infty(\Sigma,\sigma)\hookrightarrow\mathrm{BCMO}_p(\Sigma,\sigma)\quad\textit{continuously.} \tag{5.35}$$

Lastly, the following inclusion holds

$$\mathrm{BCMO}_p(\Sigma,\sigma)\subseteq L^1\Big(\Sigma,\frac{\sigma(x)}{1+|x|^{n-1+\varepsilon}}\Big)\cap L^p_{\mathrm{loc}}(\Sigma,\sigma)\quad\textit{for all }\varepsilon\in(0,\infty). \tag{5.36}$$

Proof Fix a reference point $x_0\in\Sigma$, and a dilation factor 2_* as in Definition 5.1. To verify the embedding claimed in (5.32) pick $q\in(1,\infty)$ with $q>p$. Select $f\in\mathrm{BCMO}_q(\Sigma,\sigma)$ along with a radius $R\geq 2_*$. Using Hölder's inequality, and recalling (5.1) we estimate

$$
\begin{aligned}
&\left(\fint_{\Delta(x_0,R)} |f - f_{\Delta(x_0,R)}|^p \, d\sigma\right)^{1/p} \\
&\leq \frac{\sigma\big(\Delta(x_0,R)\big)^{(1/p)(1-p/q)}}{\sigma\big(\Delta(x_0,R)\big)^{1/p}} \left(\int_{\Delta(x_0,R)} |f - f_{\Delta(x_0,R)}|^{p(q/p)} \, d\sigma\right)^{(1/p)(p/q)} \\
&= \left(\fint_{\Delta(x_0,R)} |f - f_{\Delta(x_0,R)}|^q \, d\sigma\right)^{1/q} \leq \|f\|_{\mathrm{BCMO}_q(\Sigma,\sigma)}.
\end{aligned}
\tag{5.37}
$$

The arbitrariness of $R \geq 2_*$ then entails

$$
\|f\|_{\mathrm{BCMO}_p(\Sigma,\sigma)} \leq \|f\|_{\mathrm{BCMO}_q(\Sigma,\sigma)} < \infty. \tag{5.38}
$$

This proves the embedding in (5.32).

Next, pick a function $f \in \mathrm{B}^p(\Sigma, \sigma)$. Thanks to Proposition 3.1, for every $R \geq 2_*$ we have

$$
\begin{aligned}
\left(\fint_{\Delta(x_0,R)} |f - f_{\Delta(x_0,R)}|^p \, d\sigma\right)^{1/p} &\leq 2\left(\fint_{\Delta(x_0,R)} |f|^p \, d\sigma\right)^{1/p} \\
&\leq C\|f\|_{\mathrm{B}^p(\Sigma,\sigma)}
\end{aligned}
\tag{5.39}
$$

for some constant $C \in (0, \infty)$ depending only on n, p, and the Ahlfors regularity constants of Σ. After taking the supremum over all $R \geq 2_*$ in (5.39), we obtain

$$
\|f\|_{\mathrm{BCMO}_p(\Sigma,\sigma)} \leq C\|f\|_{\mathrm{B}^p(\Sigma,\sigma)} < \infty \tag{5.40}
$$

thus proving (5.33).

To see that the embedding (5.34) holds, observe that (2.36) implies that for every function $f \in L^1_{\mathrm{loc}}(\Sigma, \sigma)$ we have

$$
\|f\|_{\mathrm{BCMO}_p(\Sigma,\sigma)} \leq \sup_{\Delta \subseteq \Sigma} \left(\fint_{\Delta} |f - f_\Delta|^p \, d\sigma\right)^{1/p} \approx \|f\|_{\mathrm{BMO}(\Sigma,\sigma)}, \tag{5.41}
$$

where the supremum is taken over all surface balls on Σ, and with implicit proportionality constants independent of f. As a consequence of this and the fact that $L^\infty(\Sigma, \sigma)$ embeds continuously into $\mathrm{BMO}(\Sigma, \sigma)$, it follows that (5.35) also holds.

There remains to show the inclusion in (5.36). To this end, fix an arbitrary function $f \in \mathrm{BCMO}_p(\Sigma, \sigma)$. Pick a surface ball $\Delta \subseteq \Sigma$, and select a sufficiently large radius $R \geq 2_*$ such that $\Delta \subseteq \Delta(x_0, R)$. Then

$$\begin{aligned}
&\Big(\int_{\Delta}|f|^p\,d\sigma\Big)^{1/p}\\
&\le\Bigg(\int_{\Delta(x_0,R)}|f-f_{\Delta(x_0,R)}|^p\,d\sigma\Big)^{1/p}+\sigma\big(\Delta(x_0,R)\big)^{1/p}|f_{\Delta(x_0,R)}|\\
&\le\sigma\big(\Delta(x_0,R)\big)^{1/p}\Big(⨍_{\Delta(x_0,R)}|f-f_{\Delta(x_0,R)}|^p\,d\sigma\Big)^{1/p}\\
&\quad+\sigma\big(\Delta(x_0,R)\big)^{1/p}⨍_{\Delta(x_0,R)}|f|\,d\sigma\\
&\le\sigma\big(\Delta(x_0,R)\big)^{1/p}\Big(\|f\|_{\mathrm{BCMO}_p(\Sigma,\sigma)}+⨍_{\Delta(x_0,R)}|f|\,d\sigma\Big)<\infty.
\end{aligned}\tag{5.42}$$

Hence $f\in L^p_{\rm loc}(\Sigma,\sigma)$, as desired.

Next, continue to assume that $f\in\mathrm{BCMO}_p(\Sigma,\sigma)$. Select $\varepsilon\in(0,\infty)$, along with a radius $R\ge 2_*$. Decompose f as

$$\begin{aligned}
f&=(f-f_{\Delta(x_0,R)})\mathbb{1}_{\Delta(x_0,R)}+(f-f_{\Delta(x_0,R)})\mathbb{1}_{\Sigma\setminus\Delta(x_0,R)}+f_{\Delta(x_0,R)}\\
&=:f_1+f_2+f_3.
\end{aligned}\tag{5.43}$$

Since $L^1\big(\Sigma,\frac{\sigma(x)}{1+|x|^{n-1+\varepsilon}}\big)$ contains constant functions, it is visible that

$$f_3\in L^1\Big(\Sigma,\frac{\sigma(x)}{1+|x|^{n-1+\varepsilon}}\Big).\tag{5.44}$$

Regarding the function f_1, use (5.1) and Hölder's inequality to estimate

$$\begin{aligned}
\int_\Sigma|f_1(x)|\frac{d\sigma(x)}{1+|x|^{n-1+\varepsilon}}&=\int_{\Delta(x_0,R)}|f-f_{\Delta(x_0,R)}|\frac{d\sigma(x)}{1+|x|^{n-1+\varepsilon}}\\
&\le\int_{\Delta(x_0,R)}|f-f_{\Delta(x_0,R)}|\,d\sigma\\
&\le\sigma\big(\Delta(x_0,R)\big)\Big(⨍_{\Delta(x_0,R)}|f-f_{\Delta(x_0,R)}|^p\,d\sigma\Big)^{1/p}\\
&\le\sigma\big(\Delta(x_0,R)\big)\|f\|_{\mathrm{BCMO}_p(\Sigma,\sigma)}<\infty.
\end{aligned}\tag{5.45}$$

As such

$$f_1\in L^1\Big(\Sigma,\frac{\sigma(x)}{1+|x|^{n-1+\varepsilon}}\Big).\tag{5.46}$$

To proceed, observe that $1+|x|^{n-1+\varepsilon}\approx 1+|x-x_0|^{n-1+\varepsilon}$ uniformly for $x\in\mathbb{R}^n$ (since these expressions are continuous with respect to x and have the same grow at infinity). Thanks to this and the Ahlfors regularity of Σ, we may compute

$$
\begin{aligned}
&\int_{\Sigma} |f_2(x)| \frac{d\sigma(x)}{1+|x|^{n-1+\varepsilon}} \\
&= \int_{\Sigma\setminus\Delta(x_0,R)} |f(x) - f_{\Delta(x_0,R)}| \frac{d\sigma(x)}{1+|x|^{n-1+\varepsilon}} \\
&\le C\sum_{j=0}^{\infty} \int_{\Delta(x_0,2^{j+1}R)\setminus\Delta(x_0,2^jR)} |f(x) - f_{\Delta(x_0,R)}| \frac{d\sigma(x)}{1+|x-x_0|^{n-1+\varepsilon}} \\
&\le C\sum_{j=0}^{\infty} \frac{1}{(2^jR)^{n-1+\varepsilon}} \int_{\Delta(x_0,2^{j+1}R)\setminus\Delta(x_0,2^jR)} |f - f_{\Delta(x_0,R)}|\, d\sigma \\
&\le C\sum_{j=0}^{\infty} \frac{1}{(2^jR)^{n-1+\varepsilon}} \int_{\Delta(x_0,2^{j+1}R)} |f - f_{\Delta(x_0,2^{j+1}R)}|\, d\sigma \\
&\quad + C\sum_{j=0}^{\infty} \frac{1}{(2^jR)^{n-1+\varepsilon}} \sigma\big(\Delta(x_0,2^{j+1}R)\big) |f_{\Delta(x_0,R)} - f_{\Delta(x_0,2^{j+1}R)}| \\
&\le C\sum_{j=0}^{\infty} \frac{1}{(2^jR)^{n-1+\varepsilon}} \int_{\Delta(x_0,2^{j+1}R)} |f - f_{\Delta(x_0,2^{j+1}R)}|\, d\sigma \\
&\quad + C\sum_{j=0}^{\infty} \frac{1}{(2^jR)^{\varepsilon}} |f_{\Delta(x_0,R)} - f_{\Delta(x_0,2^{j+1}R)}| \\
&=: \mathrm{I} + \mathrm{II},
\end{aligned} \tag{5.47}
$$

for some constant $C \in (0,\infty)$ depending on n, and Ahlfors regularity constants of Σ. Use the Ahlfors regularity of Σ, Hölder's inequality, and the fact that $R \ge 2_*$ to further estimate I as

$$
\begin{aligned}
\mathrm{I} &= C\sum_{j=0}^{\infty} \frac{1}{(2^jR)^{n-1+\varepsilon}} \int_{\Delta(x_0,2^{j+1}R)} |f - f_{\Delta(x_0,2^{j+1}R)}|\, d\sigma \\
&\le C\sum_{j=0}^{\infty} \frac{1}{(2^jR)^{\varepsilon}} ⨍_{\Delta(x_0,2^{j+1}R)} |f - f_{\Delta(x_0,2^{j+1}R)}|\, d\sigma \\
&\le C\sum_{j=0}^{\infty} \frac{1}{(2^jR)^{\varepsilon}} \Big(⨍_{\Delta(x_0,2^{j+1}R)} |f - f_{\Delta(x_0,2^{j+1}R)}|^p\, d\sigma \Big)^{1/p} \\
&\le \frac{C}{2_*^{\varepsilon}} \Big(\sum_{j=0}^{\infty} \frac{1}{2^{j\varepsilon}} \Big) \|f\|_{\mathrm{BCMO}_p(\Sigma,\sigma)} = C\|f\|_{\mathrm{BCMO}_p(\Sigma,\sigma)},
\end{aligned} \tag{5.48}
$$

for some constant $C \in (0,\infty)$ depending only on n, and the Ahlfors regularity character of Σ. To deal with II from (5.47), use a telescoping sum argument, the doubling character of σ, Hölder's inequality, and $R \geq 2_*$ to bound

$$
\begin{aligned}
\mathrm{II} &= C\sum_{j=0}^{\infty}\frac{1}{(2^jR)^{\varepsilon}}|f_{\Delta(x_0,R)} - f_{\Delta(x_0,2^{j+1}R)}| \\
&\leq \frac{C}{2_*^{\varepsilon}}\sum_{j=0}^{\infty}\frac{1}{2^{j\varepsilon}}\sum_{k=0}^{j}|f_{\Delta(x_0,2^kR)} - f_{\Delta(x_0,2^{k+1}R)}| \\
&\leq C\sum_{j=0}^{\infty}\frac{1}{2^{j\varepsilon}}\sum_{k=0}^{j}\fint_{\Delta(x_0,2^kR)}|f - f_{\Delta(x_0,2^{k+1}R)}|\,d\sigma \\
&\leq C\sum_{j=0}^{\infty}\frac{1}{2^{j\varepsilon}}\sum_{k=0}^{j}\fint_{\Delta(x_0,2^{k+1}R)}|f - f_{\Delta(x_0,2^{k+1}R)}|\,d\sigma \\
&\leq C\sum_{j=0}^{\infty}\frac{1}{2^{j\varepsilon}}\sum_{k=0}^{j}\Big(\fint_{\Delta(x_0,2^{k+1}R)}|f - f_{\Delta(x_0,2^{k+1}R)}|^p\,d\sigma\Big)^{1/p} \\
&\leq C\Big(\sum_{j=0}^{\infty}\frac{j}{2^{j\varepsilon}}\Big)\|f\|_{\mathrm{BCMO}_p(\Sigma,\sigma)} = C\|f\|_{\mathrm{BCMO}_p(\Sigma,\sigma)} < \infty,
\end{aligned}
\tag{5.49}
$$

for some constant $C \in (0,\infty)$ depending on n, and the Ahlfors regularity character of Σ. All together, (5.47)-(5.49) entail

$$
f_2 \in L^1\Big(\Sigma, \frac{\sigma(x)}{1+|x|^{n-1+\varepsilon}}\Big), \tag{5.50}
$$

which in concert with (5.43), (5.44), and (5.46) we finally obtain that f belongs to $L^1\big(\Sigma, \frac{\sigma(x)}{1+|x|^{n-1+\varepsilon}}\big)$. This establishes the inclusion (5.36), thus concluding the proof of Proposition 5.4. □

5.2 Duality between HA^p and BCMO_p

Recall the celebrated duality result by C. Fefferman and E. Stein [FeSt72] which establishes that the dual of the Hardy space H^1 is the John-Nirenberg space BMO. In the same spirit, in Theorem 5.1 we prove that in the context of Ahlfors regular sets the dual of the Beurling-Hardy space HA^p is the space $\mathrm{BCMO}_{p'}$, where $1/p + 1/p' = 1$. Establishing this requires some preparations.

To begin, the lemma recorded below is implied by [GHA-II, Lemma 4.6.9, p. 190].

Lemma 5.2 *Let $\Sigma \subseteq \mathbb{R}^n$ be a closed Ahlfors regular set, and abbreviate $\sigma := \mathcal{H}^{n-1}\lfloor\Sigma$. If a function $f \in L^1_{\rm loc}(\Sigma, \sigma)$ satisfies*

$$\int_\Sigma f a \, d\sigma = 0 \ \textit{for every } a \in L^\infty_{\rm comp}(\Sigma, \sigma) \textit{ with } \int_\Sigma a \, d\sigma = 0, \tag{5.51}$$

then necessarily f is a constant on Σ.

It is also useful to recall that on a finite measure space the following L^p-norm characterization holds.

Lemma 5.3 *Let $(X, \mathfrak{M}, \mu)$ be a finite measure space. Select an integrability exponent $p \in (1, \infty)$ and denote by $p' \in (1, \infty)$ its Hölder conjugate exponent. Then for every $f \in L^p(X, \mu)$ with $\int_X f \, d\mu = 0$ one has*

$$\begin{aligned} &\|f\|_{L^p(X,\mu)} \\ &\quad \approx \sup\left\{\left|\int_X fg \, d\mu\right| : g \in L^{p'}(X, \mu), \int_X g \, d\mu = 0, \ \|g\|_{L^{p'}(X,\mu)} \le 1\right\}. \end{aligned} \tag{5.52}$$

Proof Thanks to Hölder's inequality it is clear that the supremum in the right-hand side of (5.52) is less than or equal to $\|f\|_{L^p(X,\mu)}$. To prove the reverse inequality, fix some $\varepsilon \in (0, \infty)$. Then Riesz Representation Theorem implies that there exists $h \in L^{p'}(X, \mu)$ with $\|h\|_{L^{p'}(X,\mu)} \le 1$ satisfying

$$\|f\|_{L^p(X,\mu)} - \varepsilon \le \left|\int_X fh \, d\mu\right|. \tag{5.53}$$

Since (X, μ) is a finite measure space, it follows that h is absolutely integrable in X. As such

$$h_X := \fint_X h \, d\mu \ \text{ is a well-defined number} \tag{5.54}$$

and the function $h - h_X$ has integral zero. Using the fact that $\|h\|_{L^{p'}(X,\mu)} \le 1$, and Hölder's inequality we have

$$\begin{aligned} \|h - h_X\|_{L^{p'}(X,\mu)} &\le \|h\|_{L^{p'}(X,\mu)} + |h_X| \mu(X)^{1/p'} \\ &\le \|h\|_{L^{p'}(X,\mu)} + \mu(X)^{1/p'} \fint_X |h| \, d\mu \le 2\|h\|_{L^{p'}(X,\mu)} \le 2. \end{aligned} \tag{5.55}$$

Setting $\widetilde{h} := \frac{1}{2}(h - h_X)$, the above considerations yield

$$\widetilde{h} \in L^{p'}(X, \mu), \quad \int_X \widetilde{h} \, d\mu = 0, \quad \text{and} \quad \|\widetilde{h}\|_{L^{p'}(X,\mu)} \le 1. \tag{5.56}$$

By (5.53) and the vanishing moment of f, we obtain

$$\|f\|_{L^p(X,\mu)} - \varepsilon \leq \Big|\int_X fh\,d\mu\Big| = \Big|\int_X f(h-h_X)\,d\mu\Big| = 2\Big|\int_X f\widetilde{h}\,d\mu\Big|$$

$$\leq 2\sup\Big\{\Big|\int_X fg\,d\mu\Big| : g \in L^{p'}(X,\mu),\ \int_X g\,d\mu = 0,\ \|g\|_{L^{p'}(X,\mu)} \leq 1\Big\}. \quad (5.57)$$

Since this holds for every $\varepsilon \in (0,\infty)$, we finally obtain the right-pointing inequality of (5.52). □

Another useful abstract measure theoretic lemma is recorded below.

Lemma 5.4 *Consider $(X,\mathfrak{M},\mu)$ a finite measure space. Fix an integrability exponent $p \in (1,\infty)$, and let $p' \in (1,\infty)$ be its Hölder conjugate exponent. Also, denote by $L_0^{p'}(X,\mu)$ the collection of all functions $f \in L^{p'}(X,\mu)$ satisfying $\int_X f\,d\mu = 0$. Select a function $g \in L^p(X,\mu)$ and assume that*

$$\int_X fg\,d\mu = 0 \quad \text{for all } f \in L_0^{p'}(X,\mu). \quad (5.58)$$

Then g is a constant μ-a.e. on X.

Proof If $\mu(X) = 0$, then there is nothing to prove, since in this case every function in $L^p(X,\mu)$ is zero μ-a.e on X. Assume $0 < \mu(X) < \infty$, and fix an arbitrary function $f \in L^{p'}(X,\mu)$. We claim that

$$f - ⨍_X f\,d\mu \quad \text{belongs to } L_0^{p'}(X,\mu). \quad (5.59)$$

To verify this, first observe that

$$\int_X \Big(f - ⨍_X f\,d\mu\Big)\,d\mu = \int_X f\,d\mu - \Big(⨍_X f\,d\mu\Big)\mu(X) = 0. \quad (5.60)$$

Second, use Hölder's inequality to estimate

$$\begin{aligned}\Big(\int_X \Big|f - ⨍_X f\,d\mu\Big|^p d\mu\Big)^{1/p} &\leq \Big(\int_X |f|^p\,d\mu\Big)^{1/p} + \Big(⨍_X |f|\,d\mu\Big)\mu(X)^{1/p} \\ &\leq \Big(\int_X |f|^p\,d\mu\Big)^{1/p} + \Big(⨍_X |f|^p\,d\mu\Big)^{1/p}\mu(X)^{1/p} \\ &= 2\|f\|_{L^p(X,\mu)} < \infty. \end{aligned} \quad (5.61)$$

As such, (5.60) and (5.61) proves (5.59).

In light of (5.59) and (5.58) we may now compute

$$\begin{aligned}\int_X f\Big(g - \fint_X g\,d\mu\Big)\,d\mu &= \int_X fg\,d\mu - \int_X f\Big(\fint_X g\,d\mu\Big)\,d\mu \\ &= \int_X fg\,d\mu - \Big(\fint_X f\,d\mu\Big)\Big(\int_X g\,d\mu\Big) \\ &= \int_X \Big(f - \fint_X f\,d\mu\Big) g\,d\mu = 0.\end{aligned} \tag{5.62}$$

Since $f \in L^{p'}(X, \mu)$ was arbitrary to begin with, Riesz Representation Theorem yields

$$\Big\| g - \fint_X g \Big\|_{L^p(X,\mu)} = 0 \tag{5.63}$$

hence $g = \fint_X g\,d\mu$ at μ-a.e. point on X. This concludes the proof of Lemma 5.4. □

We are now prepared to state and prove the main result of this section, namely the duality result between HA^p and $\mathrm{BCMO}^{p'}$.

Theorem 5.1 *Consider an unbounded closed set $\Sigma \subseteq \mathbb{R}^n$ which is Ahlfors regular, and abbreviate $\sigma := \mathcal{H}^{n-1}\lfloor\Sigma$. Having fixed an integrability exponent $p \in (1, \infty)$, denote by $p' \in (1, \infty)$ its Hölder conjugate exponent. Then the dual of $\mathrm{HA}^p(\Sigma, \sigma)$ is $\widetilde{\mathrm{BCMO}}_{p'}(\Sigma, \sigma)$, i.e.,*

$$\big(\mathrm{HA}^p(\Sigma, \sigma)\big)^* = \widetilde{\mathrm{BCMO}}_{p'}(\Sigma, \sigma). \tag{5.64}$$

The nature of this identification is understood as follows. Fix a reference point $x_0 \in \Sigma$ and consider the assignment

$$\begin{aligned}&\widetilde{\mathrm{BCMO}}_{p'}(\Sigma, \sigma) \ni [f] \mapsto \Phi f := \Lambda_{[f]} \in \big(\mathrm{HA}^p(\Sigma, \sigma)\big)^* \\ &\textit{defined as} \quad \langle \Lambda_{[f]}, g\rangle := \lim_{N\to\infty} \sum_{j=0}^{N} \lambda_j \int_\Sigma f a_j\,d\sigma\end{aligned} \tag{5.65}$$

whenever

$g \in \mathrm{HA}^p(\Sigma, \sigma)$, $\{\lambda_j\}_{j\in\mathbb{N}_0} \in \ell^1(\mathbb{N}_0)$, and $\{a_j\}_{j\in\mathbb{N}_0}$ is a sequence of x_0-central $(1, p)$-atoms with the property that $g = \sum_{j=0}^{\infty} \lambda_j a_j$ in $\mathrm{HA}^p(\Sigma, \sigma)$. (5.66)

Then $\Lambda_{[f]}$ is meaningfully defined and the assignment (5.65)-(5.66) *is a well-defined bounded linear isomorphism.*

Proof To start, pick a reference point $x_0 \in \Sigma$ and a dilation factor 2_* as in Definition 5.1. Fix an arbitrary function $f \in \mathrm{BCMO}_{p'}(\Sigma, \sigma)$. Select an x_0-central $(1, p)$-atom $a : \Sigma \to \mathbb{R}$, and denote by $R \geq 2_*$ the radius associated with a in the sense of Definition 4.3. Since $a \in L^p_{\mathrm{comp}}(\Sigma, \sigma)$ and $f \in L^{p'}_{\mathrm{loc}}(\Sigma, \sigma)$ (cf. Proposition 5.4), Hölder's inequality implies that $fa \in L^1(\Sigma, \sigma)$. In particular, the integral pairing between f and a is meaningfully defined. Moreover, using the vanishing moment and normalization properties of a (cf. (4.29) and (4.30)), Hölder's inequality, and the Ahlfors regularity of Σ we have

$$\begin{aligned}\Big|\int_\Sigma f a\,d\sigma\Big| &= \Big|\int_\Sigma (f - f_{\Delta(x_0,R)}) a\,d\sigma\Big| \\ &\leq \Big(\int_{\Delta(x_0,R)} |f - f_{\Delta(x_0,R)}|^{p'}\,d\sigma\Big)^{1/p'} \|a\|_{L^p(\Sigma,\sigma)} \\ &\leq R^{-(n-1)/p'}\Big(\int_{\Delta(x_0,R)} |f - f_{\Delta(x_0,R)}|^{p'}\,d\sigma\Big)^{1/p'} \\ &\leq C\Big(\fint_{\Delta(x_0,R)} |f(x) - f_{\Delta(x_0,R)}|^{p'}\,d\sigma(x)\Big)^{1/p'} \\ &\leq C\|f\|_{\mathrm{BCMO}_{p'}(\Sigma,\sigma)}, \end{aligned} \tag{5.67}$$

for some constant $C \in (0,\infty)$ depending only on n, p, and the Ahlfors regularity of Σ.

To show that the functional $\Lambda_{[f]}$ is meaningfully defined, we need to verify that for any $g \in \mathrm{HA}^p(\Sigma,\sigma)$ as in (5.66) the limit in (5.65) exists and is independent of the atomic decomposition of g. To this end, select one such function $g \in \mathrm{HA}^p(\Sigma,\sigma)$. Given $M, N \in \mathbb{N}$ with $M > N$, use (5.67) to estimate

$$\begin{aligned}\Big|\sum_{j=0}^{M} \lambda_j \int_\Sigma f a_j\,d\sigma - \sum_{j=0}^{N} \lambda_j \int_\Sigma f a_j\,d\sigma\Big| &= \Big|\sum_{j=N+1}^{M} \lambda_j \int_\Sigma f a_j\,d\sigma\Big| \\ &\leq \sum_{j=N+1}^{M} |\lambda_j| \Big|\int_\Sigma f a_j\,d\sigma\Big| \\ &\leq C\Big(\sum_{j=N+1}^{\infty} |\lambda_j|\Big)\|f\|_{\mathrm{BCMO}_{p'}(\Sigma,\sigma)}. \end{aligned} \tag{5.68}$$

Since $\{\lambda_j\}_{j\in\mathbb{N}_0} \in \ell^1(\mathbb{N}_0)$, we see that the right-hand side of (5.68) approaches zero as $N \to \infty$. Hence, the sequence $\Big\{\sum_{j=0}^{N} \lambda_j \int_\Sigma f a_j\,d\sigma\Big\}_{N\in\mathbb{N}_0}$ is Cauchy in $\mathbb{C}$, which in turn implies that the limit in (5.65) exists.

Now, select $g \in \mathrm{HA}^p_{\mathrm{comp}}(\Sigma,\sigma)$, i.e., g has compact support and belongs to $\mathrm{HA}^p(\Sigma,\sigma)$ (cf. (4.156)). Thanks to Theorem 4.3, there exist some finite sequence of x_0-central $(1,p)$-atoms $\{a_j\}_{0\leq j\leq N}$, and a finite sequence $\{\lambda_j\}_{0\leq j\leq N} \subseteq \mathbb{R}$, for some $N \in \mathbb{N}$, such that

$$g = \sum_{j=0}^{N} \lambda_j a_j \quad \text{pointwise on } \Sigma \text{ and} \quad \sum_{j=0}^{N} |\lambda_j| \leq C\|g\|_{\mathrm{HA}^p(\Sigma,\sigma)} \tag{5.69}$$

for some constant $C \in (0, \infty)$ independent of g. Based on these and (5.67) we may estimate

$$\begin{aligned} \Big| \int_\Sigma f g \, d\sigma \Big| &\leq \sum_{j=0}^{N} |\lambda_j| \Big| \int_\Sigma f a_j \, d\sigma \Big| \\ &\leq C \|f\|_{\mathrm{BCMO}_{p'}(\Sigma,\sigma)} \sum_{j=0}^{N} |\lambda_j| \\ &\leq C \|f\|_{\mathrm{BCMO}_{p'}(\Sigma,\sigma)} \|g\|_{\mathrm{HA}^p(\Sigma,\sigma)}, \end{aligned} \tag{5.70}$$

for some constant $C \in (0, \infty)$ depending only on n, p, and the Ahlfors regularity of Σ. This shows that the mapping

$$\begin{gathered} \mathrm{HA}^p_{\mathrm{comp}}(\Sigma, \sigma) \ni g \longmapsto \Lambda_f(g) := \int_\Sigma f g \, d\sigma \in \mathbb{C} \\ \text{is a well-defined, linear, and bounded functional,} \\ \text{with norm } \|\Lambda_f\|_{(\mathrm{HA}^p_{\mathrm{comp}}(\Sigma,\sigma))^*} \leq C \|f\|_{\mathrm{BCMO}_{p'}(\Sigma,\sigma)}. \end{gathered} \tag{5.71}$$

Since $\mathrm{HA}^p_{\mathrm{comp}}(\Sigma, \sigma)$ is a dense subspace of $\mathrm{HA}^p(\Sigma, \sigma)$ (cf. Remark 4.5), standard functional analysis ensures that

$$\begin{gathered} \Lambda_f \text{ extends uniquely (by density) to a continuous linear functional} \\ \widetilde{\Lambda}_f \in \big(\mathrm{HA}^p(\Sigma, \sigma)\big)^* \text{ with norm } \|\widetilde{\Lambda}_f\|_{(\mathrm{HA}^p(\Sigma,\sigma))^*} \leq C \|f\|_{\mathrm{BCMO}_{p'}(\Sigma,\sigma)}. \end{gathered} \tag{5.72}$$

Let us now analyze the manner in which $\widetilde{\Lambda}_f$ acts on arbitrary functions from $\mathrm{HA}^p(\Sigma, \sigma)$. Pick a function $g \in \mathrm{HA}^p(\Sigma, \sigma)$ as in (5.66). Then the continuity and linearity of $\widetilde{\Lambda}_f$ on $\mathrm{HA}^p(\Sigma, \sigma)$ (see (5.72)) allows us to compute

$$\begin{aligned} \widetilde{\Lambda}_f(g) &= \widetilde{\Lambda}_f\Big(\lim_{N\to\infty} \sum_{j=0}^{N} \lambda_j a_j \Big) = \lim_{N\to\infty} \widetilde{\Lambda}_f\Big(\sum_{j=0}^{N} \lambda_j a_j \Big) = \lim_{N\to\infty} \sum_{j=0}^{N} \lambda_j \widetilde{\Lambda}_f(a_j) \\ &= \lim_{N\to\infty} \sum_{j=0}^{N} \lambda_j \Lambda_f(a_j) = \lim_{N\to\infty} \sum_{j=0}^{N} \lambda_j \int_\Sigma f a_j \, d\sigma. \end{aligned} \tag{5.73}$$

In turn, from this we see that $\Lambda_{[f]}$ defined in (5.65) is actually the functional $\widetilde{\Lambda}_f$ introduced in (5.72), i.e.,

$$\Lambda_{[f]} = \widetilde{\Lambda}_f \quad \text{as functionals in } \big(\mathrm{HA}^p(\Sigma, \sigma)\big)^*. \tag{5.74}$$

This ensures that $\Lambda_{[f]}$ is meaningfully defined.

Pressing on, we show that the mapping Φ defined in (5.65)-(5.66) is surjective. Establishing this requires some preparation. For each $k \in \mathbb{N}$, abbreviate

$$L^p_0\big(\Delta(x_0,2^k_*)\big) := \Big\{f \in L^p\big(\Delta(x_0,2^k_*),\sigma\big) : \int_{\Delta(x_0,2^k_*)} f\,d\sigma = 0\Big\} \tag{5.75}$$

and equip this with the norm $\|\cdot\|_{L^p(\Delta(x_0,2^k_*),\sigma)}$. For every $f \in L^p_0\big(\Delta(x_0,2^k_*)\big)$ define $E_k f$ at each point $x \in \Sigma$ as

$$E_k f(x) := \begin{cases} f(x) & \text{if } x \in \Delta(x_0,2^k_*), \\ 0 & \text{if } x \in \Sigma\backslash\Delta(x_0,2^k_*). \end{cases} \tag{5.76}$$

We claim that for each $k \in \mathbb{N}$ and every $f \in L^p_0\big(\Delta(x_0,2^k_*)\big)$ the function $E_k f$ is a constant multiple of an x_0-central $(1,p)$-atom. Indeed, assume $f \in L^p_0\big(\Delta(x_0,2^k_*)\big)$ is not identically zero on $\Delta(x_0,2^k_*)$, otherwise the conclusion is trivially satisfied. Then bearing in mind (5.75)-(5.76) the function

$$a := \alpha_k \cdot E_k f \quad \text{on } \Sigma \text{ where } \quad \alpha_k := \frac{2_*^{-k(n-1)/p'}}{\|f\|_{L^p(\Delta(x_0,2^k_*),\sigma)}} \tag{5.77}$$

satisfies

$$\begin{aligned} &\operatorname{supp} a \subseteq \Delta(x_0,2^k_*), \\ &\|a\|_{L^p(\Sigma,\sigma)} = \alpha_k\|E_k f\|_{L^p(\Sigma,\sigma)} = \alpha_k\|f\|_{L^p(\Delta(x_0,2^k_*),\sigma)} = 2_*^{-k(n-1)/p'}, \\ &\int_\Sigma a\,d\sigma = \alpha_k \int_{\Delta(x_0,2^k_*)} f\,d\sigma = 0. \end{aligned} \tag{5.78}$$

Hence, a is an x_0-central $(1,p)$-atom on Σ. Moreover, in light of Theorem 4.1 there exists some constant $C \in (0,\infty)$ with the property that

$$\begin{aligned} \|E_k f\|_{\mathrm{HA}^p(\Sigma,\sigma)} &= 2_*^{k(n-1)/p'}\|f\|_{L^p(\Delta(x_0,2^k_*),\sigma)}\|a\|_{\mathrm{HA}^p(\Sigma,\sigma)} \\ &\leq C 2_*^{k(n-1)/p'}\|f\|_{L^p(\Delta(x_0,2^k_*),\sigma)} \end{aligned} \tag{5.79}$$

for some constant $C \in (0,\infty)$ depending only on n, p, and the Ahlfors regularity constants of Σ. In particular, (5.79) yields

$$\begin{aligned} &E_k : L^p_0\big(\Delta(x_0,2^k_*)\big) \to \mathrm{HA}^p(\Sigma,\sigma) \text{ is a bounded linear operator with} \\ &\text{norm } \|E_k\|_{L^p_0(\Delta(x_0,2^k_*))\to \mathrm{HA}^p(\Sigma,\sigma)} \leq C 2_*^{k(n-1)/p'}. \end{aligned} \tag{5.80}$$

To proceed, fix an arbitrary linear functional $\Lambda \in \big(\mathrm{HA}^p(\Sigma,\sigma)\big)^*$. Thanks to (5.80), for each $k \in \mathbb{N}$ the mapping

$$\begin{aligned} &\Lambda \circ E_k : L^p_0\big(\Delta(x_0,2^k_*)\big) \to \mathbb{C} \text{ is well defined, linear, and bounded, with} \\ &\text{norm } \|\Lambda \circ E_k\|_{(L^p_0(\Delta(x_0,2^k_*)))^*} \leq C 2_*^{k(n-1)/p'}\|\Lambda\|_{(\mathrm{HA}^p(\Sigma,\sigma))^*} \end{aligned} \tag{5.81}$$

Since $L_0^p(\Delta(x_0, 2_*^k))$ is a linear subspace of $L^p(\Delta(x_0, 2_*^k), \sigma)$, the Hahn-Banach Theorem ensures that

$$\begin{aligned}&\Lambda \circ E_k \text{ has a continuous linear extension } \widetilde{\Lambda}_k \in \big(L^p(\Delta(x_0, 2_*^k), \sigma)\big)^* \\ &\quad \text{with norm } \|\widetilde{\Lambda}_k\|_{(L^p(\Delta(x_0,2_*^k),\sigma))^*} \leq C 2_*^{k(n-1)/p'} \|\Lambda\|_{(\mathrm{HA}^p(\Sigma,\sigma))^*}.\end{aligned} \tag{5.82}$$

By the Riesz Representation Theorem, for every $k \in \mathbb{N}$ there exists a unique function $f_k \in L^{p'}(\Delta(x_0, 2_*^k), \sigma)$ with the property that

$$\begin{aligned}\widetilde{\Lambda}_k(g) &= \int_{\Delta(x_0,2_*^k)} f_k g \, d\sigma \quad \text{for all } g \in L^p(\Delta(x_0, 2_*^k), \sigma) \\ \text{and} \quad \|f_k\|_{L^{p'}(\Delta(x_0,2_*^k),\sigma)} &= \|\widetilde{\Lambda}_k\|_{(L^p(\Delta(x_0,2_*^k),\sigma))^*}.\end{aligned} \tag{5.83}$$

In particular, whenever $k \in \mathbb{N}$ and $g \in L_0^p(\Delta(x_0, 2_*^k))$ we have

$$(\Lambda \circ E_k)(g) = \widetilde{\Lambda}_k(g) = \int_{\Delta(x_0,2_*^k)} f_k g \, d\sigma. \tag{5.84}$$

Next, we show that the functions f_{k+1} and f_k are compatible (modulo constants) on $\Delta(x_0, 2_*^k)$. More specifically, for each $k \in \mathbb{N}$ there exists a constant $c_k \in \mathbb{C}$ with the property that

$$f_{k+1}\big|_{\Delta(x_0,2_*^k)} = f_k + c_k \quad \text{at } \sigma\text{-a.e. point on } \Sigma. \tag{5.85}$$

Indeed, having fixed a function $g \in L_0^p(\Delta(x_0, 2_*^k))$, denote by $\widetilde{g}$ its extension by zero to $\Delta(x_0, 2_*^{k+1})$. Then $\widetilde{g} \in L_0^p(\Delta(x_0, 2_*^{k+1}))$. From this, (5.76), (5.82), and (5.84) we obtain the following implications:

$$\begin{aligned}E_{k+1}\widetilde{g} = E_k g \quad &\Longrightarrow \quad (\Lambda \circ E_{k+1})\widetilde{g} = (\Lambda \circ E_k) g \\ &\Longrightarrow \quad \int_{\Delta(x_0,2_*^{k+1})} f_{k+1}\widetilde{g} \, d\sigma = \int_{\Delta(x_0,2_*^k)} f_k g \, d\sigma \\ &\Longrightarrow \quad \int_{\Delta(x_0,2_*^k)} \big(f_{k+1}\big|_{\Delta(x_0,2_*^k)}\big) g \, d\sigma = \int_{\Delta(x_0,2_*^k)} f_k g \, d\sigma \\ &\Longrightarrow \quad \int_{\Delta(x_0,2_*^k)} \big(f_{k+1}\big|_{\Delta(x_0,2_*^k)} - f_k\big) g \, d\sigma = 0.\end{aligned} \tag{5.86}$$

Since this holds for every $g \in L_0^p(\Delta(x_0, 2_*^k))$, Lemma 5.4 implies that for each $k \in \mathbb{N}$ there exists $c_k \in \mathbb{C}$ for which $f_{k+1}\big|_{\Delta(x_0,2_*^k)} - f_k = c_k$ at σ-a.e. point on Σ. This justifies the validity of (5.85).

Pressing on, we inductively define

$$\widehat{f_1} := f_1 \quad \text{on } \Delta(x_0, 2_*), \text{ and}$$
$$\widehat{f_{k+1}} := f_{k+1} - \sum_{j=1}^{k} c_j \quad \text{on } \Delta(x_0, 2_*^{k+1}) \text{ for all } k \in \mathbb{N}. \tag{5.87}$$

In particular, since f_k belongs to $L^{p'}\big(\Delta(x_0, 2_*^k), \sigma\big)$ for each $k \in \mathbb{N}$ (cf. (5.83)), the function $\widehat{f_k}$ is σ-measurable on $\Delta(x_0, 2_*^k)$ and

$$\widehat{f_k} \in L^{p'}\big(\Delta(x_0, 2_*^k), \sigma\big) \quad \text{for all } k \in \mathbb{N}. \tag{5.88}$$

Observe that (5.85) along with (5.87) imply the following compatibility properties

$$\widehat{f_1}\big|_{\Delta(x_0, 2_*)} = f_1 \tag{5.89}$$

and

$$\begin{aligned} \widehat{f_{k+1}}\big|_{\Delta(x_0, 2_*^k)} &= f_{k+1}\big|_{\Delta(x_0, 2_*^k)} - \sum_{j=1}^{k} c_j \\ &= f_k + c_k - \sum_{j=1}^{k} c_j = f_k - \sum_{j=1}^{k-1} c_j = \widehat{f_k} \end{aligned} \tag{5.90}$$

for each $k \in \mathbb{N}$ with $k \geq 2$. Now, consider the function $f : \Sigma \to \mathbb{R}$ defined at each $x \in \Sigma$ by

$$f(x) := \widehat{f_k}(x) \text{ if } x \in \Delta(x_0, 2_*^k) \text{ for some } k \in \mathbb{N}. \tag{5.91}$$

We claim that f is well defined, σ-measurable, and locally integrable on Σ. First, thanks to (5.89)-(5.90) and the fact that $\Sigma = \cup_{k \in \mathbb{N}} \Delta(x_0, 2_*^k)$, it follows that f is well defined. Second, to see that f is σ-measurable, use that $\widehat{f_k}$ is σ-measurable for all $k \in \mathbb{N}_0$, and that whenever $O \subseteq \mathbb{R}$ is an open set we have

$$f^{-1}(O) = \bigcup_{k \in \mathbb{N}} \Delta(x_0, 2_*^k) \cap \widehat{f_k}^{-1}(O). \tag{5.92}$$

Third, to verify that $f \in L^1_{\mathrm{loc}}(\Sigma, \sigma)$, we rely on (5.88), (5.91), along with

$$\int_{\Delta(x_0, 2_*^k)} |f|^{p'} \, d\sigma = \int_{\Delta(x_0, 2_*^k)} |\widehat{f_k}|^{p'} \, d\sigma < \infty \quad \text{for all } k \in \mathbb{N}, \tag{5.93}$$

to arrive at the conclusion that

$$f \in L^{p'}_{\mathrm{loc}}(\Sigma, \sigma) \subseteq L^1_{\mathrm{loc}}(\Sigma, \sigma). \tag{5.94}$$

In addition to these properties, we claim that f belongs to $\mathrm{BCMO}_{p'}(\Sigma, \sigma)$. To verify this claim, fix a finite radius $R \geq 2_*$. Then (5.94) guarantees that

$$
\begin{aligned}
&(f - f_{\Delta(x_0,R)})\big|_{\Delta(x_0,R)} \in L^{p'}\big(\Delta(x_0,R),\sigma\big)\\
&\text{and} \quad \int_{\Delta(x_0,R)} (f - f_{\Delta(x_0,R)})\big|_{\Delta(x_0,R)}\, d\sigma = 0.
\end{aligned}
\tag{5.95}
$$

Fix a function $g \in L^p\big(\Delta(x_0,R),\sigma\big)$ satisfying the following properties:

$$
\int_{\Delta(x_0,R)} g\, d\sigma = 0 \ \text{ and } \ \|g\|_{L^p(\Delta(x_0,R),\sigma)} \leq 1. \tag{5.96}
$$

Choose $k \in \mathbb{N}$ satisfying

$$
2_*^k \leq R < 2_*^{k+1}. \tag{5.97}
$$

If we let $\widetilde{g}$ be the extension by zero of g to $\Delta(x_0, 2_*^{k+1})$, then

$$
\widetilde{g} \in L^p\big(\Delta(x_0,2_*^{k+1}),\sigma\big), \quad \int_{\Delta(x_0,2_*^{k+1})} \widetilde{g}\, d\sigma = 0, \quad \|\widetilde{g}\|_{L^p(\Delta(x_0,2_*^{k+1}),\sigma)} \leq 1. \tag{5.98}
$$

From this, (5.81), (5.84), (5.87), (5.91), (5.97), the vanishing moment of g, and the Ahlfors regularity of Σ we see that

$$
\begin{aligned}
\Big|\int_{\Delta(x_0,R)} &(f - f_{\Delta(x_0,R)})\big|_{\Delta(x_0,R)} g\, d\sigma\Big| = \Big|\int_{\Delta(x_0,R)} f\big|_{\Delta(x_0,R)} g\, d\sigma\Big|\\
&= \Big|\int_{\Delta(x_0,2_*^{k+1})} f\big|_{\Delta(x_0,2_*^{k+1})} \widetilde{g}\, d\sigma\Big|\\
&= \Big|\int_{\Delta(x_0,2_*^{k+1})} \widehat{f}_{k+1} \widetilde{g}\, d\sigma\Big|\\
&= \Big|\int_{\Delta(x_0,2_*^{k+1})} \Big(f_{k+1} - \sum_{j=1}^{k} c_j\Big) \widetilde{g}\, d\sigma\Big|\\
&= \Big|\int_{\Delta(x_0,2_*^{k+1})} f_{k+1} \widetilde{g}\, d\sigma\Big|\\
&= |(\Lambda \circ E_{k+1})\widetilde{g}|\\
&\leq C 2_*^{k(n-1)/p'} \|\Lambda\|_{(\mathrm{HA}^p(\Sigma,\sigma))^*} \|\widetilde{g}\|_{L^p(\Delta(x_0,2_*^{k+1}),\sigma)}\\
&\leq C R^{(n-1)/p'} \|\Lambda\|_{(\mathrm{HA}^p(\Sigma,\sigma))^*}\\
&\leq C \sigma\big(\Delta(x_0,R)\big)^{1/p'} \|\Lambda\|_{(\mathrm{HA}^p(\Sigma,\sigma))^*}
\end{aligned}
\tag{5.99}
$$

for some constant $C \in (0,\infty)$ depending only on n, p, and the Ahlfors regularity of Σ. Thanks to Lemma 5.3, and (5.95), if we take the supremum over all such functions g in the left-hand side of (5.99) we find that

$$\Big(⨍_{\Delta(x_0,R)} |f - f_{\Delta(x_0,R)}|^{p'}\,d\sigma\Big)^{1/p'} \leq C\|\Lambda\|_{(\mathrm{HA}^p(\Sigma,\sigma))^*}. \tag{5.100}$$

Since this holds for every $R \geq 2_*$, we finally obtain

$$\|f\|_{\mathrm{BCMO}_{p'}(\Sigma,\sigma)} \leq C\|\Lambda\|_{(\mathrm{HA}^p(\Sigma,\sigma))^*} < \infty, \tag{5.101}$$

thus showing that $f \in \mathrm{BCMO}_{p'}(\Sigma,\sigma)$.

Having established this membership, we proceed by showing that the action of Λ on functions from $\mathrm{HA}^p(\Sigma,\sigma)_{\mathrm{comp}}$ is given via integration against f. With this goal in mind, pick an arbitrary function $g \in \mathrm{HA}^p_{\mathrm{comp}}(\Sigma,\sigma)$, and select $k \in \mathbb{N}$ such that $\operatorname{supp} g \subseteq \Delta(x_0, 2_*^k)$. Thanks to Corollary 4.2 we have

$$g\big|_{\Delta(x_0,2_*^k)} \in L^p\big(\Delta(x_0,2_*^k),\sigma\big) \quad \text{and} \quad \int_{\Delta(x_0,2_*^k)} g\big|_{\Delta(x_0,2_*^k)}\,d\sigma = \int_\Sigma g\,d\sigma = 0. \tag{5.102}$$

Bearing this in mind, along with (5.76), (5.84), (5.87), and (5.91), we compute

$$\begin{aligned}\Lambda(g) &= \Lambda\big(E_k g\big|_{\Delta(x_0,2_*^k)}\big) = (\Lambda\circ E_k)\big(g\big|_{\Delta(x_0,2_*^k)}\big) = \widetilde{\Lambda}_k\big(g\big|_{\Delta(x_0,2_*^k)}\big)\\ &= \int_{\Delta(x_0,2_*^k)} f_k\cdot g\big|_{\Delta(x_0,2_*^k)}\,d\sigma = \int_{\Delta(x_0,2_*^k)} \widehat{f}_k\cdot g\big|_{\Delta(x_0,2_*^k)}\,d\sigma \\ &= \int_\Sigma fg\,d\sigma.\end{aligned} \tag{5.103}$$

Hence,

$$\Lambda(g) = \int_\Sigma fg\,d\sigma \quad \text{for all } g \in \mathrm{HA}^p_{\mathrm{comp}}(\Sigma,\sigma). \tag{5.104}$$

From this, the density of $\mathrm{HA}^p_{\mathrm{comp}}(\Sigma,\sigma)$ in $\mathrm{HA}^p(\Sigma,\sigma)$ (cf. Remark 4.5), recalling (5.71)-(5.72), and (5.74) we conclude that

$$\Lambda = \Lambda_{[f]} \quad \text{as functionals on } \mathrm{HA}^p(\Sigma,\sigma). \tag{5.105}$$

Collectively, (5.72), (5.101), and (5.105) entail

$$\|\Lambda\|_{(\mathrm{HA}^p(\Sigma,\sigma))^*} \approx \|f\|_{\mathrm{BCMO}_{p'}(\Sigma,\sigma)}, \tag{5.106}$$

with implicit proportionality constants independent of f.

Ultimately, our analysis so far (cf. (5.67)-(5.106)) ensures that the mapping

$$\begin{gathered}\mathrm{BCMO}_{p'}(\Sigma,\sigma) \ni f \longmapsto \Phi f := \widetilde{\Lambda}_f = \Lambda_{[f]} \in \big(\mathrm{HA}^p(\Sigma,\sigma)\big)^*\\ \text{is well defined, linear, bounded, and surjective.}\end{gathered} \tag{5.107}$$

Let us compute the kernel of Φ. On the one hand, observe that Φ maps constants into the zero functional on $\mathrm{HA}^p(\Sigma,\sigma)$, since functions from $\mathrm{HA}^p_{\mathrm{comp}}(\Sigma,\sigma)$ have integral zero (see Corollary 4.2). On the other hand, assume $f \in \mathrm{BCMO}_{p'}(\Sigma, \sigma)$ satisfies $\widetilde{\Lambda}_f = 0$ in $\big(\mathrm{HA}^p(\Sigma, \sigma)\big)^*$. Fix $k \in \mathbb{N}$ and an arbitrary function $g \in L^p_0\big(\Delta(x_0, 2^k_*)\big)$. Then in particular $E_k g \in \mathrm{HA}^p_{\mathrm{comp}}(\Sigma, \sigma)$ (see (5.76)-(5.78)), and from (5.71)-(5.72) we have

$$\int_{\Delta(x_0,2^k_*)} f\big|_{\Delta(x_0,2^k_*)} g\, d\sigma = \int_\Sigma f \cdot (E_k g)\, d\sigma = \widetilde{\Lambda}_f(E_k g) = 0. \tag{5.108}$$

Since $g \in L^p_0\big(\Delta(x_0, 2^k_*)\big)$ was taken arbitrary, Lemma 5.4 implies that f is equal to a constant σ-a.e. on $\Delta(x_0, 2^k_*)$. Because this holds for all $k \in \mathbb{N}$ we may conclude that f is a constant on Σ. Hence $\ker \Phi = \mathbb{C}$. In conclusion, viewing Φ in (5.107) as a group homomorphism, the First Isomorphism Theorem implies that

$$\widetilde{\mathrm{BCMO}}_{p'}(\Sigma, \sigma) = \mathrm{BCMO}_{p'}(\Sigma, \sigma)/\mathbb{C} \cong \big(\mathrm{HA}^p(\Sigma, \sigma)\big)^*. \tag{5.109}$$

This finishes the proof of Theorem 5.1. □

An immediate consequence of Theorem 5.1 regarding the completeness of $\widetilde{\mathrm{BCMO}}_p$ is presented below.

Corollary 5.1 *Assume $\Sigma \subseteq \mathbb{R}^n$ is a closed unbounded Ahlfors regular set, and abbreviate $\sigma := \mathcal{H}^{n-1}\lfloor\Sigma$. Then for every integrability exponent $p \in (1, \infty)$ it follows that $\widetilde{\mathrm{BCMO}}_p(\Sigma, \sigma)$ when equipped with the norm $\|\cdot\|_{\widetilde{\mathrm{BCMO}}_p(\Sigma,\sigma)}$ (defined in* (5.6)) *is a Banach space.*

Proof The completeness of $\widetilde{\mathrm{BCMO}}_p(\Sigma, \sigma)$ follows from the duality result recorded in Theorem 5.1 along with the fact that Beurling-Hardy spaces are complete (cf. Proposition 4.3). □

The following corollary provides a useful characterization of the BCMO_p-seminorm in terms of integral parings with central atoms.

Corollary 5.2 *Consider a closed unbounded Ahlfors regular set $\Sigma \subseteq \mathbb{R}^n$, and abbreviate $\sigma := \mathcal{H}^{n-1}\lfloor\Sigma$. Fix an integrability exponent $p \in (1, \infty)$, and let $p' \in (1, \infty)$ be its Hölder conjugate exponent. Then for every $f \in L^1_{\mathrm{loc}}(\Sigma, \sigma)$ one has*

$$\|f\|_{\mathrm{BCMO}_{p'}(\Sigma,\sigma)} \approx \sup\Big\{\Big|\int_\Sigma f a\, d\sigma\Big| : a \text{ is an } x_0\text{-central } (1, p)\text{-atom}\Big\}, \tag{5.110}$$

with implicit proportionality constants depending exclusively on n, p, and the Ahlfors regularity character of Σ.

Proof Start fixing a function $f \in L^1_{\mathrm{loc}}(\Sigma, \sigma)$. Given an x_0-central $(1, p)$-atom a on Σ, the computations made in (5.67) show that

$$\Big|\int_\Sigma f a\, d\sigma\Big| \le C\|f\|_{\mathrm{BCMO}_{p'}(\Sigma,\sigma)}, \tag{5.111}$$

for some constant $C \in (0,\infty)$ depending exclusively on n, p, and the Ahlfors regularity constants of Σ. In turn, this gives

$$\sup\Big\{\Big|\int_\Sigma fa\,d\sigma\Big| : a \text{ is an } x_0\text{-central } (1,p)\text{-atom}\Big\} \leq C\|f\|_{\mathrm{BCMO}_{p'}(\Sigma,\sigma)}. \tag{5.112}$$

To prove the opposite inequality, pick $\varepsilon \in (0,\infty)$ and a finite radius $R \geq 2_*$. Observe that $(f - f_{\Delta(x_0,R)})\big|_{\Delta(x_0,R)} \in L^{p'}\big(\Delta(x_0,R),\sigma\big)$ (as implied by Proposition 5.4), and

$$\int_{\Delta(x_0,R)} (f - f_{\Delta(x_0,R)})\,d\sigma = 0. \tag{5.113}$$

In light of Lemma 5.3 there exists some $h \in L^p\big(\Delta(x_0,R),\sigma\big)$ with $\int_{\Delta(x_0,R)} h\,d\sigma = 0$ and $\|h\|_{L^p(\Delta(x_0,R),\sigma)} \leq 1$ such that

$$\Big(\int_{\Delta(x_0,R)} |f - f_{\Delta(x_0,R)}|^{p'}\,d\sigma\Big)^{1/p'} - \varepsilon \leq \Big|\int_{\Delta(x_0,R)} (f - f_{\Delta(x_0,R)})\big|_{\Delta(x_0,R)}\cdot h\,d\sigma\Big|. \tag{5.114}$$

Denote by $\widetilde{h}$ the extension by zero of h to the entire Σ. Based on the properties of h mentioned above, it is evident that the function $\widetilde{a} := R^{-(n-1)/p'}\cdot\widetilde{h}$ satisfies

$$\operatorname{supp}\widetilde{a} \subseteq \Delta(x_0,R), \quad \|\widetilde{a}\|_{L^p(\Sigma,\sigma)} \leq R^{-(n-1)/p'}, \quad \int_\Sigma \widetilde{a}\,d\sigma = 0. \tag{5.115}$$

In particular, this shows that

$$\widetilde{a} := R^{-(n-1)/p'}\cdot\widetilde{h} \quad \text{is an } x_0\text{-central } (1,p)\text{-atom on } \Sigma. \tag{5.116}$$

Thanks to this, (5.114), (5.115), and the Ahlfors regularity of Σ, estimate

$$\begin{aligned}
&\Big(\int_{\Delta(x_0,R)} |f - f_{\Delta(x_0,R)}|^{p'}\,d\sigma\Big)^{1/p'} - \varepsilon \leq \Big|\int_{\Delta(x_0,R)} (f - f_{\Delta(x_0,R)})\big|_{\Delta(x_0,R)}\cdot h\,d\sigma\Big| \\
&\quad = \Big|\int_{\Delta(x_0,R)} f\big|_{\Delta(x_0,R)}\cdot h\,d\sigma\Big| = \Big|\int_{\Delta(x_0,R)} f\widetilde{h}\,d\sigma\Big| = R^{(n-1)/p'}\Big|\int_\Sigma f\widetilde{a}\,d\sigma\Big| \\
&\quad \leq C\sigma\big(\Delta(x_0,R)\big)^{1/p'}\sup\Big\{\Big|\int_\Sigma fa\,d\sigma\Big| : a \text{ in an } x_0\text{-central } (1,p)\text{-atom}\Big\},
\end{aligned} \tag{5.117}$$

for some constant $C \in (0,\infty)$ depending only on the Ahlfors regularity of Σ (in particular independent of ε and R). Since this holds for every $\varepsilon \in (0,\infty)$, we find that

$$\begin{aligned}
&\Big(⨍_{\Delta(x_0,R)} |f - f_{\Delta(x_0,R)}|^{p'}\,d\sigma\Big)^{1/p'} \\
&\qquad \leq C\cdot\sup\Big\{\Big|\int_\Sigma fa\,d\sigma\Big| : a \text{ in an } x_0\text{-central } (1,p)\text{-atom}\Big\}.
\end{aligned} \tag{5.118}$$

Moreover, from the arbitrariness of $R \geq 2_*$ this entails

$$\|f\|_{\mathrm{BCMO}_{p'}(\Sigma,\sigma)} \leq C \cdot \sup\Big\{\Big|\int_\Sigma f a \, d\sigma\Big| : a \text{ in an } x_0\text{-central } (1,p)\text{-atom}\Big\}, \tag{5.119}$$

thus justifying the right-pointing inequality in (5.110). The proof of Corollary 5.2 is therefore complete. □

In the corollary below we discuss a weak-$*$ convergence result in BCMO_p.

Corollary 5.3 *Consider a closed unbounded Ahlfors regular set $\Sigma \subseteq \mathbb{R}^n$, and abbreviate $\sigma := \mathcal{H}^{n-1}\lfloor\Sigma$. Pick an integrability exponent $p \in (1,\infty)$ and let $p' \in (1,\infty)$ be its Hölder conjugate exponent. Fix an arbitrary function $f \in L^1_{\mathrm{loc}}(\Sigma,\sigma)$, and assume that there exists a sequence of functions $\{f_j\}_{j\in\mathbb{N}} \subseteq L^1_{\mathrm{loc}}(\Sigma,\sigma)$ with the following properties:*

$$\begin{gathered}\sup_{j\in\mathbb{N}} \|f_j\|_{\mathrm{BCMO}_p(\Sigma,\sigma)} < \infty \quad \textit{and} \\ f_j \longrightarrow f \quad \textit{in } L^1_{\mathrm{loc}}(\Sigma,\sigma) \textit{ as } j \to \infty.\end{gathered} \tag{5.120}$$

Then $f \in \mathrm{BCMO}_p(\Sigma,\sigma)$ and

$$[f_j] \to [f] \quad \textit{as } j \to \infty \textit{ in } \widetilde{\mathrm{BCMO}}_p(\Sigma,\sigma) \textit{ in the weak-}*\textit{ topology}, \tag{5.121}$$

in the sense that

$$\big\langle [f_j], g\big\rangle \to \big\langle [f], g\big\rangle \quad \textit{as } j \to \infty \textit{ for every } g \in \mathrm{HA}^{p'}(\Sigma,\sigma), \tag{5.122}$$

where $\langle\cdot,\cdot\rangle$ is the duality pairing between $\widetilde{\mathrm{BCMO}}_p(\Sigma,\sigma)$ and $\mathrm{HA}^{p'}(\Sigma,\sigma)$ understood in the sense of Theorem 5.1.

Proof We argue by contradiction. Assume that the weak-$*$ convergence (5.122) fails. Then there exists a function $g \in \mathrm{HA}^{p'}(\Sigma,\sigma)$ along with a strictly increasing sequence $\{j_k\}_{k\in\mathbb{N}} \subseteq \mathbb{N}$ with the property that

$$\text{any subsequence of } \big\{\big\langle [f_{j_k}], g\big\rangle\big\}_{k\in\mathbb{N}} \text{ does not converge to } \big\langle [f], g\big\rangle. \tag{5.123}$$

Notice that the first condition in (5.120) ensures that $\big\{[f_{j_k}]\big\}_{k\in\mathbb{N}}$ is a bounded sequence in $\widetilde{\mathrm{BCMO}}_p(\Sigma,\sigma)$. Thanks to the duality $\widetilde{\mathrm{BCMO}}_p(\Sigma,\sigma) = \big(\mathrm{HA}^{p'}(\Sigma,\sigma)\big)^*$ (cf. Theorem 5.1) and the separability of $\mathrm{HA}^{p'}(\Sigma,\sigma)$ (cf. Theorem 4.4), the Sequential Banach-Alaoglu Theorem guarantees the existence of a subsequence $\big\{[f_{j_{k_\ell}}]\big\}_{\ell\in\mathbb{N}}$ of $\big\{[f_{j_k}]\big\}_{k\in\mathbb{N}}$ along with some function $\widetilde{f} \in \mathrm{BCMO}_p(\Sigma,\sigma)$ for which

$$[f_{j_{k_\ell}}] \to [\widetilde{f}] \quad \text{weak-}* \text{ in } \widetilde{\mathrm{BCMO}}_p(\Sigma,\sigma) \text{ as } \ell \to \infty. \tag{5.124}$$

We claim that $[f] = [\widetilde{f}]$ in $\widetilde{\mathrm{BCMO}}_p(\Sigma,\sigma)$. To justify this claim, select an arbitrary function $a \in L^\infty_{\mathrm{comp}}(\Sigma,\sigma)$ with $\int_\Sigma a\,d\sigma = 0$. In particular, $a \in L^{p'}_{\mathrm{comp}}(\Sigma,\sigma)$, and thus by Corollary 4.2 we have $a \in \mathrm{HA}^{p'}_{\mathrm{comp}}(\Sigma,\sigma)$. Furthermore,

$$\int_\Sigma fa\,d\sigma = \lim_{\ell\to\infty}\int_\Sigma f_{j_{k_\ell}} a\,d\sigma = \lim_{\ell\to\infty}\big\langle [f_{j_{k_\ell}}], a\big\rangle = \big\langle [\widetilde{f}], a\big\rangle = \int_\Sigma \widetilde{f} a\,d\sigma. \tag{5.125}$$

The first equality of (5.125) is a consequence of the second line in (5.120), and the fact that $a \in L^\infty_{\mathrm{comp}}(\Sigma,\sigma)$. The second and fourth equalities are implied from the fact that elements from $\widetilde{\mathrm{BCMO}}_p(\Sigma,\sigma)$, viewed as functionals from $\big(\mathrm{HA}^{p'}(\Sigma,\sigma)\big)^*$, act on functions from $\mathrm{HA}^{p'}_{\mathrm{comp}}(\Sigma,\sigma)$ via integration (cf. Theorem 5.1). The third equality follows from (5.124). In turn, from (5.125) we see that

$$\int_\Sigma (f-\widetilde{f})a\,d\sigma = 0. \tag{5.126}$$

Since this holds for every $a \in L^\infty_{\mathrm{comp}}(\Sigma,\sigma)$ with $\int_\Sigma a\,d\sigma = 0$, Lemma 5.2 guarantees that $f - \widetilde{f}$ is constant on Σ. Hence

$$[f] = [\widetilde{f}] \text{ in } \widetilde{\mathrm{BCMO}}_p(\Sigma,\sigma). \tag{5.127}$$

Finally, (5.124) and (5.127) imply that $\big\langle [f_{j_{k_\ell}}], g\big\rangle \to \big\langle [f], g\big\rangle$ as $\ell \to \infty$, a contradiction with (5.123). This finishes the proof of Corollary 5.3. □

A related result regarding weak-$*$ convergence on BCMO_p spaces is recorded below.

Corollary 5.4 *Consider a closed unbounded Ahlfors regular set $\Sigma \subseteq \mathbb{R}^n$, and abbreviate $\sigma := \mathcal{H}^{n-1}\lfloor\Sigma$. Pick an integrability exponent $p \in (1,\infty)$, and let $p' \in (1,\infty)$ be its Hölder conjugate exponent. Fix an arbitrary real-valued function $f \in \mathrm{BCMO}_p(\Sigma,\sigma)$, and for each $N \in \mathbb{N}$ define $f_N : \Sigma \to \mathbb{R}$ by setting*

$$f_N(x) := \begin{cases} N & \text{if } f(x) > N, \\ f(x) & \text{if } |f(x)| \le N, \\ -N & \text{if } f(x) < -N, \end{cases} \tag{5.128}$$

for each $x \in \Sigma$. Then $[f_N] \to [f]$ weak-$*$ in $\widetilde{\mathrm{BCMO}}_p(\Sigma,\sigma)$ as $N \to \infty$. More precisely, for every $g \in \mathrm{HA}^{p'}(\Sigma,\sigma)$ one has

$$\lim_{N\to\infty}\big\langle [f_N], g\big\rangle = \big\langle [f], g\big\rangle \tag{5.129}$$

where $\langle\,\cdot\,,\,\cdot\,\rangle$ is the duality pairing between $\widetilde{\mathrm{BCMO}}_p(\Sigma,\sigma)$ and $\mathrm{HA}^{p'}(\Sigma,\sigma)$ understood in the sense of Theorem 5.1.

Proof To begin, a direct computation shows that

$$f_N = \min\big\{\max\{f,-N\},N\big\} \quad \text{for all } N \in \mathbb{N}. \tag{5.130}$$

Hence, using (5.17)-(5.18) in Proposition 5.2 along with the fact that $\|\cdot\|_{\mathrm{BCMO}_p(\Sigma,\sigma)}$ is a seminorm (cf. (5.3)), we have

$$\sup_{N\in\mathbb{N}} \|f_N\|_{\mathrm{BCMO}_p(\Sigma,\sigma)} \le \frac{9}{4}\|f\|_{\mathrm{BCMO}_p(\Sigma,\sigma)} < \infty. \tag{5.131}$$

Next, we claim that

$$f_N \to f \ \text{ in } L^1_{\mathrm{loc}}(\Sigma,\sigma) \text{ as } N \to \infty. \tag{5.132}$$

Indeed, it is visible from (5.128) that for each $x \in \Sigma$ and $N \in \mathbb{N}$ sufficiently large, $f_N(x)$ becomes equal to $f(x)$. Hence, $f_N \to f$ pointwise everywhere on Σ as $N \to \infty$. Also, $|f_N| \le |f|$ pointwise everywhere on Σ for every $N \in \mathbb{N}$. Since $f \in L^1_{\mathrm{loc}}(\Sigma,\sigma)$, Lebesgue's Dominated Convergence Theorem implies $f_N \to f$ in $L^1_{\mathrm{loc}}(\Sigma,\sigma)$ as $N \to \infty$. This justifies (5.132). Granted this and (5.131), the weak-$*$ convergence claimed in (5.129) now follows from Corollary 5.3. The proof of Corollary 5.4 is therefore complete. □

The following result provides a useful way to describe the duality pairing between BCMO_p and $\mathrm{HA}^{p'}$ as an integral pairing. This is remarkable, since the duality result described in Theorem 5.1 is not merely realized via integration.

Corollary 5.5 *Consider a closed unbounded Ahlfors regular set $\Sigma \subseteq \mathbb{R}^n$, and abbreviate $\sigma := \mathcal{H}^{n-1}\lfloor\Sigma$. Pick an integrability exponent $p \in (1,\infty)$, and let $p' \in (1,\infty)$ be its Hölder conjugate exponent. Select two arbitrary functions $f \in \mathrm{BCMO}_p(\Sigma,\sigma)$ and $g \in \mathrm{HA}^{p'}(\Sigma,\sigma)$, and assume*

$$\int_\Sigma |fg|\,d\sigma < \infty. \tag{5.133}$$

Then the equivalence class $[f] \in \widetilde{\mathrm{BCMO}}_p(\Sigma,\sigma)$, viewed as a continuous linear functional from $\big(\mathrm{HA}^{p'}(\Sigma,\sigma)\big)^*$ (in the sense of Theorem 5.1), acts on g via integration against f, i.e.,

$$\big\langle [f], g\big\rangle = \int_\Sigma fg\,d\sigma. \tag{5.134}$$

Proof For each $N \in \mathbb{N}$, consider f_N as in (5.128). Hence, the sequence $\{f_N\}_{N\in\mathbb{N}}$ satisfies

$$\begin{aligned} &f_N \in \mathrm{BCMO}_p(\Sigma,\sigma) \ \text{ for all } N \in \mathbb{N}, \\ &\lim_{N\to\infty} [f_N] = [f] \ \text{ weak-}* \text{ in } \mathrm{BCMO}_p(\Sigma,\sigma), \\ &\text{and } \lim_{N\to\infty} f_N g = fg \ \text{ in } L^1(\Sigma,\sigma). \end{aligned} \tag{5.135}$$

The first property in (5.135) is implied by (5.131). The second property follows from Corollary 5.4. The third property of (5.135) is implied by Lebesgue's Dominated Convergence Theorem, which is applicable thanks to (5.133), the fact that $f_N g \to fg$ pointwise σ-a.e. on Σ as $N \to \infty$, and that $|f_N g| \leq |fg|$ at σ-a.e. point on Σ for all $N \in \mathbb{N}$.

Next, consider an atomic decomposition of g (cf. Theorem 4.2) as

$$g = \sum_{j=1}^{\infty} \lambda_j a_j \text{ with convergence in } L^1(\Sigma, \sigma) \text{ and in } \mathrm{HA}^{p'}(\Sigma, \sigma) \tag{5.136}$$

for some sequence of x_0-central $(1, p)$-atoms $\{a_j\}_{j\in\mathbb{N}}$, and some numerical sequence $\{\lambda_j\}_{j\in\mathbb{N}} \in \ell^1(\mathbb{N})$. For each $M \in \mathbb{N}$ abbreviate

$$g_M := \sum_{j=1}^{M} \lambda_j a_j. \tag{5.137}$$

Then

$$\begin{aligned} \big\langle [f], g \big\rangle &= \lim_{N\to\infty} \big\langle [f_N], g \big\rangle = \lim_{N\to\infty} \lim_{M\to\infty} \big\langle [f_N], g_M \big\rangle \\ &= \lim_{N\to\infty} \Big(\lim_{M\to\infty} \int_\Sigma f_N g_M \, d\sigma \Big) = \lim_{N\to\infty} \int_\Sigma f_N g \, d\sigma = \int_\Sigma f g \, d\sigma. \end{aligned} \tag{5.138}$$

The first equality follows from the weak-$*$ convergence in (5.135). The second equality is implied from the fact that $[f_N]$ is a continuous linear functional from $\big(\mathrm{HA}^{p'}(\Sigma, \sigma)\big)^*$ (cf. Theorem 5.1), and that $g_M \to g$ in $\mathrm{HA}^{p'}(\Sigma, \sigma)$ as $M \to \infty$ (cf. (5.136)). The third equality is valid since the manner in which $[f_N]$, viewed as a continuous linear functional on $\mathrm{HA}^{p'}(\Sigma, \sigma)$, acts on central atoms is via integration against f_N (cf. Theorem 5.1). The second to last equality follows directly from the convergence in $L^1(\Sigma, \sigma)$ in (5.136), along with the fact that f_N is a bounded function for every $N \in \mathbb{N}$. The last equality is guaranteed by the property in the third line of (5.135). This justifies (5.138) and finishes the proof of Corollary 5.5. □

As an important application of Corollary 5.5, in the following result we establish that central molecules pair well (in an integral sense) with functions from BCMO_p.

Corollary 5.6 *Assume $\Sigma \subseteq \mathbb{R}^n$ is a closed unbounded Ahlfors regular set, and abbreviate $\sigma := \mathcal{H}^{n-1}\lfloor\Sigma$. Pick an integrability exponent $p \in (1, \infty)$, and denote by $p' \in (1, \infty)$ its Hölder conjugate exponent. Fix a reference point $x_0 \in \Sigma$ and a dilation factor 2_* as in Definition 5.1. Also, select a parameter*

$$\theta \in \big((n-1)(p'-1), \infty\big). \tag{5.139}$$

Then for every function $f \in \mathrm{BCMO}_p(\Sigma, \sigma)$ and every x_0-central $(1, p', \theta)$-molecule M on Σ it follows that

$$\int_\Sigma |fM|\,d\sigma < \infty. \tag{5.140}$$

In particular, one has

$$\big\langle [f], M \big\rangle = \int_\Sigma fM\,d\sigma \tag{5.141}$$

where $\langle \cdot\,, \cdot \rangle$ *denotes the duality pairing between* $\widetilde{\mathrm{BCMO}}_p(\Sigma, \sigma)$ *and* $\mathrm{HA}^{p'}(\Sigma, \sigma)$ *(in the sense described in Theorem 5.1).*

Proof Fix a function $f \in \mathrm{BCMO}_p(\Sigma, \sigma)$ along with an x_0-central $(1, p', \theta)$-molecule M on Σ. Then there exists a radius $R \geq 2_*$ for which M satisfies (4.189)-(4.191) in Definition 4.4 (with p replaced by p'). With the goal of establishing (5.140) in mind, we first observe that

$$\int_\Sigma |fM|\,d\sigma \leq \int_\Sigma |f - f_{\Delta(x_0,2R)}||M|\,d\sigma + |f_{\Delta(x_0,2R)}| \int_\Sigma |M|\,d\sigma. \tag{5.142}$$

Since $f \in L^1_{\mathrm{loc}}(\Sigma, \sigma)$ (see Definition 5.1), and M is an absolutely integrable function on Σ (see Remark 4.7) it is clear that

$$|f_{\Delta(x_0,2R)}| \int_\Sigma |M|\,d\sigma < \infty. \tag{5.143}$$

To deal with the first term in the right-hand side of (5.142), we write

$$\begin{aligned}
&\int_\Sigma |f - f_{\Delta(x_0,2R)}||M|\,d\sigma \\
&= \int_{\Delta(x_0,2R)} |f - f_{\Delta(x_0,2R)}||M|\,d\sigma + \int_{\Sigma\setminus\Delta(x_0,2R)} |f - f_{\Delta(x_0,2R)}||M|\,d\sigma \\
&=: \mathrm{I}_1 + \mathrm{I}_2.
\end{aligned} \tag{5.144}$$

Using Hölder's inequality, the normalization property of M (cf. (4.189) with p replaced by p'), and the Ahlfors regularity of Σ, we estimate I_1 as

$$\begin{aligned}
\mathrm{I}_1 &\leq \Big(\int_{\Delta(x_0,2R)} |f - f_{\Delta(x_0,2R)}|^p\,d\sigma\Big)^{1/p} \|M\|_{L^{p'}(\Sigma,\sigma)} \\
&\leq R^{-(n-1)/p}\Big(\int_{\Delta(x_0,2R)} |f - f_{\Delta(x_0,2R)}|^p\,d\sigma\Big)^{1/p} \\
&\leq C\Big(⨍_{\Delta(x_0,2R)} |f - f_{\Delta(x_0,2R)}|^p\,d\sigma\Big)^{1/p} \leq C\|f\|_{\mathrm{BCMO}_p(\Sigma,\sigma)} < \infty,
\end{aligned} \tag{5.145}$$

for some $C \in (0, \infty)$ depending only on n, p, and the Ahlfors regularity constants of Σ.

Next, to estimate I_2 from (5.144), we observe that

$$
\begin{aligned}
\mathrm{I}_2 &= \sum_{j=1}^{\infty}\int_{\Delta(x_0,2^{j+1}R)\setminus\Delta(x_0,2^jR)}|f-f_{\Delta(x_0,2R)}||M|\,d\sigma\\
&\leq \sum_{j=1}^{\infty}\int_{\Delta(x_0,2^{j+1}R)\setminus\Delta(x_0,2^jR)}|f(x)-f_{\Delta(x_0,2R)}||M(x)|\frac{|x-x_0|^{\theta/p'}}{(2^jR)^{\theta/p'}}\,d\sigma(x)\\
&\leq \sum_{j=1}^{\infty}\int_{\Delta(x_0,2^{j+1}R)\setminus\Delta(x_0,2^jR)}|f(x)-f_{\Delta(x_0,2^{j+1}R)}||M(x)|\frac{|x-x_0|^{\theta/p'}}{(2^jR)^{\theta/p'}}\,d\sigma(x)\\
&\quad+\sum_{j=1}^{\infty}|f_{\Delta(x_0,2^{j+1}R)}-f_{\Delta(x_0,2R)}|\int_{\Delta(x_0,2^{j+1}R)\setminus\Delta(x_0,2^jR)}|M(x)|\frac{|x-x_0|^{\theta/p'}}{(2^jR)^{\theta/p'}}\,d\sigma(x)\\
&=: \mathrm{S}_1+\mathrm{S}_2.
\end{aligned}
\tag{5.146}
$$

To bound S_1 we rely on Hölder's inequality, the decay property of M (cf. (4.190) with p replaced by p'), and the Ahlfors regularity of Σ, to obtain

$$
\begin{aligned}
\mathrm{S}_1 &\leq \sum_{j=1}^{\infty}\frac{1}{(2^jR)^{\theta/p'}}\Big(\int_{\Delta(x_0,2^{j+1}R)}|f-f_{\Delta(x_0,2^{j+1}R)}|^p\,d\sigma\Big)^{1/p}\times\\
&\qquad\times\Big(\int_{\Sigma}|M(x)|^{p'}|x-x_0|^{\theta}\,d\sigma(x)\Big)^{1/p'}\\
&\leq \sum_{j=1}^{\infty}\frac{R^{-(n-1)/p+\theta/p'}}{(2^jR)^{\theta/p'}}\Big(\int_{\Delta(x_0,2^{j+1}R)}|f-f_{\Delta(x_0,2^{j+1}R)}|^p\,d\sigma\Big)^{1/p}\\
&\leq C\sum_{j=1}^{\infty}\frac{(2^{j+1}R)^{-(n-1)/p}}{(2^j)^{\theta/p'-(n-1)/p}}\Big(\int_{\Delta(x_0,2^{j+1}R)}|f-f_{\Delta(x_0,2^{j+1}R)}|^p\,d\sigma\Big)^{1/p}\\
&\leq C\sum_{j=1}^{\infty}\frac{1}{2^{j[\theta/p'-(n-1)/p]}}\Big(\fint_{\Delta(x_0,2^{j+1}R)}|f-f_{\Delta(x_0,2^{j+1}R)}|^p\,d\sigma\Big)^{1/p}\\
&\leq C\Big(\sum_{j=1}^{\infty}\frac{1}{2^{j[\theta/p'-(n-1)/p]}}\Big)\|f\|_{\mathrm{BCMO}_p(\Sigma,\sigma)}<\infty,
\end{aligned}
\tag{5.147}
$$

for some constant $C\in(0,\infty)$ depending only on n, p, and the Ahlfors regularity of Σ. Notice that the series in the last line of (5.147) is convergent, since $\theta>(n-1)(p'-1)$ implies $\theta/p'-(n-1)/p>0$.

Moving on, use Hölder's inequality, a telescoping sum argument (much as in (5.49)), the decay property of M (cf. (4.190)), the Ahlfors regularity of Σ, and the fact that $\theta/p'-(n-1)/p>0$ (as implied by the choice of θ), to estimate S_2 in (5.146) by

$$
\begin{aligned}
S_2 &= \sum_{j=1}^{\infty} |f_{\Delta(x_0,2^{j+1}R)} - f_{\Delta(x_0,2R)}| \int_{\Delta(x_0,2^{j+1}R)\setminus\Delta(x_0,2^jR)} |M(x)| \frac{|x-x_0|^{\theta/p'}}{(2^jR)^{\theta/p'}}\, d\sigma(x) \\
&\leq \sum_{j=1}^{\infty} \frac{1}{(2^jR)^{\theta/p'}} \Big(\sum_{k=1}^{j} |f_{\Delta(x_0,2^{k+1}R)} - f_{\Delta(x_0,2^kR)}| \Big) \\
&\qquad \times \Big(\int_{\Sigma} |M(x)|^{p'} |x-x_0|^{\theta}\, d\sigma(x) \Big)^{1/p'} \sigma\big(\Delta(x_0, 2^{j+1}R)\big)^{1/p} \\
&\leq C \sum_{j=1}^{\infty} \frac{j}{(2^jR)^{\theta/p'}} \|f\|_{\mathrm{BCMO}_p(\Sigma,\sigma)} R^{-(n-1)/p+\theta/p'} (2^jR)^{(n-1)/p} \\
&= C \Big(\sum_{j=1}^{\infty} \frac{j}{2^{j[\theta/p'-(n-1)/p]}} \Big) \|f\|_{\mathrm{BCMO}_p(\Sigma,\sigma)} < \infty
\end{aligned}
\tag{5.148}
$$

for some constant $C \in (0,\infty)$ depending only on n, p, θ, and the Ahlfors regularity character of Σ. All together, (5.146), (5.147), and (5.148) ensure the existence of some constant $C \in (0,\infty)$ depending only on n, p, θ, and the Ahlfors regularity constants of Σ such that

$$
I_2 \leq C \|f\|_{\mathrm{BCMO}_p(\Sigma,\sigma)}. \tag{5.149}
$$

As such, this, (5.142), (5.143), (5.144), and (5.145), ultimately prove (5.140). The very last claim regarding (5.141) follows from what we have established so far, along with the fact that $M \in \mathrm{HA}^{p'}(\Sigma,\sigma)$ (cf. Theorem 4.4), and Corollary 5.5. □

Chapter 6

Calderón-Zygmund Theory on Beurling-Hardy Spaces and BCMO_p Spaces

In this chapter, we take up the task of developing a Calderón-Zygmund theory for singular integral operators on uniformly rectifiable sets acting on Beurling-Hardy spaces and on spaces of functions of bounded central mean oscillations. The body of results presented here is applicable to singular integral operators of layer potential type associated with weakly elliptic systems, which will play a crucial role when dealing with boundary problems in subsequent chapters.

6.1 Singular Integral Operators on Beurling-Hardy Spaces

We begin by recalling a result from [MMMi^3, Proposition 3.4, p. 179] regarding the boundedness of convolution-type singular integral operators, nontangential maximal function estimates, and jump-formulas in the context of Lebesgue spaces.

Proposition 6.1 *Suppose $\Sigma \subseteq \mathbb{R}^n$ is a closed uniformly rectifiable set and abbreviate $\sigma := \mathcal{H}^{n-1}\lfloor\Sigma$. Assume $N = N(n) \in \mathbb{N}$ is a sufficiently large integer and consider a complex-valued function $k \in \mathscr{C}^N\big(\mathbb{R}^n \setminus \{0\}\big)$ which is odd and positive homogeneous of degree $1-n$. Also, fix an integrability exponent $p \in (1,\infty)$. In this setting, for each $f \in L^1\big(\Sigma, \frac{\sigma(x)}{1+|x|^{n-1}}\big)$ define*

$$T_\varepsilon f(x) := \int_{\substack{y\in\Sigma \\ |x-y|>\varepsilon}} k(x-y)f(y)\,d\sigma(y) \textit{ for all } x \in \Sigma \textit{ and } \varepsilon > 0, \tag{6.1}$$

$$T_{\max} f(x) := \sup_{\varepsilon>0} |T_\varepsilon f(x)| \quad \textit{for each} \quad x \in \Sigma, \tag{6.2}$$

$$Tf(x) := \lim_{\varepsilon\to 0^+} T_\varepsilon f(x) \quad \textit{for } \sigma\textit{-a.e.} \quad x \in \Sigma. \tag{6.3}$$

© The Author(s), under exclusive license to Springer Nature Switzerland AG 2026
M. Mitrea and P. Takemura, *Singular Integrals, Herz-Type Function Spaces, and Boundary Problems*, Progress in Mathematics 362,
https://doi.org/10.1007/978-3-032-12516-3_6

Then there exists some $C \in (0, \infty)$ *which depends exclusively on* n, p, *and the* UR *constants of* Σ *with the property that for each* $f \in L^p(\Sigma, \sigma)$ *one has*

$$\|T_{\max} f\|_{L^p(\Sigma,\sigma)} \leq C\Big(\sum_{|\alpha|\leq N} \sup_{S^{n-1}} |\partial^\alpha k|\Big) \|f\|_{L^p(\Sigma,\sigma)}. \tag{6.4}$$

In particular,

the truncated integral operators $T_\varepsilon : L^p(\Sigma, \sigma) \to L^p(\Sigma, \sigma)$ *are well defined, linear, and bounded in a uniform fashion with respect to the truncation parameter* $\varepsilon > 0$. (6.5)

Moreover, for each function $f \in L^1\big(\Sigma, \frac{\sigma(x)}{1+|x|^{n-1}}\big)$ *the limit defining* $Tf(x)$ *in* (6.3) *exists at* σ*-a.e.* $x \in \Sigma$ *and the operator*

$$T : L^p(\Sigma, \sigma) \longrightarrow L^p(\Sigma, \sigma) \tag{6.6}$$

is well defined, linear, and bounded. Corresponding to $p = 1$, *it follows that the operator*

$$T : L^1(\Sigma, \sigma) \longrightarrow L^{1,\infty}(\Sigma, \sigma) \tag{6.7}$$

is well defined, linear, and bounded.

Finally, assume $\Omega \subseteq \mathbb{R}^n$ *is an open set such that* $\partial\Omega$ *is a* UR *set and abbreviate* $\sigma := \mathcal{H}^{n-1}\lfloor\partial\Omega$. *Fix an aperture parameter* $\kappa \in (0, \infty)$. *With the integral kernel* k *as before, for each* $f \in L^1\big(\Sigma, \frac{\sigma(x)}{1+|x|^{n-1}}\big)$ *define*

$$\mathcal{T} f(x) := \int_{\partial\Omega} k(x - y) f(y)\, d\sigma(y) \quad \textit{for each} \quad x \in \Omega. \tag{6.8}$$

Then there exists some $C \in (0, \infty)$ *which depends exclusively on* n, p, *and the* UR *constants of* $\partial\Omega$ *with the property that for each* $f \in L^p(\partial\Omega, \sigma)$ *one has*

$$\|\mathcal{N}_\kappa(\mathcal{T} f)\|_{L^p(\partial\Omega,\sigma)} \leq C\Big(\sum_{|\alpha|\leq N} \sup_{S^{n-1}} |\partial^\alpha k|\Big) \|f\|_{L^p(\partial\Omega,\sigma)}. \tag{6.9}$$

Corresponding to $p = 1$, *for every* $f \in L^1(\partial\Omega, \sigma)$ *it follows that*

$$\|\mathcal{N}_\kappa(\mathcal{T} f)\|_{L^{1,\infty}(\partial\Omega,\sigma)} \leq C\, \|f\|_{L^1(\partial\Omega,\sigma)}. \tag{6.10}$$

Also, for each function $f \in L^1\big(\partial\Omega, \frac{\sigma(x)}{1+|x|^{n-1}}\big)$ *one has the jump-formula*

$$\Big(\mathcal{T} f\Big|_{\partial\Omega}^{\kappa-\text{n.t.}}\Big)(x) = \frac{1}{2i}\widehat{k}\big(\nu(x)\big) f(x) + (Tf)(x) \quad \textit{at } \sigma\textit{-a.e.} \ \ x \in \partial_*\Omega, \tag{6.11}$$

where $\widehat{k}$ denotes the Fourier transform of k. In particular, the jump-formula (6.11) *is valid for each function $f \in L^p(\partial\Omega, \sigma)$.*

The result recorded below establishes the boundedness from HA^p into A^p of boundary-to-boundary convolution-type singular integral operators as in (6.3).

Theorem 6.1 *Assume $\Sigma \subseteq \mathbb{R}^n$ is a closed unbounded uniformly rectifiable set, and abbreviate $\sigma := \mathcal{H}^{n-1}\lfloor\Sigma$. For some $N = N(n) \in \mathbb{N}$ sufficiently large, consider a kernel $k \in \mathscr{C}^N(\mathbb{R}^n \setminus \{0\})$ that is odd, and positive homogeneous of degree $1-n$. Recall the boundary-to-boundary convolution-type integral operator T associated with k as in* (6.3). *Then whenever $p \in (1, \infty)$ the operator*

$$T : \mathrm{HA}^p(\Sigma, \sigma) \longrightarrow \mathrm{A}^p(\Sigma, \sigma) \quad \text{is well defined, linear, and bounded.} \tag{6.12}$$

Proof In light of the embedding from (4.15) and Proposition 6.1, the action of T on functions from $\mathrm{HA}^p(\Sigma, \sigma)$ is well defined. Fix a reference point $x_0 \in \Sigma$ and a dilation factor 2_* as in Definition 3.1. Select an x_0-central $(1, p)$-atom a on Σ, and consider a finite radius $R \geq 2_*$ associated with a (in the sense of Definition 4.3). We claim that there exists some $C \in (0, \infty)$ depending only on n, p, and the Ahlfors regularity constants of Σ (in particular, it is independent of a) with the property that

$$\|Ta\|_{\mathrm{A}^p(\Sigma,\sigma)} \leq C. \tag{6.13}$$

To justify this claim, pick the unique integer $J \in \mathbb{N}$ for which

$$2_*^J \leq R < 2_*^{J+1}. \tag{6.14}$$

Upon recalling (3.2), write

$$\begin{aligned} \|Ta\|_{\mathrm{A}^p(\Sigma,\sigma)} &= \sum_{j=0}^{J+1} 2_*^{j(n-1)/p'} \|(Ta) \cdot \mathbb{1}_{C_j}\|_{L^p(\Sigma,\sigma)} \\ &\quad + \sum_{j=J+2}^{\infty} 2_*^{j(n-1)/p'} \|(Ta) \cdot \mathbb{1}_{C_j}\|_{L^p(\Sigma,\sigma)} \\ &=: \mathrm{S}_1 + \mathrm{S}_2 \end{aligned} \tag{6.15}$$

where $p' \in (1, \infty)$ is the Hölder conjugate exponent of p. Taking into account (6.14), the boundedness of T on $L^p(\Sigma, \sigma)$ (cf. Theorem 6.1), and the normalization property of a (see (4.29)), we estimate S_1 by

$$
\begin{aligned}
\mathrm{S}_1 &= \sum_{j=0}^{J+1} 2_*^{j(n-1)/p'} \|(Ta)\cdot \mathbb{1}_{C_j}\|_{L^p(\Sigma,\sigma)} \le C \sum_{j=0}^{J+1} 2_*^{j(n-1)/p'} \|a\|_{L^p(\Sigma,\sigma)} \\
&\le C \sum_{j=0}^{J+1} 2_*^{j(n-1)/p'} R^{-(n-1)/p'} \le C \sum_{j=0}^{J+1} 2_*^{j(n-1)/p'} 2_*^{-J(n-1)/p'} \\
&= C \sum_{j=0}^{J+1} 2_*^{(j-J)(n-1)/p'} \le C \sum_{\ell=0}^{\infty} 2_*^{-\ell(n-1)/p'} < \infty, \qquad (6.16)
\end{aligned}
$$

for some $C \in (0,\infty)$ depending only on n, p, the kernel k, and the UR constants of Σ.

To deal with S_2 from (6.15), fix $j \in \mathbb{N}$ such that $j \ge J+2$. Then, bearing in mind the definition of $\{C_j\}_{j\in\mathbb{N}_0}$ in (3.1) and (6.14) we see that

$$
|x-z| \approx |x-x_0| \text{ uniformly with respect } x \in C_j \text{ and } z \in B(x_0,R). \qquad (6.17)
$$

In particular, this ensures that there is enough separation between C_j and $\Delta(x_0,R)$. Since $\operatorname{supp} a \subseteq \Delta(x_0,R)$, this and the vanishing moment of a (cf. (4.30)) permit us to express Ta at each point $x \in C_j$ as

$$
\begin{aligned}
Ta(x) &= \int_\Sigma k(x-y)a(y)\,d\sigma(y) \\
&= \int_\Sigma \big(k(x-y)-k(x-x_0)\big)a(y)\,d\sigma(y). \qquad (6.18)
\end{aligned}
$$

Also, the fact that k is sufficiently smooth in $\mathbb{R}^n\setminus\{0\}$, odd, and positive homogeneous of degree $1-n$ yield

$$
|\nabla k(z)| \le \frac{C}{|z|^n} \quad \text{for all } z \in \mathbb{R}^n \setminus \{0\}, \qquad (6.19)
$$

for some constant $C \in (0,\infty)$ depending only on n and k. Use this, (6.17), the normalization property of a (cf. (4.29)), the Ahlfors regularity of Σ, the Mean Value Theorem, and Hölder's inequality, to estimate (6.18) at each $x \in C_j$ as

$$\begin{aligned}
|Ta(x)| &\le \int_{\Delta(x_0,R)} |k(x-y)-k(x-x_0)||a(y)|\,d\sigma(y) \\
&\le \int_{\Delta(x_0,R)} \Big(\sup_{\xi\in[y,x_0]} |\nabla k(x-\xi)|\Big)|y-x_0||a(y)|\,d\sigma(y) \\
&\le R\int_{\Delta(x_0,R)} \Big(\sup_{\xi\in[y,x_0]} \frac{C}{|x-\xi|^n}\Big)|a(y)|\,d\sigma(y) \\
&\le C\frac{R}{|x-x_0|^n}\int_{\Delta(x_0,R)} |a|\,d\sigma \\
&\le C\frac{R}{|x-x_0|^n}\|a\|_{L^p(\Sigma,\sigma)}\cdot\sigma\big(\Delta(x_0,R)\big)^{1/p'} \\
&\le C\frac{R}{2_*^{jn}}R^{-(n-1)/p'}R^{(n-1)/p'} = C\frac{R}{2_*^{jn}}
\end{aligned} \tag{6.20}$$

for some constant $C\in(0,\infty)$ depending only on n, p, the kernel k, and the Ahlfors regularity of Σ. This, (6.14), and the Ahlfors regularity of Σ then yield

$$\begin{aligned}
\|(Ta)\cdot\mathbb{1}_{C_j}\|_{L^p(\Sigma,\sigma)} &= \Big(\int_{C_j}|Ta|^p\,d\sigma\Big)^{1/p} \le C\frac{R}{2_*^{jn}}\sigma(C_j)^{1/p} \\
&\le C\frac{R}{2_*^{jn}}\sigma\big(\Delta(x_0,2_*^{j+1})\big)^{1/p} \le C\frac{2_*^J}{2_*^{jn}}2_*^{j(n-1)/p}
\end{aligned} \tag{6.21}$$

for some constant $C=C(n,p,k,\Sigma)\in(0,\infty)$ and for every $j\in\mathbb{N}_0$ with $j\ge J+2$. In light of this, we bound S_2 from (6.15) as follows:

$$\begin{aligned}
\mathrm{S}_2 &\le C\sum_{j=J+2}^{\infty} 2_*^{j(n-1)/p'}2_*^{-jn}2_*^J2_*^{j(n-1)/p} \\
&= C\sum_{j=J+2}^{\infty} 2_*^{J-j} \le C\sum_{\ell=0}^{\infty}2_*^{-\ell} < \infty.
\end{aligned} \tag{6.22}$$

All together, estimates (6.15), (6.16), and (6.22) justify the claim made in (6.13).

To proceed, select an arbitrary function $f\in\mathrm{HA}^p(\Sigma,\sigma)$, and use Theorem 4.2 to decompose f as

$$f=\sum_{j=0}^{\infty}\lambda_j a_j \quad \text{with convergence in } L^1(\Sigma,\sigma), \tag{6.23}$$

for some sequence of x_0-central $(1,p)$-atoms $\{a_j\}_{j\in\mathbb{N}_0}$ and some numerical sequence $\{\lambda_j\}_{j\in\mathbb{N}_0}\in\ell^1(\mathbb{N}_0)$. Without loss of generality, assume this decomposition is quasi-optimal, i.e.,

$$\sum_{j=0}^{\infty} |\lambda_j| \le 2\|f\|_{\mathrm{HA}^p(\Sigma,\sigma)}. \tag{6.24}$$

For each $m \in \mathbb{N}_0$ abbreviate

$$f_m := \sum_{j=0}^{m} \lambda_j a_j. \tag{6.25}$$

With this piece of notation, the weak-type $(1, 1)$ boundedness of T (cf. Theorem 6.1) together with (6.23) give

$$Tf_m \to Tf \ \text{ in } L^{1,\infty}(\Sigma, \sigma) \text{ as } m \to \infty. \tag{6.26}$$

Hence, restricting to a subsequence if necessary,

$$Tf_m \to Tf \ \text{ pointwise } \sigma\text{-a.e. on } \Sigma \text{ as } m \to \infty. \tag{6.27}$$

We next claim that $\{Tf_m\}_{m\in\mathbb{N}}$ is a Cauchy sequence in $\mathrm{A}^p(\Sigma, \sigma)$. To justify this, use (6.13) and (6.25), so that whenever $m, \ell \in \mathbb{N}$ are such that $\ell \ge m$ we estimate

$$\|Tf_\ell - Tf_m\|_{\mathrm{A}^p(\Sigma,\sigma)} = \sum_{j=m+1}^{\ell} |\lambda_j| \|Ta_j\|_{\mathrm{A}^p(\Sigma,\sigma)} \le C \sum_{j=m+1}^{\infty} |\lambda_j|. \tag{6.28}$$

Since $\{\lambda_j\}_{j\in\mathbb{N}_0} \in \ell^1(\mathbb{N}_0)$, if we let $m \to \infty$ then the right-hand side tends to zero. As such, $\{Tf_m\}_{m\in\mathbb{N}}$ is a Cauchy sequence in $\mathrm{A}^p(\Sigma, \sigma)$. The completeness of $\mathrm{A}^p(\Sigma, \sigma)$ (cf. Theorem 3.9) then ensures the existence of some function $g \in \mathrm{A}^p(\Sigma, \sigma)$ for which

$$Tf_m \to g \ \text{ in } \mathrm{A}^p(\Sigma, \sigma) \text{ as } m \to \infty. \tag{6.29}$$

Thanks to (3.29) in Proposition 3.3, the latter convergence also holds in $L^1(\Sigma, \sigma)$. Restricting to a subsubsequence if necessary, we can say that

$$Tf_m \to g \ \text{ pointwise } \sigma\text{-a.e. on } \Sigma \text{ as } m \to \infty, \tag{6.30}$$

which in concert with (6.27) implies

$$Tf = g \ \text{ pointwise } \sigma\text{-a.e. on } \Sigma. \tag{6.31}$$

Collectively, (6.13), (6.24), (6.25), (6.29), and (6.31) entail

$$\begin{aligned}\|Tf\|_{\mathrm{A}^p(\Sigma,\sigma)} &= \lim_{m\to\infty} \|Tf_m\|_{\mathrm{A}^p(\Sigma,\sigma)} \le \lim_{m\to\infty} \sum_{j=0}^{m} |\lambda_j| \|Ta_j\|_{\mathrm{A}^p(\Sigma,\sigma)} \\ &\le C \sum_{j=0}^{\infty} |\lambda_j| \le 2C\|f\|_{\mathrm{HA}^p(\Sigma,\sigma)},\end{aligned} \tag{6.32}$$

for some $C \in (0, \infty)$ depending only on n, p, the kernel k, and the UR constants of Σ. This ultimately concludes the proof of Theorem 6.1. □

In the following result, we establish nontangential maximal function estimates for boundary-to-domain singular integral operators of convolution-type in the context of Beurling spaces and their associated Hardy spaces.

Theorem 6.2 *Assume $\Omega \subseteq \mathbb{R}^n$ is a uniformly rectifiable domain with unbounded boundary, and abbreviate $\sigma := \mathcal{H}^{n-1}\lfloor\partial\Omega$. Select a sufficiently large integer $N = N(n) \in \mathbb{N}$, and consider a kernel $k \in \mathscr{C}^N(\mathbb{R}^n \setminus \{0\})$ that is odd, and positive homogeneous of degree $1 - n$. Recall the boundary-to-domain convolution-type integral operator $\mathcal{T}$ associated with k as in* (6.8). *Finally, pick an integrability exponent $p \in (1, \infty)$ and an aperture parameter $\kappa \in (0, \infty)$.*

Then there exists some $C \in (0, \infty)$ depending only on n, p, κ, the kernel k, and the UR *constants of $\partial\Omega$ with the property that for every $f \in \mathrm{HA}^p(\partial\Omega, \sigma)$ one has*

$$\|\mathcal{N}_\kappa(\mathcal{T} f)\|_{\mathrm{A}^p(\partial\Omega,\sigma)} \le C\|f\|_{\mathrm{HA}^p(\partial\Omega,\sigma)}. \tag{6.33}$$

Proof Fix a reference point $x_0 \in \partial\Omega$, and a dilation factor $2_* > (C_{\partial\Omega}/c_{\partial\Omega})^{1/(n-1)}$, where $C_{\partial\Omega}, c_{\partial\Omega} \in (0, \infty)$ denote the upper and lower Ahlfors regularity constants of $\partial\Omega$, respectively. Consider an x_0-central $(1, p)$-atom $a : \partial\Omega \to \mathbb{R}$ and let $R \ge 2_*$ be a radius associated with a (in the sense of Definition 4.3). We claim that there exists some $C \in (0, \infty)$ depending only on n, p, κ, and the UR constants of $\partial\Omega$ with the property that

$$\|\mathcal{N}_\kappa(\mathcal{T} a)\|_{\mathrm{A}^p(\partial\Omega,\sigma)} \le C. \tag{6.34}$$

To verify this claim, fix $\rho \in (0, \infty)$ satisfying

$$\rho > 2 + \kappa, \tag{6.35}$$

and select the unique integer $N \in \mathbb{N}$ for which

$$2_*^N \le \rho R < 2_*^{N+1}. \tag{6.36}$$

Next, with $p' \in (1, \infty)$ denoting the Hölder conjugate exponent of p, write

$$\begin{aligned} \|\mathcal{N}_\kappa(\mathcal{T} a)\|_{\mathrm{A}^p(\partial\Omega,\sigma)} &= \sum_{j=0}^{N} 2_*^{j(n-1)/p'} \|\mathcal{N}_\kappa(\mathcal{T} a) \cdot \mathbb{1}_{C_j}\|_{L^p(\partial\Omega,\sigma)} \\ &\quad + \sum_{j=N+1}^{\infty} 2_*^{j(n-1)/p'} \|\mathcal{N}_\kappa(\mathcal{T} a) \cdot \mathbb{1}_{C_j}\|_{L^p(\partial\Omega,\sigma)} \\ &=: \mathrm{I} + \mathrm{II}. \end{aligned} \tag{6.37}$$

To estimate I we first use the Cotlar-type inequality from [GHA-III, Theorem 2.4.1, pp. 382–390] to bound (with $T_{\max}$ as in (6.2), and $\mathcal{M}_{\partial\Omega}$ denoting the Hardy-

Littlewood maximal operator on $\partial\Omega$)

$$\big[\mathcal{N}_\kappa(\mathcal{T}a)\big](x) \leq (T_{\max}a)(x) + C\cdot(\mathcal{M}_{\partial\Omega}a)(x) \quad \text{for all } x \in \partial\Omega \tag{6.38}$$

and for some constant $C = C(n, \kappa, \Sigma) \in (0, \infty)$. From this, the boundedness of $T_{\max}$ on $L^p(\partial\Omega, \sigma)$ (cf. Proposition 6.1), the boundedness of $\mathcal{M}_{\partial\Omega}$ on $L^p(\partial\Omega, \sigma)$, and the normalization condition of a (cf. (4.29)) we estimate

$$\begin{aligned}\|\mathcal{N}_\kappa(\mathcal{T}a)\|_{L^p(\partial\Omega,\sigma)} &\leq \|T_{\max}a\|_{L^p(\partial\Omega,\sigma)} + C\|\mathcal{M}_{\partial\Omega}a\|_{L^p(\partial\Omega,\sigma)} \\ &\leq C\|a\|_{L^p(\partial\Omega,\sigma)} \leq CR^{-(n-1)/p'}\end{aligned} \tag{6.39}$$

for some $C \in (0, \infty)$ depending only on n, p, κ, and the UR constants of $\partial\Omega$. Granted this and (6.36), we estimate the term I from (6.37) as

$$\begin{aligned} \mathrm{I} &= \sum_{j=0}^{N} 2_*^{j(n-1)/p'}\|\mathcal{N}_\kappa(\mathcal{T}a)\cdot \mathbb{1}_{C_j}\|_{L^p(\partial\Omega,\sigma)} \leq C\sum_{j=0}^{N} 2_*^{j(n-1)/p'}R^{-(n-1)/p'} \\ &\leq C\sum_{j=0}^{N} 2_*^{(j-N)(n-1)/p'} \leq C\sum_{\ell=0}^{\infty} 2_*^{-\ell(n-1)/p'} < \infty, \end{aligned} \tag{6.40}$$

for some constant $C = C(n, p, \kappa, \partial\Omega) \in (0, \infty)$.

To deal with II from (6.37), select $j \in \mathbb{N}_0$ with $j \geq N + 1$. Fix some $x \in C_j$ along with some $z \in \Gamma_\kappa(x)$. In particular, x and z satisfy

$$|x - x_0| \leq |x - z| + |z - x_0| < (\kappa + 2)|z - x_0|. \tag{6.41}$$

Also, keeping in mind (6.36), whenever $\xi \in B(x_0, R)$ we have

$$|\xi - x_0| < R \leq \frac{2_*^{N+1}}{\rho} \leq \frac{2_*^j}{\rho} \leq \frac{|x - x_0|}{\rho}, \tag{6.42}$$

As such, (6.41)-(6.42) imply

$$|z - \xi| \geq |z - x_0| - |\xi - x_0| > c|x - x_0| \quad \text{for all } \xi \in B(x_0, R), \tag{6.43}$$

where $c := \frac{1}{\kappa+2} - \frac{1}{\rho}$ is a positive constant thanks to (6.35). Moreover, notice that the smoothness and homogeneity of degree $1 - n$ of k imply

$$|\nabla k(y)| \leq \frac{1}{|y|^n}\sup_{\omega\in S^{n-1}}|\nabla k(\omega)| \quad \text{for all } y \in \mathbb{R}^n \setminus \{0\}. \tag{6.44}$$

Availing ourselves of (6.43)-(6.44), use the vanishing moment and normalization properties of a (cf. (4.29)-(4.30)), the Mean Value Theorem, Hölder's inequality, and the Ahlfors regularity of $\partial\Omega$ to estimate

$$
\begin{aligned}
|\mathcal{T}a(z)| &= \Big|\int_{\partial\Omega} k(z-y)a(y)\,d\sigma(y)\Big| \\
&= \Big|\int_{\partial\Omega} \big(k(z-y)-k(z-x_0)\big)a(y)\,d\sigma(y)\Big| \\
&\le \int_{\Delta(x_0,R)} |k(z-y)-k(z-x_0)||a(y)|\,d\sigma(y) \\
&\le \int_{\Delta(x_0,R)} \Big(\sup_{\xi\in[y,x_0]} \frac{C}{|z-\xi|^n}\Big)|y-x_0||a(y)|\,d\sigma(y) \\
&\le \frac{CR}{|x-x_0|^n}\int_{\Delta(x_0,R)} |a|\,d\sigma \\
&\le \frac{CR}{|x-x_0|^n}\|a\|_{L^p(\partial\Omega,\sigma)}\sigma\big(\Delta(x_0,R)\big)^{1/p'} \\
&\le \frac{CR}{|x-x_0|^n}
\end{aligned}
\tag{6.45}
$$

for some constant $C \in (0,\infty)$ depending only on n, p, κ, the kernel k, and the Ahlfors regularity of $\partial\Omega$. Since $z \in \Gamma_\kappa(x)$ was taken arbitrary to begin with, we conclude that

$$
\mathcal{N}_\kappa(\mathcal{T}a)(x) \le \frac{CR}{|x-x_0|^n} \quad \text{for every } x \in C_j. \tag{6.46}
$$

Based on this, (6.36), and the Ahlfors regularity of $\partial\Omega$, estimate

$$
\begin{aligned}
\|\mathcal{N}_\kappa(\mathcal{T}a)\cdot \mathbb{1}_{C_j}\|_{L^p(\partial\Omega,\sigma)} &\le CR\Big(\int_{C_j}\frac{d\sigma(x)}{|x-x_0|^{np}}\Big)^{1/p} \\
&\le C\frac{2_*^N}{2_*^{jn}}\sigma\big(\Delta(x_0,2_*^{j+1})\big)^{1/p} \\
&\le C\frac{2_*^N}{2_*^{jn}}2_*^{j(n-1)/p} = C\frac{2_*^{-j(n-1)/p'}}{2_*^{j-N}}
\end{aligned}
\tag{6.47}
$$

for some constant $C \in (0,\infty)$ depending exclusively on n, p, κ, the kernel k, and the Ahlfors regularity character of $\partial\Omega$ (in particular, independent of a). With estimate (6.47) in hand, we can now bound II from (6.37) as

$$
\begin{aligned}
\mathrm{II} &= \sum_{j=N+1}^{\infty} 2_*^{j(n-1)/p'}\|\mathcal{N}_\kappa(\mathcal{T}a)\cdot \mathbb{1}_{C_j}\|_{L^p(\partial\Omega,\sigma)} \\
&\le \sum_{j=N+1}^{\infty}\frac{C}{2_*^{j-N}} = \sum_{\ell=1}^{\infty}\frac{C}{2_*^{\ell}} < \infty
\end{aligned}
\tag{6.48}
$$

with $C \in (0, \infty)$ of same nature as in (6.47). Putting it all together, (6.37), (6.40), and (6.48) imply (6.34).

Having established (6.34), we are now in a position to prove (6.33). To this end, fix an arbitrary function $f \in \mathrm{HA}^p(\partial\Omega, \sigma)$ with atomic decomposition (cf. Theorem 4.2)

$$f = \sum_{j=0}^{\infty} \lambda_j a_j \quad \text{in } L^1(\partial\Omega, \sigma), \tag{6.49}$$

for some sequence $\{a_j\}_{j\in\mathbb{N}_0}$ of x_0-central $(1, p)$-atoms and some numerical sequence $\{\lambda_j\}_{j\in\mathbb{N}_0} \in \ell^1(\mathbb{N}_0)$. Also, assume that such a decomposition is quasi-optimal, i.e.,

$$\sum_{j=0}^{\infty} |\lambda_j| \leq 2\|f\|_{\mathrm{HA}^p(\partial\Omega,\sigma)}. \tag{6.50}$$

For each $m \in \mathbb{N}_0$, abbreviate

$$f_m := \sum_{j=0}^{m} \lambda_j a_j. \tag{6.51}$$

Being mindful of (6.10), and that $\mathrm{HA}^p(\partial\Omega, \sigma)$ embeds continuously into $L^1(\partial\Omega, \sigma)$ (see Proposition 4.2), for each $m \in \mathbb{N}_0$ we have

$$\begin{aligned} \|\mathcal{N}_\kappa(\mathcal{T}f) - \mathcal{N}_\kappa(\mathcal{T}f_m)\|_{L^{1,\infty}(\partial\Omega,\sigma)} &\leq \|\mathcal{N}_\kappa\big(\mathcal{T}(f - f_m)\big)\|_{L^{1,\infty}(\partial\Omega,\sigma)} \\ &\leq C\|f - f_m\|_{L^1(\partial\Omega,\sigma)}. \end{aligned} \tag{6.52}$$

Thanks to (6.49) and (6.51), if we let $m \to \infty$ then the term in the right-hand side of (6.52) approaches zero. As such, $\mathcal{N}_\kappa(\mathcal{T}f_m) \to \mathcal{N}_\kappa(\mathcal{T}f)$ in $L^{1,\infty}(\partial\Omega, \sigma)$ as $m \to \infty$. Restricting to a subsequence if necessary, it follows that

$$\mathcal{N}_\kappa(\mathcal{T}f_m) \to \mathcal{N}_\kappa(\mathcal{T}f) \quad \text{at } \sigma\text{-a.e. point on } \partial\Omega \text{ as } m \to \infty. \tag{6.53}$$

Going further, we claim the sequence $\big\{\mathcal{N}_\kappa(\mathcal{T}f_m)\big\}_{m\in\mathbb{N}_0}$ is Cauchy in $\mathrm{A}^p(\partial\Omega, \sigma)$. Indeed, select $\ell, m \in \mathbb{N}_0$ with $m \geq \ell$, and use (6.34) to bound

$$\begin{aligned} &\|\mathcal{N}_\kappa(\mathcal{T}f_m) - \mathcal{N}_\kappa(\mathcal{T}f_\ell)\|_{\mathrm{A}^p(\partial\Omega,\sigma)} \leq \|\mathcal{N}_\kappa\big(\mathcal{T}(f_m - f_\ell)\big)\|_{\mathrm{A}^p(\partial\Omega,\sigma)} \\ &\quad = \Big\|\mathcal{N}_\kappa\Big(\sum_{j=\ell+1}^{m} \lambda_j \mathcal{T}a_j\Big)\Big\|_{\mathrm{A}^p(\partial\Omega,\sigma)} \leq \sum_{j=\ell+1}^{m} |\lambda_j| \cdot \|\mathcal{N}_\kappa(\mathcal{T}a_j)\|_{\mathrm{A}^p(\partial\Omega,\sigma)} \\ &\quad \leq C \sum_{j=\ell+1}^{\infty} |\lambda_j|. \end{aligned} \tag{6.54}$$

Taking into account the membership $\{\lambda_j\}_{j\in\mathbb{N}_0} \in \ell^1(\mathbb{N}_0)$, if we let $\ell \to \infty$ in (6.54) then the right-hand side tends to zero. Hence, $\{\mathcal{N}_\kappa(\mathcal{T} f_m)\}_{m\in\mathbb{N}_0}$ is a Cauchy sequence in $\mathrm{A}^p(\partial\Omega, \sigma)$, as desired. Based on the completeness of $\mathrm{A}^p(\partial\Omega, \sigma)$ (cf. Proposition 3.9),

$$\begin{gathered}\text{there exists } g \in \mathrm{A}^p(\partial\Omega,\sigma) \text{ such that}\\ \mathcal{N}_\kappa(\mathcal{T} f_m) \to g \ \text{ in } \mathrm{A}^p(\partial\Omega,\sigma) \text{ as } m \to \infty.\end{gathered} \tag{6.55}$$

Because $\mathrm{A}^p(\partial\Omega, \sigma)$ is continuously embedded in $L^1(\partial\Omega, \sigma)$ (cf. Proposition 3.3), this convergence also holds in $L^1(\partial\Omega, \sigma)$. Hence, restricting to a subsubsequence if necessary we can write

$$\mathcal{N}_\kappa(\mathcal{T} f_m) \to g \ \text{ at } \sigma\text{-a.e. point on } \partial\Omega \text{ as } m \to \infty, \tag{6.56}$$

which together with (6.53) gives

$$\mathcal{N}_\kappa(\mathcal{T} f) = g \ \text{ at } \sigma\text{-a.e. point on } \partial\Omega. \tag{6.57}$$

Collectively, (6.34), (6.50), (6.51), (6.55), and (6.57) permit us to estimate

$$\begin{aligned}\|\mathcal{N}_\kappa(\mathcal{T} f)\|_{\mathrm{A}^p(\partial\Omega,\sigma)} &= \lim_{m\to\infty} \|\mathcal{N}_\kappa(\mathcal{T} f_m)\|_{\mathrm{A}^p(\partial\Omega,\sigma)}\\ &\le \lim_{m\to\infty} \sum_{j=0}^{m} |\lambda_j| \|\mathcal{N}_\kappa(\mathcal{T} a_j)\|_{\mathrm{A}^p(\partial\Omega,\sigma)}\\ &\le C \lim_{m\to\infty} \sum_{j=0}^{m} |\lambda_j| \le C \sum_{j=0}^{\infty} |\lambda_j| \le 2C\|f\|_{\mathrm{HA}^p(\partial\Omega,\sigma)},\end{aligned} \tag{6.58}$$

for some $C \in (0, \infty)$ depending only on n, p, κ, the kernel k, and the UR constants of $\partial\Omega$. This ultimately establishes (6.33), thus concluding the proof of Theorem 6.2. □

The size of a function, measured using the nontangential maximal function, on Beurling algebras is independent of the aperture parameter. Specifically, we have the following result.

Proposition 6.2 *Consider an Ahlfors regular domain $\Omega \subseteq \mathbb{R}^n$ with unbounded boundary, and abbreviate $\sigma := \mathcal{H}^{n-1}\lfloor\partial\Omega$. Fix an integrability exponent $p \in (1, \infty)$ and two aperture parameters $\kappa, \widetilde{\kappa} \in (0, \infty)$. Then for every $\mathcal{L}^n$-measurable function $u : \Omega \to \mathbb{C}$ it follows that*

$$\|\mathcal{N}_\kappa u\|_{\mathrm{A}^p(\partial\Omega,\sigma)} \approx \|\mathcal{N}_{\widetilde{\kappa}} u\|_{\mathrm{A}^p(\partial\Omega,\sigma)} \ \text{ and } \ \|\mathcal{N}_\kappa u\|_{\mathrm{B}^p(\partial\Omega,\sigma)} \approx \|\mathcal{N}_{\widetilde{\kappa}} u\|_{\mathrm{B}^p(\partial\Omega,\sigma)}, \tag{6.59}$$

with proportionality constants depending only on n, p, κ, $\widetilde{\kappa}$, and the Ahlfors regularity character of $\partial\Omega$.

Proof Without loss of generality, assume $\widetilde{\kappa} \geq \kappa$, and pick an exponent $s \in (0, 1)$. Then thanks to [GHA-I, Theorem 8.4.6, pp. 704–705] and the lattice properties of $\mathrm{A}^p(\partial\Omega, \sigma)$ and $\mathrm{B}^p(\partial\Omega, \sigma)$ (cf. Remark 3.3) we have

$$\|\mathcal{N}_\kappa u\|_{\mathrm{A}^p(\partial\Omega,\sigma)} \leq \|\mathcal{N}_{\widetilde{\kappa}} u\|_{\mathrm{A}^p(\partial\Omega,\sigma)} \leq C\|\mathcal{M}_{\partial\Omega,s}(\mathcal{N}_\kappa u)\|_{\mathrm{A}^p(\partial\Omega,\sigma)} \tag{6.60}$$

and

$$\|\mathcal{N}_\kappa u\|_{\mathrm{B}^p(\partial\Omega,\sigma)} \leq \|\mathcal{N}_{\widetilde{\kappa}} u\|_{\mathrm{B}^p(\partial\Omega,\sigma)} \leq C\|\mathcal{M}_{\partial\Omega}(\mathcal{N}_\kappa u)\|_{\mathrm{B}^p(\partial\Omega,\sigma)}, \tag{6.61}$$

for some $C \in (0, \infty)$ depending only on n, κ, $\widetilde{\kappa}$, and the Ahlfors regularity constants of $\partial\Omega$, where $\mathcal{M}_{\partial\Omega,s}$ is the L^s-based Hardy-Littlewood maximal operator on $\partial\Omega$ (cf. (2.28)), and $\mathcal{M}_{\partial\Omega}$ is the Hardy-Littlewood maximal operator on $\partial\Omega$ (cf. (2.26)). As such, the boundedness of $\mathcal{M}_{\partial\Omega,s}$ on $\mathrm{A}^p(\partial\Omega, \sigma)$ (cf. Theorem 3.3), and the boundedness of $\mathcal{M}_{\partial\Omega}$ on $\mathrm{B}^p(\partial\Omega, \sigma)$ (cf. Theorem 3.1) readily imply (6.59). □

We shift our attention to a special class of singular integral operators whose integral kernel involves a vector-valued function that is divergence-free. This special algebraic structure of these integral kernels interplays favorably with the cancellation properties of functions from Beurling-Hardy spaces. To be specific, for some sufficiently large integer $N = N(n) \in \mathbb{N}$ consider a kernel $\vec{k} \in [\mathscr{C}^N(\mathbb{R}^n \setminus \{0\})]^n$, that is odd, positive homogeneous of degree $1 - n$, and divergence-free, i.e., for all $x \in \mathbb{R}^n \setminus \{0\}$ and all scalars $\lambda \in (0, \infty)$, the vector-valued function $\vec{k}$ satisfies

$$\begin{aligned} \vec{k}(-x) = -\vec{k}(x), \quad \vec{k}(\lambda x) = \lambda^{1-n}\vec{k}(x), \\ \text{and} \quad \operatorname{div}\vec{k} = 0 \quad \text{in } \mathbb{R}^n \setminus \{0\}. \end{aligned} \tag{6.62}$$

Consider a UR domain $\Omega \subseteq \mathbb{R}^n$. Abbreviate $\sigma := \mathcal{H}^{n-1}\lfloor\partial\Omega$ and denote by ν the geometric measure theoretic outward unit normal to Ω. In this setting, following [GHA-IV, (5.1.5)-(5.1.6), p. 532], for each function $f \in L^1\big(\partial\Omega, \frac{\sigma(x)}{1+|x|^{n-1}}\big)$, define the `Generalized Double Layer Potential` T acting on f according to

$$(Tf)(x) := \lim_{\varepsilon\to 0^+} \int_{\substack{y\in\partial\Omega \\ |x-y|>\varepsilon}} \langle \nu(y), \vec{k}(y-x)\rangle f(y)\, d\sigma(y) \quad \text{for } \sigma\text{-a.e. } x \in \partial\Omega, \tag{6.63}$$

and its formal transpose $T^{\#}$ acting on $f \in L^1\big(\partial\Omega, \frac{\sigma(x)}{1+|x|^{n-1}}\big)$ as

$$(T^{\#}f)(x) := \lim_{\varepsilon\to 0^+} \int_{\substack{y\in\partial\Omega \\ |x-y|>\varepsilon}} \langle \nu(x), \vec{k}(x-y)\rangle f(y)\, d\sigma(y) \quad \text{for } \sigma\text{-a.e. } x \in \partial\Omega. \tag{6.64}$$

The existence of these principal value limits is guaranteed by work done in [GHA-III, Theorem 2.3.2, pp. 348–357] (see also [GHA-IV, Theorem 5.1.1, pp. 532–550]). To offer an example, consider vector-valued function $\vec{k}_\Delta$ defined as

$$\vec{k}_\Delta(z) := \frac{1}{\omega_{n-1}} \frac{z}{|z|^n} \quad \text{for all } z \in \mathbb{R}^n \setminus \{0\}. \tag{6.65}$$

Then $\vec{k}_\Delta$ satisfies (6.62), and this choice yields gives rise to the **boundary-to-boundary Harmonic double layer potentials**:

$$(K_\Delta f)(x) := \lim_{\varepsilon\to 0^+} \frac{1}{\omega_{n-1}} \int\limits_{\substack{y\in\partial\Omega \\ |x-y|>\varepsilon}} \frac{\langle \nu(y), y-x\rangle}{|x-y|^n} f(y)\, d\sigma(y) \quad \text{for } \sigma\text{-a.e. } x \in \partial\Omega, \tag{6.66}$$

and its formal transpose

$$(K^\#_\Delta f)(x) := \lim_{\varepsilon\to 0^+} \frac{1}{\omega_{n-1}} \int\limits_{\substack{y\in\partial\Omega \\ |x-y|>\varepsilon}} \frac{\langle \nu(x), x-y\rangle}{|x-y|^n} f(y)\, d\sigma(y) \quad \text{for } \sigma\text{-a.e. } x \in \partial\Omega. \tag{6.67}$$

Let us comment on the boundedness of the generalized double layer potentials on L^p with $1 < p < \infty$. Consider $\vec{k} = (k_1, \dots, k_n) \in [\mathscr{C}^N(\mathbb{R}^n \setminus \{0\})]^n$ as before (i.e., satisfying (6.62)), and write the components of the geometric measure theoretic unit normal to the UR domain Ω as $\nu = (\nu_1, \dots, \nu_n)$. For each $j \in \{1, \dots, n\}$, denote by M_{ν_j} the operator of pointwise multiplication by ν_j, and let T_{k_j} be the convolution-type integral operator associated with the kernel k_j as in (6.3). Then for each function $f \in L^1\big(\partial\Omega, \frac{\sigma(x)}{1+|x|^{n-1}}\big)$, the action T and $T^\#$ on f (defined in (6.63)-(6.64)) can be written as

$$Tf = \sum_{j=1}^n (T_{k_j} M_{\nu_j}) f \quad \text{and} \quad T^\# f = \sum_{j=1}^n (M_{\nu_j} T_{k_j}) f. \tag{6.68}$$

Since $\nu_j \in L^\infty(\partial\Omega, \sigma)$, it is clear that M_{ν_j} is a bounded operator on $L^p(\partial\Omega, \sigma)$. Hence, because T_{k_j} is also bounded on $L^p(\partial\Omega, \sigma)$ (cf. Proposition 6.1), from (6.68) we see that

$$\begin{gathered} T : L^p(\partial\Omega, \sigma) \to L^p(\partial\Omega, \sigma) \quad \text{and} \quad T^\# : L^p(\partial\Omega, \sigma) \to L^p(\partial\Omega, \sigma) \\ \text{are well-defined, linear, and bounded operators.} \end{gathered} \tag{6.69}$$

Next, we prove that the generalized double layer potentials $T^\#$ maps HA^p into itself in a bounded fashion.

Theorem 6.3 *Assume $\Omega \subseteq \mathbb{R}^n$ is a uniformly rectifiable domain with unbounded boundary. Abbreviate $\sigma := \mathcal{H}^{n-1}\lfloor\partial\Omega$ and denote by ν the geometric measure theoretic outward unit normal to Ω. For some sufficiently large $N = N(n) \in \mathbb{N}$, pick a kernel $\vec{k} \in \big[\mathscr{C}^N(\mathbb{R}^n \setminus \{0\})\big]^n$ that is odd, positive homogeneous of degree $1-n$, and divergence-free, i.e.,* $\operatorname{div}\vec{k} = 0$ *in* $\mathbb{R}^n \setminus \{0\}$. *Recall the generalized double layer potential $T^\#$ associated with $\vec{k}$ as in* (6.64). *Then, for any integrability exponent $p \in (1, \infty)$, it follows that*

$$T^{\#}: \mathrm{HA}^p(\partial\Omega,\sigma) \longrightarrow \mathrm{HA}^p(\partial\Omega,\sigma) \tag{6.70}$$
is a well-defined, linear, and bounded operator.

Proof Fix an integrability exponent $p \in (1,\infty)$. Notice that the action of $T^{\#}$ on functions from $\mathrm{HA}^p(\partial\Omega,\sigma)$ is well defined, thanks to (4.15). Select an arbitrary x_0-central $(1,p)$-atom $a : \partial\Omega \to \mathbb{R}$, and letting $R \geq 2_*$ be a radius associated with a (as in Definition 4.3). Having fixed $\theta \in (0,\infty)$, with $(n-1)(p-1) < \theta < n(p-1)+1$, we claim that there exists some constant $C_\theta \in (0,\infty)$, independent of a, for which the function

$$T^{\#}a \text{ is a fixed } C_\theta\text{-multiple of an } x_0\text{-central } (1,p,\theta)\text{-molecule.} \tag{6.71}$$

To justify this claim, we first observe that the boundedness of $T^{\#}$ on $L^p(\partial\Omega,\sigma)$ (cf. (6.69)) along with the normalization property of a (see (4.29)) yield

$$\|T^{\#}a\|_{L^p(\partial\Omega,\sigma)} \leq C\|a\|_{L^p(\partial\Omega,\sigma)} \leq CR^{-(n-1)/p'}, \tag{6.72}$$

for some $C \in (0,\infty)$ depending only on n, p, and the UR constants of $\partial\Omega$, where $p' \in (1,\infty)$ denotes the Hölder conjugate exponent of p.

The next goal is to show that $T^{\#}a$ satisfies the following decay property:

$$\Big(\int_{\partial\Omega} |(T^{\#}a)(x)|^p |x-x_0|^\theta\, d\sigma(x)\Big)^{1/p} \leq CR^{-(n-1)/p'+\theta/p}, \tag{6.73}$$

for some constant $C \in (0,\infty)$ independent of a. To verify this, estimate

$$\begin{aligned} &\Big(\int_{\partial\Omega} |(T^{\#}a)(x)|^p |x-x_0|^\theta\, d\sigma(x)\Big)^{1/p} \\ &\quad\leq \Big(\int_{\substack{x\in\partial\Omega \\ |x-x_0|<2R}} |(T^{\#}a)(x)|^p |x-x_0|^\theta\, d\sigma(x)\Big)^{1/p} \\ &\qquad + \Big(\int_{\substack{x\in\partial\Omega \\ |x-x_0|\geq 2R}} |(T^{\#}a)(x)|^p |x-x_0|^\theta\, d\sigma(x)\Big)^{1/p} =: \mathrm{I} + \mathrm{II}. \end{aligned} \tag{6.74}$$

To deal with I, use (6.72) to bound

$$\begin{aligned} \mathrm{I} &= \Big(\int_{\substack{x\in\partial\Omega \\ |x-x_0|<2R}} |(T^{\#}a)(x)|^p |x-x_0|^\theta\, d\sigma(x)\Big)^{1/p} \\ &\leq (2R)^{\theta/p}\|T^{\#}a\|_{L^p(\partial\Omega,\sigma)} \leq CR^{-(n-1)/p'+\theta/p}, \end{aligned} \tag{6.75}$$

for some $C \in (0,\infty)$ depending exclusively on n, p, θ, and the UR constants of $\partial\Omega$.

To obtain a similar bound for II from (6.74), observe that

$$|x-z| \approx |x-x_0| \text{ uniformly with respect to } x \in \partial\Omega\setminus B(x_0,2R) \text{ and } z \in B(x_0,R). \tag{6.76}$$

In view of this, the vanishing moment property of a (cf. (4.30)), and bearing in mind that $\operatorname{supp} a \subseteq \Delta(x_0, R)$ (cf. (4.28)), whenever $x \in \partial\Omega \backslash B(x_0, 2R)$ we may write

$$(T^{\#}a)(x) = \int_{\partial\Omega} \big\langle \nu(x), \vec{k}(x-y)\big\rangle a(y)\, d\sigma(y)$$

$$= \int_{\Delta(x_0,R)} \big\langle \nu(x), \vec{k}(x-y) - \vec{k}(x-x_0)\big\rangle a(y)\, d\sigma(y). \tag{6.77}$$

Notice that the smoothness and the homogeneity of $\vec{k}$ yield

$$\big|\nabla\vec{k}(z)\big| \leq \frac{1}{|z|^n}\Big(\sup_{S^{n-1}} |\nabla\vec{k}|\Big) \quad \text{for all } z \in \mathbb{R}^n \setminus \{0\}. \tag{6.78}$$

Using (6.76), (6.77), (6.78), Hölder's inequality, the Mean Value Theorem, that $|\nu| = 1$ at σ-a.e. point on $\partial\Omega$, the normalization property of a (cf. (4.29)), and the Ahlfors regularity of $\partial\Omega$ it follows that

$$\begin{aligned}
|T^{\#}a(x)| &\leq \int_{\Delta(x_0,R)} \big|\big\langle \nu(x), \vec{k}(x-y) - \vec{k}(x-x_0)\big\rangle\big| |a(y)|\, d\sigma(y) \\
&\leq \int_{\Delta(x_0,R)} \big|\vec{k}(x-y) - \vec{k}(x-x_0)\big| |a(y)|\, d\sigma(y) \\
&\leq \int_{\Delta(x_0,R)} \Big(\sup_{\xi\in[y,x_0]} \big|\nabla\vec{k}(x-\xi)\big|\Big) |y-x_0| |a(y)|\, d\sigma(y) \\
&\leq CR \int_{\Delta(x_0,R)} \Big(\sup_{\xi\in[y,x_0]} \frac{1}{|x-\xi|^n}\Big) |a(y)|\, d\sigma(y), \\
&\leq \frac{CR}{|x-x_0|^n} \int_{\Delta(x_0,R)} |a|\, d\sigma \\
&\leq \frac{CR}{|x-x_0|^n} \|a\|_{L^p(\partial\Omega,\sigma)} \sigma\big(\Delta(x_0,R)\big)^{1/p'} \\
&\leq \frac{CR}{|x-x_0|^n} \|a\|_{L^p(\partial\Omega,\sigma)} R^{(n-1)/p'} \\
&\leq \frac{CR}{|x-x_0|^n},
\end{aligned} \tag{6.79}$$

for some constant $C \in (0, \infty)$ depending only on n, p, and the Ahlfors regularity character of $\partial\Omega$. In view of this, and the fact that we have $\theta < n(p-1)+1$, which is further equivalent to $\frac{\theta}{p} - n + \frac{n-1}{p} < 0$, we estimate II from (6.74) as follows:

$$
\begin{aligned}
\mathrm{II} &= \Big(\int_{\substack{x\in\partial\Omega \\ |x-x_0|\geq 2R}} |(T^{\#}a)(x)|^p |x-x_0|^{\theta}\, d\sigma(x) \Big)^{1/p} \\
&\leq CR \Big(\int_{\substack{x\in\partial\Omega \\ |x-x_0|\geq 2R}} \frac{|x-x_0|^{\theta}}{|x-x_0|^{np}}\, d\sigma(x) \Big)^{1/p} \\
&\leq CR \sum_{j=1}^{\infty} \Big(\int_{\substack{x\in\partial\Omega \\ 2^j R\leq |x-x_0|\leq 2^{j+1}R}} \frac{|x-x_0|^{\theta}}{|x-x_0|^{np}}\, d\sigma(x) \Big)^{1/p} \\
&\leq CR \sum_{j=1}^{\infty} \frac{(2^{j+1}R)^{\theta/p}}{(2^j R)^n} \sigma\big(\Delta(x_0, 2^{j+1}R)\big)^{1/p} \\
&\leq CR^{\theta/p-(n-1)} \sum_{j=0}^{\infty} (2^j)^{\theta/p-n} (2^{j+1}R)^{(n-1)/p} \\
&= CR^{-(n-1)/p'+\theta/p} \sum_{j=0}^{\infty} 2^{j[\theta/p-n+(n-1)/p]} < \infty. \qquad (6.80)
\end{aligned}
$$

Putting it all together, (6.74), (6.75), and (6.80) guarantee that there exists some constant $C \in (0,\infty)$ depending exclusively on n, p, θ, and the Ahlfors regularity character of $\partial\Omega$ (in particular, independent of the atom a) for which (6.73) holds.

Recalling that we are in the process of showing the claim made in (6.71), we are left with proving the vanishing moment property of $T^{\#}a$. To this end, introduce the (boundary-to-domain) convolution-type integral operator $\mathcal{W}$ associated with $\vec{k}$ acting on each function $f \in L^1\big(\partial\Omega, \frac{\sigma(x)}{1+|x|^{n-1}}\big)$ according to

$$
(\mathcal{W}f)(x) := \int_{\partial\Omega} \vec{k}(x-y) f(y)\, d\sigma(y) \quad \text{for all } x \in \Omega. \qquad (6.81)
$$

By the properties of the integral kernel $\vec{k}$, for each $f \in L^1\big(\partial\Omega, \frac{\sigma(x)}{1+|x|^{n-1}}\big)$ the function $\mathcal{W}f$ satisfies

$$
\begin{aligned}
&\mathcal{W}f \in \big[\mathscr{C}^N(\Omega)\big]^n, \ \operatorname{div}(\mathcal{W}f) = 0 \ \text{ in } \Omega, \text{ and at } \sigma\text{-a.e. } x \in \partial\Omega \text{ we have} \\
&(\mathcal{W}f)\Big|_{\partial\Omega}^{\kappa-\mathrm{n.t.}}(x) = \frac{1}{2i}\widehat{\vec{k}}\big(\nu(x)\big) f(x) + \lim_{\varepsilon\to 0^+} \int_{\substack{y\in\partial\Omega \\ |x-y|>\varepsilon}} \vec{k}(x-y) f(y)\, d\sigma(y), \qquad (6.82)
\end{aligned}
$$

where the last property follows from the jump-formula (6.11) recorded in Proposition 6.1. Furthermore, it is relevant to observe that the components of the vector-valued operator $\mathcal{W}$ are themselves integral operators of the form described in (6.8). As such, having fixed an aperture parameter $\kappa \in (0,\infty)$, Theorem 6.2 entails

$$
\mathcal{N}_\kappa(\mathcal{W}f) \in \mathrm{A}^p(\partial\Omega,\sigma) \quad \text{for all } f \in \mathrm{HA}^p(\partial\Omega,\sigma). \qquad (6.83)
$$

For each $j \in \{1, \dots, n\}$, denote by k_j, the j-th component of $\vec{k}$. The assumptions made on $\vec{k}$ yield that k_j belongs to $\mathscr{C}^N(\mathbb{R}^n \setminus \{0\})$, is odd, and positive homogeneous of degree $1-n$. As such, this puts k_j in $\mathcal{D}'(\mathbb{R}^n)$ for each $j \in \{1, \dots, n\}$, which in light of [Mi18, Example 4.71, p. 151, and Theorem 4.72, p. 152] allows us to conclude that

$$\partial_j k_j = \Big(\int_{S^{n-1}} k_j(\omega)\omega_j \, d\mathcal{H}^{n-1}(\omega) \Big) \delta_0 + \text{p.v.}(\partial_j k_j) \tag{6.84}$$

for each $j \in \{1, \dots, n\}$. Above, the partial derivative in the left-hand side is taken in the distributional sense, δ_0 denotes the Dirac distribution with mass at the origin in $\mathbb{R}^n$, and $\text{p.v.}(\partial_j k_j)$ is the principal value distribution associated with the function $\partial_j k_j \in \mathscr{C}^{N-1}(\mathbb{R}^n \setminus \{0\})$ according to

$$\langle \text{p.v.}(\partial_j k_j), \varphi \rangle := \lim_{\varepsilon \to 0^+} \int_{\substack{x \in \mathbb{R}^n \\ |x| > \varepsilon}} (\partial_j k_j)(x) \varphi(x) \, d\mathcal{L}^n(x) \tag{6.85}$$

for all $\varphi \in \mathscr{C}_0^\infty(\mathbb{R}^n)$, and for each $j \in \{1, \dots, n\}$. Summing up over $j \in \{1, \dots, n\}$ in (6.84), and bearing in mind that $\vec{k}$ is divergence-free, we find that

$$\begin{aligned} \operatorname{div} \vec{k} &= \Big(\int_{S^{n-1}} \langle \vec{k}(\omega), \omega \rangle \, d\mathcal{H}^{n-1}(\omega) \Big) \delta_0 + \text{p.v.}(\operatorname{div} \vec{k}) \\ &= \Big(\int_{S^{n-1}} \langle \vec{k}(\omega), \omega \rangle \, d\mathcal{H}^{n-1}(\omega) \Big) \delta_0 \quad \text{in } \mathcal{D}'(\mathbb{R}^n). \end{aligned} \tag{6.86}$$

Taking the Fourier transform of both sides of (6.86), using that $\widehat{\delta_0} = 1$ in $\mathcal{D}'(\mathbb{R}^n)$, together with

$$\widehat{(\operatorname{div} \vec{k})}(\xi) = \sum_{j=1}^{n} -i\xi_j \widehat{k_j}(\xi) = -i\xi \cdot \widehat{\vec{k}}(\xi) \quad \text{for all } \xi \in \mathbb{R}^n \setminus \{0\}, \tag{6.87}$$

we obtain

$$-i\xi \cdot \widehat{\vec{k}}(\xi) = \int_{S^{n-1}} \langle \vec{k}(\omega), \omega \rangle \, d\mathcal{H}^{n-1}(\omega) \quad \text{for every } \xi \in \mathbb{R}^n \setminus \{0\}. \tag{6.88}$$

For σ-a.e. $x \in \partial\Omega$, choosing $\xi := \nu(x) \in \mathbb{R}^n \setminus \{0\}$ implies

$$-i\nu(x) \cdot \widehat{\vec{k}}\big(\nu(x)\big) = \int_{S^{n-1}} \langle \vec{k}(\omega), \omega \rangle \, d\mathcal{H}^{n-1}(\omega). \tag{6.89}$$

Combining this formula with the jump-relation recorded in (6.82) implies that for every function $f \in L^1\big(\partial\Omega, \frac{\sigma(x)}{1+|x|^{n-1}}\big)$ we have

$$\begin{aligned}\big\langle \nu(x), (\mathcal{W} f)\big|^{\kappa-\text{n.t.}}_{\partial\Omega}(x)\big\rangle &= \frac{1}{2i}\nu(x)\cdot\widehat{\vec{k}}\big(\nu(x)\big)f(x)\\ &\quad+\lim_{\varepsilon\to 0^+}\int_{\substack{y\in\partial\Omega\\|x-y|>\varepsilon}}\big\langle \nu(x), \vec{k}(x-y)\big\rangle f(y)\,d\sigma(y)\\ &= \frac{1}{2}\Big(\int_{S^{n-1}}\big\langle\vec{k}(\omega),\omega\big\rangle\,d\mathcal{H}^{n-1}(\omega)\Big)f(x)+(T^{\#}f)(x)\end{aligned} \tag{6.90}$$

for σ-a.e. $x\in\partial\Omega$. Abbreviating $\vartheta := \frac{1}{2}\int_{S^{n-1}}\big\langle\vec{k}(\omega),\omega\big\rangle\,d\mathcal{H}^{n-1}(\omega)\in\mathbb{C}$, and specializing (6.90) to the choice $f := a$ (our x_0-central $(1,p)$-atom initially fixed) shows that we may express $T^{\#}a$ as

$$(T^{\#}a)(x) = \vartheta a(x) + \big\langle \nu(x), (\mathcal{W} a)\big|^{\kappa-\text{n.t.}}_{\partial\Omega}(x)\big\rangle \quad\text{for } \sigma\text{-a.e. } x\in\partial\Omega. \tag{6.91}$$

Upon recalling (6.83), and that $\mathrm{A}^p(\partial\Omega,\sigma)\subseteq L^1(\partial\Omega,\sigma)$ (cf. Proposition 3.3), for every aperture parameter $\kappa\in(0,\infty)$ we have $\mathcal{N}_\kappa(\mathcal{W} a)\in L^1(\partial\Omega,\sigma)$. Granted this, (6.91), the fact that $\mathrm{div}(\mathcal{W} a)=0$ in Ω (see (6.82)), the vanishing moment property of a (see (4.30)), and the Divergence Theorem from [GHA-I, Theorem 1.2.1, p. 18], we arrive at

$$\int_{\partial\Omega} T^{\#}a\,d\sigma = \int_{\partial\Omega}\big\langle \nu, (\mathcal{W} a)\big|^{\kappa-\text{n.t.}}_{\partial\Omega}\big\rangle\,d\sigma = \int_{\Omega}\mathrm{div}(\mathcal{W} a)\,d\mathcal{L}^n = 0. \tag{6.92}$$

Putting together this with (6.72) and (6.73) establishes (6.71).

As a consequence of (6.71), Theorem 4.4 ensures the existence of some constants $C\in(0,\infty)$ depending only on n, p, θ, and the Ahlfors regularity character of $\partial\Omega$ with the property that

$$\|T^{\#}a\|_{\mathrm{HA}^p(\partial\Omega,\sigma)}\le C \tag{6.93}$$

for every x_0-central $(1,p)$-atom a on $\partial\Omega$.

Moving on, the goal is to prove that the operator $T^{\#}$ maps $\mathrm{HA}^p(\partial\Omega,\sigma)$ into itself in a bounded fashion. Fix an arbitrary function $f\in\mathrm{HA}^p(\partial\Omega,\sigma)$, and use Theorem 4.2 to write

$$f = \sum_{j=0}^{\infty}\lambda_j a_j \quad\text{in } L^1(\partial\Omega,\sigma), \tag{6.94}$$

for some sequence of x_0-central $(1,p)$-atoms $\{a_j\}_{j\in\mathbb{N}_0}$ and some numerical sequence $\{\lambda_j\}_{j\in\mathbb{N}_0}\in\ell^1(\mathbb{N}_0)$. Moreover, assume that this atomic decomposition is quasi-optimal, in the sense that

$$\sum_{j=0}^{\infty}|\lambda_j|\le 2\|f\|_{\mathrm{HA}^p(\partial\Omega,\sigma)}. \tag{6.95}$$

Abbreviate

$$f_m := \sum_{j=0}^{m} \lambda_j a_j \quad \text{for each } m \in \mathbb{N}_0. \tag{6.96}$$

Since the operator $T^{\#}$ is of weak-type $(1,1)$ (cf. Proposition 6.1), the convergence (6.94) in $L^1(\partial\Omega, \sigma)$ yields

$$T^{\#} f_m \to T^{\#} f \quad \text{in } L^{1,\infty}(\partial\Omega, \sigma) \text{ as } m \to \infty. \tag{6.97}$$

Restricting to a subsequence if necessary we can say that

$$T^{\#} f_m \to T^{\#} f \quad \text{at } \sigma\text{-a.e. point on } \partial\Omega \text{ as } m \to \infty. \tag{6.98}$$

Being mindful of (6.93)-(6.96), for each $m, \ell \in \mathbb{N}_0$ with $m \geq \ell$ we have

$$\begin{aligned} \|T^{\#} f_m - T^{\#} f_\ell\|_{\mathrm{HA}^p(\partial\Omega,\sigma)} &= \|T^{\#}(f_m - f_\ell)\|_{\mathrm{HA}^p(\partial\Omega,\sigma)} \\ &\leq \sum_{j=\ell+1}^{m} |\lambda_j| \|T^{\#} a_j\|_{\mathrm{HA}^p(\partial\Omega,\sigma)} \\ &\leq C \sum_{j=\ell+1}^{\infty} |\lambda_j|. \end{aligned} \tag{6.99}$$

Letting $\ell \to \infty$ and recalling $\{\lambda_j\}_{j\in\mathbb{N}_0} \in \ell^1(\mathbb{N}_0)$, this implies that $\{T^{\#} f_m\}_{m\in\mathbb{N}_0}$ is a Cauchy sequence in $\mathrm{HA}^p(\partial\Omega, \sigma)$. From the completeness of $\mathrm{HA}^p(\partial\Omega, \sigma)$ (cf. Theorem 4.3), there exists some function $g \in \mathrm{HA}^p(\partial\Omega, \sigma)$ for which

$$T^{\#} f_m \to g \quad \text{in } \mathrm{HA}^p(\partial\Omega, \sigma) \text{ as } m \to \infty. \tag{6.100}$$

The fact that $\mathrm{HA}^p(\Sigma, \sigma) \hookrightarrow L^1(\partial\Omega, \sigma)$ continuously (cf. Theorem 4.2, item *(b)*) ensures that the convergence in (6.100) also holds in $L^1(\partial\Omega, \sigma)$. Restricting to a subsubsequence if necessary,

$$T^{\#} f_m \to g \quad \text{at } \sigma\text{-a.e. point on } \partial\Omega \text{ as } m \to \infty, \tag{6.101}$$

which in concert with (6.98) yields $T^{\#} f = g$ pointwise σ-a.e. on $\partial\Omega$. Collectively, from (6.93), (6.95), (6.98), (6.100), and (6.101) we arrive at the conclusion that

$$\begin{aligned} \|T^{\#} f\|_{\mathrm{HA}^p(\partial\Omega,\sigma)} &= \|g\|_{\mathrm{HA}^p(\partial\Omega,\sigma)} = \lim_{m\to\infty} \|T^{\#} f_m\|_{\mathrm{HA}^p(\partial\Omega,\sigma)} \\ &\leq \lim_{m\to\infty} \sum_{j=0}^{m} |\lambda_j| \|T^{\#} a_j\|_{\mathrm{HA}^p(\partial\Omega,\sigma)} \\ &\leq C \sum_{j=0}^{\infty} |\lambda_j| \leq 2C \|f\|_{\mathrm{HA}^p(\partial\Omega,\sigma)}, \end{aligned} \tag{6.102}$$

for some $C \in (0, \infty)$ depending only on n, p, the kernel $\vec{k}$, and the UR constants of $\partial\Omega$. This completes the proof of Theorem 6.3. □

Our next goal is to obtain geometrically sensitive operator norm estimates for singular integral operators of chord-dot-normal type acting on Beurling-Hardy spaces. To set the stage, consider a UR domain $\Omega \subseteq \mathbb{R}^n$, abbreviate $\sigma := \mathcal{H}^{n-1}\lfloor\partial\Omega$, and denote by ν the geometric measure theoretic outward unit normal to Ω. Given $N = N(n) \in \mathbb{N}$ a sufficiently large integer, select a complex-valued function $k \in \mathscr{C}^N(\mathbb{R}^n \setminus \{0\})$ that is even and positive homogeneous of degree $-n$. Following [GHA-IV, (5.2.136), p. 624] introduce the `chord-dot-normal singular integral operator` T associated with k acting on each $f \in L^1\big(\partial\Omega, \frac{\sigma(x)}{1+|x|^{n-1}}\big)$, for σ-a.e. $x \in \partial\Omega$, according to

$$Tf(x) := \lim_{\varepsilon\to 0^+} \int_{\substack{y\in\partial\Omega \\ |x-y|>\varepsilon}} \langle \nu(y), x-y\rangle k(x-y) f(y)\, d\sigma(y). \tag{6.103}$$

Also consider its formal transpose $T^{\#}$ (cf. [GHA-IV, (5.2.137), p. 624]), whose action on each $f \in L^1\big(\partial\Omega, \frac{\sigma(x)}{1+|x|^{n-1}}\big)$ is defined, for σ-a.e. $x \in \partial\Omega$, as

$$T^{\#} f(x) := \lim_{\varepsilon\to 0^+} \int_{\substack{y\in\partial\Omega \\ |x-y|>\varepsilon}} \langle \nu(x), y-x\rangle k(x-y) f(y)\, d\sigma(y). \tag{6.104}$$

It is of interest to bring in the `modified chord-dot-normal singular integral operator` $T_{\rm mod}$ (cf. [GHA-IV, (5.2.188), p. 632]) whose action on each function $f \in L^1\big(\partial\Omega, \frac{\sigma(x)}{1+|x|^{n}}\big)$ is given by

$$T_{\rm mod} f(x) := \lim_{\varepsilon\to 0^+} \int_{\partial\Omega} \big\langle \nu(y), \vec{k}_\varepsilon(x-y) - \vec{k}_1(-y)\big\rangle f(y)\, d\sigma(y) \tag{6.105}$$

at σ-a.e. $x \in \partial\Omega$, where

$$\vec{k}_\varepsilon := z \cdot k(z) \cdot \mathbf{1}_{\mathbb{R}^n\setminus\overline{B(0,\varepsilon)}}(z) \quad \text{for all } \varepsilon \in (0,\infty) \text{ and } z \in \mathbb{R}^n. \tag{6.106}$$

Work done in [GHA-IV, Theorem 5.2.2, pp. 624–641] ensures that these operators are meaningfully defined in the sense that the limits in (6.103), (6.104), and (6.105) exist σ-a.e. on $\partial\Omega$. Moreover, the operators T and $T_{\rm mod}$ differ by a constant when acting on functions from $L^1\big(\partial\Omega, \frac{\sigma(x)}{1+|x|^{n-1}}\big)$. Specifically, whenever $f \in L^1\big(\partial\Omega, \frac{\sigma(x)}{1+|x|^{n-1}}\big)$ it follows that

$$C_f := T_{\rm mod} f - Tf \quad \text{is a constant on } \partial\Omega. \tag{6.107}$$

It important to mention that [GHA-IV, (5.2.191), p. 632] yields

$$\begin{array}{c} T_{\rm mod} \text{ maps constants functions on } \partial\Omega \\ \text{into constants functions on } \partial\Omega. \end{array} \tag{6.108}$$

We next recall a fundamental property of the operators $T_{\rm mod}$ and $T^{\#}$. Given integrability exponents $p, p' \in (1,\infty)$ such that $1/p + 1/p' = 1$, then whenever

$$f \in L^{p'}_{\rm loc}(\partial\Omega, \sigma) \cap L^1\big(\partial\Omega, \tfrac{\sigma(x)}{1+|x|^n}\big)$$
$$\text{and } \; g \in L^p_{\rm comp}(\partial\Omega, \sigma) \; \text{ with } \; \int_{\partial\Omega} g\, d\sigma = 0, \tag{6.109}$$

it follows that (cf. [GHA-IV, (5.2.192), p. 632])

$$\int_{\partial\Omega} |T_{\rm mod} f| \cdot |g|\, d\sigma < \infty, \quad \int_{\partial\Omega} |f| \cdot |T^{\#} g|\, d\sigma < \infty,$$
$$\text{and } \; \int_{\partial\Omega} (T_{\rm mod} f) \cdot g\, d\sigma = \int_{\partial\Omega} f \cdot (T^{\#} g)\, d\sigma. \tag{6.110}$$

In particular, these properties permit us to obtain the following lemma.

Lemma 6.1 *Consider a* UR *domain* $\Omega \subseteq \mathbb{R}^n$*, and abbreviate* $\sigma := \mathcal{H}^{n-1}\lfloor\partial\Omega$*. For some sufficiently large integer* $N = N(n) \in \mathbb{N}$*, select a kernel* $k \in \mathscr{C}^N(\mathbb{R}^n \setminus \{0\})$ *that is even and positive of degree* $-n$*. Recall the chord-dot-normal singular integral operator* $T^{\#}$ *associated with* k *as in* (6.104)*, as well as the operator* $T_{\rm mod}$ *as in* (6.105)*. Also, pick an integrability exponent* $p \in (1,\infty)$*. Fix an arbitrary function* $a \in L^p_{\rm comp}(\partial\Omega, \sigma)$ *such that* $\int_{\partial\Omega} a\, d\sigma = 0$*. Then the function* $T^{\#}a$ *is absolutely integrable on* $\partial\Omega$ *and satisfies a vanishing moment condition, i.e.,*

$$\int_{\partial\Omega} |T^{\#} a|\, d\sigma < \infty \;\; \textit{and} \;\; \int_{\partial\Omega} T^{\#} a\, d\sigma = 0. \tag{6.111}$$

Proof This is a direct consequence of the vanishing moment of a, together with (6.108), (6.109), and (6.110) presently used with with $f := 1$ and $g := a$. □

Before stating the next result, make one more definition, following [GHA-V, (4.1.11)-(4.1.22), pp. 182–183]. With e denoting the base of natural logarithms, for each $m \in \mathbb{N}_0$ and $t \in [0,\infty)$ define

$$t^{\langle 0\rangle} := 1 \tag{6.112}$$

and, whenever $m \geq 1$,

$$t^{\langle m\rangle} := \begin{cases} 0 & \text{if } t = 0, \\ t \cdot \underbrace{\ln\Big(\cdots \ln\big(\ln(1/t)\big)\cdots\Big)}_{m \text{ natural logarithms}} & \text{if } 0 < t \leq ({}^m e)^{-1}, \\ ({}^m e)^{-1} & \text{if } t > ({}^m e)^{-1}, \end{cases} \tag{6.113}$$

where ${}^m e$ is the m-th tetration of e (involving m copies of e, combined by exponentiation), i.e.,

$$
{}^{m}e := \underbrace{e^{e^{\cdot^{\cdot^{\cdot^{e}}}}}}_{m \text{ copies of } e}, \quad \text{the } m\text{-th fold exponentiation of } e. \tag{6.114}
$$

We also agree to set ${}^{0}e := 1$. It is worth mentioning that

$$
\begin{gathered}
t^{\langle m\rangle} \leq t^{\langle m-1\rangle} \leq \cdots \leq t^{\langle 1\rangle} \leq \big(e^{\,\varepsilon^{-1}}/\varepsilon\big) \cdot t^{1-\varepsilon} \\
\text{for each } \ t \in [0,\infty), \ \ m \in \mathbb{N}, \ \ \varepsilon \in (0,1),
\end{gathered} \tag{6.115}
$$

and

$$
t \leq \max\{1, ({}^{m}e)t\} \cdot t^{\langle m\rangle} \quad \text{for all } \ t \in [0,\infty) \ \text{ and } \ m \in \mathbb{N}_0. \tag{6.116}
$$

In particular, (6.115) entails

$$
\lim_{t\to 0^+} t^{\langle m\rangle} = 0 \quad \text{for all } m \in \mathbb{N}. \tag{6.117}
$$

We now present the main result of this chapter, which establishes operator norm estimates for singular integral operators of chord-dot-normal type which are sensitive to the geometry of the underlying domain. This result is of fundamental importance for this work, as it opens the door for employing layer potential techniques to treat boundary problems on Beurling-Hardy space in subsequent chapters.

Theorem 6.4 *Assume $\Omega \subseteq \mathbb{R}^n$ is a* UR *domain with unbounded boundary, and abbreviate $\sigma := \mathcal{H}^{n-1}\lfloor\partial\Omega$. Denote by ν the geometric measure theoretic outward unit normal to Ω. For some sufficiently large integer $N = N(n) \in \mathbb{N}$, consider a kernel $k \in \mathscr{C}^N(\mathbb{R}^n \setminus \{0\})$ that is even and positive homogeneous of degree $-n$. Recall the chord-dot-normal singular integral operator $T^{\#}$ associated with k as in* (6.104).

Then for each integrability exponent $p \in (1,\infty)$ the operator

$$
\begin{gathered}
T^{\#} : \mathrm{HA}^p(\partial\Omega,\sigma) \longrightarrow \mathrm{HA}^p(\partial\Omega,\sigma) \\
\textit{is well defined, linear, and bounded.}
\end{gathered} \tag{6.118}
$$

Moreover, for each $m \in \mathbb{N}$ there exists some $C_m \in (0,\infty)$ depending exclusively on m, n, p, the kernel k, and the UR *constants of $\partial\Omega$ with the property that*

$$
\|T^{\#}\|_{\mathrm{HA}^p(\partial\Omega,\sigma)\to\mathrm{HA}^p(\partial\Omega,\sigma)} \leq C_m \cdot \|\nu\|^{\langle m\rangle}_{[\mathrm{BMO}(\partial\Omega,\sigma)]^n}. \tag{6.119}
$$

Proof Fix some $p \in (1,\infty)$ and $m \in \mathbb{N}$. The claim made in (6.118) is justified by Theorem 6.3 and the fact that chord-dot-normal singular integral operators are themselves generalized double layers (see [GHA-IV, Example 5.1.7, pp. 556–557]). Pick a reference point $x_0 \in \partial\Omega$ along with a dilation factor $2_* > (C_{\partial\Omega}/c_{\partial\Omega})^{1/(n-1)}$ as in Definition 3.1, where $C_{\partial\Omega}, c_{\partial\Omega} \in (0,\infty)$ denote the upper and lower Ahlfors regularity constants of $\partial\Omega$, respectively. Select an x_0-central $(1,p)$-atom a on $\partial\Omega$, i.e., a is a σ-measurable function on $\partial\Omega$ such that for some $R \in [2_*,\infty)$ the following properties hold:

$$\operatorname{supp} a \subseteq \Delta(x_0, R), \tag{6.120}$$

$$\|a\|_{L^p(\Sigma,\sigma)} \leq R^{(n-1)(1/p-1)}, \tag{6.121}$$

$$\int_\Sigma a \, d\sigma = 0. \tag{6.122}$$

Fix a parameter $\varepsilon \in \big(0, \frac{1}{n-1}\big)$. We claim that there exists some $C \in (0, \infty)$ depending only on n, m, p, the kernel k, and the UR constants of $\partial\Omega$ such that

$$\begin{gathered}\text{the function } M := T^\# a \text{ is a } C \cdot \|\nu\|^{\langle m\rangle}_{[\mathrm{BMO}(\partial\Omega,\sigma)]^n}\text{-multiple of an}\\ x_0\text{-central } (1,p)_\varepsilon\text{-molecule on } \partial\Omega \text{ associated with the}\\ \text{radius } 2R \text{ (in the sense of Definition 4.5).}\end{gathered} \tag{6.123}$$

To justify this claim, we observe that the membership $a \in L^p(\partial\Omega, \sigma)$ (cf. (6.121)) and [MMMi[3], Corollary 4.2, p. 292] implies that $M = T^\# a$ is meaningfully defined in a pointwise sense and, as a function, belongs to $L^p(\partial\Omega, \sigma)$. Moreover, there exists some $C \in (0, \infty)$ depending only on n, m, p, the kernel k, and the UR constants of $\partial\Omega$ for which

$$\begin{aligned}\|M\|_{L^p(\partial\Omega,\sigma)} = \|T^\# a\|_{L^p(\partial\Omega,\sigma)} &\leq C \cdot \|\nu\|^{\langle m\rangle}_{[\mathrm{BMO}(\partial\Omega,\sigma)]^n} \|a\|_{L^p(\partial\Omega,\sigma)}\\ &\leq C \cdot \|\nu\|^{\langle m\rangle}_{[\mathrm{BMO}(\partial\Omega,\sigma)]^n} R^{(n-1)(1/p-1)},\end{aligned} \tag{6.124}$$

where the last inequality follows from (6.121). In particular, this implies

$$\Big(\int_{\Delta(x_0,2R)} |M|^p \, d\sigma\Big)^{1/p} \leq C \cdot \|\nu\|^{\langle m\rangle}_{[\mathrm{BMO}(\partial\Omega,\sigma)]^n} R^{(n-1)(1/p-1)}. \tag{6.125}$$

To proceed, notice that

$$\begin{gathered}|x - z| \approx |x - x_0| \text{ uniformly with respect to}\\ x \in \partial\Omega \setminus B(x_0, 2R) \text{ and } z \in B(x_0, R).\end{gathered} \tag{6.126}$$

Thanks to this, there is enough separation between $\partial\Omega \setminus B(x_0, 2R)$ and the support of a (which is contained in $B(x_0, R) \cap \partial\Omega$; see (6.120)). As such, being mindful of (6.104) and the vanishing moment property of a (cf. (6.122)), for σ-a.e. $x \in \partial\Omega \setminus B(x_0, 2R)$ we may express $M(x) = (T^\# a)(x)$ as

$$M(x) = -\int_{\Delta(x_0,R)} \Big\{\big\langle\nu(x), x - y\big\rangle k(x - y) - \big\langle\nu(x), x - x_0\big\rangle k(x - x_0)\Big\} a(y) \, d\sigma(y). \tag{6.127}$$

For each $x \in \partial\Omega \setminus B(x_0, 2R)$, introduce

$$M_\Delta(x)$$
$$:= -\int_{\Delta(x_0,R)} \Big\{\big\langle \nu_{\Delta(x_0,R)}, x-y\big\rangle k(x-y) - \big\langle \nu_{\Delta(x_0,R)}, x-x_0\big\rangle k(x-x_0)\Big\} a(y)\,d\sigma(y), \tag{6.128}$$

where $\nu_{\Delta(x_0,R)} := \fint_{\Delta(x_0,R)} \nu\,d\sigma$. Since a belongs to $L^1_{\rm comp}(\partial\Omega,\sigma)$ and k is sufficiently smooth in $\mathbb{R}^n\setminus\{0\}$ (hence bounded on any compact set away from the origin), it is clear that M_Δ is well defined via an absolutely convergent integral. Fix an arbitrary point $x\in\partial\Omega\backslash B(x_0,2R)$ and consider the function

$$F_x(y) := \big\langle \nu(x) - \nu_{\Delta(x_0,R)}, x-y\big\rangle k(x-y) \quad \text{for each } y\in B(x_0,R). \tag{6.129}$$

By design, $F_x\in\mathscr{C}^N\big(B(x_0,R)\big)$. Bearing in mind this and (6.127)-(6.128), write

$$M(x) - M_\Delta(x) = -\int_{\Delta(x_0,R)} \big(F_x(y) - F_x(x_0)\big)a(y)\,d\sigma(y). \tag{6.130}$$

Our first goal here is to estimate $M - M_\Delta$ on $\partial\Omega\backslash B(x_0,2R)$. To this end, notice that the homogeneity of degree $-n$ of the function k yields

$$|k(z)| \le \frac{1}{|z|^n}\sup_{\omega\in S^{n-1}}|k(\omega)| \ \text{ and } \ |\nabla k(z)| \le \frac{1}{|z|^{n+1}}\sup_{\omega\in S^{n-1}}|\nabla k(\omega)| \tag{6.131}$$

for all $z\in\mathbb{R}^n\setminus\{0\}$. Thanks to these and the definition of F_x (see (6.129)), a direct computation shows that

$$|\nabla F_x(y)| \le C\frac{|\nu(x) - \nu_{\Delta(x_0,R)}|}{|x-y|^n} \quad \text{for all } y\in B(x_0,R), \tag{6.132}$$

for some constant $C\in(0,\infty)$ depending on n and the kernel k. Employing the Mean Value Theorem, keeping in mind (6.132) and (6.126), for each $y\in B(x_0,R)$ we estimate

$$\begin{aligned} |F_x(y) - F_x(x_0)| &\le \sup_{\xi\in[y,x_0]} \big|\nabla F_x(\xi)\big|\cdot|y-x_0| \\ &\le CR \sup_{\xi\in[y,x_0]} \frac{|\nu(x) - \nu_{\Delta(x_0,R)}|}{|x-\xi|^n} \\ &\le CR\frac{|\nu(x) - \nu_{\Delta(x_0,R)}|}{|x-x_0|^n}. \end{aligned} \tag{6.133}$$

Use this, (6.130), Hölder's inequality, the Ahlfors regularity of $\partial\Omega$, and the normalization property of a (see (6.121)) to bound, for σ-a.e. $x\in\partial\Omega\backslash B(x_0,2R)$,

$$
\begin{aligned}
|M(x) - M_\Delta(x)| &= CR\frac{|\nu(x) - \nu_{\Delta(x_0,R)}|}{|x - x_0|^n}\int_{\Delta(x_0,R)} |a|\,d\sigma \\
&\le CR\frac{|\nu(x) - \nu_{\Delta(x_0,R)}|}{|x - x_0|^n}\sigma\big(\Delta(x_0,R)\big)^{1-1/p}\|a\|_{L^p(\partial\Omega,\sigma)} \\
&\le CR\frac{|\nu(x) - \nu_{\Delta(x_0,R)}|}{|x - x_0|^n},
\end{aligned}
\tag{6.134}
$$

for some $C \in (0, \infty)$ depending only on k, n, p, and the Ahlfors regularity of $\partial\Omega$.

To proceed, recall the annuli introduced on (4.198). Availing ourselves of the estimate in (6.134), for every $\ell \in \mathbb{N}$ we have

$$
\begin{aligned}
&\Big(\int_{A_\ell(x_0,2R)} |M - M_\Delta|^p\,d\sigma\Big)^{1/p} \\
&\le CR\Big(\int_{A_\ell(x_0,2R)} \frac{|\nu(x) - \nu_{\Delta(x_0,R)}|^p}{|x - x_0|^{np}}\,d\sigma(x)\Big)^{1/p} \\
&\le \frac{CR}{(2^\ell R)^n}\Big(\int_{A_\ell(x_0,2R)} |\nu(x) - \nu_{\Delta(x_0,R)}|^p\,d\sigma(x)\Big)^{1/p}.
\end{aligned}
\tag{6.135}
$$

Taking into account that $\|\cdot\|_{\mathrm{BMO}_p(\partial\Omega,\sigma)} \approx \|\cdot\|_{\mathrm{BMO}(\partial\Omega,\sigma)}$ (cf. (2.36)) and the Ahlfors regularity of $\partial\Omega$, we proceed to estimate the integral in the right-hand side of (6.135) as

$$
\begin{aligned}
&\Big(\int_{A_\ell(x_0,2R)} |\nu - \nu_{\Delta(x_0,R)}|^p\,d\sigma\Big)^{1/p} \\
&\le C(2^\ell R)^{(n-1)/p}\Big(\fint_{\Delta(x_0,2^{\ell+1}R)} |\nu - \nu_{\Delta(x_0,R)}|^p\,d\sigma\Big)^{1/p} \\
&\le C(2^\ell R)^{(n-1)/p}\Big(\fint_{\Delta(x_0,2^{\ell+1}R)} |\nu - \nu_{\Delta(x_0,2^{\ell+1}R)}|^p\,d\sigma\Big)^{1/p} \\
&\quad + C(2^\ell R)^{(n-1)/p}|\nu_{\Delta(x_0,R)} - \nu_{\Delta(x_0,2^{\ell+1}R)}| \\
&\le C(2^\ell R)^{(n-1)/p}\Big(\fint_{\Delta(x_0,2^{\ell+1}R)} |\nu - \nu_{\Delta(x_0,2^{\ell+1}R)}|^p\,d\sigma\Big)^{1/p} \\
&\quad + C(2^\ell R)^{(n-1)/p}\sum_{j=0}^{\ell}|\nu_{\Delta(x_0,2^jR)} - \nu_{\Delta(x_0,2^{j+1}R)}| \\
&\le C(2^\ell R)^{(n-1)/p}\Big(\fint_{\Delta(x_0,2^{\ell+1}R)} |\nu - \nu_{\Delta(x_0,2^{\ell+1}R)}|^p\,d\sigma\Big)^{1/p} \\
&\quad + C(2^\ell R)^{(n-1)/p}\sum_{j=0}^{\ell}\fint_{\Delta(x_0,2^{j+1}R)} |\nu - \nu_{\Delta(x_0,2^{j+1}R)}|\,d\sigma
\end{aligned}
$$

$$
\begin{aligned}
&\le C(2^\ell R)^{(n-1)/p}\|\nu\|_{[\mathrm{BMO}_p(\partial\Omega,\sigma)]^n} \\
&\quad + C(2^\ell R)^{(n-1)/p}(\ell+1)\|\nu\|_{[\mathrm{BMO}(\partial\Omega,\sigma)]^n} \\
&\le C(2^\ell R)^{(n-1)/p}\ell\|\nu\|_{[\mathrm{BMO}(\partial\Omega,\sigma)]^n},
\end{aligned}
\tag{6.136}
$$

for every $\ell \in \mathbb{N}$, and for some constant $C = C(n, p, k, \partial\Omega) \in (0, \infty)$. Together, (6.135) and (6.136) yield

$$
\begin{aligned}
&\Big(\int_{A_\ell(x_0,2R)} |M - M_\Delta|^p \, d\sigma\Big)^{1/p} \\
&\quad \le \frac{CR}{(2^\ell R)^n}(2^\ell R)^{(n-1)/p}\ell\|\nu\|_{[\mathrm{BMO}(\partial\Omega,\sigma)]^n} \\
&\quad = C\ell 2^{\ell(n-1)[1/p-1-1/(n-1)]} R^{(n-1)(1/p-1)}\|\nu\|_{[\mathrm{BMO}(\partial\Omega,\sigma)]^n}.
\end{aligned}
\tag{6.137}
$$

Moving forward, consider the function $\varphi : [0, \infty) \to \mathbb{R}$ defined according to

$$
\varphi(t) := t \cdot 2^{-t(1-\varepsilon(n-1))} \quad \text{for all } t \ge 0. \tag{6.138}
$$

Elementary calculus shows that φ is a bounded function and in fact

$$
0 \le \varphi(t) \le C_1 := \frac{2^{-1/\ln 2}}{\big(1-\varepsilon(n-1)\big)\ln 2} \quad \text{for all } t \ge 0. \tag{6.139}
$$

In particular, whenever $t := \ell \in \mathbb{N}$ this entails

$$
\varphi(\ell) = \ell \cdot 2^{-\ell[1-\varepsilon(n-1)]} \le C_1 \implies \ell \cdot 2^{-\ell} \le C_1 \cdot 2^{-\varepsilon(n-1)\ell}, \tag{6.140}
$$

which in turn gives

$$
\ell \cdot 2^{\ell(n-1)[1/p-1-1/(n-1)]} \le C_1 \cdot 2^{\ell(n-1)(1/p-1-\varepsilon)}. \tag{6.141}
$$

Combining (6.137) and (6.141) we conclude that there exists some $C \in (0, \infty)$ depending only on n, p, ε, the kernel k, and the Ahlfors regularity character of $\partial\Omega$ satisfying

$$
\begin{aligned}
&\Big(\int_{A_\ell(x_0,2R)} |M - M_\Delta|^p \, d\sigma\Big)^{1/p} \\
&\quad \le C2^{\ell(n-1)(1/p-1-\varepsilon)} R^{(n-1)(1/p-1)}\|\nu\|_{[\mathrm{BMO}(\partial\Omega,\sigma)]^n}
\end{aligned}
\tag{6.142}
$$

for every $\ell \in \mathbb{N}$.

Pressing forward, we now focus on obtaining a similar estimate for the function M_Δ (cf. (6.128)) on $\partial\Omega\setminus B(x_0, 2R)$. To this end, decompose

$$
M_\Delta(x) = M_\Delta^{(1)}(x) + M_\Delta^{(2)}(x) \quad \text{for each } x \in \partial\Omega\setminus B(x_0, 2R), \tag{6.143}
$$

where

$$M_\Delta^{(1)}(x) := -\int_{\Delta(x_0,R)} \langle \nu_{\Delta(x_0,R)}, x - x_0\rangle \big[k(x-y) - k(x-x_0)\big] a(y)\,d\sigma(y), \tag{6.144}$$

$$M_\Delta^{(2)}(x) := \int_{\Delta(x_0,R)} \langle \nu_{\Delta(x_0,R)}, y - x_0\rangle k(x-y) a(y)\,d\sigma(y), \tag{6.145}$$

for each $x \in \partial\Omega \backslash B(x_0, 2R)$.

Fix $x \in \partial\Omega\backslash B(x_0, 2R)$. Using the Mean Value Theorem, Hölder's inequality, (6.126), (6.131), the normalization property of a (cf. (6.121)), and the Ahlfors regularity of $\partial\Omega$, estimate $M_\Delta^{(1)}(x)$ as

$$\begin{aligned}
&|M_\Delta^{(1)}(x)| \\
&\leq \int_{\Delta(x_0,R)} \big|\langle \nu_{\Delta(x_0,R)}, x - x_0\rangle\big| |k(x-y) - k(x-x_0)| |a(y)|\,d\sigma(y) \\
&\leq \big|\langle \nu_{\Delta(x_0,R)}, x - x_0\rangle\big| \int_{\Delta(x_0,R)} \Big(\sup_{\xi\in[y,x_0]} |\nabla k(x-\xi)|\Big)|y-x_0||a(y)|\,d\sigma(y) \\
&\leq \big|\langle \nu_{\Delta(x_0,R)}, x - x_0\rangle\big| \frac{CR}{|x-x_0|^{n+1}} \int_{\Delta(x_0,R)} |a|\,d\sigma \\
&\leq \big|\langle \nu_{\Delta(x_0,R)}, x - x_0\rangle\big| \frac{CR}{|x-x_0|^{n+1}} \sigma\big(\Delta(x_0,R)\big)^{1-1/p} \|a\|_{L^p(\partial\Omega,\sigma)} \\
&\leq \big|\langle \nu_{\Delta(x_0,R)}, x - x_0\rangle\big| \frac{CR}{|x-x_0|^{n+1}} R^{(n-1)(1-1/p)} R^{(n-1)(1/p-1)} \\
&= \big|\langle \nu_{\Delta(x_0,R)}, x - x_0\rangle\big| \frac{CR}{|x-x_0|^{n+1}},
\end{aligned} \tag{6.146}$$

for some constant $C = C(n, p, k, \partial\Omega) \in (0, \infty)$. Now, relying on [MMMi[3], Proposition 2.15, p. 100] (currently used with $z := x_0$, $y := x_0$, $\lambda := 2^{\ell+1}$), estimates (6.146) and (6.141), and using the Ahlfors regularity of $\partial\Omega$, for each $\ell \in \mathbb{N}$ we have

$$\begin{aligned}
&\Big(\int_{A_\ell(x_0,2R)} |M_\Delta^{(1)}|^p\,d\sigma\Big)^{1/p} \\
&\quad\leq CR\Big(\int_{A_\ell(x_0,2R)} \frac{\big|\langle \nu_{\Delta(x_0,R)}, x - x_0\rangle\big|^p}{|x-x_0|^{(n+1)p}}\,d\sigma(x)\Big)^{1/p} \\
&\quad\leq \frac{CR}{(2^\ell R)^{n+1}} \sigma\big(\Delta(x_0, 2^{\ell+1}R)\big)^{1/p} \sup_{x\in\Delta(x_0,2^{\ell+1}R)} \big|\langle \nu_{\Delta(x_0,R)}, x - x_0\rangle\big|
\end{aligned}$$

$$
\begin{aligned}
&\le \frac{CR^2}{(2^\ell R)^{n+1}}(2^{\ell+1}R)^{(n-1)/p}(2+\ell)2^{\ell+1}\|\nu\|_{[\mathrm{BMO}(\partial\Omega,\sigma)]^n} \\
&\le C\ell 2^{\ell(n-1)[1/p-1-1/(n-1)]}R^{(n-1)(1/p-1)}\|\nu\|_{[\mathrm{BMO}(\partial\Omega,\sigma)]^n} \\
&\le C2^{\ell(n-1)(1/p-1-\varepsilon)}R^{(n-1)(1/p-1)}\|\nu\|_{[\mathrm{BMO}(\partial\Omega,\sigma)]^n} \qquad (6.147)
\end{aligned}
$$

for some constant $C = C(n, p, \varepsilon, k, \partial\Omega) \in (0, \infty)$.

Next, continue to assume that $x \in \partial\Omega\setminus B(x_0, 2R)$, and recall the definition of $M^{(2)}_\Delta(x)$ given in (6.145). In view of [MMMi[3], Proposition 2.15, p. 100] (presently employed with $x = x_0$, $z := x_0$, $\lambda := 1$), bearing in mind (6.126), (6.131), using Hölder's inequality, the normalization property of a (see (6.121)), and the Ahlfors regularity of $\partial\Omega$, we estimate

$$
\begin{aligned}
|M^{(2)}_\Delta(x)| &\le \int_{\Delta(x_0,R)} \big|\langle \nu_{\Delta(x_0,R)}, y-x_0\rangle\big|\cdot|k(x-y)a(y)|\,d\sigma(y) \\
&\le C\cdot\Big(\sup_{y\in\Delta(x_0,R)}\big|\langle \nu_{\Delta(x_0,R)}, y-x_0\rangle\big|\Big)\int_{\Delta(x_0,R)}\frac{|a(y)|}{|x-y|^n}\,d\sigma(y) \\
&\le CR\|\nu\|_{[\mathrm{BMO}(\partial\Omega,\sigma)]^n}\frac{1}{|x-x_0|^n}\int_{\Delta(x_0,R)}|a|\,d\sigma \\
&\le CR\|\nu\|_{[\mathrm{BMO}(\partial\Omega,\sigma)]^n}\frac{1}{|x-x_0|^n}\|a\|_{L^p(\partial\Omega,\sigma)}\sigma\big(\Delta(x_0,R)\big)^{1-1/p} \\
&\le CR\|\nu\|_{[\mathrm{BMO}(\partial\Omega,\sigma)]^n}\frac{1}{|x-x_0|^n}R^{(n-1)(1/p-1)}R^{(n-1)(1-1/p)} \\
&= CR\|\nu\|_{[\mathrm{BMO}(\partial\Omega,\sigma)]^n}\frac{1}{|x-x_0|^n}. \qquad (6.148)
\end{aligned}
$$

From this, Hölder's inequality, the Ahlfors regularity of $\partial\Omega$, and the fact that $\varepsilon < \frac{1}{n-1}$, for every $\ell \in \mathbb{N}$ we obtain

$$
\begin{aligned}
&\Big(\int_{A_\ell(x_0,2R)}|M^{(2)}_\Delta|^p\,d\sigma\Big)^{1/p} \\
&\le CR\|\nu\|_{[\mathrm{BMO}(\partial\Omega,\sigma)]^n}\Big(\int_{A_\ell(x_0,2R)}\frac{d\sigma(x)}{|x-x_0|^{np}}\Big)^{1/p} \\
&\le CR\|\nu\|_{[\mathrm{BMO}(\partial\Omega,\sigma)]^n}\frac{1}{(2^\ell R)^n}\sigma\big(\Delta(x_0,2^{\ell+1}R)\big)^{1/p} \\
&\le CR\|\nu\|_{[\mathrm{BMO}(\partial\Omega,\sigma)]^n}\frac{1}{(2^\ell R)^n}(2^{\ell+1}R)^{(n-1)/p} \\
&= C2^{\ell(n-1)[1/p-1-1/(n-1)]}R^{(n-1)(1/p-1)}\|\nu\|_{[\mathrm{BMO}(\partial\Omega,\sigma)]^n}
\end{aligned}
$$

$$\leq C 2^{\ell(n-1)(1/p-1-\varepsilon)} R^{(n-1)(1/p-1)} \|\nu\|_{[\mathrm{BMO}(\partial\Omega,\sigma)]^n} \tag{6.149}$$

for some constant $C = C(n, p, k, \partial\Omega) \in (0, \infty)$.

In concert, (6.143), (6.147), and (6.149) guarantee the existence of some constant $C = C(n, p, \varepsilon, k, \partial\Omega) \in (0, \infty)$ with the property that

$$\Big(\int_{A_\ell(x_0,2R)} |M_\Delta|^p \, d\sigma\Big)^{1/p} \leq C 2^{\ell(n-1)(1/p-1-\varepsilon)} R^{(n-1)(1/p-1)} \|\nu\|_{[\mathrm{BMO}(\partial\Omega,\sigma)]^n} \tag{6.150}$$

for each $\ell \in \mathbb{N}$. All together, (6.142) and (6.150) (along with (6.116)) imply that there exists some constant $C \in (0, \infty)$ depending exclusively on n, m, p, ε, the kernel k, and the Ahlfors regularity character of $\partial\Omega$ satisfying

$$\Big(\int_{A_\ell(x_0,2R)} |M|^p \, d\sigma\Big)^{1/p} \leq C 2^{\ell(n-1)(1/p-1-\varepsilon)} R^{(n-1)(1/p-1)} \|\nu\|^{\langle m\rangle}_{[\mathrm{BMO}(\partial\Omega,\sigma)]^n}. \tag{6.151}$$

Lastly, notice that the vanishing moment of $M = T^\# a$ follows from Lemma 6.1. This, (6.125), and (6.151), then finishes the justification of the claim made in (6.123).

Having established (6.123), invoke Theorem 4.4 to guarantee that $T^\# a$ belongs to $\mathrm{HA}^p(\partial\Omega, \sigma)$ and there exists some $C \in (0, \infty)$ independent of a for which

$$\|T^\# a\|_{\mathrm{HA}^p(\partial\Omega,\sigma)} \leq C \cdot \|\nu\|^{\langle m\rangle}_{[\mathrm{BMO}(\partial\Omega,\sigma)]^n}. \tag{6.152}$$

At this stage of the proof, there remains to show that $T^\#$ is bounded from $\mathrm{HA}^p(\partial\Omega, \sigma)$ into itself and has operator norm satisfying (6.119). To verify this, fix an arbitrary function $f \in \mathrm{HA}^p(\partial\Omega, \sigma)$, and use Theorem 4.2 to write

$$f = \sum_{j=0}^{\infty} \lambda_j a_j \quad \text{in } L^1(\partial\Omega, \sigma), \tag{6.153}$$

for some $\{a_j\}_{j\in\mathbb{N}_0}$ is a sequence of x_0-central $(1, p)$-atoms and some numerical sequence $\{\lambda_j\}_{j\in\mathbb{N}_0} \in \ell^1(\mathbb{N}_0)$. Also, make the assumption that this decomposition is quasi-optimal, i.e.,

$$\sum_{j=0}^{\infty} |\lambda_j| \leq 2\|f\|_{\mathrm{HA}^p(\partial\Omega,\sigma)}. \tag{6.154}$$

For each $r \in \mathbb{N}_0$, abbreviate

$$f_r := \sum_{j=0}^{r} \lambda_j a_j. \tag{6.155}$$

In light of the weak-type $(1, 1)$ boundedness of the operator $T^{\#}$ (cf. Proposition 6.1), (6.153), and (6.155), we have

$$T^{\#} f_r \longrightarrow T^{\#} f \quad \text{in } L^{1,\infty}(\partial\Omega, \sigma) \text{ as } r \longrightarrow \infty. \tag{6.156}$$

Hence, restricting to a subsequence if necessary

$$T^{\#} f_r \longrightarrow T^{\#} f \quad \text{at } \sigma\text{-a.e. point on } \partial\Omega \text{ as } r \longrightarrow \infty. \tag{6.157}$$

Thanks to (6.152), whenever $r, s \in \mathbb{N}_0$ and $r \geq s$, estimate

$$\begin{aligned} \|T^{\#} f_r - T^{\#} f_s\|_{\mathrm{HA}^p(\partial\Omega,\sigma)} &= \|T^{\#}(f_r - f_s)\|_{\mathrm{HA}^p(\partial\Omega,\sigma)} \\ &= \Big\| \sum_{j=s+1}^{r} \lambda_j (T^{\#} a_j) \Big\|_{\mathrm{HA}^p(\partial\Omega,\sigma)} \\ &\leq \sum_{j=s+1}^{r} |\lambda_j| \|T^{\#} a_j\|_{\mathrm{HA}^p(\partial\Omega,\sigma)} \\ &\leq C \cdot \|\nu\|^{\langle m \rangle}_{[\mathrm{BMO}(\partial\Omega,\sigma)]^n} \sum_{j=s+1}^{\infty} |\lambda_j|. \end{aligned} \tag{6.158}$$

Since $\{\lambda_j\}_{j\in\mathbb{N}_0} \in \ell^1(\mathbb{N}_0)$, if we let $s \longrightarrow \infty$ then $\sum_{j=s+1}^{\infty} |\lambda_j| \longrightarrow 0$. As such, the sequence $\{T^{\#} f_r\}_{r\in\mathbb{N}_0}$ is Cauchy in $\mathrm{HA}^p(\partial\Omega, \sigma)$. From the completeness of $\mathrm{HA}^p(\partial\Omega, \sigma)$ (cf. Theorem 4.3) there exists some $g \in \mathrm{HA}^p(\partial\Omega, \sigma)$ for which

$$T^{\#} f_r \longrightarrow g \quad \text{in } \mathrm{HA}^p(\partial\Omega, \sigma), \text{ as } r \longrightarrow \infty. \tag{6.159}$$

Taking into account that $\mathrm{HA}^p(\partial\Omega, \sigma)$ embeds continuously into $L^1(\partial\Omega, \sigma)$ (cf. item *(b)* in Theorem 4.2), this convergence also holds in $L^1(\partial\Omega, \sigma)$. As such, restricting to a subsubsequence if necessary, we have

$$T^{\#} f_r \longrightarrow g \quad \text{at } \sigma\text{-a.e. point on } \partial\Omega \text{ as } r \longrightarrow \infty. \tag{6.160}$$

A combination of this and (6.157) yields

$$T^{\#} f = g \in \mathrm{HA}^p(\partial\Omega, \sigma) \quad \text{at } \sigma\text{-a.e. point on } \partial\Omega. \tag{6.161}$$

Granted this, (6.154), and inequality (6.152), for each $r \in \mathbb{N}_0$ we may bound

$$
\begin{aligned}
\|T^{\#} f_r\|_{\mathrm{HA}^p(\partial\Omega,\sigma)} &\leq \sum_{j=0}^{r} |\lambda_j| \|T^{\#} a_j\|_{\mathrm{HA}^p(\partial\Omega,\sigma)} \\
&\leq C \cdot \|\nu\|^{\langle m\rangle}_{[\mathrm{BMO}(\partial\Omega,\sigma)]^n} \sum_{j=0}^{\infty} |\lambda_j| \\
&\leq 2C \cdot \|\nu\|^{\langle m\rangle}_{[\mathrm{BMO}(\partial\Omega,\sigma)]^n} \|f\|_{\mathrm{HA}^p(\partial\Omega,\sigma)},
\end{aligned} \tag{6.162}
$$

for some constant $C = C(n, m, p, \varepsilon, k, \partial\Omega) \in (0, \infty)$. Keeping in mind (6.159) and (6.161), if we let $r \to \infty$ in (6.162) then

$$
\|T^{\#} f\|_{\mathrm{HA}^p(\partial\Omega,\sigma)} \leq 2C \cdot \|\nu\|^{\langle m\rangle}_{[\mathrm{BMO}(\partial\Omega,\sigma)]^n} \|f\|_{\mathrm{HA}^p(\partial\Omega,\sigma)}. \tag{6.163}
$$

This ultimately concludes the proof of Theorem 6.4. □

6.2 Singular Integral Operators on BCMO_p Spaces

Our next goal is to study the action of singular integral operators on the space of functions of bounded central mean oscillations. Let $\Sigma \subseteq \mathbb{R}^n$ be a UR set, and abbreviate $\sigma := \mathcal{H}^{n-1}\lfloor\Sigma$. Having fixed a sufficiently large integer $N = N(n) \in \mathbb{N}$, consider a kernel $k \in \mathscr{C}^N(\mathbb{R}^n \setminus \{0\})$ that is odd and positive homogeneous of degree $1-n$. Recall the convolution-type integral operator T associated with k whose action of each function $f \in L^1\big(\Sigma, \frac{\sigma(x)}{1+|x|^{n-1}}\big)$ is given by

$$
(Tf)(x) := \lim_{\varepsilon\to 0^+} \int_{\substack{y\in\Sigma \\ |x-y|>\varepsilon}} k(x-y) f(y)\, d\sigma(y) \text{ for } \sigma\text{-a.e. } x \in \Sigma. \tag{6.164}
$$

Notice that the space $L^1\big(\Sigma, \frac{\sigma(x)}{1+|x|^{n-1}}\big)$ is the largest functional analytic environment that makes the truncated integrals in (6.164) absolutely convergent. Since the space $\mathrm{BCMO}_p(\Sigma, \sigma)$ is not included in $L^1\big(\Sigma, \frac{\sigma(x)}{1+|x|^{n-1}}\big)$, it is of interest to consider the modified version of T. Following [GHA-III, (2.3.31), p. 352], define the `modified convolution-type integral operator` T_{mod} acting on each $f \in L^1\big(\Sigma, \frac{\sigma(x)}{1+|x|^{n}}\big)$ according to

$$
\begin{gathered}
(T_{\mathrm{mod}} f)(x) := \lim_{\varepsilon\to 0^+} \int_{\Sigma} \big\{k_\varepsilon(x-y) - k_1(-y)\big\} f(y)\, d\sigma(y), \\
\text{at } \sigma\text{-a.e. } x \in \Sigma, \text{ where } k_\varepsilon(z) := k(z) \cdot \mathbb{1}_{\mathbb{R}^n \setminus \overline{B(0,\varepsilon)}}(z) \\
\text{for all } \varepsilon \in (0,\infty) \text{ and } z \in \mathbb{R}^n \setminus \{0\}.
\end{gathered} \tag{6.165}
$$

In relation to this operator, a result from [GHA-III, Theorem 2.3.2, pp. 348–357] ensures that

$$\begin{aligned}&T_{\rm mod}f \quad \text{is well defined for all } f\in L^1\big(\Sigma,\tfrac{\sigma(x)}{1+|x|^n}\big), \text{ and}\\ &\text{for any } p\in(1,\infty) \text{ and } f\in L^1\big(\Sigma,\tfrac{\sigma(x)}{1+|x|^n}\big)\cap L^p_{\rm loc}(\Sigma,\sigma)\\ &\quad\text{it follows that } T_{\rm mod}f\in L^p_{\rm loc}(\Sigma,\sigma)\subseteq L^1_{\rm loc}(\Sigma,\sigma).\end{aligned}\tag{6.166}$$

Moreover, the operators T and $T_{\rm mod}$ are compatible (up to additive constants) in the sense that for any function $f\in L^1\big(\Sigma,\frac{\sigma(x)}{1+|x|^{n-1}}\big)\subseteq L^1\big(\Sigma,\frac{\sigma(x)}{1+|x|^n}\big)$ one has

$$\begin{aligned}&C_f:=T_{\rm mod}f-Tf \quad \text{is a constant function on } \Sigma, \text{ and}\\ &|C_f|\le C_{\Sigma,k}\cdot\|f\|_{L^1(\Sigma,\frac{\sigma(x)}{1+|x|^{n-1}})} \text{ for some } C_{\Sigma,k}\in(0,\infty).\end{aligned}\tag{6.167}$$

We also wish to highlight that [GHA-III, Theorem 2.3.2, pp. 348–357] guarantees that if $p,p'\in(1,\infty)$ are such that $1/p+1/p'=1$, then whenever

$$\begin{aligned}&f\in L^{p'}_{\rm loc}(\Sigma,\sigma)\cap L^1\big(\Sigma,\tfrac{\sigma(x)}{1+|x|^n}\big) \text{ and}\\ &g\in L^p_{\rm comp}(\Sigma,\sigma) \text{ satisfies } \int_\Sigma g\,d\sigma=0,\end{aligned}\tag{6.168}$$

it follows that

$$\begin{aligned}&\int_\Sigma |T_{\rm mod}f|\cdot|g|\,d\sigma<\infty,\quad \int_\Sigma |f|\cdot|Tg|\,d\sigma<\infty,\\ &\text{and}\quad \int_\Sigma (T_{\rm mod}f)\cdot g\,d\sigma=-\int_\Sigma f\cdot(Tg)\,d\sigma.\end{aligned}\tag{6.169}$$

Our first result in the section is in stark contrast with the fact that the operator $T_{\rm mod}:L^\infty(\Sigma,\sigma)\to{\rm BMO}(\Sigma,\sigma)$ is well defined and bounded (cf. [GHA-III, item (7) in Theorem 2.3.2, pp. 348–357]). In fact, the result recorded below establishes that $T_{\rm mod}$ is a bounded operator from ${\rm B}^p(\Sigma,\sigma)$ into ${\rm BCMO}_p(\Sigma,\sigma)$. The reader is reminded that ${\rm B}^p(\Sigma,\sigma)$ has been introduced in Definition 3.1 and ${\rm BCMO}_p(\Sigma,\sigma)$ first appears in Definition 5.1.

Theorem 6.5 *Let $\Sigma\subseteq\mathbb{R}^n$ be an unbounded* UR *set, and abbreviate $\sigma:=\mathcal{H}^{n-1}\lfloor\Sigma$. For a sufficiently large integer $N=N(n)\in\mathbb{N}$, select a kernel $k\in\mathscr{C}^N(\mathbb{R}^n\setminus\{0\})$ that is odd, and positive homogeneous of degree $1-n$. Consider the integral operator $T_{\rm mod}$ associated with k as in* (6.165). *Finally, fix an exponent $p\in(1,\infty)$. Then the operator*

$$\begin{aligned}&T_{\rm mod}:{\rm B}^p(\Sigma,\sigma)\longrightarrow{\rm BCMO}_p(\Sigma,\sigma)\\ &\textit{is well defined, linear, and bounded.}\end{aligned}\tag{6.170}$$

In particular,

$$\begin{gathered}\textit{the operator } [T_{\mathrm{mod}}] : \mathrm{B}^p(\Sigma,\sigma) \longrightarrow \widetilde{\mathrm{BCMO}}_p(\Sigma,\sigma)\\ \textit{defined as } [T_{\mathrm{mod}}]f := [T_{\mathrm{mod}}f] \textit{ for all } f \in \mathrm{B}^p(\Sigma,\sigma)\\ \textit{is well defined, linear, and bounded.}\end{gathered}\tag{6.171}$$

Proof To begin, observe that (3.39)-(3.40) in Proposition 3.4, along with (6.166), ensure that

$$\begin{gathered}T_{\mathrm{mod}}f \quad \text{is meaningfully defined and}\\ T_{\mathrm{mod}}f \in L^1_{\mathrm{loc}}(\Sigma,\sigma), \quad \text{for all } f \in \mathrm{B}^p(\Sigma,\sigma).\end{gathered}\tag{6.172}$$

Henceforth, fix a reference point $x_0 \in \Sigma$ and a dilation factor 2_* (as in Definition 5.1). Having fixed a function $f \in \mathrm{B}^p(\Sigma,\sigma)$, and a radius $R \in [2_*,\infty)$, decompose f as

$$f = f\cdot \mathbb{1}_{\Delta(x_0,2R)} + \big(f - f\cdot \mathbb{1}_{\Delta(x_0,2R)}\big) =: f_1 + f_2.\tag{6.173}$$

It is apparent that $|f_1| \le |f|$ and $|f_2| \le 2|f|$ at σ-a.e. point on Σ. In turn, the lattice property of $\mathrm{B}^p(\Sigma,\sigma)$ (cf. Remark 3.3) implies that $f_1, f_2 \in \mathrm{B}^p(\Sigma,\sigma)$. Thus, the action of T_{mod} on f_1 and f_2 is also well defined (cf. (6.172)). Abbreviate

$$\begin{gathered}c_1 := T_{\mathrm{mod}}f_1 - Tf_1, \quad \text{and } c_2 := T_{\mathrm{mod}}f_2(x_0)\\ \text{and define } c_R := c_1 + c_2.\end{gathered}\tag{6.174}$$

We claim that c_1 and c_2 are well-defined constants on Σ. To verify this, notice that f_1 belongs to $L^p(\Sigma,\sigma)$ (cf. (3.40)), which in particular puts f_1 in $L^1\big(\Sigma, \frac{\sigma(x)}{1+|x|^{n-1}}\big)$. As such, (6.167) implies that c_1 a is well-defined number. In relation to c_2, observe that f_2 from (6.173) belongs to $L^1\big(\Sigma, \frac{\sigma(x)}{1+|x|^{n}}\big)$ (cf. (3.39) used with $\varepsilon := 1$) and is supported in $\Sigma\setminus\Delta(x_0,2R)$. Hence, we may express

$$T_{\mathrm{mod}}f_2(x_0) = \int_\Sigma \big\{k_{2R}(x_0-y) - k_1(-y)\big\} f_2(y)\,d\sigma(y).\tag{6.175}$$

In light of [GHA-III, (2.3.119), p. 368], this is an absolutely convergent integral, hence c_2 is a well-defined number, as desired.

Moving on, estimate

$$\begin{aligned}&\Big(\fint_{\Delta(x_0,R)} |T_{\mathrm{mod}}f - c_R|^p\,d\sigma\Big)^{1/p}\\ &\le \Big(\fint_{\Delta(x_0,R)} |T_{\mathrm{mod}}f_1 - c_1|^p\,d\sigma\Big)^{1/p} + \Big(\fint_{\Delta(x_0,R)} |T_{\mathrm{mod}}f_2 - c_2|^p\,d\sigma\Big)^{1/p}\\ &=: \mathrm{I} + \mathrm{II}.\end{aligned}\tag{6.176}$$

To deal with I, use (6.174), the boundedness of T on $L^p(\Sigma,\sigma)$ (cf. Theorem 6.1), the doubling character of σ (as implied by the Ahlfors regularity of Σ), together with Proposition 3.1 to write

$$
\begin{aligned}
\mathrm{I} &= \Big(\fint_{\Delta(x_0,R)} |T_{\mathrm{mod}} f_1 - c_1|^p \, d\sigma \Big)^{1/p} = \Big(\fint_{\Delta(x_0,R)} |T f_1|^p \, d\sigma \Big)^{1/p} \\
&\leq \frac{1}{\sigma\big(\Delta(x_0,R)\big)^{1/p}} \|T(f \cdot \mathbb{1}_{\Delta(x_0,2R)})\|_{L^p(\Sigma,\sigma)} \\
&\leq \frac{C}{\sigma\big(\Delta(x_0,2R)\big)^{1/p}} \|f \cdot \mathbb{1}_{\Delta(x_0,2R)}\|_{L^p(\Sigma,\sigma)} \\
&= C \Big(\fint_{\Delta(x_0,2R)} |f|^p \, d\sigma \Big)^{1/p} \\
&\leq C \|f\|_{\mathrm{B}^p(\Sigma,\sigma)}, \qquad (6.177)
\end{aligned}
$$

for some constant $C \in (0, \infty)$ depending only on n, p, and the Ahlfors regularity of Σ.

Going further, in order to obtain a similar bound for II in (6.176), we first prove that

$$
T_{\mathrm{mod}} f_2(x) = \int_\Sigma \big(k(x-y) - k_1(-y)\big) f_2(y)\, d\sigma(y) \quad \text{for every } x \in \Delta(x_0, R). \qquad (6.178)
$$

To justify this, fix $x \in \Delta(x_0, R)$, and choose $\varepsilon > 0$ sufficiently small, say

$$
0 < \varepsilon < R. \qquad (6.179)
$$

Then, since $\operatorname{supp} f_2 \subseteq \Sigma \setminus \Delta(x_0, 2R)$, we may write

$$
\begin{aligned}
&\int_\Sigma \big\{k_\varepsilon(x-y) - k_1(-y)\big\} f_2(y)\, d\sigma(y) \\
&\quad = \int\limits_{\substack{y\in\Sigma \\ |x-y|<R}} \big\{k_\varepsilon(x-y) - k_1(-y)\big\} f_2(y)\, d\sigma(y) \\
&\qquad + \int\limits_{\substack{y\in\Sigma \\ |x-y|\geq R}} \big\{k_\varepsilon(x-y) - k_1(-y)\big\} f_2(y)\, d\sigma(y) \\
&\quad = \int\limits_{\substack{y\in\Sigma \\ |x-y|<R \\ |y-x_0|\geq 2R}} \big\{k_\varepsilon(x-y) - k_1(-y)\big\} f_2(y)\, d\sigma(y) \\
&\qquad + \int\limits_{\substack{y\in\Sigma \\ |x-y|\geq R \\ |y-x_0|\geq 2R}} \big\{k_\varepsilon(x-y) - k_1(-y)\big\} f_2(y)\, d\sigma(y) =: \mathrm{I}_\varepsilon + \mathrm{II}_\varepsilon. \qquad (6.180)
\end{aligned}
$$

Being mindful of $|x - x_0| < R$, whenever $y \in \Sigma$ satisfies $|x - y| \leq R$ it follows that $|y - x_0| < 2R$. In turn, this implies

$$\mathrm{I}_\varepsilon = 0. \tag{6.181}$$

On the other hand, because $\varepsilon < R$, and $|x-x_0| < R$, if $y \in \Sigma$ is such that $|y-x_0| \geq 2R$ then

$$|x - y| \geq |y - x_0| - |x - x_0| > 2R - R = R > \varepsilon. \tag{6.182}$$

Hence, recalling the definition of k_ε in (6.165), the integral

$$\begin{aligned} \mathrm{II}_\varepsilon &= \int\limits_{\substack{y\in\Sigma \\ |x-y|\geq R \\ |y-x_0|\geq 2R}} \big\{k(x-y) - k_1(-y)\big\} f_2(y)\, d\sigma(y) \\ &= \int_\Sigma \big\{k(x-y) - k_1(-y)\big\} f_2(y)\, d\sigma(y) \end{aligned} \tag{6.183}$$

becomes independent of ε. In concert, (6.165), (6.180), (6.181), and (6.183) readily gives (6.178).

Notice that the smoothness and homogeneity of k ensure that there exists some dimensional constant $C \in (0, \infty)$ for which

$$|\nabla k(z)| \leq \frac{C}{|z|^n} \quad \text{for all} \quad z \in \mathbb{R}^n \setminus \{0\}. \tag{6.184}$$

Also, a direct computation shows that

$$|y - z| \approx |y - x_0| \quad \text{uniformly on } y \in \partial\Omega \setminus B(x_0, 2R) \text{ and } z \in B(x_0, R). \tag{6.185}$$

Availing ourselves with (6.184), (6.185), (6.175), (6.178), and employing the Mean Value Theorem, for every $x \in \Delta(x_0, R)$ we estimate II from (6.176) as

$$
\begin{aligned}
|T_{\rm mod}f_2(x)-c_2| &= |T_{\rm mod}f_2(x)-T_{\rm mod}f_2(x_0)| \\
&\leq \int\limits_{\substack{y\in\Sigma\\ |y-x_0|\geq 2R}} |k(x-y)-k(x_0-y)||f_2(y)|\,d\sigma(y) \\
&\leq \int\limits_{\substack{y\in\Sigma\\ |y-x_0|\geq 2R}} \Big(\sup_{\xi\in[x,x_0]}|\nabla k(\xi-y)|\Big)|x-x_0||f_2(y)|\,d\sigma(y) \\
&\leq C|x-x_0|\int\limits_{\substack{y\in\Sigma\\ |y-x_0|>2R}} \Big(\sup_{\xi\in[x,x_0]}\frac{1}{|\xi-y|^n}\Big)|f(y)|\,d\sigma(y), \\
&\leq CR\int\limits_{\substack{y\in\Sigma\\ |y-x_0|>2R}} \frac{|f(y)|}{|y-x_0|^n}\,d\sigma(y), \qquad (6.186)
\end{aligned}
$$

for some constant $C=C(k,n)\in(0,\infty)$. Choose $\ell\in\mathbb{N}$ with the property that

$$
2_*^{\ell}\leq 2R<2_*^{\ell+1}. \qquad (6.187)
$$

Continue to assume that $x\in\Delta(x_0,R)$, recall (3.4) and the family $\{C_j\}_{j\in\mathbb{N}_0}$ (cf. (3.1)). Use (6.187), Hölder's inequality, the Ahlfors regularity of Σ, and continue estimating (6.186) as

$$
\begin{aligned}
|T_{\rm mod}f_2(x)-c_2| &\leq C2^{\ell}\sum_{j=\ell}^{\infty}\int_{C_j}\frac{|f(y)|}{|y-x_0|^n}\,d\sigma(y)\leq C2_*^{\ell}\sum_{j=\ell}^{\infty}2_*^{-jn}\int_{C_j}|f|\,d\sigma \\
&\leq C2_*^{\ell}\sum_{j=\ell}^{\infty}2_*^{-jn}\|f\cdot\mathbb{1}_{C_j}\|_{L^p(\Sigma,\sigma)}\sigma(C_j)^{1-1/p}. \\
&\leq C2_*^{\ell}\sum_{j=\ell}^{\infty}2_*^{-jn}\|f\cdot\mathbb{1}_{C_j}\|_{L^p(\Sigma,\sigma)}2_*^{j(n-1)(1-1/p)} \\
&= C\sum_{j=\ell}^{\infty}2_*^{\ell-j}\big(2_*^{-j(n-1)/p}\|f\cdot\mathbb{1}_{C_j}\|_{L^p(\Sigma,\sigma)}\big) \\
&\leq C\Big(\sum_{j=\ell}^{\infty}2_*^{\ell-j}\Big)\|f\|_{\mathrm{B}^p(\Sigma,\sigma)}=C\|f\|_{\mathrm{B}^p(\Sigma,\sigma)}
\end{aligned} \qquad (6.188)
$$

for some constant $C=C(n,p,k,\Sigma)\in(0,\infty)$. Hence

$$
\mathrm{II}=\Big(⨍_{\Delta(x_0,R)}|T_{\rm mod}f_2(x)-c_2|^p\,d\sigma(x)\Big)^{1/p}\leq C\|f\|_{\mathrm{B}^p(\Sigma,\sigma)}, \qquad (6.189)
$$

for some constant $C \in (0, \infty)$ independent of f and $R \in [2_*, \infty)$.

Collectively, from (6.176), (6.177), (6.189), we deduce that there exists some constant $C \in (0, \infty)$, depending only on n, p, the kernel k, and the Ahlfors regularity of Σ with the property that for every radius $R \in [2_*, \infty)$, there exists some $c_R \in \mathbb{C}$ satisfying

$$\Big(\fint_{\Delta(x_0,R)} |T_{\mathrm{mod}} f(x) - c_R|^p \, d\sigma(x)\Big)^{1/p} \leq C\|f\|_{\mathrm{B}^p(\Sigma,\sigma)}. \tag{6.190}$$

Finally, being mindful of (5.7), this ultimately proves that

$$\|T_{\mathrm{mod}} f\|_{\mathrm{BCMO}_p(\Sigma,\sigma)} \leq C\|f\|_{\mathrm{B}^p(\Sigma,\sigma)}. \tag{6.191}$$

This concludes the proof of Theorem 6.5. □

It is instructive to give an alternative proof of Theorem 6.5, now taking advantage of the duality result established in Theorem 5.1 and the boundedness result obtained in Theorem 6.1.

Alternative proof of Theorem 6.5. Fix an arbitrary function $f \in \mathrm{B}^p(\Sigma, \sigma)$, along with an x_0-central $(1, p')$-atom a on Σ, where $p' \in (1, \infty)$ denotes the Hölder conjugate exponent of p. From Proposition 3.4, and Definition 4.3 we have

$$\begin{aligned} & f \in L^p_{\mathrm{loc}}(\Sigma, \sigma) \cap L^1\big(\Sigma, \tfrac{\sigma(x)}{1+|x|^n}\big), \\ \text{and } \ & a \in L^{p'}_{\mathrm{comp}}(\Sigma, \sigma) \ \text{ with } \int_\Sigma a \, d\sigma = 0. \end{aligned} \tag{6.192}$$

As such, thanks to (3.70), (6.169), Theorem 4.1, and Theorem 6.1, we may write

$$\begin{aligned} \Big|\int_\Sigma (T_{\mathrm{mod}} f) \cdot a \, d\sigma\Big| &= \Big|\int_\Sigma f \cdot (Ta) \, d\sigma\Big| \leq \int_\Sigma |f||Ta| \, d\sigma \\ &\leq \|f\|_{B^{p'}(\Sigma,\sigma)} \|Ta\|_{\mathrm{A}^p(\Sigma,\sigma)} \\ &\leq \|T\|_{\mathrm{HA}^p(\Sigma,\sigma)\to \mathrm{A}^p(\Sigma,\sigma)} \|f\|_{B^{p'}(\Sigma,\sigma)} \|a\|_{\mathrm{HA}^p(\Sigma,\sigma)} \\ &\leq C\|T\|_{\mathrm{HA}^p(\Sigma,\sigma)\to \mathrm{A}^p(\Sigma,\sigma)} \|f\|_{B^{p'}(\Sigma,\sigma)} \end{aligned} \tag{6.193}$$

for some constant $C \in (0, \infty)$ independent of f and a. After taking the supremum over all x_0-central $(1, p')$-atoms a, Corollary 5.2 implies that

$$\|T_{\mathrm{mod}} f\|_{\mathrm{BCMO}_p(\Sigma,\sigma)} \leq C\|T\|_{\mathrm{HA}^p(\Sigma,\sigma)\to \mathrm{A}^p(\Sigma,\sigma)} \|f\|_{\mathrm{B}^p(\Sigma,\sigma)}. \tag{6.194}$$

This then establishes Theorem 6.5. □

To proceed, we briefly elaborate on the modified versions of generalized double layers. Assume $\Omega \subseteq \mathbb{R}^{n-1}$ is a UR domain, and abbreviate $\sigma := \mathcal{H}^{n-1} \lfloor \partial\Omega$. Denote by ν the geometric measure theoretic outward unit normal to Ω. For some sufficiently large integer $N = N(n) \in \mathbb{N}$, select a vector-valued function $\vec{k} \in [\mathscr{C}^N(\mathbb{R}^n \setminus \{0\})]^n$, and suppose that $\vec{k}$ is odd, positive homogeneous of degree $1-n$, and divergence-free, i.e., $\operatorname{div}\vec{k} = 0$ in $\mathbb{R}^n \setminus \{0\}$. As in [GHA-IV, (5.1.4), p. 532], define the `boundary-to-domain Generalized Double Layer` $\mathcal{T}$ associated with $\vec{k}$ acting on each function $f \in L^1\big(\partial\Omega, \frac{\sigma(x)}{1+|x|^{n-1}}\big)$ via

$$\mathcal{T}f(x) := \int_{\partial\Omega} \langle \nu(y), \vec{k}(x-y)\rangle f(y)\, d\sigma(y) \quad \text{for all } x \in \Omega. \tag{6.195}$$

Also, as in [GHA-IV, (5.1.44), p. 538] introduce the `modified boundary-to-domain Generalized Double Layer` $\mathcal{T}_{\rm mod}$ acting on each $f \in L^1\big(\partial\Omega, \frac{\sigma(x)}{1+|x|^n}\big)$ according to

$$\mathcal{T}_{\rm mod}f(x) := \int_{\partial\Omega} \langle \nu(y), \vec{k}(x-y) - \vec{k}(-y)\mathbb{1}_{\mathbb{R}^n\setminus\overline{B(0,1)}}(y)\rangle f(y)\, d\sigma(y) \tag{6.196}$$

for all $x \in \Omega$. Work done in [GHA-IV, Theorem 5.1.1, pp. 532–550] ensures that the operators $\mathcal{T}$ and $\mathcal{T}_{\rm mod}$ are meaningfully defined, and they are compatible on $L^1\big(\partial\Omega, \frac{\sigma(x)}{1+|x|^{n-1}}\big)$, in the sense that

$$\begin{gathered}\text{for each } f \in L^1\big(\partial\Omega, \tfrac{\sigma(x)}{1+|x|^{n-1}}\big), \text{ the difference}\\ C_f := \mathcal{T}_{\rm mod}f - \mathcal{T}f \text{ is a constant function in } \Omega.\end{gathered} \tag{6.197}$$

As a consequence, it follows that

$$\nabla\mathcal{T}_{\rm mod}f = \nabla\mathcal{T}f \quad \text{in } \Omega \text{ for each } f \in L^1\big(\partial\Omega, \tfrac{\sigma(x)}{1+|x|^{n-1}}\big). \tag{6.198}$$

In addition, at each point $x \in \Omega$, one may express

$$\begin{gathered}\partial_\ell(\mathcal{T}_{\rm mod}f)(x) = \int_{\partial\Omega} \langle \nu(y), (\partial_\ell\vec{k})(x-y)\rangle f(y)\, d\sigma(y)\\ \text{for each } \ell \in \{1,\dots,n\}, \text{ and } f \in L^1\big(\partial\Omega, \tfrac{\sigma(x)}{1+|x|^n}\big)\end{gathered} \tag{6.199}$$

Equally important, the operator $\mathcal{T}_{\rm mod}$ has the following property:

$$\begin{gathered}\mathcal{T}_{\rm mod} \text{ maps constants function on } \partial\Omega\\ \text{into constants functions in the set } \Omega.\end{gathered} \tag{6.200}$$

We may also consider a modified version of the boundary-to-boundary generalized double layer T (introduced in (6.63)). The `modified boundary-to-boundary Generalized Double Layer` $T_{\rm mod}$ acts on each $f \in L^1\big(\partial\Omega, \frac{\sigma(x)}{1+|x|^n}\big)$ via (cf. [GHA-IV, (5.1.65)-(5.1.66), p. 541]

$$T_{\rm mod}f(x) := \lim_{\varepsilon\to 0^+} \int_{\partial\Omega} \langle \nu(y), \vec{k}_\varepsilon(x-y) - \vec{k}_1(-y)\rangle f(y)\, d\sigma(y), \tag{6.201}$$

at σ-a.e. $x \in \partial\Omega$, where

$$\vec{k}_\varepsilon := \vec{k} \cdot \mathbb{1}_{\mathbb{R}^n \setminus \overline{B(0,\varepsilon)}} \quad \text{for each } \varepsilon \in (0, \infty). \tag{6.202}$$

In relation to this, [GHA-IV, Theorem 5.1.1, pp. 532–550] guarantees that T_{mod} is a well-defined operator and it compatible with T (cf. (6.63)) in the space $L^1\big(\partial\Omega, \frac{\sigma(x)}{1+|x|^{n-1}}\big)$. Specifically, for every function $f \in L^1\big(\partial\Omega, \frac{\sigma(x)}{1+|x|^{n-1}}\big)$ it follows that

$$\text{the difference } C_f := T_{\text{mod}} f - Tf \text{ is a constant function on } \partial\Omega. \tag{6.203}$$

Furthermore,

$$\begin{gathered} T_{\text{mod}} \text{ maps constant functions on } \partial\Omega \\ \text{into constant functions on } \partial\Omega. \end{gathered} \tag{6.204}$$

Another fundamental property of T_{mod} is the fact that if $p, p' \in (1, \infty)$ are such that $1/p + 1/p' = 1$, then whenever

$$\begin{gathered} f \in L^{p'}_{\text{loc}}(\partial\Omega, \sigma) \cap L^1\big(\partial\Omega, \tfrac{\sigma(x)}{1+|x|^n}\big), \\ \text{and } \ g \in L^p_{\text{comp}}(\partial\Omega, \sigma) \ \text{ with } \int_{\partial\Omega} g \, d\sigma = 0 \end{gathered} \tag{6.205}$$

it follows that (cf. [GHA-IV, (5.1.70), p. 542]

$$\begin{gathered} \int_{\partial\Omega} |T_{\text{mod}} f| \cdot |g| \, d\sigma < \infty, \quad \int_{\partial\Omega} |f| \cdot |T^{\#} g| \, d\sigma < \infty, \\ \text{and } \int_{\partial\Omega} (T_{\text{mod}} f) \cdot g \, d\sigma = \int_{\partial\Omega} f \cdot (T^{\#} g) \, d\sigma, \end{gathered} \tag{6.206}$$

where $T^{\#}$ is the formal transpose generalized double layer defined in (6.64).

Having discussed the modified generalized double layers, we next prove that the space of functions of bounded central mean oscillations is stable under the action of T_{mod}. This is made precise in the result recorded below.

Theorem 6.6 *Suppose $\Omega \subseteq \mathbb{R}^n$ is a* UR *domain with unbounded boundary, and abbreviate $\sigma := \mathcal{H}^{n-1} \lfloor \partial\Omega$. For some sufficiently large integer $N = N(n) \in \mathbb{N}$, pick a kernel $\vec{k} \in \big[\mathscr{C}^N(\mathbb{R}^n \setminus \{0\})\big]^n$ that is odd, positive homogeneous of degree $1 - n$, and divergence-free, i.e.,* $\operatorname{div} \vec{k} = 0$ *in $\mathbb{R}^n \setminus \{0\}$. Recall the generalized double layer potential T_{mod} associated with $\vec{k}$ as in* (6.201)*. Then, for any integrability exponent $p \in (1, \infty)$, it follows that*

$$\begin{gathered} T_{\text{mod}} : \text{BCMO}_p(\partial\Omega, \sigma) \longrightarrow \text{BCMO}_p(\partial\Omega, \sigma) \\ \textit{is a well-defined, linear, and bounded operator.} \end{gathered} \tag{6.207}$$

In particular, with $[\,\cdot\,]$ *denoting the equivalence class modulo constants in the space* $\mathrm{BCMO}_p(\partial\Omega,\sigma)$*, one has that the operator*

$$\begin{gathered}[T_{\mathrm{mod}}]:\widetilde{\mathrm{BCMO}}_p(\partial\Omega,\sigma)\to\widetilde{\mathrm{BCMO}}_p(\partial\Omega,\sigma),\quad \textit{given by}\\ [T_{\mathrm{mod}}][f]:=[T_{\mathrm{mod}}f]\ \textit{for all}\ f\in\mathrm{BCMO}_p(\partial\Omega,\sigma),\\ \textit{is meaningfully defined, linear, and bounded.}\end{gathered}\tag{6.208}$$

Proof Fix an integrability exponent $p\in(1,\infty)$. Thanks to (6.166), Proposition 5.4, and (5.36) (presently used with $\varepsilon:=1$), whenever $f\in\mathrm{BCMO}_p(\partial\Omega,\sigma)$ the function $T_{\mathrm{mod}}f$ is well defined and belongs to $L^1_{\mathrm{loc}}(\Sigma,\sigma)$.

Fix an arbitrary function $f\in\mathrm{BCMO}_p(\partial\Omega,\sigma)$ along with an x_0-central $(1,p')$-atom a on $\partial\Omega$, where $p'\in(1,\infty)$ denotes the Hölder conjugate exponent of p. From (5.36) in Proposition 5.4 and Definition 4.3 we have

$$\begin{gathered}f\in L^p_{\mathrm{loc}}(\partial\Omega,\sigma)\cap L^1\big(\partial\Omega,\tfrac{\sigma(x)}{1+|x|^n}\big),\\ \text{and}\ \ a\in L^{p'}_{\mathrm{comp}}(\partial\Omega,\sigma)\ \ \text{with}\ \ \int_{\partial\Omega}a\,d\sigma=0.\end{gathered}\tag{6.209}$$

In light of (6.206) (with $g:=a$, and the roles of p and p' interchanged), we compute

$$\begin{aligned}\Big|\int_{\partial\Omega}(T_{\mathrm{mod}}f)\cdot a\,d\sigma\Big|&=\Big|\int_{\partial\Omega}f\cdot(T^{\#}a)\,d\sigma\Big|=\big|\langle[f],T^{\#}a\rangle\big|\\ &\le\|f\|_{\mathrm{BCMO}_p(\partial\Omega,\sigma)}\|T^{\#}a\|_{\mathrm{HA}^{p'}(\partial\Omega,\sigma)}\\ &\le\|f\|_{\mathrm{BCMO}_p(\partial\Omega,\sigma)}\|T^{\#}\|_{\mathrm{HA}^{p'}(\partial\Omega,\sigma)\to\mathrm{HA}^{p'}(\partial\Omega,\sigma)}\|a\|_{\mathrm{HA}^{p'}(\partial\Omega,\sigma)}\\ &\le C\|f\|_{\mathrm{BCMO}_p(\partial\Omega,\sigma)}\end{aligned}\tag{6.210}$$

for some $C\in(0,\infty)$ depending only on n, p, and the UR constants of $\partial\Omega$, where $\langle\cdot,\cdot\rangle$ denotes the duality pairing between $\widetilde{\mathrm{BCMO}}_p(\partial\Omega,\sigma)$ and $\mathrm{HA}^{p'}(\partial\Omega,\sigma)$, understood in the sense of Theorem 5.1. The second equality in (6.210) follows from Corollary 5.6 and the fact that $T^{\#}a$ is a fixed multiple of a central molecule (see (6.71)). The first inequality is guaranteed by the continuity of $[f]$ (viewed as a linear functional on $\mathrm{HA}^{p'}(\partial\Omega,\sigma)$; cf. Theorem 5.1), while the second inequality is a consequence of the boundedness of $T^{\#}$ on $\mathrm{HA}^{p'}(\partial\Omega,\sigma)$ established in Theorem 6.3. The last inequality is justified by Theorem 4.1. Since (6.210) holds for every x_0-central $(1,p')$-atom a, Corollary 5.2 yields

$$\|T_{\mathrm{mod}}\|_{\mathrm{BCMO}_p(\partial\Omega,\sigma)}\le C\|f\|_{\mathrm{BCMO}_p(\partial\Omega,\sigma)},\tag{6.211}$$

for some $C\in(0,\infty)$ depending only on n, p, and the UR constants of $\partial\Omega$. As such, this proves (6.207).

Next, recall that T_{mod} maps constants into constants (cf. (6.204)). Then whenever $f\in\mathrm{BCMO}_p(\partial\Omega,\sigma)$ and $c\in\mathbb{C}$ we have

$$[T_{\rm mod}(f+c)] = [T_{\rm mod}f + T_{\rm mod}c] = [T_{\rm mod}f] \quad \text{as equivalence classes in } \widetilde{\rm BCMO}_p(\partial\Omega,\sigma). \tag{6.212}$$

In turn, this proves that the operator $[T_{\rm mod}]$ in (6.208) is well defined. The boundedness of the operator $[T_{\rm mod}]$ follows immediately from what we have proved so far. This concludes the proof of Theorem 6.6. □

To proceed, we introduce the notion of a central Carleson measure in open sets from $\mathbb{R}^n$ having Ahlfors regular boundary.

Definition 6.1 Assume $\Omega \subseteq \mathbb{R}^n$ is an open set with an Ahlfors regular boundary, and abbreviate $\sigma := \mathcal{H}^{n-1}\lfloor\partial\Omega$. Fix a dilation factor $2_* > \big(C_{\partial\Omega}/c_{\partial\Omega}\big)^{1/(n-1)}$, where $C_{\partial\Omega}, c_{\partial\Omega} \in (0,\infty)$ denote the upper and lower Ahlfors regularity constants of $\partial\Omega$, respectively. Also, pick a reference point $x_0 \in \partial\Omega$. A Borel measure μ in Ω is called a `central Carleson measure` if there exists some constant $C \in (0,\infty)$ with the property that

$$\mu\big(B(x_0,R)\cap\Omega\big) \le C\cdot\sigma\big(\Delta(x_0,R)\big) \quad \text{for every } R \ge 2_*. \tag{6.213}$$

Define `central Carleson measure norm` (or simply CCM-`norm`) of μ in Ω as being the following quantity:

$$\|\mu\|_{\rm CCM(\Omega)} := \sup_{R\ge 2_*}\frac{\mu\big(B(x_0,R)\cap\Omega\big)}{\sigma\big(\Delta(x_0,R)\big)}. \tag{6.214}$$

The result recorded below ensures that the property of a measure μ being a central Carleson measure is independent of the choice of the reference point x_0 and the dilation factor 2_* in Definition 6.1.

Proposition 6.3 *Assume $\Omega \subseteq \mathbb{R}^n$ is an open set with an Ahlfors regular boundary, and abbreviate $\sigma := \mathcal{H}^{n-1}\lfloor\partial\Omega$. Denote by $C_{\partial\Omega}, c_{\partial\Omega} \in (0,\infty)$ the upper and lower Ahlfors regularity constants of $\partial\Omega$, respectively. Fix two reference points $x_0, x_1 \in \partial\Omega$, and two dilation factors $2_*, 2_\diamond \ge \big(C_{\partial\Omega}/c_{\partial\Omega}\big)^{1/(n-1)}$. Then for any Borel measure μ in Ω one has*

$$\sup_{R\ge 2_*}\frac{\mu\big(B(x_0,R)\cap\Omega\big)}{\sigma\big(\Delta(x_0,R)\big)} \approx \sup_{R\ge 2_\diamond}\frac{\mu\big(B(x_1,R)\cap\Omega\big)}{\sigma\big(\Delta(x_1,R)\big)}, \tag{6.215}$$

with implicit proportionality constants depending only on n, x_0, x_1, 2_*, $2_\diamond$, and the Ahlfors regularity character of $\partial\Omega$. Consequently, the central Carleson measure norm of μ (cf. (6.214)) is independent of x_0 and 2_*.

Proof Fix a Borel measure μ in Ω, and select an arbitrary radius $R \in [2_*,\infty)$. Select a parameter $\lambda \in (0,\infty)$ such that

$$\lambda \ge \max\Big\{2, \frac{2|x_0-x_1|}{2_*}, \frac{2_\diamond}{2_*}\Big\} \tag{6.216}$$

and define

$$\widetilde{R} := \lambda R \in (0, \infty). \tag{6.217}$$

It is apparent that (6.216)-(6.217) imply

$$\widetilde{R} \geq 2_\diamond. \tag{6.218}$$

Also, from the choice of λ whenever $x \in B(x_0, R) \cap \Omega$ we have

$$\begin{aligned} |x - x_1| &\leq |x - x_0| + |x_0 - x_1| < R + \frac{\lambda}{2} 2_* \\ &\leq \frac{\lambda}{2} R + \frac{\lambda}{2} R = \lambda R = \widetilde{R}. \end{aligned} \tag{6.219}$$

As such,

$$B(x_0, R) \cap \Omega \subseteq B(x_1, \widetilde{R}) \cap \Omega. \tag{6.220}$$

Collectively, (6.217), (6.218), (6.220), and the Ahlfors regularity of $\partial\Omega$, yield

$$\begin{aligned} \frac{\mu\big(B(x_0, R) \cap \Omega\big)}{\sigma\big(\Delta(x_0, R)\big)} &\leq \frac{\sigma\big(\Delta(x_1, \widetilde{R})\big)}{\sigma\big(\Delta(x_0, R)\big)} \frac{\mu\big(B(x_1, \widetilde{R}) \cap \Omega\big)}{\sigma\big(\Delta(x_1, \widetilde{R})\big)} \\ &\leq C \frac{\widetilde{R}^{n-1}}{R^{n-1}} \frac{\mu\big(B(x_1, \widetilde{R}) \cap \Omega\big)}{\sigma\big(\Delta(x_1, \widetilde{R})\big)} \\ &\leq C \lambda^{n-1} \sup_{R \geq 2_\diamond} \frac{\mu\big(B(x_1, R) \cap \Omega\big)}{\sigma\big(\Delta(x_1, R)\big)}, \end{aligned} \tag{6.221}$$

for some constant $C = C(\partial\Omega) \in (0, \infty)$. Since $R \geq 2_*$ was taken arbitrary to begin with, this ultimately gives

$$\sup_{R \geq 2_*} \frac{\mu\big(B(x_0, R) \cap \Omega\big)}{\sigma\big(\Delta(x_0, R)\big)} \leq C \cdot \sup_{R \geq 2_\diamond} \frac{\mu\big(B(x_1, R) \cap \Omega\big)}{\sigma\big(\Delta(x_1, R)\big)}, \tag{6.222}$$

for some constant $C \in (0, \infty)$ depending only on n, x_0, x_1, 2_*, $2_\diamond$, and the Ahlfors regularity character of $\partial\Omega$. To prove the reverse inequality, run the same reasoning as before now switching x_0 with x_1, and 2_* with $2_\diamond$. This finishes the proof of Proposition 6.3. □

In order to establish one of the main results of this section, we next recall a L^p-square function estimate involving convolution-type integral operators in UR domains obtained in [GHA-III, Theorem 2.4.1, pp. 382–390].

Theorem 6.7 *Assume $\Omega \subseteq \mathbb{R}^n$ is a* UR *domain, and abbreviate $\sigma := \mathcal{H}^{n-1} \lfloor \partial\Omega$. For a sufficiently large integer $N = N(n) \in \mathbb{N}$, consider a kernel $k \in \mathscr{C}^N(\mathbb{R}^n \setminus \{0\})$ that is odd and positive homogeneous of degree $1 - n$. Recall the boundary-to-domain convolution-type integral operator $\mathcal{T}$ associated with k, whose action on* *each* $f \in L^1\big(\partial\Omega, \frac{\sigma(x)}{1+|x|^{n-1}}\big)$ *is given by*

$$\mathcal{T}f(x) := \int_{\partial\Omega} k(x-y)f(y)\,d\sigma(y) \quad \textit{for all } x \in \Omega. \tag{6.223}$$

Finally, fix an integrability exponent $p \in (1,\infty)$. *Then there exists some* $C \in (0,\infty)$ *depending only on* n, p, *and the* UR *constants of* $\partial\Omega$ *such that, for every function* $f \in L^p(\partial\Omega,\sigma)$, *one has*

$$\int_\Omega |\nabla(\mathcal{T}f)|^p\,\mathrm{dist}(\cdot,\partial\Omega)^{p-1}\,d\mathcal{L}^n \le C\int_{\partial\Omega}|f|^p\,d\sigma. \tag{6.224}$$

The following result establishes a connection between central Carleson measures and functions of bounded central mean oscillations. This plays a fundamental role in the treatment of the Dirichlet Problem with boundary data taken from BCMO_p in subsequent chapters.

Theorem 6.8 *Consider a* UR *domain* $\Omega \subseteq \mathbb{R}^n$ *with unbounded boundary, and abbreviate* $\sigma := \mathcal{H}^{n-1}\lfloor\partial\Omega$. *For some sufficiently large integer* $N = N(n) \in \mathbb{N}$, *select a vector-valued kernel* $\vec{k} \in \big[\mathscr{C}^N(\mathbb{R}^n\setminus\{0\})\big]^n$ *that is odd, positive homogeneous of degree* $1-n$, *and divergence-free, i.e.,* $\mathrm{div}\,\vec{k} = 0$ *in* $\mathbb{R}^n\setminus\{0\}$. *Recall the boundary-to-domain modified generalized double layer potential* $\mathcal{T}_{\mathrm{mod}}$ *associated with* $\vec{k}$ *from* (6.196). *Fix a reference point* $x_0 \in \partial\Omega$ *and a dilation factor* 2_* *as in Definition 6.1. Finally, pick an integrability exponent* $p \in (1,\infty)$.

Then there exists some $C \in (0,\infty)$ *depending only on* n, p, *and the* UR *constants of* $\partial\Omega$ *with the property that whenever* $f \in \mathrm{BCMO}_p(\partial\Omega,\sigma)$ *one has*

$$\sup_{R\ge 2_*}\left(\frac{1}{\sigma\big(\Delta(x_0,R)\big)}\int_{B(x_0,R)\cap\Omega}|\nabla(\mathcal{T}_{\mathrm{mod}}f)|^p\,\mathrm{dist}(\cdot,\partial\Omega)^{p-1}\,d\mathcal{L}^n\right)^{1/p}$$
$$\le C\|f\|_{\mathrm{BCMO}_p(\partial\Omega,\sigma)}. \tag{6.225}$$

In particular, $|\nabla(\mathcal{T}_{\mathrm{mod}}f)|^p\,\mathrm{dist}(\cdot,\partial\Omega)^{p-1}\,d\mathcal{L}^n$ *is a central Carleson measure in* Ω.

Proof Notice that the action of $\mathcal{T}_{\mathrm{mod}}$ on any function from $\mathrm{BCMO}_p(\partial\Omega,\sigma)$ is well defined, since $\mathrm{BCMO}_p(\partial\Omega,\sigma) \subseteq L^1\big(\partial\Omega, \frac{\sigma(x)}{1+|x|^n}\big)$ (cf. Proposition 5.4).

Fix an arbitrary function $f \in \mathrm{BCMO}_p(\partial\Omega,\sigma)$, and pick a finite radius $R \ge 2_*$. Decompose f as

$$\begin{aligned} f &= \big(f - f_{\Delta(x_0,2R)}\big)\cdot \mathbb{1}_{\Delta(x_0,2R)} + \big(f - f_{\Delta(x_0,2R)}\big)\cdot \mathbb{1}_{\partial\Omega\setminus\Delta(x_0,2R)} + f_{\Delta(x_0,2R)} \\ &=: f_1 + f_2 + f_3. \end{aligned} \tag{6.226}$$

In light of the reasoning presented in (5.44)-(5.50) it follows that f_1, f_2, and f_3 belong to $L^1\big(\partial\Omega, \frac{\sigma(x)}{1+|x|^n}\big)$. Since the term f_3 is itself a constant, (6.200) ensures that

$\mathcal{T}_{\rm mod} f_3$ is a constant function in Ω. Hence

$$\nabla(\mathcal{T}_{\rm mod} f)(x) = \nabla(\mathcal{T}_{\rm mod} f_1)(x) + \nabla(\mathcal{T}_{\rm mod} f_2)(x) \quad \text{for each } x \in \Omega. \tag{6.227}$$

Bearing in mind that $\mathrm{BCMO}_p(\partial\Omega, \sigma) \subseteq L^p_{\rm loc}(\partial\Omega, \sigma)$ (cf. Proposition 5.4), we see that the function f_1 in (6.226) actually belongs to $L^p(\partial\Omega, \sigma) \subseteq L^1\big(\partial\Omega, \frac{\sigma(x)}{1+|x|^{n-1}}\big)$. Hence, (5.1), (6.198), Theorem 6.7, and the doubling character of σ entail

$$\begin{aligned}
&\int_\Omega |\nabla(\mathcal{T}_{\rm mod} f_1)(x)|^p \operatorname{dist}(x, \partial\Omega)^{p-1}\, d\mathcal{L}^n(x) \\
&\quad = \int_\Omega |\nabla(\mathcal{T} f_1)(x)|^p \operatorname{dist}(x, \partial\Omega)^{p-1}\, d\mathcal{L}^n(x) \\
&\quad \leq C \int_{\partial\Omega} |f_1|^p \, d\sigma \\
&\quad = C \int_{\Delta(x_0,2R)} |f - f_{\Delta(x_0,2R)}|^p \, d\sigma \\
&\quad \leq C\sigma\big(\Delta(x_0, R)\big) \|f\|^p_{\mathrm{BCMO}_p(\partial\Omega,\sigma)},
\end{aligned} \tag{6.228}$$

for some $C \in (0, \infty)$ depending only on n, p, and the Ahlfors regularity constants of $\partial\Omega$.

Going further, the goal is to obtain a similar type of estimate for $\nabla(\mathcal{T}_{\rm mod} f_2)$. Fix a location $x \in B(x_0, R) \cap \Omega$. Notice that if $y \in \Delta(x_0, 2^{k+1}R) \setminus \Delta(x_0, 2^k R)$ and $k \in \mathbb{N}$ then

$$|x - y| \geq |y - x_0| - |x - x_0| \geq 2^k R - R \geq 2^k R - 2^{k-1} R = 2^{k-1} R. \tag{6.229}$$

Using this, (6.199), (6.78), the definition of f_2 (cf. (6.226)), along with the Ahlfors regularity of $\partial\Omega$, and Hölder's inequality, we estimate

$$\begin{aligned}
|\nabla(\mathcal{T}_{\rm mod} f_2)(x)| &\leq \int_{\partial\Omega} |\nabla\vec{k}(x-y)||f_2(y)|\, d\sigma(y) \\
&\leq C \int_{\partial\Omega\setminus\Delta(x_0,2R)} \frac{|f(y) - f_{\Delta(x_0,2R)}|}{|x-y|^n}\, d\sigma(y) \\
&\leq C \sum_{k=1}^{\infty} \int_{\Delta(x_0,2^{k+1}R)\setminus\Delta(x_0,2^k R)} \frac{|f(y) - f_{\Delta(x_0,2R)}|}{|x-y|^n}\, d\sigma(y) \\
&\leq C \sum_{k=1}^{\infty} \frac{1}{(2^k R)^n} \int_{\Delta(x_0,2^{k+1}R)} |f - f_{\Delta(x_0,2^{k+1}R)}|\, d\sigma \\
&\quad + C \sum_{k=1}^{\infty} \frac{\sigma\big(\Delta(x_0, 2^{k+1}R)\big)}{(2^k R)^n} |f_{\Delta(x_0,2R)} - f_{\Delta(x_0,2^{k+1}R)}|
\end{aligned}$$

$$
\begin{aligned}
&\leq C \sum_{k=1}^{\infty} \frac{2^n}{2^{k+1}R} \frac{1}{(2^{k+1}R)^{n-1}} \int_{\Delta(x_0,2^{k+1}R)} |f - f_{\Delta(x_0,2^{k+1}R)}| \, d\sigma \\
&\quad + C \sum_{k=1}^{\infty} \frac{(2^k R)^{n-1}}{(2^k R)^n} \sum_{j=1}^{k} |f_{\Delta(x_0,2^j R)} - f_{\Delta(x_0,2^{j+1}R)}| \\
&\leq C \sum_{k=1}^{\infty} \frac{1}{2^k R} \fint_{\Delta(x_0,2^{k+1}R)} |f - f_{\Delta(x_0,2^{k+1}R)}| \, d\sigma \\
&\quad + C \sum_{k=1}^{\infty} \frac{1}{2^k R} \sum_{j=1}^{k} \fint_{\Delta(x_0,2^{j+1}R)} |f - f_{\Delta(x_0,2^{j+1}R)}| \, d\sigma \\
&\leq C \sum_{k=1}^{\infty} \frac{1}{2^k R} \Big(\fint_{\Delta(x_0,2^{k+1}R)} |f - f_{\Delta(x_0,2^{k+1}R)}|^p \, d\sigma \Big)^{1/p} \\
&\quad + C \sum_{k=1}^{\infty} \frac{1}{2^k R} \sum_{j=1}^{k} \Big(\fint_{\Delta(x_0,2^{j+1}R)} |f - f_{\Delta(x_0,2^{j+1}R)}|^p \, d\sigma \Big)^{1/p} \\
&\leq \frac{C}{R} \Big(\sum_{k=1}^{\infty} \frac{1}{2^k} + \sum_{k=1}^{\infty} \frac{k}{2^k} \Big) \|f\|_{\mathrm{BCMO}_p(\partial\Omega,\sigma)} \\
&\leq \frac{C}{R} \|f\|_{\mathrm{BCMO}_p(\partial\Omega,\sigma)}
\end{aligned}
\tag{6.230}
$$

for some constant $C = C(n, \partial\Omega) \in (0, \infty)$. In turn, this and the Ahlfors regularity of $\partial\Omega$ permit us to estimate

$$
\begin{aligned}
&\int_{B(x_0,R)\cap\Omega} |\nabla(\mathcal{T}_{\mathrm{mod}} f_2)(x)|^p \, \mathrm{dist}(x, \partial\Omega)^{p-1} \, d\mathcal{L}^n(x) \\
&\leq \frac{C}{R^p} \|f\|^p_{\mathrm{BCMO}_p(\partial\Omega,\sigma)} \int_{B(x_0,R)} |x - x_0|^{p-1} \, d\mathcal{L}^n(x) \\
&\leq \frac{C}{R^p} \|f\|^p_{\mathrm{BCMO}_p(\partial\Omega,\sigma)} R^{p-1} \mathcal{L}^n\big(B(x_0, R)\big) \\
&\leq C \|f\|^p_{\mathrm{BCMO}_p(\partial\Omega,\sigma)} R^{n-1} \\
&\leq C \cdot \sigma\big(\Delta(x_0, R)\big) \|f\|^p_{\mathrm{BCMO}_p(\partial\Omega,\sigma)},
\end{aligned}
\tag{6.231}
$$

for some constant $C \in (0, \infty)$ depending only on n, p, and the Ahlfors regularity of $\partial\Omega$.

In concert, (6.227), (6.228) and (6.231), imply

$$
\begin{aligned}
&\int_{B(x_0,R)\cap\Omega} |\nabla(\mathcal{T}_{\rm mod} f)|^p \operatorname{dist}(\cdot,\partial\Omega)^{p-1}\,d\mathcal{L}^n \\
&\quad \le C\int_{B(x_0,R)\cap\Omega} |\nabla(\mathcal{T}_{\rm mod} f_1)|^p \operatorname{dist}(\cdot,\partial\Omega)^{p-1}\,d\mathcal{L}^n \\
&\qquad + C\int_{B(x_0,R)\cap\Omega} |\nabla(\mathcal{T}_{\rm mod} f_2)|^p \operatorname{dist}(\cdot,\partial\Omega)^{p-1}\,d\mathcal{L}^n \\
&\quad \le C\cdot\sigma\big(\Delta(x_0,R)\big)\|f\|^p_{\mathrm{BCMO}_p(\partial\Omega,\sigma)},
\end{aligned}
\tag{6.232}
$$

for some $C \in (0,\infty)$ depending on n, p, and the UR constants of $\partial\Omega$. Since $R \ge 2_*$ was taken arbitrary to begin with, (6.232) ultimately establishes (6.225). In particular, this ensures that $|\nabla(\mathcal{T}_{\rm mod} f)|^p \operatorname{dist}(\cdot,\partial\Omega)^{p-1}\,d\mathcal{L}^n$ is a central Carleson measure in Ω. The proof of Theorem 6.8 is therefore complete. □

We now obtain some crucial geometrically sensitive operator norm estimates for chord-dot-normal singular integral operators on spaces of bounded central mean oscillations functions. The reader is advised to recall the piece of notation defined in (6.113).

Theorem 6.9 *Assume $\Omega \subseteq \mathbb{R}^n$ is a* UR *domain with unbounded boundary. Abbreviate $\sigma := \mathcal{H}^{n-1}\lfloor\partial\Omega$, and denote by ν the geometric measure theoretic outward unit normal to Ω. For a sufficiently large integer $N = N(n) \in \mathbb{N}$, consider a kernel $k \in \mathscr{C}^N(\mathbb{R}^n \setminus \{0\})$ that is even and positive homogeneous of degree $-n$, and recall the modified chord-dot-normal singular integral operator $T_{\rm mod}$ associated with k as in* (6.105).

Then for each integrability exponent $p \in (1,\infty)$ the operator

$$
\begin{gathered}
T_{\rm mod} : \mathrm{BCMO}_p(\partial\Omega,\sigma) \longrightarrow \mathrm{BCMO}_p(\partial\Omega,\sigma) \\
\text{is well-defined, linear, and bounded.}
\end{gathered}
\tag{6.233}
$$

Moreover, for each $m \in \mathbb{N}$ there exists some $C_m \in (0,\infty)$ depending only on n, m p, the kernel k, and the UR *constants of Ω with the property that*

$$
\|T_{\rm mod}\|_{\mathrm{BCMO}_p(\partial\Omega,\sigma)\to\mathrm{BCMO}_p(\partial\Omega,\sigma)} \le C_m\cdot\|\nu\|^{\langle m\rangle}_{[\mathrm{BMO}(\partial\Omega,\sigma)]^n}.
\tag{6.234}
$$

In particular, with $[\,\cdot\,]$ denoting the equivalence class modulo constants in the space $\mathrm{BCMO}_p(\partial\Omega,\sigma)$, *it follows that the operator*

$$
\begin{gathered}
[T_{\rm mod}] : \widetilde{\mathrm{BCMO}}_p(\partial\Omega,\sigma) \longrightarrow \widetilde{\mathrm{BCMO}}_p(\partial\Omega,\sigma) \ \text{ given by} \\
[T_{\rm mod}][f] := [T_{\rm mod} f] \text{ for all } f \in \mathrm{BCMO}_p(\partial\Omega,\sigma) \\
\text{is meaningfully defined, linear, and bounded,}
\end{gathered}
\tag{6.235}
$$

and, with $C_m \in (0,\infty)$ as in (6.234), *one also has*

$$\big\|[T_{\rm mod}]\big\|_{\widetilde{\mathrm{BCMO}}_p(\partial\Omega,\sigma)\to\widetilde{\mathrm{BCMO}}_p(\partial\Omega,\sigma)} \leq C_m \cdot \|\nu\|^{\langle m\rangle}_{[\mathrm{BMO}(\partial\Omega,\sigma)]^n}. \tag{6.236}$$

Proof Fix an exponent $p \in (1,\infty)$ and $m \in \mathbb{N}$. The claim made in (6.233) is seen from Theorem 6.6 and the fact that any modified chord-dot-normal singular integral operator is a modified generalized double layer (as seen from [GHA-IV, Example 5.1.7, pp. 556–557]). In light of the property recorded in (6.166), for any function f belonging to $\mathrm{BCMO}_p(\partial\Omega,\sigma)$, in particular $f \in L^1\big(\partial\Omega, \frac{\sigma(x)}{1+|x|^n}\big) \cap L^p_{\rm loc}(\Sigma,\sigma)$ (cf. (5.36) with $\varepsilon := 1$), we have that $T_{\rm mod}f$ is a well-defined function which belongs to $L^1_{\rm loc}(\Sigma,\sigma)$.

Fix an arbitrary function $f \in \mathrm{BCMO}_p(\partial\Omega,\sigma)$, along with an x_0-central $(1,p')$-atom a on $\partial\Omega$ (cf. Definition 4.3), where $p' \in (1,\infty)$ denotes the Hölder conjugate exponent of p. Then using (6.109)-(6.110) (with $g := a$ and with the roles of p and p' switched) we obtain

$$\int_{\partial\Omega} (T_{\rm mod}f)\cdot a\,d\sigma = \int_{\partial\Omega} f\cdot (T^{\#}a)\,d\sigma. \tag{6.237}$$

Following the same computations as in (6.210), now taking into consideration Theorem 6.4, we may conclude that there exists some $C_m = C(n,p,m,\partial\Omega) \in (0,\infty)$ with the property that

$$\Big|\int_{\partial\Omega} (T_{\rm mod}f)\cdot a\,d\sigma\Big| \leq C_m \cdot \|\nu\|^{\langle m\rangle}_{[\mathrm{BMO}(\partial\Omega,\sigma)]^n}\|f\|_{\mathrm{BCMO}_p(\partial\Omega,\sigma)}. \tag{6.238}$$

Taking the supremum in (6.238) over all x_0-central $(1,p')$-atoms a, Corollary 5.2 entails

$$\|T_{\rm mod}f\|_{\mathrm{BCMO}_p(\partial\Omega,\sigma)} \leq C_m \cdot \|\nu\|^{\langle m\rangle}_{[\mathrm{BMO}(\partial\Omega,\sigma)]^n}\|f\|_{\mathrm{BCMO}_p(\partial\Omega,\sigma)}. \tag{6.239}$$

This establishes the estimate in (6.234). The remaining claims made in relation to $[T_{\rm mod}]$ follow from what we have proved so far together with the same type of reasoning as in (6.212), while bearing in mind that $T_{\rm mod}$ maps constants into constants (cf. (6.108)). This concludes the proof of Theorem 6.9. □

The following result identifies $[T_{\rm mod}]$ acting on the space of functions of bounded central mean oscillations as the adjoint operator $T^{\#}$ in the context of Beurling-Hardy spaces.

Theorem 6.10 *Assume $\Omega \subseteq \mathbb{R}^n$ is a* UR *domain with unbounded boundary. Abbreviate $\sigma := \mathcal{H}^{n-1}\lfloor\partial\Omega$, and denote by ν the geometric measure theoretic outward unit normal to Ω. Pick integrability exponents $p, p' \in (1,\infty)$ such that $1/p + 1/p' = 1$. Consider $T^{\#}$ and $[T_{\rm mod}]$ the operators introduced in* (6.104) *and* (6.235)*, respectively.*

Then the adjoint operator $(T^{\#})^* : \big(\mathrm{HA}^p(\partial\Omega,\sigma)\big)^* \to \big(\mathrm{HA}^p(\partial\Omega,\sigma)\big)^*$ of $T^{\#}$, as a mapping acting from $\mathrm{HA}^p(\partial\Omega,\sigma)$ into itself (cf. Theorem 6.4), is identified with $[T_{\rm mod}] : \widetilde{\mathrm{BCMO}}_{p'}(\partial\Omega,\sigma) \to \widetilde{\mathrm{BCMO}}_{p'}(\partial\Omega,\sigma)$ in the following precise sense: For every functional $\Lambda_f \in \big(\mathrm{HA}^p(\partial\Omega,\sigma)\big)^*$ determined by a func-

tion $f \in \mathrm{BCMO}_{p'}(\partial\Omega,\sigma)$ *(in the sense of Theorem 5.1), and every function* $g \in \mathrm{HA}^p(\partial\Omega,\sigma)$ *one has*

$$
{}_{(\mathrm{HA}^p(\partial\Omega,\sigma))^*}\big\langle (T^\#)^*\Lambda_f, g\big\rangle_{\mathrm{HA}^p(\partial\Omega,\sigma)} = {}_{\widetilde{\mathrm{BCMO}}_{p'}(\partial\Omega,\sigma)}\big\langle [T_{\mathrm{mod}}f], g\big\rangle_{\mathrm{HA}^p(\partial\Omega,\sigma)}. \tag{6.240}
$$

Proof To lighten notation, we omit the subscripts indicating the spaces involved from the duality brackets. Consider a linear functional $\Lambda_f \in \big(\mathrm{HA}^p(\partial\Omega,\sigma)\big)^*$ induced by a (unique modulo constants) function $f \in \mathrm{BCMO}_{p'}(\partial\Omega,\sigma)$ in the sense described in Theorem 5.1. Given $g \in \mathrm{HA}^p(\partial\Omega,\sigma)$, use Theorem 4.2 to find an atomic decomposition as

$$
g = \lim_{N\to\infty} g_N \text{ in } \mathrm{HA}^p(\partial\Omega,\sigma), \text{ where } g_N := \sum_{j=0}^{N} \lambda_j a_j \text{ for all } N \in \mathbb{N}_0, \tag{6.241}
$$

for some numerical sequence $\{\lambda_j\}_{j\in\mathbb{N}_0}$ in $\ell^1(\mathbb{N})$, and some sequence of x_0-central $(1,p)$-atoms $\{a_j\}_{j\in\mathbb{N}_0}$. Then

$$
\begin{aligned}
\big\langle (T^\#)^*\Lambda_f, g\big\rangle &= \lim_{N\to\infty}\big\langle (T^\#)^*\Lambda_f, g_N\big\rangle = \lim_{N\to\infty}\sum_{j=0}^{N}\lambda_j\big\langle (T^\#)^*\Lambda_f, a_j\big\rangle \\
&= \lim_{N\to\infty}\sum_{j=0}^{N}\lambda_j\big\langle \Lambda_f, T^\# a_j\big\rangle = \lim_{N\to\infty}\sum_{j=0}^{N}\lambda_j\int_{\partial\Omega} f\cdot (T^\# a_j)\,d\sigma \\
&= \lim_{N\to\infty}\sum_{j=0}^{N}\lambda_j\int_{\partial\Omega} (T_{\mathrm{mod}}f)\cdot a_j\,d\sigma = \lim_{N\to\infty}\sum_{j=0}^{N}\lambda_j\big\langle [T_{\mathrm{mod}}f], a_j\big\rangle \\
&= \lim_{N\to\infty}\big\langle [T_{\mathrm{mod}}f], g_N\big\rangle = \big\langle [T_{\mathrm{mod}}f], g\big\rangle.
\end{aligned} \tag{6.242}
$$

The first equality is a consequence of the continuity of $(T^\#)^*\Lambda_f$ on $\mathrm{HA}^p(\partial\Omega,\sigma)$. The second equality and the second-to-last equality follow from (6.241). The third equality is seen from the definition of the adjoint of an operator, while the fourth equality is justified by Corollary 5.6 together with the fact that $T^\# a_j$ is a fixed multiple of a central molecule (cf. (6.123)). The fifth equality can be seen from (6.237). The sixth equality is valid since $T_{\mathrm{mod}}f$ belongs to $\mathrm{BCMO}_{p'}(\partial\Omega,\sigma)$ (cf. Theorem 6.9) and its action on central atoms is given by an integral pairing (cf. Theorem 5.1). The last equality is implied by (6.241) and the continuity of $[T_{\mathrm{mod}}f] \in \widetilde{\mathrm{BCMO}}_{p'}(\partial\Omega,\sigma)$, when viewed as a functional on $\mathrm{HA}^p(\partial\Omega,\sigma)$ (in the sense of Theorem 5.1). This concludes the proof of Theorem 6.10. □

Chapter 7

Weakly Elliptic Systems, and Layer Potentials on UR Domains

In this chapter we collect some background results pertaining to weakly elliptic systems and boundary layer potentials. The discussion presented here is based on [GHA-V, Chapter 1]. To set the stage, fix $n \in \mathbb{N}$ with $n \geq 2$, along with some $M \in \mathbb{N}$, and consider a second-order, homogeneous, constant complex coefficient, $M \times M$ system L in $\mathbb{R}^n$ which may be written as

$$L := \big(a^{\alpha\beta}_{jk}\partial_j\partial_k\big)_{1\leq\alpha,\beta\leq M} \tag{7.1}$$

for some complex numbers $a^{\alpha\beta}_{jk}$. Here and elsewhere, we adopt the summation convention over repeated indices. The (real) transpose of L is given by

$$L^\top := \big(a^{\beta\alpha}_{kj}\partial_j\partial_k\big)_{1\leq\alpha,\beta\leq M}. \tag{7.2}$$

The action of L on any vector-valued distribution $u \in (u_\beta)_{1\leq\beta\leq M}$ is defined as

$$Lu = \big(a^{\alpha\beta}_{jk}\partial_j\partial_k u_\beta\big)_{1\leq\alpha\leq M}. \tag{7.3}$$

Denote by $\mathfrak{L}$ the collection of all such systems L as in (7.1).

Introduce the collection of all coefficient tensors with constant complex entries as

$$\mathfrak{A} := \Big\{A = \big(a^{\alpha\beta}_{jk}\big)_{\substack{1\leq\alpha,\beta\leq M\\ 1\leq j,k\leq n}} : \text{ each } a^{\alpha\beta}_{jk} \text{ belongs to } \mathbb{C}\Big\}. \tag{7.4}$$

For each coefficient tensor $A = \big(a^{\alpha\beta}_{jk}\big)_{\substack{1\leq\alpha,\beta\leq M\\ 1\leq j,k\leq n}} \in \mathfrak{A}$ associate the system $L_A \in \mathfrak{L}$ according to

$$L_A := \big(a^{\alpha\beta}_{jk}\partial_j\partial_k\big)_{1\leq\alpha,\beta\leq M}. \tag{7.5}$$

With this piece of notation, consider the collection of all coefficient tensors used to represent a given system $L \in \mathfrak{L}$ as

$$\mathfrak{A}_L := \big\{A \in \mathfrak{A} : L = L_A\big\}. \tag{7.6}$$

© The Author(s), under exclusive license to Springer Nature Switzerland AG 2026
M. Mitrea and P. Takemura, *Singular Integrals, Herz-Type Function Spaces, and Boundary Problems*, Progress in Mathematics 362,
https://doi.org/10.1007/978-3-032-12516-3_7

Significantly, a system L does not uniquely determine its coefficients. Indeed, let $L \in \mathfrak{L}$ and assume $L = L_A$ for some $A = \big(a^{\alpha\beta}_{jk}\big)_{\substack{1\leq\alpha,\beta\leq M\\ 1\leq j,k\leq n}}$ belonging to $\mathfrak{A}_L$. Select complex numbers $c^{\alpha\beta}_{jk}$ satisfying $c^{\alpha\beta}_{jk} = -c^{\alpha\beta}_{kj}$ for each $j, k \in \{1,\dots,n\}$ and each $\alpha,\beta \in \{1,\dots,M\}$. If we set $\widetilde{A} := \big(\widetilde{a}^{\alpha\beta}_{jk}\big)_{\substack{1\leq\alpha,\beta\leq M\\ 1\leq j,k\leq n}}$ where $\widetilde{a}^{\alpha\beta}_{jk} := a^{\alpha\beta}_{jk} + c^{\alpha\beta}_{jk}$ then the antisymmetry in the lower indices of $\big(c^{\alpha\beta}_{jk}\big)_{\substack{1\leq\alpha,\beta\leq M\\ 1\leq j,k\leq n}}$ ensures that

$$L_{\widetilde{A}} = \big(\widetilde{a}^{\alpha\beta}_{jk}\partial_j\partial_k\big)_{1\leq\alpha,\beta\leq M} = \big(a^{\alpha\beta}_{jk}\partial_j\partial_k\big)_{1\leq\alpha,\beta\leq M} = L_A. \tag{7.7}$$

For this reason, the library $\mathfrak{A}_L$ of all coefficient tensors used to represent the system L has infinite cardinality, i.e., $\#\mathfrak{A}_L = \infty$.

Call the system L `weakly elliptic` if its `characteristic matrix`

$$L(\xi) := \big(a^{\alpha\beta}_{jk}\xi_j\xi_k\big)_{1\leq\alpha,\beta\leq M} \text{ is invertible for all } \xi \in \mathbb{R}^n \setminus \{0\}, \tag{7.8}$$

or equivalently,

$$\det\big[L(\xi)\big] \neq 0 \text{ whenever } \xi \in \mathbb{R}^n \setminus \{0\}. \tag{7.9}$$

This condition is implied by the `Legendre-Hadamard (strong) ellipticity` condition which stipulates the existence of some $c \in (0,\infty)$ satisfying

$$\begin{gathered} \mathrm{Re}\big\langle -L(\xi)\zeta, \overline{\zeta}\big\rangle \geq c|\xi|^2|\zeta|^2 \\ \text{for all } \xi \in \mathbb{R}^n \text{ and } \zeta \in \mathbb{C}^M. \end{gathered} \tag{7.10}$$

One of the key features of the class of weakly elliptic systems is the fact that they possess a (matrix-valued) canonical fundamental solution. This is summarized in following result (cf. [Mi18, Theorem 11.1, p. 393]):

Theorem 7.1 *Let L be a homogeneous, second-order, constant complex coefficient, $M \times M$ system in $\mathbb{R}^n$, which is weakly elliptic (in the sense described in* (7.8)*). Consider the $M \times M$ matrix-valued function $E = \big(E_{\alpha\beta}\big)_{1\leq\alpha,\beta\leq M}$ defined at each $x \in \mathbb{R} \setminus \{0\}$ by*[1]

$$E(x) := \begin{cases} -\dfrac{\Delta_x^{(n-1)/2}}{4(2\pi i)^{n-1}}\left\{\displaystyle\int_{S^{n-1}} |\langle x,\xi\rangle|\,[L(\xi)]^{-1}\,d\mathcal{H}^{n-1}(\xi)\right\} & \text{if } n \text{ is odd,} \\[2ex] \dfrac{\Delta_x^{(n-2)/2}}{(2\pi i)^n}\left\{\displaystyle\int_{S^{n-1}} \ln|\langle x,\xi\rangle|\,[L(\xi)]^{-1}\,d\mathcal{H}^{n-1}(\xi)\right\} & \text{if } n \text{ is even.} \end{cases} \tag{7.11}$$

In relation to the $\mathbb{C}^{M\times M}$-valued function in (7.11)*, the following properties are valid:*

[1] for each $m \in \mathbb{N}$, denote by Δ_x^m the m-fold application of the Laplacian in the variable x

1. *For each* $\alpha, \beta \in \{1, \dots, M\}$ *it follows that* $E_{\alpha\beta} \in \mathscr{C}^\infty(\mathbb{R}^n \setminus \{0\})$ *and satisfies* $E_{\alpha\beta}(x) = E_{\alpha\beta}(-x)$ *for every* $x \in \mathbb{R}^n \setminus \{0\}$.

2. *For each fixed point* $y \in \mathbb{R}^n$ *one has* $L\big[E(\cdot - y)\big] = \delta_y I_{M\times M}$ *in the sense of distributions in* $\mathbb{R}^n$, *where* $I_{M\times M}$ *is the* $M \times M$ *identity matrix and* δ_y *denotes the Dirac distribution with mass at* y *in* $\mathbb{R}^n$. *That is, using the standard Kronecker delta notation,*
$$a^{\alpha\beta}_{jk}\partial_{x_j}\partial_{x_k}\big[E_{\beta\gamma}(x-y)\big] = \delta_{\alpha\gamma}\delta_y(x), \qquad x \in \mathbb{R}^n, \tag{7.12}$$
in the sense of distributions, for every $\alpha, \gamma \in \{1, \dots, M\}$.

3. *The transpose of* E, *i.e.,* $E^\top = (E_{\beta\alpha})_{1\le\alpha,\beta\le M}$, *is a fundamental solution for the transpose system* $L^\top$. *In other words, for each fixed point* $y \in \mathbb{R}^n$ *one has* $L^\top\big[E^\top(\cdot - y)\big] = \delta_y I_{M\times M}$ *in the sense of distributions in* $\mathbb{R}^n$, *i.e.,*
$$a^{\beta\alpha}_{kj}\partial_{x_j}\partial_{x_k}\big[E_{\gamma\beta}(x-y)\big] = \delta_{\alpha\gamma}\delta_y(x), \qquad x \in \mathbb{R}^n, \tag{7.13}$$
in the sense of distributions, for every $\alpha, \gamma \in \{1, \dots, M\}$.

4. *For every multi-index* $\alpha \in \mathbb{N}_0^n$ *with* $n + |\alpha| > 2$, *the function* $\partial^\alpha E$ *is positive homogeneous of degree* $2 - n - |\alpha|$ *and there exists a constant* $C_\alpha \in (0, \infty)$ *with the property that*
$$\big|(\partial^\alpha E)(x)\big| \le C_\alpha |x|^{2-n-|\alpha|} \text{ for all } x \in \mathbb{R}^n \setminus \{0\}. \tag{7.14}$$
Finally, corresponding to $n = 2$ *and* $\alpha = (0, \dots, 0)$, *there exists* $C \in (0, \infty)$ *such that* $|E(x)| \le C\big(1 + \big|\ln|x|\big|\big)$ *for every* $x \in \mathbb{R}^2 \setminus \{0\}$.

5. *Let 'hat' denote the Fourier transform in* $\mathbb{R}^n$ *(originally defined on Schwartz functions as in (2.4), then extended to tempered distributions via duality). Then* $\widehat{E}$ *is a tempered distribution in* $\mathbb{R}^n$ *(which is positive homogeneous of degree* -2 *if* $n \ge 3$*), whose restriction to* $\mathbb{R}^n \setminus \{0\}$ *is a (matrix-valued) function of class* $\mathscr{C}^\infty$. *In fact,*
$$\widehat{E}(\xi) = \big[L(\xi)\big]^{-1} \text{ for every } \xi \in \mathbb{R}^n \setminus \{0\}, \tag{7.15}$$
More generally, given any $\gamma \in \mathbb{N}_0^n$ *it follows that the tempered distribution* $\widehat{\partial^\gamma E}$ *is a function of class* $\mathscr{C}^\infty$ *when restricted to* $\mathbb{R}^n \setminus \{0\}$ *which, regarded as such, satisfies*
$$\widehat{\partial^\gamma E}(\xi) = i^{|\gamma|}\,\xi^\gamma\,\big[L(\xi)\big]^{-1} \text{ for every } \xi \in \mathbb{R}^n \setminus \{0\}, \tag{7.16}$$
and
$$\text{if } \gamma \in \mathbb{N}_0^n \text{ then } \widehat{\partial^\gamma E} = i^{|\gamma|}\,\xi^\gamma\,\big[L(\xi)\big]^{-1} \text{ as tempered distributions in } \mathbb{R}^n \text{ when either } |\gamma| > 0 \text{ or } n \ge 3. \tag{7.17}$$

6. *Writing E_L in place of E to emphasize the dependence on L, matters may be arranged so that*

$$\begin{gathered}(E_L)^\top = E_{L^\top}, \quad \overline{(E_L)} = E_{\overline{L}}, \quad (E_L)^* = E_{L^*}, \\ \text{as well as } E_{\lambda L} = \lambda^{-1} E_L \text{ for each } \lambda \in \mathbb{C}\setminus\{0\}.\end{gathered} \tag{7.18}$$

where $\top$, $\bar{\cdot}$, $*$ denote, respectively, transposition, complex conjugation, and complex (or Hermitian) adjunction.

Example 7.1 The canonical fundamental for the Laplacian in $\mathbb{R}^n$ is defined at each $x \in \mathbb{R}^n \setminus \{0\}$ by

$$E_\Delta(x) := \begin{cases} \dfrac{1}{\omega_{n-1}(2-n)} \dfrac{1}{|x|^{n-2}}, & \text{if } n \geq 3, \\ \dfrac{1}{2\pi} \ln |x|, & \text{if } n = 2, \end{cases} \tag{7.19}$$

where, as usual, ω_{n-1} denotes the surface area of the unit sphere in $\mathbb{R}^n$ (cf. [Mi18, Section 7.1]).

The existence and properties of fundamental solutions for weakly elliptic systems (as described above) play a crucial role in the theory of boundary layer potentials. To further elaborate on this, consider a UR domain $\Omega \subseteq \mathbb{R}^n$. Set $\sigma := \mathcal{H}^{n-1}\lfloor\partial\Omega$, and denote by $\nu = (\nu_1, \dots, \nu_n)$ the geometric measure theoretic outward unit normal to Ω. Select a second-order, homogeneous, constant complex coefficient, weakly elliptic, $M \times M$ system L in $\mathbb{R}^n$, and let $E = (E_{\alpha\beta})_{1\leq\alpha,\beta\leq M}$ be its canonical fundamental solution as in Theorem 7.1. Fix a coefficient tensor $A = \big(a_{jk}^{\alpha\beta}\big)_{\substack{1\leq j,k\leq n \\ 1\leq\alpha,\beta\leq M}} \in \mathfrak{A}_L$ along with an arbitrary function

$$f = (f_\alpha)_{1\leq\alpha\leq M} \in \Big[L^1\big(\partial\Omega, \tfrac{\sigma(x)}{1+|x|^{n-1}}\big)\Big]^M. \tag{7.20}$$

Following [GHA-IV, (1.3.18), p. 27], introduce the `boundary-to-domain double layer potential` $\mathcal{D}_A$ acting on f according to

$$\mathcal{D}_A f(x) := \Big(-\int_{\partial\Omega} \nu_k(y) a_{jk}^{\beta\alpha} \left(\partial_j E_{\gamma\beta}\right)(x-y) f_\alpha(y)\, d\sigma(y)\Big)_{1\leq\gamma\leq M} \tag{7.21}$$

at each point $x \in \Omega$. It turns out that (7.20) is the most general functional analytic setting in which the integrals in (7.21) are absolutely convergent. The double layer operator $\mathcal{D}_A$ may be regarded as a mechanism for creating lots of smooth null-solutions for the given system L in Ω, in the sense that for each function f as in (7.20) we have (cf. [GHA-IV, (1.3.24), p. 28]):

$$\mathcal{D}_A f \in \big[\mathscr{C}^\infty(\Omega)\big]^M \quad \text{and} \quad L(\mathcal{D}_A f) = 0 \text{ in } \Omega. \tag{7.22}$$

Moving on, consider the `boundary-to-boundary double layer potential` K_A and its `''transpose'' double layer potential` $K^{\#}_A$, whose action on f as in (7.20) is respectively defined, at σ-a.e. point $x \in \partial\Omega$, via (cf. [GHA-IV, (1.3.68), p. 38])

$$K_A f(x) := \Big(- \lim_{\varepsilon \to 0^+} \int_{\substack{y \in \partial\Omega \\ |x-y|>\varepsilon}} \nu_k(y) a^{\beta\alpha}_{jk} (\partial_j E_{\gamma\beta})(x-y) f_\alpha(y)\, d\sigma(y) \Big)_{1 \le \gamma \le M}, \quad (7.23)$$

and (cf. [GHA-IV, (1.3.72), p. 40])

$$K^{\#}_A f(x) := \Big(\lim_{\varepsilon \to 0^+} \int_{\substack{y \in \partial\Omega \\ |x-y|>\varepsilon}} \nu_k(x) a^{\beta\alpha}_{jk} (\partial_j E_{\gamma\beta})(x-y) f_\gamma(y)\, d\sigma(y) \Big)_{1 \le \alpha \le M}. \quad (7.24)$$

The operator K_A arises naturally when studying the nontangential behavior of $\mathcal{D}_A$ near the boundary $\partial\Omega$. To be precise, given an aperture parameter $\kappa \in (0, \infty)$, and a coefficient tensor $A \in \mathfrak{A}_L$, the operator $\mathcal{D}_A$ satisfies, for every function f as in (7.20), the following jump-relation (cf. [GHA-IV, Theorem 1.5.1, pp. 86–102]):

$$(\mathcal{D}_A f)\Big|^{\kappa-\mathrm{n.t.}}_{\partial\Omega} = \big(\tfrac{1}{2} I + K_A\big) f \ \text{ at } \sigma\text{-a.e. point on } \partial\Omega. \quad (7.25)$$

Having introduced the layer potentials $\mathcal{D}_A$ and K_A (see (7.21) and (7.23)), we now consider their modified versions. Let Ω, σ, and ν be as before, as well as L and E. Given a parameter $\varepsilon \in (0, \infty)$ along with indices $\gamma, \beta \in \{1, \dots, M\}$ and $r \in \{1, \dots, n\}$ abbreviate

$$k^{(r\gamma\beta)}_{\varepsilon} := (\partial_r E_{\gamma\beta}) \cdot \mathbb{1}_{\mathbb{R}^n \setminus \overline{B(0,\varepsilon)}}. \quad (7.26)$$

Fix a coefficient tensor $A = \big(a^{\alpha\beta}_{jk}\big)_{\substack{1 \le j,k \le n \\ 1 \le \alpha,\beta \le M}} \in \mathfrak{A}_L$ and a function

$$f = (f_\alpha)_{1 \le \alpha \le M} \in \Big[L^1\big(\partial\Omega, \tfrac{\sigma(x)}{1+|x|^n}\big) \Big]^M. \quad (7.27)$$

(The reader is alerted to the change in power for the weight, compared with (7.20)). The `modified boundary-to-domain double layer potential` $\mathcal{D}_{A,\mathrm{mod}}$ (cf. [GHA-IV, (1.8.6), p. 174]) acting on f, at each $x \in \Omega$, is given by

$$\begin{aligned} &\mathcal{D}_{A,\mathrm{mod}} f(x) \\ &:= \Big(- \int_{\partial\Omega} \nu_s(y) a^{\beta\alpha}_{rs} \Big\{ (\partial_r E_{\gamma\beta})(x-y) - k^{(r\gamma\beta)}_1(-y) \Big\} f_\alpha(y)\, d\sigma(y) \Big)_{1 \le \gamma \le M} \end{aligned} \quad (7.28)$$

while the action of the `modified boundary-to-boundary double layer potential` $K_{A,\mathrm{mod}}$ (cf. [GHA-IV, (1.8.24), p. 177]) on f, at σ-a.e. $x \in \partial\Omega$, is

$$K_{A,\mathrm{mod}}f(x) := \left(- \lim_{\varepsilon \to 0^+} \int_{\partial\Omega} \nu_s(y) a_{rs}^{\beta\alpha} \Big\{ k_\varepsilon^{(r\gamma\beta)}(x-y) - k_1^{(r\gamma\beta)}(-y) \Big\} f_\alpha(y)\, d\sigma(y) \right)_{1\le\gamma\le M}. \tag{7.29}$$

A wealth of properties of $\mathcal{D}_{A,\mathrm{mod}}$ and $K_{A,\mathrm{mod}}$ have been established in [GHA-IV, Theorem 1.8.2, pp. 173–181]. For example, the modified double layer $\mathcal{D}_{A,\mathrm{mod}}$ serves as a mechanism for creating a multitude of smooth null-solutions for the system L, in the sense that whenever $f \in \big[L^1(\partial\Omega, \frac{\sigma(x)}{1+|x|^n})\big]^M$ it follows that

$$\mathcal{D}_{A,\mathrm{mod}}f \in \big[\mathscr{C}^\infty(\Omega)\big]^M \quad \text{and} \quad L(\mathcal{D}_{A,\mathrm{mod}}f) = 0 \ \text{ in } \Omega. \tag{7.30}$$

Moreover, for each function $f \in \big[L^1(\partial\Omega, \frac{\sigma(x)}{1+|x|^n})\big]^M$ the following jump-formula holds:

$$\big(\mathcal{D}_{A,\mathrm{mod}}f\big)\Big|_{\partial\Omega}^{\kappa-\mathrm{n.t.}} = \big(\tfrac{1}{2}I + K_{A,\mathrm{mod}}\big)f \ \text{ at } \sigma\text{-a.e. point on } \partial\Omega. \tag{7.31}$$

Moving on, we elaborate on the notion of distinguished coefficient tensors for a second-order weakly elliptic system L. The starting point is the following result from [GHA-V, Proposition 1.2.1, pp. 11–13].

Proposition 7.1 *Let L be a homogeneous, second-order, constant complex coefficient, weakly elliptic $M \times M$ system in $\mathbb{R}^n$, and consider the matrix-valued function defined for each $\xi \in \mathbb{R}^n \setminus \{0\}$ as*

$$\big(\mathcal{E}_{\gamma\beta}(\xi)\big)_{1\le\gamma,\beta\le M} := -\big[L(\xi)\big]^{-1} \in \mathbb{C}^{M\times M} \tag{7.32}$$

(recall that the characteristic matrix $L(\xi)$ of L has been defined in (7.8)). Also, denote by $E = (E_{\alpha\beta})_{1\le\alpha,\beta\le M}$ be the canonical fundamental solution associated with the given system L as in Theorem 7.1.

Then for each coefficient tensor $A = \big(a_{jk}^{\alpha\beta}\big)_{\substack{1\le\alpha,\beta\le M\\ 1\le j,k\le n}} \in \mathfrak{A}_L$ (cf. (7.6)) the following conditions are equivalent:

(a) For each $k, k' \in \{1, \dots, n\}$ and each $\alpha, \gamma \in \{1, \dots, M\}$ there holds

$$\big(x_{k'} a_{jk}^{\beta\alpha} - x_k a_{jk'}^{\beta\alpha}\big)(\partial_j E_{\gamma\beta})(x) = 0 \ \textit{for all} \ x = (x_i)_{1\le i\le n} \in \mathbb{R}^n \setminus \{0\}. \tag{7.33}$$

(b) For each $s, s' \in \{1, \dots, n\}$ and each $\alpha, \gamma \in \{1, \dots, M\}$, in the sense of tempered distributions in $\mathbb{R}^n$ one has

$$\Big[a_{rs}^{\beta\alpha}\partial_{\xi_{s'}} - a_{rs'}^{\beta\alpha}\partial_{\xi_s}\Big]\big[\xi_r \mathcal{E}_{\gamma\beta}(\xi)\big] = 0. \tag{7.34}$$

(c) For each $k, k' \in \{1, \dots, n\}$ and each $\alpha, \gamma \in \{1, \dots, M\}$ one has

$$(a^{\beta\alpha}_{k'k} - a^{\beta\alpha}_{kk'} + \xi_j a^{\beta\alpha}_{jk}\partial_{\xi_{k'}} - \xi_j a^{\beta\alpha}_{jk'}\partial_{\xi_k})\mathcal{E}_{\gamma\beta}(\xi) = 0 \text{ for all } \xi \in \mathbb{R}^n \setminus \{0\} \quad (7.35)$$

and also

$$\int_{S^1} (a^{\beta\alpha}_{jk}\xi_{k'} - a^{\beta\alpha}_{jk'}\xi_k)\xi_j\, \mathcal{E}_{\gamma\beta}(\xi)\, d\mathcal{H}^1(\xi) = 0 \ \text{ if } \ n = 2. \quad (7.36)$$

(d) One has

$$\begin{gathered}\xi_r\xi_j\Big[a^{\beta\alpha}_{rs'}(a^{\lambda\mu}_{sj} + a^{\lambda\mu}_{js}) - a^{\beta\alpha}_{rs}(a^{\lambda\mu}_{s'j} + a^{\lambda\mu}_{js'})\Big]\mathcal{E}_{\mu\beta}(\xi) + a^{\lambda\alpha}_{s's} - a^{\lambda\alpha}_{ss'} = 0 \\ \text{for all } \xi \in S^{n-1}, \text{ all } s, s' \in \{1, \dots, n\}, \text{ and all } \alpha, \lambda \in \{1, \dots, M\},\end{gathered} \quad (7.37)$$

with the cancellation condition

$$\begin{gathered}\int_{S^1} \Big(a^{\beta\alpha}_{rs}\xi_{s'} - a^{\beta\alpha}_{rs'}\xi_s\Big)\xi_r\mathcal{E}_{\lambda\beta}(\xi)\, d\mathcal{H}^1(\xi) = 0 \\ \text{for all } s, s' \in \{1, \dots, n\} \text{ and } \alpha, \lambda \in \{1, \dots, M\},\end{gathered} \quad (7.38)$$

additionally imposed in the case when $n = 2$.

(e) For each $\xi \in S^{n-1}$ and each $\alpha, \lambda \in \{1, \dots, M\}$,

$$\begin{gathered}\text{the expression } \big(a^{\lambda\mu}_{sj} + a^{\lambda\mu}_{js}\big)\mathcal{E}_{\mu\beta}(\xi)\xi_j\xi_r a^{\beta\alpha}_{rs'} + a^{\lambda\alpha}_{s's} \\ \text{is symmetric in the indices } s, s' \in \{1, \dots, n\},\end{gathered} \quad (7.39)$$

with the condition that for each $\alpha, \lambda \in \{1, \dots, M\}$

$$\begin{gathered}\text{the expression } \int_{S^1} a^{\beta\alpha}_{rs}\xi_{s'}\xi_r\mathcal{E}_{\lambda\beta}(\xi)\, d\mathcal{H}^1(\xi) \\ \text{is symmetric in the indices } s, s' \in \{1, 2\},\end{gathered} \quad (7.40)$$

also imposed in the case when $n = 2$.

(f) There exists a matrix-valued function

$$k = \big\{k_{\gamma\alpha}\big\}_{1 \le \gamma, \alpha \le M} : \mathbb{R}^n \setminus \{0\} \longrightarrow \mathbb{C}^{M \times M} \quad (7.41)$$

with the property that for each $\gamma, \alpha \in \{1, \dots, M\}$ and $s \in \{1, \dots, n\}$ one has

$$a^{\beta\alpha}_{rs}(\partial_r E_{\gamma\beta})(x) = x_s k_{\gamma\alpha}(x) \ \text{ for all } \ x \in \mathbb{R}^n \setminus \{0\}. \quad (7.42)$$

Distinguished coefficient tensors have been first introduced in [GHA-V, Definition 1.2.2, p.17], and this concept plays a vastly important role in the present work.

Definition 7.1 Given a second-order, weakly elliptic, homogeneous, $M \times M$ system L in $\mathbb{R}^n$, with constant complex coefficients, call

$$A = \left(a_{rs}^{\alpha\beta}\right)_{\substack{1\le r,s\le n\\ 1\le\alpha,\beta\le M}} \in \mathfrak{A}_L \tag{7.43}$$

a `distinguished coefficient tensor` for L provided any of the equivalent conditions *(a)-(f)* in Proposition 7.1 holds. Also, denote by $\mathfrak{A}_L^{\mathrm{dis}}$ the family of such distinguished coefficient tensors for L, more explicitly,

$$\mathfrak{A}_L^{\mathrm{dis}} := \Big\{A = \left(a_{rs}^{\alpha\beta}\right)_{\substack{1\le r,s\le n\\ 1\le\alpha,\beta\le M}} \in \mathfrak{A}_L : \text{condition (7.33) holds} \\ \text{for each } \ k,k' \in \{1,\dots,n\} \ \text{ and } \ \alpha,\gamma \in \{1,\dots,M\}\Big\}. \tag{7.44}$$

The significance of distinguished coefficient tensors is apparent from the following result which appears in [GHA-V, Proposition 1.3.2, pp. 21–22] linking distinguished coefficient tensors to boundary layer potentials having a chord-dot-normal structure.

Proposition 7.2 *Let L be a homogeneous, second-order, constant complex coefficient, weakly elliptic $M \times M$ system in $\mathbb{R}^n$, and suppose $A \in \mathfrak{A}_L$. Then the following statements are equivalent.*

(i) The coefficient tensor A belongs to $\mathfrak{A}_L^{\mathrm{dis}}$.

(ii) Whenever Ω is a half-space in $\mathbb{R}^n$, the boundary-to-boundary double layer potential K_A associated with A and Ω as in (7.23) *is the zero operator.*

(iii) There exists a matrix-valued function $k \in \left[\mathscr{C}^\infty(\mathbb{R}^n \setminus \{0\})\right]^{M\times M}$ which is even, positive homogeneous of degree $-n$, and with the property that for each UR *domain $\Omega \subseteq \mathbb{R}^n$ the matrix-valued integral kernel of the double layer potential operator K_A associated with A and Ω as in* (7.23) *has the chord-dot-normal structure*

$$\langle \nu(y), x-y\rangle k(x-y) \\ \textit{for each } x \in \partial\Omega \textit{ and } \mathcal{H}^{n-1}\textit{-a.e. } y \in \partial\Omega, \tag{7.45}$$

where ν is the geometric measure theoretic outward unit normal to Ω.

(iv) Whenever Ω is a half-space in $\mathbb{R}^n$, the "transpose" double layer potential $K_A^{\#}$ associated with A and Ω as in (7.24) *is the zero operator.*

(v) There exists a matrix-valued function $k^{\#} \in \left[\mathscr{C}^\infty(\mathbb{R}^n \setminus \{0\})\right]^{M\times M}$ which is even, positive homogeneous of degree $-n$, and with the property that for each UR *domain $\Omega \subseteq \mathbb{R}^n$ the (matrix-valued) integral kernel of the "transpose" double layer potential operator $K_A^{\#}$ associated with A and Ω as in* (7.24) *has the chord-dot-normal structure*

$$\langle\nu(x), y-x\rangle k^{\#}(x-y) \quad \text{for } \mathcal{H}^{n-1}\text{-a.e. } x\in\partial\Omega \text{ and each } y\in\partial\Omega, \tag{7.46}$$

where ν is the geometric measure theoretic outward unit normal to Ω.

Moreover, whenever either (hence all) of the above conditions holds, the matrices $k, k^{\#}$ in items (iii), (v) above are related to each other via $k^{\#} = k^{\top}$, where the superscript $\top$ indicates transposition.

Remark 7.1 In the context of Proposition 7.2, if L has a distinguished coefficient tensor and $A \in \mathfrak{A}_L^{\rm dis}$, then items *(iii)* and *(v)* imply that there exist two matrix-valued functions $k, k^{\#} \in [\mathscr{C}^\infty(\mathbb{R}^n\setminus\{0\})]^{M\times M}$ which are even, and positive homogeneous of degree $-n$ such that for each function $f \in [L^1(\partial\Omega, \frac{\sigma(x)}{1+|x|^{n-1}})]^M$,

$$(K_A f)(x) = \lim_{\varepsilon\to 0^+} \int_{\substack{y\in\partial\Omega\\ |x-y|>\varepsilon}} \langle\nu(y), x-y\rangle k(x-y) f(y)\, d\sigma(y), \tag{7.47}$$

and

$$(K_A^{\#} f)(x) = \lim_{\varepsilon\to 0^+} \int_{\substack{y\in\partial\Omega\\ |x-y|>\varepsilon}} \langle\nu(x), y-x\rangle k^{\#}(x-y) f(y)\, d\sigma(y) \tag{7.48}$$

at σ-a.e. point $x\in\partial\Omega$. Moreover, the modified boundary-to-boundary double layer

$$K_{A,\mathrm{mod}} \text{ (cf. (7.29)) may be thought as a } M\times M \text{ matrix operator in which each entry is of the form } T_{\mathrm{mod}} \text{ (as in (6.201)),} \tag{7.49}$$

and similarly, the modified boundary-to-domain double layer

$$\mathcal{D}_{A,\mathrm{mod}} \text{ (cf. (7.28)) may be regarded as a } M\times M \text{ matrix operator in whose entry is of the form } \mathcal{T}_{\mathrm{mod}} \text{ (as in (6.196)).} \tag{7.50}$$

In the subsequent discussion, assume $\Omega\subseteq\mathbb{R}^n$ is a UR domain, and abbreviate $\sigma := \mathcal{H}^{n-1}\lfloor\partial\Omega$. Denote by $\nu = (\nu_1,\dots,\nu_n)$ the geometric measure theoretic outward unit normal to Ω. Suppose L is a second-order, homogeneous, constant complex coefficient, $M\times M$ system in $\mathbb{R}^n$ as in (7.1), and let E be its canonical fundamental solution (in the sense of Theorem 7.1). Another integral operator of layer potential type of relevance for us is the `modified single layer potential` $\mathscr{S}_{\mathrm{mod}}$ (cf. [GHA-IV, (1.5.50), p. 95]), whose action on each function $f\in\big[L^1\big(\partial\Omega, \frac{\sigma(x)}{1+|x|^{n-1}}\big)\big]^M$ is defined as

$$\mathscr{S}_{\mathrm{mod}} f(x) := \int_{\partial\Omega} \big\{E(x-y) - E_*(y)\big\} f(y)\, d\sigma(y) \quad \text{for each } x\in\Omega, \tag{7.51}$$

where $E_* := E \cdot \mathbb{1}_{\mathbb{R}^n \setminus B(0,1)}$. As opposed to the double layer potential $\mathcal{D}_A$ (cf. (7.21)), the description of $\mathcal{S}_{\text{mod}}$ does not depend on a choice of a coefficient tensor to write the system L.

As noted in [GHA-IV, (1.5.51), p. 94], for any function $f \in \big[L^1\big(\partial\Omega, \frac{\sigma(x)}{1+|x|^{n-1}}\big)\big]^M$, and any multi-index $\alpha \in \mathbb{N}_0^n$ with $|\alpha| \geq 1$ one has

$$\partial^\alpha(\mathcal{S}_{\text{mod}} f)(x) = \int_{\partial\Omega} (\partial^\alpha E)(x-y) f(y)\, d\sigma(y) \;\; \text{for each } x \in \Omega. \tag{7.52}$$

In particular, for each f as above this yields

$$\nabla(\mathcal{S}_{\text{mod}} f)(x) = \int_{\partial\Omega} (\nabla E)(x-y) f(y)\, d\sigma(y) \;\; \text{for each } \; x \in \Omega. \tag{7.53}$$

It has been shown in [GHA-IV, Theorem 1.5.1, pp. 86–102] that the operator in (7.51) can be employed to generate a multitude of smooth null solutions for the system L. Specifically, whenever $f \in \big[L^1\big(\partial\Omega, \frac{\sigma(x)}{1+|x|^{n-1}}\big)\big]^M$ we have

$$\mathcal{S}_{\text{mod}} f \in \big[\mathscr{C}^\infty(\Omega)\big]^M \quad \text{and} \quad L(\mathcal{S}_{\text{mod}} f) = 0 \;\text{ in } \Omega. \tag{7.54}$$

Fix a coefficient tensor $A = \big(a_{jk}^{\alpha\beta}\big)_{\substack{1\leq j,k\leq n\\ 1\leq\alpha,\beta\leq M}}$, with each $a_{jk}^{\alpha\beta} \in \mathbb{C}$, along with an aperture parameter $\kappa \in (0,\infty)$. Given a function $u = (u_\beta)_{1\leq\beta\leq M} \in \big[W^{1,1}_{\text{loc}}(\Omega)\big]^M$ for which the nontangential boundary trace $(\nabla u)\big|^{\kappa-\text{n.t.}}_{\partial\Omega}$ exists σ-a.e. on $\partial\Omega$, we follow [GHA-III, (1.7.9), p. 189] and define the `conormal derivative` of u associated with A as the vector-valued function

$$\partial_\nu^A u := \Big(\nu_j a_{jk}^{\alpha\beta} (\partial_k u_\beta)\big|^{\kappa-\text{n.t.}}_{\partial\Omega}\Big)_{1\leq\alpha\leq M} \quad \text{at } \sigma\text{-a.e. point on } \partial\Omega. \tag{7.55}$$

In connection with this, [GHA-IV, Theorem 1.5.1, pp. 86–102] tells us that the modified single layer potential $\mathcal{S}_{\text{mod}}$ enjoys the following property. Under the assumption that $\Omega \subseteq \mathbb{R}^n$ is a UR domain and setting $\sigma := \mathcal{H}^{n-1}\lfloor\partial\Omega$, for each coefficient tensor $A = \big(a_{jk}^{\alpha\beta}\big)_{\substack{1\leq j,k\leq n\\ 1\leq\alpha,\beta\leq M}}$, and each function $f \in \big[L^1\big(\partial\Omega, \frac{\sigma(x)}{1+|x|^{n-1}}\big)\big]^M$ one has the jump-formula

$$\partial_\nu^A(\mathcal{S}_{\text{mod}} f) = \big(-\tfrac{1}{2}I + K^\#_{A^\top}\big) f \;\; \text{at } \sigma\text{-a.e. point on } \partial\Omega, \tag{7.56}$$

where $K^\#_{A^\top}$ is the transpose double layer potential associated with $A^\top$ (cf. (7.24)).

We conclude this chapter by recalling the boundary-to-boundary version of the modified single layer operator. Retain the setting of UR domain $\Omega \subseteq \mathbb{R}^n$, and as before $\sigma := \mathcal{H}^{n-1}\lfloor\partial\Omega$. Consider a weakly elliptic, second-order, homogeneous, constant, complex coefficient, $M \times M$ system L in $\mathbb{R}^n$, and denote by E its canonical fundamental solution (in the sense of Theorem 7.1). As in [GHA-IV, (1.5.73), p 97], we define the `modified boundary-to-boundary single layer potential operator`

$$S_{\text{mod}}f(x) := \int_{\partial\Omega} \{E(x-y) - E_*(-y)\} f(y)\, d\sigma(y) \text{ at } \sigma\text{-a.e. } x \in \partial\Omega, \quad \text{for each } f \in \Big[L^1\big(\partial\Omega, \tfrac{\sigma(x)}{1+|x|^{n-1}}\big)\Big]^M, \text{ where } E_* := E \cdot \mathbb{1}_{\mathbb{R}^n\setminus B(0,1)}. \tag{7.57}$$

Here are some properties of S_{mod}, recently established in [GHA-IV, Theorem 1.5.1, pp. 86–102], which are of interest to us. First, this operator is meaningfully defined, via an absolutely convergent integral, at σ-a.e. point in Ω. In addition,

$$S_{\text{mod}} : L^1\Big[\Big(\partial\Omega, \tfrac{\sigma(x)}{1+|x|^{n-1}}\Big)\Big]^M \to \big[L^1_{\text{loc}}(\partial\Omega, \sigma)\big]^M \tag{7.58}$$

and

$$S_{\text{mod}} : L^1\Big[\Big(\partial\Omega, \tfrac{\sigma(x)}{1+|x|^{n-1-\varepsilon}}\Big)\Big]^M \to \Big[L^1\Big(\partial\Omega, \tfrac{\sigma(x)}{1+|x|^{n}}\Big)\Big]^M, \quad \text{for each } \varepsilon > 0, \tag{7.59}$$

are well-defined, linear, and bounded operators.

Second, for each aperture parameter $\kappa \in (0, \infty)$, and each function f belonging to $\big[L^1(\partial\Omega, \frac{\sigma(x)}{1+|x|^{n-1}})\big]^M$ one has

$$\Big((\mathcal{S}_{\text{mod}}f)\Big|^{\kappa-\text{n.t.}}_{\partial\Omega}\Big)(x) = (S_{\text{mod}}f)(x) \quad \text{at } \sigma\text{-a.e. point } x \in \partial\Omega. \tag{7.60}$$

For more properties pertaining to boundary layer potentials (and their modified versions) associated with weakly elliptic second-order constant complex coefficient systems, see [GHA-IV, Theorem 1.5.1, pp. 86–102] and [GHA-IV, Theorem 1.8.2, pp. 173–181].

Chapter 8

Layer Potentials on Beurling-Hardy Spaces and BCMO_p Spaces

In this chapter the principal aim is to establish boundedness and invertibility results for boundary layer potentials associated with weakly elliptic systems in the context of Beurling-Hardy spaces HA^p and spaces of functions of bounded central mean oscillations BCMO_p. This is possible thanks to the Calderón-Zygmund theory for singular integral operators developed in Chapter 6. The invertibility results obtained here play a decisive role in the treatment of boundary value problems in AR domains and boundary data selected from HA^p spaces and BCMO_p spaces (see Chapter 9). As a preamble, we obtain some crucial geometric operator norm estimates which are sensitive to flatness for boundary-to-boundary double layer potentials associated with weakly elliptic systems that have distinguished coefficient tensors.

Theorem 8.1 *Assume $\Omega \subseteq \mathbb{R}^n$ is a* UR *domain such that $\partial\Omega$ is an unbounded set. Abbreviate $\sigma := \mathcal{H}^{n-1}\lfloor\partial\Omega$ and denote by ν the geometric measure theoretic outward unit normal to Ω. Fix some $M \in \mathbb{N}$, and let L be a homogeneous, second-order, constant complex coefficient, weakly elliptic $M \times M$ system in $\mathbb{R}^n$ for which $\mathfrak{A}_L^{\mathrm{dis}} \neq \varnothing$. Select $A \in \mathfrak{A}_L^{\mathrm{dis}}$ and consider the boundary-to-boundary transpose double layer potential $K_A^{\#}$, as well as the modified double layer $K_{A,\mathrm{mod}}$, associated with Ω and the coefficient tensor A (as in* (7.24) *and* (7.29)*, respectively). Finally, pick an integrability exponent $p \in (1,\infty)$.*

Then

$$K_A^{\#} : [\mathrm{HA}^p(\partial\Omega,\sigma)]^M \to [\mathrm{HA}^p(\partial\Omega,\sigma)]^M \quad \textit{and}$$
$$K_{A,\mathrm{mod}} : [\mathrm{BCMO}_p(\partial\Omega,\sigma)]^M \to [\mathrm{BCMO}_p(\partial\Omega,\sigma)]^M \tag{8.1}$$

are well-defined, linear, and bounded operators.

In fact, for each $m \in \mathbb{N}$ there exists some $C_m \in (0,\infty)$ depending only on m, n, A, p, and the UR *constants of Ω, with the property that*

$$\|K_A^{\#}\|_{[\mathrm{HA}^p(\partial\Omega,\sigma)]^M \to [\mathrm{HA}^p(\partial\Omega,\sigma)]^M} \leq C_m \cdot \|\nu\|_{[\mathrm{BMO}(\partial\Omega,\sigma)]^n}^{\langle m\rangle} \tag{8.2}$$

and

© The Author(s), under exclusive license to Springer Nature Switzerland AG 2026

M. Mitrea and P. Takemura, *Singular Integrals, Herz-Type Function Spaces, and Boundary Problems*, Progress in Mathematics 362,
https://doi.org/10.1007/978-3-032-12516-3_8

$$\|K_{A,\mathrm{mod}}\|_{[\mathrm{BCMO}_p(\partial\Omega,\sigma)]^M\to[\mathrm{BCMO}_p(\partial\Omega,\sigma)]^M} \leq C_m \cdot \|\nu\|^{\langle m\rangle}_{[\mathrm{BMO}(\partial\Omega,\sigma)]^n}. \tag{8.3}$$

Moreover, with $[\,\cdot\,]$ *denoting the equivalence class modulo constants in the space* $\mathrm{BCMO}_p(\partial\Omega,\sigma)$, *it follows that the operator*

$$\begin{gathered}[K_{A,\mathrm{mod}}] : \widetilde{\mathrm{BCMO}}_p(\partial\Omega,\sigma) \to \widetilde{\mathrm{BCMO}}_p(\partial\Omega,\sigma) \ \text{ given by} \\ [K_{A,\mathrm{mod}}][f] := [K_{A,\mathrm{mod}}f] \text{ for all } f \in \mathrm{BCMO}_p(\partial\Omega,\sigma) \\ \text{is meaningfully defined, linear, and bounded,}\end{gathered} \tag{8.4}$$

and, with $C_m \in (0,\infty)$ *of same nature as before, one also has*

$$\big\|[K_{A,\mathrm{mod}}]\big\|_{[\widetilde{\mathrm{BCMO}}_p(\partial\Omega,\sigma)]^M\to[\widetilde{\mathrm{BCMO}}_p(\partial\Omega,\sigma)]^M} \leq C_m \cdot \|\nu\|^{\langle m\rangle}_{[\mathrm{BMO}(\partial\Omega,\sigma)]^n}. \tag{8.5}$$

Proof In view of Proposition 7.2, and (7.48)-(7.49) in Remark 7.1, all claims made in (8.1)-(8.5) are consequences of the boundedness results recorded in Theorem 6.4 and Theorem 6.9. □

It turns out that Theorem 6.9 permits us to obtain invertibility of results on Beurling-Hardy spaces and on BCMO_p spaces in the setting of AR domains whose boundaries are "sufficiently flat". Before stating said result, we record the following abstract invertibility lemma.

Lemma 8.1 *Let* $(X,\|\cdot\|_X)$ *be a q-Banach space for some* $q \in (0,1]$, *i.e.,* $(X,\|\cdot\|_X)$ *is a quasi-Banach space with the property that for all* $x,y \in X$ *one has*

$$\|x+y\|_X^q \leq \|x\|_X^q + \|y\|_X^q. \tag{8.6}$$

Also, denote by I *the identity operator on* X, *and by* $\mathscr{B}(X)$ *the space of bounded linear operators on* X *endowed with the usual operator norm* $\|\cdot\|_{X\to X}$.

Then $\big(\mathscr{B}(X),\|\cdot\|_{X\to X}\big)$ *is a q-Banach space. Also, given a spectral parameter* $z \in \mathbb{C}\setminus\{0\}$, *and* $T \in \mathscr{B}(X)$ *with operator norm*

$$\|T\|_{X\to X} < |z|,$$

one may conclude that the operator $zI+T$ *is invertible on* X *and its inverse is given by the Neumann series*

$$(zI+T)^{-1} = \sum_{k=0}^{\infty}(-1)^k z^{-k-1}T^k,$$

with convergence in $\big(\mathscr{B}(X),\|\cdot\|_{X\to X}\big)$. *As a consequence of this and the classical Aoki-Rolewicz Theorem (see [MiMiMiMo13, Theorem 1.4, p. 5]), the same conclusions hold when* X *is a quasi-Banach space.*

We are now in a position to prove the following invertibility result on Beurling-Hardy spaces for layer potentials associated with systems possessing distinguished coefficient tensors.

Theorem 8.2 *Suppose* $\Omega \subseteq \mathbb{R}^n$ *is an Ahlfors regular domain. Set* $\sigma := \mathcal{H}^{n-1}\lfloor\partial\Omega$ *and denote by* ν *the geometric measure theoretic outward unit normal to* Ω*. Fix* $M \in \mathbb{N}$*, and let* L *be a homogeneous, second-order, constant complex coefficient, weakly elliptic* $M \times M$ *system in* $\mathbb{R}^n$*, and assume that* $\mathfrak{A}_L^{\rm dis} \neq \varnothing$*. Pick* $A \in \mathfrak{A}_L^{\rm dis}$ *and consider the boundary-to-boundary double layer potential operator* $K_A^{\#}$ *associated with* A *as in* (7.24)*. Finally, fix an integrability exponent* $p \in (1, \infty)$*, and a parameter* $\varepsilon \in (0, \infty)$*.*

Then there exists some threshold $\delta \in (0, 1)$*, depending only on* n*,* p*,* A*,* ε*, and the Ahlfors regularity constant of* $\partial\Omega$*, with the property that whenever*

$$\|\nu\|_{[\mathrm{BMO}(\partial\Omega,\sigma)]^n} < \delta \tag{8.7}$$

it follows that for each spectral parameter $z \in \mathbb{C}$ *with* $|z| \geq \varepsilon$ *the operator*

$$\begin{gathered} zI + K_A^{\#} : [\mathrm{HA}^p(\partial\Omega,\sigma)]^M \longrightarrow [\mathrm{HA}^p(\partial\Omega,\sigma)]^M \\ \text{is well defined, linear, bounded, and invertible.} \end{gathered} \tag{8.8}$$

Proof Thanks to [MMMi[3], Theorem 2.3 and Theorem 2.4, pp. 64–65], there exists some threshold $\delta_* \in (0, 1)$ depending only on n and the Ahlfors regularity constants of $\partial\Omega$ such that whenever $\|\nu\|_{[\mathrm{BMO}(\partial\Omega,\sigma)]^n} < \delta_*$ it follows that Ω is a UR domain with unbounded boundary, and UR constants solely controlled in terms of n and Ahlfors regularity character of $\partial\Omega$. In such a scenario, apply Theorem 8.1 (with $m := 1$) to ensure the existence of some constant $C_1 \in (0, \infty)$, depending only on n, p, and Ahlfors regularity constants of $\partial\Omega$, such that estimate (8.2) holds. Bearing in mind (6.113)-(6.117), choose some number $t_\varepsilon \in (0, 1/e)$ sufficiently small satisfying $t_\varepsilon \cdot \ln(1/t_\varepsilon) < \varepsilon/C_1$. Next, pick $\delta \in \big(0, \min\{t_\varepsilon, \delta_*\}\big)$. Hence, whenever $\|\nu\|_{[\mathrm{BMO}(\partial\Omega,\sigma)]^n} < \delta$ we have

$$\begin{aligned} \big\|K_A^{\#}\big\|_{[\mathrm{HA}^p(\partial\Omega,\sigma)]^M \to [\mathrm{HA}^p p(\partial\Omega,\sigma)]^M} &\leq C_1 \|\nu\|^{\langle 1\rangle}_{[\mathrm{BMO}(\partial\Omega,\sigma)]^n} \\ &\leq C_1 (t_\varepsilon)^{\langle 1\rangle} < \varepsilon. \end{aligned} \tag{8.9}$$

In view of this and the fact that $\mathrm{HA}_p(\partial\Omega,\sigma)$ is a Banach space (cf. Proposition 4.3), for every spectral parameter $z \in \mathbb{C}$ with $|z| \geq \varepsilon$, the operator $zI + K_A^{\#}$ becomes invertible on the space $[\mathrm{HA}_p(\partial\Omega,\sigma)]^M$ via a Neumann series (cf. Lemma 8.1), and its inverse is given by

$$(zI + K_A^{\#})^{-1} = \sum_{k=0}^{\infty} (-1)^k z^{-k-1} (K_A^{\#})^k, \tag{8.10}$$

with convergence in the space of bounded linear operators on $[\mathrm{HA}^p(\partial\Omega,\sigma)]^M$ (equipped with the usual operator norm). The proof of Theorem 8.2 is therefore complete. □

A result of similar flavor as before can be obtained in BCMO$_p$. Specifically, we have the following theorem.

Theorem 8.3 *Consider an Ahlfors regular domain $\Omega \subseteq \mathbb{R}^n$. Set $\sigma := \mathcal{H}^{n-1}\lfloor\partial\Omega$ and denote by ν the geometric measure theoretic outward unit normal to Ω. Also, consider a homogeneous, second-order, constant complex coefficient, weakly elliptic $M \times M$ system L in $\mathbb{R}^n$, and assume that $\mathfrak{A}_L^{\rm dis} \neq \varnothing$. Pick $A \in \mathfrak{A}_L^{\rm dis}$ and consider the modified boundary-to-boundary double layer potential operators $K_{A,\rm mod}$ associated with A as in (7.29). Finally, fix an integrability exponent $p \in (1,\infty)$, along with a parameter $\varepsilon \in (0,\infty)$.*

Then there exists some threshold $\delta \in (0,1)$, depending only on n, p, A, ε, and the Ahlfors regularity constant of $\partial\Omega$, with the property that whenever

$$\|\nu\|_{[\mathrm{BMO}(\partial\Omega,\sigma)]^n} < \delta \tag{8.11}$$

it follows that for each spectral parameter $z \in \mathbb{C}$ with $|z| \geq \varepsilon$ the operator

$$\begin{gathered} zI + [K_{A,\rm mod}] : \big[\widetilde{\mathrm{BCMO}}_p(\partial\Omega,\sigma)\big]^M \longrightarrow \big[\widetilde{\mathrm{BCMO}}_p(\partial\Omega,\sigma)\big]^M \\ \text{is well defined, linear, bounded, and invertible.} \end{gathered} \tag{8.12}$$

Proof The invertibility claimed in (8.12) may be obtained via the same type of argument presented in the proof of Theorem 8.2, now taking into consideration (8.5), and the fact that $\widetilde{\mathrm{BCMO}}_p(\partial\Omega,\sigma)$ is a Banach space (cf. Corollary 5.1). □

Remark 8.1 A more explicit version of Theorem 8.2 and Theorem 8.3, clarifying the timing of the choices of thresholds, goes as follows. Fix two numbers $n, M \in \mathbb{N}$, with $n \geq 2$, and assume L is a second-order, homogeneous, constant complex coefficient, weakly elliptic $M\times M$ system in $\mathbb{R}^n$. Assume that $\mathfrak{A}_L^{\rm dis} \neq \varnothing$, and select a distinguished coefficient tensor $A \in \mathfrak{A}_L^{\rm dis}$. Pick an integrability exponent $p \in (1,\infty)$, a constant $C_{\rm ADR} \in (0,\infty)$, and a parameter $\varepsilon \in (0,\infty)$. Under these assumptions, there exists a threshold $\delta \in (0,1)$ depending exclusively on M, n, p, ε, A, and $C_{\rm ADR}$ with the following significance: If $\Omega \subseteq \mathbb{R}^n$ is an AR domain, with Ahlfors regularity constants $C_{\partial\Omega}, c_{\partial\Omega} \in (0,\infty)$ and geometric measure theoretic outward unit normal ν satisfying

$$\max\big\{C_{\partial\Omega}, \tfrac{1}{c_{\partial\Omega}}\big\} \leq C_{\rm ADR} \quad \text{and} \quad \|\nu\|_{[\mathrm{BMO}(\partial\Omega,\sigma)]^n} < \delta, \tag{8.13}$$

then Ω is a UR domain, with $\partial\Omega$ unbounded, and UR constants controlled solely in terms of n and $C_{\rm ADR}$. In this scenario, both conclusions (8.8) and (8.12) hold true.

Chapter 9
Boundary Value Problems on HA^p Spaces and BCMO_p Spaces

The Calderón-Zygmund theory for singular integral operators on Beurling-Hardy spaces and the space of bounded central mean oscillation functions opens the door for employing boundary layer potential techniques for treating boundary problems on these scales. This chapter is devoted to studying the Neumann Problem with boundary data prescribed in HA^p and the Dirichlet Problem with boundary data prescribed in BCMO_p for second-order homogeneous constant complex coefficient weakly elliptic systems in δ-AR domains, with $\delta \in (0, 1)$ sufficiently small.

9.1 The Neumann Problem with Boundary Data in HA^p Spaces

Our main result in this section (cf. Theorem 9.4 below) describes the well-posedness of the Neumann Problem for weakly elliptic systems in δ-AR domains $\Omega \subseteq \mathbb{R}^n$, with $\delta \in (0, 1)$ sufficiently small, and boundary data in Beurling-Hardy spaces. We also establish a Fatou-type result in the context of HA^p spaces, ensuring that the latter scale is the correct space of boundary data in the context of the HA^p-Neumann Problem.

We begin by recalling a Fatou-type result from [GHA-III, Theorem 3.3.5, p. 707] which guarantees the existence of nontangential boundary traces of null-solutions for weakly elliptic systems L and their gradients.

Theorem 9.1 *Consider a* UR *domain $\Omega \subseteq \mathbb{R}^n$, and abbreviate $\sigma := \mathcal{H}^{n-1}\lfloor\partial\Omega$. Pick a second-order, homogeneous, constant complex coefficient, weakly elliptic $M \times M$ system L in $\mathbb{R}^n$. Fix some aperture parameter $\kappa \in (0, \infty)$, and some integrability exponent $p \in (1, \infty]$, and assume $u \in [\mathscr{C}^\infty(\Omega)]^M$ has the following properties:*

$$Lu = 0 \ \ in\ \Omega \ \ and \ \ \mathcal{N}_\kappa(\nabla u) \in L^p_{\mathrm{loc}}(\partial\Omega, \sigma). \tag{9.1}$$

© The Author(s), under exclusive license to Springer Nature Switzerland AG 2026
M. Mitrea and P. Takemura, *Singular Integrals, Herz-Type Function Spaces, and Boundary Problems*, Progress in Mathematics 362,
https://doi.org/10.1007/978-3-032-12516-3_9

Then the κ-nontangential trace $(\nabla u)\big|_{\partial\Omega}^{\kappa-\mathrm{n.t.}}$ exists at σ-a.e. point on $\partial\Omega$, it is actually independent of κ, and the function $(\nabla u)\big|_{\partial\Omega}^{\kappa-\mathrm{n.t.}}$ is σ-measurable and belongs to $[L^p_{\mathrm{loc}}(\partial\Omega,\sigma)]^{M\times n}$.

Furthermore, the nontangential boundary trace $u\big|_{\partial\Omega}^{\kappa-\mathrm{n.t.}}$ also exists σ-a.e. on $\partial\Omega$, is a σ-measurable function which is actually independent of the aperture parameter.

As a preamble to the next result, we recall the following definition.

Definition 9.1 Let $\Omega\subseteq\mathbb{R}^n$ be a nonempty open set with the property that

$$0<\mathcal{H}^{n-1}\big(\partial\Omega\cap B(x,r)\big)<\infty \quad \text{for each } x\in\partial\Omega \text{ and } r\in(0,\infty), \tag{9.2}$$

and abbreviate $\sigma:=\mathcal{H}^{n-1}\lfloor\partial\Omega$. For each $\mathcal{L}^n$-measurable function $u:\Omega\to\mathbb{C}$ define its $\mathfrak{P}$-`maximal function`, at each $x\in\partial\Omega$, according to

$$(\mathfrak{P}u)(x):=\sup_{0<r<2\,\mathrm{diam}(\partial\Omega)}\left\{\frac{1}{\sigma\big(\Delta(x,r)\big)}\int_{\Omega\cap B(x,r)}|u|\,d\mathcal{L}^n\right\}\in[0,\infty]. \tag{9.3}$$

The next result provides a useful estimate for the grand maximal function. See [GHA-II, Theorem 10.1.2, p. 610] for its proof.

Theorem 9.2 *Assume $\Omega\subseteq\mathbb{R}^n$ an Ahlfors regular domain. Set $\sigma:=\mathcal{H}^{n-1}\lfloor\partial\Omega$, and denote by ν the geometric measure theoretic outward unit normal to Ω. Consider a vector field $\vec{F}:\Omega\to\mathbb{C}^n$ having $\mathcal{L}^n$-measurable components and with the property that for some aperture parameter $\kappa\in(0,\infty)$ one has*

$$\mathcal{N}_\kappa\vec{F}\in L^1_{\mathrm{loc}}(\partial\Omega,\sigma) \text{ and } \vec{F}\big|_{\partial\Omega}^{\kappa-\mathrm{n.t.}} \text{ exists } \sigma\text{-a.e. on } \partial\Omega. \tag{9.4}$$

In particular, $\vec{F}\in\big[L^1_{\mathrm{loc}}(\Omega,\mathcal{L}^n)\big]^n$, and one also assumes that the divergence of $\vec{F}$, computed in the sense of distributions in Ω, is of function type, i.e., it satisfies

$$\mathrm{div}\,\vec{F}\in L^1_{\mathrm{loc}}(\Omega,\mathcal{L}^n). \tag{9.5}$$

Fix some $\gamma\in(0,1)$ along with some power

$$s\in\Big(\frac{n-1}{n-1+\gamma},\infty\Big). \tag{9.6}$$

Finally, recall that $[\,\cdot\,]^{\#}_{\gamma}$ denotes the Fefferman-Stein grand maximal function from Definition 4.1.

Then there exists some constant $C=C(\partial\Omega,\kappa,n,\gamma,s)\in(0,\infty)$ with the property that

$$\Big[\nu\cdot\Big(\vec{F}\big|_{\partial\Omega}^{\kappa-\mathrm{n.t.}}\Big)\Big]^{\#}_{\gamma}\leq C\cdot\mathcal{M}_{\partial\Omega,s}\big(\mathcal{N}_\kappa\vec{F}\big)+C\cdot\mathfrak{P}\big(\mathrm{div}\,\vec{F}\big)\quad \text{everywhere on } \partial\Omega, \tag{9.7}$$

where $\mathcal{M}_{\partial\Omega,s}$ is the L^s-based Hardy-Littlewood maximal operator on $\partial\Omega$ (cf. (2.28)).

We now present a Fatou-type result in the context of Beurling-Hardy spaces HA^p.

Theorem 9.3 *Suppose $\Omega \subseteq \mathbb{R}^n$ is a* UR *domain with unbounded boundary. Abbreviate $\sigma := \mathcal{H}^{n-1}\lfloor\partial\Omega$, and denote by ν the geometric measure theoretic outward unit normal to Ω. Consider L a second-order, homogeneous, constant complex coefficient, weakly elliptic $M \times M$ system in $\mathbb{R}^n$. Also, fix an integrability exponent $p \in (1,\infty)$, along with an aperture parameter $\kappa \in (0,\infty)$. Finally, assume $u : \Omega \to \mathbb{C}^M$ is a function satisfying:*

$$\begin{aligned} &u \in [\mathscr{C}^\infty(\Omega)]^M, \quad Lu = 0 \ \text{ in } \Omega, \\ &\text{and} \quad \mathcal{N}_\kappa(\nabla u) \in \mathrm{A}^p(\partial\Omega,\sigma). \end{aligned} \tag{9.8}$$

Then the nontangential boundary trace $(\nabla u)\big|_{\partial\Omega}^{\kappa-\text{n.t.}}$ exists at σ-a.e. point on $\partial\Omega$, and the vector-valued function $(\nabla u)\big|_{\partial\Omega}^{\kappa-\text{n.t.}}$ actually belongs to $[\mathrm{A}^p(\partial\Omega,\sigma)]^{M\times n}$.

Furthermore, for any coefficient tensor $A \in \mathfrak{A}_L$ it follows that conormal derivative $\partial_\nu^A u$ (cf. (7.55)) belongs to $[\mathrm{HA}^p(\partial\Omega,\sigma)]^M$ in a quantitative fashion, i.e., there exists some constant $C = C(\partial\Omega, A, n, p, \kappa) \in (0,\infty)$ for which

$$\|\partial_\nu^A u\|_{[\mathrm{HA}^p(\partial\Omega,\sigma)]^M} \leq C\|\mathcal{N}_\kappa(\nabla u)\|_{\mathrm{A}^p(\partial\Omega,\sigma)}. \tag{9.9}$$

Proof To get started, notice that the assumptions made in (9.8) along with the fact that $\mathrm{A}^p(\partial\Omega,\sigma) \subseteq L^p_{\text{loc}}(\partial\Omega,\sigma)$ (cf. Proposition 3.3), permit us to invoke Theorem 9.1 to guarantee that the nontangential limit $(\nabla u)\big|_{\partial\Omega}^{\kappa-\text{n.t.}}$ exists σ-a.e. on $\partial\Omega$. As such, since

$$\Big|(\nabla u)\big|_{\partial\Omega}^{\kappa-\text{n.t.}}(x)\Big| \leq \mathcal{N}_\kappa(\nabla u)(x) \ \text{ at } \sigma\text{-a.e. } x \in \partial\Omega, \tag{9.10}$$

the membership in the second line of (9.8), and the lattice property of $\mathrm{A}^p(\partial\Omega,\sigma)$ (cf. Remark 3.3) ensure that $(\nabla u)\big|_{\partial\Omega}^{\kappa-\text{n.t.}} \in \mathrm{A}^p(\partial\Omega,\sigma)$.

To verify the last claim made in the statement of the present theorem, fix a coefficient tensor $A = \big(a_{jk}^{\alpha\beta}\big)_{\substack{1\leq j,k\leq n \\ 1\leq\alpha,\beta\leq M}} \in \mathfrak{A}_L$ and write $L = \big(a_{jk}^{\alpha\beta}\partial_j\partial_k\big)_{1\leq\alpha,\beta\leq M}$. For each $\alpha \in \{1,\dots,M\}$, abbreviate

$$\vec{F}_\alpha := \big(a_{jk}^{\alpha\beta}(\partial_k u_\beta)\big)_{1\leq j\leq n} \ \text{ in } \Omega. \tag{9.11}$$

The existence of the nontangential limit $(\nabla u)\big|_{\partial\Omega}^{\kappa-\text{n.t.}}$ σ-a.e on $\partial\Omega$ (established above) implies that

$$(\vec{F}_\alpha)\big|_{\partial\Omega}^{\kappa-\text{n.t.}} \ \text{ exists } \sigma\text{-a.e on } \partial\Omega. \tag{9.12}$$

Upon recalling (7.55), for each $\alpha \in \{1,\dots,M\}$ we may express the α-th component of the conormal derivative $\partial_\nu^A u$ as

$$\big(\partial_\nu^A u\big)_\alpha = \nu_j a_{jk}^{\alpha\beta}(\partial_k u_\beta)\big|_{\partial\Omega}^{\kappa-\text{n.t.}} = \nu \cdot (\vec{F}_\alpha)\big|_{\partial\Omega}^{\kappa-\text{n.t.}}. \tag{9.13}$$

In addition, since $Lu = 0$ in Ω (cf. (9.8)) it follows that

$$\operatorname{div} \vec{F}_\alpha = a^{\alpha\beta}_{jk} \partial_j \partial_k u_\beta = (Lu)_\alpha = 0 \in L^1_{\mathrm{loc}}(\Omega, \mathcal{L}^n). \tag{9.14}$$

Notice that $\mathcal{N}_\kappa(\partial_k u_\beta) \le \mathcal{N}_\kappa(\nabla u)$ pointwise everywhere on $\partial\Omega$ for every indices $k \in \{1, \ldots, n\}$ and $\beta \in \{1, \ldots, M\}$. This, (9.11), (3.32), the fact that $\mathcal{N}_\kappa(\nabla u)$ belongs to $\mathrm{A}^p(\partial\Omega, \sigma)$ (cf. (9.8)), and the lattice property of $\mathrm{A}^p(\partial\Omega, \sigma)$ (cf. Remark 3.3) yield

$$\begin{aligned} &\mathcal{N}_\kappa \vec{F}_\alpha \in \mathrm{A}^p(\partial\Omega, \sigma) \subseteq L^1_{\mathrm{loc}}(\partial\Omega, \sigma) \quad \text{and} \\ &\|\mathcal{N}_\kappa \vec{F}_\alpha\|_{\mathrm{A}^p(\partial\Omega,\sigma)} \le C \|\mathcal{N}_\kappa(\nabla u)\|_{\mathrm{A}^p(\partial\Omega,\sigma)} \end{aligned} \tag{9.15}$$

for each $\alpha \in \{1, \ldots, M\}$, and for some constant $C \in (0, \infty)$ depending on n and A. Thanks to this, (9.12), and (9.14), Theorem 9.2 ensures that for each $\gamma \in (0, 1)$ and each $s \in \big(\frac{n-1}{n-1+\gamma}, \infty\big)$ there exists some constant $C = C(\partial\Omega, \kappa, n, \gamma, s) \in (0, \infty)$ with the property that

$$\Big[\nu \cdot \Big(\vec{F}_\alpha\Big|^{\kappa-\mathrm{n.t.}}_{\partial\Omega}\Big)\Big]^{\#}_{\gamma} \le C \cdot \mathcal{M}_{\partial\Omega,s}(\mathcal{N}_\kappa \vec{F}_\alpha) \text{ everywhere on } \partial\Omega, \tag{9.16}$$

for every $\alpha \in \{1, \ldots, M\}$. Restricting ourselves to the range $s \in \big(\frac{n-1}{n-1+\gamma}, 1\big)$, using (9.15) and the boundedness of $\mathcal{M}_{\partial\Omega,s}$ on $\mathrm{A}^p(\partial\Omega, \sigma)$ (cf. Theorem 3.3) we arrive at

$$\begin{aligned} \Big\|\nu \cdot \Big(\vec{F}_\alpha\Big|^{\kappa-\mathrm{n.t.}}_{\partial\Omega}\Big)\Big\|_{\mathrm{HA}^p(\partial\Omega,\sigma)} &= \Big\|\Big[\nu \cdot \Big(\vec{F}_\alpha\Big|^{\kappa-\mathrm{n.t.}}_{\partial\Omega}\Big)\Big]^{\#}_{\gamma}\Big\|_{\mathrm{A}^p(\partial\Omega,\sigma)} \\ &\le C \|\mathcal{N}_\kappa \vec{F}_\alpha\|_{\mathrm{A}^p(\partial\Omega,\sigma)} < \infty \end{aligned} \tag{9.17}$$

for each $\alpha \in \{1, \ldots, M\}$, and for some constant $C \in (0, \infty)$ depending only on n, p, γ, s, and the Ahlfors regularity constants of $\partial\Omega$. All together, (9.11), (9.13), (9.15), and (9.17) imply (9.9). This concludes the proof of Theorem 9.3. □

We are now in a position to state and prove our main result in this section regarding the well-posedness of the Neumann Problem with boundary data taken from Beurling Hardy spaces.

Theorem 9.4 *Assume $\Omega \subseteq \mathbb{R}^n$ is an Ahlfors regular domain. Set $\sigma := \mathcal{H}^{n-1} \lfloor \partial\Omega$ and denote by ν the geometric measure theoretic outward unit normal to Ω. Fix an integrability exponent $p \in (1, \infty)$. Suppose L is a second-order, homogeneous, constant complex coefficients, weakly elliptic $M \times M$ system in $\mathbb{R}^n$, and select a coefficient tensor $A \in \mathfrak{A}_L$. Finally, pick an aperture parameter $\kappa \in (0, \infty)$. Consider the Neumann Problem for L in Ω with boundary data in the Beurling-Hardy space $[\mathrm{HA}^p(\partial\Omega, \sigma)]^M$ formulated as follows:*

$$\begin{cases} u \in [\mathscr{C}^\infty(\Omega)]^M, \\ Lu = 0 \ \ in\ \Omega, \\ \mathcal{N}_\kappa(\nabla u) \in \mathrm{A}^p(\partial\Omega, \sigma), \\ \partial_\nu^A u = f \in [\mathrm{HA}^p(\partial\Omega, \sigma)]^M. \end{cases} \tag{9.18}$$

Then the following statements are true:

(a) [Existence, Integral Representation, and Estimates] *If* $A^\top \in \mathfrak{A}_{L^\top}^{\mathrm{dis}}$ *then there exists some threshold* $\delta \in (0,1)$ *depending only on* n, p, κ, A, *and the Ahlfors regularity constants of* $\partial\Omega$ *with the property that whenever*

$$\|\nu\|_{[\mathrm{BMO}(\partial\Omega,\sigma)]^n} < \delta \tag{9.19}$$

it follows that the Neumann Problem (9.18) *is solvable. Specifically, the operator* $-\frac{1}{2}I + K^\#_{A^\top}$ *is invertible on the Beurling-Hardy space* $[\mathrm{HA}^p(\partial\Omega,\sigma)]^M$ *and the function* $u : \Omega \longrightarrow \mathbb{C}^M$ *defined by*

$$u := \mathscr{S}_{\mathrm{mod}}\big[\big(-\tfrac{1}{2}I + K^\#_{A^\top}\big)^{-1} f\big] \ \ in\ \Omega \tag{9.20}$$

is a solution of the Neumann Problem (9.18) *and satisfies*

$$\|\mathcal{N}_\kappa(\nabla u)\|_{\mathrm{A}^p(\partial\Omega,\sigma)} \approx \|f\|_{[\mathrm{HA}^p(\partial\Omega,\sigma)]^M} \tag{9.21}$$

with implicit proportionality constants independent of f.

(b) [Uniqueness (modulo constants)] *If* $A \in \mathfrak{A}_L^{\mathrm{dis}}$ *then there exists some threshold* $\delta \in (0,1)$ *depending only on* n, p, κ, A, *and the Ahlfors regularity constants of* $\partial\Omega$ *with the property that if* $\|\nu\|_{[\mathrm{BMO}(\partial\Omega,\sigma)]^n} < \delta$ *then any two solutions of the Neumann Problem* (9.18) *differ by a constant in* $\mathbb{C}^M$.

(c) [Well-Posedness] *If* $A \in \mathfrak{A}_L^{\mathrm{dis}}$ *and* $A^\top \in \mathfrak{A}_{L^\top}^{\mathrm{dis}}$ *then there exists some threshold* $\delta \in (0,1)$ *depending only on* n, p, κ, A, *and the Ahlfors regularity constants of* $\partial\Omega$ *with the property that whenever* $\|\nu\|_{[\mathrm{BMO}(\partial\Omega,\sigma)]^n} < \delta$ *it follows that the Neumann Problem* (9.18) *has a solution that is unique modulo constants, and satisfies* (9.21).

Proof First assume that $\mathfrak{A}_{L^\top}^{\mathrm{dis}} \neq \varnothing$, and $A^\top \in \mathfrak{A}_{L^\top}^{\mathrm{dis}}$. Then Theorem 8.2 (currently used with $\varepsilon := 1/2$, and $z := -1/2$) ensures that there exists some $\delta \in (0,1)$ for which whenever $\|\nu\|_{[\mathrm{BMO}(\partial\Omega,\sigma)]^n} < \delta$, it follows that $-\frac{1}{2}I + K^\#_{A^\top}$ is a bounded and invertible operator on $[\mathrm{HA}^p(\partial\Omega,\sigma)]^M$. As such, the function u from (9.20) is meaningfully defined, and thanks to (7.54) and (7.56) it satisfies

$$u \in [\mathscr{C}^\infty(\Omega)]^M, \quad Lu = 0 \ \text{ in } \Omega, \text{ and } \ \partial_\nu^A u = f \ \text{ pointwise } \sigma\text{-a.e. on } \partial\Omega. \tag{9.22}$$

Additionally, (7.53), Theorem 7.1, and Theorem 6.2 imply

$$\begin{aligned}\|\mathcal{N}_\kappa(\nabla u)\|_{\mathrm{A}^p(\partial\Omega,\sigma)} &\leq C\big\|\big(-\tfrac{1}{2}I + K^{\#}_{A^\top}\big)^{-1} f\big\|_{[\mathrm{HA}^p(\partial\Omega,\sigma)]^M} \\ &\leq C\|f\|_{[\mathrm{HA}^p(\partial\Omega,\sigma)]^M},\end{aligned} \tag{9.23}$$

for some constant $C \in (0, \infty)$ depending only on n, p, A, and the Ahlfors regularity character of $\partial\Omega$. Hence (9.22)-(9.23) show that the function u from (9.20) is indeed a solution of the Neumann Problem (9.18). The estimates in (9.21) can be seen from (9.22), (9.23), and (9.9). This finishes the treatment of item *(a)*.

To deal with the uniqueness (modulo constants) claimed in *(b)*, assume $A \in \mathfrak{A}_L^{\mathrm{dis}}$, and consider u a solution of the homogeneous Neumann Problem (9.18) (corresponding to the boundary datum $f \equiv 0$). Bearing in mind that $\mathrm{A}^p(\partial\Omega,\sigma) \hookrightarrow L^p(\Sigma,\sigma)$ continuously (cf. Proposition 3.3), such a solution u satisfies

$$\begin{cases} u \in [\mathscr{C}^\infty(\Omega)]^M, \quad Lu = 0 \ \text{ in } \Omega, \\ \mathcal{N}_\kappa(\nabla u) \in L^p(\partial\Omega,\sigma), \\ \partial_\nu^A u = 0 \quad \text{pointwise } \sigma\text{-a.e. on } \partial\Omega. \end{cases} \tag{9.24}$$

Hence, [MMMi³, Theorem 6.17, p. 396] ensures that u is a constant function in Ω. This concludes the justification of *(b)*. Lastly, the claim made in *(c)* is a consequence of *(a)* and *(b)*. This completes the proof of Theorem 9.4. □

<u>`Comment 1.`</u> We wish to highlight that the boundary condition in (9.18) is naturally formulated, thanks to our Fatou-type result recorded in Theorem 9.3.

<u>`Comment 2.`</u> Recall that corresponding to the end point case $p = 1$, the Beurling space $A^1(\partial\Omega,\sigma)$ coincides with the Lebesgue space $L^1(\partial\Omega,\sigma)$, and $\mathrm{HA}^1(\partial\Omega,\sigma)$ coincides with the (standard) Hardy space $H^1(\partial\Omega,\sigma)$ (cf. (3.1) and (4.1)). In this context, the Neumann Problem (9.18) reads as

$$\begin{cases} u \in [\mathscr{C}^\infty(\Omega)]^M, \quad Lu = 0 \ \text{ in } \Omega, \\ \mathcal{N}_\kappa(\nabla u) \in L^1(\partial\Omega,\sigma), \\ \partial_\nu^A u = f \in [H^1(\partial\Omega,\sigma)]^M. \end{cases} \tag{9.25}$$

Such a problem has been treated recently in [GHA-V, Theorem 8.4.25, p. 843], where similar conclusions as in Theorem 9.3 have been obtained. Based on this, the well-posedness result presented in Theorem 9.3 may be regarded as a natural extension of (9.25), where the boundary data is now taken from Beurling-Hardy spaces, which constitute a nested family of scales included in $H^1(\partial\Omega,\sigma)$ (cf. (4.11), (4.12)). We also wish to mention that in the context of Lipschitz domains, and $\mathscr{C}^1$ domains, the Neumann Problem for the Laplacian on $H^1(\partial\Omega,\sigma)$ formulated in (9.25) has been solved in [DaKe87] and [FaKe81].

<u>`Comment 3.`</u> In view of the well-posedness of (9.25) (commented above), it is worth to record an alternative approach to the uniqueness of the Neumann Problem (9.18).

Specifically, taking into account the inclusion $\mathrm{A}^p(\partial\Omega,\sigma) \subseteq L^1(\partial\Omega,\sigma)$ (cf. (3.29)), it follows that whenever u is a solutions to the homogeneous Neumann Problem (9.18) (corresponding to $f \equiv 0$), we have

$$\begin{cases} u \in [\mathscr{C}^\infty(\Omega)]^M, \quad Lu = 0 \text{ in } \Omega, \\ \mathcal{N}_\kappa(\nabla u) \in L^1(\partial\Omega,\sigma), \\ \partial_\nu^A u = 0 \text{ pointwise } \sigma\text{-a.e. on } \partial\Omega. \end{cases} \tag{9.26}$$

As such, since (4.12) is well-posed (cf. [GHA-V, Theorem 8.4.25, p. 843]), this necessarily yields that u is a constant function in Ω.

9.2 The Dirichlet Problem with Boundary Data in BCMO_p Spaces

Theorem 9.5 below establishes the solvability of the Dirichlet Problem in δ-AR domains, with $\delta \in (0,1)$ sufficiently small, for second-order weakly elliptic systems L satisfying $\mathfrak{A}_L^{\mathrm{dis}} \neq \varnothing$ when the boundary data is prescribed in the space of functions of bounded central p-mean oscillations, and the size condition is formulated in terms of central Carleson measures (cf. Definition 6.1).

Theorem 9.5 *Assume $\Omega \subseteq \mathbb{R}^n$ is an Ahlfors regular domain. Set $\sigma := \mathcal{H}^{n-1}\lfloor\partial\Omega$, and denote by ν the geometric measure theoretic outward unit normal to Ω. Pick an integrability exponent $p \in (1,\infty)$. Suppose L is a second-order, homogeneous, constant complex coefficients, weakly elliptic $M \times M$ system in $\mathbb{R}^n$, and select an aperture parameter $\kappa \in (0,\infty)$. Formulate the Dirichlet Problem for L in Ω with boundary data in $[\mathrm{BCMO}_p(\partial\Omega,\sigma)]^M$ as follows:*

$$\begin{cases} u \in [\mathscr{C}^\infty(\Omega)]^M, \quad Lu = 0 \text{ in } \Omega, \\ |\nabla u|^p \operatorname{dist}(\cdot,\partial\Omega)^{p-1}\, d\mathcal{L}^n \text{ is a central Carleson measure in } \Omega, \\ u\big|_{\partial\Omega}^{\kappa-\mathrm{n.t.}} = f \in \big[\mathrm{BCMO}_p(\partial\Omega,\sigma)\big]^M. \end{cases} \tag{9.27}$$

Then whenever $\mathfrak{A}_L^{\mathrm{dis}} \neq \varnothing$ there exists a threshold $\delta \in (0,1)$ depending on n, M, p, A, and the Ahlfors regularity constants of $\partial\Omega$ with the property that if

$$\|\nu\|_{[\mathrm{BMO}(\partial\Omega,\sigma)]^n} < \delta \tag{9.28}$$

it follows that the Dirichlet Problem (9.27) is solvable. More specifically, if $A \in \mathfrak{A}_L^{\mathrm{dis}}$ and $\big[K_{A,\mathrm{mod}}\big]$ acts on $[\widetilde{\mathrm{BCMO}}_p(\partial\Omega,\sigma)]^M$ as in (8.4), then the operator $\frac{1}{2}I + [K_{A,\mathrm{mod}}]$ is invertible on $[\widetilde{\mathrm{BCMO}}_p(\partial\Omega,\sigma)]^M$ and a solution of the Dirichlet Problem (9.27) is given by the function $u : \Omega \to \mathbb{C}^M$ defined as

$$u := \mathcal{D}_{A,\mathrm{mod}}g + c \quad in\ \Omega \tag{9.29}$$

for a suitable constant $c \in \mathbb{C}^M$ and a choice of

$$g \in \big[\mathrm{BCMO}_p(\Sigma, \sigma)\big]^M \ \ such\ that\ \ [g] = \Big(\tfrac{1}{2}I + \big[K_{A,\mathrm{mod}}\big]\Big)^{-1}[f]. \tag{9.30}$$

Moreover, this solution satisfies

$$\sup_{R \geq 2_*} \left(\frac{1}{\sigma\big(\Delta(x_0, R)\big)} \int_{B(x_0,R)\cap\Omega} |\nabla u|^p \,\mathrm{dist}(\cdot, \partial\Omega)^{p-1}\, d\mathcal{L}^n\right)^{1/p} \leq C\|f\|_{[\mathrm{BCMO}_p(\partial\Omega,\sigma)]^M} \tag{9.31}$$

some constant $C \in (0, \infty)$ independent of f.

Proof of Theorem 9.5 Thanks to Theorem 8.3 (currently used with $\varepsilon = z := 1/2$), there exists $\delta \in (0, 1)$ depending exclusively on n, M, p, A, and the Ahlfors regularity constants of $\partial\Omega$ such that if $\|\nu\|_{[\mathrm{BMO}(\partial\Omega,\sigma)]^n} < \delta$ then the operator $\frac{1}{2}I + [K_{A,\mathrm{mod}}]$ is invertible on $\big[\widetilde{\mathrm{BCMO}}_p(\partial\Omega, \sigma)\big]^M$. Hence, there exists a function $g \in \big[\mathrm{BCMO}_p(\partial\Omega, \sigma)\big]^M$ with the property that

$$\big[\big(\tfrac{1}{2}I + K_{A,\mathrm{mod}}\big)g\big] = \big(\tfrac{1}{2}I + [K_{A,\mathrm{mod}}]\big)[g] = [f], \tag{9.32}$$

as equivalence classes (modulo constants) in $\big[\widetilde{\mathrm{BCMO}}_p(\partial\Omega, \sigma)\big]^M$. As such, there exists some constant $c \in \mathbb{C}^M$ satisfying

$$\big(\tfrac{1}{2}I + K_{A,\mathrm{mod}}\big)g = f - c. \tag{9.33}$$

Thanks to Proposition 5.4 (see (5.36)), the function u given in (9.29) is meaningfully defined. Moreover, (7.30), (7.31), and (9.33), entail

$$u \in [\mathscr{C}^\infty(\Omega)]^M \text{ and } Lu = 0 \text{ in } \Omega \tag{9.34}$$

and

$$u\Big|_{\partial\Omega}^{\kappa-\mathrm{n.t.}} = (\mathcal{D}_{A,\mathrm{mod}}g)\Big|_{\partial\Omega}^{\kappa-\mathrm{n.t.}} + c = \big(\tfrac{1}{2}I + K_{A,\mathrm{mod}}\big)g + c = f. \tag{9.35}$$

Finally, bearing in mind that $A \in \mathfrak{A}_L^{\mathrm{dis}}$, along with (9.29) and (7.50), Theorem 6.8 ensures that

$$|\nabla u|^p \,\mathrm{dist}(\cdot, \partial\Omega)^{p-1}\, d\mathcal{L}^n \text{ is a central Carleson measure in } \Omega \text{ satisfying estimate (9.31).} \tag{9.36}$$

This concludes the proof of Theorem 9.5. □

We close this chapter with some comments in relation to the Dirichlet Problem formulated in the above theorem.

<u>`Comment 1.`</u> The boundary problem in (9.27) is of same nature as the Dirichlet Problem with boundary data in the John-Nirenberg space $\mathrm{BMO}(\partial\Omega, \sigma)$, namely

$$\begin{cases} u \in [\mathscr{C}^\infty(\Omega)]^M, \quad Lu = 0 \ \text{ in } \Omega, \\ |\nabla u|^2 \operatorname{dist}(\cdot, \partial\Omega)\, d\mathcal{L}^n \ \text{ is a Carleson measure in } \Omega, \\ u\big|_{\partial\Omega}^{\kappa-\mathrm{n.t.}} = f \in [\mathrm{BMO}(\partial\Omega, \sigma)]^M. \end{cases} \tag{9.37}$$

The solvability of this problem has been established in [GHA-V, Theorem 8.4.33, p. 857]. It is of interest to observe that actually there is a more flexible version of (9.37), in which the size condition involves an L^p-styled Carleson measure condition:

$$\begin{cases} u \in [\mathscr{C}^\infty(\Omega)]^M, \quad Lu = 0 \ \text{ in } \Omega, \\ |\nabla u|^p \operatorname{dist}(\cdot, \partial\Omega)^{p-1}\, d\mathcal{L}^n \ \text{ is a Carleson,} \\ \text{measure in } \Omega \text{ for each } p \in (1, \infty), \\ u\big|_{\partial\Omega}^{\kappa-\mathrm{n.t.}} = f \in [\mathrm{BMO}(\partial\Omega, \sigma)]^M, \end{cases} \tag{9.38}$$

which also has been proved to be solvable (cf. [GHA-V, Theorem 8.4.33, p. 857]). What makes this flexibility possible is the fact that the John-Nirenberg space $\mathrm{BMO}(\partial\Omega, \sigma)$ can be characterized in terms of L^p-styled mean oscillations (as a consequence of the classical John-Nirenberg inequality). In contrast with the formulation of our Dirichlet Problem (9.27), the exponent p appearing in the (central) Carleson measure condition is tied up with the exponent p intrinsic to the space $\mathrm{BCMO}_p(\partial\Omega, \sigma)$. This is due to the fact that the space $\mathrm{BCMO}_p(\partial\Omega, \sigma)$ fundamentally depends on p (cf. [ChLa89, Proposition 5.2] and [AlLaGu00, Proposition 3.3]).

Comment 2. In the context of Lipschitz domains, and $\mathscr{C}^1$ domains, the Dirichlet Problem (9.37) corresponding to the Laplacian, i.e., $L = \Delta$, has been treated in [FaKe81] and [FaNe80].

Chapter 10
Herz Spaces of First Generation

The main aim of this chapter is to study Herz spaces of first generation $\mathrm{A}_{p,q}$ in the setting of unbounded Ahlfors regular sets. We establish a wealth of fundamental properties for this scale of spaces, such as embeddings, natural versions of Lebesgue's Monotone Convergence Theorem and Lebesgue's Dominated Convergence Theorem, completeness, density, duality results, and boundedness of the Hardy-Littlewood maximal operator. Significantly, we find that Herz spaces of first generation fall within the scope of Generalized Banach Function Spaces (in the sense of [GHA-II]). This allows us to obtain a powerful extrapolation result on $\mathrm{A}_{p,q}$ (see Theorem 10.3), which will play a crucial role in the treatment of singular integral operators on these spaces.

10.1 Definitions and Basic Properties

The goal here is to explore the functional analytic properties of first-generation Herz spaces. We begin with the following definition.

Definition 10.1 Assume $\Sigma \subseteq \mathbb{R}^n$ is a closed unbounded Ahlfors regular set, abbreviate $\sigma := \mathcal{H}^{n-1}\lfloor\Sigma$ and pick integrability exponents $p, q \in (0, \infty)$. Denote by $c_\Sigma, C_\Sigma \in (0, \infty)$ the respective lower and upper Ahlfors regularity constants of Σ. Having fixed a reference point $x_0 \in \Sigma$ and a dilation factor $2_* > (C_\Sigma/c_\Sigma)^{1/(n-1)}$, define the family $\{C_k\}_{k\in\mathbb{N}_0}$ as in (3.1).

For each σ-measurable function $f : \Sigma \to \mathbb{R}$, introduce

$$\|f\|_{\mathrm{A}_{p,q}(\Sigma,\sigma)} := \Big\{\sum_{k=0}^{\infty} \big(2_*^{k(n-1)(1/q-1/p)}\|f\cdot\mathbb{1}_{C_k}\|_{L^p(\Sigma,\sigma)}\big)^q\Big\}^{1/q} \tag{10.1}$$

and

© The Author(s), under exclusive license to Springer Nature Switzerland AG 2026
M. Mitrea and P. Takemura, *Singular Integrals, Herz-Type Function Spaces, and Boundary Problems*, Progress in Mathematics 362,
https://doi.org/10.1007/978-3-032-12516-3_10

$$\|f\|_{\mathrm{B}_{p,q}(\Sigma,\sigma)} := \sup_{k\in\mathbb{N}_0} \left\{2_*^{-k(n-1)(1/q+1/p-1)} \|f\cdot \mathbb{1}_{C_k}\|_{L^p(\Sigma,\sigma)}\right\}. \tag{10.2}$$

Define the `Herz space of first generation` $\mathrm{A}_{p,q}(\Sigma,\sigma)$ as the collection

$$\mathrm{A}_{p,q}(\Sigma,\sigma) := \left\{f : \Sigma \to \mathbb{R} : f \text{ is } \sigma\text{-measurable and } \|f\|_{\mathrm{A}_{p,q}(\Sigma,\sigma)} < \infty\right\}, \tag{10.3}$$

as well as the space $\mathrm{B}_{p,q}(\Sigma,\sigma)$ by

$$\mathrm{B}_{p,q}(\Sigma,\sigma) := \left\{f : \Sigma \to \mathbb{R} : f \text{ is } \sigma\text{-measurable and } \|f\|_{\mathrm{B}_{p,q}(\Sigma,\sigma)} < \infty\right\}. \tag{10.4}$$

We wish to single out that Beurling space A^p (studied in Section 3) arises when $q = 1$, i.e., from (3.2) and (10.1) we see that

$$\mathrm{A}_{p,1}(\Sigma,\sigma) = \mathrm{A}^p(\Sigma,\sigma) \text{ for all } p \in (0,\infty), \tag{10.5}$$

while from (10.2) and (3.4) we conclude that

$$\mathrm{B}_{p,1}(\Sigma,\sigma) = \mathrm{B}^p(\Sigma,\sigma) \text{ for all } p \in (0,\infty). \tag{10.6}$$

As such, the Herz scale is a generalization of Beurling spaces. Furthermore, the family of Herz spaces is a natural extension of the Lebesgue scale, as it is visible from (10.1) that

$$\mathrm{A}_{p,p}(\Sigma,\sigma) = L^p(\Sigma,\sigma) \text{ for all } p \in (0,\infty). \tag{10.7}$$

Observe that the spaces $\mathrm{A}_{p,q}(\Sigma,\sigma)$ and $\mathrm{B}_{p,q}(\Sigma,\sigma)$ are lattices. Specifically,

> if f, g are σ-measurable functions on Σ such that $|f| \leq |g|$ at σ-a.e. point on Σ, and $g \in \mathrm{A}_{p,q}(\Sigma,\sigma)$ then $f \in \mathrm{A}_{p,q}(\Sigma,\sigma)$ and $\|f\|_{\mathrm{A}_{p,q}(\Sigma,\sigma)} \leq \|g\|_{\mathrm{A}_{p,q}(\Sigma,\sigma)}$, (10.8)

and similarly for $\mathrm{B}_{p,q}(\Sigma,\sigma)$.

A relevant observation about the nature of the quasinorm defined in (10.1) is that for any σ-measurable function f defined on Σ, for every $j \in \mathbb{N}_0$ and every $p, q \in (0,\infty)$, we have

$$\begin{aligned}
&\|f\cdot \mathbb{1}_{C_j}\|_{L^p(\Sigma,\sigma)} \\
&= 2_*^{-j(n-1)(1/q-1/p)} \left[\left(2_*^{j(n-1)(1/q-1/p)} \|f\cdot \mathbb{1}_{C_j}\|_{L^p(\Sigma,\sigma)}\right)^q\right]^{1/q} \\
&\leq 2_*^{-j(n-1)(1/q-1/p)} \left\{\sum_{k=0}^{\infty} \left(2_*^{k(n-1)(1/q-1/p)} \|f\cdot \mathbb{1}_{C_k}\|_{L^p(\Sigma,\sigma)}\right)^q\right\}^{1/q} \\
&= 2_*^{-j(n-1)(1/q-1/p)} \|f\|_{\mathrm{A}_{p,q}(\Sigma,\sigma)}.
\end{aligned} \tag{10.9}$$

Next, it is useful to record the manner in which $\|\cdot\|_{A_{p,q}}$ dilates with respect to p and q. Precisely, it follows from definitions (cf. (10.1)) that whenever f is a σ-measurable function on Σ, and $\theta \in (0,\infty)$, the following identity holds:

$$\|f\|^{1/\theta}_{A_{p/\theta,q/\theta}(\Sigma,\sigma)} = \||f|^{1/\theta}\|_{A_{p,q}(\Sigma,\sigma)} \quad \text{for all } p,q \in (0,\infty). \tag{10.10}$$

In particular, for $\theta := q$, this and (10.5) imply

$$\|f\|^{1/q}_{A^{p/q}(\Sigma,\sigma)} = \||f|^{1/q}\|_{A_{p,q}(\Sigma,\sigma)} \quad \text{for all } p,q \in (0,\infty). \tag{10.11}$$

We proceed by providing examples of functions that belong to Herz spaces of first generation.

Example 10.1 Consider a closed unbounded Ahlfors regular set $\Sigma \subseteq \mathbb{R}^n$, and abbreviate $\sigma := \mathcal{H}^{n-1}\lfloor\Sigma$. Fix a reference point $x_0 \in \Sigma$, a parameter $r \in \mathbb{R}$, and define the function $\varphi : \Sigma \to \mathbb{R}$ by

$$\varphi(x) := \frac{1}{1+|x-x_0|^r} \quad \text{for each } x \in \Sigma. \tag{10.12}$$

Then whenever $p,q \in (1,\infty)$ and $r > (n-1)/q$ one has $\varphi \in A_{p,q}(\Sigma,\sigma)$. To prove this, consider a dilation factor 2_* as in Definition 10.1, and bring in $\|\cdot\|_{A_{p,q}(\Sigma,\sigma)}$ defined in relation to x_0 and 2_*. By the Ahlfors regularity of Σ, we estimate

$$\begin{aligned}
\|\varphi\|^q_{A_{p,q}(\Sigma,\sigma)} &= \sum_{k=0}^{\infty} 2_*^{k(n-1)(1/q-1/p)q}\Big(\int_{C_k}\frac{d\sigma(x)}{(1+|x-x_0|^r)^p}\Big)^{q/p}\\
&\le \sigma(C_0)^{q/p} + \sum_{k=1}^{\infty} 2_*^{k(n-1)(1-q/p)}\Big(\int_{C_k}\frac{d\sigma(x)}{|x-x_0|^{rp}}\Big)^{q/p}\\
&\le \sigma(C_0)^{q/p} + \sum_{k=1}^{\infty} 2_*^{k(n-1)(1-q/p)}\frac{\sigma(C_k)^{q/p}}{2_*^{krq}}\\
&\le \sigma(C_0)^{q/p} + C\sum_{k=1}^{\infty} 2_*^{k(n-1)(1-q/p)}\frac{2_*^{k(n-1)q/p}}{2_*^{krq}}\\
&= \sigma(C_0)^{q/p} + C\sum_{k=1}^{\infty} 2_*^{k(n-1-rq)} < \infty,
\end{aligned} \tag{10.13}$$

since $n-1-rq<0$ and $2_*>1$. Hence, $\varphi \in A_{p,q}(\Sigma,\sigma)$. In particular, for r replaced by $n-1$ or n, the corresponding function φ in (10.12) belongs to $A_{p,q}(\Sigma,\sigma)$ for all $p,q \in (1,\infty)$.

Next we provide a norm characterization for the space $B_{p,q}$.

Proposition 10.1 *Consider a closed unbounded Ahlfors regular set* $\Sigma \subseteq \mathbb{R}^n$, *and abbreviate* $\sigma := \mathcal{H}^{n-1}\lfloor\Sigma$. *Fix a reference point* $x_0 \in \Sigma$, *and select a dilation factor*

2_* *as in Definition 10.1. Also, pick exponents* $p \in (1, \infty)$ *and* $q \in (0, 1]$. *Then for every* σ*-measurable function* f *defined on* Σ *one has*

$$\|f\|_{\mathrm{B}_{p,q}(\Sigma,\sigma)} \approx \sup_{R \geq 2_*} \Big\{ R^{(n-1)(1-1/q)} \Big(\fint_{\Delta(x_0,R)} |f|^p \, d\sigma \Big)^{1/p} \Big\}, \tag{10.14}$$

with implicit proportionality constants independent of f.

Proof To begin, fix a finite radius $R \geq 2_*$, and select $j \in \mathbb{N}$ such that

$$2_*^j \leq R < 2_*^{j+1}. \tag{10.15}$$

Based on this, (10.2), the Ahlfors regularity of Σ, and the fact that $0 < q \leq 1 < p$ (which implies $1/q + 1/p - 1 > 0$), we may bound

$$\begin{aligned}
& R^{(n-1)(1-1/q)} \Big(\fint_{\Delta(x_0,R)} |f|^p \, d\sigma \Big)^{1/p} \\
& \leq C R^{(n-1)(1-1/q)} \Big(\frac{1}{R^{n-1}} \int_{\Delta(x_0,R)} |f|^p \, d\sigma \Big)^{1/p} \\
& \leq C \frac{2_*^{j(n-1)(1-1/q)}}{2_*^{j(n-1)/p}} \sum_{k=0}^{j} \|f \cdot \mathbb{1}_{C_k}\|_{L^p(\Sigma,\sigma)} \\
& \leq C \sum_{k=0}^{j} 2_*^{(k-j)(n-1)(1/q+1/p-1)} \big(2_*^{-k(n-1)(1/q+1/p-1)} \|f \cdot \mathbb{1}_{C_k}\|_{L^p(\Sigma,\sigma)} \big) \\
& \leq C \Big(\sum_{k=0}^{j} 2_*^{(k-j)(n-1)(1/q+1/p-1)} \Big) \|f\|_{\mathrm{B}_{p,q}(\Sigma,\sigma)} \\
& \leq C \Big(\sum_{\ell=0}^{\infty} 2_*^{-\ell(n-1)(1/q+1/p-1)} \Big) \|f\|_{\mathrm{B}_{p,q}(\Sigma,\sigma)}
\end{aligned} \tag{10.16}$$

for some constant $C = C(n, p, q, \Sigma) \in (0, \infty)$. After taking the supremum over $R \geq 2_*$ in the right-hand side of (10.16), we arrive at

$$\sup_{R \geq 2_*} \Big\{ R^{(n-1)(1-1/q)} \Big(\fint_{\Delta(x_0,R)} |f|^p \, d\sigma \Big) \Big\} \leq C \|f\|_{\mathrm{B}_{p,q}(\Sigma,\sigma)}. \tag{10.17}$$

In the opposite direction, upon recalling (3.1) and using the Ahlfors regularity of Σ, for each $k \in \mathbb{N}_0$ we may estimate

$$
\begin{aligned}
&2_*^{-k(n-1)(1/q+1/p-1)}\|f\cdot\mathbb{1}_{C_k}\|_{L^p(\Sigma,\sigma)}\\
&\quad\le C2_*^{k(n-1)(1-1/q)}\sigma\big(\Delta(x_0,2_*^{k+1})\big)^{-1/p}\Big(\int_{\Delta(x_0,2_*^{k+1})}|f|^p\,d\sigma\Big)^{1/p}\\
&\quad= C2_*^{(k+1)(n-1)(1-1/q)}\Big(⨍_{\Delta(x_0,2_*^{k+1})}|f|^p\,d\sigma\Big)^{1/p}\\
&\quad\le C\cdot\sup_{R\ge 2_*}\Big\{R^{(n-1)(1-1/q)}\Big(⨍_{\Delta(x_0,R)}|f|^p\,d\sigma\Big)^{1/p}\Big\},
\end{aligned}
\tag{10.18}
$$

for some constant $C=C(n,p,q,\Sigma)\in(0,\infty)$. Taking the supremum over $k\in\mathbb{N}_0$ in the right-hand side of (10.18) then yields

$$
\|f\|_{\mathrm{B}_{p,q}(\Sigma,\sigma)}\le C\cdot\sup_{R\ge 2_*}\Big\{R^{(n-1)(1-1/q)}\Big(⨍_{\Delta(x_0,R)}|f|^p\,d\sigma\Big)^{1/p}\Big\}. \tag{10.19}
$$

All together, (10.17) and (10.19) imply (10.14). This finishes the proof of Proposition 10.1. □

The result recorded below features a Hölder-type inequality in the context of firs-generation Herz spaces.

Proposition 10.2 *Assume $\Sigma\subseteq\mathbb{R}^n$ is a closed unbounded Ahlfors regular set, and abbreviate $\sigma:=\mathcal{H}^{n-1}\lfloor\Sigma$. Fix two σ-measurable functions f and g defined on Σ. Then the following statements are true:*

(a) Given two exponents $p,q\in(1,\infty)$, and denoting by p',q' their respective Hölder conjugate exponents, one has

$$
\int_\Sigma|fg|\,d\sigma\le\|f\|_{\mathrm{A}_{p,q}(\Sigma,\sigma)}\|g\|_{\mathrm{A}_{p',q'}(\Sigma,\sigma)}. \tag{10.20}
$$

(b) For every two exponents $p\in(1,\infty)$ and $q\in(0,1]$ one has

$$
\int_\Sigma|fg|\,d\sigma\le\|f\|_{\mathrm{A}_{p,q}(\Sigma,\sigma)}\|g\|_{\mathrm{B}_{p',q}(\Sigma,\sigma)}, \tag{10.21}
$$

where p' denotes the Hölder conjugate exponent of p.

Proof To justify the claim made in *(a)*, use the ordinary version of Hölder's inequality to estimate

$$
\begin{aligned}
&\int_{\Sigma} |fg|\, d\sigma \\
&= \sum_{k=0}^{\infty} \int_{C_k} |fg|\, d\sigma \leq \sum_{k=0}^{\infty} \|f \cdot \mathbb{1}_{C_k}\|_{L^p(\Sigma,\sigma)} \|g \cdot \mathbb{1}_{C_k}\|_{L^{p'}(\Sigma,\sigma)} \\
&= \sum_{k=0}^{\infty} \big(2_*^{k(n-1)(1/q-1/p)} \|f \cdot \mathbb{1}_{C_k}\|_{L^p(\Sigma,\sigma)}\big) \big(2_*^{k(n-1)(1/q'-1/p')} \|g \cdot \mathbb{1}_{C_k}\|_{L^{p'}(\Sigma,\sigma)}\big) \\
&\leq \Big\{ \sum_{k=0}^{\infty} \big(2_*^{k(n-1)(1/q-1/p)} \|f \cdot \mathbb{1}_{C_k}\|_{L^p(\Sigma,\sigma)}\big)^q \Big\}^{1/q} \\
&\quad \times \Big\{ \sum_{k=0}^{\infty} \big(2_*^{k(n-1)(1/q'-1/p')} \|g \cdot \mathbb{1}_{C_k}\|_{L^{p'}(\Sigma,\sigma)}\big)^{q'} \Big\}^{1/q'} \\
&= \|f\|_{\mathrm{A}_{p,q}(\Sigma,\sigma)} \|g\|_{\mathrm{A}_{p',q'}(\Sigma,\sigma)}.
\end{aligned} \tag{10.22}
$$

As regards item *(b)*, we once again use the standard version of Hölder's inequality along with the fact that $0 < q \leq 1$ to obtain

$$
\begin{aligned}
\int_{\Sigma} |fg|\, d\sigma &= \sum_{k=0}^{\infty} \int_{C_k} |fg|\, d\sigma \leq \sum_{k=0}^{\infty} \|f \cdot \mathbb{1}_{C_k}\|_{L^p(\Sigma,\sigma)} \|g \cdot \mathbb{1}_{C_k}\|_{L^{p'}(\Sigma,\sigma)} \\
&\leq \sum_{k=0}^{\infty} \big(2_*^{k(n-1)(1/q-1/p)} \|f \cdot \mathbb{1}_{C_k}\|_{L^p(\Sigma,\sigma)}\big) \\
&\quad \times \big(2_*^{-k(n-1)(1/q+1/p'-1)} \|g \cdot \mathbb{1}_{C_k}\|_{L^{p'}(\Sigma,\sigma)}\big) \\
&\leq \Big\{ \sum_{k=0}^{\infty} \big(2_*^{k(n-1)(1/q-1/p)} \|f \cdot \mathbb{1}_{C_k}\|_{L^p(\Sigma,\sigma)}\big)^q \Big\}^{1/q} \\
&\qquad \times \sup_{k \in \mathbb{N}_0} \big(2_*^{-k(n-1)(1/q+1/p'-1)} \|g \cdot \mathbb{1}_{C_k}\|_{L^{p'}(\Sigma,\sigma)}\big) \\
&= \|f\|_{\mathrm{A}_{p,q}(\Sigma,\sigma)} \|g\|_{\mathrm{B}_{p',q}(\Sigma,\sigma)}.
\end{aligned} \tag{10.23}
$$

This ultimately concludes the proof of Proposition 10.2. □

The following proposition contains a number of useful embedding results involving first-generation Herz spaces.

Proposition 10.3 *Assume $\Sigma \subseteq \mathbb{R}^n$ is a closed unbounded Ahlfors regular set, and abbreviate $\sigma := \mathcal{H}^{n-1}\lfloor\Sigma$. Also, fix two exponents $p, q \in (0, \infty)$. Then the following statements are true:*

(a) If $p_1, p_2 \in (0, \infty)$, and $p_1 \leq p_2$, one has the continuous embedding

$$\mathrm{A}_{p_2,q}(\Sigma,\sigma) \hookrightarrow \mathrm{A}_{p_1,q}(\Sigma,\sigma), \tag{10.24}$$

i.e., the scale of Herz spaces $\mathrm{A}_{p,q}(\Sigma,\sigma)$ *is monotonically decreasing in the exponent* p.

(b) If $q_1, q_2 \in (0,\infty)$, *and* $q_1 \leq q_2$, *then the following embedding is continuous*

$$\mathrm{A}_{p,q_1}(\Sigma,\sigma) \hookrightarrow \mathrm{A}_{p,q_2}(\Sigma,\sigma), \tag{10.25}$$

i.e., the scale of Herz spaces $\mathrm{A}_{p,q}(\Sigma,\sigma)$ *is monotonically increasing in the exponent* q. *In particular, when* $0 < q \leq 1$, *it follows that*

$$\mathrm{A}_{p,q}(\Sigma,\sigma) \hookrightarrow \mathrm{A}_{p,1}(\Sigma,\sigma) = \mathrm{A}^p(\Sigma,\sigma) \quad \textit{continuously.} \tag{10.26}$$

(c) Whenever $0 < q \leq p < \infty$, *it follows that*

$$\mathrm{A}_{p,q}(\Sigma,\sigma) \hookrightarrow L^q(\Sigma,\sigma) \quad \textit{continuously,} \tag{10.27}$$

and

$$\mathrm{A}_{p,q}(\Sigma,\sigma) \hookrightarrow L^p(\Sigma,\sigma) \quad \textit{continuously.} \tag{10.28}$$

In particular, if $r \in (0,\infty)$ *is such that* $q \leq r \leq p$, *then*

$$\mathrm{A}_{p,q}(\Sigma,\sigma) \hookrightarrow L^r(\Sigma,\sigma) \quad \textit{continuously.} \tag{10.29}$$

(d) Whenever $0 < p \leq q < \infty$, *one has*

$$L^q(\Sigma,\sigma) \hookrightarrow \mathrm{A}_{p,q}(\Sigma,\sigma) \quad \textit{continuously,} \tag{10.30}$$

and

$$L^p(\Sigma,\sigma) \hookrightarrow \mathrm{A}_{p,q}(\Sigma,\sigma) \quad \textit{continuously.} \tag{10.31}$$

(e) If $p \in [1,\infty)$, *one has the continuous embedding*

$$\mathrm{A}_{p,q}(\Sigma,\sigma) \hookrightarrow L^1\Big(\Sigma, \frac{\sigma(x)}{1+|x|^{n-1}}\Big). \tag{10.32}$$

(f) The following sequence of inclusions hold

$$L^\infty_{\mathrm{comp}}(\Sigma,\sigma) \subseteq \mathrm{A}_{p,q}(\Sigma,\sigma) \subseteq L^p_{\mathrm{loc}}(\Sigma,\sigma). \tag{10.33}$$

As a consequence, one may also conclude that

$$\bigcup_{\substack{1<p<\infty\\ 0<q<\infty}} \mathrm{A}_{p,q}(\Sigma,\sigma) \subseteq L^1_{\mathrm{loc}}(\Sigma,\sigma). \tag{10.34}$$

(g) Assume $p, q \in (1, \infty)$ and select $\tau \in \mathbb{R}$ such that

$$\frac{-(n-1)}{p} < \tau < (n-1)\Big(\frac{1}{q} - \frac{1}{p}\Big) \tag{10.35}$$

Then, having fixed a reference point $x_0 \in \Sigma$, it follows that the function

$$|\cdot - x_0|^{\tau p} \quad \text{belongs to the Muckenhoupt class } A_p(\Sigma, \sigma) \tag{10.36}$$

and the following embedding is well defined and continuous

$$\mathrm{A}_{p,q}(\Sigma, \sigma) \hookrightarrow L^p(\Sigma, \sigma) + L^p\big(\Sigma, |x - x_0|^{\tau p}\sigma(x)\big). \tag{10.37}$$

Proof Throughout, let $x_0 \in \Sigma$ be a reference point and fix a dilation factor 2_* as in Definition 10.1. Also, bring in the family of sets $\{C_k\}_{k\in\mathbb{N}_0}$ as in (3.1). First, consider the claim made in *(a)*, and pick some function $f \in \mathrm{A}_{p_2,q}(\Sigma, \sigma)$. Then the same type of computations as those made in (3.33) (with $p := p_1$ and $q := p_2$) presently give that for each $k \in \mathbb{N}_0$ we have

$$\begin{aligned} &\|f \cdot \mathbb{1}_{C_k}\|_{L^{p_1}(\Sigma,\sigma)} \leq C 2_*^{k(n-1)(1/p_1 - 1/p_2)} \|f \cdot \mathbb{1}_{C_k}\|_{L^{p_2}(\Sigma,\sigma)} \\ &\Rightarrow 2_*^{k(n-1)(1/q - 1/p_1)} \|f \cdot \mathbb{1}_{C_k}\|_{L^{p_1}(\Sigma,\sigma)} \leq C 2_*^{k(n-1)(1/q - 1/p_2)} \|f \cdot \mathbb{1}_{C_k}\|_{L^{p_2}(\Sigma,\sigma)} \end{aligned} \tag{10.38}$$

for some constant $C = C(n, p_1, p_2, \Sigma) \in (0, \infty)$. In turn, this and (10.1) readily imply

$$\|f\|_{\mathrm{A}_{p_1,q}(\Sigma,\sigma)} \leq C\|f\|_{\mathrm{A}_{p_2,q}(\Sigma,\sigma)} < \infty, \tag{10.39}$$

which ultimately proves (10.24).

In relation to the claim made in *(b)*, pick a function $f \in \mathrm{A}_{p,q_1}(\Sigma, \sigma)$. Since $q_1 \leq q_2$, we have

$$\begin{aligned} \|f\|_{\mathrm{A}_{p,q_2}(\Sigma,\sigma)} &= \Big\{\sum_{k=0}^{\infty} \big(2_*^{k(n-1)(1/q_2 - 1/p)} \|f \cdot \mathbb{1}_{C_k}\|_{L^p(\Sigma,\sigma)}\big)^{q_2}\Big\}^{(q_1/q_2)q_1} \\ &\leq \Big\{\sum_{k=0}^{\infty} \big(2_*^{k(n-1)(1/q_1 - 1/p)} \|f \cdot \mathbb{1}_{C_k}\|_{L^p(\Sigma,\sigma)}\big)^{q_1}\Big\}^{1/q_1} \\ &= \|f\|_{\mathrm{A}_{p,q_1}(\Sigma,\sigma)} < \infty. \end{aligned} \tag{10.40}$$

This justifies (10.25). The embedding in (10.26) is a direct consequence of (10.25) (for $q_1 := q$ and $q_2 := 1$), and (10.5).

Next, assume $q \leq p$. Bearing in mind (10.7), the embedding in (10.27) is a consequence of (10.24) used with $p_1 := q$ and $p_2 := p$, while (10.28) follows from (10.25) with $q_1 := q$ and $q_2 := p$. Having established (10.27)-(10.28), a interpolation argument readily implies (10.29). This takes care of the claim made in item *(c)*.

Regarding the embeddings claimed in item *(d)*, suppose $p \leq q$. Thanks to (10.7), use (10.24) with $p_1 := p$ and $p_2 := q$ to obtain (10.30), while (10.31) follows from (10.25) with $q_1 := p$ and $q_2 := q$.

Moving on, consider the claim made in item *(e)*. Make the assumption that $1 \leq p < \infty$. In the case when $q \leq 1$, the embedding in (10.32) follows from (10.26) and (3.31) in Proposition 3.3. Assume next that $1 < q < \infty$. Since $p \geq 1$, by the monotonicity property from item *(a)*, it suffices to show (10.32) for $p = 1$. To this end, select an arbitrary function $f \in \mathrm{A}_{1,q}(\Sigma, \sigma)$. Using Hölder's inequality, along with the fact that $1 + |x|^{n-1} \approx 1 + |x - x_0|^{n-1}$ uniformly for $x \in \Sigma$, we estimate

$$
\begin{aligned}
\int_\Sigma |f(x)| \frac{d\sigma(x)}{1 + |x|^{n-1}} &\leq C \int_\Sigma |f(x)| \frac{d\sigma(x)}{1 + |x - x_0|^{n-1}} \\
&= C \sum_{k=0}^{\infty} \int_{C_k} |f(x)| \frac{d\sigma(x)}{1 + |x - x_0|^{n-1}} \\
&\leq C \|f \cdot \mathbb{1}_{C_0}\|_{L^1(\Sigma,\sigma)} + \sum_{k=1}^{\infty} \int_{C_k} |f(x)| \frac{d\sigma(x)}{|x - x_0|^{n-1}} \\
&\leq C \|f \cdot \mathbb{1}_{C_0}\|_{L^1(\Sigma,\sigma)} + C \sum_{k=1}^{\infty} 2_*^{-k(n-1)} \|f \cdot \mathbb{1}_{C_k}\|_{L^1(\Sigma,\sigma)} \\
&= C \sum_{k=0}^{\infty} \big(2_*^{k(n-1)(1/q-1)} \|f \cdot \mathbb{1}_{C_k}\|_{L^1(\Sigma,\sigma)}\big) 2_*^{-k(n-1)/q} \\
&\leq \Big\{ \sum_{k=0}^{\infty} \big(2_*^{k(n-1)(1/q-1)} \|f \cdot \mathbb{1}_{C_k}\|_{L^1(\Sigma,\sigma)}\big)^q \Big\}^{1/q} \times \Big\{ \sum_{k=0}^{\infty} 2_*^{-k(n-1)q'/q} \Big\}^{1/q'} \\
&= C \|f\|_{\mathrm{A}_{1,q}(\Sigma,\sigma)}.
\end{aligned}
\tag{10.41}
$$

This justifies the claim made in item *(e)*.

The inclusions in (10.33) are clear (10.1) and from the fact that $\{C_k\}_{k \in \mathbb{N}_0}$ is a disjoint family of sets covering Σ. Furthermore, (10.34) is a direct consequence of the second inclusion in (10.33) and (10.24). The claim in item *(f)* is therefore established.

Finally, consider the claim made in item *(g)*. Observe that (10.35) ensures that τ satisfies

$$
-(n-1) < \tau p < (n-1)\Big(\frac{p}{q} - 1\Big) < (n-1)(p-1). \tag{10.42}
$$

As such, (2.70) readily implies (10.36). Next, fix a function $f \in \mathrm{A}_{p,q}(\Sigma, \sigma)$, and decompose

$$
f = f \cdot \mathbb{1}_{\Delta(x_0, 2_*)} + f \cdot \mathbb{1}_{\Sigma \setminus \Delta(x_0, 2_*)} =: f_1 + f_2. \tag{10.43}
$$

From (10.1) and item *(f)* it follows that $f_1 \in L^p(\Sigma, \sigma)$ and

$$\|f_1\|_{L^p(\Sigma,\sigma)} = \|f \cdot \mathbb{1}_{C_0}\|_{L^p(\Sigma,\sigma)} \leq \|f\|_{\mathrm{A}_{p,q}(\Sigma,\sigma)}. \tag{10.44}$$

Now, using Hölder's inequality and the fact that $\tau - (n-1)(1/q - 1/p) < 0$ we find that

$$\begin{aligned}
&\|f_2\|_{L^p(\Sigma,|x-x_0|^{\tau p}\sigma(x))} \\
&= \left(\int_{\Sigma \setminus \Delta(x_0,2_*)} |f(x)|^p |x-x_0|^{\tau p}\, d\sigma(x) \right)^{1/p} \\
&\leq \sum_{k=1}^{\infty} \left(\int_{\Delta(x_0,2_*^{k+1}) \setminus \Delta(x_0,2_*^k)} |f(x)|^p |x-x_0|^{\tau p}\, d\sigma(x) \right)^{1/p} \\
&\leq \sum_{k=1}^{\infty} 2_*^{k\tau} \left(\int_{C_k} |f(x)|^p\, d\sigma(x) \right)^{1/p} \\
&= \sum_{k=1}^{\infty} 2_*^{k[\tau-(n-1)(1/q-1/p)]} \big(2_*^{k(n-1)(1/q-1/p)} \|f \cdot \mathbb{1}_{C_k}\|_{L^p(\Sigma,\sigma)} \big) \\
&\leq \Big\{ \sum_{k=0}^{\infty} 2_*^{k[\tau-(n-1)(1/q-1/p)]q'} \Big\}^{1/q'} \\
&\quad \times \Big\{ \sum_{k=0}^{\infty} \big(2_*^{k(n-1)(1/q-1/p)} \|f \cdot \mathbb{1}_{C_k}\|_{L^p(\Sigma,\sigma)} \big)^q \Big\}^{1/q} \\
&= C\|f\|_{\mathrm{A}_{p,q}(\Sigma,\sigma)},
\end{aligned} \tag{10.45}$$

for some constant $C = C(n, p, q, \tau) \in (0, \infty)$. This, (10.44), and (10.43) show that f belongs to $L^p(\Sigma, \sigma) + L^p\big(\Sigma, |x - x_0|^{\tau p}\sigma(x)\big)$ and

$$\begin{aligned}
&\|f\|_{L^p(\Sigma,\sigma)+L^p(\Sigma,|x-x_0|^{\tau p}\sigma(x))} \\
&= \inf_{\substack{f=g+h \\ g\in L^p(\Sigma,\sigma) \\ h\in L^p(\Sigma,|x-x_0|^{\tau p}\sigma(x))}} \big\{ \|g\|_{L^p(\Sigma,\sigma)} + \|h\|_{L^p(\Sigma,|x-x_0|^{\tau p}\sigma(x))} \big\} \\
&\leq \|f_1\|_{L^p(\Sigma,\sigma)} + \|f_2\|_{L^p(\Sigma,|x-x_0|^{\tau p}\sigma(x))} \\
&\leq C\|f\|_{\mathrm{A}_{p,q}(\Sigma,\sigma)}.
\end{aligned} \tag{10.46}$$

This concludes the proof of the claims made in item *(g)*. As such, the proof of Proposition 10.3 is now complete. □

It is useful to have a version of Lebesgue's Dominated Convergence Theorem in the setting of Herz spaces. Such a result reads as follows.

Proposition 10.4 *Consider a closed unbounded Ahlfors regular set $\Sigma \subseteq \mathbb{R}^n$, and abbreviate $\sigma := \mathcal{H}^{n-1}\lfloor\Sigma$. Let $\{f_j\}_{j\in\mathbb{N}}$ be a sequence of σ-measurable functions on Σ such that for some σ-measurable function $f : \Sigma \to \mathbb{R}$ one has*

$$f_j \to f \ \text{ at } \sigma\text{-a.e. point on } \Sigma \text{ as } j \to \infty. \tag{10.47}$$

Having fixed two exponents $p, q \in (0, \infty)$, assume there exists $g \in \mathrm{A}_{p,q}(\Sigma, \sigma)$ for which

$$|f_j(x)| \leq g(x) \ \text{ at } \sigma\text{-a.e. } x \in \Sigma \text{ and for all } j \in \mathbb{N}. \tag{10.48}$$

Then

$$\|f_j - f\|_{\mathrm{A}_{p,q}(\Sigma,\sigma)} \to 0 \ \text{ as } \ j \to \infty. \tag{10.49}$$

Proof To begin, bring in the family of sets $\{C_k\}_{k\in\mathbb{N}_0}$ as in (3.1), and notice that the membership $g \in \mathrm{A}_{p,q}(\Sigma, \sigma)$ implies

$$g \cdot \mathbb{1}_{C_k} \in L^p(\Sigma, \sigma) \quad \text{for every } k \in \mathbb{N}_0. \tag{10.50}$$

Hence, from this, (10.47), (10.48), and the classical version of Lebesgue's Dominated Convergence Theorem we have

$$\|(f_j - f)\mathbb{1}_{C_k}\|_{L^p(\Sigma,\sigma)} \to 0 \ \text{ as } j \to \infty, \text{ for all } k \in \mathbb{N}_0. \tag{10.51}$$

In turn, this permits us to conclude that, having fixed $\varepsilon \in (0, \infty)$, for each $k \in \mathbb{N}_0$ there exists $J_k \in \mathbb{N}$ such that

$$\|(f_j - f) \cdot \mathbb{1}_{C_k}\|_{L^p(\Sigma,\sigma)} < \frac{\varepsilon^{1/q}}{2_*^{k(n-1)(1/q-1/p)}2^{(k+1)/q}}, \quad \text{for all} \quad j \geq J_k. \tag{10.52}$$

Next, as $g \in \mathrm{A}_{p,q}(\Sigma, \sigma)$, there exists some $N \in \mathbb{N}$ satisfying

$$\sum_{k=N+1}^{\infty} \big(2_*^{k(n-1)(1/q-1/p)}\|g\|_{L^p(C_k,\sigma)}\big)^q < \varepsilon. \tag{10.53}$$

Collectively, since $|f| \leq g$ pointwise σ-a.e. on Σ (as implied by (10.47)-(10.48)), (10.52), and (10.53), ensure that whenever $j \in \mathbb{N}$ with $j \geq \max\{J_0, \ldots, J_N\}$ we have

$$
\begin{aligned}
&\|f_j - f\|^q_{\mathrm{A}_{p,q}(\Sigma,\sigma)} \\
&\quad = \sum_{k=0}^{N} \big(2_*^{k(n-1)(1/q-1/p)} \|(f_j - f)\cdot \mathbb{1}_{C_k}\|_{L^p(C_k,\sigma)}\big)^q \\
&\qquad + \sum_{k=N+1}^{\infty} \big(2_*^{k(n-1)(1/q-1/p)} \|(f_j - f)\cdot \mathbb{1}_{C_k}\|_{L^p(C_k,\sigma)}\big)^q \\
&\quad \le \sum_{k=0}^{N} \frac{\varepsilon}{2^{k+1}} + 2^q \sum_{k=N+1}^{\infty} \big[2_*^{k(n-1)(1/q-1/p)} (\|f_j\|_{L^p(\Sigma,\sigma)} + \|f\|_{L^p(\Sigma,\sigma)})\big]^q \\
&\quad \le \sum_{k=0}^{\infty} \frac{\varepsilon}{2^{k+1}} + 2^q \sum_{k=N+1}^{\infty} \big(2_*^{k(n-1)(1/q-1/p)} \|g\|_{L^p(C_k,\sigma)}\big)^q \\
&\quad < (1+2^q)\varepsilon.
\end{aligned}
\tag{10.54}
$$

Since $\varepsilon \in (0,\infty)$ is arbitrary, this ultimately justifies the convergence in (10.49), thus concluding the proof of Proposition 10.4. □

We now present a version of Lebesgue's Monotone Convergence Theorem in the context of first-generation Herz spaces.

Proposition 10.5 *Suppose $\Sigma \subseteq \mathbb{R}^n$ is a closed unbounded Ahlfors regular set, and abbreviate $\sigma := \mathcal{H}^{n-1}\lfloor\Sigma$. Assume $\{f_j\}_{j\in\mathbb{N}}$ is a sequence of non-negative σ-measurable functions on Σ for which $f_j \nearrow f$ pointwise σ-a.e. on Σ as $j \to \infty$, for some non-negative σ-measurable function f on Σ. Then whenever $p, q \in (0,\infty)$ it follows that*

$$
\|f_j\|_{\mathrm{A}_{p,q}(\Sigma,\sigma)} \nearrow \|f\|_{\mathrm{A}_{p,q}(\Sigma,\sigma)} \quad \text{as } j \to \infty. \tag{10.55}
$$

Proof Since $\{f_j\}_{j\in\mathbb{N}}$ increases to f pointwise σ-a.e. on Σ, and $\mathrm{A}_{p,q}(\Sigma,\sigma)$ is a lattice (cf. (10.8)) we have

$$
\|f_j\|_{\mathrm{A}_{p,q}(\Sigma,\sigma)} \le \|f_{j+1}\|_{\mathrm{A}_{p,q}(\Sigma,\sigma)} \le \|f\|_{\mathrm{A}_{p,q}(\Sigma,\sigma)}, \quad \text{for all } j \in \mathbb{N}. \tag{10.56}
$$

This further gives

$$
\limsup_{j\to\infty} \|f_j\|_{\mathrm{A}_{p,q}(\Sigma,\sigma)} \le \|f\|_{\mathrm{A}_{p,q}(\Sigma,\sigma)}. \tag{10.57}
$$

Now, having fixed some $N \in \mathbb{N}$, use the standard version of Lebesgue's Monotone Convergence Theorem to write

$$\begin{aligned}
&\Big\{\sum_{k=0}^{N}\big(2_*^{k(n-1)(1/q-1/p)}\|f\cdot\mathbb{1}_{C_k}\|_{L^p(\Sigma,\sigma)}\big)^q\Big\}^{1/q}\\
&\quad\le\liminf_{j\to\infty}\Big\{\sum_{k=0}^{N}\big(2_*^{k(n-1)(1/q-1/p)}\|f_j\cdot\mathbb{1}_{C_k}\|_{L^p(\Sigma,\sigma)}\big)^q\Big\}^{1/q}\\
&\quad\le\liminf_{j\to\infty}\|f_j\|_{\mathrm{A}_{p,q}(\Sigma,\sigma)}. \qquad (10.58)
\end{aligned}$$

Letting $N\to\infty$ in the right-hand side of (10.58) yields

$$\|f\|_{\mathrm{A}_{p,q}(\Sigma,\sigma)}\le\liminf_{j\to\infty}\|f_j\|_{\mathrm{A}_{p,q}(\Sigma,\sigma)}. \qquad (10.59)$$

Together, (10.57) and (10.59) imply (10.55). This finishes the proof of Theorem 10.5. □

Our next result contains the completeness of first-generation Herz spaces.

Proposition 10.6 *Consider a closed unbounded Ahlfors regular set $\Sigma\subseteq\mathbb{R}^n$, and abbreviate $\sigma:=\mathcal{H}^{n-1}\lfloor\Sigma$. Then for every $p,q\in(0,\infty)$ it follows that the Herz space $\mathrm{A}_{p,q}(\Sigma,\sigma)$ is complete when equipped with $\|\cdot\|_{\mathrm{A}_{p,q}(\Sigma,\sigma)}$. Consequently, $\mathrm{A}_{p,q}(\Sigma,\sigma)$ is a Banach space when $p,q\in(1,\infty)$, and a Quasi-Banach space when $q\in(0,1]$ and $p\in(0,\infty)$.*

Proof This follows via the same type of argument as in the proof of Proposition 3.9, naturally adapted to the quasinorm of $\mathrm{A}_{p,q}(\Sigma,\sigma)$. □

In the proposition below, we identify certain dense subspaces of $\mathrm{A}_{p,q}$.

Proposition 10.7 *Consider $\Sigma\subseteq\mathbb{R}^n$ a closed unbounded Ahlfors regular set, and abbreviate $\sigma:=\mathcal{H}^{n-1}\lfloor\Sigma$. Also, select two exponents $p\in(1,\infty)$ and $q\in(0,\infty)$. Then the following statements are true:*

(a) The space $\mathrm{Lip}_c(\Sigma)$, consisting of Lipschitz functions with compact support on Σ, is a dense subspace of $\mathrm{A}_{p,q}(\Sigma,\sigma)$.

(b) The collection $\big\{\phi|_\Sigma:\phi\in\mathscr{C}_0^\infty(\mathbb{R}^n)\big\}$ is a dense subspace of $\mathrm{A}_{p,q}(\Sigma,\sigma)$.

Proof We first prove the claim made in item *(a)*. Fix a reference point $x_0\in\Sigma$ and a dilation factor 2_* as in Definition 10.1. Once more, bring in the family of sets $\{C_k\}_{k\in\mathbb{N}_0}$ as in (3.1). Having fixed a function $f\in\mathrm{A}_{p,q}(\Sigma,\sigma)$, the observation made in (10.9) yields $f\big|_{C_k}\in L^p(C_k,\sigma)$, for every $k\in\mathbb{N}_0$. Since for each $k\in\mathbb{N}_0$ the space $\mathrm{Lip}_c(C_k)$ is dense in $L^p(C_k,\sigma)$ (cf. [GHA-I, Proposition 3.7.1, p. 279]) there exists a sequence $\big\{\varphi_j^{(k)}\big\}_{j\in\mathbb{N}}$ in $\mathrm{Lip}_c(C_k)$ with the property that

$$\varphi_j^{(k)}\to f\big|_{C_k}\ \text{ in }\ L^p(C_k,\sigma),\ \text{ as }\ j\to\infty. \qquad (10.60)$$

Fix $\varepsilon > 0$. Then given $k \in \mathbb{N}_0$, there exists $J_k \in \mathbb{N}$ satisfying

$$\|\varphi^{(k)}_{J_k} - f|_{C_k}\|_{L^p(C_k,\sigma)} < \frac{\varepsilon^{1/q}}{2_*^{k(n-1)(1/q-1/p)}2^{(k+1)/q}}. \tag{10.61}$$

To proceed denote by $\widetilde{\varphi}^{(k)}_{J_k}$ the extension by zero of $\varphi^{(k)}_{J_k}$ to the entire Σ, and notice that $\widetilde{\varphi}^{(k)}_{J_k}$ is a Lipschitz function supported in C_k. For each $\ell \in \mathbb{N}_0$ define the function $\psi_\ell : \Sigma \to \mathbb{R}$ by setting

$$\psi_\ell := \sum_{k=0}^{\ell} \widetilde{\varphi}^{(k)}_{J_k}. \tag{10.62}$$

By design, each ψ_ℓ is supported in $\Delta(x_0, 2_*^{\ell+1})$, and being a finite sum of Lipschitz functions on Σ, we therefore have $\{\psi_\ell\}_{\ell\in\mathbb{N}} \subseteq \mathrm{Lip}_c(\Sigma)$. In addition, observe that for all $k, \ell \in \mathbb{N}_0$ the function ψ_ℓ satisfies

$$\psi_\ell\big|_{C_k} = \begin{cases} \varphi^{(k)}_{J_k} & \text{if } 0 \le k \le \ell, \\ 0 & \text{if } k \ge \ell+1. \end{cases} \tag{10.63}$$

Now, bearing in mind (10.1), and that f belongs to $\mathrm{A}_{p,q}(\Sigma,\sigma)$, we can guarantee the existence of some number $N \in \mathbb{N}$ satisfying

$$\sum_{k=\ell+1}^{\infty} \big(2_*^{k(n-1)(1/q-1/p)}\|f\|_{L^p(C_k,\sigma)}\big)^q < \varepsilon \quad \text{for all } \ell \ge N. \tag{10.64}$$

Hence, whenever $\ell \in \mathbb{N}_0$, and $\ell \ge N$, (10.61), (10.63), and (10.64) imply

$$\begin{aligned}
\|\psi_\ell - f\|^q_{\mathrm{A}_{p,q}(\Sigma,\sigma)} &= \sum_{k=0}^{\ell} \big(2_*^{k(n-1)(1/q-1/p)}\|\psi_\ell - f\|_{L^p(C_k,\sigma)}\big)^q \\
&\quad + \sum_{k=\ell+1}^{\infty} \big(2_*^{k(n-1)(1/q-1/p)}\|\psi_\ell - f\|_{L^p(C_k,\sigma)}\big)^q \\
&= \sum_{k=0}^{\ell} \big(2_*^{k(n-1)(1/q-1/p)}\|\varphi^{(k)}_{J_k} - f|_{C_k}\|_{L^p(C_k,\sigma)}\big)^q \\
&\quad + \sum_{k=\ell+1}^{\infty} \big(2_*^{k(n-1)(1/q-1/p)}\|f\|_{L^p(C_k,\sigma)}\big)^q \\
&< \sum_{k=0}^{\ell} \frac{\varepsilon}{2^{k+1}} + \varepsilon \le 2\varepsilon.
\end{aligned} \tag{10.65}$$

Since $\varepsilon > 0$ was taken arbitrary to begin with, (10.65) shows that $\psi_\ell \to f$ in $\mathrm{A}_{p,q}(\Sigma, \sigma)$ as $\ell \to \infty$. This establishes the density of $\mathrm{Lip}_c(\Sigma)$ on $\mathrm{A}_{p,q}(\Sigma, \sigma)$, as claimed in item *(a)*.

Regarding the claim made in item *(b)*, pick an arbitrary function $f \in \mathrm{A}_{p,q}(\Sigma, \sigma)$. Thanks to *(a)* and [GHA-I, (3.7.3), p. 279], given $\varepsilon > 0$ we can find $\Psi \in \mathrm{Lip}_c(\mathbb{R}^n)$ for which

$$\|f - \Psi|_\Sigma\|_{\mathrm{A}_{p,q}(\Sigma,\sigma)} < \varepsilon. \tag{10.66}$$

Via a mollifier argument, construct a sequence $\{\phi_t\}_{1<t<1}$ of functions in $\mathscr{C}_0^\infty(\mathbb{R}^n)$ satisfying

$$\begin{aligned} &\operatorname{supp}\phi_t \subseteq K \ \text{ for some compact set } K \subseteq \mathbb{R}^n, \text{ for all } t \in (0,1), \\ &\quad \text{and such that } \sup_{x \in \mathbb{R}^n} |\phi_t(x) - \Psi(x)| \to 0 \ \text{ as } t \to 0^+. \end{aligned} \tag{10.67}$$

It is therefore possible to select $t_0 \in (0, 1)$ small enough so that

$$\sup_{\mathbb{R}^n} |\phi_{t_0} - \Psi| < \varepsilon. \tag{10.68}$$

Hence, $\phi_{t_0}\big|_\Sigma \in \mathrm{Lip}_c(\Sigma)$ and thanks to (10.66) and (10.68) we may estimate

$$\begin{aligned} \|f - \phi_{t_0}|_\Sigma\|_{\mathrm{A}_{p,q}(\Sigma,\sigma)} &\leq \|f - \Psi|_\Sigma\|_{\mathrm{A}_{p,q}(\Sigma,\sigma)} + \|\phi_{t_0}|_\Sigma - \Psi|_\Sigma\|_{\mathrm{A}_{p,q}(\Sigma,\sigma)} \\ &< \varepsilon + \big(\sup_{\mathbb{R}^n} |\phi_{t_0} - \Psi|\big)\|\mathbb{1}_{K\cap\Sigma}\|_{\mathrm{A}_{p,q}(\Sigma,\sigma)} \\ &< C \cdot \varepsilon \end{aligned} \tag{10.69}$$

for some constant $C = C(\Psi, K, n, p, q, \Sigma) \in (0, \infty)$. This establishes the claim made in item *(b)*, and ultimately finishes the proof of Proposition 10.7. □

The following proposition provides the separability of Herz spaces.

Proposition 10.8 *Consider $\Sigma \subseteq R^n$ a closed unbounded Ahlfors regular set, and abbreviate $\sigma := \mathcal{H}^{n-1}\lfloor\Sigma$. Then for any given exponents $p \in (1, \infty)$ and $q \in (0, \infty)$ the space $\mathrm{A}_{p,q}(\Sigma, \sigma)$ is separable.*

Proof Fix $p \in (1, \infty)$ and $q \in (0, \infty)$. Recall the family $\{C_k\}_{k\in\mathbb{N}_0}$ defined in (3.1). Since for every $k \in \mathbb{N}_0$ the space $L^p(C_k, \sigma)$ is separable (see [GHA-I, Lemma 3.6.4, p. 278]), there exists a countable dense subset $\mathscr{F}_k$ of $L^p(C_k, \sigma)$. For each $k \in \mathbb{N}_0$, denote by $\widetilde{\mathscr{F}_k}$ the collection of all extensions by zero (to the entire Σ) of functions from $\mathscr{F}_k$. In particular $\widetilde{\mathscr{F}_k}$ is also a countable subset $\mathrm{A}_{p,q}(\Sigma, \sigma)$. Hence, letting

$$\widetilde{\mathscr{F}} := \text{the collection of all finite sums of functions from } \cup_{k\in\mathbb{N}_0}\widetilde{\mathscr{F}_k}, \tag{10.70}$$

it follows that $\widetilde{\mathscr{F}}$ is a countable subset of $\mathrm{A}_{p,q}(\Sigma, \sigma)$.

We claim that

$$\widetilde{\mathscr{F}} \text{ is dense in } \mathrm{A}_{p,q}(\Sigma,\sigma). \tag{10.71}$$

To justify this, pick an arbitrary function $f \in \mathrm{A}_{p,q}(\Sigma,\sigma)$, and fix $\varepsilon \in (0,\infty)$. Then, whenever $k \in \mathbb{N}_0$, the function $f\big|_{C_k}$ belongs to $L^p(C_k,\sigma)$ (as a consequence of (10.9)). As such, there exists some $\varphi_k \in \mathscr{F}_k$ for which

$$\big\|f\big|_{C_k} - \varphi_k\big\|_{L^p(C_k,\sigma)} < \frac{\varepsilon}{2^k 2_*^{k(n-1)(1/q-1/p)}}. \tag{10.72}$$

Moreover, if $\widetilde{\varphi}_k$ is the extension of φ_k by zero to Σ, then $\widetilde{\varphi}_k \in \widetilde{\mathscr{F}}_k$. In light of (10.70), we see that

$$\Phi_N := \sum_{k=0}^{N} \widetilde{\varphi}_k \quad \text{belongs to } \widetilde{\mathscr{F}} \quad \text{for all } N \in \mathbb{N}_0. \tag{10.73}$$

From this, and keeping in mind that each $\widetilde{\varphi}_k$ is supported in C_k, we conclude that the functions defined in (10.73) enjoy the following properties:

$$\begin{aligned} &\Phi_N\big|_{C_k} = \varphi_k \quad \text{for all } k, N \in \mathbb{N}_0 \text{ with } k \le N, \text{ while} \\ &\Phi_N \cdot \mathbb{1}_{C_k} = 0 \quad \text{for all } k, N \in \mathbb{N}_0 \text{ with } k \ge N+1. \end{aligned} \tag{10.74}$$

To proceed, observe that the membership of f into $\mathrm{A}_{p,q}(\Sigma,\sigma)$ implies the existence of some $\widetilde{N} \in \mathbb{N}$ satisfying

$$\Big\{\sum_{k=N+1}^{\infty} \big(2_*^{k(n-1)(1/q-1/p)}\|f \cdot \mathbb{1}_{C_k}\|_{L^p(\Sigma,\sigma)}\Big\}^{1/q} < \varepsilon \quad \text{for all } N \ge \widetilde{N}. \tag{10.75}$$

Hence, if $N \ge \widetilde{N}$, then (10.72)-(10.75) allow us to estimate

$$\begin{aligned} \|f - \Phi_N\|^q_{\mathrm{A}_{p,q}(\Sigma,\sigma)} &= \sum_{k=0}^{N} \big(2_*^{k(n-1)(1/q-1/p)}\|f\big|_{C_k} - \varphi_k\|_{L^p(C_k,\sigma)}\big)^q \\ &\quad + \sum_{k=N+1}^{\infty} \big(2_*^{k(n-1)(1/q-1/p)}\|f \cdot \mathbb{1}_{C_k}\|_{L^p(\Sigma,\sigma)}\big)^q \\ &\le \sum_{k=0}^{N} \frac{\varepsilon^q}{2^{kq}} + \varepsilon^q \le C \cdot \varepsilon^q, \end{aligned} \tag{10.76}$$

for some constant $C = C(q) \in (0,\infty)$. Since $\varepsilon > 0$ was taken arbitrary to begin with, (10.76) shows that $\Phi_N \to f$ in $\mathrm{A}_{p,q}(\Sigma,\sigma)$ as $N \to \infty$, thus concluding the proof of (10.71). This ultimately shows the separability of $\mathrm{A}_{p,q}(\Sigma,\sigma)$. $\square$

10.2 Duality Theory

The goal here is to identify the dual spaces of first-generation Herz spaces.

Proposition 10.9 *Assume $\Sigma \subseteq \mathbb{R}^n$ is a closed unbounded Ahlfors regular set, and abbreviate $\sigma := \mathcal{H}^{n-1}\lfloor\Sigma$. Use the symbol $(\cdot)'$ to denote Hölder conjugation of exponents. Then the following statements are valid:*

(a) If $p, q \in (1, \infty)$ then the dual of $\mathrm{A}_{p,q}(\Sigma, \sigma)$ is $\mathrm{A}_{p',q'}(\Sigma, \sigma)$, i.e.,

$$\big(\mathrm{A}_{p,q}(\Sigma, \sigma)\big)^* = \mathrm{A}_{p',q'}(\Sigma, \sigma). \tag{10.77}$$

Specifically, the mapping

$$\mathrm{A}_{p',q'}(\Sigma, \sigma) \ni f \longmapsto \Phi f := \Lambda_f \in \big(\mathrm{A}_{p,q}(\Sigma, \sigma)\big)^* \tag{10.78}$$

where

$$\Lambda_f(g) := \int_\Sigma fg\, d\sigma \quad \textit{for all } g \in \mathrm{A}_{p,q}(\Sigma, \sigma) \tag{10.79}$$

is a well-defined, linear, isometric isomorphism.

(b) If $p \in (1, \infty)$ and $q \in (0, 1]$ then the dual of $\mathrm{A}_{p,q}(\Sigma, \sigma)$ is $\mathrm{B}_{p',q}(\Sigma, \sigma)$, i.e.,

$$\big(\mathrm{A}_{p,q}(\Sigma, \sigma)\big)^* = \mathrm{B}_{p',q}(\Sigma, \sigma). \tag{10.80}$$

Concretely, the assignment

$$\mathrm{B}_{p',q}(\Sigma, \sigma) \ni f \longmapsto \Psi f := \Lambda_f \in \big(\mathrm{A}_{p,q}(\Sigma, \sigma)\big)^* \tag{10.81}$$

where

$$\Lambda_f(g) := \int_\Sigma fg\, d\sigma \quad \textit{for all } g \in \mathrm{A}_{p,q}(\Sigma, \sigma) \tag{10.82}$$

is a well-defined, linear, isometric isomorphism.

Proof We first prove the claim made in item *(a)*.

<u>`Well-definiteness of Φ.`</u> Fix a function $f \in \mathrm{A}_{p',q'}(\Sigma, \sigma)$. In light of Herz-based Hölder's inequality recorded in Proposition 10.2, the integral defining Λ_f in (10.79) is absolutely convergent and for every $g \in \mathrm{A}_{p,q}(\Sigma, \sigma)$ we have

$$|\Lambda_f(g)| \leq \int_\Sigma |fg|\, d\sigma \leq \|f\|_{\mathrm{A}_{p',q'}(\Sigma,\sigma)} \|g\|_{\mathrm{A}_{p,q}(\Sigma,\sigma)}. \tag{10.83}$$

As such, Λ_f is well defined, linear, and continuous. This implies that the mapping

$$\mathrm{A}_{p,q}(\Sigma, \sigma) \ni g \longmapsto \Lambda_f(g) = \int_\Sigma fg\, d\sigma \quad \text{is a bounded} \\ \text{linear functional with norm } \|\Lambda_f\|_{(\mathrm{A}_{p,q}(\Sigma,\sigma))^*} \leq \|f\|_{\mathrm{A}_{p',q'}(\Sigma,\sigma)}. \tag{10.84}$$

Surjectivity of Φ. Our next goal is to prove that every continuous linear functional on $\mathrm{A}_{p,q}(\Sigma,\sigma)$ is of the form Λ_f (as in (10.84)) for some $f \in \mathrm{A}_{p',q'}(\Sigma,\sigma)$. Establishing this requires some preparation. Recall $\{C_k\}_{k\in\mathbb{N}_0}$ from (3.1), and consider the sequence of Banach spaces $\mathscr{A}_\Sigma := \{L^p(C_k,\sigma)\}_{k\in\mathbb{N}_0}$ and its dual sequence $\mathscr{A}^*_\Sigma := \{L^{p'}(C_k,\sigma)\}_{k\in\mathbb{N}_0}$. Define the sequence spaces

$$\ell_q(\mathscr{A}_\Sigma) := \Big\{F = \{f_k\}_{k\in\mathbb{N}_0} \in \prod_{k\in\mathbb{N}_0} L^p(C_k,\sigma)\,; \quad \|F\|_{\ell_q(\mathscr{A}_\Sigma)} := \Big(\sum_{k=0}^{\infty} \|f_k\|^q_{L^p(C_k,\sigma)}\Big)^{1/q} < \infty\Big\}, \tag{10.85}$$

and

$$\ell_{q'}(\mathscr{A}^*_\Sigma) := \Big\{F = \{f_k\}_{k\in\mathbb{N}_0} \in \prod_{k\in\mathbb{N}_0} L^{p'}(C_k,\sigma) : \quad \|F\|_{\ell_{q'}(\mathscr{A}^*_\Sigma)} := \Big(\sum_{k=0}^{\infty} \|f_k\|^{q'}_{L^{p'}(C_k,\sigma)}\Big)^{1/q'} < \infty\Big\}. \tag{10.86}$$

These are Banach spaces and, the dual of $\ell_q(\mathscr{A}_\Sigma)$ is $\ell_{q'}(\mathscr{A}^*_\Sigma)$ (cf. [Tr95, Lemma p. 68]), i.e.,

$$\big(\ell_q(\mathscr{A}_\Sigma)\big)^* = \ell_{q'}(\mathscr{A}^*_\Sigma). \tag{10.87}$$

Specifically, the mapping

$$\ell_{q'}(\mathscr{A}^*_\Sigma) \ni G \longmapsto {}_{\ell_q(\mathscr{A}_\Sigma)}(\cdot, G)_{\ell_{q'}(\mathscr{A}^*_\Sigma)} \in \big(\ell_q(\mathscr{A}_\Sigma)\big)^* \tag{10.88}$$

is a well-defined isomorphism, where

$${}_{\ell_q(\mathscr{A}_\Sigma)}(F, G)_{\ell_{q'}(\mathscr{A}^*_\Sigma)} = \sum_{k=0}^{\infty} \int_{C_k} f_k g_k \, d\sigma \tag{10.89}$$

whenever $F = \{f_k\}_{k\in\mathbb{N}_0} \in \ell_q(\mathscr{A}_\Sigma)$ and $G = \{g_k\}_{k\in\mathbb{N}_0} \in \ell_{q'}(\mathscr{A}^*_\Sigma)$.

Next, for each σ-measurable function f on Σ define

$$\Theta f := \Big\{2_*^{k(n-1)(1/q-1/p)} \cdot f\big|_{C_k}\Big\}_{k\in\mathbb{N}_0} \quad \text{for all } f \in \mathrm{A}_{p,q}(\Sigma,\sigma). \tag{10.90}$$

Then for every $f \in \mathrm{A}_{p,q}(\Sigma,\sigma)$ we have

$$\|\Theta f\|_{\ell_q(\mathscr{A}_\Sigma)} = \Big\{\sum_{k=0}^{\infty} \big(2_*^{k(n-1)(1/q-1/p)} \|f|_{C_k}\|_{L^p(C_k,\sigma)}\big)^q\Big\}^{1/q} = \|f\|_{\mathrm{A}_{p,q}(\Sigma,\sigma)}. \tag{10.91}$$

Hence, the operator

$$\Theta : \mathrm{A}_{p,q}(\Sigma, \sigma) \to \ell_q(\mathscr{A}_\Sigma) \text{ is a well-defined linear isometry, hence a bounded operator.} \tag{10.92}$$

This further implies that

$$\Theta : \mathrm{A}_{p,q}(\Sigma, \sigma) \to \Theta\big(\mathrm{A}_{p,q}(\Sigma, \sigma)\big) \text{ is a bounded isometric isomorphism onto its image.} \tag{10.93}$$

Denoting by $\widetilde{\Theta}$ the inverse of this isometry, we see that

$$\widetilde{\Theta} : \Theta\big(\mathrm{A}_{p,q}(\Sigma, \sigma)\big) \to \mathrm{A}_{p,q}(\Sigma, \sigma) \text{ is a continuous linear operator satisfying } \widetilde{\Theta}(\Theta f) = f \text{ for all } f \in \mathrm{A}_{p,q}(\Sigma, \sigma). \tag{10.94}$$

We claim that $\Theta\big(\mathrm{A}_{p,q}(\Sigma, \sigma)\big)$ is a closed subspace of $\ell_q(\mathscr{A}_\Sigma)$. To justify this, assume $\{f_j\}_{j\in\mathbb{N}_0}$ is a sequence in $\mathrm{A}_{p,q}(\Sigma, \sigma)$ such that there exists some $F \in \ell_q(\mathscr{A}_\Sigma)$ satisfying

$$\Theta f_j \longrightarrow F \text{ in } \ell_q(\mathscr{A}_\Sigma) \text{ as } j \to \infty. \tag{10.95}$$

In particular, $\{\Theta f_j\}_{j\in\mathbb{N}_0}$ is a Cauchy sequence in $\ell_q(\mathscr{A}_\Sigma)$. This and the fact that Θ is a linear isometry (cf. (10.93)) yield

$$\|f_j - f_\ell\|_{\mathrm{A}_{p,q}(\Sigma,\sigma)} = \|\Theta f_j - \Theta f_\ell\|_{\ell_q(\mathscr{A}_\Sigma)} \quad \text{for every } j, \ell \in \mathbb{N}. \tag{10.96}$$

Hence $\{f_j\}_{j\in\mathbb{N}_0}$ is a Cauchy sequence in $\mathrm{A}_{p,q}(\Sigma, \sigma)$. Since $\mathrm{A}_{p,q}(\Sigma, \sigma)$ is complete (cf. Proposition 10.6), there exists some $f \in \mathrm{A}_{p,q}(\Sigma, \sigma)$ such that $f_j \to f$ in $\mathrm{A}_{p,q}(\Sigma, \sigma)$ as $j \to \infty$. Since Θ is a continuous mapping, it follows that $\Theta f_j \to \Theta f$ in $\ell_q(\mathscr{A}_\Sigma)$ as $j \to \infty$. In concert with (10.95) we obtain $F = \Theta f$. This establishes that $\Theta\big(\mathrm{A}_{p,q}(\Sigma, \sigma)\big)$ is a closed subspace of $\ell_q(\mathscr{A}_\Sigma)$. Upon recalling that $\ell_q(\mathscr{A}_\Sigma)$ is a Banach space, we see that

$$\Big(\Theta\big(\mathrm{A}_{p,q}(\Sigma, \sigma)\big), \|\cdot\|_{\ell_q(\mathscr{A}_\Sigma)}\Big) \text{ is a Banach space.} \tag{10.97}$$

To proceed, pick an arbitrary functional $\Lambda \in \big(\mathrm{A}_{p,q}(\Sigma, \sigma)\big)^*$. In view of the continuity of Λ and $\widetilde{\Theta}$ (cf. (10.94)), the mapping

$$\Lambda \circ \widetilde{\Theta} : \Theta\big(\mathrm{A}_{p,q}(\Sigma, \sigma)\big) \to \mathbb{C} \text{ is a continuous linear functional.} \tag{10.98}$$

Thanks to the Hahn-Banach Theorem, $\Lambda \circ \widetilde{\Theta}$ extends to a continuous linear functional $\widetilde{\Lambda} : \ell_q(\mathscr{A}_\Sigma) \to \mathbb{C}$ with the same operator norm as $\Lambda \circ \widetilde{\Theta}$, i.e.,

$$\widetilde{\Lambda} \in \big(\ell_q(\mathscr{A}_\Sigma)\big)^* \text{ with } \widetilde{\Lambda}\big|_{\Theta(\mathrm{A}_{p,q}(\Sigma,\sigma))} = \Lambda \circ \widetilde{\Theta} \text{ in } \Theta\big(\mathrm{A}_{p,q}(\Sigma, \sigma)\big) \text{ and } \|\widetilde{\Lambda}\|_{(\ell_q(\mathscr{A}_\Sigma))^*} = \|\Lambda \circ \widetilde{\Theta}\|_{[\Theta(\mathrm{A}_{p,q}(\Sigma,\sigma))]^*}. \tag{10.99}$$

In particular, (10.87)-(10.89) guarantee the existence of some $F = \{f_k\}_{k\in\mathbb{N}_0}$ in $\ell_{q'}(\mathscr{A}^*_\Sigma)$ with the property that

$$\begin{aligned}\widetilde{\Lambda}(G) &= \sum_{k=0}^{\infty}\int_{C_k} f_k g_k\, d\sigma \quad \text{for each } G = \{g_k\}_{k\in\mathbb{N}_0} \in \ell_q(\mathscr{A}_\Sigma),\\ \text{and}\quad \|\widetilde{\Lambda}\|_{(\ell_q(\mathscr{A}_\Sigma))^*} &= \|F\|_{\ell_{q'}(\mathscr{A}^*_\Sigma)} = \Big(\sum_{k=0}^{\infty}\|f_k\|^{q'}_{L^{p'}(C_k,\sigma)}\Big)^{1/q'}.\end{aligned} \tag{10.100}$$

Based on this, (10.90), (10.94), and (10.99), for each $g \in \mathrm{A}_{p,q}(\Sigma,\sigma)$ we write

$$\begin{aligned}\Lambda(g) &= (\Lambda\circ\widetilde{\Theta})(\Theta g) = \widetilde{\Lambda}(\Theta g) = \widetilde{\Lambda}\Big(\big\{2_*^{k(n-1)(1/q-1/p)}\cdot g\big|_{C_k}\big\}_{k\in\mathbb{N}_0}\Big)\\ &= \sum_{k=0}^{\infty} 2_*^{k(n-1)(1/q-1/p)}\int_{C_k} f_k\big(g\big|_{C_k}\big)\, d\sigma.\end{aligned} \tag{10.101}$$

Next, for each $k \in \mathbb{N}_0$ denote by $\widetilde{f_k}$ the extension by zero of f_k to Σ. Then since $\{C_k\}_{k\in\mathbb{N}_0}$ is a disjoint family of sets covering Σ (cf. (3.1)) and each $\widetilde{f_k}$ is supported in C_k, the function $f : \Sigma \to \mathbb{C}$ given by

$$\begin{aligned}f &:= \sum_{k=0}^{\infty} 2_*^{k(n-1)(1/q-1/p)}\widetilde{f_k} \text{ is well defined and satisfies}\\ f &= 2_*^{k(n-1)(1/q-1/p)}\widetilde{f_k} \text{ on } C_k \text{ for every } k \in \mathbb{N}_0.\end{aligned} \tag{10.102}$$

Together, (10.102) and (10.100) imply that

$$\begin{aligned}\|f\|_{\mathrm{A}_{p',q'}(\Sigma,\sigma)} &= \Big\{\sum_{k=0}^{\infty}\big(2_*^{k(n-1)(1/q'-1/p')}\|f\|_{L^{p'}(C_k,\sigma)}\big)^{q'}\Big\}^{1/q'}\\ &= \Big\{\sum_{k=0}^{\infty}\big(2_*^{k(n-1)(1/q'-1/p')}\|2_*^{k(n-1)(1/q-1/p)}f_k\|_{L^{p'}(C_k,\sigma)}\big)^{q'}\Big\}^{1/q'}\\ &= \Big\{\sum_{k=0}^{\infty}\|f_k\|^{q'}_{L^{p'}(C_k,\sigma)}\Big\}^{1/q'} = \|F\|_{\ell_{q'}(\mathscr{A}^*_\Sigma)} < \infty.\end{aligned} \tag{10.103}$$

This establishes that $f \in \mathrm{A}_{p',q'}(\Sigma,\sigma)$.

Finally, we claim that Λ coincides with the functional Λ_f associated with f defined in (10.102). To verify this, pick $g \in \mathrm{A}_{p,q}(\Sigma,\sigma)$. Recall (10.101), the definition of f in (10.102), and use Lebesgue's Dominated Convergence Theorem to compute

$$\Lambda(g) = \sum_{k=0}^{\infty} 2_*^{k(n-1)(1/q-1/p)} \int_{C_k} f_k \cdot \big(g\big|_{C_k}\big)\, d\sigma$$

$$= \sum_{k=0}^{\infty} 2_*^{k(n-1)(1/q-1/p)} \int_{\Sigma} \widetilde{f_k} \cdot g\, d\sigma$$

$$= \lim_{N\to\infty} \int_{\Sigma} \Big(\sum_{k=0}^{N} 2_*^{k(n-1)(1/q-1/p)} \widetilde{f_k}\Big) g\, d\sigma = \int_{\Sigma} fg\, d\sigma = \Lambda_f(g). \qquad (10.104)$$

Hence $\Lambda = \Lambda_f$ as functionals in $\big(\mathrm{A}_{p,q}(\Sigma,\sigma)\big)^*$. This proves that the mapping Φ in (10.78) is surjective. Moreover, on account of (10.91), (10.93), (10.94), (10.99), (10.100), and (10.103) we have

$$\begin{aligned}
\|\Lambda_f\|_{(\mathrm{A}_{p,q}(\Sigma,\sigma))^*} &= \|\Lambda\|_{(\mathrm{A}_{p,q}(\Sigma,\sigma))^*} = \sup_{\substack{g\in \mathrm{A}_{p,q}(\Sigma,\sigma)\\ \|g\|_{\mathrm{A}_{p,q}(\Sigma,\sigma)}\leq 1}} |\Lambda(g)| \\
&= \sup_{\substack{g\in \mathrm{A}_{p,q}(\Sigma,\sigma)\\ \|g\|_{\mathrm{A}_{p,q}(\Sigma,\sigma)}\leq 1}} |(\Lambda\circ\widetilde{\Theta})(\Theta g)| \\
&= \sup_{\substack{h\in \Theta(\mathrm{A}_{p,q}(\Sigma,\sigma))\\ \|h\|_{\ell_q(\mathscr{A}_\Sigma)}\leq 1}} |(\Lambda\circ\widetilde{\Theta})(h)| = \|\Lambda\circ\widetilde{\Theta}\|_{[\Theta(\mathrm{A}_{p,q}(\Sigma,\sigma))]^*} \\
&= \|\widetilde{\Lambda}\|_{(\ell_q(\mathscr{A}_\Sigma))^*} = \|F\|_{\ell_{q'}(\mathscr{A}_\Sigma^*)} = \|f\|_{\mathrm{A}_{p',q'}(\Sigma,\sigma)}
\end{aligned} \qquad (10.105)$$

proving that the mapping in (10.78)-(10.79) is an isometry.

<u>Injectivity of Φ.</u> To show that Φ is injective, pick $f \in \mathrm{A}_{p',q'}(\Sigma,\sigma)$ and assume $\Lambda_f = 0$ as a functional in $\big(\mathrm{A}_{p,q}(\Sigma,\sigma)\big)^*$, i.e.,

$$\int_{\Sigma} fg\, d\sigma = 0 \quad \text{for all } g \in \mathrm{A}_{p,q}(\Sigma,\sigma). \qquad (10.106)$$

Notice that for every $g \in L^p(\Sigma,\sigma)$ and $k \in \mathbb{N}_0$ we have $g \cdot \mathbb{1}_{C_k} \in \mathrm{A}_{p,q}(\Sigma,\sigma)$. This and (10.106) imply that

$$\int_{\Sigma} (f\cdot \mathbb{1}_{C_k})\cdot g\, d\sigma = \int_{\Sigma} f\cdot (g\cdot \mathbb{1}_{C_k})\, d\sigma = 0 \qquad (10.107)$$
$$\text{for all } g \in L^p(\Sigma,\sigma) \text{ and } k \in \mathbb{N}_0$$

Hence, by Riesz Representation Theorem, we find that $f = 0$ at σ-a.e. point on C_k, for each $k \in \mathbb{N}_0$. Since $\{C_k\}_{k\in\mathbb{N}_0}$ covers Σ, this ultimately gives that $f = 0$ at σ-a.e. point on Σ. This establishes that Φ is injective, which ultimately concludes the proof of the duality result claimed in item *(a)*.

Pressing forward, consider the claim made in item *(b)*.

Well-definiteness of Ψ. Given $f \in \mathrm{B}_{p',q}(\Sigma, \sigma)$, the Hölder-type inequality in item *(b)* of Proposition 10.2 gives

$$\int_\Sigma |fg|\, d\sigma \leq \|f\|_{\mathrm{B}_{p',q}(\Sigma,\sigma)} \|g\|_{\mathrm{A}_{p,q}(\Sigma,\sigma)} \quad \text{for all } g \in \mathrm{A}_{p,q}(\Sigma, \sigma). \tag{10.108}$$

Hence, for each $f \in \mathrm{B}_{p',q}(\Sigma, \sigma)$ the mapping

$$\begin{aligned} &\mathrm{A}_{p,q}(\Sigma, \sigma) \ni g \longmapsto \Lambda_f(g) := \int_\Sigma fg\, d\sigma \quad \text{is a continuous} \\ &\text{linear functional with norm } \|\Lambda_f\|_{(\mathrm{A}_{p,q}(\Sigma,\sigma))^*} \leq \|f\|_{\mathrm{B}_{p',q}(\Sigma,\sigma)}. \end{aligned} \tag{10.109}$$

This ensures that the mapping Ψ given in (10.81) is well defined and linear.

Surjectivity of Ψ. For each $j \in \mathbb{N}_0$, denote by E_j the operator of extension by zero of function on C_j to the entire Σ (cf. (3.103)). Then the same type of reasoning used in (3.105), (3.104), and (3.106), now bearing in mind (10.1), shows that

$$\begin{aligned} &E_j : L^p(C_j, \sigma) \longrightarrow \mathrm{A}_{p,q}(\Sigma, \sigma) \text{ is a bounded linear operator} \\ &\text{with norm } \|E_j\|_{L^p(C_j,\sigma)\to \mathrm{A}_{p,q}(\Sigma,\sigma)} = 2_*^{j(n-1)(1/q-1/p)} \text{ for all } j \in \mathbb{N}_0. \end{aligned} \tag{10.110}$$

Fix an arbitrary linear functional $\Lambda \in \big(\mathrm{A}_{p,q}(\Sigma, \sigma)\big)^*$. Then (10.110) imply that for every $j \in \mathbb{N}_0$ we have

$$\begin{aligned} &\Lambda \circ E_j : L^p(C_j, \sigma) \longrightarrow \mathbb{R} \text{ is a well-defined,} \\ &\text{continuous, and linear functional, with norm} \\ &\|\Lambda \circ E_j\|_{(L^p(C_k,\sigma))^*} \leq 2_*^{j(n-1)(1/q-1/p)} \|\Lambda\|_{(\mathrm{A}_{p,q}(\Sigma,\sigma))^*}. \end{aligned} \tag{10.111}$$

By Riesz Representation Theorem, for each $j \in \mathbb{N}_0$ there exists $f_j \in L^{p'}(C_j, \sigma)$ satisfying

$$\begin{aligned} &(\Lambda \circ E_j)(g) = \int_{C_j} g f_j\, d\sigma \quad \text{for all } g \in L^p(C_j, \sigma) \\ &\text{and} \quad \|\Lambda \circ E_j\|_{(L^p(C_j,\sigma))^*} = \|f_j\|_{L^{p'}(C_j,\sigma)}. \end{aligned} \tag{10.112}$$

Bearing in mind that for each $j \in \mathbb{N}_0$ the function $E_j f_j$ is supported on C_j, and that $\{C_j\}_{j\in\mathbb{N}_0}$ is a disjoint family of subsets of Σ, it follows that the function $f : \Sigma \to \mathbb{R}$ given by

$$f := \sum_{j=0}^{\infty} E_j f_j \tag{10.113}$$

is well defined and σ-measurable. In particular, it is clear that $f \cdot \mathbb{1}_{C_k} = E_k f_k$ for each $k \in \mathbb{N}_0$. Thanks to this, (10.112), and (10.111) we may now estimate

$$\begin{aligned}
\|f\|_{\mathrm{B}_{p',q}(\Sigma,\sigma)} &= \sup_{k\in\mathbb{N}_0}\left\{2_*^{-k(n-1)(1/q-1/p)}\|E_k f_k\|_{L^{p'}(C_k,\sigma)}\right\} \\
&= \sup_{k\in\mathbb{N}_0}\left\{2_*^{-k(n-1)(1/q-1/p)}\|f_k\|_{L^{p'}(C_k,\sigma)}\right\} \\
&= \sup_{k\in\mathbb{N}_0}\left\{2_*^{-k(n-1)(1/q-1/p)}\|\Lambda\circ E_k\|_{(\mathrm{A}_{p,q}(\Sigma,\sigma))^*}\right\} \\
&\leq \|\Lambda\|_{(\mathrm{A}_{p,q}(\Sigma,\sigma))^*} < \infty. \qquad (10.114)
\end{aligned}$$

This establishes the membership $f \in \mathrm{B}_{p',q}(\Sigma,\sigma)$.

At this stage, we claim that Λ coincides with the functional Λ_f associated with f as in (10.113). To see that this is the case, pick a function $g \in \mathrm{A}_{p,q}(\Sigma,\sigma)$ and decompose it as

$$g = \sum_{k=0}^{\infty} g \cdot \mathbb{1}_{C_k} \quad \text{pointwise } \sigma\text{-a.e. on } \Sigma. \qquad (10.115)$$

From Proposition 10.4, the series in (10.115) also converges in $\mathrm{A}_{p,q}(\Sigma,\sigma)$. From this, the continuity of the linear functional Λ, (10.112), and Lebesgue's Dominated Convergence Theorem we compute

$$\begin{aligned}
\Lambda(g) &= \sum_{k=0}^{\infty}\Lambda(g\cdot\mathbb{1}_{C_k}) = \sum_{k=0}^{\infty}(\Lambda\circ E_k)\big(g\big|_{C_k}\big) \\
&= \sum_{k=0}^{\infty}\int_{C_k} f_k\cdot\big(g\big|_{C_k}\big)\,d\sigma = \sum_{k=0}^{\infty}\int_{\Sigma}(E_k f_k)\cdot g\,d\sigma \\
&= \lim_{N\to\infty}\int_{\Sigma}\Big(\sum_{k=0}^{N} E_k f_k\Big)\cdot g\,d\sigma = \int_{\Sigma} fg\,d\sigma = \Lambda_f(g). \qquad (10.116)
\end{aligned}$$

Hence, $\Lambda = \Lambda_f$, which shows that Ψ is a surjective mapping. Let us also note here that, as a consequence of this, (10.109), and (10.114), we have

$$\|\Lambda\|_{(\mathrm{A}_{p,q}(\Sigma,\sigma))^*} = \|\Lambda_f\|_{(\mathrm{A}_{p,q}(\Sigma,\sigma))^*} = \|f\|_{\mathrm{B}_{p',q}(\Sigma,\sigma)}. \qquad (10.117)$$

Injectivity of Ψ. To justify the injectivity of Ψ, let $f \in \mathrm{B}_{p',q}(\Sigma,\sigma)$ be such that $\Lambda_f = 0$ as a functional in $\big(\mathrm{A}_{p,q}(\Sigma,\sigma)\big)^*$. This means that

$$\int_{\Sigma} fg\,d\sigma = 0 \quad \text{for all } g \in \mathrm{A}_{p,q}(\Sigma,\sigma). \qquad (10.118)$$

Via the same reasoning as in (10.107) we may then conclude that $f = 0$ pointwise σ-a.e. on Σ. Hence Ψ is injective, as desired. As a consequence of this and the surjectivity of Ψ it follows that for every continuous linear functional $\Lambda \in \big(\mathrm{A}_{p,q}(\Sigma,\sigma)\big)^*$ there exists a unique function $f \in \mathrm{B}_{p',q}(\Sigma,\sigma)$ such that $\Lambda = \Lambda_f$. This and (10.117) then entail

$$\|\Psi f\|_{(\mathrm{A}_{p,q}(\Sigma,\sigma))^*} = \|f\|_{\mathrm{B}_{p',q}(\Sigma,\sigma)} \quad \text{for all } f \in \mathrm{B}_{p',q}(\Sigma,\sigma). \tag{10.119}$$

This establishes the claim made in item *(b)*, which ultimately finishes the proof of Proposition 10.9. □

As a consequence of the previous duality result, we have the following norm characterizations.

Corollary 10.1 *Suppose $\Sigma \subseteq \mathbb{R}^n$ is a closed unbounded Ahlfors regular set, and abbreviate $\sigma := \mathcal{H}^{n-1}\lfloor\Sigma$. Pick two exponents $p \in (1,\infty)$ and $q \in (0,\infty)$. Also, fix a σ-measurable function f on Σ. Then the following statements are valid:*

(a) If $q \in (1,\infty)$ then f belongs to $\mathrm{A}_{p,q}(\Sigma,\sigma)$ if and only if

$$\sup_{\substack{g\in \mathrm{A}_{p',q'}(\Sigma,\sigma)\\ \|g\|_{\mathrm{A}_{p',q'}(\Sigma,\sigma)}\leq 1}} \int |fg|\,d\sigma < \infty \tag{10.120}$$

where $p', q' \in (1,\infty)$ denote the Hölder conjugate exponent of p and q, respectively. In fact, whenever $q \in (1,\infty)$ one has

$$\|f\|_{\mathrm{A}_{p,q}(\Sigma,\sigma)} = \sup_{\substack{g\in \mathrm{A}_{p',q'}(\Sigma,\sigma)\\ \|g\|_{\mathrm{A}_{p',q'}(\Sigma,\sigma)}\leq 1}} \left|\int fg\,d\sigma\right| = \sup_{\substack{g\in \mathrm{A}_{p',q'}(\Sigma,\sigma)\\ \|g\|_{\mathrm{A}_{p',q'}(\Sigma,\sigma)}\leq 1}} \int |fg|\,d\sigma. \tag{10.121}$$

(b) If $q \in (0,1]$ then f belongs to $\mathrm{B}_{p,q}(\Sigma,\sigma)$ if and only if

$$\sup_{\substack{g\in \mathrm{A}_{p',q}(\Sigma,\sigma)\\ \|g\|_{\mathrm{A}_{p',q}(\Sigma,\sigma)}\leq 1}} \int |fg|\,d\sigma < \infty \tag{10.122}$$

where $p' \in (1,\infty)$ denotes the Hölder conjugate exponent of p. In fact, whenever $q \in (0,1]$ one has

$$\|f\|_{\mathrm{B}_{p,q}(\Sigma,\sigma)} = \sup_{\substack{g\in \mathrm{A}_{p',q}(\Sigma,\sigma)\\ \|g\|_{\mathrm{A}_{p',q}(\Sigma,\sigma)}\leq 1}} \left|\int fg\,d\sigma\right| = \sup_{\substack{g\in \mathrm{A}_{p',q}(\Sigma,\sigma)\\ \|g\|_{\mathrm{A}_{p',q}(\Sigma,\sigma)}\leq 1}} \int |fg|\,d\sigma. \tag{10.123}$$

Proof Consider the claim made in item *(a)*. In one direction, if $f \in \mathrm{A}_{p,q}(\Sigma,\sigma)$ then the finiteness of the supremum in (10.120) is a consequence of (10.20). To deal with the opposite direction, assume that (10.120) holds. Then this implies that the mapping $\Lambda_f : \mathrm{A}_{p',q'}(\Sigma,\sigma) \to \mathbb{C}$ given by

$$\Lambda_f(g) = \int_\Sigma fg\,d\sigma \quad \text{for all } g \in \mathrm{A}_{p',q'}(\Sigma,\sigma) \tag{10.124}$$

is well defined (as an absolutely convergent integral) and linear. Moreover, thanks to (10.20) we see that $\Lambda_f \in \big(\mathrm{A}_{p',q'}(\Sigma,\sigma)\big)^*$. As such, item *(a)* in Proposition 10.9 implies that $f \in \mathrm{A}_{p,q}(\Sigma,\sigma)$, and (10.20) allows us to estimate

$$\begin{aligned}\|f\|_{\mathrm{A}_{p,q}(\Sigma,\sigma)} = \|\Lambda_f\|_{(\mathrm{A}_{p',q'}(\Sigma,\sigma))^*} &= \sup_{\substack{g\in\mathrm{A}_{p',q'}(\Sigma,\sigma)\\ \|g\|_{\mathrm{A}_{p',q'}(\Sigma,\sigma)}\leq 1}} \Big|\int fg\,d\sigma\Big| \\ &\leq \sup_{\substack{g\in\mathrm{A}_{p',q'}(\Sigma,\sigma)\\ \|g\|_{\mathrm{A}_{p',q'}(\Sigma,\sigma)}\leq 1}} \int |fg|\,d\sigma \leq \|f\|_{\mathrm{A}_{p,q}(\Sigma,\sigma)}.\end{aligned} \tag{10.125}$$

The claims in (10.121) are implicit in what we have proved so far, and this finishes the treatment of item *(a)*. The proof of item *(b)* follows via a similar argument, now bearing in mind (10.21) and item *(b)* in Proposition 10.9. □

10.3 Boundedness of the Hardy-Littlewood Maximal Operator

In the result recorded below we establish the boundedness of the Hardy-Littlewood maximal operator, as well as its L^s-based version, on first-generation Herz spaces.

Theorem 10.1 *Assume $\Sigma \subseteq \mathbb{R}^n$ is a closed unbounded Ahlfors regular set, and abbreviate $\sigma := \mathcal{H}^{n-1}\lfloor\Sigma$. Select two exponents $p \in (1,\infty)$ and $q \in (0,\infty)$. Then the following statements are valid:*

(a) If $0 < q \leq p < \infty$ then for every $s \in (0,q)$ it follows that

$$\begin{gathered}\mathcal{M}_{\Sigma,s} : \mathrm{A}_{p,q}(\Sigma,\sigma) \to \mathrm{A}_{p,q}(\Sigma,\sigma) \\ \text{is well defined, sublinear, and bounded,}\end{gathered} \tag{10.126}$$

where $\mathcal{M}_{\Sigma,s}$ is the L^s-based Hardy-Littlewood maximal operator on Σ (cf. (2.28)).

(b) If $1 < p,q < \infty$ then

$$\begin{gathered}\mathcal{M}_{\Sigma} : \mathrm{A}_{p,q}(\Sigma,\sigma) \to \mathrm{A}_{p,q}(\Sigma,\sigma) \\ \text{is well defined, sublinear, and bounded,}\end{gathered} \tag{10.127}$$

where $\mathcal{M}_{\Sigma}$ is the Hardy-Littlewood maximal operator on Σ (cf. (2.26)).

Proof Consider first the claim made in *(a)*. Suppose $0 < q \leq p < \infty$, and select some $s \in (0,q)$. Fix an arbitrary function $f \in \mathrm{A}_{p,q}(\Sigma,\sigma)$. Then thanks to (10.11) we have

$$\||f|^q\|_{A^{p/q}(\Sigma,\sigma)} = \|f\|^q_{A_{p,q}(\Sigma,\sigma)} < \infty \quad \text{and} \quad |f|^q \in A^{p/q}(\Sigma,\sigma). \tag{10.128}$$

Upon recalling (2.28), it is visible that

$$\mathcal{M}_{\Sigma,s}f = \big[\mathcal{M}_{\Sigma,s/q}(|f|^q)\big]^{1/q} \quad \text{everywhere on } \Sigma. \tag{10.129}$$

Since $0 < s/q < 1$ (as implied by the choice of s) and $1 \leq p/q < \infty$, it follows that $\mathcal{M}_{\Sigma,s/q}$ is bounded on $A^{p/q}(\Sigma,\sigma)$ (cf. Theorem 3.3). In light of this, (10.128), (10.129), and (10.11) there exists some $C = C(n,p,q,s,\Sigma) \in (0,\infty)$ with the property that

$$\begin{aligned}\|\mathcal{M}_{\Sigma,s}f\|_{A_{p,q}(\Sigma,\sigma)} &= \Big\|\big[\mathcal{M}_{\Sigma,s/q}(|f|^q)\big]^{1/q}\Big\|_{A_{p,q}(\Sigma,\sigma)} = \|\mathcal{M}_{\Sigma,s/q}(|f|^q)\|^{1/q}_{A^{p/q}(\Sigma,\sigma)} \\ &\leq C\||f|^q\|^{1/q}_{A^{p/q}(\Sigma,\sigma)} = C\|f\|_{A_{p,q}(\Sigma,\sigma)}.\end{aligned} \tag{10.130}$$

This finishes the treatment of item *(a)*.

Next, consider the claim made in item *(b)*. Corresponding to the case when we have $1 < q \leq p < \infty$, the boundedness in (10.127) follows from what we have proved so far, for the choice $s := 1$. Henceforth, suppose $1 < p < q < \infty$, and fix a parameter $s \in (0,1)$. Then $1 < q' < p' < \infty$, where p' and q' are the Hölder conjugate exponents of p and q, respectively. In particular, the current item *(a)* implies that the Hardy-Littlewood maximal operator $\mathcal{M}_\Sigma$ is bounded on $A_{p',q'}(\Sigma,\sigma)$. Thanks to this, (10.10), (10.20), and the Fefferman-Stein maximal inequality (cf. Theorem 2.2) for each $f \in A_{p/s,q/s}(\Sigma,\sigma)$, and $g \in A_{p',q'}(\Sigma,\sigma)$ with $\|g\|_{A_{p',q'}(\Sigma,\sigma)} \leq 1$ we estimate

$$\begin{aligned}\int_\Sigma (\mathcal{M}_\Sigma f)^{1/s} \cdot |g|\, d\sigma &= \int_\Sigma \big[\mathcal{M}_{\Sigma,s}(|f|^{1/s})\big] \cdot |g|\, d\sigma \\ &\leq \int_\Sigma |f|^{1/s} \cdot (\mathcal{M}_\Sigma g)\, d\sigma \\ &\leq C\||f|^{1/s}\|_{A_{p,q}(\Sigma,\sigma)}\|\mathcal{M}_\Sigma g\|_{A_{p',q'}(\Sigma,\sigma)} \\ &\leq C\||f|^{1/s}\|_{A_{p,q}(\Sigma,\sigma)}\|g\|_{A_{p',q'}(\Sigma,\sigma)} \\ &\leq C\|f\|^{1/s}_{A_{p/s,q/s}(\Sigma,\sigma)} < \infty,\end{aligned} \tag{10.131}$$

for some constant $C = C(n,p,q,s,\Sigma) \in (0,\infty)$. Since this holds for all such functions g, item *(a)* in Corollary 10.1 ensures that for every $f \in A_{p/s,q/s}(\Sigma,\sigma)$ fixed the function $(\mathcal{M}_\Sigma f)^{1/s}$ belongs to $A_{p,q}(\Sigma,\sigma)$ and, bearing in mind (10.10), we have

$$\begin{aligned}\|\mathcal{M}_\Sigma f\|^{1/s}_{A_{p/s,q/s}(\Sigma,\sigma)} &= \|(\mathcal{M}_\Sigma f)^{1/s}\|_{A_{p,q}(\Sigma,\sigma)}\\ &= \sup_{\substack{g\in A_{p',q'}(\Sigma,\sigma)\\ \|g\|_{A_{p',q'}(\Sigma,\sigma)}\leq 1}} \int_\Sigma (\mathcal{M}_\Sigma f)^{1/s}\cdot|g|\,d\sigma\\ &\leq C\|f\|^{1/s}_{A_{p/s,q/s}(\Sigma,\sigma)},\end{aligned} \tag{10.132}$$

for some constant $C = C(n, p, q, s, \Sigma) \in (0, \infty)$. This further implies that

$$\|\mathcal{M}_\Sigma f\|_{A_{p/s,q/s}(\Sigma,\sigma)} \leq C\|f\|_{A_{p/s,q/s}(\Sigma,\sigma)} \tag{10.133}$$

for every function $f \in A_{p/s,q/s}(\Sigma,\sigma)$, and for some constant $C \in (0,\infty)$ of same nature as before.

At this stage, consider arbitrary exponents $\widetilde{p},\widetilde{q} \in (1,\infty)$, with $1 < \widetilde{p} < \widetilde{q} < \infty$. Select a parameter $s \in (1/\widetilde{p}, 1)$, and define $p := s\widetilde{p}$ and $q := s\widetilde{q}$. Notice that $1 < p < q < \infty$. Thus, for these specific choices of p, q, and s, estimate (10.132) reads

$$\|\mathcal{M}_\Sigma f\|_{A_{\widetilde{p},\widetilde{q}}(\Sigma,\sigma)} \leq C\|f\|_{A_{\widetilde{p},\widetilde{q}}(\Sigma,\sigma)}, \tag{10.134}$$

for every $f \in A_{\widetilde{p},\widetilde{q}}(\Sigma,\sigma)$. This justifies the claim made in *(b)*, which ultimately concludes the proof of Theorem 10.1. □

10.4 Generalized Banach Function Spaces and Extrapolation Theory

In this section we briefly elaborate on the theory of Generalized Banach Function Spaces (GBFS), following the treatment in [GHA-II, Chapter 5]. We recall a powerful extrapolation result in the context of GBFS with the purpose of obtaining a similar result for first-generation Herz spaces (in Section 10.5). We begin with the following definition.

Definition 10.2 Let $(X,\mathfrak{M},\mu)$ be an arbitrary measure space and set

$$\mathscr{M}(X,\mu) := \{f : X \to \mathbb{R} : f \text{ is } \mu\text{-measurable}\}, \tag{10.135}$$

$$\mathscr{M}_+(X,\mu) := \{f \in \mathscr{M}(X,\mu) : f \geq 0\}. \tag{10.136}$$

A mapping $\rho : \mathscr{M}_+(X,\mu) \to [0,\infty]$ is called a `function norm` provided, for all functions $f, g \in \mathscr{M}_+(X,\mu)$, the following properties are satisfied:

(P1) $\rho(f) = 0$ if and only if $f = 0$ μ-a.e. in X, for each $\lambda \in [0,\infty)$ it follows that $\rho(\lambda f) = \lambda\rho(f)$, and $\rho(f + g) \leq \rho(f) + \rho(g)$;

(P2) if $f \leq g$ μ-a.e. in X then $\rho(f) \leq \rho(g)$;

(P3) if $\{f_k\}_{k\in\mathbb{N}} \subseteq \mathscr{M}_+(X,\mu)$ is such that f_k increases to f pointwise μ-a.e. in X as $k \to \infty$, then $\rho(f_k)$ increases to $\rho(f)$ as $k \to \infty$;

(P4) there exists a collection of sets $\{Y_j\}_{j\in\mathbb{N}} \subseteq \mathfrak{M}$ such that $X = \cup_{j\in\mathbb{N}} Y_j$, and $\rho(\mathbb{1}_{Y_j}) < \infty$, for all $j \in \mathbb{N}$;

(P5) there exists a collection of sets $\{Z_j\}_{j\in\mathbb{N}} \subseteq \mathfrak{M}$ such that $X = \cup_{j\in\mathbb{N}} Z_j$, and with the property that for each $j \in \mathbb{N}$ there exists a constant $c_j \in (0,\infty)$ satisfying $\int_{Z_j} f\, d\mu \leq c_j \rho(f)$.

Moreover, if ρ is a function norm as described above, the set

$$\mathbb{X} := \{f \in \mathscr{M}(X,\mu) : \rho(|f|) < \infty\} \tag{10.137}$$

is called a `Generalized Banach Function Space` (GBFS) on (X,μ) . In such a scenario, for each $f \in \mathscr{M}(X,\mu)$ define

$$\|f\|_{\mathbb{X}} := \begin{cases} \rho(|f|), & \text{if } f \in \mathbb{X} \\ \infty, & \text{if } f \notin \mathbb{X}. \end{cases} \tag{10.138}$$

Remark 10.1 With the piece of notation introduced in (10.138), we may recast (10.137) as

$$\mathbb{X} = \{f \in \mathscr{M}(X,\mu) : \|f\|_{\mathbb{X}} < \infty\}. \tag{10.139}$$

Remark 10.2 If conditions (P4) and (P5) from Definition 10.2 are replaced, respectively, by

(P4') For every μ-measurable set $E \subseteq X$ with $\mu(E) < \infty$, one has $\rho(\mathbb{1}_E) < \infty$. (10.140)

and

(P5') For every μ-measurable set $E \subseteq X$ with $\mu(E) < \infty$, there exists a constant $C_E \in (0,\infty)$ (independent of f) such that $\int_E |f|\, d\mu \leq C_E \|f\|$. (10.141)

then the resulting space $\mathbb{X}$ defined as in (10.137), with function norm ρ satisfying (P1)-(P3), (P4'), and (P5'), is called a `Classical Banach Function Space`.

Remark 10.3 We wish to highlight that, in relation to properties (P4) in Definition 10.2, matters can be arranged so that the family $\{Y_j\}_{j\in\mathbb{N}}$ satisfies either of the following properties

(a) $\{Y_j\}_{j\in\mathbb{N}}$ covers X up to a set of μ-measure zero, i.e.

$$\mu\Big(X \setminus \cup_{j\in\mathbb{N}} Y_j\Big) = 0, \tag{10.142}$$

(b) $\{Y_j\}_{j\in\mathbb{N}}$ is an increasing family of sets in X, i.e.,

$$Y_j \subseteq Y_{j+1} \quad \text{for all } j \in \mathbb{N}, \tag{10.143}$$

(c) $\{Y_j\}_{j\in\mathbb{N}}$ is a family of pairwise disjoint sets, i.e.,

$$Y_j \cap Y_k = \varnothing \quad \text{for all } j, k \in \mathbb{N}. \tag{10.144}$$

Similar properties may be also considered for $\{Z_j\}_{j\in\mathbb{N}}$ in the context of (P5) in Definition 10.2.

Let us now recall the associated space of a Generalized Banach Function Space.

Definition 10.3 Suppose $(X, \mathfrak{M}, \mu)$ is a measure space and consider ρ a function norm (in the sense of Definition 10.2). The associated norm of ρ is the function $\rho' : \mathscr{M}_+(X, \mu) \to [0, \infty]$ defined by

$$\rho'(g) := \sup\Big\{\int_X fg\,d\mu : f \in \mathscr{M}_+(X, \mu),\ \rho(f) \leq 1\Big\} \text{ for all } g \in \mathscr{M}_+(X, \mu). \tag{10.145}$$

With this piece of notation, the Generalized Banach Function Space on (X, μ) defined in relation to ρ' as in Definition 10.2, namely

$$\mathbb{X}' := \{g \in \mathscr{M}(X, \mu) : \rho'(|g|) < \infty\}, \tag{10.146}$$

is called the associated space (a.k.a. Köthe dual) of $\mathbb{X}$.

Remark 10.4 It has been shown in [GHA-II, Proposition 5.1.10, p. 224] that the associated norm ρ' is function norm (in the sense of Definition 10.2). As such, the associated space $\mathbb{X}'$ is indeed a Generalized Banach Function Space on (X, μ).

Remark 10.5 Implementing (10.138) with ρ replaced by ρ', the formula (10.145) gives

$$\|g\|_{\mathbb{X}'} = \sup\Big\{\int_X |fg|\,d\mu : f \in \mathbb{X},\ \|f\|_{\mathbb{X}} \leq 1\Big\} \text{ for all } g \in \mathscr{M}(X, \mu). \tag{10.147}$$

Moreover, much as in (10.139), we may write

$$\mathbb{X}' := \{g \in \mathscr{M}(X, \mu) : \|g\|_{\mathbb{X}'} < \infty\}. \tag{10.148}$$

Remark 10.6 In the context of Definition 10.3, [GHA-II, Proposition 5.1.12, p. 226] ensures that whenever f and g are μ-measurable functions defined on X one has the following Hölder-type inequality:

$$\int_X |fg|\,d\mu \leq \|f\|_{\mathbb{X}}\|g\|_{\mathbb{X}'}. \tag{10.149}$$

It turns out that the second associated space of a Generalized Banach Function Space coincides with the original space. More specifically, a version of the Lorentz-Luxemburg theorem holds in the context of GBFS's (cf. [GHA-II, Proposition 5.1.14, pp. 226–227]).

Proposition 10.10 *Consider $\mathbb{X}$ a Generalized Banach Function Space on an arbitrary measure space $(X, \mathfrak{M}, \mu)$. Assume that μ is sigma-finite. Then $\mathbb{X}'' = \mathbb{X}$ isometrically, i.e., $\mathbb{X}$ coincides with its second associated space $\mathbb{X}'' := (\mathbb{X}')'$ and*

$$\|f\|_{\mathbb{X}} = \|f\|_{\mathbb{X}''} = \sup\Big\{\int_X |fg|\,d\mu : g \in \mathbb{X}',\ \|g\|_{\mathbb{X}'} \leq 1\Big\} \quad \textit{for every } f \in \mathbb{X}. \tag{10.150}$$

A remarkable result obtained in [GHA-II] is an extrapolation result obtained on Generalized Banach Function Spaces on Ahlfors regular sets. In essence, this states that estimates at the level of Muckenhoupt weighted Lebesgue spaces yield similar estimates at the level of Generalized Banach Function Spaces in the aforementioned geometric setting. More precisely, the following result is implied by [GHA-II, Proposition 5.2.1, pp. 223–224].

Theorem 10.2 *Suppose $\Sigma \subseteq \mathbb{R}^n$ is a closed Ahlfors regular set, and abbreviate $\sigma := \mathcal{H}^{n-1}\lfloor\Sigma$. Consider a Generalized Banach Function Space $\mathbb{X}$, and let $\mathbb{X}'$ be its associated space. With $\mathcal{M}_\Sigma$ denoting the Hardy-Littlewood maximal operator on Σ, make the assumption that*

$$\begin{gathered}\mathcal{M}_\Sigma : \mathbb{X} \to \mathbb{X} \quad \textit{and} \quad \mathcal{M}_\Sigma : \mathbb{X}' \to \mathbb{X}' \\ \textit{and are well-defined bounded mappings.}\end{gathered} \tag{10.151}$$

Then there exists some constant $C_\Sigma \in (0, \infty)$, depending only on the dimension n and the Ahlfors regularity constants of Σ with the following significance. Having fixed some integrability exponent $p_0 \in (1, \infty)$, introduce

$$\begin{aligned} W_{\mathbb{X},p_0} &:= C_\Sigma^{p_0}\|\mathcal{M}_\Sigma\|_{\mathbb{X}\to\mathbb{X}}^{p_0-1} \cdot \|\mathcal{M}_\Sigma\|_{\mathbb{X}'\to\mathbb{X}'} \\ \textit{and} \quad W_{\mathbb{X}',p_0} &:= C_\Sigma^{p_0}\|\mathcal{M}_\Sigma\|_{\mathbb{X}'\to\mathbb{X}'}^{p_0-1} \cdot \|\mathcal{M}_\Sigma\|_{\mathbb{X}\to\mathbb{X}}. \end{aligned} \tag{10.152}$$

Also, suppose f, g are two σ-measurable real-valued functions defined on Σ with the property that for each Muckenhoupt weight $w \in A_{p_0}(\Sigma, \sigma)$ with $[w]_{A_{p_0}} \leq W_{\mathbb{X},p_0}$ one has

$$\|f\|_{L^{p_0}(\Sigma,w)} \leq C_w\|g\|_{L^{p_0}(\Sigma,w)}, \tag{10.153}$$

for some constant $C_w \in (0, \infty)$ which depends only on n, Σ, p_0, and w. Then

$$\|f\|_{\mathbb{X}} \leq 2^{2-1/p_0}\Bigg(\sup_{\substack{w\in A_{p_0}(\Sigma,\sigma)\\ [w]_{A_{p_0}}\leq W_{\mathbb{X},p_0}}} C_w\Bigg)\|g\|_{\mathbb{X}}. \tag{10.154}$$

In addition to this, if (10.153) *holds for every Muckenhoupt weight* $w \in A_{p_0}(\Sigma, \sigma)$ *with* $[w]_{A_{p_0}} \leq W_{\mathbb{X}', p_0}$ *then*

$$\|f\|_{\mathbb{X}'} \leq 2^{2-1/p_0} \left(\sup_{\substack{w \in A_{p_0}(\Sigma,\sigma) \\ [w]_{A_{p_0}} \leq W_{\mathbb{X}',p_0}}} C_w \right) \|g\|_{\mathbb{X}'}. \tag{10.155}$$

10.5 Extrapolation Results for Herz Spaces $\mathrm{A}_{p,q}$

The main result of this section is an extrapolation theorem in the context of first-generation Herz spaces. We start by showing that $\mathrm{A}_{p,q}$ spaces fit within the framework of Generalized Banach Function Spaces. Specifically, we have the following proposition.

Proposition 10.11 *Consider* $\Sigma \subseteq \mathbb{R}^n$ *a closed unbounded Ahlfors regular set, and abbreviate* $\sigma := \mathcal{H}^{n-1}\lfloor\Sigma$. *Fix two exponents* $p, q \in (1, \infty)$. *Then* $\|\cdot\|_{\mathrm{A}_{p,q}(\Sigma,\sigma)}$ (*cf.* (10.1)) *is a function norm on* (Σ, σ) (*in the sense of Definition 10.2*). *In particular, the Herz space* $\mathrm{A}_{p,q}(\Sigma, \sigma)$ *is a Generalized Banach Function Space on* (Σ, σ).

Proof It is evident from (10.1) and (10.8) that $\|\cdot\|_{\mathrm{A}_{p,q}(\Sigma,\sigma)}$ satisfies properties (P1)-(P2) in Definition 10.2. Moreover, Proposition 10.5 readily imply that $\|\cdot\|_{\mathrm{A}_{p,q}(\Sigma,\sigma)}$ satisfies the monotonicity property (P3) in Definition 10.2.

Next, recall the collection of annuli $\{C_k\}_{k\in\mathbb{N}_0}$ (cf. (3.1)). In particular, this family covers Σ, and

$$\|\mathbb{1}_{C_k}\|_{\mathrm{A}_{p,q}(\Sigma,\sigma)} = 2_*^{k(n-1)(1/q-1/p)} \sigma(C_k)^{1/p} < \infty \quad \text{for all } k \in \mathbb{N}_0. \tag{10.156}$$

As such, $\|\cdot\|_{\mathrm{A}_{p,q}(\Sigma,\sigma)}$ satisfies (P4) in Definition 10.2 in relation to $\{C_k\}_{k\in\mathbb{N}_0}$.

Lastly, for each σ-measurable function f defined on Σ, and each $k \in \mathbb{N}_0$, we make use of (10.20) to estimate

$$\int_{C_k} f\, d\sigma \leq \|\mathbb{1}_{C_k}\|_{\mathrm{A}_{p',q'}(\Sigma,\sigma)} \|f\|_{\mathrm{A}_{p,q}(\Sigma,\sigma)} = c_k \|f\|_{\mathrm{A}_{p,q}(\Sigma,\sigma)}, \tag{10.157}$$

where $p', q' \in (1, \infty)$ denotes the Hölder conjugate exponent of p and q, respectively, and $c_k := \|\mathbb{1}_{C_k}\|_{\mathrm{A}_{p',q'}(\Sigma,\sigma)}$ which, according to (10.156), is a finite positive number. Hence, $\|\cdot\|_{\mathrm{A}_{p,q}(\Sigma,\sigma)}$ also satisfies (P5) with respect to $\{C_k\}_{k\in\mathbb{N}_0}$. This finally shows that $\|\cdot\|_{\mathrm{A}_{p,q}(\Sigma,\sigma)}$ is a function norm, and therefore $\mathrm{A}_{p,q}(\Sigma, \sigma)$ is in fact a Generalized Banach Function Space on (Σ, σ). □

Remark 10.7 We wish to recall from [GHA-II, Proposition 5.1.8, p. 222] that every Generalized Banach Function Space is complete. As such, this and Proposition 10.11 reprove that, in the regime $p, q \in (1, \infty)$, the Herz space $\mathrm{A}_{p,q}(\Sigma, \sigma)$ is Banach.

Remark 10.8 Notice that if either $p \in (0,1)$ or $q \in (0,1)$, then the Herz space $\mathrm{A}_{p,q}(\Sigma,\sigma)$ fails to be a Generalized Banach Function Space, simply because in this regime $\|\cdot\|_{\mathrm{A}_{p,q}(\Sigma,\sigma)}$ is only a quasinorm.

Remark 10.9 It is worth pointing out that if $p,q \in (1,\infty)$ and $1 < q < p < \infty$, the Herz space $\mathrm{A}_{p,q}(\Sigma,\sigma)$ fails to be Classical Banach Function Spaces (in the sense described in Remark 10.2). To justify this, we show that $\|\cdot\|_{\mathrm{A}_{p,q}}$ does not satisfy condition (P4') (cf. (10.140)). The goal is to find

$$\begin{aligned}&\text{a } \sigma\text{-measurable subset } E \subseteq \Sigma \text{ such that } \sigma(E) < \infty\\ &\text{and } \|\mathbb{1}_E\|_{\mathrm{A}_{p,q}(\Sigma,\sigma)} = \infty.\end{aligned} \tag{10.158}$$

First observe that whenever $E \subseteq \Sigma$ is a σ-measurable set we have

$$\begin{aligned}\|\mathbb{1}_E\|^q_{\mathrm{A}_{p,q}(\Sigma,\sigma)} &= \sum_{k=0}^{\infty} 2_*^{k(n-1)(1-q/p)} \|\mathbb{1}_E \cdot \mathbb{1}_{C_k}\|^q_{L^p(\Sigma,\sigma)}\\ &= \sum_{k=0}^{\infty} 2_*^{k(n-1)(1/q-1/p)} \sigma(E \cap C_k)^{q/p}.\end{aligned} \tag{10.159}$$

Thanks to this, finding E that satisfy (10.158) is reduced to finding a sequence of non-negative numbers $\{\lambda_k\}_{k\in\mathbb{N}_0}$ such that

$$\sum_{k=0}^{\infty} \lambda_k < \infty \quad \text{and} \quad \sum_{k=0}^{\infty} 2_*^{k(n-1)(1-q/p)} \lambda_k^{q/p} = \infty. \tag{10.160}$$

To this end, since $p/q - 1 > 0$, fix a parameter $\gamma \in (0,\infty)$ satisfying $0 < \gamma < p/q - 1$. For each $k \in \mathbb{N}_0$ define

$$\lambda_k := 2_*^{-k(n-1)\gamma}. \tag{10.161}$$

Then, because $2_* > 1$, we see that the first condition on (10.160) is satisfied for this choice of sequence $\{\lambda_k\}_{k\in\mathbb{N}_0}$. Moreover, since the condition $\gamma < p/q - 1$ is equivalent to $1 - (q/p)(1+\gamma) > 0$, we may compute

$$\sum_{k=0}^{\infty} 2_*^{k(n-1)(1-q/p)} \lambda_k^{q/p} = \sum_{k=0}^{\infty} 2_*^{k(n-1)[1-(q/p)(1+\gamma)]} = \infty. \tag{10.162}$$

Hence, the sequence $\{\lambda_k\}_{k\in\mathbb{N}_0}$ defined in (10.161) also satisfies the second condition in (10.160). This ultimately shows that $\mathrm{A}_{p,q}(\Sigma,\sigma)$ fails to be Classical Banach Function Space, whenever $p,q \in (1,\infty)$ with $1 < q < p < \infty$, and explains why we need to resort to the more inclusive category of Generalized Banach Function Spaces introduced in [GHA-II].

The next order of business is to identify the associated spaces of first-generation Herz spaces.

Proposition 10.12 *Suppose $\Sigma \subseteq \mathbb{R}^n$ is a closed unbounded Ahlfors regular set, and abbreviate $\sigma := \mathcal{H}^{n-1}\lfloor\Sigma$. Select two integrability exponents $p, q \in (1, \infty)$, and denote by $p', q' \in (1, \infty)$ the Hölder conjugate exponents of p and q, respectively. Then for every σ-measurable function f defined on Σ, one has*

$$\|f\|_{\mathrm{A}_{p',q'}(\Sigma,\sigma)} = \sup\Big\{\int_\Sigma |fg|\,d\sigma : g \in \mathrm{A}_{p,q}(\Sigma,\sigma) \text{ with } \|g\|_{\mathrm{A}_{p,q}(\Sigma,\sigma)} \leq 1\Big\}. \tag{10.163}$$

As a consequence, the associated space of the Herz space $\mathrm{A}_{p,q}(\Sigma,\sigma)$ may be identified as

$$\big(\mathrm{A}_{p,q}(\Sigma,\sigma)\big)' = \mathrm{A}_{p',q'}(\Sigma,\sigma). \tag{10.164}$$

Proof That the supremum in (10.163) is less than or equal to $\|f\|_{\mathrm{A}_{p',q'}(\Sigma,\sigma)}$ follows immediately from item *(a)* in Proposition 10.2. To prove of the opposite inequality, select $\eta \in (0, \infty)$ with the property that

$$0 < \eta < \|f\|_{\mathrm{A}_{p',q'}(\Sigma,\sigma)} = \Big\|\big\{2_*^{k(n-1)(1/q'-1/p')}\|f \cdot \mathbb{1}_{C_k}\|_{L^{p'}(\Sigma,\sigma)}\big\}_{k\in\mathbb{N}_0}\Big\|_{\ell^{q'}(\mathbb{N}_0)}. \tag{10.165}$$

Via duality, there exists some sequence $\{b_k\}_{k\in\mathbb{N}_0} \subseteq \mathbb{R}$ for which

$$\|\{b_k\}_{k\in\mathbb{N}_0}\|_{\ell^q(\mathbb{N}_0)} \leq 1 \text{ and } \eta \leq \sum_{k=0}^{\infty}\big[2_*^{k(n-1)(1/q'-1/p')}\|f \cdot \mathbb{1}_{C_k}\|_{L^{p'}(\Sigma,\sigma)}\big]|b_k|. \tag{10.166}$$

Next, for each $k \in \mathbb{N}_0$, select $\zeta_k \in (0, \infty)$ satisfying

$$0 < \zeta_k < \|f \cdot \mathbb{1}_{C_k}\|_{L^{p'}(\Sigma,\sigma)}. \tag{10.167}$$

Again, by duality, this ensures that for each $k \in \mathbb{N}_0$ there exists some $g_k \in L^p(\Sigma, \sigma)$ such that

$$\|g_k\|_{L^p(\Sigma,\sigma)} \leq 1 \text{ and } \zeta_k \leq \int_\Sigma |f \cdot \mathbb{1}_{C_k}| \cdot |g_k|\,d\sigma. \tag{10.168}$$

Define the function

$$\widetilde{g} := \sum_{\ell=0}^{\infty} 2_*^{\ell(n-1)(1/q'-1/p')}b_\ell(g_\ell \cdot \mathbb{1}_{C_\ell}) \text{ on } \Sigma. \tag{10.169}$$

By design, g is a σ-measurable function which is finite σ-a.e. on Σ. Bearing in mind (10.166), (10.168), and (10.169), we may estimate

$$
\begin{aligned}
&\|\widetilde{g}\|_{\mathrm{A}_{p,q}(\Sigma,\sigma)} \\
&= \Big\{\sum_{k=0}^{\infty}\big(2_*^{k(n-1)(1/q-1/p)}\|\widetilde{g}\cdot \mathbb{1}_{C_k}\|_{L^p(\Sigma,\sigma)}\big)^q\Big\}^{1/q} \\
&= \Big\{\sum_{k=0}^{\infty}\big(2_*^{k(n-1)(1/q-1/p)}2_*^{k(n-1)(1/q'-1/p')}|b_k|\|g_k\cdot \mathbb{1}_{C_k}\|_{L^p(\Sigma,\sigma)}\big)^q\Big\}^{1/q} \\
&\leq \|\{b_k\}_{k\in\mathbb{N}_0}\|_{\ell^q(\mathbb{N}_0)} \leq 1.
\end{aligned}
\tag{10.170}
$$

Hence, $\widetilde{g} \in \mathrm{A}_{p,q}(\Sigma,\sigma)$ and $\|\widetilde{g}\|_{\mathrm{A}_{p,q}(\Sigma,\sigma)} \leq 1$. Furthermore, thanks to (10.168)-(10.169), we have

$$
\begin{aligned}
\int_\Sigma |f\widetilde{g}|\,d\sigma &= \sum_{k=0}^{\infty} 2_*^{\ell(n-1)(1/q'-1/p')}|b_k|\int_{C_k}|f\cdot g_k| \\
&\geq \sum_{k=0}^{\infty} 2_*^{\ell(n-1)(1/q'-1/p')}|b_k|\zeta_k.
\end{aligned}
\tag{10.171}
$$

Letting $\zeta_k \nearrow \|f\cdot \mathbb{1}_{C_k}\|_{L^{p'}(\Sigma,\sigma)}$ for each $k \in \mathbb{N}_0$, using Lebesgue's Monotone Convergence Theorem for series, and (10.166) then leads to the conclusion that

$$
\int_\Sigma |f\widetilde{g}|\,d\sigma \geq \sum_{k=0}^{\infty} 2_*^{\ell(n-1)(1/q'-1/p')}|b_k|\|f\cdot \mathbb{1}_{C_k}\|_{L^{p'}(\Sigma,\sigma)} \geq \eta. \tag{10.172}
$$

Since this holds for every η as in (10.165), we can pass to the limit $\eta \nearrow \|f\|_{\mathrm{A}_{p',q'}(\Sigma,\sigma)}$ to obtain

$$
\int_\Sigma |f\widetilde{g}|\,d\sigma \geq \|f\|_{\mathrm{A}_{p',q'}(\Sigma,\sigma)}. \tag{10.173}
$$

In concert with (10.170), this gives

$$
\begin{aligned}
&\|f\|_{\mathrm{A}_{p',q'}(\Sigma,\sigma)} \\
&\quad\leq \sup\Big\{\int_\Sigma |fg|\,d\sigma : g \in \mathrm{A}_{p,q}(\Sigma,\sigma),\ \text{with } \|g\|_{\mathrm{A}_{p,q}(\Sigma,\sigma)} \leq 1\Big\}.
\end{aligned}
\tag{10.174}
$$

This concludes the justification of (10.163). Finally, (10.164) is a direct consequence of (10.163), and (10.147)-(10.148). The proof of Proposition 10.12 is now complete. □

The relevance of viewing Herz spaces as Generalized Banach Function Spaces is most apparent from the following extrapolation result.

Theorem 10.3 *Suppose $\Sigma \subseteq \mathbb{R}^n$ is a closed unbounded Ahlfors regular set and abbreviate $\sigma := \mathcal{H}^{n-1}\lfloor\Sigma$. Also, denote by $\mathcal{M}_\Sigma$ the Hardy-Littlewood maximal operator on Σ (cf. (2.26)). Then there exists some $C_\Sigma \in (0,\infty)$ depending only on the dimension n and the Ahlfors regularity constants of Σ with the following* *significance. Having fixed integrability exponents $p, q, p_0 \in (1, \infty)$, introduce the constant*

$$W_{p,q,p_0} := C_\Sigma^{p_0} \|\mathcal{M}_\Sigma\|^{p_0-1}_{\mathrm{A}_{p,q}(\Sigma,\sigma)\to\mathrm{A}_{p,q}(\Sigma,\sigma)} \cdot \|\mathcal{M}_\Sigma\|_{\mathrm{A}_{p',q'}(\Sigma,\sigma)\to\mathrm{A}_{p',q'}(\Sigma,\sigma)}, \tag{10.175}$$

which is finite thanks to item (b) in Theorem 10.1, where $p', q' \in (1, \infty)$ denote the Hölder conjugate exponent of p and q, respectively. Also, assume that f and g are two σ-measurable functions defined on Σ with the property that for each Muckenhoupt weight $w \in A_{p_0}(\Sigma, \sigma)$ with $[w]_{A_{p_0}(\Sigma,\sigma)} \leq W_{p,q,p_0}$ one has

$$\|f\|_{L^{p_0}(\Sigma,w)} \leq C_w \|g\|_{L^{p_0}(\Sigma,w)}, \tag{10.176}$$

for some constant $C_w \in (0, \infty)$ which depends only on Σ, n, p_0, and $[w]_{A_{p_0}}$. Then

$$\|f\|_{\mathrm{A}_{p,q}(\Sigma,\sigma)} \leq 2^{2-1/p_0} \Bigg(\sup_{\substack{w\in A_p(\Sigma,\sigma) \\ [w]_{A_p} \leq W_{p,q,p_0}}} C_w \Bigg) \|g\|_{\mathrm{A}_{p,q}(\Sigma,\sigma)}. \tag{10.177}$$

Proof Thanks to Proposition 10.11, Proposition 10.12, and the boundedness of the Hardy-Littlewood maximal operator $\mathcal{M}_\Sigma$ on $\mathrm{A}_{p,q}(\Sigma, \sigma)$ (cf. item *b)* in Theorem 10.1), all claims in Theorem 10.3 are consequences of the extrapolation result recorded in Theorem 10.2. □

Chapter 11
Herz-type Hardy Spaces of First Generation

The central aim of this chapter is to introduce the scale of first-generation Herz-type Hardy spaces $\mathrm{HA}_{p,q}$ in the setting of unbounded Ahlfors regular sets. Significantly, we develop an atomic and molecular theory for this scale of spaces, which plays a fundamental role in the treatments of singular integral operators on these spaces.

11.1 Definitions and Basic Properties

The Hardy spaces associated with first-generation Herz spaces are defined by demanding the membership of the Fefferman-Stein grand maximal function to $\mathrm{A}_{p,q}$. More precisely, we make the following definition.

Definition 11.1 Assume $\Sigma \subseteq \mathbb{R}^n$ is a closed unbounded Ahlfors regular set, and abbreviate $\sigma := \mathcal{H}^{n-1}\lfloor\Sigma$. Consider two integrability exponents $p, q \in (0, \infty)$, and a background parameter $\gamma \in (0, 1)$. The `Herz-type Hardy space` is defined as

$$\mathrm{HA}_{p,q}(\Sigma, \sigma) := \big\{ f \in \big(\mathrm{Lip}_c(\Sigma)\big)' : f^{\#}_{\gamma} \in \mathrm{A}_{p,q}(\Sigma, \sigma) \big\}. \tag{11.1}$$

Equip this space with the quasinorm

$$\|f\|_{\mathrm{HA}_{p,q}(\Sigma,\sigma)} := \|f^{\#}_{\gamma}\|_{\mathrm{A}_{p,q}(\Sigma,\sigma)} \text{ for every } f \in \mathrm{HA}_{p,q}(\Sigma, \sigma). \tag{11.2}$$

Since in the case when $p = q$ the Herz scale $\mathrm{A}_{p,p}$ coincides with the Lebesgue scale L^p (cf. (10.7)), we see from (11.1) that

$$\mathrm{HA}_{p,p}(\Sigma, \sigma) = H^p(\Sigma, \sigma) \text{ for all } p \in \big(\tfrac{n-1}{n}, \infty\big), \tag{11.3}$$

where $H^p(\Sigma, \sigma)$ is the (Lebesgue-based) Hardy space on Σ (from [GHA-II, Definition 4.2.1]. As such, the scale of Herz-type Hardy spaces may be regarded as an extension of standard Hardy spaces. Moreover, in the case when $q = 1$, we obtain

© The Author(s), under exclusive license to Springer Nature Switzerland AG 2026
M. Mitrea and P. Takemura, *Singular Integrals, Herz-Type Function Spaces, and Boundary Problems*, Progress in Mathematics 362,
https://doi.org/10.1007/978-3-032-12516-3_11

$$\mathrm{HA}_{p,1}(\Sigma,\sigma)=\mathrm{HA}^p(\Sigma,\sigma). \tag{11.4}$$

This is a simple consequence of (10.5) and (11.1). Hence, Herz-type Hardy spaces also generalize the scale of Beurling-Hardy spaces.

Upon recalling (10.1), observe that that whenever $p\in(1,\infty)$ and $q\in(0,1]$, the quasinorm (11.2) is a q-norm, since for all $f,g\in\mathrm{HA}_{p,q}(\Sigma,\sigma)$ we have

$$\|f+g\|^q_{\mathrm{HA}_{p,q}(\Sigma,\sigma)}\leq\|f\|^q_{\mathrm{HA}_{p,q}(\Sigma,\sigma)}+\|g\|^q_{\mathrm{HA}_{p,q}(\Sigma,\sigma)}. \tag{11.5}$$

Remark 11.1 Perhaps a more accurate piece of notation for our brand of Hardy space introduced above would be $\mathrm{H}_\gamma\mathrm{A}_{p,q}$ to emphasize a natural dependence on the parameter γ. This being said, in Theorem 11.1 and Theorem 11.2 we are going to show that $\mathrm{H}_\gamma\mathrm{A}_{p,q}$ is independent of γ as long as this parameter belongs to $\big((n-1)(1/q-1),1\big)$. For this reason, we may drop the dependence on γ and simply write $\mathrm{HA}_{p,q}$.

In the following result we provide alternative characterizations of the Herz-type Hardy spaces introduced above.

Proposition 11.1 *Consider a closed unbounded Ahlfors regular set $\Sigma\subseteq\mathbb{R}^n$, and abbreviate $\sigma:=\mathcal{H}^{n-1}\lfloor\Sigma$. Pick two exponents $p\in[1,\infty)$, and $q\in(0,\infty)$, and a parameter $\gamma\in(0,1)$. Then the following statements are equivalent.*

(a) $f\in\mathrm{HA}_{p,q}(\Sigma,\sigma)$.

(b) $f\in L^p_{\mathrm{loc}}(\Sigma,\sigma)$ *and* $f^\#_\gamma\in\mathrm{A}_{p,q}(\Sigma,\sigma)$.

(c) $f\in L^1_{\mathrm{loc}}(\Sigma,\sigma)$ *and* $f^\#_\gamma\in\mathrm{A}_{p,q}(\Sigma,\sigma)$.

(d) $f\in\mathrm{A}_{p,q}(\Sigma,\sigma)$ *and* $f^\#_\gamma\in\mathrm{A}_{p,q}(\Sigma,\sigma)$.

Proof The implication *(a)*⇒*(b)* follows from (11.1), item *(f)* in Proposition 10.3, and item *(c)* in Proposition 4.1. The fact that *(b)*⇒*(c)* is a consequence $L^p_{\mathrm{loc}}(\Sigma,\sigma)$ being included $L^1_{\mathrm{loc}}(\Sigma,\sigma)$. Moreover, *(c)*⇒*(d)* is obtained from item *(a)* from Proposition 4.1 along with the lattice property of $\mathrm{A}_{p,q}(\Sigma,\sigma)$ (cf. (10.8)). Lastly, the veracity of *(d)*⇒*(a)* is seen from the inclusions $\mathrm{A}_{p,q}(\Sigma,\sigma)\subseteq L^1_{\mathrm{loc}}(\Sigma,\sigma)\subseteq(\mathrm{Lip}_c(\Sigma))'$. □

In the regime $p\in[1,\infty)$, and $q\in(0,\infty)$, Proposition 11.1 permits us to obtain the following description for Herz-type Hardy spaces

$$\begin{aligned}\mathrm{HA}_{p,q}(\Sigma,\sigma)&=\big\{f\in L^p_{\mathrm{loc}}(\Sigma,\sigma):f^\#_\gamma\in\mathrm{A}_{p,q}(\Sigma,\sigma)\big\}\\&=\big\{f\in L^1_{\mathrm{loc}}(\Sigma,\sigma):f^\#_\gamma\in\mathrm{A}_{p,q}(\Sigma,\sigma)\big\}\\&=\big\{f\in\mathrm{A}_{p,q}(\Sigma,\sigma):f^\#_\gamma\in\mathrm{A}_{p,q}(\Sigma,\sigma)\big\}.\end{aligned} \tag{11.6}$$

Hence, the elements of $\mathrm{HA}_{p,q}(\Sigma,\sigma)$ are genuine functions in the aforementioned regime.

Remark 11.2 If matters are specialized to the case when $0 < q \leq 1 < p < \infty$, we obtain

$$\begin{aligned} \mathrm{HA}_{p,q}(\Sigma,\sigma) &= \{f \in L^1(\Sigma,\sigma) : f_\gamma^\# \in \mathrm{A}_{p,q}(\Sigma,\sigma)\} \\ &= \{f \in L^p(\Sigma,\sigma) : f_\gamma^\# \in \mathrm{A}_{p,q}(\Sigma,\sigma)\}. \end{aligned} \tag{11.7}$$

This is a consequence of property *(c)* from Proposition 4.1, and item *(c)* in Proposition 10.3 (used with $r := 1$). In particular,

$$\begin{aligned} &\mathrm{HA}_{p,q}(\Sigma,\sigma) = \{f \in L^r(\Sigma,\sigma) : f_\gamma^\# \in \mathrm{A}_{p,q}(\Sigma,\sigma)\} \\ &\quad \text{whenever } 0 < q \leq 1 < p < \infty \text{ and } 1 \leq r \leq p. \end{aligned} \tag{11.8}$$

Remarkably, much as in the classical theory of Hardy spaces, the phenomenon regarding the coincidence of the Hardy spaces H^p and the Lebesgue spaces L^p, in the regime $p \in (1, \infty)$ has a natural counterpart in the context of Herz-type Hardy spaces $\mathrm{HA}_{p,q}$ for a certain range of p and q. This is made precise in the following result.

Proposition 11.2 *Suppose $\Sigma \subseteq \mathbb{R}^n$ is a closed unbounded Ahlfors regular set, and abbreviate $\sigma := \mathcal{H}^{n-1}\lfloor\Sigma$. Then for any two exponents $p, q \in (1, \infty)$ it follows that*

$$\mathrm{HA}_{p,q}(\Sigma,\sigma) = \mathrm{A}_{p,q}(\Sigma,\sigma) \quad \textit{with equivalent norms.} \tag{11.9}$$

Proof The inclusion $\mathrm{HA}_{p,q}(\Sigma,\sigma) \subseteq \mathrm{A}_{p,q}(\Sigma,\sigma)$ is a consequence of Proposition 11.1. Moreover, item *(a)* in Proposition 4.1, the lattice property of $\mathrm{A}_{p,q}(\Sigma,\sigma)$ (cf. (10.8)), and the inclusion $\mathrm{A}_{p,q}(\Sigma,\sigma) \subseteq L^1_{\mathrm{loc}}(\Sigma,\sigma)$ (cf. item *(f)* in Proposition 10.3) imply that there exists a constant $c \in (0, \infty)$ such that whenever $f \in \mathrm{A}_{p,q}(\Sigma,\sigma)$ we have $|f| \leq c \cdot f_\gamma^\#$ at σ-a.e. point on Σ. Therefore,

$$\|f\|_{\mathrm{A}_{p,q}(\Sigma,\sigma)} \leq c\|f_\gamma^\#\|_{\mathrm{A}_{p,q}(\Sigma,\sigma)} = c\|f\|_{\mathrm{HA}_{p,q}(\Sigma,\sigma)}. \tag{11.10}$$

On the other hand, pick $f \in \mathrm{A}_{p,q}(\Sigma,\sigma)$, which in particular puts f in $L^1_{\mathrm{loc}}(\Sigma,\sigma)$ (cf. item *(f)* in Proposition 10.3). Thanks to (4.7), and the boundedness of the Hardy-Littlewood maximal operator on $\mathrm{A}_{p,q}(\Sigma,\sigma)$ (cf. Theorem 10.1) we may estimate

$$\begin{aligned} \|f\|_{\mathrm{HA}_{p,q}(\Sigma,\sigma)} &= \|f_\gamma^\#\|_{\mathrm{A}_{p,q}(\Sigma,\sigma)} \leq \|\mathcal{M}_\Sigma f\|_{\mathrm{A}_{p,q}(\Sigma,\sigma)} \\ &\leq C\|f\|_{\mathrm{A}_{p,q}(\Sigma,\sigma)} < \infty, \end{aligned} \tag{11.11}$$

for some constant $C \in (0, \infty)$ independent of f. This then implies $f \in \mathrm{HA}_{p,q}(\Sigma,\sigma)$, which in turn concludes the proof of $\mathrm{HA}_{p,q}(\Sigma,\sigma) = \mathrm{A}_{p,q}(\Sigma,\sigma)$ as vector spaces. The fact that the norms $\|\cdot\|_{\mathrm{HA}_{p,q}(\Sigma,\sigma)}$ and $\|\cdot\|_{\mathrm{A}_{p,q}(\Sigma,\sigma)}$ are equivalent follows from (11.10) and (11.11). □

In view of Proposition 11.2, we focus on developing the theory of Herz-type Hardy spaces in the regime $q \in (0, 1]$, and $p \in (1, \infty)$. For these ranges we have the following embedding results.

Proposition 11.3 *Consider a closed unbounded Ahlfors regular set* $\Sigma \subseteq \mathbb{R}^n$, *and abbreviate* $\sigma := \mathcal{H}^{n-1}\lfloor\Sigma$. *Also, fix two exponents* $p \in (1,\infty)$ *and* $q \in (0,1]$. *In this setting, one has the following embeddings. First, if* $p_1, p_2 \in (0,\infty)$ *and* $p_1 \leq p_2$, *then*

$$\mathrm{HA}_{p_2,q}(\Sigma,\sigma) \hookrightarrow \mathrm{HA}_{p_1,q}(\Sigma,\sigma) \quad \textit{continuously,} \tag{11.12}$$

i.e., the scale of Herz-type spaces $\mathrm{HA}_{p,q}(\Sigma,\sigma)$ *is monotonically decreasing in the exponent* p. *Consequently,*

$$\mathrm{HA}_{p,q}(\Sigma,\sigma) \hookrightarrow H^q(\Sigma,\sigma) \quad \textit{continuously.} \tag{11.13}$$

Second, if $q_1, q_2 \in (0,\infty)$ *and* $q_1 \leq q_2$, *then*

$$\mathrm{HA}_{p,q_1}(\Sigma,\sigma) \hookrightarrow \mathrm{HA}_{p,q_2}(\Sigma,\sigma) \quad \textit{continuously,} \tag{11.14}$$

i.e., the scale of Herz-type spaces $\mathrm{HA}_{p,q}(\Sigma,\sigma)$ *is monotonically increasing in the exponent* q. *In particular,*

$$\mathrm{HA}_{p,q}(\Sigma,\sigma) \hookrightarrow \mathrm{HA}^p(\Sigma,\sigma) \quad \textit{continuously.} \tag{11.15}$$

Additionally, one has

$$\mathrm{HA}_{p,q}(\Sigma,\sigma) \hookrightarrow \mathrm{A}_{p,q}(\Sigma,\sigma) \quad \textit{continuously.} \tag{11.16}$$

As a consequence, whenever $1 \leq r \leq p$ *one also has the continuous embeddings:*

$$\mathrm{HA}_{p,q}(\Sigma,\sigma) \hookrightarrow L^r(\Sigma,\sigma), \tag{11.17}$$

and

$$\mathrm{HA}_{p,q}(\Sigma,\sigma) \hookrightarrow L^1\Big(\Sigma, \frac{\sigma(x)}{1+|x|^{n-1}}\Big). \tag{11.18}$$

Lastly, the following inclusions hold:

$$\bigcup_{p<p_1<\infty} \mathrm{HA}_{p_1,q}(\Sigma,\sigma) \subseteq \mathrm{HA}_{p,q}(\Sigma,\sigma) \subseteq L^p_{\mathrm{loc}}(\Sigma,\sigma) \subseteq L^1_{\mathrm{loc}}(\Sigma,\sigma). \tag{11.19}$$

Proof The embeddings claimed in (11.12) and (11.15) are consequences of items *(a)* and *(b)* in Proposition 10.3, together with (11.1)-(11.2). As such, (11.13) can be obtained from (11.3), and (11.12) used with $p_2 := p$ and $p_1 = q$. Also, taking $q_1 := q$ and $q_2 := 1$ in (11.14), and recalling (11.4), yields the continuous embedding (11.15).

Next, the validity of (11.16) can be seen from the second line in (11.6), item *(c)* in Proposition 4.1, and the lattice property of $\mathrm{A}_{p,q}(\Sigma,\sigma)$ (cf. (10.8)). A combination of (10.29) and (11.16) readily gives (11.17). Thanks to (11.17), and bearing in mind that $L^p(\Sigma,\sigma) \hookrightarrow L^1(\Sigma, \frac{\sigma(x)}{1+|x|^{n-1}})$ continuously, we also obtain the embedding (11.18). Finally, the inclusions (11.19) can be seen from (11.12), and (11.6). □

Let us now prove the completeness of Herz-type Hardy spaces.

Proposition 11.4 *Assume* $\Sigma \subseteq \mathbb{R}^n$ *is a closed unbounded Ahlfors regular set, and abbreviate* $\sigma := \mathcal{H}^{n-1}\lfloor\Sigma$. *Select two exponents* $p \in (1, \infty)$, *and* $q \in (0, \infty)$. *Then the Herz-type Hardy space* $\mathrm{HA}_{p,q}(\Sigma, \sigma)$ *is complete when equipped with* $\|\cdot\|_{\mathrm{HA}_{p,q}(\Sigma,\sigma)}$. *In particular,* $\mathrm{HA}_{p,q}(\Sigma, \sigma)$ *is a Banach space in the range* $q \in [1, \infty)$, *and a quasi-Banach space whenever* $q \in (0, 1)$.

Proof In the case when $q \in (1, \infty)$, the completeness of $\mathrm{HA}_{p,q}(\Sigma, \sigma)$ is a consequence of the fact that this space coincides with $\mathrm{A}_{p,q}(\Sigma, \sigma)$ (cf. Proposition 11.2) along with the completeness of $\mathrm{A}_{p,q}(\Sigma, \sigma)$ proved in Proposition 10.6. Also, if $q = 1$ then $\mathrm{HA}_{p,1}(\Sigma, \sigma)$ coincides with the Beurling-Hardy space $\mathrm{HA}^p(\Sigma, \sigma)$, which according to Proposition 4.3 is complete. In the case when $q \in (0, 1)$, the completeness of $\mathrm{HA}_{p,q}(\Sigma, \sigma)$ can be verified by using the same type of argument as in the proof of Proposition 4.3, while now bearing in mind (11.16) and (11.19) from Proposition 11.3. □

11.2 Atomic Theory in First-Generation Herz-type Hardy Spaces

Our main result in this section establishes an atomic characterization for first-generation Herz-type Hardy spaces $\mathrm{HA}_{p,q}$ (cf. Theorem 11.2) which is of fundamental importance to our work. We begin by introducing the notion of a central (q, p)-atom in the setting of Ahlfors regular sets.

Definition 11.2 Suppose $\Sigma \subseteq \mathbb{R}^n$ is a closed unbounded Ahlfors regular set, and abbreviate $\sigma := \mathcal{H}^{n-1}\lfloor\Sigma$. Denote by $c_\Sigma, C_\Sigma \in (0, \infty)$ the lower and upper Ahlfors regularity constants of Σ, respectively. Consider a reference point $x_0 \in \Sigma$ along with a dilation factor $2_* > (C_\Sigma/c_\Sigma)^{1/(n-1)}$. Fix two exponents $p \in (1, \infty)$ and $q \in (0, \infty)$. A σ-measurable function $a : \Sigma \to \mathbb{R}$ is called a x_0-`central` (q, p)-`atom` provided there exists a finite radius $R \geq 2_*$ such that the following three properties are true:

(1) Localization Condition:

$$\operatorname{supp} a \subseteq \Delta(x_0, R). \tag{11.20}$$

(2) Normalization Condition:

$$\|a\|_{L^p(\Sigma,\sigma)} \leq R^{(n-1)(1/p-1/q)}. \tag{11.21}$$

(3) Vanishing Moment Condition:

$$\int_\Sigma a\, d\sigma = 0. \tag{11.22}$$

Remark 11.3 Notice that the localization and normalization properties (11.20)-(11.21) imply that a is an absolutely integrable function on Σ. As such, the integral in (11.22) is meaningfully defined.

The purpose of the following result is to establish the existence of a constant that bounds the $\mathrm{HA}_{p,q}$-quasinorm of every central (q,p)-atom.

Theorem 11.1 *Assume $\Sigma \subseteq \mathbb{R}^n$ is a closed unbounded Ahlfors regular set, and abbreviate $\sigma := \mathcal{H}^{n-1}\lfloor\Sigma$. Fix a reference point $x_0 \in \Sigma$. Select two exponents $p \in (1,\infty)$ and $q \in \big(\frac{n-1}{n},1\big]$, along with a parameter $\gamma \in \big((n-1)(1/q-1),1\big)$. Then there exists a constant $C \in (0,\infty)$ depending only on p, q, n, γ, and the Ahlfors regularity constants of Σ, with the property that for every x_0-central (q,p)-atom $a : \Sigma \to \mathbb{R}$ one has*

$$\|a\|_{\mathrm{HA}_{p,q}(\Sigma,\sigma)} = \|a^{\#}_{\gamma}\|_{\mathrm{A}_{p,q}(\Sigma,\sigma)} \leq C. \tag{11.23}$$

Proof The proof is similar to that of Theorem 4.1. As such, here we only emphasize the main differences. Let 2_* be as in Definition 11.2, and fix an x_0-central (q,p)-atom a on Σ. Denote by $R \geq 2_*$ a radius associated with a in the sense of Definition 11.2. Select $\ell \in \mathbb{N}$ the unique integer for which

$$2_*^{\ell} \leq R < 2_*^{\ell+1}. \tag{11.24}$$

Recall the family of annuli $\{C_k\}_{k\in\mathbb{N}_0}$ from (3.1). Having fixed a number $k \in \mathbb{N}_0$ and $x \in C_k$, pick a bump function $\psi \in \mathcal{B}^{\gamma}_{x}(\Sigma)$. Employing the same type of computations as in (4.38), now bearing in mind the normalization condition (11.21) of a, yields

$$|\langle a,\psi\rangle| \leq C\frac{2_*^{(\ell-k)\gamma}}{2_*^{k(n-1)}}2_*^{\ell(n-1)(1-1/q)} \quad \text{whenever } k \geq \ell+2, \tag{11.25}$$

for some constant $C \in (0,\infty)$ depending only on n, γ, and the Ahlfors regularity character of Σ. Taking the supremum over all $\psi \in \mathcal{B}^{\gamma}_{x}(\Sigma)$ in (11.25) gives

$$a^{\#}_{\gamma}(x) \leq C\frac{2_*^{(\ell-k)\gamma}}{2_*^{k(n-1)}}2_*^{\ell(n-1)(1-1/q)} \quad \text{for all } x \in C_k \text{ and } k \in \mathbb{N}_0 \text{ with } k \geq \ell+2. \tag{11.26}$$

Keeping this in mind and using the Ahlfors regularity of Σ, for all $k \in \mathbb{N}_0$ with $k \geq \ell+2$ we have

$$\begin{aligned}\|a^{\#}_{\gamma}\cdot\mathbb{1}_{C_k}\|_{L^p(\Sigma,\sigma)} &\leq C\frac{2_*^{(\ell-k)\gamma}}{2_*^{k(n-1)}}2_*^{\ell(n-1)(1-1/q)}\sigma(C_k)^{1/p}\\ &\leq C\frac{2_*^{(\ell-k)\gamma}}{2_*^{k(n-1)}}2_*^{\ell(n-1)(1-1/q)}2_*^{k(n-1)/p},\end{aligned} \tag{11.27}$$

for some constant $C = C(n, p, \gamma, \Sigma) \in (0, \infty)$. This and $(n-1)(1/q-1) < \gamma < 1$ (which forces $1 - 1/q + \gamma/(n-1) > 0$) allow us to write

$$
\begin{aligned}
&\sum_{k=\ell+2}^{\infty} \big(2_*^{k(n-1)(1/q-1/p)} \|a_\gamma^\# \cdot \mathbb{1}_{C_k}\|_{L^p(\Sigma,\sigma)}\big)^q \\
&\quad \leq \sum_{k=\ell+2}^{\infty} \Big(2_*^{k(n-1)(1/q-1/p)} \frac{2_*^{(\ell-k)\gamma}}{2_*^{k(n-1)}} 2_*^{\ell(n-1)(1-1/q)} 2_*^{k(n-1)/p}\Big)^q \\
&\quad = C \sum_{k=\ell+2}^{\infty} \big(2_*^{k(n-1)(1/q-1)} 2_*^{(\ell-k)\gamma} 2_*^{\ell(n-1)(1-1/q)}\big)^q \\
&\quad = C \sum_{k=\ell+2}^{\infty} 2_*^{(\ell-k)(n-1)[1-1/q+\gamma/(n-1)]} \\
&\quad \leq C \sum_{j=2}^{\infty} 2_*^{-j(n-1)[1-1/q+\gamma/(n-1)]} < \infty,
\end{aligned}
\tag{11.28}
$$

for some constant $C = C(n, p, q, \gamma, \Sigma) \in (0, \infty)$.

Pressing on, using estimate (4.7), the boundedness of the Hardy-Littlewood maximal operator $\mathcal{M}_\Sigma$ on $L^p(\Sigma, \sigma)$, the normalization property of a (cf. (11.21)), and (11.24), we may estimate

$$
\begin{aligned}
&\sum_{k=0}^{\ell+1}\big(2_*^{k(n-1)(1/q-1/p)}\|a_\gamma^\# \cdot \mathbb{1}_{C_k}\|_{L^p(\Sigma,\sigma)}\big)^q \\
&\qquad \le \|\mathcal{M}_\Sigma a\|_{L^p(\Sigma,\sigma)}\sum_{k=0}^{\ell+1} 2_*^{k(n-1)(1/q-1/p)q} \\
&\qquad \le C\|a\|_{L^p(\Sigma,\sigma)}\sum_{k=0}^{\ell+1} 2_*^{k(n-1)(1/q-1/p)q} \\
&\qquad \le CR^{(n-1)(1/p-1/q)}\sum_{k=0}^{\ell+1} 2_*^{k(n-1)(1/q-1/p)q} \\
&\qquad \le C2_*^{\ell(n-1)(1/p-1/q)}\sum_{k=0}^{\ell+1} 2_*^{k(n-1)(1/q-1/p)q} \\
&\qquad = C\sum_{k=0}^{\ell+1} 2_*^{(k-\ell)(n-1)(1/q-1/p)q} \\
&\qquad \le C\sum_{j=0}^{\infty} 2_*^{-j(n-1)(1/q-1/p)q} < \infty,
\end{aligned}
\tag{11.29}
$$

for some constant $C = C(n, p, q, \gamma, \Sigma) \in (0, \infty)$.

Together, estimates (11.28) and (11.29) ensure the existence of some $C \in (0, \infty)$ depending only on n, p, q, γ, and the Ahlfors regularity constants of Σ with the property that

$$
\begin{aligned}
\|a\|^q_{\mathrm{HA}_{p,q}(\Sigma,\sigma)} &= \|a_\gamma^\#\|^q_{\mathrm{A}_{p,q}(\Sigma,\sigma)} \\
&= \sum_{k=0}^{\ell+1}\big(2_*^{k(n-1)(1/q-1/p)}\|a_\gamma^\# \cdot \mathbb{1}_{C_k}\|_{L^p(\Sigma,\sigma)}\big)^q \\
&\quad + \sum_{k=\ell+2}^{\infty}\big(2_*^{k(n-1)(1/q-1/p)}\|a_\gamma^\# \cdot \mathbb{1}_{C_k}\|_{L^p(\Sigma,\sigma)}\big)^q \le C.
\end{aligned}
\tag{11.30}
$$

This readily implies (11.23), and concludes the proof of Theorem 11.1. □

We are now in a position to state and prove the main result of this section, namely, the atomic characterization for first-generation Herz-type Hardy spaces.

Theorem 11.2 *Suppose $\Sigma \subseteq \mathbb{R}^n$ is a closed unbounded Ahlfors regular set, and abbreviate $\sigma := \mathcal{H}^{n-1}\lfloor\Sigma$. Fix a reference point $x_0 \in \Sigma$, along with two exponents $p \in (1, \infty)$ and $q \in \big(\frac{n-1}{n}, 1\big]$. Then, in this context, the following statements are true.*

(a) *For every* $f \in \mathrm{HA}_{p,q}(\Sigma, \sigma)$ *there exist a sequence* x_0*-central* (q, p)*-atoms* $\{a_k\}_{k\in\mathbb{N}_0}$ *and a numerical sequence* $\{\lambda_k\}_{k\in\mathbb{N}_0} \subseteq \mathbb{R}$ *such that*

$$f = \sum_{k=0}^{\infty} \lambda_k a_k \tag{11.31}$$

with simultaneous convergence in $L^r(\Sigma, \sigma)$ *for every* $r \in [1, p]$, *and in* $\mathrm{HA}_{p,q}(\Sigma, \sigma)$. *Moreover, there exists a constant* $C \in (0, \infty)$ *depending only* n, p, q, *and the Ahlfors regularity character of* Σ *with the property that*

$$\Big(\sum_{k=0}^{\infty} |\lambda_k|^q\Big)^{1/q} \leq C\|f\|_{\mathrm{HA}_{p,q}(\Sigma,\sigma)}. \tag{11.32}$$

(b) *Conversely, assume* $f : \Sigma \to \mathbb{R}$ *is a* σ*-measurable function for which there exists some sequence of* x_0*-central* (q, p)*-atoms* $\{a_k\}_{k\in\mathbb{N}_0}$ *and a numerical sequence* $\{\lambda_k\}_{k\in\mathbb{N}_0} \in \ell^q(\mathbb{N}_0)$ *satisfying*

$$f = \sum_{k=0}^{\infty} \lambda_k a_k \ \text{ pointwise } \sigma\text{-a.e. on } \Sigma. \tag{11.33}$$

Then $f \in \mathrm{HA}_{p,q}(\Sigma, \sigma)$ *quantitatively, i.e., there exists a constant* $C \in (0, \infty)$ *depending only* n, p, q, *and the Ahlfors regularity character of* Σ *such that*

$$\|f\|_{\mathrm{HA}_{p,q}(\Sigma,\sigma)} \leq C\Big(\sum_{k=0}^{\infty} |\lambda_k|^q\Big)^{1/q}. \tag{11.34}$$

(c) *For every* $f \in \mathrm{HA}_{p,q}(\Sigma, \sigma)$ *and* $r \in [1, p]$ *one has*

$$\|f\|_{\mathrm{HA}_{p,q}(\Sigma,\sigma)} \approx \inf\Big\{\Big(\sum_{k=0}^{\infty} |\lambda_k|^q\Big)^{1/q} : f = \sum_{k=0}^{\infty} \lambda_k a_k \text{ in } L^r(\Sigma, \sigma)\Big\}, \tag{11.35}$$

where the infimum is taken over all atomic representations of f, *and with implicit proportionality constants depending only on* Σ, n, p, q, *and* r.

Proof The approach used to prove the present theorem parallels that of Theorem 4.2. For this reason, here we only highlight the main differences. Throughout, fix a background parameter $\gamma \in (0, 1)$ satisfying

$$(n - 1)(1/q - 1) < \gamma < 1. \tag{11.36}$$

Consider first the claim made in *(a)*. Recall the construction of the smooth partition of unit $\{\phi_k\}_{k\in\mathbb{N}_0}$ made in (4.50)-(4.78), where

$$\phi_k = \frac{\psi_k}{\sum_{j=0}^{\infty} \psi_j} \quad \text{in } \mathbb{R}^n \text{ for each } k \in \mathbb{N}_0, \tag{11.37}$$

and the function ψ_k is as in (4.53). On account of (4.79)-(4.86) the following properties have been obtained:

$$\begin{aligned} &0 \le \phi_k \le 1 \ \text{ in } \mathbb{R}^n, \ \ \operatorname{supp}\big(\phi_k\big|_\Sigma\big) \subseteq \widetilde{A}_k \ \text{ for all } k \in \mathbb{N}_0, \\ &\text{and } \sum_{k=0}^{\infty} \phi_k = 1 \ \text{ at every point in } \mathbb{R}^n. \end{aligned} \tag{11.38}$$

Next, fix an arbitrary function $f \in \mathrm{HA}_{p,q}(\Sigma, \sigma)$. The property recorded in the second line of (11.38) yields

$$f = \sum_{k=0}^{\infty} f \cdot \big(\phi_k\big|_\Sigma\big) \ \text{ pointwise } \sigma\text{-a.e. on } \Sigma. \tag{11.39}$$

Since we have the pointwise domination

$$\Big|\sum_{k=0}^{N} f \cdot \big(\phi_k\big|_\Sigma\big)\Big| \le |f| \sum_{k=0}^{N} \big(\phi_k\big|_\Sigma\big) \le |f| \quad \text{for every } N \in \mathbb{N}. \tag{11.40}$$

and the membership $f \in L^r(\Sigma, \sigma)$ for every $r \in [1, p]$ (cf. (11.8)), Lebesgue's Dominated Convergence Theorem implies

$$f = \sum_{k=0}^{\infty} f \cdot \big(\phi_k\big|_\Sigma\big) \ \text{ in } L^r(\Sigma, \sigma) \text{ for every } r \in [1, p]. \tag{11.41}$$

Retaining the same piece of notation introduced in (4.92) and (4.114), the computations employed in (4.91)-(4.93) and (4.105)-(4.115) in our current scenario entail the following writing for f:

$$f = \sum_{k=0}^{\infty} g_k + \sum_{k=0}^{\infty} h_k \ \text{ pointwise } \sigma\text{-a.e. on } \Sigma. \tag{11.42}$$

In light of (11.41), this convergence also holds in $L^r(\Sigma, \sigma)$ whenever $1 \le r \le p$. Consider the functions g_k and h_k (defined on (4.92) and (4.114), respectively), and recall from (4.93)-(4.95) and (4.116)-(4.135) that we have obtained the following properties:

$$
\begin{aligned}
&\operatorname{supp} g_k \subseteq \Delta(x_0, 2_*^{k+2}) \text{ and } \int_\Sigma g_k \, d\sigma = 0 \text{ for each } k \in \mathbb{N}_0, \\
&\|g_0\|_{L^p(\Sigma,\sigma)} \leq 2 \sum_{j=0}^{1} \|f \cdot \mathbb{1}_{C_j}\|_{L^p(\Sigma,\sigma)}, \\
&\|g_k\|_{L^p(\Sigma,\sigma)} \leq 2 \sum_{j=k-1}^{k+1} \|f \cdot \mathbb{1}_{C_j}\|_{L^p(\Sigma,\sigma)} \text{ for each } k \in \mathbb{N},
\end{aligned}
\tag{11.43}
$$

while

$$
\begin{aligned}
&\operatorname{supp} h_k \subseteq \Delta(x_0, 2_*^{k+3}) \text{ and } \int_\Sigma h_k \, d\sigma = 0 \text{ for each } k \in \mathbb{N}_0, \\
&\|h_0\|_{L^p(\Sigma,\sigma)} \leq \widetilde{C} \sum_{j=0}^{2} \|f_\gamma^\# \cdot \mathbb{1}_{C_j}\|_{L^p(\Sigma,\sigma)}, \\
&\|h_k\|_{L^p(\Sigma,\sigma)} \leq \widetilde{C} \sum_{j=k-1}^{k+2} \|f_\gamma^\# \cdot \mathbb{1}_{C_j}\|_{L^p(\Sigma,\sigma)} \text{ for each } k \in \mathbb{N}.
\end{aligned}
\tag{11.44}
$$

Next, recall the set of indices $\mathcal{J}_1$ and $\mathcal{J}_2$ as in (4.96) and (4.136), and for each $k \in \mathbb{N}_0$ define the functions $a_k, b_k : \Sigma \to \mathbb{R}$ by setting

$$
\begin{aligned}
a_0 &:= \begin{cases} \Big(2(2_*^2)^{(n-1)(1/q-1/p)} \sum_{\ell=0}^{1} \|f \cdot \mathbb{1}_{C_\ell}\|_{L^p(\Sigma,\sigma)}\Big)^{-1} g_0 & \text{if } 0 \in \mathcal{J}_1, \\ 0 & \text{if } 0 \notin \mathcal{J}_1, \end{cases} \\
a_k &:= \begin{cases} \Big(2(2_*^{k+2})^{(n-1)(1/q-1/p)} \sum_{\ell=k-1}^{k+1} \|f \cdot \mathbb{1}_{C_\ell}\|_{L^p(\Sigma,\sigma)}\Big)^{-1} g_k & \text{if } k \in \mathcal{J}_1, \\ 0 & \text{if } k \notin \mathcal{J}_1, \end{cases}
\end{aligned}
\tag{11.45}
$$

for $k \in \mathbb{N}$, and

$$
\begin{aligned}
b_0 &:= \begin{cases} \Big(\widetilde{C}(2_*^3)^{(n-1)(1/q-1/p)} \sum_{\ell=0}^{2} \|f_\gamma^\# \cdot \mathbb{1}_{C_\ell}\|_{L^p(\Sigma,\sigma)}\Big)^{-1} h_0 & \text{if } 0 \in \mathcal{J}_2, \\ 0 & \text{if } 0 \notin \mathcal{J}_2, \end{cases} \\
b_k &:= \begin{cases} \Big(\widetilde{C}(2_*^{k+3})^{(n-1)(1/q-1/p)} \sum_{\ell=k-1}^{k+2} \|f_\gamma^\# \cdot \mathbb{1}_{C_\ell}\|_{L^p(\Sigma,\sigma)}\Big)^{-1} h_k & \text{if } k \in \mathcal{J}_2, \\ 0 & \text{if } k \notin \mathcal{J}_2, \end{cases}
\end{aligned}
\tag{11.46}
$$

for $k \in \mathbb{N}$. The properties listed in (11.43)-(11.44) then ensure that for every $k \in \mathbb{N}_0$ the functions a_k and b_k satisfy

$$\begin{cases} \operatorname{supp} a_k \subseteq \Delta(x_0, 2_*^{k+2}), \\ \|a_k\|_{L^p(\Sigma,\sigma)} \leq (2_*^{k+2})^{(n-1)(1/p-1/q)}, \\ \int_\Sigma a_k \, d\sigma = 0, \end{cases} \tag{11.47}$$

and

$$\begin{cases} \operatorname{supp} b_k \subseteq \Delta(x_0, 2_*^{k+3}), \\ \|b_k\|_{L^p(\Sigma,\sigma)} \leq (2_*^{k+3})^{(n-1)(1/p-1/q)}, \\ \int_\Sigma b_k \, d\sigma = 0. \end{cases} \tag{11.48}$$

Hence, a_k and b_k are x_0-central (q, p)-atoms on Σ for each $k \in \mathbb{N}_0$.

Pressing on, define sequences $\{\lambda_k\}_{k\in\mathbb{N}_0}$ and $\{\gamma_k\}_{k\in\mathbb{N}_0}$ by setting

$$\begin{aligned} \lambda_0 &:= \begin{cases} 2(2_*^2)^{(n-1)(1/q-1/p)} \sum_{\ell=0}^{1} \|f \cdot \mathbb{1}_{C_\ell}\|_{L^p(\Sigma,\sigma)} & \text{if } 0 \in \mathcal{J}_1, \\ 0 & \text{if } 0 \notin \mathcal{J}_1, \end{cases} \\ \lambda_k &:= \begin{cases} 2(2_*^{k+2})^{(n-1)(1/q-1/p)} \sum_{\ell=k-1}^{k+1} \|f \cdot \mathbb{1}_{C_\ell}\|_{L^p(\Sigma,\sigma)} & \text{if } k \in \mathcal{J}_1, \\ 0 & \text{if } k \notin \mathcal{J}_1, \end{cases} \end{aligned} \tag{11.49}$$

for $k \in \mathbb{N}$, and

$$\begin{aligned} \gamma_0 &:= \begin{cases} \widetilde{C}(2_*^3)^{(n-1)(1/q-1/p)} \sum_{\ell=0}^{2} \|f_\gamma^{\#} \mathbb{1}_{C_\ell}\|_{L^p(\Sigma,\sigma)} & \text{if } 0 \in \mathcal{J}_2, \\ 0 & \text{if } 0 \notin \mathcal{J}_2, \end{cases} \\ \gamma_k &:= \begin{cases} \widetilde{C}(2_*^{k+3})^{(n-1)(1/q-1/p)} \sum_{\ell=k-1}^{k+2} \|f_\gamma^{\#} \mathbb{1}_{C_\ell}\|_{L^p(\Sigma,\sigma)} & \text{if } k \in \mathcal{J}_2, \\ 0 & \text{if } k \notin \mathcal{J}_2, \end{cases} \end{aligned} \tag{11.50}$$

for $k \in \mathbb{N}$. In relation to these, we observe that $\{\lambda_k\}_{k\in\mathbb{N}_0}$ and $\{\gamma_k\}_{k\in\mathbb{N}_0}$ belong to $\ell^q(\mathbb{N}_0)$, since by design

$$\begin{aligned} \sum_{k=0}^{\infty} |\lambda_k|^q &\leq C \sum_{\ell=0}^{1} \|f \cdot \mathbb{1}_{C_\ell}\|_{L^p(\Sigma,\sigma)}^q \\ &\quad + C \sum_{k=1}^{\infty} \sum_{\ell=k-1}^{k+1} 2_*^{k(n-1)(1/q-1/p)q} \|f \cdot \mathbb{1}_{C_\ell}\|_{L^p(\Sigma,\sigma)}^q, \\ &\leq C \sum_{j=0}^{\infty} \left(2_*^{j(n-1)(1/q-1/p)} \|f \cdot \mathbb{1}_{C_j}\|_{L^p(\Sigma,\sigma)}\right)^q \\ &= C\|f\|_{\mathrm{A}_{p,q}(\Sigma,\sigma)}^q \leq C\|f\|_{\mathrm{HA}_{p,q}(\Sigma,\sigma)}^q < \infty, \end{aligned} \tag{11.51}$$

while

$$\sum_{k=0}^{\infty}|\gamma_k|^q \leq C\sum_{\ell=0}^{2}\|f_\gamma^{\#}\mathbb{1}_{C_\ell}\|^q_{L^p(\Sigma,\sigma)}$$
$$+ C\sum_{k=1}^{\infty}\sum_{\ell=k-1}^{k+2}\big(2_*^{(k+3)(n-1)(1/q-1/p)}\|f_\gamma^{\#}\cdot\mathbb{1}_{C_\ell}\|_{L^p(\Sigma,\sigma)}\big)^q$$
$$= C\sum_{k=0}^{\infty}\big(2_*^{k(n-1)(1/q-1/p)}\|f_\gamma^{\#}\cdot\mathbb{1}_{C_k}\|_{L^p(\Sigma,\sigma)}\big)^q$$
$$= C\|f_\gamma^{\#}\|^q_{\mathrm{A}_{p,q}(\Sigma,\sigma)} = C\|f\|^q_{\mathrm{HA}_{p,q}(\Sigma,\sigma)} < \infty, \tag{11.52}$$

for some constant $C = C(n, p, q, \Sigma) \in (0, \infty)$.

Collectively, (11.42), (11.45), (11.46), (11.49), and (11.50) give the following atomic decomposition of f:

$$f = \sum_{k=0}^{\infty}\lambda_k a_k + \sum_{k=0}^{\infty}\gamma_k b_k \tag{11.53}$$

with convergence in $L^r(\Sigma,\sigma)$ for each $r \in [1, p]$. After relabeling, this yields

$$f = \sum_{k=0}^{\infty}\widetilde{\lambda}_k\widetilde{a}_k \quad \text{in } L^r(\Sigma,\sigma), \text{ for each } r \in [1, p], \tag{11.54}$$

where $\{\widetilde{a}_k\}_{k\in\mathbb{N}_0}$ is a sequence of x_0-central (q, p)-atoms, and $\{\widetilde{\lambda}_k\}_{k\in\mathbb{N}_0} \in \ell^q(\mathbb{N}_0)$ are obtained by interlacing $\{a_k\}_{k\in\mathbb{N}_0}$ with $\{b_k\}_{k\in\mathbb{N}_0}$, and $\{\lambda_k\}_{k\in\mathbb{N}_0}$ with $\{\gamma_k\}_{k\in\mathbb{N}_0}$ as described in (4.147). Based on this and (11.51)-(11.52), we also have

$$\Big(\sum_{k=0}^{\infty}|\widetilde{\lambda}_k|^q\Big)^{1/q} \leq C\|f\|_{\mathrm{HA}_{p,q}(\Sigma,\sigma)}. \tag{11.55}$$

At this stage, we are left with proving that the series in (11.54) converges to f in $\mathrm{HA}_{p,q}(\Sigma,\sigma)$. To verify this, fix $N, M \in \mathbb{N}_0$, with $M \geq N$. Using the fact that $\|\cdot\|_{\mathrm{HA}_{p,q}(\Sigma,\sigma)}$ is a q-norm (cf. (11.54)), along with Theorem 11.1, there exists a constant $C \in (0, \infty)$ depending only on n, p, q, γ, and the Ahlfors regularity of Σ such that

$$\Big\| \sum_{k=0}^{M} \widetilde{\lambda}_k \widetilde{a}_k - \sum_{k=0}^{N} \widetilde{\lambda}_k \widetilde{a}_k \Big\|^q_{\mathrm{HA}_{p,q}(\Sigma,\sigma)} = \Big\| \sum_{k=N+1}^{M} \widetilde{\lambda}_j \widetilde{a}_k \Big\|^q_{\mathrm{HA}_{p,q}(\Sigma,\sigma)}$$

$$\le \sum_{k=N+1}^{M} |\widetilde{\lambda}_k|^q \|\widetilde{a}_k\|^q_{\mathrm{HA}_{p,q}(\Sigma,\sigma)}$$

$$\le C \sum_{k=N+1}^{\infty} |\widetilde{\lambda}_k|^q. \tag{11.56}$$

Since $\{\widetilde{\lambda}_k\}_{k\in\mathbb{N}_0} \in \ell^q(\mathbb{N}_0)$ we have $\sum_{k=N+1}^{\infty} |\widetilde{\lambda}_k|^q \to 0$ as $N \to \infty$. This implies that $\big\{\sum_{k=0}^{N} \lambda_k a_k\big\}_{N\in\mathbb{N}_0}$ is a Cauchy sequence in $\mathrm{HA}_{p,q}(\Sigma,\sigma)$. From the completeness of $\mathrm{HA}_{p,q}(\Sigma,\sigma)$ (cf. Theorem 11.4), there exists some $\widetilde{f} \in \mathrm{HA}_{p,q}(\Sigma,\sigma)$ satisfying

$$\widetilde{f} = \sum_{k=0}^{\infty} \widetilde{\lambda}_k \widetilde{a}_k \ \text{ in } \mathrm{HA}_{p,q}(\Sigma,\sigma). \tag{11.57}$$

Bearing in mind that $\mathrm{HA}_{p,q}(\Sigma,\sigma) \hookrightarrow L^r(\Sigma,\sigma)$ continuously, for each $r \in [1,p]$ (cf. Proposition 11.3) the convergence (11.57) also holds in $L^r(\Sigma,\sigma)$, which in concert with (11.54) implies $\widetilde{f} = f$ at σ-a.e. point on Σ. Hence, the series in (11.54) converges to f in $\mathrm{HA}_{p,q}(\Sigma,\sigma)$, as desired. This concludes the proof of item *(a)*.

We now turn our attention to the claim made in item *(b)*. Consider a σ-measurable function f on Σ such that

$$f = \sum_{j=0}^{\infty} \lambda_j a_j \ \text{ pointwise } \sigma\text{-a.e. on } \Sigma, \tag{11.58}$$

for some sequence $\{a_k\}_{k\in\mathbb{N}_0}$ of x_0-central (q,p)-atoms (in the sense of Definition 11.2) and some numerical sequence $\{\lambda_k\}_{k\in\mathbb{N}_0} \in \ell^q(\mathbb{N}_0)$. The membership of f to $\mathrm{HA}_{p,q}(\Sigma,\sigma)$ can be obtained via the same type of reasoning used to prove item *(b)* from Theorem 4.2, now bearing in mind (11.56), Theorem 11.1, Proposition 11.3 (used with $r := 1$), and (11.5). Moreover, thanks to (11.5) and Theorem 11.1 we have

$$\|f\|^q_{\mathrm{HA}_{p,q}(\Sigma,\sigma)} \le \Big\| \sum_{j=0}^{\infty} \lambda_j a_j \Big\|^q_{\mathrm{HA}_{p,q}(\Sigma,\sigma)}$$

$$\le \sum_{j=0}^{\infty} |\lambda_j|^q \|a_j\|^q_{\mathrm{HA}_{p,q}(\Sigma,\sigma)}$$

$$\le C \sum_{j=0}^{\infty} |\lambda_j|^q, \tag{11.59}$$

for some constant $C \in (0, \infty)$ independent of f. This finishes the proof of the claim made in *(b)*. The claim made item *(c)* is a direct consequence of items *(a)* and *(b)*. The proof of Theorem 11.2 is therefore complete. □

Remark 11.4 Notice that the quantity in the right-hand side of (11.35) is independent of the background parameter $\gamma \in \big((n-1)(1/q-1), 1\big)$. As such, the atomic characterization of $\mathrm{HA}_{p,q}(\Sigma, \sigma)$ obtained in Theorem 11.2 implies that $\mathrm{HA}_{p,q}(\Sigma, \sigma)$ is actually independent of such a parameter.

It turns out that functions from $\mathrm{HA}_{p,q}$ satisfy a vanishing moment condition.

Corollary 11.1 *Consider a closed unbounded Ahlfors regular set $\Sigma \subseteq \mathbb{R}^n$, and abbreviate $\sigma := \mathcal{H}^{n-1}\lfloor\Sigma$. Then given two exponents $p \in (1, \infty)$ and $q \in \big(\frac{n-1}{n}, 1\big]$, it follows that for every $f \in \mathrm{HA}_{p,q}(\Sigma, \sigma)$ one has $f \in L^1(\Sigma, \sigma)$ and*

$$\int_\Sigma f \, d\sigma = 0. \tag{11.60}$$

Proof Thanks to item *(a)* from Theorem 11.2, the vanishing moment condition of central (q, p)-atoms (cf. Definition 11.2), and (11.7), any $f \in \mathrm{HA}_{p,q}(\Sigma, \sigma)$ has an atomic decomposition as in (11.31) with convergence in $L^1(\Sigma, \sigma)$. As such, we have

$$\int_\Sigma f \, d\sigma = \sum_{j=0}^{\infty} \lambda_j \int_\Sigma a_j \, d\sigma = 0, \tag{11.61}$$

as desired. □

We now record a description of the class of compactly supported functions from $\mathrm{HA}_{p,q}$.

Corollary 11.2 *Consider a closed unbounded Ahlfors regular set $\Sigma \subseteq \mathbb{R}^n$ and abbreviate $\sigma := \mathcal{H}^{n-1}\lfloor\Sigma$. Having fixed $p \in (1, \infty)$, and $q \in \big(\frac{n-1}{n}, 1\big]$, introduce*

$$\mathrm{HA}_{p,q,\mathrm{comp}}(\Sigma, \sigma) := \big\{f \in \mathrm{HA}_{p,q}(\Sigma, \sigma) : f \textit{ is compactly supported}\big\}. \tag{11.62}$$

Finally, fix a reference point $x_0 \in \Sigma$. Then given a σ-measurable function $f : \Sigma \to \mathbb{R}$, the following conditions are equivalent:

(a) f belongs to $\mathrm{HA}_{p,q,\mathrm{comp}}(\Sigma, \sigma)$;

(b) f belongs to $L^p_{\mathrm{comp}}(\Sigma, \sigma)$ and has integral zero;

(c) f is a finite linear combination of x_0-central (q, p)-atoms;

(d) f is a multiple of an x_0-central (q, p)-atom.

Proof The equivalence between items *(a)*, *(b)*, *(c)*, and *(d)* can be obtained via a similar type of argument as that used in the proof of Corollary 4.2, now bearing in mind Theorem 11.1, Theorem 11.2, Corollary 11.1, and Proposition 11.3. □

Remark 11.5 To summarize, the previous corollary establishes that

$$\begin{aligned} \mathrm{HA}_{p,q,\mathrm{comp}}(\Sigma,\sigma) &= \Big\{f \in L^p_{\mathrm{comp}}(\Sigma,\sigma) : \int_\Sigma f\, d\sigma = 0\Big\} \\ &= \big\{\text{multiples of } x_0\text{-central } (q,p)\text{-atoms}\big\}. \end{aligned} \tag{11.63}$$

From this, and Theorem 11.2 we see that $\mathrm{HA}_{p,q,\mathrm{comp}}(\Sigma,\sigma)$ is a dense subspace of $\mathrm{HA}_{p,q}(\Sigma,\sigma)$.

It is useful to record a quantitative atomic description for $\mathrm{HA}_{p,q,\mathrm{comp}}$.

Theorem 11.3 *Assume $\Sigma \subseteq \mathbb{R}^n$ is a closed unbounded Ahlfors regular set, and abbreviate $\sigma := \mathcal{H}^{n-1}\lfloor\Sigma$. Fix two exponents $p \in (1,\infty)$ and $q \in \big(\frac{n-1}{n},1\big]$, along with a reference point $x_0 \in \Sigma$. Then there exists a constant $C \in (0,\infty)$ depending only on n, p, q, and the Ahlfors regularity constants of Σ such that for every $f \in \mathrm{HA}_{p,q,\mathrm{comp}}(\Sigma,\sigma)$ there exist a finite sequence of x_0-central (q,p)-atoms $\{a_k\}_{0\le k\le N}$, and a finite numerical sequence $\{\lambda_k\}_{0\le k\le N}$, for some $N \in \mathbb{N}$, with the property that*

$$\begin{aligned} & f = \sum_{k=0}^{N} \lambda_k a_k \quad \textit{pointwise } \sigma\textit{-a.e. on } \Sigma \\ & \textit{and} \quad \Big(\sum_{k=0}^{N} |\lambda_k|^q\Big)^{1/q} \le C\|f\|_{\mathrm{HA}_{p,q}(\Sigma,\sigma)}. \end{aligned} \tag{11.64}$$

Proof Given that the proof of Theorem 11.2 builds upon that of Theorem 4.2, the finite atomic decomposition claimed for $\mathrm{HA}_{p,q,\mathrm{comp}}(\Sigma,\sigma)$ can be deduced via the same type of reasoning as presented in the proof of Theorem 4.3. □

It is of relevance to obtain the separability of first-generation Herz-type Hardy spaces.

Proposition 11.5 *Suppose $\Sigma \subseteq \mathbb{R}^n$ is a closed unbounded Ahlfors regular set, and abbreviate $\sigma := \mathcal{H}^{n-1}\lfloor\Sigma$. Then whenever $p \in (1,\infty)$ and $q \in \big(\frac{n-1}{n},1\big]$, the Herz-type Hardy space $\mathrm{HA}_{p,q}(\Sigma,\sigma)$ is separable.*

Proof The separability of the Herz-type Hardy space $\mathrm{HA}_{p,q}(\Sigma,\sigma)$ may be obtained by making natural alterations in the proof of Theorem 4.4, now bearing in mind Theorem 11.1, Theorem 11.2, and Definition 11.2. □

11.3 Molecules for First-Generation Herz-type Hardy Spaces

The goal here is to develop a molecular theory for first-generation Herz-type Hardy spaces on unbounded Ahlfors regular sets. The notion of central molecules introduced below can be thought of as a weakened version of the central (q, p)-atoms introduced in Definition 11.2, where now the localization condition is replaced by a decay condition. The precise definition is given below.

Definition 11.3 Assume $\Sigma \subseteq \mathbb{R}^n$ is a closed unbounded Ahlfors regular set, and abbreviate $\sigma := \mathcal{H}^{n-1}\lfloor\Sigma$. Denote by $C_\Sigma, c_\Sigma \in (0, \infty)$ the upper and lower Ahlfors regularity constants of Σ, respectively. Select a reference point $x_0 \in \Sigma$ and a dilation factor $2_* > (C_\Sigma/c_\Sigma)^{1/(n-1)}$. Pick two exponents $p \in (1, \infty)$ and $q \in \big(\frac{n-1}{n}, 1\big]$, and fix parameter $\theta \in \big((n-1)(p/q-1), \infty\big)$. A σ-measurable function $M : \Sigma \to \mathbb{R}$ is called an x_0-`central` (q, p, θ)-`molecule` provided there exists a radius $R \in [2_*, \infty)$ for which the following three properties hold:

(1) Normalization Condition:

$$\|M\|_{L^p(\Sigma,\sigma)} \leq R^{(n-1)(1/p-1/q)}. \tag{11.65}$$

(2) Decay Condition:

$$\Big(\int_\Sigma |M(x)|^p |x - x_0|^\theta \, d\sigma(x)\Big)^{1/p} \leq R^{(n-1)(1/p-1/q)+\theta/p}. \tag{11.66}$$

(3) Vanishing Moment Condition:

$$\int_\Sigma M \, d\sigma = 0. \tag{11.67}$$

Remark 11.6 Following the same type of computations as in Remark 4.7, while keeping in mind that $(n-1)(p/q-1) < \theta < \infty$, we have that the normalization condition (11.65) and decay condition (11.66) imply that the function M belongs to $L^1(\Sigma, \sigma)$. Hence the integral in (11.67) is meaningful.

Remark 11.7 In the regime

$$1 < p < \infty, \quad \tfrac{n-1}{n} < q \leq 1, \quad \text{and} \quad (n-1)(p/q-1) < \theta < \infty, \tag{11.68}$$

we claim that if M is an x_0-central (q, p, θ)-molecule on Σ, then M is also an x_0-central $(1, p, \theta)$-molecule on Σ (in the sense of Definition 4.4). Indeed, since $q \leq 1$ we have

$$(n-1)(p-1) < (n-1)(p/q-1) < \theta < \infty. \tag{11.69}$$

Based on this, whenever M is an x_0-central (q, p, θ)-molecule on Σ and $R \geq 2_*$ is such that (11.65)-(11.66) are satisfied, then

$$\|M\|_{L^p(\Sigma,\sigma)} \leq R^{(n-1)(1/p-1/q)} \leq R^{(n-1)(1/p-1)} \tag{11.70}$$

and

$$\begin{aligned}\Big(\int_\Sigma |M(x)|^p |x-x_0|^\theta \, d\sigma(x)\Big)^{1/p} &\leq R^{(n-1)(1/p-1/q)+\theta/p} \\ &\leq R^{(n-1)(1/p-1)+\theta/p}.\end{aligned} \tag{11.71}$$

This shows that M satisfies the normalization and decay conditions in (4.189)-(4.190), which together with the vanishing moment condition (11.67) it follows that M is an x_0-central $(1, p, \theta)$-molecule.

It will be useful to introduce a second related version of central molecules in the context of Herz-type Hardy spaces.

Definition 11.4 Suppose $\Sigma \subseteq \mathbb{R}^n$ is a closed unbounded Ahlfors regular set, and abbreviate $\sigma := \mathcal{H}^{n-1}\lfloor\Sigma$. Denote by $C_\Sigma, c_\Sigma \in (0,\infty)$ the upper and lower Ahlfors regularity constants of Σ, respectively. Fix a reference point $x_0 \in \Sigma$ and a dilation factor $2_* > (C_\Sigma/c_\Sigma)^{1/(n-1)}$. Pick two exponents $p \in (1,\infty)$ and $q \in \big(\frac{n-1}{n}, 1\big]$, along with a parameter $\varepsilon \in (1/q-1,\infty)$. A σ-measurable function $M : \Sigma \to \mathbb{R}$ is called an x_0-`central` $(q,p)_\varepsilon$-`molecule` provided there exists a radius $R \in [2_*,\infty)$ with the property that for each $k \in \mathbb{N}_0$ one has

$$\Big(\int_{A_k(x_0,R)} |M|^p \, d\sigma\Big)^{1/p} \leq 2^{k(n-1)(1/p-1-\varepsilon)} \cdot R^{(n-1)(1/p-1/q)}, \tag{11.72}$$

where

$$A_k(x_0,R) := \begin{cases} \Delta(x_0,R) & \text{if } k=0, \\ \Delta(x_0,2^kR)\setminus\Delta(x_0,2^{k-1}R) & \text{if } k\geq 1, \end{cases} \tag{11.73}$$

and

$$\int_\Sigma M \, d\sigma = 0. \tag{11.74}$$

Remark 11.8 Similar computations to those in (4.200) ensure that whenever M is a σ-measurable function satisfying (11.72) then $M \in L^1(\Sigma,\sigma)$. As such, the vanishing moment property in (11.74) is meaningfully formulated.

We now establish that both definitions of central molecules introduced in Definition 11.3 and Definition 11.4 are compatible up to multiplicative constants.

Proposition 11.6 *Assume $\Sigma \subseteq \mathbb{R}^n$ is a closed unbounded Ahlfors regular set, and abbreviate $\sigma := \mathcal{H}^{n-1}\lfloor\Sigma$. Select two exponents $p \in (1,\infty)$ and $q \in \big(\frac{n-1}{n}, 1\big]$, and fix a reference point $x_0 \in \Sigma$. Then the following statements hold:*

(a) For every $\varepsilon \in (1/q-1,\infty)$ there exists $\theta = \theta(\varepsilon) \in \big((n-1)(p/q-1),\infty\big)$ and a constant $C_\varepsilon \in (0,\infty)$ depending only on the n, p, q, and ε, such that every x_0-central $(q,p)_\varepsilon$-molecule (in the sense of Definition 11.4) is a C_ε-multiple of an x_0-central (q,p,θ)-molecule (in the sense of Definition 11.3).

(b) In the opposite direction, for every $\theta \in \big((n-1)(p/q-1), \infty\big)$ *there exists some* $\varepsilon = \varepsilon(\theta) \in (1/q - 1, \infty)$*, and a constant* $C_\theta \in (0, \infty)$ *depending exclusively on n, p, q, and* θ*, such that every* x_0*-central* (q, p, θ)*-molecule (in the sense of Definition 11.3) is a* C_θ*-multiple of an* x_0*-central* $(q, p)_\varepsilon$*-molecule (in the sense of Definition 11.4).*

Proof To prove the claim made in *(a)*, fix an x_0-central $(q, p)_\varepsilon$-molecule $M : \Sigma \to \mathbb{R}$ in the sense of Definition 11.4. It is visible that the condition $\varepsilon > 1/q - 1$ implies

$$(n-1)\Big(\frac{p}{q} - 1\Big) < (n-1)\Big(\varepsilon + 1 - \frac{1}{p}\Big)p. \tag{11.75}$$

Then, choose some $\theta \in (0, \infty)$ satisfying

$$(n-1)\Big(\frac{p}{q} - 1\Big) < \theta < (n-1)\Big(\varepsilon + 1 - \frac{1}{p}\Big)p. \tag{11.76}$$

Reasoning as in (4.203)-(4.206), this time bearing in mind (11.76), (11.72), (11.73), and the fact that $1/p - 1 - \varepsilon < 0$ (as implied by the choices of ε and q), it follows that there exists some constant $C_\varepsilon \in (0, \infty)$ with the property that $C_\varepsilon^{-1} \cdot M$ an x_0-central (q, p, θ)-molecule, as desired.

To justify the claim in *(b)*, assume M is an x_0-central (q, p, θ)-molecule on Σ in the sense of Definition 11.3. Select

$$\varepsilon := \frac{\theta}{p(n-1)} + \frac{1}{p} - 1 \in (0, \infty). \tag{11.77}$$

Since $(n-1)(p/q - 1) < \theta < \infty$, this choice of ε satisfies

$$\varepsilon > \frac{1}{q} - 1 \quad \text{and} \quad -\frac{\theta}{p} = (n-1)\Big(\frac{1}{p} - 1 - \varepsilon\Big). \tag{11.78}$$

Thanks to this, the normalization condition in (11.65), and the decay condition in (11.66), the same type of computations as in (4.208)-(4.209) ensure that there exists some constant $C_\theta \in (0, \infty)$ such that $C_\theta^{-1} \cdot M$ is an x_0-central $(q, p)_\varepsilon$-molecule on Σ, as wanted. □

It turns out central (q, p, θ)-molecules belong to first-generation Herz-type Hardy space $\mathrm{HA}_{p,q}$, and there exists a constant that bounds the $\mathrm{HA}_{p,q}$-quasinorm of every central (q, p, θ)-molecule.

Theorem 11.4 *Assume* $\Sigma \subseteq \mathbb{R}^n$ *is a closed unbounded Ahlfors regular set, and abbreviate* $\sigma := \mathcal{H}^{n-1}\lfloor\Sigma$*. Pick two exponents* $p \in (1, \infty)$ *and* $q \in \big(\frac{n-1}{n}, 1\big]$*, along with a parameter* $\theta \in \big((n-1)(p/q-1), \infty\big)$*. Also, fix a reference point* $x_0 \in \Sigma$*. Then there exists some constant* $C \in (0, \infty)$ *depending exclusively on n, p, q,* θ*, and the Ahlfors regularity character of* Σ*, with the property that for every* x_0*-central* (q, p, θ)*-molecule M on* Σ *(in the sense of Definition 11.3) one has*

$$\|M\|_{\mathrm{HA}_{p,q}(\Sigma,\sigma)} \leq C. \tag{11.79}$$

Moreover, the same type of conclusions hold for x_0-central $(q,p)_\varepsilon$-molecules (in the sense of Definition 11.4) for all $\varepsilon \in (1/q-1,\infty)$.

Proof The approach used to prove (11.79) is similar to that used to establish Theorem 4.4. Thus, here we only emphasize the key differences. To get started, fix an arbitrary x_0-central (q,p,θ)-molecule M on Σ. The analysis carried out in (4.211)-(4.221) (which is actually independent of the nature of the central molecule) leads to the following writing

$$M\cdot \mathbb{1}_{\Delta_N} = \sum_{k=0}^{N}\alpha_k + \sum_{k=0}^{N}\beta_k - \frac{\mathbb{1}_{A_N}}{\sigma(A_N)}\int_{\Sigma\setminus\Delta_N} M\,d\sigma \quad \text{for every } N\in\mathbb{N}, \tag{11.80}$$

where for each $k\in\mathbb{N}_0$ the functions $\alpha_k,\beta_k:\Sigma\to\mathbb{R}$ are defined according to

$$\alpha_k := (M - M_{A_k})\mathbb{1}_{A_k} \quad \text{and} \tag{11.81}$$

$$\beta_k := \Big(\frac{\mathbb{1}_{A_{k+1}}}{\sigma(A_{k+1})} - \frac{\mathbb{1}_{A_k}}{\sigma(A_k)}\Big)\int_{\Sigma\setminus\Delta_k} M\,d\sigma. \tag{11.82}$$

Availing ourselves with the computations made in (4.222), (4.223), and (4.224), we see that for each $k\in\mathbb{N}_0$ the function α_k satisfy

$$\begin{aligned} &\operatorname{supp}\alpha_k \subseteq \Delta(x_0, 2_*^k R), \quad \int_\Sigma \alpha_k\,d\sigma = 0, \\ &\text{and} \quad \|\alpha_k\|_{L^p(\Sigma,\sigma)} \leq 2\Big(\int_{A_k}|M|^p\,d\sigma\Big)^{1/p}. \end{aligned} \tag{11.83}$$

For $k=0$, use the normalization condition of M (cf. (11.65)) to bound

$$\begin{aligned} \|\alpha_0\|_{L^p(\Sigma,\sigma)} &\leq 2\Big(\int_{A_0}|M|^p\,d\sigma\Big)^{1/p} \\ &\leq 2\Big(\int_\Sigma |M|^p\,d\sigma\Big)^{1/p} \leq 2R^{(n-1)(1/p-1/q)}. \end{aligned} \tag{11.84}$$

Corresponding to $k\in\mathbb{N}$, use the decay condition of M (cf. (11.66)), and recall the definition of A_j in (4.213), to obtain

$$
\begin{aligned}
\|\alpha_k\|_{L^p(\Sigma,\sigma)} &\leq 2\Big(\int_{A_k} |M|^p\,d\sigma\Big)^{1/p} \\
&\leq 2\Big(\int_{A_k} |M(x)|^p |x-x_0|^\theta (2_*^{k-1}R)^{-\theta}\,d\sigma(x)\Big)^{1/p} \\
&\leq 2(2_*^{k-1}R)^{-\theta/p}\Big(\int_\Sigma |M(x)|^p |x-x_0|^\theta\,d\sigma(x)\Big)^{1/p} \\
&\leq 2(2_*^{k-1}R)^{-\theta/p} R^{(n-1)(1/p-1/q)+\theta/p} \\
&= C 2_*^{-k\theta/p} R^{(n-1)(1/p-1/q)}
\end{aligned}
\tag{11.85}
$$

for some constant $C = C(\Sigma, p, \theta) \in (0, \infty)$. In concert, (11.84) and (11.85) imply that there exists $C' = C'(\Sigma, p, \theta) \in (0, \infty)$ such that

$$
\|\alpha_k\|_{L^p(\Sigma,\sigma)} \leq C' 2_*^{-k\theta/p} R^{(n-1)(1/p-1/q)} \quad \text{for all } k \in \mathbb{N}_0. \tag{11.86}
$$

For each $k \in \mathbb{N}_0$ define the function $a_k : \Sigma \to \mathbb{R}$ by setting

$$
a_k := \lambda_k^{-1}\alpha_k, \quad \text{with } \lambda_k := C' 2_*^{-k[\theta/p+(n-1)(1/p-1/q)]} \in (0, \infty). \tag{11.87}
$$

Thanks to (11.83), (11.86), and (11.87), for every $k \in \mathbb{N}_0$ the function a_k satisfies:

$$
\begin{gathered}
\operatorname{supp} a_k \subseteq \Delta(x_0, 2_*^k R), \quad \int_\Sigma a_k\,d\sigma = 0, \\
\text{and} \quad \|a_k\|_{L^p(\Sigma,\sigma)} \leq (2_*^k R)^{(n-1)(1/p-1/q)}.
\end{gathered}
\tag{11.88}
$$

Hence, a_k is an x_0-central (q, p)-atom on Σ for all $k \in \mathbb{N}_0$. Furthermore, bearing in mind that $\theta > (n-1)(p/q-1)$, which in particular yields $\theta/p+(n-1)(1/p-1/q) > 0$, the sequence $\{\lambda_k\}_{k\in\mathbb{N}_0}$ defined in (11.87) satisfies

$$
\sum_{j=0}^{\infty} |\lambda_j|^q = C' \sum_{j=0}^{\infty} 2_*^{-j[\theta/p+(n-1)(1/p-1/q)]q} < \infty. \tag{11.89}
$$

Moving forward, having fixed an arbitrary $k \in \mathbb{N}_0$, we turn our attention to the function β_k defined in (11.82). It is visible from (11.82) that we have

$$
\operatorname{supp} \beta_k \subseteq \Delta(x_0, 2^{k+1}R) \quad \text{and} \quad \int_\Sigma \beta_k\,d\sigma = 0. \tag{11.90}
$$

Moreover, combining the computations made in (4.233)-(4.234) with the decay property (11.66) of M, and bearing in mind that

$$
\theta > (n-1)(p/q-1) \geq (n-1)(p-1) \implies (n-1)/p' - \theta/p < 0, \tag{11.91}
$$

where $p' \in (1, \infty)$ denotes the Hölder conjugate exponent of p, we may estimate

$$\begin{aligned}\|\beta_k\|_{L^p(\Sigma,\sigma)} &\leq C(2_*^{k+1}R)^{(n-1)(1/p-1)}\Big(\int_\Sigma |M(x)|^p|x-x_0|^\theta\,d\sigma(x)\Big)^{1/p}\\ &\quad\times\Big(\int_{\Sigma\setminus\Delta_k}|x-x_0|^{-\theta p'/p}\,d\sigma(x)\Big)^{1/p'}\\ &\leq C(2_*^{k+1}R)^{(n-1)(1/p-1)}R^{(n-1)(1/p-1/q)+\theta/p}\\ &\quad\times\Big(\int_{\Sigma\setminus\Delta_k}|x-x_0|^{-\theta p'/p}\,d\sigma(x)\Big)^{1/p'}\\ &\leq C(2_*^{k+1}R)^{(n-1)(1/p-1)}R^{(n-1)(1/p-1/q)+\theta/p}\\ &\quad\times(2_*^{k+1}R)^{(n-1)(1-1/p)-\theta/p}\sum_{\ell=0}^{\infty}2^{\ell[(n-1)/p'-\theta/p]}\\ &= C''R^{(n-1)(1/p-1/q)}(2_*^{k+1})^{-\theta/p}\end{aligned} \tag{11.92}$$

for some $C''=C''(\Sigma,\theta,n,p)\in(0,\infty)$. For each $k\in\mathbb{N}_0$, define $b_k:\Sigma\to\mathbb{R}$ as

$$b_k:=\widetilde{\lambda}_k^{-1}\beta_k,\quad\text{where } \widetilde{\lambda}_k:=C''2_*^{(k+1)(n-1)(1/q-1/p)}2_*^{-(k+1)\theta/p}\in(0,\infty). \tag{11.93}$$

Based on (11.90) and (11.92), we see that for each $k\in\mathbb{N}_0$ the function b_k has the following properties

$$\begin{aligned}&\operatorname{supp} b_k\subseteq\Delta(x_0,2_*^{k+1}R),\qquad \int_\Sigma b_k\,d\sigma=0,\\ &\text{and}\quad \|b_k\|_{L^p(\Sigma,\sigma)}\leq(2_*^{k+1}R)^{(n-1)(1/p-1/q)}.\end{aligned} \tag{11.94}$$

Hence, b_k is an x_0-central (q,p)-atom on Σ for every $k\in\mathbb{N}_0$. Also, since the choice of θ yields $(n-1)(1/q-1/p)-\theta/p<0$, we see that the sequence $\{\widetilde{\lambda}_j\}_{j\in\mathbb{N}_0}$ defined in (11.93) satisfies

$$\sum_{k=0}^{\infty}|\widetilde{\lambda}_k|^q=C''\sum_{k=0}^{\infty}2_*^{(k+1)[(n-1)(1/q-1/p)-\theta/p]q}<\infty. \tag{11.95}$$

At this stage of the proof, we may employ the same reasoning as in (4.239)-(4.242), bearing in mind the absolute integrability of M (see Remark 11.6), as well as (11.87) and (11.93), to deduce the following atomic decomposition for M:

$$M=\sum_{k=0}^{\infty}\lambda_k a_k+\sum_{k=0}^{\infty}\widetilde{\lambda}_k b_k\ \text{ pointwise } \sigma\text{-a.e. on } \Sigma. \tag{11.96}$$

Thanks to item *(b)* in Theorem 11.2, we find that M belongs to $\mathrm{HA}_{p,q}(\Sigma,\sigma)$ and

$$\|M\|^q_{\mathrm{HA}_{p,q}(\Sigma,\sigma)} \leq C\sum_{j=0}^{\infty}|\lambda_j|^q + C\sum_{j=0}^{\infty}|\widetilde{\lambda}_j|^q, \tag{11.97}$$

for some constant $C \in (0,\infty)$ independent of M. Notice that the sequences $\{\lambda_k\}_{k\in\mathbb{N}_0}$ and $\{\widetilde{\lambda}_k\}_{k\in\mathbb{N}_0}$ (defined in (11.87) and (11.93), respectively) depend exclusively on n, p, q, θ, and the Ahlfors regularity of Σ, hence so does the quantity in the right-hand side of (11.97). This establishes (11.79).

The very last claim regarding (11.79) when M is an x_0-central $(q,p)_\varepsilon$-molecule, for each $\varepsilon \in (1/q-1,\infty)$, is a consequence of what we have proved so far, together with with the fact that every x_0-central $(q,p)_\varepsilon$-molecule is a fixed constant multiple of an x_0-central (q,p,θ)-molecule (cf. *(a)* in Proposition 11.6). This finishes the proof of Theorem 11.4. □

Chapter 12
Functions of (p,q)-Bounded Central Mean Oscillations

In this chapter we introduce and study the space of functions of (p,q)-bounded central mean oscillations $\mathrm{BCMO}_{p,q}$ on unbounded Ahlfors regular sets, which extends the scale of bounded central p-mean oscillations functions BCMO_p discussed in Chapter 5. We begin by establishing some basic functional analytic properties of $\mathrm{BCMO}_{p,q}$, with the ultimate goal of proving that the dual of first-generation Herz-type Hardy spaces $\mathrm{HA}_{p,q}$ coincides with $\mathrm{BCMO}_{p',q}$ modulo constants.

12.1 Definitions and Basic Properties

Functions of (p,q)-bounded central mean oscillations, in the context of unbounded Ahlfors regular sets, are defined by demanding that L^p-styled central mean oscillations have power-growth control. A precise definition is recorded below.

Definition 12.1 Consider $\Sigma \subseteq \mathbb{R}^n$ a closed unbounded Ahlfors regular set, and abbreviate $\sigma := \mathcal{H}^{n-1}\lfloor\Sigma$. Denote by $c_\Sigma, C_\Sigma \in (0,\infty)$ the lower and upper Ahlfors regularity constants of Σ, respectively, and fix a dilation factor $2_* > (C_\Sigma/c_\Sigma)^{1/(n-1)}$. Select a reference point $x_0 \in \Sigma$, and pick two exponents $p,q \in (0,\infty)$. For each function $f \in L^1_{\mathrm{loc}}(\Sigma,\sigma)$ define

$$\|f\|_{\mathrm{BCMO}_{p,q}(\Sigma,\sigma)} := \sup_{R\geq 2_*}\left\{R^{(n-1)(1-1/q)}\left(⨍_{\Delta(x_0,R)}|f - f_{\Delta(x_0,R)}|^p\,d\sigma\right)^{1/p}\right\} \in [0,\infty] \tag{12.1}$$

where $f_{\Delta(x_0,R)} = ⨍_{\Delta(x_0,R)} f\,d\sigma$. The `space of functions of` (p,q)`-bounded central mean oscillations` on Σ is defined as

$$\mathrm{BCMO}_{p,q}(\Sigma,\sigma) := \left\{f \in L^1_{\mathrm{loc}}(\Sigma,\sigma) : \|f\|_{\mathrm{BCMO}_{p,q}(\Sigma,\sigma)} < \infty\right\}. \tag{12.2}$$

© The Author(s), under exclusive license to Springer Nature Switzerland AG 2026
M. Mitrea and P. Takemura, *Singular Integrals, Herz-Type Function Spaces, and Boundary Problems*, Progress in Mathematics 362,
https://doi.org/10.1007/978-3-032-12516-3_12

Corresponding to the case when $q = 1$, the space $\mathrm{BCMO}_{p,1}(\Sigma,\sigma)$ coincides with the space of functions of bounded central p-mean oscillations $\mathrm{BCMO}_p(\Sigma,\sigma)$ (cf. Definition 5.1), i.e.,

$$\mathrm{BCMO}_{p,1}(\Sigma,\sigma) = \mathrm{BCMO}_p(\Sigma,\sigma). \tag{12.3}$$

Notice that $\|\cdot\|_{\mathrm{BCMO}_{p,q}(\Sigma,\sigma)}$ (cf. (12.2)) defines a seminorm on $\mathrm{BCMO}_{p,q}(\Sigma,\sigma)$ since for every $f \in L^1_{\mathrm{loc}}(\Sigma,\sigma)$ we have

$$\|f\|_{\mathrm{BCMO}_{p,q}(\Sigma,\sigma)} = 0 \quad \text{if and only if} \quad f \text{ is a constant function on } \Sigma. \tag{12.4}$$

Because of this, we may mod out constant functions from $\mathrm{BCMO}_{p,q}(\Sigma,\sigma)$. Precisely, given two functions $f, g \in L^1_{\mathrm{loc}}(\Sigma,\sigma)$, consider the equivalence relation $\sim$ given by:

$$f \sim g \iff f - g \text{ is a constant.} \tag{12.5}$$

For each function $f \in \mathrm{BCMO}_{p,q}(\Sigma,\sigma)$, denote by $[f]$ the equivalence modulo constants of f. Introduce

$$\begin{aligned} \widetilde{\mathrm{BCMO}}_{p,q}(\Sigma,\sigma) &:= \mathrm{BCMO}_{p,q}(\Sigma,\sigma)/\sim \\ &= \big\{[f] : f \in \mathrm{BCMO}_{p,q}(\Sigma,\sigma)\big\}, \end{aligned} \tag{12.6}$$

and equip this space with

$$\|[f]\|_{\widetilde{\mathrm{BCMO}}_{p,q}(\Sigma,\sigma)} := \|f\|_{\mathrm{BCMO}_{p,q}(\Sigma,\sigma)} \quad \text{for all } f \in \mathrm{BCMO}_{p,q}(\Sigma,\sigma). \tag{12.7}$$

In this scenario, $\big(\widetilde{\mathrm{BCMO}}_{p,q}(\Sigma,\sigma), \|\cdot\|_{\widetilde{\mathrm{BCMO}}_{p,q}(\Sigma,\sigma)}\big)$ becomes a genuine normed space.

Much as in Proposition 5.1, here we establish a natural criterion for the membership to $\mathrm{BCMO}_{p,q}$. The precise statement is given below.

Proposition 12.1 *Consider a closed unbounded Ahlfors regular set $\Sigma \subseteq \mathbb{R}^n$, and abbreviate $\sigma := \mathcal{H}^{n-1}\lfloor\Sigma$. Fix a reference point $x_0 \in \Sigma$, and a dilation factor 2_* as in Definition 5.1. Select two exponents $p \in (1,\infty)$ and $q \in (0,\infty)$. Then for every function $f \in L^1_{\mathrm{loc}}(\Sigma,\sigma)$ one has*

$$\sup_{R\geq 2_*}\inf_{c\in\mathbb{C}}\left\{R^{(n-1)(1-1/q)}\Big(\fint_{\Delta(x_0,R)}|f-c|^p\,d\sigma\Big)^{1/p}\right\} \approx \|f\|_{\mathrm{BCMO}_{p,q}(\Sigma,\sigma)} \tag{12.8}$$

with implicit proportionality constants independent of f. In particular, a given function $f \in L^1_{\mathrm{loc}}(\Sigma,\sigma)$ belongs to the space $\mathrm{BCMO}_{p,q}(\Sigma,\sigma)$ *if and only if*

$$\sup_{R\geq 2_*}\inf_{c\in\mathbb{C}}\left\{R^{(n-1)(1-1/q)}\Big(\fint_{\Delta(x_0,R)}|f-c|^p\,d\sigma\Big)^{1/p}\right\} < \infty. \tag{12.9}$$

Proof Select a function $f \in L^1_{\mathrm{loc}}(\Sigma,\sigma)$. For each radius $R \geq 2_*$ we have

$$
\begin{aligned}
&\inf_{c\in\mathbb{C}}\Big\{R^{(n-1)(1-1/q)}\Big(\fint_{\Delta(x_0,R)}|f-c|^p\,d\sigma\Big)^{1/p}\Big\}\\
&\quad\le\Big\{R^{(n-1)(1-1/q)}\Big(\fint_{\Delta(x_0,R)}|f-f_{\Delta(x_0,R)}|^p\,d\sigma\Big)^{1/p}\Big\}\\
&\quad\le\|f\|_{\mathrm{BCMO}_{p,q}(\Sigma,\sigma)}<\infty.
\end{aligned}
\tag{12.10}
$$

After taking the supremum in the left-hand side of (12.10) we obtain the right-pointing in (12.8).

To verify the opposite inequality, fix an arbitrary radius $R \geq 2_*$ and a number $c \in \mathbb{C}$. The same type of computations made in (5.10)-(5.11) yield

$$
\begin{aligned}
&R^{(n-1)(1-1/q)}\Big(\fint_{\Delta(x_0,R)}|f-f_{\Delta(x_0,R)}|^p\,d\sigma\Big)^{1/p}\\
&\qquad\le 2R^{(n-1)(1-1/q)}\Big(\fint_{\Delta(x_0,R)}|f-c|^p\,d\sigma\Big)^{1/p}.
\end{aligned}
\tag{12.11}
$$

Since $R \geq 2_*$ and $c \in \mathbb{C}$ are arbitrary, we see from (12.1) that

$$
\begin{aligned}
&\|f\|_{\mathrm{BCMO}_{p,q}(\Sigma,\sigma)}\\
&\qquad\le 2\sup_{R\ge 2_*}\inf_{c\in\mathbb{C}}\Big\{R^{(n-1)(1-1/q)}\Big(\fint_{\Delta(x_0,R)}|f-c|^p\,d\sigma\Big)^{1/p}\Big\}.
\end{aligned}
\tag{12.12}
$$

This establishes the left-pointing inequality in (12.8), concluding the proof of Proposition 12.8. □

We collect some basic properties of functions of (p,q)-bounded central mean oscillations in the proposition recorded below.

Proposition 12.2 *Assume $\Sigma \subseteq \mathbb{R}^n$ is a closed unbounded Ahlfors regular set, and abbreviate $\sigma := \mathcal{H}^{n-1}\lfloor\Sigma$. Fix two exponents $p \in (1,\infty)$ and $q \in (0,\infty)$. Then for every $f, g \in L^1_{\mathrm{loc}}(\Sigma,\sigma)$ one has*

$$\big\||f|\big\|_{\mathrm{BCMO}_{p,q}(\Sigma,\sigma)}\le 2\|f\|_{\mathrm{BCMO}_{p,q}(\Sigma,\sigma)},\tag{12.13}$$

$$\big\|\max\{f,g\}\big\|_{\mathrm{BCMO}_{p,q}(\Sigma,\sigma)}\le\frac{3}{2}\Big(\|f\|_{\mathrm{BCMO}_{p,q}(\Sigma,\sigma)}+\|g\|_{\mathrm{BCMO}_{p,q}(\Sigma,\sigma)}\Big),\tag{12.14}$$

$$\big\|\min\{f,g\}\big\|_{\mathrm{BCMO}_{p,q}(\Sigma,\sigma)}\le\frac{3}{2}\Big(\|f\|_{\mathrm{BCMO}_{p,q}(\Sigma,\sigma)}+\|g\|_{\mathrm{BCMO}_{p,q}(\Sigma,\sigma)}\Big).\tag{12.15}$$

In particular, whenever $f, g \in \mathrm{BCMO}_{p,q}(\Sigma,\sigma)$ it follows that

$$|f|,\ \max\{f,g\},\ \min\{f,g\}\in\mathrm{BCMO}_{p,q}(\Sigma,\sigma).\tag{12.16}$$

Proof The proof of all claims follows by naturally adapting the reasoning presented in the proof of Proposition 5.2. □

It turns out that membership to the space $\mathrm{BCMO}_{p,q}$ is actually independent of the reference point x_0 and the dilation factor 2_* appearing in Definition 12.1. A precise statement of this is given below.

Proposition 12.3 *Suppose $\Sigma \subseteq \mathbb{R}^n$ is a closed unbounded Ahlfors regular set, and abbreviate $\sigma := \mathcal{H}^{n-1}\lfloor\Sigma$. Denote by $c_\Sigma, C_\Sigma \in (0,\infty)$ the lower and upper Ahlfors regularity constants of Σ, respectively. Select two reference points $x_0, x_1 \in \Sigma$, and two dilation factors $2_*, 2_\diamond > (C_\Sigma/c_\Sigma)^{1/(n-1)}$. Fix two exponents $p \in (1,\infty)$ and $q \in (0,\infty)$. Then for every function $f \in L^1_{\mathrm{loc}}(\Sigma,\sigma)$ one has*

$$\begin{aligned} &\sup_{R\geq 2_*}\Big\{R^{(n-1)(1-1/q)}\Big(\fint_{\Delta(x_0,R)}|f-f_{\Delta(x_0,R)}|^p\,d\sigma\Big)^{1/p}\Big\} \\ &\qquad\approx \sup_{R\geq 2_\diamond}\Big\{R^{(n-1)(1-1/q)}\Big(\fint_{\Delta(x_1,R)}|f-f_{\Delta(x_1,R)}|^p\,d\sigma\Big)^{1/p}\Big\}, \end{aligned} \tag{12.17}$$

with implicit proportionality constants independent of the function f. As a consequence, the $\mathrm{BCMO}_{p,q}$-seminorm in (12.1) is independent of the choices of x_0 and 2_*.

Proof Select a function $f \in L^1_{\mathrm{loc}}(\Sigma,\sigma)$, and fix a radius $R \in [2_*,\infty)$. Pick $\lambda \in (0,\infty)$ satisfying

$$\lambda > \max\Big\{2, \frac{2}{2_*}|x_0 - x_1|, \frac{2_\diamond}{2_*}\Big\}, \tag{12.18}$$

and set $\widetilde{R} := \lambda R \in (0,\infty)$. Then $\widetilde{R} \geq 2_\diamond$ and, much as in (5.26)-(5.27) we have

$$\Delta(x_0,R) \subseteq \Delta(x_1,\widetilde{R}). \tag{12.19}$$

Using to this, and following the same type of computations as in (5.29), we may bound

$$\begin{aligned} &R^{(n-1)(1-1/q)}\Big(\fint_{\Delta(x_0,R)}|f-f_{\Delta(x_1,\widetilde{R})}|^p\,d\sigma\Big)^{1/p} \\ &\quad\leq C\frac{\lambda^{(n-1)/p}}{\lambda^{(n-1)(1-1/q)}}\widetilde{R}^{(n-1)(1-1/q)}\Big(\fint_{\Delta(x_1,\widetilde{R})}|f-f_{\Delta(x_1,\widetilde{R})}|^p\,d\sigma\Big)^{1/p} \\ &\quad\leq C\frac{\lambda^{(n-1)/p}}{\lambda^{(n-1)(1-1/q)}}\sup_{r\geq 2_\diamond}\Big\{r^{(n-1)(1-1/q)}\Big(\fint_{\Delta(x_1,r)}|f-f_{\Delta(x_1,r)}|^p\,d\sigma\Big)^{1/p}\Big\}. \end{aligned} \tag{12.20}$$

Also, a similar type of reasoning as the one used in (5.30) yields

$$
\begin{aligned}
&R^{(n-1)(1-1/q)}|f_{\Delta(x_0,R)} - f_{\Delta(x_1,\widetilde{R})}| \\
&\quad \le C\frac{\lambda^{n-1}}{\lambda^{(n-1)(1-1/q)}}\widetilde{R}^{(n-1)(1-1/q)}\Big(⨍_{\Delta(x_1,\widetilde{R})} |f - f_{\Delta(x_1,\widetilde{R})}|^p\,d\sigma\Big)^{1/p} \\
&\quad \le C\frac{\lambda^{n-1}}{\lambda^{(n-1)(1-1/q)}}\sup_{r\ge 2_\diamond}\Big\{r^{(n-1)(1-1/q)}\Big(⨍_{\Delta(x_1,r)} |f - f_{\Delta(x_1,r)}|^p\,d\sigma\Big)^{1/p}\Big\}.
\end{aligned}
\tag{12.21}
$$

Combining (12.20) and (12.21) implies

$$
\begin{aligned}
&R^{(n-1)(1-1/q)}\Big(⨍_{\Delta(x_0,R)} |f - f_{\Delta(x_0,R)}|^p\,d\sigma\Big)^{1/p} \\
&\quad \le R^{(n-1)(1-1/q)}\Big(⨍_{\Delta(x_0,R)} |f - f_{\Delta(x_1,\widetilde{R})}|^p\,d\sigma\Big)^{1/p} \\
&\qquad + R^{(n-1)(1-1/q)}|f_{\Delta(x_0,R)} - f_{\Delta(x_1,\widetilde{R})}| \\
&\quad \le C\cdot\sup_{r\ge 2_\diamond}\Big\{r^{(n-1)(1-1/q)}\Big(⨍_{\Delta(x_1,r)} |f - f_{\Delta(x_1,r)}|^p\,d\sigma\Big)^{1/p}\Big\},
\end{aligned}
\tag{12.22}
$$

for some $C = C(n, p, q, x_0, x_1, 2_*, 2_\diamond, \Sigma) \in (0, \infty)$. After taking the supremum over $R \in [2_*, \infty)$ in the left-hand side of (12.22) we arrive at

$$
\begin{aligned}
&\sup_{R\ge 2_*}\Big\{R^{(n-1)(1-1/q)}\Big(⨍_{\Delta(x_0,R)} |f - f_{\Delta(x_0,R)}|^p\,d\sigma\Big)^{1/p}\Big\} \\
&\quad \le C\cdot\sup_{R\ge 2_\diamond}\Big\{R^{(n-1)(1-1/q)}\Big(⨍_{\Delta(x_1,R)} |f - f_{\Delta(x_1,R)}|^p\,d\sigma\Big)^{1/p}\Big\},
\end{aligned}
\tag{12.23}
$$

for some $C \in (0, \infty)$ of same nature as in (12.22). This shows the right-pointing inequality in (12.17). The left-pointing inequality in (12.17) follows from the same argument as before now with x_1 in place of x_0, and $2_\diamond$ in place of 2_*. This concludes the proof of (12.17). □

Proposition 12.4 *Assume $\Sigma \subseteq \mathbb{R}^n$ is a closed unbounded Ahlfors regular set, and abbreviate $\sigma := \mathcal{H}^{n-1}\lfloor\Sigma$. Pick two exponents $p \in (1, \infty)$, and $q \in (0, \infty)$. In this setting, the following statements are true:*

(a) If $p_1, p_2 \in (1, \infty)$, and $p_1 \le p_2$, then

$$
\mathrm{BCMO}_{p_2,q}(\Sigma, \sigma) \hookrightarrow \mathrm{BCMO}_{p_1,q}(\Sigma, \sigma) \quad \textit{continuously,} \tag{12.24}
$$

i.e., the scale of spaces $\mathrm{BCMO}_{p,q}(\Sigma, \sigma)$ is monotonically decreasing in p.

(b) Whenever $q_1, q_2 \in (0, \infty)$ are such that $q_1 \le q_2$, one has

$$
\mathrm{BCMO}_{p,q_2}(\Sigma, \sigma) \hookrightarrow \mathrm{BCMO}_{p,q_1}(\Sigma, \sigma) \quad \textit{continuously,} \tag{12.25}
$$

i.e., the scale of spaces $\mathrm{BCMO}_{p,q}(\Sigma,\sigma)$ *is monotonically decreasing in* q. *In particular, in the regime* $0 < q \le 1$, *it follows that*

$$\mathrm{BCMO}_p(\Sigma,\sigma) \hookrightarrow \mathrm{BCMO}_{p,q}(\Sigma,\sigma) \quad \textit{continuously.} \tag{12.26}$$

As a consequence, whenever $0 < q \le 1$, *one also has the continuous embeddings*

$$\begin{aligned} \mathrm{BMO}(\Sigma,\sigma) &\hookrightarrow \mathrm{BCMO}_{p,q}(\Sigma,\sigma) \quad \textit{and} \\ L^\infty(\Sigma,\sigma) &\hookrightarrow \mathrm{BCMO}_{p,q}(\Sigma,\sigma). \end{aligned} \tag{12.27}$$

(c) If $0 < q \le 1$, *one has*

$$\mathrm{B}_{p,q}(\Sigma,\sigma) \hookrightarrow \mathrm{BCMO}_{p,q}(\Sigma,\sigma) \quad \textit{continuously,} \tag{12.28}$$

where $\mathrm{B}_{p,q}(\Sigma,\sigma)$ *is the space defined in* (10.4). *In addition, in relation to the (homogeneous) Hölder space* $\dot{\mathscr{C}}^{(n-1)(1/q-1)}(\Sigma)$, *one has*

$$\dot{\mathscr{C}}^{(n-1)(1/q-1)}(\Sigma) \hookrightarrow \mathrm{BCMO}_{p,q}(\Sigma,\sigma) \quad \textit{continuously.} \tag{12.29}$$

(d) Whenever $0 < q \le 1$ *and* $(n-1)(\frac{1}{q}-1) < \varepsilon < \infty$, *the following inclusion holds*

$$\mathrm{BCMO}_{p,q}(\Sigma,\sigma) \subseteq L^1\Big(\Sigma, \frac{\sigma(x)}{1+|x|^{n-1+\varepsilon}}\Big). \tag{12.30}$$

Consequently, if $\frac{n-1}{n} < q \le 1$ *then*

$$\mathrm{BCMO}_{p,q}(\Sigma,\sigma) \subseteq L^1\Big(\Sigma, \frac{\sigma(x)}{1+|x|^{n}}\Big). \tag{12.31}$$

(e) One has the following inclusions

$$\bigcup_{0<q<\infty} \mathrm{BCMO}_{p,q}(\Sigma,\sigma) \subseteq L^p_{\mathrm{loc}}(\Sigma,\sigma) \subseteq L^1_{\mathrm{loc}}(\Sigma,\sigma). \tag{12.32}$$

Proof Throughout, fix reference point $x_0 \in \Sigma$, and a dilation factor $2_* \in (0,\infty)$ as in Definition 12.1. Consider first the claim made in *(a)*. Assume $p_1 < p_2$, otherwise (12.24) is trivial. Then for every $f \in L^1_{\mathrm{loc}}(\Sigma,\sigma)$ and $R \ge 2_*$, Hölder's inequality gives

$$\begin{aligned} &R^{(n-1)(1-1/q)}\Big(⨍_{\Delta(x_0,R)} |f - f_{\Delta(x_0,R)}|^{p_1}\,d\sigma\Big)^{1/p_1} \\ &\le R^{(n-1)(1-1/q)}\Big(⨍_{\Delta(x_0,R)} |f - f_{\Delta(x_0,R)}|^{p_2}\,d\sigma\Big)^{1/p_2}, \end{aligned} \tag{12.33}$$

which further entails $\|f\|_{\mathrm{BCMO}_{p_1,q}(\Sigma,\sigma)} \le \|f\|_{\mathrm{BCMO}_{p_2,q}(\Sigma,\sigma)}$. This justifies (12.24).

In relation to the claim made in *(b)*, observe that whenever $q_1 \leq q_2$ and $R \geq 2_*$ we have $R^{(n-1)(1-1/q_1)} \leq R^{(n-1)(1-1/q_2)}$. Hence for each $f \in L^1_{\rm loc}(\Sigma, \sigma)$ we have

$$\begin{aligned} R^{(n-1)(1-1/q_1)} \Big(\fint_{\Delta(x_0,R)} |f - f_{\Delta(x_0,R)}|^p \, d\sigma \Big)^{1/p} \\ \leq R^{(n-1)(1-1/q_2)} \Big(\fint_{\Delta(x_0,R)} |f - f_{\Delta(x_0,R)}|^p \, d\sigma \Big)^{1/p}. \end{aligned} \tag{12.34}$$

This readily implies that $\|f\|_{\mathrm{BCMO}_{p,q_1}(\Sigma,\sigma)} \leq \|f\|_{\mathrm{BCMO}_{p,q_2}(\Sigma,\sigma)}$, thus proving (12.25). In particular, when $0 < q \leq 1$, then (12.26) is a consequence of (12.3) and (12.25) with $q_1 := q$ and $q_2 := 1$. As such, the embeddings in (12.27) follows from (12.26), (5.34), and (5.35). This completes the treatment of item *(b)*.

To justify the embeddings claimed in item *(c)*, fix an exponent $q \in (0, 1]$ and a finite number $R \geq 2_*$. Then for every $f \in \mathrm{B}_{p,q}(\Sigma, \sigma)$, use Hölder's inequality and Proposition 10.1 to estimate

$$\begin{aligned} R^{(n-1)(1-1/q)} \Big(\fint_{\Delta(x_0,R)} |f - f_{\Delta(x_0,R)}|^p \, d\sigma \Big)^{1/p} \\ \leq 2R^{(n-1)(1-1/q)} \Big(\fint_{\Delta(x_0,R)} |f|^p \, d\sigma \Big)^{1/p} \\ \leq C\|f\|_{\mathrm{B}_{p,q}(\Sigma,\sigma)} < \infty, \end{aligned} \tag{12.35}$$

for some constant $C = C(n, p, q, \Sigma) \in (0, \infty)$. Since this holds for all finite radii $R \geq 2_*$, (12.35) ensures that

$$\|f\|_{\mathrm{BCMO}_{p,q}(\Sigma,\sigma)} \leq C\|f\|_{\mathrm{B}_{p,q}(\Sigma,\sigma)}, \tag{12.36}$$

thus proving (12.28). Regarding the embedding in (12.29), continue to assume that $q \in (0, 1]$. Given a function $f \in \dot{\mathscr{C}}^{(n-1)(1/q-1)}(\Sigma)$, we have

$$\|f\|_{\dot{\mathscr{C}}^{(n-1)(1/q-1)}(\Sigma)} \approx \sup_{\substack{x\in\Sigma \\ r>0}} \Big\{ r^{-(n-1)(1/q-1)} \Big(\fint_{\Delta(x,r)} |f - f_{\Delta(x,r)}|^p \Big)^{1/p} \Big\} \tag{12.37}$$

(cf. [GHA-I, Proposition 7.4.7, p. 588]) with proportionality constants depending only on n, q, and the Ahlfors regularity constants of Σ. In turn, this and (12.2) give

$$\|f\|_{\mathrm{BCMO}_{p,q}(\Sigma,\sigma)} \leq \sup_{\substack{x\in\Sigma \\ r>0}} \Big\{ r^{-(n-1)(1/q-1)} \Big(\fint_{\Delta(x,r)} |f - f_{\Delta(x,r)}|^p \Big)^{1/p} \Big\} < \infty. \tag{12.38}$$

This establishes (12.29), hence finishing the justification of item *(c)*.

Pressing on, we now show the embeddings claimed in item *(d)*. Assume $0 < q < 1$, as in the case when $q = 1$ the embedding (12.30) follows from (5.36). Pick an arbitrary function $f \in \mathrm{BCMO}_{p,q}(\Sigma, \sigma)$. Having fixed a finite radius $R \geq 2_*$, decompose f as in (5.43), i.e., write

$$\begin{aligned} f &= (f - f_{\Delta(x_0,R)})\mathbb{1}_{\Delta(x_0,R)} + (f - f_{\Delta(x_0,R)})\mathbb{1}_{\Sigma\setminus\Delta(x_0,R)} + f_{\Delta(x_0,R)} \\ &=: f_1 + f_2 + f_3. \end{aligned} \tag{12.39}$$

Since $L^1\big(\Sigma, \frac{\sigma(x)}{1+|x|^{n-1+\varepsilon}}\big)$ contains constants functions, it is immediate that

$$f_3 \in L^1\Big(\Sigma, \frac{\sigma(x)}{1+|x|^{n-1+\varepsilon}}\Big). \tag{12.40}$$

To see that f_1 belongs to $L^1\big(\Sigma, \frac{\sigma(x)}{1+|x|^{n-1+\varepsilon}}\big)$, use Hölder's inequality, and the Ahlfors regularity of Σ to bound

$$\begin{aligned} &\int_\Sigma \frac{|f_1(x)|}{1+|x|^{n-1+\varepsilon}}\,d\sigma(x) \\ &\quad\le \int_{\Delta(x_0,R)} |f - f_{\Delta(x_0,R)}|\,d\sigma \\ &\quad\le \frac{\sigma\big(\Delta(x_0,R)\big)}{R^{(n-1)(1-1/q)}}\Big\{R^{(n-1)(1-1/q)}\Big(⨍_{\Delta(x_0,R)} |f - f_{\Delta(x_0,R)}|^p\,d\sigma\Big)^{1/p}\Big\} \\ &\quad\le CR^{(n-1)/q}\|f\|_{\mathrm{BCMO}_{p,q}(\Sigma,\sigma)} < \infty, \end{aligned} \tag{12.41}$$

for some constant $C \in (0,\infty)$ depending on the Ahlfors regularity of Σ. In order to prove the membership $f_2 \in L^1\big(\Sigma, \frac{\sigma(x)}{1+|x|^{n-1+\varepsilon}}\big)$, employ the same type of computations as in (5.47) to deduce

$$\begin{aligned} &\int_\Sigma \frac{|f_2(x)|}{1+|x|^{n-1+\varepsilon}}\,d\sigma(x) \\ &\qquad\le C\sum_{j=0}^{\infty} \frac{1}{(2^jR)^{n-1+\varepsilon}} \int_{\Delta(x_0,2^{j+1}R)} |f - f_{\Delta(x_0,2^{j+1}R)}|\,d\sigma \\ &\qquad\quad + C\sum_{j=0}^{\infty} \frac{1}{(2^jR)^{\varepsilon}} |f_{\Delta(x_0,R)} - f_{\Delta(x_0,2^{j+1}R)}| \\ &\qquad =: \mathrm{I} + \mathrm{II}, \end{aligned} \tag{12.42}$$

for some $C = C(n,\Sigma) \in (0,\infty)$. To deal with I in (12.42), we rely on Hölder's inequality, the Ahlfors regularity of Σ, (12.1), and the fact that $\varepsilon > (n-1)(\frac{1}{q}-1)$, to bound

$$
\begin{aligned}
\mathrm{I} &\le C\sum_{j=0}^{\infty}\frac{(2^{j+1}R)^{(n-1)(1/q-1)}}{(2^jR)^{n-1+\varepsilon}}\sigma\big(\Delta(x_0,2^{j+1}R)\big)\\
&\quad\times\Big\{(2^{j+1}R)^{(n-1)(1-1/q)}\Big(⨍_{\Delta(x_0,2^{j+1}R)}|f-f_{\Delta(x_0,2^{j+1}R)}|^p\Big)^{1/p}\Big\}\\
&\le C\Big(\sum_{j=0}^{\infty}\frac{(2^jR)^{(n-1)(1/q-1)}}{(2^jR)^{n-1+\varepsilon}}(2^jR)^{n-1}\Big)\|f\|_{\mathrm{BCMO}_{p,q}(\Sigma,\sigma)}\\
&= CR^{(n-1)[1/q-1-\varepsilon/(n-1)]}\Big(\sum_{j=0}^{\infty}2^{j(n-1)[1/q-1-\varepsilon/(n-1)]}\Big)\|f\|_{\mathrm{BCMO}_{p,q}(\Sigma,\sigma)}<\infty,
\end{aligned}
\tag{12.43}
$$

for some constant $C=C(n,\Sigma)\in(0,\infty)$.

As regards II from (12.42), we rework the computations from (5.49) up to the second to last line (as to obtain the first inequality displayed below), to estimate

$$
\begin{aligned}
\mathrm{II} &\le C\sum_{j=0}^{\infty}\frac{1}{2^{j\varepsilon}}\sum_{k=0}^{j}\Big(⨍_{\Delta(x_0,2^{k+1}R)}|f-f_{\Delta(x_0,2^{k+1}R)}|^p\,d\sigma\Big)^{1/p}\\
&\le C\Big(\sum_{j=0}^{\infty}\frac{1}{2^{j\varepsilon}}\sum_{k=0}^{j}(2^{k+1}R)^{(n-1)(1/q-1)}\Big)\|f\|_{\mathrm{BCMO}_{p,q}(\Sigma,\sigma)}\\
&\le C\Big(\sum_{j=0}^{\infty}\frac{1}{2^{j\varepsilon}}\sum_{k=0}^{j}(2^{k+1}R)^{(n-1)(1/q-1)}\Big)\|f\|_{\mathrm{BCMO}_{p,q}(\Sigma,\sigma)}\\
&= C\Big(\sum_{j=0}^{\infty}2^{j(n-1)[1/q-1-\varepsilon/(n-1)]}\sum_{k=0}^{j}2^{(k-j)(n-1)(1/q-1)}\Big)\\
&\quad\times R^{(n-1)(1/q-1)}\|f\|_{\mathrm{BCMO}_{p,q}(\Sigma,\sigma)}\\
&\le C\Big(\sum_{j=0}^{\infty}2^{j(n-1)[1/q-1-\varepsilon/(n-1)]}\Big)\Big(\sum_{\ell=0}^{\infty}2^{-\ell(n-1)(1/q-1)}\Big)\\
&\quad\times R^{(n-1)(1/q-1)}\|f\|_{\mathrm{BCMO}_{p,q}(\Sigma,\sigma)}\\
&\le CR^{(n-1)(1/q-1)}\|f\|_{\mathrm{BCMO}_{p,q}(\Sigma,\sigma)}<\infty,
\end{aligned}
\tag{12.44}
$$

for some $C=C(n,q,\varepsilon,\Sigma)\in(0,\infty)$, since $0<q<1$ and $\varepsilon>(n-1)(\frac{1}{q}-1)$. This shows that f_2 belongs to $L^1\big(\Sigma,\frac{\sigma(x)}{1+|x|^{n-1+\varepsilon}}\big)$. Collectively, (12.39), (12.40), (12.41), (12.42), (12.43), and (12.44) ultimately imply that f belongs to $L^1\big(\Sigma,\frac{\sigma(x)}{1+|x|^{n-1+\varepsilon}}\big)$. The inclusion (12.31) is a consequence of (12.30) used with $\varepsilon:=1$. This completes the proof of item *(d)*.

Regarding the inclusions claimed in item *(e)*, pick some $f \in \mathrm{BCMO}_{p,q}(\Sigma,\sigma)$ and observe that for every $R \geq 2_*$ we have

$$\begin{aligned}
&\Big(\int_{\Delta(x_0,R)} |f|^p\,d\sigma\Big)^{1/p}\\
&\quad \leq C\Big(\int_{\Delta(x_0,R)} |f - f_{\Delta(x_0,R)}|^p\,d\sigma\Big)^{1/p} + C|f_{\Delta(x_0,R)}|\sigma\big(\Delta(x_0,\widetilde{R})\big)^{1/p}\\
&\quad = CR^{(n-1)(1/q-1)}\sigma\big(\Delta(x_0,R)\big)^{1/p}\|f\|_{\mathrm{BCMO}_{p,q}(\Sigma,\sigma)}\\
&\qquad + C|f_{\Delta(x_0,R)}|\sigma\big(\Delta(x_0,R)\big)^{1/p}\\
&\quad < \infty.
\end{aligned} \tag{12.45}$$

This ensures that $f \in L^p_{\mathrm{loc}}(\Sigma,\sigma) \subseteq L^1_{\mathrm{loc}}(\Sigma,\sigma)$. The proof of Proposition 12.4 is now complete. □

12.2 Duality between $\mathrm{HA}_{p,q}$ and $\mathrm{BCMO}_{p,q}$

In the following theorem, we identify the dual of the first-generation Herz-type Hardy space $\mathrm{HA}_{p,q}$ as the space of (p',q)-bounded central mean oscillations functions $\mathrm{BCMO}_{p',q}$ modulo constants.

Theorem 12.1 *Assume $\Sigma \subseteq \mathbb{R}^n$ is a closed unbounded Ahlfors regular set, and abbreviate $\sigma := \mathcal{H}^{n-1}\lfloor\Sigma$. Given two exponents $p \in (1,\infty)$ and $q \in \big(\frac{n-1}{n},1\big]$, denote by $p' \in (1,\infty)$ the Hölder conjugate exponent of p. Then the dual of $\mathrm{HA}_{p,q}(\Sigma,\sigma)$ is $\widetilde{\mathrm{BCMO}}_{p',q}(\Sigma,\sigma)$, i.e.,*

$$\big(\mathrm{HA}_{p,q}(\Sigma,\sigma)\big)^* = \widetilde{\mathrm{BCMO}}_{p',q}(\Sigma,\sigma). \tag{12.46}$$

The specific nature of this identification is as follows. Fix a reference point $x_0 \in \Sigma$ and consider the assignment

$$\begin{aligned}
&\widetilde{\mathrm{BCMO}}_{p',q}(\Sigma,\sigma) \ni [f] \mapsto \Phi f := \Lambda_{[f]} \in \big(\mathrm{HA}_{p,q}(\Sigma,\sigma)\big)^*\\
&\qquad \text{defined as} \quad \langle\Lambda_{[f]}, g\rangle := \lim_{N\to\infty}\sum_{j=0}^{N}\lambda_j\int_\Sigma f a_j\,d\sigma
\end{aligned} \tag{12.47}$$

whenever

$$\begin{aligned}
&g \in \mathrm{HA}_{p,q}(\Sigma,\sigma),\ \{\lambda_j\}_{j\in\mathbb{N}_0} \in \ell^q(\mathbb{N}_0), \text{ and } \{a_j\}_{j\in\mathbb{N}_0} \text{ is a sequence}\\
&\text{of } x_0\text{-central } (q,p)\text{-atoms with the property that } g = \textstyle\sum_{j=0}^\infty \lambda_j a_j \text{ in}\\
&\mathrm{HA}_{p,q}(\Sigma,\sigma).
\end{aligned} \tag{12.48}$$

Then $\Lambda_{[f]}$ *is meaningfully defined, and the assignment* (12.47)-(12.48) *is a well-defined bounded linear isomorphism.*

Proof The identification claimed in (12.46) can be obtained by making natural alterations in the reasoning used to establish Theorem 5.1, now bearing in mind Definition 11.2, Definition 12.1, Theorem 11.2, Theorem 11.3, Proposition 12.4, Theorem 11.1, Corollary 11.1, Corollary 11.2, and Remark 11.5. □

The completeness of $\widetilde{\mathrm{BCMO}}_{p,q}(\Sigma,\sigma)$ can be obtained via duality.

Corollary 12.1 *Assume* $\Sigma \subseteq \mathbb{R}^n$ *is a closed unbounded Ahlfors regular set, and abbreviate* $\sigma := \mathcal{H}^{n-1}\lfloor\Sigma$. *Then for every* $p \in (1,\infty)$ *and* $q \in \left(\frac{n-1}{n}, 1\right]$ *the space* $\widetilde{\mathrm{BCMO}}_{p,q}(\Sigma,\sigma)$ *is Banach when equipped with the norm* $\|\cdot\|_{\widetilde{\mathrm{BCMO}}_{p,q}(\Sigma,\sigma)}$ *(cf.* (12.7)).

Proof The completeness of $\widetilde{\mathrm{BCMO}}_{p,q}(\Sigma,\sigma)$ is a consequence of the duality result recorded in Theorem 12.1 and the completeness of Herz-type Hardy spaces obtained in Proposition 11.4. □

We have the following description for the seminorm $\|\cdot\|_{\mathrm{BCMO}_{p,q}(\Sigma,\sigma)}$ in terms of integral pairings with central (q,p)-atoms.

Corollary 12.2 *Consider a closed unbounded Ahlfors regular set* $\Sigma \subseteq \mathbb{R}^n$, *and abbreviate* $\sigma := \mathcal{H}^{n-1}\lfloor\Sigma$. *Select two exponents* $p \in (1,\infty)$ *and* $q \in \left(\frac{n-1}{n}, 1\right]$, *and denote by* $p' \in (1,\infty)$ *the Hölder conjugate exponent of* p. *Then for every function* $f \in L^1_{\mathrm{loc}}(\Sigma,\sigma)$ *one has*

$$\|f\|_{\mathrm{BCMO}_{p,q}(\Sigma,\sigma)} \approx \sup\left\{\left|\int_\Sigma f a\,d\sigma\right| : a \text{ is an } x_0\text{-central } (q,p')\text{-atom on } \Sigma\right\}, \tag{12.49}$$

with implicit proportionality constants depending only on p, q, *and the Ahlfors regularity character of* Σ.

Proof This follows by running a similar argument as in the proof of Corollary 5.2, now keeping in mind Definition 11.2, Definition 12.1, Proposition 12.4, and Lemma 5.3. □

In the spirit of Corollary 5.3 we also have the following result regarding weak-$*$ convergence of sequences in $\mathrm{BCMO}_{p,q}$.

Corollary 12.3 *Consider a closed unbounded Ahlfors regular set* $\Sigma \subseteq \mathbb{R}^n$, *and abbreviate* $\sigma := \mathcal{H}^{n-1}\lfloor\Sigma$. *Pick two exponents* $p \in (1,\infty)$ *and* $q \in \left(\frac{n-1}{n}, 1\right]$, *and denote by* $p' \in (1,\infty)$ *the Hölder conjugate exponent of* p. *Select a function* $f \in L^1_{\mathrm{loc}}(\Sigma,\sigma)$, *and assume there exists a sequence of functions* $\{f_j\}_{j\in\mathbb{N}} \subseteq L^1_{\mathrm{loc}}(\Sigma,\sigma)$ *with the following properties*

$$\sup_{j \in \mathbb{N}} \|f_j\|_{\mathrm{BCMO}_{p,q}(\Sigma,\sigma)} < \infty \text{ and} \qquad (12.50)$$
$$f_j \longrightarrow f \text{ in } L^1_{\mathrm{loc}}(\Sigma, \sigma) \text{ as } j \to \infty.$$

Then $f \in \mathrm{BCMO}_{p,q}(\Sigma, \sigma)$ *and*

$$[f_j] \to [f] \text{ as } j \to \infty \text{ in } \widetilde{\mathrm{BCMO}}_{p,q}(\Sigma, \sigma) \text{ in the weak-}* \text{ topology,} \qquad (12.51)$$

in the sense that

$$\langle [f_j], g \rangle \to \langle [f], g \rangle \text{ as } j \to \infty, \text{ for every } g \in \mathrm{HA}_{p',q}(\Sigma, \sigma) \qquad (12.52)$$

where $\langle \cdot, \cdot \rangle$ *denotes the duality pairing between* $\mathrm{BCMO}_{p,q}(\Sigma, \sigma)$ *and* $\mathrm{HA}_{p',q}(\Sigma, \sigma)$ *understood in the sense of Theorem 12.1.*

Proof Natural modifications in the argument presented in the proof of Corollary 5.3, together with the separability of $\mathrm{HA}_{p',q}(\Sigma, \sigma)$ (cf. Theorem 11.5), the duality result proved in Theorem 12.1, and Lemma 5.2, give the weak-$*$ convergence claimed in (12.51). □

A version of Corollary 5.4 regarding weak-$*$ convergence is also valid in the context of $\mathrm{BCMO}_{p,q}$ spaces. More specifically, we have the following result.

Corollary 12.4 *Assume* $\Sigma \subseteq \mathbb{R}^n$ *is a closed unbounded Ahlfors regular set, and abbreviate* $\sigma := \mathcal{H}^{n-1} \lfloor \Sigma$. *Select two exponents* $p \in (1, \infty)$ *and* $q \in \left(\frac{n-1}{n}, 1\right]$, *and denote by* $p' \in (1, \infty)$ *the Hölder conjugate exponent of* p. *Given a function* $f \in \mathrm{BCMO}_{p,q}(\Sigma, \sigma)$, *for each* $N \in \mathbb{N}$ *define*

$$f_N(x) := \begin{cases} N, & \text{if } f(x) > N, \\ f(x), & \text{if } |f(x)| \le N, \\ -N, & \text{if } f(x) < -N, \end{cases} \qquad (12.53)$$

for each $x \in \Sigma$. *Then* $[f_N] \to [f]$ *weak-*$*$ *in* $\widetilde{\mathrm{BCMO}}_{p,q}(\Sigma, \sigma)$ *as* $N \to \infty$. *Specifically, for every* $g \in \mathrm{HA}^{p'}(\Sigma, \sigma)$ *one has*

$$\lim_{N \to \infty} \langle [f_N], g \rangle = \langle [f], g \rangle \qquad (12.54)$$

where $\langle \cdot, \cdot \rangle$ *is the duality pairing between* $\mathrm{BCMO}_{p,q}(\Sigma, \sigma)$ *and* $\mathrm{HA}_{p',q}(\Sigma, \sigma)$ *understood in the sense of Theorem 12.1.*

Proof The reasoning presented in the proof of Corollary 5.4, combined with Proposition 12.2 and Corollary 12.3, yield the desired weak-$*$ convergence. □

It turns out that if a $\mathrm{BCMO}_{p,q}$ function pairs well (in an integral sense) with functions from $\mathrm{HA}_{p',q}$ then the corresponding duality pairing is realized via an integral pairing.

Corollary 12.5 *Consider a closed unbounded Ahlfors regular set $\Sigma \subseteq \mathbb{R}^n$, and abbreviate $\sigma := \mathcal{H}^{n-1}\lfloor\Sigma$. Select integrability exponents $p \in (1,\infty)$ and $q \in \big(\frac{n-1}{n}, 1\big]$, and denote by $p' \in (1,\infty)$ the Hölder conjugate exponent of p. Pick two functions $f \in \mathrm{BCMO}_{p,q}(\Sigma,\sigma)$ and $g \in \mathrm{HA}_{p',q}(\Sigma,\sigma)$, and assume*

$$\int_\Sigma |fg|\, d\sigma < \infty. \tag{12.55}$$

Then the equivalence class $[f] \in \widetilde{\mathrm{BCMO}}_{p,q}(\Sigma,\sigma)$, viewed as a continuous linear function from $\big(\mathrm{HA}_{p',q}(\Sigma,\sigma)\big)^*$ (understood in the sense of Theorem 12.1), acts on g via integration against f, i.e.,

$$\langle [f], g\rangle = \int_\Sigma fg\, d\sigma. \tag{12.56}$$

Proof The proof is similar to that of Corollary 5.5, now with $\mathrm{BCMO}_{p,q}(\Sigma,\sigma)$ in place of $\mathrm{BCMO}_p(\Sigma,\sigma)$, this time using Corollary 12.4, the duality result recorded in Theorem 12.1, and the atomic decomposition of $\mathrm{HA}_{p,q}(\Sigma,\sigma)$ established in Theorem 11.2. □

In the following corollary, we show that central (q,p')-molecules from $\mathrm{HA}_{p',q}$ pair well, in an integral sense, with functions from $\mathrm{BCMO}_{p,q}$.

Corollary 12.6 *Suppose $\Sigma \subseteq \mathbb{R}^n$ is a closed unbounded Ahlfors regular set, and abbreviate $\sigma := \mathcal{H}^{n-1}\lfloor\Sigma$. Select two exponents $p \in (1,\infty)$ and $q \in \big(\frac{n-1}{n}, 1]$, and denote by $p' \in (1,\infty)$ the Hölder conjugate exponent of p. Pick a reference point $x_0 \in \Sigma$ and a dilation factor 2_* as in Definition 12.1. Also, fix a parameter*

$$\theta \in \big((n-1)(p'/q - 1), \infty\big). \tag{12.57}$$

Then for every function $f \in \mathrm{BCMO}_{p,q}(\Sigma,\sigma)$ and every x_0-central (q,p',θ)-molecule M on Σ it follows that

$$\int_\Sigma |fM|\, d\sigma < \infty. \tag{12.58}$$

In particular, one has

$$\big\langle [f], M\big\rangle = \int_\Sigma fM\, d\sigma \tag{12.59}$$

where $\big\langle \cdot\,,\,\cdot\big\rangle$ denotes the duality pairing between $\widetilde{\mathrm{BCMO}}_{p,q}(\Sigma,\sigma)$ and $\mathrm{HA}_{p',q}(\Sigma,\sigma)$ (in the sense described in Theorem 12.1).

Proof The proof parallels the computations made in the proof of Corollary 12.5, now bearing in mind (11.65)-(11.67) in Definition 11.3, as well as (12.1) in Definition 12.1, and the fact that $\theta/p'(n-1) + 1/p' - 1/q > 0$ (as implied by the choice of θ). □

Chapter 13
Calderón-Zygmund Theory on First-Generation Herz-type Spaces

Having established the functional analytic framework in Chapters 10, 11, and 12, here we take up the task of developing a comprehensive Calderón-Zygmund theory for singular integral operators on uniformly rectifiable sets in the context of first-generation Herz spaces, and other related scales. This is applicable to SIO's of layer potential type associated with weakly elliptic systems, and provides the groundwork for employing boundary integral methods for the treatment of boundary value problems in Chapter 16. In Section 13.1 we focus on obtaining boundedness results for integral operators on first-generation Herz spaces $\mathrm{A}_{p,q}$. In Section 13.2 we treat singular integral operators on first-generation Herz-type Hardy spaces $\mathrm{HA}_{p,q}$. In Section 13.3 we study modified singular integral operators in relation to functions of (p,q)-bounded central mean oscillations, and (p,q)-central Carleson measures.

13.1 Singular Integral Operators on Herz Spaces $\mathrm{A}_{p,q}$

Our first result concerns convolution-type singular integral operators, nontangential estimates, and jump-formulas in the context of first-generation Herz spaces.

Theorem 13.1 *Assume $\Sigma \subseteq \mathbb{R}^n$ is a closed unbounded uniformly rectifiable set, and abbreviate $\sigma := \mathcal{H}^{n-1}\lfloor\Sigma$. For some sufficiently large integer $N = N(n) \in \mathbb{N}$, consider a complex-valued kernel $k \in \mathscr{C}^N\big(\mathbb{R}^n \setminus \{0\}\big)$ which is odd and positive homogeneous of degree $1-n$. In this setting, for each function $f \in L^1\big(\Sigma, \frac{\sigma(x)}{1+|x|^{n-1}}\big)$ define*

$$T_\varepsilon f(x) := \int\limits_{\substack{y\in\Sigma\\ |x-y|>\varepsilon}} k(x-y)f(y)\,d\sigma(y) \ \textit{for all } x \in \Sigma \textit{ and } \varepsilon > 0, \tag{13.1}$$

$$T_{\max} f(x) := \sup_{\varepsilon>0} |T_\varepsilon f(x)| \ \textit{ for each } \ x \in \Sigma, \tag{13.2}$$

© The Author(s), under exclusive license to Springer Nature Switzerland AG 2026

M. Mitrea and P. Takemura, *Singular Integrals, Herz-Type Function Spaces, and Boundary Problems*, Progress in Mathematics 362,
https://doi.org/10.1007/978-3-032-12516-3_13

$$Tf(x) := \lim_{\varepsilon \to 0^+} T_\varepsilon f(x) \text{ for } \sigma\text{-a.e. } x \in \Sigma. \tag{13.3}$$

Finally, fix two exponents $p, q \in (1, \infty)$.

Then there exists some $C \in (0, \infty)$ which depends exclusively on n, p, q, and the UR *constants of Σ with the property that for each $f \in \mathrm{A}_{p,q}(\Sigma, \sigma)$ one has*

$$\|T_{\max} f\|_{\mathrm{A}_{p,q}(\Sigma,\sigma)} \leq C\Big(\sum_{|\alpha| \leq N} \sup_{S^{n-1}} |\partial^\alpha k|\Big) \|f\|_{\mathrm{A}_{p,q}(\Sigma,\sigma)}. \tag{13.4}$$

Moreover, the operator

$$T : \mathrm{A}_{p,q}(\Sigma, \sigma) \longrightarrow \mathrm{A}_{p,q}(\Sigma, \sigma) \tag{13.5}$$

is well defined, linear, and bounded.

Next, assume $\Omega \subseteq \mathbb{R}^n$ is an open set such that $\partial\Omega$ is a uniformly rectifiable set, and abbreviate $\sigma := \mathcal{H}^{n-1} \lfloor \partial\Omega$. With the integral kernel k as before, for each function $f \in L^1\big(\partial\Omega, \frac{\sigma(x)}{1+|x|^{n-1}}\big)$ define

$$\mathcal{T} f(x) := \int_{\partial\Omega} k(x-y) f(y)\, d\sigma(y) \text{ for each } x \in \Omega. \tag{13.6}$$

Then for each aperture parameter $\kappa \in (0, \infty)$ there exists some $C \in (0, \infty)$ depending only on n, p, q, κ, and the UR *constants of $\partial\Omega$ with the property that for each function $f \in \mathrm{A}_{p,q}(\partial\Omega, \sigma)$ one has*

$$\|\mathcal{N}_\kappa(\mathcal{T} f)\|_{\mathrm{A}_{p,q}(\partial\Omega,\sigma)} \leq C\Big(\sum_{|\alpha| \leq N} \sup_{S^{n-1}} |\partial^\alpha k|\Big) \|f\|_{\mathrm{A}_{p,q}(\partial\Omega,\sigma)}. \tag{13.7}$$

Finally, for each function $f \in \mathrm{A}_{p,q}(\partial\Omega, \sigma)$ the following jump-formula holds:

$$\Big(\mathcal{T} f\Big|_{\partial\Omega}^{\kappa-\mathrm{n.t.}}\Big)(x) = \frac{1}{2i}\widehat{k}\big(\nu(x)\big) f(x) + (Tf)(x) \text{ at } \sigma\text{-a.e. } x \in \partial_*\Omega, \tag{13.8}$$

where $\widehat{k}$ denotes the Fourier transform (cf. (2.4)) *of k.*

Proof The boundedness properties claimed in (13.4), (13.5), and (13.7) are consequences of the extrapolation result recorded in Theorem 10.3 together with [MMMi[3], Proposition 3.4, p. 179]. Moreover, the jump-formula displayed in (13.8) follows from [GHA-III, Theorem 2.5.1, p. 407] and the fact that $\mathrm{A}_{p,q}(\Sigma, \sigma)$ embeds continuously into $L^1\big(\Sigma, \frac{\sigma(x)}{1+|x|^{n-1}}\big)$ (cf. item *(e)* in Proposition 10.3). □

The next result describes operator norm estimates for chord-dot-normal singular integral operators acting on first-generation Herz spaces which are sensitive to the geometry of the underlying UR domain. This is a key result when dealing with boundary value problems with data in $\mathrm{A}_{p,q}$ spaces in subsequent chapters. Recall the piece of notation defined in (6.113).

Theorem 13.2 *Assume $\Omega \subseteq \mathbb{R}^n$ is a* UR *domain with unbounded boundary, and abbreviate $\sigma := \mathcal{H}^{n-1}\lfloor\partial\Omega$. Denote by ν the geometric measure theoretic outward unit normal to Ω. For some sufficiently large integer $N = N(n) \in \mathbb{N}$, consider a kernel $k \in \mathscr{C}^N(\mathbb{R}^n \setminus \{0\})$ that is even and positive homogeneous of degree $-n$. Recall the chord-dot-normal singular integral operators T and $T^{\#}$ associated with k as in* (6.103) *and* (6.104)*, respectively. Finally, fix two exponents $p, q \in (1, \infty)$.*

Then the operators

$$T : [\mathrm{A}_{p,q}(\partial\Omega, \sigma)]^M \to [\mathrm{A}_{p,q}(\partial\Omega, \sigma)]^M$$
$$T^{\#} : [\mathrm{A}_{p,q}(\partial\Omega, \sigma)]^M \to [\mathrm{A}_{p,q}(\partial\Omega, \sigma)]^M \tag{13.9}$$

are well defined, linear, and bounded. Moreover, for each $m \in \mathbb{N}$ there exits some $C_m \in (0, \infty)$ depending exclusively on m, n, p, q, the kernel k, and the UR *constants of $\partial\Omega$, with the property that*

$$\|T\|_{\mathrm{A}_{p,q}(\partial\Omega,\sigma)\to\mathrm{A}_{p,q}(\partial\Omega,\sigma)} \leq C_m \cdot \|\nu\|^{\langle m\rangle}_{[\mathrm{BMO}(\partial\Omega,\sigma)]^n}, \tag{13.10}$$

$$\|T^{\#}\|_{\mathrm{A}_{p,q}(\partial\Omega,\sigma)\to\mathrm{A}_{p,q}(\partial\Omega,\sigma)} \leq C_m \cdot \|\nu\|^{\langle m\rangle}_{[\mathrm{BMO}(\partial\Omega,\sigma)]^n}. \tag{13.11}$$

Proof The fact that the operators in (13.9) are well defined and bounded, with operator norms satisfying (13.10) and (13.11), is a consequence of the extrapolation result recorded in Theorem 10.3 along with the boundedness result from [MMMi[3], Corollary 4.2, p. 292]. □

The goal now is to obtain boundedness results for commutators acting on first-generation Herz spaces.

Theorem 13.3 *Consider a closed unbounded Ahlfors regular set $\Sigma \subseteq \mathbb{R}^n$, and abbreviate $\sigma := \mathcal{H}^{n-1}\lfloor\Sigma$. Pick an integrability exponent $p_0 \in (1, \infty)$ along with a non-decreasing function $\Phi : (0, \infty) \to (0, \infty)$. Assume T is a bounded linear operator on $L^{p_0}(\Sigma, w)$ for every weight $w \in A_{p_0}(\Sigma, \sigma)$, with operator norm controlled by $\Phi([w]_{A_{p_0}})$. Finally, fix two exponents $p, q \in (1, \infty)$ and a complex-valued function $b \in L^\infty(\Sigma, \sigma)$.*

Then the following inclusion holds

$$\mathrm{A}_{p,q}(\Sigma, \sigma) \subseteq \bigcup_{w \in A_{p_0}(\Sigma,\sigma)} L^{p_0}(\Sigma, w), \tag{13.12}$$

and $[M_b, T]$ maps $\mathrm{A}_{p,q}(\Sigma, \sigma)$ into itself, where $[M_b, T]$ is the commutator of T and the operator M_b of pointwise multiplication on $\mathrm{A}_{p,q}(\Sigma, \sigma)$ by the function b, i.e.,

$$[M_b, T]f := b(Tf) - T(bf) \quad \textit{for each } f \in \mathrm{A}_{p,q}(\Sigma, \sigma). \tag{13.13}$$

Moreover, there exist constants $C_1, C \in (0, \infty)$ depending only on n, p_0, p, q, and the Ahlfors regularity of Σ such that

$$\|[M_b, T]\|_{A_{p,q}(\Sigma,\sigma)\to A_{p,q}(\Sigma,\sigma)} \leq C_1 \cdot \Phi(C)\|b\|_{\mathrm{BMO}(\Sigma,\sigma)}, \tag{13.14}$$

Proof All claims are consequences of [MMMi³, Theorem 4.3, p. 294], [GHA-II, Corollary 5.2.2, p. 240], and the extrapolation result recorded in Theorem 10.3. □

13.2 Singular Integral Operators on Herz-type Hardy Spaces $\mathrm{HA}_{p,q}$

This section contains results pertaining to the boundedness of singular integral operators acting on first-generation Herz-type Hardy spaces defined on uniformly rectifiable sets. A crucial role is played by the atomic and molecular theory developed in Chapter 11. We begin by studying singular integral operators of convolution-type acting from $\mathrm{HA}_{p,q}$ spaces.

Theorem 13.4 *Consider a closed unbounded uniformly rectifiable set $\Sigma \subseteq \mathbb{R}^n$, and abbreviate $\sigma := \mathcal{H}^{n-1}\lfloor\Sigma$. For some sufficiently large integer $N = N(n) \in \mathbb{N}$, select a kernel $k \in \mathscr{C}^N(\mathbb{R}^n \setminus \{0\})$ that is odd and positive homogeneous of degree $1-n$. Recall the boundary-to-boundary convolution-type integral operator T associated with k as in* (13.3)*. Finally, fix two exponents $p \in (1,\infty)$ and $q \in \big(\frac{n-1}{n}, 1\big]$. Then, in this context, the operator*

$$\begin{gathered} T : \mathrm{HA}_{p,q}(\Sigma,\sigma) \to \mathrm{A}_{p,q}(\Sigma,\sigma) \\ \textit{is well defined, linear, and bounded.} \end{gathered} \tag{13.15}$$

Proof Notice that the action of T on functions from $\mathrm{HA}_{p,q}(\Sigma,\sigma)$ is well defined, thanks to the fact $\mathrm{HA}_{p,q}(\Sigma,\sigma) \hookrightarrow L^1\big(\Sigma, \frac{\sigma(x)}{1+|x|^{n-1}}\big)$ (cf. Proposition 11.3). The proof of the boundedness claimed in (13.15) mirrors that of Theorem 6.1. As such, here we only emphasize the key differences in our current functional analytic setting. Fix some reference point $x_0 \in \Sigma$ and a dilation factor 2_* as in Definition 10.1. Consider an x_0-central (q,p)-atom a on Σ, and let $R \geq 2_*$ be a finite radius associated with a (as in Definition 11.2). The first claim we make is that there exists some $C \in (0,\infty)$ depending solely on n, p, q, and the Ahlfors regularity constants of Σ with the property that

$$\|Ta\|_{\mathrm{A}_{p,q}(\Sigma,\sigma)} \leq C. \tag{13.16}$$

To verify this, pick the unique integer $J \in \mathbb{N}$ for which

$$2_*^J \leq R < 2_*^{J+1}, \tag{13.17}$$

and write

$$
\begin{aligned}
\|Ta\|^q_{\mathrm{A}_{p,q}(\Sigma,\sigma)} &= \sum_{j=0}^{J+1} \big(2_*^{j(n-1)(1/q-1/p)}\|(Ta)\cdot \mathbb{1}_{C_j}\|_{L^p(\Sigma,\sigma)}\big)^q \\
&\quad + \sum_{j=J+2}^{\infty} \big(2_*^{j(n-1)(1/q-1/p)}\|(Ta)\cdot \mathbb{1}_{C_j}\|_{L^p(\Sigma,\sigma)}\big)^q \\
&=: \mathrm{S}_1 + \mathrm{S}_2. \qquad (13.18)
\end{aligned}
$$

In light of the boundedness of T on $L^p(\Sigma,\sigma)$ (cf. Theorem 6.1), the normalization property of a (cf. (11.21)), (13.17), and the fact that $q \le 1 < p$, we may estimate S_1 as follows:

$$
\begin{aligned}
\mathrm{S}_1 &= \sum_{j=0}^{J+1} \big(2_*^{j(n-1)(1/q-1/p)}\|(Ta)\cdot \mathbb{1}_{C_j}\|_{L^p(\Sigma,\sigma)}\big)^q \\
&\le C\sum_{j=0}^{J+1} \big(2_*^{j(n-1)(1/q-1/p)}\|a\|_{L^p(\Sigma,\sigma)}\big)^q \\
&\le C\sum_{j=0}^{J+1} \big(2_*^{j(n-1)(1/q-1/p)}R^{(n-1)(1/p-1/q)}\big)^q \\
&\le C\sum_{j=0}^{J+1} \big(2_*^{j(n-1)(1/q-1/p)}2_*^{J(n-1)(1/p-1/q)}\big)^q \\
&\le C\sum_{\ell=0}^{\infty} 2_*^{-\ell(n-1)(1/q-1/p)q} < \infty, \qquad (13.19)
\end{aligned}
$$

for some constant $C = C(n,p,q,k,\Sigma) \in (0,\infty)$ independent of the atom a.

Next, fix some $j \in \mathbb{N}_0$ such that $j \ge J+2$, and some $x \in C_j$. Following the same type of reasoning as in (6.17)-(6.18), employing the same type of computations as in (6.20), now bearing in mind the vanishing moment and normalization properties of a (cf. (11.22)-(11.21)), we presently obtain

$$
\begin{aligned}
|Ta(x)| &\le \int_{\Delta(x_0,R)} |k(x-y)-k(x-x_0)||a(y)|\,d\sigma(y) \\
&\le C\frac{R}{|x-x_0|^n}\|a\|_{L^p(\Sigma,\sigma)}\sigma\big(\Delta(x_0,R)\big)^{1-1/p} \\
&\le C\frac{R}{2_*^{jn}}R^{(n-1)(1/p-1/q)}R^{(n-1)(1-1/p)} \\
&= C\frac{R}{2_*^{jn}}R^{(n-1)(1-1/q)} \qquad (13.20)
\end{aligned}
$$

for some constant $C \in (0, \infty)$ depending only on n, p, the kernel k, and the Ahlfors regularity of Σ. Since this holds for all $x \in C_j$, the Ahlfors regularity of Σ and (13.17) entail

$$\begin{aligned}\|(Ta)\cdot \mathbb{1}_{C_j}\|_{L^p(\Sigma,\sigma)} &= \Big(\int_{C_j} |Ta|^p\,d\sigma\Big)^{1/p}\\ &\leq C\frac{R}{2_*^{jn}}R^{(n-1)(1-1/q)}\sigma\big(\Delta(x_0, 2_*^{j+1})\big)^{1/p}\\ &\leq C\frac{2_*^J}{2_*^{jn}}2_*^{J(n-1)(1-1/q)}2_*^{j(n-1)/p}\\ &= C2_*^{-jn}2_*^{J(n-1)[n/(n-1)-1/q]}2_*^{j(n-1)/p}\end{aligned} \tag{13.21}$$

for some constant $C = C(n, p, k, \Sigma) \in (0, \infty)$. Having established (13.21), and recalling that $q > (n-1)/n$, we estimate S_2 from (13.18) as follows:

$$\begin{aligned}\mathrm{S}_2 &\leq C\sum_{j=J+2}^{\infty}\big(2_*^{j(n-1)(1/q-1/p)}2_*^{-jn}2_*^{J(n-1)[n/(n-1)-1/q]}2_*^{j(n-1)/p}\big)^q\\ &= C\sum_{j=J+2}^{\infty}2_*^{(j-J)(n-1)[1/q-n/(n-1)]q}\\ &\leq C\sum_{\ell=0}^{\infty}2_*^{\ell(n-1)[1/q-n/(n-1)]q} < \infty.\end{aligned} \tag{13.22}$$

Collectively, (13.18), (13.19), and (13.22) imply the validity of estimate (13.16).

Pressing forward, select an arbitrary function $f \in \mathrm{HA}_{p,q}(\Sigma, \sigma)$ and use Theorem 11.2 to write

$$f = \sum_{j=0}^{\infty}\lambda_j a_j \ \text{ in } L^1(\Sigma, \sigma) \tag{13.23}$$

for some sequence of x_0-central (q, p)-atoms $\{a_j\}_{j\in\mathbb{N}_0}$, and some numerical sequence $\{\lambda_j\}_{j\in\mathbb{N}_0} \in \ell^q(\mathbb{N}_0)$. Assume this atomic decomposition is quasi-optimal, i.e.,

$$\Big(\sum_{j=0}^{\infty}|\lambda_j|^q\Big)^{1/q} \leq 2\|f\|_{\mathrm{HA}_{p,q}(\Sigma,\sigma)}. \tag{13.24}$$

For each $m \in \mathbb{N}_0$ abbreviate

$$f_m := \sum_{j=0}^{m}\lambda_j a_j. \tag{13.25}$$

In light of weak-type $(1, 1)$ boundedness of T (cf. Theorem 6.1), and the convergence in $L^1(\Sigma, \sigma)$ from (13.23), we have

$$Tf_m \to Tf \quad \text{in } L^{1,\infty}(\Sigma, \sigma) \text{ as } m \to \infty. \tag{13.26}$$

As such, restricting to a subsequence if necessary, it follows that

$$Tf_m \to Tf \quad \text{pointwise } \sigma\text{-a.e. on } \Sigma \text{ as } m \to \infty. \tag{13.27}$$

We now show that the sequence $\{Tf_m\}_{m\in\mathbb{N}}$ is Cauchy in $\mathrm{A}_{p,q}(\Sigma, \sigma)$. Indeed, bearing in mind that $q \le 1$, and employing (13.16) and (13.25), for each $m, \ell \in \mathbb{N}$ with $\ell > m$ we bound

$$\|Tf_\ell - Tf_m\|^q_{\mathrm{A}_{p,q}(\Sigma,\sigma)} = \sum_{j=m+1}^{\ell} |\lambda_j|^q \|Ta_j\|^q_{\mathrm{A}_{p,q}(\Sigma,\sigma)} \le C \sum_{j=m+1}^{\infty} |\lambda_j|^q. \tag{13.28}$$

Because $\{\lambda_j\}_{j\in\mathbb{N}_0} \in \ell^q(\mathbb{N}_0)$, letting $m \to \infty$ we see that the right-hand side tends to zero. Hence, $\{Tf_m\}_{m\in\mathbb{N}}$ is a Cauchy sequence in $\mathrm{A}_{p,q}(\Sigma, \sigma)$. From the completeness of $\mathrm{A}_{p,q}(\Sigma, \sigma)$ (cf. Proposition 10.6), there exists some function $g \in \mathrm{A}_{p,q}(\Sigma, \sigma)$ for which

$$Tf_m \to g \quad \text{in } \mathrm{A}_{p,q}(\Sigma, \sigma) \text{ as } m \to \infty. \tag{13.29}$$

Thanks to Proposition 10.3 this convergence also holds in $L^1(\Sigma, \sigma)$. Hence, restricting to a subsubsequence if necessary we can say that

$$Tf_m \to g \quad \text{pointwise } \sigma\text{-a.e. on } \Sigma \text{ as } m \to \infty. \tag{13.30}$$

In concert with (13.27) this implies

$$Tf = g \quad \text{pointwise } \sigma\text{-a.e. on } \Sigma. \tag{13.31}$$

From (13.24), (13.25), (13.29), and (13.31) we conclude that

$$\begin{aligned} \|Tf\|^q_{\mathrm{A}_{p,q}(\Sigma,\sigma)} &= \lim_{m\to\infty} \|Tf_m\|^q_{\mathrm{A}_{p,q}(\Sigma,\sigma)} \le \lim_{m\to\infty} \sum_{j=0}^{m} |\lambda_j|^q \|Ta_j\|^q_{\mathrm{A}_{p,q}(\Sigma,\sigma)} \\ &\le C \sum_{j=0}^{\infty} |\lambda_j|^q \le C\|f\|^q_{\mathrm{HA}_{p,q}(\Sigma,\sigma)} \end{aligned} \tag{13.32}$$

for some $C \in (0, \infty)$ depending only on n, p, q, the kernel k, and the Ahlfors regularity constants of Σ. This finishes the proof of Theorem 13.4. □

Our next result concerns nontangential maximal function estimates involving convolution-type integral operators in the context of first-generation Herz-type Hardy spaces.

Theorem 13.5 *Assume $\Omega \subseteq \mathbb{R}^n$ is a* UR *domain with unbounded boundary, and abbreviate $\sigma := \mathcal{H}^{n-1}\lfloor\partial\Omega$. For some sufficiently large integer $N = N(n) \in \mathbb{N}$, consider a kernel $k \in \mathscr{C}^N(\mathbb{R}^n\setminus\{0\})$ that is odd and positive homogeneous of degree*

$1-n$. Recall the boundary-to-domain convolution-type integral operator $\mathcal{T}$ associated with k as in (13.6)*. Finally, fix two exponents $p \in (1, \infty)$ and $q \in \big(\frac{n-1}{n}, 1\big]$, along with an aperture parameter $\kappa \in (0, \infty)$.*

Then there exists some $C \in (0, \infty)$ depending exclusively on n, p, q, κ, and the UR *constants of $\partial\Omega$, with the property that for every $f \in \mathrm{HA}_{p,q}(\partial\Omega, \sigma)$ one has*

$$\|\mathcal{N}_\kappa(\mathcal{T}f)\|_{\mathrm{A}_{p,q}(\partial\Omega,\sigma)} \leq C\|f\|_{\mathrm{HA}_{p,q}(\partial\Omega,\sigma)}. \tag{13.33}$$

Proof Fix a reference point $x_0 \in \partial\Omega$, and a dilation factor $2_* > (C_{\partial\Omega}/c_{\partial\Omega})^{1/(n-1)}$, where $C_{\partial\Omega}, c_{\partial\Omega} \in (0, \infty)$ are the upper and lower Ahlfors regularity constants of $\partial\Omega$, respectively. The proof is similar to that of Theorem 6.2. With this in mind, here we only highlight the main differences.

Fix an x_0-central (q, p)-atom $a : \partial\Omega \to \mathbb{R}$, and let $R \geq 2_*$ be a radius associated with a (as in Definition 11.2). We claim that there exists some constant $C \in (0, \infty)$ depending only on n, p, q, κ, and the Ahlfors regularity of $\partial\Omega$ (in particular, independent of a) with the property that

$$\|\mathcal{N}_\kappa(\mathcal{T}a)\|_{\mathrm{A}_{p,q}(\partial\Omega,\sigma)} \leq C. \tag{13.34}$$

To justify this, select a number $\rho \in (0, \infty)$ satisfying

$$\rho > 2 + \kappa, \tag{13.35}$$

and choose the unique integer $N \in \mathbb{N}$ satisfying

$$2_*^N \leq \rho R < 2_*^{N+1}. \tag{13.36}$$

Next, write

$$\begin{aligned}\|\mathcal{N}_\kappa(\mathcal{T}a)\|^q_{\mathrm{A}_{p,q}(\partial\Omega,\sigma)} &= \sum_{j=0}^{N} \big(2_*^{j(n-1)(1/q-1/p)}\|\mathcal{N}_\kappa(\mathcal{T}a)\cdot \mathbb{1}_{C_j}\|_{L^p(\partial\Omega,\sigma)}\big)^q \\ &\quad + \sum_{j=N+1}^{\infty} \big(2_*^{j(n-1)(1/q-1/p)}\|\mathcal{N}_\kappa(\mathcal{T}a)\cdot \mathbb{1}_{C_j}\|_{L^p(\partial\Omega,\sigma)}\big)^q \\ &=: \mathrm{I} + \mathrm{II}.\end{aligned} \tag{13.37}$$

To deal with I, recall the Cotlar-type inequality recorded in (6.38), then use the same argument as in (6.39), and the normalization of a (cf. (11.21)) to estimate

$$\begin{aligned}\|\mathcal{N}_\kappa(\mathcal{T}a)\|_{L^p(\partial\Omega,\sigma)} &\leq \|T_{\max}a\|_{L^p(\partial\Omega,\sigma)} + C\|\mathcal{M}_{\partial\Omega}a\|_{L^p(\partial\Omega,\sigma)} \\ &\leq C\|a\|_{L^p(\partial\Omega,\sigma)} \leq CR^{(n-1)(1/p-1/q)},\end{aligned} \tag{13.38}$$

for some constant $C \in (0, \infty)$ independent of a. Granted this and (13.36), we bound the term I in (13.37) as

$$\begin{aligned} \mathrm{I} &\leq \sum_{j=0}^{N} \big(2_*^{j(n-1)(1/q-1/p)} R^{(n-1)(1/p-1/q)}\big)^q \leq C \sum_{j=0}^{N} 2_*^{(j-N)(n-1)(1/q-1/p)q} \\ &\leq C \sum_{\ell=0}^{\infty} 2_*^{-\ell(n-1)(1/q-1/p)q} < \infty, \end{aligned} \tag{13.39}$$

for some $C = C(\partial\Omega, \kappa, n, p) \in (0, \infty)$.

To deal with II in (13.37) fix some $j \in \mathbb{N}$, with $j \geq N+1$, along with some $x \in C_j$. Based on the analysis carried out in (6.41)-(6.45), together with the Ahlfors regularity of $\partial\Omega$ and the normalization property of a (cf. (11.21)), for every $z \in \Gamma_\kappa(x)$ we have:

$$\begin{aligned} |(\mathcal{T}a)(z)| &\leq \frac{CR}{|x-x_0|^n} \|a\|_{L^p(\partial\Omega,\sigma)} \sigma\big(\Delta(x_0, R)\big)^{1-1/p} \\ &\leq \frac{CR}{|x-x_0|^n} R^{(n-1)(1/p-1/q)} R^{(n-1)(1-1/p)} \\ &= \frac{CR}{|x-x_0|^n} R^{(n-1)(1-1/q)}, \end{aligned} \tag{13.40}$$

for some constant $C = C(n, p, q, \kappa, k, \partial\Omega) \in (0, \infty)$. After taking the supremum over all $z \in \Gamma_\kappa(x)$ we find that

$$\mathcal{N}_\kappa(\mathcal{T}a)(x) \leq \frac{CR}{|x-x_0|^n} R^{(n-1)(1-1/q)} \quad \text{for all } x \in C_j. \tag{13.41}$$

Based on this, (13.36), and the Ahlfors regularity of $\partial\Omega$, we may estimate

$$\begin{aligned} \|\mathcal{N}_\kappa(\mathcal{T}a) \cdot \mathbb{1}_{C_j}\|_{L^p(\partial\Omega,\sigma)} &\leq CRR^{(n-1)(1-1/q)} \Big(\int_{C_j} \frac{d\sigma(x)}{|x-x_0|^{np}} \Big)^{1/p} \\ &\leq C \frac{2_*^N}{2_*^{jn}} 2_*^{N(n-1)(1-1/q)} \sigma\big(\Delta(x_0, 2_*^{j+1})\big)^{1/p} \\ &\leq C \frac{2_*^{j(n-1)/p}}{2_*^{jn}} 2_*^N 2_*^{N(n-1)(1-1/q)}, \end{aligned} \tag{13.42}$$

for some $C = C(\partial\Omega, n, p, q, k, \kappa) \in (0, \infty)$. Keeping this in mind and that $q > \frac{n-1}{n}$ (which further implies that $q/(n-1) + q - 1 > 0$), we estimate II from (13.37) as

$$
\begin{aligned}
\mathrm{II} &= \sum_{j=N+1}^{\infty} \left(2_*^{j(n-1)(1/q-1/p)} \|\mathcal{N}_\kappa(\mathcal{T}a)\cdot \mathbb{1}_{C_j}\|_{L^p(\partial\Omega,\sigma)}\right)^q \\
&\leq C \sum_{j=N+1}^{\infty} 2_*^{j(n-1)(1-q/p)} \frac{2_*^{j(n-1)q/p}}{2_*^{jnq}} 2_*^{Nq} 2_*^{N(n-1)(q-1)} \\
&= C \sum_{j=N+1}^{\infty} \frac{2_*^{j(n-1)}}{2_*^{j(n-1)q} 2_*^{jq}} 2_*^{Nq} 2_*^{N(n-1)(q-1)} \\
&= C \sum_{j=N+1}^{\infty} 2_*^{(N-j)q} 2_*^{(N-j)(n-1)(q-1)} \\
&= C \sum_{\ell=1}^{\infty} 2_*^{-\ell(n-1)[q/(n-1)+q-1]} < \infty.
\end{aligned} \tag{13.43}
$$

At this stage, (13.34) follows from a combination of (13.37), (13.39), and (13.43).

In order to verify estimate (13.33), fix an arbitrary function $f \in \mathrm{HA}_{p,q}(\partial\Omega,\sigma)$ and use Theorem 11.2 to decompose f into central atoms as

$$
f = \sum_{j=0}^{\infty} \lambda_j a_j \quad \text{in } L^1(\partial\Omega,\sigma), \tag{13.44}
$$

for some sequence of x_0-central (q,p)-atoms $\{a_j\}_{j\in\mathbb{N}_0}$ and some numerical sequence $\{\lambda_j\}_{j\in\mathbb{N}_0} \in \ell^q(\mathbb{N}_0)$. Also, without loss of generality, assume that such a decomposition is quasi-optimal in the sense that

$$
\Big(\sum_{j=0}^{\infty} |\lambda_j|^q\Big)^{1/q} \leq 2\|f\|_{\mathrm{HA}_{p,q}(\partial\Omega,\sigma)}. \tag{13.45}
$$

For each $m \in \mathbb{N}_0$, abbreviate

$$
f_m := \sum_{j=0}^{m} \lambda_j a_j. \tag{13.46}
$$

Based on the same type of reasoning as the one presented in (6.52)-(6.57), now using that $\mathrm{HA}_{p,q}(\partial\Omega,\sigma)$ continuously embeds into $L^1(\partial\Omega,\sigma)$ (cf. Proposition 11.3), the completeness of $\mathrm{A}_{p,q}(\partial\Omega,\sigma)$ (cf. Proposition 10.6), together with (13.34), (13.45), and the fact that $\{\lambda_j\}_{j\in\mathbb{N}_0} \in \ell^q(\mathbb{N}_0)$ we obtain

$$
\mathcal{N}_\kappa(\mathcal{T} f_m) \to \mathcal{N}_\kappa(\mathcal{T} f) \quad \text{in } \mathrm{A}_{p,q}(\partial\Omega,\sigma) \text{ as } m \to \infty \tag{13.47}
$$

and

$$
\begin{aligned}
\|\mathcal{N}_\kappa(\mathcal{T} f)\|^q_{\mathrm{A}_{p,q}(\partial\Omega,\sigma)} &= \lim_{m\to\infty} \|\mathcal{N}_\kappa(\mathcal{T} f_m)\|^q_{\mathrm{A}_{p,q}(\partial\Omega,\sigma)} \\
&\leq \lim_{m\to\infty} \sum_{j=0}^{m} |\lambda_j|^q \cdot \|\mathcal{N}_\kappa(\mathcal{T} a_j)\|^q_{\mathrm{A}_{p,q}(\partial\Omega,\sigma)} \\
&\leq C \sum_{j=0}^{\infty} |\lambda_j|^q \leq C\|f\|^q_{\mathrm{HA}_{p,q}(\partial\Omega,\sigma)}.
\end{aligned} \tag{13.48}
$$

for some constant $C = C(\partial\Omega, n, p, q, \kappa) \in (0,\infty)$ independent of f. This establishes (13.33), thus concluding the proof of Theorem 13.5. □

Our next result concerns the boundedness of generalized double layer potentials on Herz-type Hardy spaces.

Theorem 13.6 *Assume $\Omega \subseteq \mathbb{R}^n$ is a uniformly rectifiable domain with unbounded boundary. Abbreviate $\sigma := \mathcal{H}^{n-1}\lfloor\partial\Omega$, and denote by ν the geometric measure theoretic outward unit normal to Ω. For some sufficiently large integer $N = N(n) \in \mathbb{N}$, select a vector-valued kernel $\vec{k} \in \big[\mathscr{C}^N(\mathbb{R}^n\setminus\{0\})\big]^n$ that is odd, positive homogeneous of degree $1-n$, and divergence-free, i.e., $\operatorname{div}\vec{k} = 0$ in $\mathbb{R}^n\setminus\{0\}$. Recall the generalized double layer potential $T^\#$ associated with $\vec{k}$ as in (6.64). Then, whenever $p \in (1,\infty)$ and $q \in \big(\frac{n-1}{n}, 1\big]$, it follows that*

$$
\begin{gathered}
T^\# : \mathrm{HA}_{p,q}(\partial\Omega,\sigma) \to \mathrm{HA}_{p,q}(\partial\Omega,\sigma) \\
\textit{is a well-defined, linear, and bounded operator.}
\end{gathered} \tag{13.49}
$$

Proof Select some reference point $x_0 \in \Sigma$ and a dilation factor 2_* as in Definition 10.1. The justification of the boundedness claimed in (13.49) follows the same type of argument used in the proof of Theorem 6.3. For this reason, here we may only emphasize the key differences. Fix an arbitrary x_0-central (q,p)-atom a on $\partial\Omega$, and let $R \geq 2_*$ be the radius associated with a (in the sense of Definition 11.2). Fix a parameter $\theta \in (0,\infty)$ satisfying

$$
(n-1)(p/q-1) < \theta < n(p-1)+1. \tag{13.50}
$$

Such a choice of θ is possible since $q > \frac{n-1}{n}$. We claim that there exists some constant $C_\theta \in (0,\infty)$, independent of a, such that the function

$$
T^\# a \text{ is a } C_\theta\text{-multiple of an } x_0\text{-central } (q,p,\theta)\text{-molecule on } \partial\Omega. \tag{13.51}
$$

To verify this, we begin by noticing that the normalization property of a (cf. (11.21)), and the boundedness of $T^\#$ on $L^p(\partial\Omega,\sigma)$ (cf. (6.69)) yield

$$
\|T^\# a\|_{L^p(\partial\Omega,\sigma)} \leq C\|a\|_{L^p(\partial\Omega,\sigma)} \leq CR^{(n-1)(1/p-1/q)}, \tag{13.52}
$$

for some $C = C(n,p,k,\partial\Omega) \in (0,\infty)$. Next, we prove that there exists some constant $C \in (0,\infty)$, independent of a, for which

$$\Big(\int_{\partial\Omega}|(T^{\#}a)(x)|^p|x-x_0|^\theta\,d\sigma(x)\Big)^{1/p}\leq CR^{(n-1)(1/p-1/q)+\theta/p}. \tag{13.53}$$

Indeed, begin by spitting the above integral as

$$\begin{aligned}\Big(\int_{\partial\Omega}&|(T^{\#}a)(x)|^p|x-x_0|^\theta\,d\sigma(x)\Big)^{1/p}\\ &\leq\Big(\int_{\substack{x\in\partial\Omega\\|x-x_0|<2R}}|(T^{\#}a)(x)|^p|x-x_0|^\theta\,d\sigma(x)\Big)^{1/p}\\ &\quad+\Big(\int_{\substack{x\in\partial\Omega\\|x-x_0|\geq 2R}}|(T^{\#}a)(x)|^p|x-x_0|^\theta\,d\sigma(x)\Big)^{1/p}\\ &=:\mathrm{I}+\mathrm{II}.\end{aligned} \tag{13.54}$$

Using (13.52), we estimate I as

$$\begin{aligned}\mathrm{I}&=\Big(\int_{\substack{x\in\partial\Omega\\|x-x_0|<2R}}|(T^{\#}a)(x)|^p|x-x_0|^\theta\,d\sigma(x)\Big)^{1/p}\\ &\leq(2R)^{\theta/p}\|T^{\#}a\|_{L^p(\partial\Omega,\sigma)}\leq CR^{(n-1)(1/p-1/q)+\theta/p},\end{aligned} \tag{13.55}$$

for some $C\in(0,\infty)$ depending exclusively on n, p, θ, and the UR constants of $\partial\Omega$. Next, in view of the normalization property of a (cf. (11.21)), and the Ahlfors regularity of $\partial\Omega$, the same type of reasoning as in (6.76)-(6.79) produces

$$\begin{aligned}|T^{\#}a(x)|&\leq\frac{CR}{|x-x_0|^n}\|a\|_{L^p(\partial\Omega,\sigma)}\sigma\big(\Delta(x_0,R)\big)^{1-1/p}\\ &\leq\frac{CR}{|x-x_0|^n}R^{(n-1)(1-1/q)},\end{aligned} \tag{13.56}$$

for every $x\in\partial\Omega\setminus B(x_0,2R)$, and for some constant $C=C(n,p,\partial\Omega)\in(0,\infty)$. Thanks to this, and the upper bound for θ in (13.50), we estimate II from (13.54) as follows:

$$
\begin{aligned}
\mathrm{II} &= \Big(\int_{\substack{x \in \partial\Omega \\ |x-x_0| \geq 2R}} |(T^{\#}a)(x)|^p |x-x_0|^{\theta}\, d\sigma(x) \Big)^{1/p} \\
&\leq CRR^{(n-1)(1-1/q)} \Big(\int_{\substack{x \in \partial\Omega \\ |x-x_0| \geq 2R}} \frac{|x-x_0|^{\theta}}{|x-x_0|^{np}}\, d\sigma(x) \Big)^{1/p} \\
&\leq CRR^{(n-1)(1-1/q)} \sum_{j=1}^{\infty} \Big(\int_{\substack{x \in \partial\Omega \\ 2^j R \leq |x-x_0| \leq 2^{j+1} R}} \frac{|x-x_0|^{\theta}}{|x-x_0|^{np}}\, d\sigma(x) \Big)^{1/p} \\
&\leq CRR^{(n-1)(1-1/q)} \sum_{j=1}^{\infty} \frac{(2^{j+1}R)^{\theta/p}}{(2^j R)^n} \sigma\big(\Delta(x_0, 2^{j+1}R)\big)^{1/p} \\
&\leq C \frac{RR^{(n-1)(1-1/q)} R^{\theta/p}}{R^n} \sum_{j=0}^{\infty} (2^j)^{\theta/p - n} (2^{j+1}R)^{(n-1)/p} \\
&= CR^{(n-1)(1/p-1/q)+\theta/p} \sum_{j=0}^{\infty} 2^{j[\theta/p - n + (n-1)/p]} < \infty.
\end{aligned} \tag{13.57}
$$

In concert, (13.54), (13.55), (13.57) imply the existence of some $C \in (0, \infty)$ depending only on n, p, q, θ, and the Ahlfors regularity constants of $\partial\Omega$ such that (13.53) holds.

Going further, by running the same argument as in (6.81)-(6.92), now keeping in mind Theorem 13.5 and item *(c)* in Proposition 10.3 (used with $r := 1$), we conclude that

$$
\int_{\partial\Omega} T^{\#}a\, d\sigma = 0. \tag{13.58}
$$

As such, the claim made in (13.51) is a consequence of (13.52), (13.53), (13.58).

Having established (13.51), we may invoke Theorem 11.4 to ensure the existence of some constant $C \in (0, \infty)$ for which

$$
\|T^{\#}a\|_{\mathrm{HA}_{p,q}(\partial\Omega,\sigma)} \leq C \tag{13.59}
$$

for every x_0-central (q, p)-atom a on $\partial\Omega$.

As a next step, select an arbitrary function $f \in \mathrm{HA}_{p,q}(\partial\Omega, \sigma)$, and use Theorem 11.2 to decompose f as

$$
f = \sum_{j=0}^{\infty} \lambda_j a_j \ \text{ in } L^1(\partial\Omega, \sigma), \tag{13.60}
$$

where $\{\lambda_j\}_{j \in \mathbb{N}_0} \in \ell^q(\mathbb{N}_0)$, and $\{a_j\}_{j \in \mathbb{N}_0}$ is some sequence of x_0-central (q, p)-atoms. Assume that this atomic decomposition is quasi-optimal, i.e.,

$$\sum_{j=0}^{\infty} |\lambda_j| \leq 2\|f\|_{\mathrm{HA}_{p,q}(\partial\Omega,\sigma)}. \tag{13.61}$$

For each $m \in \mathbb{N}_0$ abbreviate

$$f_m := \sum_{j=0}^{m} \lambda_j a_j. \tag{13.62}$$

Then the same type of argument as in (6.97)-(6.101), now using the continuous embedding $\mathrm{HA}_{p,q}(\partial\Omega,\sigma) \hookrightarrow L^1(\partial\Omega,\sigma)$ (cf. Proposition 11.3), the fact that $\mathrm{HA}_{p,q}(\partial\Omega,\sigma)$ is a quasi-Banach space (cf. Proposition 11.4), along with (13.59), (13.61), (11.5), and that $\{\lambda_j\}_{j\in\mathbb{N}_0} \in \ell^q(\mathbb{N}_0)$, allows us to conclude that

$$\lim_{m\to\infty} T^{\#} f_m = T^{\#} f \quad \text{in } \mathrm{HA}_{p,q}(\partial\Omega,\sigma) \tag{13.63}$$

and

$$\begin{aligned}\|T^{\#} f\|^q_{\mathrm{HA}_{p,q}(\partial\Omega,\sigma)} &= \lim_{m\to\infty} \|T^{\#} f_m\|^q_{\mathrm{HA}_{p,q}(\partial\Omega,\sigma)} \\ &\leq \lim_{m\to\infty} \sum_{j=0}^{m} |\lambda_j|^q \|T^{\#} a_j\|^q_{\mathrm{HA}_{p,q}(\partial\Omega,\sigma)} \\ &\leq C \sum_{j=0}^{\infty} |\lambda_j|^q \leq 2C\|f\|^q_{\mathrm{HA}_{p,q}(\partial\Omega,\sigma)}\end{aligned} \tag{13.64}$$

for some $C \in (0,\infty)$ depending only on n, p, q, the kernel $\vec{k}$ and the UR constants of $\partial\Omega$. This ultimately proves Theorem 13.6. □

We next establish geometric operator norm estimates (which are sensitive to "flatness") for chord-dot-normal singular integral operators in the context of first-generation Herz-type Hardy spaces. This is a key result that opens the door for employing boundary layer methods in the treatment of boundary value problems with data in $\mathrm{HA}_{p,q}$. Recall the piece of notation introduced in (6.113).

Theorem 13.7 *Assume $\Omega \subseteq \mathbb{R}^n$ is a* UR *domain with unbounded boundary. Abbreviate $\sigma := \mathcal{H}^{n-1}\lfloor\partial\Omega$, and denote by ν the geometric measure theoretic outward unit normal to Ω. For a sufficiently large integer $N = N(n) \in \mathbb{N}$, select a kernel $k \in \mathscr{C}^N(\mathbb{R}^n \setminus \{0\})$ that is even, and positive homogeneous of degree $-n$. Recall the chord-dot-normal singular integral operator $T^{\#}$ associated with k as in* (6.104). *Also, fix two exponents $p \in (1,\infty)$ and $q \in \big(\frac{n-1}{n}, 1\big]$.*

Then

$$\begin{gathered} T^{\#} : \mathrm{HA}_{p,q}(\partial\Omega,\sigma) \to \mathrm{HA}_{p,q}(\partial\Omega,\sigma) \\ \text{is well defined, linear, and bounded.}\end{gathered} \tag{13.65}$$

Moreover, for each $m \in \mathbb{N}$ there exits some $C_m \in (0,\infty)$ depending exclusively on m, n, p, q, the kernel k, and the UR *constants of $\partial\Omega$ with the property that*

$$\|T^{\#}\|_{\mathrm{HA}_{p,q}(\partial\Omega,\sigma)\to\mathrm{HA}_{p,q}(\partial\Omega,\sigma)} \leq C_m \cdot \|\nu\|^{\langle m\rangle}_{[\mathrm{BMO}(\partial\Omega,\sigma)]^n}. \tag{13.66}$$

Proof The claim made in (13.65) is a consequence of Theorem 13.6 together with the fact that chord-dot-normal singular integral operators are themselves generalized double layers (see [GHA-IV, Example 5.1.7, pp. 556–557]). Select a reference point $x_0 \in \partial\Omega$ and a dilation factor $2_* > (C_{\partial\Omega}/c_{\partial\Omega})^{1/(n-1)}$, where $C_{\partial\Omega}, c_{\partial\Omega} \in (0,\infty)$ are the upper and lower Ahlfors regularity constants of $\partial\Omega$, respectively. To establish the operator norm estimate claimed in (13.66), we revisit the proof of Theorem 6.4. The core of the matter is to show the existence of some $C \in (0,\infty)$ depending exclusively on n, m, p, q, the kernel k, and the UR constants of $\partial\Omega$ with the property that whenever $\varepsilon \in \big(\frac{1}{q} - 1, \frac{1}{n-1}\big)$, and a is an x_0-central (q,p)-atom on $\partial\Omega$, with $R \geq 2_*$ associated with a as in Definition 11.2,

$$\begin{gathered}\text{the function } M := T^{\#}a \text{ is a } C\cdot\|\nu\|^{\langle m\rangle}_{[\mathrm{BMO}(\partial\Omega,\sigma)]^n}\text{-multiple of an}\\ x_0\text{-central } (q,p)_\varepsilon\text{-molecule on } \partial\Omega \text{ associated with the}\\ \text{radius } 2R \text{ (in the sense of Definition 11.4).}\end{gathered} \tag{13.67}$$

To get started, employ the same argument as in (6.124)-(6.125), now taking into account the normalization of a (cf. (11.21)) to obtain

$$\Big(\int_{\Delta(x_0,2R)} |M|^p\,d\sigma\Big)^{1/p} \leq C\cdot\|\nu\|^{\langle m\rangle}_{[\mathrm{BMO}(\partial\Omega,\sigma)]^n} R^{(n-1)(1/p-1/q)}. \tag{13.68}$$

Next, using the vanishing moment property of the atom a (cf. (11.22)), and taking into account that $\operatorname{supp} a \subseteq \Delta(x_0,R)$, for σ-a.e. $x \in \partial\Omega\setminus B(x_0,2R)$ we may express

$$M(x) = -\int\limits_{\Delta(x_0,R)} \Big\{\big\langle\nu(x), x-y\big\rangle k(x-y) - \big\langle\nu(x), x-x_0\big\rangle k(x-x_0)\Big\} a(y)\,d\sigma(y). \tag{13.69}$$

Retaining the same piece of notation as in (6.128), and using the reasoning presented in (6.129)-(6.134) together with the normalization of a (cf. (11.21)), we may deduce

$$\begin{aligned}|M(x) - M_\Delta(x)| &\leq CR\frac{|\nu(x)-\nu_{\Delta(x_0,R)}|}{|x-x_0|^n}\sigma\big(\Delta(x_0,R)\big)^{1-1/p}\|a\|_{L^p(\partial\Omega,\sigma)}\\ &\leq CR\frac{|\nu(x)-\nu_{\Delta(x_0,R)}|}{|x-x_0|^n}R^{(n-1)(1-1/p)}R^{(n-1)(1/p-1/q)}\\ &\leq CR\frac{|\nu(x)-\nu_{\Delta(x_0,R)}|}{|x-x_0|^n}R^{(n-1)(1-1/q)},\end{aligned} \tag{13.70}$$

for σ-a.e. $x \in \partial\Omega\setminus B(x_0,2R)$, where $C \in (0,\infty)$ depends only on n, p, the kernel k, and the Ahlfors regularity character of $\partial\Omega$.

Recall the annulus $A_\ell(x_0,R)$ introduced in (4.198) for each $\ell \in \mathbb{N}$. Taking into consideration (13.70), and using the same approach as in (6.135)-(6.142) we presently obtain

$$\Big(\int_{A_\ell(x_0,2R)} |M - M_\Delta|^p \, d\sigma\Big)^{1/p} \leq C 2^{\ell(n-1)(1/p-1-\varepsilon)} R^{(n-1)(1/p-1/q)} \|\nu\|_{[\mathrm{BMO}(\partial\Omega,\sigma)]^n} \tag{13.71}$$

for all $\ell \in \mathbb{N}$, and for some $C = C(n, p, k, \varepsilon, \partial\Omega) \in (0, \infty)$.

To proceed, with $M_\Delta^{(1)}$ and $M_\Delta^{(2)}$ as in (6.144)-(6.145), we have the following decomposition

$$M_\Delta(x) = M_\Delta^{(1)}(x) + M_\Delta^{(2)}(x) \text{ for each } x \in \partial\Omega\backslash B(x_0, 2R). \tag{13.72}$$

Availing ourselves of the computations made in (6.146), and using the normalization of a (cf. (11.21)), for each $x \in \partial\Omega\backslash B(x_0, 2R)$ it follows that

$$\big|M_\Delta^{(1)}(x)\big| = \big|\langle \nu_{\Delta(x_0,R)}, x - x_0\rangle\big| \frac{CR}{|x - x_0|^{n+1}} R^{(n-1)(1-1/q)} \tag{13.73}$$

for some constant $C = C(n, p, k, \partial\Omega) \in (0, \infty)$ independent of the atom a. Now, based on this, and the same type of reasoning as in (6.147), for every $\ell \in \mathbb{N}$ it follows that

$$\Big(\int_{A_\ell(x_0,2R)} |M_\Delta^{(1)}|^p \, d\sigma\Big)^{1/p} \leq C 2^{\ell(n-1)(1/p-1-\varepsilon)} R^{(n-1)(1/p-1/q)} \|\nu\|_{[\mathrm{BMO}(\partial\Omega,\sigma)]^n}, \tag{13.74}$$

for some constant $C = C(n, p, k, \varepsilon, \partial\Omega) \in (0, \infty)$ independent of the atom a. In relation to the term $M_\Delta^{(2)}(x)$ in (13.72), we may take advantage of the computations made in (6.148), combined with the normalization of a (cf. (11.21)), to bound

$$|M_\Delta^{(2)}(x)| \leq CR \cdot R^{(n-1)(1-1/q)} \|\nu\|_{[\mathrm{BMO}(\partial\Omega,\sigma)]^n} \frac{1}{|x - x_0|^n}, \tag{13.75}$$

for every $x \in \partial\Omega\backslash B(x_0, 2R)$. Thanks to this, a quick look at (6.149) reveals that for every $\ell \in \mathbb{N}$ we have

$$\Big(\int_{A_\ell(x_0,2R)} |M_\Delta^{(2)}|^p \, d\sigma\Big)^{1/p} \leq C 2^{\ell(n-1)(1/p-1-\varepsilon)} R^{(n-1)(1/p-1/q)} \|\nu\|_{[\mathrm{BMO}(\partial\Omega,\sigma)]^n} \tag{13.76}$$

for some constant $C = C(n, p, k, \partial\Omega) \in (0, \infty)$ (in particular, C is independent of the atom a). Collectively, (13.72), (13.74) and (13.76) ensure that there exists some constant $C \in (0, \infty)$ depending exclusively on n, p, q, k, ε, and the Ahlfors regularity character of $\partial\Omega$ with the property that, for each $\ell \in \mathbb{N}$ we have

$$\Big(\int_{A_\ell(x_0,2R)} |M_\Delta|^p\,d\sigma\Big)^{1/p} \leq C2^{\ell(n-1)(1/p-1-\varepsilon)}R^{(n-1)(1/p-1/q)}\|\nu\|_{[\mathrm{BMO}(\partial\Omega,\sigma)]^n}. \tag{13.77}$$

All together, estimates (13.71) and (13.77) imply the existence of some constant $C = (n, p, k, \varepsilon, \partial\Omega) \in (0,\infty)$ satisfying

$$\Big(\int_{A_\ell(x_0,2R)} |M|^p\,d\sigma\Big)^{1/p} \leq C2^{\ell(n-1)(1/p-1-\varepsilon)}R^{(n-1)(1/p-1/q)}\|\nu\|_{[\mathrm{BMO}(\partial\Omega,\sigma)]^n}. \tag{13.78}$$

Lastly, observe that the vanishing moment property of M follows from Lemma 6.1. This, (13.68), and (13.78) finally prove the claim in (13.67).

Once we have established (13.67), invoke Theorem 11.4 to guarantee that there exists some constant $C = C(n, p, q, \varepsilon, k, \partial\Omega) \in (0,\infty)$ such that for every x_0-central (q, p)-atom a on $\partial\Omega$ we have $T^\# a \in \mathrm{HA}_{p,q}(\Sigma,\sigma)$ and

$$\|T^\# a\|_{\mathrm{HA}_{p,q}(\partial\Omega,\sigma)} \leq C\cdot\|\nu\|^{\langle m\rangle}_{[\mathrm{BMO}(\partial\Omega,\sigma)]^n}. \tag{13.79}$$

At this stage, we are left with proving that $T^\#$ is bounded on $\mathrm{HA}_{p,q}(\partial\Omega,\sigma)$ with an operator norm satisfying (13.66). To this end, fix a function $f \in \mathrm{HA}_{p,q}(\partial\Omega,\sigma)$, and use Theorem 11.2 to write

$$f = \sum_{j=0}^{\infty}\lambda_j a_j \ \text{ in } L^1(\partial\Omega,\sigma), \tag{13.80}$$

where $\{\lambda_j\}_{j\in\mathbb{N}_0} \in \ell^q(\mathbb{N}_0)$, and $\{a_j\}_{j\in\mathbb{N}_0}$ is a sequence of x_0-central (q, p)-atoms. Also, make the assumption that this decomposition is quasi-optimal, i.e.,

$$\Big(\sum_{j=0}^{\infty}|\lambda_j|^q\Big)^{1/q} \leq 2\|f\|_{\mathrm{HA}_{p,q}(\partial\Omega,\sigma)}. \tag{13.81}$$

For each $N \in \mathbb{N}_0$, abbreviate

$$f_N := \sum_{j=0}^{N}\lambda_j a_j. \tag{13.82}$$

Relying on the same type of argument as in (6.156)–(6.161), now employing (11.5), (13.79), (13.81), the completeness of $\mathrm{HA}_{p,q}(\partial\Omega,\sigma)$ (cf. Theorem 11.4), and using that $\mathrm{HA}_{p,q}(\Sigma,\sigma)$ embeds continuously into $L^1(\partial\Omega,\sigma)$ (cf. Proposition 11.3), we may conclude that

$$T^\# f_N \to T^\# f \ \text{ in } \mathrm{HA}_{p,q}(\partial\Omega,\sigma) \text{ as } N\to\infty, \tag{13.83}$$

and

$$
\begin{aligned}
\|T^{\#} f\|^q_{\mathrm{HA}_{p,q}(\partial\Omega,\sigma)} &= \lim_{N\to\infty} \|T^{\#} f_N\|^q_{\mathrm{HA}_{p,q}(\partial\Omega,\sigma)} \\
&\leq \lim_{N\to\infty} \sum_{j=0}^{N} |\lambda_j|^q \|T^{\#} a_j\|^q_{\mathrm{HA}_{p,q}(\partial\Omega,\sigma)} \\
&\leq C \cdot \big(\|\nu\|^{\langle m\rangle}_{[\mathrm{BMO}(\partial\Omega,\sigma)]^n}\big)^q \sum_{j=0}^{\infty} |\lambda_j|^q \\
&\leq 2^q C \cdot \big(\|\nu\|^{\langle m\rangle}_{[\mathrm{BMO}(\partial\Omega,\sigma)]^n}\big)^q \|f\|^q_{\mathrm{HA}_{p,q}(\partial\Omega,\sigma)}.
\end{aligned}
\tag{13.84}
$$

This concludes the proof of Theorem 13.7. □

13.3 Singular Integral Operators on $\mathrm{BCMO}_{p,q}$ Spaces

Here we obtain boundedness results for modified singular integral operators acting on, or from, spaces of functions of (p,q)-bounded central mean oscillations, and establish certain central Carleson measure estimates. To get started, the result recorded below gives the boundedness of modified convolution-type integral operators from $\mathrm{B}_{p,q}$ (cf. Definition 10.1) into $\mathrm{BCMO}_{p,q}$ (cf. Definition 12.1).

Theorem 13.8 *Let $\Sigma \subseteq \mathbb{R}^n$ be an unbounded* UR *set, and abbreviate $\sigma := \mathcal{H}^{n-1}\lfloor\Sigma$. For some sufficiently large $N = N(n) \in \mathbb{N}$, select a kernel $k \in \mathscr{C}^N(\mathbb{R}^n \setminus \{0\})$ that is odd, and positive homogeneous of degree $1-n$. Denote $k_\varepsilon := k \cdot \mathbb{1}_{\mathbb{R}^n\setminus\overline{B(0,\varepsilon)}}$ for each $\varepsilon \in (0,\infty)$, and recall the modified boundary-to-boundary convolution type integral operator T_{mod} whose action on each $f \in L^1\big(\Sigma, \frac{\sigma(x)}{1+|x|^n}\big)$ is given by*

$$
(T_{\mathrm{mod}} f)(x) := \lim_{\varepsilon\to 0^+} \int_\Sigma \big(k_\varepsilon(x-y) - k_1(-y)\big) f(y)\, d\sigma(y) \quad \text{at } \sigma\text{-a.e. } x \in \Sigma. \tag{13.85}
$$

Finally, fix two exponents $p \in (1,\infty)$ and $q \in \big(\frac{n-1}{n}, 1\big]$. Then the operator

$$
\begin{gathered}
T_{\mathrm{mod}} : \mathrm{B}_{p,q}(\Sigma,\sigma) \longrightarrow \mathrm{BCMO}_{p,q}(\Sigma,\sigma) \\
\text{is well defined, linear, and bounded.}
\end{gathered}
\tag{13.86}
$$

In particular,

$$
\begin{gathered}
\text{the operator } \ [T_{\mathrm{mod}}] : \mathrm{B}_{p,q}(\Sigma,\sigma) \longrightarrow \widetilde{\mathrm{BCMO}}_{p,q}(\Sigma,\sigma) \\
\text{defined as } [T_{\mathrm{mod}}]f := [T_{\mathrm{mod}} f] \text{ for each } f \in \mathrm{B}_{p,q}(\Sigma,\sigma) \\
\text{is well defined, linear, and bounded.}
\end{gathered}
\tag{13.87}
$$

Proof Notice that the action of T_{mod} on functions from $\mathrm{B}_{p,q}(\Sigma,\sigma)$ is meaningful, thanks to (6.166) and items *(c)-(d)* from Proposition 12.4. Fix a function $f \in \mathrm{B}_{p,q}(\Sigma,\sigma)$ along with an x_0-central (q,p')-atom a on Σ (in the sense of Definition 11.2), where $p' \in (1,\infty)$ denotes the Hölder conjugate exponent of p. Thanks to items *(c)-(e)* in Proposition 12.4, we have that

$$\begin{gathered} f \in L^p_{\text{loc}}(\Sigma,\sigma) \cap L^1\big(\Sigma, \tfrac{\sigma(x)}{1+|x|^n}\big) \\ \text{and } \ a \in L^{p'}_{\text{comp}}(\Sigma,\sigma) \ \text{ with } \ \int_\Sigma a\,d\sigma = 0. \end{gathered} \tag{13.88}$$

Hence, the formula in (6.169), the boundedness result recorded in Theorem 13.4, the Hölder-type inequality from item *(b)* in Proposition 10.2, and Theorem 11.1 yield

$$\begin{aligned} \Big|\int_\Sigma (T_{\text{mod}} f)\cdot a\,d\sigma\Big| &= \Big|\int_\Sigma f\cdot (Ta)\,d\sigma\Big| \leq \|f\|_{\mathrm{B}_{p,q}(\Sigma,\sigma)}\|Ta\|_{\mathrm{A}_{p',q}(\Sigma,\sigma)} \\ &\leq C\|a\|_{\mathrm{HA}_{p',q}(\Sigma,\sigma)}\|f\|_{\mathrm{B}_{p,q}(\Sigma,\sigma)} \leq C\|f\|_{\mathrm{B}_{p,q}(\Sigma,\sigma)}, \end{aligned} \tag{13.89}$$

for some constant $C \in (0,\infty)$ independent of a and f. After taking the supremum over all x_0-central (q,p')-atoms in the right-hand side of (13.89), and invoking Corollary 12.2 we conclude that there exists some $C \in (0,\infty)$ depending exclusively on n, p, q, the kernel k, and the UR constants of Σ with the property that

$$\|T_{\text{mod}} f\|_{\mathrm{BCMO}_{p,q}(\Sigma,\sigma)} \leq C\|f\|_{\mathrm{B}_{p,q}(\Sigma,\sigma)}. \tag{13.90}$$

In turn, this proves the boundedness claimed in (13.86). The claim made in (13.87) follows from what we have proved so far. This ultimately establishes Theorem 13.8.□

We now establish the boundedness of (modified) generalized double layer potentials on the space $\mathrm{BCMO}_{p,q}$ (cf. Definition 12.1).

Theorem 13.9 *Suppose* $\Omega \subseteq \mathbb{R}^n$ *is a* UR *domain with unbounded boundary, and abbreviate* $\sigma := \mathcal{H}^{n-1}\lfloor\partial\Omega$. *Denote by* ν *the geometric measure theoretic outward unit normal to* Ω. *For some sufficiently large integer* $N = N(n) \in \mathbb{N}$, *pick a vector-valued kernel* $\vec{k} \in \big[\mathscr{C}^N(\mathbb{R}^n \setminus \{0\})\big]^n$ *that is odd, positive homogeneous of degree* $1-n$, *and divergence-free, i.e.,* $\operatorname{div}\vec{k} = 0$ *in* $\mathbb{R}^n \setminus \{0\}$. *Recall the generalized double layer potential* T_{mod} *associated with* $\vec{k}$ *as in* (6.201). *Then, for any two exponents* $p \in (1,\infty)$ *and* $q \in \big(\tfrac{n-1}{n}, 1\big]$, *it follows that*

$$\begin{gathered} T_{\text{mod}} : \mathrm{BCMO}_{p,q}(\partial\Omega,\sigma) \to \mathrm{BCMO}_{p,q}(\partial\Omega,\sigma) \\ \textit{is a well-defined, linear, and bounded operator.} \end{gathered} \tag{13.91}$$

In particular, denoting by $[\,\cdot\,]$ *the equivalence class modulo constants in the space* $\mathrm{BCMO}_{p,q}(\partial\Omega,\sigma)$, *one also has that the operator*

$$[T_{\text{mod}}] : \widetilde{\mathrm{BCMO}}_{p,q}(\partial\Omega,\sigma) \to \widetilde{\mathrm{BCMO}}_{p,q}(\partial\Omega,\sigma) \quad \textit{given by}$$
$$[T_{\text{mod}}][f] := [T_{\text{mod}} f] \textit{ for all } f \in \mathrm{BCMO}_{p,q}(\partial\Omega,\sigma), \tag{13.92}$$

is meaningfully defined, linear, and bounded.

Proof The proof follows the same type of argument used to justify Theorem 6.6, now taking into account Proposition 12.4, Definition 11.2, (6.169), the duality result recorded in Theorem 12.1, Corollary 12.2, Corollary 12.5, Corollary 12.6, the boundedness result in Theorem 13.6, as well as (13.51). □

Next, we introduce a new class of central Carleson measures which generalizes that considered in Definition 6.1.

Definition 13.1 Assume $\Omega \subseteq \mathbb{R}^n$ is an open set with an Ahlfors regular boundary, and abbreviate $\sigma := \mathcal{H}^{n-1}\lfloor\partial\Omega$. Denote by $C_{\partial\Omega}, c_{\partial\Omega} \in (0,\infty)$ the upper and lower Ahlfors regularity constants of $\partial\Omega$, respectively. Fix a reference point $x_0 \in \partial\Omega$, and a dilation factor $2_* > \big(C_{\partial\Omega}/c_{\partial\Omega}\big)^{1/(n-1)}$. Select two exponents $p,q \in (0,\infty)$. A Borel measure μ in Ω is called a (p,q)-`central Carleson measure` in Ω provided there exists a constant $C \in (0,\infty)$ with the property that

$$\mu\big(B(x_0,R)\cap\Omega\big)^{1/p} \leq C\cdot\sigma\big(\Delta(x_0,R)\big)^{1/p+1/q-1} \quad \text{for every } R \geq 2_*. \tag{13.93}$$

Define the (p,q)-`central Carleson measure norm` (or simply $\mathrm{CCM}_{p,q}$-`norm`) of μ in Ω as the following quantity:

$$\|\mu\|_{\mathrm{CCM}_{p,q}(\Omega)} := \sup_{R\geq 2_*} \frac{\mu\big(B(x_0,R)\cap\Omega\big)^{1/p}}{\sigma\big(\Delta(x_0,R)\big)^{1/p+1/q-1}}. \tag{13.94}$$

Remark 13.1 In the context of Definition 13.1, condition (13.93) corresponding to $q := 1$ is equivalent to (6.213). Hence, the class of $(1,p)$-central Carleson measures coincides with the class of central Carleson measures (in the sense of Definition 6.1).

Remark 13.2 We wish to emphasize here that condition (13.93) is actually independent of the choice of a reference point $x_0 \in \partial\Omega$, and the dilation factor $2_* > \big(C_{\partial\Omega}/c_{\partial\Omega}\big)^{1/(n-1)}$. This can be verified by reasoning as in the proof of Proposition 6.3.

It turns out that one can use functions of (p,q)-bounded central mean oscillations to construct (p,q)-central Carleson measures via (generalized) double layer potentials. This is made precise in the result recorded below.

Theorem 13.10 *Consider a* UR *domain* $\Omega \subseteq \mathbb{R}^n$ *with unbounded boundary, and abbreviate* $\sigma := \mathcal{H}^{n-1}\lfloor\partial\Omega$. *For some sufficiently large integer* $N = N(n) \in \mathbb{N}$, *select a vector-valued kernel* $\vec{k} \in \big[\mathscr{C}^N(\mathbb{R}^n\setminus\{0\})\big]^n$ *that is odd, positive homogeneous of degree* $1-n$, *and divergence-free, i.e.,* $\operatorname{div}\vec{k} = 0$ *in* $\mathbb{R}^n\setminus\{0\}$. *Recall the boundary-to-domain modified generalized double layer potential* $\mathcal{T}_{\text{mod}}$ *associated with* $\vec{k}$ *as in*

(6.196). *Fix a reference point* $x_0 \in \partial\Omega$, *and a dilation factor* 2_* *as in Definition 13.1. Finally, pick two exponents* $p \in (1, \infty)$ *and* $q \in \left(\frac{n-1}{n}, 1\right]$.

Then there exists some $C \in (0, \infty)$ *depending only on* n, p, *and the Ahlfors regularity constants of* $\partial\Omega$ *with the property that whenever* $f \in \mathrm{BCMO}_{p,q}(\partial\Omega, \sigma)$ *one has*

$$\left(\int_{B(x_0,R)\cap\Omega} |\nabla(\mathcal{T}_{\mathrm{mod}} f)|^p \operatorname{dist}(\cdot, \partial\Omega)^{p-1}\, d\mathcal{L}^n \right)^{1/p} \leq C \cdot \sigma\big(\Delta(x_0, R)\big)^{1/p+1/q-1} \|f\|_{\mathrm{BCMO}_{p,q}(\partial\Omega,\sigma)} \quad (13.95)$$

for every radius $R \geq 2_*$. *In particular,* $|\nabla(\mathcal{T}_{\mathrm{mod}} f)|^p \operatorname{dist}(\cdot, \partial\Omega)^{p-1}\, d\mathcal{L}^n$ *is a* (p, q)-*central Carleson measure in* Ω.

Proof Observe that the action of $\mathcal{T}_{\mathrm{mod}}$ on functions from $\mathrm{BCMO}_{p,q}(\partial\Omega, \sigma)$ is well defined thanks to item *(d)* in Proposition 12.4. The case $q = 1$ of estimate (13.95) is implied by Theorem 6.8, since $\mathrm{BCMO}_{p,1}$ coincides with BCMO_p. The proof of (13.95) in the regime $q \in \left(\frac{n-1}{n}, 1\right)$ follows by a similar argument as the one presented in the proof of Theorem 6.8. For this reason, here we revisit only those aspects of said proof that change significantly in our current functional analytic setting.

Fix a function $f \in \mathrm{BCMO}_{p,q}(\partial\Omega, \sigma)$ and a finite radius $R \geq 2_*$. Decompose f as

$$\begin{aligned} f &= \big(f - f_{\Delta(x_0,2R)}\big) \cdot \mathbb{1}_{\Delta(x_0,2R)} + \big(f - f_{\Delta(x_0,2R)}\big) \cdot \mathbb{1}_{\partial\Omega\setminus\Delta(x_0,2R)} + f_{\Delta(x_0,2R)} \\ &=: f_1 + f_2 + f_3. \end{aligned} \quad (13.96)$$

Taking advantage of the computations made in (12.40)(12.44), it follows that f_1, f_2, and f_3 belong to $L^1\big(\partial\Omega, \frac{\sigma(x)}{1+|x|^n}\big)$. Since the term f_3 is itself a constant, (6.200) ensures that $\mathcal{T}_{\mathrm{mod}} f_3$ is a constant function in Ω. Hence,

$$\nabla(\mathcal{T}_{\mathrm{mod}} f)(x) = \nabla(\mathcal{T}_{\mathrm{mod}} f_1)(x) + \nabla(\mathcal{T}_{\mathrm{mod}} f_2)(x) \quad \text{for each } x \in \Omega. \quad (13.97)$$

Notice that $f_1 \in L^p(\partial\Omega, \sigma) \subseteq L^1\big(\partial\Omega, \frac{\sigma(x)}{1+|x|^{n-1}}\big)$, since $\mathrm{BCMO}_{p,q}(\partial\Omega, \sigma)$ is included in $L^p_{\mathrm{loc}}(\partial\Omega, \sigma)$ (cf. item *(e)* in Proposition 12.4). Hence, using (6.198), Theorem 6.7, and the Ahlfors regularity of $\partial\Omega$ we obtain

$$
\begin{aligned}
&\int_{\Omega} |\nabla(\mathcal{T}_{\text{mod}} f_1)(x)|^p \operatorname{dist}(x, \partial\Omega)^{p-1}\, d\mathcal{L}^n(x) \\
&= \int_{\Omega} |\nabla(\mathcal{T} f_1)(x)|^p \operatorname{dist}(x, \partial\Omega)^{p-1}\, d\mathcal{L}^n(x) \\
&\leq C \int_{\Delta(x_0,2R)} |f - f_{\Delta(x_0,2R)}|^p \, d\sigma \\
&= C \frac{\sigma(\Delta(x_0, 2R))}{(2R)^{(n-1)(1-1/q)p}} \Big\{ (2R)^{(n-1)(1-1/q)} \Big(\fint_{\Delta(x_0,2R)} |f - f_{\Delta(x_0,2R)}|^p \, d\sigma \Big)^{1/p} \Big\}^p \\
&\leq C \sigma\big(\Delta(x_0, R)\big)^{1+p(1/q-1)} \|f\|^p_{\mathrm{BCMO}_{p,q}(\partial\Omega,\sigma)}
\end{aligned}
\tag{13.98}
$$

for some constant $C = C(n, p, q, \partial\Omega) \in (0, \infty)$.

Next, fix an arbitrary point $x \in B(x_0, R) \cap \Omega$. Availing ourselves of the analysis carried out from (6.229) up to the penultimate line in (6.230), and using the fact that $q \in \big(\frac{n-1}{n}, 1\big]$, we may deduce

$$
\begin{aligned}
|\nabla(\mathcal{T}_{\text{mod}} f_2)(x)| &\leq C \sum_{k=1}^{\infty} \frac{1}{2^k R} \Big(\fint_{\Delta(x_0,2^{k+1}R)} |f - f_{\Delta(x_0,2^{k+1}R)}|^p \, d\sigma \Big)^{1/p} \\
&\quad + C \sum_{k=1}^{\infty} \frac{1}{2^k R} \sum_{j=1}^{k} \Big(\fint_{\Delta(x_0,2^{j+1}R)} |f - f_{\Delta(x_0,2^{j+1}R)}|^p \, d\sigma \Big)^{1/p} \\
&\leq C \Big(\sum_{k=1}^{\infty} \frac{(2^{k+1}R)^{(n-1)(1/q-1)}}{2^k R} \Big) \|f\|_{\mathrm{BCMO}_{p,q}(\partial\Omega,\sigma)} \\
&\quad + C \Big(\sum_{k=1}^{\infty} \frac{1}{2^k R} \sum_{j=1}^{k} (2^{j+1}R)^{(n-1)(1/q-1)} \Big) \|f\|_{\mathrm{BCMO}_{p,q}(\partial\Omega,\sigma)} \\
&\leq C \frac{R^{(n-1)(1/q-1)}}{R} \Big(\sum_{k=1}^{\infty} 2^{k(n-1)[1/q-n/(n-1)]} \Big) \|f\|_{\mathrm{BCMO}_{p,q}(\partial\Omega,\sigma)}
\end{aligned}
\tag{13.99}
$$

for some $C \in (0, \infty)$ depending only on n, q, and the Ahlfors regularity constants of $\partial\Omega$. Since this holds for every $x \in B(x_0, R) \cap \Omega$, it follows that

$$
\begin{aligned}
&\int_{B(x_0,R)\cap\Omega} |\nabla(\mathcal{T}_{\mathrm{mod}} f_2)(x)|^p \,\mathrm{dist}(x,\partial\Omega)^{p-1}\, d\mathcal{L}^n(x)\\
&\leq C\frac{R^{(n-1)(1/q-1)p}}{R^p}\|f\|^p_{\mathrm{BCMO}_{p,q}(\partial\Omega,\sigma)}\int_{B(x_0,R)}|x-x_0|^{p-1}\,d\mathcal{L}^n(x)\\
&\leq C\frac{R^{(n-1)(1/q-1)p}}{R^p}\|f\|^p_{\mathrm{BCMO}_{p,q}(\partial\Omega,\sigma)}\cdot R^{p-1}\cdot\mathcal{L}^n\big(B(x_0,R)\big)\\
&\leq CR^{(n-1)[1+p(1/q-1)]}\|f\|^p_{\mathrm{BCMO}_{p,q}(\partial\Omega,\sigma)}\\
&\leq C\sigma\big(\Delta(x_0,R)\big)^{1+p(1/q-1)}\|f\|^p_{\mathrm{BCMO}_{p,q}(\partial\Omega,\sigma)}.
\end{aligned}
\tag{13.100}
$$

In concert, (13.97), (13.98), and (13.100) finally give

$$
\begin{aligned}
&\Big(\int_{B(x_0,R)\cap\Omega}|\nabla(\mathcal{T}_{\mathrm{mod}} f)|^p\,\mathrm{dist}(\cdot,\partial\Omega)^{p-1}\,d\mathcal{L}^n\Big)^{1/p}\\
&\quad\leq C\Big(\int_{B(x_0,R)\cap\Omega}|\nabla(\mathcal{T}_{\mathrm{mod}} f_1)|^p\,\mathrm{dist}(\cdot,\partial\Omega)^{p-1}\,d\mathcal{L}^n\Big)^{1/p}\\
&\qquad+C\Big(\int_{B(x_0,R)\cap\Omega}|\nabla(\mathcal{T}_{\mathrm{mod}} f_2)|^p\,\mathrm{dist}(\cdot,\partial\Omega)^{p-1}\,d\mathcal{L}^n\Big)^{1/p}\\
&\quad\leq C\sigma\big(\Delta(x_0,R)\big)^{1/p+1/q-1}\|f\|_{\mathrm{BCMO}_{p,q}(\partial\Omega,\sigma)},
\end{aligned}
\tag{13.101}
$$

for some $C\in(0,\infty)$ depending only on n, p, q, and the Ahlfors regularity constants of $\partial\Omega$. This ultimately establishes (13.95), thus finishing the proof of Theorem 13.10.□

Next, we obtain a boundedness result on the space of functions of (p,q)-bounded central mean oscillations functions for the class of singular integral operators whose kernel has a chord-dot-normal structure. Most importantly, the result recorded below establishes geometric operator norm estimates for such a class of SIO's in the context of $\mathrm{BCMO}_{p,q}$ spaces. The reader is advised to recall the piece of notation introduced in (6.113).

Theorem 13.11 *Assume $\Omega\subseteq\mathbb{R}^n$ is a* UR *domain with unbounded boundary. Abbreviate $\sigma:=\mathcal{H}^{n-1}\lfloor\partial\Omega$ and denote by ν the geometric measure theoretic outward unit normal to Ω. For a sufficiently large integer $N=N(n)\in\mathbb{N}$, pick a kernel $k\in\mathscr{C}^N(\mathbb{R}^n\setminus\{0\})$ that is even, and positive homogeneous of degree $-n$, and recall the modified chord-dot-normal singular integral operator T_{mod} associated with k as in* (6.105). *Also, select two exponents $p\in(1,\infty)$ and $q\in\big(\frac{n-1}{n},1\big]$.*

Then for each $m\in\mathbb{N}$ there exists some $C_m\in(0,\infty)$ depending only on m, n, p, q, the kernel k, and the UR *constants of Ω with the property that*

$$
\|T_{\mathrm{mod}}\|_{\mathrm{BCMO}_{p,q}(\partial\Omega,\sigma)\to\mathrm{BCMO}_{p,q}(\partial\Omega,\sigma)}\leq C_m\cdot\|\nu\|^{\langle m\rangle}_{[\mathrm{BMO}(\partial\Omega,\sigma)]^n}.
\tag{13.102}
$$

Also, if $[\,\cdot\,]$ *denotes the equivalence class modulo constants in* $\mathrm{BCMO}_{p,q}(\partial\Omega,\sigma)$, *then the operator*

$$\begin{gathered}[T_{\mathrm{mod}}]:\widetilde{\mathrm{BCMO}}_{p,q}(\partial\Omega,\sigma)\to\widetilde{\mathrm{BCMO}}_{p,q}(\partial\Omega,\sigma)\ \textit{given by}\\ [T_{\mathrm{mod}}][f]:=[T_{\mathrm{mod}}f]\ \textit{for all}\ f\in\mathrm{BCMO}_{p,q}(\partial\Omega,\sigma)\\ \textit{is meaningfully defined, linear, and bounded.}\end{gathered}\tag{13.103}$$

Moreover, with $C_m\in(0,\infty)$ *as in* (13.102), *one also has*

$$\big\|[T_{\mathrm{mod}}]\big\|_{\widetilde{\mathrm{BCMO}}_{p,q}(\partial\Omega,\sigma)\to\widetilde{\mathrm{BCMO}}_{p,q}(\partial\Omega,\sigma)}\leq C_m\cdot\|\nu\|^{\langle m\rangle}_{[\mathrm{BMO}(\partial\Omega,\sigma)]^n}.\tag{13.104}$$

Proof Observe that the action of T_{mod} on functions from $\mathrm{BCMO}_{p,q}(\partial\Omega,\sigma)$ is meaningfully defined thanks to (6.166) and the fact that $\mathrm{BCMO}_{p,q}(\partial\Omega,\sigma)$ is included in $L^1\big(\partial\Omega,\frac{\sigma(x)}{1+|x|^n}\big)$ (cf. Proposition 12.4). Fix a function $f\in\mathrm{BCMO}_{p,q}(\partial\Omega,\sigma)$ and an x_0-central (q,p')-atom a on $\partial\Omega$, where $p'\in(1,\infty)$ denotes the Hölder conjugate exponent of p. In particular,

$$\begin{gathered}f\in L^p_{\mathrm{loc}}(\partial\Omega,\sigma)\cap L^1\big(\partial\Omega,\tfrac{\sigma(x)}{1+|x|^n}\big)\\ \text{and}\ \ a\in L^{p'}_{\mathrm{comp}}(\partial\Omega,\sigma)\ \ \text{with}\ \ \int_{\partial\Omega}a\,d\sigma=0.\end{gathered}\tag{13.105}$$

Hence, (6.169) entails that for each $m\in\mathbb{N}$ we have

$$\begin{aligned}\Big|\int_{\partial\Omega}(T_{\mathrm{mod}}f)\cdot a\,d\sigma\Big|&=\Big|\int_{\partial\Omega}f\cdot(T^{\#}a)\,d\sigma\Big|=\big|\langle[f],T^{\#}a\rangle\big|\\ &\leq\|f\|_{\mathrm{BCMO}_{p,q}(\partial\Omega,\sigma)}\|T^{\#}a\|_{\mathrm{HA}_{p',q}(\partial\Omega,\sigma)}\\ &\leq\|f\|_{\mathrm{BCMO}_{p,q}(\partial\Omega,\sigma)}\|T^{\#}\|_{\mathrm{HA}_{p',q}(\partial\Omega,\sigma)\to\mathrm{HA}_{p',q}(\partial\Omega,\sigma)}\|a\|_{\mathrm{HA}_{p',q}(\partial\Omega,\sigma)}\\ &\leq C_m\|\nu\|^{\langle m\rangle}_{[\mathrm{BMO}(\partial\Omega,\sigma)]^n}\|f\|_{\mathrm{BCMO}_{p,q}(\partial\Omega,\sigma)},\end{aligned}\tag{13.106}$$

for some constant $C_m=C(n,p,q,m,\partial\Omega)\in(0,\infty)$. The second equality in (13.106) follows from the fact that $T^{\#}a$ is a fixed multiple of an x_0-central molecule on $\mathrm{HA}_{p',q}(\partial\Omega,\sigma)$ (cf. (13.67)) along with (12.59) from Corollary 12.6 and Proposition 11.6. The first inequality is a consequence of the continuity of $[f]$ viewed as a linear functional acting on $\mathrm{HA}_{p',q}(\partial\Omega,\sigma)$ (cf. Theorem 12.1), while the second inequality follows from the boundedness of $T^{\#}$ on $\mathrm{HA}_{p',q}(\partial\Omega,\sigma)$ (cf. Theorem 13.7). The third inequality is justified by (13.66) and Theorem 11.1. Taking the supremum in (6.238) over all central (q,p')-atoms, Corollary 12.2 yields

$$\|T_{\mathrm{mod}}f\|_{\mathrm{BCMO}_{p,q}(\partial\Omega,\sigma)}\leq C_m\|\nu\|^{\langle m\rangle}_{[\mathrm{BMO}(\partial\Omega,\sigma)]^n}\|f\|_{\mathrm{BCMO}_{p,q}(\partial\Omega,\sigma)}.\tag{13.107}$$

This ultimately proves (13.102). The well-definiteness of the operator $[T_{\mathrm{mod}}]$ claimed in (13.103) is seen from what we have proved so far and the fact that T_{mod} maps constant functions into constant functions (cf. (6.108)). Finally, (13.104) is a direct consequence of (13.102)–(13.103). This concludes the proof of Theorem 13.11. □

Chapter 14
Herz-based Sobolev Spaces of First Generation

In this chapter, we shall primarily be focused on studying Herz-based Sobolev spaces (of first generation) on boundaries of Ahlfors regular domains, along with their homogeneous versions. Here, the Calderón-Zygmund theory on Herz spaces $\mathrm{A}_{p,q}$ developed in Chapter 13 plays an important role, as it allows us to develop a functional analytic theory for the aforementioned scales of Sobolev spaces. The consideration of these scales opens the door for treating boundary value problems in which the boundary data, hence also the solution, exhibit additional regularity.

To get started, recall from [GHA-II, Definition 11.1.2, p. 684] the following notion of weak tangential derivatives.

Definition 14.1 Consider $\Omega \subseteq \mathbb{R}^n$ an Ahlfors regular domain, let $\nu := (\nu_1, \dots, \nu_n)$ denote its geometric measure theoretic outward unit normal, and set $\sigma := \mathcal{H}^{n-1} \lfloor \partial\Omega$. Given a locally integrable function f defined on $\partial\Omega$, and indices $j, k \in \{1, \dots, n\}$, one says that the `weak tangential derivative` of f exists in $L^1_{\mathrm{loc}}(\partial\Omega, \sigma)$ if there exists some function $f_{jk} \in L^1_{\mathrm{loc}}(\partial\Omega, \sigma)$ for which

$$\int_{\partial\Omega} (\partial_{\tau_{jk}} \varphi) f \, d\sigma = - \int_{\partial\Omega} \varphi f_{jk} \, d\sigma \quad \text{for all } \varphi \in \mathscr{C}^\infty_0(\mathbb{R}^n), \tag{14.1}$$

where

$$\partial_{\tau_{jk}} \varphi := \nu_j (\partial_k \varphi)\big|_{\partial\Omega} - \nu_k (\partial_j \varphi)\big|_{\partial\Omega} \quad \text{for each } j, k \in \{1, \dots, n\}. \tag{14.2}$$

In this case, for each $j, k \in \{1, \dots, n\}$, denote

$$\partial_{\tau_{jk}} f := f_{jk} \tag{14.3}$$

so (14.1) becomes the integration by parts formula on the boundary

$$\int_{\partial\Omega} (\partial_{\tau_{jk}} \varphi) f \, d\sigma = - \int_{\partial\Omega} \varphi (\partial_{\tau_{jk}} f) \, d\sigma \quad \text{for all } \varphi \in \mathscr{C}^\infty_0(\mathbb{R}^n). \tag{14.4}$$

© The Author(s), under exclusive license to Springer Nature Switzerland AG 2026
M. Mitrea and P. Takemura, *Singular Integrals, Herz-Type Function Spaces, and Boundary Problems*, Progress in Mathematics 362,
https://doi.org/10.1007/978-3-032-12516-3_14

Moreover, the tangential gradient of f is defined as (with the summation convention over repeated indices in effect)

$$\nabla_{\tan} f := \big(\nu_k \partial_{\tau_{jk}} f\big)_{1\leq j\leq n}. \tag{14.5}$$

For each $p \in [1,\infty]$, denote the local Sobolev space by

$$L^p_{1,\mathrm{loc}}(\partial\Omega,\sigma) := \Big\{ f \in L^p_{\mathrm{loc}}(\partial\Omega,\sigma) : \partial_{\tau_{jk}} f \in L^p_{\mathrm{loc}}(\partial\Omega,\sigma) \\ \text{for each } j,k \in \{1,\dots,n\}\Big\}. \tag{14.6}$$

In the definition below, we introduce inhomogeneous Sobolev spaces associated with first-generation Herz spaces on boundaries of Ahlfors regular domains.

Definition 14.2 Assume $\Omega \subseteq \mathbb{R}^n$ is an Ahlfors regular domain with unbounded boundary, and abbreviate $\sigma := \mathcal{H}^{n-1}\lfloor\partial\Omega$. Fix two exponents $p,q \in (1,\infty)$, and define the Herz-based inhomogeneous Sobolev space by

$$\mathrm{A}^1_{p,q}(\partial\Omega,\sigma) := \Big\{ f \in \mathrm{A}_{p,q}(\partial\Omega,\sigma) : \partial_{\tau_{jk}} f \in \mathrm{A}_{p,q}(\partial\Omega,\sigma) \\ \text{for each } \ j,k \in \{1,\dots,n\}\Big\}. \tag{14.7}$$

Equip this space with the norm

$$\|f\|_{\mathrm{A}^1_{p,q}(\partial\Omega,\sigma)} := \|f\|_{\mathrm{A}_{p,q}(\partial\Omega,\sigma)} + \sum_{j,k=1}^{n} \|\partial_{\tau_{jk}} f\|_{\mathrm{A}_{p,q}(\partial\Omega,\sigma)} \\ \text{for each function } f \in \mathrm{A}^1_{p,q}(\partial\Omega,\sigma). \tag{14.8}$$

Work done in [GHA-II, Proposition 11.4.2, p. 729] shows that whenever $\Omega \subseteq \mathbb{R}^n$ is a uniformly rectifiable domain, with $\sigma := \mathcal{H}^{n-1}\lfloor\partial\Omega$, and $p \in (1,\infty)$ then

$$|\nabla_{\tan} f(x)| \approx \sum_{j,k=1}^{n} |\partial_{\tau_{jk}} f(x)| \ \text{ uniformly for} \\ f \in L^p_{1,\mathrm{loc}}(\partial\Omega,\sigma) \ \text{ and } \sigma\text{-a.e. } x \in \partial\Omega. \tag{14.9}$$

The comparability (14.9) and the fact that $\mathrm{A}_{p,q}(\partial\Omega,\sigma) \subseteq L^p_{\mathrm{loc}}(\partial\Omega,\sigma)$ (cf. item *(f)* in Proposition 10.3) then imply that for each $p,q \in (1,\infty)$ and $f \in \mathrm{A}^1_{p,q}(\partial\Omega,\sigma)$ we have

$$\|\nabla_{\tan} f\|_{[\mathrm{A}_{p,q}(\partial\Omega,\sigma)]^n} \approx \sum_{j,k=1}^{n} \|\partial_{\tau_{jk}} f\|_{\mathrm{A}_{p,q}(\partial\Omega,\sigma)}. \tag{14.10}$$

In particular, this and (14.8) entail

$$\|f\|_{\mathrm{A}^1_{p,q}(\partial\Omega,\sigma)} \approx \|f\|_{\mathrm{A}_{p,q}(\partial\Omega,\sigma)} + \|\nabla_{\tan} f\|_{[\mathrm{A}_{p,q}(\partial\Omega,\sigma)]^n} \tag{14.11}$$

uniformly for $f \in \mathrm{A}^1_{p,q}(\partial\Omega, \sigma)$.

It turns out that the Herz-based Sobolev space is a Banach space.

Proposition 14.1 *Consider an Ahlfors regular domain $\Omega \subseteq \mathbb{R}^n$ with unbounded boundary, and abbreviate $\sigma := \mathcal{H}^{n-1}\lfloor\partial\Omega$. Fix two exponents $p, q \in (1, \infty)$. Then the Herz-based Sobolev space $\mathrm{A}^1_{p,q}(\partial\Omega, \sigma)$ is a Banach when equipped with the norm $\|\cdot\|_{\mathrm{A}^1_{p,q}(\partial\Omega,\sigma)}$.*

Proof Consider an arbitrary Cauchy sequence $\{f_m\}_{m\in\mathbb{N}}$ in $\mathrm{A}^1_{p,q}(\partial\Omega, \sigma)$. Then for each fixed $\varepsilon \in (0, \infty)$ there exists $M \in \mathbb{N}$ such that for every $m, \ell \in \mathbb{N}$, with $m, \ell \geq M$, we have

$$
\begin{aligned}
\|f_m - f_\ell\|_{\mathrm{A}_{p,q}(\partial\Omega,\sigma)} &+ \sum_{j,k=1}^{n} \|\partial_{\tau_{jk}} f_m - \partial_{\tau_{jk}} f_\ell\|_{\mathrm{A}_{p,q}(\partial\Omega,\sigma)} \\
&= \|f_m - f_\ell\|_{\mathrm{A}^1_{p,q}(\partial\Omega,\sigma)} < \varepsilon.
\end{aligned}
\tag{14.12}
$$

In particular, this implies that $\{f_m\}_{m\in\mathbb{N}}$ and $\{\partial_{\tau_{jk}} f_m\}_{m\in\mathbb{N}}$ are Cauchy sequences in $\mathrm{A}_{p,q}(\partial\Omega, \sigma)$, for each $j, k \in \{1, \dots, n\}$. From the fact that $\mathrm{A}_{p,q}(\partial\Omega, \sigma)$ is a Banach space (cf. Proposition 10.6), for each $j, k \in \{1, \dots, n\}$ we conclude that

$$
\begin{gathered}
\text{there exist functions } f, \widetilde{f}_{jk} \in \mathrm{A}_{p,q}(\partial\Omega, \sigma) \text{ with the property that} \\
\lim_{m\to\infty} \|f_m - f\|_{\mathrm{A}_{p,q}(\partial\Omega,\sigma)} = 0 \text{ and } \lim_{m\to\infty} \|\partial_{\tau_{jk}} f_m - \widetilde{f}_{jk}\|_{\mathrm{A}_{p,q}(\partial\Omega,\sigma)} = 0.
\end{gathered}
\tag{14.13}
$$

To proceed, fix some $m \in \mathbb{N}$, two indices $j, k \in \{1, \dots, n\}$, and an arbitrary function $\varphi \in \mathscr{C}^\infty_0(\mathbb{R}^n)$. Bearing in mind (14.4), we have

$$
\int_{\partial\Omega} (\partial_{\tau_{jk}}\varphi) f_m \, d\sigma = -\int_{\partial\Omega} \varphi(\partial_{\tau_{jk}} f_m) \, d\sigma.
\tag{14.14}
$$

Because φ is compactly supported in $\mathbb{R}^n$, we have $\mathbb{1}_{\operatorname{supp}\varphi\cap\partial\Omega} \in \mathrm{A}_{p',q'}(\partial\Omega, \sigma)$ by (10.33), where $p', q' \in (1, \infty)$ denotes the Hölder conjugate exponent of p and q, respectively. Hence, since $\partial_{\tau_{jk}}\varphi$ is itself a bounded function on $\partial\Omega$, use the brand of Hölder inequality recorded in (10.20) to estimate

$$
\begin{aligned}
&\Big| \int_{\partial\Omega} (\partial_{\tau_{jk}}\varphi) f_m \, d\sigma - \int_{\partial\Omega} (\partial_{\tau_{jk}}\varphi) f \, d\sigma \Big| \\
&\quad \leq \int_{\operatorname{supp}\varphi\cap\partial\Omega} |\partial_{\tau_{jk}}\varphi| |f_m - f| \, d\sigma \\
&\quad \leq \big(\sup_{\partial\Omega} |\partial_{\tau_{jk}}\varphi|\big) \int_{\operatorname{supp}\varphi\cap\partial\Omega} |f_m - f| \, d\sigma \\
&\quad \leq \big(\sup_{\partial\Omega} |\partial_{\tau_{jk}}\varphi|\big) \|\mathbb{1}_{\operatorname{supp}\varphi\cap\partial\Omega}\|_{\mathrm{A}_{p',q'}(\partial\Omega,\sigma)} \|f_m - f\|_{\mathrm{A}_{p,q}(\partial\Omega,\sigma)}.
\end{aligned}
\tag{14.15}
$$

This and the first convergence in the second line of (14.13) ultimately imply

$$\lim_{m\to\infty}\int_{\partial\Omega}(\partial_{\tau_{jk}}\varphi)f_m\,d\sigma=\int_{\partial\Omega}(\partial_{\tau_{jk}}\varphi)f\,d\sigma. \tag{14.16}$$

Furthermore, since φ is smooth and compactly supported (hence bounded in $\mathbb{R}^n$), we may once again appeal to the Hölder inequality from (10.20) to estimate

$$\begin{aligned}
&\Big|\int_{\partial\Omega}\varphi(\partial_{\tau_{jk}}f_m)\,d\sigma-\int_{\partial\Omega}\varphi\widetilde{f}_{jk}\,d\sigma\Big|\\
&\quad\le\int_{\operatorname{supp}\varphi\cap\partial\Omega}|\varphi||\partial_{\tau_{jk}}f_m-\widetilde{f}_{jk}|\,d\sigma\\
&\quad\le\Big(\sup_{\partial\Omega}|\varphi|\Big)\int_{\operatorname{supp}\varphi\cap\partial\Omega}|\partial_{\tau_{jk}}f_m-\widetilde{f}_{jk}|\,d\sigma\\
&\quad\le\Big(\sup_{\partial\Omega}|\varphi|\Big)\|\mathbb{1}_{\operatorname{supp}\varphi\cap\partial\Omega}\|_{\mathrm{A}_{p',q'}(\partial\Omega,\sigma)}\|\partial_{\tau_{jk}}f_m-\widetilde{f}_{jk}\|_{\mathrm{A}_{p,q}(\partial\Omega,\sigma)}.
\end{aligned} \tag{14.17}$$

Keeping in mind the second convergence in the second line of (14.13), this entails

$$\lim_{m\to\infty}\int_{\partial\Omega}\varphi(\partial_{\tau_{jk}}f_m)\,d\sigma=\int_{\partial\Omega}\varphi\widetilde{f}_{jk}\,d\sigma. \tag{14.18}$$

Collectively, (14.14), (14.16), and (14.18) give

$$\int_{\partial\Omega}(\partial_{\tau_{jk}}\varphi)f\,d\sigma=-\int_{\partial\Omega}\varphi\widetilde{f}_{jk}\,d\sigma. \tag{14.19}$$

for every $j,k\in\{1,\dots,n\}$, and $\varphi\in\mathscr{C}_0^\infty(\mathbb{R}^n)$. This yields the existence of the weak tangential derivative of f and, in fact, shows that

$$\partial_{\tau_{jk}}f=\widetilde{f}_{jk}\in\mathrm{A}_{p,q}(\partial\Omega,\sigma)\ \text{ for all } j,k\in\{1,\dots,n\}. \tag{14.20}$$

In concert with (14.13), this establishes the membership $f\in\mathrm{A}^1_{p,q}(\partial\Omega,\sigma)$. Furthermore, if we let $\ell\to\infty$ in (14.12), then the convergences in (14.13), along with (14.20), yield

$$\begin{aligned}
\|f_m-f\|_{\mathrm{A}^1_{p,q}(\partial\Omega,\sigma)}&=\|f_m-f\|_{\mathrm{A}_{p,q}(\partial\Omega,\sigma)}\\
&\quad+\sum_{j,k=1}^{n}\|\partial_{\tau_{jk}}f_m-\partial_{\tau_{jk}}f\|_{\mathrm{A}_{p,q}(\partial\Omega,\sigma)}\le\varepsilon,
\end{aligned} \tag{14.21}$$

for every $m\ge M$. Hence

$$\lim_{m\to\infty}\|f_m-f\|_{\mathrm{A}^1_{p,q}(\partial\Omega,\sigma)}=0, \tag{14.22}$$

which ultimately proves that $\mathrm{A}^1_{p,q}(\partial\Omega,\sigma)$ is Banach, as desired. □

Here is a formal definition of what we call Herz-based homogeneous Sobolev spaces on boundaries of Ahlfors regular domains.

Definition 14.3 Consider an Ahlfors regular domain $\Omega \subseteq \mathbb{R}^n$ with unbounded boundary, and set $\sigma := \mathcal{H}^{n-1}\lfloor\partial\Omega$. Pick two integrability exponents $p \in (1,\infty)$ and $q \in (0,\infty)$. Define the `Herz-based homogeneous Sobolev space` (of order one on $\partial\Omega$) as

$$\dot{\mathrm{A}}^1_{p,q}(\partial\Omega,\sigma) := \Big\{ f \in L^1\big(\partial\Omega, \tfrac{\sigma(x)}{1+|x|^n}\big) \cap L^p_{\mathrm{loc}}(\partial\Omega,\sigma) : \\ \partial_{\tau_{jk}} f \in \mathrm{A}_{p,q}(\partial\Omega,\sigma) \text{ for each } j,k \in \{1,\dots,n\}\Big\}, \tag{14.23}$$

and equip this space with the seminorm

$$\|f\|_{\dot{\mathrm{A}}^1_{p,q}(\partial\Omega,\sigma)} := \sum_{j,k=1}^{n} \|\partial_{\tau_{jk}} f\|_{\mathrm{A}_{p,q}(\partial\Omega,\sigma)} \text{ for all } f \in \dot{\mathrm{A}}^1_{p,q}(\partial\Omega,\sigma). \tag{14.24}$$

From Definition 14.2, items *(e)* and *(f)* in Proposition 10.3, and the fact that the space $L^1\big(\partial\Omega, \frac{\sigma(x)}{1+|x|^{n-1}}\big)$ is continuously embedded into $L^1\big(\partial\Omega, \frac{\sigma(x)}{1+|x|^n}\big)$, we have that the following embedding is well defined and continuous:

$$\mathrm{A}^1_{p,q}(\partial\Omega,\sigma) \hookrightarrow \dot{\mathrm{A}}^1_{p,q}(\partial\Omega,\sigma). \tag{14.25}$$

The following result provides a membership criterion to a "large" weighted Lebesgue space (in which we wish to eventually place $\dot{\mathrm{A}}^1_{p,q}$).

Lemma 14.1 *Consider a closed unbounded Ahlfors regular set $\Sigma \subseteq \mathbb{R}^n$ and abbreviate $\sigma := \mathcal{H}^{n-1}\lfloor\Sigma$. Pick a reference point $x_0 \in \Sigma$, and a dilation factor $2_* > 1$. Select two integrability exponents $p \in (1,\infty)$ and $q \in (0,\infty)$. Having fixed a function $f \in L^1_{\mathrm{loc}}(\Sigma,\sigma)$, assume that*

$$C_* := \sup_{k\in\mathbb{N}_0} \frac{2_*^{k(n-1)(1/q-1/p)}}{2_*^k}\Big(\int_{\Delta_k} |f - f_{\Delta_k}|^p \, d\sigma\Big)^{1/p} < \infty, \tag{14.26}$$

where, for each $k \in \mathbb{N}_0$,

$$\Delta_k := \Delta(x_0, 2_*^{k+1}) \quad \textit{and} \quad f_{\Delta_k} := \fint_{\Delta_k} f \, d\sigma. \tag{14.27}$$

Then there exists some constant $C = C(\Sigma,n,p,q) \in (0,\infty)$ with the property that for every $k \in \mathbb{N}_0$ one has

$$\int_\Sigma \frac{|f(x) - f_{\Delta_k}|}{(2_*^{k+1} + |x-x_0|)^n} \, d\sigma(x) \le \frac{C \cdot C_*}{\sigma(\Delta_k)^{1/q}}. \tag{14.28}$$

In particular, the function f *belongs to* $L^1\big(\Sigma, \frac{\sigma(x)}{1+|x|^n}\big)$.

Some remarks concerning the above result are in order.

Remark 14.1 Lemma 14.1 may be regarded as a weightless, flexible, and weaker version of [MMMi³, Lemma 2.16, p. 150]. In fact, corresponding to the choice $p = q$, condition (14.26) simply reads:

$$\sup_{k\in\mathbb{N}_0} \frac{1}{2_*^k}\Big(\int_{\Delta_k} |f - f_{\Delta_k}|^p \, d\sigma\Big)^{1/p} < \infty. \tag{14.29}$$

This is weaker than the demand made in [MMMi³, Lemma 2.16, p. 150], namely

$$\sup_{r>0} \frac{1}{r}\Big(\int_{\Delta(x_0,r)} |f - f_{\Delta(x_0,r)}|^p \, d\sigma\Big)^{1/p} < \infty. \tag{14.30}$$

In particular, we cannot directly appeal to [MMMi³, Lemma 2.16, p. 150] to derive conclusions about the nature of functions satisfying (14.26). This being said, our proof is inspired by the approach in [MMMi³].

Here is the proof of Lemma 14.1.

Proof of Lemma 14.1 Let us start by observing that from the Ahlfors regularity of Σ, Hölder's inequality, and (14.26), for each $j \in \mathbb{N}_0$, we have

$$\begin{aligned} |f_{\Delta_{j+1}} - f_{\Delta_j}| &\leq \fint_{\Delta_j} |f - f_{\Delta_{j+1}}| \, d\sigma \leq C \fint_{\Delta_{j+1}} |f - f_{\Delta_{j+1}}| \, d\sigma \\ &\leq C\Big(\fint_{\Delta_{j+1}} |f - f_{\Delta_{j+1}}|^p \, d\sigma\Big)^{1/p} \\ &\leq C 2_*^{-j(n-1)/p}\Big(\frac{2_*^j}{2_*^{j(n-1)(1/q-1/p)}} C_*\Big) \\ &\leq \frac{C \cdot C_* \cdot 2_*^j}{2_*^{j(n-1)q}}, \end{aligned} \tag{14.31}$$

for some constant $C \in (0, \infty)$ depending only on n, p, q, and Σ.

Having fixed a number $k \in \mathbb{N}_0$, write

$$\begin{aligned} \int_\Sigma \frac{|f(x) - f_{\Delta_k}|}{(2_*^{k+1} + |x - x_0|)^n} \, d\sigma(x) &= \int_{\Delta_k} \frac{|f(x) - f_{\Delta_k}|}{(2_*^{k+1} + |x - x_0|)^n} \, d\sigma(x) \\ &\quad + \sum_{\ell=k+1}^{\infty} \int_{\Delta_\ell \setminus \Delta_{\ell-1}} \frac{|f(x) - f_{\Delta_k}|}{(2_*^{k+1} + |x - x_0|)^n} \, d\sigma(x) \\ &=: \mathrm{S}_1 + \mathrm{S}_2. \end{aligned} \tag{14.32}$$

First, thanks to the Ahlfors regularity of Σ, Hölder's inequality, and (14.26), we may estimate S_1 as

$$
\begin{aligned}
S_1 &\le \frac{1}{2_*^{n(k+1)}} \int_{\Delta_k} |f - f_{\Delta_k}|\, d\sigma \le \frac{1}{2_*^{n(k+1)}} \Big(\int_{\Delta_k} |f - f_{\Delta_k}|^p\, d\sigma \Big)^{1/p} \cdot \sigma(\Delta_k)^{1-1/p} \\
&\le \frac{C}{2_*^{nk}} \Big(\frac{2_*^k}{2_*^{k(n-1)(1/q-1/p)}} C_* \Big) 2_*^{k(n-1)(1-1/p)} = \frac{C \cdot C_*}{2_*^{k(n-1)/q}}.
\end{aligned}
\tag{14.33}
$$

Next, to deal with S_2 from (14.32), fix $\ell \in \mathbb{N}_0$ with $\ell \ge k+1$. Then, using Hölder's inequality, the Ahlfors regularity of Σ, (14.26), and (14.31), we have

$$
\begin{aligned}
&\int_{\Delta_\ell \setminus \Delta_{\ell-1}} \frac{|f(x) - f_{\Delta_k}|}{(2_*^{k+1} + |x - x_0|)^n}\, d\sigma(x) \\
&\quad \le \int_{\Delta_\ell \setminus \Delta_{\ell-1}} \frac{|f(x) - f_{\Delta_k}|}{|x - x_0|^n}\, d\sigma(x) \\
&\quad \le \frac{1}{2_*^{\ell n}} \int_{\Delta_\ell} |f - f_{\Delta_k}|\, d\sigma \\
&\quad \le \frac{1}{2_*^{\ell n}} \int_{\Delta_\ell} |f - f_{\Delta_\ell}|\, d\sigma + \frac{\sigma(\Delta_\ell)}{2_*^{\ell n}} |f_{\Delta_\ell} - f_{\Delta_k}| \\
&\quad \le \frac{\sigma(\Delta_\ell)^{1-1/p}}{2_*^{\ell n}} \Big(\int_{\Delta_\ell} |f - f_{\Delta_\ell}|^p\, d\sigma \Big)^{1/p} + C \frac{2_*^{\ell(n-1)}}{2_*^{\ell n}} \sum_{j=k}^{\ell-1} |f_{\Delta_{j+1}} - f_{\Delta_j}| \\
&\quad \le \frac{C \cdot 2_*^{\ell(n-1)(1-1/p)}}{2_*^{\ell n}} \Big(\frac{2_*^\ell \cdot C_*}{2_*^{\ell(n-1)(1/q-1/p)}} \Big) + \frac{C \cdot C_*}{2_*^\ell} \sum_{j=k}^{\ell-1} \frac{2_*^j}{2_*^{j(n-1)q}} \\
&\quad = \frac{C \cdot C_*}{2_*^{\ell(n-1)/q}} + C \cdot C_* \sum_{j=k+1}^{\ell} \frac{2_*^{j-\ell}}{2_*^{j(n-1)q}},
\end{aligned}
\tag{14.34}
$$

for some constant $C = C(n, p, q, \Sigma) \in (0, \infty)$. With (14.34) in hand, use Fubini's Theorem for series to estimate S_2 in (14.32) as

$$
\begin{aligned}
S_2 &\le C \cdot C_*\left\{\sum_{\ell=k+1}^{\infty}\frac{1}{2_*^{\ell(n-1)/q}}+\sum_{\ell=k+1}^{\infty}\sum_{j=k+1}^{\ell}\frac{2_*^{j-\ell}}{2_*^{j(n-1)q}}\right\}\\
&= C \cdot C_*\left\{\frac{1}{2_*^{k(n-1)/q}}\sum_{\ell=k+1}^{\infty}\frac{1}{2_*^{(\ell-k)(n-1)/q}}+\sum_{j=k+1}^{\infty}\frac{1}{2_*^{j(n-1)q}}\Big(\sum_{\ell=j}^{\infty}2_*^{j-\ell}\Big)\right\}\\
&\le C \cdot C_*\left\{\frac{1}{2_*^{k(n-1)/q}}\sum_{\ell=1}^{\infty}\frac{1}{2_*^{\ell(n-1)/q}}+\frac{1}{2_*^{k(n-1)/q}}\sum_{j=1}^{\infty}\frac{1}{2_*^{j(n-1)q}}\sum_{\ell=0}^{\infty}2_*^{-\ell}\right\}\\
&= \frac{C \cdot C_*}{2_*^{k(n-1)/q}},
\end{aligned}
\tag{14.35}
$$

for some constant $C \in (0, \infty)$ independent of f. Putting it all together, (14.32), (14.33), (14.35), and the Ahlfors regularity of Σ readily imply (14.28).

Regarding the very last claim made in the statement of Lemma 14.1, observe that since $2_* > 1$, for each $x \in \Sigma$ we have

$$
\begin{aligned}
(2_* + |x - x_0|)^n &\le (2_* + |x_0| + |x|)^n \le (2_* + |x_0|)^n (1 + |x|)^n\\
&\le 2^n (2_* + |x_0|)^n (1 + |x|^n).
\end{aligned}
\tag{14.36}
$$

From this and (14.28) (used with $k := 0$) we obtain

$$
\begin{aligned}
\int_\Sigma \frac{|f(x)|}{1+|x|^n}\,d\sigma(x) &\le \int_\Sigma \frac{|f(x) - f_{\Delta_0}|}{1+|x|^n}\,d\sigma(x) + |f_{\Delta_0}|\int_\Sigma \frac{d\sigma(x)}{1+|x|^n}\\
&\le C\int_\Sigma \frac{|f - f_{\Delta_0}|}{(2_* + |x - x_0|)^n}\,d\sigma(x) + |f_{\Delta_0}|\int_\Sigma \frac{d\sigma(x)}{1+|x|^n}\\
&< \infty.
\end{aligned}
\tag{14.37}
$$

Hence $f \in L^1\big(\Sigma, \frac{\sigma(x)}{1+|x|^n}\big)$, as desired. This ultimately proves Lemma 14.1. □

Remark 14.2 As pointed out in Remark 14.1, in the special case $p = q$ condition (14.26) reduces to (14.29). In such a scenario, conclusion (14.28) of Lemma 14.1 simply reads:

$$
\int_\Sigma \frac{|f - f_{\Delta_k}|}{(2_*^{k+1} + |x - x_0|)^n}\,d\sigma(x) \le \frac{C \cdot C_*}{\sigma(\Delta_k)^{1/p}}. \tag{14.38}
$$

We continue by giving a more complete and streamlined description of the Herz-based homogeneous Sobolev spaces on boundaries of two-sided NTA domains.

Theorem 14.1 *Consider $\Omega \subseteq \mathbb{R}^n$ a two-sided* NTA *domain, suppose $\partial\Omega$ is an unbounded Ahlfors regular set, and abbreviate $\sigma := \mathcal{H}^{n-1}\lfloor\partial\Omega$. Select a reference point $x_0 \in \partial\Omega$, and with $C_{\partial\Omega}, c_{\partial\Omega} \in (0, \infty)$ denoting the upper and lower* AR *constants of $\partial\Omega$, respectively, pick a dilation factor $2_* > (C_{\partial\Omega}/c_{\partial\Omega})^{1/(n-1)}$. Lastly, given two exponents $p \in (1, \infty)$ and $q \in (0, \infty)$, assume that*

$$f \in L^1_{\mathrm{loc}}(\partial\Omega,\sigma) \ \textit{satisfies} \ \partial_{\tau_{jk}} f \in \mathrm{A}_{p,q}(\partial\Omega,\sigma) \quad \textit{for each} \ j,k \in \{1,\dots,n\}. \tag{14.39}$$

Then the following statements are equivalent:

(a) The function f belongs to $L^1\big(\partial\Omega, \frac{\sigma(x)}{1+|x|^n}\big)$.

(b) With the same piece of notation as (14.27)*, there exists some $C \in (0,\infty)$ depending only on n, p, q, x_0, and the Ahlfors regularity constants of $\partial\Omega$ with the property that:*

(b1) Whenever $0 < q \leq p$ one has

$$\sup_{\ell\in\mathbb{N}_0} \frac{1}{2_*^\ell}\Big(\int_{\Delta_\ell} |f - f_{\Delta_\ell}|^p \, d\sigma\Big)^{1/p} \leq C \sum_{j,k=1}^n \|\partial_{\tau_{jk}} f\|_{\mathrm{A}_{p,q}(\partial\Omega,\sigma)}. \tag{14.40}$$

(b2) Whenever $1 < p < q$ one has

$$\sup_{\ell\in\mathbb{N}_0} \frac{2_*^{\ell(n-1)(1/q-1/p)}}{2_*^\ell}\Big(\int_{\Delta_\ell} |f - f_{\Delta_\ell}|^p \, d\sigma\Big)^{1/p} \leq C \sum_{j,k=1}^n \|\partial_{\tau_{jk}} f\|_{\mathrm{A}_{p,q}(\partial\Omega,\sigma)}. \tag{14.41}$$

(c) Retaining the piece of notation introduced in (14.27)*, for each $\ell \in \mathbb{N}_0$ there exists some $C_\ell \in (0,\infty)$ depending exclusively on ℓ, n, p, q, x_0, and the Ahlfors regularity constants of $\partial\Omega$ satisfying:*

(c1) Whenever $0 < q \leq p$ one has

$$\int_{\partial\Omega} \frac{|f(x) - f_{\Delta_\ell}|}{1+|x|^n} \, d\sigma(x) \leq \frac{C_\ell}{\sigma(\Delta_\ell)^{1/p}} \sum_{j,k=1}^n \|\partial_{\tau_{jk}} f\|_{\mathrm{A}_{p,q}(\partial\Omega,\sigma)}. \tag{14.42}$$

(c2) Whenever $1 < p < q$ one has

$$\int_{\partial\Omega} \frac{|f(x) - f_{\Delta_\ell}|}{1+|x|^n} \, d\sigma(x) \leq \frac{C_\ell}{\sigma(\Delta_\ell)^{1/q}} \sum_{j,k=1}^n \|\partial_{\tau_{jk}} f\|_{\mathrm{A}_{p,q}(\partial\Omega,\sigma)}. \tag{14.43}$$

(d) The function f belongs to $\dot{\mathrm{A}}^1_{p,q}(\partial\Omega,\sigma)$.

Proof Consider the implication *(a)⇒(b)*. Assume $f \in L^1\big(\partial\Omega, \frac{\sigma(x)}{1+|x|^n}\big)$, and denote

$$\Omega_+ := \Omega \quad \text{and} \quad \Omega_- := \mathbb{R}^n \backslash \overline{\Omega}. \tag{14.44}$$

With this piece of notation, for each point $x \in \Omega_\pm$ define

$$u_\pm(x) := \frac{1}{\omega_{n-1}} \int_{\partial\Omega} \Big\{ \frac{\langle \nu(y), y-x\rangle}{|x-y|^n} - \frac{\langle \nu(y), y\rangle}{|y|^n} \cdot \mathbb{1}_{\mathbb{R}^n \setminus B(0,1)}(y) \Big\} f(y)\, d\sigma(y), \tag{14.45}$$

where ν is the geometric measure theoretic outward unit normal to $\Omega = \Omega_+$. In particular, $-\nu$ is the geometric measure theoretic outward unit normal to Ω_- (cf. [GHA-I, item *(6)* in Lemma 5.10.9, pp 466–467]). Hence, we can recast (14.45) as

$$\begin{aligned} u_+(x) &= \mathcal{D}^+_{\Delta,\mathrm{mod}} f(x) \ \text{ for all } x \in \Omega_+, \ \text{ and} \\ u_-(x) &= -\mathcal{D}^-_{\Delta,\mathrm{mod}} f(x) \ \text{ for all } x \in \Omega_-, \end{aligned} \tag{14.46}$$

where $\mathcal{D}^\pm_{\Delta,\mathrm{mod}}$ are the boundary-to-domain harmonic double layer potential canonically associated with $\Omega_\pm$ (cf. [GHA-IV, (1.8.50), p. 182]). From [MMMi[3], Theorem 3.2, p. 173], we know that $u_\pm \in \mathscr{C}^\infty(\Omega_\pm)$ and (with the summation convention over repeated indices in effect)

$$(\partial_\ell u_\pm)(x) = \pm \int_{\partial\Omega} (\partial_r E_\Delta)(y-x)(\partial_{\tau_{\ell r}} f)(y)\, d\sigma(y), \tag{14.47}$$

for every $x \in \Omega_\pm$ and $\ell \in \{1, \dots, n\}$, where E_Δ is the canonical fundamental solution for the Laplacian Δ in $\mathbb{R}^n$ (cf. (7.19)). Thanks to the nontangential maximal function estimate recorded in Theorem 13.1, along with (14.39) we may therefore bound

$$\|\mathcal{N}_\kappa(\nabla u_\pm)\|_{\mathrm{A}_{p,q}(\partial\Omega,\sigma)} \le C \sum_{j,k=1}^n \|\partial_{\tau_{jk}} f\|_{\mathrm{A}_{p,q}(\partial\Omega,\sigma)} < \infty. \tag{14.48}$$

Moreover, (14.46) and the jump-formula recorded in [MMMi[3], Theorem 3.2, (2), p. 175] yield the following writing for f

$$f = u_+\big|^{\kappa-\mathrm{n.t.}}_{\partial\Omega} - u_-\big|^{\kappa-\mathrm{n.t.}}_{\partial\Omega} \ \text{ pointwise } \sigma\text{-a.e. on } \partial\Omega. \tag{14.49}$$

Next, in light of (14.48) the function

$$\begin{aligned} &g := \mathcal{N}_\kappa(\nabla u_+) + \mathcal{N}_\kappa(\nabla u_-) \ \text{ belongs to } \mathrm{A}_{p,q}(\partial\Omega,\sigma) \\ &\text{and has norm } \|g\|_{\mathrm{A}_{p,q}(\partial\Omega,\sigma)} \le C \sum_{j,k=1}^n \|\partial_{\tau_{jk}} f\|_{\mathrm{A}_{p,q}(\partial\Omega,\sigma)}. \end{aligned} \tag{14.50}$$

In particular, (14.49)-(14.50) permit us to invoke [MMMi[3], Proposition 2.24, p. 152] to guarantee the existence of some constant $C \in (0, \infty)$ depending only on the geometry of Ω for which

$$|f(x) - f(y)| \le C|x-y|\big[g(x) + g(y)\big] \ \text{ for } \sigma\text{-a.e. } x, y \in \partial\Omega. \tag{14.51}$$

Once these properties have been obtained, we proceed by fixing $\ell \in \mathbb{N}_0$ and estimating

$$
\begin{aligned}
\Big(\int_{\Delta_\ell} |f(x) - f_{\Delta_\ell}|^p \, d\sigma(x)\Big)^{1/p}
&= \Big(\int_{\Delta_\ell} \Big|f(x) - \fint_{\Delta_\ell} f(y)\, d\sigma(y)\Big|^p \, d\sigma(x)\Big)^{1/p} \\
&\leq \Big(\int_{\Delta_\ell} \Big(\fint_{\Delta_\ell} |f(x) - f(y)|\, d\sigma(y)\Big)^p \, d\sigma(x)\Big)^{1/p} \\
&\leq C\Big(\int_{\Delta_\ell} \Big(\fint_{\Delta_\ell} |x-y|\big[g(x) + g(y)\big]\, d\sigma(y)\Big)^p \, d\sigma(x)\Big)^{1/p} \\
&\leq C \cdot 2_*^{\ell+1}\Big(\int_{\Delta_\ell} \Big(\fint_{\Delta_\ell} \big[g(x) + g(y)\big]\, d\sigma(y)\Big)^p \, d\sigma(x)\Big)^{1/p} \\
&= C \cdot 2_*^{\ell+1}\Big(\int_{\Delta_\ell} \Big(g(x) + \fint_{\Delta_\ell} g(y)\, d\sigma(y)\Big)^p \, d\sigma(x)\Big)^{1/p} \\
&\leq C \cdot 2_*^{\ell+1}\Big(\int_{\Delta_\ell} g(x)^p \, d\sigma(x)\Big)^{1/p} + C \cdot 2_*^{\ell+1} \sigma(\Delta_\ell)^{1/p} \fint_{\Delta_\ell} g(y)\, d\sigma(y) \\
&\leq C \cdot 2_*^{\ell+1}\Big(\int_{\Delta_\ell} g^p \, d\sigma\Big)^{1/p},
\end{aligned}
\tag{14.52}
$$

for some constant $C = C(\Omega) \in (0, \infty)$, where in the second inequality we have used (14.51), while the third inequality follows from the fact that $|x - y| \leq 2 \cdot 2_*^{\ell+1}$ for all $x, y \in \Delta_\ell$, and the final step is justified by Hölder's inequality.

In the diagonal case $p = q$, from (14.50) and (14.52) we see that

$$
\begin{aligned}
\frac{1}{2_*^{\ell}}\Big(\int_{\Delta_\ell} |f(x) - f_{\Delta_\ell}|^p \, d\sigma(x)\Big)^{1/p} &\leq C\|g\|_{L^p(\partial\Omega,\sigma)} \\
&\leq C \sum_{j,k=1}^{n} \|\partial_{\tau_{jk}} f\|_{L^p(\partial\Omega,\sigma)}
\end{aligned}
\tag{14.53}
$$

for every $\ell \in \mathbb{N}_0$. In turn, after taking the supremum over $\ell \in \mathbb{N}_0$ in (14.53) we obtain (14.40) (in the case when $p = q$).

Let us now consider the case when $0 < q < p$. In particular, $1/q - 1/p > 0$. Based on this, (14.50), and (14.52) we may estimate, for some $C = C(\partial\Omega, n, p, q) \in (0, \infty)$,

$$
\begin{aligned}
\frac{1}{2_*^{\ell}}\Big(\int_{\Delta_\ell}|f(x)-f_{\Delta_\ell}|^p\,d\sigma(x)\Big)^{1/p}
&\leq C\sum_{k=0}^{\ell}\|g\cdot \mathbb{1}_{C_k}\|_{L^p(\partial\Omega,\sigma)}\\
&= C\sum_{k=0}^{\ell}2_*^{-k(n-1)(1/q-1/p)}2_*^{k(n-1)(1/q-1/p)}\|g\cdot \mathbb{1}_{C_k}\|_{L^p(\partial\Omega,\sigma)}\\
&\leq C\Big(\sum_{k=0}^{\ell}2_*^{-k(n-1)(1/q-1/p)}\Big)\|g\|_{\mathrm{A}_{p,q}(\partial\Omega,\sigma)}\\
&\leq C\Big(\sum_{k=0}^{\infty}2_*^{-k(n-1)(1/q-1/p)}\Big)\|g\|_{\mathrm{A}_{p,q}(\partial\Omega,\sigma)}\\
&\leq C\sum_{j,k=1}^{n}\|\partial_{\tau_{jk}}f\|_{\mathrm{A}_{p,q}(\partial\Omega,\sigma)}<\infty, \qquad (14.54)
\end{aligned}
$$

for all $\ell\in\mathbb{N}_0$. After taking the supremum over all $\ell\in\mathbb{N}_0$, we arrive at estimate (14.40) from subitem *(b1)*.

Next, assume $1<p<q$, hence $1/q-1/p<0$. From the second inequality in (14.54), and keeping in mind (14.50), we see that for some constant $C\in(0,\infty)$ depending only n, p, q, and the Ahlfors regularity constants of $\partial\Omega$,

$$
\begin{aligned}
\frac{2_*^{\ell(n-1)(1/q-1/p)}}{2_*^{\ell}}\Big(\int_{\Delta_\ell}|f(x)-f_{\Delta_\ell}|^p\,d\sigma(x)\Big)^{1/p}
&\leq C\cdot 2_*^{\ell(n-1)(1/q-1/p)}\Big(\sum_{k=0}^{\ell}2_*^{-k(n-1)(1/q-1/p)}\Big)\|g\|_{\mathrm{A}_{p,q}(\partial\Omega,\sigma)}\\
&\leq C\cdot\Big(\sum_{k=0}^{\ell}2_*^{(\ell-k)(n-1)(1/q-1/p)}\Big)\|g\|_{\mathrm{A}_{p,q}(\partial\Omega,\sigma)}\\
&\leq C\Big(\sum_{k=0}^{\infty}2_*^{k(n-1)(1/q-1/p)}\Big)\|g\|_{\mathrm{A}_{p,q}(\partial\Omega,\sigma)}\\
&\leq C\sum_{j,k=1}^{n}\|\partial_{\tau_{jk}}f\|_{\mathrm{A}_{p,q}(\partial\Omega,\sigma)}<\infty, \qquad (14.55)
\end{aligned}
$$

for every $\ell\in\mathbb{N}_0$. Taking the supremum over $\ell\in\mathbb{N}_0$ in (14.55) we arrive at estimate (14.41) from subitem *(b2)*. This completes the treatment of *(a)⟹(b)*.

We now prove the implication *(b)⟹(c)*. Thanks to Lemma 14.1 and Remark 14.2, the assumptions made in *(b)* yield

$$\int_{\Sigma} \frac{|f(x) - f_{\Delta_\ell}|}{(2_*^{\ell+1} + |x - x_0|)^n} \, d\sigma(x) \leq \frac{C \cdot C_*}{\sigma(\Delta_\ell)^{1/p}} \quad \text{whenever } 0 < q \leq p, \tag{14.56}$$

and

$$\int_{\Sigma} \frac{|f(x) - f_{\Delta_\ell}|}{(2_*^{\ell+1} + |x - x_0|)^n} \, d\sigma(x) \leq \frac{C \cdot C_*}{\sigma(\Delta_\ell)^{1/q}} \quad \text{whenever } 1 < p < q, \tag{14.57}$$

for some constant $C = C(\partial\Omega, n, p, q) \in (0, \infty)$, and for every $\ell \in \mathbb{N}_0$. Having fix $\ell \in \mathbb{N}_0$, observe that since $2_* > 1$, we have

$$2_*^{\ell+1} + |x - x_0| \leq 2_*^{\ell+1} + |x| + |x_0| \leq (2_*^{\ell+1} + |x_0|)(1 + |x|) \tag{14.58}$$

which in turn implies

$$(2_*^{\ell+1} + |x - x_0|)^n \leq 2^n (2_*^{\ell+1} + |x_0|)^n (1 + |x|^n). \tag{14.59}$$

This and estimate (14.56) (respectively (14.57)) then readily imply (14.42) (respectively (14.43)). Therefore *(b)⟹(c)* is established.

Regarding the implication *(c)⟹(a)*, we make use of (14.42) and (14.43) for the choice $\ell := 1$ to write

$$\begin{aligned} \int_{\partial\Omega} \frac{|f(x)|}{1 + |x|^n} \, d\sigma(x) &\leq \int_{\partial\Omega} \frac{|f(x) - f_{\Delta_0}|}{1 + |x|^n} \, d\sigma(x) + |f_{\Delta_0}| \int_{\partial\Omega} \frac{d\sigma(x)}{1 + |x|^n} \\ &\leq C \sum_{j,k=1}^{n} \|\partial_{\tau_{jk}} f\|_{\mathrm{A}_{p,q}(\partial\Omega, \sigma)} + c_f < \infty, \end{aligned} \tag{14.60}$$

and conclude that f belongs to $L^1\big(\partial\Omega, \frac{\sigma(x)}{1+|x|^n}\big)$, as desired. This finishes the proof of the equivalences *(a)⟺(b)⟺(c)*.

Furthermore, *(d)⟹(a)* is clear from (14.23) in Definition 14.3, while *(b)⟹(d)* is visible from the equivalence *(a)⟺(b)*, (14.39), and the fact that conditions *(b1)-(b2)* imply $f \in L^p_{\mathrm{loc}}(\partial\Omega, \sigma)$. This ultimately finishes the proof of Theorem 14.1. □

Notice that $\dot{\mathrm{A}}^1_{p,q}(\partial\Omega, \sigma)$ contains all constant functions on $\partial\Omega$, and their seminorm are zero. For this reason, we shall occasionally mod out constants from $\dot{\mathrm{A}}^1_{p,q}(\partial\Omega, \sigma)$. More precisely, given $f, g \in L^1\big(\partial\Omega, \frac{\sigma(x)}{1+|x|^n}\big) \cap L^p_{\mathrm{loc}}(\partial\Omega, \sigma)$, consider the equivalence relation

$$f \sim g \quad \overset{\text{def}}{\Longleftrightarrow} \quad f - g \ \text{ is a constant.} \tag{14.61}$$

Denoting by $[f]$ the equivalence class modulo constants of f, set

$$\|[f]\|_{\dot{\mathrm{A}}^1_{p,q}(\partial\Omega, \sigma)/\sim} := \|f\|_{\dot{\mathrm{A}}^1_{p,q}(\partial\Omega, \sigma)} \quad \text{for every } f \in \dot{\mathrm{A}}^1_{p,q}(\partial\Omega, \sigma). \tag{14.62}$$

It is therefore of interest to establish when this becomes a genuine norm on the quotient space

$$\dot{\mathrm{A}}^1_{p,q}(\partial\Omega,\sigma)/\!\sim\, = \big\{[f] : f \in \dot{\mathrm{A}}^1_{p,q}(\partial\Omega,\sigma)\big\}. \tag{14.63}$$

The result displayed below establishes the completeness of (14.63) equipped with (14.62).

Proposition 14.2 *Suppose* $\Omega \subseteq \mathbb{R}^n$ *is a two-sided* NTA *domain with* $\partial\Omega$ *being an unbounded Ahlfors regular set, and abbreviate* $\sigma := \mathcal{H}^{n-1}\lfloor\partial\Omega$. *Select two integrability exponents* $p \in (1,\infty)$ *and* $q \in (0,\infty)$.

Then (14.62) *is a genuine norm (respectively a quasinorm) on* $\dot{\mathrm{A}}^1_{p,q}(\partial\Omega,\sigma)/\!\sim$ *for* $q \geq 1$ *(respectively for* $0 < q < 1$*). Moreover,* $\dot{\mathrm{A}}^1_{p,q}(\partial\Omega,\sigma)/\!\sim$ *is a Banach space (respectively a quasi-Banach space), for* $q \geq 1$ *(respectively for* $0 < q < 1$*), when equipped with* $\|\cdot\|_{\dot{\mathrm{A}}^1_{p,q}(\partial\Omega,\sigma)/\sim}$ *(cf.* (14.62)*).*

Proof The fact that (14.62) is a norm (respectively quasinorm) on $\dot{\mathrm{A}}^1_{p,q}(\partial\Omega,\sigma)/\!\sim$ for $q \geq 1$ (respectively for $0 < q < 1$) follows from (14.40) and (14.41).

Consider a sequence $\{f_\alpha\}_{\alpha\in\mathbb{N}} \subseteq \dot{\mathrm{A}}^1_{p,q}(\partial\Omega,\sigma)$ such that, at the level of equivalence classes modulo constants, $\{[f_\alpha]\}_{\alpha\in\mathbb{N}}$ is a Cauchy sequence in $\dot{\mathrm{A}}^1_{p,q}(\partial\Omega,\sigma)/\!\sim$. Then having fixed an $\varepsilon > 0$ there exists some $N \in \mathbb{N}$ such that for all $\alpha,\beta > N$ we have

$$\sum_{j,k=1}^{n} \|\partial_{\tau_{jk}} f_\alpha - \partial_{\tau_{jk}} f_\beta\|_{\mathrm{A}_{p,q}(\partial\Omega,\sigma)} = \|f_\alpha - f_\beta\|_{\dot{\mathrm{A}}^1_{p,q}(\partial\Omega,\sigma)} < \varepsilon. \tag{14.64}$$

In particular, this implies that, for each $j,k \in \{1,\dots,n\}$, the sequence $\{\partial_{\tau_{jk}} f_\alpha\}$ is Cauchy in the space $\mathrm{A}_{p,q}(\partial\Omega,\sigma)$ which is complete (cf. Proposition 10.6). As such, for each $j,k \in \{1,\dots,n\}$ there exists a function $g_{jk} \in \mathrm{A}_{p,q}(\partial\Omega,\sigma)$ with the property that

$$\partial_{\tau_{jk}} f_\alpha \to g_{jk} \ \text{ in } \mathrm{A}_{p,q}(\partial\Omega,\sigma) \text{ as } \alpha \to \infty. \tag{14.65}$$

Next, pick a reference point $x_0 \in \partial\Omega$ and a dilation factor $2_* > \big(C_{\partial\Omega}/c_{\partial\Omega}\big)^{1/(n-1)}$ (as is Definition 10.1). For each $\ell \in \mathbb{N}_0$ abbreviate

$$\Delta_\ell := \Delta(x_0, 2_*^{\ell+1}) \quad \text{and} \quad f_{\alpha,\Delta_\ell} := \fint_{\Delta_\ell} f_\alpha \, d\sigma \quad \text{for all } \alpha \in \mathbb{N}. \tag{14.66}$$

In light of Theorem 14.1 item *(c)*, for each $\ell \in \mathbb{N}$ there exists some finite constant $C_\ell > 0$ satisfying

$$\|(f_\alpha - f_{\alpha,\Delta_\ell}) - (f_\beta - f_{\beta,\Delta_\ell})\|_{L^1(\partial\Omega,\frac{\sigma(x)}{1+|x|^n})}$$
$$\leq \begin{cases} \dfrac{C_\ell}{\sigma(\Delta_\ell)^{1/p}} \displaystyle\sum_{j,k=1}^{n} \|\partial_{\tau_{jk}} f_\alpha - \partial_{\tau_{jk}} f_\beta\|_{\mathrm{A}_{p,q}(\partial\Omega,\sigma)} & \text{if } 0 < q \leq p, \\ \dfrac{C_\ell}{\sigma(\Delta_\ell)^{1/q}} \displaystyle\sum_{j,k=1}^{n} \|\partial_{\tau_{jk}} f_\alpha - \partial_{\tau_{jk}} f_\beta\|_{\mathrm{A}_{p,q}(\partial\Omega,\sigma)} & \text{if } 1 < p < q, \end{cases} \tag{14.67}$$

for every $\alpha, \beta \in \mathbb{N}$. Thanks to the convergence recorded in (14.65), this implies that for each fixed $\ell \in \mathbb{N}_0$ the sequence $\{f_\alpha - f_{\alpha,\Delta_\ell}\}_{\alpha\in\mathbb{N}}$ is Cauchy in the Banach space $L^1(\partial\Omega, \frac{\sigma(x)}{1+|x|^n})$. Hence, for each $\ell \in \mathbb{N}_0$ there exists a function $h_\ell \in L^1(\partial\Omega, \frac{\sigma(x)}{1+|x|^n})$ such that

$$f_\alpha - f_{\alpha,\Delta_\ell} \to h_\ell \ \text{ in } L^1(\partial\Omega, \tfrac{\sigma(x)}{1+|x|^n}) \text{ as } \alpha \to \infty. \tag{14.68}$$

Furthermore, from Theorem 14.1 item *(b)*, there exists some constant $C \in (0,\infty)$ with the property that whenever $\ell \in \mathbb{N}_0$ we have

$$\Big(\int_{\Delta_\ell} |(f_\alpha - f_{\alpha,\Delta_\ell}) - (f_\beta - f_{\beta,\Delta_\ell})|^p \, d\sigma\Big)^{1/p}$$
$$\leq \begin{cases} 2_*^\ell \displaystyle\sum_{j,k=1}^{n} \|\partial_{\tau_{jk}} f_\alpha - \partial_{\tau_{jk}} f_\beta\|_{\mathrm{A}_{p,q}(\partial\Omega,\sigma)} & \text{if } 0 < q \leq p, \\ \dfrac{2_*^\ell}{2_*^{\ell(n-1)(1/q-1/p)}} \displaystyle\sum_{j,k=1}^{n} \|\partial_{\tau_{jk}} f_\alpha - \partial_{\tau_{jk}} f_\beta\|_{\mathrm{A}_{p,q}(\partial\Omega,\sigma)} & \text{if } 1 < p < q. \end{cases} \tag{14.69}$$

for every $\alpha, \beta \in \mathbb{N}$. Bearing in mind (14.65), this guarantees that for each $\ell \in \mathbb{N}_0$, the sequence $\{f_\alpha\big|_{\Delta_\ell} - f_{\alpha,\Delta_\ell}\}_{\alpha\in\mathbb{N}}$ is Cauchy in the Banach space $L^p(\Delta_\ell, \sigma)$. As such, for each $\ell \in \mathbb{N}_0$ there exists some function $\zeta_\ell \in L^p(\Delta_\ell, \sigma)$ for which

$$f_\alpha\big|_{\Delta_\ell} - f_{\alpha,\Delta_\ell} \to \zeta_\ell \ \text{ in } L^p(\Delta_\ell, \sigma) \text{ as } \alpha \to \infty. \tag{14.70}$$

For each $\ell \in \mathbb{N}_0$ and $\alpha \in \mathbb{N}$ we use Hölder inequality to estimate

$$\begin{aligned} &\|h_\ell\big|_{\Delta_\ell} - \zeta_\ell\|_{L^1(\Delta_\ell,\sigma)} \\ &\quad\leq \|(f_\alpha\big|_{\Delta_\ell} - f_{\alpha,\Delta_\ell}) - h_\ell\big|_{\Delta_\ell}\|_{L^1(\Delta_\ell,\sigma)} + \|(f_\alpha\big|_{\Delta_\ell} - f_{\alpha,\Delta_\ell}) - \zeta_\ell\|_{L^1(\Delta_\ell,\sigma)} \\ &\quad\leq C_\ell \|(f_\alpha - f_{\alpha,\Delta_\ell}) - h_\ell\|_{L^1(\partial\Omega,\frac{\sigma(x)}{1+|x|^n})} \\ &\qquad + C_{\ell,p}\|(f_\alpha\big|_{\Delta_\ell} - f_{\alpha,\Delta_\ell}) - \zeta_\ell\|_{L^p(\Delta_\ell,\sigma)}. \end{aligned} \tag{14.71}$$

In concert with (14.68) and (14.70), this gives

$$h_\ell\big|_{\Delta_\ell} = \zeta_\ell \ \text{ pointwise } \sigma\text{-a.e. on } \partial\Omega, \text{ for all } \ell \in \mathbb{N}_0. \tag{14.72}$$

From the convergence in (14.68) we also see that for every $\ell, \ell' \in \mathbb{N}_0$ we have

$$(f_{\alpha,\Delta_\ell} - f_{\alpha,\Delta_{\ell'}}) \longrightarrow (h_\ell - h_{\ell'}) \text{ in } L^1\big(\partial\Omega, \tfrac{\sigma(x)}{1+|x|^n}\big) \text{ as } \alpha \to \infty. \tag{14.73}$$

In particular, this forces $h_\ell - h_{\ell'}$ to be a constant, say $h_\ell - h_{\ell'} \equiv c_{\ell,\ell'} \in \mathbb{R}$, on $\partial\Omega$. Therefore, (14.72) implies that for every $\ell, \ell' \in \mathbb{N}_0$ we have

$$\begin{aligned}\Big(\int_{\Delta_{\ell'}} |h_\ell|^p\, d\sigma\Big)^{1/p} &\leq \Big(\int_{\Delta_{\ell'}} |h_\ell - h_{\ell'}|^p\, d\sigma\Big)^{1/p} + \Big(\int_{\Delta_{\ell'}} |h_{\ell'}|^p\, d\sigma\Big)^{1/p} \\ &= c_{\ell,\ell'} \cdot \sigma(\Delta_\ell)^{1/p} + \Big(\int_{\Delta_{\ell'}} |\zeta_{\ell'}|^p\, d\sigma\Big)^{1/p} < \infty.\end{aligned} \tag{14.74}$$

Ultimately, this proves that $h_\ell \in L^p_{\rm loc}(\partial\Omega, \sigma)$, for every $\ell \in \mathbb{N}_0$. This and (14.68) then show that

$$h_\ell \in L^1\big(\partial\Omega, \tfrac{\sigma(x)}{1+|x|^n}\big) \cap L^p_{\rm loc}(\partial\Omega, \sigma) \text{ for all } \ell \in \mathbb{N}_0. \tag{14.75}$$

To proceed, abbreviate

$$h := h_1 \text{ and } c_\alpha := f_{\alpha,\Delta_1} \in \mathbb{C} \text{ for all } \alpha \in \mathbb{N}. \tag{14.76}$$

Then with this piece of notation, (14.68) and (14.75) read as follows:

$$\begin{aligned}&h \in L^1\big(\partial\Omega, \tfrac{\sigma(x)}{1+|x|^n}\big) \cap L^p_{\rm loc}(\partial\Omega, \sigma) \text{ and} \\ &(f_\alpha - c_\alpha) \longrightarrow h \text{ in } L^1\big(\partial\Omega, \tfrac{\sigma(x)}{1+|x|^n}\big) \text{ as } \alpha \to \infty.\end{aligned} \tag{14.77}$$

Based on this, (14.1), (14.2), (14.65), and Lebesgue's Dominated Convergence Theorem, for each $\varphi \in \mathscr{C}_0^\infty(\mathbb{R}^n)$ and $j, k \in \{1, \dots, n\}$ we may compute

$$\begin{aligned}\int_{\partial\Omega} h(\partial_{\tau_{jk}}\varphi)\, d\sigma &= \lim_{\alpha\to\infty} \int_{\partial\Omega} (f_\alpha - c_\alpha)(\partial_{\tau_{jk}}\varphi)\, d\sigma \\ &= -\lim_{\alpha\to\infty} \int_{\partial\Omega} [\partial_{\tau_{jk}}(f_\alpha - c_\alpha)]\varphi\, d\sigma \\ &= -\lim_{\alpha\to\infty} \int_{\partial\Omega} (\partial_{\tau_{jk}} f_\alpha)\varphi\, d\sigma = -\int_{\partial\Omega} g_{jk}\varphi\, d\sigma,\end{aligned} \tag{14.78}$$

since tangential derivatives annihilates constants. Hence

$$\partial_{\tau_{jk}} h = g_{jk} \in \mathrm{A}_{p,q}(\partial\Omega, \sigma) \text{ for all } j, k \in \{1, \dots, n\}, \tag{14.79}$$

which in concert with (14.77) implies

$$h \in \dot{\mathrm{A}}_{p,q}(\partial\Omega, \sigma). \tag{14.80}$$

Finally, letting $\beta \to \infty$ in (14.64) and using (14.65) together with (14.79) we obtain

$$\sum_{j,k=1}^{n} \|\partial_{\tau_{jk}} f_\alpha - \partial_{\tau_{jk}} h\|_{A_{p,q}(\partial\Omega,\sigma)} = \|f_\alpha - h\|_{\dot{A}^1_{p,q}(\partial\Omega,\sigma)} \le \varepsilon \quad \text{for all } \alpha > N. \tag{14.81}$$

Since $\varepsilon > 0$ was taken arbitrary to begin with, (14.81) shows that $[f_\alpha] \to [h]$ in $\dot{A}_{p,q}(\partial\Omega,\sigma)/\sim$ as $\alpha \to \infty$, thus concluding the proof of Proposition 14.2. □

14.1 Tangential Derivatives on Herz Spaces of First Generation

The goal here is to study nontangential boundary traces for functions satisfying certain size conditions related to first-generation Herz spaces, and obtain estimates involving weak tangential derivatives. We begin with the following result which describes the independence of the membership of the nontangential maximal operator to Herz spaces on the aperture parameter.

Proposition 14.3 *Let $\Omega \subseteq \mathbb{R}^n$ be an Ahlfors regular domain with unbounded boundary. Abbreviate $\sigma := \mathcal{H}^{n-1}\lfloor\partial\Omega$, and select two exponents $p \in (1,\infty)$ and $q \in (0,\infty)$. Assume that $u : \Omega \to \mathbb{C}$ is a Lebesgue measurable function and fix two aperture parameters $\kappa, \widetilde{\kappa} \in (0,\infty)$. Then*

$$\|\mathcal{N}_\kappa u\|_{A_{p,q}(\partial\Omega,\sigma)} \approx \|\mathcal{N}_{\widetilde{\kappa}} u\|_{A_{p,q}(\partial\Omega,\sigma)}, \tag{14.82}$$

with implicit proportionality constants independent of u.

Proof Without loss of generality, assume $\kappa \ge \widetilde{\kappa}$. Thanks to [GHA-I, Theorem 8.4.6, p. 704], for every parameter $s \in (0,\infty)$, there exists some constant $C_s \in (0,\infty)$, such that

$$0 \le \mathcal{N}_\kappa u \le \mathcal{N}_{\widetilde{\kappa}} u \le C \cdot \mathcal{M}_{\partial\Omega,s}(\mathcal{N}_\kappa u) \quad \text{everywhere on } \partial\Omega. \tag{14.83}$$

Granted this, (14.82) readily follows from the lattice property of $A_{p,q}(\partial\Omega,\sigma)$ (cf. (10.8)), and the boundedness of the Hardy-Littlewood maximal operator. Specifically, from Theorem 10.1 we know that $\mathcal{M}_{\partial\Omega,s}$ is bounded on $A_{p,q}(\Sigma,\sigma)$ if $0 < q \le p$ and $0 < s < q$, and $\mathcal{M}_{\partial\Omega}$ is bounded on $A_{p,q}(\Sigma,\sigma)$ if $p, q \in (1,\infty)$. □

We continue by recording the following lemma regarding the membership of nontangential boundary traces to Herz spaces.

Lemma 14.2 *Consider an Ahlfors regular domain $\Omega \subseteq \mathbb{R}^n$, and set $\sigma := \mathcal{H}^{n-1}\lfloor\partial\Omega$. Fix integrability exponents $p \in (1,\infty)$ and $q \in (0,\infty)$, along with an aperture parameter $\kappa \in (0,\infty)$. Assume $u : \Omega \to \mathbb{R}$ is a Lebesgue measurable function such that*

$$\begin{gathered} u\big|^{\kappa-\mathrm{n.t.}}_{\partial\Omega} \text{ exists } \sigma\text{-a.e. on } \partial\Omega, \text{ and} \\ \mathcal{N}^\delta_\kappa u \in A_{p,q}(\partial\Omega,\sigma) \text{ for some } \delta > 0. \end{gathered} \tag{14.84}$$

Then $u\big|_{\partial\Omega}^{\kappa-\mathrm{n.t.}}$ *is a* σ*-measurable function on* $\partial\Omega$ *that belongs to* $\mathrm{A}_{p,q}(\partial\Omega,\sigma)$ *and*

$$\mathcal{N}_\kappa^\varepsilon u \to u\big|_{\partial\Omega}^{\kappa-\mathrm{n.t.}} \quad \textit{in } \mathrm{A}_{p,q}(\partial\Omega,\sigma) \textit{ as } \varepsilon \to 0^+. \tag{14.85}$$

Proof As in [GHA-I, §8.8], define $A_\kappa(\partial\Omega) := \{x \in \partial\Omega : x \in \overline{\Gamma_\kappa(x)}\}$. Since Ω is an Ahlfors regular domain, we may invoke [GHA-I, Proposition 8.8.4, p. 780] to conclude that

$$\sigma\big(\partial\Omega\backslash A_\kappa(\partial\Omega)\big) = \sigma\big(\partial\Omega\backslash\partial_*\Omega\big) + \sigma\big(\partial_*\Omega\backslash A_\kappa(\partial\Omega)\big) = 0. \tag{14.86}$$

Hence, in light of [GHA-I, Lemma 8.9.4. p. 791] (used with $\mathcal{U} := \partial\Omega$) we have that

$$u\big|_{\partial\Omega}^{\kappa-\mathrm{n.t.}} \text{ is a } \sigma\text{-measurable function on } \partial\Omega. \tag{14.87}$$

Moreover, since

$$\Big|u\big|_{\partial\Omega}^{\kappa-\mathrm{n.t.}}\Big| \leq \mathcal{N}_\kappa^\delta u \in \mathrm{A}_{p,q}(\partial\Omega,\sigma) \quad \text{for } \sigma\text{-a.e. } x \in \partial\Omega, \tag{14.88}$$

(see [GHA-I, (8.9.8), p. 788]) the lattice property of Herz spaces (cf. (10.8)) guarantees the membership of $u\big|_{\partial\Omega}^{\kappa-\mathrm{n.t.}}$ to $\mathrm{A}_{p,q}(\partial\Omega,\sigma)$.

Having established these properties, we follow the proof of [GHA-I, Lemma 8.9.4. p. 791] to justify the following pointwise convergence:

$$\lim_{\varepsilon\to 0^+} \big(\mathcal{N}_\kappa^\varepsilon u\big)(x) = \Big|u\big|_{\partial\Omega}^{\kappa-\mathrm{n.t.}}(x)\Big| \quad \text{at } \sigma\text{-a.e. } x \in \partial\Omega. \tag{14.89}$$

Notice that for every $\varepsilon \in (0,\delta)$ we have

$$0 \leq \mathcal{N}_\kappa^\varepsilon u \leq \mathcal{N}_\kappa^\delta u \in \mathrm{A}_{p,q}(\partial\Omega,\sigma). \tag{14.90}$$

This together with Proposition 10.4 and (14.89) imply the validity of the convergence in (14.85), thus concluding the proof of Lemma 14.2. □

It is also useful to note that the truncated nontangential maximal function of a continuous and compactly supported function is actually bounded and has compact support.

Lemma 14.3 *Let* $\Omega \subseteq \mathbb{R}^n$ *be an Ahlfors regular domain, and set* $\sigma := \mathcal{H}^{n-1}\lfloor\partial\Omega$. *Also, fix an aperture parameter* $\kappa \in (0,\infty)$. *Then for every* $\varphi \in \mathscr{C}_0^0(\mathbb{R}^n)$ *the function* $\mathcal{N}_\kappa\varphi$ *belongs to* $L^\infty_{\mathrm{comp}}(\partial\Omega,\sigma)$.

Proof For each $E \subseteq \Omega$, define

$$\pi_k(E) := \big\{x \in \partial\Omega : \Gamma_\kappa(x) \cap E \neq \varnothing\big\}. \tag{14.91}$$

Pick a function $\varphi \in \mathscr{C}_0^0(\mathbb{R}^n)$. Fix a reference point $x_0 \in \partial\Omega$ and choose $R > 0$ large enough so that $\operatorname{supp}\varphi \subseteq B(x_0,R)$. Then

$$\operatorname{supp}(\mathcal{N}_\kappa\varphi) \subseteq \pi_k(\operatorname{supp}\varphi) \subseteq \pi_k\big(B(x_0,R)\cap\Omega\big) \subseteq \Delta\big(x_0,(2+\kappa)R\big), \tag{14.92}$$

with the last inclusion coming from [GHA-I, (8.1.17), p. 675]. Since we also have that the function $\mathcal{N}_\kappa\phi$ is σ-measurable on $\partial\Omega$ (see [GHA-I, Proposition 8.2.3, pp. 684–685]), and $\sup_{\partial\Omega}|\mathcal{N}_\kappa\phi| \leq \sup_{\overline{\Omega}}|\varphi| < \infty$ we ultimately see that $\mathcal{N}_\kappa\varphi$ belongs to $L^\infty_{\rm comp}(\partial\Omega,\sigma)$. □

The next result sheds light on the membership of tangential derivatives of nontangential boundary traces to Herz spaces.

Theorem 14.2 *Consider an Ahlfors regular domain $\Omega \subseteq \mathbb{R}^n$ with unbounded boundary, and set $\sigma := \mathcal{H}^{n-1}\lfloor\partial\Omega$. Select an aperture parameter $\kappa \in (0,\infty)$, and two integrability exponents $p,q \in (1,\infty)$. Pick a function $u \in W^{1,1}_{\rm loc}(\Omega)$ and assume that for some $\delta > 0$ one has*

$$\begin{gathered} u\big|^{\kappa-\rm n.t.}_{\partial\Omega}(x) \ \text{ exists for } \sigma\text{-a.e. } x \in \partial\Omega, \\ \mathcal{N}^\delta_\kappa u \in L^p_{\rm loc}(\partial\Omega,\sigma), \ \text{ and } \ \mathcal{N}^\delta_\kappa(\nabla u) \in \mathrm{A}_{p,q}(\partial\Omega,\sigma). \end{gathered} \tag{14.93}$$

Then for any $\kappa' \in (0,\infty)$ the nontangential trace $u\big|^{\kappa'-\rm n.t.}_{\partial\Omega}$ exists σ-a.e. on $\partial\Omega$, is actually independent of κ', and

$$u\big|^{\kappa'-\rm n.t.}_{\partial\Omega} \in L^p_{1,\rm loc}(\partial\Omega,\sigma). \tag{14.94}$$

Moreover, with the dependence of κ' dropped, this function satisfies

$$\begin{gathered} \partial_{\tau_{jk}}\big(u\big|^{\rm n.t.}_{\partial\Omega}\big) \in \mathrm{A}_{p,q}(\partial\Omega,\sigma) \ \text{ for all } j,k \in \{1,\dots,n\}, \text{ and} \\ \sum_{j,k=1}^n \big\|\partial_{\tau_{jk}}\big(u\big|^{\rm n.t.}_{\partial\Omega}\big)\big\|_{\mathrm{A}_{p,q}(\partial\Omega,\sigma)} \leq C\|\mathcal{N}^\delta_\kappa(\nabla u)\|_{\mathrm{A}_{p,q}(\partial\Omega,\sigma)}, \end{gathered} \tag{14.95}$$

for some constant $C \in (0,\infty)$ independent of u.

Proof In view of the fact that $\mathcal{N}^\delta_\kappa(\nabla u) \in \mathrm{A}_{p,q}(\partial\Omega,\sigma) \subseteq L^p_{\rm loc}(\partial\Omega,\sigma)$ (see item *(g)* in Proposition 10.3) together with the other assumptions made in (14.93), we may invoke [GHA-II, Corollary 11.4.7, p. 717] to obtain the existence of $u\big|^{\kappa'-\rm n.t.}_{\partial\Omega}$ at σ-a.e. point on $\partial\Omega$, the independence of κ', and the membership in (14.94).

From now on we may drop the dependence on κ', and abbreviate

$$f := u\big|^{\rm n.t.}_{\partial\Omega} \ \text{ on } \partial\Omega. \tag{14.96}$$

Fix indices $j,k \in \{1,\dots,n\}$, and pick a test function $\varphi \in \mathscr{C}^\infty_0(\mathbb{R}^n)$ satisfying

$$\operatorname{supp}\varphi \subseteq \{x \in \mathbb{R}^n : \operatorname{dist}(x,\partial\Omega) < \delta\}. \tag{14.97}$$

For each $\varepsilon \in (0,\infty)$, define $O_\varepsilon := \{x \in \Omega : \operatorname{dist}(x,\partial\Omega) < \varepsilon\}$. Following the proof of [GHA-II, Proposition 11.3.4, pp. 712–716] we may conclude that

$$\Big|\int_{\partial\Omega} f(\partial_{\tau_{jk}}\varphi)\,d\sigma\Big| \leq C\int_{O_\varepsilon} |\nabla u|\{\varepsilon^{-1}|\varphi| + |\nabla\varphi|\}\,d\mathcal{L}^n \quad \text{for all } \varepsilon > 0 \tag{14.98}$$

and

$$\lim_{\varepsilon\to 0^+}\int_{O_\varepsilon} |\nabla u||\nabla\varphi|\,d\mathcal{L}^n = 0. \tag{14.99}$$

In addition, having fixed $\varepsilon \in (0,\delta)$, [GHA-II, Proposition 8.6.10, p. 745] along with item *(a)* in Proposition 10.2 imply

$$\begin{aligned}
\varepsilon^{-1}\int_{O_\varepsilon} |\nabla u||\varphi|\,d\mathcal{L}^n &\leq C\big\|\mathcal{N}_\kappa^\varepsilon\big(|\nabla u||\varphi|\big)\big\|_{L^1(\partial\Omega,\sigma)} \\
&\leq C\big\|\mathcal{N}_\kappa^\varepsilon(\nabla u)\cdot\mathcal{N}_\kappa^\varepsilon\varphi\big\|_{L^1(\partial\Omega,\sigma)} \\
&\leq C\big\|\mathcal{N}_\kappa^\delta(\nabla u)\cdot\mathcal{N}_\kappa^\varepsilon\varphi\big\|_{L^1(\partial\Omega,\sigma)} \\
&\leq C\big\|\mathcal{N}_\kappa^\delta(\nabla u)\big\|_{\mathrm{A}_{p,q}(\partial\Omega,\sigma)}\cdot\big\|\mathcal{N}_\kappa^\varepsilon\varphi\big\|_{\mathrm{A}_{p',q'}(\partial\Omega,\sigma)}.
\end{aligned} \tag{14.100}$$

In view of Lemma 14.3 and item *(f)* in Proposition 10.3, we see that the function $\mathcal{N}_\kappa\varphi$ belongs to $L^\infty_{\mathrm{comp}}(\partial\Omega,\sigma) \subseteq \mathrm{A}_{p',q'}(\partial\Omega,\sigma)$. Hence, Lemma 14.2 gives

$$\lim_{\varepsilon\to 0^+}\big\|\mathcal{N}_\kappa^\varepsilon\varphi\big\|_{\mathrm{A}_{p',q'}(\partial,\Omega,\sigma)} = \big\|\varphi\big|_{\partial\Omega}\big\|_{\mathrm{A}_{p',q'}(\partial\Omega,\sigma)}. \tag{14.101}$$

This, (14.98), and (14.100) then entail

$$\begin{aligned}
\Big|\int_{\partial\Omega} f\cdot(\partial_{\tau_{jk}}\varphi)\,d\sigma\Big| &\leq C\big\|\mathcal{N}_\kappa^\delta(\nabla u)\big\|_{\mathrm{A}_{p,q}(\partial\Omega,\sigma)}\big\|\mathcal{N}_\kappa^\varepsilon\varphi\big\|_{\mathrm{A}_{p',q'}(\partial\Omega,\sigma)} \\
&\quad + C\int_{O_\varepsilon}|\nabla u||\nabla\varphi|\,d\mathcal{L}^n.
\end{aligned} \tag{14.102}$$

After passing to the limit $\varepsilon \to 0^+$, keeping in mind (14.99) and (14.101), we arrive at

$$\Big|\int_{\partial\Omega} f\cdot(\partial_{\tau_{jk}}\varphi)\,d\sigma\Big| \leq C\big\|\mathcal{N}_\kappa^\delta(\nabla u)\big\|_{\mathrm{A}_{p,q}(\partial\Omega,\sigma)}\big\|\varphi\big|_{\partial\Omega}\big\|_{\mathrm{A}_{p',q'}(\partial\Omega,\sigma)}, \tag{14.103}$$

for some constant $C = C(\partial\Omega,\kappa,p,q) \in (0,\infty)$. Since this inequality involves only the behavior of φ near the boundary, the assumption made in (14.97) becomes redundant. Hence (14.103) is valid for every $\varphi \in \mathscr{C}_0^\infty(\mathbb{R}^n)$, with constant $C \in (0,\infty)$ depending only on $\partial\Omega$, κ, p, q.

To proceed, recall from Proposition 10.7 that

$$\big\{\varphi\big|_{\partial\Omega} : \varphi \in \mathscr{C}_0^\infty(\mathbb{R}^n)\big\} \text{ is a dense subspace of } \mathrm{A}_{p,q}(\partial\Omega,\sigma). \tag{14.104}$$

Continue to assume that $j, k \in \{1, \dots, n\}$ are fixed, and consider the functional

$$\begin{aligned}\Lambda_{jk} : \big\{\phi|_{\partial\Omega} : \phi \in \mathscr{C}_0^\infty(\mathbb{R}^n)\big\} \to \mathbb{R} \\ \text{given by } \Lambda_{jk}\big(\phi|_{\partial\Omega}\big) := \int_{\partial\Omega} f(\partial_{\tau_{jk}}\varphi)\, d\sigma.\end{aligned} \tag{14.105}$$

Hence, thanks to [GHA-II, Lemma 11.1.1, p. 683], Λ_{jk} is a well-defined linear functional, while (14.103) yields the boundedness of Λ_{jk} in the norm of $\mathrm{A}_{p',q'}(\partial\Omega, \sigma)$. As such, from (14.104) and item *(a)* in Proposition 10.9 it follows that Λ_{jk} extends to a bounded linear function $\widetilde{\Lambda}_{jk}$ in $\big(\mathrm{A}_{p',q'}(\partial\Omega, \sigma)\big)^* = \mathrm{A}_{p,q}(\partial\Omega, \sigma)$ with norm

$$\|\widetilde{\Lambda}_{jk}\|_{(\mathrm{A}_{p',q'}(\partial\Omega,\sigma))^*} \leq C\|\mathcal{N}_\kappa^\delta(\nabla u)\|_{\mathrm{A}_{p,q}(\partial\Omega,\sigma)}. \tag{14.106}$$

In particular, there exists some function $h_{jk} \in \mathrm{A}_{p,q}(\partial\Omega, \sigma)$ for which

$$\begin{aligned}\Lambda_{jk}\big(\phi|_{\partial\Omega}\big) = \widetilde{\Lambda}_{jk}\big(\phi|_{\partial\Omega}\big) = \int_{\partial\Omega} h_{jk}\varphi\, d\sigma \quad \text{for all } \varphi \in \mathscr{C}_0^\infty(\mathbb{R}^n), \\ \text{and } \|h_{jk}\|_{\mathrm{A}_{p,q}(\partial\Omega,\sigma)} = \|\widetilde{\Lambda}_{jk}\|_{(\mathrm{A}_{p',q'}(\partial\Omega,\sigma))^*}.\end{aligned} \tag{14.107}$$

From this and (14.105) we arrive at

$$\int_{\partial\Omega} f(\partial_{\tau_{jk}}\varphi)\, d\sigma = \int_{\partial\Omega} h_{jk}\varphi\, d\sigma \quad \text{for all } \varphi \in \mathscr{C}_0^\infty(\mathbb{R}^n), \tag{14.108}$$

and therefore, bearing in mind (14.1) and (14.3), we obtain

$$\partial_{\tau_{jk}} f = -h_{jk} \in \mathrm{A}_{p,q}(\partial\Omega, \sigma). \tag{14.109}$$

Putting it all together, since $j, k \in \{1, \dots, n\}$ was taken arbitrary to begin with, (14.96), (14.106), (14.107), and (14.109) imply the claim made in (14.95). This finishes the proof of Theorem 14.2. □

The following result provides estimates for tangential derivatives of nontangential boundary traces for functions with a size condition on the gradient, measured on Herz spaces.

Theorem 14.3 *Consider $\Omega \subseteq \mathbb{R}^n$ an* NTA *domain with $\partial\Omega$ being an unbounded Ahlfors regular set, and abbreviate $\sigma := \mathcal{H}^{n-1}\lfloor\partial\Omega$. Pick a function $u \in \mathscr{C}^1(\Omega)$ and assume that for some $p, q \in (1, \infty)$ and some aperture parameter $\kappa \in (0, \infty)$ one has*

$$\mathcal{N}_\kappa(\nabla u) \in \mathrm{A}_{p,q}(\partial\Omega, \sigma). \tag{14.110}$$

Then the nontangential trace $u\big|_{\partial\Omega}^{\kappa-\mathrm{n.t.}}$ exists σ-a.e. on $\partial\Omega$, and as a function it belongs to $\dot{\mathrm{A}}^1_{p,q}(\partial\Omega, \sigma)$ is a quantitative fashion, i.e., there exists some $C = C(p, q, \partial\Omega) \in (0, \infty)$, with the property that

$$\Big\| u\big|_{\partial\Omega}^{\kappa-\text{n.t.}} \Big\|_{\dot{\mathrm{A}}^1_{p,q}(\partial\Omega,\sigma)} \leq C\big\|\mathcal{N}_\kappa(\nabla u)\big\|_{\mathrm{A}_{p,q}(\partial\Omega,\sigma)}. \tag{14.111}$$

Proof To get started, observe that the membership displayed in (14.110), the fact that $\mathrm{A}_{p,q}(\partial\Omega,\sigma) \subseteq L^p_{\mathrm{loc}}(\partial\Omega,\sigma)$ (cf. item *(f)* in Proposition 10.3) along with [MMMi[3], Proposition 2.24, pp. 152–153] ensure that

$$\begin{aligned} &u\big|_{\partial\Omega}^{\kappa-\text{n.t.}} \text{ exists for } \sigma\text{-a.e. point on } \partial\Omega, \text{ and} \\ &\text{the function } u\big|_{\partial\Omega}^{\kappa-\text{n.t.}} \text{ is } \sigma\text{-measurable on } \partial\Omega. \end{aligned} \tag{14.112}$$

Moreover, having fixed a reference point $x_0 \in \partial\Omega$, we know from [MMMi[3], Proposition 2.24, pp. 152–153] that there exists an aperture parameter $\widetilde{\kappa} \in (0,\infty)$ along with some constant $C \in (0,\infty)$ independent of u, such that for every $\delta \in (0,\infty)$ one can find a compact set $K_\delta \subseteq \Omega$ with the property that

$$\mathcal{N}_\kappa^\delta u(x) \leq C\delta \cdot \mathcal{N}_{\widetilde{\kappa}}(\nabla u)(x) + \sup_{K_\delta}|u| \quad \text{for each } x \in \Delta(x_0,\delta). \tag{14.113}$$

Thanks to Proposition 14.3, and (14.110), we have $\mathcal{N}_{\widetilde{\kappa}}(\nabla u) \in \mathrm{A}_{p,q}(\partial\Omega,\sigma)$. This, (14.113), and item *(f)* in Proposition 10.3 imply that $\mathcal{N}_\kappa^\delta u$ belongs to $L^p_{\mathrm{loc}}(\partial\Omega,\sigma)$. In addition, Theorem 14.2 yields

$$\begin{aligned} &u\big|_{\partial\Omega}^{\kappa-\text{n.t.}} \in L^p_{\mathrm{loc}}(\partial\Omega,\sigma), \quad \partial_{\tau_{jk}}\Big(u\big|_{\partial\Omega}^{\kappa-\text{n.t.}}\Big) \in \mathrm{A}_{p,q}(\partial\Omega,\sigma) \ \text{ for all } j,k \in \{1,\dots,n\}, \\ &\qquad \text{and } \sum_{j,k=1}^n \Big\|\partial_{\tau_{jk}}\Big(u\big|_{\partial\Omega}^{\kappa-\text{n.t.}}\Big)\Big\|_{\mathrm{A}_{p,q}(\partial\Omega,\sigma)} \leq C\|\mathcal{N}_\kappa(\nabla u)\|_{\mathrm{A}_{p,q}(\partial\Omega,\sigma)}. \end{aligned} \tag{14.114}$$

Once the properties described in the previous paragraph have been obtained, we are left with proving the membership of $u\big|_{\partial\Omega}^{\kappa-\text{n.t.}}$ to $L^1\big(\partial\Omega, \frac{\sigma(x)}{1+|x|^n}\big)$. For simplicity, abbreviate

$$f := u\big|_{\partial\Omega}^{\kappa-\text{n.t.}} \quad \text{and} \quad g := \mathcal{N}_\kappa(\nabla u). \tag{14.115}$$

With these pieces of notation, in view of [MMMi[3], (2.610), p. 153] we have

$$|f(x) - f(y)| \leq C[g(x) + g(y)] \quad \text{for } \sigma\text{-a.e. } x,y \in \partial\Omega, \tag{14.116}$$

for some constant $C = C(\Omega,\kappa) \in (0,\infty)$.

Consider a dilation factor $2_* > (C_{\partial\Omega}/c_{\partial\Omega})^{1/(n-1)}$, where $C_{\partial\Omega}, c_{\partial\Omega} \in (0,\infty)$ denotes the upper and lower Ahlfors regularity constants of $\partial\Omega$. We may then follow the proof of Theorem 14.1 (see (14.52)-(14.55)) to obtain the following estimates

$$\sup_{\ell\in\mathbb{N}_0} \frac{1}{2_*^\ell}\Big(\int_{\Delta_\ell} |f - f_{\Delta_\ell}|^p \, d\sigma\Big)^{1/p} \leq C\|g\|_{\mathrm{A}_{p,q}(\partial\Omega,\sigma)} \ \text{ if } 1 < q \leq p, \tag{14.117}$$

and

$$\sup_{\ell\in\mathbb{N}_0}\frac{2_*^{\ell(n-1)(1/q-1/p)}}{2_*^{\ell}}\Big(\int_{\Delta_\ell}|f-f_{\Delta_\ell}|^p\,d\sigma\Big)^{1/p}\leq C\|g\|_{\mathrm{A}_{p,q}(\partial\Omega,\sigma)}\ \text{ if } 1<p<q, \tag{14.118}$$

for some constant $C = C(n, p, q, \partial\Omega) \in (0, \infty)$, where

$$\Delta_\ell := \Delta(x_0, 2_*^{\ell+1}) \quad\text{and}\quad f_{\Delta_\ell} := \fint_{\Delta_\ell} f\,d\sigma \quad\text{for each } \ell\in\mathbb{N}_0. \tag{14.119}$$

Collectively, (14.117)-(14.118), and Lemma 14.1 imply that $f \in L^1\big(\partial\Omega, \frac{\sigma(x)}{1+|x|^n}\big)$. From this, (14.114), and (14.115) we come to the conclusion that

$$\Big\|u\Big|_{\partial\Omega}^{\kappa-\mathrm{n.t.}}\Big\|_{\dot{\mathrm{A}}^1_{p,q}(\partial\Omega,\sigma)}\leq C\big\|\mathcal{N}_\kappa(\nabla u)\big\|_{\mathrm{A}_{p,q}(\partial\Omega,\sigma)}, \tag{14.120}$$

ultimately finishing the proof of Theorem 14.3. □

Chapter 15
Layer Potentials on Herz-type Spaces of First Generation, and Invertibility Results

In this chapter we shall specialize the Calderón-Zygmund theory developed in Chapter 13 to boundary layer potentials in UR domains associated with second-order, homogeneous, constant complex coefficient, weakly elliptic systems. This is done in the context of first-generation Herz spaces, Herz-type Hardy spaces, and spaces of (p, q)-bounded central mean oscillations. The invertibility results obtained here play a decisive role later on in establishing solvability results for boundary problems in the aforementioned function spaces.

15.1 Layer Potentials on Herz Spaces $\mathbf{A}_{p,q}$

We begin by recalling a result from [GHA-IV, Proposition 1.5.6, pp. 132–133] regarding the manner in which tangential derivatives commute with layer potentials.

Proposition 15.1 *Suppose $\Omega \subseteq \mathbb{R}^n$ is a* UR *domain. Abbreviate $\sigma := \mathcal{H}^{n-1}\lfloor\partial\Omega$ and denote by $\nu = (\nu_1, \dots, \nu_n)$ the geometric measure theoretic outward unit normal to Ω. Assume L is a homogeneous, second-order, constant complex coefficient, weakly elliptic $M \times M$ system in $\mathbb{R}^n$ and let $E = (E_{\alpha\beta})_{1\leq\alpha,\beta\leq M}$ be its canonical matrix-valued fundamental solution* (*cf. Proposition 7.1*)*. Also, pick a coefficient tensor*

$$A = \big(a_{jk}^{\alpha\beta}\big)_{\substack{1\leq\alpha,\beta\leq M\\ 1\leq j,k\leq n}} \in \mathfrak{A}_L \tag{15.1}$$

and bring in K_A (*cf.* (7.23))*, the boundary-to-boundary double layer potential operator associated with Ω and A. In addition, for each $j, k \in \{1, \dots, n\}$ define the singular integral operator U_{jk} acting on each given matrix-valued function $F = (F_{\alpha s})_{\substack{1\leq\alpha\leq M\\ 1\leq s\leq n}}$ with entries belonging to $L^1\big(\partial\Omega, \frac{\sigma(x)}{1+|x|^{n-1}}\big)$ according to $U_{jk}F = \big((U_{jk}F)_\gamma\big)_{1\leq\gamma\leq M}$ where, for each index $\gamma \in \{1, \dots, M\}$,*

© The Author(s), under exclusive license to Springer Nature Switzerland AG 2026
M. Mitrea and P. Takemura, *Singular Integrals, Herz-Type Function Spaces, and Boundary Problems*, Progress in Mathematics 362,
https://doi.org/10.1007/978-3-032-12516-3_15

$$
\begin{aligned}
&(U_{jk}F)_\gamma(x)\\
&:= -\lim_{\varepsilon\to 0^+}\int\limits_{\substack{y\in\partial\Omega\\|x-y|>\varepsilon}}[\nu_k(x)-\nu_k(y)]\nu_j(y)a^{\beta\alpha}_{rs}(\partial_r E_{\gamma\beta})(x-y)F_{\alpha s}(y)\,d\sigma(y)\\
&\quad+\lim_{\varepsilon\to 0^+}\int\limits_{\substack{y\in\partial\Omega\\|x-y|>\varepsilon}}[\nu_j(x)-\nu_j(y)]\nu_k(y)a^{\beta\alpha}_{rs}(\partial_r E_{\gamma\beta})(x-y)F_{\alpha s}(y)\,d\sigma(y)\\
&\quad+\lim_{\varepsilon\to 0^+}\int\limits_{\substack{y\in\partial\Omega\\|x-y|>\varepsilon}}[\nu_k(y)-\nu_k(x)]\nu_s(y)a^{\beta\alpha}_{rs}(\partial_r E_{\gamma\beta})(x-y)F_{\alpha j}(y)\,d\sigma(y)\\
&\quad-\lim_{\varepsilon\to 0^+}\int\limits_{\substack{y\in\partial\Omega\\|x-y|>\varepsilon}}[\nu_j(y)-\nu_j(x)]\nu_s(y)a^{\beta\alpha}_{rs}(\partial_r E_{\gamma\beta})(x-y)F_{\alpha k}(y)\,d\sigma(y)
\end{aligned}
\tag{15.2}
$$

at σ-a.e. point $x\in\partial\Omega$. Finally, fix some integrability exponents $p,q\in(1,\infty]$ and consider a function

$$
\begin{aligned}
f&\in\Big[L^1\big(\partial\Omega,\tfrac{\sigma(x)}{1+|x|^{n-1}}\big)\cap L^p_{\rm loc}(\partial\Omega,\sigma)\Big]^M \quad\textit{with the property that}\\
\partial_{\tau_{jk}}f&\in\Big[L^1\big(\partial\Omega,\tfrac{\sigma(x)}{1+|x|^{n-1}}\big)\cap L^q_{\rm loc}(\partial\Omega,\sigma)\Big]^M \quad\textit{for all } j,k\in\{1,\dots,n\}.
\end{aligned}
\tag{15.3}
$$

Then for each $j,k\in\{1,\dots,n\}$ one has

$$
\partial_{\tau_{jk}}(K_A f)=K_A(\partial_{\tau_{jk}}f)+U_{jk}(\nabla_{\tan}f)\quad\textit{at } \sigma\textit{-a.e. point on }\ \partial\Omega,
\tag{15.4}
$$

where $\nabla_{\tan}f$ is regarded as the $M\times n$ matrix-valued function $F=(F_{\alpha s})_{\substack{1\leq\alpha\leq M\\1\leq s\leq n}}$ whose entry $F_{\alpha s}$ is the s-th component of $\nabla_{\tan}f_\alpha$.

Our next result contains crucially important geometrically sensitive operator norm estimates for boundary layer potentials acting on first-generation Herz spaces. The reader is advised to recall the piece of notation introduced in (6.113).

Theorem 15.1 *Consider a* UR *domain $\Omega\subseteq\mathbb{R}^n$, such that $\partial\Omega$ is an unbounded set. Abbreviate $\sigma:=\mathcal{H}^{n-1}\lfloor\partial\Omega$, and denote by ν the geometric measure theoretic outward unit normal to Ω. Also, let L be a second-order, homogeneous, constant complex coefficient, weakly elliptic $M\times M$ system in $\mathbb{R}^n$ with the property that $\mathfrak{A}^{\rm dis}_L\neq\varnothing$. Pick a coefficient tensor $A\in\mathfrak{A}^{\rm dis}_L$, and recall the boundary-to-boundary double layer potentials K_A and $K^\#_A$ defined in* (7.23) *and* (7.24)*. Lastly, select two exponents $p,q\in(1,\infty)$.*

Then

$$
\begin{aligned}
K_A &: [\mathrm{A}_{p,q}(\partial\Omega,\sigma)]^M \to [\mathrm{A}_{p,q}(\partial\Omega,\sigma)]^M \\
K_A &: [\mathrm{A}^1_{p,q}(\partial\Omega,\sigma)]^M \to [\mathrm{A}^1_{p,q}(\partial\Omega,\sigma)]^M \\
K^{\#}_A &: [\mathrm{A}_{p,q}(\partial\Omega,\sigma)]^M \to [\mathrm{A}_{p,q}(\partial\Omega,\sigma)]^M
\end{aligned}
\tag{15.5}
$$

are well defined, linear, and bounded. In fact, for every $m \in \mathbb{N}$ there exists some $C_m \in (0,\infty)$ depending only on n, p, q, A, and the UR *constants of $\partial\Omega$ such that*

$$
\|K_A\|_{[\mathrm{A}_{p,q}(\partial\Omega,\sigma)]^M \to [\mathrm{A}_{p,q}(\partial\Omega,\sigma)]^M} \le C_m \cdot \|\nu\|^{\langle m\rangle}_{[\mathrm{BMO}(\partial\Omega,\sigma)]^n}, \tag{15.6}
$$

$$
\|K_A\|_{[\mathrm{A}^1_{p,q}(\partial\Omega,\sigma)]^M \to [\mathrm{A}^1_{p,q}(\partial\Omega,\sigma)]^M} \le C_m \cdot \|\nu\|^{\langle m\rangle}_{[\mathrm{BMO}(\partial\Omega,\sigma)]^n}, \tag{15.7}
$$

$$
\|K^{\#}_A\|_{[\mathrm{A}_{p,q}(\partial\Omega,\sigma)]^M \to [\mathrm{A}_{p,q}(\partial\Omega,\sigma)]^M} \le C_m \cdot \|\nu\|^{\langle m\rangle}_{[\mathrm{BMO}(\partial\Omega,\sigma)]^n}. \tag{15.8}
$$

Proof If A is a distinguished coefficient tensor for L, then the kernels of K_A and $K^{\#}_A$ have a chord-dot-normal structure (cf. Remark 7.1). As such, estimates (15.6) are implied by Theorem 13.2. As regards estimate (15.7), start by picking a vector-valued function $f \in [\mathrm{A}^1_{p,q}(\partial\Omega,\sigma)]^M$. Recall the operator U_{jk} defined in (15.2), which is a sum of commutators. Then, keeping in mind (15.4) from Proposition 15.1, Theorem 13.3, (15.6), (14.3), (14.5), and (14.11), we estimate

$$
\begin{aligned}
&\|K_A f\|_{[\mathrm{A}^1_{p,q}(\partial\Omega,\sigma)]^M} \\
&\quad = \|K_A f\|_{[\mathrm{A}_{p,q}(\partial\Omega,\sigma)]^M} + \sum_{j,k=1}^{n} \|\partial_{\tau_{jk}}(K_A f)\|_{[\mathrm{A}_{p,q}(\partial\Omega,\sigma)]^M} \\
&\quad \le \|K_A f\|_{[\mathrm{A}_{p,q}(\partial\Omega,\sigma)]^M} \\
&\qquad + \sum_{j,k=1}^{n} \big(\|K_A(\partial_{\tau_{jk}} f)\|_{[\mathrm{A}_{p,q}(\partial\Omega,\sigma)]^M} + \|U_{jk}(\nabla_{\tan} f)\|_{[\mathrm{A}_{p,q}(\partial\Omega,\sigma)]^M}\big) \\
&\quad \le C_m \|\nu\|^{\langle m\rangle}_{[\mathrm{BMO}(\partial\Omega,\sigma)]^n} \|f\|_{[\mathrm{A}_{p,q}(\partial\Omega,\sigma)]^M} \\
&\qquad + C_m \|\nu\|^{\langle m\rangle}_{[\mathrm{BMO}(\partial\Omega,\sigma)]^n} \sum_{j,k=1}^{n} \|\partial_{\tau_{jk}} f\|_{[\mathrm{A}_{p,q}(\partial\Omega,\sigma)]^M} \\
&\qquad\quad + C \|\nu\|_{[\mathrm{BMO}(\partial\Omega,\sigma)]^n} \|\nabla_{\tan} f\|_{[\mathrm{A}_{p,q}(\partial\Omega,\sigma)]^M} \\
&\quad \le C \|\nu\|^{\langle m\rangle}_{[\mathrm{BMO}(\partial\Omega,\sigma)]^n} \|f\|_{[\mathrm{A}^1_{p,q}(\partial\Omega,\sigma)]^M},
\end{aligned}
\tag{15.9}
$$

for some $C \in (0,\infty)$ depending only on m, n, p, q, A, and the UR constants of $\partial\Omega$. This concludes the proof of Theorem 15.1. □

In turn, the operator norm estimates obtained in Theorem 15.1 allow us to obtain the following fundamental invertibility result.

Theorem 15.2 *Assume $\Omega \subseteq \mathbb{R}^n$ is an Ahlfors regular domain. Set $\sigma := \mathcal{H}^{n-1}\lfloor\partial\Omega$ and denote by ν the geometric measure theoretic outward unit normal to Ω. Also,* *let L be a homogeneous, second-order, constant complex coefficient, weakly elliptic $M \times M$ system in $\mathbb{R}^n$, and assume that $\mathfrak{A}_L^{\rm dis} \neq \varnothing$. Pick $A \in \mathfrak{A}_L^{\rm dis}$ and consider the boundary-to-boundary double layer potential operators K_A and $K_A^{\#}$ associated with A as in (7.23) and (7.24). Lastly, fix two exponents $p, q \in (1, \infty)$, along with a parameter $\varepsilon \in (0, \infty)$.*

Then there exists some threshold $\delta \in (0, 1)$, depending only on n, p, q, A, ε, and the Ahlfors regularity constant of $\partial\Omega$, with the property that whenever

$$\|\nu\|_{[\mathrm{BMO}(\partial\Omega,\sigma)]^n} < \delta \tag{15.10}$$

it follows that for each spectral parameter $z \in \mathbb{C}$ with $|z| \geq \varepsilon$ the following operators are well defined, linear, bounded, and invertible:

$$\begin{aligned} zI + K_A &: [\mathrm{A}_{p,q}(\partial\Omega,\sigma)]^M \longrightarrow [\mathrm{A}_{p,q}(\partial\Omega,\sigma)]^M, \\ zI + K_A &: [\mathrm{A}^1_{p,q}(\partial\Omega,\sigma)]^M \longrightarrow [\mathrm{A}^1_{p,q}(\partial\Omega,\sigma)]^M, \\ zI + K_A^{\#} &: [\mathrm{A}_{p,q}(\partial\Omega,\sigma)]^M \longrightarrow [\mathrm{A}_{p,q}(\partial\Omega,\sigma)]^M. \end{aligned} \tag{15.11}$$

Proof The invertibility of the operators in (15.11) is obtained by following the same type of argument used to prove Theorem 8.3, now keeping in mind the boundedness result obtained in Theorem 15.1, and the fact that the spaces $\mathrm{A}_{p,q}(\partial\Omega,\sigma)$ and $\mathrm{A}^1_{p,q}(\partial\Omega,\sigma)$ are Banach (cf. Proposition 10.6 and Proposition 14.1). □

Remark 15.1 Estimates on Muckenhoupt weighted Lebesgue spaces for the tangential gradients of functions belonging to hybrid Lebesgue-based Sobolev spaces have been established in [MMMi[3], Remark 4.20, p. 315]. The argument presented there mostly relies on mapping and invertibility properties of layer potentials. Following a similar template, and availing ourselves of Theorem 15.1, Theorem 13.1, and Proposition 13.3 it follows that, in the context of Theorem 15.2, there exists some constant $C \in (0, \infty)$ with the property that

$$\begin{aligned} &\text{if } f \in [\mathrm{A}^1_{p,q}(\partial\Omega,\sigma)]^M \text{ and } g := (zI + K_A)^{-1} f \in [\mathrm{A}^1_{p,q}(\partial\Omega,\sigma)]^M \\ &\quad \text{then } \|\nabla_{\tan} g\|_{[\mathrm{A}^1_{p,q}(\partial\Omega,\sigma)]^{n\times M}} \leq C \|\nabla_{\tan} f\|_{[\mathrm{A}^1_{p,q}(\partial\Omega,\sigma)]^{n\times M}}. \end{aligned} \tag{15.12}$$

As a preamble to the invertibility result in Theorem 15.4, here we establish the boundedness of boundary layer potentials acting on Herz-based homogeneous Sobolev spaces.

Theorem 15.3 *Consider $\Omega \subseteq \mathbb{R}^n$ a two-sided* NTA *domain such that $\partial\Omega$ is an unbounded Ahlfors regular set, and abbreviate $\sigma := \mathcal{H}^{n-1}\lfloor\partial\Omega$. Assume L is a second-order, homogeneous, constant complex coefficient, weakly elliptic $M \times M$ system in $\mathbb{R}^n$. Select a coefficient tensor $A \in \mathfrak{A}_L$ and recall the boundary-to-domain modified double layer potential $K_{A,\mathrm{mod}}$ associated with A as in (7.29).*

Then for all exponents $p, q \in (1, \infty)$ the operator

$$
\begin{gathered}
K_{A,\text{mod}} : \big[\dot{\mathrm{A}}^1_{p,q}(\partial\Omega,\sigma)\big]^M \to \big[\dot{\mathrm{A}}^1_{p,q}(\partial\Omega,\sigma)\big]^M \\
\textit{is well defined, linear, and bounded.}
\end{gathered}
\tag{15.13}
$$

Moreover, at the level of equivalence classes modulo constants, it follows that

$$
\begin{gathered}
[K_{A,\text{mod}}] : \big[\dot{\mathrm{A}}^1_{p,q}(\partial\Omega,\sigma)/\sim\big]^M \to \big[\dot{\mathrm{A}}^1_{p,q}(\partial\Omega,\sigma)/\sim\big]^M \ \textit{given by} \\
[K_{A,\text{mod}}][f] := [K_{A,\text{mod}}f] \textit{ for all } f \in \big[\dot{\mathrm{A}}^1_{p,q}(\partial\Omega,\sigma)\big]^M \\
\textit{is well defined, linear, and bounded.}
\end{gathered}
\tag{15.14}
$$

Proof The assumption that Ω is an NTA domain with $\partial\Omega$ an Ahlfors regular set guarantees that Ω is a UR domain (cf. [MMMi[3], (2.48), p. 38]). Fix an arbitrary vector-valued function $f \in \big[\dot{\mathrm{A}}^1_{p,q}(\partial\Omega,\sigma)\big]^M$ and define

$$
u := \mathcal{D}_{A,\text{mod}}f \ \text{ in } \Omega. \tag{15.15}
$$

In light of [MMMi[3], Theorem 3.2, p. 173], this function u enjoys the following properties:

$$
u \in [\mathscr{C}^\infty(\mathbb{R}^n)]^M \ \text{ and } \ u\big|^{\kappa-\text{n.t.}}_{\partial\Omega} = (\tfrac{1}{2}I + K_{A,\text{mod}})f \ \text{ pointwise } \sigma\text{-a.e. on } \partial\Omega. \tag{15.16}
$$

Also, from [MMMi[3], (3.57), p. 175] together with the nontangential maximal function estimate (13.7) recorded in Theorem 13.1 we see that

$$
\mathcal{N}_\kappa(\nabla u) \in \mathrm{A}_{p,q}(\partial\Omega,\sigma) \ \text{ and } \ \|\mathcal{N}_\kappa(\nabla u)\|_{\mathrm{A}_{p,q}(\partial\Omega,\sigma)} \le C\|f\|_{[\dot{\mathrm{A}}^1_{p,q}(\partial\Omega,\sigma)]^M} \tag{15.17}
$$

for some constant $C \in (0,\infty)$ independent of u. This and Theorem 14.3 (see in particular (14.111)) further imply

$$
u\big|^{\kappa-\text{n.t.}}_{\partial\Omega} \in \big[\dot{\mathrm{A}}^1_{p,q}(\partial\Omega,\sigma)\big]^M \ \text{ and } \ \Big\|u\big|^{\kappa-\text{n.t.}}_{\partial\Omega}\Big\|_{[\dot{\mathrm{A}}^1_{p,q}(\partial\Omega,\sigma)]^M} \le C\|f\|_{[\dot{\mathrm{A}}^1_{p,q}(\partial\Omega,\sigma)]^M}. \tag{15.18}
$$

From (15.18) and (15.16) we therefore deduce that

$$
K_{A,\text{mod}}f = u\big|^{\kappa-\text{n.t.}}_{\partial\Omega} - \tfrac{1}{2}f \ \text{ belongs to } \ [\dot{\mathrm{A}}^1_{p,q}(\partial\Omega,\sigma)]^M \tag{15.19}
$$

and

$$
\begin{aligned}
\|K_{A,\text{mod}}f\|_{[\dot{\mathrm{A}}^1_{p,q}(\partial\Omega,\sigma)]^M} &\le \Big\|u\big|^{\kappa-\text{n.t.}}_{\partial\Omega}\Big\|_{[\dot{\mathrm{A}}^1_{p,q}(\partial\Omega,\sigma)]^M} + \tfrac{1}{2}\|f\|_{[\dot{\mathrm{A}}^1_{p,q}(\partial\Omega,\sigma)]^M} \\
&\le C\|f\|_{[\dot{\mathrm{A}}^1_{p,q}(\partial\Omega,\sigma)]^M}
\end{aligned}
\tag{15.20}
$$

for some $C \in (0,\infty)$ independent of f. This concludes the proof of (15.13). The very last claim made in (15.14) is a consequence of (15.13) and the fact that $K_{A,\text{mod}}$ maps $\mathbb{C}^M$-constants into $\mathbb{C}^M$-constants (see [MMMi[3], (3.62), p. 176]). □

In the following result, we specialize the boundedness results obtained in (15.14) for modified double layers associated with weakly elliptic systems which have a distinguished coefficient tensor. Crucially in this case we obtain operator norm estimates that are sensitive to the geometry of the underlying domain. The reader is advised to recall the piece of notation defined in (6.113).

Theorem 15.4 *Let $\Omega \subseteq \mathbb{R}^n$ be a two-sided* NTA *domain, and suppose $\partial\Omega$ an unbounded Ahlfors regular set. Abbreviate $\sigma := \mathcal{H}^{n-1}\lfloor\partial\Omega$, and denote by ν the geometric measure theoretic outward unit normal to Ω. Consider L a second-order, homogeneous, constant complex coefficient, weakly elliptic $M \times M$ system in $\mathbb{R}^n$. Assume $\mathfrak{A}_L^{\mathrm{dis}} \neq \varnothing$ and select $A \in \mathfrak{A}_L^{\mathrm{dis}}$. Fix two exponents $p, q \in (1,\infty)$ and recall the operator $[K_{A,\mathrm{mod}}]$ associated with A as in* (15.14)*.*

Then for each $m \in \mathbb{N}$, there exists some $C_m \in (0,\infty)$ depending on n, m, p, q, A, and the Ahlfors regularity constants of $\partial\Omega$, with the property that

$$\big\|[K_{A,\mathrm{mod}}]\big\|_{[\dot{\mathrm{A}}^1_{p,q}(\partial\Omega,\sigma)/\sim]^M\to[\dot{\mathrm{A}}^1_{p,q}(\partial\Omega,\sigma)/\sim]^M} \leq C_m\|\nu\|^{\langle m\rangle}_{[\mathrm{BMO}(\partial\Omega,\sigma)]^n}. \tag{15.21}$$

Proof First, recall from (2.18), (2.23), and (2.24) that Ω being a two-sided NTA domain implies that Ω satisfies a two-sided local John condition, which further implies that Ω satisfies a two-sided corkscrew condition. Since $\partial\Omega$ is an Ahlfors regular set, we conclude that Ω is a UR domain.

Fix a function $f \in \big[\dot{\mathrm{A}}^1_{p,q}(\partial\Omega,\sigma)/\sim\big]^M$. Then based on Theorem 15.3, (15.4) in Proposition 15.1, Theorem 13.3, (15.6), (14.3), (14.5), and (14.11), we may estimate

$$\begin{aligned}
\big\|[K_{A,\mathrm{mod}}][f]\big\|_{[\dot{\mathrm{A}}^1_{p,q}(\partial\Omega,\sigma)/\sim]^M} &= \big\|[K_{A,\mathrm{mod}}f]\big\|_{[\dot{\mathrm{A}}^1_{p,q}(\partial\Omega,\sigma)]^M}\\
&= \sum_{j,k=1}^n \|\partial_{\tau_{jk}}(K_{A,\mathrm{mod}}f)\|_{[\mathrm{A}_{p,q}(\partial\Omega,\sigma)]^M}\\
&\leq \sum_{j,k=1}^n \|K_A(\partial_{\tau_{jk}}f)\|_{[\mathrm{A}_{p,q}(\partial\Omega,\sigma)]^M}\\
&\quad + \sum_{j,k=1}^n \|U_{jk}(\nabla_{\tan}f)\|_{[\mathrm{A}_{p,q}(\partial\Omega,\sigma)]^M}\\
&\leq C_m\|\nu\|^{\langle m\rangle}_{[\mathrm{BMO}(\partial\Omega,\sigma)]^n}\sum_{j,k=1}^n \|\partial_{\tau_{jk}}f\|_{[\mathrm{A}_{p,q}(\partial\Omega,\sigma)]^M}\\
&\quad + C\|\nu\|_{[\mathrm{BMO}(\partial\Omega,\sigma)]^n}\|\nabla_{\tan}f\|_{[\mathrm{A}_{p,q}(\partial\Omega,\sigma)]^M}\\
&\leq C_m\|\nu\|^{\langle m\rangle}_{[\mathrm{BMO}(\partial\Omega,\sigma)]^n}\big\|[f]\big\|_{[\dot{\mathrm{A}}^1_{p,q}(\partial\Omega,\sigma)/\sim]^M},
\end{aligned} \tag{15.22}$$

proving (15.21). □

The operator norm estimate of the modified boundary-to-boundary double layer potential obtained in Theorem 15.4 permits us to deduce the following basic invertibility result.

Theorem 15.5 *Let $\Omega \subseteq \mathbb{R}^n$ be an Ahlfors regular domain. Set $\sigma := \mathcal{H}^{n-1}\lfloor\partial\Omega$ and denote by ν the geometric measure theoretic outward unit normal to Ω. Also, let L be a homogeneous, second-order, constant complex coefficient, weakly elliptic $M \times M$ system in $\mathbb{R}^n$, and assume that $\mathfrak{A}_L^{\rm dis} \neq \varnothing$. Fix two exponents $p, q \in (1, \infty)$. Pick $A \in \mathfrak{A}_L^{\rm dis}$ and recall the operator $[K_{A,\rm mod}]$ associated with A as in (15.14). Also, select a parameter $\varepsilon \in (0, \infty)$.*

Then there exists a threshold $\delta \in (0, 1)$ depending exclusively on n, p, q, A, ε, and the Ahlfors regularity constants of $\partial\Omega$, with the property that whenever

$$\|\nu\|_{[\mathrm{BMO}(\partial\Omega,\sigma)]^n} < \delta \tag{15.23}$$

it follows that for each spectral parameter $z \in \mathbb{C}$ with $|z| \geq \varepsilon$ the operator

$$\begin{gathered} zI + [K_{A,\rm mod}] : [\dot{\mathrm{A}}^1_{p,q}(\partial\Omega,\sigma)/\sim]^M \longrightarrow [\dot{\mathrm{A}}^1_{p,q}(\partial\Omega,\sigma)/\sim]^M \\ \text{is well defined, linear, bounded, and invertible.} \end{gathered} \tag{15.24}$$

Proof Thanks to [MMMi[3], Theorem 2.3, and Theorem 2.4, pp. 64–65], there exists a sufficiently small threshold $\delta \in (0, 1)$ for which whenever $\Omega \subseteq \mathbb{R}^n$ is an Ahlfors regular domain satisfying (15.23), it follows that Ω is a two-sided NTA domain with unbounded topological boundary. In such a scenario, Ω is also a UR domain with UR constants solely controlled in terms of n and the Ahlfors regularity constants of $\partial\Omega$. As such, in light of (15.21), the invertibility result claimed in (15.24) is obtained via a Neumann series argument (cf. Lemma 8.1), much as in the proof of Theorem 8.2, while keeping in mind that $\dot{\mathrm{A}}^1_{p,q}(\partial\Omega,\sigma)/\sim$ is a Banach space (cf. Proposition 14.2). □

15.2 Layer Potentials on Herz-type Hardy Spaces $\mathrm{HA}_{p,q}$

Here we obtain operator norm estimates, which are sensitive to the geometry of the underlying domain, for transpose double layer potentials in the context of first-generation Herz-type Hardy spaces. More precisely, we have the following result, which plays an important role in Theorem 15.7.

Theorem 15.6 *Let $\Omega \subseteq \mathbb{R}^n$ be a* UR *domain such that $\partial\Omega$ is an unbounded set. Abbreviate $\sigma := \mathcal{H}^{n-1}\lfloor\partial\Omega$ and denote by ν the geometric measure theoretic outward unit normal to Ω. Consider L a second-order, homogeneous, constant complex coefficient, weakly elliptic $M \times M$ system in $\mathbb{R}^n$ for which $\mathfrak{A}_L^{\rm dis} \neq \varnothing$. Pick $A \in \mathfrak{A}_L^{\rm dis}$ and consider the boundary-to-boundary double layer potential $K_A^{\#}$ associated with*

Ω and the coefficient tensor A as in (7.24). Lastly, fix two exponents $p \in (1, \infty)$ and $q \in \big(\frac{n-1}{n}, 1\big]$.

Then the operator

$$
\begin{gathered}
K_A^{\#} : [\mathrm{HA}_{p,q}(\partial\Omega, \sigma)]^M \to [\mathrm{HA}_{p,q}(\partial\Omega, \sigma)]^M \\
\text{is well defined, linear, and bounded,}
\end{gathered}
\tag{15.25}
$$

and for each $m \in \mathbb{N}$, there exists some $C_m \in (0, \infty)$ depending only on m, n, A, p, q, and the UR *constants of Ω, with the property that*

$$
\|K_A^{\#}\|_{[\mathrm{HA}_{p,q}(\partial\Omega,\sigma)]^M \to [\mathrm{HA}_{p,q}(\partial\Omega,\sigma)]^M} \leq C_m \cdot \|\nu\|^{\langle m \rangle}_{[\mathrm{BMO}(\partial\Omega,\sigma)]^n}. \tag{15.26}
$$

Proof The assumption $A \in \mathfrak{A}_L^{\mathrm{dis}}$ ensures that the integral kernel of the layer potential $K_A^{\#}$ is of chord-dot-normal type (cf. Proposition 7.2, and (7.48)). As such, the claims made in (15.25)-(15.26) become direct consequences of Theorem 13.7. □

As a consequence of Theorem 15.6 we now establish a key invertibility result for double layer potentials associated with weakly elliptic systems that possess a distinguished coefficient tensor.

Theorem 15.7 *Assume $\Omega \subseteq \mathbb{R}^n$ is an Ahlfors regular domain. Set $\sigma := \mathcal{H}^{n-1} \lfloor \partial\Omega$ and denote by ν the geometric measure theoretic outward unit normal to Ω. Also, let L be a homogeneous, second-order, constant complex coefficient, weakly elliptic $M \times M$ system in $\mathbb{R}^n$, and suppose that $\mathfrak{A}_L^{\mathrm{dis}} \neq \varnothing$. Pick $A \in \mathfrak{A}_L^{\mathrm{dis}}$ and recall the boundary-to-boundary double layer potential operator $K_A^{\#}$ associated with A as in (7.24). Finally, fix two exponents $p \in (1, \infty)$ and $q \in \big(\frac{n-1}{n}, 1\big]$, along with a parameter $\varepsilon \in (0, \infty)$.*

Then there exists some threshold $\delta \in (0, 1)$, depending only on n, p, q, A, ε, and the Ahlfors regularity constant of $\partial\Omega$, with the property that whenever

$$
\|\nu\|_{[\mathrm{BMO}(\partial\Omega,\sigma)]^n} < \delta \tag{15.27}
$$

it follows that for each spectral parameter $z \in \mathbb{C}$ with $|z| \geq \varepsilon$ the operator

$$
\begin{gathered}
zI + K_A^{\#} : [\mathrm{HA}_{p,q}(\partial\Omega, \sigma)]^M \longrightarrow [\mathrm{HA}_{p,q}(\partial\Omega, \sigma)]^M \\
\text{is well defined, linear, bounded, and invertible.}
\end{gathered}
\tag{15.28}
$$

Proof Thanks to Theorem 15.6 and the completeness result recorded in Proposition 11.4, the invertibility result claimed in (15.28) is obtained via a Neumann series argument (cf. Lemma 8.1), using the same type of reasoning as the one presented in the proof of Theorem 8.2. □

15.3 Layer Potentials on $\mathrm{BCMO}_{p,q}$

The boundedness result in Theorem 13.11 permits us to deduce geometrically operator norm estimates for modified boundary-to-boundary layer potentials associated with systems that possess distinguished coefficient tensors.

Theorem 15.8 *Let $\Omega \subseteq \mathbb{R}^n$ be a* UR *domain such that $\partial\Omega$ is an unbounded set. Abbreviate $\sigma := \mathcal{H}^{n-1}\lfloor\partial\Omega$ and denote by ν the geometric measure theoretic outward unit normal to Ω. Consider L a second-order, homogeneous, constant complex coefficient, weakly elliptic $M \times M$ system in $\mathbb{R}^n$ for which $\mathfrak{A}_L^{\mathrm{dis}} \neq \varnothing$. Pick $A \in \mathfrak{A}_L^{\mathrm{dis}}$ and consider the modified boundary-to-boundary double layer potential operators $K_{A,\mathrm{mod}}$ associated with A as in* (7.29).

Then

$$\begin{gathered} K_{A,\mathrm{mod}} : [\mathrm{BCMO}_{p,q}(\partial\Omega,\sigma)]^M \to [\mathrm{BCMO}_{p,q}(\partial\Omega,\sigma)]^M \\ \text{is a well-defined, linear, and bounded operator.} \end{gathered} \tag{15.29}$$

In fact, for each $m \in \mathbb{N}$ there exists some $C_m \in (0,\infty)$ depending only on m, n, A, p, q, and the UR *constants of Ω with the property that*

$$\|K_{A,\mathrm{mod}}\|_{[\mathrm{BCMO}_{p,q}(\partial\Omega,\sigma)]^M \to [\mathrm{BCMO}_{p,q}(\partial\Omega,\sigma)]^M} \leq C_m \cdot \|\nu\|^{\langle m\rangle}_{[\mathrm{BMO}(\partial\Omega,\sigma)]^n}. \tag{15.30}$$

Moreover, with $[\,\cdot\,]$ denoting the equivalence class modulo constants in the space $\mathrm{BCMO}_{p,q}(\partial\Omega,\sigma)$, it follows that the operator

$$\begin{gathered} [K_{A,\mathrm{mod}}] : \widetilde{\mathrm{BCMO}}_{p,q}(\partial\Omega,\sigma) \to \widetilde{\mathrm{BCMO}}_{p,q}(\partial\Omega,\sigma) \ \text{ given by} \\ [K_{A,\mathrm{mod}}][f] := [K_{A,\mathrm{mod}} f] \text{ for all } f \in \mathrm{BCMO}_{p,q}(\partial\Omega,\sigma) \\ \text{is meaningfully defined, linear, and bounded,} \end{gathered} \tag{15.31}$$

and, with $C_m \in (0,\infty)$ of same nature as before, one also has

$$\begin{aligned} \big\|[K_{A,\mathrm{mod}}]\big\|&_{[\widetilde{\mathrm{BCMO}}_{p,q}(\partial\Omega,\sigma)]^M \to [\widetilde{\mathrm{BCMO}}_{p,q}(\partial\Omega,\sigma)]^M} \\ &\leq C_m \cdot \|\nu\|^{\langle m\rangle}_{[\mathrm{BMO}(\partial\Omega,\sigma)]^n}. \end{aligned} \tag{15.32}$$

Proof Since $A \in \mathfrak{A}_L^{\mathrm{dis}}$, the operator $K_{A,\mathrm{mod}}$ may be thought as a $M \times M$ matrix operator whose entries are modified chord-dot-normal singular integral operators (as in (6.201)) Granted this, Theorem 13.11 readily implies (15.30), (15.31), and (15.32). □

The nature of the estimate proved in (15.32) opens the door for obtaining the following invertibility result on the space of (p,q)-bounded central mean oscillations functions.

Theorem 15.9 *Assume $\Omega \subseteq \mathbb{R}^n$ is an Ahlfors regular domain. Set $\sigma := \mathcal{H}^{n-1}\lfloor\partial\Omega$ and denote by ν the geometric measure theoretic outward unit normal to Ω. Also, let L be a homogeneous, second-order, constant complex coefficient, weakly elliptic $M \times M$ system in $\mathbb{R}^n$, and suppose that $\mathfrak{A}_L^{\rm dis} \neq \emptyset$. Pick $A \in \mathfrak{A}_L^{\rm dis}$ and consider the modified boundary-to-boundary double layer potential operators $K_{A,\rm mod}$ associated with A as in (7.29). Finally, fix two exponents $p \in (1, \infty)$ and $q \in \big(\frac{n-1}{n}, 1\big]$, along with a parameter $\varepsilon \in (0, \infty)$.*

Then there exists some threshold $\delta \in (0, 1)$, depending only on n, p, q, A, ε, and the Ahlfors regularity constant of $\partial\Omega$, with the property that whenever

$$\|\nu\|_{[\mathrm{BMO}(\partial\Omega,\sigma)]^n} < \delta \tag{15.33}$$

it follows that for each spectral parameter $z \in \mathbb{C}$ with $|z| \geq \varepsilon$ the operator

$$zI + [K_{A,\rm mod}] : \big[\widetilde{\mathrm{BCMO}}_{p,q}(\partial\Omega, \sigma)\big]^M \longrightarrow \big[\widetilde{\mathrm{BCMO}}_{p,q}(\partial\Omega, \sigma)\big]^M \tag{15.34}$$

is well defined, linear, bounded, and invertible.

Proof A combination of the argument presented in the proof of Theorem 8.2, along with Theorem 15.8 and the completeness of $\widetilde{\mathrm{BCMO}}_{p,q}(\partial\Omega, \sigma)$ (cf. Corollary 12.1), implies that the operator in (15.34) is invertible via a Neumann series. □

Remark 15.2 We conclude this chapter by observing that a similar interpretation to the one detailed in Remark 8.1, regarding the timing of the choices of thresholds, applies to the statements of Theorem 15.2, Theorem 15.5, Theorem 15.7, and Theorem 15.9.

Chapter 16

Boundary Problems on First-Generation Herz-type Spaces

The machinery developed in Chapters 13, 14, and 15 paves the way for solving boundary value problems in Ahlfors regular domains for weakly elliptic systems, and boundary condition formulated in terms of first-generation Herz-type spaces. Here we obtain solvability and well-posedness results for the Dirichlet, Neumann, Regularity, and Transmission Problems with boundary data in $\mathrm{A}_{p,q}$, the Neumann Problem with boundary data in $\mathrm{HA}_{p,q}$, and the Dirichlet Problem with boundary data in $\mathrm{BCMO}_{p,q}$.

To set the stage, we recall Theorem 16.1 below from [GHA-III, Theorem 4.4.1, pp. 799–801] regarding Poisson integral representation formulas. Stating such a result requires the following definition.

Definition 16.1 A set $\Omega \subseteq \mathbb{R}^n$ is called `globally pathwise nontangentially accessible` provided Ω is an open, nonempty, proper subset of $\mathbb{R}^n$ such that:

$$\begin{array}{l}\text{given any } \kappa > 0 \text{ there exist } \widetilde{\kappa} \geq \kappa \text{ along with } c \in [1, \infty) \text{ such that} \\ \sigma\text{-a.e. point } x \in \partial\Omega \text{ has the property that any } y \in \Gamma_\kappa(x) \text{ may be} \\ \text{joined by a rectifiable curve } \gamma_{x,y} \text{ such that } \gamma_{x,y} \setminus \{x\} \subseteq \Gamma_{\widetilde{\kappa}}(x) \\ \text{and whose length is } \leq c|x-y|.\end{array} \tag{16.1}$$

It is relevant to highlight that (cf. [GHA-I, (8.9.180), p. 814])

$$\text{any one-sided NTA domain with unbounded boundary is a globally pathwise nontangentially accessible set.} \tag{16.2}$$

We are now ready to state the Poisson integral formulas mentioned earlier.

Theorem 16.1 *Let Ω (with $n \in \mathbb{N}$ and $n \geq 2$) be an open nonempty proper subset of $\mathbb{R}^n$ which is globally nontangentially accessible, and such that $\partial\Omega$ is unbounded and Ahlfors regular. Abbreviate $\sigma := \mathcal{H}^{n-1}\lfloor\partial\Omega$ and denote by $\nu = (\nu_1, \dots, \nu_n)$ the geometric measure theoretic outward unit normal to Ω. Next, suppose L is a weakly elliptic, homogeneous, constant complex coefficient, second-order, $M \times M$ system in*

© The Author(s), under exclusive license to Springer Nature Switzerland AG 2026
M. Mitrea and P. Takemura, *Singular Integrals, Herz-Type Function Spaces, and Boundary Problems*, Progress in Mathematics 362,
https://doi.org/10.1007/978-3-032-12516-3_16

$\mathbb{R}^n$. Fix an aperture parameter $\kappa \in (0, \infty)$, along with an arbitrary point $x_\diamond \in \Omega$, and suppose $0 < \rho < \frac{1}{4}\operatorname{dist}(x_\diamond, \partial\Omega)$. Finally, define $K := \overline{B(x_\diamond, \rho)}$.

Then there exists some $\widetilde{\kappa} > 0$, which depends only on Ω and κ, with the following significance. Assume G is a matrix-valued function satisfying

$$\begin{cases} G^{x_\diamond} = \big(G^{x_\diamond}_{\alpha\beta}\big)_{1\leq\alpha,\beta\leq M} \in \big[L^1_{\rm loc}(\Omega, \mathcal{L}^n)\big]^{M\times M}, \\ \big(L^\top G^{x_\diamond}_{\bullet\beta}\big)_\alpha = -\delta_{x_\diamond}\delta_{\alpha\beta} \ \textit{in} \ \big[\mathcal{D}'(\Omega)\big]^M \ \textit{for all} \ \alpha, \beta \in \{1, \dots, M\}, \\ \big(\nabla G^{x_\diamond}\big)\Big|^{\widetilde{\kappa}-{\rm n.t.}}_{\partial\Omega} \ \textit{exists (in } \mathbb{C}^{n\cdot M^2}\textit{) at } \sigma\textit{-a.e. point on } \partial\Omega, \\ G^{x_\diamond}\Big|^{\widetilde{\kappa}-{\rm n.t.}}_{\partial\Omega} = 0 \in \mathbb{C}^{M\times M} \ \textit{at } \sigma\textit{-a.e. point on } \partial\Omega, \\ \mathcal{N}^{\Omega\setminus K}_{\widetilde{\kappa}}(\nabla G^{x_\diamond}) < \infty \ \textit{at } \sigma\textit{-a.e. point on } \partial\Omega \end{cases} \tag{16.3}$$

(where $G^{x_\diamond}_{\bullet\beta}$ is the β-th column of the $M \times M$ matrix G), and assume $u = (u_\beta)_{1\leq\beta\leq M}$ is a $\mathbb{C}^M$-valued function in Ω satisfying

$$\begin{cases} u \in \big[\mathscr{C}^\infty(\Omega)\big]^M, \quad Lu = 0 \ \textit{in} \ \Omega, \\ u\Big|^{\kappa-{\rm n.t.}}_{\partial\Omega} \ \textit{exists at } \sigma\textit{-a.e. point on } \partial\Omega, \\ \mathcal{N}_\kappa u < \infty \ \textit{at } \sigma\textit{-a.e. point on } \partial\Omega, \\ \displaystyle\int_{\partial\Omega} \mathcal{N}_\kappa u \cdot \mathcal{N}^{\Omega\setminus K}_{\widetilde{\kappa}}(\nabla G^{x_\diamond})\, d\sigma < \infty. \end{cases} \tag{16.4}$$

Then for any choice of a coefficient tensor $A = \big(a^{\alpha\beta}_{rs}\big)_{\substack{1\leq r,s\leq n \\ 1\leq\alpha,\beta\leq M}} \in \mathfrak{A}_L$ one has the Poisson integral representation formula

$$u(x_\diamond) = \left(-\int_{\partial_*\Omega} \Big\langle u\Big|^{\kappa-{\rm n.t.}}_{\partial\Omega}, \partial^{A^\top}_\nu G^{x_\diamond}_{\bullet\beta}\Big\rangle\, d\sigma \right)_{1\leq\beta\leq M} \tag{16.5}$$

where $\partial^{A^\top}_\nu$ stands for the conormal derivative associated with $A^\top$, acting on the columns of the matrix-valued function G according to

$$\partial^{A^\top}_\nu G^{x_\diamond}_{\bullet\beta} := \Big(\nu_r a^{\gamma\alpha}_{sr}\big(\partial_s G^{x_\diamond}_{\gamma\beta}\big)\Big|^{\widetilde{\kappa}-{\rm n.t.}}_{\partial\Omega}\Big)_{1\leq\alpha\leq M} \quad \textit{at } \sigma\textit{-a.e. point on } \partial_*\Omega, \tag{16.6}$$

for each $\beta \in \{1, \dots, M\}$.

A fundamental property of boundary-to-domain double layer potentials is recorded below (cf. [GHA-IV, Lemma 1.3.2, pp. 29–30] for a proof).

Proposition 16.1 *Let $\Omega \subseteq \mathbb{R}^n$ be an Ahlfors regular domain. Define $\sigma := \mathcal{H}^{n-1}\lfloor\partial\Omega$, and denote by $\nu = (\nu_1, \dots, \nu_n)$ the geometric measure theoretic outward unit normal to Ω. Also, for some $M \in \mathbb{N}$, consider a weakly elliptic, homogeneous, constant* *(complex) coefficient, second-order, $M \times M$ system L in $\mathbb{R}^n$, written for some choice of a coefficient tensor $A = \big(a_{rs}^{\alpha\beta}\big)_{\substack{1\le r,s\le n\\ 1\le\alpha,\beta\le M}}$. Finally, associate with A and Ω the double layer potential operator $\mathcal{D}_A$ as in* (7.21), *and consider a function*

$$f = (f_\alpha)_{1\le\alpha\le M} \in \Big[L^1\Big(\partial\Omega, \frac{\sigma(x)}{1+|x|^{n-1}}\Big)\Big]^M \text{ with the property that}$$

$$\partial_{\tau_{jk}} f_\alpha \in L^1\Big(\partial\Omega, \frac{\sigma(x)}{1+|x|^{n-1}}\Big) \text{ for } j,k \in \{1,\dots,n\} \text{ and } \alpha \in \{1,\dots,M\}. \tag{16.7}$$

Then for each index $\ell \in \{1,\dots,n\}$ and each point $x \in \Omega$ one has

$$\partial_\ell(\mathcal{D}_A f)(x) = \Bigg(\int_{\partial\Omega} a_{rs}^{\beta\alpha}(\partial_r E_{\gamma\beta})(x-y)(\partial_{\tau_{\ell s}} f_\alpha)(y)\, d\sigma(y)\Bigg)_{1\le\gamma\le M}. \tag{16.8}$$

In particular, for each aperture parameter $\kappa > 0$, the nontangential boundary trace

$$\big(\nabla\mathcal{D}_A f\big)\Big|_{\partial\Omega}^{\kappa-\mathrm{n.t.}} \text{ exists (in } \mathbb{C}^{n\times M}\text{) at } \sigma\text{-a.e. point on } \partial\Omega. \tag{16.9}$$

Let us prove an auxiliary lemma regarding estimates for fundamental solutions of second-order weakly elliptic systems.

Lemma 16.1 *Consider $a, b \in \mathbb{R}^n$, with $a \neq b$, and select a radius $r \in (0, |a-b|/2)$. Let L be a second-order, homogeneous, constant complex coefficient, weakly elliptic $M \times M$ system in $\mathbb{R}^n$, and denote by E its canonical matrix-valued fundamental solution (in the sense of Theorem 7.1). Then for every multi-index $\alpha \in \mathbb{N}_0^n$ there exists a constant $C \in (0,\infty)$ depending only on a, b, r, n, α, and L with the property that for every $x \in \mathbb{R}^n \setminus \big(B(a,r) \cup B(b,r)\big)$ one has*

$$|\partial^\alpha E(x-a) - \partial^\alpha E(x-b)| \le \frac{C}{1+|x|^{n-1+|\alpha|}}. \tag{16.10}$$

Proof This is a straightforward consequence of the Mean Value Theorem and the estimates for the fundamental solution from Theorem 7.1. □

16.1 The Dirichlet Problem with Boundary Data in $\mathrm{A}_{p,q}$

We are now ready to state and prove the well-posedness of the Dirichlet Problem for weakly elliptic systems on Ahlfors regular domains and boundary data taken from first-generation Herz spaces.

Theorem 16.2 *Consider an Ahlfors regular domain $\Omega \subseteq \mathbb{R}^n$, set $\sigma := \mathcal{H}^{n-1}\lfloor\partial\Omega$, and denote by ν its geometric measure theoretic outward unit normal. Let L be a second-order, homogeneous, constant complex coefficient, weakly elliptic $M \times M$ system in $\mathbb{R}^n$, for some $M \in \mathbb{N}$. Select two exponents $p, q \in (1, \infty)$ along with an aperture parameter $\kappa \in (0, \infty)$. In this context, formulate the Dirichlet Problem for L in Ω with boundary data in $[\mathrm{A}_{p,q}(\partial\Omega, \sigma)]^M$ as follows:*

$$\begin{cases} u \in [\mathscr{C}^\infty(\Omega)]^M, \\ Lu = 0 \ \ in\ \Omega, \\ \mathcal{N}_\kappa u \in \mathrm{A}_{p,q}(\partial\Omega, \sigma), \\ u\big|_{\partial\Omega}^{\kappa-\mathrm{n.t.}} = f \in [\mathrm{A}_{p,q}(\partial\Omega, \sigma)]^M. \end{cases} \tag{16.11}$$

Then, in relation to this, the following statements are true:

(a) [Existence, Integral Representation, and Estimates] *If $\mathfrak{A}_L^{\mathrm{dis}} \neq \varnothing$ then there exists a threshold $\delta \in (0, 1)$ depending on n, M, p, q, L and the Ahlfors regularity constants of $\partial\Omega$ with the property that if*

$$\|\nu\|_{[\mathrm{BMO}(\partial\Omega,\sigma)]^n} < \delta \tag{16.12}$$

it follows that the Dirichlet Problem (16.11) *is solvable. Specifically, if $A \in \mathfrak{A}_L^{\mathrm{dis}}$ then the operator $\frac{1}{2}I + K_A$ is invertible on $[\mathrm{A}_{p,q}(\partial\Omega, \sigma)]^M$, and the function $u : \Omega \to \mathbb{C}^M$ defined by*

$$u := \mathcal{D}_A\big[\big(\tfrac{1}{2}I + K_A\big)^{-1} f\big] \ \ in\ \Omega \tag{16.13}$$

is a solution of the Dirichlet Problem (16.11), *and satisfies*

$$\big\|\mathcal{N}_\kappa u\big\|_{\mathrm{A}_{p,q}(\partial\Omega,\sigma)} \approx \|f\|_{[\mathrm{A}_{p,q}(\partial\Omega,\sigma)]^M}, \tag{16.14}$$

with implicit proportionality constants independent of the boundary datum f.

(b) [Uniqueness] *Whenever $\mathfrak{A}_{L^\top}^{\mathrm{dis}} \neq \varnothing$, there exists a threshold $\delta \in (0, 1)$ depending on n, M, p, q, L, and the Ahlfors regularity constants of $\partial\Omega$ with the property that whenever $\|\nu\|_{[\mathrm{BMO}(\partial\Omega,\sigma)]^n} < \delta$ the Dirichlet Problem* (16.11) *admits at most one solution.*

(c) [Well-Posedness] *If $\mathfrak{A}_L^{\mathrm{dis}} \neq \varnothing$ and $\mathfrak{A}_{L^\top}^{\mathrm{dis}} \neq \varnothing$, then there exists a threshold $\delta \in (0, 1)$ depending on n, M, p, q, L, and the Ahlfors regularity constants of $\partial\Omega$ with the property that whenever $\|\nu\|_{[\mathrm{BMO}(\partial\Omega,\sigma)]^n} < \delta$ the Dirichlet Problem* (16.11) *is well-posed (i.e., it is uniquely solvable and the solution satisfies the naturally accompanying estimates from* (16.14)).

Proof Consider the claim made in item *(a)*. The existence of a threshold $\delta \in (0, 1)$ depending on n, M, p, q, A and the Ahlfors regularity constants of $\partial\Omega$ for which whenever $\|\nu\|_{[\mathrm{BMO}(\partial\Omega,\sigma)]^n} < \delta$ the operator $\frac{1}{2}I + K_A$ is invertible on $[\mathrm{A}_{p,q}(\partial\Omega,\sigma)]^M$ follows from Theorem 15.2 (presently used with $\varepsilon := 1/2$, and $z := 1/2$). Thanks to this invertibility result, the function u given in (16.13) is well defined. The fact that this is a solution of the problem (16.11) is a consequence of (13.7), (7.22), (7.25), along with the fact that $\mathrm{A}_{p,q}(\partial\Omega,\sigma) \hookrightarrow L^1\big(\partial\Omega, \frac{\sigma(x)}{1+|x|^{n-1}}\big)$ continuously (cf. item *(e)* from Proposition 10.3). To justify (16.14) observe that the inverse $(\frac{1}{2}I + K_A)^{-1}$ is bounded on $[\mathrm{A}_{p,q}(\partial\Omega,\sigma)]^M$, as a consequence of the Open Mapping Theorem (in our case this inverse is obtained via a Neumann series; see Theorem 15.2). Granted this, we have

$$\begin{aligned}\|\mathcal{N}_\kappa u\|_{\mathrm{A}_{p,q}(\partial\Omega,\sigma)} &\leq C\|(\tfrac{1}{2}I + K_A)^{-1} f\|_{[\mathrm{A}_{p,q}(\partial\Omega,\sigma)]^M} \\ &\leq C\|f\|_{[\mathrm{A}_{p,q}(\partial\Omega,\sigma)]^M}\end{aligned} \tag{16.15}$$

for some constant $C \in (0,\infty)$ independent of f. To establish the opposite inequality, notice that $|f| = \big|u\big|_{\partial\Omega}^{\kappa-\mathrm{n.t.}}\big| \leq \mathcal{N}_\kappa u$ at σ-a.e. point on $\partial\Omega$. Hence, by the lattice property of $\mathrm{A}_{p,q}(\partial\Omega,\sigma)$ from (10.8), we have $\|f\|_{[\mathrm{A}_{p,q}(\partial\Omega,\sigma)]^M} \leq \|\mathcal{N}_\kappa u\|_{\mathrm{A}_{p,q}(\partial\Omega,\sigma)}$. This completes the treatment of the claim in *(a)*.

Let us now turn to the uniqueness claimed in item *(b)*. Choose a coefficient tensor $A \in \mathfrak{A}_L$ such that $A^\top \in \mathfrak{A}^{\mathrm{dis}}_{L^\top}$. Thanks to Theorem 15.2 (written with L replaced by $L^\top$, A replaced by $A^\top$, and $\varepsilon := 1/2$) we can ensure the existence of a threshold $\delta \in (0,1)$ depending on M, n, p, q, A, and the Ahlfors regularity constants of $\partial\Omega$ with the property that whenever $\|\nu\|_{[\mathrm{BMO}(\partial\Omega,\sigma)]^n} < \delta$ the operator

$$\tfrac{1}{2}I + K_{A^\top} : [\mathrm{A}^1_{p',q'}(\partial\Omega,\sigma)]^M \to [\mathrm{A}^1_{p',q'}(\partial\Omega,\sigma)]^M \text{ is invertible,} \tag{16.16}$$

where $p', q' \in (1,\infty)$ denote the Hölder conjugate exponents of p and q, respectively.

By decreasing $\delta \in (0,1)$ if necessary, we may guarantee that Ω is an NTA domain with unbounded boundary (cf. [MMMi^3, Theorem 2.3, pp. 64–65]), which in turn implies that Ω is globally pathwise nontangentially connected (cf. (16.2)). Next, fix an arbitrary point $x_\diamond \in \Omega$ along with some $x_* \in \mathbb{R}^n \setminus \overline{\Omega}$, and let $\widetilde{\kappa} \in (0,\infty)$ be as in Theorem 16.1. Also, pick a radius $\rho \in \big(0, \frac{1}{4}\operatorname{dist}(x_\diamond, \partial\Omega)\big)$. Having fixed an index $\beta \in \{1,\dots,M\}$, define the function

$$f^{(\beta)}(x) := (f^{(\beta)}_\alpha)_{1\leq\alpha\leq M} := \big(E_{\beta\alpha}(x - x_\diamond) - E_{\beta\alpha}(x - x_*)\big)_{1\leq\alpha\leq M} \tag{16.17}$$

for each $x \in \partial\Omega$, where $E := (E_{\alpha\beta})_{1\leq\alpha,\beta\leq M}$ denotes the canonical matrix-valued fundamental solution of the system L (as in Theorem 7.1). In relation to this, we claim that

$$f^{(\beta)} \in [\mathrm{A}^1_{p',q'}(\partial\Omega,\sigma)]^M. \tag{16.18}$$

Indeed, by decreasing ρ if necessary, matters may be arranged so that

$$\partial\Omega \cap \big(B(x_\diamond,\rho) \cup B(x_*,\rho)\big) = \varnothing. \tag{16.19}$$

Thus, keeping in mind (16.17), we invoke Lemma 16.1 (by taking $a := x_\diamond$, $b := x_*$, and $r := \rho$) to obtain

$$|f^{(\beta)}(x)| \leq \frac{C}{1+|x|^{n-1}} \quad \text{and} \quad |\partial_j f^{(\beta)}(x)| \leq \frac{C}{1+|x|^n} \tag{16.20}$$

for every $x \in \partial\Omega$, and $j \in \{1, \ldots, n\}$, and for some constant $C \in (0, \infty)$ depending only on n, $x_\diamond$, x_*, ρ, L. Since the functions in the right-hand side of (16.20) belong to $\mathrm{A}_{p',q'}(\partial\Omega, \sigma)$ (cf. Example 10.1), the lattice property described in (10.8), and (14.2) imply that the membership in (16.18) holds.

Going further, thanks to (16.18) and (16.16), the function $v_\beta : \Omega \to \mathbb{C}^M$ given by

$$v_\beta := (v_{\beta\alpha})_{1\leq\alpha\leq M} := \mathcal{D}_{A^\top}\Big[\big(\tfrac{1}{2}I + K_{A^\top}\big)^{-1} f^{(\beta)}\Big] \tag{16.21}$$

is well defined, and satisfies

$$v_\beta \in [\mathscr{C}^\infty(\Omega)]^M, \quad L^\top v_\beta = 0 \text{ in } \Omega, \tag{16.22}$$

$$\mathcal{N}_{\widetilde{\kappa}} v_\beta \in \mathrm{A}_{p',q'}(\partial\Omega, \sigma), \quad \mathcal{N}_{\widetilde{\kappa}}(\nabla v_\beta) \in \mathrm{A}_{p',q'}(\partial\Omega, \sigma), \tag{16.23}$$

$$\text{and } v_\beta\Big|_{\partial\Omega}^{\widetilde{\kappa}-\mathrm{n.t.}} = f^{(\beta)} \text{ at } \sigma\text{-a.e. point on } \partial\Omega. \tag{16.24}$$

Indeed, the properties listed in (16.22) and (16.24) are consequences of (7.22) and (7.25). The memberships in (16.23) follow from (16.18), Theorem 13.1, and Proposition 16.1. In particular, thanks to (16.9) we also have

$$(\nabla v_\beta)\Big|_{\partial\Omega}^{\widetilde{\kappa}-\mathrm{n.t.}} \text{ exists (in } \mathbb{C}^{n\times M}) \text{ at } \sigma\text{-a.e. point on } \partial\Omega. \tag{16.25}$$

To proceed, for each $\alpha, \beta \in \{1, \ldots, M\}$ introduce the function

$$G^{x_\diamond}_{\alpha\beta} := v_{\beta\alpha}(x) - \big(E_{\beta\alpha}(x - x_\diamond) - E_{\beta\alpha}(x - x_*)\big) \quad \text{for all } x \in \Omega\backslash\{x_\diamond\}. \tag{16.26}$$

Define $G^{x_\diamond} := \big(G^{x_\diamond}_{\alpha\beta}\big)_{1\leq\alpha,\beta\leq M}$ a.e. in Ω, which in particular is a locally integrable function in Ω with respect to the n-dimensional Lebesgue measure $\mathcal{L}^n$. Moreover, thanks to item *(c)* in Theorem 7.1 and (16.17), the function $G^{x_\diamond}$ enjoys the following properties:

$$\begin{gathered} L^\top G^{x_\diamond} = \delta_{x_\diamond} I_{M\times M} \text{ in } [\mathcal{D}'(\Omega)]^{M\times M}, \\ G^{x_\diamond}\Big|_{\partial\Omega}^{\widetilde{\kappa}-\mathrm{n.t.}} = 0 \text{ at } \sigma\text{-a.e. point on } \partial\Omega, \text{ and} \\ (\nabla G^{x_\diamond})\Big|_{\partial\Omega}^{\widetilde{\kappa}-\mathrm{n.t.}} \text{ exists at } \sigma\text{-a.e. point on } \partial\Omega. \end{gathered} \tag{16.27}$$

Also, if $v := (v_{\beta\alpha})_{1\leq\alpha,\beta\leq M}$ then based on item *(d)* in Theorem 7.1 we have

$$\begin{aligned} \big(\mathcal{N}^{\Omega\backslash K}_{\widetilde{\kappa}} G^{x_\diamond}\big)(x) &\leq (\mathcal{N}_{\widetilde{\kappa}} v)(x) + C(1 + |x - x_0|)^{1-n}, \\ \big(\mathcal{N}^{\Omega\backslash K}_{\widetilde{\kappa}} (\nabla G^{x_\diamond})\big)(x) &\leq \big(\mathcal{N}_{\widetilde{\kappa}}(\nabla v)\big)(x) + C(1 + |x - x_0|)^{-n}, \end{aligned} \tag{16.28}$$

for each $x \in \partial\Omega$, and for some constant $C \in (0, \infty)$ independent of x. From this, (16.23), and Example 10.1 we find that

$$\mathcal{N}_{\widetilde{\kappa}}^{\Omega \setminus K}(\nabla G^{x_\diamond}) \in A_{p',q'}(\partial\Omega, \sigma). \tag{16.29}$$

Now, suppose $u = (u_\beta)_{1 \leq \beta \leq M} : \Omega \to \mathbb{C}^M$ is a function satisfying

$$\begin{cases} u \in [\mathscr{C}^\infty(\Omega)]^M, \quad Lu = 0 \text{ in } \Omega, \\ u\big|_{\partial\Omega}^{\kappa-\text{n.t.}} \text{ exists } \sigma\text{-a.e. on } \partial\Omega, \\ \text{and } \mathcal{N}_\kappa u \in A_{p,q}(\partial\Omega, \sigma). \end{cases} \tag{16.30}$$

Then the condition in the third line of (16.30), (16.29), and item *(a)* in Proposition 10.2 yield

$$\begin{aligned} &\int_{\partial\Omega} \mathcal{N}_\kappa u \cdot \mathcal{N}_{\widetilde{\kappa}}^{\Omega \setminus K}(\nabla G^{x_\diamond})\, d\sigma \\ &\quad \leq \|\mathcal{N}_\kappa u\|_{A_{p,q}(\partial\Omega,\sigma)} \|\mathcal{N}_{\widetilde{\kappa}}^{\Omega \setminus K}(\nabla G^{x_\diamond})\|_{A_{p',q'}(\partial\Omega,\sigma)} < \infty. \end{aligned} \tag{16.31}$$

Therefore, in the current scenario we may invoke Theorem 16.1 in order to obtain the Poisson integral representation formula for u as in (16.5). In particular, whenever the function u vanishes on the boundary, i.e., $u\big|_{\partial\Omega}^{\kappa-\text{n.t.}} = 0$ at σ-a.e. point on $\partial\Omega$, we conclude from this that $u(x_\diamond) = 0$. Since $x_\diamond \in \Omega$ was taken arbitrarily, then necessarily u is identically zero in Ω. This concludes the proof of uniqueness of the Dirichlet Problem (16.11) claimed in *(b)*.

Ultimately, the claim made in *(c)* is a consequence of items *(a)* and *(b)*. The proof of Theorem 16.2 is now complete. □

Remark 16.1 Here is an alternative proof of the uniqueness of the Dirichlet Problem (16.11). Work under the same assumptions in Theorem 16.2 and place ourselves in the context of item *(b)*. Assume u is a solution of the homogeneous version of (16.11), i.e., corresponding to the boundary datum $f \equiv 0$. Select $\tau \in \mathbb{R}$ satisfying

$$\frac{-(n-1)}{p} < \tau < (n-1)\Big(\frac{1}{q} - \frac{1}{p}\Big) \tag{16.32}$$

and fix a reference point $x_0 \in \partial\Omega$. Based on item *(g)* in Proposition 10.3, we see that such a function u has the property that

$$\begin{cases} u \in [\mathscr{C}^\infty(\Omega)]^M, \quad Lu = 0 \text{ in } \Omega, \\ \mathcal{N}_\kappa u \in L^p(\partial\Omega, \sigma) + L^p\big(\partial\Omega, |x - x_0|^{\tau p}\sigma(x)\big), \\ u\big|_{\partial\Omega}^{\kappa-\text{n.t.}} = 0 \text{ at } \sigma\text{-a.e. point on } \partial\Omega. \end{cases} \tag{16.33}$$

Since the function $|\cdot - x_0|^{\tau p}$ belongs to the Muckenhoupt class $A_p(\partial\Omega, \sigma)$ (cf. (10.36)), we may invoke [MMMi³, Theorem 6.3, p. 376] to guarantee that $u \equiv 0$ in Ω.

16.2 Regularity Problems with Boundary Data in $\mathrm{A}_{p,q}$

Our next goal is to establish well-posedness results for a variety of Regularity Problems in the context of first-generation Herz-based Sobolev spaces. We begin by dealing with the Inhomogeneous Regularity Problem.

Theorem 16.3 *Let $\Omega \subseteq \mathbb{R}^n$ be an Ahlfors regular domain, and set $\sigma := \mathcal{H}^{n-1}\lfloor\partial\Omega$. Denote by ν the geometric measure theoretic outward unit normal to Ω. Select two exponents $p, q \in (1, \infty)$ along with an aperture parameter $\kappa \in (0, \infty)$. Let L be a second-order, homogeneous, constant complex coefficient, weakly elliptic, $M \times M$ system in $\mathbb{R}^n$, and consider the Inhomogeneous Regularity Problem for L in Ω with boundary data in $[\mathrm{A}^1_{p,q}(\partial\Omega, \sigma)]^M$ formulated as follows:*

$$\begin{cases} u \in \big[\mathscr{C}^\infty(\Omega)\big]^M, \\ Lu = 0 \ \ in\ \Omega, \\ \mathcal{N}_\kappa u,\ \mathcal{N}_\kappa(\nabla u) \in \mathrm{A}_{p,q}(\partial\Omega, \sigma), \\ u\big|^{\kappa-\mathrm{n.t.}}_{\partial\Omega} = f \in [\mathrm{A}^1_{p,q}(\partial\Omega, \sigma)]^M. \end{cases} \tag{16.34}$$

In relation to this, the following statements are valid:

(a) [Existence, Estimates, and Integral Representation] *If $\mathfrak{A}^{\mathrm{dis}}_L \neq \varnothing$ then there exists $\delta \in (0, 1)$ depending only on n, p, q, L, and the Ahlfors regularity of $\partial\Omega$ with the property that whenever*

$$\|\nu\|_{[\mathrm{BMO}(\partial\Omega,\sigma)]^n} < \delta \tag{16.35}$$

it follows that the Inhomogeneous Regularity Problem (16.34) *is solvable. More precisely, if $A \in \mathfrak{A}^{\mathrm{dis}}_L$ then the operator $\frac{1}{2}I + K_A$ is invertible on $[\mathrm{A}^1_{p,q}(\partial\Omega, \sigma)]^M$ and the function $u : \Omega \to \mathbb{C}^M$ defined as*

$$u := \mathcal{D}_A\big[\big(\tfrac{1}{2}I + K_A\big)^{-1} f\big] \ \ in\ \Omega \tag{16.36}$$

is a solution of the Inhomogeneous Regularity Problem (16.34)*, which satisfies*

$$\|\mathcal{N}_\kappa u\|_{\mathrm{A}_{p,q}(\partial\Omega,\sigma)} \approx \|f\|_{[\mathrm{A}_{p,q}(\partial\Omega,\sigma)]^M} \tag{16.37}$$

and

$$\|\mathcal{N}_\kappa(\nabla u)\|_{\mathrm{A}_{p,q}(\partial\Omega,\sigma)} \approx \|\nabla_{\tan} f\|_{[\mathrm{A}_{p,q}(\partial\Omega,\sigma)]^{n\times M}}. \tag{16.38}$$

In particular, one has

$$\|\mathcal{N}_\kappa u\|_{\mathrm{A}_{p,q}(\partial\Omega,\sigma)} + \|\mathcal{N}_\kappa(\nabla u)\|_{\mathrm{A}_{p,q}(\partial\Omega,\sigma)} \approx \|f\|_{[\mathrm{A}^1_{p,q}(\partial\Omega,\sigma)]^M}. \tag{16.39}$$

(b) [Uniqueness] *If* $\mathfrak{A}^{\mathrm{dis}}_{L^\top} \neq \varnothing$, *then there exists a threshold* $\delta \in (0,1)$ *depending on* n, M, p, q, L, *and the Ahlfors regularity constants of* $\partial\Omega$ *with the property that whenever* $\|\nu\|_{[\mathrm{BMO}(\partial\Omega,\sigma)]^n} < \delta$ *it follows that the Inhomogeneous Regularity Problem* (16.34) *admits at most one solution.*

(c) [Well-Posedness] *If* $\mathfrak{A}^{\mathrm{dis}}_{L} \neq \varnothing$ *and* $\mathfrak{A}^{\mathrm{dis}}_{L^\top} \neq \varnothing$ *it follows that there exists* $\delta \in (0,1)$ *depending only on* n, p, q, L, *and the Ahlfors regularity of* $\partial\Omega$ *with the property that whenever* $\|\nu\|_{[\mathrm{BMO}(\partial\Omega,\sigma)]^n} < \delta$ *the Inhomogeneous Regularity Problem* (16.34) *is uniquely solvable.*

Proof We only need to show that the function u from (16.36) satisfies

$$\mathcal{N}_\kappa(\nabla u) \in \mathrm{A}_{p,q}(\partial\Omega,\sigma) \tag{16.40}$$

in a quantitative fashion as in (16.38), since the all remaining claims made in items *(a)-(c)* are consequences of Theorem 15.2 and Theorem 16.2. To justify (16.40), abbreviate $g := (\frac{1}{2}I + K_A)^{-1} f \in [\mathrm{A}^1_{p,q}(\partial\Omega,\sigma)]^M$. Since $\mathrm{A}_{p,q}(\partial\Omega,\sigma)$ embeds continuously into $L^1\big(\partial\Omega, \frac{\sigma(x)}{1+|x|^{n-1}}\big)$ (cf. item *(e)* in Proposition 10.3), this membership guarantees that the function $g = (g_\alpha)_{1\leq\alpha\leq M}$ and all its tangential derivatives $\partial_{\tau_{jk}} g$ belong to $\big[L^1\big(\partial\Omega, \frac{\sigma(x)}{1+|x|^{n-1}}\big)\big]^M$. As such, for each $\ell \in \{1,\dots,n\}$ Proposition 16.1 yields

$$\begin{aligned}(\partial_\ell u)(x) &= \partial_\ell(\mathcal{D}_A g)(x)\\ &= \Big(\int_{\partial\Omega} a^{\beta\alpha}_{rs}(\partial_r E_{\gamma\beta})(x-y)(\partial_{\tau_{\ell s}} g_\alpha)(y)\,d\sigma(y)\Big)_{1\leq\gamma\leq M}\end{aligned} \tag{16.41}$$

for each $x \in \Omega$. Thanks to the smoothness, homogeneity, and parity properties of the canonical fundamental solution $E = (E_{\gamma\beta})_{1\leq\gamma,\beta\leq M}$ (cf. Theorem 7.1), along with the nontangential estimates from Theorem 13.1, (14.10), and (15.12) we see that

$$\|\mathcal{N}_\kappa(\nabla u)\|_{\mathrm{A}_{p,q}(\partial\Omega,\sigma)} \leq C\|\nabla_{\tan} g\|_{[\mathrm{A}_{p,q}(\partial\Omega,\sigma)]^{n\times M}} \leq C\|\nabla_{\tan} f\|_{[\mathrm{A}_{p,q}(\partial\Omega,\sigma)]^{n\times M}} \tag{16.42}$$

for some constant $C \in (0,\infty)$ that depends only on n, M, p, q, A, and the Ahlfors regularity character of $\partial\Omega$. In turn, this establishes the membership (16.40) and the right-pointing inequality in (16.38). The left-point inequality in (16.38) follows from (14.10) and Theorem 14.2. This completes the proof of Theorem 16.3. □

Moving on, we formulate and solve the Homogeneous Regularity Problem with boundary data taken from Herz-based homogeneous Sobolev spaces $\dot{\mathrm{A}}^1_{p,q}$.

Theorem 16.4 *Let $\Omega \subseteq \mathbb{R}^n$ be an Ahlfors regular domain, set $\sigma := \mathcal{H}^{n-1}\lfloor\partial\Omega$, and denote by ν the geometric measure theoretic outward unit normal to Ω. Select two exponents $p, q \in (1, \infty)$, and fix an aperture parameter $\kappa \in (0, \infty)$. Given a homogeneous, second-order, constant complex coefficient, weakly elliptic $M \times M$ system L in $\mathbb{R}^n$, formulate the Homogeneous Regularity Problem for L in Ω with boundary data in $[\dot{\mathrm{A}}^1_{p,q}(\partial\Omega, \sigma)]^M$ as follows:*

$$\begin{cases} u \in \big[\mathscr{C}^\infty(\Omega)\big]^M, \\ Lu = 0 \ \text{ in } \Omega, \\ \mathcal{N}_\kappa(\nabla u) \in \mathrm{A}_{p,q}(\partial\Omega, \sigma), \\ u\big|_{\partial\Omega}^{\kappa-\mathrm{n.t.}} = f \in \big[\dot{\mathrm{A}}^1_{p,q}(\partial\Omega, \sigma)\big]^M. \end{cases} \tag{16.43}$$

Then the following statements are true:

(a) [Existence, Integral Representation, and Estimates] *If $\mathfrak{A}^{\mathrm{dis}}_L \neq \varnothing$, then there exists a small threshold $\delta \in (0, 1)$ which depends only on n, p, q, L, and the Ahlfors regularity constants of $\partial\Omega$ with the property that whenever*

$$\|\nu\|_{[\mathrm{BMO}(\partial\Omega,\sigma)]^n} < \delta \tag{16.44}$$

it follows that the Homogeneous Regularity Problem (16.43) *is solvable. More specifically, if $A \in \mathfrak{A}^{\mathrm{dis}}_L$ and $\big[K_{A,\mathrm{mod}}\big]$ acts on $\big[\dot{\mathrm{A}}^1_{p,q}(\partial\Omega, \sigma)/\!\sim\big]^M$ as in* (15.14)*, then the operator*

$$\tfrac{1}{2}I + \big[K_{A,\mathrm{mod}}\big] : \big[\dot{\mathrm{A}}^1_{p,q}(\partial\Omega, \sigma)/\!\sim\big]^M \longrightarrow \big[\dot{\mathrm{A}}^1_{p,q}(\partial\Omega, \sigma)/\!\sim\big]^M \tag{16.45}$$

is an isomorphism, and the Homogeneous Regularity Problem (16.43) *is solved by* (*cf.* (7.28))

$$u := \mathcal{D}_{A,\mathrm{mod}}\, g + c \ \text{ in } \Omega \tag{16.46}$$

for a suitable constant $c \in \mathbb{C}^M$ and some function

$$g \in \big[\dot{\mathrm{A}}^1_{p,q}(\partial\Omega, \sigma)\big]^M \ \text{ satisfying } \ [g] = \Big(\tfrac{1}{2}I + \big[K_{A,\mathrm{mod}}\big]\Big)^{-1}[f]. \tag{16.47}$$

Moreover, for this solution one has

$$\|\mathcal{N}_\kappa(\nabla u)\|_{\mathrm{A}_{p,q}(\partial\Omega,\sigma)} \approx \|\nabla_{\tan} f\|_{[\mathrm{A}_{p,q}(\partial\Omega,\sigma)]^{n\times M}} \tag{16.48}$$

with implicit proportionality constants independent of f.

(b) [Uniqueness] *Whenever $\mathfrak{A}^{\mathrm{dis}}_{L^\top} \neq \varnothing$, there exists $\delta \in (0, 1)$ depending only on n, p, q, L, and the Ahlfors regularity of $\partial\Omega$ such that if $\|\nu\|_{[\mathrm{BMO}(\partial\Omega,\sigma)]^n} < \delta$ then the Homogeneous Regularity Problem* (16.43) *has a unique solution.*

(c) [Well-Posedness] *If* $\mathfrak{A}_L^{\rm dis} \neq \varnothing$ *and* $\mathfrak{A}_{L^\top}^{\rm dis} \neq \varnothing$ *it follows that there exists* $\delta \in (0,1)$ *depending only on* n, p, q, L, *and the Ahlfors regularity of* $\partial\Omega$ *with the property that if* $\|\nu\|_{[\mathrm{BMO}(\partial\Omega,\sigma)]^n} < \delta$ *then the Homogeneous Regularity Problem* (16.43) *is uniquely solvable.*

Proof The invertibility claimed in item *(a)* follows from Theorem 15.5 (presently used with $z := 1/2$ and $\varepsilon := 1/2$). Hence, there exists $g \in \big[\mathring{\mathrm{A}}^1_{p,q}(\partial\Omega,\sigma)\big]^M$ for which

$$\big[\tfrac{1}{2}g + K_{A,\mathrm{mod}}g\big] = \Big(\tfrac{1}{2}I + \big[K_{A,\mathrm{mod}}\big]\Big)[g] = [f]. \tag{16.49}$$

As such, one can choose some constant $c \in \mathbb{C}^M$ with the property that

$$\tfrac{1}{2}g + K_{A,\mathrm{mod}}g = f - c. \tag{16.50}$$

Thanks to this, (7.30), (7.31), and the nontangential estimate (13.7) recorded in Theorem 13.1 it follows that the function u from (16.46) is a solution of the Homogeneous Regularity Problem (16.43), and there exists $C = C(n, p, q, \kappa, A, \partial\Omega) \in (0,\infty)$ for which

$$\begin{aligned} \|\mathcal{N}_\kappa(\nabla u)\|_{\mathrm{A}_{p,q}(\partial\Omega,\sigma)} &\leq C \sum_{j,k=1}^{n} \|\partial_{\tau_{jk}} f\|_{[\mathrm{A}_{p,q}(\partial\Omega,\sigma)]^M} \\ &\approx C\|\nabla_{\tan} f\|_{[\mathrm{A}_{p,q}(\partial\Omega,\sigma)]^{n\times M}}, \end{aligned} \tag{16.51}$$

where the last comparability follows from (14.10). The reverse inequality in (16.51) is implied by Theorem 14.3, (14.10), and (14.24). This finishes the treatment of existence, integral representation, and estimates claimed in item *(a)*.

The uniqueness claimed in item *(b)* can be dealt with via the same type of argument presented in the proof of [MMMi[3], Theorem 6.8, pp. 384–390] with the Herz space $\mathrm{A}_{p,q}(\partial\Omega,\sigma)$ playing the role of the Muckenhoupt weighted Lebesgue space $L^p(\partial\Omega, w)$, while keeping in mind Theorem 14.3, the invertibility result recorded in Theorem 15.2, and the continuous embedding $\mathrm{A}_{p,q}(\partial\Omega,\sigma) \hookrightarrow L^1\big(\partial\Omega, \frac{\sigma(x)}{1+|x|^{n-1}}\big)$ (cf. item *(e)* in Proposition 10.3).

Lastly, the well-posedness statement in item *(c)* is a direct consequence of items *(a)* and *(b)*. This ultimately establishes Theorem 16.4. □

16.3 The Neumann Problem with Boundary Data in $\mathrm{A}_{p,q}$

The goal here is to establish the well-posedness of the Neumann Problem with boundary data taken from first-generation Herz spaces $\mathrm{A}_{p,q}$. We begin by obtaining the following a Fatou-type result.

Proposition 16.2 *Consider* $\Omega \subseteq \mathbb{R}^n$ *(with* $n \in \mathbb{N}$ *with* $n \geq 2$*) a* UR *domain, abbreviate* $\sigma := \mathcal{H}^{n-1}\lfloor\partial\Omega$, *and denote by* ν *the geometric measure theoretic outward unit* *normal to Ω. Let L be a second-order, homogeneous, constant complex coefficient, weakly elliptic $M \times M$ system in $\mathbb{R}^n$, and fix a coefficient tensor $A \in \mathfrak{A}_L$. Pick two exponents $p \in (1,\infty)$ and $q \in (0,\infty)$, and assume $u : \Omega \to \mathbb{C}^M$ is a function satisfying*

$$\begin{aligned} & u \in \big[\mathscr{C}^\infty(\Omega)\big]^M, \ \ Lu = 0 \text{ in } \Omega, \\ & \text{and } \ \mathcal{N}_\kappa(\nabla u) \in \mathrm{A}_{p,q}(\partial\Omega, \sigma). \end{aligned} \tag{16.52}$$

Then the nontangential boundary trace $(\nabla u)\big|^{\kappa-\mathrm{n.t.}}_{\partial\Omega}$ exists (in $\mathbb{C}^{M\times n}$) at σ-a.e. point on $\partial\Omega$ and, in a quantitative fashion,

$$\begin{aligned} & (\nabla u)\big|^{\kappa-\mathrm{n.t.}}_{\partial\Omega} \text{ belongs to } [\mathrm{A}_{p,q}(\partial\Omega,\sigma)]^{M\times n} \\ & \text{while } \partial^A_\nu u \text{ belongs to } [\mathrm{A}_{p,q}(\partial\Omega,\sigma)]^{M}, \end{aligned} \tag{16.53}$$

i.e., there exists a constant $C \in (0,\infty)$ independent of u for which

$$\Big\|(\nabla u)\big|^{\kappa-\mathrm{n.t.}}_{\partial\Omega}\Big\|_{[\mathrm{A}_{p,q}(\partial\Omega,\sigma)]^{M\times n}} \leq C\|\mathcal{N}_\kappa(\nabla u)\|_{\mathrm{A}_{p,q}(\partial\Omega,\sigma)} \tag{16.54}$$

$$\text{and } \ \|\partial^A_\nu u\|_{[\mathrm{A}_{p,q}(\partial\Omega,\sigma)]^{M}} \leq C\|\mathcal{N}_\kappa(\nabla u)\|_{\mathrm{A}_{p,q}(\partial\Omega,\sigma)}. \tag{16.55}$$

Proof A combination of Theorem 9.1, the inclusions recorded in item *(f)* from Proposition 10.3, the lattice property of $\mathrm{A}_{p,q}(\partial\Omega,\sigma)$ (see (10.8)), and (7.55) gives the desired conclusions. □

We are now in a position to state and prove the well-posedness result for the Neumann Problem in the context of first-generation Herz spaces announced earlier.

Theorem 16.5 *Let $\Omega \subseteq \mathbb{R}^n$ be an Ahlfors regular domain, and set $\sigma := \mathcal{H}^{n-1}\lfloor\partial\Omega$. Denote by ν the geometric measure theoretic outward unit normal to Ω. Pick two exponents $p, q \in (1,\infty)$ along with an aperture parameter $\kappa \in (0,\infty)$. Let L be a second-order, homogeneous, constant complex coefficient, weakly elliptic $M \times M$ system in $\mathbb{R}^n$, and select a coefficient tensor $A \in \mathfrak{A}_L$. Formulate the Neumann Problem for L in Ω with boundary data in $[\mathrm{A}_{p,q}(\partial\Omega,\sigma)]^M$ as follows:*

$$\begin{cases} u \in [\mathscr{C}^\infty(\Omega)]^M, \\ Lu = 0 \ \text{ in } \Omega, \\ \mathcal{N}_\kappa(\nabla u) \in \mathrm{A}_{p,q}(\partial\Omega,\sigma), \\ \partial^A_\nu u = f \in [\mathrm{A}_{p,q}(\partial\Omega,\sigma)]^M. \end{cases} \tag{16.56}$$

In relation to this, the following statements are valid:

(a) [Existence, Integral Representation, and Estimates] *If $\mathfrak{A}^{\mathrm{dis}}_{L^\top} \neq \varnothing$ and $A^\top \in \mathfrak{A}^{\mathrm{dis}}_{L^\top}$, there exists a threshold $\delta \in (0,1)$ depending on n, M, p, q, A, and the Ahlfors regularity constants of $\partial\Omega$ with the property that whenever*

$$\|\nu\|_{[\mathrm{BMO}(\partial\Omega,\sigma)]^n} < \delta \tag{16.57}$$

it follows that the Neumann Problem (16.56) *is solvable. Concretely, the operator* $-\frac{1}{2}I + K^{\#}_{A^\top}$ *is invertible on* $[\mathrm{A}_{p,q}(\partial\Omega,\sigma)]^M$, *and the function* $u : \Omega \to \mathbb{C}^M$ *defined as*

$$u := \mathscr{S}_{\mathrm{mod}}\big[\big(-\tfrac{1}{2}I + K^{\#}_{A^\top}\big)^{-1} f\big] \ \ in\ \Omega \tag{16.58}$$

is a solution of the Neumann Problem (16.56) *which satisfies*

$$\big\|\mathcal{N}_\kappa(\nabla u)\big\|_{\mathrm{A}_{p,q}(\partial\Omega,\sigma)} \approx \|f\|_{[\mathrm{A}_{p,q}(\partial\Omega,\sigma)]^M} \tag{16.59}$$

with implicit proportionality constants independent of f.

(b) [Uniqueness (modulo constants)] *Whenever* $A \in \mathfrak{A}^{\mathrm{dis}}_L$ *there exists some threshold* $\delta \in (0,1)$ *depending on* n, M, p, q, A, *and the Ahlfors regularity constants of* $\partial\Omega$ *with the property that if* $\|\nu\|_{[\mathrm{BMO}(\partial\Omega,\sigma)]^n} < \delta$ *then any two solutions of the Neumann Problem* (16.56) *differ by a* $\mathbb{C}^M$*-constant.*

(c) [Well-Posedness] *If* $A \in \mathfrak{A}^{\mathrm{dis}}_L$ *and* $A^\top \in \mathfrak{A}^{\mathrm{dis}}_{L^\top}$ *then there exists some threshold* $\delta \in (0,1)$ *depending on* n, M, p, q, L, *and the Ahlfors regularity constants of* $\partial\Omega$ *with the property that whenever* $\|\nu\|_{[\mathrm{BMO}(\partial\Omega,\sigma)]^n} < \delta$, *the Neumann Problem* (16.56) *has a solution that is unique modulo constants, and satisfies* (16.59).

Proof We first deal with the claim made in item *(a)*. The invertibility of the operator $-\frac{1}{2}I + K^{\#}_{A^\top}$ on $[\mathrm{A}_{p,q}(\partial\Omega,\sigma)]^M$ follows from Theorem 15.2 (presently used with A replaced by $A^\top$, $\varepsilon := 1/2$, and $z := 1/2$), assuming $\delta \in (0,1)$ is sufficiently small. Thanks to said invertibility, the function u given in (16.58) is well defined. In view of (7.51)-(7.56), along with (13.7) and the fact that $\mathrm{A}_{p,q}(\partial\Omega,\sigma) \hookrightarrow L^1\big(\partial\Omega, \frac{\sigma(x)}{1+|x|^{n-1}}\big)$ continuously (cf. item *(e)* in Proposition 10.3), it follows that u is indeed a solution of (16.56) and satisfies

$$\begin{aligned}\|\mathcal{N}_\kappa(\nabla u)\|_{\mathrm{A}_{p,q}(\partial\Omega,\sigma)} &\leq C\big\|\big(-\tfrac{1}{2}I + K^{\#}_{A^\top}\big)^{-1} f\big\|_{\mathrm{A}_{p,q}(\partial\Omega,\sigma)} \\ &\leq C\|f\|_{\mathrm{A}_{p,q}(\partial\Omega,\sigma)},\end{aligned} \tag{16.60}$$

for some constant $C \in (0,\infty)$ independent of u. The reverse inequality follows from (16.55). This establishes the veracity of the claim made in *(a)*.

To verify the claim made in item *(b)*, we may employ the same approach used in the proof of uniqueness of the Neumann Problem with boundary data in Muckenhoupt weighted Lebesgue spaces from [MMMi[3], Theorem 6.1, pp. 396–404]. The argument presented there is largely independent of the underlying functional analytic context, and the only aspect we should pay attention to is the validity of the estimate in [MMMi[3], (6.131), p. 400] in the current setting of Herz spaces. Hence, thanks to Theorem 15.1, and Theorem 13.3, the version of said estimate do hold in our functional analytic setting and this finishes the treatment of the claim made in *(b)*.

Ultimately, the well-posedness claimed in *(c)* is a consequence of *(a)* and *(b)*. The proof of Theorem 16.5 is now complete. □

Remark 16.2 Let us present an alternative proof of the uniqueness (modulo constants) of the Neumann Problem (16.56). Make the same assumptions as in Theorem 16.5, and work in the setting of item *(b)*. Consider a solution u of the homogeneous version of the Neumann Problem (16.56) (i.e., corresponding to the boundary datum $f \equiv 0$). Pick $\tau \in \mathbb{R}$ satisfying

$$\frac{-(n-1)}{p} < \tau < (n-1)\Big(\frac{1}{q} - \frac{1}{p}\Big) \tag{16.61}$$

and fix a reference point $x_0 \in \partial\Omega$. Then item *(g)* in Proposition 10.3 implies that u is also a solution of

$$\begin{cases} u \in [\mathscr{C}^\infty(\Omega)]^M, \quad Lu = 0 \ \text{ in } \Omega, \\ \mathcal{N}_\kappa(\nabla u) \in L^p(\partial\Omega, \sigma) + L^p\big(\partial\Omega, |x - x_0|^{\tau p}\sigma(x)\big), \\ \partial_\nu^A u = 0 \ \text{ pointwise } \sigma\text{-a.e. on } \partial\Omega. \end{cases} \tag{16.62}$$

Bearing in mind that $|\cdot - x_0|^{\tau p}$ belongs to the Muckenhoupt class $A_p(\partial\Omega, \sigma)$ (cf. (10.36)), work done in [MMMi[3], Theorem 6.14, pp. 407–408] ensures that u is a $\mathbb{C}^M$-constant in Ω, as desired.

16.4 The Transmission Problem with Boundary Data in $\mathbf{A}_{p,q}$

The goal now is to obtain a well-posedness result for the Transmission Problem formulated in the context of first-generation Herz spaces. A key aspect of this boundary problem is that one now seeks two functions defined on either side of an interface, and the boundary data match their nontangential traces and conormal derivatives suitably coupled.

Theorem 16.6 *domain such that $\partial\Omega$ is an unbounded Ahlfors regular set. Abbreviate $\sigma := \mathcal{H}^{n-1}\lfloor\partial\Omega$, denote by ν the geometric measure theoretic outward unit normal to Ω, and define*

$$\Omega_+ := \Omega \ \text{ and } \ \Omega_- := \mathbb{R}^n \setminus \overline{\Omega}. \tag{16.63}$$

Next, pick a second-order, homogeneous, constant complex coefficient, weakly elliptic, $M \times M$ system L in $\mathbb{R}^n$. Also, fix two exponents $p, q \in (1, \infty)$. Having selected some coefficient tensor $A \in \mathfrak{A}_L$, along with an aperture parameter $\kappa \in (0, \infty)$ and a transmission parameter $\eta \in \mathbb{C}$, consider the following Transmission Problem:

$$\begin{cases} u^{\pm} \in [\mathscr{C}^{\infty}(\Omega)]^M, \quad Lu^{\pm} = 0 \ \ in\ \Omega_{\pm}, \\ \mathcal{N}_{\kappa}(\nabla u^{\pm}) \in \mathrm{A}_{p,q}(\partial\Omega, \sigma), \\ u^{+}\big|_{\partial\Omega}^{\kappa-\mathrm{n.t.}} - u^{-}\big|_{\partial\Omega}^{\kappa-\mathrm{n.t.}} = f \in [\dot{\mathrm{A}}^1_{p,q}(\partial\Omega, \sigma)]^M, \\ \partial^A_{\nu} u^{+} - \eta \cdot \partial^A_{\nu} u^{-} = g \in [\mathrm{A}_{p,q}(\partial\Omega, \sigma)]^M. \end{cases} \tag{16.64}$$

Lastly, suppose that either

$$A^{\top} \in \mathfrak{A}^{\mathrm{dis}}_{L^{\top}} \ \ and\ \eta \in \mathbb{C} \setminus \{-1\} \tag{16.65}$$

or

$$A \in \mathfrak{A}^{\mathrm{dis}}_{L} \ \ and\ \eta \in \mathbb{C} \setminus \{0, -1\}. \tag{16.66}$$

Then for each $m \in \mathbb{N}$ there exists some threshold $\delta \in (0, 1)$, depending solely on n, M, A, p, q, m, the two-sided NTA *constants of Ω, and the Ahlfors regularity constants of $\partial\Omega$ with the property that whenever*

$$\left|\tfrac{1-\eta}{1+\eta}\right| \cdot \|\nu\|^{\langle m \rangle}_{[\mathrm{BMO}(\partial\Omega, \sigma)]^n} < \delta \tag{16.67}$$

it follows that the Transmission Problem (16.64) *is well posed. Specifically, there exists a pair of functions u^+ and u^- solving* (16.64)*, which are unique modulo additive constants from $\mathbb{C}^M$, and satisfy*

$$\|\mathcal{N}_{\kappa}(\nabla u^{\pm})\|_{\mathrm{A}_{p,q}(\partial\Omega, \sigma)} \leq C\big(\|f\|_{[\dot{\mathrm{A}}^1_{p,q}(\partial\Omega, \sigma)]^M} + \|g\|_{[\mathrm{A}_{p,q}(\partial\Omega, \sigma)]^M}\big), \tag{16.68}$$

for some constant $C \in (0, \infty)$ independent of the boundary data f and g.

Proof Bearing in mind that the Herz space $\mathrm{A}_{p,q}(\partial\Omega, \sigma)$ is a Generalized Banach Function Space on $\partial\Omega$ (cf. Proposition 10.11) such that the Hardy-Littlewood maximal operator $\mathcal{M}_{\partial\Omega}$ is bounded (cf. Theorem 10.1) both on $\mathrm{A}_{p,q}(\partial\Omega, \sigma)$ and on its associated space $\big(\mathrm{A}_{p,q}(\partial\Omega, \sigma)\big)' = \mathrm{A}_{p',q'}(\partial\Omega, \sigma)$, where $p', q' \in (1, \infty)$ are the Hölder conjugate exponents of p and q, respectively (cf. Proposition 10.12), it follows that the well posedness of (16.64) is a consequence of [GHA-V, Theorem 8.4.44, p. 877]. □

16.5 The Neumann Problem with Boundary Data in $\mathrm{HA}_{p,q}$

The principal aim here is to establish the well-posedness of the Neumann Problem for weakly elliptic systems with boundary data in first-generation Herz-type Hardy spaces. As a preamble, the result recorded below describes a useful Fatou-type result in the context of $\mathrm{HA}_{p,q}$ spaces.

Theorem 16.7 *Assume $\Omega \subseteq \mathbb{R}^n$ is a* UR *domain with unbounded boundary. Abbreviate $\sigma := \mathcal{H}^{n-1}\lfloor\partial\Omega$, and denote by ν the geometric measure theoretic outward unit normal to Ω. Consider L a second-order, homogeneous, constant complex coefficient, weakly elliptic $M \times M$ system in $\mathbb{R}^n$. Also, pick two exponents $p \in (1, \infty)$ and $q \in \big(\frac{n-1}{n}, 1\big]$, along with an aperture parameter $\kappa \in (0, \infty)$. Suppose $u : \Omega \to \mathbb{C}^M$ is a function satisfying*

$$u \in [\mathscr{C}^\infty(\Omega)]^M, \quad Lu = 0 \ \text{ in } \Omega, \qquad \text{and} \quad \mathcal{N}_\kappa(\nabla u) \in \mathrm{A}_{p,q}(\partial\Omega, \sigma). \tag{16.69}$$

Then the nontangential boundary trace $(\nabla u)\big|_{\partial\Omega}^{\kappa-\mathrm{n.t.}}$ exists at σ-a.e. point on $\partial\Omega$, and the vector-valued function $(\nabla u)\big|_{\partial\Omega}^{\kappa-\mathrm{n.t.}}$ belongs to $[\mathrm{A}_{p,q}(\partial\Omega, \sigma)]^{M\times n}$.

Furthermore, for any choice of a coefficient tensor $A \in \mathfrak{A}_L$ it follows that the conormal derivative $\partial_\nu^A u$ (cf. (7.55)) belongs to $[\mathrm{HA}_{p,q}(\partial\Omega, \sigma)]^M$ quantitatively, i.e., there exists a constant $C = C(\partial\Omega, A, n, p, q, \kappa) \in (0, \infty)$ with the property that

$$\|\partial_\nu^A u\|_{[\mathrm{HA}_{p,q}(\partial\Omega,\sigma)]^M} \leq C\|\mathcal{N}_\kappa(\nabla u)\|_{\mathrm{A}_{p,q}(\partial\Omega,\sigma)}. \tag{16.70}$$

Proof Notice that the existence of the nontangential limit $(\nabla u)\big|_{\partial\Omega}^{\kappa-\mathrm{n.t.}}$ and its membership to $[\mathrm{A}_{p,q}(\partial\Omega, \sigma)]^{M\times n}$ follow from Proposition 16.2. To verify (16.70), select a coefficient tensor $A = \big(a_{jk}^{\alpha\beta}\big)_{\substack{1\leq j,k\leq n\\ 1\leq\alpha,\beta\leq M}} \in \mathfrak{A}_L$, and for each $\alpha \in \{1, \dots, M\}$ abbreviate $\vec{F}_\alpha := \big(a_{jk}^{\alpha\beta}(\partial_k u_\beta)\big)_{1\leq j\leq n}$. Using the same type of reasoning as in (9.13)-(9.15), now taking into account (16.69) and item *(f)* in Proposition 10.3, for each $\alpha \in \{1, \dots, M\}$ we have

$$\big(\partial_\nu^A u\big)_\alpha = \nu \cdot (\vec{F}_\alpha)\big|_{\partial\Omega}^{\kappa-\mathrm{n.t.}}, \quad \operatorname{div} \vec{F}_\alpha = 0 \in L^1_{\mathrm{loc}}(\Omega, \mathcal{L}^n), \tag{16.71}$$

and

$$\mathcal{N}_\kappa \vec{F}_\alpha \in \mathrm{A}_{p,q}(\partial\Omega, \sigma) \subseteq L^1_{\mathrm{loc}}(\partial\Omega, \sigma) \quad \text{and} \qquad \|\mathcal{N}_\kappa \vec{F}_\alpha\|_{\mathrm{A}_{p,q}(\partial\Omega,\sigma)} \leq C\|\mathcal{N}_\kappa(\nabla u)\|_{\mathrm{A}_{p,q}(\partial\Omega,\sigma)}. \tag{16.72}$$

Moreover, thanks to Theorem 9.2, whenever $\gamma, s \in \mathbb{R}$ are such that

$$(n-1)(\tfrac{1}{q} - 1) < \gamma < 1 \quad \text{and} \quad \tfrac{n-1}{n-1+\gamma} < s < q \tag{16.73}$$

there exists some constant $C = C(\partial\Omega, \kappa, n, \gamma, s) \in (0, \infty)$ with the property that

$$\Big[\nu \cdot \Big(\vec{F}_\alpha\big|_{\partial\Omega}^{\kappa-\mathrm{n.t.}}\Big)\Big]_\gamma^\# \leq C \cdot \mathcal{M}_{\partial\Omega,s}(\mathcal{N}_\kappa \vec{F}_\alpha) \quad \text{everywhere on } \partial\Omega, \tag{16.74}$$

for each $\alpha \in \{1, \dots, M\}$. From this, (16.71), and the boundedness of the L^s-based Hardy-Littlewood maximal operator $\mathcal{M}_{\partial\Omega,s}$ on $\mathrm{A}_{p,q}(\partial\Omega, \sigma)$ (cf. item *(a)* in Theorem 10.1) we obtain

$$
\begin{aligned}
\big\|(\partial_\nu^A u)_\alpha\big\|_{\mathrm{HA}_{p,q}(\partial\Omega,\sigma)} &= \Big\|\nu\cdot\big(\vec{F}_\alpha\big|_{\partial\Omega}^{\kappa-\mathrm{n.t.}}\big)\Big\|_{\mathrm{HA}_{p,q}(\partial\Omega,\sigma)} \\
&= \Big\|\Big[\nu\cdot\big(\vec{F}_\alpha\big|_{\partial\Omega}^{\kappa-\mathrm{n.t.}}\big)\Big]_\gamma^{\#}\Big\|_{\mathrm{A}_{p,q}(\partial\Omega,\sigma)} \\
&\leq C\|\mathcal{N}_\kappa\vec{F}_\alpha\|_{\mathrm{A}_{p,q}(\partial\Omega,\sigma)} < \infty, \qquad (16.75)
\end{aligned}
$$

for each $\alpha\in\{1,\dots,M\}$, and for a constant $C=C(\partial\Omega,\kappa,n,p,q,\gamma,s)\in(0,\infty)$. This finishes the justification of (16.70), which ultimately establishes Theorem 16.7. □

We are now ready to formulate and solve the Neumann Problem on Herz-type Hardy spaces advertised earlier.

Theorem 16.8 *Assume $\Omega\subseteq\mathbb{R}^n$ is an Ahlfors regular domain. Set $\sigma:=\mathcal{H}^{n-1}\lfloor\partial\Omega$ and denote by ν the geometric measure theoretic outward unit normal to Ω. Fix two exponents $p\in(1,\infty)$ and $q\in\big(\frac{n-1}{n},1\big]$. Suppose L is a second-order, homogeneous, constant complex coefficients, weakly elliptic $M\times M$ system in $\mathbb{R}^n$, and fix a coefficient tensor $A\in\mathfrak{A}_L$. Finally, select an aperture parameter $\kappa\in(0,\infty)$. Formulate the Neumann Problem for L in Ω with boundary data in the Herz-type Hardy space $[\mathrm{HA}_{p,q}(\partial\Omega,\sigma)]^M$ as follows:*

$$
\begin{cases}
u\in[\mathscr{C}^\infty(\Omega)]^M,\\
Lu=0 \ \text{ in } \Omega,\\
\mathcal{N}_\kappa(\nabla u)\in \mathrm{A}_{p,q}(\partial\Omega,\sigma),\\
\partial_\nu^A u=f\in[\mathrm{HA}_{p,q}(\partial\Omega,\sigma)]^M.
\end{cases}
\qquad (16.76)
$$

Regarding this problem, the following statements are valid:

(a) [Existence, Integral Representation, and Estimates] *If $A^\top\in\mathfrak{A}_{L^\top}^{\mathrm{dis}}$ then there exists some threshold $\delta\in(0,1)$ depending only on n, p, q, κ, A, and the Ahlfors regularity constants of $\partial\Omega$ with the property that whenever*

$$\|\nu\|_{[\mathrm{BMO}(\partial\Omega,\sigma)]^n}<\delta \qquad (16.77)$$

it follows that the Neumann Problem (16.76) *is solvable. Specifically, the operator $-\frac{1}{2}I+K_{A^\top}^{\#}$ is invertible on the Herz-type Hardy space $[\mathrm{HA}_{p,q}(\partial\Omega,\sigma)]^M$, and the function $u:\Omega\to\mathbb{C}^M$ defined by*

$$u:=\mathscr{S}_{\mathrm{mod}}\big[\big(-\tfrac{1}{2}I+K_{A^\top}^{\#}\big)^{-1}f\big] \ \text{ in } \Omega \qquad (16.78)$$

is a solution of the Neumann Problem (16.76) *and satisfies*

$$\|\mathcal{N}_\kappa(\nabla u)\|_{\mathrm{A}_{p,q}(\partial\Omega,\sigma)}\approx\|f\|_{[\mathrm{HA}_{p,q}(\partial\Omega,\sigma)]^M}, \qquad (16.79)$$

with implicit proportionality constants independent of the boundary datum f.

(b) [Uniqueness modulo constants] *If $A \in \mathfrak{A}_L^{\mathrm{dis}}$ then there exists $\delta \in (0,1)$ depending only on n, p, q, κ, A, and the Ahlfors regularity constants of $\partial\Omega$ with the property that whenever $\|\nu\|_{[\mathrm{BMO}(\partial\Omega,\sigma)]^n} < \delta$ it follows that any two solutions of* (16.76) *differ by a constant from $\mathbb{C}^M$.*

(c) [Well-Posedness] *If $\mathfrak{A}_L^{\mathrm{dis}} \neq \varnothing$ and $\mathfrak{A}_{L^\top}^{\mathrm{dis}} \neq \varnothing$ there exists a threshold $\delta \in (0,1)$ depending only on n, p, q, κ, A, and the Ahlfors regularity constants of $\partial\Omega$, with the property that whenever $\|\nu\|_{[\mathrm{BMO}(\partial\Omega,\sigma)]^n} < \delta$ then the Neumann Problem* (16.76) *is solvable, its solution is unique modulo $\mathbb{C}^n$-constants and satisfies* (9.21).

Proof To get started, make the assumption that $\mathfrak{A}_{L^\top}^{\mathrm{dis}} \neq \varnothing$, and $A^\top \in \mathfrak{A}_{L^\top}^{\mathrm{dis}}$. In light of Theorem 15.7 (presently used with $\varepsilon := 1/2$, and $z := -1/2$) there exists some $\delta \in (0,1)$ for which whenever $\|\nu\|_{[\mathrm{BMO}(\partial\Omega,\sigma)]^n} < \delta$ the operator $-\frac{1}{2}I + K^\#_{A^\top}$ is bounded and invertible on $[\mathrm{HA}_{p,q}(\partial\Omega,\sigma)]^M$. As such, the function u introduced in (9.20) is well defined, and thanks to (7.54) and (7.56) it satisfies

$$u \in [\mathscr{C}^\infty(\Omega)]^M, \quad Lu = 0 \ \text{ in } \Omega, \text{ and } \ \partial^A_\nu u = f \ \text{ pointwise } \sigma\text{-a.e. on } \partial\Omega. \tag{16.80}$$

Additionally, from (7.53), Theorem 7.1, and Theorem 13.5 we obtain

$$\begin{aligned}\|\mathcal{N}_\kappa(\nabla u)\|_{\mathrm{A}_{p,q}(\partial\Omega,\sigma)} &\leq C\big\|\big(-\tfrac{1}{2}I + K^\#_{A^\top}\big)^{-1} f\big\|_{[\mathrm{HA}_{p,q}(\partial\Omega,\sigma)]^M} \\ &\leq C\|f\|_{[\mathrm{HA}_{p,q}(\partial\Omega,\sigma)]^M}\end{aligned} \tag{16.81}$$

for some $C \in (0,\infty)$ depending only on n, p, q, A, and the Ahlfors regularity character of $\partial\Omega$. Together, (16.80) and (16.81) imply that the function u defined in (16.78) is indeed a solution of the Neumann Problem (16.76). The estimates in (9.21) follows from (16.80), (16.81), and (16.70). This establishes the claim in item *(a)*.

Regarding the uniqueness claimed in item *(b)*, suppose $A \in \mathfrak{A}_L^{\mathrm{dis}}$ and assume u is a solution of the homogeneous Neumann Problem (16.76) (corresponding to the boundary datum $f \equiv 0$). Upon recalling that $\mathrm{A}_{p,q}(\partial\Omega,\sigma) \subseteq L^p(\partial\Omega,\sigma)$ (cf. item *(c)* in Proposition 10.3), the function u under consideration satisfies

$$\begin{cases} u \in [\mathscr{C}^\infty(\Omega)]^M, \quad Lu = 0 \ \text{ in } \Omega, \\ \mathcal{N}_\kappa(\nabla u) \in L^p(\partial\Omega,\sigma), \\ \partial^A_\nu u = 0 \ \text{ pointwise } \sigma\text{-a.e. on } \partial\Omega. \end{cases} \tag{16.82}$$

Hence, [MMMi³, Theorem 6.17, p. 396] ensures that u is a constant function in Ω. This establishes the claim made in item *(b)*. Lastly, the claim made in *(c)* is a consequence of *(a)* and *(b)*. This concludes the proof of Theorem 16.8. □

Comment 1. In view of the Fatou-type result described in Theorem 16.7 the boundary condition involving the conormal derivative $\partial_\nu^A u$ in (16.76) is meaningfully formulated.

Comment 2. In the context of Theorem 16.8, notice that corresponding to the case $q = 1$ the Neumann Problem formulated in (16.76) reduces to the Neumann Problem with boundary data in the Beurling-Hardy space $\mathrm{HA}^p(\partial\Omega, \sigma)$. As such, Theorem 16.8 recovers our earlier result obtained in Theorem 9.4.

16.6 The Dirichlet Problem with Boundary Data in $\mathrm{BCMO}_{p,q}$

Our next result concerns the solvability of the Dirichlet Problem with boundary data in the space of (p, q)-bounded central mean oscillations functions, and size condition formulated in terms of (p, q)-central Carleson measures.

Theorem 16.9 *Assume $\Omega \subseteq \mathbb{R}^n$ is an Ahlfors regular domain. Set $\sigma := \mathcal{H}^{n-1}\lfloor\partial\Omega$, and denote by ν the geometric measure theoretic outward unit normal to Ω. Pick two exponents $p \in (1, \infty)$ and $q \in \big(\frac{n-1}{n}, 1\big]$. Suppose L is a second-order, homogeneous, constant complex coefficients, weakly elliptic $M \times M$ system in $\mathbb{R}^n$, and select an aperture parameter $\kappa \in (0, \infty)$. Formulate the Dirichlet Problem for L in Ω with boundary data taken from $[\mathrm{BCMO}_{p,q}(\partial\Omega, \sigma)]^M$ as:*

$$\begin{cases} u \in [\mathscr{C}^\infty(\Omega)]^M, \quad Lu = 0 \ \text{ in } \Omega, \\ |\nabla u|^p \operatorname{dist}(\cdot, \partial\Omega)^{p-1}\, d\mathcal{L}^n \ \text{ is a } (p,q)\text{-central Carleson measure in } \Omega, \\ u\big|_{\partial\Omega}^{\kappa-\text{n.t.}} = f \in \big[\mathrm{BCMO}_{p,q}(\partial\Omega, \sigma)\big]^M. \end{cases} \tag{16.83}$$

Then whenever $\mathfrak{A}_L^{\mathrm{dis}} \neq \varnothing$ there exists a threshold $\delta \in (0, 1)$ depending on n, M, p, q, A, and the Ahlfors regularity constants of $\partial\Omega$ with the property that whenever

$$\|\nu\|_{[\mathrm{BMO}(\partial\Omega,\sigma)]^n} < \delta \tag{16.84}$$

it follows that the Dirichlet Problem is solvable. More transparently, whenever $A \in \mathfrak{A}_L^{\mathrm{dis}}$ and $\big[K_{A,\mathrm{mod}}\big]$ acts on $\big[\widetilde{\mathrm{BCMO}}_{p,q}(\partial\Omega, \sigma)\big]^M$ as in (15.31), then the operator $\frac{1}{2}I + [K_{A,\mathrm{mod}}]$ is invertible on $\big[\widetilde{\mathrm{BCMO}}_{p,q}(\partial\Omega, \sigma)\big]^M$ and the Dirichlet Problem (16.83) is solved by the function $u : \Omega \to \mathbb{C}^M$ defined as

$$u := \mathcal{D}_{A,\mathrm{mod}} g + c \ \text{ in } \Omega \tag{16.85}$$

for a suitable constant $c \in \mathbb{C}^M$ and some choice of

$$g \in \big[\mathrm{BCMO}_{p,q}(\partial\Omega, \sigma)\big]^M \ \text{ such that } \ [g] = \Big(\tfrac{1}{2}I + \big[K_{A,\mathrm{mod}}\big]\Big)^{-1}[f]. \tag{16.86}$$

Moreover, this solution satisfies

$$
\sup_{R \geq 2_*} \left\{ R^{(n-1)(1-1/q)} \left(\frac{1}{\sigma(\Delta(x_0, R))} \int_{B(x_0,R)\cap\Omega} |\nabla u|^p \operatorname{dist}(\cdot, \partial\Omega)^{p-1}\, d\mathcal{L}^n \right)^{1/p} \right\}
\leq C\|f\|_{[\mathrm{BCMO}_{p,q}(\partial\Omega,\sigma)]^M}, \tag{16.87}
$$

some constant $C \in (0, \infty)$ independent of f.

Proof The validity of all claims follows by combining the reasoning presented in the proof of Theorem 9.5, the embeddings established in Proposition 12.4, the (p, q)-central Carleson measure estimates from Theorem 13.10, and the invertibility result recorded in Theorem 15.9. □

Comment 1. In relation to the boundary value problem formulated in (16.83), observe that corresponding to the case $q = 1$ Theorem 16.9 recovers the solvability result recorded in Theorem 9.5.

Comment 2. The Dirichlet Problem (16.83) is closely related to the Dirichlet Problem with boundary data in the (homogeneous) Hölder space $\dot{\mathscr{C}}^{(n-1)(1/q-1)}(\partial\Omega)$, namely

$$
\begin{cases} u \in \left[\dot{\mathscr{C}}^{(n-1)(1/q-1)}(\overline{\Omega}) \cap \mathscr{C}^\infty(\Omega)\right]^M, & Lu = 0 \text{ in } \Omega, \\ u\big|_{\partial\Omega} = f \in \left[\dot{\mathscr{C}}^{(n-1)(1/q-1)}(\partial\Omega)\right]^M. \end{cases} \tag{16.88}
$$

The latter has been solved in [GHA-V, Theorem 8.4.36, pp. 862–863], where it was shown that any solution u of this problem has the property that

$$
\sup_{\substack{x\in\partial\Omega \\ R>0}} \left\{ R^{(n-1)(1-1/q)} \left(\frac{1}{\sigma(\Delta(x, R))} \int_{B(x,R)\cap\Omega} |\nabla u|^p \operatorname{dist}(\cdot, \partial\Omega)^{p-1}\, d\mathcal{L}^n \right)^{1/p} \right\}
\leq C\|f\|_{[\dot{\mathscr{C}}^{(n-1)(1/q-1)}(\partial\Omega)]^M}. \tag{16.89}
$$

Since $\dot{\mathscr{C}}^{(n-1)(1/q-1)}(\partial\Omega)$ continuously embeds into $\mathrm{BCMO}_{p,q}(\partial\Omega, \sigma)$ (cf. item *(c)* in Proposition 12.4), this should be compared with the estimate obtained in (16.87).

Chapter 17
Measure Theoretic Herz Spaces

The topics treated in this chapter are as follows. In Section 17.1 we introduce the (homogeneous and inhomogeneous) measure theoretic Herz spaces and establish some of their basic functional analytic properties, while in Section 17.2 we show that this scale of spaces is stable under the real method of interpolation. In Section 17.3, we study second-generation Herz spaces in the context of unbounded Ahlfors regular sets. In Section 17.4 we prove a real interpolation result on $\dot{\mathrm{K}}^{\alpha}_{p,q}$ spaces, and derive embedding results as well as a characterization of $\dot{\mathrm{K}}^{\alpha}_{p,q}$ in terms of weighted Lebesgue spaces. In Section 17.5 we deduce the boundedness of the Hardy-Littlewood maximal operator on $\dot{\mathrm{K}}^{\alpha}_{p,q}$. Finally, in Section 17.6 and Section 17.7 we introduce new classes of Herz-type spaces, which we refer to as Geometric Herz spaces, and Herz spaces associated with Generalized Banach Function Spaces.

17.1 Definitions and Basic Properties

To get started, consider an arbitrary measure space $(\mathfrak{X}, \mathfrak{M}, \mu)$. Assume there exists some collection of sets $\mathscr{C} := \{C_k\}_{k\in\mathbb{Z}} \subseteq \mathfrak{M}$ satisfying

$$\begin{aligned} &\mu\Big(\mathfrak{X}\setminus \textstyle\bigcup_{k\in\mathbb{Z}} C_k\Big) = 0, \quad \mu(C_k) > 0 \quad \text{for all } k \in \mathbb{Z}, \\ &\text{and} \quad C_k \cap C_j = \varnothing \quad \text{for every } k, j \in \mathbb{Z} \text{ with } k \neq j. \end{aligned} \tag{17.1}$$

Fix a dilation factor $2_* > 1$, along with two exponents $p, q \in (0, \infty]$ and an index $\alpha \in \mathbb{R}$. For each μ-measurable function $f : \mathfrak{X} \to \mathbb{R}$ introduce

$$\|f\|_{\dot{\mathrm{K}}^{\alpha}_{p,q}(\mathfrak{X},\mu,\mathscr{C})} := \begin{cases} \Big[\displaystyle\sum_{k=-\infty}^{\infty} \big(2_*^{k\alpha}\|f\cdot \mathbb{1}_{C_k}\|_{L^p(\mathfrak{X},\mu)}\big)^q\Big]^{1/q} & \text{if } q<\infty, \\ \displaystyle\sup_{k\in\mathbb{Z}} \big\{2_*^{k\alpha}\|f\cdot \mathbb{1}_{C_k}\|_{L^p(\mathfrak{X},\mu)}\big\} & \text{if } q=\infty. \end{cases} \tag{17.2}$$

© The Author(s), under exclusive license to Springer Nature Switzerland AG 2026

M. Mitrea and P. Takemura, *Singular Integrals, Herz-Type Function Spaces, and Boundary Problems*, Progress in Mathematics 362,
https://doi.org/10.1007/978-3-032-12516-3_17

Define the homogeneous measure theoretic Herz space on $\mathfrak{X}$ associated with $\mathscr{C}$ as

$$\begin{aligned} &\dot{\mathrm{K}}^{\alpha}_{p,q}(\mathfrak{X},\mu,\mathscr{C}) \\ &\quad := \Big\{ f : \mathfrak{X} \to \mathbb{R} : f \text{ is } \mu\text{-measurable and } \|f\|_{\dot{\mathrm{K}}^{\alpha}_{p,q}(\mathfrak{X},\mu,\mathscr{C})} < \infty \Big\}. \end{aligned} \tag{17.3}$$

It is visible from (17.2) that if $p, q \in [1, \infty]$ and $\alpha \in \mathbb{R}$ then $\|\cdot\|_{\dot{\mathrm{K}}^{\alpha}_{p,q}(\mathfrak{X},\mu,\mathscr{C})}$ is a norm on $\dot{\mathrm{K}}^{\alpha}_{p,q}(\mathfrak{X},\mu,\mathscr{C})$, while

$$\begin{aligned} &\text{whenever } p \in [1,\infty],\ q \in (0,1), \text{ and } \alpha \in \mathbb{R}, \text{ the quasinorm} \\ &\|\cdot\|_{\dot{\mathrm{K}}^{\alpha}_{p,q}(\mathfrak{X},\mu,\mathscr{C})} \text{ is a } q\text{-norm, i.e., for every } \mu\text{-measurable functions } f, g \\ &\text{on } \mathfrak{X} \text{ one has } \|f+g\|^{q}_{\dot{\mathrm{K}}^{\alpha}_{p,q}(\mathfrak{X},\mu,\mathscr{C})} \le \|f\|^{q}_{\dot{\mathrm{K}}^{\alpha}_{p,q}(\mathfrak{X},\mu,\mathscr{C})} + \|g\|^{q}_{\dot{\mathrm{K}}^{\alpha}_{p,q}(\mathfrak{X},\mu,\mathscr{C})}. \end{aligned} \tag{17.4}$$

Moreover, we also see from (17.2) that the space $\dot{\mathrm{K}}^{\alpha}_{p,q}(\mathfrak{X},\mu,\mathscr{C})$ satisfies the lattice property, i.e.,

$$\begin{aligned} &\text{if } f, g : \mathfrak{X} \to \mathbb{R} \text{ are two } \mu\text{-measurable functions with the property} \\ &\text{that } |g| \le |f| \text{ at } \mu\text{-a.e. point on } \mathfrak{X} \text{ and } f \in \dot{\mathrm{K}}^{\alpha}_{p,q}(\mathfrak{X},\mu,\mathscr{C}), \text{ then } g \text{ also} \\ &\text{belongs to } \dot{\mathrm{K}}^{\alpha}_{p,q}(\mathfrak{X},\mu,\mathscr{C}), \text{ and } \|g\|_{\dot{\mathrm{K}}^{\alpha}_{p,q}(\mathfrak{X},\mu,\mathscr{C})} \le \|f\|_{\dot{\mathrm{K}}^{\alpha}_{p,q}(\mathfrak{X},\mu,\mathscr{C})}. \end{aligned} \tag{17.5}$$

Consequently, it follows that

$$\begin{aligned} &\text{the operator of pointwise multiplication by a given} \\ &\text{function } b \in L^{\infty}(\mathfrak{X},\mu) \text{ is a bounded mapping} \\ &\text{from } \dot{\mathrm{K}}^{\alpha}_{p,q}(\mathfrak{X},\mu,\mathscr{C}) \text{ into itself, with operator norm} \\ &\le \|b\|_{L^{\infty}(\mathfrak{X},\mu)}. \end{aligned} \tag{17.6}$$

Another useful property of the quasinorm on Herz spaces is the manner in which it "dilates" in relation to the parameters p, q, and α. Specifically, given $s \in (0,\infty)$, for each μ-measurable function $f : \mathfrak{X} \to \mathbb{R}$ one has

$$\big\| |f|^{s} \big\|_{\dot{\mathrm{K}}^{\alpha}_{p,q}(\mathfrak{X},\mu,\mathscr{C})} = \|f\|^{s}_{\dot{\mathrm{K}}^{\alpha/s}_{ps,qs}(\mathfrak{X},\mu,\mathscr{C})}. \tag{17.7}$$

It is relevant to observe that in the diagonal case $p = q \in (0,\infty)$, the homogeneous Herz space $\dot{\mathrm{K}}^{\alpha}_{p,q}(\mathfrak{X},\mu,\mathscr{C})$ coincides with a weighted Lebesgue space. More precisely, if we set

$$\dot{w} := \sum_{k=-\infty}^{\infty} 2^{k}_{*} \cdot \mathbb{1}_{C_k} \quad \text{on } \mathfrak{X}, \tag{17.8}$$

then from (17.2) we see that

$$\dot{\mathrm{K}}^{\alpha}_{p,p}(\mathfrak{X},\mu,\mathscr{C}) = L^{p}\big(\mathfrak{X}, \dot{w}^{\alpha p}\mu\big) \quad \text{for each } p \in (0,\infty) \text{ and } \alpha \in \mathbb{R}. \tag{17.9}$$

To proceed, we introduce the inhomogeneous version of measure theoretic Herz spaces. Continue to assume that $(\mathfrak{X}, \mathfrak{M}, \mu)$ is a generic measure space, and now suppose there exists a collection of sets $\widetilde{\mathscr{C}} := \{\widetilde{C}_k\}_{k\in\mathbb{N}_0} \subseteq \mathfrak{M}$ such that

$$\begin{aligned} \mathfrak{X} = \textstyle\bigcup_{k\in\mathbb{N}_0} \widetilde{C}_k, \quad \mu(\widetilde{C}_k) > 0 \quad \text{for all } k \in \mathbb{N}_0 \\ \text{and} \quad \widetilde{C}_k \cap \widetilde{C}_j = \varnothing \quad \text{for every } k, j \in \mathbb{Z}, \text{ with } k \neq j. \end{aligned} \tag{17.10}$$

Select a dilation factor $2_* > 1$, two exponents $p, q \in (0, \infty]$, and an index $\alpha \in \mathbb{R}$. For each μ-measurable function $f : \mathfrak{X} \to \mathbb{R}$ define

$$\|f\|_{\mathrm{K}^\alpha_{p,q}(\mathfrak{X},\mu,\widetilde{\mathscr{C}})} := \begin{cases} \Big[\displaystyle\sum_{k=0}^{\infty} \big(2_*^{k\alpha}\|f \cdot \mathbb{1}_{\widetilde{C}_k}\|_{L^p(\mathfrak{X},\mu)}\big)^q\Big]^{1/q} & \text{if } q < \infty, \\ \displaystyle\sup_{k\in\mathbb{N}_0} \big\{2_*^{k\alpha}\|f \cdot \mathbb{1}_{\widetilde{C}_k}\|_{L^p(\mathfrak{X},\mu)}\big\} & \text{if } q = \infty. \end{cases} \tag{17.11}$$

Introduce the `inhomogeneous measure theoretic Herz space` on $\mathfrak{X}$ associated with $\widetilde{\mathscr{C}}$ as

$$\begin{aligned} &\mathrm{K}^\alpha_{p,q}(\mathfrak{X}, \mu, \widetilde{\mathscr{C}}) \\ &\quad := \big\{f : \mathfrak{X} \to \mathbb{R} : f \text{ is } \mu\text{-measurable and } \|f\|_{\mathrm{K}^\alpha_{p,q}(\mathfrak{X},\mu,\widetilde{\mathscr{C}})} < \infty\big\}. \end{aligned} \tag{17.12}$$

From (17.11) we see that whenever $p, q \in [1, \infty]$ and $\alpha \in \mathbb{R}$ then $\|\cdot\|_{\mathrm{K}^\alpha_{p,q}(\mathfrak{X},\mu,\widetilde{\mathscr{C}})}$ is a norm on $\mathrm{K}^\alpha_{p,q}(\mathfrak{X}, \mu, \widetilde{\mathscr{C}})$, while

> whenever $p \in [1, \infty]$, $q \in (0, 1)$, and $\alpha \in \mathbb{R}$, the quasinorm $\|\cdot\|_{\mathrm{K}^\alpha_{p,q}(\mathfrak{X},\mu,\widetilde{\mathscr{C}})}$ is a q-norm, i.e., for every μ-measurable functions f, g on $\mathfrak{X}$ one has $\|f + g\|^q_{\mathrm{K}^\alpha_{p,q}(\mathfrak{X},\mu,\widetilde{\mathscr{C}})} \leq \|f\|^q_{\mathrm{K}^\alpha_{p,q}(\mathfrak{X},\mu,\widetilde{\mathscr{C}})} + \|g\|^q_{\mathrm{K}^\alpha_{p,q}(\mathfrak{X},\mu,\widetilde{\mathscr{C}})}$. (17.13)

It is also clear from definitions that $\mathrm{K}^\alpha_{p,q}(\mathfrak{X}, \mu, \widetilde{\mathscr{C}})$ is a lattice, i.e.,

> if $f, g : \mathfrak{X} \to \mathbb{R}$ are two μ-measurable functions with the property that $|g| \leq |f|$ at μ-a.e. point on $\mathfrak{X}$ and $f \in \mathrm{K}^\alpha_{p,q}(\mathfrak{X}, \mu, \widetilde{\mathscr{C}})$, then g also belongs to $\mathrm{K}^\alpha_{p,q}(\mathfrak{X}, \mu, \widetilde{\mathscr{C}})$, and $\|g\|_{\mathrm{K}^\alpha_{p,q}(\mathfrak{X},\mu,\widetilde{\mathscr{C}})} \leq \|f\|_{\mathrm{K}^\alpha_{p,q}(\mathfrak{X},\mu,\widetilde{\mathscr{C}})}$. (17.14)

In particular, this implies

> the operator of pointwise multiplication by a given function $b \in L^\infty(\mathfrak{X}, \mu)$ is a bounded mapping from $\mathrm{K}^\alpha_{p,q}(\mathfrak{X}, \mu, \widetilde{\mathscr{C}})$ into itself, with operator norm $\leq \|b\|_{L^\infty(\mathfrak{X},\mu)}$. (17.15)

It is clear from (17.11) that for every $s \in (0, \infty)$, and for each μ-measurable function $f : \mathfrak{X} \to \mathbb{R}$ one has

$$\| |f|^s \|_{\mathrm{K}^{\alpha}_{p,q}(\mathfrak{X},\mu,\widetilde{\mathscr{C}})} = \|f\|^s_{\mathrm{K}^{\alpha/s}_{ps,qs}(\mathfrak{X},\mu,\widetilde{\mathscr{C}})}. \tag{17.16}$$

In addition, we can see from (17.11) that whenever $\alpha_0, \alpha_1 \in \mathbb{R}$ and $\alpha_0 \leq \alpha_1$ the following continuous embedding holds

$$\mathrm{K}^{\alpha_1}_{p,q}(\mathfrak{X},\mu,\widetilde{\mathscr{C}}) \hookrightarrow \mathrm{K}^{\alpha_0}_{p,q}(\mathfrak{X},\mu,\widetilde{\mathscr{C}}). \tag{17.17}$$

It is useful to notice that in the diagonal case $p = q \in (0,\infty)$, the space $\mathrm{K}^{\alpha}_{p,q}$ coincides with a weighted Lebesgue spaces. More precisely, if we introduce

$$w := \sum_{k=0}^{\infty} 2^k_* \cdot \mathbb{1}_{\widetilde{C}_k} \quad \text{on } \mathfrak{X} \tag{17.18}$$

then we can see from (17.11) that

$$\mathrm{K}^{\alpha}_{p,p}(\mathfrak{X},\mu,\widetilde{\mathscr{C}}) = L^p\big(\mathfrak{X}, w^{\alpha p}\mu\big) \quad \text{for each } p \in (0,\infty) \text{ and } \alpha \in \mathbb{R}. \tag{17.19}$$

Corresponding to the case $\alpha = 0$, the identifications (17.19) and (17.9) yield

$$\mathrm{K}^{0}_{p,p}(\mathfrak{X},\mu,\widetilde{\mathscr{C}}) = \dot{\mathrm{K}}^{0}_{p,p}(\mathfrak{X},\mu,\mathscr{C}) = L^p(\mathfrak{X},\mu) \quad \text{for each } p \in (0,\infty). \tag{17.20}$$

Our first result in this section provides a Hölder-type inequality in the context of measure theoretic Herz spaces.

Proposition 17.1 *Consider a measure space $(\mathfrak{X},\mathfrak{M},\mu)$ along with a collection of sets $\mathscr{C} = \{C_k\}_{k\in\mathbb{Z}} \subseteq \mathfrak{M}$ satisfying* (17.1)*. Select two μ-measurable functions f and g defined on $\mathfrak{X}$. Then the following statements are true.*

(a) Having fixed two exponents $p, q \in [1,\infty)$, and an index $\alpha \in \mathbb{R}$, it follows that

$$\int_{\mathfrak{X}} |fg|\, d\mu \leq \|f\|_{\dot{\mathrm{K}}^{\alpha}_{p,q}(\mathfrak{X},\mu,\mathscr{C})} \|g\|_{\dot{\mathrm{K}}^{-\alpha}_{p',q'}(\mathfrak{X},\mu,\mathscr{C})} \tag{17.21}$$

where $p', q' \in (1,\infty]$ denote the Hölder conjugate exponents of p and q, respectively.

(b) Having fixed two exponents $p \in [1,\infty)$ and $q \in (0,1]$, along with an index $\alpha \in \mathbb{R}$, it follows that

$$\int_{\mathfrak{X}} |fg|\, d\mu \leq \|f\|_{\dot{\mathrm{K}}^{\alpha}_{p,q}(\mathfrak{X},\mu,\mathscr{C})} \|g\|_{\dot{\mathrm{K}}^{-\alpha}_{p',\infty}(\mathfrak{X},\mu,\mathscr{C})} \tag{17.22}$$

where $p' \in (1,\infty]$ denotes the Hölder conjugate exponent of p.

Similar estimates as in (17.21)-(17.22) *hold in the context of inhomogeneous measure theoretic Herz spaces.*

Proof To justify the claim made in *(a)*, use Hölder's inequality to estimate

$$
\begin{aligned}
\int_{\mathfrak{X}} |fg|\, d\mu &= \sum_{k=-\infty}^{\infty} \int_{C_k} |fg|\, d\mu \leq \sum_{k=-\infty}^{\infty} \|f \cdot \mathbb{1}_{C_k}\|_{L^p(\mathfrak{X},\mu)} \|g \cdot \mathbb{1}_{C_k}\|_{L^{p'}(\mathfrak{X},\mu)} \\
&= \sum_{k=-\infty}^{\infty} \big(2_*^{k\alpha} \|f \cdot \mathbb{1}_k\|_{L^p(\mathfrak{X},\mu)}\big)\big(2_*^{-k\alpha} \|g \cdot \mathbb{1}_k\|_{L^{p'}(\mathfrak{X},\mu)}\big) \\
&\leq \Big[\sum_{k=-\infty}^{\infty} \big(2_*^{k\alpha} \|f \cdot \mathbb{1}_{C_k}\|_{L^p(\mathfrak{X},\mu)}\big)^q\Big]^{1/q} \\
&\quad \times \Big[\sum_{k=-\infty}^{\infty} \big(2_*^{-k\alpha} \|g \cdot \mathbb{1}_{C_k}\|_{L^{p'}(\mathfrak{X},\mu)}\big)^{q'}\Big]^{1/q'} \\
&= \|f\|_{\dot{\mathrm{K}}^{\alpha}_{p,q}(\mathfrak{X},\mu,\mathscr{C})} \|g\|_{\dot{\mathrm{K}}^{-\alpha}_{p',q'}(\mathfrak{X},\mu,\mathscr{C})}.
\end{aligned}
\tag{17.23}
$$

Regarding the claim made in item *(b)*, use Hölder's inequality along with the fact that $0 < q \leq 1$ to obtain

$$
\begin{aligned}
\int_{\mathfrak{X}} |fg|\, d\mu &= \sum_{k=-\infty}^{\infty} \int_{C_k} |fg|\, d\mu \leq \sum_{k=-\infty}^{\infty} \|f \cdot \mathbb{1}_k\|_{L^p(\mathfrak{X},\mu)} \|g \cdot \mathbb{1}_k\|_{L^{p'}(\mathfrak{X},\mu)} \\
&= \sum_{k=-\infty}^{\infty} \big(2_*^{k\alpha} \|f \cdot \mathbb{1}_k\|_{L^p(\mathfrak{X},\mu)}\big)\big(2_*^{-k\alpha} \|g \cdot \mathbb{1}_k\|_{L^{p'}(\mathfrak{X},\mu)}\big) \\
&\leq \Big[\sum_{k=-\infty}^{\infty} \big(2_*^{-k\alpha} \|f \cdot \mathbb{1}_k\|_{L^p(\mathfrak{X},\mu)}\big)\Big]^{q(1/q)} \\
&\quad \times \sup_{k\in\mathbb{Z}} \big(2_*^{-k\alpha} \|g \cdot \mathbb{1}_k\|_{L^{p'}(\mathfrak{X},\mu)}\big) \\
&\leq \|f\|_{\dot{\mathrm{K}}^{\alpha}_{p,q}(\mathfrak{X},\mu,\mathscr{C})} \|g\|_{\dot{\mathrm{K}}^{\alpha}_{p',\infty}(\mathfrak{X},\mu,\mathscr{C})}.
\end{aligned}
\tag{17.24}
$$

The same reasoning as in (17.23)-(17.24) produce similar estimates in the context of inhomogeneous measure theoretic Herz spaces. This concludes the proof of Proposition 17.1. □

The next goal is to obtain a version of Lebesgue's Monotone Convergence result in the context of measure theoretic Herz spaces.

Proposition 17.2 *Consider a measure space $(\mathfrak{X}, \mathfrak{M}, \mu)$ along with a collection of sets $\mathscr{C} = \{C_k\}_{k\in\mathbb{Z}} \subseteq \mathfrak{M}$ satisfying* (17.1)*. Let $\{f_j\}_{j\in\mathbb{N}}$ be a sequence of non-negative μ-measurable functions on $\mathfrak{X}$ such that $f_j \nearrow f$ pointwise μ-a.e. on $\mathfrak{X}$ as $j \to \infty$, for some non-negative μ-measurable function f on $\mathfrak{X}$. Select two exponents $p \in (0, \infty)$ and $q \in (0, \infty]$, along with an index $\alpha \in \mathbb{R}$. Then*

$$\|f_j\|_{\dot{\mathrm{K}}^{\alpha}_{p,q}(\mathfrak{X},\mu,\mathscr{C})} \nearrow \|f\|_{\dot{\mathrm{K}}^{\alpha}_{p,q}(\mathfrak{X},\mu,\mathscr{C})} \quad as\ j \to \infty. \tag{17.25}$$

The same type of conclusion holds for inhomogeneous measure theoretic Herz spaces.

Proof Bearing in mind the format of the quasinorm defined in (17.2) and the lattice property in (17.5), the validity of the convergence in (17.25) follows from a similar reasoning to that used in the proof of Proposition 10.5. □

The next result establishes the completeness of measure theoretic Herz spaces.

Proposition 17.3 *Consider a measure space $(\mathfrak{X}, \mathfrak{M}, \mu)$, along with a collection $\mathscr{C} = \{C_k\}_{k\in\mathbb{Z}} \subseteq \mathfrak{M}$ satisfying* (17.1). *Then whenever $p, q \in (0, \infty]$ the space $\dot{\mathrm{K}}^{\alpha}_{p,q}(\mathfrak{X}, \mu, \mathscr{C})$ is complete when equipped with $\|\cdot\|_{\dot{\mathrm{K}}^{\alpha}_{p,q}(\mathfrak{X},\mu,\mathscr{C})}$. In particular, in the regime $p, q \in [1, \infty]$ (respectively $p \in (0, 1)$ or $q \in (0, 1)$), one may conclude that $\big(\dot{\mathrm{K}}^{\alpha}_{p,q}(\mathfrak{X}, \mu, \mathscr{C}), \|\cdot\|_{\dot{\mathrm{K}}^{\alpha}_{p,q}(\mathfrak{X},\mu,\mathscr{C})}\big)$ is a Banach space (respectively quasi-Banach space). The same results hold for the inhomogeneous Herz spaces.*

Proof The proof of the completeness of the space $\dot{\mathrm{K}}^{\alpha}_{p,q}(\mathfrak{X}, \mu, \mathscr{C})$ follows via a similar type of argument as the one presented in the proof of Proposition 3.9. □

We also have the following duality result in the context of measure theoretic Herz spaces.

Proposition 17.4 *Assume $(\mathfrak{X}, \mathfrak{M}, \mu)$ is a measure space, and let $\mathscr{C} = \{C_k\}_{k\in\mathbb{Z}} \subseteq \mathfrak{M}$ be a collection satisfying* (17.1). *Select two exponents $p \in (1, \infty)$ and $q \in (0, \infty)$, and an index $\alpha \in \mathbb{R}$. Then the following statements are valid.*

(a) Whenever $1 < q < \infty$ it follows that the dual of $\mathrm{K}^{\alpha}_{p,q}(\mathfrak{X}, \mu, \mathscr{C})$ is $\mathrm{K}^{-\alpha}_{p',q'}(\mathfrak{X}, \mu, \mathscr{C})$, i.e.,

$$\big(\mathrm{K}^{\alpha}_{p,q}(\mathfrak{X}, \mu, \mathscr{C})\big)^* = \mathrm{K}^{-\alpha}_{p',q'}(\mathfrak{X}, \mu, \mathscr{C}), \tag{17.26}$$

where $p', q' \in (1, \infty)$ denote the Hölder conjugate exponents of p and q, respectively. Specifically, the mapping

$$\dot{\mathrm{K}}^{-\alpha}_{p',q'}(\mathfrak{X}, \mu, \mathscr{C}) \ni f \longmapsto \Phi f := \Lambda_f \in \big(\dot{\mathrm{K}}^{\alpha}_{p,q}(\mathfrak{X}, \mu, \mathscr{C})\big)^* \tag{17.27}$$

where

$$\Lambda_f(g) := \int_{\mathfrak{X}} fg\, d\mu \quad \text{for all } g \in \dot{\mathrm{K}}^{\alpha}_{p,q}(\mathfrak{X}, \mu, \mathscr{C}) \tag{17.28}$$

is a well-defined, linear, continuous and isometric isomorphism.

(b) Whenever $0 < q \le 1$ it follows that the dual of $\mathrm{K}^{\alpha}_{p,q}(\mathfrak{X}, \mu, \mathscr{C})$ is $\mathrm{K}^{-\alpha}_{p',\infty}(\mathfrak{X}, \mu, \mathscr{C})$, i.e.,

$$\big(\dot{\mathrm{K}}^{\alpha}_{p,q}(\mathfrak{X}, \mu, \mathscr{C})\big)^* = \dot{\mathrm{K}}^{-\alpha}_{p',\infty}(\mathfrak{X}, \mu, \mathscr{C}). \tag{17.29}$$

where $p' \in (1, \infty)$ is the Hölder conjugate exponent of p. Specifically, the mapping

$$\dot{\mathrm{K}}^{-\alpha}_{p',\infty}(\mathfrak{X},\mu,\mathscr{C}) \ni f \longmapsto \Psi f := \Lambda_f \in \big(\dot{\mathrm{K}}^{\alpha}_{p,q}(\mathfrak{X},\mu,\mathscr{C})\big)^* \tag{17.30}$$

where

$$\Lambda_f(g) := \int_{\mathfrak{X}} fg\,d\mu \quad \textit{for all } g \in \dot{\mathrm{K}}^{\alpha}_{p,q}(\mathfrak{X},\mu,\mathscr{C}) \tag{17.31}$$

is well defined, linear, and an isometric isomorphism.

Proof Based on the format of the quasinorm defined in (17.2) and Proposition 17.1, the duality results claimed in items *(a)* and *(b)* may be obtained via a similar approach as that used to establish the duality result recorded in Proposition 10.9. □

We next show that measure theoretic Herz spaces fit within the framework of Generalized Banach Function Spaces (in the sense described in Definition 10.2).

Proposition 17.5 *Consider a measure space* $(\mathfrak{X},\mathfrak{M},\mu)$ *and let* $\mathscr{C} = \{C_k\}_{k\in\mathbb{Z}} \subseteq \mathfrak{M}$ *be a collection satisfying* (17.1). *Assume there exists a family of sets* $\{B_j\}_{j\in\mathbb{N}} \subseteq \mathfrak{M}$ *satisfying*

$$\mathfrak{X} = \bigcup_{j\in\mathbb{N}} B_j \quad \textit{and} \quad 0 < \mu(B_j) < \infty \quad \textit{for all } j \in \mathbb{N}. \tag{17.32}$$

Select two exponents $p, q \in (1,\infty)$, *along an index* $\alpha \in \mathbb{R}$. *Then* $\|\cdot\|_{\dot{\mathrm{K}}^{\alpha}_{p,q}(\mathfrak{X},\mu,\mathscr{C})}$ *is a function norm (in the sense of Definition 10.2). In particular, the homogeneous measure theoretic Herz space* $\dot{\mathrm{K}}^{\alpha}_{p,q}(\mathfrak{X},\mu,\mathscr{C})$ *is a Generalized Banach Function Space on* $\mathfrak{X}$. *Similar conclusions hold in the context of inhomogeneous measure theoretic Herz spaces.*

Proof It is clear from (17.2) that the function $\|\cdot\|_{\dot{\mathrm{K}}^{\alpha}_{p,q}(\mathfrak{X},\mu,\mathscr{C})}$ satisfies properties (P1) and (P2) in Definition 10.2. Moreover, the fact that $\|\cdot\|_{\dot{\mathrm{K}}^{\alpha}_{p,q}(\mathfrak{X},\mu,\mathscr{C})}$ satisfies (P3) in Definition 10.2 follows from Proposition 17.25. To verify property (P4) in Definition 10.2, define

$$U_{j,k} := B_j \cap C_k \quad \text{for each } j \in \mathbb{N} \text{ and } k \in \mathbb{Z}. \tag{17.33}$$

From this, (17.1), and (17.32) we see that

$$\mu\Bigg(\mathfrak{X} \setminus \bigcup_{\substack{j\in\mathbb{N}\\ k\in\mathbb{Z}}} U_{j,k}\Bigg) = 0. \tag{17.34}$$

Moreover, based on (17.33), (17.2), and (17.32), for every $j \in \mathbb{N}$ and $k \in \mathbb{Z}$ we have

$$\|\mathbb{1}_{U_{j,k}}\|_{\dot{\mathrm{K}}^{\alpha}_{p,q}(\mathfrak{X},\mu,\mathscr{C})} = 2^{k\alpha}_{*}\|\mathbb{1}_{B_j\cap C_k}\|_{L^p(\mathfrak{X},\mu)} \le 2^{k\alpha}_{*}\mu(B_j)^{1/p} < \infty. \tag{17.35}$$

This shows that $\|\cdot\|_{\dot{\mathrm{K}}^{\alpha}_{p,q}(\mathfrak{X},\mu,\mathscr{C})}$ satisfies (P4) in Definition 10.2 relative to the countable family of μ-measurable sets $\{U_{j,k}\}_{\substack{j\in\mathbb{N}\\ k\in\mathbb{Z}}}$. Additionally, in relation to this collection of sets, for every non-negative μ-measurable function f on $\mathfrak{X}$, and every $j \in \mathbb{N}$ and $k \in \mathbb{Z}$, use Hölder's inequality to estimate

$$\begin{aligned}\int_{U_{j,k}} f\,d\mu = \int_{B_j\cap C_k} f\,d\mu &\le 2_*^{\alpha k}\|f\cdot \mathbb{1}_{B_j\cap C_k}\|_{L^p(\mathfrak{X},\mu)}\cdot 2_*^{-\alpha k}\|\mathbb{1}_{B_j\cap C_k}\|_{L^{p'}(\mathfrak{X},\mu)}\\ &\le 2_*^{\alpha k}\|f\cdot \mathbb{1}_{C_k}\|_{L^p(\mathfrak{X},\mu)}\cdot 2_*^{-\alpha k}\|\mathbb{1}_{B_j}\|_{L^{p'}(\mathfrak{X},\mu)}\\ &\le 2_*^{-\alpha k}\mu(B_j)^{1/p'}\|f\|_{\dot{\mathrm{K}}^{\alpha}_{p,q}(\mathfrak{X},\mu,\mathscr{C})},\end{aligned} \tag{17.36}$$

where $p' \in (1,\infty)$ denotes the Hölder conjugate exponent of p. Since $\mu(B_j) < \infty$ (cf. (17.32)), this ensures that $\|\cdot\|_{\dot{\mathrm{K}}^{\alpha}_{p,q}(\mathfrak{X},\mu,\mathscr{C})}$ satisfies (P5) in Definition 10.2 with respect to the collection $\{U_{j,k}\}_{\substack{j\in\mathbb{N}\\ k\in\mathbb{Z}}}$. All together, we conclude that $\|\cdot\|_{\dot{\mathrm{K}}^{\alpha}_{p,q}(\mathfrak{X},\mu,\mathscr{C})}$ is a function norm on $\mathfrak{X}$. Finally, this guarantees that $\dot{\mathrm{K}}^{\alpha}_{p,q}(\mathfrak{X},\mu,\mathscr{C})$ is a Generalized Banach Function Space on $(\mathfrak{X},\mu)$. □

Having established the GBFS character of measure theoretic Herz spaces, we may now identify their associated spaces.

Proposition 17.6 *Consider a measure space $(\mathfrak{X},\mathfrak{M},\mu)$ along with a countable collection of sets $\mathscr{C} = \{C_k\}_{k\in\mathbb{Z}} \subseteq \mathfrak{M}$ satisfying* (17.1)*. Fix two exponents $p, q \in (1,\infty)$, along with an index $\alpha \in \mathbb{R}$. Denote by $p', q' \in (1,\infty)$ the Hölder conjugate exponents of p and q, respectively. Then for every μ-measurable function f defined on $\mathfrak{X}$ one has*

$$\|f\|_{\dot{\mathrm{K}}^{-\alpha}_{p',q'}(\mathfrak{X},\mu,\mathscr{C})} = \sup\Big\{\int_{\mathfrak{X}}|fg|\,d\mu : g\in \dot{\mathrm{K}}^{\alpha}_{p,q}(\mathfrak{X},\mu,\mathscr{C}),\ \textit{with}\ \|g\|_{\dot{\mathrm{K}}^{\alpha}_{p,q}(\mathfrak{X},\mu,\mathscr{C})}\le 1\Big\}. \tag{17.37}$$

In particular, whenever a family of sets $\{B_j\}_{j\in\mathbb{N}} \subseteq \mathfrak{M}$ satisfying (17.32) *exists, it follows that the associated space of $\dot{\mathrm{K}}^{\alpha}_{p,q}(\mathfrak{X},\mu,\mathscr{C})$ is $\dot{\mathrm{K}}^{-\alpha}_{p',q'}(\mathfrak{X},\mu,\mathscr{C})$, i.e.,*

$$\big(\dot{\mathrm{K}}^{\alpha}_{p,q}(\mathfrak{X},\mu,\mathscr{C})\big)' = \dot{\mathrm{K}}^{-\alpha}_{p',q'}(\mathfrak{X},\mu,\mathscr{C}). \tag{17.38}$$

Lastly, similar conclusions hold in the context of inhomogeneous measure theoretic Herz spaces.

Proof The right-pointing inequality on (17.37) is a consequence of item *(a)* in Proposition 17.1. To justify the opposite inequality, we adapt the reasoning used to prove Proposition 10.12. First, notice that we can express

$$\|f\|_{\dot{\mathrm{K}}^{-\alpha}_{p',q'}(\mathfrak{X},\mu,\mathscr{C})} = \Big\|\big\{2_*^{-k\alpha}\|f\cdot\mathbb{1}_{C_k}\|_{L^{p'}(\mathfrak{X},\mu)}\big\}_{k\in\mathbb{Z}}\Big\|_{\ell^{q'}(\mathbb{Z})}. \tag{17.39}$$

Select $\eta \in (0,\infty)$ satisfying

$$0 < \eta < \|f\|_{\dot{\mathrm{K}}^{-\alpha}_{p',q'}(\mathfrak{X},\mu,\mathscr{C})} = \left\| \left\{ 2_*^{-k\alpha} \|f \cdot \mathbb{1}_{C_k}\|_{L^{p'}(\mathfrak{X},\mu)} \right\}_{k\in\mathbb{Z}} \right\|_{\ell^{q'}(\mathbb{Z})}. \tag{17.40}$$

Via (Köthe) duality, there exists some sequence $\{b_k\}_{k\in\mathbb{Z}} \subseteq \mathbb{R}$ such that

$$\|\{b_k\}_{k\in\mathbb{Z}}\|_{\ell^q(\mathbb{Z})} \leq 1 \quad \text{and} \quad \eta \leq \sum_{k=-\infty}^{\infty} \left[2_*^{-k\alpha} \|f \cdot \mathbb{1}_{C_k}\|_{L^{p'}(\mathfrak{X},\mu)}\right] |b_k|. \tag{17.41}$$

For each $k \in \mathbb{Z}$, pick $\zeta_k \in (0, \infty)$ satisfying

$$0 < \zeta_k < \|f \cdot \mathbb{1}_{C_k}\|_{L^{p'}(\mathfrak{X},\mu)}. \tag{17.42}$$

Again, via (Köthe) duality, this implies that for each $k \in \mathbb{Z}$ there exists a function $g_k \in L^p(\mathfrak{X}, \mu)$ for which

$$\|g_k\|_{L^p(\mathfrak{X},\mu)} \leq 1 \quad \text{and} \quad \zeta_k \leq \int_{\mathfrak{X}} |f \cdot \mathbb{1}_{C_k}| \cdot |g_k| \, d\mu. \tag{17.43}$$

Next, define

$$\widetilde{g} := \sum_{\ell=-\infty}^{\infty} 2_*^{-\ell\alpha} b_\ell (g_\ell \cdot \mathbb{1}_{C_\ell}) \;\text{ on } \mathfrak{X}. \tag{17.44}$$

It is evident that g is a μ-measurable function and finite μ-a.e. on $\mathfrak{X}$. Thanks to (17.41), (17.43), and (17.44), we may bound

$$\begin{aligned} \|\widetilde{g}\|_{\dot{\mathrm{K}}^{\alpha}_{p,q}(\mathfrak{X},\mu,\mathscr{C})} &= \left\{ \sum_{k=-\infty}^{\infty} \left(2_*^{k\alpha} \|\widetilde{g} \cdot \mathbb{1}_{C_k}\|_{L^p(\mathfrak{X},\mu)}\right)^q \right\}^{1/q} \\ &= \left\{ \sum_{k=-\infty}^{\infty} \left(2_*^{k\alpha} 2_*^{-k\alpha} |b_k| \|g_k \cdot \mathbb{1}_{C_k}\|_{L^p(\mathfrak{X},\mu)}\right)^q \right\}^{1/q} \\ &\leq \|\{b_k\}_{k\in\mathbb{Z}}\|_{\ell^q(\mathbb{Z})} \leq 1. \end{aligned} \tag{17.45}$$

As such, $\widetilde{g} \in \dot{\mathrm{K}}^{\alpha}_{p,q}(\mathfrak{X}, \mu, \mathscr{C})$ and $\|\widetilde{g}\|_{\dot{\mathrm{K}}^{\alpha}_{p,q}(\mathfrak{X},\mu,\mathscr{C})} \leq 1$. In addition, (17.43) and (17.44) entail

$$\int_{\mathfrak{X}} |f\widetilde{g}| \, d\mu = \sum_{k=-\infty}^{\infty} 2_*^{-\ell\alpha} |b_k| \int_{C_k} |f \cdot g_k| \, d\mu \geq \sum_{k=-\infty}^{\infty} 2_*^{-\ell\alpha} |b_k| \zeta_k. \tag{17.46}$$

Since this holds for every ζ_k as in (17.42), let $\zeta_k \nearrow \|f \cdot \mathbb{1}_{C_k}\|_{L^{p'}(\mathfrak{X},\mu)}$ for every $k \in \mathbb{N}_0$, use Lebesgue's Monotone Convergence Theorem for series, and (17.41) to obtain

$$\int_{\mathfrak{X}} |f\widetilde{g}| \, d\mu \geq \sum_{k=0}^{\infty} 2_*^{-\ell\alpha} |b_k| \|f \cdot \mathbb{1}_{C_k}\|_{L^{p'}(\mathfrak{X},\mu)} \geq \eta. \tag{17.47}$$

Because this holds for each η satisfying (10.165), we may let $\eta \nearrow \|f\|_{\dot{K}^{-\alpha}_{p',q'}(\mathfrak{X},\mu,\mathscr{C})}$ which in turn yields

$$\int_{\mathfrak{X}} |f\widetilde{g}|\,d\mu \geq \|f\|_{\dot{K}^{-\alpha}_{p',q'}(\mathfrak{X},\mu,\mathscr{C})}. \tag{17.48}$$

Hence, based on (17.45) we arrive at

$$\|f\|_{\dot{K}^{-\alpha}_{p',q'}(\mathfrak{X},\mu,\mathscr{C})} \leq \sup\Big\{\int_{\mathfrak{X}} |fg|\,d\mu : g \in \dot{K}^{\alpha}_{p,q}(\mathfrak{X},\mu,\mathscr{C}),\ \text{with } \|g\|_{\dot{K}^{\alpha}_{p,q}(\mathfrak{X},\mu,\mathscr{C})} \leq 1\Big\}. \tag{17.49}$$

This finishes the proof of (17.39). Ultimately, the identification claimed in (17.38) is a consequence of Proposition 17.5 and (17.39). □

17.2 Real Interpolation of Measure Theoretic Herz Spaces

In this section we identify the intermediate spaces, obtained by the real method of interpolation, between measure theoretic Herz spaces. Stating said result requires some preliminaries.

To set the stage, consider $(X_0, \|\cdot\|_{X_0})$ and $(X_1, \|\cdot\|_{X_1})$ two quasi-normed spaces that are `compatible`, i.e., X_0 and X_1 embed continuously into a Hausdorff topological vector space $\mathcal{X}$. In this scenario, the algebraic sum $X_0 + X_1$ is a quasi-normed space when equipped with

$$\|x\|_{X_0+X_1} := \inf_{\substack{x=x_0+x_1\\ x_0\in X_0,\ x_1\in X_1}} \big(\|x_0\|_{X_0} + \|x_1\|_{X_1}\big) \quad \text{for all } x \in X_0 + X_1. \tag{17.50}$$

Also, equip the intersection $X_0 \cap X_1$ with the quasinorm

$$\|x\|_{X_0\cap X_1} := \max\big\{\|x\|_{X_0}, \|x\|_{X_1}\big\} \quad \text{for all } x \in X_0 \cap X_1. \tag{17.51}$$

Fix a parameter $\theta \in (0,1)$ along with some $q \in (0,\infty]$. For each $t \in (0,\infty)$ and $x \in X_0 + X_1$ introduce the K-`functional` associated with X_0 and X_1 as

$$K(t,x,X_0,X_1) := \inf_{\substack{x=x_0+x_1\\ x_0\in X_0,\ x_1\in X_1}} \big(\|x_0\|_{X_0} + t\|x_1\|_{X_1}\big). \tag{17.52}$$

Note that for each $x \in X_0 + X_1$ we have

$$K(t,x,X_0,X_1) = tK(t^{-1},x,X_0,X_1) \quad \text{for each } t \in (0,\infty) \tag{17.53}$$

$$(0,\infty) \ni t \mapsto K(t,x,X_0,X_1) \quad \text{is non-decreasing, and} \tag{17.54}$$

$$\tfrac{1}{2}K(2t,x,X_0,X_1) \le K(t,x,X_0,X_1) \le K(2t,x,X_0,X_1) \quad \text{for each } t \in (0,\infty). \tag{17.55}$$

For each $x \in X_0 + X_1$, define the quasinorm

$$\|x\|_{(X_0,X_1)_{\theta,q}} := \begin{cases} \left(\displaystyle\int_0^\infty \left[t^{-\theta}K(t,x,X_0,X_1)\right]^q \frac{dt}{t} \right)^{1/q} & \text{if } 0 < q < \infty, \\ \sup\limits_{t>0} \left[t^{-\theta}K(t,x,X_0,X_1)\right] & \text{if } q = \infty, \end{cases} \tag{17.56}$$

and define the `intermediate space` (for the real method of interpolation) between X_0 and X_1 as

$$(X_0,X_1)_{\theta,q} := \left\{x \in X_0 + X_1 : \|x\|_{(X_0,X_1)_{\theta,q}} < \infty\right\}. \tag{17.57}$$

In this context, it is relevant to observe that

$$(X_0,X_1)_{\theta,q} = (X_1,X_0)_{1-\theta,q} \quad \text{and} \tag{17.58}$$

$$(X_0,X_1)_{\theta,q} \hookrightarrow X_0 + X_1 \quad \text{continuously.} \tag{17.59}$$

Going further, given a sigma-finite measure space $(\mathfrak{X},\mathfrak{M},\mu)$, denote by $L^0(\mathfrak{X},\mu)$ the space of all μ-measurable functions which are finite μ-a.e. on $\mathfrak{X}$. Equip this space with the topology of `convergence in measure on sets of finite measure`, i.e., a sequence $\{f_j\}_{j\in\mathbb{N}} \subseteq L^0(\mathfrak{X},\mu)$ converges to some $f \in L^0(\mathfrak{X},\mu)$ provided

$$\begin{aligned} &\text{for all } A \in \mathfrak{M}, \text{ with } \mu(A) < \infty, \text{ and for all } \varepsilon > 0 \implies \\ &\lim_{j\to\infty} \mu\Big(\big\{x \in A : |f_j(x) - f(x)| > \varepsilon\big\}\Big) = 0. \end{aligned} \tag{17.60}$$

This turns $L^0(\mathfrak{X},\mu)$ into a Hausdorff topological vector space (see [GHA-I, Lemma 3.2.1, p. 251]).

The following lemma provides an embedding result for Measure Theoretic Herz spaces:

Lemma 17.1 *Consider a measure space $(\mathfrak{X},\mathfrak{M},\mu)$, along with a collection of sets $\mathscr{C} = \{C_k\}_{k\in\mathbb{Z}} \subseteq \mathfrak{M}$ satisfying* (17.1). *Fix two exponents $p \in (0,\infty)$ and $q \in (0,\infty]$, and an index $\alpha \in \mathbb{R}$. Then*

$$\dot{\mathrm{K}}^{\alpha}_{p,q}(\mathfrak{X},\mu,\mathscr{C}) \hookrightarrow L^0(\mathfrak{X},\mu) \quad \textit{continuously.} \tag{17.61}$$

In particular, any two homogeneous measure theoretic Herz spaces are compatible. The same type of conclusions hold in the context of inhomogeneous measure theoretic Herz spaces.

Proof Consider a sequence of functions $\{f_j\}_{j\in\mathbb{N}} \subseteq \dot{\mathrm{K}}^{\alpha}_{p,q}(\mathfrak{X},\mu,\mathscr{C})$ such that

$$\lim_{j\to\infty} f_j = f \ \text{ in } \ \dot{\mathrm{K}}^{\alpha}_{p,q}(\mathfrak{X},\mu,\mathscr{C}) \ \text{ for some } f \in \dot{\mathrm{K}}^{\alpha}_{p,q}(\mathfrak{X},\mu,\mathscr{C}). \tag{17.62}$$

Fix $\varepsilon > 0$, and $A \in \mathfrak{M}$ with $\mu(A) < \infty$. For each $j \in \mathbb{N}$ and $k \in \mathbb{Z}$, abbreviate

$$A_k^{(j)} := \big\{x \in A \cap C_k : |f_j(x) - f(x)| > \varepsilon\big\} \subseteq C_k. \tag{17.63}$$

Let $j \in \mathbb{N}$ be fixed, and $N \in \mathbb{N}$ be arbitrary. Keeping in mind that $\{C_k\}_{k\in\mathbb{Z}}$ cover $\mathfrak{X}$ up to a set of μ-measure zero (see (17.1)), estimate

$$\begin{aligned}
&\mu\Big(\{x \in A : |f_j(x) - f(x)| > \varepsilon\}\Big)\\
&\quad\le \sum_{k=-\infty}^{\infty} \mu\Big(\{x \in A \cap C_k : |f_j(x) - f(x)| > \varepsilon\}\Big)\\
&\quad= \sum_{k=-\infty}^{\infty} \mu\big(A_k^{(j)}\big) = \sum_{k=-N}^{N} \mu\big(A_k^{(j)}\big) + \sum_{\substack{k\in\mathbb{Z}\\ |k|>N}} \mu\big(A_k^{(j)}\big)\\
&\quad= \sum_{k=-N}^{N} \mu\big(A_k^{(j)}\big) + \mu\Bigg(\bigcup_{\substack{k\in\mathbb{Z}\\ |k|>N}} A_k^{(j)}\Bigg)\\
&\quad\le \sum_{k=-N}^{N} \mu\big(A_k^{(j)}\big) + \mu\Bigg(A \setminus \bigcup_{\substack{k\in\mathbb{Z}\\ |k|\le N}} C_k\Bigg).
\end{aligned} \tag{17.64}$$

To proceed, fix an arbitrary $\rho \in (0,\infty)$. By design of the family $\{C_k\}_{k\in\mathbb{Z}}$ (see (17.1)), we can find some $\widetilde{N} \in \mathbb{N}$ for which

$$\mu\Bigg(A \setminus \bigcup_{\substack{k\in\mathbb{Z}\\ |k|\le\widetilde{N}}} C_k\Bigg) < \frac{\rho}{2}. \tag{17.65}$$

Keeping in mind (17.2) and (17.63), for each $k \in \mathbb{Z}$ with $|k| \le \widetilde{N}$, and $j \in \mathbb{N}$ we have

$$\begin{aligned}
\varepsilon \cdot \mu\big(A_k^{(j)}\big)^{1/p} &= \Big(\int_{A_k^{(j)}} \varepsilon^p \, d\mu\Big)^{1/p} \le \Big(\int_{A_k^{(j)}} |f_j - f|^p \, d\mu\Big)^{1/p}\\
&\le \Big(\int_{C_k} |f_j - f|^p \, d\mu\Big)^{1/p} = \|(f_j - f)\cdot \mathbb{1}_{C_k}\|_{L^p(\mathfrak{X},\mu)}\\
&\le 2_*^{-\alpha k}\|f_j - f\|_{\dot{\mathrm{K}}^{\alpha}_{p,q}(\mathfrak{X},\mu,\mathscr{C})},
\end{aligned} \tag{17.66}$$

which further implies

$$\mu\big(A_k^{(j)}\big) \le \varepsilon^{-p} 2_*^{-\alpha k p} \|f_j - f\|^p_{\dot{\mathrm{K}}^{\alpha}_{p,q}(\mathfrak{X},\mu,\mathscr{C})}$$
$$\le \varepsilon^{-p}\Big(\max_{|k|\le \widetilde{N}} \{2_*^{-\alpha k p}\}\Big)\|f_j - f\|^p_{\dot{\mathrm{K}}^{\alpha}_{p,q}(\mathfrak{X},\mu,\mathscr{C})}. \tag{17.67}$$

In light of (17.62), choose $j_0 \in \mathbb{N}$ sufficiently large so that for all $j \in \mathbb{N}$, with $j \ge j_0$, we have

$$\varepsilon^{-p}\Big(\max_{|k|\le \widetilde{N}} \{2_*^{-\alpha k p}\}\Big)\|f_j - f\|^p_{\dot{\mathrm{K}}^{\alpha}_{p,q}(\mathfrak{X},\mu,\widetilde{\mathscr{C}})} \le \frac{\rho}{2(2\widetilde{N}+1)}. \tag{17.68}$$

Taking $N := \widetilde{N}$ in (17.64), use (17.65)-(17.68) to finally obtain

$$\mu\Big(\{x \in A : |f_j(x) - f(x)| > \varepsilon\}\Big) \le \sum_{k=-\widetilde{N}}^{\widetilde{N}} \frac{\rho}{2(2\widetilde{N}+1)} + \frac{\rho}{2} = \rho \tag{17.69}$$

for all $j \ge j_0$. Since $\rho > 0$ was arbitrary to begin with, this ultimately proves that $f_j \to f$ as $j \to \infty$ in the topology of $L^0(\mathfrak{X},\mu)$. This then establishes the continuous embedding in (17.61). By making natural alterations to the reasoning above, the same continuous embedding holds in the context of inhomogeneous measure theoretic Herz spaces. This finishes the proof of Lemma 17.1. □

The following result establishes that the scale of measure theoretic Herz spaces is stable under the real method of interpolation. This plays a crucial role in subsequent developments.

Theorem 17.1 *Consider a measure space $(\mathfrak{X},\mathfrak{M},\mu)$ and let $\mathscr{C} = \{C_k\}_{k\in\mathbb{Z}} \subseteq \mathfrak{M}$ be a collection of sets satisfying* (17.1)*. Fix exponents $p \in (0,\infty)$ and $q_0, q_1, q \in (0,\infty]$ along with two indices $\alpha_0, \alpha_1 \in \mathbb{R}$ such that $\alpha_0 \ne \alpha_1$. Then whenever $\theta \in (0,1)$ and $\alpha := (1-\theta)\alpha_0 + \theta\alpha_1$ it follows that*

$$\big(\dot{\mathrm{K}}^{\alpha_0}_{p,q_0}(\mathfrak{X},\mu,\mathscr{C}),\, \dot{\mathrm{K}}^{\alpha_1}_{p,q_1}(\mathfrak{X},\mu,\mathscr{C})\big)_{\theta,q} = \dot{\mathrm{K}}^{\alpha}_{p,q}(\mathfrak{X},\mu,\mathscr{C}). \tag{17.70}$$

Proof Notice that the intermediate space in the left-hand side of (17.58) is meaningfully defined, thanks to Lemma 17.1. Also, in light of the first line in (17.58), it suffices to show (17.70) under the assumption that $\alpha_0 < \alpha_1$. The crux of the matter is verifying the following identification:

$$\big(\dot{\mathrm{K}}^{\alpha_0}_{p,1}(\mathfrak{X},\mu,\mathscr{C}),\, \dot{\mathrm{K}}^{\alpha_1}_{p,1}(\mathfrak{X},\mu,\mathscr{C})\big)_{\theta,q} = \dot{\mathrm{K}}^{\alpha}_{p,q}(\mathfrak{X},\mu,\mathscr{C}). \tag{17.71}$$

Indeed, assume momentarily that (17.71) holds. Select two indices $\beta_1, \beta_2 \in \mathbb{R}$ such that $-\infty < \beta_0 < \alpha_0 < \alpha_1 < \beta_1 < \infty$, and choose $\theta_0, \theta_1 \in (0,1)$ satisfying

$$\alpha_0 = (1-\theta_0)\beta_0 + \theta_0\beta_1 \quad \text{and} \quad \alpha_1 = (1-\theta_1)\beta_0 + \theta_1\beta_1. \tag{17.72}$$

According to (17.71) we then have

$$\begin{aligned}\big(\dot{\mathrm{K}}^{\beta_0}_{p,1}(\mathfrak{X},\mu,\mathscr{C}),\,\dot{\mathrm{K}}^{\beta_1}_{p,1}(\mathfrak{X},\mu,\mathscr{C})\big)_{\theta_0,q_0} &= \dot{\mathrm{K}}^{\alpha_0}_{p,q_0}(\mathfrak{X},\mu,\mathscr{C}),\\ \big(\dot{\mathrm{K}}^{\beta_0}_{p,1}(\mathfrak{X},\mu,\mathscr{C}),\,\dot{\mathrm{K}}^{\beta_1}_{p,1}(\mathfrak{X},\mu,\mathscr{C})\big)_{\theta_1,q_1} &= \dot{\mathrm{K}}^{\alpha_1}_{p,q_1}(\mathfrak{X},\mu,\mathscr{C})\end{aligned} \tag{17.73}$$

Denote by $\widetilde{\theta} := (1-\theta)\theta_0+\theta\theta_1 \in (0,1)$, and observe that $\alpha = (1-\widetilde{\theta})\beta_0+\widetilde{\theta}\beta_1$. Granted these, (17.71), and (17.73), invoke the Reiteration Theorem [GHA-II, Theorem 1.3.9, p. 36] to obtain

$$\begin{aligned}&\big(\dot{\mathrm{K}}^{\alpha_0}_{p,q_0}(\mathfrak{X},\mu,\mathscr{C}),\,\dot{\mathrm{K}}^{\alpha_1}_{p,q_1}(\mathfrak{X},\mu,\mathscr{C})\big)_{\theta,q}\\ &\quad= \Big(\big(\dot{\mathrm{K}}^{\beta_0}_{p,1}(\mathfrak{X},\mu,\mathscr{C}),\,\dot{\mathrm{K}}^{\beta_1}_{p,1}(\mathfrak{X},\mu,\mathscr{C})\big)_{\theta_0,q_0},\\ &\qquad\qquad \big(\dot{\mathrm{K}}^{\beta_0}_{p,1}(\mathfrak{X},\mu,\mathscr{C}),\,\dot{\mathrm{K}}^{\beta_1}_{p,1}(\mathfrak{X},\mu,\mathscr{C})\big)_{\theta_1,q_1}\Big)_{\theta,q}\\ &\quad= \big(\dot{\mathrm{K}}^{\beta_0}_{p,1}(\mathfrak{X},\mu,\mathscr{C}),\,\dot{\mathrm{K}}^{\beta_1}_{p,1}(\mathfrak{X},\mu,\mathscr{C})\big)_{(1-\theta)\theta_0+\theta\theta_1,q}\\ &\quad= \dot{\mathrm{K}}^{(1-\widetilde{\theta})\beta_0+\widetilde{\theta}\beta_1}_{p,q}(\mathfrak{X},\mu,\mathscr{C})\\ &\quad= \dot{\mathrm{K}}^{\alpha}_{p,q}(\mathfrak{X},\mu,\mathscr{C}).\end{aligned} \tag{17.74}$$

This justifies (17.70).

We are left with verifying (17.96). Fix an arbitrary function $f \in \dot{\mathrm{K}}^{\alpha}_{p,q}(\mathfrak{X},\mu,\mathscr{C})$, and make the following observation. Having fixed $j \in \mathbb{Z}$, decompose f as

$$f = \sum_{k=j+1}^{\infty} f\cdot\mathbb{1}_{C_k} + \sum_{k=-\infty}^{j} f\cdot\mathbb{1}_{C_k} =: f_0^{(j)} + f_1^{(j)}. \tag{17.75}$$

In this regard, we claim that

$$f_0^{(j)} \in \dot{\mathrm{K}}^{\alpha_0}_{p,1}(\mathfrak{X},\mu,\mathscr{C}) \quad\text{and}\quad f_1^{(j)} \in \dot{\mathrm{K}}^{\alpha_1}_{p,1}(\mathfrak{X},\mu,\mathscr{C}). \tag{17.76}$$

To deduce the first membership in (17.76), recall that we are working under the assumption that $\alpha_0-\alpha_1<0$, and use Hölder's inequality to estimate

$$\begin{aligned}&\|f_0^{(j)}\|_{\dot{\mathrm{K}}^{\alpha_0}_{p,1}(\mathfrak{X},\mu,\mathscr{C})}\\ &\quad= \Big\|\sum_{k=j+1}^{\infty} f\cdot\mathbb{1}_{C_k}\Big\|_{\dot{\mathrm{K}}^{\alpha_0}_{p,1}(\mathfrak{X},\mu,\mathscr{C})} = \sum_{\ell=j+1}^{\infty} 2_*^{\ell\alpha_0}\|f\cdot\mathbb{1}_{C_\ell}\|_{L^p(\mathfrak{X},\mu)}\\ &\quad= \sum_{\ell=j+1}^{\infty} \big(2_*^{\ell\alpha}\|f\cdot\mathbb{1}_{C_\ell}\|_{L^p(\mathfrak{X},\mu)}\big)2_*^{\ell\theta(\alpha_0-\alpha_1)}\end{aligned}$$

$$\leq \begin{cases} \Big(\sum\limits_{\ell=-\infty}^{\infty} (2_*^{\ell\alpha}\|f\cdot \mathbb{1}_{C_\ell}\|_{L^p(\mathfrak{X},\mu)})^q\Big)^{1/q}\Big(\sum\limits_{\ell=j+1}^{\infty} 2_*^{\ell\theta(\alpha_0-\alpha_1)q'}\Big)^{1/q'} & \text{if } q>1 \\ \Big(\sum\limits_{\ell=-\infty}^{\infty} (2_*^{\ell\alpha}\|f\cdot \mathbb{1}_{C_\ell}\|_{L^p(\mathfrak{X},\mu)})^q\Big)^{1/q}\Big(\sup\limits_{\ell\geq j+1} 2_*^{\ell\theta(\alpha_0-\alpha_1)}\Big) & \text{if } q\leq 1, \end{cases}$$

$$\leq C 2_*^{j\theta(\alpha_0-\alpha_1)}\|f\|_{\dot{\mathrm{K}}^{\alpha}_{p,q}(\mathfrak{X},\mu,\mathscr{C})} < \infty \tag{17.77}$$

for some constant $C = C(2_*, q, \alpha_0, \alpha_1, \theta) \in (0,\infty)$. In relation to the second membership in (17.76), once again use Hölder's inequality, and the fact that $\alpha_1 - \alpha_0 > 0$, to obtain

$$\begin{aligned}
&\|f_1^{(j)}\|_{\dot{\mathrm{K}}^{\alpha_1}_{p,1}(\mathfrak{X},\mu,\mathscr{C})} \\
&= \Big\|\sum_{k=-\infty}^{j} f\cdot \mathbb{1}_{C_k}\Big\|_{\dot{\mathrm{K}}^{\alpha_1}_{p,1}(\mathfrak{X},\mu,\mathscr{C})} = \sum_{\ell=-\infty}^{j} 2_*^{\ell\alpha_1}\|f\cdot \mathbb{1}_{C_\ell}\|_{L^p(\mathfrak{X},\mu)} \\
&= \sum_{\ell=-\infty}^{j} (2_*^{\ell\alpha}\|f\cdot \mathbb{1}_{C_\ell}\|_{L^p(\mathfrak{X},\mu)}) 2_*^{\ell(1-\theta)(\alpha_1-\alpha_0)} \\
&\leq \begin{cases} \Big(\sum\limits_{\ell=-\infty}^{\infty} (2_*^{\ell\alpha}\|f\cdot \mathbb{1}_{C_\ell}\|_{L^p(\mathfrak{X},\mu)})^q\Big)^{1/q}\Big(\sum\limits_{\ell=-\infty}^{j} 2_*^{\ell(1-\theta)(\alpha_1-\alpha_0)q'}\Big)^{1/q'} & \text{if } q>1 \\ \Big(\sum\limits_{\ell=-\infty}^{\infty} (2_*^{\ell\alpha}\|f\cdot \mathbb{1}_{C_\ell}\|_{L^p(\mathfrak{X},\mu)})^q\Big)^{1/q}\Big(\sup\limits_{\ell\leq j} 2_*^{\ell(1-\theta)(\alpha_1-\alpha_0)}\Big) & \text{if } q\leq 1 \end{cases} \\
&\leq C 2_*^{j(1-\theta)(\alpha_1-\alpha_0)}\|f\|_{\dot{\mathrm{K}}^{\alpha}_{p,q}(\mathfrak{X},\mu,\mathscr{C})} < \infty,
\end{aligned} \tag{17.78}$$

for some constant $C = C(2_*, \alpha_0, \alpha_1, q, \theta) \in (0,\infty)$. This establishes (17.76), which implies $f \in \dot{\mathrm{K}}^{\alpha_0}_{p,1}(\mathfrak{X},\mu,\mathscr{C}) + \dot{\mathrm{K}}^{\alpha_1}_{p,1}(\mathfrak{X},\mu,\mathscr{C})$.

For simplicity, for each $t \in (0,\infty)$ abbreviate

$$\begin{aligned}
K(t,f) &:= K\big(t, f, \dot{\mathrm{K}}^{\alpha_0}_{p,1}(\mathfrak{X},\mu,\mathscr{C}), \dot{\mathrm{K}}^{\alpha_1}_{p,1}(\mathfrak{X},\mu,\mathscr{C})\big) \\
&= \inf_{\substack{f=f_0+f_1 \\ f_0\in\dot{\mathrm{K}}^{\alpha_0}_{p,1}(\mathfrak{X},\mu,\mathscr{C}), \\ f_1\in\dot{\mathrm{K}}^{\alpha_1}_{p,1}(\mathfrak{X},\mu,\mathscr{C})}} \Big\{\|f_0\|_{\dot{\mathrm{K}}^{\alpha_0}_{p,1}(\mathfrak{X},\mu,\mathscr{C})} + t\|f_1\|_{\dot{\mathrm{K}}^{\alpha_1}_{p,1}(\mathfrak{X},\mu,\mathscr{C})}\Big\},
\end{aligned} \tag{17.79}$$

and set

$$\tau := \frac{-(\alpha-\alpha_0)}{\theta} = \frac{\alpha-\alpha_1}{1-\theta} \in (-\infty,0). \tag{17.80}$$

We will prove the present theorem for $q \in (0,\infty)$, as the case $q = \infty$ follows via a similar reasoning. Keeping in mind (17.75), (17.76), (17.80), and that the function

$K(\cdot, f)$ defined in (17.79) is non-decreasing on $(0, \infty)$ (cf. (17.54)), estimate

$$\begin{aligned}
&\left(\int_0^\infty \left[t^{-\theta}K(t,f)\right]^q \frac{dt}{t}\right)^{1/q} = \left(\sum_{j=-\infty}^{\infty} \int_{2_*^{(j+1)\tau}}^{2_*^{j\tau}} \left[t^{-\theta}K(t,f)\right]^q \frac{dt}{t}\right)^{1/q} \\
&\leq C\left(\sum_{j=\infty}^{\infty} 2_*^{-j\theta\tau q} K(2_*^{j\tau}, f)^q \int_{2_*^{(j+1)\iota}}^{2_*^{j\tau}} \frac{dt}{t}\right)^{1/q} \\
&= C\left(\sum_{j=-\infty}^{\infty} 2_*^{-j\theta\tau q} K(2_*^{j\tau}, f)^q\right)^{1/q} \\
&\leq C\left\{\sum_{j=-\infty}^{\infty} 2_*^{-j\theta\tau q}\left(\|f_0^{(j)}\|_{\dot{\mathrm{K}}^{\alpha_0}_{p,1}(\mathfrak{X},\mu,\mathscr{C})} + 2_*^{j\tau}\|f_1^{(j)}\|_{\dot{\mathrm{K}}^{\alpha_1}_{p,1}(\mathfrak{X},\mu,\mathscr{C})}\right)^q\right\}^{1/q} \\
&\leq C\left\{\sum_{j=-\infty}^{\infty} 2_*^{-j\theta\tau q}\left\|\sum_{k=j+1}^{\infty} f\cdot \mathbb{1}_{C_k}\right\|^q_{\dot{\mathrm{K}}^{\alpha_0}_{p,1}(\mathfrak{X},\mu,\mathscr{C})}\right\}^{1/q} \\
&\quad + C\left\{\sum_{j=-\infty}^{\infty} 2_*^{j\tau q(1-\theta)}\left\|\sum_{k=-\infty}^{j} f\cdot \mathbb{1}_{C_k}\right\|^q_{\dot{\mathrm{K}}^{\alpha_1}_{p,1}(\mathfrak{X},\mu,\mathscr{C})}\right\}^{1/q} \\
&= C\left\{\sum_{j=-\infty}^{\infty}\left(\sum_{\ell=j+1}^{\infty} 2_*^{(j-\ell)(\alpha-\alpha_0)} 2_*^{\ell\alpha}\|f\cdot \mathbb{1}_{C_\ell}\|_{L^p(\mathfrak{X},\mu)}\right)^q\right\}^{1/q} \\
&\quad + C\left\{\sum_{j=-\infty}^{\infty}\left(\sum_{\ell=-\infty}^{j} 2_*^{(\ell-j)(\alpha_1-\alpha)} 2_*^{\ell\alpha}\|f\cdot \mathbb{1}_{C_\ell}\|_{L^p(\mathfrak{X},\mu)}\right)^q\right\}^{1/q} =: \mathrm{I} + \mathrm{II}.
\end{aligned} \tag{17.81}$$

Let us bound I. In the case when $0 < q \leq 1$, use Fubini's Theorem for series and the fact that $\alpha - \alpha_0 > 0$ to further obtain

$$\begin{aligned}
\mathrm{I} &\leq C\left\{\sum_{j=-\infty}^{\infty}\sum_{\ell=j+1}^{\infty} 2_*^{(j-\ell)(\alpha-\alpha_0)q} 2_*^{\ell\alpha q}\|f\cdot \mathbb{1}_{C_\ell}\|^q_{L^p(\mathfrak{X},\mu)}\right\}^{1/q} \\
&= C\left\{\sum_{\ell=-\infty}^{\infty} 2_*^{\ell\alpha q}\|f\cdot \mathbb{1}_{C_\ell}\|^q_{L^p(\mathfrak{X},\mu)}\sum_{j=-\infty}^{\ell-1} 2_*^{(j-\ell)(\alpha-\alpha_0)q}\right\}^{1/q} \\
&\leq C\left\{\sum_{\ell=-\infty}^{\infty} 2_*^{\ell\alpha q}\|f\cdot \mathbb{1}_{C_\ell}\|^q_{L^p(\mathfrak{X},\mu)}\right\}^{1/q} = C\|f\|_{\dot{\mathrm{K}}^{\alpha}_{p,q}(\mathfrak{X},\mu,\mathscr{C})}
\end{aligned} \tag{17.82}$$

for some constant $C = C(2_*, \alpha, \alpha_0, q) \in (0, \infty)$. When $1 < q < \infty$, denote by $q' \in (1, \infty)$ the Hölder conjugate exponent of q, employ Hölder's inequality, Fubini's Theorem for series, along with $\alpha - \alpha_0 > 0$ to estimate I as

$$
\begin{aligned}
\mathrm{I} &\leq \Bigg\{ \sum_{j=-\infty}^{\infty} \Big(\sum_{\ell=j+1}^{\infty} 2_*^{(j-\ell)(\alpha-\alpha_0)q/2} 2_*^{\ell\alpha q} \| f \cdot \mathbb{1}_{C_\ell} \|_{L^p(\mathfrak{X},\mu)}^q \Big) \times \\
&\qquad \times \Big(\sum_{\ell=j+1}^{\infty} 2_*^{(j-\ell)(\alpha-\alpha_0)q'/2} \Big)^{q/q'} \Bigg\}^{1/q} \\
&\leq C \Bigg\{ \sum_{j=-\infty}^{\infty} \sum_{\ell=j+1}^{\infty} 2_*^{(j-\ell)(\alpha-\alpha_0)q/2} 2_*^{\ell\alpha q} \| f \cdot \mathbb{1}_{C_\ell} \|_{L^p(\mathfrak{X},\mu)}^q \Bigg\}^{1/q} \\
&= C \Bigg\{ \sum_{\ell=-\infty}^{\infty} 2_*^{\ell\alpha q} \| f \cdot \mathbb{1}_{C_\ell} \|_{L^p(\mathfrak{X},\mu)}^q \sum_{j=-\infty}^{\ell-1} 2_*^{(j-\ell)(\alpha-\alpha_0)q/2} \Bigg\}^{1/q} \\
&\leq C \Bigg\{ \sum_{\ell=-\infty}^{\infty} 2_*^{\ell\alpha q} \| f \cdot \mathbb{1}_{C_\ell} \|_{L^p(\mathfrak{X},\mu)}^q \Bigg\}^{1/q} = C \| f \|_{\dot{\mathrm{K}}^{\alpha}_{p,q}(\mathfrak{X},\mu,\mathscr{C})}
\end{aligned}
\tag{17.83}
$$

for some constant $C = C(2_*, \alpha, \alpha_0, q) \in (0, \infty)$.

We now deal with the term II in (17.81). Consider first the case when $0 < q \leq 1$. Then using Fubini's Theorem for series, and the fact that $\alpha_1 - \alpha > 0$, we have

$$
\begin{aligned}
\mathrm{II} &\leq C \Bigg\{ \sum_{j=-\infty}^{\infty} \sum_{\ell=-\infty}^{j} 2_*^{(\ell-j)(\alpha_1-\alpha)q} 2_*^{\ell\alpha q} \| f \cdot \mathbb{1}_{C_\ell} \|_{L^p(\mathfrak{X},\mu)}^q \Bigg\}^{1/q} \\
&= C \Bigg\{ \sum_{\ell=-\infty}^{\infty} 2_*^{\ell\alpha q} \| f \cdot \mathbb{1}_{C_\ell} \|_{L^p(\mathfrak{X},\mu)}^q \sum_{j=\ell}^{\infty} 2_*^{(\ell-j)(\alpha_1-\alpha)q} \Bigg\}^{1/q} \\
&\leq C \Bigg\{ \sum_{\ell=-\infty}^{\infty} 2_*^{\ell\alpha q} \| f \cdot \mathbb{1}_{C_\ell} \|_{L^p(\mathfrak{X},\mu)}^q \Bigg\}^{1/q} = C \| f \|_{\dot{\mathrm{K}}^{\alpha}_{p,q}(\mathfrak{X},\mu,\mathscr{C})}
\end{aligned}
\tag{17.84}
$$

for some $C = C(2_*, \alpha, \alpha_1, q) \in (0, \infty)$. Corresponding to $1 < q < \infty$, denote by $q' \in (1, \infty)$ the Hölder conjugate exponent of q, use Hölder's inequality, Fubini's Theorem for series, and the fact that $\alpha_1 - \alpha > 0$, to estimate

$$
\mathrm{II} \leq C \Bigg\{ \sum_{j=-\infty}^{\infty} \Big(\sum_{\ell=-\infty}^{j} 2_*^{(\ell-j)(\alpha_1-\alpha)q/2} 2_*^{\ell\alpha q} \| f \cdot \mathbb{1}_{C_\ell} \|_{L^p(\mathfrak{X},\mu)}^q \Big) \times
$$

$$\times \Big(\sum_{\ell=-\infty}^{j} 2_*^{(\ell-j)(\alpha_1-\alpha)q'/2} \Big)^{q/q'} \Bigg\}^{1/q}$$

$$\leq C \Bigg\{ \sum_{j=-\infty}^{\infty} \sum_{\ell=-\infty}^{j} 2_*^{(\ell-j)(\alpha_1-\alpha)q/2} 2_*^{\ell\alpha q} \|f \cdot \mathbb{1}_{C_\ell}\|_{L^p(\mathfrak{X},\mu)}^q \Bigg\}^{1/q}$$

$$= C \Bigg\{ \sum_{\ell=-\infty}^{\infty} 2_*^{\ell\alpha q} \|f \cdot \mathbb{1}_{C_\ell}\|_{L^p(\mathfrak{X},\mu)}^q \sum_{j=\ell}^{\infty} 2_*^{(\ell-j)(\alpha_1-\alpha)q/2} \Bigg\}^{1/q}$$

$$\leq C \Bigg\{ \sum_{\ell=-\infty}^{\infty} 2_*^{\ell\alpha q} \|f \cdot \mathbb{1}_{C_\ell}\|_{L^p(\mathfrak{X},\mu)}^q \Bigg\}^{1/q} = C\|f\|_{\dot{\mathrm{K}}^{\alpha}_{p,q}(\mathfrak{X},\mu,\mathscr{C})} \tag{17.85}$$

for some $C = C(2_*, \alpha, \alpha_1, q) \in (0, \infty)$. Collectively, (17.81), (17.82), (17.83), (17.84), and (17.85) imply that there exits some constant $C \in (0, \infty)$ depending only on 2_*, α, α_0, α_1, and q with the property that

$$\Big(\int_0^\infty \big[t^{-\theta} K(t, f)\big]^q \frac{dt}{t} \Big)^{1/q} \leq C\|f\|_{\dot{\mathrm{K}}^{\alpha}_{p,q}(\mathfrak{X},\mu,\mathscr{C})} < \infty, \tag{17.86}$$

which in concert with (17.56), (17.57), and (17.79) ultimately proves the continuous embedding:

$$\dot{\mathrm{K}}^{\alpha}_{p,q}(\mathfrak{X}, \mu, \mathscr{C}) \hookrightarrow \big(\dot{\mathrm{K}}^{\alpha_0}_{p,1}(\mathfrak{X}, \mu, \mathscr{C}),\ \dot{\mathrm{K}}^{\alpha_1}_{p,1}(\mathfrak{X}, \mu, \mathscr{C})\big)_{\theta,q}. \tag{17.87}$$

Let us now deal with the reverse inclusion. To this end, fix an arbitrary function $f \in \big(\dot{\mathrm{K}}^{\alpha_0}_{p,1}(\mathfrak{X}, \mu, \mathscr{C}),\ \dot{\mathrm{K}}^{\alpha_1}_{p,1}(\mathfrak{X}, \mu, \mathscr{C})\big)_{\theta,q}$ and decompose f as

$$f = f_0 + f_1, \quad \text{for some } f_0 \in \dot{\mathrm{K}}^{\alpha_0}_{p,1}(\mathfrak{X}, \mu, \mathscr{C}) \text{ and } f_1 \in \dot{\mathrm{K}}^{\alpha_1}_{p,1}(\mathfrak{X}, \mu, \mathscr{C}). \tag{17.88}$$

Mindful of this and (17.2), for each $j \in \mathbb{Z}$ estimate

$$\begin{aligned}
&2_*^{j\alpha}\|f \cdot \mathbb{1}_{C_j}\|_{L^p(\mathfrak{X},\mu)} \\
&\quad \leq 2_*^{j\alpha}\|f_0 \cdot \mathbb{1}_{C_j}\|_{L^p(\mathfrak{X},\mu)} + 2_*^{j\alpha}\|f_1 \cdot \mathbb{1}_{C_j}\|_{L^p(\mathfrak{X},\mu)} \\
&\quad = 2_*^{j(\alpha-\alpha_0)}\Big(2_*^{j\alpha_0}\|f_0 \cdot \mathbb{1}_{C_j}\|_{L^p(\mathfrak{X},\mu)} + 2_*^{j(\alpha_0-\alpha_1)} 2_*^{j\alpha_1}\|f_1 \cdot \mathbb{1}_{C_j}\|_{L^p(\mathfrak{X},\mu)}\Big) \\
&\quad \leq C 2_*^{j(\alpha-\alpha_0)}\Big(\|f_0\|_{\dot{\mathrm{K}}^{\alpha_0}_{p,1}(\mathfrak{X},\mu,\mathscr{C})} + 2_*^{(j+1)(\alpha_0-\alpha_1)}\|f_1\|_{\dot{\mathrm{K}}^{\alpha_1}_{p,1}(\mathfrak{X},\mu,\mathscr{C})}\Big),
\end{aligned} \tag{17.89}$$

for some constant $C = C(2_*, \alpha_0, \alpha_1) \in (0, \infty)$. Taking the infimum in the right-hand side over all possible writings of f as in (17.88), and recalling (17.79), we arrive at

$$2_*^{j\alpha}\|f \cdot \mathbb{1}_{C_j}\|_{L^p(\mathfrak{X},\mu)} \leq C 2_*^{j(\alpha-\alpha_0)} \cdot K\big(2_*^{(j+1)(\alpha_0-\alpha_1)}, f\big) \quad \text{for all } j \in \mathbb{Z}. \tag{17.90}$$

Moreover, recalling that $\alpha_0 < \alpha_1$ and (17.80), we have $(\alpha_0 - \alpha_1)/\tau > 0$. As such, for all $j \in \mathbb{Z}$ and $t \in \big(2_*^{(j+1)\tau}, 2_*^{j\tau}\big)$ it follows that

$$\begin{gathered} t \approx 2_*^{j\tau} \quad \text{uniformly on } j \text{ and} \\ 2_*^{(j+1)(\alpha_0-\alpha_1)} < t^{(\alpha_0-\alpha_1)/\tau} < 2_*^{j(\alpha_0-\alpha_1)}. \end{gathered} \tag{17.91}$$

This and the monotonicity of $K(\cdot, f)$ (cf. (17.54)) then yield

$$\begin{aligned} &\int_{2_*^{(j+1)\tau}}^{2_*^{j\tau}} \big[t^{(\alpha-\alpha_0)/\tau} K\big(t^{(\alpha_0-\alpha_1)/\tau}, f\big)\big]^q \frac{dt}{t} \\ &\qquad \geq C\big[2_*^{j(\alpha-\alpha_0)} K\big(2_*^{(j+1)(\alpha_0-\alpha_1)}, f\big)\big]^q \end{aligned} \tag{17.92}$$

for some constant $C = C(2_*, \alpha_0, \alpha_1, \theta) \in (0, \infty)$. In turn, use (17.90) and (17.92) to estimate

$$\begin{aligned} \|f\|_{\dot{\mathrm{K}}^{\alpha}_{p,q}(\mathfrak{X},\mu,\mathscr{C})} &= \Big\{ \sum_{j=-\infty}^{\infty} \big(2_*^{j\alpha} \|f \cdot \mathbb{1}_{C_k}\|_{L^p(\mathfrak{X},\mu)}\big)^q \Big\}^{1/q} \\ &\leq C\Big\{ \sum_{j=-\infty}^{\infty} \big[2_*^{j(\alpha-\alpha_0)} K\big(2_*^{(j+1)(\alpha_0-\alpha_1)}, f\big)\big]^q \Big\}^{1/q} \\ &\leq C\left\{ \sum_{j=-\infty}^{\infty} \int_{2_*^{(j+1)\tau}}^{2_*^{j\tau}} \big[t^{(\alpha-\alpha_0)/\tau} K\big(t^{(\alpha_0-\alpha_1)/\tau}, f\big)\big]^q \frac{dt}{t} \right\}^{1/q} \\ &= C\left\{ \int_0^{\infty} \big[s^{-\theta} K(s, f)\big]^q \frac{ds}{s} \right\}^{1/q} \\ &= C\|f\|_{\big(\dot{\mathrm{K}}^{\alpha_0}_{p,1}(\mathfrak{X},\mu,\mathscr{C}),\, \dot{\mathrm{K}}^{\alpha_1}_{p,1}(\mathfrak{X},\mu,\mathscr{C})\big)_{\theta,q}} < \infty, \end{aligned} \tag{17.93}$$

where in the penultimate equality we made the change of variable $s := t^{(\alpha_0-\alpha_1)/\tau}$ (which implies $ds/s \approx dt/t$), and used the fact that $-\theta = (\alpha - \alpha_0)/(\alpha_0 - \alpha_1)$. This shows that the following embedding is continuous

$$\big(\dot{\mathrm{K}}^{\alpha_0}_{p,1}(\mathfrak{X}, \mu, \mathscr{C}),\, \dot{\mathrm{K}}^{\alpha_1}_{p,1}(\mathfrak{X}, \mu, \mathscr{C})\big)_{\theta,q} \hookrightarrow \dot{\mathrm{K}}^{\alpha}_{p,q}(\mathfrak{X}, \mu, \mathscr{C}). \tag{17.94}$$

This and (17.87) justify (17.71), which ultimately proves Theorem 17.1. □

We have a similar interpolation result in the context of inhomogeneous measure theoretic Herz spaces.

Theorem 17.2 *Consider a measure space $(\mathfrak{X}, \mathfrak{M}, \mu)$ and let $\widetilde{\mathscr{C}} = \{\widetilde{C}_k\}_{k\in\mathbb{Z}} \subseteq \mathfrak{M}$ be a collection of sets satisfying* (17.10). *Fix exponents $p \in (0, \infty)$ and $q_0, q_1, q \in (0, \infty]$ along with two indices $\alpha_0, \alpha_1 \in \mathbb{R}$ such that $\alpha_0 \neq \alpha_1$. Then whenever $\theta \in (0, 1)$ and*

$\alpha := (1-\theta)\alpha_0 + \theta\alpha_1$ *it follows that*

$$\big(\mathrm{K}^{\alpha_0}_{p,q_0}(\mathfrak{X},\mu,\widetilde{\mathscr{C}}),\ \mathrm{K}^{\alpha_1}_{p,q_1}(\mathfrak{X},\mu,\widetilde{\mathscr{C}})\big)_{\theta,q} = \mathrm{K}^{\alpha}_{p,q}(\mathfrak{X},\mu,\widetilde{\mathscr{C}}). \tag{17.95}$$

Proof In light of Lemma 17.1, the intermediate space in the left-hand side of (17.95) is meaningful. By similar reasons as in (17.71)-(17.74), we may work under the assumption that $\alpha_0 < \alpha_1$, and the key step to prove (17.95) is verifying the identification

$$\big(\mathrm{K}^{\alpha_0}_{p,1}(\mathfrak{X},\mu,\widetilde{\mathscr{C}}),\ \mathrm{K}^{\alpha_1}_{p,1}(\mathfrak{X},\mu,\widetilde{\mathscr{C}})\big)_{\theta,q} = \mathrm{K}^{\alpha}_{p,q}(\mathfrak{X},\mu,\widetilde{\mathscr{C}}). \tag{17.96}$$

Recall the continuous embedding (17.17). Hence, [GHA-II, Lemma 1.3.4, p. 30] (presently used with $t_* := 1$) ensures that for every $f \in \mathrm{K}^{\alpha_0}_{p,1}(\mathfrak{X},\mu,\widetilde{\mathscr{C}}) + \mathrm{K}^{\alpha_1}_{p,1}(\mathfrak{X},\mu,\widetilde{\mathscr{C}})$ the following equivalence holds:

$$\|f\|_{\left(\mathrm{K}^{\alpha_0}_{p,1}(\mathfrak{X},\mu,\widetilde{\mathscr{C}}),\,\mathrm{K}^{\alpha_1}_{p,1}(\mathfrak{X},\mu,\widetilde{\mathscr{C}})\right)_{\theta,q}} \approx \begin{cases} \left(\displaystyle\int_0^1 \big[t^{-\theta}K(t,f)\big]^q \frac{dt}{t}\right)^{1/q} & \text{if } q<\infty, \\ \displaystyle\sup_{t\in(0,1)} \big[t^{-\theta}K(t,f)\big] & \text{if } q=\infty, \end{cases} \tag{17.97}$$

where for every $t \in (0,\infty)$ we abbreviate

$$\begin{aligned} K(t,f) &:= K\big(t,f;\mathrm{K}^{\alpha_0}_{p,1}(\mathfrak{X},\mu,\widetilde{\mathscr{C}}),\mathrm{K}^{\alpha_1}_{p,1}(\mathfrak{X},\mu,\widetilde{\mathscr{C}})\big) \\ &= \inf_{\substack{f=f_0+f_1 \\ f_0\in\mathrm{K}^{\alpha_0}_{p,1}(\mathfrak{X},\mu,\widetilde{\mathscr{C}}), \\ f_1\in\mathrm{K}^{\alpha_1}_{p,1}(\mathfrak{X},\mu,\widetilde{\mathscr{C}})}} \Big\{\|f_0\|_{\mathrm{K}^{\alpha_0}_{p,1}(\mathfrak{X},\mu,\widetilde{\mathscr{C}})} + t\|f_1\|_{\mathrm{K}^{\alpha_1}_{p,1}(\mathfrak{X},\mu,\widetilde{\mathscr{C}})}\Big\}. \end{aligned} \tag{17.98}$$

In order to show (17.96), fix an arbitrary function $f \in \mathrm{K}^{\alpha}_{p,q}(\mathfrak{X},\mu,\widetilde{\mathscr{C}})$. Given $j \in \mathbb{N}_0$, decompose f as

$$f = \sum_{k=j+1}^{\infty} f\cdot\mathbb{1}_{\widetilde{C}_k} + \sum_{k=0}^{j} f\cdot\mathbb{1}_{\widetilde{C}_k} =: f_0^{(j)} + f_1^{(j)}. \tag{17.99}$$

Following the same type of computations as in (17.77)-(17.78) we may ensure that

$$f_0^{(j)} \in \mathrm{K}^{\alpha_0}_{p,1}(\mathfrak{X},\mu,\widetilde{\mathscr{C}}) \quad\text{and}\quad f_1^{(j)} \in \mathrm{K}^{\alpha_1}_{p,1}(\mathfrak{X},\mu,\widetilde{\mathscr{C}}). \tag{17.100}$$

In particular, this implies $f \in \mathrm{K}^{\alpha_0}_{p,1}(\mathfrak{X},\mu,\widetilde{\mathscr{C}}) + \mathrm{K}^{\alpha_1}_{p,1}(\mathfrak{X},\mu,\widetilde{\mathscr{C}})$. To proceed, set

$$\tau := \frac{-(\alpha-\alpha_0)}{\theta} = \frac{\alpha-\alpha_1}{1-\theta} \in (-\infty,0). \tag{17.101}$$

Throughout, assume that $0 < q < \infty$, as the proof for $q = \infty$ follows via a similar reasoning. Keeping in mind (17.99), (17.100), (17.101), and making natural alterations in the computations presented in (17.81) we may estimate

$$\begin{aligned}
\Big(\int_0^1 [t^{-\theta}K(t,f)]^q \frac{dt}{t}\Big)^{1/q} &= \Big(\sum_{j=0}^{\infty}\int_{2_*^{(j+1)\tau}}^{2_*^{j\tau}} [t^{-\theta}K(t,f)]^q \frac{dt}{t}\Big)^{1/q} \\
&= C\Big\{\sum_{j=0}^{\infty} 2_*^{-j\theta\tau q}\Big(\sum_{\ell=j+1}^{\infty} 2_*^{\ell\alpha_0}\|f\cdot \mathbb{1}_{\widetilde{C}_\ell}\|_{L^p(\mathfrak{X},\mu)}\Big)^q\Big\}^{1/q} \\
&\quad + C\Big\{\sum_{j=0}^{\infty} 2_*^{j\tau q(1-\theta)}\Big(\sum_{\ell=0}^{j} 2_*^{\ell\alpha_1}\|f\cdot \mathbb{1}_{\widetilde{C}_\ell}\|_{L^p(\mathfrak{X},\mu)}\Big)^q\Big\}^{1/q} \\
&=: \mathrm{I} + \mathrm{II}. \qquad (17.102)
\end{aligned}$$

At this stage, the same type of argument as in (17.82), (17.83), (17.84), and (17.85), ensures the existence of some constant $C \in (0,\infty)$, depending only on α, α_0, α_1, and q, with the property that

$$\mathrm{I} \le C\|f\|_{\mathrm{K}^{\alpha}_{p,q}(\mathfrak{X},\mu,\widetilde{\mathscr{C}})} \quad\text{and}\quad \mathrm{II} \le C\|f\|_{\mathrm{K}^{\alpha}_{p,q}(\mathfrak{X},\mu,\widetilde{\mathscr{C}})} \qquad (17.103)$$

which, in concert with (17.97) and (17.102), entails

$$\|f\|_{\left(\mathrm{K}^{\alpha_0}_{p,1}(\mathfrak{X},\mu,\widetilde{\mathscr{C}}),\,\mathrm{K}^{\alpha_1}_{p,1}(\mathfrak{X},\mu,\widetilde{\mathscr{C}})\right)_{\theta,q}} \le C\|f\|_{\mathrm{K}^{\alpha}_{p,q}(\mathfrak{X},\mu,\widetilde{\mathscr{C}})} < \infty. \qquad (17.104)$$

Hence,

$$\mathrm{K}^{\alpha}_{p,q}(\mathfrak{X},\mu,\widetilde{\mathscr{C}}) \hookrightarrow \big(\mathrm{K}^{\alpha_0}_{p,1}(\mathfrak{X},\mu,\widetilde{\mathscr{C}}),\,\mathrm{K}^{\alpha_1}_{p,1}(\mathfrak{X},\mu,\widetilde{\mathscr{C}})\big)_{\theta,q} \text{ continuously.} \qquad (17.105)$$

In the opposite direction, fix $f \in \big(\mathrm{K}^{\alpha_0}_{p,1}(\mathfrak{X},\mu,\widetilde{\mathscr{C}}),\,\mathrm{K}^{\alpha_1}_{p,1}(\mathfrak{X},\mu,\widetilde{\mathscr{C}})\big)_{\theta,q}$ arbitrary and decompose it as

$$f = f_0 + f_1, \ \text{ where } f_0 \in \mathrm{K}^{\alpha_0}_{p,1}(\mathfrak{X},\mu,\widetilde{\mathscr{C}}) \text{ and } f_1 \in \mathrm{K}^{\alpha_1}_{p,1}(\mathfrak{X},\mu,\widetilde{\mathscr{C}}). \qquad (17.106)$$

Then much as in (17.89), for each $j \in \mathbb{N}_0$ estimate

$$\begin{aligned}
&2_*^{j\alpha}\|f\cdot \mathbb{1}_{\widetilde{C}_j}\|_{L^p(\mathfrak{X},\mu)} \\
&\quad \le C2_*^{j(\alpha-\alpha_0)}\Big(\|f_0\|_{\mathrm{K}^{\alpha_0}_{p,1}(\mathfrak{X},\mu,\widetilde{\mathscr{C}})} + 2_*^{(j+1)(\alpha_0-\alpha_1)}\|f_1\|_{\mathrm{K}^{\alpha_1}_{p,1}(\mathfrak{X},\mu,\widetilde{\mathscr{C}})}\Big), \qquad (17.107)
\end{aligned}$$

for some constant $C = C(2_*,\alpha_0,\alpha_1) \in (0,\infty)$. Taking the infimum in the right-hand side over all decomposition of f as in (17.106) then yields

$$2_*^{j\alpha}\|f\cdot \mathbb{1}_{\widetilde{C}_j}\|_{L^p(\mathfrak{X},\mu)} \le C2_*^{j(\alpha-\alpha_0)}\cdot K\big(2_*^{(j+1)(\alpha_0-\alpha_1)}, f\big) \ \text{ for all } j \in \mathbb{N}_0. \qquad (17.108)$$

Furthermore, because $(\alpha_0-\alpha_1)/\tau>0$, for every $j\in\mathbb{N}_0$ and $t\in\big(2_*^{(j+1)\tau},2_*^{j\tau}\big)$ we see that $t\approx 2_*^{j\tau}$ uniformly on j and $2_*^{(j+1)(\alpha_0-\alpha_1)}<t^{(\alpha_0-\alpha_1)/\tau}<2_*^{j(\alpha_0-\alpha_1)}$. Based on this, and the monotonicity of $K(\cdot,f)$ (cf. (17.54)), it follows that

$$\int_{2_*^{(j+1)\tau}}^{2_*^{j\tau}}\big[t^{(\alpha-\alpha_0)/\tau}K\big(t^{(\alpha_0-\alpha_1)/\tau},f\big)\big]^q\,\frac{dt}{t}$$
$$\geq C\big[2_*^{j(\alpha-\alpha_0)}K\big(2^{(j+1)(\alpha_0-\alpha_1)},f\big)\big]^q, \tag{17.109}$$

for some constant $C=C(2_*,\alpha_0,\alpha_1,\theta)\in(0,\infty)$. In light of this, (17.108), and (17.97) we may deduce

$$\begin{aligned}
\|f\|_{\mathrm{K}^\alpha_{p,q}(\mathfrak{X},\mu,\widetilde{\mathscr{C}})}&=\Big\{\sum_{j=0}^{\infty}\big(2_*^{j\alpha}\|f\cdot\mathbb{1}_{\widetilde{C}_k}\|_{L^p(\mathfrak{X},\mu)}\big)^q\Big\}^{1/q}\\
&\leq C\Big\{\sum_{j=0}^{\infty}\big[2_*^{j(\alpha-\alpha_0)}K\big(2_*^{(j+1)(\alpha_0-\alpha_1)},f\big)\big]^q\Big\}^{1/q}\\
&\leq C\left\{\sum_{j=0}^{\infty}\int_{2_*^{(j+1)\tau}}^{2_*^{j\tau}}\big[t^{(\alpha-\alpha_0)/\tau}K\big(t^{(\alpha_0-\alpha_1)/\tau},f\big)\big]^q\,\frac{dt}{t}\right\}^{1/q}\\
&=C\left\{\int_0^1\big[t^{(\alpha-\alpha_0)/\tau}K\big(t^{(\alpha_0-\alpha_1)/\tau},f\big)\big]^q\,\frac{dt}{t}\right\}^{1/q}\\
&=C\left\{\int_0^1\big[s^{-\theta}K(s,f)\big]^q\,\frac{ds}{s}\right\}^{1/q}\\
&\leq C\|f\|_{\left(\mathrm{K}^{\alpha_0}_{p,1}(\mathfrak{X},\mu,\widetilde{\mathscr{C}}),\,\mathrm{K}^{\alpha_1}_{p,1}(\mathfrak{X},\mu,\widetilde{\mathscr{C}})\right)_{\theta,q}},
\end{aligned} \tag{17.110}$$

where in the second-to-last equality we made the change of variable $s:=t^{(\alpha_0-\alpha_1)/\tau}$ (which implies $ds/s\approx dt/t$), and used the fact that $-\theta=(\alpha-\alpha_0)/(\alpha_0-\alpha_1)$. Hence

$$\big(\mathrm{K}^{\alpha_0}_{p,1}(\mathfrak{X},\mu,\widetilde{\mathscr{C}}),\,\mathrm{K}^{\alpha_1}_{p,1}(\mathfrak{X},\mu,\widetilde{\mathscr{C}})\big)_{\theta,q}\hookrightarrow\mathrm{K}^{\alpha}_{p,q}(\mathfrak{X},\mu,\widetilde{\mathscr{C}})\ \text{ continuously.} \tag{17.111}$$

This concludes the proof Theorem 17.2. □

17.3 Herz Spaces of Second Generation on Ahlfors Regular Sets

In this section we introduce the second-generation Herz spaces in the setting of unbounded Ahlfors regular sets. To set the stage, assume $\Sigma\subseteq\mathbb{R}^n$ is a closed unbounded Ahlfors regular set, and abbreviate $\sigma:=\mathcal{H}^{n-1}\lfloor\Sigma$. Fix a reference point

$x_0 \in \Sigma$ along with a dilation factor $2_* > (C_\Sigma/c_\Sigma)^{1/(n-1)}$, where $C_\Sigma, c_\Sigma \in (0,\infty)$ denote the upper and lower Ahlfors regularity constants of Σ, respectively. For each $k \in \mathbb{Z}$ set

$$C_k := \Delta(x_0, 2_*^{k+1}) \setminus \Delta(x_0, 2_*^k), \tag{17.112}$$

and

$$\widetilde{C}_0 := \Delta(x_0, 2_*) \quad \text{and} \quad \widetilde{C}_k := C_k \quad \text{for every } k \in \mathbb{N}. \tag{17.113}$$

Select two exponents $p, q \in (0,\infty]$, along with an index $\alpha \in \mathbb{R}$. For each σ-measurable function f on Σ, introduce

$$\|f\|_{\dot{\mathrm{K}}^{\alpha}_{p,q}(\Sigma,\sigma)} := \begin{cases} \Big[\sum\limits_{k=-\infty}^{\infty} \big(2_*^{k\alpha}\|f\cdot\mathbb{1}_{C_k}\|_{L^p(\Sigma,\sigma)}\big)^q\Big]^{1/q} & \text{if } q<\infty,\\ \sup\limits_{k\in\mathbb{Z}} \big\{2_*^{k\alpha}\|f\cdot\mathbb{1}_{C_k}\|_{L^p(\Sigma,\sigma)}\big\} & \text{if } q=\infty, \end{cases} \tag{17.114}$$

and

$$\|f\|_{\mathrm{K}^{\alpha}_{p,q}(\Sigma,\sigma)} := \begin{cases} \Big[\sum\limits_{k=0}^{\infty} \big(2_*^{k\alpha}\|f\cdot\mathbb{1}_{\widetilde{C}_k}\|_{L^p(\Sigma,\sigma)}\big)^q\Big]^{1/q} & \text{if } q<\infty,\\ \sup\limits_{k\in\mathbb{N}_0} \big\{2_*^{k\alpha}\|f\cdot\mathbb{1}_{\widetilde{C}_k}\|_{L^p(\Sigma,\sigma)}\big\} & \text{if } q=\infty. \end{cases} \tag{17.115}$$

Define the `homogeneous Herz space` on Σ as

$$\dot{\mathrm{K}}^{\alpha}_{p,q}(\Sigma,\sigma) := \Big\{f:\Sigma\to\mathbb{R} : f \text{ is } \sigma\text{-measurable and } \|f\|_{\dot{\mathrm{K}}^{\alpha}_{p,q}(\Sigma,\sigma)}<\infty\Big\}, \tag{17.116}$$

and the `inhomogeneous Herz space` on Σ

$$\mathrm{K}^{\alpha}_{p,q}(\Sigma,\sigma) := \Big\{f:\Sigma\to\mathbb{R} : f \text{ is } \sigma\text{-measurable and } \|f\|_{\mathrm{K}^{\alpha}_{p,q}(\Sigma,\sigma)}<\infty\Big\}. \tag{17.117}$$

Observe that $\mathscr{C} := \{C_k\}_{k\in\mathbb{Z}}$ and $\widetilde{\mathscr{C}} := \{\widetilde{C}_k\}_{k\in\mathbb{N}_0}$ are collections of σ-measurable subsets of Σ satisfying

$$\begin{aligned} &\sigma\Big(\Sigma\setminus \textstyle\bigcup_{k\in\mathbb{Z}} C_k\Big)=0, \quad \sigma(C_k)>0 \quad \text{for all } k\in\mathbb{Z},\\ &\text{and} \quad C_k\cap C_j=\varnothing \quad \text{for every } k,j\in\mathbb{Z}, \text{ with } k\neq j \end{aligned} \tag{17.118}$$

and

$$\begin{aligned} &\Sigma = \textstyle\bigcup_{k\in\mathbb{N}_0} \widetilde{C}_k \quad \sigma(\widetilde{C}_k)>0 \quad \text{for all } k\in\mathbb{N}_0,\\ &\text{and} \quad \widetilde{C}_k\cap\widetilde{C}_j=\varnothing \quad \text{for every } k,j\in\mathbb{N}_0, \text{ with } k\neq j. \end{aligned} \tag{17.119}$$

Upon recalling (17.1)-(17.3) and (17.10)-(17.12), we see that

$$\dot{K}^{\alpha}_{p,q}(\Sigma,\sigma) = \dot{K}^{\alpha}_{p,q}(\Sigma,\sigma,\mathscr{C}) \quad \text{and} \quad K^{\alpha}_{p,q}(\Sigma,\sigma) = K^{\alpha}_{p,q}(\Sigma,\sigma,\widetilde{\mathscr{C}}), \tag{17.120}$$

i.e., the homogeneous and inhomogeneous Herz spaces defined in (17.116)-(17.117) fit within the framework on measure theoretic Herz spaces discussed in Section 17.1. Moreover, the choice $\alpha := (n-1)(1/q - 1/p)$ gives rise to the first-generation Herz spaces $A_{p,q}$ (introduced in Section 10), i.e.,

$$K^{(n-1)(1/q-1/p)}_{p,q}(\Sigma,\sigma) = A_{p,q}(\Sigma,\sigma). \tag{17.121}$$

In the following discussion, continue to assume that $\Sigma \subseteq \mathbb{R}^n$ is a closed unbounded Ahlfors regular set, and that $\sigma := \mathcal{H}^{n-1}\lfloor\Sigma$. Here we record some useful functional analytic properties of homogeneous and inhomogeneous Herz spaces, which are obtained by specializing our results from Section 17.1 to the current setting of Ahlfors regular sets. In this regard, we wish to point out that Proposition 17.3 ensures that whenever $p, q \in (0,\infty]$ and $\alpha \in \mathbb{R}$ the spaces

$$\big(\dot{K}^{\alpha}_{p,q}(\Sigma,\sigma), \|\cdot\|_{\dot{K}^{\alpha}_{p,q}(\Sigma,\sigma)}\big) \text{ and } \big(K^{\alpha}_{p,q}(\Sigma,\sigma), \|\cdot\|_{K^{\alpha}_{p,q}(\Sigma,\sigma)}\big) \text{ are complete.} \tag{17.122}$$

In the regime $p, q \in (1,\infty)$, Proposition 17.5 (used presently with $B_j := \Delta(x_0, j)$, for each $j \in \mathbb{N}$) ensures that

$$\begin{gathered}\text{the spaces } \dot{K}^{\alpha}_{p,q}(\Sigma,\sigma) \text{ and } K^{\alpha}_{p,q}(\Sigma,\sigma)\\ \text{are Generalized Banach Function Spaces on } (\Sigma,\sigma).\end{gathered} \tag{17.123}$$

Moreover, Proposition 17.6 permits us to identify their associated spaces as

$$\big(\dot{K}^{\alpha}_{p,q}(\Sigma,\sigma)\big)' = \dot{K}^{-\alpha}_{p',q'}(\Sigma,\sigma) \quad \text{and} \quad \big(K^{\alpha}_{p,q}(\Sigma,\sigma)\big)' = K^{-\alpha}_{p',q'}(\Sigma,\sigma) \tag{17.124}$$

where $p', q' \in (1,\infty)$ denote the Hölder conjugate exponents of p and q, respectively.

It is pertinent to highlight that in the diagonal case $p = q \in (0,\infty)$, Herz spaces coincides (as quasi-normed spaces) with weighted Lebesgue spaces:

$$\dot{K}^{\alpha}_{p,p}(\Sigma,\sigma) = L^p\big(\Sigma, |x-x_0|^{\alpha p}\sigma(x)\big) \tag{17.125}$$

$$\text{and } K^{\alpha}_{p,p}(\Sigma,\sigma) = L^p\big(\Sigma, (1+|x-x_0|)^{\alpha p}\sigma(x)\big). \tag{17.126}$$

These identifications follow from (17.112)-(17.115), along with the fact that

$$|x-x_0| \approx 2^k_* \quad \text{uniformly for } x \in C_k \text{ and } k \in \mathbb{Z}, \tag{17.127}$$

and

$$1 + |x-x_0| \approx 2^k_* \quad \text{uniformly for } x \in \widetilde{C}_k \text{ and } k \in \mathbb{N}_0. \tag{17.128}$$

Corresponding to the case when $\alpha = 0$, we see from (17.125)-(17.126) that

$$\dot{K}^{0}_{p,q}(\Sigma,\sigma) = K^{0}_{p,q}(\Sigma,\sigma) = L^p(\Sigma,\sigma). \tag{17.129}$$

Below, we record some useful embeddings for homogeneous Herz spaces.

Proposition 17.7 *Assume $\Sigma \subseteq \mathbb{R}^n$ is a closed unbounded Ahlfors regular, and abbreviate $\sigma := \mathcal{H}^{n-1}\lfloor\Sigma$. Select two exponents $p \in (0, \infty)$ and $q \in (0, \infty]$, along with an index $\alpha \in \mathbb{R}$. Then the following statements are true.*

(a) If $q_1, q_2 \in (0, \infty]$ and $q_1 \leq q_2$, then

$$\dot{\mathrm{K}}^{\alpha}_{p,q_1}(\Sigma, \sigma) \hookrightarrow \dot{\mathrm{K}}^{\alpha}_{p,q_2}(\Sigma, \sigma) \quad \textit{continuously,} \tag{17.130}$$

i.e., the scale of homogeneous Herz spaces $\dot{\mathrm{K}}^{\alpha}_{p,q}(\Sigma, \sigma)$ is monotonically increasing with respect to q.

(b) If either

$$q \in (0, \infty] \;\; \textit{and} \;\; \alpha < 0, \quad \textit{or} \quad p \in (1, \infty), \;\; q \in (0, 1], \;\; \textit{and} \;\; \alpha \leq 0, \tag{17.131}$$

then

$$\dot{\mathrm{K}}^{\alpha}_{p,q}(\Sigma, \sigma) \subseteq L^p_{\mathrm{loc}}(\Sigma, \sigma). \tag{17.132}$$

(c) Assume that $p \in (1, \infty)$, and denote by $p' := p/(p-1) \in (1, \infty)$ its Hölder conjugate exponent. Then whenever $\alpha < \frac{n-1}{p'}$, it follows that

$$\dot{\mathrm{K}}^{\alpha}_{p,q}(\Sigma, \sigma) \subseteq L^1_{\mathrm{loc}}(\Sigma, \sigma). \tag{17.133}$$

(d) If $\alpha > \frac{-(n-1)}{p}$, then

$$L^\infty_{\mathrm{comp}}(\Sigma, \sigma) \subseteq \dot{\mathrm{K}}^{\alpha}_{p,q}(\Sigma, \sigma). \tag{17.134}$$

(e) If $p \in (1, \infty)$ and $\frac{-(n-1)}{p} < \alpha < \frac{n-1}{p'}$, then

$$L^\infty_{\mathrm{comp}}(\Sigma, \sigma) \subseteq \dot{\mathrm{K}}^{\alpha}_{p,q}(\Sigma, \sigma) \subseteq L^1_{\mathrm{loc}}(\Sigma, \sigma). \tag{17.135}$$

Proof Select a reference point $x_0 \in \Sigma$ and pick a dilation factor $2_* > (C_\Sigma/c_\Sigma)^{1/(n-1)}$, where $C_\Sigma, c_\Sigma \in (0, \infty)$ denote the upper and lower Ahlfors regularity constants of Σ, respectively. The claim made in item *(a)* follows from (17.114) and the fact that $q_1 \leq q_2$ implies $\ell^{q_1}(\mathbb{Z}) \hookrightarrow \ell^{q_2}(\mathbb{Z})$ continuously.

To justify the inclusion claimed in *(b)*, fix a function $f \in \dot{\mathrm{K}}^{\alpha}_{p,q}(\Sigma, \sigma)$, and a compact subset $K \subseteq \Sigma$. Based on (17.112) and (17.118), there exists $N_K \in \mathbb{Z}$ such that

$$K \setminus \{x_0\} \subseteq \bigcup_{k=-\infty}^{N_K} C_k. \tag{17.136}$$

Consider first the case when $q \in (0, \infty]$ and $\alpha < 0$. Then

$$\begin{aligned}\int_K |f|^p\,d\sigma &\le \sum_{k=-\infty}^{N_K} \big(2_*^{k\alpha}\|f\cdot \mathbb{1}_{C_k}\|_{L^p(\Sigma,\sigma)}\big)^p 2_*^{-k\alpha p}\\ &\le \Big(\sum_{k=-\infty}^{N_K} 2_*^{-k\alpha p}\Big)\|f\|^p_{\dot{\mathrm{K}}^\alpha_{p,q}(\Sigma,\sigma)} < \infty,\end{aligned} \tag{17.137}$$

which shows that $f \in L^p_{\mathrm{loc}}(\Sigma,\sigma)$. In the scenario when $p \in (1,\infty)$ and $q \in (0,1]$, with $\alpha \le 0$, compute

$$\begin{aligned}\Big(\int_K |f|^p\,d\sigma\Big)^{1/p} &\le \sum_{k=-\infty}^{N_K} 2_*^{k\alpha}\|f\cdot \mathbb{1}_{C_k}\|_{L^p(\Sigma,\sigma)} 2_*^{-k\alpha}\\ &\le \Big(\sum_{k=-\infty}^{N_K} \big(2_*^{k\alpha}\|f\cdot \mathbb{1}_{C_k}\|_{L^p(\Sigma,\sigma)}\big)^q 2_*^{-k\alpha q}\Big)^{1/q}\\ &\le \Big(\sup_{k\le N_K} 2_*^{-k\alpha q}\Big)\Big(\sum_{k=-\infty}^{N_K} \big(2_*^{k\alpha}\|f\cdot \mathbb{1}_{C_k}\|_{L^p(\Sigma,\sigma)}\big)^q\Big)^{1/q} < \infty,\end{aligned} \tag{17.138}$$

which once again gives $f \in L^p_{\mathrm{loc}}(\Sigma,\sigma)$.

Moving on, consider the claim made in *(c)*. Then with f and K as before, use Hölder's inequality and the Ahlfors regularity of Σ to estimate

$$\begin{aligned}\int_K |f|\,d\sigma &\le \sum_{k=-\infty}^{N_K} \int_{C_k} |f|\,d\sigma\\ &\le \sum_{k=-\infty}^{N_K} \|f\cdot \mathbb{1}_{C_k}\|_{L^p(\Sigma,\sigma)}\sigma(C_k)^{1/p'}\\ &\le \sum_{k=-\infty}^{N_K} \big(2_*^{k\alpha}\|f\cdot \mathbb{1}_{C_k}\|_{L^p(\Sigma,\sigma)}\big)2_*^{k[(n-1)/p'-\alpha]}\\ &\le \Big(\sum_{k=-\infty}^{N_K} 2_*^{k[(n-1)/p'-\alpha]}\Big)\|f\|_{\dot{\mathrm{K}}^\alpha_{p,q}(\Sigma,\sigma)}.\end{aligned} \tag{17.139}$$

Thanks to the fact that $\alpha < \frac{n-1}{p'}$, this series is convergent. As such, this proves the membership $f \in L^1_{\mathrm{loc}}(\Sigma,\sigma)$.

Next, assume that $\alpha > \frac{-(n-1)}{p}$ and pick a function $f \in L^\infty_{\mathrm{comp}}(\Sigma,\sigma)$. Consider a sufficiently large integer $N_f \in \mathbb{N}$ so that $\operatorname{supp} f \subseteq \Delta(x_0, 2_*^{N_f})$. Using the Ahlfors regularity of Σ we see that

$$\begin{aligned}
\|f\|_{\dot{\mathrm{K}}^{\alpha}_{p,q}(\Sigma,\sigma)} &= \Big(\sum_{k=-\infty}^{N_f} \big(2_*^{k\alpha}\|f\cdot \mathbb{1}_{C_k}\|_{L^p(\Sigma,\sigma)}\big)^q\Big)^{1/q} \\
&\leq \|f\|_{L^\infty(\Sigma,\sigma)}\Big(\sum_{k=-\infty}^{N_f} \big(2_*^{k\alpha}\|\mathbb{1}_{C_k}\|_{L^p(\Sigma,\sigma)}\big)^q\Big)^{1/q} \\
&= \|f\|_{L^\infty(\Sigma,\sigma)}\Big(\sum_{k=-\infty}^{N_f} 2_*^{k\alpha q}\sigma(C_k)^{q/p}\Big)^{1/q} \\
&\leq \|f\|_{L^\infty(\Sigma,\sigma)}\Big(\sum_{k=-\infty}^{N_f} 2_*^{k[\alpha+(n-1)/p]q}\Big)^{1/q}. \qquad (17.140)
\end{aligned}$$

Notice that this series is convergent, since $\alpha + \frac{n-1}{p} > 0$. Hence, $f \in \dot{\mathrm{K}}^{\alpha}_{p,q}(\Sigma,\sigma)$ as desired. This completes the treatment of the claim made in *(d)*. Lastly, the inclusions claimed in *(e)* are direct consequences of items *(c)* and *(d)*. This finishes the proof of Proposition 17.7. □

Regarding the scale of inhomogeneous Herz spaces, we have the following embedding result.

Proposition 17.8 *Assume $\Sigma \subseteq \mathbb{R}^n$ is a closed unbounded Ahlfors regular, and abbreviate $\sigma := \mathcal{H}^{n-1}\lfloor\Sigma$. Select two exponents $p \in (0,\infty)$ and $q \in (0,\infty]$, along with an index $\alpha \in \mathbb{R}$. Then the following statements are true.*

(a) If $q_1, q_2 \in (0,\infty]$ and $q_1 \leq q_2$, then

$$\mathrm{K}^{\alpha}_{p,q_1}(\Sigma,\sigma) \hookrightarrow \mathrm{K}^{\alpha}_{p,q_2}(\Sigma,\sigma) \quad \textit{continuously}, \qquad (17.141)$$

i.e., the scale of inhomogeneous Herz spaces $\mathrm{K}^{\alpha}_{p,q}(\Sigma,\sigma)$ is monotonically increasing with respect to q.

(b) If $\alpha_1, \alpha_2 \in (0,\infty]$ and $\alpha_1 \leq \alpha_2$, then

$$\mathrm{K}^{\alpha_2}_{p,q}(\Sigma,\sigma) \hookrightarrow \mathrm{K}^{\alpha_1}_{p,q}(\Sigma,\sigma) \quad \textit{continuously}, \qquad (17.142)$$

i.e., the scale of inhomogeneous Herz spaces $\mathrm{K}^{\alpha}_{p,q}(\Sigma,\sigma)$ is monotonically decreasing with respect to α.

(c) The following inclusions hold

$$L^\infty_{\mathrm{comp}}(\Sigma,\sigma) \subseteq \mathrm{K}^{\alpha}_{p,q}(\Sigma,\sigma) \subseteq L^p_{\mathrm{loc}}(\Sigma,\sigma). \qquad (17.143)$$

In particular, whenever $p \in [1,\infty)$ it follows that

$$L^\infty_{\mathrm{comp}}(\Sigma,\sigma) \subseteq \mathrm{K}^{\alpha}_{p,q}(\Sigma,\sigma) \subseteq L^1_{\mathrm{loc}}(\Sigma,\sigma). \qquad (17.144)$$

(d) Suppose that $p \in (1, \infty)$. If either

$$q \in (1, \infty] \text{ and } \alpha > 0, \quad \text{or} \quad q \in (0, 1] \text{ and } \alpha \geq 0, \tag{17.145}$$

then

$$\mathrm{K}^{\alpha}_{p,q}(\Sigma, \sigma) \hookrightarrow L^p(\Sigma, \sigma) \quad \text{continuously.} \tag{17.146}$$

(e) Assume that $p \in (1, \infty)$, and denote by $p' := p/(p-1) \in (1, \infty)$ its Hölder conjugate exponent. If either

$$q \in (1, \infty] \text{ and } \alpha > \tfrac{n-1}{p'}, \quad \text{or} \quad q \in (0, 1] \text{ and } \alpha \geq \tfrac{n-1}{p'}, \tag{17.147}$$

then

$$\mathrm{K}^{\alpha}_{p,q}(\Sigma, \sigma) \hookrightarrow L^1(\Sigma, \sigma) \quad \text{continuously.} \tag{17.148}$$

As a consequence of this and item (d), it follows that

$$\mathrm{K}^{\alpha}_{p,q}(\Sigma, \sigma) \hookrightarrow L^r(\Sigma, \sigma) \quad \text{continuously, whenever } 1 \leq r \leq p. \tag{17.149}$$

Proof The embedding in *(a)* is visible from (17.115) and the fact whenever $q_1 \leq q_2$ we have $\ell^{q_1}(\mathbb{N}_0) \hookrightarrow \ell^{q_2}(\mathbb{N}_0)$. Moreover, the embedding claimed in *(b)* follows from (17.115) together with the observation that $2_*^{k\alpha_1} \leq 2_*^{k\alpha_2}$ for all $k \in \mathbb{N}_0$.

Next, consider the claim made in *(c)*. Pick a function $f \in L^{\infty}_{\mathrm{comp}}(\Sigma, \sigma)$, and select $N_f \in \mathbb{N}_0$ with the property that $\operatorname{supp} f \subseteq \Delta(x_0, 2_*^{N_f})$. Then

$$\begin{aligned} \|f\|_{\mathrm{K}^{\alpha}_{p,q}(\Sigma,\sigma)} &= \Big(\sum_{k=0}^{N_f} \big(2_*^{k\alpha} \|f \cdot \mathbb{1}_{C_k}\|_{L^p(\Sigma,\sigma)} \big)^q \Big)^{1/q} \\ &\leq \|f\|_{L^{\infty}(\Sigma,\sigma)} \Big(\sum_{k=0}^{N_f} 2_*^{k\alpha q} \Big)^{1/q} < \infty. \end{aligned} \tag{17.150}$$

This gives $f \in \mathrm{K}^{\alpha}_{p,q}(\Sigma, \sigma)$, as desired. To establish the second inclusion in (17.143), select a function $f \in \mathrm{K}^{\alpha}_{p,q}(\Sigma, \sigma)$. Given a compact set $K \subseteq \Sigma$, select $N_K \in \mathbb{N}$ such that

$$K \subseteq \bigcup_{k=0}^{N_K} \widetilde{C}_k, \tag{17.151}$$

where the collection $\{\widetilde{C}_k\}_{k \in \mathbb{N}_0}$ is as in (17.113) (which in particular satisfies (17.119)). Based on this and (17.115), we have

$$
\begin{aligned}
\Big(\int_K |f|^p\,d\sigma\Big)^{1/p} &\leq C\sum_{k=0}^{N_K}\|f\cdot\mathbb{1}_{\widetilde{C}_k}\|_{L^p(\Sigma,\sigma)} \\
&\leq C\Big(\sum_{k=0}^{N_K}2_*^{-k\alpha}\Big)\|f\|_{\mathrm{K}^{\alpha}_{p,q}(\Sigma,\sigma)}<\infty.
\end{aligned}
\tag{17.152}
$$

As such, $f \in L^p_{\text{loc}}(\Sigma, \sigma)$. This proves the inclusions in (17.143). The inclusion in (17.145) is a direct consequence of (17.143).

The next goal is to prove the embedding claimed in *(d)*. Pick an arbitrary function $f \in \mathrm{K}^{\alpha}_{p,q}(\Sigma, \sigma)$ and estimate

$$
\|f\|_{L^p(\Sigma,\sigma)} \leq \sum_{k=0}^{\infty}\|f\cdot\mathbb{1}_{\widetilde{C}_k}\|_{L^p(\Sigma,\sigma)} = \sum_{k=0}^{\infty}\big(2_*^{k\alpha}\|f\cdot\mathbb{1}_{\widetilde{C}_k}\|_{L^p(\Sigma,\sigma)}\big)2_*^{-k\alpha}. \tag{17.153}
$$

In the case when $q \in (1, \infty]$ and $\alpha > 0$, use Hölder's inequality to further obtain

$$
\begin{aligned}
\|f\|_{L^p(\Sigma,\sigma)} &\leq \Big(\sum_{k=0}^{\infty}\big(2_*^{k\alpha}\|f\cdot\mathbb{1}_{\widetilde{C}_k}\|_{L^p(\Sigma,\sigma)}\big)^q\Big)^{1/q}\Big(\sum_{k=0}^{\infty}2_*^{-k\alpha q'}\Big)^{1/q'} \\
&= \Big(\sum_{k=0}^{\infty}2_*^{-k\alpha q'}\Big)^{1/q'}\|f\|_{\mathrm{K}^{\alpha}_{p,q}(\Sigma,\sigma)}<\infty,
\end{aligned}
\tag{17.154}
$$

where $q' \in [1, \infty)$ denotes the Hölder conjugate exponent of q. In the scenario when $q \in (0, 1]$ and $\alpha \geq 0$, we observe that $2_*^{-k\alpha} \leq 1$ for all $k \in \mathbb{N}_0$. Based on this, estimate (17.153) as

$$
\|f\|_{L^p(\Sigma,\sigma)} \leq \Big(\sum_{k=0}^{\infty}\big(2_*^{k\alpha}\|f\cdot\mathbb{1}_{\widetilde{C}_k}\|_{L^p(\Sigma,\sigma)}\big)^q\Big)^{1/q} = \|f\|_{\mathrm{K}^{\alpha}_{p,q}(\Sigma,\sigma)}<\infty. \tag{17.155}
$$

Hence, (17.154)-(17.155) show that in either of the cases (17.145), the function f belongs to $L^p(\Sigma, \sigma)$. This then justifies (17.146).

Lastly, consider the claim made in *(e)*. Pick a function $f \in \mathrm{K}^{\alpha}_{p,q}(\Sigma, \sigma)$. Using Hölder's inequality, and the Ahlfors regularity of Σ, we bound

$$
\begin{aligned}
\int_\Sigma |f|\,d\sigma &= \sum_{k=0}^{\infty}\int_{\widetilde{C}_k}|f|\,d\sigma \leq \sum_{k=0}^{\infty}\|f\cdot\mathbb{1}_{\widetilde{C}_k}\|_{L^p(\Sigma,\sigma)}\sigma(\widetilde{C}_k)^{1/p'} \\
&\leq C\sum_{k=0}^{\infty}\big(2_*^{k\alpha}\|f\cdot\mathbb{1}_{\widetilde{C}_k}\|_{L^p(\Sigma,\sigma)}\big)2_*^{-k[\alpha-(n-1)/p']},
\end{aligned}
\tag{17.156}
$$

for some constant $C = C(\Sigma, p) \in (0, \infty)$. When $q \in (1, \infty)$ and $\alpha > (n-1)/p'$, use Hölder's inequality and the fact that $\alpha - (n-1)/p' > 0$ to bound (17.156) as

$$\int_{\Sigma} |f| \, d\sigma \le C\Big(\sum_{k=0}^{\infty} (2_*^{k\alpha} \|f \cdot \mathbb{1}_{C_k}\|_{L^p(\Sigma,\sigma)})^q \Big)^{1/q} \Big(\sum_{k=0}^{\infty} 2_*^{-k[\alpha-(n-1)/p']q'} \Big)^{1/q'}$$
$$= C\|f\|_{\mathrm{K}^{\alpha}_{p,q}(\Sigma,\sigma)} < \infty, \tag{17.157}$$

for some constant $C = C(\Sigma, p, q, \alpha) \in (0, \infty)$. Corresponding to the case $q \in (0, 1]$ and $\alpha \ge (n-1)/p'$, we may observe that $2_*^{-k[\alpha-(n-1)/p']} \le 1$ for all $k \in \mathbb{N}_0$. Thanks to this, we estimate (17.156) by

$$\int_{\Sigma} |f| \, d\sigma \le C\Big(\sum_{k=0}^{\infty} (2_*^{k\alpha} \|f \cdot \mathbb{1}_{C_k}\|_{L^p(\Sigma,\sigma)})^q \Big)^{1/q} = C\|f\|_{\mathrm{K}^{\alpha}_{p,q}(\Sigma,\sigma)}, \tag{17.158}$$

for some $C = C(\Sigma, p) \in (0, \infty)$. Therefore, in either case, (17.157)-(17.158) ensure that $f \in L^1(\Sigma, \sigma)$. This concludes the proof of (17.148). At this stage, the continuous embedding claimed in (17.149) is a direct consequence of (17.146), (17.148), and an interpolation argument for Lebesgue spaces. □

We conclude by presenting a version of Lebesgue's Dominated Convergence Theorem in the context of second-generation Herz spaces.

Proposition 17.9 *Consider a closed unbounded Ahlfors regular set $\Sigma \subseteq \mathbb{R}^n$, and abbreviate $\sigma := \mathcal{H}^{n-1}\lfloor\Sigma$. Let $\{f_j\}_{j\in\mathbb{N}}$ be a sequence of σ-measurable functions on Σ such that for some σ-measurable function $f : \Sigma \to \mathbb{R}$ one has*

$$f_j \to f \quad \text{at } \sigma\text{-a.e. point on } \Sigma \text{ as } j \to \infty. \tag{17.159}$$

Having fixed two exponents $p, q \in (0, \infty)$, assume there exists $g \in \dot{\mathrm{K}}^{\alpha}_{p,q}(\Sigma, \sigma)$ for which

$$|f_j(x)| \le g(x) \quad \text{at } \sigma\text{-a.e. } x \in \Sigma \text{ and for all } j \in \mathbb{N}. \tag{17.160}$$

Then

$$\|f_j - f\|_{\dot{\mathrm{K}}^{\alpha}_{p,q}(\Sigma,\sigma)} \to 0 \quad \text{as} \quad j \to \infty. \tag{17.161}$$

A similar result also holds in the context of inhomogeneous Herz spaces.

Proof The proof of the current result follows by naturally adapting the argument used to establish Proposition 10.4. □

17.4 Real Interpolation of Herz Spaces of Second Generation

In this section we specialize the interpolation result obtained in Theorem 17.1 in the setting of Ahlfors regular set, and deduce some embedding and boundedness results for second-generation Herz spaces.

Theorem 17.3 *Consider a closed unbounded Ahlfors regular set* $\Sigma \subseteq \mathbb{R}^n$*, and abbreviate* $\sigma := \mathcal{H}^{n-1}\lfloor\Sigma$. *Select exponents* $p \in (0,\infty)$ *and* $q_0, q_1, q \in (0,\infty]$ *along with two indices* $\alpha_0, \alpha_1 \in \mathbb{R}$ *such that* $\alpha_0 \neq \alpha_1$. *Then whenever* $\theta \in (0,1)$ *and* $\alpha := (1-\theta)\alpha_0 + \theta\alpha_1$ *it follows that*

$$\Big(\dot{\mathrm{K}}^{\alpha_0}_{p,q_0}(\Sigma,\sigma),\, \dot{\mathrm{K}}^{\alpha_1}_{p,q_1}(\Sigma,\sigma)\Big)_{\theta,q} = \dot{\mathrm{K}}^{\alpha}_{p,q}(\Sigma,\sigma) \tag{17.162}$$

and

$$\Big(\mathrm{K}^{\alpha_0}_{p,q_0}(\Sigma,\sigma),\, \mathrm{K}^{\alpha_1}_{p,q_1}(\Sigma,\sigma)\Big)_{\theta,q} = \mathrm{K}^{\alpha}_{p,q}(\Sigma,\sigma). \tag{17.163}$$

Proof Bearing in mind (17.118), (17.119), and (17.120), all claims are consequences of Theorem 17.1. □

The flexibility for the choices of indices q_0, q_1 in Theorem 17.3 and the identification recorded in (17.125) allow us to characterize Herz spaces as intermediate spaces (obtained via the real method of interpolation) between weighted Lebesgue spaces. This is made precise in the following corollary.

Corollary 17.1 *Consider a closed unbounded Ahlfors regular set* $\Sigma \subseteq \mathbb{R}^n$*, and abbreviate* $\sigma := \mathcal{H}^{n-1}\lfloor\Sigma$. *Select a reference point* $x_0 \in \Sigma$. *Fix two exponents* $p \in (0,\infty)$ *and* $q \in (0,\infty]$*, along with an index* $\alpha \in \mathbb{R}$. *Then whenever* $a, b \in \mathbb{R}$ *are such that* $-\infty < a < \alpha p < b < \infty$*, there exists* $\theta \in (0,1)$ *for which*

$$\dot{\mathrm{K}}^{\alpha}_{p,q}(\Sigma,\sigma) = \Big(L^p\big(\Sigma, |x-x_0|^a\sigma(x)\big),\, L^p\big(\Sigma, |x-x_0|^b\sigma(x)\big)\Big)_{\theta,q} \tag{17.164}$$

and

$$\mathrm{K}^{\alpha}_{p,q}(\Sigma,\sigma) = \Big(L^p\big(\Sigma, (1+|x-x_0|)^a\sigma(x)\big),\, L^p\big(\Sigma, (1+|x-x_0|)^b\sigma(x)\big)\Big)_{\theta,q}. \tag{17.165}$$

In particular, the following continuous embeddings are valid:

$$\dot{\mathrm{K}}^{\alpha}_{p,q}(\Sigma,\sigma) \hookrightarrow L^p\big(\Sigma, |x-x_0|^a\sigma(x)\big) + L^p\big(\Sigma, |x-x_0|^b\sigma(x)\big), \tag{17.166}$$

and

$$\mathrm{K}^{\alpha}_{p,q}(\Sigma,\sigma) \hookrightarrow L^p\big(\Sigma, (1+|x-x_0|)^a\sigma(x)\big) + L^p\big(\Sigma, (1+|x-x_0|)^b\sigma(x)\big). \tag{17.167}$$

Proof Select two numbers $a, b \in \mathbb{R}$ satisfying $-\infty < a < \alpha p < b < \infty$. To justify (17.164), choose $q_0 = q_1 := p$, $\alpha_0 := a/p$, and $\alpha_1 := b/p$. Also, pick $\theta \in (0,1)$ such $\alpha = (1-\theta)(a/p) + \theta(b/p)$, which is possible thanks to the way a and b have been selected. For these choices, use (17.125) together with (17.162) to obtain

$$\begin{aligned}\dot{\mathrm{K}}^{\alpha}_{p,q}(\Sigma,\sigma) &= \Big(\dot{\mathrm{K}}^{a/p}_{p,p}(\Sigma,\sigma),\, \dot{\mathrm{K}}^{b/p}_{p,p}(\Sigma,\sigma)\Big)_{\theta,q} \\ &= \Big(L^p\big(\Sigma, |x-x_0|^a\sigma(x)\big),\, L^p\big(\Sigma, |x-x_0|^b\sigma(x)\big)\Big)_{\theta,q}.\end{aligned}$$

This proves (17.164). As a consequence of this and (17.59) we obtain the continuous embedding claimed in (17.166). Via a similar reasoning, the claims corresponding to the inhomogeneous Herz space $\mathrm{K}^{\alpha}_{p,q}(\Sigma,\sigma)$ also hold. □

The relevance of identifying homogeneous and inhomogeneous Herz spaces as intermediate spaces (obtained via the real method of interpolation) between weighted Lebesgue spaces is that matters can be arranged so that the weights belong to some Muckenhoupt class. This is made precise in the follows remark.

Remark 17.1 In the regime

$$1 < p < \infty, \quad 0 < q \leq \infty, \quad \text{and} \quad \tfrac{-(n-1)}{p} < \alpha < \tfrac{n-1}{p'}, \tag{17.168}$$

where $p' := p/(p-1) \in (1,\infty)$ is the Hölder conjugate exponent of p, we can choose numbers $a, b \in \mathbb{R}$ satisfying

$$-(n-1) < a < \alpha p < b < (p-1)(n-1), \tag{17.169}$$

This, (2.70), and (2.71) imply that the functions

$$\begin{gathered}|\cdot - x_0|^a, \quad |\cdot - x_0|^b, \quad (1+|\cdot - x_0|)^a, \quad \text{and} \quad (1+|\cdot - x_0|)^b \\ \text{belong to the Muckenhoupt class } A_p(\Sigma,\sigma).\end{gathered} \tag{17.170}$$

As such, we conclude that

> whenever (17.168) holds, the Herz spaces $\dot{\mathrm{K}}^{\alpha}_{p,q}(\Sigma,\sigma)$ and $\mathrm{K}^{\alpha}_{p,q}(\Sigma,\sigma)$ are intermediate spaces (for the real method of interpolation) between Muckenhoupt weighted Lebesgue spaces as in (17.164)-(17.165). (17.171)

In this context, thanks to (2.84) and Corollary 17.1, whenever (17.168) holds the following inclusions hold:

$$\begin{gathered}\dot{\mathrm{K}}^{\alpha}_{p,q}(\Sigma,\sigma) \subseteq \textstyle\bigcup_{1<r<p} L^r_{\mathrm{loc}}(\Sigma,\sigma) \subseteq L^1_{\mathrm{loc}}(\Sigma,\sigma) \quad \text{and} \\ \mathrm{K}^{\alpha}_{p,q}(\Sigma,\sigma) \subseteq \textstyle\bigcup_{1<r<p} L^r_{\mathrm{loc}}(\Sigma,\sigma) \subseteq L^1_{\mathrm{loc}}(\Sigma,\sigma).\end{gathered} \tag{17.172}$$

Another useful embedding for second-generation Herz spaces is recorded below.

Corollary 17.2 *Consider a closed unbounded Ahlfors regular set $\Sigma \subseteq \mathbb{R}^n$, and abbreviate $\sigma := \mathcal{H}^{n-1}\lfloor\Sigma$. Fix two exponents $p \in (1,\infty)$ and $q \in (0,\infty]$, along with an index $\alpha \in \mathbb{R}$. Denote by $p' := p/(p-1) \in (1,\infty)$ the Hölder conjugate exponent of p. Then whenever $\frac{-(n-1)}{p} < \alpha < \frac{n-1}{p'}$ one has*

$$\dot{\mathrm{K}}^{\alpha}_{p,q}(\Sigma,\sigma) \hookrightarrow L^1\Big(\Sigma, \frac{\sigma(x)}{1+|x|^{n-1}}\Big) \quad \textit{continuously}, \tag{17.173}$$

and

$$\mathrm{K}^{\alpha}_{p,q}(\Sigma,\sigma) \hookrightarrow L^1\Big(\Sigma, \frac{\sigma(x)}{1+|x|^{n-1}}\Big) \quad \textit{continuously.} \tag{17.174}$$

Proof Since $\frac{-(n-1)}{p} < \alpha < \frac{n-1}{p'}$ we can pick numbers $a, b \in \mathbb{R}$ satisfying (17.169)-(17.170). As such, a combination of (2.83), (17.166), and (17.167) implies the embeddings (17.173)-(17.174). □

Next, we notice that in general the scale of Herz spaces (of first generation) $\mathrm{A}_{p,q}$ is not stable under the real method interpolation. More specifically, we have the following result.

Corollary 17.3 *Consider a closed unbounded Ahlfors regular set $\Sigma \subseteq \mathbb{R}^n$, and abbreviate $\sigma := \mathcal{H}^{n-1}\lfloor\Sigma$. Select exponents $p \in (0,\infty)$, and $q_0, q_1, q \in (0,\infty)$, with $q_0 \neq q_1$, and an index $\theta \in (0,1)$. Then the following identification holds:*

$$\big(\mathrm{A}_{p,q_0}(\Sigma,\sigma),\, \mathrm{A}_{p,q_1}(\Sigma,\sigma)\big)_{\theta,q} = \mathrm{K}^{(n-1)[(1-\theta)/q_0+\theta/q_1-1/p]}_{p,q}(\Sigma,\sigma). \tag{17.175}$$

In particular, for the choice $q := \big((1-\theta)/q_0+\theta/q_1\big)^{-1} \in (0,\infty)$ one has

$$\big(\mathrm{A}_{p,q_0}(\Sigma,\sigma),\, \mathrm{A}_{p,q_1}(\Sigma,\sigma)\big)_{\theta,q} = \mathrm{A}_{p,q}(\Sigma,\sigma). \tag{17.176}$$

Proof The equality displayed in (17.175) is a direct consequence of Theorem 17.3 and the fact that

$$\mathrm{A}_{p,q_j}(\Sigma,\sigma) = \mathrm{K}^{(n-1)(1/q_j-1/p)}_{p,q_j}(\Sigma,\sigma) \quad \text{for each } j \in \{0,1\}. \tag{17.177}$$

Also, (17.176) can be seen from (17.175) written with $q := \big((1-\theta)/q_0+\theta/q_1\big)^{-1}$.□

17.5 Boundedness of the Hardy-Littlewood Maximal Operator

We are now in a position to prove the boundedness of the Hardy-Littlewood maximal operator on second-generation Herz spaces.

Theorem 17.4 *Consider a closed unbounded Ahlfors regular set $\Sigma \subseteq \mathbb{R}^n$, and abbreviate $\sigma := \mathcal{H}^{n-1}\lfloor\Sigma$. Select two exponents $p \in (1,\infty)$ and $q \in (0,\infty]$, and an index $\alpha \in \mathbb{R}$ satisfying*

$$\frac{-(n-1)}{p} < \alpha < \frac{(n-1)}{p'}, \tag{17.178}$$

where $p' \in (1,\infty)$ denotes the Hölder conjugate exponent of p. Then the Hardy-Littlewood maximal operator on Σ (cf. (2.26))

$$\mathcal{M}_\Sigma : \dot{\mathrm{K}}^{\alpha}_{p,q}(\Sigma,\sigma) \to \dot{\mathrm{K}}^{\alpha}_{p,q}(\Sigma,\sigma) \quad \textit{is well defined, sub-linear, and bounded.} \tag{17.179}$$

Corresponding to the inhomogeneous case, it follows that

$$\mathcal{M}_\Sigma : \mathrm{K}^{\alpha}_{p,q}(\Sigma,\sigma) \to \mathrm{K}^{\alpha}_{p,q}(\Sigma,\sigma) \quad \text{is well defined, sub-linear, and bounded.} \tag{17.180}$$

Proof Since α satisfies (17.178), we may select $a, b \in \mathbb{R}$ satisfying (17.169). In this context, (2.44), (2.45), and (17.170), together with Corollary 17.1 and [GHA-II, Proposition 1.3.7, pp. 33–34], imply the boundedness claimed in (17.179)-(17.180).□

Remark 17.2 It is useful to notice that under the same assumptions as in Theorem 17.4, one may also conclude that

$$\begin{gathered}\mathcal{M}_\Sigma : \dot{\mathrm{K}}^{-\alpha}_{p',q}(\Sigma,\sigma) \to \dot{\mathrm{K}}^{-\alpha}_{p',q}(\Sigma,\sigma)\\ \text{and}\quad \mathcal{M}_\Sigma : \mathrm{K}^{-\alpha}_{p',q}(\Sigma,\sigma) \to \mathrm{K}^{-\alpha}_{p',q}(\Sigma,\sigma)\\ \text{are well defined, sub-linear, and bounded.}\end{gathered} \tag{17.181}$$

This is a consequence of (17.179) used with p replaced by p', and α replaced by $-\alpha$, bearing in mind that the condition (17.178) is equivalent to $\frac{-(n-1)}{p'} < -\alpha < \frac{(n-1)}{p}$.

The following result may be regarded as an improvement of Theorem 17.4.

Theorem 17.5 *Consider a closed unbounded Ahlfors regular set $\Sigma \subseteq \mathbb{R}^n$, and abbreviate $\sigma := \mathcal{H}^{n-1}\lfloor\Sigma$. Select two exponents $p \in (0,\infty)$ and $q \in (0,\infty]$, a parameter $s \in (0,\infty)$ such that $p/s > 1$, and an index $\alpha \in \mathbb{R}$ satisfying*

$$\frac{-(n-1)}{p} < \alpha < (n-1)\Big(\frac{1}{s} - \frac{1}{p}\Big). \tag{17.182}$$

Then the L^s-based Hardy-Littlewood maximal operator on Σ (cf. (2.28))

$$\mathcal{M}_{\Sigma,s} : \dot{\mathrm{K}}^{\alpha}_{p,q}(\Sigma,\sigma) \to \dot{\mathrm{K}}^{\alpha}_{p,q}(\Sigma,\sigma) \quad \text{is well defined, sub-linear, and bounded.} \tag{17.183}$$

Corresponding to the inhomogeneous case, it follows that

$$\mathcal{M}_{\Sigma,s} : \mathrm{K}^{\alpha}_{p,q}(\Sigma,\sigma) \to \mathrm{K}^{\alpha}_{p,q}(\Sigma,\sigma) \quad \text{is well defined, sub-linear, and bounded.} \tag{17.184}$$

Proof Observe that condition (17.182) is equivalent to

$$\frac{-(n-1)}{p/s} < \alpha s < (n-1)\Big(1 - \frac{1}{p/s}\Big). \tag{17.185}$$

Since $p/s > 1$, Theorem 17.4 (used with α replaced by αs) guarantees that $\mathcal{M}_\Sigma$ is bounded on $\dot{\mathrm{K}}^{\alpha s}_{p/s,q/s}(\Sigma,\sigma)$. Granted this and (17.7), for each $f \in \dot{\mathrm{K}}^{\alpha}_{p,q}(\Sigma,\sigma)$ we have

$$\begin{aligned}\|\mathcal{M}_{\Sigma,s}f\|_{\dot{\mathrm{K}}^{\alpha}_{p,q}(\Sigma,\sigma)} &= \big\|(\mathcal{M}_{\Sigma}|f|^{s})^{1/s}\big\|_{\dot{\mathrm{K}}^{\alpha}_{p,q}(\Sigma,\sigma)} = \big\|\mathcal{M}_{\Sigma}|f|^{s}\big\|^{1/s}_{\dot{\mathrm{K}}^{\alpha s}_{p/s,q/s}(\Sigma,\sigma)} \\ &\leq C\big\||f|^{s}\big\|^{1/s}_{\dot{\mathrm{K}}^{\alpha s}_{p/s,q/s}(\Sigma,\sigma)} = C\|f\|_{\dot{\mathrm{K}}^{\alpha}_{p,q}(\Sigma,\sigma)}.\end{aligned} \tag{17.186}$$

This proves the claim made in (17.183). Via a similar argument, now bearing in mind (17.16), we may also obtain the boundedness claimed in (17.184). □

It turns out that measuring the size of a function using nontangential maximal functions on second-generation Herz spaces is independent of the aperture parameter. A detailed statement is contained in the following proposition.

Proposition 17.10 *Assume $\Omega \subseteq \mathbb{R}^n$ is an Ahlfors regular domain with $\partial\Omega$ an unbounded set, and abbreviate $\sigma := \mathcal{H}^{n-1}\lfloor\partial\Omega$. Select exponents $p \in (0,\infty)$ and $q \in (0,\infty]$, along with an index $\alpha > -(n-1)/p$. Finally, fix two aperture parameters $\kappa, \widetilde{\kappa} \in (0,\infty)$.*

Then whenever $u : \Omega \to \mathbb{R}$ is an $\mathcal{L}^n$-measurable function, one has

$$\begin{aligned}&\|\mathcal{N}_{\kappa}u\|_{\dot{\mathrm{K}}^{\alpha}_{p,q}(\partial\Omega,\sigma)} \approx \|\mathcal{N}_{\widetilde{\kappa}}u\|_{\dot{\mathrm{K}}^{\alpha}_{p,q}(\partial\Omega,\sigma)} \\ \text{and} \quad &\|\mathcal{N}_{\kappa}u\|_{\mathrm{K}^{\alpha}_{p,q}(\partial\Omega,\sigma)} \approx \|\mathcal{N}_{\widetilde{\kappa}}u\|_{\mathrm{K}^{\alpha}_{p,q}(\partial\Omega,\sigma)},\end{aligned} \tag{17.187}$$

with implicit proportionality constants depending only on n, p, q, α, κ, $\widetilde{\kappa}$, and the Ahlfors regularity character of $\partial\Omega$.

Proof Without loss of generality, assume $\widetilde{\kappa} \geq \kappa$. Choose $s \in (0,p)$ sufficiently small such that

$$\frac{-(n-1)}{p} < \alpha < (n-1)\Big(\frac{1}{s} - \frac{1}{p}\Big). \tag{17.188}$$

In this regime, we may employ (17.5) and [GHA-I, Theorem 8.4.6, pp. 704–705] to obtain

$$\|\mathcal{N}_{\kappa}u\|_{\dot{\mathrm{K}}^{\alpha}_{p,q}(\partial\Omega,\sigma)} \leq \|\mathcal{N}_{\widetilde{\kappa}}u\|_{\dot{\mathrm{K}}^{\alpha}_{p,q}(\partial\Omega,\sigma)} \leq C\|\mathcal{M}_{\partial\Omega,s}(\mathcal{N}_{\kappa}u)\|_{\dot{\mathrm{K}}^{\alpha}_{p,q}(\partial\Omega,\sigma)}, \tag{17.189}$$

for some constant $C \in (0,\infty)$ depending only on n, κ, $\widetilde{\kappa}$, and the Ahlfors regularity constants of $\partial\Omega$. This and the boundedness of the L^s-based Hardy-Littlewood maximal operator $\mathcal{M}_{\partial\Omega,s}$ on $\dot{\mathrm{K}}^{\alpha}_{p,q}(\partial\Omega,\sigma)$ (cf. Theorem 17.5) imply the first proportionality in (17.187). The same type of argument also holds in the context of inhomogeneous Herz spaces $\mathrm{K}^{\alpha}_{p,q}(\partial\Omega,\sigma)$. □

17.6 A New Class: Geometric Herz Spaces

In this section we introduce a new brand of Herz-type spaces, which we shall refer to as Geometric Herz spaces, which generalizes $\mathrm{K}^{\alpha}_{p,q}(\Sigma,\sigma)$, in that now the functions into consideration are going to be localized on "tubular" regions determined by a given fixed closed subset of an Ahlfors regular set.

To elaborate, assume $\Sigma \subseteq \mathbb{R}^n$ is a closed unbounded Ahlfors regular set, and abbreviate $\sigma := \mathcal{H}^{n-1}\lfloor\Sigma$. With $C_\Sigma, c_\Sigma \in (0,\infty)$ denoting the upper and lower Ahlfors regularity constants of Σ, respectively, select a dilation factor $2_* > \big(C_\Sigma/c_\Sigma\big)^{1/(n-1)}$. Consider a d-Ahlfors regular set $E \subseteq \Sigma$ for some $d \in [0, n-1)$, i.e., E is a closed subset of Σ, and has the property that

$$\mathcal{H}^d\big(B(x,R)\cap E\big) \approx R^d \quad \text{uniformly for } x \in E \text{ and } R \in \big(0, 2\mathrm{diam}(E)\big), \tag{17.190}$$

where $\mathcal{H}^d$ denotes the d-dimensional Hausdorff measure in $\mathbb{R}^n$. Define the tubular regions determined by E as

$$\begin{aligned} C_k^E &:= \big\{x \in \Sigma : 2_*^k < \mathrm{dist}(x,E) \le 2_*^{k+1}\big\} \quad \text{for each } k \in \mathbb{Z}, \\ \widetilde{C}_0^E &:= \big\{x \in \Sigma : \mathrm{dist}(x,E) \le 2_*\big\} \quad \text{and} \quad \widetilde{C}_k^E := C_k^E \quad \text{for each } k \in \mathbb{N}. \end{aligned} \tag{17.191}$$

Make the following non-degeneracy assumption

$$\sigma(C_k^E) > 0 \quad \text{each } k \in \mathbb{Z}. \tag{17.192}$$

Since $d < n-1$, it is evident from definitions that

$$\sigma\Bigg(\Sigma \setminus \bigcup_{k\in\mathbb{Z}} C_k^E\Bigg) = \sigma(E) = \mathcal{H}^{n-1}(E) = 0, \tag{17.193}$$

i.e., $\{C_k^E\}_{k\in\mathbb{Z}}$ covers the entire Σ up to a set of $\mathcal{H}^{n-1}$-measure zero. Moreover,

$$\Sigma = \bigcup_{k\in\mathbb{N}_0} \widetilde{C}_k^E. \tag{17.194}$$

Having selected indices $p, q \in (0,\infty]$, and $\alpha \in \mathbb{R}$, for each σ-measurable function f defined on Σ, introduce the quasinorms:

$$\|f\|_{\dot{\mathrm{K}}^{\alpha}_{p,q}(\Sigma,\sigma;E)} := \begin{cases} \Bigg[\displaystyle\sum_{k=-\infty}^{\infty} \big(2_*^{k\alpha}\|f\cdot \mathbb{1}_{C_k^E}\|_{L^p(\Sigma,\sigma)}\big)^q\Bigg]^{1/q} & \text{if } q<\infty, \\ \displaystyle\sup_{k\in\mathbb{Z}} \big\{2_*^{k\alpha}\|f\cdot \mathbb{1}_{C_k^E}\|_{L^p(\Sigma,\sigma)}\big\} & \text{if } q=\infty, \end{cases} \tag{17.195}$$

and

$$\|f\|_{\mathrm{K}^{\alpha}_{p,q}(\Sigma,\sigma;E)} := \begin{cases} \Bigg[\displaystyle\sum_{k=0}^{\infty} \big(2_*^{k\alpha}\|f\cdot \mathbb{1}_{\widetilde{C}_k^E}\|_{L^p(\Sigma,\sigma)}\big)^q\Bigg]^{1/q} & \text{if } q<\infty, \\ \displaystyle\sup_{k\in\mathbb{N}_0} \big\{2_*^{k\alpha}\|f\cdot \mathbb{1}_{\widetilde{C}_k^E}\|_{L^p(\Sigma,\sigma)}\big\} & \text{if } q=\infty. \end{cases} \tag{17.196}$$

With these pieces of notation, define the `homogeneous Geometric Herz space` on Σ associated with E as

$$\begin{aligned}&\dot{\mathrm{K}}^{\alpha}_{p,q}(\Sigma,\sigma;E)\\&\quad:=\big\{f:\Sigma\to\mathbb{R}:f\text{ is }\sigma\text{-measurable and }\|f\|_{\dot{\mathrm{K}}^{\alpha}_{p,q}(\Sigma,\sigma;E)}<\infty\big\},\end{aligned}\tag{17.197}$$

and the `inhomogeneous Geometric Herz space` on Σ associated with E as

$$\begin{aligned}&\mathrm{K}^{\alpha}_{p,q}(\Sigma,\sigma;E)\\&\quad:=\big\{f:\Sigma\to\mathbb{R}:f\text{ is }\sigma\text{-measurable and }\|f\|_{\mathrm{K}^{\alpha}_{p,q}(\Sigma,\sigma;E)}<\infty\big\}.\end{aligned}\tag{17.198}$$

We shall refer to E as the `reference set` to the corresponding Geometric Herz spaces $\dot{\mathrm{K}}^{\alpha}_{p,q}(\Sigma,\sigma;E)$ and $\mathrm{K}^{\alpha}_{p,q}(\Sigma,\sigma;E)$.

It is of relevance to note that this brand of Geometric Herz spaces fall within the framework of measure theoretic Herz spaces introduced in (17.3) and (17.12). As such, all the basic properties obtained in Section 17.1 hold in the context of Geometric Herz spaces. For instance, Proposition 17.3 entails that

the spaces $\dot{\mathrm{K}}^{\alpha}_{p,q}(\Sigma,\sigma;E)$ and $\mathrm{K}^{\alpha}_{p,q}(\Sigma,\sigma;E)$ are quasi-Banach spaces, whenever $p,q\in(0,\infty]$ and $\alpha\in\mathbb{R}$. (17.199)

In particular, they are genuine Banach spaces in the regime $p,q\in[1,\infty]$.

Being mindful of the fact that $\Sigma=\cup_{j\in\mathbb{N}}\Delta(z_0,j)$, for any fixed reference point $z_0\in\Sigma$, and that $0<\sigma\big(\Delta(z_0,j)\big)<\infty$, Proposition 17.5 ensures that

for every indices $p,q\in(1,\infty)$ and $\alpha\in\mathbb{R}$, the spaces $\dot{\mathrm{K}}^{\alpha}_{p,q}(\Sigma,\sigma;E)$ and $\mathrm{K}^{\alpha}_{p,q}(\Sigma,\sigma;E)$ are Generalized Banach Function Spaces on (Σ,σ) (in the sense of Definition 10.2), (17.200)

and, thanks to Proposition 17.6, their associated space may be identified as

$$\begin{aligned}\big(\dot{\mathrm{K}}^{\alpha}_{p,q}(\Sigma,\sigma;E)\big)'&=\dot{\mathrm{K}}^{-\alpha}_{p',q'}(\Sigma,\sigma;E)\\\text{and }\big(\mathrm{K}^{\alpha}_{p,q}(\Sigma,\sigma;E)\big)'&=\mathrm{K}^{-\alpha}_{p',q'}(\Sigma,\sigma;E),\end{aligned}\tag{17.201}$$

where $p',q'\in(1,\infty)$ are the Hölder conjugate exponents of p and q, respectively.

Another relevant feature of Geometric Herz spaces is that corresponding to the diagonal case $p=q\in(0,\infty)$ they coincide with certain weighted Lebesgue spaces. Specifically, given $p\in(0,\infty)$ and $\alpha\in\mathbb{R}$ we have

$$\dot{\mathrm{K}}^{\alpha}_{p,p}(\Sigma,\sigma;E)=L^p\big(\Sigma,\operatorname{dist}(x,E)^{\alpha p}\sigma(x)\big)\tag{17.202}$$

$$\text{and }\mathrm{K}^{\alpha}_{p,p}(\Sigma,\sigma;E)=L^p\Big(\Sigma,\big(1+\operatorname{dist}(x,E)\big)^{\alpha p}\sigma(x)\Big).\tag{17.203}$$

These identifications follow from (17.192), (17.193), (17.195), and (17.196), together with the fact that

$$\operatorname{dist}(x, E)^{\alpha p} \approx 2_*^{k\alpha} \quad \text{uniformly for } x \in C_k^E \text{ and } k \in \mathbb{Z}, \tag{17.204}$$

and

$$\big(1 + \operatorname{dist}(x, E)\big)^{\alpha p} \approx 2_*^{k\alpha} \quad \text{uniformly for } x \in \widetilde{C}_k^E \text{ and for } k \in \mathbb{N}_0. \tag{17.205}$$

To proceed, we provide a wealth of examples of d-Ahlfors regular sets that can be used as reference sets for Geometric Herz spaces.

Example 17.1 Consider a closed unbounded Ahlfors regular set $\Sigma \subseteq \mathbb{R}^n$, and abbreviate $\sigma := \mathcal{H}^{n-1}\lfloor\Sigma$. Pick a reference point $x_0 \in \Sigma$. Then the set $E := \{x_0\} \subseteq \Sigma$ is a 0-Ahlfors regular set, and the corresponding family of regions $\{C_k^E\}_{k\in\mathbb{Z}}$ as in (17.191) become the family of annuli $\{C_k\}_{k\in\mathbb{Z}}$ defined in (17.112). In this case, it is visible from (17.191), (17.197), and (17.198) that the homogeneous and inhomogeneous Geometric Herz spaces associated with the singleton $\{x_0\}$ coincide with the standard homogeneous and inhomogeneous Herz spaces defined in (17.116) and (17.117), respectively:

$$\dot{\mathrm{K}}^{\alpha}_{p,q}\big(\Sigma, \sigma; \{x_0\}\big) = \dot{\mathrm{K}}^{\alpha}_{p,q}(\Sigma, \sigma) \quad \text{and} \quad \mathrm{K}^{\alpha}_{p,q}\big(\Sigma, \sigma; \{x_0\}\big) = \mathrm{K}^{\alpha}_{p,q}(\Sigma, \sigma). \tag{17.206}$$

More generally, given $N \in \mathbb{N}$ and arbitrary distinct points $x_1, \dots, x_N \in \Sigma$, we can also choose the 0-Ahlfors regular set $E := \{x_1, \dots, x_N\} \subseteq \Sigma$. For this choice, we refer to the Geometric Herz space associated with $\{x_1, \dots, x_N\}$ as N-`centered Herz space` on Σ.

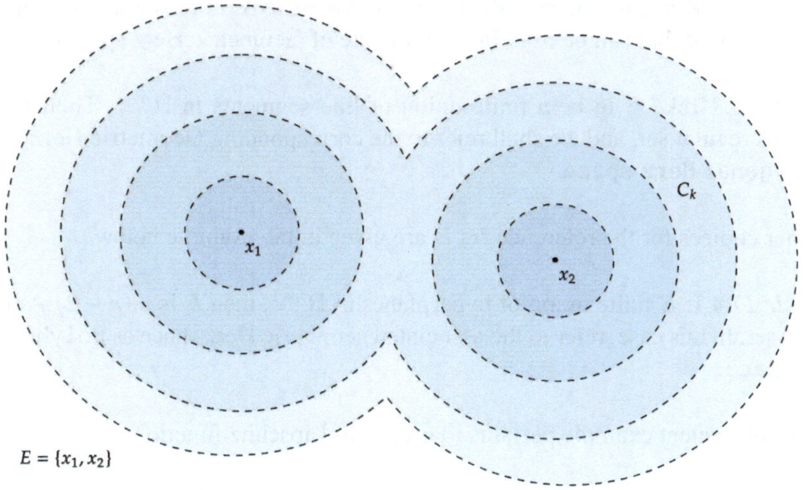

Fig. 17.1: Example of a 0-AR set in $\mathbb{R}^2$ and its tubular neighborhoods.

For the next few examples, we consider the Euclidean setting $\Sigma := \mathbb{R}^{n-1}$, a scenario in which $\mathcal{H}^{n-1} \equiv \mathcal{L}^{n-1}$ in $\mathbb{R}^{n-1}$, and assume $n \geq 3$.

Example 17.2 Choose E to be any linear subspace of $\mathbb{R}^{n-1}$ with dimension $d \in \mathbb{N}$ and $1 \leq d < n-1$. Then E is a d-AR set. This choice leads to a new class of Herz spaces, which we refer to as `tubular Herz spaces` in $\mathbb{R}^{n-1}$.

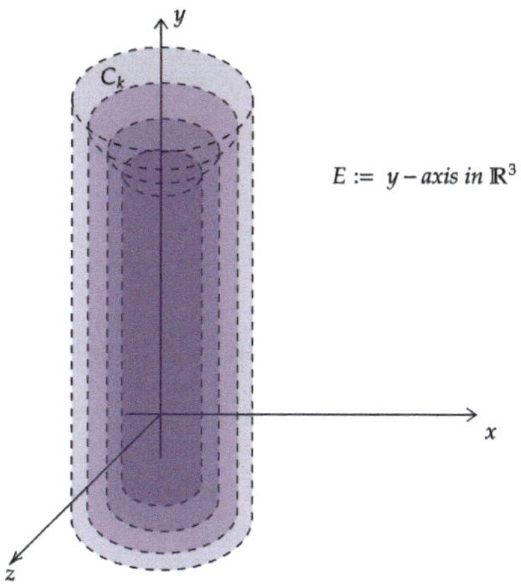

Fig. 17.2: Example of a 1-AR set in $\mathbb{R}^3$ and its tubular neighborhoods.

We continue to present non-standard choices of d-Ahlfors regular sets in the Euclidean space that can be used in the template of Geometric Herz spaces.

Example 17.3 Take E to be a finite union of line segments in $\mathbb{R}^{n-1}$. Then E is a 1-Ahlfors regular set, and we shall refer to the corresponding Geometric Herz space as `polygonal Herz space`.

Other choices for the reference set E are given in the example below.

Example 17.4 If E finite union of hyperplanes in $\mathbb{R}^{n-1}$, then E is a $(n-2)$-Ahlfors regular set. In this case, refer to the associated geometric Herz space as `polyhedral Herz space`.

The subsequent example pertains to graphs of Lipschitz functions.

Example 17.5 Consider a Lipschitz function $\phi : \mathbb{R}^d \to \mathbb{R}$, for some $d \in \mathbb{N}$ with $1 \leq d < n-1$. Choosing $E := \{(x, \phi(x)) : x \in \mathbb{R}^d\}$, i.e., the graph of ϕ, it follows that E is a d-Ahlfors regular set in $\mathbb{R}^{n-1}$. This gives rise to another family of geometric Herz spaces, which we refer to as `rough-tubular Herz spaces`.

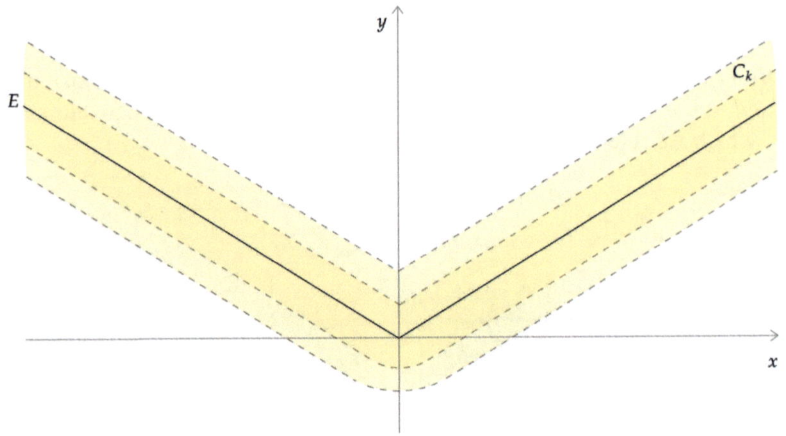

Fig. 17.3: Graph of a Lipschitz function, and its tubular neighborhoods.

Our last example involves a fractal, which highlights the degree of flexibility that the framework of Geometric Herz spaces provides.

Example 17.6 The von Koch snowflake in the plane $\mathbb{R}^2$ is a $\frac{\ln(4)}{\ln(3)}$-Ahlfors regular set, and its Hausdorff dimension is $\frac{\ln(4)}{\ln(3)} < 2$. Choosing E to be the von Koch snowflake makes E a suitable reference set for geometric Herz spaces. In such a case, we call the resulting space the `snowflake Herz space`.

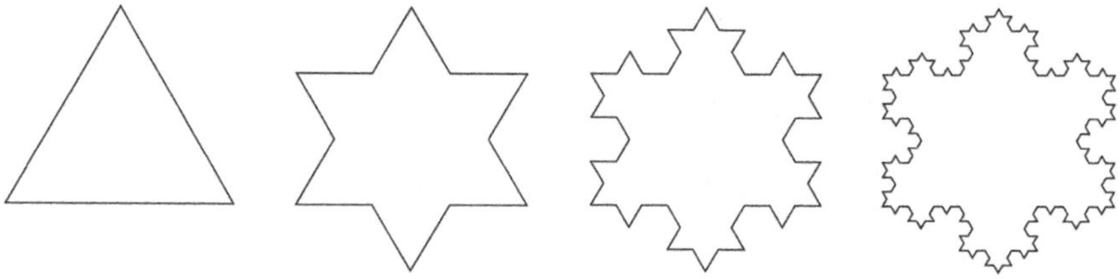

Fig. 17.4: First four iterations of the von Koch snowflake.

Remarkably, the fact that Geometric Herz spaces fall within the scope of measure theoretic Herz spaces allow us to obtain the following interpolation result. Succinctly put, the scale of Geometric Herz spaces is stable under the real method of interpolation.

Proposition 17.11 *Consider a closed unbounded Ahlfors regular set* $\Sigma \subseteq \mathbb{R}^n$, *and abbreviate* $\sigma := \mathcal{H}^{n-1}\lfloor\Sigma$. *Fix some dimension* $d \in [0, n-1)$ *and let* $E \subseteq \Sigma$ *be a* d*-Ahlfors regular set. Select exponents* $p \in (0, \infty)$ *and* $q_0, q_1, q \in (0, \infty]$ *along with two indices* $\alpha_0, \alpha_1 \in \mathbb{R}$ *such that* $\alpha_0 \neq \alpha_1$. *Then whenever* $\theta \in (0, 1)$ *and* $\alpha := (1-\theta)\alpha_0 + \theta\alpha_1$ *it follows that*

$$\left(\dot{\mathrm{K}}^{\alpha_0}_{p,q_0}(\Sigma, \sigma; E),\, \dot{\mathrm{K}}^{\alpha_1}_{p,q_1}(\Sigma, \sigma; E)\right)_{\theta,q} = \dot{\mathrm{K}}^{\alpha}_{p,q}(\Sigma, \sigma; E) \tag{17.207}$$

and

$$\left(\mathrm{K}^{\alpha_0}_{p,q_0}(\Sigma, \sigma; E),\, \mathrm{K}^{\alpha_1}_{p,q_1}(\Sigma, \sigma; E)\right)_{\theta,q} = \mathrm{K}^{\alpha}_{p,q}(\Sigma, \sigma; E). \tag{17.208}$$

Proof Recall that Lemma 17.1 implies that any two Geometric Herz spaces are compatible. As such, (17.207)-(17.208) are direct consequences of Theorem 17.1.□

The above proposition together with the identifications recorded in (17.202)-(17.203) permit us to obtain the following characterization of Geometric Herz spaces in terms of intermediate spaces between weighted Lebesgue spaces. Such a result may be regarded as an improvement of Corollary 17.1.

Corollary 17.4 *Suppose* $\Sigma \subseteq \mathbb{R}^n$ *is a closed unbounded Ahlfors regular set, and abbreviate* $\sigma := \mathcal{H}^{n-1}\lfloor\Sigma$. *Fix some* d-AR *set* $E \subseteq \Sigma$, *with* $d \in [0, n-1)$. *Pick two exponents* $p \in (0, \infty)$ *and* $q \in (0, \infty]$, *along with an index* $\alpha \in \mathbb{R}$. *Then whenever* $a, b \in \mathbb{R}$ *are such that* $-\infty < a < \alpha p < b < \infty$, *there exists* $\theta \in (0, 1)$ *for which*

$$\dot{\mathrm{K}}^{\alpha}_{p,q}(\Sigma, \sigma; E) = \left(L^p\big(\Sigma, \mathrm{dist}(x, E)^a \sigma(x)\big),\, L^p\big(\Sigma, \mathrm{dist}(x, E)^b \sigma(x)\big)\right)_{\theta,q} \tag{17.209}$$

and

$$\mathrm{K}^{\alpha}_{p,q}(\Sigma,\sigma;E) \\ = \Big(L^p\big(\Sigma,(1+\mathrm{dist}(x,E))^a\sigma(x)\big),\ L^p\big(\Sigma,(1+\mathrm{dist}(x,E))^b\sigma(x)\big)\Big)_{\theta,q}. \tag{17.210}$$

Consequently, the following continuous embeddings are valid:

$$\dot{\mathrm{K}}^{\alpha}_{p,q}(\Sigma,\sigma;E) \hookrightarrow L^p\big(\Sigma,\mathrm{dist}(x,E)^a\sigma(x)\big) + L^p\big(\Sigma,\mathrm{dist}(x,E)^b\sigma(x)\big) \tag{17.211}$$

and

$$\mathrm{K}^{\alpha}_{p,q}(\Sigma,\sigma;E) \hookrightarrow L^p\big(\Sigma,(1+\mathrm{dist}(x,E))^a\sigma(x)\big) + L^p\big(\Sigma,(1+\mathrm{dist}(x,E))^b\sigma(x)\big). \tag{17.212}$$

Proof Thanks to (17.202)-(17.203) and Proposition 17.11, all claims may be obtained via a similar reasoning to that used in the proof of Corollary 17.1. □

Remark 17.3 The importance of the characterization obtained in Corollary 17.4 lies in the fact that for a certain range of α's we can actually identify Geometric Herz spaces as intermediate spaces between Muckenhoupt weighted Lebesgue spaces. More precisely, consider indices $p, q, \alpha \in \mathbb{R}$ satisfying

$$1 < p < \infty, \quad 0 < q \le \infty, \quad \text{and} \quad \tfrac{-(n-1-d)}{p} < \alpha < \tfrac{(n-1-d)(p-1)}{p}. \tag{17.213}$$

In this regime, choose $a, b \in \mathbb{R}$ such that

$$-(n-1-d) < a < \alpha p < b < (n-1-d)(p-1). \tag{17.214}$$

Hence, thanks to Proposition 2.4 and Remark 2.1, it follows that the functions

$$\begin{gathered}\mathrm{dist}(\cdot,E)^a,\ \mathrm{dist}(\cdot,E)^b,\ \big(1+\mathrm{dist}(\cdot,E)\big)^a,\ \text{and}\ \big(1+\mathrm{dist}(\cdot,E)\big)^b \\ \text{belong to the Muckenhoupt class } A_p(\Sigma,\sigma).\end{gathered} \tag{17.215}$$

This observation permits us to conclude that

> in the regime (17.213), the Geometric Herz spaces $\dot{\mathrm{K}}^{\alpha}_{p,q}(\Sigma,\sigma;E)$ and $\mathrm{K}^{\alpha}_{p,q}(\Sigma,\sigma;E)$ are intermediate spaces (for the real method of interpolation) between Muckenhoupt weighted Lebesgue spaces as in (17.209)-(17.210). (17.216)

As a consequence of (17.216), we have the following embedding result for Geometric Herz spaces.

Corollary 17.5 *Let $\Sigma \subseteq \mathbb{R}^n$ be a closed unbounded Ahlfors regular set, and abbreviate $\sigma := \mathcal{H}^{n-1}\lfloor\Sigma$. Select a d-Ahlfors regular set $E \subseteq \Sigma$, for some $d \in [0, n-1)$. Fix two exponents $p \in (1,\infty)$ and $q \in (0,\infty]$, along with an index $\alpha \in \mathbb{R}$ satisfying*

$$\frac{-(n-1-d)}{p} < \alpha < \frac{(n-1-d)(p-1)}{p}. \tag{17.217}$$

Then the following embeddings are well defined and continuous

$$\dot{\mathrm{K}}^{\alpha}_{p,q}(\Sigma, \sigma; E) \hookrightarrow L^1\Big(\Sigma, \frac{\sigma(x)}{1+|x|^{n-1}}\Big) \tag{17.218}$$

and

$$\mathrm{K}^{\alpha}_{p,q}(\Sigma, \sigma; E) \hookrightarrow L^1\Big(\Sigma, \frac{\sigma(x)}{1+|x|^{n-1}}\Big). \tag{17.219}$$

Moreover, the following inclusions are valid:

$$\begin{aligned} \dot{\mathrm{K}}^{\alpha}_{p,q}(\Sigma, \sigma; E) &\subseteq \textstyle\bigcup_{1<r<p} L^r_{\mathrm{loc}}(\Sigma, \sigma) \subseteq L^1_{\mathrm{loc}}(\Sigma, \sigma) \quad \textit{and} \\ \mathrm{K}^{\alpha}_{p,q}(\Sigma, \sigma; E) &\subseteq \textstyle\bigcup_{1<r<p} L^r_{\mathrm{loc}}(\Sigma, \sigma) \subseteq L^1_{\mathrm{loc}}(\Sigma, \sigma). \end{aligned} \tag{17.220}$$

Proof All claims may be obtained via Corollary 17.4, (17.215)-(17.216), and (2.83)-(2.84). □

Having identified Geometric Herz spaces as intermediate spaces between Muckenhoupt weighted Lebesgue spaces, we next obtain a boundedness result for the Hardy-Littlewood maximal operator acting on these scale.

Proposition 17.12 *Consider a closed unbounded Ahlfors regular set $\Sigma \subseteq \mathbb{R}^n$, and abbreviate $\sigma := \mathcal{H}^{n-1}\lfloor\Sigma$. Pick a d-AR set $E \subseteq \Sigma$ for some $d \in [0, n-1)$. Select two exponents $p \in (1, \infty)$ and $q \in (0, \infty]$, along with an index $\alpha \in \mathbb{R}$ satisfying*

$$\frac{-(n-1-d)}{p} < \alpha < \frac{(n-1-d)(p-1)}{p}. \tag{17.221}$$

Then the Hardy-Littlewood maximal operator on Σ (cf. (2.26))

$$\begin{gathered} \mathcal{M}_\Sigma : \dot{\mathrm{K}}^{\alpha}_{p,q}(\Sigma, \sigma; E) \to \dot{\mathrm{K}}^{\alpha}_{p,q}(\Sigma, \sigma; E) \\ \textit{is well defined, sub-linear, and bounded.} \end{gathered} \tag{17.222}$$

Corresponding to the inhomogeneous case, it follows that

$$\begin{gathered} \mathcal{M}_\Sigma : \mathrm{K}^{\alpha}_{p,q}(\Sigma, \sigma; E) \to \mathrm{K}^{\alpha}_{p,q}(\Sigma, \sigma; E) \\ \textit{is well defined, sub-linear, and bounded.} \end{gathered} \tag{17.223}$$

Proof Having selected $a, b \in \mathbb{R}$ satisfying (17.214), we see that (17.215) is valid. Thanks to (2.44), Hardy-Littlewood maximal operator

$$\begin{gathered} \mathcal{M}_\Sigma \text{ is bounded on } L^p\big(\Sigma, \mathrm{dist}(x,E)^a\sigma(x)\big),\ \ L^p\big(\Sigma, \mathrm{dist}(x,E)^b\sigma(x)\big) \\ L^p\big(\Sigma, (1+\mathrm{dist}(x,E))^a\sigma(x)\big), \text{ and } L^p\big(\Sigma, (1+\mathrm{dist}(x,E))^b\sigma(x)\big). \end{gathered} \tag{17.224}$$

As such, the boundedness claimed in (17.222)-(17.223) follows from these, (17.209)-(17.210), and [GHA-II, Proposition 1.3.7, pp. 33–34]. □

Remark 17.4 In the context of Proposition 17.12 we may also conclude that

$$\begin{aligned} &\mathcal{M}_\Sigma : \dot{\mathrm{K}}^{-\alpha}_{p',q}(\Sigma,\sigma;E) \to \dot{\mathrm{K}}^{-\alpha}_{p',q}(\Sigma,\sigma;E) \\ \text{and}\quad &\mathcal{M}_\Sigma : \mathrm{K}^{-\alpha}_{p',q}(\Sigma,\sigma;E) \to \mathrm{K}^{-\alpha}_{p',q}(\Sigma,\sigma;E) \\ &\text{are well defined, sub-linear, and bounded,} \end{aligned} \tag{17.225}$$

where $p' := p/(p-1) \in (1,\infty)$ denotes the Hölder conjugate exponent of p. Indeed, this follows from the (17.222) and (17.223), used with p' in place of p, and $-\alpha$ in place of α, and the fact that (17.221) is equivalent to

$$\frac{-(n-1-d)}{p'} < -\alpha < \frac{(n-1-d)(p'-1)}{p'}. \tag{17.226}$$

17.7 A New Class: Herz Spaces associated with Generalized Banach Function Spaces.

In this section, we introduce Herz-type spaces associated with Generalized Banach Function Spaces, and we show that these spaces are stable under the real method of interpolation.

To set the stage, consider $\Sigma \subseteq \mathbb{R}^n$ a closed unbounded Ahlfors regular set, and abbreviate $\sigma := \mathcal{H}^{n-1}\lfloor\Sigma$. Suppose $\mathbb{X}$ is a Generalized Banach Function Space on Σ (in the sense of Definition 10.2) equipped with a norm $\|\cdot\|_{\mathbb{X}}$. Fix a reference point $x_0 \in \Sigma$, and a dilation factor $2_* > \big(C_\Sigma/c_\Sigma\big)^{1/(n-1)}$, where $C_\Sigma, c_\Sigma \in (0,\infty)$ denote the upper and lower Ahlfors regularity constants of Σ, respectively. Recall the families $\{C_k\}_{k\in\mathbb{Z}}$ and $\{\widetilde{C}_k\}_{k\in\mathbb{N}_0}$ from (17.112)-(17.113). Select indices $q \in (0,\infty]$, and $\alpha \in \mathbb{R}$. For each σ-measurable function f on Σ introduce the quasinorms

$$\|f\|_{\dot{\mathrm{K}}^{\alpha}_{\mathbb{X},q}(\Sigma,\sigma)} := \begin{cases} \Big[\displaystyle\sum_{k=-\infty}^{\infty} \big(2_*^{k\alpha}\|f\cdot \mathbb{1}_{C_k}\|_{\mathbb{X}}\big)^q\Big]^{1/q} & \text{if } q<\infty, \\ \displaystyle\sup_{k\in\mathbb{Z}} \big\{2_*^{k\alpha}\|f\cdot \mathbb{1}_{C_k}\|_{\mathbb{X}}\big\} & \text{if } q=\infty, \end{cases} \tag{17.227}$$

and

$$\|f\|_{\mathrm{K}^{\alpha}_{\mathbb{X},q}(\Sigma,\sigma)} := \begin{cases} \Big[\displaystyle\sum_{k=0}^{\infty} \big(2_*^{k\alpha}\|f\cdot \mathbb{1}_{\widetilde{C}_k}\|_{\mathbb{X}}\big)^q\Big]^{1/q} & \text{if } q<\infty, \\ \displaystyle\sup_{k\in\mathbb{N}_0} \big\{2_*^{k\alpha}\|f\cdot \mathbb{1}_{\widetilde{C}_k}\|_{\mathbb{X}}\big\} & \text{if } q=\infty. \end{cases} \tag{17.228}$$

Define the homogeneous Herz space associated with $\mathbb{X}$ as

$$\dot{\mathrm{K}}^{\alpha}_{\mathbb{X},q}(\Sigma,\sigma) := \big\{f:\Sigma\to\mathbb{R} : f \text{ is } \sigma\text{-measurable and } \|f\|_{\dot{\mathrm{K}}^{\alpha}_{\mathbb{X},q}(\Sigma,\sigma)} < \infty\big\}, \tag{17.229}$$

and the `inhomogeneous Herz space associated with` $\mathbb{X}$ as

$$\mathrm{K}^{\alpha}_{\mathbb{X},q}(\Sigma,\sigma) := \big\{f:\Sigma\to\mathbb{R} : f \text{ is } \sigma\text{-measurable and } \|f\|_{\mathrm{K}^{\alpha}_{\mathbb{X},q}(\Sigma,\sigma)} < \infty\big\}. \tag{17.230}$$

It is visible from these definitions that $\|\cdot\|_{\dot{\mathrm{K}}^{\alpha}_{\mathbb{X},q}(\Sigma,\sigma)}$ and $\|\cdot\|_{\mathrm{K}^{\alpha}_{\mathbb{X},q}(\Sigma,\sigma)}$ are quasinorms. In fact, when $q\in[1,\infty]$ these are genuine norms.

Based on the generalized Hölder inequality in (10.149), similar computations as in (17.23)-(17.24) ensure that if f and g are σ-measurable functions on Σ, then

$$\int_\Sigma |fg|\,d\sigma \le \|f\|_{\dot{\mathrm{K}}^{\alpha}_{\mathbb{X},q}(\Sigma,\sigma)}\|g\|_{\dot{\mathrm{K}}^{-\alpha}_{\mathbb{X}',q'}(\Sigma,\sigma)} \quad \text{whenever } q,q'\in(1,\infty) \text{ satisfies } 1/q+1/q'=1, \tag{17.231}$$

and

$$\int_\Sigma |fg|\,d\sigma \le \|f\|_{\dot{\mathrm{K}}^{\alpha}_{\mathbb{X},q}(\Sigma,\sigma)}\|g\|_{\dot{\mathrm{K}}^{-\alpha}_{\mathbb{X}',\infty}(\Sigma,\sigma)} \quad \text{whenever } q\in(0,1]. \tag{17.232}$$

In the proposition below we show that, in the regime $q\in(1,\infty)$, Herz spaces associated with a GBFS are actually Generalized Banach Function Spaces.

Proposition 17.13 *Assume $\Sigma\subseteq\mathbb{R}^n$ is a closed unbounded Ahlfors regular set, and abbreviate $\sigma:=\mathcal{H}^{n-1}\lfloor\Sigma$. Let $\mathbb{X}$ be a Generalized Banach Function Space on (Σ,σ), and select indices $q\in(1,\infty)$ and $\alpha\in\mathbb{R}$. Then $\|\cdot\|_{\dot{\mathrm{K}}^{\alpha}_{\mathbb{X},q}(\Sigma,\sigma)}$ is a function norm on Σ (in the sense of Definition 10.2). Consequently, $\dot{\mathrm{K}}^{\alpha}_{\mathbb{X},q}(\Sigma,\sigma)$ is a Generalized Banach Function Space on (Σ,σ). The same conclusions hold for the inhomogeneous version $\mathrm{K}^{\alpha}_{\mathbb{X},q}(\Sigma,\sigma)$.*

Proof It is apparent from (17.227) that $\|\cdot\|_{\dot{\mathrm{K}}^{\alpha}_{\mathbb{X},q}(\Sigma,\sigma)}$ satisfies properties (P1) and (P2) from Definition 10.2. To verify (P3), consider a sequence $\{f_j\}_{j\in\mathbb{N}}$ of σ-measurable functions defined on Σ, and suppose $f_j\nearrow f$ at σ-a.e. point on Σ, as $j\to\infty$, for some σ-measurable function f defined on Σ. Hence, for each $k\in\mathbb{Z}$, we have that $f_j\cdot\mathbb{1}_{C_k}\nearrow f\cdot\mathbb{1}_{C_k}$ at σ-a.e. point on Σ as $j\to\infty$, for every $k\in\mathbb{Z}$. This and the fact that $\|\cdot\|_{\mathbb{X}}$ satisfies (P3) from Definition 10.2 imply

$$2_*^{k\alpha}\|f_j\cdot\mathbb{1}_{C_k}\|_{\mathbb{X}}\nearrow 2_*^{k\alpha}\|f\cdot\mathbb{1}_{C_k}\|_{\mathbb{X}} \quad \text{as } j\to\infty, \text{ for each } k\in\mathbb{Z}. \tag{17.233}$$

Using Lebesgue's Monotone Convergence Theorem for series, we find that

$$\Big[\sum_{k=-\infty}^{\infty}\big(2_*^{k\alpha}\|f_j\cdot\mathbb{1}_{C_k}\|_{\mathbb{X}}\big)^q\Big]^{1/q}\nearrow\Big[\sum_{k=-\infty}^{\infty}\big(2_*^{k\alpha}\|f\cdot\mathbb{1}_{C_k}\|_{\mathbb{X}}\big)^q\Big]^{1/q} \quad \text{as } j\to\infty, \tag{17.234}$$

and therefore $\|f_j\|_{\dot{K}^{\alpha}_{\mathbb{X},q}(\Sigma,\sigma)} \nearrow \|f\|_{\dot{K}^{\alpha}_{\mathbb{X},q}(\Sigma,\sigma)}$ as $j \to \infty$. This proves that the norm $\|\cdot\|_{\dot{K}^{\alpha}_{\mathbb{X},q}(\Sigma,\sigma)}$ satisfies (P3) from Definition 10.2.

Pressing forward, because $\|\cdot\|_{\mathbb{X}}$ and its associated function norm $\|\cdot\|_{\mathbb{X}'}$ satisfy (P4) from Definition 10.2, there exist families of σ-measurable sets $\{Y_j\}_{j\in\mathbb{N}}$ and $\{Z_j\}_{j\in\mathbb{N}}$ that cover the entire Σ, i.e.,

$$\Sigma = \bigcup_{j\in\mathbb{N}} Y_j = \bigcup_{j\in\mathbb{N}} Z_j, \tag{17.235}$$

and, for every $j \in \mathbb{N}$ they satisfy

$$\|\mathbb{1}_{Y_j}\|_{\mathbb{X}} < \infty \quad \text{and} \quad \|\mathbb{1}_{Z_j}\|_{\mathbb{X}'} < \infty. \tag{17.236}$$

For each $j \in \mathbb{N}$ and $\ell \in \mathbb{Z}$ define

$$U_{j,\ell} := Y_j \cap C_\ell \quad \text{and} \quad V_{j,\ell} := Z_j \cap C_\ell \tag{17.237}$$

which are σ-measurable sets. From (17.118) and (17.235) we see that

$$\sigma\Bigg(\Sigma \setminus \bigcup_{\substack{j\in\mathbb{N}\\ \ell\in\mathbb{Z}}} U_{j,\ell}\Bigg) = \sigma\Bigg(\Sigma \setminus \bigcup_{\substack{j\in\mathbb{N}\\ \ell\in\mathbb{Z}}} V_{j,\ell}\Bigg) = \sigma\big(\{x_0\}\big) = 0, \tag{17.238}$$

that is, both families $\{U_{j,\ell}\}_{\substack{j\in\mathbb{N}\\ \ell\in\mathbb{Z}}}$ and $\{V_{j,\ell}\}_{\substack{j\in\mathbb{N}\\ \ell\in\mathbb{Z}}}$ cover Σ up to a set of measure zero (with respect to σ). Bearing in mind (17.118), (17.237), and the first property in (17.236), for $j \in \mathbb{N}$ and $\ell \in \mathbb{Z}$ we have

$$\begin{aligned} \|\mathbb{1}_{U_{j,\ell}}\|_{\dot{K}^{\alpha}_{\mathbb{X},q}(\Sigma,\sigma)} &= \Big[\sum_{k=-\infty}^{\infty} \big(2_*^{k\alpha}\|\mathbb{1}_{U_{j,\ell}} \cdot \mathbb{1}_{C_k}\|_{\mathbb{X}}\big)^q\Big]^{1/q} \\ &= 2_*^{\ell\alpha}\|\mathbb{1}_{Y_j\cap C_\ell}\|_{\mathbb{X}} \le 2_*^{\ell\alpha}\|\mathbb{1}_{Y_j}\|_{\mathbb{X}} < \infty. \end{aligned} \tag{17.239}$$

This proves that $\|\cdot\|_{\dot{K}^{\alpha}_{\mathbb{X},q}(\Sigma,\sigma)}$ satisfies (P4) from Definition 10.2 relative to the family $\{U_{j,\ell}\}_{\substack{j\in\mathbb{N}\\ \ell\in\mathbb{Z}}}$ (see also (a) in Remark 10.3).

To proceed, fix an arbitrary σ-measurable function f defined on Σ. Using (10.149), while bearing in mind (17.227), the definition of $V_{j,\ell}$ in (17.237), and the second property in (17.236), for each $j \in \mathbb{N}$ and $\ell \in \mathbb{Z}$ we compute

$$\begin{aligned} \int_{V_{j,\ell}} |f|\, d\sigma &= \int_{Z_j} |f \cdot \mathbb{1}_{C_\ell}|\, d\sigma \le \|f \cdot \mathbb{1}_{C_\ell}\|_{\mathbb{X}}\|\mathbb{1}_{Z_j}\|_{\mathbb{X}'} \\ &\le 2_*^{-\ell\alpha}\|f\|_{\dot{K}^{\alpha}_{\mathbb{X},q}(\Sigma,\sigma)}\|\mathbb{1}_{Z_j}\|_{\mathbb{X}'} := c_{j,\ell}\|f\|_{\dot{K}^{\alpha}_{\mathbb{X},q}(\Sigma,\sigma)}, \end{aligned} \tag{17.240}$$

with $c_{j,\ell} := 2_*^{-\ell\alpha}\|\mathbb{1}_{Z_j}\|_{\mathbb{X}'} \in (0,\infty)$. This proves that $\|\cdot\|_{\dot{\mathrm{K}}^{\alpha}_{\mathbb{X},q}(\Sigma,\sigma)}$ satisfies (P5) from Definition 10.2 relative to the family $\{V_{j,\ell}\}_{\substack{j\in\mathbb{N}\\ \ell\in\mathbb{Z}}}$. This establishes that $\|\cdot\|_{\dot{\mathrm{K}}^{\alpha}_{\mathbb{X},q}(\Sigma,\sigma)}$ is a function norm on Σ, thus completing the proof of Proposition 17.13. □

Having established that Herz spaces associated with GBFS's are themselves Generalized Banach Function Spaces, the next order of business is to identify their associated spaces. A concrete statement is as follows.

Proposition 17.14 *Consider an unbounded Ahlfors regular set $\Sigma \subseteq \mathbb{R}^n$, and abbreviate $\sigma := \mathcal{H}^{n-1}\lfloor\Sigma$. Suppose $\mathbb{X}$ is a* GBFS *over Σ. Select indices $q \in (1,\infty)$ and $\alpha \in \mathbb{R}$. Denote by $q' \in (1,\infty)$ the Hölder conjugate exponent of q, and $\mathbb{X}'$ the associated space of $\mathbb{X}$ (in the sense of Definition 10.3). Then for each σ-measurable function f defined on Σ one has*

$$\|f\|_{\dot{\mathrm{K}}^{-\alpha}_{\mathbb{X}',q'}(\Sigma,\sigma)} = \sup\Big\{\int_\Sigma |fg|\,d\sigma : g \in \dot{\mathrm{K}}^{\alpha}_{\mathbb{X},q}(\Sigma,\sigma) \text{ with } \|g\|_{\dot{\mathrm{K}}^{\alpha}_{\mathbb{X},q}(\Sigma,\sigma)} \le 1\Big\}. \tag{17.241}$$

Consequently, the associated space of $\dot{\mathrm{K}}^{\alpha}_{\mathbb{X},q}(\Sigma,\sigma)$ is $\dot{\mathrm{K}}^{-\alpha}_{\mathbb{X}',q'}(\Sigma,\sigma)$, i.e.,

$$\big(\dot{\mathrm{K}}^{\alpha}_{\mathbb{X},q}(\Sigma,\sigma)\big)' = \dot{\mathrm{K}}^{-\alpha}_{\mathbb{X}',q'}(\Sigma,\sigma). \tag{17.242}$$

Proof The veracity of (17.241) can be established via an analogous proof to that of Proposition 17.6 (when $\mathfrak{X} := \Sigma$ and $\mu := \sigma$), which involves replacing the Lebesgue space $L^p(\mathfrak{X},\mu)$ by $\mathbb{X}$, and $L^{p'}(\mathfrak{X},\mu)$ by $\mathbb{X}'$, and now bearing in mind (17.231). The identification (17.242) is then a direct consequence of (17.241) and (10.146). □

Next, we show that Herz spaces associated with Generalized Banach Function Spaces are embedded into the space of measurable functions. This is made precise in the following proposition.

Proposition 17.15 *Consider a closed unbounded Ahlfors regular set $\Sigma \subseteq \mathbb{R}^n$, and abbreviate $\sigma := \mathcal{H}^{n-1}\lfloor\Sigma$. Let $\mathbb{X}$ be a Generalized Banach Function Space on (Σ,σ). With $\{C_k\}_{k\in\mathbb{Z}}$ as in* (17.112)*, assume that*

$$\mathbb{1}_{C_k} \in \mathbb{X}' \quad \text{for all } k \in \mathbb{Z}, \tag{17.243}$$

where $\mathbb{X}'$ is the associated space of $\mathbb{X}$. Fix two indices $q \in (0,\infty]$ and $\alpha \in \mathbb{R}$. Then the following embedding is continuous:

$$\dot{\mathrm{K}}^{\alpha}_{\mathbb{X},q}(\Sigma,\sigma) \hookrightarrow L^0(\Sigma,\sigma), \tag{17.244}$$

where $L^0(\Sigma,\sigma)$ denotes the space of all σ-measurable functions which are finite σ-a.e. on Σ equipped with the topology of convergence in measure on sets of finite measure (in the sense described in (17.60)*). The same continuous embedding holds for the inhomogeneous version $\mathrm{K}^{\alpha}_{\mathbb{X},q}(\Sigma,\sigma)$.*

Proof The proof of (17.244) is analogous to that of Lemma 17.1. For this reason, we shall only emphasize those aspects that cannot be directly applied to our current functional analytic setting. To get started, consider a sequence of functions $\{f_j\}_{j\in\mathbb{X}}$ in $\dot{\mathrm{K}}^{\alpha}_{\mathbb{X},q}(\Sigma,\sigma)$ and suppose that

$$\lim_{j\to\infty} f_j = f \quad \text{in } \dot{\mathrm{K}}^{\alpha}_{\mathbb{X},q}(\Sigma,\sigma), \tag{17.245}$$

for some $f \in \dot{\mathrm{K}}^{\alpha}_{\mathbb{X},q}(\Sigma,\sigma)$. Having fixed $\varepsilon \in (0,\infty)$, and a σ-measurable set $A \subseteq \Sigma$ with $\sigma(A) < \infty$, for each $j \in \mathbb{N}$ and $k \in \mathbb{Z}$ define

$$A_k^{(j)} := \big\{x \in A \cap C_k : |f_j(x) - f(x)| > \varepsilon\big\} \subseteq C_k. \tag{17.246}$$

Next, fix some $j \in \mathbb{N}$ and pick an arbitrary $\rho \in (0,\infty)$. Following the same type of reasoning as in (17.64)-(17.65), there exists some $\widetilde{N} \in \mathbb{N}$ for which

$$\begin{aligned}\sigma\Big(\{x \in A : |f_j(x) - f(x)| > \varepsilon\}\Big) &\leq \sum_{k=-\widetilde{N}}^{\widetilde{N}} \sigma\big(A_k^{(j)}\big) + \sigma\Bigg(A \setminus \bigcup_{\substack{k\in\mathbb{Z}\\ |k|\leq\widetilde{N}}} C_k\Bigg)\\ &< \sum_{k=-\widetilde{N}}^{\widetilde{N}} \sigma\big(A_k^{(j)}\big) + \frac{\rho}{2}.\end{aligned} \tag{17.247}$$

Being mindful of (17.243), (17.246), (17.227), and (10.149), for each $k \in \mathbb{Z}$ with $|k| < \widetilde{N}$ we estimate

$$\begin{aligned}\varepsilon\cdot\sigma\big(A_k^{(j)}\big) &= \int_{A_k^{(j)}} \varepsilon\, d\sigma \leq \int_{A_k^{(j)}} |f - f_j|\, d\sigma \leq \int_{C_k} |f - f_j|\, d\sigma\\ &\leq \|(f - f_j)\cdot \mathbb{1}_{C_k}\|_{\mathbb{X}} \cdot \|\mathbb{1}_{C_k}\|_{\mathbb{X}'} \leq 2_*^{-k\alpha}\|f - f_j\|_{\mathrm{K}^{\alpha}_{\mathbb{X},q}(\Sigma,\sigma)} \cdot \|\mathbb{1}_{C_k}\|_{\mathbb{X}'}\\ &\leq \Bigg(\max_{\substack{k\in\mathbb{Z}\\ |k|<\widetilde{N}}} \{2_*^{-k\alpha}\|\mathbb{1}_{C_k}\|_{\mathbb{X}'}\}\Bigg)\cdot \|f - f_j\|_{\mathrm{K}^{\alpha}_{\mathbb{X},q}(\Sigma,\sigma)} < \infty.\end{aligned} \tag{17.248}$$

Thanks to (17.245), there exists some $j_0 \in \mathbb{N}$ sufficiently large such that whenever $j \in \mathbb{N}$ and $j \geq j_0$ we have

$$\sigma\big(A_k^{(j)}\big) \leq \varepsilon^{-1}\Bigg(\max_{\substack{k\in\mathbb{Z}\\ |k|<\widetilde{N}}} \{2_*^{-k\alpha}\|\mathbb{1}_{C_k}\|_{\mathbb{X}'}\}\Bigg)\cdot\|f - f_j\|_{\mathrm{K}^{\alpha}_{\mathbb{X},q}(\Sigma,\sigma)} < \frac{\rho}{2(2\widetilde{N}+1)}. \tag{17.249}$$

In concert, this and (17.247) entail

$$\sigma\Big(\{x \in A : |f_j(x) - f(x)| > \varepsilon\}\Big) < \sum_{k=-\widetilde{N}}^{\widetilde{N}} \frac{\rho}{2(2\widetilde{N}+1)} + \frac{\rho}{2} = \rho. \tag{17.250}$$

Ultimately, since $\rho \in (0, \infty)$ was taken arbitrary, we find that $f_j \to f$ in $L^0(\Sigma, \sigma)$ as $j \to \infty$. This proves that the embedding (17.244) is continuous, concluding the proof of Proposition 17.15. $\square$

An inspection of the proof of the interpolation result recorded in Theorem 17.1 reveals the Lebesgue space L^p in the definition of standard Herz spaces $\dot{\mathrm{K}}^{\alpha}_{p,q}$ plays only an incidental role. As such, the argument presented in the proof of this theorem is rather flexible and allows us to obtain the following interpolation result for Herz spaces associated with Generalized Banach Function Spaces.

Theorem 17.6 *Let $\Sigma \subseteq \mathbb{R}^n$ be a closed unbounded Ahlfors regular set, and abbreviate $\sigma := \mathcal{H}^{n-1}\lfloor\Sigma$. Consider a Generalized Banach Function Space $\mathbb{X}$ on Σ and assume that* (17.243) *holds. Select indices $q_0, q_1, q \in (0, \infty]$ along with $\alpha_0, \alpha_1 \in \mathbb{R}$ such that $\alpha_0 \neq \alpha_1$. Then whenever $\theta \in (0, 1)$ and $\alpha := (1 - \theta)\alpha_0 + \theta\alpha_1$ it follows that*

$$\Big(\dot{\mathrm{K}}^{\alpha_0}_{\mathbb{X},q_0}(\Sigma, \sigma), \dot{\mathrm{K}}^{\alpha_1}_{\mathbb{X},q_1}(\Sigma, \sigma)\Big)_{\theta,q} = \dot{\mathrm{K}}^{\alpha}_{\mathbb{X},q}(\Sigma, \sigma). \tag{17.251}$$

The same type of conclusion holds for the inhomogeneous version of Herz spaces associated with $\mathbb{X}$.

Proof Notice that Proposition 17.15 ensures that the spaces $\dot{\mathrm{K}}^{\alpha_0}_{\mathbb{X},q_0}(\Sigma, \sigma)$ and $\dot{\mathrm{K}}^{\alpha_1}_{\mathbb{X},q_1}(\Sigma, \sigma)$ are compatible. As such, the intermediate space in the left-hand side of (17.251) is meaningfully defined. The identification (17.251) may be established by running the same argument used in the proof of Theorem 17.1 (when $\mathfrak{X} := \Sigma$ and $\mu := \sigma$) by replacing the Lebesgue space $L^p(\mathfrak{X}, \mu)$ by the Generalized Banach Function Space $\mathbb{X}$. $\square$

Remark 17.5 Theorem 17.6 may be regarded as a purely Harmonic Analysis flavored result which establishes that the scale of Herz spaces associated with a GBFS is stable under the real method of interpolation.

Remark 17.6 For the choice $\mathbb{X} := L^p(\Sigma, \sigma)$, with $p \in (1, \infty)$, Proposition 17.15 recovers Theorem 17.3.

Remark 17.7 For each choice of a Generalized Banach Function Space $\mathbb{X}$, Theorem 17.6 produces a new result. Many examples are discussed in Section 23.5.

Chapter 18

Calderón-Zygmund Theory on Herz Spaces of Second Generation

This chapter is primarily focused on developing a Calderón-Zygmund theory for second-generation Herz spaces on uniformly rectifiable sets. At the core of our approach one finds the interpolation result described in Theorem 17.3, the characterization obtained in Corollary 17.1, along with recently established results in the context of Muckenhoupt weighted Lebesgue spaces in [MMMi^3]. The theory presented here is applicable to boundary layer potentials associated with weakly elliptic systems, which will be crucial in the treatment of boundary value problems on second-generation Herz spaces in subsequent chapters.

18.1 Singular Integral Operators on Second-Generation Herz Spaces

We start by discussing the boundedness of convolution-type singular integral operators acting on second-generation Herz spaces.

Theorem 18.1 *Assume $\Sigma \subseteq \mathbb{R}^n$ is a closed unbounded uniformly rectifiable set, and abbreviate $\sigma := \mathcal{H}^{n-1}\lfloor\Sigma$. For some sufficiently large integer $N = N(n) \in \mathbb{N}$, consider a complex-valued function $k \in \mathscr{C}^N(\mathbb{R}^n \setminus \{0\})$ which is odd and positive homogeneous of degree $1-n$. In this setting, for each function $f \in L^1\big(\Sigma, \frac{\sigma(x)}{1+|x|^{n-1}}\big)$ define*

$$T_\varepsilon f(x) := \int_{\substack{y\in\Sigma \\ |x-y|>\varepsilon}} k(x-y)f(y)\,d\sigma(y) \text{ for all } x \in \Sigma \text{ and } \varepsilon > 0, \tag{18.1}$$

$$T_{\max} f(x) := \sup_{\varepsilon>0} |T_\varepsilon f(x)| \text{ for each } x \in \Sigma, \tag{18.2}$$

$$Tf(x) := \lim_{\varepsilon\to 0^+} T_\varepsilon f(x) \text{ for } \sigma\text{-a.e. } x \in \Sigma. \tag{18.3}$$

© The Author(s), under exclusive license to Springer Nature Switzerland AG 2026
M. Mitrea and P. Takemura, *Singular Integrals, Herz-Type Function Spaces, and Boundary Problems*, Progress in Mathematics 362,
https://doi.org/10.1007/978-3-032-12516-3_18

Finally, select two exponents $p \in (1, \infty)$ *and* $q \in (0, \infty]$, *along with an index* $\alpha \in \mathbb{R}$ *satisfying*

$$-\frac{n-1}{p} < \alpha < \frac{n-1}{p'} \tag{18.4}$$

where $p' \in (1, \infty)$ *denotes the Hölder conjugate exponent of* p.

Then there exists $C \in (0, \infty)$ *depending exclusively on* n, p, q, α, *and the* UR *constants of* Σ *with the property that for each function* $f \in \dot{\mathrm{K}}^{\alpha}_{p,q}(\Sigma, \sigma)$ *one has*

$$\|T_{\max} f\|_{\dot{\mathrm{K}}^{\alpha}_{p,q}(\Sigma,\sigma)} \leq C\Big(\sum_{|\alpha| \leq N} \sup_{S^{n-1}} |\partial^{\alpha} k|\Big) \|f\|_{\dot{\mathrm{K}}^{\alpha}_{p,q}(\Sigma,\sigma)}. \tag{18.5}$$

Moreover, the principal value singular integral operator

$$T : \dot{\mathrm{K}}^{\alpha}_{p,q}(\Sigma, \sigma) \longrightarrow \dot{\mathrm{K}}^{\alpha}_{p,q}(\Sigma, \sigma) \tag{18.6}$$

is well defined, linear, and bounded.

Next, assume that $\Omega \subseteq \mathbb{R}^n$ *is an open set such that* $\partial\Omega$ *is an unbounded uniformly rectifiable set, abbreviate* $\sigma := \mathcal{H}^{n-1} \lfloor \partial\Omega$. *With the integral kernel* k *as before, for each function* $f \in L^1\big(\partial\Omega, \frac{\sigma(x)}{1+|x|^{n-1}}\big)$ *define*

$$\mathcal{T} f(x) := \int_{\partial\Omega} k(x-y) f(y)\, d\sigma(y) \quad \textit{for each} \quad x \in \Omega. \tag{18.7}$$

Then for each aperture parameter $\kappa \in (0, \infty)$ *there exists some* $C \in (0, \infty)$ *depending only on* n, p, q, *and the* UR *constants of* $\partial\Omega$ *with the property that for each* $f \in \dot{\mathrm{K}}^{\alpha}_{p,q}(\partial\Omega, \sigma)$ *one has*

$$\|\mathcal{N}_{\kappa}(\mathcal{T} f)\|_{\dot{\mathrm{K}}^{\alpha}_{p,q}(\partial\Omega,\sigma)} \leq C\Big(\sum_{|\alpha| \leq N} \sup_{S^{n-1}} |\partial^{\alpha} k|\Big) \|f\|_{\dot{\mathrm{K}}^{\alpha}_{p,q}(\partial\Omega,\sigma)}. \tag{18.8}$$

Furthermore, for each function $f \in \dot{\mathrm{K}}^{\alpha}_{p,q}(\partial\Omega, \sigma)$ *the following jump-formula holds*

$$\Big(\mathcal{T} f\Big|^{\kappa-\text{n.t.}}_{\partial\Omega}\Big)(x) = \frac{1}{2i}\widehat{k}\big(\nu(x)\big) f(x) + (T f)(x) \quad \textit{at } \sigma\textit{-a.e. } x \in \partial_*\Omega, \tag{18.9}$$

where $\widehat{k}$ *denotes the Fourier transform (cf.* (2.4)) *of* k. *Lastly, similar conclusions pertaining to mapping properties, boundedness, nontangential maximal function estimates, and jump-formulas hold in the context of inhomogeneous Herz spaces* $\mathrm{K}^{\alpha}_{p,q}$.

Proof All claims may be obtained by combining Corollary 17.1, Remark 17.1, Corollary 17.2, [MMMi³, Proposition 3.4, pp. 179–180], and [GHA-II, Proposition 1.3.7, pp. 33–34]. □

Remark 18.1 Relying on the characterization established in Corollary 17.4, the observations made in (17.214)-(17.216), the embeddings from Corollary 17.5, together with [MMMi³, Proposition 3.4, pp. 179–180], and [GHA-II, Proposition 1.3.7, pp. 33–34], a similar result to that in Theorem 18.1 may be obtained in the context of geometric Herz spaces. Specifically, in the setting of Theorem 18.1, consider a d-Ahlfors regular subset E of Σ for some $d \in [0, n-1)$, where p, q, and α are indices satisfying

$$1 < p < \infty, \quad 0 < q \leq \infty, \quad \text{and} \quad \tfrac{-(n-1-d)}{p} < \alpha < \tfrac{(n-1-d)(p-1)}{p}. \tag{18.10}$$

In this regime, the same type of conclusions (as in Theorem 18.1) regarding mapping properties, boundedness of singular integrals, nontangential maximal function estimates, and jump-formulas hold in the context of the geometric Herz spaces $\dot{\mathrm{K}}^{\alpha}_{p,q}(\Sigma, \sigma; E)$ and $\mathrm{K}^{\alpha}_{p,q}(\Sigma, \sigma; E)$.

The next goal is to obtain geometrically sensitive operator norm estimates for singular integral operators of chord-dot-normal type in the context of second-generation Herz spaces. The key feature of the result recorded below is that the norms of these operators are controlled in terms of the BMO seminorm of the outward unit normal to the underlying domain. This is crucial to the present work, as it paves the way for treating boundary value problems with boundary data taken from second-generation Herz spaces. The reader is advised to recall the piece of notation defined in (6.113).

Theorem 18.2 *Assume $\Omega \subseteq \mathbb{R}^n$ is a* UR *domain with unbounded boundary, and abbreviate $\sigma := \mathcal{H}^{n-1}\lfloor\partial\Omega$. Denote by ν the geometric measure theoretic outward unit normal to Ω. For some sufficiently large integer $N = N(n) \in \mathbb{N}$, consider a kernel $k \in \mathscr{C}^N(\mathbb{R}^n \setminus \{0\})$ that is even and positive homogeneous of degree $-n$. Recall chord-dot-normal singular integral operators T and $T^{\#}$ associated with k as in* (6.103)-(6.104). *Finally, fix two exponents exponents $p \in (1, \infty)$ and $q \in (0, \infty]$, along with an index $\alpha \in \mathbb{R}$ satisfying*

$$-\frac{n-1}{p} < \alpha < \frac{n-1}{p'} \tag{18.11}$$

where $p' \in (1, \infty)$ denotes the Hölder conjugate exponent of p.

Then the operators

$$\begin{aligned} T &: \dot{\mathrm{K}}^{\alpha}_{p,q}(\partial\Omega, \sigma) \to \dot{\mathrm{K}}^{\alpha}_{p,q}(\partial\Omega, \sigma), \\ T^{\#} &: \dot{\mathrm{K}}^{\alpha}_{p,q}(\partial\Omega, \sigma) \to \dot{\mathrm{K}}^{\alpha}_{p,q}(\partial\Omega, \sigma), \\ T &: \mathrm{K}^{\alpha}_{p,q}(\partial\Omega, \sigma) \to \mathrm{K}^{\alpha}_{p,q}(\partial\Omega, \sigma), \\ T^{\#} &: \mathrm{K}^{\alpha}_{p,q}(\partial\Omega, \sigma) \to \mathrm{K}^{\alpha}_{p,q}(\partial\Omega, \sigma), \end{aligned} \tag{18.12}$$

are well defined, linear, and bounded. Moreover, for each $m \in \mathbb{N}$ there exits some $C_m \in (0, \infty)$ depending exclusively on m, n, p, q, α, the kernel k, and the UR *constants of $\partial\Omega$ with the property that*

$$\|T\|_{\dot{\mathrm{K}}^{\alpha}_{p,q}(\partial\Omega,\sigma)\to\dot{\mathrm{K}}^{\alpha}_{p,q}(\partial\Omega,\sigma)} \leq C_m \cdot \|\nu\|^{\langle m\rangle}_{[\mathrm{BMO}(\partial\Omega,\sigma)]^n}, \tag{18.13}$$

$$\|T^{\#}\|_{\dot{\mathrm{K}}^{\alpha}_{p,q}(\partial\Omega,\sigma)\to\dot{\mathrm{K}}^{\alpha}_{p,q}(\partial\Omega,\sigma)} \leq C_m \cdot \|\nu\|^{\langle m\rangle}_{[\mathrm{BMO}(\partial\Omega,\sigma)]^n}. \tag{18.14}$$

Corresponding to the inhomogeneous Herz spaces, one also has

$$\|T\|_{\mathrm{K}^{\alpha}_{p,q}(\partial\Omega,\sigma)\to\mathrm{K}^{\alpha}_{p,q}(\partial\Omega,\sigma)} \leq C_m \cdot \|\nu\|^{\langle m\rangle}_{[\mathrm{BMO}(\partial\Omega,\sigma)]^n}, \tag{18.15}$$

$$\|T^{\#}\|_{\mathrm{K}^{\alpha}_{p,q}(\partial\Omega,\sigma)\to\mathrm{K}^{\alpha}_{p,q}(\partial\Omega,\sigma)} \leq C_m \cdot \|\nu\|^{\langle m\rangle}_{[\mathrm{BMO}(\partial\Omega,\sigma)]^n}. \tag{18.16}$$

Proof A combination of Corollary 17.1, Remark 17.1, Corollary 17.2, [MMMi[3], Corollary 4.2, pp. 292–293], and [GHA-II, Proposition 1.3.7, pp. 33–34] implies all estimates in (18.13)-(18.16). □

Remark 18.2 In light of Corollary 17.4, the observations made in (17.214)-(17.216), the embeddings obtained in Corollary 17.5, [MMMi[3], Corollary 4.2, pp. 292–293], and [GHA-II, Proposition 1.3.7, pp. 33–34], the same type of operator norm estimates as in (18.13)-(18.16) hold in the context of homogeneous and inhomogeneous Geometric Herz spaces. Specifically, in the setting of Theorem 18.2, assume that $E \subseteq \partial\Omega$ is a d-AR for some $d \in [0, n-1)$. Select indices p, q, and α satisfying

$$1 < p < \infty, \quad 0 < q \leq \infty, \quad \text{and} \quad \tfrac{-(n-1-d)}{p} < \alpha < \tfrac{(n-1-d)(p-1)}{p}. \tag{18.17}$$

Then the operators

$$\begin{aligned} T &: \dot{\mathrm{K}}^{\alpha}_{p,q}(\partial\Omega,\sigma;E) \to \dot{\mathrm{K}}^{\alpha}_{p,q}(\partial\Omega,\sigma;E),\\ T^{\#} &: \dot{\mathrm{K}}^{\alpha}_{p,q}(\partial\Omega,\sigma;E) \to \dot{\mathrm{K}}^{\alpha}_{p,q}(\partial\Omega,\sigma;E),\\ T &: \mathrm{K}^{\alpha}_{p,q}(\partial\Omega,\sigma;E) \to \mathrm{K}^{\alpha}_{p,q}(\partial\Omega,\sigma;E),\\ T^{\#} &: \mathrm{K}^{\alpha}_{p,q}(\partial\Omega,\sigma;E) \to \mathrm{K}^{\alpha}_{p,q}(\partial\Omega,\sigma;E), \end{aligned} \tag{18.18}$$

are well defined, linear, and bounded. Moreover, given $m \in \mathbb{N}$ there exits some constant $C_m \in (0,\infty)$ depending exclusively on m, n, p, q, α, d, the d-Ahlfors regularity of E, the kernel k, and the UR character of $\partial\Omega$, with the property that

$$\|T\|_{\dot{\mathrm{K}}^{\alpha}_{p,q}(\partial\Omega,\sigma;E)\to\dot{\mathrm{K}}^{\alpha}_{p,q}(\partial\Omega,\sigma;E)} \leq C_m \cdot \|\nu\|^{\langle m\rangle}_{[\mathrm{BMO}(\partial\Omega,\sigma)]^n}, \tag{18.19}$$

$$\|T^{\#}\|_{\dot{\mathrm{K}}^{\alpha}_{p,q}(\partial\Omega,\sigma;E)\to\dot{\mathrm{K}}^{\alpha}_{p,q}(\partial\Omega,\sigma;E)} \leq C_m \cdot \|\nu\|^{\langle m\rangle}_{[\mathrm{BMO}(\partial\Omega,\sigma)]^n}, \tag{18.20}$$

$$\|T\|_{\mathrm{K}^{\alpha}_{p,q}(\partial\Omega,\sigma;E)\to\mathrm{K}^{\alpha}_{p,q}(\partial\Omega,\sigma;E)} \leq C_m \cdot \|\nu\|^{\langle m\rangle}_{[\mathrm{BMO}(\partial\Omega,\sigma)]^n}, \tag{18.21}$$

$$\|T^{\#}\|_{\mathrm{K}^{\alpha}_{p,q}(\partial\Omega,\sigma;E)\to\mathrm{K}^{\alpha}_{p,q}(\partial\Omega,\sigma;E)} \leq C_m \cdot \|\nu\|^{\langle m\rangle}_{[\mathrm{BMO}(\partial\Omega,\sigma)]^n}. \tag{18.22}$$

It is also of interest to obtain bounds for commutators on second-generation Herz spaces.

Theorem 18.3 *Assume $\Sigma \subseteq \mathbb{R}^n$ is a closed unbounded Ahlfors regular set, and abbreviate $\sigma := \mathcal{H}^{n-1}\lfloor\Sigma$. Fix some integrability exponent $p_0 \in (1,\infty)$ along with a non-decreasing function $\Phi : (0,\infty) \to (0,\infty)$. Suppose T is a bounded linear operator on $L^{p_0}(\Sigma, w)$ for every weight $w \in A_{p_0}(\Sigma,\sigma)$, with operator norm controlled by $\Phi([w]_{A_{p_0}})$. Select exponents $p \in (1,\infty)$ and $q \in (0,\infty]$, along with an index $\alpha \in \mathbb{R}$ satisfying*

$$-\frac{n-1}{p} < \alpha < \frac{n-1}{p'} \tag{18.23}$$

where $p' \in (1,\infty)$ denotes the Hölder conjugate exponent of p. Finally, pick a complex-valued function $b \in L^\infty(\Sigma,\sigma)$.

Then

$$\begin{gathered}[M_b, T] : \dot{\mathrm{K}}^{\alpha}_{p,q}(\Sigma,\sigma) \to \dot{\mathrm{K}}^{\alpha}_{p,q}(\Sigma,\sigma) \\ \text{is well defined, linear, and bounded,}\end{gathered} \tag{18.24}$$

where $[M_b, T]$ is the commutator of T and the operator M_b of pointwise multiplication on $\dot{\mathrm{K}}^{\alpha}_{p,q}(\Sigma,\sigma)$ by the function b, i.e.,

$$[M_b, T]f = b(Tf) - T(bf) \quad \text{for each } f \in \dot{\mathrm{K}}^{\alpha}_{p,q}(\Sigma,\sigma). \tag{18.25}$$

Moreover, there exist constants $C_1, C \in (0,\infty)$ depending only on n, p_0, p, q, α, and the Ahlfors regularity character of Σ with the property

$$\|[M_b, T]\|_{\dot{\mathrm{K}}^{\alpha}_{p,q}(\Sigma,\sigma)\to\dot{\mathrm{K}}^{\alpha}_{p,q}(\Sigma,\sigma)} \leq C_1 \cdot \Phi(C)\|b\|_{\mathrm{BMO}(\Sigma,\sigma)}. \tag{18.26}$$

Similar conclusions pertaining to mapping properties and estimates for commutators hold in the context of inhomogeneous Herz spaces $\mathrm{K}^{\alpha}_{p,q}(\Sigma,\sigma)$.

Proof The claims made in (18.24) and (18.26) follow from Corollary 17.1, Remark 17.1, together with [MMMi[3], Theorem 4.3, p. 294] and [GHA-II, Proposition 1.3.7, pp. 33–34]. □

Remark 18.3 Commutator estimates of the sort recorded in (18.26) hold in the context of Geometric Herz spaces. Precisely, retain the same background assumptions as in Theorem 18.3. Suppose $E \subseteq \Sigma$ is a d-Ahlfors regular set for some $d \in [0, n-1)$. Select indices p, q, and α satisfying

$$1 < p < \infty, \quad 0 < q \leq \infty, \quad \text{and} \quad \tfrac{-(n-1-d)}{p} < \alpha < \tfrac{(n-1-d)(p-1)}{p}. \tag{18.27}$$

Then, in this context, the commutators

$$\begin{aligned}&[M_b, T] : \dot{\mathrm{K}}^{\alpha}_{p,q}(\Sigma,\sigma;E) \to \dot{\mathrm{K}}^{\alpha}_{p,q}(\Sigma,\sigma;E),\\ &[M_b, T] : \mathrm{K}^{\alpha}_{p,q}(\Sigma,\sigma;E) \to \mathrm{K}^{\alpha}_{p,q}(\Sigma,\sigma;E)\end{aligned} \tag{18.28}$$

are well defined, linear, and bounded. Moreover, there exist constants $C_1, C \in (0,\infty)$ depending exclusively on n, p_0, p, q, α, d, the d-Ahlfors regularity of E, and the Ahlfors regularity of Σ with the property that

$$\begin{aligned} \|[M_b, T]\|_{\dot{\mathrm{K}}^{\alpha}_{p,q}(\Sigma,\sigma;E)\to\dot{\mathrm{K}}^{\alpha}_{p,q}(\Sigma,\sigma;E)} &\leq C_1\cdot\Phi(C)\|b\|_{\mathrm{BMO}(\Sigma,\sigma)}, \\ \|[M_b, T]\|_{\mathrm{K}^{\alpha}_{p,q}(\Sigma,\sigma;E)\to\mathrm{K}^{\alpha}_{p,q}(\Sigma,\sigma;E)} &\leq C_1\cdot\Phi(C)\|b\|_{\mathrm{BMO}(\Sigma,\sigma)}. \end{aligned} \tag{18.29}$$

Indeed, this follows from Corollary 17.4, the observations made in (17.214)-(17.216), along with [MMMi[3], Theorem 4.3, p. 294], and [GHA-II, Proposition 1.3.7, pp. 33–34].

18.2 Layer Potentials on Second-Generation Herz Spaces

Here we specialize the results obtained in the previous section to singular integral operators of layer potential type. The result recorded below provides operator norm estimates (which are sensitive to geometry) for double layer potentials associated with weakly elliptic systems that possess distinguished coefficient tensors.

Theorem 18.4 *Assume $\Omega \subseteq \mathbb{R}^n$ is a* UR *domain with unbounded boundary. Abbreviate $\sigma := \mathcal{H}^{n-1}\lfloor\partial\Omega$, and denote by ν the geometric measure theoretic outward unit normal to Ω. Consider a second-order, homogeneous, constant complex coefficient, weakly-elliptic $M \times M$ system L in $\mathbb{R}^n$, for some $M \in \mathbb{N}$, with the property that $\mathfrak{A}^{\mathrm{dis}}_L \neq \emptyset$. Pick a coefficient tensor $A \in \mathfrak{A}^{\mathrm{dis}}_L$, and recall the boundary-to-boundary double layer potentials K_A and $K^{\#}_A$ defined in* (7.23) *and* (7.24)*, respectively. Finally, select two exponents $p \in (1,\infty)$ and $q \in (0,\infty]$ along with an index $\alpha \in \mathbb{R}$ satisfying*

$$-\frac{n-1}{p} < \alpha < \frac{n-1}{p'} \tag{18.30}$$

where $p' \in (1,\infty)$ denotes the Hölder conjugate exponent of p.

Then

$$\begin{aligned} K_A &: [\dot{\mathrm{K}}^{\alpha}_{p,q}(\partial\Omega,\sigma)]^M \to [\dot{\mathrm{K}}_{p,q}(\partial\Omega,\sigma)]^M, \\ K^{\#}_A &: [\dot{\mathrm{K}}^{\alpha}_{p,q}(\partial\Omega,\sigma)]^M \to [\dot{\mathrm{K}}_{p,q}(\partial\Omega,\sigma)]^M, \\ K_A &: [\mathrm{K}^{\alpha}_{p,q}(\partial\Omega,\sigma)]^M \to [\mathrm{K}_{p,q}(\partial\Omega,\sigma)]^M, \\ K^{\#}_A &: [\mathrm{K}^{\alpha}_{p,q}(\partial\Omega,\sigma)]^M \to [\mathrm{K}_{p,q}(\partial\Omega,\sigma)]^M, \end{aligned} \tag{18.31}$$

are well-defined, linear, and bounded operators. In fact, for every $m \in \mathbb{N}$ there exists some $C_m \in (0,\infty)$ depending only on n, p, q, α, A, and the UR *constants of $\partial\Omega$ with the property that*

$$\|K_A\|_{[\dot{\mathrm{K}}^{\alpha}_{p,q}(\partial\Omega,\sigma)]^M\to[\dot{\mathrm{K}}_{p,q}(\partial\Omega,\sigma)]^M} \leq C_m\cdot\|\nu\|^{\langle m\rangle}_{[\mathrm{BMO}(\partial\Omega,\sigma)]^n}, \tag{18.32}$$

$$\|K_A^\#\|_{[\dot{\mathrm{K}}^{\alpha}_{p,q}(\partial\Omega,\sigma)]^M\to[\dot{\mathrm{K}}^{\alpha}_{p,q}(\partial\Omega,\sigma)]^M}\leq C_m\cdot\|\nu\|^{\langle m\rangle}_{[\mathrm{BMO}(\partial\Omega,\sigma)]^n}. \tag{18.33}$$

Corresponding to the homogeneous case, it follows that

$$\|K_A\|_{[\mathrm{K}^{\alpha}_{p,q}(\partial\Omega,\sigma)]^M\to[\mathrm{K}_{p,q}(\partial\Omega,\sigma)]^M}\leq C_m\cdot\|\nu\|^{\langle m\rangle}_{[\mathrm{BMO}(\partial\Omega,\sigma)]^n}, \tag{18.34}$$

$$\|K_A^\#\|_{[\mathrm{K}^{\alpha}_{p,q}(\partial\Omega,\sigma)]^M\to[\mathrm{K}^{\alpha}_{p,q}(\partial\Omega,\sigma)]^M}\leq C_m\cdot\|\nu\|^{\langle m\rangle}_{[\mathrm{BMO}(\partial\Omega,\sigma)]^n}. \tag{18.35}$$

Proof Since A is a distinguished coefficient tensor for L, the integral kernels of the boundary-to-boundary double layer potentials K_A and $K_A^\#$ have a chord-dot-normal structure (cf. Remark 7.1). As such, the claims made in (18.31), as well as estimates (18.32)-(18.35), follow from Theorem 18.2. □

Remark 18.4 Relying on (18.19)-(18.22) from Remark 18.2, we may also obtain the same type of geometric operator norm estimates for boundary layer potentials in the context of Geometric Herz spaces. Specifically, work in same setting of Theorem 18.4, and let $E\subseteq\partial\Omega$ be a d-AR set, for some $d\in[0,n-1)$. Select indices p, q, and α satisfying

$$1<p<\infty,\quad 0<q\leq\infty,\quad\text{and}\quad \frac{-(n-1-d)}{p}<\alpha<\frac{(n-1-d)(p-1)}{p}. \tag{18.36}$$

Then

$$\begin{aligned}
K_A &:[\dot{\mathrm{K}}^{\alpha}_{p,q}(\partial\Omega,\sigma;E)]^M\to[\dot{\mathrm{K}}_{p,q}(\partial\Omega,\sigma;E)]^M,\\
K_A^\# &:[\dot{\mathrm{K}}^{\alpha}_{p,q}(\partial\Omega,\sigma;E)]^M\to[\dot{\mathrm{K}}_{p,q}(\partial\Omega,\sigma;E)]^M,\\
K_A &:[\mathrm{K}^{\alpha}_{p,q}(\partial\Omega,\sigma;E)]^M\to[\mathrm{K}_{p,q}(\partial\Omega,\sigma;E)]^M,\\
K_A^\# &:[\mathrm{K}^{\alpha}_{p,q}(\partial\Omega,\sigma;E)]^M\to[\mathrm{K}_{p,q}(\partial\Omega,\sigma;E)]^M,
\end{aligned} \tag{18.37}$$

are well-defined, linear, and bounded operators. Moreover, for every $m\in\mathbb{N}$ there exists some $C_m\in(0,\infty)$ depending only on n, p, q, α, A, d, the d-Ahlfors regularity constants of E, and the UR constants of $\partial\Omega$ with the property that

$$\|K_A\|_{[\dot{\mathrm{K}}^{\alpha}_{p,q}(\partial\Omega,\sigma;E)]^M\to[\dot{\mathrm{K}}_{p,q}(\partial\Omega,\sigma;E)]^M}\leq C_m\cdot\|\nu\|^{\langle m\rangle}_{[\mathrm{BMO}(\partial\Omega,\sigma)]^n}, \tag{18.38}$$

$$\|K_A^\#\|_{[\dot{\mathrm{K}}^{\alpha}_{p,q}(\partial\Omega,\sigma;E)]^M\to[\dot{\mathrm{K}}^{\alpha}_{p,q}(\partial\Omega,\sigma;E)]^M}\leq C_m\cdot\|\nu\|^{\langle m\rangle}_{[\mathrm{BMO}(\partial\Omega,\sigma)]^n}, \tag{18.39}$$

$$\|K_A\|_{[\mathrm{K}^{\alpha}_{p,q}(\partial\Omega,\sigma;E)]^M\to[\mathrm{K}_{p,q}(\partial\Omega,\sigma;E)]^M}\leq C_m\cdot\|\nu\|^{\langle m\rangle}_{[\mathrm{BMO}(\partial\Omega,\sigma)]^n}, \tag{18.40}$$

$$\|K_A^\#\|_{[\mathrm{K}^{\alpha}_{p,q}(\partial\Omega,\sigma;E)]^M\to[\mathrm{K}^{\alpha}_{p,q}(\partial\Omega,\sigma;E)]^M}\leq C_m\cdot\|\nu\|^{\langle m\rangle}_{[\mathrm{BMO}(\partial\Omega,\sigma)]^n}. \tag{18.41}$$

We are now in a position to state and prove the following basic invertibility result.

Theorem 18.5 *Suppose $\Omega\subseteq\mathbb{R}^n$ is an Ahlfors regular domain. Set $\sigma:=\mathcal{H}^{n-1}\lfloor\partial\Omega$ and denote by ν the geometric measure theoretic outward unit normal to Ω. Also,* *let L be a homogeneous, second-order, constant complex coefficient, weakly elliptic $M \times M$ system in $\mathbb{R}^n$, and assume $\mathfrak{A}_L^{\text{dis}} \neq \varnothing$. Pick $A \in \mathfrak{A}_L^{\text{dis}}$ and consider the boundary-to-boundary double layers K_A and $K_A^{\#}$ associated with A as in* (7.23) *and* (7.24)*, respectively. Select two exponents $p \in (1, \infty)$ and $q \in (0, \infty]$, along with an index $\alpha \in \mathbb{R}$ satisfying*

$$-\frac{n-1}{p} < \alpha < \frac{n-1}{p'} \tag{18.42}$$

where $p' \in (1, \infty)$ denotes the Hölder conjugate exponent of p. Lastly, fix some number $\varepsilon \in (0, \infty)$.

Then there exists a threshold $\delta \in (0, 1)$, depending exclusively on n, p, q, α, ε, A, and the Ahlfors regularity character of $\partial\Omega$, with the property that whenever

$$\|\nu\|_{[\mathrm{BMO}(\partial\Omega,\sigma)]^n} < \delta \tag{18.43}$$

it follows that for each spectral parameter $z \in \mathbb{C}$ with $|z| \geq \varepsilon$ the following operators are well defined, linear, bounded, and invertible:

$$\begin{aligned} zI + K_A &: [\dot{\mathrm{K}}^{\alpha}_{p,q}(\partial\Omega,\sigma)]^M \longrightarrow [\dot{\mathrm{K}}^{\alpha}_{p,q}(\partial\Omega,\sigma)]^M, \\ zI + K_A^{\#} &: [\dot{\mathrm{K}}^{\alpha}_{p,q}(\partial\Omega,\sigma)]^M \longrightarrow [\dot{\mathrm{K}}^{\alpha}_{p,q}(\partial\Omega,\sigma)]^M. \end{aligned} \tag{18.44}$$

Similar conclusions hold for the inhomogeneous Herz spaces.

Proof Thanks to Theorem 18.4, the invertibility of the operators in (18.44) follows via a Neumann series argument (cf. Lemma 8.1) much as in the proof of Theorem 8.2. □

Remark 18.5 In view of (18.38)-(18.41), invertibility results of the sort described in Theorem 18.5 are also valid in the context of Geometric Herz spaces $\dot{\mathrm{K}}^{\alpha}_{p,q}(\partial\Omega,\sigma;E)$ and $\mathrm{K}^{\alpha}_{p,q}(\partial\Omega,\sigma;E)$ associated with a d-Ahlfors regular set $E \subseteq \partial\Omega$, for some $d \in [0, n-1)$, provided p, q, and α satisfy (18.17).

Chapter 19
Boundary Value Problems on Herz Spaces of Second Generation

Here we make use of the Calderón-Zygmund theory for singular integral operators developed in Chapter 18 as a means of solving boundary value problems such as Dirichlet, Neumann, Regularity, and Transmission Problems for weakly elliptic systems in δ-AR domains with boundary data selected from homogeneous and inhomogeneous Herz spaces of second generation, as well as from geometric Herz spaces.

19.1 The Dirichlet Problem with Boundary Data in $\dot{\mathrm{K}}^{\alpha}_{p,q}$

We begin by dealing with the Dirichlet Problem with boundary data prescribed in second-generation Herz spaces.

Theorem 19.1 *Let $\Omega \subseteq \mathbb{R}^n$ be an Ahlfors regular domain, and set $\sigma := \mathcal{H}^{n-1}\lfloor\partial\Omega$. Denote by ν the geometric measure theoretic outward unit normal to Ω. Pick two exponents $p \in (1,\infty)$ and $q \in (0,\infty]$ along with an index $\alpha \in \mathbb{R}$ satisfying*

$$-\frac{n-1}{p} < \alpha < \frac{n-1}{p'} \tag{19.1}$$

where $p' \in (1,\infty)$ denotes the Hölder conjugate exponent of p. Also, fix an aperture parameter $\kappa \in (0,\infty)$. Let L be a second-order, homogeneous, constant complex coefficient, weakly elliptic $M \times M$ system in $\mathbb{R}^n$, and formulate the Dirichlet Problem for L in Ω with boundary data in the Herz space $[\dot{\mathrm{K}}^{\alpha}_{p,q}(\partial\Omega,\sigma)]^M$ as follows:

© The Author(s), under exclusive license to Springer Nature Switzerland AG 2026

M. Mitrea and P. Takemura, *Singular Integrals, Herz-Type Function Spaces, and Boundary Problems*, Progress in Mathematics 362,
https://doi.org/10.1007/978-3-032-12516-3_19

$$\begin{cases} u \in [\mathscr{C}^\infty(\Omega)]^M, \\ Lu = 0 \ \ in\ \Omega, \\ \mathcal{N}_\kappa u \in \dot{\mathrm{K}}^{\alpha}_{p,q}(\partial\Omega, \sigma), \\ u\big|_{\partial\Omega}^{\kappa-\mathrm{n.t.}} = f \in [\dot{\mathrm{K}}^{\alpha}_{p,q}(\partial\Omega, \sigma)]^M. \end{cases} \tag{19.2}$$

Regarding this problem, the following statements are true:

(a) [Existence, Integral Representation, and Estimates] *If* $\mathfrak{A}^{\mathrm{dis}}_L \neq \varnothing$, *there exists a threshold* $\delta \in (0,1)$ *depending on* n, M, p, q, α, L, *and the Ahlfors regularity constants of* $\partial\Omega$ *with the property that if*

$$\|\nu\|_{[\mathrm{BMO}(\partial\Omega,\sigma)]^n} < \delta \tag{19.3}$$

then the Dirichlet Problem (19.2) *is solvable. Concretely, if* $A \in \mathfrak{A}^{\mathrm{dis}}_L$ *then the operator* $\frac{1}{2}I + K_A$ *is invertible on* $[\dot{\mathrm{K}}^{\alpha}_{p,q}(\partial\Omega,\sigma)]^M$, *the function* $u : \Omega \to \mathbb{C}^M$ *given by*

$$u := \mathcal{D}_A\big[\big(\tfrac{1}{2}I + K_A\big)^{-1} f\big] \ \ in\ \Omega \tag{19.4}$$

is a solution of the Dirichlet Problem (19.2), *and satisfies*

$$\big\|\mathcal{N}_\kappa u\big\|_{\dot{\mathrm{K}}^{\alpha}_{p,q}(\partial\Omega,\sigma)} \approx \|f\|_{[\dot{\mathrm{K}}^{\alpha}_{p,q}(\partial\Omega,\sigma)]^M} \tag{19.5}$$

with implicit proportionality constants independent of the boundary datum f.

(b) [Uniqueness] *Whenever* $\mathfrak{A}^{\mathrm{dis}}_{L^\top} \neq \varnothing$, *there exists a threshold* $\delta \in (0,1)$ *depending on* n, M, p, q, α, L, *and the Ahlfors regularity constants of* $\partial\Omega$ *with the property that whenever* $\|\nu\|_{[\mathrm{BMO}(\partial\Omega,\sigma)]^n} < \delta$ *the Dirichlet Problem* (19.2) *admits at most one solution.*

(c) [Well-Posedness] *If* $\mathfrak{A}^{\mathrm{dis}}_L \neq \varnothing$ *and* $\mathfrak{A}^{\mathrm{dis}}_{L^\top} \neq \varnothing$, *then there exists a threshold* $\delta \in (0,1)$ *depending on* n, M, p, q, α, L, *and the Ahlfors regularity constants of* $\partial\Omega$ *with the property that whenever* $\|\nu\|_{[\mathrm{BMO}(\partial\Omega,\sigma)]^n} < \delta$ *the Dirichlet Problem* (19.2) *is well posed (i.e., it is uniquely solvable and the solution satisfies the naturally accompanying estimates formulated in* (19.5)).

Lastly, the same type of conclusions regarding existence, integral representations, estimates, uniqueness, and well-posedness, hold in the context of inhomogeneous Herz spaces $\mathrm{K}^{\alpha}_{p,q}(\partial\Omega,\sigma)$.

Proof The claim made in *(a)* can be obtained via the same approach used in the treatment of item *(a)* in Theorem 16.2, but now bearing in mind the invertibility result recorded in Theorem 18.5, and the embedding from Corollary 17.2.

To establish the uniqueness claimed in *(b)*, we rely on a well-posedness result for the Dirichlet Problem with boundary data belonging to sums of Muckenhoupt weighted Lebesgue spaces from [MMMi[3]]. More specifically, assume that u is a solution of the homogeneous Dirichlet Problem (19.2), i.e., corresponding to the boundary datum $f \equiv 0$. Since $-\frac{n-1}{p} < \alpha < \frac{n-1}{p'}$ (cf. (19.1)) we can select two numbers $a, b \in \mathbb{R}$ such that

$$-(n-1) < b < \alpha p < a < (p-1)(n-1). \tag{19.6}$$

Hence, as pointed out in (17.170), the functions

$$|\cdot - x_0|^a \text{ and } |\cdot - x_0|^b \text{ belong to the Muckenhoupt class } A_p(\partial\Omega, \sigma). \tag{19.7}$$

Granted this, and bearing in mind Corollary 17.1, our null-solution u of (19.2) satisfies

$$\begin{cases} u \in [\mathscr{C}^\infty(\Omega)]^M, \quad Lu = 0 \text{ in } \Omega, \\ \mathcal{N}_\kappa u \in L^p\big(\partial\Omega, |x - x_0|^a \sigma(x)\big) + L^p\big(\partial\Omega, |x - x_0|^b \sigma(x)\big), \\ u\Big|_{\partial\Omega}^{\kappa-\mathrm{n.t.}} = 0 \text{ at } \sigma\text{-a.e. point on } \partial\Omega. \end{cases}$$

At this stage, we may invoke [MMMi[3], Theorem 6.3, p. 376] to ensure that $u \equiv 0$ in Ω, thus completing the treatment of the claim made in *(b)*. Ultimately, the well-posedness claimed in *(c)* is a direct consequence of *(a)* and *(b)*, thus completing the proof of Theorem 19.1. □

Remark 19.1 The approach used to establish Theorem 19.1 relies on mapping properties of layer potentials of Herz spaces, and embeddings obtained in Corollary 17.1. Given that analogous results to these are valid for Geometric Herz spaces (see Corollary 17.4, Remark 17.3, and Remark 18.5) the same type of argument used in the proof of Theorem 19.1 can be used to treat the Dirichlet problem with boundary data taken from Geometric Herz spaces formulated as:

$$\begin{cases} u \in [\mathscr{C}^\infty(\Omega)]^M, \\ Lu = 0 \text{ in } \Omega, \\ \mathcal{N}_\kappa u \in \dot{\mathrm{K}}^{\alpha}_{p,q}(\partial\Omega, \sigma; E), \\ u\Big|_{\partial\Omega}^{\kappa-\mathrm{n.t.}} = f \in [\dot{\mathrm{K}}^{\alpha}_{p,q}(\partial\Omega, \sigma; E)]^M, \end{cases} \tag{19.8}$$

Specifically, given $E \subseteq \partial\Omega$ a d-AR set for some $d \in [0, n-1)$, and indices $p \in (1, \infty)$, $q \in (0, \infty]$, and $\alpha \in \mathbb{R}$ such that $\frac{-(n-1-d)}{p} < \alpha < \frac{(n-1-d)(p-1)}{p}$, then (19.8) admits a solution in the setting of item *(a)* in Theorem 19.1, while uniqueness is valid in the context of item *(b)* in Theorem 19.1.

19.2 Regularity Problems with Boundary Data in $\mathrm{K}^{\alpha}_{p,q}$

Our next result concerns the well-posedness of the inhomogeneous Regularity Problem with boundary data taken from second-generation Herz spaces. To set the stage we make the following definition.

Definition 19.1 Consider an Ahlfors regular domain $\Omega \subseteq \mathbb{R}^n$ with unbounded boundary, and set $\sigma := \mathcal{H}^{n-1}\lfloor\partial\Omega$. Fix two exponents $p, q \in (1, \infty)$, along with an index $\alpha \in \mathbb{R}$ satisfying

$$-\frac{n-1}{p} < \alpha < \frac{n-1}{p'}. \tag{19.9}$$

define the $\dot{\mathrm{K}}^{\alpha}_{p,q}$-`styled Sobolev space` on $\partial\Omega$ as

$$\dot{\mathrm{K}}^{\alpha,1}_{p,q}(\partial\Omega,\sigma) := \Big\{ f \in \dot{\mathrm{K}}^{\alpha}_{p,q}(\partial\Omega,\sigma) : \partial_{\tau_{jk}} f \in \dot{\mathrm{K}}^{\alpha}_{p,q}(\partial\Omega,\sigma) \quad \text{for each } j,k \in \{1,\dots,n\} \Big\}. \tag{19.10}$$

Equip this space with the norm

$$\|f\|_{\dot{\mathrm{K}}^{\alpha,1}_{p,q}(\partial\Omega,\sigma)} := \|f\|_{\dot{\mathrm{K}}^{\alpha}_{p,q}(\partial\Omega,\sigma)} + \textstyle\sum_{j,k=1}^{n} \|\partial_{\tau_{jk}} f\|_{\dot{\mathrm{K}}^{\alpha}_{p,q}(\partial\Omega,\sigma)} \quad \text{for every } f \in \dot{\mathrm{K}}^{\alpha,1}_{p,q}(\partial\Omega,\sigma). \tag{19.11}$$

Finally, define $\mathrm{K}^{\alpha,1}_{p,q}(\partial\Omega,\sigma)$ the $\mathrm{K}^{\alpha}_{p,q}$-`styled Sobolev space` on $\partial\Omega$ in a similar fashion, now with the inhomogeneous Herz space $\mathrm{K}^{\alpha}_{p,q}(\partial\Omega,\sigma)$ playing the role of $\dot{\mathrm{K}}^{\alpha}_{p,q}(\partial\Omega,\sigma)$.

We are now in a position to state and prove the result announced earlier.

Theorem 19.2 *Let $\Omega \subseteq \mathbb{R}^n$ be an Ahlfors regular domain, and set $\sigma := \mathcal{H}^{n-1}\lfloor\partial\Omega$. Denote by ν the geometric measure theoretic outward unit normal to Ω, and fix an aperture parameter $\kappa \in (0,\infty)$. Also select two exponents $p, q \in (1,\infty)$, along with an index $\alpha \in \mathbb{R}$ satisfying*

$$-\frac{n-1}{p} < \alpha < \frac{n-1}{p'}. \tag{19.12}$$

Lastly, consider a second-order, homogeneous, constant complex coefficient, weakly elliptic, $M\times M$ system L in $\mathbb{R}^n$, and formulate the Inhomogeneous Regularity Problem for L in Ω with boundary data in the Herz-based Sobolev space $[\dot{\mathrm{K}}^{\alpha,1}_{p,q}(\partial\Omega,\sigma)]^M$ as:

$$\begin{cases} u \in \big[\mathscr{C}^\infty(\Omega)\big]^M, \\ Lu = 0 \ \text{ in } \Omega, \\ \mathcal{N}_\kappa u,\ \mathcal{N}_\kappa(\nabla u) \in \dot{\mathrm{K}}^{\alpha}_{p,q}(\partial\Omega,\sigma), \\ u\big|^{\kappa-\mathrm{n.t.}}_{\partial\Omega} = f \in \big[\dot{\mathrm{K}}^{\alpha,1}_{p,q}(\partial\Omega,\sigma)\big]^M. \end{cases} \tag{19.13}$$

Then the following statements are true:

(*a*) [Existence, Estimates, and Integral Representation] *If* $\mathfrak{A}^{\mathrm{dis}}_{L} \neq \emptyset$ *then there exists* $\delta \in (0,1)$ *that depends only on* n, p, q, α, L, *and the Ahlfors regularity of* $\partial\Omega$ *with the property that whenever*

$$\|\nu\|_{[\mathrm{BMO}(\partial\Omega,\sigma)]^n} < \delta \tag{19.14}$$

then the Inhomogeneous Regularity Problem (19.13) *is solvable. More specifically, if* $A \in \mathfrak{A}^{\mathrm{dis}}_{L}$ *then the operator* $\frac{1}{2}I + K_A$ *is invertible on* $[\dot{\mathrm{K}}^{\alpha,1}_{p,q}(\partial\Omega,\sigma)]^M$ *and the function* $u : \Omega \longrightarrow \mathbb{C}^M$ *defined as*

$$u := \mathcal{D}_A\big[\big(\tfrac{1}{2}I + K_A\big)^{-1} f\big] \ \text{ in } \Omega \tag{19.15}$$

is a solution of the Inhomogeneous Regularity Problem (19.13), *which satisfies*

$$\|\mathcal{N}_\kappa u\|_{\dot{\mathrm{K}}^{\alpha}_{p,q}(\partial\Omega,\sigma)} \approx \|f\|_{[\dot{\mathrm{K}}^{\alpha}_{p,q}(\partial\Omega,\sigma)]^M} \tag{19.16}$$

and

$$\|\mathcal{N}_\kappa u\|_{\dot{\mathrm{K}}^{\alpha}_{p,q}(\partial\Omega,\sigma)} + \|\mathcal{N}_\kappa(\nabla u)\|_{\dot{\mathrm{K}}^{\alpha}_{p,q}(\partial\Omega,\sigma)} \approx \|f\|_{[\dot{\mathrm{K}}^{\alpha,1}_{p,q}(\partial\Omega,\sigma)]^M} \tag{19.17}$$

with implicit proportionality constants independent of f.

(*b*) [Uniqueness] *If* $\mathfrak{A}^{\mathrm{dis}}_{L^\top} \neq \emptyset$ *then there exists a threshold* $\delta \in (0,1)$ *depending on* n, M, p, q, α, L, *and the Ahlfors regularity constants of* $\partial\Omega$ *with the property that whenever* $\|\nu\|_{[\mathrm{BMO}(\partial\Omega,\sigma)]^n} < \delta$ *the Inhomogeneous Regularity Problem* (19.13) *admits at most one solution.*

(*c*) [Well-Posedness] *If* $\mathfrak{A}^{\mathrm{dis}}_{L} \neq \emptyset$ *and* $\mathfrak{A}^{\mathrm{dis}}_{L^\top} \neq \emptyset$ *it follows that there exists some* $\delta \in (0,1)$ *depending only on* n, p, q, α, L, *and the Ahlfors regularity constants of* $\partial\Omega$ *with the property that whenever* $\|\nu\|_{[\mathrm{BMO}(\partial\Omega,\sigma)]^n} < \delta$ *the Inhomogeneous Regularity Problem* (19.13) *is uniquely solvable.*

The same type of conclusions regarding existence, uniqueness, and estimates hold for the inhomogeneous Herz space $\mathrm{K}^{\alpha}_{p,q}(\partial\Omega,\sigma)$ *and its Sobolev counterpart* $\mathrm{K}^{\alpha,1}_{p,q}(\partial\Omega,\sigma)$ *(cf. Definition 19.1).*

Proof Thanks to Theorem 17.4, Remark 17.2, Proposition 17.6, and the fact that $\dot{\mathrm{K}}^{\alpha}_{p,q}(\partial\Omega,\sigma)$ is a Generalized Banach Function Space (cf. (17.123)), the veracity of all claims *(a)*-*(c)* follows from [GHA-V, Theorem 8.4.41, pp. 871–872]. □

Remark 19.2 Via a similar approach, we may obtain analogous conclusions in relation to the Inhomogeneous Regularity Problem formulated in the context of Geometric Herz spaces. More precisely, given $E \subseteq \partial\Omega$ a d-AR set for some $d \in [0, n-1)$, along with indices $p, q \in (1,\infty)$ and $\alpha \in \mathbb{R}$ satisfying

$$\frac{-(n-1-d)}{p} < \alpha < \frac{(n-1-d)(p-1)}{p} \tag{19.18}$$

define the Geometric Herz-based Sobolev space on $\partial\Omega$ according to

$$\dot{\mathrm{K}}^{\alpha,1}_{p,q}(\partial\Omega,\sigma;E) := \Big\{ f \in \dot{\mathrm{K}}^{\alpha}_{p,q}(\partial\Omega,\sigma;E) : \partial_{\tau_{jk}} f \in \dot{\mathrm{K}}^{\alpha}_{p,q}(\partial\Omega,\sigma;E) \\ \text{for each } j,k \in \{1,\dots,n\} \Big\}, \tag{19.19}$$

and equip this space with the norm

$$\|f\|_{\dot{\mathrm{K}}^{\alpha,1}_{p,q}(\partial\Omega,\sigma;E)} := \|f\|_{\dot{\mathrm{K}}^{\alpha}_{p,q}(\partial\Omega,\sigma;E)} + \textstyle\sum_{j,k=1}^{n} \|\partial_{\tau_{jk}} f\|_{\dot{\mathrm{K}}^{\alpha}_{p,q}(\partial\Omega,\sigma;E)} \\ \text{for every } f \in \dot{\mathrm{K}}^{\alpha,1}_{p,q}(\partial\Omega,\sigma;E). \tag{19.20}$$

In light of (17.200), (17.201), Proposition 17.12, and Remark 17.4, we can now employ [GHA-V, Theorem 8.4.41, pp. 871–872] to ensure that the Inhomogeneous Regularity Problem with boundary data in Geometric Herz-based Sobolev spaces formulated as

$$\begin{cases} u \in \big[\mathscr{C}^\infty(\Omega)\big]^M, \\ Lu = 0 \ \text{ in } \Omega, \\ \mathcal{N}_\kappa u,\ \mathcal{N}_\kappa(\nabla u) \in \dot{\mathrm{K}}^{\alpha}_{p,q}(\partial\Omega,\sigma;E), \\ u\big|_{\partial\Omega}^{\kappa-\mathrm{n.t.}} = f \in \big[\dot{\mathrm{K}}^{\alpha,1}_{p,q}(\partial\Omega,\sigma;E)\big]^M, \end{cases} \tag{19.21}$$

is solvable in the context of item *(a)* in Theorem 19.2, and has a unique solution a in the setting of item *(b)* in Theorem 19.2.

Next, we discuss the Homogeneous Regularity Problem in the context of second-generation Herz spaces. To facilitate its formulation, we first make the following definition.

Definition 19.2 Assume Ω is an Ahlfors regular domain with unbounded boundary, and abbreviate $\sigma := \mathcal{H}^{n-1}\lfloor\Sigma$. Select two exponents $p,q \in (0,\infty]$ along with an index $\alpha \in \mathbb{R}$. Introduce the $\dot{\mathrm{K}}^{\alpha}_{p,q}$-based homogeneous Sobolev space on $\partial\Omega$ as

$$\dot{\mathrm{K}}^{\alpha,(1)}_{p,q}(\partial\Omega,\sigma) := \Big\{ f \in L^1\big(\partial\Omega, \tfrac{\sigma(x)}{1+|x|^n}\big) \cap \dot{\mathrm{K}}^{\alpha}_{p,q,\mathrm{loc}}(\partial\Omega,\sigma) : \\ \partial_{\tau_{jk}} f \in \dot{\mathrm{K}}^{\alpha}_{p,q}(\partial\Omega,\sigma) \ \text{ for } \ j,k \in \{1,\dots,n\} \Big\}, \tag{19.22}$$

where the membership $f \in \dot{\mathrm{K}}^{\alpha}_{p,q,\mathrm{loc}}(\partial\Omega,\sigma)$ means $f \cdot \mathbb{1}_\Delta \in \dot{\mathrm{K}}^{\alpha}_{p,q}(\partial\Omega,\sigma)$ for every surface ball $\Delta \subseteq \partial\Omega$. Equip this space with the seminorm

$$\|f\|_{\dot{\mathrm{K}}^{\alpha,(1)}_{p,q}(\partial\Omega,\sigma)} := \sum_{j,k=1}^{n} \|\partial_{\tau_{jk}} f\|_{\dot{\mathrm{K}}^{\alpha}_{p,q}(\partial\Omega,\sigma)} \quad \text{for each } f \in \dot{\mathrm{K}}^{\alpha,(1)}_{p,q}(\partial\Omega,\sigma). \tag{19.23}$$

Finally, in a similar fashion introduce the $\mathrm{K}^{\alpha}_{p,q}$-based homogeneous Sobolev space on $\partial\Omega$, denoted by $\mathrm{K}^{\alpha,(1)}_{p,q}(\partial\Omega,\sigma)$, where now the inhomogeneous Herz space $\mathrm{K}^{\alpha}_{p,q}(\partial\Omega,\sigma)$ play the role of $\dot{\mathrm{K}}^{\alpha}_{p,q}(\partial\Omega,\sigma)$.

We are now in a position to treat the Homogeneous Regularity Problem.

Theorem 19.3 *Consider $\Omega \subseteq \mathbb{R}^n$ an Ahlfors regular domain, set $\sigma := \mathcal{H}^{n-1}\lfloor\partial\Omega$, and denote by ν the geometric measure theoretic outward unit normal to Ω. Select two exponents $p, q \in (1,\infty)$ along with an index $\alpha \in \mathbb{R}$ satisfying*

$$-\frac{n-1}{p} < \alpha < \frac{n-1}{p'} \tag{19.24}$$

where $p' \in (1,\infty)$ denotes the Hölder conjugate exponent of p. Given L a second-order, homogeneous, constant complex coefficient, weakly elliptic, $M \times M$ system in $\mathbb{R}^n$, and an aperture parameter $\kappa \in (0,\infty)$, consider the Homogeneous Regularity Problem for L in Ω with boundary data in $[\dot{\mathrm{K}}^{\alpha,(1)}_{p,q}(\partial\Omega,\sigma)]^M$ formulated as follows:

$$\begin{cases} u \in [\mathscr{C}^\infty(\Omega)]^M, \\ Lu = 0 \ \text{ in } \Omega, \\ \mathcal{N}_\kappa(\nabla u) \in \dot{\mathrm{K}}^{\alpha}_{p,q}(\partial\Omega,\sigma), \\ u\big|^{\kappa-\mathrm{n.t.}}_{\partial\Omega} = f \in \big[\dot{\mathrm{K}}^{\alpha,(1)}_{p,q}(\partial\Omega,\sigma)\big]^M. \end{cases} \tag{19.25}$$

In relation to this, the following statements are true:

(a) [Existence, Estimates, and Integral Representation] *If $\mathfrak{A}^{\mathrm{dis}}_L \neq \varnothing$ then there exists some $\delta \in (0,1)$ depending on n, M, L, p, q, α, and the Ahlfors regularity constants of $\partial\Omega$ with the property that whenever*

$$\|\nu\|_{[\mathrm{BMO}(\partial\Omega,\sigma)]^n} < \delta \tag{19.26}$$

then the Homogeneous Regularity Problem (19.25) *is solvable. More transparently, the modified boundary-to-boundary single layer potential S_{mod} from* (7.57) *induces a well-defined linear isomorphism*

$$\begin{gathered} \big[S_{\mathrm{mod}}\big] : \big[\dot{\mathrm{K}}^{\alpha}_{p,q}(\partial\Omega,\sigma)\big]^M \longrightarrow \big[\dot{\mathrm{K}}^{\alpha,(1)}_{p,q}(\partial\Omega,\sigma)/\sim\big]^M \\ \text{defined as } \ \big[S_{\mathrm{mod}}\big]g := \big[S_{\mathrm{mod}}g\big] \ \text{ for all } \ g \in \big[\dot{\mathrm{K}}^{\alpha}_{p,q}(\partial\Omega,\sigma)\big]^M. \end{gathered} \tag{19.27}$$

Moreover, if $[f] \in \big[\dot{\mathrm{K}}^{\alpha,(1)}_{p,q}(\partial\Omega,\sigma)/\sim\big]^M$ denotes the equivalence class (modulo constants) of the boundary datum f and, the function $g \in \big[\dot{\mathrm{K}}^{\alpha}_{p,q}(\partial\Omega,\sigma)\big]^M$ is such that $\big[S_{\mathrm{mod}}\big]g = [f]$, then there exists some constant $c \in \mathbb{C}^M$ for which the function $u : \Omega \to \mathbb{C}^M$ given by

$$u := \mathscr{S}_{\rm mod} g + c \ \ in\ \Omega \tag{19.28}$$

is a solution of the Homogeneous Regularity Problem (19.25)*, which satisfies*

$$\|\mathcal{N}_\kappa(\nabla u)\|_{\dot{\mathrm{K}}^\alpha_{p,q}(\partial\Omega,\sigma)} \approx \|\nabla_{\tan} f\|_{[\dot{\mathrm{K}}^\alpha_{p,q}(\partial\Omega,\sigma)]^{n\times M}} \tag{19.29}$$

with implicit proportionality constants independent of f*.*

Alternatively, for each coefficient tensor $A \in \mathfrak{A}^{\rm dis}_L$*, if* $\big[K_{A,\rm mod}\big][g] := \big[K_{A,\rm mod} g\big]$ *for each* $[g] \in \big[\dot{\mathrm{K}}^{\alpha,(1)}_{p,q}(\partial\Omega,\sigma)/\sim\big]^M$*, where* $K_{A,\rm mod}$ *is as in* (7.29)*, then the operator*

$$\tfrac{1}{2}I + \big[K_{A,\rm mod}\big] : \big[\dot{\mathrm{K}}^{\alpha,(1)}_{p,q}(\partial\Omega,\sigma)/\sim\big]^M \longrightarrow \big[\dot{\mathrm{K}}^{\alpha,(1)}_{p,q}(\partial\Omega,\sigma)/\sim\big]^M \tag{19.30}$$

is an isomorphism, and the Homogeneous Regularity Problem (19.25) *may also be solved as (cf.* (7.28))

$$u := \mathcal{D}_{A,\rm mod} h + c \ \ in\ \Omega, \tag{19.31}$$

for some constant $c \in \mathbb{C}^M$ *and some function*

$$h \in \big[\dot{\mathrm{K}}^{\alpha,(1)}_{p,q}(\partial\Omega,\sigma)\big]^M \ \ satisfying\ \ [h] = \Big(\tfrac{1}{2}I + \big[K_{A,\rm mod}\big]\Big)^{-1}[f]. \tag{19.32}$$

Furthermore, the solution u from (19.31) *continues to satisfy* (19.29)*.*

(b) [Uniqueness] *Whenever* $\mathfrak{A}^{\rm dis}_{L^\top} \neq \varnothing$*, there exists* $\delta \in (0,1)$ *depending only on* n*,* L*,* p*,* q*,* α*, and the Ahlfors regularity constants of* $\partial\Omega$ *such that if* $\|\nu\|_{[\mathrm{BMO}(\partial\Omega,\sigma)]^n} < \delta$ *then the Homogeneous Regularity Problem* (19.25) *has at most one solution.*

(c) [Well-Posedness and Additional Regularity] *If* $\mathfrak{A}^{\rm dis}_L \neq \varnothing$ *and* $\mathfrak{A}^{\rm dis}_{L^\top} \neq \varnothing$*, then there exists a threshold* $\delta \in (0,1)$ *depending exclusively on* n*,* L*,* p*,* q*,* α*, and the Ahlfors regularity constants of* $\partial\Omega$ *with the property that if* $\|\nu\|_{[\mathrm{BMO}(\partial\Omega,\sigma)]^n} < \delta$*, it follows that the Homogeneous Regularity Problem* (19.25) *is well posed. In addition, the unique solution u of* (19.25) *satisfies*

$$\mathcal{N}_\kappa u \in \dot{\mathrm{K}}^\alpha_{p,q}(\partial\Omega,\sigma) \iff f \in \big[\dot{\mathrm{K}}^{\alpha,1}_{p,q}(\partial\Omega,\sigma)\big]^M \tag{19.33}$$

where $\dot{\mathrm{K}}^{\alpha,1}_{p,q}(\partial\Omega,\sigma)$ *denotes the Herz-based Sobolev space of order one on* $\partial\Omega$ *defined as in* (19.10)*.*

Furthermore, similar conclusions regarding existence, estimates, integral representations, mapping properties, uniqueness, regularity, and well-posedness also hold in the context of inhomogeneous Herz spaces $\mathrm{K}^\alpha_{p,q}(\partial\Omega,\sigma)$*.*

Proof Bearing in mind Theorem 17.4, Remark 17.2 Proposition 17.6, and the fact that $\dot{\mathrm{K}}^{\alpha}_{p,q}(\partial\Omega, \sigma)$ is a Generalized Banach Function Space (cf. (17.123)), all claims may be obtained via [GHA-V, Theorem 8.4.42, pp. 873–875]. □

Remark 19.3 For the same reasons as in Remark 19.2, similar solvability and uniqueness conclusions may be obtained in relation to the Homogeneous Regularity Problem with boundary data in Geometric Herz spaces:

$$\begin{cases} u \in [\mathscr{C}^{\infty}(\Omega)]^M, \\ Lu = 0 \ \text{ in } \Omega, \\ \mathcal{N}_{\kappa}(\nabla u) \in \dot{\mathrm{K}}^{\alpha}_{p,q}(\partial\Omega, \sigma; E), \\ u\big|^{\kappa-\text{n.t.}}_{\partial\Omega} = f \in \big[\dot{\mathrm{K}}^{\alpha,(1)}_{p,q}(\partial\Omega, \sigma; E)\big]^M, \end{cases} \tag{19.34}$$

with $E \subseteq \partial\Omega$ being d-Ahlfors regular set, and indices p, q, α satisfying

$$1 < p < \infty, \quad 1 < q < \infty, \quad \text{and} \quad \tfrac{-(n-1-d)}{p} < \alpha < \tfrac{(n-1-d)(p-1)}{p}. \tag{19.35}$$

Here, the space $\dot{\mathrm{K}}^{\alpha,(1)}_{p,q}(\partial\Omega, \sigma; E)$ is the Geometric Herz-based homogeneous Sobolev space defined in an analogous fashion to (19.22).

19.3 The Neumann Problem with Boundary Data in $\mathrm{K}^{\alpha}_{p,q}$

Our next goal is to formulate and solve the Neumann Problem for weakly elliptic systems in δ-AR domains with boundary data in homogeneous and inhomogeneous second-generation Herz spaces. We first present a Fatou-type result in the context of $\dot{\mathrm{K}}^{\alpha}_{p,q}$ spaces.

Proposition 19.1 *Assume $\Omega \subseteq \mathbb{R}^n$ is an Ahlfors regular domain with unbounded boundary. Set $\sigma := \mathcal{H}^{n-1}\lfloor\partial\Omega$ and denote by ν the geometric measure theoretic outward unit normal to Ω. Let L be a second-order, homogeneous, constant complex coefficient, weakly elliptic $M\times M$ system in $\mathbb{R}^n$, and pick a coefficient tensor $A \in \mathfrak{A}_L$. Select two exponents $p \in (1, \infty)$ and $q \in (0, \infty]$, along with an index $\alpha \in \mathbb{R}$ satisfying*

$$-\frac{n-1}{p} < \alpha < \frac{n-1}{p'}, \tag{19.36}$$

with $p' \in (1, \infty)$ denoting the Hölder conjugate exponent of p. Fix an aperture parameter $\kappa \in (0, \infty)$, and assume $u : \Omega \to \mathbb{C}^M$ is a function satisfying:

$$\begin{aligned} &u \in \big[\mathscr{C}^{\infty}(\Omega)\big]^M, \ Lu = 0 \text{ in } \Omega, \\ &\text{and } \mathcal{N}_{\kappa}(\nabla u) \in \dot{\mathrm{K}}^{\alpha}_{p,q}(\partial\Omega, \sigma). \end{aligned} \tag{19.37}$$

Then the nontangential boundary trace $(\nabla u)\big|_{\partial\Omega}^{\kappa-\mathrm{n.t.}}$ *exists (in* $\mathbb{C}^{M\times n}$*) at* σ*-a.e. point on* $\partial\Omega$ *and, in a quantitative fashion,*

$$\begin{aligned}&(\nabla u)\big|_{\partial\Omega}^{\kappa-\mathrm{n.t.}} \text{ belongs to } [\dot{\mathrm{K}}^{\alpha}_{p,q}(\partial\Omega,\sigma)]^{M\times n}\\ &\text{while } \partial^A_\nu u \text{ belongs to } [\dot{\mathrm{K}}^{\alpha}_{p,q}(\partial\Omega,\sigma)]^{M},\end{aligned} \tag{19.38}$$

i.e., there exists a constant $C\in(0,\infty)$ *independent of* u *for which*

$$\big\|(\nabla u)\big|_{\partial\Omega}^{\kappa-\mathrm{n.t.}}\big\|_{[\dot{\mathrm{K}}^{\alpha}_{p,q}(\partial\Omega,\sigma)]^{M\times n}} \le C\|\mathcal{N}_\kappa(\nabla u)\|_{\dot{\mathrm{K}}^{\alpha}_{p,q}(\partial\Omega,\sigma)} \tag{19.39}$$

$$\text{and}\quad \|\partial^A_\nu u\|_{[\dot{\mathrm{K}}^{\alpha}_{p,q}(\partial\Omega,\sigma)]^{n}} \le C\|\mathcal{N}_\kappa(\nabla u)\|_{\dot{\mathrm{K}}^{\alpha}_{p,q}(\partial\Omega,\sigma)}. \tag{19.40}$$

Similar conclusions hold in the context of inhomogeneous Herz spaces $\mathrm{K}^{\alpha}_{p,q}(\partial\Omega,\sigma)$*.*

Proof All conclusions follow from Theorem 9.1, the inclusions in (17.172), and the lattice property of $\dot{\mathrm{K}}^{\alpha}_{p,q}(\partial\Omega,\sigma)$ (see (17.5)). □

We are now ready to state and prove the well-posedness of the Neumann Problem announced earlier.

Theorem 19.4 *Assume* $\Omega\subseteq\mathbb{R}^n$ *is an Ahlfors regular domain, set* $\sigma:=\mathcal{H}^{n-1}\lfloor\partial\Omega$*, and denote by* ν *the geometric measure theoretic outward unit normal to* Ω*. Select two exponents* $p\in(1,\infty)$ *and* $q\in(0,\infty]$*, along with an index* $\alpha\in\mathbb{R}$ *satisfying*

$$-\frac{n-1}{p}<\alpha<\frac{n-1}{p'}, \tag{19.41}$$

with $p'\in(1,\infty)$ *denotes the Hölder conjugate exponent of* p*. Consider a second-order, homogeneous, constant complex coefficient, weakly elliptic* $M\times M$ *system* L *in* $\mathbb{R}^n$*, and select a coefficient tensor* $A\in\mathfrak{A}_L$*. Having fixed an aperture parameter* $\kappa\in(0,\infty)$*, formulate the Neumann Problem for* L *in* Ω *with boundary data in the Herz space* $[\dot{\mathrm{K}}^{\alpha}_{p,q}(\partial\Omega,\sigma)]^M$ *as follows:*

$$\begin{cases} u\in[\mathscr{C}^\infty(\Omega)]^M,\\ Lu=0 \ \text{ in } \Omega,\\ \mathcal{N}_\kappa(\nabla u)\in\dot{\mathrm{K}}^{\alpha}_{p,q}(\partial\Omega,\sigma),\\ \partial^A_\nu u=f\in[\dot{\mathrm{K}}^{\alpha}_{p,q}(\partial\Omega,\sigma)]^M.\end{cases} \tag{19.42}$$

Then the following statements are valid:

(a) [Existence, Integral Representation, and Estimates] *If* $A^\top\in\mathfrak{A}^{\mathrm{dis}}_{L^\top}$ *then there exists some* $\delta\in(0,1)$ *depending only on* n*,* M*,* p*,* q*,* α*,* A*, and the Ahlfors regularity constants of* $\partial\Omega$ *with the property that whenever*

$$\|\nu\|_{[\mathrm{BMO}(\partial\Omega,\sigma)]^n} < \delta \tag{19.43}$$

the Neumann Problem (19.42) *is solvable. Concretely, the operator* $-\frac{1}{2}I + K^{\#}_{A^{\top}}$ *is invertible on* $[\dot{\mathrm{K}}^{\alpha}_{p,q}(\partial\Omega,\sigma)]^M$ *and the function* $u : \Omega \to \mathbb{C}^M$ *given by*

$$u := \mathscr{S}_{\mathrm{mod}}\big[\big(-\tfrac{1}{2}I + K^{\#}_{A^{\top}}\big)^{-1} f\big] \ \ in\ \Omega \tag{19.44}$$

is a solution of the Neumann Problem (19.42) *with the property that*

$$\big\|\mathcal{N}_{\kappa}(\nabla u)\big\|_{\dot{\mathrm{K}}^{\alpha}_{p,q}(\partial\Omega,\sigma)} \approx \|f\|_{[\dot{\mathrm{K}}^{\alpha}_{p,q}(\partial\Omega,\sigma)]^M} \tag{19.45}$$

with implicit proportionality constants independent of f.

(b) [Uniqueness (modulo constants)] *Whenever* $A \in \mathfrak{A}^{\mathrm{dis}}_{L}$, *there exists* $\delta \in (0,1)$ *depending on n, M, p, q, α, A, and the Ahlfors regularity constants of* $\partial\Omega$ *with the property that if* $\|\nu\|_{[\mathrm{BMO}(\partial\Omega,\sigma)]^n} < \delta$ *then any two solutions of the Neumann Problem* (19.42) *differ by a constant in* $\mathbb{C}^M$.

(c) [Well-Posedness] *If* $A \in \mathfrak{A}^{\mathrm{dis}}_{L}$ *and* $A^{\top} \in \mathfrak{A}^{\mathrm{dis}}_{L^{\top}}$ *then there exists* $\delta \in (0,1)$ *depending on n, M, p, q, α, L, and the Ahlfors regularity constants of* $\partial\Omega$ *with the property that whenever* $\|\nu\|_{[\mathrm{BMO}(\partial\Omega,\sigma)]^n} < \delta$ *the Neumann Problem* (19.42) *has a solution that is unique modulo constants, and satisfies* (19.45).

Proof The veracity of the claims made in *(a)* follows from the same type of argument as in Theorem 16.5, being mindful of Theorem 18.5, Proposition 19.1, and Corollary 17.2.

The justify the uniqueness (modulo constants) claimed in item *(b)*, suppose that u is a solution of the homogeneous version of the Neumann Problem (19.42) (i.e., corresponding to the boundary datum $f \equiv 0$). Bearing in mind the range of α specified in (19.41), choose two numbers $a, b \in \mathbb{R}$ satisfying

$$-(n-1) < b < \alpha p < a < (p-1)(n-1). \tag{19.46}$$

As such, (17.170) ensures that the functions

$$|\cdot - x_0|^a \ \text{ and } \ |\cdot - x_0|^b \ \text{ belong to the Muckenhoupt class } A_p(\partial\Omega,\sigma). \tag{19.47}$$

Thanks to this, and Corollary 17.1, we may conclude that the null-solution u under consideration satisfies

$$\begin{cases} u \in [\mathscr{C}^{\infty}(\Omega)]^M, \quad Lu = 0 \ \text{ in } \Omega, \\ \mathcal{N}_{\kappa} u \in L^p\big(\partial\Omega, |x-x_0|^a\sigma(x)\big) + L^p\big(\partial\Omega, |x-x_0|^b\sigma(x)\big), \\ \partial^A_{\nu} u = 0 \ \text{ at } \sigma\text{-a.e. point on } \partial\Omega. \end{cases}$$

Work done in [MMMi[3], Theorem 6.14, p. 407] concerning the well-posedness of the Neumann Problem on sums of Muckenhoupt weighted Lebesgue spaces then implies that u is a constant function in Ω. This completes the treatment of item *(b)*. Ultimately, the well-posedness claimed in *(c)* is a direct consequence of *(a)*-*(b)*. This concludes the proof of Theorem 19.4. □

Remark 19.4 Based on Corollary 17.1, (17.169)-(17.171), Remark 18.5, and Remark 18.4, a well-posedness result as the one described in Theorem 19.4 also holds for the Neumann Problem with boundary data in Geometric Herz spaces:

$$\begin{cases} u \in [\mathscr{C}^\infty(\Omega)]^M, \\ Lu = 0 \ \text{ in } \Omega, \\ \mathcal{N}_\kappa(\nabla u) \in \dot{\mathrm{K}}^\alpha_{p,q}(\partial\Omega, \sigma; E), \\ \partial^A_\nu u = f \in [\dot{\mathrm{K}}^\alpha_{p,q}(\partial\Omega, \sigma; E)]^M \end{cases} \tag{19.48}$$

with $E \subseteq \partial\Omega$ a d-AR set for some $d \in [0, n-1)$, and indices $p \in (1,\infty)$, $q \in (0,\infty]$, and $\alpha \in \mathbb{R}$ satisfying $\frac{-(n-1-d)}{p} < \alpha < \frac{(n-1-d)(p-1)}{p}$.

19.4 The Transmission Problem with Boundary Data in $\mathrm{K}^\alpha_{p,q}$

Finally, we formulate and solve the Transmission Problem for weakly elliptic systems in δ-AR domains and boundary data in second-generation Herz spaces.

Theorem 19.5 *Select $n \in \mathbb{N}$ with $n \geq 2$ and suppose that $\Omega \subseteq \mathbb{R}^n$ is a two-sided* NTA *domain such that $\partial\Omega$ is an unbounded Ahlfors regular set. Abbreviate $\sigma := \mathcal{H}^{n-1}\lfloor\partial\Omega$, denote by ν the geometric measure theoretic outward unit normal to Ω, and set*

$$\Omega_+ := \Omega \ \text{ and } \ \Omega_- := \mathbb{R}^n \setminus \Omega. \tag{19.49}$$

Next, pick a second-order, homogeneous, constant complex coefficient, weakly elliptic, $M \times M$ system L in $\mathbb{R}^n$. Also, fix two exponents $p, q \in (1,\infty)$, along with an index $\alpha \in \mathbb{R}$ satisfying

$$-\frac{n-1}{p} < \alpha < \frac{n-1}{p'}, \tag{19.50}$$

where $p' \in (1,\infty)$ denote the Hölder conjugate exponent of p. Having selected some coefficient tensor $A \in \mathfrak{A}_L$, an aperture parameter $\kappa \in (0,\infty)$, and a transmission parameter $\eta \in \mathbb{C}$, consider the following Transmission Problem:

$$\begin{cases} u^\pm \in [\mathscr{C}^\infty(\Omega)]^M, \ \ Lu^\pm = 0 \ \text{ in } \Omega_\pm, \\ \mathcal{N}_\kappa(\nabla u^\pm) \in \dot{\mathrm{K}}^\alpha_{p,q}(\partial\Omega, \sigma), \\ u^+\big|^{\kappa-\mathrm{n.t.}}_{\partial\Omega} - u^-\big|^{\kappa-\mathrm{n.t.}}_{\partial\Omega} = f \in [\dot{\mathrm{K}}^{\alpha,(1)}_{p,q}(\partial\Omega, \sigma)]^M, \\ \partial^A_\nu u^+ - \eta \cdot \partial^A_\nu u^- = g \in [\dot{\mathrm{K}}^\alpha_{p,q}(\partial\Omega, \sigma)]^M. \end{cases} \tag{19.51}$$

Lastly, suppose that either

$$A^{\top} \in \mathfrak{A}^{\mathrm{dis}}_{L^{\top}} \text{ and } \eta \in \mathbb{C} \setminus \{-1\} \tag{19.52}$$

or

$$A \in \mathfrak{A}^{\mathrm{dis}}_{L} \text{ and } \eta \in \mathbb{C} \setminus \{0, -1\}. \tag{19.53}$$

Then for each $m \in \mathbb{N}$ there exists some threshold $\delta \in (0,1)$, depending solely on n, M, A, p, q, α, the two-sided NTA *constants of Ω, and the Ahlfors regularity constants of $\partial\Omega$ with the property that whenever*

$$\left|\tfrac{1-\eta}{1+\eta}\right| \cdot \|\nu\|^{\langle m\rangle}_{[\mathrm{BMO}(\partial\Omega,\sigma)]^n} < \delta \tag{19.54}$$

it follows that the Transmission Problem (16.64) *is well posed. Specifically, there exists a pair of functions u^+ and u^- solving the Transmission Problem* (16.64)*, which are unique modulo additive constants from $\mathbb{C}^M$, and satisfy*

$$\|\mathcal{N}_\kappa(\nabla u^{\pm})\|_{\dot{\mathrm{K}}^{\alpha}_{p,q}(\partial\Omega,\sigma)} \le C\big(\|f\|_{[\dot{\mathrm{K}}^{\alpha,(1)}_{p,q}(\partial\Omega,\sigma)]^M} + \|g\|_{[\dot{\mathrm{K}}^{\alpha}_{p,q}(\partial\Omega,\sigma)]^M}\big) \tag{19.55}$$

for some constant $C \in (0,\infty)$ independent of the boundary data f and g.

Finally, similar conclusions regarding well-posedness of the Transmission Problem hold in the context of inhomogeneous Herz spaces $\mathrm{K}^{\alpha}_{p,q}(\partial\Omega,\sigma)$.

Proof Thanks to Theorem 17.4, Proposition 17.6, and the fact that $\dot{\mathrm{K}}^{\alpha}_{p,q}(\partial\Omega,\sigma)$ is a Generalized Banach Function Space (cf. (17.123)), the well posedness of (19.51) follows from [GHA-V, Theorem 8.4.44, p. 877]. □

Remark 19.5 For similar reasons as in the past (see Remark 19.3), a well-posedness result of the sort described Theorem 19.5 may also be obtained for the Transmission Problem with boundary data taken from Geometric Herz spaces:

$$\begin{cases} u^{\pm} \in [\mathscr{C}^\infty(\Omega)]^M, \;\; Lu^{\pm} = 0 \text{ in } \Omega_{\pm}, \\ \mathcal{N}_\kappa(\nabla u^{\pm}) \in \dot{\mathrm{K}}^{\alpha}_{p,q}(\partial\Omega,\sigma;E), \\ u^{+}\big|^{\kappa-\mathrm{n.t.}}_{\partial\Omega} - u^{-}\big|^{\kappa-\mathrm{n.t.}}_{\partial\Omega} = f \in [\dot{\mathrm{K}}^{\alpha,(1)}_{p,q}(\partial\Omega,\sigma;E)]^M, \\ \partial^A_\nu u^+ - \eta\cdot\partial^A_\nu u^- = g \in [\dot{\mathrm{K}}^{\alpha}_{p,q}(\partial\Omega,\sigma;E)]^M, \end{cases} \tag{19.56}$$

with $E \subseteq \partial\Omega$ being a d-Ahlfors regular set, $d \in [0, n-1)$, and indices p, q, α satisfying

$$1 < p < \infty, \quad 1 < q < \infty, \quad \text{and} \quad \tfrac{-(n-1-d)}{p} < \alpha < \tfrac{(n-1-d)(p-1)}{p}. \tag{19.57}$$

Chapter 20

Inhomogeneous Herz-type Hardy Spaces of Second Generation

In this chapter we introduce and study the scale of second-generation inhomogeneous Herz-type Hardy spaces in the setting of unbounded Ahlfors regular sets. We develop a functional analytic framework for this brand of Hardy spaces, that includes atomic and molecular theory. We start by making the following definition. The reader is advised to recall the Fefferman-Stein grand maximal function introduced in Definition 4.1.

Definition 20.1 Consider a closed unbounded Ahlfors regular set $\Sigma \subseteq \mathbb{R}^n$, and abbreviate $\sigma := \mathcal{H}^{n-1}\lfloor\Sigma$. Having selected numbers p, q, α, γ satisfying

$$1 < p < \infty, \quad 0 < q \leq \infty, \quad \alpha \in \mathbb{R}, \quad \text{and} \quad 0 < \gamma < 1, \tag{20.1}$$

define the `inhomogeneous Herz-type Hardy space` on Σ as

$$\mathrm{HK}^{\alpha}_{p,q}(\Sigma,\sigma) := \big\{ f \in \big(\mathrm{Lip}_c(\Sigma)\big)' : f^{\#}_{\gamma} \in \mathrm{K}^{\alpha}_{p,q}(\Sigma,\sigma) \big\}, \tag{20.2}$$

and equip this space with the quasinorm

$$\|f\|_{\mathrm{HK}^{\alpha}_{p,q}(\Sigma,\sigma)} := \|f^{\#}_{\gamma}\|_{\mathrm{K}^{\alpha}_{p,q}(\Sigma,\sigma)} \quad \text{for each} \quad f \in \mathrm{HK}^{\alpha}_{p,q}(\Sigma,\sigma). \tag{20.3}$$

Notice that in the case when $0 < q \leq 1$, the quasinorm (20.3) is a q-norm, i.e., for all functions $f, g \in \mathrm{HK}^{\alpha}_{p,q}(\Sigma,\sigma)$ one has

$$\|f+g\|^{q}_{\mathrm{HK}^{\alpha}_{p,q}(\Sigma,\sigma)} \leq \|f\|^{q}_{\mathrm{HK}^{\alpha}_{p,q}(\Sigma,\sigma)} + \|g\|^{q}_{\mathrm{HK}^{\alpha}_{p,q}(\Sigma,\sigma)}. \tag{20.4}$$

A comment about the nature of the elements in $\mathrm{HK}^{\alpha}_{p,q}(\Sigma,\sigma)$ is in order. Upon recalling properties *(a)* and *(c)* from Proposition 4.1, and in view of the fact that the Herz space $\mathrm{K}^{\alpha}_{p,q}(\Sigma,\sigma)$ consists of functions from $L^{p}_{\mathrm{loc}}(\Sigma,\sigma) \subseteq L^{1}_{\mathrm{loc}}(\Sigma,\sigma)$ (cf. (17.143)-(17.144)), the elements of inhomogeneous Herz-type Hardy spaces are genuine functions:

© The Author(s), under exclusive license to Springer Nature Switzerland AG 2026
M. Mitrea and P. Takemura, *Singular Integrals, Herz-Type Function Spaces, and Boundary Problems*, Progress in Mathematics 362,
https://doi.org/10.1007/978-3-032-12516-3_20

$$\begin{aligned}\mathrm{HK}^{\alpha}_{p,q}(\Sigma,\sigma) &= \big\{f \in L^1_{\mathrm{loc}}(\Sigma,\sigma) : f^{\#}_{\gamma} \in \mathrm{K}^{\alpha}_{p,q}(\Sigma,\sigma)\big\} \\ &= \big\{f \in L^p_{\mathrm{loc}}(\Sigma,\sigma) : f^{\#}_{\gamma} \in \mathrm{K}^{\alpha}_{p,q}(\Sigma,\sigma)\big\},\end{aligned} \tag{20.5}$$

whenever p, q, α, γ are as in (20.1). With this in hand, we may invoke item *(a)* in Proposition 4.1 and the lattice property of $\mathrm{K}^{\alpha}_{p,q}(\Sigma,\sigma)$ (cf. (17.5)) to conclude that

$$\begin{gathered}\mathrm{HK}^{\alpha}_{p,q}(\Sigma,\sigma) \hookrightarrow \mathrm{K}^{\alpha}_{p,q}(\Sigma,\sigma) \text{ continuously,} \\ \text{if } \ 1 < p < \infty, \quad 0 < q \leq \infty, \quad \alpha \in \mathbb{R}.\end{gathered} \tag{20.6}$$

Recall from item *(d)* in Proposition 17.8 that

$$\begin{gathered}\mathrm{K}^{\alpha}_{p,q}(\Sigma,\sigma) \hookrightarrow L^p(\Sigma,\sigma) \text{ continuously} \\ \text{if } \ 1 < p < \infty, \quad 0 < q \leq \infty, \quad 0 < \alpha < \infty.\end{gathered} \tag{20.7}$$

As such, item *(c)* in Proposition 4.1 allows us to describe the inhomogeneous Herz-type Hardy space as

$$\begin{gathered}\mathrm{HK}^{\alpha}_{p,q}(\Sigma,\sigma) = \big\{f \in L^p(\Sigma,\sigma) : f^{\#}_{\gamma} \in \mathrm{K}^{\alpha}_{p,q}(\Sigma,\sigma)\big\}, \\ \text{if } \ 1 < p < \infty, \quad 0 < q \leq \infty, \quad 0 < \alpha < \infty, \quad \text{and} \quad 0 < \gamma < 1.\end{gathered} \tag{20.8}$$

From (20.6) and (20.7) we also see that

$$\begin{gathered}\mathrm{HK}^{\alpha}_{p,q}(\Sigma,\sigma) \hookrightarrow L^p(\Sigma,\sigma) \text{ continuously} \\ \text{if } \ 1 < p < \infty, \quad 0 < q \leq \infty, \quad 0 < \alpha < \infty.\end{gathered} \tag{20.9}$$

In particular, since $L^p(\Sigma,\sigma) \hookrightarrow L^1\big(\Sigma, \frac{\sigma(x)}{1+|x|^{n-1}}\big)$ continuously if $p \in (1,\infty)$, we also have

$$\begin{gathered}\mathrm{HK}^{\alpha}_{p,q}(\Sigma,\sigma) \hookrightarrow L^1\big(\Sigma, \tfrac{\sigma(x)}{1+|x|^{n-1}}\big) \text{ continuously,} \\ \text{whenever } \ 1 < p < \infty, \quad 0 < q \leq \infty, \quad \text{and} \quad 0 < \alpha < \infty.\end{gathered} \tag{20.10}$$

Recall from (17.149) that

$$\begin{gathered}\mathrm{K}^{\alpha}_{p,q}(\Sigma,\sigma) \hookrightarrow L^r(\Sigma,\sigma) \text{ continuously} \\ \text{if } \ 1 < p < \infty, \ \ 0 < q \leq 1, \ \ \tfrac{n-1}{p'} \leq \alpha < \infty, \ \ 1 \leq r \leq p\end{gathered} \tag{20.11}$$

where $p' \in (1,\infty)$ is the Hölder conjugate exponent of p. In concert with items *(a)* and *(c)* in Proposition 4.1, this yields the following quantitative description

$$\begin{gathered}\mathrm{HK}^{\alpha}_{p,q}(\Sigma,\sigma) = \big\{f \in L^r(\Sigma,\sigma) : f^{\#}_{\gamma} \in \mathrm{K}^{\alpha}_{p,q}(\Sigma,\sigma)\big\} \ \text{ if } 1 \leq r \leq p, \\ \text{assuming } \ 1 < p < \infty, \ \ 0 < q \leq 1, \ \ \text{and } \ \tfrac{n-1}{p'} \leq \alpha < \infty.\end{gathered} \tag{20.12}$$

In particular, this further entails

$$\begin{aligned}&\mathrm{HK}^{\alpha}_{p,q}(\Sigma,\sigma)\hookrightarrow L^r(\Sigma,\sigma)\ \text{ continuously},\\ &\text{whenever }\ 1<p<\infty,\ \ 0<q\leq 1,\ \ \tfrac{n-1}{p'}\leq\alpha<\infty,\ \ 1\leq r\leq p.\end{aligned}\tag{20.13}$$

Having discussed some embeddings for Herz-type Hardy spaces, we proceed by showing that $\mathrm{HK}^{\alpha}_{p,q}(\Sigma,\sigma)$ actually coincides with $\mathrm{K}^{\alpha}_{p,q}(\Sigma,\sigma)$ for a certain range of α's. This is made precise in the following proposition.

Proposition 20.1 *Let $\Sigma\subseteq\mathbb{R}^n$ be a closed unbounded Ahlfors regular set, and abbreviate $\sigma:=\mathcal{H}^{n-1}\lfloor\Sigma$. Then, with equivalent quasinorms,*

$$\begin{aligned}&\mathrm{HK}^{\alpha}_{p,q}(\Sigma,\sigma)=\mathrm{K}^{\alpha}_{p,q}(\Sigma,\sigma)\ \ \textit{whenever}\\ &1<p<\infty,\ \ 0<q\leq\infty\ ,-\tfrac{n-1}{p}<\alpha<\tfrac{n-1}{p'},\end{aligned}\tag{20.14}$$

where $p'\in(1,\infty)$ is the Hölder conjugate exponent of p.

Proof The proof follows by the same type of argument used to establish Proposition 11.2, keeping in mind (20.5), item *(c)* in Proposition 4.1, and the lattice property of $\mathrm{K}^{\alpha}_{p,q}(\Sigma,\sigma)$ (cf. (17.5)), along with the choice of α in (20.14), the boundedness of the Hardy-Littlewood maximal operator on $\mathrm{K}^{\alpha}_{p,q}(\Sigma,\sigma)$ (cf. Theorem (17.4)), and (4.7). □

Remark 20.1 The above result should be regarded as a generalization of Proposition 11.2. Indeed, if $p\in(1,\infty)$, $q\in(0,\infty)$, then choosing $\alpha:=(n-1)(1/q-1/p)$ leads to $\mathrm{HK}^{(n-1)(1/q-1/p)}_{p,q}(\Sigma,\sigma)=\mathrm{HA}_{p,q}(\Sigma,\sigma)$. Also, observe that condition on α in (20.14) forces $1<q<\infty$, which agrees with the range given in Proposition 11.2.

The completeness of Herz-type Hardy spaces $\mathrm{HK}^{\alpha}_{p,q}$ makes the object of our next proposition.

Proposition 20.2 *Assume Σ is a closed unbounded Ahlfors regular set, and abbreviate $\sigma:=\mathcal{H}^{n-1}\lfloor\Sigma$. Then whenever $p\in(1,\infty)$, $q\in(0,\infty]$, and $\alpha\in\mathbb{R}$, the space $\mathrm{HK}^{\alpha}_{p,q}(\Sigma,\sigma)$ is complete when equipped with the quasinorm $\|\cdot\|_{\mathrm{HK}^{\alpha}_{p,q}(\Sigma,\sigma)}$.*

Proof The proof of the current result follows from a similar reasoning as the one used to establish Proposition 4.3, taking into account the continuous embeddings (20.6) and (20.5), as well as Lemma 4.1. □

20.1 Atomic Theory in Inhomogeneous Herz-type Hardy Spaces

This section contains an atomic characterization of inhomogeneous Herz-type Hardy spaces $\mathrm{HK}^{\alpha}_{p,q}$ in the context of unbounded Ahlfors regular sets. This plays a crucial role in subsequent developments. Stating said result requires the following definition of central (α,p)-atoms.

Definition 20.2 Assume $\Sigma \subseteq \mathbb{R}^n$ is a closed unbounded Ahlfors regular set, and abbreviate $\sigma := \mathcal{H}^{n-1}\lfloor\Sigma$. Denote by $c_\Sigma, C_\Sigma \in (0,\infty)$ the lower and upper Ahlfors regularity constants of Σ, respectively. Fix a dilation factor $2_* > (C_\Sigma/c_\Sigma)^{1/(n-1)}$ along with a reference point $x_0 \in \Sigma$. Pick an integrability exponent $p \in (1,\infty)$ and an index $\alpha \in \mathbb{R}$. A σ-measurable function $a : \Sigma \to \mathbb{R}$ is called an x_0-`central` (α, p)-`atom` on Σ if there exists a finite radius $R \geq 2_*$ for which the following three properties hold:

(1) Localization Condition:

$$\operatorname{supp} a \subseteq \Delta(x_0, R). \tag{20.15}$$

(2) Normalization Condition:

$$\|a\|_{L^p(\Sigma,\sigma)} \leq R^{-\alpha}. \tag{20.16}$$

(3) Vanishing Moment Condition:

$$\int_\Sigma a\, d\sigma = 0. \tag{20.17}$$

Remark 20.2 Observe that conditions (20.15)-(20.16) guarantees the function a is absolutely integrable on Σ. As such, the vanishing moment condition (20.17) is meaningfully formulated.

Remark 20.3 It is worth to highlight that central atoms are independent of the reference point x_0 and the dilation factor 2_* in the following sense. Consider two reference points $x_0, x_1 \in \Sigma$ and two dilation factors $2_*, 2_\diamond > (C_\Sigma/c_\Sigma)^{1/(n-1)}$. We claim that there exists a constant $c \in (0,\infty)$ depending only on x_0, x_1, 2_*, $2_\diamond$, α, such that whenever a is an x_0-central (α, p)-atom on Σ associated with a radius $R \geq 2_*$ as in Definition 20.2, then $\widetilde{a} := ca$ on Σ satisfies

$$\operatorname{supp} \widetilde{a} \subseteq \Delta(x_1, \widetilde{R}), \quad \|\widetilde{a}\|_{L^p(\Sigma,\sigma)} \leq \widetilde{R}^{-\alpha}, \quad \text{and} \quad \int_\Sigma \widetilde{a}\, d\sigma = 0, \tag{20.18}$$

for some $\widetilde{R} \geq 2_\diamond$. In other words, $\widetilde{a}$ is an x_1-central (α, p)-atom (in the sense of Definition 20.2 written with x_1 in place of x_0, and $2_\diamond$ in place of 2_*). To verify this claim, fix a an x_0-central (α, p)-atom on Σ, and let $R \geq 2_*$ be a radius associated with a as in Definition 20.2. Fix a scalar $\lambda \in (0,\infty)$ such that

$$\lambda \geq 2 \cdot \max\left\{1, \frac{2_\diamond}{2_*}, \frac{|x_0 - x_1|}{2_*}\right\}, \tag{20.19}$$

and set $\widetilde{R} = \lambda R$. It is visible that $\widetilde{R} \geq \lambda 2_* \geq 2_\diamond$. Also, $\Delta(x_0, R) \subseteq \Delta(x_1, \widetilde{R})$, since whenever $z \in \Sigma$ and $|z - x_0| < R$ we have

$$|z - x_1| \leq |z - x_0| + |x_0 - x_1| < R + \frac{\lambda}{2} 2_* \leq \frac{\lambda}{2} R + \frac{\lambda}{2} R = \widetilde{R}. \tag{20.20}$$

Thanks to these and (20.15)-(20.17), if we set $c := \lambda^{-\alpha} \in (0,\infty)$ and $\widetilde{a} := ca$ on Σ then

$$\begin{aligned}&\operatorname{supp}\widetilde{a} \subseteq \Delta(x_0,R) \subseteq \Delta(x_1,\widetilde{R}),\\ &\|\widetilde{a}\|_{L^p(\Sigma,\sigma)} = \lambda^{-\alpha}\|a\|_{L^p(\Sigma,\sigma)} \leq (\lambda R)^{-\alpha} = \widetilde{R}^{-\alpha},\\ &\text{and} \quad \int_\Sigma \widetilde{a}\,d\sigma = \lambda^{-\alpha}\int_\Sigma a\,d\sigma = 0,\end{aligned} \tag{20.21}$$

thus establishing (20.18).

Before presenting the atomic characterization of inhomogeneous Herz-type Hardy spaces, we first show that every x_0-central (α,p)-atom belongs $\mathrm{HK}^{\alpha}_{p,q}$.

Proposition 20.3 *Assume $\Sigma \subseteq \mathbb{R}^n$ is a closed unbounded Ahlfors regular set, and abbreviate $\sigma := \mathcal{H}^{n-1}\lfloor\Sigma$. Fix a reference point $x_0 \in \Sigma$. Select two exponents $p \in (1,\infty)$, and $q \in (0,\infty]$, along with indices $\alpha,\gamma \in \mathbb{R}$ satisfying*

$$\tfrac{n-1}{p'} \leq \alpha < \tfrac{n-1}{p'} + 1 \quad \textit{and} \quad \alpha - \tfrac{n-1}{p'} < \gamma < 1 \tag{20.22}$$

where $p' \in (1,\infty)$ denote the Hölder conjugate exponent of p. Then there exists a constant $C \in (0,\infty)$ depending exclusively on p, q, α, γ, and the Ahlfors regularity constants of Σ with the property that for every x_0-central (α,p)-atom a on Σ one has

$$\|a\|_{\mathrm{HK}^{\alpha}_{p,q}(\Sigma,\sigma)} = \|a^{\#}_{\gamma}\|_{\mathrm{K}^{\alpha}_{p,q}(\Sigma,\sigma)} \leq C. \tag{20.23}$$

Proof The proof mirrors that of Theorem 4.1. For this reason, here we only emphasize the main differences. Fix a dilation factor 2_* as in Definition 20.2. Pick $a : \Sigma \to \mathbb{R}$ an x_0-central (α,p)-atom, and let $R \in [2_*,\infty)$ be a radius associated with a (in the sense of Definition 20.2). Consider $\ell \in \mathbb{N}$ the unique integer satisfying

$$2_*^{\ell} \leq R < 2_*^{\ell+1}. \tag{20.24}$$

A combination of the computations made in (4.37)-(4.39) with normalization condition (20.16) yields

$$\begin{aligned}&a^{\#}_{\gamma}(x) \leq C 2_*^{(\ell-k)\gamma} 2_*^{\ell[(n-1)/p'-\alpha]} 2_*^{-k(n-1)} \quad \text{for all } x \in \widetilde{C}_k\\ &\text{and every } k \in \mathbb{N}_0 \text{ such that } k \geq \ell+2,\end{aligned} \tag{20.25}$$

for some constant $C = C(\Sigma,n,p,\gamma) \in (0,\infty)$, where $\{\widetilde{C}_k\}_{k\in\mathbb{N}_0}$ is the family of annuli as in (17.113). Thanks to this, and the Ahlfors regularity of Σ, whenever $q < \infty$ we may estimate

$$
\begin{aligned}
&\Big(\sum_{k=\ell+2}^{\infty}\big(2_*^{k\alpha}\|a_\gamma^{\#}\cdot\mathbb{1}_{\widetilde{C}_k}\|_{L^p(\Sigma,\sigma)}\big)^q\Big)^{1/q}\\
&\quad\le C\Big(\sum_{k=\ell+2}^{\infty}\big(2_*^{k\alpha}2_*^{(\ell-k)\gamma}2_*^{\ell[(n-1)/p'-\alpha]}2_*^{-k(n-1)}\sigma(\widetilde{C}_k)^{1/p}\big)^q\Big)^{1/q}\\
&\quad\le C\Big(\sum_{k=\ell+2}^{\infty}\big(2_*^{k\alpha}2_*^{(\ell-k)\gamma}2_*^{\ell[(n-1)/p'-\alpha]}2_*^{-k(n-1)(1-1/p)}\big)^q\Big)^{1/q}\\
&\quad= C\Big(\sum_{k=\ell+2}^{\infty}\big(2_*^{k\alpha}2_*^{(\ell-k)\gamma}2_*^{\ell[(n-1)/p'-\alpha]}2_*^{-k(n-1)/p'}\big)^q\Big)^{1/q}\\
&\quad= C\Big(\sum_{k=\ell+2}^{\infty}2_*^{(\ell-k)[(n-1)/p'-\alpha+\gamma]q}\Big)^{1/q}\\
&\quad\le C\Big(\sum_{j=0}^{\infty}2_*^{-j[(n-1)/p'-\alpha+\gamma]q}\Big)^{1/q}<\infty
\end{aligned}
\tag{20.26}
$$

for some constant $C = C(\Sigma, n, p, q, \alpha, \gamma) \in (0, \infty)$. The convergence of the series in (20.26) follows from the fact that $\gamma > \alpha - (n-1)/p'$ (see (20.22)). Corresponding to $q = \infty$, estimate (20.26) reads as

$$
\sup_{k\ge\ell+2}\big(2_*^{k\alpha}\|a_\gamma^{\#}\cdot\mathbb{1}_{\widetilde{C}_k}\|_{L^p(\Sigma,\sigma)}\big)\le C\cdot\sup_{j\in\mathbb{N}_0}2_*^{-j[(n-1)/p'-\alpha+\gamma]}<\infty. \tag{20.27}
$$

Next, thanks to (20.24), the absolute integrability of a (cf. Remark 20.2), the normalization condition (20.16) of a, and boundedness of the Hardy-Littlewood maximal operator $\mathcal{M}_\Sigma$ on $L^p(\Sigma, \sigma)$, and (4.7), whenever $q < \infty$ we have

$$
\begin{aligned}
&\Big(\sum_{k=0}^{\ell+1}\big(2_*^{k\alpha}\|a_\gamma^{\#}\cdot\mathbb{1}_{\widetilde{C}_k}\|_{L^p(\Sigma,\sigma)}\big)^q\Big)^{1/q}\le\|\mathcal{M}_\Sigma a\|_{L^p(\Sigma,\sigma)}\Big(\sum_{k=0}^{\ell+1}2_*^{k\alpha q}\Big)^{1/q}\\
&\quad\le C\|a\|_{L^p(\Sigma,\sigma)}\Big(\sum_{k=0}^{\ell+1}2_*^{k\alpha q}\Big)^{1/q}\le CR^{-\alpha}\Big(\sum_{k=0}^{\ell+1}2_*^{k\alpha q}\Big)^{1/q}\\
&\quad\le C\Big(\sum_{k=0}^{\ell+1}2_*^{(k-\ell)\alpha q}\Big)^{1/q}\le C\Big(\sum_{j=0}^{\infty}2_*^{-j\alpha q}\Big)^{1/q}<\infty,
\end{aligned}
\tag{20.28}
$$

for some constant $C = C(\Sigma, n, p, q, \alpha) \in (0, \infty)$. It is straightforward to adapt this to the case when $q = \infty$ to obtain

$$
\sup_{0\le k\le\ell+1}\big(2_*^{k\alpha}\|a_\gamma^{\#}\cdot\mathbb{1}_{\widetilde{C}_k}\|_{L^p(\Sigma,\sigma)}\big)\le C. \tag{20.29}
$$

Having established these, (20.23) follows immediately from (20.3), (17.115), (20.26), (20.27), (20.28), and (20.29). This finishes the proof of Proposition 20.3. □

We are now well prepared to prove an atomic characterization for inhomogeneous Herz-type Hardy spaces $\mathrm{HK}^{\alpha}_{p,q}$ in the setting of unbounded Ahlfors regular sets. As alluded to before, this plays a central role in the theory developed here.

Theorem 20.1 *Assume $\Sigma \subseteq \mathbb{R}^n$ is a closed unbounded Ahlfors regular set, and abbreviate $\sigma := \mathcal{H}^{n-1}\lfloor\Sigma$. Fix a reference point $x_0 \in \Sigma$. Select exponents $p \in (1, \infty)$ and $q \in (0, \infty]$, along with an index $\alpha \in \mathbb{R}$ such that*

$$\frac{n-1}{p'} \leq \alpha < \frac{n-1}{p'} + 1, \tag{20.30}$$

where $p' \in (1, \infty)$ denotes the Hölder conjugate exponent of p. Then the following statements are valid.

(a) For every $f \in \mathrm{HK}^{\alpha}_{p,q}(\Sigma, \sigma)$ there exist a sequence $\{a_k\}_{k\in\mathbb{N}_0}$ of x_0-central (α, p)-atoms and a numerical sequence $\{\lambda_k\}_{k\in\mathbb{N}_0} \subseteq \mathbb{R}$ such that

$$f = \sum_{k=0}^{\infty} \lambda_k a_k \tag{20.31}$$

with convergence in $L^p(\Sigma, \sigma)$. In the particular case when $0 < q \leq 1$, this series converges to f simultaneously in $\mathrm{HK}^{\alpha}_{p,q}(\Sigma, \sigma)$ and in $L^r(\Sigma, \sigma)$ whenever $1 \leq r \leq p$. Moreover, there exists a constant $C \in (0, \infty)$ depending only n, p, q, α, and the Ahlfors regularity character of Σ with the property that

$$\Big(\sum_{k=0}^{\infty} |\lambda_k|^q\Big)^{1/q} \leq C\|f\|_{\mathrm{HK}^{\alpha}_{p,q}(\Sigma,\sigma)}. \tag{20.32}$$

In the case when $q = \infty$, this estimate is understood in the ℓ^∞-norm. Lastly, matters can be arranged so that each central atom a_k is associated with $\Delta(x_0, 2^{k+1}_*)$ in the sense of Definition 20.2, i.e., a_k satisfies localization and normalization conditions relative to $\Delta(x_0, 2^{k+1}_*)$ for each $k \in \mathbb{N}_0$.

(b) Conversely, assume $f : \Sigma \to \mathbb{R}$ is a σ-measurable function with the property that there exists a sequence of x_0-central (α, p)-atoms $\{a_k\}_{k\in\mathbb{N}_0}$ and a numerical sequence $\{\lambda_k\}_{k\in\mathbb{N}_0} \in \ell^q(\mathbb{N}_0)$ satisfying

$$f = \sum_{k=0}^{\infty} \lambda_k a_k \ \textit{ pointwise } \sigma\textit{-a.e. on } \Sigma. \tag{20.33}$$

If $0 < q \leq 1$ then it follows that $f \in \mathrm{HK}^{\alpha}_{p,q}(\Sigma, \sigma)$ in a quantitative fashion, i.e., and there exists some constant $C \in (0, \infty)$ depending only n, p, q, α, and the Ahlfors regularity character of Σ with the property that

$$\|f\|_{\mathrm{HK}^{\alpha}_{p,q}(\Sigma,\sigma)} \leq C\Big(\sum_{k=0}^{\infty}|\lambda_k|^q\Big)^{1/q}. \tag{20.34}$$

If $1 < q \leq \infty$ then under the extra assumption that the series in (20.33) converges to f in $L^1_{\mathrm{loc}}(\Sigma, \sigma)$ and that a_k satisfies localization and normalization conditions (as in Definition 20.2) relative to the surface ball $\Delta(x_0, 2^{k+1}_*)$ for each $k \in \mathbb{N}_0$, it follows that $f \in \mathrm{HK}^{\alpha}_{p,q}(\Sigma, \sigma)$ quantitatively as in (20.34). When $q = \infty$, estimate (20.34) is naturally understood in the ℓ^∞-norm.

(c) For every $f \in \mathrm{HK}^{\alpha}_{p,q}(\Sigma, \sigma)$ one has

$$\|f\|_{\mathrm{HK}^{\alpha}_{p,q}(\Sigma,\sigma)} \approx \inf\Big\{\Big(\sum_{k=0}^{\infty}|\lambda_k|^q\Big)^{1/q} : f = \sum_{k=0}^{\infty}\lambda_k a_k \text{ in } L^p(\Sigma,\sigma)\Big\} \tag{20.35}$$

with implicit proportionality constants depending only on Σ, n, p, q. In the case when $1 < q \leq \infty$, this infimum is taken over all atomic representations of f as in (20.31) with a_k satisfying localization and normalization conditions (as in Definition 20.2) relative to $\Delta(x_0, 2^{k+1}_*)$ for each $k \in \mathbb{N}_0$. Corresponding to the case $q = \infty$, the norm of the numerical sequence in (20.35) is taken to be ℓ^∞.

Proof Throughout, fix a parameter $\gamma \in \mathbb{R}$ satisfying

$$\alpha - \tfrac{n-1}{p'} < \gamma < 1. \tag{20.36}$$

To deal with the claim made in item *(a)*, we rely on the same type of argument used to establish item *(a)* from Theorem 4.2. Recall the construction of the smooth partition of unit $\{\phi_k\}_{k\in\mathbb{N}_0}$ from (4.50)-(4.78) given by

$$\phi_k = \frac{\psi_k}{\sum_{k=0}^{\infty}\psi_k} \quad \text{in } \mathbb{R}^n, \text{ for each } k \in \mathbb{N}_0, \tag{20.37}$$

where for each $k \in \mathbb{N}_0$ the function ψ_k is as in (4.53). In particular, in (4.79)-(4.86) we have obtained the following properties

$$\begin{gathered} 0 \leq \phi_k \leq 1 \;\text{ in } \mathbb{R}^n, \;\; \mathrm{supp}\,\big(\phi_k\big|_{\Sigma}\big) \subseteq \widetilde{A}_k \;\text{ for all } k \in \mathbb{N}_0, \\ \text{and } \sum_{k=0}^{\infty}\phi_k = 1 \;\text{ at every point in } \mathbb{R}^n, \end{gathered} \tag{20.38}$$

Fix an arbitrary function $f \in \mathrm{HK}^{\alpha}_{p,q}(\Sigma, \sigma)$. Thanks to the third property in (20.38), we have

$$f = \sum_{k=0}^{\infty} f \cdot \big(\phi_k\big|_{\Sigma}\big) \;\text{ pointwise } \sigma\text{-a.e. on } \Sigma. \tag{20.39}$$

In view of this, the fact that $f \in L^p(\Sigma, \sigma)$ (cf. (20.8)), and

$$\Big|\sum_{k=0}^{N} f\cdot\big(\phi_k\big|_{\Sigma}\big)\Big| \leq |f| \quad \text{at } \sigma\text{-a.e. point on } \Sigma \text{ and for all } N\in\mathbb{N}_0 \tag{20.40}$$

(as implied by the third property in (20.38)), Lebesgue's Dominated Convergence Theorem gives

$$f=\sum_{k=0}^{\infty} f\big(\phi_k\big|_{\Sigma}\big) \quad \text{in } L^p(\Sigma,\sigma). \tag{20.41}$$

For each $k\in\mathbb{N}_0$, recall the functions g_k and h_k defined in (4.92) and (4.114), respectively. Then the same type of reasoning as in (4.91)-(4.93) and (4.105)-(4.115) in our current context produces the following writing of f:

$$f=\sum_{k=0}^{\infty} f\big(\phi_k\big|_{\Sigma}\big)=\sum_{k=0}^{\infty} g_k+\sum_{k=0}^{\infty} h_k \quad \text{pointwise } \sigma\text{-a.e. on } \Sigma. \tag{20.42}$$

In particular, thanks to (20.41), this convergence holds in $L^p(\Sigma,\sigma)$. Based on (4.93)-(4.95) and (4.116)-(4.135), and recalling $\{\widetilde{C}_k\}_{k\in\mathbb{N}_0}$ the family of annuli defined in (17.113), we have

$$\begin{aligned}
&\operatorname{supp} g_k\subseteq \Delta(x_0,2_*^{k+2}) \quad \text{and} \quad \int_{\Sigma} g_k\,d\sigma=0 \quad \text{for each } k\in\mathbb{N}_0,\\
&\|g_0\|_{L^p(\Sigma,\sigma)}\leq 2\sum_{j=0}^{1}\|f\cdot\mathbb{1}_{\widetilde{C}_j}\|_{L^p(\Sigma,\sigma)},\\
&\|g_k\|_{L^p(\Sigma,\sigma)}\leq 2\sum_{j=k-1}^{k+1}\|f\cdot\mathbb{1}_{\widetilde{C}_j}\|_{L^p(\Sigma,\sigma)} \quad \text{for each } k\in\mathbb{N},
\end{aligned} \tag{20.43}$$

while

$$\begin{aligned}
&\operatorname{supp} h_k\subseteq \Delta(x_0,2_*^{k+3}) \quad \text{and} \quad \int_{\Sigma} h_k\,d\sigma=0 \quad \text{for each } k\in\mathbb{N}_0,\\
&\|h_0\|_{L^p(\Sigma,\sigma)}\leq \widetilde{C}\sum_{j=0}^{2}\|f^{\#}_{\gamma}\cdot\mathbb{1}_{\widetilde{C}_j}\|_{L^p(\Sigma,\sigma)},\\
&\|h_k\|_{L^p(\Sigma,\sigma)}\leq \widetilde{C}\sum_{j=k-1}^{k+2}\|f^{\#}_{\gamma}\cdot\mathbb{1}_{\widetilde{C}_j}\|_{L^p(\Sigma,\sigma)} \quad \text{for each } k\in\mathbb{N}.
\end{aligned} \tag{20.44}$$

Considering the set of indices $\mathcal{J}_1$ and $\mathcal{J}_2$ as in (4.96) and (4.136), for each $k\in\mathbb{N}_0$ introduce the functions $a_k,b_k:\Sigma\to\mathbb{R}$ by setting

$$a_0 := \begin{cases} \Big(2(2_*^2)^\alpha \sum_{\ell=0}^{1} \|f\cdot \mathbb{1}_{\widetilde{C}_\ell}\|_{L^p(\Sigma,\sigma)}\Big)^{-1} g_0 & \text{if } 0 \in \mathcal{J}_1,\\ 0 & \text{if } 0 \notin \mathcal{J}_1,\end{cases}$$

$$a_k := \begin{cases} \Big(2(2_*^{k+2})^\alpha \sum_{\ell=k-1}^{k+1} \|f\cdot \mathbb{1}_{\widetilde{C}_\ell}\|_{L^p(\Sigma,\sigma)}\Big)^{-1} g_k & \text{if } k \in \mathcal{J}_1,\\ 0 & \text{if } k \notin \mathcal{J}_1,\end{cases} \quad (\text{for } k \in \mathbb{N}) \tag{20.45}$$

and

$$b_0 := \begin{cases} \Big(\widetilde{C}(2_*^3)^\alpha \sum_{\ell=0}^{2} \|f_\gamma^\# \cdot \mathbb{1}_{\widetilde{C}_\ell}\|_{L^p(\Sigma,\sigma)}\Big)^{-1} h_0 & \text{if } 0 \in \mathcal{J}_2,\\ 0 & \text{if } 0 \notin \mathcal{J}_2,\end{cases}$$

$$b_k := \begin{cases} \Big(\widetilde{C}(2_*^{k+3})^\alpha \sum_{\ell=k-1}^{k+2} \|f_\gamma^\# \cdot \mathbb{1}_{\widetilde{C}_\ell}\|_{L^p(\Sigma,\sigma)}\Big)^{-1} h_k & \text{if } k \in \mathcal{J}_2,\\ 0 & \text{if } k \notin \mathcal{J}_2,\end{cases} \quad (\text{for } k \in \mathbb{N}). \tag{20.46}$$

Based on these and the properties recorded in (20.43)-(20.44), for each $k \in \mathbb{N}_0$ we have

$$\operatorname{supp} a_k \subseteq \Delta(x_0, 2_*^{k+2}), \quad \|a_k\|_{L^p(\Sigma,\sigma)} \leq (2_*^{k+2})^{-\alpha}, \quad \int_\Sigma a_k\, d\sigma = 0, \tag{20.47}$$

and

$$\operatorname{supp} b_k \subseteq \Delta(x_0, 2_*^{k+3}), \quad \|b_k\|_{L^p(\Sigma,\sigma)} \leq (2_*^{k+3})^{-\alpha}, \quad \int_\Sigma b_k\, d\sigma = 0. \tag{20.48}$$

As such, for each $k \in \mathbb{N}_0$, the functions a_k and b_k are x_0-central (α, p)-atoms on Σ.

Moving on, define the sequences $\{\lambda_k\}_{k\in\mathbb{N}_0}$ and $\{\gamma_k\}_{k\in\mathbb{N}_0}$ as

$$\lambda_0 := \begin{cases} 2(2_*^2)^\alpha \sum_{\ell=0}^{1} \|f\cdot \mathbb{1}_{\widetilde{C}_\ell}\|_{L^p(\Sigma,\sigma)} & \text{if } 0 \in \mathcal{J}_1,\\ 0 & \text{if } 0 \notin \mathcal{J}_1,\end{cases}$$

$$\lambda_k := \begin{cases} 2(2_*^{k+2})^\alpha \sum_{\ell=k-1}^{k+1} \|f\cdot \mathbb{1}_{\widetilde{C}_\ell}\|_{L^p(\Sigma,\sigma)} & \text{if } k \in \mathcal{J}_1,\\ 0 & \text{if } k \notin \mathcal{J}_1,\end{cases} \quad (\text{for } k \in \mathbb{N}) \tag{20.49}$$

and

$$\gamma_0 := \begin{cases} \widetilde{C}(2_*^3)^\alpha \sum_{\ell=0}^{2} \|f_\gamma^\# \cdot \mathbb{1}_{\widetilde{C}_\ell}\|_{L^p(\Sigma,\sigma)} & \text{if } 0 \in \mathcal{J}_2,\\ 0 & \text{if } 0 \notin \mathcal{J}_2,\end{cases}$$

$$\gamma_k := \begin{cases} \widetilde{C}(2_*^{k+3})^\alpha \sum_{\ell=k-1}^{k+2} \|f_\gamma^\# \cdot \mathbb{1}_{\widetilde{C}_\ell}\|_{L^p(\Sigma,\sigma)} & \text{if } k \in \mathcal{J}_2,\\ 0 & \text{if } k \notin \mathcal{J}_2,\end{cases} \quad (\text{for } k \in \mathbb{N}). \tag{20.50}$$

We claim that the sequence $\{\lambda_k\}_{k\in\mathbb{N}_0}$ belongs to $\ell^q(\mathbb{N}_0)$. Indeed, being mindful of (20.49) and assuming $q<\infty$, estimate

$$\begin{aligned}\sum_{k=0}^{\infty}|\lambda_k|^q &\leq C\sum_{\ell-0}^{1}\|f\cdot\mathbb{1}_{\widetilde{C}_\ell}\|^q_{L^p(\Sigma,\sigma)}+C\sum_{k=1}^{\infty}\sum_{\ell=k-1}^{k+1}(2_*^{k\alpha q}\|f\cdot\mathbb{1}_{\widetilde{C}_\ell}\|_{L^p(\Sigma,\sigma)})^q,\\ &\leq C\sum_{j=0}^{\infty}(2_*^{j\alpha}\|f\cdot\mathbb{1}_{\widetilde{C}_j}\|_{L^p(\Sigma,\sigma)})^q\\ &= C\|f\|^q_{\mathrm{K}^\alpha_{p,q}(\Sigma,\sigma)}\leq C\|f\|^q_{\mathrm{HK}^\alpha_{p,q}(\Sigma,\sigma)}\end{aligned} \tag{20.51}$$

for some constant $C=C(\Sigma,n,p,q,\alpha)\in(0,\infty)$, where the last inequality follows from the fact that $\mathrm{HK}^\alpha_{p,q}(\Sigma,\sigma)\hookrightarrow\mathrm{K}^\alpha_{p,q}(\Sigma,\sigma)$ continuously (see (20.6)). Similarly, corresponding to $q=\infty$ we have

$$\sup_{k\in\mathbb{N}_0}|\lambda_k|\leq C\|f\|_{\mathrm{HK}^\alpha_{p,\infty}(\Sigma,\sigma)}. \tag{20.52}$$

All together, this argument implies that $\{\lambda_k\}_{k\in\mathbb{N}_0}\in\ell^q(\mathbb{N}_0)$, as desired.

Next, we claim that the sequence $\{\gamma_k\}_{k\in\mathbb{N}_0}$ defined in (20.50) also belongs to $\ell^q(\mathbb{N}_0)$. To justify this, whenever $q<\infty$ compute

$$\begin{aligned}\sum_{k=0}^{\infty}|\gamma_k|^q &\leq C\sum_{\ell=0}^{2}\|f_\gamma^\#\cdot\mathbb{1}_{\widetilde{C}_\ell}\|^q_{L^p(\Sigma,\sigma)}\\ &\quad+C\sum_{k=1}^{\infty}\sum_{\ell=k-1}^{k+2}(2_*^{(k+3)\alpha}\|f_\gamma^\#\cdot\mathbb{1}_{\widetilde{C}_\ell}\|_{L^p(\Sigma,\sigma)})^q\\ &\leq C\sum_{j=0}^{\infty}(2_*^{j\alpha}\|f_\gamma^\#\cdot\mathbb{1}_{\widetilde{C}_j}\|_{L^p(\Sigma,\sigma)})^q=C\|f_\gamma^\#\|^q_{\mathrm{K}^\alpha_{p,q}(\Sigma,\sigma)}\\ &= C\|f\|^q_{\mathrm{HK}^\alpha_{p,q}(\Sigma,\sigma)},\end{aligned} \tag{20.53}$$

for some constant $C=C(\Sigma,n,p,q,\alpha)\in(0,\infty)$. Corresponding to $q=\infty$ similar computations yield

$$\sup_{k\in\mathbb{N}_0}|\gamma_k|\leq C\|f\|_{\mathrm{HK}^\alpha_{p,\infty}(\Sigma,\sigma)}. \tag{20.54}$$

Hence $\{\gamma_k\}_{k\in\mathbb{N}_0}\in\ell^q(\mathbb{N}_0)$, as wanted.

Bearing in mind (20.45), (20.46), (20.49), and (20.50) we may recast (20.42) as

$$f=\sum_{k=0}^{\infty}\lambda_k a_k+\sum_{k=0}^{\infty}\gamma_k b_k\quad\text{in } L^p(\Sigma,\sigma). \tag{20.55}$$

To express f as a single series requires some adjustments. Specifically, keeping in mind $\mathcal{J}_1$ and $\mathcal{J}_2$ (cf. (4.96) and (4.136)), for each $k \in \mathbb{N}_0$ consider

$$\mu_k := 2_*^{\alpha}|\lambda_k| + |\gamma_k| \in [0, \infty), \tag{20.56}$$

and define $c_k : \Sigma \to \mathbb{R}$ by setting

$$c_k := \begin{cases} \mu_k^{-1}(\lambda_k a_k + \gamma_k b_k) & \text{if } k \in \mathcal{J}_1 \cup \mathcal{J}_2, \\ 0 & \text{if } k \notin \mathcal{J}_1 \cup \mathcal{J}_2. \end{cases} \tag{20.57}$$

With this piece of notation, we can express f from (20.55) as

$$f = \sum_{k=0}^{\infty} \mu_k c_k \quad \text{in } L^p(\Sigma, \sigma). \tag{20.58}$$

It is visible from (20.51), (20.53), and (20.56) that $\{\mu_k\}_{k\in\mathbb{N}_0}$ belongs to $\ell^q(\mathbb{N}_0)$, and thanks to (20.51)-(20.54) it satisfies

$$\Big(\sum_{k=0}^{\infty} |\mu_k|^q\Big)^{1/q} \leq C\|f\|_{\mathrm{HK}^{\alpha}_{p,q}(\Sigma,\sigma)}, \tag{20.59}$$

with the understanding that the term in the left side is written as an ℓ^∞-norm in the case when $q = \infty$.

Moreover, recalling (20.47) and (20.48), for each $k \in \mathbb{N}_0$ the function c_k (cf. (20.57)) satisfies

$$\operatorname{supp} c_k \subseteq \Delta(x_0, 2_*^{k+3}), \qquad \int_\Sigma c_k \, d\sigma = 0, \tag{20.60}$$

and

$$\|c_k\|_{L^p(\Sigma,\sigma)} \leq |\mu_k|^{-1}\big[|\lambda_k|(2_*^{k+2})^{-\alpha} + |\gamma_k|(2_*^{k+3})^{-\alpha}\big] = (2_*^{k+3})^{-\alpha}. \tag{20.61}$$

As such, for each $k \in \mathbb{N}_0$ the function c_k is an x_0-central (α, p)-atom on Σ. Hence, (20.58) and (20.59) give the desired atomic decomposition for f as in (20.31).

Next we show that in the regime $0 < q \leq 1$ the series (20.58) also converges to f in $\mathrm{HK}^{\alpha}_{p,q}(\Sigma, \sigma)$ and in $L^r(\Sigma, \sigma)$ whenever $1 \leq r \leq p$. Indeed, in the case when $0 < q \leq 1$, recall from (20.13) that $\mathrm{HK}^{\alpha}_{p,q}(\Sigma, \sigma) \hookrightarrow L^r(\Sigma, \sigma)$ for each $r \in [1, p]$. In light of this, Lebesgue's Dominated Convergence Theorem entails that the series in (20.41) convergence in $L^r(\Sigma, \sigma)$ for all $r \in [1, p]$. Running the same reasoning as before (which, starting with the series in (20.41) and (20.42), now being convergent in $L^r(\Sigma, \sigma)$, shows that (20.55) now converges in $L^r(\Sigma, \sigma)$) yields that the series (20.58) converges in $L^r(\Sigma, \sigma)$ for every $r \in [1, p]$. Next, convergence in $\mathrm{HK}^{\alpha}_{p,q}(\Sigma, \sigma)$ follows by an argument analogous to the one presented in (11.56)-(11.57), bearing in mind (20.4), the completeness of $\mathrm{HK}^{\alpha}_{p,q}(\Sigma, \sigma)$ (cf.

Proposition 20.2), and that $\mathrm{HK}^{\alpha}_{p,q}(\Sigma,\sigma)$ embeds continuously into $L^1(\Sigma,\sigma)$ (see (20.13) currently used with $r := 1$).

To verify the very last claim made in item *(a)* we relabel the sequences $\{c_k\}_{k\in\mathbb{N}_0}$ and $\{\mu_k\}_{k\in\mathbb{N}_0}$ as

$$\begin{aligned} &\widetilde{c}_0 = \widetilde{c}_1 := 0, \quad \widetilde{c}_{k+2} := c_k \ \text{ for } k \geq 0, \text{ and} \\ &\widetilde{\mu}_0 = \widetilde{\mu}_1 := 0, \quad \widetilde{\mu}_{k+2} = \mu_k \ \text{ for } k \geq 0. \end{aligned} \tag{20.62}$$

With this in hand, (20.58), (20.60), and (20.61) ensure that $\{\widetilde{\mu}_k\}_{k\in\mathbb{N}_0} \in \ell^q(\mathbb{N}_0)$, and for every $k \in \mathbb{N}_0$ the function $\widetilde{c}_k$ is an x_0-central (α,p)-atom supported in $\Delta(x_0, 2_*^{k+1})$ which satisfies the L^p-normalization condition (20.16) relative to this surface ball, and

$$f = \sum_{k=0}^{\infty} \widetilde{\mu}_k \widetilde{c}_k \ \text{ in } L^p(\Sigma,\sigma). \tag{20.63}$$

Ultimately, this completes the proof of item *(a)*.

To justify the claim made in *(b)*, select a σ-measurable function $f : \Sigma \to \mathbb{R}$, and assume f has an atomic decomposition as in (20.33). To show that f belongs to $\mathrm{HK}^{\alpha}_{p,q}(\Sigma,\sigma)$ in the case when $0 < q \leq 1$, employ the same type of reasoning as in (11.58)-(11.59), keeping in mind (20.4), (20.13), the completeness of $\mathrm{HK}^{\alpha}_{p,q}(\Sigma,\sigma)$ (cf. Proposition 20.2), and Proposition 20.3. Now, suppose $1 < q \leq \infty$ and make the extra assumption that the sequence of x_0-central (α,p)-atoms $\{a_j\}_{j\in\mathbb{N}_0}$ is such that

$$\operatorname{supp} a_j \subseteq \Delta(x_0, 2_*^{j+1}), \quad \|a_j\|_{L^p(\Sigma,\sigma)} \leq 2_*^{-(j+1)\alpha}, \quad \int_{\Sigma} a_j \, d\sigma = 0, \tag{20.64}$$

for each $j \in \mathbb{N}_0$. Also assume that the series in (20.33) converges to f in $L^1_{\mathrm{loc}}(\Sigma,\sigma)$. Then $f \in L^1_{\mathrm{loc}}(\Sigma,\sigma)$, and thanks to Lemma 4.1 it follows that

$$\langle f,\psi\rangle = \sum_{k=0}^{\infty} \lambda_k \langle a_k,\psi\rangle \ \text{ for each } \psi \in \mathrm{Lip}_c(\Sigma). \tag{20.65}$$

Then whenever $x \in \Sigma$ and $\psi \in \mathscr{B}^{\gamma}_x(\Sigma) \subseteq \mathrm{Lip}_c(\Sigma)$, this implies

$$|\langle f,\psi\rangle| \leq \sum_{k=0}^{\infty} |\lambda_k||\langle a_k,\psi\rangle| \leq \sum_{k=0}^{\infty} |\lambda_k|(a_k)^{\#}_{\gamma}(x). \tag{20.66}$$

After taking the supremum over all bump functions $\psi \in \mathscr{B}^{\gamma}_x(\Sigma)$ we arrive at

$$f^{\#}_{\gamma}(x) \leq \sum_{k=0}^{\infty} |\lambda_k|(a_k)^{\#}_{\gamma}(x) \ \text{ for every } x \in \Sigma. \tag{20.67}$$

To proceed, consider the case when $q < \infty$. Upon recalling (17.113), (17.115), and (20.3), use (20.67) and (4.7) to write

$$
\begin{aligned}
\|f\|^q_{\mathrm{HK}^\alpha_{p,q}(\Sigma,\sigma)} &= \|f^\#_\gamma\|^q_{\mathrm{K}^\alpha_{p,q}(\Sigma,\sigma)} \\
&\leq \sum_{k=0}^{\infty} 2_*^{k\alpha q}\Big(\sum_{j=0}^{\infty}|\lambda_j|\|(a_j)^\#_\gamma\cdot\mathbb{1}_{\widetilde{C}_k}\|_{L^p(\Sigma,\sigma)}\Big)^q \\
&\leq \sum_{k=0}^{2} 2_*^{k\alpha q}\Big(\sum_{j=0}^{\infty}|\lambda_j|\|(a_j)^\#_\gamma\cdot\mathbb{1}_{\widetilde{C}_k}\|_{L^p(\Sigma,\sigma)}\Big)^q \\
&\quad + C\sum_{k=3}^{\infty} 2_*^{k\alpha q}\Big(\sum_{j=0}^{k-3}|\lambda_j|\|(a_j)^\#_\gamma\cdot\mathbb{1}_{\widetilde{C}_k}\|_{L^p(\Sigma,\sigma)}\Big)^q \\
&\quad + C\sum_{k=3}^{\infty} 2_*^{k\alpha q}\Big(\sum_{j=k-2}^{\infty}|\lambda_j|\|(a_j)^\#_\gamma\cdot\mathbb{1}_{\widetilde{C}_k}\|_{L^p(\Sigma,\sigma)}\Big)^q \\
&\leq \sum_{k=0}^{2} 2_*^{k\alpha q}\Big(\sum_{j=0}^{\infty}|\lambda_j|\|\mathcal{M}_\Sigma a_j\|_{L^p(\Sigma,\sigma)}\Big)^q \\
&\quad + C\sum_{k=3}^{\infty} 2_*^{k\alpha q}\Big(\sum_{j=0}^{k-3}|\lambda_j|\|(a_j)^\#_\gamma\cdot\mathbb{1}_{\widetilde{C}_k}\|_{L^p(\Sigma,\sigma)}\Big)^q \\
&\quad + C\sum_{k=3}^{\infty} 2_*^{k\alpha q}\Big(\sum_{j=k-2}^{\infty}|\lambda_j|\|\mathcal{M}_\Sigma a_j\|_{L^p(\Sigma,\sigma)}\Big)^q \\
&=: \mathrm{I}_1 + \mathrm{I}_2 + \mathrm{I}_3.
\end{aligned} \tag{20.68}
$$

In light of the boundedness of the Hardy-Littlewood maximal operator $\mathcal{M}_\Sigma$ on $L^p(\Sigma,\sigma)$, Hölder's inequality, the normalization condition in (20.64), and the fact that $\alpha > 0$, we may bound the term I_1 as

$$
\begin{aligned}
\mathrm{I}_1 &\leq C\sum_{k=0}^{2} 2_*^{k\alpha q}\Big(\sum_{j=0}^{\infty}|\lambda_j|\|a_j\|_{L^p(\Sigma,\sigma)}\Big)^q \leq C\sum_{k=0}^{2} 2_*^{k\alpha q}\Big(\sum_{j=0}^{\infty}|\lambda_j|2_*^{-j\alpha}\Big)^q \\
&\leq C\sum_{k=0}^{2} 2_*^{k\alpha q}\Big(\sum_{j=0}^{\infty}|\lambda_j|^q\Big)\Big(\sum_{j=0}^{\infty}2_*^{-j\alpha q'}\Big)^{q/q'} \leq C\sum_{j=0}^{\infty}|\lambda_j|^q,
\end{aligned} \tag{20.69}
$$

for some constant $C = C(\Sigma, p, q, \alpha) \in (0,\infty)$, where $q' \in (1,\infty)$ is the Hölder conjugate exponent of q. To deal with I_3 from (20.68), use the boundedness of the Hardy-Littlewood maximal operator $\mathcal{M}_\Sigma$ on $L^p(\Sigma,\sigma)$, the normalization condition in (20.64), Hölder's inequality, Fubini's Theorem for series, and the fact that $\alpha > 0$ to estimate

$$\begin{aligned}
\mathrm{I}_3 &\le C\sum_{k=3}^{\infty} 2_*^{k\alpha q}\Big(\sum_{j=k-2}^{\infty}|\lambda_j|\|a_j\|_{L^p(\Sigma,\sigma)}\Big)^q \le C\sum_{k=3}^{\infty} 2_*^{k\alpha q}\Big(\sum_{j=k-2}^{\infty}|\lambda_j|2_*^{-j\alpha}\Big)^q \\
&\le C\sum_{k=3}^{\infty} 2_*^{k\alpha q}\Big(\sum_{j=k-2}^{\infty}|\lambda_j|^q 2_*^{-j\alpha q/2}\Big)\Big(\sum_{j=k-2}^{\infty} 2_*^{-j\alpha q'/2}\Big)^{q/q'} \\
&\le C\sum_{k=3}^{\infty} 2_*^{k\alpha q}\Big(\sum_{j=k-2}^{\infty}|\lambda_j|^q 2_*^{-j\alpha q/2}\Big)2_*^{-k\alpha q/2} \\
&= C\sum_{k=3}^{\infty} 2_*^{k\alpha q/2}\Big(\sum_{j=k-2}^{\infty}|\lambda_j|^q 2_*^{-j\alpha q/2}\Big) = C\sum_{j=1}^{\infty}|\lambda_j|^q 2_*^{-j\alpha q/2}\Big(\sum_{k=3}^{j+2} 2_*^{k\alpha q/2}\Big) \\
&\le C\sum_{j=0}^{\infty}|\lambda_j|^q\Big(\sum_{k=0}^{j+2} 2_*^{(k-j)\alpha q/2}\Big) \le C\sum_{j=0}^{\infty}|\lambda_j|^q,
\end{aligned} \tag{20.70}$$

for some constant $C = C(\Sigma, p, q, \alpha) \in (0, \infty)$.

To further analyze the term I_2 appearing in (20.68), we now employ (20.24)-(20.25) (presently used with the choices $a := a_j$ and $\ell := j + 1$) which ensures that whenever the integers $k, j \in \mathbb{N}_0$ are such that $k \ge j + 3 = (j + 1) + 2$, the central atom a_j satisfies

$$(a_j)^{\#}_{\gamma}(x) \le C 2_*^{(j-k)\gamma} 2_*^{j[(n-1)/p'-\alpha]} 2_*^{-k(n-1)} \quad \text{for each } x \in \widetilde{C}_k. \tag{20.71}$$

In light of this, the Ahlfors regularity of Σ, Hölder's inequality, Fubini's Theorem for series, and the fact that $\gamma + (n - 1)/p' - \alpha > 0$ (as implied by (20.36)), we may estimate

$$
\begin{aligned}
\mathrm{I}_2 &\leq C \sum_{k=3}^{\infty} 2_*^{k\alpha q} \Big(\sum_{j=0}^{k-3} |\lambda_j| \|(a_j)_\gamma^\# \cdot \mathbb{1}_{\widetilde{C}_k}\|_{L^p(\Sigma,\sigma)} \Big)^q \\
&\leq C \sum_{k=3}^{\infty} 2_*^{k\alpha q} \Big(\sum_{j=0}^{k-3} |\lambda_j| 2_*^{(j-k)\gamma} 2_*^{j[(n-1)/p'-\alpha]} 2_*^{-k(n-1)} \sigma(\widetilde{C}_k)^{1/p} \Big)^q \\
&\leq C \sum_{k=3}^{\infty} 2_*^{k\alpha q} \Big(\sum_{j=0}^{k-3} |\lambda_j| 2_*^{(j-k)\gamma} 2_*^{j[(n-1)/p'-\alpha]} 2_*^{-k(n-1)(1-1/p)} \Big)^q \\
&= C \sum_{k=3}^{\infty} 2_*^{k\alpha q} \Big(\sum_{j=0}^{k-3} |\lambda_j| 2_*^{(j-k)\gamma} 2_*^{j[(n-1)/p'-\alpha]} 2_*^{-k(n-1)/p'} \Big)^q \\
&= C \sum_{k=3}^{\infty} \Big(\sum_{j=0}^{k-3} |\lambda_j| 2_*^{(j-k)[\gamma+(n-1)/p'-\alpha]} \Big)^q \\
&\leq C \sum_{k=3}^{\infty} \Big(\sum_{j=0}^{k-3} |\lambda_j|^q 2_*^{(j-k)[\gamma+(n-1)/p'-\alpha]q/2} \Big) \Big(\sum_{j=0}^{k-3} 2_*^{(j-k)[\gamma+(n-1)/p'-\alpha]q'/2} \Big)^{q/q'} \\
&\leq C \sum_{k=3}^{\infty} \Big(\sum_{j=0}^{k-3} |\lambda_j|^q 2_*^{(j-k)[\gamma+(n-1)/p'-\alpha]q/2} \Big) \\
&= C \sum_{j=0}^{\infty} |\lambda_j|^q \Big(\sum_{k=j+3}^{\infty} 2_*^{(j-k)[\gamma+(n-1)/p'-\alpha]q/2} \Big) \\
&\leq C \sum_{j=0}^{\infty} |\lambda_j|^q,
\end{aligned} \tag{20.72}
$$

for some constant $C = C(\Sigma, n, p, q, \gamma) \in (0, \infty)$. All together, estimates (20.68), (20.69), (20.70), and (20.72) guarantee the existence of a constant $C \in (0, \infty)$ depending only on p, q, α n, γ, and the Ahlfors regularity character of Σ with the property that

$$
\|f\|_{\mathrm{HK}^{\alpha}_{p,q}(\Sigma,\sigma)} \leq C \Big(\sum_{j=0}^{\infty} |\lambda_j|^q \Big)^{1/q}. \tag{20.73}
$$

This establishes the membership $f \in \mathrm{HK}^{\alpha}_{p,q}(\Sigma, \sigma)$. Furthermore, the computations (20.68), (20.69), (20.70), and (20.72) can be naturally adapted to the case $q = \infty$. This finishes the treatment of the claims made in item *(b)*. Ultimately, the claim made in item *(c)* is a direct consequence of *(a)* and *(b)*. This concludes the proof of Theorem 20.1. □

Remark 20.4 Recall that if $\alpha := (n-1)(1/q - 1/p)$ then (17.121) yields

$$\mathrm{HK}_{p,q}^{(n-1)(1/q-1/p)}(\Sigma,\sigma)=\mathrm{HA}_{p,q}(\Sigma,\sigma). \tag{20.74}$$

Also, observe that for this choice of α, the condition

$$\tfrac{n-1}{p'}\le\alpha<\tfrac{n-1}{p'}+1 \quad\text{and}\quad \alpha-\tfrac{n-1}{p'}<\gamma<1 \tag{20.75}$$

forces q to satisfy

$$\tfrac{n-1}{n}<q\le 1 \quad\text{and}\quad (n-1)(\tfrac{1}{q}-1)<\gamma<1. \tag{20.76}$$

In turn, these are precisely the ranges for which we have developed the theory of first-generation Herz-type Hardy spaces $\mathrm{HA}_{p,q}(\Sigma,\sigma)$ (see for instance Theorem 11.1 and Theorem 11.2).

Remark 20.5 Observe that the right-hand side in (20.35) is independent of the background parameter γ in (20.36). As such, the atomic characterization described in Theorem 20.1 ensures that the space $\mathrm{HK}^{\alpha}_{p,q}$ is actually independent of γ.

Remark 20.6 It is worth mentioning that the atomic description of $\mathrm{HK}^{\alpha}_{p,q}$ obtained in Theorem 20.1 combined with Remark 20.3 imply that the definition of $\mathrm{HK}^{\alpha}_{p,q}$ is independent of the reference point x_0 and the dilation factor 2_*.

20.2 Molecules for Inhomogeneous Herz-type Hardy Spaces

In this section we develop a molecular theory for the scale of inhomogeneous Herz-type Hardy Spaces $\mathrm{HK}^{\alpha}_{p,q}$ in the setting of unbounded Ahlfors regular sets. The notion of a central molecule introduced here may be regarded as a weaker version of central (α,p)-atoms (in the sense of Definition 20.2) in which the localization condition of central atoms is replaced by a decay condition (better understood in an integral sense). We begin by making the following definition.

Definition 20.3 Assume $\Sigma\subseteq\mathbb{R}^n$ is a closed unbounded Ahlfors regular set, and abbreviate $\sigma:=\mathcal{H}^{n-1}\lfloor\Sigma$. Denote by $c_\Sigma, C_\Sigma\in(0,\infty)$ the lower and upper Ahlfors regularity constants of Σ, respectively. Fix a reference point $x_0\in\Sigma$ along with a dilation factor $2_*>(C_\Sigma/c_\Sigma)^{1/(n-1)}$. Also, select an exponent $p\in(1,\infty)$ and denote by $p'\in(1,\infty)$ its Hölder conjugate exponent. Select indices $\alpha,\theta\in\mathbb{R}$ satisfying

$$\tfrac{n-1}{p'}\le\alpha<\infty \quad\text{and}\quad \alpha p<\theta<\infty. \tag{20.77}$$

A σ-measurable function $M:\Sigma\to\mathbb{R}$ is called an x_0-`central` (α,p,θ)-`molecule` provided there exists a radius $R\in[2_*,\infty)$ such that the following properties hold:

(1) Normalization Condition:

$$\|M\|_{L^p(\Sigma,\sigma)}\le R^{-\alpha}. \tag{20.78}$$

(2) Decay Condition:

$$\Big(\int_{\Sigma}|M(x)|^p|x-x_0|^{\theta}\,d\sigma(x)\Big)^{1/p}\leq R^{-\alpha+\theta/p}. \tag{20.79}$$

(3) Vanishing Moment Condition:

$$\int_{\Sigma}M\,d\sigma=0. \tag{20.80}$$

Remark 20.7 The normalization condition (20.78) and decay condition (20.79) imply that M is an absolutely integrable function on Σ. This follows from the same type of computations made in (4.192)-(4.193), now using that $\alpha\geq(n-1)/p'$ and $\theta>\alpha p$. As such, the vanishing moment condition in (20.80) is meaningfully formulated.

It is relevant to introduce a second notion of central molecules in the second-generation Herz-type Hardy spaces, of the sort described below.

Definition 20.4 Assume $\Sigma\subseteq\mathbb{R}^n$ is a closed unbounded Ahlfors regular set, and abbreviate $\sigma:=\mathcal{H}^{n-1}\lfloor\Sigma$. Denote by $c_\Sigma, C_\Sigma\in(0,\infty)$ the lower and upper Ahlfors regularity constants of Σ, respectively. Fix a reference point $x_0\in\Sigma$ along with a dilation factor $2_*>(C_\Sigma/c_\Sigma)^{1/(n-1)}$. Select an exponent $p\in(1,\infty)$ and denote by $p'\in(1,\infty)$ its Hölder conjugate exponent. Fix two indices $\alpha,\varepsilon\in\mathbb{R}$ satisfying

$$\tfrac{n-1}{p'}\leq\alpha<\infty\quad\text{and}\quad\tfrac{\alpha}{n-1}-\tfrac{1}{p'}<\varepsilon<\infty. \tag{20.81}$$

A σ-measurable function $M:\Sigma\to\mathbb{R}$ is called an x_0-`central` $(\alpha,p)_\varepsilon$-`molecule` provided there exists a radius $R\in[2_*,\infty)$ with the property that for each $k\in\mathbb{N}_0$ one has

$$\Big(\int_{A_k(x_0,R)}|M|^p\,d\sigma\Big)^{1/p}\leq 2^{k(n-1)(1/p-1-\varepsilon)}\cdot R^{-\alpha}, \tag{20.82}$$

where

$$A_k(x_0,R):=\begin{cases}\Delta(x_0,R) & \text{if } k=0,\\ \Delta(x_0,2^kR)\backslash\Delta(x_0,2^{k-1}R) & \text{if } k\geq 1,\end{cases} \tag{20.83}$$

and

$$\int_{\Sigma}M\,d\sigma=0. \tag{20.84}$$

Remark 20.8 Much as in Remark 4.9, condition (20.82) implies that M is an absolutely integrable function on Σ. This guarantees that the integral in (20.84) is well defined.

The two notions of central molecules introduced in Definition 20.3 and Definition 20.4 are compatible in the sense made precise in our next proposition.

Proposition 20.4 *Assume $\Sigma \subseteq \mathbb{R}^n$ is a closed unbounded Ahlfors regular set, and abbreviate $\sigma := \mathcal{H}^{n-1}\lfloor\Sigma$. Fix a reference point $x_0 \in \Sigma$. Select two exponents $p \in (1,\infty)$ and $q \in (0,\infty]$, along with an index $\alpha \in \mathbb{R}$ such that*

$$\tfrac{n-1}{p'} \leq \alpha < \infty \tag{20.85}$$

where $p' \in (1,\infty)$ denotes the Hölder conjugate exponent of p. Then the following statements are valid:

(a) For every $\varepsilon \in \big(\alpha/(n-1) - 1/p', \infty\big)$, there exists $\theta = \theta(\varepsilon) \in (\alpha p, \infty)$ and a constant $C_\varepsilon \in (0,\infty)$ depending on n, p, q, α, and ε with the property that every x_0-central $(\alpha, p)_\varepsilon$-molecule (in the sense of Definition 20.4) is a C_ε-multiple of an x_0-central (α, p, θ)-molecule (in the sense of Definition 20.3).

(b) Conversely, for every $\theta \in (\alpha p, \infty)$ there exists $\varepsilon \in \big(\alpha/(n-1) - 1/p', \infty\big)$ and a constant $C_\theta \in (0,\infty)$ depending on n, p, q, α, and θ such that every x_0-central (α, p, θ)-molecule (in the sense of Definition 20.3) is a C_θ-multiple of an x_0-central $(\alpha, p)_\varepsilon$-molecule (in the sense of Definition 20.4).

Proof To verify the claim made in item *(a)* notice that if $\varepsilon > \alpha/(n-1) - 1/p'$ then we can choose $\theta \in (0,\infty)$ satisfying

$$\alpha p < \theta < (n-1)\Big(\varepsilon + 1 - \frac{1}{p}\Big)p. \tag{20.86}$$

In turn, using (20.82), and reasoning as in (4.203)-(4.206) yields the validity of the claim made in item *(a)*.

To establish the claim made in item *(b)*, observe that since $\theta > \alpha p$, then the choice

$$\varepsilon := \frac{\theta}{p(n-1)} + \frac{1}{p} - 1 \tag{20.87}$$

satisfies

$$\varepsilon > \frac{\alpha}{n-1} - \frac{1}{p'} \quad \text{and} \quad -\frac{\theta}{p} = (n-1)\Big(\frac{1}{p} - 1 - \varepsilon\Big). \tag{20.88}$$

Granted this, the veracity of *(b)* can be obtained via the same type of argument as (4.208)-(4.209), now bearing in mind (20.78)-(20.79). □

In the following result we prove that central (α, p, θ)-molecules belong to the Herz-type Hardy space $\mathrm{HK}^{\alpha}_{p,q}$.

Theorem 20.2 *Assume $\Sigma \subseteq \mathbb{R}^n$ is a closed unbounded Ahlfors regular set, and abbreviate $\sigma := \mathcal{H}^{n-1}\lfloor\Sigma$. Pick two exponents $p \in (1,\infty)$ and $q \in (0,\infty]$ along with an index $\alpha \in \mathbb{R}$ such that*

$$\tfrac{n-1}{p'} \leq \alpha < \tfrac{n-1}{p'} + 1, \tag{20.89}$$

where $p' \in (1,\infty)$ denotes the Hölder conjugate exponent of p. Also, select a parameter $\theta \in (\alpha p, \infty)$. Finally, fix a reference point $x_0 \in \Sigma$. Then there exists some *constant $C \in (0, \infty)$ depending exclusively on n, p, q, α, θ, and the Ahlfors regularity character of Σ with the property that for every x_0-central (α, p, θ)-molecule M on Σ (in the sense of Definition 20.3) one has*

$$\|M\|_{\mathrm{HK}^{\alpha}_{p,q}(\Sigma,\sigma)} \leq C. \tag{20.90}$$

Moreover, the same type of conclusions hold for x_0-central $(\alpha, p)_{\varepsilon}$-molecules (in the sense of Definition 20.4) for all $\varepsilon \in \big(\alpha/(n-1) - 1/p', \infty\big)$.

Proof Consider a dilation factor 2_* as in Definition 20.3. Fix an arbitrary x_0-central (α, p, θ)-molecule $M : \Sigma \to \mathbb{R}$, and denote by $R \geq 2_*$ the radius associated with M in the sense of Definition 20.3. Select the unique integer $m \in \mathbb{N}$ for which

$$2_*^m \leq R < 2_*^{m+1}. \tag{20.91}$$

Denote by $\Delta_{-1} := \varnothing$, and for each for each $j \in \mathbb{N}_0$ define

$$\begin{aligned} \Delta_j &:= \Delta(x_0, 2_*^{j+m}), \\ A_j &:= \Delta_j \setminus \Delta_{j-1}, \\ M_{A_j} &:= ⨍_{A_j} M \, d\sigma. \end{aligned} \tag{20.92}$$

Notice that the absolute integrability of M (cf. Remark 11.6) guarantees the quantities M_{A_j} are meaningfully defined. Being mindful of (20.92) and the choice of 2_*, a quick inspection on the computations made in (4.215)-(4.216) reveals that

$$\sigma(A_j) \approx (2_*^{j+m})^{n-1} \quad \text{for each } j \in \mathbb{N}_0 \tag{20.93}$$

with proportionality constants depending only on n and the Ahlfors regularity of Σ.

Also, the same type of argument used in (4.217)-(4.221) yields

$$M \cdot \mathbb{1}_{\Delta_N} = \sum_{j=0}^{N} \alpha_j + \sum_{j=0}^{N-1} \beta_j - \frac{\mathbb{1}_{A_N}}{\sigma(A_N)} \int_{\Sigma \setminus \Delta_N} M \, d\sigma \quad \text{for every } N \in \mathbb{N}. \tag{20.94}$$

where α_j and β_j are defined as in (4.219)-(4.220) in relation to the sets Δ_j and A_j from (20.92), i.e.,

$$\alpha_j := (M - M_{A_j})\mathbb{1}_{A_j} \quad \text{and} \tag{20.95}$$

$$\beta_j := \Big(\frac{\mathbb{1}_{A_{j+1}}}{\sigma(A_{j+1})} - \frac{\mathbb{1}_{A_j}}{\sigma(A_j)}\Big) \int_{\Sigma \setminus \Delta_j} M \, d\sigma. \tag{20.96}$$

By making natural alterations in (4.222)-(4.227), and in (4.231)-(4.235), now bearing in mind (20.78)-(20.80), (20.91), (20.92), as well as $\alpha \geq (n-1)/p'$ and $\theta > \alpha p$, we may conclude that for every $j \in \mathbb{N}_0$ the functions α_j and β_j satisfy the following properties:

$$\operatorname{supp}\alpha_j \subseteq \Delta(x_0, 2_*^{j+m}), \quad \|\alpha_j\|_{L^p(\Sigma,\sigma)} \leq \widetilde{C}2_*^{-j\theta/p}2_*^{-m\alpha}, \quad \int_\Sigma \alpha_j \, d\sigma = 0, \tag{20.97}$$

and

$$\operatorname{supp}\beta_j \subseteq \Delta(x_0, 2_*^{j+m+1}), \quad \|\beta_j\|_{L^p(\Sigma,\sigma)} \leq \widetilde{C}2_*^{-(j+1)\theta/p}2_*^{-m\alpha}, \quad \int_\Sigma \beta_j \, d\sigma = 0, \tag{20.98}$$

for some constant $\widetilde{C} \in (0, \infty)$ depending only on n, p, θ, and the Ahlfors regularity character of Σ. For each $j \in \mathbb{N}_0$, introduce coefficients $\lambda_j, \widetilde{\lambda}_j \in \mathbb{R}$ by setting

$$\lambda_j := \widetilde{C}2_*^{j(\alpha-\theta/p)} \quad \text{and} \quad \widetilde{\lambda}_j := \widetilde{C}2_*^{(j+1)(\alpha-\theta/p)}. \tag{20.99}$$

For each $j \in \mathbb{N}_0$, consider the functions $a_j, b_j : \Sigma \to \mathbb{R}$ defined as

$$a_j := \lambda_j^{-1}\alpha_j \quad \text{and} \quad b_j := \widetilde{\lambda}_j^{-1}\beta_j. \tag{20.100}$$

Based on this, (20.97), (20.98), and (20.99) we have that for each $j \in \mathbb{N}_0$ the functions a_j and b_j satisfy

$$\operatorname{supp} a_j \subseteq \Delta(x_0, 2_*^{j+m}), \quad \|a_j\|_{L^p(\Sigma,\sigma)} \leq (2_*^{j+m})^{-\alpha}, \quad \int_\Sigma a_j \, d\sigma = 0, \tag{20.101}$$

and

$$\operatorname{supp} b_j \subseteq \Delta(x_0, 2_*^{j+m+1}), \quad \|b_j\|_{L^p(\Sigma,\sigma)} \leq (2_*^{j+m+1})^{-\alpha}, \quad \int_\Sigma b_j \, d\sigma = 0. \tag{20.102}$$

Hence, a_j and b_j are x_0-central (α, p)-atoms for all $j \in \mathbb{N}_0$. Also, thanks to the fact that $\theta > \alpha p$ it is visible from (20.99) that the sequences

$$\{\lambda_j\}_{j\in\mathbb{N}_0} \text{ and } \{\widetilde{\lambda}_j\}_{j\in\mathbb{N}_0} \text{ belong to } \ell^q(\mathbb{N}_0). \tag{20.103}$$

In light of (20.94), (20.100), the same exact reasoning presented in (4.239)-(4.242) entails

$$M = \sum_{j=0}^{\infty} \lambda_j a_j + \sum_{j=0}^{\infty} \widetilde{\lambda}_j b_j \quad \text{pointwise } \sigma\text{-a.e. on } \Sigma. \tag{20.104}$$

We claim that this series also converges to M in $L^1(\Sigma, \sigma)$. To justify this, we first observe that since M is an absolutely integrable function on Σ (cf. Remark 20.7), Lebesgue's Dominated Convergence Theorem gives

$$M \cdot \mathbb{1}_{\Delta_N} \to M \quad \text{in } L^1(\Sigma, \sigma) \text{ as } N \to \infty. \tag{20.105}$$

Additionally, for each $N \in \mathbb{N}$ we have

$$\left\| \frac{\mathbb{1}_{A_N}}{\sigma(A_N)} \int_{\Sigma \setminus \Delta_N} M \, d\sigma \right\|_{L^1(\Sigma,\sigma)} \leq \frac{\|\mathbb{1}_{A_N}\|_{L^1(\Sigma,\sigma)}}{\sigma(A_N)} \int_{\Sigma \setminus \Delta_N} |M| \, d\sigma$$
$$= \int_\Sigma |M \cdot \mathbb{1}_{\Sigma \setminus \Delta_N}| \, d\sigma. \tag{20.106}$$

Since $M \cdot \mathbb{1}_{\Sigma \setminus \Delta_N} \to 0$ pointwise everywhere on Σ as $N \to \infty$, and $M \in L^1(\Sigma, \sigma)$ (cf. Remark 20.7), Lebesgue's Dominated Convergence Theorem entails

$$\int_\Sigma |M \cdot \mathbb{1}_{\Sigma \setminus \Delta_N}| \, d\sigma \to 0 \quad \text{as } N \to \infty. \tag{20.107}$$

Granted this, letting $N \to \infty$ in (20.106) yields

$$\frac{\mathbb{1}_{A_N}}{\sigma(A_N)} \int_{\Sigma \setminus \Delta_N} M \, d\sigma \to 0 \quad \text{in } L^1(\Sigma, \sigma). \tag{20.108}$$

Hence, letting $N \to \infty$ in (20.94), using (20.100), (20.105), and (20.108) we obtain the desired convergence in $L^1(\Sigma, \sigma)$, namely

$$M = \sum_{j=0}^{\infty} \lambda_j a_j + \sum_{j=0}^{\infty} \widetilde{\lambda}_j b_j \quad \text{in } L^1(\Sigma, \sigma). \tag{20.109}$$

Moving on, we combine these two series to obtain an atomic decomposition of M as a single series (as described in Theorem 20.1). To this end, we rely on a similar type of argument as the one in (20.56)-(20.58). More specifically, for each $j \in \mathbb{N}_0$ define

$$\gamma_j := 2_*^\alpha |\lambda_j| + |\widetilde{\lambda}_j| + 2^{-j} \in (0, \infty) \quad \text{and} \quad c_j := \gamma_j^{-1}\big(\lambda_j a_j + \widetilde{\lambda}_j b_j\big) \ \text{ on } \Sigma \tag{20.110}$$

Thanks to (20.103), the sequence $\{\gamma_j\}_{j \in \mathbb{N}_0}$ belongs to $\ell^q(\mathbb{N}_0)$. Moreover, in light of the properties listed in (20.101)-(20.102) we have

$$\operatorname{supp} c_j \subseteq \Delta(x_0, 2_*^{j+m+1}), \qquad \int_\Sigma c_j \, d\sigma = 0, \tag{20.111}$$

and

$$\|c_j\|_{L^p(\Sigma,\sigma)} \leq |\gamma_j|^{-1}\big(|\lambda_j| \|a_j\|_{L^p(\Sigma,\sigma)} + |\widetilde{\lambda}_j| \|b_j\|_{L^p(\Sigma,\sigma)}\big)$$
$$\leq \frac{1}{2_*^\alpha |\lambda_j| + |\widetilde{\lambda}_j| + 2^{-j}} \big(|\lambda_j| (2_*^{j+m})^{-\alpha} + |\widetilde{\lambda}_j| (2_*^{j+m+1})^{-\alpha}\big)$$
$$= \frac{2_*^\alpha |\lambda_j| + |\widetilde{\lambda}_j|}{2_*^\alpha |\lambda_j| + |\widetilde{\lambda}_j| + 2^{-j}} (2_*^{j+m+1})^{-\alpha}$$
$$< \big(2_*^{j+m+1}\big)^{-\alpha}. \tag{20.112}$$

Hence, for each $j \in \mathbb{N}_0$ the function c_j is an x_0-central (α, p)-atom on Σ that satisfies localization and normalization conditions like (20.15)-(20.16) relative to the surface ball $\Delta(x_0, 2_*^{j+m+1})$. A combination of this, (20.104), (20.109), and (20.110) yield the following atomic decomposition:

$$M = \sum_{j=0}^{\infty}(\lambda_j a_j + \widetilde{\lambda}_j b_j) = \sum_{j=0}^{\infty}\gamma_j \cdot \gamma_j^{-1}(\lambda_j a_j + \widetilde{\lambda}_j b_j) = \sum_{j=0}^{\infty}\gamma_j c_j \tag{20.113}$$

with convergence both in $L^1(\Sigma, \sigma)$ and pointwise σ-a.e. on Σ.

At this stage, we may relabel the sequences $\{\gamma_j\}_{j\in\mathbb{N}_0}$ and $\{c_j\}_{j\in\mathbb{N}_0}$ as $\{\widetilde{\gamma}_j\}_{j\in\mathbb{N}_0}$ and $\{\widetilde{c}_j\}_{j\in\mathbb{N}_0}$ by setting

$$\begin{aligned} &\widetilde{\gamma}_0 = \cdots = \widetilde{\gamma}_{m-1} := 0, \quad \text{and} \quad \widetilde{\gamma}_{j+m} := \gamma_j, \quad \text{for } j \geq 0 \\ &\widetilde{c}_0 = \cdots = \widetilde{c}_{m-1} := 0, \quad \text{and} \quad \widetilde{c}_{j+m} := c_j, \quad \text{for } j \geq 0. \end{aligned} \tag{20.114}$$

Observe that (20.103) and (20.110) guarantees $\{\widetilde{\gamma}_j\}_{j\in\mathbb{N}_0} \in \ell^q(\mathbb{N}_0)$, while (20.111)-(20.112) imply that $\widetilde{c}_j$ is an x_0-central (α, p)-atom supported in $\Delta(x_0, 2_*^{j+1})$ (in the sense of Definition 20.2) for each $j \in \mathbb{N}_0$. Also, (20.113) and (20.114) imply

$$M = \sum_{j=0}^{\infty}\widetilde{\gamma}_j\widetilde{c}_j \quad \text{pointwise } \sigma\text{-a.e. on } \Sigma \text{ and in } L^1(\Sigma, \sigma). \tag{20.115}$$

With this in hand, invoke item *(b)* from Theorem 20.1 to ensure that M belongs to $\mathrm{HK}^{\alpha}_{p,q}(\Sigma, \sigma)$ quantitatively, i.e.,

$$\begin{aligned} \|M\|_{\mathrm{HK}^{\alpha}_{p,q}(\Sigma,\sigma)} &\leq C\Big(\sum_{j=0}^{\infty}|\widetilde{\gamma}_j|^q\Big)^{1/q} \\ &\leq C\Big(\sum_{j=0}^{\infty}|\lambda_j|^q\Big)^{1/q} + C\Big(\sum_{j=0}^{\infty}|\widetilde{\lambda}_j|^q\Big)^{1/q} + C\Big(\sum_{j=0}^{\infty}\frac{1}{2^{jq}}\Big)^{1/q} < \infty, \end{aligned} \tag{20.116}$$

for some constant $C \in (0, \infty)$ depending only on n, α, p, q, α, and the Ahlfors regularity character of Σ. From the definition of the sequences $\{\lambda_j\}_{j\in\mathbb{N}_0}$ and $\{\widetilde{\lambda}_j\}_{j\in\mathbb{N}_0}$ (cf. (20.99)), it is apparent that these are independent of M, and in fact they depend only on n, p, q, θ, and the Ahlfors regularity of Σ. As such, (20.90) follows from (20.116).

The very last claim regarding estimate (20.90) for x_0-central $(\alpha, p)_\varepsilon$-molecules follows from what we have proved so far together with Proposition 20.4. This ultimately finishes the proof of Theorem 20.2. □

Chapter 21

Singular Integral Operators on Herz-type Hardy Spaces of Second Generation

The goal here is to develop a Calderón-Zygmund theory for singular integral operators on UR sets in the context of second-generation Herz-type Hardy spaces. This plays a crucial role in subsequent chapters as it allows us to employ boundary layer methods for treating boundary value problems with boundary data in $\mathrm{HK}^{\alpha}_{p,q}$. We begin by studying principal value convolution-type SIO's on $\mathrm{HK}^{\alpha}_{p,q}$ spaces.

Theorem 21.1 *Consider a closed unbounded uniformly rectifiable set $\Sigma \subseteq \mathbb{R}^n$, and abbreviate $\sigma := \mathcal{H}^{n-1}\lfloor\Sigma$. For some sufficiently large integer $N = N(n) \in \mathbb{N}$, select a kernel $k \in \mathscr{C}^N(\mathbb{R}^n \setminus \{0\})$ that is odd and positive homogeneous of degree $1-n$. Recall the convolution-type integral operator T associated with k as in* (18.3)*. Also, select two exponents $p \in (1,\infty)$ and $q \in (0,\infty]$, along with an index $\alpha \in \mathbb{R}$ satisfying*

$$\frac{n-1}{p'} \leq \alpha < \frac{n-1}{p'} + 1 \tag{21.1}$$

where $p' \in (1,\infty)$ is the Hölder conjugate exponent of p. Then, in this setting, the operator

$$T : \mathrm{HK}^{\alpha}_{p,q}(\Sigma,\sigma) \longrightarrow \mathrm{K}^{\alpha}_{p,q}(\Sigma,\sigma) \tag{21.2}$$

is well defined, linear, and bounded.

Proof Consider first the case when $0 < q \leq 1$. In this scenario, the proof of the boundedness claimed in (21.2) mirrors that of Theorem 6.1. For this reason, here we only emphasize the main differences. Fix a reference point $x_0 \in \Sigma$ and a dilation factor 2_* (as in Definition 20.2). Pick an x_0-central (α,p)-atom a on Σ, and let $R \geq 2_*$ be a finite radius associated with a as in Definition 20.2. In relation to this, we claim that there exists some constant $C \in (0,\infty)$ depending only on n, p, q, α, and the Ahlfors regularity of Σ (in particular, independent of a), with the property that

$$\|Ta\|_{\mathrm{K}^{\alpha}_{p,q}(\Sigma,\sigma)} \leq C. \tag{21.3}$$

To justify this, let $J \in \mathbb{N}$ be the unique integer for which

$$2_*^J \leq R < 2_*^{J+1}. \tag{21.4}$$

© The Author(s), under exclusive license to Springer Nature Switzerland AG 2026
M. Mitrea and P. Takemura, *Singular Integrals, Herz-Type Function Spaces, and Boundary Problems*, Progress in Mathematics 362,
https://doi.org/10.1007/978-3-032-12516-3_21

Bearing in mind (17.115) and the family of annuli $\{\widetilde{C}_k\}_{k\in\mathbb{N}_0}$ defined in (17.113), write

$$\begin{aligned}\|Ta\|^q_{\mathrm{K}^\alpha_{p,q}(\Sigma,\sigma)} &= \sum_{j=0}^{J+1}\big(2_*^{j\alpha}\|(Ta)\cdot\mathbb{1}_{\widetilde{C}_j}\|_{L^p(\Sigma,\sigma)}\big)^q \\ &\quad+\sum_{j=J+2}^{\infty}\big(2_*^{j\alpha}\|(Ta)\cdot\mathbb{1}_{\widetilde{C}_j}\|_{L^p(\Sigma,\sigma)}\big)^q \\ &=: \mathrm{S}_1+\mathrm{S}_2. \end{aligned}\tag{21.5}$$

Thanks to the boundedness of T on $L^p(\Sigma,\sigma)$ (cf. Theorem 6.1), the normalization condition of a (cf. (20.16)), together with (21.4) and the fact that $\alpha\geq(n-1)/p'>0$, we estimate S_1 as

$$\begin{aligned}\mathrm{S}_1 &= \sum_{j=0}^{J+1}\big(2_*^{j\alpha}\|(Ta)\cdot\mathbb{1}_{\widetilde{C}_j}\|_{L^p(\Sigma,\sigma)}\big)^q \leq C\sum_{j=0}^{J+1}\big(2_*^{j\alpha}\|a\|_{L^p(\Sigma,\sigma)}\big)^q \\ &\leq C\sum_{j=0}^{J+1}\big(2_*^{j\alpha}R^{-\alpha}\big)^q \leq C\sum_{j=0}^{J+1}2_*^{(j-J)\alpha q}\leq C\sum_{\ell=0}^{\infty}2_*^{-\ell\alpha q}<\infty \end{aligned}\tag{21.6}$$

for some constant $C=C(n,p,q,\alpha,k,\Sigma)\in(0,\infty)$.

To bound S_2 from (21.5), fix some $j\in\mathbb{N}_0$ such that $j\geq J+2$ along with a location $x\in\widetilde{C}_j$. Availing ourselves with the computations made in (6.17)-(6.18), and using the vanishing moment property of the atom a, we express

$$\begin{aligned}Ta(x) &= \int_\Sigma k(x-y)a(y)\,d\sigma(y) \\ &= \int_\Sigma \big(k(x-y)-k(x-x_0)\big)a(y)\,d\sigma(y). \end{aligned}\tag{21.7}$$

Following the computations made in (6.20), now taking into account the localization and normalization properties of a (cf. (20.15)-(20.16)) and (21.4), estimate

$$\begin{aligned}|Ta(x)| &\leq \int_{\Delta(x_0,R)}|k(x-y)-k(x-x_0)||a(y)|\,d\sigma(y) \\ &\leq C\frac{R}{|x-x_0|^n}\|a\|_{L^p(\Sigma,\sigma)}\sigma\big(\Delta(x_0,R)\big)^{1/p'} \\ &\leq C\frac{R}{2_*^{jn}}R^{-\alpha}R^{(n-1)/p'} \\ &\leq C\frac{2_*^J}{2_*^{jn}}2_*^{J[(n-1)/p'-\alpha]} \end{aligned}\tag{21.8}$$

for some constant $C = C(n, p, k, \alpha, \Sigma) \in (0, \infty)$. Thanks to this and the Ahlfors regularity of Σ, we have

$$\begin{aligned} \|(Ta) \cdot \mathbb{1}_{\widetilde{C}_j}\|_{L^p(\Sigma,\sigma)} &= \Big(\int_{\widetilde{C}_j} |Ta|^p \, d\sigma \Big)^{1/p} \\ &\leq C \frac{2_*^J}{2_*^{jn}} 2_*^{J[(n-1)/p' - \alpha]} \sigma(\widetilde{C}_j)^{1/p} \\ &\leq C 2_*^{-jn} 2_*^{J[(n-1)/p' - \alpha + 1]} 2_*^{j(n-1)/p}, \end{aligned} \tag{21.9}$$

for some constant $C = C(n, p, \alpha, k, \Sigma) \in (0, \infty)$. Having established this, recall that $\alpha < \frac{n-1}{p'} + 1$ (cf. (21.1)), and proceed to estimate S_2 from (21.5) as

$$\begin{aligned} S_2 &\leq C \sum_{j=J+2}^{\infty} \big(2_*^{j\alpha} 2_*^{-jn} 2_*^{J[(n-1)/p' - \alpha + 1]} 2_*^{j(n-1)/p} \big)^q \\ &= C \sum_{j=J+2}^{\infty} 2_*^{(j-J)[(n-1)/p' - \alpha + 1]q} = C \sum_{\ell=0}^{\infty} 2_*^{-\ell[(n-1)/p' - \alpha + 1]q} < \infty. \end{aligned} \tag{21.10}$$

Notice that the constants in the right-hand side of estimate (21.6) and (21.10) depend exclusively on n, p, q, α, the kernel k, and the Ahlfors regularity of Σ. As such, this implies the claim made in (21.3).

To proceed, continue to assume that $0 < q \leq 1$. Fix a function $f \in \mathrm{HK}^{\alpha}_{p,q}(\Sigma, \sigma)$. Then item *(a)* from Theorem 20.1 ensures the existence of a sequence $\{\lambda_j\}_{j \in \mathbb{N}_0}$ in $\ell^q(\mathbb{N}_0)$ and a sequence of x_0-central (α, p)-atoms $\{a_j\}_{j \in \mathbb{N}_0}$ for which

$$f = \sum_{j=0}^{\infty} \lambda_j a_j \ \text{ in } L^1(\Sigma, \sigma). \tag{21.11}$$

Using this, the conclusion (21.2) in the case when $0 < q \leq 1$ can be obtained via the same reasoning as in (13.25)-(13.32), bearing in mind the continuous embedding $\mathrm{K}^{\alpha}_{p,q}(\Sigma, \sigma) \hookrightarrow L^1(\Sigma, \sigma)$ (cf. item *(e)* in Proposition 17.8), and the completeness of $\mathrm{K}^{\alpha}_{p,q}(\Sigma, \sigma)$ (cf. (17.122)).

Pressing forward, consider the case $1 < q < \infty$, and denote by $q' \in (1, \infty)$ its Hölder conjugate exponent. Select a function $f \in \mathrm{HK}^{\alpha}_{p,q}(\Sigma, \sigma)$ and use Theorem 20.1 *(a)* to obtain

$$f = \sum_{j=0}^{\infty} \lambda_j a_j \ \text{ in } L^p(\Sigma, \sigma) \tag{21.12}$$

for some sequence of x_0-central (α, p)-atoms $\{a_j\}_{j \in \mathbb{N}_0}$ and some numerical sequence $\{\lambda_j\}_{j \in \mathbb{N}_0}$ belonging to $\ell^q(\mathbb{N}_0)$. Assume that for each $j \in \mathbb{N}_0$, the central atom a_j satisfies localization and normalization properties (as in (20.15)-(20.16)) relative to the ball $\Delta(x_0, 2_*^{j+1})$, i.e.,

$$\operatorname{supp} a_j \subseteq \Delta(x_0, 2_*^{j+1}) \quad \text{and} \quad \|a_j\|_{L^p(\Sigma,\sigma)} \le \left(2_*^{j+1}\right)^{-\alpha}. \tag{21.13}$$

Also, suppose that the atomic decomposition (21.12) is quasi-optimal, in the sense that

$$\Big(\sum_{j=0}^{\infty} |\lambda_j|^q\Big)^{1/q} \le 2\|f\|_{\mathrm{HK}^{\alpha}_{p,q}(\Sigma,\sigma)}. \tag{21.14}$$

The boundedness of T on $L^p(\Sigma, \sigma)$ (cf. Theorem 6.1) together with the convergence (21.12) yield

$$Tf = \sum_{j=0}^{\infty} \lambda_j T a_j \quad \text{in } L^p(\Sigma, \sigma). \tag{21.15}$$

Hence,

$$\begin{aligned}
\|Tf\|^q_{\mathrm{K}^{\alpha}_{p,q}(\Sigma,\sigma)} &= \sum_{\ell=0}^{\infty} \big(2_*^{\ell\alpha}\|(Tf)\cdot \mathbb{1}_{\widetilde{C}_\ell}\|_{L^p(\Sigma,\sigma)}\big)^q \\
&\le \sum_{\ell=0}^{\infty} 2_*^{\ell\alpha q}\Big(\sum_{j=0}^{\infty} |\lambda_j|\|(Ta_j)\cdot \mathbb{1}_{\widetilde{C}_\ell}\|_{L^p(\Sigma,\sigma)}\Big)^q \\
&\le \sum_{\ell=0}^{2} 2_*^{\ell\alpha q}\Big(\sum_{j=0}^{\infty} |\lambda_j|\|(Ta_j)\cdot \mathbb{1}_{\widetilde{C}_\ell}\|_{L^p(\Sigma,\sigma)}\Big)^q \\
&\quad + C\sum_{\ell=3}^{\infty} 2_*^{\ell\alpha q}\Big(\sum_{j=0}^{\ell-3} |\lambda_j|\|(Ta_j)\cdot \mathbb{1}_{\widetilde{C}_\ell}\|_{L^p(\Sigma,\sigma)}\Big)^q \\
&\quad + C\sum_{\ell=3}^{\infty} 2_*^{\ell\alpha q}\Big(\sum_{j=\ell-2}^{\infty} |\lambda_j|\|(Ta_j)\cdot \mathbb{1}_{\widetilde{C}_\ell}\|_{L^p(\Sigma,\sigma)}\Big)^q \\
&=: \mathrm{I}_1 + \mathrm{I}_2 + \mathrm{I}_3.
\end{aligned} \tag{21.16}$$

In light of the boundedness of T in $L^p(\Sigma, \sigma)$ (cf. Proposition 6.1), the normalization property in (21.13), Hölder's inequality, we estimate I_1 as

$$\begin{aligned}
\mathrm{I}_1 &\le C\sum_{\ell=0}^{2} 2_*^{\ell\alpha q}\Big(\sum_{j=0}^{\infty} |\lambda_j|\|a_j\|_{L^p(\Sigma,\sigma)}\Big)^q \le C\Big(\sum_{j=0}^{\infty} |\lambda_j| 2_*^{-j\alpha}\Big)^q \\
&\le C\Big(\sum_{j=0}^{\infty} |\lambda_j|^q\Big)\Big(\sum_{j=0}^{\infty} 2_*^{-j\alpha q'}\Big)^{q/q'} = C\sum_{j=0}^{\infty} |\lambda_j|^q,
\end{aligned} \tag{21.17}$$

for some constant $C = C(n, p, q, \alpha, k, \Sigma) \in (0, \infty)$.

To deal with the term I_3 from (21.16), use Fubini's Theorem for series, and the boundedness of T in $L^p(\Sigma, \sigma)$ (cf. Proposition 6.1), the normalization of central atoms as in (21.13), together with Hölder's inequality to obtain

$$
\begin{aligned}
\mathrm{I}_3 &\leq C\sum_{\ell=3}^{\infty} 2_*^{\ell\alpha q}\Big(\sum_{j=\ell-2}^{\infty}|\lambda_j|\|a_j\|_{L^p(\Sigma,\sigma)}\Big)^q \leq C\sum_{\ell=3}^{\infty} 2_*^{\ell\alpha q}\Big(\sum_{j=\ell-2}^{\infty}|\lambda_j|2_*^{-j\alpha}\Big)^q \\
&\leq C\sum_{\ell=3}^{\infty} 2_*^{\ell\alpha q}\Big(\sum_{j-\ell-2}^{\infty}2_*^{-j\alpha q'/2}\Big)^{q/q'}\Big(\sum_{j=\ell-2}^{\infty}|\lambda_j|^q 2_*^{-j\alpha q/2}\Big) \\
&\leq C\sum_{\ell=3}^{\infty} 2_*^{\ell\alpha q}2_*^{-\ell\alpha q/2}\Big(\sum_{j=\ell-2}^{\infty}|\lambda_j|^q 2_*^{-j\alpha q/2}\Big) \\
&= C\sum_{\ell=3}^{\infty} 2_*^{\ell\alpha q/2}\Big(\sum_{j=\ell-2}^{\infty}|\lambda_j|^q 2_*^{-j\alpha q/2}\Big) \\
&= C\sum_{j=1}^{\infty}|\lambda_j|^q\Big(\sum_{\ell=3}^{j+2}2_*^{(\ell-j)\alpha q/2}\Big) \leq C\sum_{j=0}^{\infty}|\lambda_j|^q
\end{aligned}
\tag{21.18}
$$

for some constant $C \in (0,\infty)$ of same nature as in (21.17).

The next goal is to estimate I_2 from (21.16). To this end, fix $\ell, j \in \mathbb{N}_0$ such that $\ell \geq 3$ and $j \leq \ell - 3$. Recalling that $\operatorname{supp} a_j \subseteq \Delta(x_0, 2_*^{j+1})$, and relying on (21.9) (currently used with $a := a_j$, $R := 2_*^{j+1}$, $j := \ell$, and $J := j+1$) we have

$$
\begin{aligned}
\|(Ta_j)\cdot \mathbb{1}_{\widetilde{C}_\ell}\|_{L^p(\Sigma,\sigma)} &= \Big(\int_{\widetilde{C}_\ell}|Ta_j|\,d\sigma\Big)^{1/p} \\
&\leq C2_*^{-\ell n}2_*^{(j+1)[(n-1)/p'-\alpha+1]}2_*^{\ell(n-1)/p} \\
&= C2_*^{-\ell[(n-1)/p'+1]}2_*^{j[(n-1)/p'-\alpha+1]}
\end{aligned}
\tag{21.19}
$$

for some constant $C \in (0,\infty)$ depending only on n, p, q, α, the kernel k, and the Ahlfors regularity of Σ. Using Hölder's inequality along with Fubini's Theorem and $\alpha < \frac{n-1}{p'} + 1$, we bound I_2 as follows

$$
\begin{aligned}
\mathrm{I}_2 &\le C \sum_{\ell=3}^{\infty} 2_*^{\ell\alpha q} \Big(\sum_{j=0}^{\ell-3} |\lambda_j| 2_*^{-\ell[(n-1)/p'+1]} 2_*^{j[(n-1)/p'-\alpha+1]} \Big)^q \\
&= C \sum_{\ell=3}^{\infty} \Big(\sum_{j=0}^{\ell-3} |\lambda_j| 2_*^{(j-\ell)[(n-1)/p'-\alpha+1]} \Big)^q \\
&\le C \sum_{\ell=3}^{\infty} \Big(\sum_{j=0}^{\ell-3} |\lambda_j|^q 2_*^{(j-\ell)[(n-1)/p'-\alpha+1]q/2} \Big) \Big(\sum_{j=0}^{\ell-3} 2_*^{(j-\ell)[(n-1)/p'-\alpha+1]q'/2} \Big)^{q/q'} \\
&\le C \sum_{\ell=3}^{\infty} \Big(\sum_{j=0}^{\ell-3} |\lambda_j|^q 2_*^{(j-\ell)[(n-1)/p'-\alpha+1]q/2} \Big) \Big(\sum_{m=0}^{\infty} 2_*^{-m[(n-1)/p'-\alpha+1]q'/2} \Big)^{q/q'} \\
&= C \sum_{\ell=3}^{\infty} \Big(\sum_{j=0}^{\ell-3} |\lambda_j|^q 2_*^{(j-\ell)[(n-1)/p'-\alpha+1]q/2} \Big) \\
&= C \sum_{j=0}^{\infty} |\lambda_j|^q \Big(\sum_{\ell=j+3}^{\infty} 2_*^{(j-\ell)[(n-1)/p'-\alpha+1]q/2} \Big) \\
&\le C \sum_{j=0}^{\infty} |\lambda_j|^q \Big(\sum_{m=0}^{\infty} 2_*^{-m[(n-1)/p'-\alpha+1]q/2} \Big) \\
&= C \sum_{j=0}^{\infty} |\lambda_j|^q,
\end{aligned}
\tag{21.20}
$$

for some constant $C = C(n, p, q, \alpha, k, \Sigma) \in (0, \infty)$. All together, (21.14), (21.16), (21.17), (21.18), and (21.20) ensure that there exists some $C \in (0, \infty)$ depending only on n, p, q, α, the kernel k, and the Ahlfors regularity character of Σ with the property that

$$
\|Tf\|_{\mathrm{K}^{\alpha}_{p,q}(\Sigma,\sigma)} \le C\|f\|_{\mathrm{HK}^{\alpha}_{p,q}(\Sigma,\sigma)}. \tag{21.21}
$$

This proves the boundedness of T claimed in (21.2), in the case when $1 < q < \infty$. Corresponding to the case when $q = \infty$, the claim in (21.2) follows via a similar argument as before replacing the ℓ^q-norm by the ℓ^∞-norm. This ultimately concludes the proof of Theorem 21.1. □

Our next goal is to establish nontangential maximal function estimates for boundary-to-domain integral operators of convolution-type acting from $\mathrm{HK}^{\alpha}_{p,q}$.

Theorem 21.2 *Assume $\Omega \subseteq \mathbb{R}^n$ is a* UR *domain with unbounded boundary, and abbreviate $\sigma := \mathcal{H}^{n-1}\lfloor\partial\Omega$. For some sufficiently large integer $N = N(n) \in \mathbb{N}$, consider a kernel $k \in \mathscr{C}^N(\mathbb{R}^n \setminus \{0\})$ that is odd, and positive homogeneous of degree $1-n$. Recall the boundary-to-domain convolution-type integral operator $\mathcal{T}$ associated with k as in* (18.7). *Fix an aperture parameter $\kappa \in (0, \infty)$. Finally, pick* *two exponents* $p \in (1, \infty)$ *and* $q \in (0, \infty]$, *along with an index* $\alpha \in \mathbb{R}$ *satisfying*

$$\frac{n-1}{p'} \leq \alpha < \frac{n-1}{p'} + 1 \tag{21.22}$$

where $p' \in (1, \infty)$ *is the Hölder conjugate exponent of* p.

Then there exists some $C \in (0, \infty)$ *depending exclusively on,* n, p, q, α, κ, *the kernel* k, *and the* UR *constants of* $\partial\Omega$ *such that for every* $f \in \mathrm{HK}^{\alpha}_{p,q}(\partial\Omega, \sigma)$ *one has*

$$\|\mathcal{N}_\kappa(\mathcal{T} f)\|_{\mathrm{K}^{\alpha}_{p,q}(\partial\Omega,\sigma)} \leq C\|f\|_{\mathrm{HK}^{\alpha}_{p,q}(\partial\Omega,\sigma)}. \tag{21.23}$$

Proof Fix a reference point $x_0 \in \partial\Omega$ and a dilation factor $2_* > (C_{\partial\Omega}/c_{\partial\Omega})^{1/(n-1)}$, where $C_{\partial\Omega}, c_{\partial\Omega} \in (0, \infty)$ denote the upper and lower Ahlfors regularity constants of $\partial\Omega$, respectively. Consider first the case when $0 < q \leq 1$. In this scenario, the proof of (21.23) follows the same type of argument used to prove Theorem 6.2. As such, here we highlight the key differences. Select an x_0-central (α, p)-atom $a : \partial\Omega \to \mathbb{R}$ and let $R \geq 2_*$ be a radius associated with a (as in Definition 20.2). Much as in (6.34), the key step is to verify the existence of a constant $C \in (0, \infty)$ depending only on n, p, q, α, κ, and the UR constants of $\partial\Omega$ (in particular, independent of a) with the property that

$$\|\mathcal{N}_\kappa(\mathcal{T} a)\|_{\mathrm{K}^{\alpha}_{p,q}(\partial\Omega,\sigma)} \leq C. \tag{21.24}$$

To justify this claim, choose $\rho \in (0, \infty)$ such that

$$\rho > 2 + \kappa, \tag{21.25}$$

and let $N \in \mathbb{N}$ be the unique integer satisfying

$$2_*^N \leq \rho R < 2_*^{N+1}. \tag{21.26}$$

Also, upon recalling (17.115) and the family of annuli $\{\widetilde{C}_k\}_{k\in\mathbb{N}_0}$ introduced in (17.113), write

$$\begin{aligned}\|\mathcal{N}_\kappa(\mathcal{T} a)\|^q_{\mathrm{K}^{\alpha}_{p,q}(\partial\Omega,\sigma)} &= \sum_{j=0}^{N} \big(2_*^{j\alpha}\|\mathcal{N}_\kappa(\mathcal{T} a)\cdot \mathbb{1}_{\widetilde{C}_j}\|_{L^p(\partial\Omega,\sigma)}\big)^q \\ &\quad + \sum_{j=N+1}^{\infty} \big(2_*^{j\alpha}\|\mathcal{N}_\kappa(\mathcal{T} a)\cdot \mathbb{1}_{\widetilde{C}_j}\|_{L^p(\partial\Omega,\sigma)}\big)^q \\ &=: \mathrm{I} + \mathrm{II}.\end{aligned} \tag{21.27}$$

To deal with I, we apply the same type of reasoning as in (6.38)-(6.39), now using the normalization condition of a (see (20.16)), to bound

$$\|\mathcal{N}_\kappa(\mathcal{T} a)\|_{L^p(\partial\Omega,\sigma)} \leq C R^{-\alpha} \tag{21.28}$$

for some $C \in (0, \infty)$ that depends only on n, p, κ, and the UR constants of $\partial\Omega$. Based on this, and (21.26), estimate the term I from (21.27) by

$$\mathrm{I} \leq C \sum_{j=0}^{N} 2_*^{j\alpha q} R^{-\alpha q} \leq C \sum_{j=0}^{N} 2_*^{(j-N)\alpha q} \leq C \sum_{\ell=0}^{\infty} 2_*^{-\ell\alpha q} \tag{21.29}$$

for some constant $C \in (0, \infty)$ depending only on n, p, κ, and the Ahlfors regularity character of $\partial\Omega$. Notice that this series converges since $\alpha \geq (n-1)/p' > 0$ (cf. (21.22)).

The next goal is to obtain a similar estimate for the term II from (21.27). Being mindful of the normalization property of a (see (20.16)), and using an analogous argument to the one in (6.41)-(6.46), we obtain the following pointwise bound:

$$\begin{gathered} \mathcal{N}_\kappa(\mathcal{T}a)(x) \leq \frac{CR}{|x-x_0|^n} R^{(n-1)/p'-\alpha} \\ \text{for each } x \in \widetilde{C}_j \text{ and } j \in \mathbb{N}_0 \text{ with } j \geq N+1, \end{gathered} \tag{21.30}$$

for some constant $C \in (0, \infty)$ depending only on n, p, κ, the kernel k, and the Ahlfors regularity character of $\partial\Omega$. Granted this, the Ahlfors regularity of $\partial\Omega$, and (21.26), whenever $j \in \mathbb{N}_0$ with $j \geq N+1$ we have

$$\begin{aligned} \|\mathcal{N}_\kappa(\mathcal{T}a) \cdot \mathbb{1}_{\widetilde{C}_j}\|_{L^p(\partial\Omega,\sigma)} &\leq C R^{(n-1)/p'+1-\alpha} \Big(\int_{\widetilde{C}_j} \frac{d\sigma(x)}{|x-x_0|^{np}} \Big)^{1/p} \\ &\leq C 2_*^{N[(n-1)/p'+1-\alpha]} \frac{1}{2_*^{jn}} \sigma\big(\Delta(x_0, 2_*^{j+1})\big)^{1/p} \\ &\leq C 2_*^{N[(n-1)/p'+1-\alpha]} \frac{1}{2_*^{jn}} 2_*^{j(n-1)/p} \\ &= C 2_*^{N[(n-1)/p'+1-\alpha]} 2_*^{-j[(n-1)/p'+1]}, \end{aligned} \tag{21.31}$$

for some constant $C \in (0, \infty)$ of same nature as in (21.30). From this, we estimate II from (21.27) as

$$\begin{aligned} \mathrm{II} &\leq \sum_{j=N+1}^{\infty} 2_*^{j\alpha q} 2_*^{N[(n-1)/p'+1-\alpha]q} 2_*^{-j[(n-1)/p'+1]q} \\ &= C \sum_{j=N+1}^{\infty} 2_*^{(N-j)[(n-1)/p'+1-\alpha]q} = C \sum_{\ell=1}^{\infty} 2_*^{-\ell[(n-1)/p'+1-\alpha]q}, \end{aligned} \tag{21.32}$$

for some $C \in (0, \infty)$ depending only on n, p, κ, the kernel k, and the Ahlfors regularity constants of $\partial\Omega$. Since $\alpha < (n-1)/p' + 1$ (cf. (21.22)), this series is convergent. Collectively, (21.24) follows from (21.29) and (21.32). At this stage, the validity of estimate (21.23) in the regime $0 < q \leq 1$ can be obtained via a a similar reasoning as in (6.49)-(6.58), now relying on (20.4), (20.13) (presently used with $r := 1$), estimate (21.24), the atomic characterization described in Theorem 20.1, and the completeness of $\mathrm{K}^{\alpha}_{p,q}(\partial\Omega,\sigma)$ (cf. Proposition 20.2).

Moving forward, we shift our focus to the case $1 < q < \infty$. Fix an arbitrary function $f \in \mathrm{HK}^{\alpha}_{p,q}(\partial\Omega,\sigma)$, and use item *(a)* in Theorem 20.1 to write

$$f = \sum_{j=1}^{\infty} \lambda_j a_j \quad \text{in } L^p(\partial\Omega,\sigma), \tag{21.33}$$

for some numerical sequence $\{\lambda_j\}_{j\in\mathbb{N}_0}$ belonging to $\ell^q(\mathbb{N}_0)$, and some sequence $\{a_j\}_{j\in\mathbb{N}_0}$ of x_0-central (α,p)-atoms. Suppose that for each $j \in \mathbb{N}_0$ the central atom a_j satisfies localization and normalization properties (cf. (20.15)-(20.16)) relative to the surface ball $\Delta(x_0, 2_*^{j+1})$, i.e.,

$$\operatorname{supp} a_j \subseteq \Delta(x_0, 2_*^{j+1}) \quad \text{and} \quad \|a_j\|_{L^p(\partial\Omega,\sigma)} \leq \big(2_*^{j+1}\big)^{-\alpha}. \tag{21.34}$$

Moreover, make the assumption that the atomic decomposition (21.12) is quasi-optimal, that is

$$\Big(\sum_{j=0}^{\infty} |\lambda_j|^q\Big)^{1/q} \leq 2\|f\|_{\mathrm{HK}^{\alpha}_{p,q}(\partial\Omega,\sigma)}. \tag{21.35}$$

Notice that the nontangential maximal estimate recorded in (6.9) along with the size condition of a_j (cf. (21.34)) imply

$$\|\mathcal{N}_\kappa(\mathcal{T}a_j)\|_{L^p(\partial\Omega,\sigma)} \leq C\|a_j\|_{L^p(\partial\Omega,\sigma)} \leq C2_*^{-j\alpha} \quad \text{for each } j \in \mathbb{N}_0, \tag{21.36}$$

for some constant $C \in (0,\infty)$ depending only on n, p, and the UR character of $\partial\Omega$. This and Hölder's inequality yield

$$\begin{aligned}\sum_{j=0}^{\infty} |\lambda_j|\|\mathcal{N}_\kappa(\mathcal{T}a_j)\|_{L^p(\partial\Omega,\sigma)} &\leq C\sum_{j=0}^{\infty} |\lambda_j| 2_*^{-j\alpha}\\ &\leq C\Big(\sum_{j=0}^{\infty} |\lambda_j|^q\Big)^{1/q}\Big(\sum_{j=0}^{\infty} 2_*^{-j\alpha q'}\Big)^{1/q'} < \infty,\end{aligned} \tag{21.37}$$

where $q' \in (1,\infty)$ denotes the Hölder conjugate exponent of q. Hence,

$$\sum_{j=0}^{\infty} \lambda_j \cdot \mathcal{N}_\kappa(\mathcal{T}a_j) \quad \text{converges in } L^p(\partial\Omega,\sigma). \tag{21.38}$$

In turn this permit us to estimate

$$
\begin{aligned}
\|\mathcal{N}_\kappa(\mathcal{T}f)\|^q_{\mathrm{K}^\alpha_{p,q}(\partial\Omega,\sigma)} &= \sum_{\ell=0}^{\infty}\big(2_*^{\ell\alpha}\|\mathcal{N}_\kappa(\mathcal{T}f)\cdot\mathbb{1}_{\widetilde{C}_\ell}\|_{L^p(\partial\Omega,\sigma)}\big)^q \\
&\le \sum_{\ell=0}^{\infty}2_*^{\ell\alpha q}\Big(\sum_{j=0}^{\infty}|\lambda_j|\|\mathcal{N}_\kappa(\mathcal{T}a_j)\cdot\mathbb{1}_{\widetilde{C}_\ell}\|_{L^p(\partial\Omega,\sigma)}\Big)^q \\
&\le \sum_{\ell=0}^{2}2_*^{\ell\alpha q}\Big(\sum_{j=0}^{\infty}|\lambda_j|\|\mathcal{N}_\kappa(\mathcal{T}a_j)\|_{L^p(\partial\Omega,\sigma)}\Big)^q \\
&\quad + \sum_{\ell=3}^{\infty}2_*^{\ell\alpha q}\Big(\sum_{j=0}^{\ell-3}|\lambda_j|\|\mathcal{N}_\kappa(\mathcal{T}a_j)\cdot\mathbb{1}_{\widetilde{C}_\ell}\|_{L^p(\partial\Omega,\sigma)}\Big)^q \\
&\quad + \sum_{\ell=3}^{\infty}2_*^{\ell\alpha q}\Big(\sum_{j=\ell-2}^{\infty}|\lambda_j|\|\mathcal{N}_\kappa(\mathcal{T}a_j)\|_{L^p(\partial\Omega,\sigma)}\Big)^q \\
&=: \mathrm{I}+\mathrm{II}+\mathrm{III}.
\end{aligned}
\tag{21.39}
$$

To deal with I, use (21.37) and (21.35) to write

$$
\mathrm{I}\le C\sum_{\ell=0}^{2}2_*^{\ell\alpha q}\Big(\sum_{j=0}^{\infty}|\lambda_j|^q\Big)\le C\sum_{j=0}^{\infty}|\lambda_j|^q\le C\|f\|^q_{\mathrm{HK}^\alpha_{p,q}(\partial\Omega,\sigma)},
\tag{21.40}
$$

for some constant $C=C(n,p,q,\alpha,k,\partial\Omega)\in(0,\infty)$.

Keeping in min (21.35) and (21.36), employing Hölder's inequality together with the same type of computations as in (21.18), we bound III from (21.39) as

$$
\mathrm{III}\le C\sum_{\ell=3}^{\infty}2_*^{\ell\alpha q}\Big(\sum_{j=\ell-2}^{\infty}|\lambda_j|2_*^{-j\alpha}\Big)^q\le C\sum_{j=0}^{\infty}|\lambda_j|^q\le C\|f\|^q_{\mathrm{HK}^\alpha_{p,q}(\partial\Omega,\sigma)},
\tag{21.41}
$$

for some constant $C\in(0,\infty)$ depending only on n, p, q, α, κ, and the UR character of $\partial\Omega$.

To deal with II from (21.39) we may assume without loss of generality that (cf. Remark 20.3)

$$
2_*\ge 2(C_{\partial\Omega}/c_{\partial\Omega})^{1/(n-1)}\ge 2\quad\text{and}\quad 2_*^2>2+\kappa.
\tag{21.42}
$$

Select two numbers $j,\ell\in\mathbb{N}$ such that $\ell\ge 3$ and $j+3\le\ell$. Fix some $x\in\widetilde{C}_\ell$ along with $z\in\Gamma_\kappa(x)$. In particular, x and z satisfy

$$
|x-x_0|\le|x-z|+|z-x_0|<(\kappa+2)|z-x_0|.
\tag{21.43}
$$

Also, keeping in mind that $j+3\le\ell$, whenever $\xi\in B(x_0,2_*^{j+1})$ we have

$$|\xi - x_0| < 2_*^{j+1} \leq 2_*^{\ell-2} \leq 2_*^{-2}|x - x_0|. \tag{21.44}$$

As such, (21.43)-(21.44) entail

$$|z - \xi| \geq |z - x_0| - |\xi - x_0| > c|x - x_0| \quad \text{for every } \xi \in B(x_0, 2_*^{j+1}), \tag{21.45}$$

where $c := \frac{1}{\kappa+2} - \frac{1}{2_*^2}$ is a positive quantity, thanks to (21.42). This, (21.34), (21.42), and the same type of argument as in (6.44)-(6.45), allow us to estimate

$$\begin{aligned} |(\mathcal{T} a_j)(z)| &\leq \frac{C 2_*^j}{|x - x_0|^n} \|a\|_{L^p(\partial\Omega,\sigma)} \sigma\big(\Delta(x_0, 2_*^{j+1})\big)^{1/p'} \\ &\leq \frac{C 2_*^j}{|x - x_0|^n} (2_*^{j+1})^{-\alpha} 2_*^{(j+1)(n-1)/p'} \\ &= \frac{C 2_*^{j[(n-1)/p'+1-\alpha]}}{|x - x_0|^n}, \end{aligned} \tag{21.46}$$

for some constant $C \in (0, \infty)$ depending only on n, p, κ, the kernel k, and the Ahlfors regularity of $\partial\Omega$. After taking the supremum over $z \in \Gamma_\kappa(x)$, we arrive at

$$\mathcal{N}_\kappa(\mathcal{T} a_j)(x) \leq \frac{C 2_*^{j[(n-1)/p'+1-\alpha]}}{|x - x_0|^n} \quad \text{for all } x \in \widetilde{C}_\ell. \tag{21.47}$$

Based on this and the reasoning presented in (21.31), for every $j, \ell \in \mathbb{N}_0$ such that $\ell \geq 3$ and $j + 3 \leq \ell$ it follows that

$$\|\mathcal{N}_\kappa(\mathcal{T} a_j) \cdot \mathbb{1}_{C_\ell}\|_{L^p(\partial\Omega,\sigma)} \leq C 2_*^{j[(n-1)/p'+1-\alpha]} 2_*^{-\ell[(n-1)/p'+1]} \tag{21.48}$$

for some $C \in (0, \infty)$ of same nature as in (21.46). Having established this, and keeping in mind (21.35), we may follow the same type of computations as in (21.20) to derive

$$\mathrm{II} \leq C \sum_{j=0}^{\infty} |\lambda_j|^q \leq C \|f\|^q_{\mathrm{HK}^\alpha_{p,q}(\partial\Omega,\sigma)}, \tag{21.49}$$

for some constant $C \in (0, \infty)$ depending exclusively on n, p, q, α, κ, the kernel k, and the UR character of $\partial\Omega$. Collectively, estimates (21.40), (21.41), and (21.49) readily imply (21.23), in the case when $1 < q < \infty$. The validity of (21.23) in the case when $q = \infty$ can be obtained via an analogous argument as before. This ultimately proves Theorem 21.2. □

We next establish a boundedness result for generalized double layer potentials on inhomogeneous Herz-type Hardy spaces $\mathrm{HK}^\alpha_{p,q}$.

Theorem 21.3 *Assume $\Omega \subseteq \mathbb{R}^n$ is a uniformly rectifiable domain with unbounded boundary. Abbreviate $\sigma := \mathcal{H}^{n-1}\lfloor\partial\Omega$, and denote by ν the geometric measure theoretic outward unit normal to Ω. For some sufficiently large integer $N = N(n) \in \mathbb{N}$,* *select a vector-valued kernel* $\vec{k} \in \left[\mathscr{C}^N(\mathbb{R}^n \setminus \{0\})\right]^n$ *that is odd, positive homogeneous of degree* $1-n$, *and divergence-free, i.e.,* $\operatorname{div}\vec{k} = 0$ *in* $\mathbb{R}^n \setminus \{0\}$. *Recall the generalized double layer potential* $T^{\#}$ *associated with* $\vec{k}$ *as in (6.64). Select two exponents* $p \in (1,\infty)$ *and* $q \in (0,1]$ *along with an index* $\alpha \in \mathbb{R}$ *satisfying*

$$\frac{n-1}{p'} \leq \alpha < \frac{n-1}{p'} + 1 \tag{21.50}$$

where $p' := p/(p-1) \in (1,\infty)$ *denotes the Hölder conjugate exponent of* p. *Then*

$$T^{\#} : \mathrm{HK}^{\alpha}_{p,q}(\partial\Omega,\sigma) \to \mathrm{HK}^{\alpha}_{p,q}(\partial\Omega,\sigma) \tag{21.51}$$

is a well-defined, linear, and bounded operator.

Proof Select a reference point $x_0 \in \partial\Omega$ and a dilation factor $2_* > (C_{\partial\Omega}/c_{\partial\Omega})^{1/(n-1)}$, where $C_{\partial\Omega}, c_{\partial\Omega} \in (0,\infty)$ denote the upper and lower Ahlfors regularity constants of $\partial\Omega$, respectively. The justification of the boundedness claimed in (13.49) mirrors that used in the proof of Theorem 6.3. Hence, we focus only on the main differences. Select an arbitrary x_0-central (α,p)-atom a on $\partial\Omega$, and denote by $R \geq 2_*$ a finite radius associated with a (in the sense of Definition 20.2). Fix a parameter $\theta \in (0,\infty)$ satisfying

$$\alpha p < \theta < n(p-1) + 1, \tag{21.52}$$

which is possible since $\alpha < \frac{n-1}{p'} + 1$ (cf. (21.50)). We claim that there exists some constant $C_\theta \in (0,\infty)$, which is independent of the atom a, with the property that the function

$$T^{\#}a \text{ is a } C_\theta\text{-multiple of an } x_0\text{-central } (\alpha,p,\theta)\text{-molecule on } \partial\Omega. \tag{21.53}$$

To justify this, observe that the normalization property of a (cf. (20.16)), and the boundedness of $T^{\#}$ on $L^p(\partial\Omega,\sigma)$ (cf. (6.69)) imply

$$\|T^{\#}a\|_{L^p(\partial\Omega,\sigma)} \leq C\|a\|_{L^p(\partial\Omega,\sigma)} \leq CR^{-\alpha}, \tag{21.54}$$

for some $C = C(n,p,k,\partial\Omega) \in (0,\infty)$. Based on this, we estimate

$$\begin{aligned}\Big(\int_{\substack{x\in\partial\Omega\\ |x-x_0|<2R}} &|(T^{\#}a)(x)|^p|x-x_0|^\theta\,d\sigma(x)\Big)^{1/p}\\ &\leq (2R)^{\theta/p}\|T^{\#}a\|_{L^p(\partial\Omega,\sigma)} \leq CR^{-\alpha+\theta/p},\end{aligned} \tag{21.55}$$

for some $C = C(n,p,k,\theta,\partial\Omega) \in (0,\infty)$. Moreover, the same type of computations as in (6.76)-(6.79), the Ahlfors regularity of $\partial\Omega$, and the normalization of a (cf. (20.16)) yield

$$
\begin{aligned}
|T^{\#}a(x)| &\leq \frac{CR}{|x-x_0|^n}\|a\|_{L^p(\partial\Omega,\sigma)}\sigma\big(\Delta(x_0,R)\big)^{1/p'} \\
&\leq \frac{CR}{|x-x_0|^n}R^{(n-1)/p'-\alpha},
\end{aligned}
\tag{21.56}
$$

whenever $x \in \partial\Omega \setminus B(x_0, 2R)$, and for some constant $C = C(n, p, \partial\Omega) \in (0, \infty)$. Hence, combining (21.56) and the computations made in (6.80) gives

$$
\begin{aligned}
&\Big(\int_{\substack{x\in\partial\Omega\\ |x-x_0|\geq 2R}} |(T^{\#}a)(x)|^p|x-x_0|^\theta\,d\sigma(x)\Big)^{1/p} \\
&\leq CRR^{(n-1)/p'-\alpha}\sum_{j=1}^{\infty}\frac{(2^{j+1}R)^{\theta/p}}{(2^jR)^n}\sigma\big(\Delta(x_0,2^{j+1}R)\big)^{1/p} \\
&\leq C\frac{RR^{(n-1)/p'-\alpha}R^{\theta/p}}{R^n}\sum_{j=0}^{\infty}(2^j)^{\theta/p-n}(2^{j+1}R)^{(n-1)/p} \\
&= CR^{-\alpha+\theta/p}\sum_{j=0}^{\infty}2^{j[\theta/p-n+(n-1)/p]} < \infty,
\end{aligned}
\tag{21.57}
$$

for some $C = C(n, p, \partial\Omega) \in (0, \infty)$. The finiteness of this series follows from the upper bound for θ in (21.52). All together, (21.55) and (21.57) ensure the existence of some $C \in (0, \infty)$ depending exclusively on n, p, q, α, θ, the kernel $\vec{k}$, and the UR character of $\partial\Omega$, with the property that

$$
\begin{aligned}
&\Big(\int_{\partial\Omega} |(T^{\#}a)(x)|^p|x-x_0|^\theta\,d\sigma(x)\Big)^{1/p} \\
&\quad\leq \Big(\int_{\substack{x\in\partial\Omega\\ |x-x_0|<2R}} |(T^{\#}a)(x)|^p|x-x_0|^\theta\,d\sigma(x)\Big)^{1/p} \\
&\quad\quad+ \Big(\int_{\substack{x\in\partial\Omega\\ |x-x_0|\geq 2R}} |(T^{\#}a)(x)|^p|x-x_0|^\theta\,d\sigma(x)\Big)^{1/p} \\
&\quad\leq CR^{-\alpha+\theta/p}.
\end{aligned}
\tag{21.58}
$$

At this stage, if we now take into account that $\mathrm{K}^{\alpha}_{p,q}(\partial\Omega,\sigma) \subseteq L^1(\partial\Omega,\sigma)$ (cf. (17.148)) and the nontangential maximal estimate obtained in Theorem 21.2, then the same type of reasoning as in (6.81)-(6.92) ensures that

$$
\int_{\partial\Omega} T^{\#}a\,d\sigma = 0. \tag{21.59}
$$

All together, (21.59), (21.54), (21.58) establish the claim made in (21.53). Since we are in the regime $0 < q \leq 1$, the boundedness of T on $\mathrm{HK}^{\alpha}_{p,q}(\partial\Omega,\sigma)$ can be obtained by running the same type of argument as in (13.60)-(13.64), bearing in mind the atomic characterization from Theorem 20.1, the continuous embedding $\mathrm{HK}^{\alpha}_{p,q}(\partial\Omega,\sigma) \hookrightarrow L^1(\partial\Omega,\sigma)$ (cf. (20.13) with $r := 1$), and the completeness of $\mathrm{HK}^{\alpha}_{p,q}(\partial\Omega,\sigma)$ (cf. Proposition 20.2), and (20.4). This finishes the proof of Theorem 21.3. □

In the next result we obtain geometric estimates for singular integral operators of chord-dot-normal type on Herz-type Hardy spaces of second generation which play crucial role in the present work. The reader is advised to recall the piece of notation introduced in (6.113).

Theorem 21.4 *Assume* $\Omega \subseteq \mathbb{R}^n$ *is a* UR *domain with unbounded boundary, and abbreviate* $\sigma := \mathcal{H}^{n-1}\lfloor\partial\Omega$. *Denote by* ν *the geometric measure theoretic outward unit normal to* Ω. *For some sufficiently large integer* $N = N(n) \in \mathbb{N}$, *select a kernel* $k \in \mathscr{C}^N(\mathbb{R}\setminus\{0\})$ *that is even and positive homogeneous of degree* $-n$. *Recall the chord-dot-normal singular integral operator* $T^{\#}$ *associated with* k *as in* (6.104). *Finally, select indices* $p, q, \alpha \in \mathbb{R}$ *such that*

$$1 < p < \infty, \quad 0 < q \leq 1, \quad \tfrac{n-1}{p'} \leq \alpha < \tfrac{n-1}{p'} + 1 \tag{21.60}$$

where $p' \in (1,\infty)$ *denotes the Hölder conjugate exponent of* p.

Then

$$\begin{gathered} T^{\#} : \mathrm{HK}^{\alpha}_{p,q}(\partial\Omega,\sigma) \to \mathrm{HK}^{\alpha}_{p,q}(\partial\Omega,\sigma) \\ \text{is well defined, linear, and bounded.} \end{gathered} \tag{21.61}$$

Moreover, whenever $m \in \mathbb{N}$ *there exits a constant* $C_m \in (0,\infty)$ *depending exclusively on* m, n, p, q, α, *the kernel* k, *and the* UR *constants of* Ω *with the property that*

$$\|T^{\#}\|_{\mathrm{HK}^{\alpha}_{p,q}(\partial\Omega,\sigma)\to\mathrm{HK}^{\alpha}_{p,q}(\partial\Omega,\sigma)} \leq C_m \cdot \|\nu\|^{\langle m\rangle}_{[\mathrm{BMO}(\partial\Omega,\sigma)]^n}. \tag{21.62}$$

Proof The claim made in (21.61) follows from Theorem 21.3 and the fact that chord-dot-normal singular integral operators are themselves generalized double layers [GHA-IV, Example 5.1.7, p.. 556–557]. Choose some $m \in \mathbb{N}$. Fix a reference point $x_0 \in \partial\Omega$ and a dilation factor $2_* > (C_{\partial\Omega}/c_{\partial\Omega})^{1/(n-1)}$, where $C_{\partial\Omega}, c_{\partial\Omega} \in (0,\infty)$ denote the upper and lower Ahlfors regularity constants of $\partial\Omega$, respectively. The proof of (21.62) relies on the same type of argument used in the proof of Theorem 6.4. The crux of the matter lies in showing the existence of some $C \in (0,\infty)$ depending only on n, m, p, α, the kernel k, and the UR constants of $\partial\Omega$ such that whenever $\varepsilon \in (0,\infty)$ satisfies $\frac{\alpha}{n-1} - \frac{1}{p'} < \varepsilon < \frac{1}{n-1}$, and a is an x_0-central (α, p)-atom on $\partial\Omega$, with a finite radius $R \geq 2_*$ associated with a as in Definition 20.2, then

$$\begin{gathered} \text{the function } M := T^{\#}a \text{ is a } C \cdot \|\nu\|^{\langle m\rangle}_{[\mathrm{BMO}(\partial\Omega,\sigma)]^n}\text{-multiple of} \\ \text{an } x_0\text{-central } (\alpha,p)_\varepsilon\text{-molecule on } \partial\Omega \text{ associated with} \\ \text{the radius } 2R \text{ (in the sense of Definition 20.4).} \end{gathered} \tag{21.63}$$

To begin, fix one such x_0-central (α, p)-atom a on $\partial\Omega$. The reasoning in (6.124)-(6.125) combined with the normalization property of a (see (20.16)) yield

$$\Big(\int_{\Delta(x_0,2R)} |M|^p \, d\sigma\Big)^{1/p} \le C \cdot R^{-\alpha} \|\nu\|^{\langle m \rangle}_{[\mathrm{BMO}(\partial\Omega,\sigma)]^n}. \tag{21.64}$$

Next, being mindful of (6.126) and the vanishing moment property of a (see (20.17)) we write

$$M(x) = -\int_{\Delta(x_0,R)} \Big\{\langle \nu(x), x-y\rangle k(x-y) - \langle \nu(x), x-x_0\rangle k(x-x_0)\Big\} a(y)\, d\sigma(y), \tag{21.65}$$

for σ-a.e. $x \in \partial\Omega \backslash B(x_0, 2R)$. Recalling the piece of notation introduced in (6.128), using the same line of reasoning as in (6.129)-(6.134) along with the normalization property of a (see (20.16)) we deduce that for σ-a.e. $x \in \partial\Omega\backslash B(x_0, 2R)$ we have

$$\begin{aligned} |M(x) - M_\Delta(x)| &\le CR \frac{|\nu(x) - \nu_{\Delta(x_0,R)}|}{|x - x_0|^n} \sigma\big(\Delta(x_0, R)\big)^{1/p'} \|a\|_{L^p(\partial\Omega,\sigma)} \\ &\le CR \frac{|\nu(x) - \nu_{\Delta(x_0,R)}|}{|x - x_0|^n} R^{(n-1)/p'-\alpha}, \end{aligned} \tag{21.66}$$

for some $C \in (0, \infty)$ depending only on n, p, the kernel k, and the Ahlfors regularity constants of $\partial\Omega$. Granted this, the definition of annulus $A_\ell(x_0, 2R)$ (as in (4.198)), and following a similar approach as in (6.135)-(6.142), we see that there exists some $C = C(n, p, k, \varepsilon, \partial\Omega) \in (0, \infty)$ such that

$$\Big(\int_{A_\ell(x_0,2R)} |M - M_\Delta|^p \, d\sigma\Big)^{1/p} \le C 2^{\ell(n-1)(1/p-1-\varepsilon)} R^{-\alpha} \|\nu\|_{[\mathrm{BMO}(\partial\Omega,\sigma)]^n}, \tag{21.67}$$

for each $\ell \in \mathbb{N}$.

To obtain a similar estimate for M_Δ, recall that

$$M_\Delta(x) = M^{(1)}_\Delta(x) + M^{(2)}_\Delta(x) \quad \text{for each } x \in \partial\Omega\backslash B(x_0, 2R), \tag{21.68}$$

where $M^{(1)}_\Delta$ and $M^{(2)}_\Delta$ are as in (6.144)-(6.145). Taking into consideration the normalization of a (see (20.16)), a quick inspection on (6.146) gives that for each $x \in \partial\Omega\backslash B(x_0, 2R)$ we have

$$|M^{(1)}_\Delta(x)| \le \big|\langle \nu_{\Delta(x_0,R)}, x - x_0\rangle\big| \frac{CR}{|x - x_0|^{n+1}} R^{(n-1)/p'-\alpha}, \tag{21.69}$$

for some constant $C \in (0, \infty)$ depending only on n, p, the kernel k, and the Ahlfors regularity of $\partial\Omega$. With this in hand, we appeal to the reasoning in (6.147) to obtain

$$\Big(\int_{A_\ell(x_0,2R)}|M_\Delta^{(1)}|^p\,d\sigma\Big)^{1/p}\le C2^{\ell(n-1)(1/p-1-\varepsilon)}R^{-\alpha}\|\nu\|_{[\mathrm{BMO}(\partial\Omega,\sigma)]^n},\qquad(21.70)$$

for all $\ell\in\mathbb{N}$. Next, in view of the normalization condition of a (see (20.16)), the computations in (6.148) entail that for every $x\in\partial\Omega\setminus B(x_0,2R)$ we have

$$|M_\Delta^{(2)}(x)|\le CR\cdot R^{(n-1)/p'-\alpha}\|\nu\|_{[\mathrm{BMO}(\partial\Omega,\sigma)]^n}\frac{1}{|x-x_0|^n},\qquad(21.71)$$

for some constant $C\in(0,\infty)$ depending only on n, p, the kernel k, and the Ahlfors regularity of $\partial\Omega$. Availing ourselves of (21.71), the calculations presented in (6.149) presently give

$$\Big(\int_{A_\ell(x_0,2R)}|M_\Delta^{(2)}|^p\,d\sigma\Big)^{1/p}\le C2^{\ell(n-1)(1/p-1-\varepsilon)}R^{-\alpha}\|\nu\|_{[\mathrm{BMO}(\partial\Omega,\sigma)]^n},\qquad(21.72)$$

for each $\ell\in\mathbb{N}$. Together, (21.68), (21.70), and (21.72) show that, for every $\ell\in\mathbb{N}$ we have

$$\Big(\int_{A_\ell(x_0,2R)}|M_\Delta|^p\,d\sigma\Big)^{1/p}\le C2^{\ell(n-1)(1/p-1-\varepsilon)}R^{-\alpha}\|\nu\|_{[\mathrm{BMO}(\partial\Omega,\sigma)]^n}.\qquad(21.73)$$

Lastly, observe that the vanishing moment of M follows from Lemma 6.1. The latter property, (21.73), (21.67), (21.64), and Definition 20.4 finally prove the claim made in (21.63).

Once (21.63) has been established, we may invoke Theorem 20.2 to conclude that whenever $\frac{\alpha}{n-1}-\frac{1}{p'}<\varepsilon<\frac{1}{n-1}$ there exists some $C_m\in(0,\infty)$, that depends exclusively on n, m, p, q, α, the kernel k, and the UR constants of $\partial\Omega$, with the property that for every x_0-central (α,p)-atom a on $\partial\Omega$ we have

$$T^{\#}a\in \mathrm{HK}^\alpha_{p,q}(\partial\Omega,\sigma)\ \text{ and }\ \|T^{\#}a\|_{\mathrm{HK}^\alpha_{p,q}(\partial\Omega,\sigma)}\le C_m\cdot\|\nu\|^{\langle m\rangle}_{[\mathrm{BMO}(\partial\Omega,\sigma)]^n}.\qquad(21.74)$$

Finally, using the line of reasoning presented in (6.153)-(6.163), now being mindful of Theorem 20.1, the continuous embedding $\mathrm{HK}^\alpha_{p,q}(\partial\Omega,\sigma)\hookrightarrow L^1(\partial\Omega,\sigma)$ (cf. (20.13) used with $r:=1$), Proposition 20.2, as well as (20.4) and the fact that $0<q\le 1$, we conclude that $T^{\#}$ is a bounded linear operator from $\mathrm{HK}^\alpha_{p,q}(\partial\Omega,\sigma)$ into itself with operator norm satisfying (21.62). This completes the proof of Theorem 21.4. □

Chapter 22

Boundary Problems on Second-Generation Herz-type Hardy Spaces

The Calderón-Zygmund theory for singular integral operators on inhomogeneous Herz-type Hardy spaces of second generation discussed in Chapter 21 is robust and can be applied to boundary layer potentials associated with weakly elliptic systems. Our main objective here is to employ this as a means of treating the Neumann Problem for weakly elliptic systems in Ahlfors regular domains with boundary data in second-generation Herz-type Hardy spaces.

22.1 Boundedness of Layer Potentials and Invertibility Results

Our first result concerns geometric operator norm estimates for double layer potentials associated with weakly elliptic systems that have distinguished coefficient tensors.

Theorem 22.1 *Let $\Omega \subseteq \mathbb{R}^n$ be a* UR *domain such that $\partial\Omega$ is an unbounded set. Abbreviate $\sigma := \mathcal{H}^{n-1}\lfloor\partial\Omega$ and denote by ν the geometric measure theoretic outward unit normal to Ω. Fix $M \in \mathbb{N}$, and consider a second-order, homogeneous, constant complex coefficient, weakly elliptic $M \times M$ system L in $\mathbb{R}^n$ for which $\mathfrak{A}_L^{\rm dis} \neq \varnothing$. Pick $A \in \mathfrak{A}_L^{\rm dis}$ and consider the boundary-to-boundary transpose double layer potential $K_A^{\#}$ associated with Ω and the coefficient tensor A as in* (7.24). *Lastly, select indices $p, q, \alpha \in \mathbb{R}$ satisfying*

$$1 < p < \infty, \quad 0 < q \leq 1, \quad \tfrac{n-1}{p'} \leq \alpha < \tfrac{n-1}{p'} + 1. \tag{22.1}$$

Then the operator

$$\begin{gathered} K_A^{\#} : [\mathrm{HK}^{\alpha}_{p,q}(\partial\Omega,\sigma)]^M \to [\mathrm{HK}^{\alpha}_{p,q}(\partial\Omega,\sigma)]^M \\ \textit{is well defined, linear, and bounded,} \end{gathered} \tag{22.2}$$

© The Author(s), under exclusive license to Springer Nature Switzerland AG 2026
M. Mitrea and P. Takemura, *Singular Integrals, Herz-Type Function Spaces, and Boundary Problems*, Progress in Mathematics 362,
https://doi.org/10.1007/978-3-032-12516-3_22

and for each $m \in \mathbb{N}$, *there exists some* $C_m \in (0,\infty)$ *depending only on* m, n, A, p, q, α, *and the* UR *constants of* Ω, *with the property that*

$$\|K_A^{\#}\|_{[\mathrm{HK}^{\alpha}_{p,q}(\partial\Omega,\sigma)]^M\to[\mathrm{HK}^{\alpha}_{p,q}(\partial\Omega,\sigma)]^M} \leq C_m \cdot \|\nu\|^{\langle m\rangle}_{[\mathrm{BMO}(\partial\Omega,\sigma)]^n}. \tag{22.3}$$

Proof Since $A \in \mathfrak{A}_L^{\mathrm{dis}}$, the integral kernel of layer potential $K_A^{\#}$ becomes of chord-dot-normal type (cf. Proposition 7.2, and (7.48)). As such, the claim made in (22.2) follow from Theorem 21.4, while estimate (15.26) also follow from Theorem 21.4.□

The operator norm estimate obtained in the previous theorem permits us to derive an invertibility result on the space $\mathrm{HK}^{\alpha}_{p,q}$ when the underlying domain has a "sufficiently flat" boundary.

Theorem 22.2 *Assume* $\Omega \subseteq \mathbb{R}^n$ *is an Ahlfors regular domain. Set* $\sigma := \mathcal{H}^{n-1}\lfloor\partial\Omega$ *and denote by* ν *the geometric measure theoretic outward unit normal to* Ω. *Also, let* L *be a homogeneous, second-order, constant complex coefficient, weakly elliptic* $M\times M$ *system in* $\mathbb{R}^n$, *and suppose that* $\mathfrak{A}_L^{\mathrm{dis}} \neq \varnothing$. *Pick* $A \in \mathfrak{A}_L^{\mathrm{dis}}$ *and recall the boundary-to-boundary transpose double layer potential operator* $K_A^{\#}$ *associated with* A *as in* (7.24). *Furthermore, select indices* $p, q, \alpha \in \mathbb{R}$ *satisfying*

$$1 < p < \infty, \quad 0 < q \leq 1, \quad \tfrac{n-1}{p'} \leq \alpha < \tfrac{n-1}{p'} + 1. \tag{22.4}$$

Finally, fix a parameter $\varepsilon \in (0,\infty)$.

Then there exists some threshold $\delta \in (0,1)$, *depending only on* n, p, q, α, A, ε, *and the Ahlfors regularity constant of* $\partial\Omega$, *with the property that whenever*

$$\|\nu\|_{[\mathrm{BMO}(\partial\Omega,\sigma)]^n} < \delta \tag{22.5}$$

it follows that for each spectral parameter $z \in \mathbb{C}$ *with* $|z| \geq \varepsilon$ *the operator*

$$\begin{gathered} zI + K_A^{\#} : [\mathrm{HK}^{\alpha}_{p,q}(\partial\Omega,\sigma)]^M \longrightarrow [\mathrm{HK}^{\alpha}_{p,q}(\partial\Omega,\sigma)]^M \\ \text{is well defined, linear, bounded, and invertible.} \end{gathered} \tag{22.6}$$

Proof In light of the boundedness result recorded in Theorem 22.2, the completeness of the space $\mathrm{HK}^{\alpha}_{p,q}(\partial\Omega,\sigma)$ (cf. Proposition 20.2), and the fact that $\|\cdot\|_{\mathrm{HK}^{\alpha}_{p,q}(\partial\Omega,\sigma)}$ is a q-norm on $\mathrm{HK}^{\alpha}_{p,q}(\partial\Omega,\sigma)$ (cf. (20.4)), the invertibility claimed in (22.6) can be obtained using a Neumann series argument (cf. Lemma 8.1), reasoning in an analogous fashion to the proof of Theorem 8.2. □

Remark 22.1 We close by noting that a similar interpretation to the one detailed in Remark 8.1 applies to the statement of Theorem 22.2.

22.2 The Neumann Problem with Boundary Data in $\mathbf{HK}^{\alpha}_{p,q}$

In this section, we treat the Neumann Problem with boundary data prescribed in second-generation Herz-type Hardy spaces $\mathrm{HK}^{\alpha}_{p,q}$. We begin by studying the properties of null solutions of weakly elliptic systems whose nontangential maximal functions belong to $\mathrm{K}^{\alpha}_{p,q}$, with the goal of bringing forth the following Fatou-type result:

Theorem 22.3 *Assume $\Omega \subseteq \mathbb{R}^n$ is a UR domain with unbounded boundary. Abbreviate $\sigma := \mathcal{H}^{n-1}\lfloor\partial\Omega$, and denote by ν the geometric measure theoretic outward unit normal to Ω. Consider L a second-order, homogeneous, constant complex coefficient, weakly elliptic $M \times M$ system in $\mathbb{R}^n$. Also, select indices $p, q, \alpha \in \mathbb{R}$ satisfying*

$$1 < p < \infty, \;\; 0 < q \leq 1, \;\; \text{and} \;\; \tfrac{n-1}{p'} \leq \alpha < 1 + \tfrac{n-1}{p'} \tag{22.7}$$

where $p' \in (1, \infty)$ denotes the Hölder conjugate exponent of p. Having fixed an aperture parameter $\kappa \in (0, \infty)$, suppose $u : \Omega \to \mathbb{C}^M$ is a vector-valued function satisfying

$$\begin{aligned} & u \in [\mathscr{C}^\infty(\Omega)]^M, \quad Lu = 0 \;\; \text{in } \Omega, \\ & \text{and} \;\; \mathcal{N}_\kappa(\nabla u) \in \mathrm{K}^{\alpha}_{p,q}(\partial\Omega, \sigma). \end{aligned} \tag{22.8}$$

Then the nontangential boundary trace $(\nabla u)\big|^{\kappa-\mathrm{n.t.}}_{\partial\Omega}$ exists σ-a.e. on $\partial\Omega$ and the vector-valued function $(\nabla u)\big|^{\kappa-\mathrm{n.t.}}_{\partial\Omega}$ belongs to $[\mathrm{K}^{\alpha}_{p,q}(\partial\Omega, \sigma)]^{M\times n}$.

Moreover, for any coefficient tensor $A \in \mathfrak{A}_L$ it follows that the conormal derivative $\partial^A_\nu u$ (cf. (7.55)) belongs to $[\mathrm{HK}^{\alpha}_{p,q}(\partial\Omega, \sigma)]^M$ in a quantitative fashion, i.e., there exists some constant $C = C(\partial\Omega, A, n, p, q, \alpha, \kappa) \in (0, \infty)$ for which

$$\|\partial^A_\nu u\|_{[\mathrm{HK}^{\alpha}_{p,q}(\partial\Omega,\sigma)]^M} \leq C\|\mathcal{N}_\kappa(\nabla u)\|_{\mathrm{K}^{\alpha}_{p,q}(\partial\Omega,\sigma)}. \tag{22.9}$$

Proof Observe that the properties listed in (22.8), and the fact that $\mathrm{K}^{\alpha}_{p,q}(\partial\Omega, \sigma)$ is included in $L^p_{\mathrm{loc}}(\partial\Omega, \sigma)$ (cf. item *(c)* in Proposition 17.8), allow us to use Theorem 9.1 to ensure the existence of the nontangential limit $(\nabla u)\big|^{\kappa-\mathrm{n.t.}}_{\partial\Omega}$ at σ-a.e. point on $\partial\Omega$. Since this is pointwise controlled by $\mathcal{N}_\kappa(\nabla u) \in \mathrm{K}^{\alpha}_{p,q}(\partial\Omega, \sigma)$, the lattice property of $\mathrm{K}^{\alpha}_{p,q}(\partial\Omega, \sigma)$ (cf. (17.5)) implies that $(\nabla u)\big|^{\kappa-\mathrm{n.t.}}_{\partial\Omega}$ belongs to $[\mathrm{K}^{\alpha}_{p,q}(\partial\Omega, \sigma)]^{M\times n}$, as desired.

Using the same type of reasoning presented in (9.13)-(9.15), keeping in mind (22.8), the lattice property of $\mathrm{K}^{\alpha}_{p,q}(\partial\Omega, \sigma)$ (cf. (17.5)), (17.144), and denoting by

$$\vec{F}_\gamma := \big(a^{\gamma\beta}_{jk}(\partial_k u_\beta)\big)_{1\leq j\leq n} \;\; \text{in } \Omega \text{ for each } \gamma \in \{1, \dots, M\}, \tag{22.10}$$

we obtain that

$$\operatorname{div} \vec{F}_\gamma = 0 \in L^1_{\mathrm{loc}}(\partial\Omega, \sigma) \;\; \text{and} \;\; \mathcal{N}_\kappa \vec{F}_\gamma \in \mathrm{K}^{\alpha}_{p,q}(\partial\Omega, \sigma) \subseteq L^1_{\mathrm{loc}}(\partial\Omega, \sigma) \tag{22.11}$$

for each $\gamma \in \{1, \dots, M\}$. Thanks to these and Theorem 9.2, whenever $\widetilde{\gamma} \in (0,1)$ and $s \in \big(\frac{n-1}{n-1+\widetilde{\gamma}}, \infty\big)$ we have

$$\Big[\nu \cdot \big(\vec{F}_\gamma\big|_{\partial\Omega}^{\kappa-\mathrm{n.t.}}\big)\Big]^{\#}_{\widetilde{\gamma}} \leq C\mathcal{M}_{\partial\Omega,s}\big(\mathcal{N}_\kappa \vec{F}_\gamma\big) \quad \text{everywhere on } \partial\Omega, \tag{22.12}$$

for some constant $C = C(\partial\Omega, n, \kappa, s, \widetilde{\gamma}) \in (0, \infty)$.

Since $\alpha < \frac{n-1}{p'} + 1$, choose $\widetilde{\gamma}$ such that

$$\alpha - \tfrac{n-1}{p'} < \widetilde{\gamma} < 1. \tag{22.13}$$

In turn, this permits us to select s satisfying

$$\frac{n-1}{n-1+\widetilde{\gamma}} < s < \min\Big\{p, \Big(\frac{\alpha}{n-1} + \frac{1}{p}\Big)^{-1}\Big\}. \tag{22.14}$$

In particular, this and the lower bound for α in (22.7) give

$$\frac{-(n-1)}{p} < \alpha < (n-1)\Big(\frac{1}{s} - \frac{1}{p}\Big). \tag{22.15}$$

As such, Theorem 17.5 ensures that the L^s-based Hardy-Littlewood maximal operator $\mathcal{M}_{\partial\Omega,s}$ is bounded on $\mathrm{K}^{\alpha}_{p,q}(\partial\Omega, \sigma)$. Granted this and the membership $\mathcal{N}_\kappa \vec{F}_\gamma \in \mathrm{K}^{\alpha}_{p,q}(\partial\Omega, \sigma)$ (cf. (22.11)), we may conclude from (22.12) that

$$\begin{aligned}\Big\|\nu \cdot \big(\vec{F}_\gamma\big|_{\partial\Omega}^{\kappa-\mathrm{n.t.}}\big)\Big\|_{\mathrm{HK}^{\alpha}_{p,q}(\partial\Omega,\sigma)} &= \Big\|\Big[\nu \cdot \big(\vec{F}_\gamma\big|_{\partial\Omega}^{\kappa-\mathrm{n.t.}}\big)\Big]^{\#}_{\widetilde{\gamma}}\Big\|_{\mathrm{K}^{\alpha}_{p,q}(\partial\Omega,\sigma)} \\ &\leq C\|\mathcal{N}_\kappa \vec{F}_\gamma\|_{\mathrm{K}^{\alpha}_{p,q}(\partial\Omega,\sigma)} < \infty,\end{aligned} \tag{22.16}$$

for each $\gamma \in \{1, \dots, M\}$, and for some $C \in (0, \infty)$ depending exclusively on n, p, q, α, $\widetilde{\gamma}$, s, and the Ahlfors regularity constants of $\partial\Omega$. Being mindful of (22.10) and the fact that $(\partial^A_\nu u)_\gamma = \nu \cdot (\vec{F}_\gamma)\big|_{\partial\Omega}^{\kappa-\mathrm{n.t.}}$ (cf. (9.13)), estimate (22.16) readily implies (22.9). This completes the proof of Theorem 22.3. □

We are now in a position to solve the Neumann Problem for weakly elliptic systems in δ-AR domains with boundary data in second-generation Herz-type Hardy spaces.

Theorem 22.4 *Assume $\Omega \subseteq \mathbb{R}^n$ is an Ahlfors regular domain. Set $\sigma := \mathcal{H}^{n-1}\lfloor\partial\Omega$ and denote by ν the geometric measure theoretic outward unit normal to Ω. Select indices $p, q, \alpha \in \mathbb{R}$ satisfying*

$$1 < p < \infty, \quad 0 < q \leq 1, \quad \text{and} \quad \tfrac{n-1}{p'} \leq \alpha < 1 + \tfrac{n-1}{p'}. \tag{22.17}$$

Suppose L is a second-order, homogeneous, constant complex coefficients, weakly elliptic $M \times M$ system in $\mathbb{R}^n$, and fix a coefficient tensor $A \in \mathfrak{A}_L$. Finally, pick an aperture parameter $\kappa \in (0, \infty)$. Consider the Neumann Problem for L in Ω with *boundary data in the second-generation Herz-type Hardy space* $[\mathrm{HK}^{\alpha}_{p,q}(\partial\Omega,\sigma)]^M$ *formulated as follows:*

$$\begin{cases} u \in [\mathscr{C}^\infty(\Omega)]^M, \\ Lu = 0 \ \text{ in } \Omega, \\ \mathcal{N}_\kappa(\nabla u) \in \mathrm{K}^{\alpha}_{p,q}(\partial\Omega,\sigma), \\ \partial^A_\nu u = f \in [\mathrm{HK}^{\alpha}_{p,q}(\partial\Omega,\sigma)]^M. \end{cases} \tag{22.18}$$

Then the following statements are valid:

(a) [Existence, Integral Representation, and Estimates] *If* $A^\top \in \mathfrak{A}^{\text{dis}}_{L^\top}$ *then there exists some threshold* $\delta \in (0,1)$ *depending only on* n, p, q, α, κ, A, *and the Ahlfors regularity constants of* $\partial\Omega$ *with the property that if*

$$\|\nu\|_{[\mathrm{BMO}(\partial\Omega,\sigma)]^n} < \delta \tag{22.19}$$

it follows that the Neumann Problem formulated in (22.18) *is solvable. Concretely, the operator* $-\frac{1}{2}I + K^{\#}_{A^\top}$ *is invertible on the spaces* $[\mathrm{HK}^{\alpha}_{p,q}(\partial\Omega,\sigma)]^M$ *and the function* $u : \Omega \to \mathbb{C}^M$ *defined by*

$$u := \mathscr{S}_{\text{mod}}\big[\big(-\tfrac{1}{2}I + K^{\#}_{A^\top}\big)^{-1} f\big] \ \text{ in } \Omega \tag{22.20}$$

is a solution of the Neumann Problem (22.18) *and satisfies*

$$\|\mathcal{N}_\kappa(\nabla u)\|_{\mathrm{K}^{\alpha}_{p,q}(\partial\Omega,\sigma)} \approx \|f\|_{[\mathrm{HK}^{\alpha}_{p,q}(\partial\Omega,\sigma)]^M} \tag{22.21}$$

with implicit proportionality constants independent of the boundary datum f.

(b) [Uniqueness modulo constants] *If* $A \in \mathfrak{A}^{\text{dis}}_{L}$ *then there exists a threshold* $\delta \in (0,1)$ *depending only on* n, p, q, α, κ, A, *and the Ahlfors regularity constants of* $\partial\Omega$ *with the property that whenever* $\|\nu\|_{[\mathrm{BMO}(\partial\Omega,\sigma)]^n} < \delta$ *it follows that any two solutions of* (22.18) *differ by a constant in* $\mathbb{C}^M$.

(c) [Well-Posedness] *If* $A \in \mathfrak{A}^{\text{dis}}_{L}$ *and* $A^\top \in \mathfrak{A}^{\text{dis}}_{L^\top}$ *then there exists* $\delta \in (0,1)$ *depending only on* n, p, q, α, κ, A, *and the Ahlfors regularity constants of* $\partial\Omega$ *such that whenever* $\|\nu\|_{[\mathrm{BMO}(\partial\Omega,\sigma)]^n} < \delta$ *it follows that the Neumann Problem* (22.18) *is solvable, its solution is unique modulo constants, and satisfies* (22.21).

Proof The validity of all claims may be justified via the same type of argument used to establish Theorem 16.8, now bearing in mind the invertibility result recorded in Theorem 22.2, the nontangential maximal function estimates from Theorem 21.2, along with estimate (22.9) and the embedding in (20.9). □

Chapter 23

A New Class of Herz-type Spaces, Singular Integrals, and Boundary Problems

The topics treated in this chapter are as follows. In Section 23.1 we introduce a new class of Herz-type spaces, referred to as Composite Herz spaces, and explore some of their basic properties. In Section 23.2 we develop a duality theory for this new brand of Herz spaces. In Section 23.3 the notions of A_∞ indices and compliancy of sequence (quasi)-norms are introduced. These novel concepts play a fundamental role in the development of the theory of Composite Herz spaces. Sections 23.4 and 23.5 contain a wealth of examples of sequence spaces and function spaces that can be used in the blueprint of Composite Herz spaces. In Section 23.6 we study sublinear operators on Composite Herz spaces, such as the Hardy-Littlewood maximal operator, and prove some key boundedness results. In Section 23.7 we introduce and study a subspace of Composite Herz spaces, called vanishing Composite Herz space, which provides a natural environment in which the Hardy-Littlewood maximal operator is bounded. In Section 23.8 we develop a Calderón-Zygmund theory for singular integral operators on uniformly rectifiable sets in the context of Composite Herz spaces. This includes boundedness results for singular integral operators of convolution-type and geometrically sensitive operator norm estimates for singular integral operators of chord-dot-normal type. In Section 23.9 we provide an ample list of examples of existing Herz-type spaces, and produce new ones based on our template for Composite Herz spaces. In Section 23.10 we define a new brand of spaces, which we call vanishing Morrey-based Herz spaces, and establish connections with vanishing Composite Herz spaces. Section 23.11 contains a boundedness result for a certain class of fractional sublinear operators, while in Section 23.12 an extrapolation result for Composite Herz spaces is established. This body of work culminates with the solution of the Dirichlet, Neumann, Regularity, and Transmission Problems for weakly elliptic systems in δ-AR domains with boundary data in Composite Herz spaces, in Section 23.13

© The Author(s), under exclusive license to Springer Nature Switzerland AG 2026
M. Mitrea and P. Takemura, *Singular Integrals, Herz-Type Function Spaces, and Boundary Problems*, Progress in Mathematics 362,
https://doi.org/10.1007/978-3-032-12516-3_23

23.1 Composite Herz Spaces

Motivated by the goal of generalizing and unifying the large family of Herz-type spaces, here we introduce a new class of spaces which we refer to as Composite Herz spaces. This requires some preliminary work. We begin by defining the notion of a Generalized Quasi-Banach Function Space.

Definition 23.1 Consider $(X, \mathfrak{M}, \mu)$ a measure space, and let $\mathscr{M}(X, \mu)$ and $\mathscr{M}_+(X, \mu)$ be as in (10.135) and (10.136). A mapping $\rho : \mathscr{M}_+(X, \mu) \to [0, \infty]$ is called a `function quasi-norm` if there exists some constant $K \in [1, \infty)$ such that for all $f, g \in \mathscr{M}_+(X, \mu)$ the following properties hold:

(Q1) $\rho(f) = 0$ if and only if $f = 0$ μ-a.e. in X, for each $\lambda \in [0, \infty)$ it follows that $\rho(\lambda f) = \lambda \rho(f)$, and $\rho(f + g) \leq K\big(\rho(f) + \rho(g)\big)$;

(Q2) if $f \leq g$ pointwise μ-a.e. in X then $\rho(f) \leq \rho(g)$;

(Q3) if $\{f_k\}_{k\in\mathbb{N}} \subseteq \mathscr{M}_+(X, \mu)$ is such that $f_k \nearrow f$ pointwise μ-a.e. in X as $k \to \infty$, then $\rho(f_k) \nearrow \rho(f)$ as $k \to \infty$;

(Q4) there exists a collection of sets $\{Y_j\}_{j\in\mathbb{N}} \subseteq \mathfrak{M}$ such that $X = \cup_{j\in\mathbb{N}} Y_j$, and $\rho(\mathbb{1}_{Y_j}) < \infty$ for all $j \in \mathbb{N}$;

(Q5) there exists a collection of sets $\{Z_j\}_{j\in\mathbb{N}} \subseteq \mathfrak{M}$ satisfying $X = \cup_{j\in\mathbb{N}} Z_j$, and such that for every $j \in \mathbb{N}$ there exists some constant $c_j \in (0, \infty)$ for which $\int_{Z_j} f \, d\mu \leq c_j \rho(f)$.

The smallest constant $K \in [1, \infty)$ for which the quasi-triangle inequality in (Q1) holds is called the `modulus of concavity` and it is denoted by $K_{\mathbb{X}}$.

If ρ is a function quasinorm as above, for each $f \in \mathscr{M}(X, \mu)$ introduce

$$\|f\|_{\mathbb{X}} := \rho(|f|) \in [0, \infty]. \tag{23.1}$$

With this piece of notation, the set

$$\mathbb{X} := \{f \in \mathscr{M}(X, \mu) : \|f\|_{\mathbb{X}} < \infty\} \tag{23.2}$$

is called a `Generalized Quasi-Banach Function Space` (GQBFS) on (X, μ). When $X \subseteq \mathbb{Z}$ is countable and μ is the counting measure, refer to $\mathbb{X}$ as a `Generalized Quasi-Banach Sequence Space` (GQBSS) on (X, μ). Moreover, when the modulus of concavity $K_{\mathbb{X}} = 1$, refer to $\mathbb{X}$ as a `Generalized Banach Sequence Space` (GQBSS) on (X, μ). In this case, $\mathscr{M}(X, \mu) = \mathbb{R}^X$ is the collection of all sequences in $\mathbb{R}$ indexed by X.

In relation to each function quasinorm ρ, the function $\rho' : \mathscr{M}_+(X, \mu) \to [0, \infty]$ as in (10.145) will still be called the associated norm of ρ. Denoting by

$$\|g\|_{\mathbb{X}'} := \rho'(|g|) = \sup\left\{\int_X |fg|\,d\mu : f \in \mathbb{X},\ \ \|f\|_{\mathbb{X}} < 1\right\} \text{ for all } g \in \mathcal{M}(\Sigma, \sigma), \tag{23.3}$$

call the collection $\mathbb{X}'$ defined as in (10.148) the `associated space` of the Generalized Quasi-Banach Function Space $\mathbb{X}$. When $\mathbb{X}$ is a Generalized Quasi-Banach Sequence space, for each $a = \{a_j\}_{j\in X} \in \mathbb{R}^X$ formula (23.3) reads as

$$\|a\|_{\mathbb{X}'} = \sup\left\{\sum_{j\in X} |a_j b_j| : b = \{b_j\}_{j\in X} \in \mathbb{X},\ \ \|b\|_{\mathbb{X}} < 1\right\}. \tag{23.4}$$

Remark 23.1 The notion of Generalized Quasi-Banach Function Spaces is a natural generalization of Generalized Banach Function Spaces (introduced in [GHA-II, Chapter 5]; see also Definition 10.2). The only difference here is that now the function ρ satisfies a quasi-triangle inequality.

Remark 23.2 In the context of Definition 23.1, if $X = \mathbb{Z}$ (respectively $X = \mathbb{N}_0$) we often refer to the corresponding set $\mathbb{X}$ (cf. (23.2)) as Generalized Quasi-Banach Sequence Space on $\mathbb{Z}$ (respectively $X = \mathbb{N}_0$), with the understanding that $\mathbb{Z}$ (respectively $X = \mathbb{N}_0$) is naturally equipped with the counting measure.

Remark 23.3 It is pertinent to remark that, as seen from (Q2) in Definition 23.1 and (23.1), for any two functions $f, g \in \mathcal{M}(X, \mu)$ we have

$$f = g \quad \text{pointwise } \mu\text{-a.e. in } X \quad \Longrightarrow \quad \|f\|_{\mathbb{X}} = \|g\|_{\mathbb{X}}. \tag{23.5}$$

Moreover, it is visible from (Q2) in Definition 23.1 that

$$\text{if } |g| \le |f| \text{ at } \mu\text{-a.e. in } X, \text{ and } f \in \mathbb{X}, \text{ then } g \text{ also belong to } \mathbb{X} \text{ and } \|g\|_{\mathbb{X}} \le \|f\|_{\mathbb{X}}. \tag{23.6}$$

In particular, this entails that

the operator of pointwise multiplication by some function $b \in L^\infty(X, \mu)$ is a bounded mapping from $\mathbb{X}$ into itself, and has operator norm $\le \|b\|_{L^\infty(X,\mu)}$. (23.7)

Remark 23.4 In the survey paper [LoNi24], the authors introduce Quasi-Banach Function Spaces as in (23.2) by demanding, beforehand, that $\mathbb{X}$ is complete and is associated with a function quasinorm ρ satisfying only properties (Q1), (Q2), together with the following saturation property:

$$\begin{gathered}\text{for every } E \in \mathfrak{M}, \text{ with } \mu(E) > 0, \text{ there exists } F \subseteq E \\ \text{a } \mu\text{-measurable set such that } \mu(F) > 0 \text{ and } \rho(\mathbb{1}_F) < \infty.\end{gathered} \tag{23.8}$$

It turns out that this is equivalent to (Q3) (cf. [LoNi24, Proposition 2.5, p. 254]). It has also been shown in [LoNi24, Proposition 2.2, p. 252] that a quasi-normed space $\mathbb{X}$ is complete if and only if $\mathbb{X}$ satisfies the Riesz-Fischer property, i.e., if $\{f_k\}_{k\in\mathbb{N}} \subseteq \mathbb{X}$, and $\sum_{k=1}^{\infty} K_{\mathbb{X}}^k \|f_k\|_{\mathbb{X}} < \infty$, where $K_{\mathbb{X}} \in [1, \infty)$ is the modulus of concavity of $\|\cdot\|_{\mathbb{X}}$, then $\sum_{k=1}^{\infty} f_k$ converges in $\mathbb{X}$, and $\|\sum_{k=1}^{\infty} f_k\|_{\mathbb{X}} \leq K_{\mathbb{X}} \sum_{k=1}^{\infty} K_{\mathbb{X}}^k \|f_k\|_{\mathbb{X}}$. (23.9)

It has been pointed out in [LoNi24, Remark 2.1, p. 251] that this condition is implied by the Fatou Property (Q3) in Definition 23.1. As such, it follows that

Generalized Quasi-Banach Function Spaces (in the sense of Definition 23.1) are complete, hence genuine quasi-Banach spaces. (23.10)

In relation to the notion of Generalized Banach Function Space, we make the following remarks.

Remark 23.5 In the context of Definition 23.1, observe that if the modulus of concavity $K_{\mathbb{X}}$ equals 1, then $\mathbb{X}$ becomes a Generalized Banach Function Space on (X, μ) (in the sense of Definition 10.2). Moreover, a version of the generalized Hölder inequality (10.149) also holds in the case when $\mathbb{X}$ is a Generalized Quasi-Banach Function Space. Specifically, for every $f, g \in \mathscr{M}(X, \mu)$ we have

$$\int_X |fg| \, d\mu \leq \|f\|_{\mathbb{X}} \|g\|_{\mathbb{X}'}. \tag{23.11}$$

This follows from the same reasoning presented in the proof of [GHA-II, Proposition 5.1.12, p. 226], as its validity is independent of the triangle inequality. In particular, whenever $\mathbb{X}$ is a Generalized Quasi-Banach Sequence Space, for any two sequences $a = \{a_j\}_{j\in X}$ and $b = \{b_j\}_{j\in X}$ in $\mathbb{R}^X$ inequality (23.11) reads as

$$\sum_{j\in X} |a_j b_j| \leq \|a\|_{\mathbb{X}} \|b\|_{\mathbb{X}'}. \tag{23.12}$$

Remark 23.6 Notice that if ρ' is the associated norm of a function quasinorm ρ, then ρ' satisfies a genuine triangle inequality. This follows immediately from standard properties of integrals. As such, this observation combined with the proof of [GHA-II, Proposition 5.1.10, p. 223] ensures that if ρ is a function quasinorm on (X, μ), then its associated norm ρ' is actually a function norm on (X, μ). Hence, if $\mathbb{X}$ is a Generalized Quasi-Banach Function Space, the associated space $\mathbb{X}'$ is a Generalized Banach Function Space.

Remark 23.7 It is worth mentioning that if $\mathcal{J} \subseteq \mathbb{Z}$ is a countable set, μ is the counting measure on $\mathcal{J}$, and $\mathbb{X}$ is a Generalized Quasi-Banach Sequence Space on $(\mathcal{J}, \mu)$, then for each $\ell \in \mathcal{J}$ the sequence $\mathbb{1}_{\{\ell\}} = \{\delta_{\ell k}\}_{k\in\mathcal{J}}$ belongs to $\mathbb{X}$, where $\delta_{\ell k}$ denotes the Kronecker delta. Indeed, let $\{Y_j\}_{j\in\mathbb{N}}$ be a partition of $\mathcal{J}$ satisfying property (Q4) in Definition 23.1 for the function quasinorm $\|\cdot\|_{\mathbb{X}}$. Then whenever $\ell \in \mathcal{J}$ there exists some $j_\ell \in \mathbb{N}$ for which $\{\ell\} \subseteq Y_{j_\ell}$. This and (Q2) in Definition 23.1 imply

$$\mathbb{1}_{\{\ell\}} \leq \mathbb{1}_{Y_{j_\ell}} \implies \|\mathbb{1}_{\{\ell\}}\|_{\mathbb{X}} \leq \|\mathbb{1}_{Y_{j_\ell}}\|_{\mathbb{X}} < \infty. \tag{23.13}$$

Hence, $\mathbb{1}_{\{\ell\}} \in \mathbb{X}$ for all $\ell \in \mathbb{Z}$, as desired. The same type of reasoning can be used to show that $\mathbb{1}_{\{\ell\}} \in \mathbb{X}'$ for all $\ell \in \mathbb{Z}$.

The purpose of the following result is to give a more streamlined description of Generalized Quasi-Banach Sequence Spaces.

Proposition 23.1 *Let $\mathcal{J} \subsetneq \mathbb{Z}$ an infinite set, and denote by μ the counting measure on $\mathcal{J}$. Consider $\mathbb{R}^{\mathcal{J}}$ (respectively $\mathbb{R}_+^{\mathcal{J}}$) the collection of all sequences in $\mathbb{R}$ (respectively $\mathbb{R}_+$) indexed by $\mathcal{J}$, i.e., $\mathscr{M}(\mathcal{J},\mu) = \mathbb{R}^{\mathcal{J}}$ and $\mathscr{M}_+(\mathcal{J},\mu) = \mathbb{R}_+^{\mathcal{J}}$. Then $\rho : \mathscr{M}_+(\mathcal{J},\mu) \to [0,\infty)$ is a function quasinorm if and only if there exists $K \in [1,\infty)$ such that for every two sequences $a = \{a_j\}_{j\in\mathcal{J}}$ and $b = \{b_j\}_{j\in\mathcal{J}}$ in $\mathbb{R}_+^{\mathcal{J}}$ the following properties hold:*

(Q1’) *$\rho(a) = 0$ if and only if $a_j = 0$ for all $j \in \mathcal{J}$, for each $\lambda \in [0,\infty)$ one has $\rho(\lambda a) = \lambda\rho(a)$, and $\rho(a+b) \leq K\big(\rho(a)+\rho(b)\big)$;*

(Q2’) *if $a_j \leq b_j$ for all $j \in \mathcal{J}$ then $\rho(a) \leq \rho(b)$;*

(Q3’) *whenever $k \in \mathbb{N}$ and $a^{(k)} = \{a_j^{(k)}\}_{j\in\mathcal{J}}$ is a sequence of non-negative numbers such that $a_j^{(k)} \nearrow a_j$ as $k \to \infty$ for each $j \in \mathcal{J}$, then $\rho(a^{(k)}) \nearrow \rho(a)$ as $k \to \infty$;*

(Q4’) *for all $j \in \mathcal{J}$ one has $\rho(\mathbb{1}_{\{j\}}) < \infty$;*

(Q5’) *for every $j \in \mathcal{J}$ there exists a constant $c_j \in (0,\infty)$ satisfying $a_j \leq c_j\rho(a)$.*

Proof It is visible that properties (Q1’)-(Q3’) are just a rephrasing of (Q1)-(Q3) in terms of sequences and the counting measure. To see that (Q4’) implies (Q4), consider the collection $\{Y_j\}_{j\in\mathcal{J}}$ defined as $Y_j := \{j\}$ for all $j \in \mathcal{J}$. Reindexing it by $\mathbb{N}$ in place of $\mathcal{J}$ the does the job. Also, it is apparent that (Q5’) implies (Q5) for $Z_j := Y_j$ for all $j \in \mathcal{J}$.

In the opposite direction, that (Q4) implies (Q4’) can be seem by the same reasoning as in Remark 23.7. Lastly, assume that ρ satisfies (Q5). Then there exists a partition $\{Z_k\}_{k\in\mathbb{N}}$ of $\mathcal{J}$ with the property that for each $k \in \mathbb{N}$ there exists some $c_k \in (0,\infty)$ satisfying

$$\sum_{\ell\in Z_k} a_\ell \leq c_k\rho(a). \tag{23.14}$$

In particular, for each $j \in \mathcal{J}$ there exists some $k_j \in \mathbb{N}$ for which $j \in Z_{k_j}$. Hence

$$a_j \leq \sum_{\ell\in Z_{k_j}} a_\ell \leq c_{k_j}\rho(a), \tag{23.15}$$

which shows that ρ satisfies (Q5’). This concludes the proof of Proposition 23.1. □

Next, we introduce the Composite Herz spaces.

Definition 23.2 Assume $\Sigma \subseteq \mathbb{R}^n$ is a closed unbounded Ahlfors regular set, and abbreviate $\sigma := \mathcal{H}^{n-1}\lfloor\Sigma$. Denote by $c_\Sigma, C_\Sigma \in (0,\infty)$ the lower and upper Ahlfors regularity constant of Σ, respectively. Fix a dilation factor $2_* > (C_\Sigma/c_\Sigma)^{1/(n-1)}$ along with a reference point $x_0 \in \Sigma$. Define

$$C_k := \Delta(x_0, 2_*^{k+1})\backslash\Delta(x_0, 2_*^k) \quad \text{for all } k \in \mathbb{Z}, \tag{23.16}$$

and

$$\widetilde{C}_0 := \Delta(x_0, 2_*) \quad \text{and} \quad \widetilde{C}_k := C_k \quad \text{for all } k \in \mathbb{N}. \tag{23.17}$$

Let $\mathbb{X}$ be a Generalized Banach Function Space on (Σ, σ). Suppose that $\mathbb{Y}$ is a Generalized Quasi-Banach Sequence Space on $(\mathbb{Z}, \mu)$, where μ denotes the counting measure on $\mathbb{Z}$. For each function $f \in \mathscr{M}(\Sigma, \sigma)$ define

$$\|f\|_{\dot{\mathrm{K}}(\Sigma;\mathbb{X};\mathbb{Y})} := \Big\|\big\{\|f \cdot \mathbb{1}_{C_k}\|_{\mathbb{X}}\big\}_{k\in\mathbb{Z}}\Big\|_{\mathbb{Y}} \in [0,\infty]. \tag{23.18}$$

Introduce the `homogeneous Composite Herz Space` associated with $\mathbb{X}$ and $\mathbb{Y}$ as

$$\dot{\mathrm{K}}(\Sigma;\mathbb{X};\mathbb{Y}) := \big\{f \in \mathscr{M}(\Sigma,\sigma) : \|f\|_{\dot{\mathrm{K}}(\Sigma;\mathbb{X};\mathbb{Y})} < \infty\big\}. \tag{23.19}$$

With μ still denoting the counting measure, suppose now that $\mathbb{Y}$ is a Generalized Quasi-Banach Sequence Space on $(\mathbb{N}_0, \mu)$. For every function $f \in \mathscr{M}(\Sigma, \sigma)$ let

$$\|f\|_{\mathrm{K}(\Sigma;\mathbb{X};\mathbb{Y})} := \Big\|\big\{\|f \cdot \mathbb{1}_{\widetilde{C}_k}\|_{\mathbb{X}}\big\}_{k\in\mathbb{N}_0}\Big\|_{\mathbb{Y}} \in [0,\infty]. \tag{23.20}$$

Introduce the `inhomogeneous Composite Herz Space` associated with $\mathbb{X}$ and $\mathbb{Y}$ as

$$\mathrm{K}(\Sigma;\mathbb{X};\mathbb{Y}) := \big\{f \in \mathscr{M}(\Sigma,\sigma) : \|f\|_{\mathrm{K}(\Sigma;\mathbb{X};\mathbb{Y})} < \infty\big\}. \tag{23.21}$$

Call $\mathbb{X}$ and $\mathbb{Y}$ the `base spaces` for $\dot{\mathrm{K}}(\Sigma;\mathbb{X};\mathbb{Y})$ and $\mathrm{K}(\Sigma;\mathbb{X};\mathbb{Y})$.

For the following two remarks, work under the same geometric setting as in Definition 23.2.

Remark 23.8 We claim that for every function $f \in \dot{\mathrm{K}}(\Sigma;\mathbb{X};\mathbb{Y})$ one has

$$f \cdot \mathbb{1}_{C_\ell} \in \mathbb{X} \quad \text{for all } \ell \in \mathbb{Z}. \tag{23.22}$$

To justify this claim, use the generalized Hölder inequality on $\mathbb{Y}$ (cf. (23.12)) along with the fact that $\mathbb{1}_\ell = \{\delta_{k\ell}\}_{k\in\mathbb{Z}} \in \mathbb{Y}'$ (see (Q4') in Proposition 23.1) and (23.18) to obtain

$$\begin{aligned}\|f\cdot \mathbb{1}_{C_\ell}\|_{\mathbb{X}} &= \sum_{k=-\infty}^{\infty} \|f\cdot \mathbb{1}_{C_k}\|_{\mathbb{X}}\cdot \delta_{k\ell} \leq \Big\|\big\{\|f\cdot \mathbb{1}_{C_k}\|_{\mathbb{X}}\big\}_{k\in\mathbb{Z}}\Big\|_{\mathbb{Y}} \|\{\delta_{k\ell}\}_{k\in\mathbb{Z}}\|_{\mathbb{Y}'} \\ &= \|f\|_{\dot{\mathrm{K}}(\Sigma;\mathbb{X};\mathbb{Y})} \|\mathbb{1}_{\{\ell\}}\|_{\mathbb{Y}'} < \infty. \end{aligned} \tag{23.23}$$

Remark 23.9 It is clear from definitions that $\dot{\mathrm{K}}(\Sigma;\mathbb{X};\mathbb{Y})$ satisfies the lattice property, i.e.,

$$\text{if } f,g:\Sigma\to\mathbb{R} \text{ are two } \sigma\text{-measurable functions with the property that } |g|\leq|f| \text{ at } \sigma\text{-a.e. point on } \Sigma \text{ and } f\in\dot{\mathrm{K}}(\Sigma;\mathbb{X};\mathbb{Y}), \text{ then } g \text{ also belongs to } \dot{\mathrm{K}}(\Sigma;\mathbb{X};\mathbb{Y}), \text{ and } \|g\|_{\dot{\mathrm{K}}(\Sigma;\mathbb{X};\mathbb{Y})} \leq \|f\|_{\dot{\mathrm{K}}(\Sigma;\mathbb{X};\mathbb{Y})}. \tag{23.24}$$

For later use, another relevant property is that

$$\text{the operator of pointwise multiplication by a given function } b\in L^\infty(\Sigma,\sigma) \text{ is a bounded mapping from } \dot{\mathrm{K}}(\Sigma;\mathbb{X};\mathbb{Y}) \text{ into itself, with operator norm } \leq \|b\|_{L^\infty(\Sigma,\sigma)}. \tag{23.25}$$

The observations made in Remarks 23.8 and 23.9 are also valid in the context of inhomogeneous Composite Herz spaces.

One of our main goals in this section is to establish that Composite Herz spaces are Generalized Quasi-Banach Function Spaces, as well as to identify their associated spaces. We begin by providing a generalized Hölder inequality on Composite Herz spaces.

Proposition 23.2 *Suppose $\Sigma\subseteq\mathbb{R}^n$ is a closed unbounded Ahlfors regular set, and abbreviate $\sigma:=\mathcal{H}^{n-1}\lfloor\Sigma$. Let $\mathbb{X}$ be a Generalized Banach Function Space on (Σ,σ), and $\mathbb{Y}$ a Generalized Quasi-Banach Sequence Space on $(\mathbb{Z},\mu)$, where μ denotes the counting measure. Denote by $\mathbb{X}'$ and $\mathbb{Y}'$ the associated spaces of $\mathbb{X}$ and $\mathbb{Y}$, respectively. Then for every $f,g\in\mathscr{M}(\Sigma,\sigma)$ one has*

$$\int_\Sigma |fg|\,d\sigma \leq \|f\|_{\dot{\mathrm{K}}(\Sigma;\mathbb{X};\mathbb{Y})}\|g\|_{\dot{\mathrm{K}}(\Sigma;\mathbb{X}';\mathbb{Y}')}. \tag{23.26}$$

The same estimate holds in the context of inhomogeneous Composite Herz spaces.

Proof Fix two σ-measurable functions f and g defined on Σ. Using the generalized Hölder inequality on $\mathbb{X}$ (cf. (23.11)) and on $\mathbb{Y}$ (cf. (23.12)), we may estimate

$$\begin{aligned}\int_\Sigma |fg|\,d\sigma &= \sum_{k=-\infty}^{\infty}\int_{C_k}|fg|\,d\sigma \leq \sum_{k=-\infty}^{\infty}\|f\cdot\mathbb{1}_{C_k}\|_{\mathbb{X}}\cdot\|g\cdot\mathbb{1}_{C_k}\|_{\mathbb{X}'} \\ &\leq \Big\|\big\{\|f\cdot\mathbb{1}_{C_k}\|_{\mathbb{X}}\big\}_{k\in\mathbb{Z}}\Big\|_{\mathbb{Y}}\cdot\Big\|\big\{\|g\cdot\mathbb{1}_{C_k}\|_{\mathbb{X}'}\big\}_{k\in\mathbb{Z}}\Big\|_{\mathbb{Y}'} \\ &= \|f\|_{\dot{\mathrm{K}}(\Sigma;\mathbb{X};\mathbb{Y})}\|g\|_{\dot{\mathrm{K}}(\Sigma;\mathbb{X}';\mathbb{Y}')}. \end{aligned} \tag{23.27}$$

This establishes the validity of (23.26), and the same is true for inhomogeneous Composite Herz spaces. □

It is useful to record a version of Lebesgue's Monotone Convergence Theorem on Composite Herz spaces.

Proposition 23.3 *Consider a closed unbounded Ahlfors regular set $\Sigma \subseteq \mathbb{R}^n$, and abbreviate $\sigma := \mathcal{H}^{n-1}\lfloor\Sigma$. Assume $\mathbb{X}$ is a Generalized Banach Function Space on (Σ, σ), and $\mathbb{Y}$ be a Generalized Quasi-Banach Sequence Space on $(\mathbb{Z}, \mu)$, where μ denotes the counting measure. Assume $\{f_j\}_{j\in\mathbb{N}}$ is a sequence of non-negative σ-measurable functions on Σ that increases to some $f \in \mathcal{M}_+(\Sigma, \sigma)$ at σ-a.e. point on Σ as $j \to \infty$. Then*

$$\|f_j\|_{\dot{\mathbb{K}}(\Sigma;\mathbb{X};\mathbb{Y})} \nearrow \|f\|_{\dot{\mathbb{K}}(\Sigma;\mathbb{X};\mathbb{Y})} \quad \textit{as } j \to \infty. \tag{23.28}$$

The same convergence holds for inhomogeneous Composite Herz spaces.

Proof The convergence assumption made over $\{f_j\}_{j\in\mathbb{N}}$ implies that for each $k \in \mathbb{Z}$ we have $f_j \cdot \mathbb{1}_{C_k} \nearrow f \cdot \mathbb{1}_{C_k}$ as $j \to \infty$ at σ-a.e. point on Σ, where $\{C_k\}_{k\in\mathbb{Z}}$ is defined as in (23.16). In turn, since $\|\cdot\|_{\mathbb{X}}$ is a function norm (see (P3) from Definition 10.2), for each $k \in \mathbb{Z}$ it follows that $\|f_j \cdot \mathbb{1}_{C_k}\|_{\mathbb{X}} \nearrow \|f \cdot \mathbb{1}_{C_k}\|_{\mathbb{X}}$ as $j \to \infty$. Using (Q3') from Proposition 23.1 we then obtain

$$\Big\|\big\{\|f_j \cdot \mathbb{1}_{C_k}\|_{\mathbb{X}}\big\}_{k\in\mathbb{Z}}\Big\|_{\mathbb{Y}} \nearrow \Big\|\big\{\|f \cdot \mathbb{1}_{C_k}\|_{\mathbb{X}}\big\}_{k\in\mathbb{Z}}\Big\|_{\mathbb{Y}} \quad \text{as } j \to \infty. \tag{23.29}$$

This and (23.18) readily imply (23.28). □

We now establish that Composite Herz spaces are themselves Generalized Quasi-Banach Function Spaces.

Proposition 23.4 *Consider $\Sigma \subseteq \mathbb{R}^n$ a closed unbounded Ahlfors regular set, and abbreviate $\sigma := \mathcal{H}^{n-1}\lfloor\Sigma$. Let $\mathbb{X}$ be a Generalized Banach Function Space on (Σ, σ), and $\mathbb{Y}$ a Generalized Quasi-Banach Sequence Space on $(\mathbb{Z}, \mu)$, where μ is the counting measure. Also, denote by $K_{\mathbb{Y}} \in [1, \infty)$ the modulus of concavity of the quasinorm $\|\cdot\|_{\mathbb{Y}}$.*

Then $\|\cdot\|_{\dot{\mathbb{K}}(\Sigma;\mathbb{X};\mathbb{Y})}$ is a function quasinorm (in the sense of Definition 23.1) with modulus of concavity $\leq K_{\mathbb{Y}}$. Consequently, $\big(\dot{\mathbb{K}}(\Sigma; \mathbb{X}; \mathbb{Y}), \|\cdot\|_{\dot{\mathbb{K}}(\Sigma;\mathbb{X};\mathbb{Y})}\big)$ is a Generalized Quasi-Banach Function Space on (Σ, σ). In particular, whenever $\mathbb{Y}$ is a Generalized Banach Sequence Space on $(\mathbb{Z}, \mu)$, then $\big(\dot{\mathbb{K}}(\Sigma; \mathbb{X}; \mathbb{Y}), \|\cdot\|_{\dot{\mathbb{K}}(\Sigma;\mathbb{X};\mathbb{Y})}\big)$ is actually a Generalized Banach Function Space on (Σ, σ). The same conclusions hold in the context of inhomogeneous Composite Herz spaces.

Proof The fact that the function $\|\cdot\|_{\dot{\mathbb{K}}(\Sigma;\mathbb{X};\mathbb{Y})}$ satisfies properties (Q1)-(Q2) in Definition 23.1 follows immediately from (23.18), and the fact that $\|\cdot\|_{\mathbb{X}}$ is a function norm and $\|\cdot\|_{\mathbb{Y}}$ is a function quasinorm. It is visible from (23.18) and definitions that $\|\cdot\|_{\dot{\mathbb{K}}(\Sigma;\mathbb{X};\mathbb{Y})}$ has modulus of concavity $\leq K_{\mathbb{Y}}$. From Proposition 23.3 it follows that $\|\cdot\|_{\dot{\mathbb{K}}(\Sigma;\mathbb{X};\mathbb{Y})}$ also satisfies property (Q3) in Definition 23.1.

Next, we verify that $\|\cdot\|_{\dot{\mathbb{K}}(\Sigma;\mathbb{X};\mathbb{Y})}$ satisfies property (Q4) in Definition 23.1. To this end, consider the family of annuli $\{C_\ell\}_{\ell\in\mathbb{Z}}$ as in Definition 23.2. Being $\|\cdot\|_{\mathbb{X}}$ a function norm on Σ (in the sense of Definition 10.2), let $\{Y_j\}_{j\in\mathbb{N}}$ be a collection of σ-measurable subsets on Σ such that

$$\Sigma = \bigcup_{j\in\mathbb{N}} Y_j \quad \text{and} \quad \|\mathbb{1}_{Y_j}\|_{\mathbb{X}} < \infty. \tag{23.30}$$

For each $j \in \mathbb{N}$ and $\ell \in \mathbb{Z}$ define

$$U_{j,\ell} := Y_j \cap C_\ell \subseteq Y_j. \tag{23.31}$$

It is clear from this that

$$\mathbb{1}_{U_{j,\ell}} \leq \mathbb{1}_{Y_j} \quad \text{for each } j \in \mathbb{N} \text{ and } \ell \in \mathbb{Z}. \tag{23.32}$$

From (23.16) and (23.30) we see that $\big\{U_{j,\ell}\big\}_{\substack{j\in\mathbb{N}\\ \ell\in\mathbb{Z}}}$ is a countable collection of σ-measurable sets that covers Σ up to a set of measure zero, i.e.,

$$\sigma\Big(\Sigma \setminus \big(\cup_{\substack{j\in\mathbb{N}\\ \ell\in\mathbb{Z}}} U_{j,\ell}\big)\Big) = \sigma(\{x_0\}) = 0. \tag{23.33}$$

Moreover, thanks to (23.16), (23.18), Remark 23.7, (23.30), (23.31), (23.32), and the lattice property of $\|\cdot\|_{\mathbb{X}}$ (cf. (P3) in Definition 10.2), we have

$$\begin{aligned} \|\mathbb{1}_{U_{j,\ell}}\|_{\dot{\mathbb{K}}(\Sigma;\mathbb{X};\mathbb{Y})} &= \Big\|\big\{\|\mathbb{1}_{U_{j,\ell}} \cdot \mathbb{1}_{C_k}\|_{\mathbb{X}}\big\}_{k\in\mathbb{Z}}\Big\|_{\mathbb{Y}} = \Big\|\big\{\|\mathbb{1}_{U_{j,\ell}}\|_{\mathbb{X}}\delta_{\ell k}\big\}_{k\in\mathbb{Z}}\Big\|_{\mathbb{Y}} \\ &= \|\mathbb{1}_{U_{j,\ell}}\|_{\mathbb{X}} \cdot \|\mathbb{1}_{\{\ell\}}\|_{\mathbb{Y}} \leq \|\mathbb{1}_{Y_j}\|_{\mathbb{X}} \cdot \|\mathbb{1}_{\{\ell\}}\|_{\mathbb{Y}} < \infty, \end{aligned} \tag{23.34}$$

for every $j \in \mathbb{N}$ and $\ell \in \mathbb{Z}$. As such, this shows that $\|\cdot\|_{\dot{\mathbb{K}}(\Sigma;\mathbb{X};\mathbb{Y})}$ satisfies (Q4) from Definition 23.1 with respect to the collection $\big\{U_{j,\ell}\big\}_{\substack{j\in\mathbb{N}\\ \ell\in\mathbb{Z}}}$.

We next show that $\|\cdot\|_{\dot{\mathbb{K}}(\Sigma;\mathbb{X};\mathbb{Y})}$ satisfies (Q5) in Definition 23.1. With this goal in mind, let $\{Z_j\}_{j\in\mathbb{N}}$ be a collection of σ-measurable subsets on Σ such that

$$\Sigma = \bigcup_{j\in\mathbb{N}} Z_j \quad \text{and} \quad \|\mathbb{1}_{Z_j}\|_{\mathbb{X}'} < \infty \tag{23.35}$$

(its existence is ensured from the fact that $\|\cdot\|_{\mathbb{X}'}$ is a function norm; cf. Remark 10.4 and Remark 10.5). For each $j \in \mathbb{N}$ and $\ell \in \mathbb{Z}$ introduce

$$V_{j,\ell} := Z_j \cap C_\ell. \tag{23.36}$$

Much as before, $\{V_{j,\ell}\}_{\substack{j\in\mathbb{N}\\ \ell\in\mathbb{Z}}}$ is a countable collection of σ-measurable sets that covers Σ up to a set of σ-measure zero.

Fix a σ-measurable function f on Σ. Using the generalized Hölder inequality on $\mathbb{X}$ (cf. (23.11)) and on $\mathbb{Y}$ (cf. (23.12)), being mindful of (23.36), for each $j \in \mathbb{N}$ and $\ell \in \mathbb{Z}$ we estimate

$$
\begin{aligned}
\int_{V_{j,\ell}} f\,d\sigma &= \int_{\Sigma}(f\cdot\mathbb{1}_{C_\ell})\cdot\mathbb{1}_{Z_j}\,d\sigma \le \|f\cdot\mathbb{1}_{C_\ell}\|_{\mathbb{X}}\cdot\|\mathbb{1}_{Z_j}\|_{\mathbb{X}'}\\
&= \|\mathbb{1}_{Z_j}\|_{\mathbb{X}'}\sum_{k=-\infty}^{\infty}\|f\cdot\mathbb{1}_{C_\ell}\|_{\mathbb{X}}\cdot\delta_{\ell k}\\
&\le \|\mathbb{1}_{Z_j}\|_{\mathbb{X}'}\cdot\Big\|\big\{\|f\cdot\mathbb{1}_{C_\ell}\|_{\mathbb{X}}\cdot\delta_{\ell k}\big\}_{k\in\mathbb{Z}}\Big\|_{\mathbb{Y}}\cdot\Big\|\big\{\delta_{\ell k}\big\}_{k\in\mathbb{Z}}\Big\|_{\mathbb{Y}'}\\
&\le \|\mathbb{1}_{Z_j}\|_{\mathbb{X}'}\cdot\|\mathbb{1}_{\{\ell\}}\|_{\mathbb{Y}'}\cdot\Big\|\big\{\|f\cdot\mathbb{1}_{C_k}\|_{\mathbb{X}}\big\}_{k\in\mathbb{Z}}\Big\|_{\mathbb{Y}}\\
&= \|\mathbb{1}_{Z_j}\|_{\mathbb{X}'}\cdot\|\mathbb{1}_{\{\ell\}}\|_{\mathbb{Y}'}\cdot\|f\|_{\dot{\mathbb{K}}(\Sigma;\mathbb{X};\mathbb{Y})}.
\end{aligned}
\tag{23.37}
$$

Based on (23.35) and Remark 23.7, we have that $c_{j,\ell} := \|\mathbb{1}_{Z_j}\|_{\mathbb{X}'}\cdot\|\mathbb{1}_{\{\ell\}}\|_{\mathbb{Y}'} \in (0,\infty)$ for every $j \in \mathbb{N}$ and $\ell \in \mathbb{Z}$. Hence, $\|\cdot\|_{\dot{\mathbb{K}}(\Sigma;\mathbb{X};\mathbb{Y})}$ satisfies (Q5) in Definition 23.1 with respect to $\{V_{j,\ell}\}_{\substack{j\in\mathbb{N}\\ \ell\in\mathbb{Z}}}$. This ultimately proves that $\|\cdot\|_{\dot{\mathbb{K}}(\Sigma;\mathbb{X};\mathbb{Y})}$ is a function quasinorm on Σ, thus establishing that the Composite Herz space $\dot{\mathbb{K}}(\Sigma;\mathbb{X};\mathbb{Y})$, when equipped with $\|\cdot\|_{\dot{\mathbb{K}}(\Sigma;\mathbb{X};\mathbb{Y})}$, is a Generalized Quasi-Banach Function Space on (Σ,σ). This completes the proof of Proposition 23.4. □

As a consequence of Proposition 23.4 and Remark 23.4, we have the completeness of Composite Herz spaces.

Corollary 23.1 *Assume $\Sigma \subseteq \mathbb{R}^n$ in a closed unbounded Ahlfors regular set, and abbreviate $\sigma := \mathcal{H}^{n-1}\lfloor\Sigma$. Let $\mathbb{X}$ be a Generalized Banach Function Space on (Σ,σ), and $\mathbb{Y}$ a Generalized Quasi-Banach Sequence Space on $(\mathbb{Z},\mu)$ where μ is the counting measure.*

Then the Composite Herz space $\dot{\mathbb{K}}(\Sigma;\mathbb{X};\mathbb{Y})$ is complete when equipped with the quasinorm $\|\cdot\|_{\dot{\mathbb{K}}(\Sigma;\mathbb{X};\mathbb{Y})}$, hence a quasi-Banach space. In particular, when $\mathbb{Y}$ is a Generalized Banach Sequence Space on $(\mathbb{Z},\mu)$, then $\dot{\mathbb{K}}(\Sigma;\mathbb{X};\mathbb{Y})$ is a genuine Banach space. The same conclusions hold for inhomogeneous Composite Herz spaces.

Proof The completeness of $\big(\dot{\mathbb{K}}(\Sigma;\mathbb{X};\mathbb{Y}),\|\cdot\|_{\dot{\mathbb{K}}(\Sigma;\mathbb{X};\mathbb{Y})}\big)$ readily follows from Proposition 23.4 and the fact that Generalized Quasi-Banach Function Spaces are quasi-Banach spaces (cf. (23.10)). □

23.2 Duality and Associated Spaces

Here we discuss the duality theory of Composite Herz spaces. We begin by identifying their associated space.

Proposition 23.5 *Consider a closed unbounded Ahlfors regular set* $\Sigma \subseteq \mathbb{R}^n$, *and abbreviate* $\sigma := \mathcal{H}^{n-1}\lfloor\Sigma$. *Assume* $\mathbb{X}$ *is a Generalized Banach Function Space on* (Σ, σ) *and, with* μ *denoting the counting measure on* $\mathbb{Z}$, *assume* $\mathbb{Y}$ *is a Generalized Quasi-Banach Sequence Space on* $(\mathbb{Z}, \mu)$. *Then for every function* $f \in \mathscr{M}(\Sigma, \sigma)$ *one has*

$$\|f\|_{\dot{\mathrm{K}}(\Sigma;\mathbb{X}';\mathbb{Y}')} = \sup\Big\{\int_\Sigma |fg|\,d\sigma : g \in \dot{\mathrm{K}}(\Sigma;\mathbb{X};\mathbb{Y}) \text{ and } \|g\|_{\dot{\mathrm{K}}(\Sigma;\mathbb{X};\mathbb{Y})} \leq 1\Big\} \tag{23.38}$$

where $\mathbb{X}'$ *and* $\mathbb{Y}'$ *denote the associated spaces of* $\mathbb{X}$ *and* $\mathbb{Y}$, *respectively. Consequently, the associated space of* $\dot{\mathrm{K}}(\Sigma;\mathbb{X};\mathbb{Y})$ *is identified as*

$$\big(\dot{\mathrm{K}}(\Sigma;\mathbb{X};\mathbb{Y})\big)' = \dot{\mathrm{K}}(\Sigma;\mathbb{X}';\mathbb{Y}'). \tag{23.39}$$

The same type of conclusions hold for the inhomogeneous Composite Herz spaces.

Proof Fix a σ-measurable function f on Σ, and assume f is not identically zero, otherwise (23.38) is trivially satisfied. Thanks to Proposition 23.2, we see that the supremum in the right-hand side of (23.38) is less than or equal to $\|f\|_{\dot{\mathrm{K}}(\Sigma;\mathbb{X}';\mathbb{Y}')}$. To prove the opposite inequality, pick $\eta \in (0,\infty)$ such that

$$0 < \eta < \|f\|_{\dot{\mathrm{K}}(\Sigma;\mathbb{X}';\mathbb{Y}')} = \Big\|\{\|f\cdot \mathbb{1}_{C_k}\|_{\mathbb{X}'}\}_{k\in\mathbb{Z}}\Big\|_{\mathbb{Y}'} \tag{23.40}$$

where $\{C_k\}_{k\in\mathbb{Z}}$ is as in Definition 23.2. From this and Definition 10.3, there exists some sequence $\{b_k\}_{k\in\mathbb{Z}} \in \mathbb{Y}$ such that

$$\big\|\{b_k\}_{k\in\mathbb{Z}}\big\|_{\mathbb{Y}} \leq 1 \text{ and } \eta < \sum_{k=-\infty}^{\infty} \|f\cdot \mathbb{1}_{C_k}\|_{\mathbb{X}'}|b_k| \tag{23.41}$$

For each $k \in \mathbb{Z}$ choose $\zeta_k \in (0,\infty)$ satisfying

$$\zeta_k < \|f\cdot \mathbb{1}_{C_k}\|_{\mathbb{X}'}. \tag{23.42}$$

Then, bearing in mind Definition 10.3, for each $k \in \mathbb{Z}$ there exists $g_k \in \mathbb{X}$ for which

$$\|g_k\|_{\mathbb{X}} \leq 1 \text{ and } \zeta_k < \int_{C_k} |fg_k|\,d\sigma. \tag{23.43}$$

Define the function $\widetilde{g} : \Sigma \to \mathbb{R}$ by setting

$$\widetilde{g} := \sum_{\ell=-\infty}^{\infty} b_\ell g_\ell \cdot \mathbb{1}_{C_\ell}. \tag{23.44}$$

Since $\{C_k\}_{k\in\mathbb{Z}}$ (as in (23.16)) is a disjoint family of σ-measurable sets that cover Σ up to a set of σ-measure zero, $\widetilde{g}$ is well defined σ-a.e. on Σ and has the property that $\widetilde{g} \cdot \mathbb{1}_{C_k} = b_k g_k \cdot \mathbb{1}_{C_k}$ for every $k \in \mathbb{Z}$. Based on this, (23.41), and (23.43), we estimate

$$\begin{aligned}
\|\widetilde{g}\|_{\dot{\mathbb{K}}(\Sigma;\mathbb{X};\mathbb{Y})} &= \Big\|\big\{\|\widetilde{g}\cdot\mathbb{1}_{C_k}\|_{\mathbb{X}}\big\}_{k\in\mathbb{Z}}\Big\|_{\mathbb{Y}} = \Big\|\big\{|b_k|\|g_k\cdot\mathbb{1}_{C_k}\|_{\mathbb{X}}\big\}_{k\in\mathbb{Z}}\Big\|_{\mathbb{Y}} \\
&\le \Big\|\big\{|b_k|\|g_k\|_{\mathbb{X}}\big\}_{k\in\mathbb{Z}}\Big\|_{\mathbb{Y}} \le \big\|\{|b_k|\}_{k\in\mathbb{Z}}\big\|_{\mathbb{Y}} \\
&= \big\|\{b_k\}_{k\in\mathbb{Z}}\big\|_{\mathbb{Y}} \le 1.
\end{aligned} \tag{23.45}$$

Hence $\widetilde{g} \in \dot{\mathbb{K}}(\Sigma;\mathbb{X};\mathbb{Y})$ and $\|\widetilde{g}\|_{\dot{\mathbb{K}}(\Sigma;\mathbb{X};\mathbb{Y})} \le 1$. Bearing in mind (23.44) and (23.43), estimate

$$\int_\Sigma |f\widetilde{g}|\,d\sigma = \sum_{k=-\infty}^{\infty} \int_{C_k} |f\widetilde{g}|\,d\sigma = \sum_{k=-\infty}^{\infty} |b_k| \int_{C_k} |f g_k|\,d\sigma \ge \sum_{k=-\infty}^{\infty} |b_k|\zeta_k. \tag{23.46}$$

Since this holds for every ζ_k as in (23.42), we may send $\zeta_k \nearrow \|f\cdot\mathbb{1}_{C_k}\|_{\mathbb{X}'}$ for each $k \in \mathbb{Z}$ and use (23.41) to obtain

$$\int_\Sigma |f\widetilde{g}|\,d\sigma \ge \sum_{k=-\infty}^{\infty} |b_k|\|f\cdot\mathbb{1}_{C_k}\|_{\mathbb{X}'} > \eta. \tag{23.47}$$

By the arbitrariness of η as in (23.40), if we let $\eta \nearrow \|f\|_{\dot{\mathbb{K}}(\Sigma;\mathbb{X}';\mathbb{Y}')}$ in (23.47) we find that

$$\begin{aligned}
\|f\|_{\dot{\mathbb{K}}(\Sigma;\mathbb{X}';\mathbb{Y}')} &\le \int_\Sigma |f\widetilde{g}|\,d\sigma \\
&\le \sup\Big\{\int_\Sigma |fg|\,d\sigma : g \in \dot{\mathbb{K}}(\Sigma;\mathbb{X};\mathbb{Y}) \text{ and } \|g\|_{\dot{\mathbb{K}}(\Sigma;\mathbb{X};\mathbb{Y})} \le 1\Big\}.
\end{aligned} \tag{23.48}$$

This establishes (23.38). Finally, the identification in (23.39) is a consequence of what we proved so far along with (23.3). This finishes the proof of Proposition 23.5.□

In order to study the (topological) dual of Composite Herz spaces, recall the following notion of order continuity.

Definition 23.3 Consider $(X, \mathfrak{M}, \mu)$ a measure space, and let $\mathbb{X}$ be a Generalized Quasi-Banach Function Space on (X, μ). The space $\mathbb{X}$ is said to be `order continuous` if for every sequence of functions $\{f_k\}_{k\in\mathbb{N}} \subseteq \mathbb{X}$ satisfying $f_k \searrow 0$ pointwise μ-a.e. on X as $k \to \infty$ one has $\|f_k\|_{\mathbb{X}} \searrow 0$ as $k \to \infty$.

It turns out that if the base spaces $\mathbb{X}$ and $\mathbb{Y}$ are order continuous, then the corresponding Composite Herz space is also order continuous. This is made precise in the following result.

Proposition 23.6 *Consider $\Sigma \subseteq \mathbb{R}^n$ a closed unbounded Ahlfors regular set, and abbreviate $\sigma := \mathcal{H}^{n-1}\lfloor\Sigma$. Assume $\mathbb{X}$ is a Generalized Banach Function Space on (Σ, σ) and $\mathbb{Y}$ is a Generalized Quasi-Banach Sequence Space on $(\mathbb{Z}, \mu)$, where μ denotes the counting measure. If $\mathbb{X}$ and $\mathbb{Y}$ are order continuous, then $\dot{\mathbb{K}}(\Sigma; \mathbb{X}; \mathbb{Y})$ is also order continuous. The same type of conclusion holds for inhomogeneous Composite Herz spaces.*

Proof Consider a sequence of functions $\{f_j\}_{j\in\mathbb{N}} \subseteq \dot{\mathbb{K}}(\Sigma; \mathbb{X}; \mathbb{Y})$ such that $f_j \searrow 0$ pointwise σ-a.e. on Σ as $j \to \infty$. From this and Remark 23.8, for each $k \in \mathbb{Z}$ we have $\{f_j \cdot \mathbb{1}_{C_k}\}_{j\in\mathbb{N}} \subseteq \mathbb{X}$ and

$$f_j \cdot \mathbb{1}_{C_k} \searrow 0 \text{ pointwise } \sigma\text{-a.e. on } \Sigma \text{ as } j \to \infty. \tag{23.49}$$

Since $\mathbb{X}$ is order continuous, this further implies

$$\|f_j \cdot \mathbb{1}_{C_k}\|_{\mathbb{X}} \searrow 0 \quad \text{as } j \to \infty \text{ for each } k \in \mathbb{Z}. \tag{23.50}$$

This convergence is now understood as a pointwise convergence of sequences indexed by $\mathbb{Z}$, i.e.,

$$\big\{\|f_j \cdot \mathbb{1}_{C_k}\|_{\mathbb{X}}\big\}_{k\in\mathbb{Z}} \searrow \mathbf{0} \quad \text{pointwise on } \mathbb{Z} \text{ as } j \to \infty, \tag{23.51}$$

where $\mathbf{0}$ denotes the sequence (indexed by $\mathbb{Z}$) whose all terms equal zero. The fact that $\mathbb{Y}$ is order continuous along with (23.18) give

$$\|f_j\|_{\dot{\mathbb{K}}(\Sigma;\mathbb{X};\mathbb{Y})} = \Big\|\big\{\|f_j \cdot \mathbb{1}_{C_k}\|_{\mathbb{X}}\big\}_{k\in\mathbb{Z}}\Big\|_{\mathbb{Y}} \searrow 0 \quad \text{as } j \to \infty. \tag{23.52}$$

Hence $\dot{\mathbb{K}}(\Sigma; \mathbb{X}; \mathbb{Y})$ is order continuous, as desired. □

Remark 23.10 It has been shown in [LoNi24, Proposition 3.15, p. 262] that if $\mathbb{X}$ is a Generalized Quasi-Banach Function Space on a measure space (X, μ) one has

$$\mathbb{X}' = \mathbb{X}^*, \text{ whenever } \mathbb{X} \text{ is order continuous.} \tag{23.53}$$

Based on this, we have the following consequence of Proposition 23.5:

Corollary 23.2 *Assume $\Sigma \subseteq \mathbb{R}^n$ is closed unbounded Ahlfors regular set, and abbreviate $\sigma := \mathcal{H}^{n-1}\lfloor\Sigma$. Let $\mathbb{X}$ be a Generalized Banach Function Space on (Σ, σ) and $\mathbb{Y}$ a Generalized Quasi-Banach Sequence Space on $(\mathbb{Z}, \mu)$, where μ denotes the counting measure. Then whenever $\mathbb{X}$ and $\mathbb{Y}$ are order continuous, the topological dual of $\dot{\mathbb{K}}(\Sigma; \mathbb{X}; \mathbb{Y})$ is identified as*

$$\big(\dot{\mathbb{K}}(\Sigma; \mathbb{X}; \mathbb{Y})\big)^* = \dot{\mathbb{K}}(\Sigma; \mathbb{X}^*; \mathbb{Y}^*) \tag{23.54}$$

where $\mathbb{X}^*$ *and* $\mathbb{Y}^*$ *denote the topological dual spaces of* $\mathbb{X}$ *and* $\mathbb{Y}$, *respectively.*

Proof From Proposition 23.5, Proposition 23.6, and (23.53), we see that

$$\big(\dot{\mathrm{K}}(\Sigma;\mathbb{X};\mathbb{Y})\big)^* = \big(\dot{\mathrm{K}}(\Sigma;\mathbb{X};\mathbb{Y})\big)' = \dot{\mathrm{K}}(\Sigma;\mathbb{X}';\mathbb{Y}') = \dot{\mathrm{K}}(\Sigma;\mathbb{X}^*;\mathbb{Y}^*). \tag{23.55}$$

This establishes (23.54). □

23.3 A_∞ Index and Compliancy

In this section, we introduce the notion of A_∞ index of a Generalized Banach Function Space $\mathbb{X}$, and the notion of compliancy of sequence quasinorms. These novel concepts play a fundamental role in the development of the theory of Composite Herz spaces in subsequent sections.

We begin with the following lemma which establishes that the norm of a Generalized Banach Function Space satisfies an A_∞ condition.

Lemma 23.1 *Assume* $\Sigma \subseteq \mathbb{R}^n$ *is a closed set which is Ahlfors regular, and abbreviate* $\sigma := \mathcal{H}^{n-1}\lfloor\Sigma$. *Consider* $\mathbb{X}$ *a Generalized Banach Function Space on* (Σ, σ), *and with* $\mathbb{X}'$ *denoting its associated space, suppose*

$$\mathcal{M}_\Sigma : \mathbb{X}' \to \mathbb{X}' \ \textit{ is well defined and bounded,} \tag{23.56}$$

where $\mathcal{M}_\Sigma$ *denotes the Hardy-Littlewood maximal operator on* Σ.

Then there exists a constant $C \in (0,\infty)$ *and an exponent* $\tau \in (0,1)$ *that depend only on n, the operator norm of* $\mathcal{M}_\Sigma$ *on* $\mathbb{X}'$, *and the Ahlfors regularity constants of* Σ *with the property that for every surface ball* $\Delta \subseteq \Sigma$ *and every* σ*-measurable set* $E \subseteq \Delta$ *one has*

$$\frac{\|\mathbb{1}_E\|_{\mathbb{X}}}{\|\mathbb{1}_\Delta\|_{\mathbb{X}}} \le C\left(\frac{\sigma(E)}{\sigma(\Delta)}\right)^\tau. \tag{23.57}$$

The proof of this lemma is based on the Rubio de Francia algorithm, and is inspired by the proof presented in [IzNo16, Lemma 6, p. 8], which is done in the context of (classical) Banach Function spaces in the Euclidean setting. As such, our result naturally extends the techniques used there to the context of GBFS on Ahlfors regular sets.

Proof of Lemma 23.1 Denote by $A := \|\mathcal{M}_\Sigma\|_{\mathbb{X}'\to\mathbb{X}'} \in (0,\infty)$. Fix an arbitrary function $g \in \mathbb{X}'$, and assume g is not identically zero on Σ. Introduce

$$(Rg)(x) := \sum_{k=0}^{\infty} \frac{(\mathcal{M}_\Sigma)^k g(x)}{(2A)^k} \quad \text{at } \sigma\text{-a.e. } x \in \Sigma, \tag{23.58}$$

where for each $k \in \mathbb{N}_0$ we set

$$(\mathcal{M}_\Sigma)^k g := \begin{cases} |g| & \text{if } k = 0, \\ \mathcal{M}_\Sigma g & \text{if } k = 1, \\ \mathcal{M}_\Sigma\big[(\mathcal{M}_\Sigma)^{k-1} g\big] & \text{if } k \geq 2. \end{cases} \tag{23.59}$$

We claim that the function Rg is well defined σ-a.e. on Σ. Indeed, using the boundedness of $\mathcal{M}_\Sigma$ on $\mathbb{X}'$ we see that

$$\begin{aligned} \sum_{k=0}^{\infty} \frac{\|(\mathcal{M}_\Sigma)^k g\|_{\mathbb{X}'}}{(2A)^k} &\leq \Big(\sum_{k=0}^{\infty} \frac{\|\mathcal{M}_\Sigma\|^k_{\mathbb{X}'\to\mathbb{X}'}}{(2A)^k} \Big) \|g\|_{\mathbb{X}'} \\ &= \Big(\sum_{k=0}^{\infty} \frac{1}{2^k} \Big) \|g\|_{\mathbb{X}'} = 2\|g\|_{\mathbb{X}'} < \infty. \end{aligned} \tag{23.60}$$

Then the series in (23.58) is absolutely convergent in $\mathbb{X}'$. From [GHA-II, Proposition 5.1.8, p. 222], it follows that said series converges in $\mathbb{X}'$. Since $\mathbb{X}' \subseteq L^1_{\rm loc}(\Sigma, \sigma)$ (cf. [GHA-II, Lemma 5.2.5, p. 245]) we then find that the series in (23.58) is finite σ-a.e. on Σ. Hence, Rg is well defined and finite σ-a.e. on Σ.

Next, we claim that the function Rg (cf. (23.58)) satisfies the following properties:

$$\begin{aligned} &|g(x)| \leq (Rg)(x) \ \text{ at } \sigma\text{-a.e. } x \in \Sigma, \ \ \|Rg\|_{\mathbb{X}'} \leq 2\|g\|_{\mathbb{X}'}, \\ &\text{and } \ \mathcal{M}_\Sigma(Rg) \leq 2A(Rg) \ \text{ pointwise everywhere on } \Sigma. \end{aligned} \tag{23.61}$$

Indeed, the first property in (23.61) is apparent from (23.59) and the fact that $|g|$ is a participant of the series (23.58). The second property in (23.61) can be seen from (23.58) and (23.60). The third property in (23.61) follows from (23.58) along with

$$\mathcal{M}_\Sigma(Rg) \leq \sum_{k=0}^{\infty} \frac{(\mathcal{M}_\Sigma)^{k+1} g(x)}{(2A)^k} = 2A \sum_{k=0}^{\infty} \frac{(\mathcal{M}_\Sigma)^{k+1} g}{(2A)^{k+1}} \leq 2A(Rg) \tag{23.62}$$

pointwise everywhere Σ. Notice that the first inequality in (23.62) is valid since $\mathcal{M}_\Sigma$ commutes with an infinite series (as implied by its very design). In particular, this and (2.42) ensure that

$$Rg \in A_1(\Sigma, \sigma) \ \text{ and } \ [Rg]_{A_1} \leq 2A. \tag{23.63}$$

By the Reverse Hölder inequality for A_1 weights (cf. item *(6)* in Proposition 2.3), there exists some constant $C \in (0, \infty)$ and some exponent $\tau \in (0, 1)$, both depending exclusively on A, n, and the Ahlfors regularity constants of Σ, with the property that for every surface ball $\Delta \subseteq \Sigma$ and every σ-measurable subset $E \subseteq \Delta$ we have

$$\frac{Rg(E)}{Rg(\Delta)} \leq C \left(\frac{\sigma(E)}{\sigma(\Delta)} \right)^{\tau}. \tag{23.64}$$

This, the first and second properties in (23.61), along with the generalized Hölder inequality (10.149) permit us to estimate

$$\begin{aligned}\int_{\Sigma} |\mathbb{1}_E \cdot g|\, d\sigma &\le \int_E Rg\, d\sigma = Rg(E) \le C\left(\frac{\sigma(E)}{\sigma(\Delta)}\right)^{\tau} Rg(\Delta) \\ &= C\left(\frac{\sigma(E)}{\sigma(\Delta)}\right)^{\tau} \int_{\Delta} Rg\, d\sigma \le C\left(\frac{\sigma(E)}{\sigma(\Delta)}\right)^{\tau} \|Rg\|_{\mathbb{X}'} \|\mathbb{1}_{\Delta}\|_{\mathbb{X}} \\ &\le C\left(\frac{\sigma(E)}{\sigma(\Delta)}\right)^{\tau} \|g\|_{\mathbb{X}'} \|\mathbb{1}_{\Delta}\|_{\mathbb{X}}.\end{aligned} \tag{23.65}$$

Specializing this inequality for all functions $g \in \mathbb{X}'$ with $\|g\|_{\mathbb{X}'} \le 1$, and then taking the supremum over all such functions (being mindful of (10.147)) we find that

$$\|\mathbb{1}_E\|_{\mathbb{X}} \le C\left(\frac{\sigma(E)}{\sigma(\Delta)}\right)^{\tau} \|\mathbb{1}_{\Delta}\|_{\mathbb{X}}. \tag{23.66}$$

This implies (23.57), which ultimately concludes the proof of Lemma 23.1. □

Remark 23.11 Notice that the if (23.57) is valid for some constant $C \in (0, \infty)$ and some exponent $\tau \in (0, 1)$, then necessarily $C \ge 1$.

The above lemma suggests making the following basic definition:

Definition 23.4 Consider a closed set $\Sigma \subseteq \mathbb{R}$ which is Ahlfors regular, and abbreviate $\sigma := \mathcal{H}^{n-1}\lfloor\Sigma$. Let $\mathbb{X}$ be a Generalized Banach Function Space on (Σ, σ). Define the A_∞ **index** of $\mathbb{X}$ by setting

$$\Upsilon_{\mathbb{X}} := \sup\big\{\tau \in (0, \infty) : \text{ (23.57) holds for some } C \in (0, \infty)\big\}. \tag{23.67}$$

Remark 23.12 Notice that Lemma 23.1 ensures that the set over which we take the supremum in (23.67) is non empty, as long as $\mathcal{M}_\Sigma$ is bounded on $\mathbb{X}'$. Also, observe that $\Upsilon_{\mathbb{X}} \in (0, \infty]$.

Throughout, we employ the following basic convention:

Remark 23.13 Work in the context of Definition 23.4. Consider an arbitrary Generalized Banach Function Space $\mathbb{X}$ on (Σ, σ) with A_∞ index $\Upsilon_{\mathbb{X}}$. For a number $\tau \in \mathbb{R}$ we write $\tau \precsim \Upsilon_{\mathbb{X}}$ provided $\tau < \Upsilon_{\mathbb{X}}$ if the supremum in (23.67) is infinite or not attained, and $\tau \le \Upsilon_{\mathbb{X}}$ if the supremum is finite and attained. With this piece of notation, observe that whenever $0 < \tau \precsim \Upsilon_{\mathbb{X}}$, there exists some constant $C \in (0, \infty)$ such that (23.57) holds. This can be easily seen from (23.67), (23.57), and the fact that whenever $\Delta \subseteq \Sigma$ is a surface ball and $E \subseteq \Delta$ is a σ-measurable set, we have $\frac{\sigma(E)}{\sigma(\Delta)} \le 1$.

Another useful lemma is recorded below.

Lemma 23.2 *Consider Σ a closed Ahlfors regular set, and abbreviate $\sigma := \mathcal{H}^{n-1}\lfloor\Sigma$. Let $\mathbb{X}$ be a Generalized Banach Function Space, and with $\mathbb{X}'$ denoting its associated space, assume*

$$\mathcal{M}_\Sigma : \mathbb{X}' \to \mathbb{X}' \text{ is well defined and bounded.} \tag{23.68}$$

Then there exists some constant $C \in (0, \infty)$ depending only on n, $\|\mathcal{M}_\Sigma\|_{\mathbb{X}'\to\mathbb{X}'}$, and the Ahlfors regularity character of Σ such that for every surface ball $\Delta \subseteq \Sigma$ one has

$$\sigma(\Delta) \leq \|\mathbb{1}_\Delta\|_{\mathbb{X}} \cdot \|\mathbb{1}_\Delta\|_{\mathbb{X}'} \leq C \cdot \sigma(\Delta). \tag{23.69}$$

Proof To get started, fix an arbitrary surface ball $\Delta \subseteq \Sigma$. Thanks to the generalized Hölder inequality in Remark 10.6 we have

$$\sigma(\Delta) = \int_\Delta \mathbb{1}_\Delta \cdot \mathbb{1}_\Delta \, d\sigma \leq \|\mathbb{1}_\Delta\|_{\mathbb{X}} \|\mathbb{1}_\Delta\|_{\mathbb{X}'}. \tag{23.70}$$

This justifies the first inequality in (23.69). In the opposite direction, for each function $f \in L^1_{\text{loc}}(\Sigma, \sigma)$ introduce

$$P_\Delta f := \left(⨍_\Delta |f| \, d\sigma \right) \cdot \mathbb{1}_\Delta \text{ on } \Sigma. \tag{23.71}$$

Upon recalling the non-centered Hardy-Littlewood maximal operator M_Σ (see (2.25)), along with (2.27), we see that for every $f \in L^1_{\text{loc}}(\Sigma, \sigma)$ we have

$$0 \leq P_\Delta f \leq M_\Sigma f \leq C \cdot \mathcal{M}_\Sigma f \text{ everywhere on } \Sigma, \tag{23.72}$$

for some constant $C \in (0, \infty)$ depending only on n and the Ahlfors regularity of Σ. Thanks to this and (23.68), for every function $f \in \mathbb{X}'$ estimate

$$\begin{aligned} \frac{1}{\sigma(\Delta)} \left(\int_\Sigma |f \cdot \mathbb{1}_\Delta| \, d\sigma \right) \cdot \|\mathbb{1}_\Delta\|_{\mathbb{X}'} &= \left(⨍_\Delta |f| \, d\sigma \right) \cdot \|\mathbb{1}_\Delta\|_{\mathbb{X}'} \\ &= \|P_\Delta f\|_{\mathbb{X}'} \leq C \|\mathcal{M}_\Sigma f\|_{\mathbb{X}'} \leq C \|f\|_{\mathbb{X}'} \end{aligned} \tag{23.73}$$

for some constant $C \in (0, \infty)$ depending only on n, $\|\mathcal{M}_\Sigma\|_{\mathbb{X}'\to\mathbb{X}'}$, and the Ahlfors regularity character of Σ. In particular, for every function $f \in \mathbb{X}'$ with $\|f\|_{\mathbb{X}'} \leq 1$ this further entails

$$\left(\int_\Sigma |f \cdot \mathbb{1}_\Delta| \, d\sigma \right) \cdot \|\mathbb{1}_\Delta\|_{\mathbb{X}'} \leq C \cdot \sigma(\Delta). \tag{23.74}$$

After taking the supremum over all such functions f, and recalling Proposition 10.10, we finally obtain (23.69). This concludes the proof of Lemma 23.2. □

Remark 23.14 Notice that in the proof of Lemma 23.2 we are implicitly using the fact that $\mathbb{X}' \subseteq L^1_{\rm loc}(\Sigma, \sigma)$, as ensured by the assumption (23.68) (see also [GHA-II, Lemma 5.2.5, p. 245]).

Much of the success of our development of a comprehensive theory for Composite Herz spaces is due to our notion of compliancy, which we introduce next.

Definition 23.5 Fix $\eta, \eta' \in (0, \infty)$ and a number $2_* \in (1, \infty)$. Denote by $\mathbb{R}_+^{\mathbb{Z}}$ the collection of all sequences in $\mathbb{R}_+$ indexed over $\mathbb{Z}$. A function $\|\cdot\| : \mathbb{R}_+^{\mathbb{Z}} \to [0, \infty]$ is said to be (η, η')-`compliant` if there exists some $C \in (0, \infty)$, called the compliancy constant of $\|\cdot\|$, such that for every sequence $\{a_\ell\}_{\ell\in\mathbb{Z}} \in \mathbb{R}_+^{\mathbb{Z}}$ one has

(C1) Boundedness of forward shift:

$$\|\{a_{\ell+1}\}_{\ell\in\mathbb{Z}}\| \leq C\|\{a_\ell\}_{\ell\in\mathbb{Z}}\|, \tag{23.75}$$

(C2) Boundedness of backward shift:

$$\|\{a_{\ell-1}\}_{\ell\in\mathbb{Z}}\| \leq C\|\{a_\ell\}_{\ell\in\mathbb{Z}}\|, \tag{23.76}$$

(C3) Boundedness of upper tail of weighted series:

$$\Big\|\Big\{\sum_{\ell=k}^{\infty} 2_*^{(k-\ell)\eta} a_\ell\Big\}_{k\in\mathbb{Z}}\Big\| \leq C\|\{a_\ell\}_{\ell\in\mathbb{Z}}\|, \tag{23.77}$$

(C4) Boundedness of lower tail of weighted series:

$$\Big\|\Big\{\sum_{\ell=-\infty}^{k} 2_*^{(\ell-k)\eta'} a_\ell\Big\}_{k\in\mathbb{Z}}\Big\| \leq C\|\{a_\ell\}_{\ell\in\mathbb{Z}}\|. \tag{23.78}$$

The subsequent lemma ensures that (η, η')-compliancy is a monotonically increasing property with respect to η and η'.

Lemma 23.3 *Select numbers $\eta_1, \eta_1', \eta_2, \eta_2' \in (0, \infty)$ with $0 < \eta_1 < \eta_2 < \infty$ and $0 < \eta_1' < \eta_2' < \infty$. Consider a function $\|\cdot\| : \mathbb{R}_+^{\mathbb{Z}} \to [0, \infty]$ that satisfies the lattice property, i.e.,*

$$\begin{gathered}\text{if } \{a_\ell\}_{\ell\in\mathbb{Z}}, \{b_\ell\}_{\ell\in\mathbb{Z}} \in \mathbb{R}_+^{\mathbb{Z}} \text{ are such that } a_\ell \leq b_\ell \\ \text{for all } \ell \in \mathbb{Z} \text{ then } \|\{a_\ell\}_{\ell\in\mathbb{Z}}\| \leq \|\{b_\ell\}_{\ell\in\mathbb{Z}}\|.\end{gathered} \tag{23.79}$$

Then whenever $\|\cdot\|$ is (η_1, η_1')-compliant, it follows that $\|\cdot\|$ is (η_2, η_2')-compliant.

Proof Notice that conditions (C1)-(C2) appearing in Definition 23.5 are independent of the exponents η_2, η_2'. As such, we only need to verify conditions (C3)-(C4) in

Definition 23.5 (currently written with $\eta := \eta_2$ and $\eta' := \eta_2'$). With this goal in mind, fix a number $2_* \in (1, \infty)$ and select a sequence $\{a_\ell\}_{\ell\in\mathbb{Z}} \subseteq \mathbb{R}_+^{\mathbb{Z}}$. Since $\eta_1 < \eta_2$ and $\eta_1' < \eta_2'$ we have

$$2_*^{(k-\ell)\eta_2} a_\ell \leq 2_*^{(k-\ell)\eta_1} a_\ell \text{ whenever } k, \ell \in \mathbb{Z} \text{ with } k \geq \ell \tag{23.80}$$

and

$$2_*^{(\ell-k)\eta_2'} a_\ell \leq 2_*^{(\ell-k)\eta_1'} a_\ell \text{ whenever } k, \ell \in \mathbb{Z} \text{ with } k \leq \ell. \tag{23.81}$$

For every $k \in \mathbb{Z}$ these yield

$$\sum_{\ell=k}^{\infty} 2_*^{(k-\ell)\eta_2} a_\ell \leq \sum_{\ell=k}^{\infty} 2_*^{(k-\ell)\eta_1} a_\ell \text{ and } \sum_{\ell=-\infty}^{k} 2_*^{(\ell-k)\eta_2'} a_\ell \leq \sum_{\ell=-\infty}^{k} 2_*^{(\ell-k)\eta_1'} a_\ell. \tag{23.82}$$

Together with (23.79), and the fact that $\|\cdot\|$ satisfies (C3)-(C4) in Definition 23.5 with $\eta := \eta_1$ and $\eta' := \eta_1'$, we find that

$$\left\|\left\{\sum_{\ell=k}^{\infty} 2_*^{(k-\ell)\eta_2} a_\ell\right\}_{k\in\mathbb{Z}}\right\| \leq \left\|\left\{\sum_{\ell=k}^{\infty} 2_*^{(k-\ell)\eta_1} a_\ell\right\}_{k\in\mathbb{Z}}\right\| \leq C\|\{a_\ell\}_{\ell\in\mathbb{Z}}\| \tag{23.83}$$

and

$$\left\|\left\{\sum_{\ell=-\infty}^{k} 2_*^{(\ell-k)\eta_2'} a_\ell\right\}_{k\in\mathbb{Z}}\right\| \leq \left\|\left\{\sum_{\ell=-\infty}^{k} 2_*^{(\ell-k)\eta_1'} a_\ell\right\}_{k\in\mathbb{Z}}\right\| \leq C\|\{a_\ell\}_{\ell\in\mathbb{Z}}\| \tag{23.84}$$

for some $C \in (0, \infty)$. This establishes that $\|\cdot\|$ is (η_2, η_2')-compliant, as desired. □

The purpose of the following lemma is to obtain an alternative description of compliancy in the case when the function $\|\cdot\|$ satisfies a lattice property and a quasi-triangle inequality.

Proposition 23.7 *Fix two exponents $\eta, \eta' \in (0, \infty)$, pick a number $2_* \in (1, \infty)$, and consider a function $\|\cdot\| : \mathbb{R}_+^{\mathbb{Z}} \to [0, \infty]$. Then the following statements are true:*

(a) Assume $\|\cdot\|$ satisfies the lattice property (23.79) *and is (η, η')-compliant. Then there exists $C \in (0, \infty)$ such that for each sequence $\{a_\ell\}_{\ell\in\mathbb{Z}} \subseteq \mathbb{R}_+$ one has*

$$\begin{aligned} &\left\|\left\{\sum_{\ell=k+2}^{\infty} 2_*^{(k-\ell)\eta} a_\ell\right\}_{k\in\mathbb{Z}}\right\| \leq C\|\{a_\ell\}_{\ell\in\mathbb{Z}}\| \text{ and} \\ &\left\|\left\{\sum_{\ell=-\infty}^{k-2} 2_*^{(\ell-k)\eta'} a_\ell\right\}_{k\in\mathbb{Z}}\right\| \leq C\|\{a_\ell\}_{\ell\in\mathbb{Z}}\|. \end{aligned} \tag{23.85}$$

(b) Conversely, suppose $\|\cdot\|$ is positive homogeneous and satisfies a quasi-triangle inequality, i.e., there exists some $K \in [1, \infty)$ such that for every $a, b \in \mathbb{R}_+^{\mathbb{Z}}$ one

has

$$\|\lambda a\| = \lambda\|a\| \quad \textit{for all } \lambda \in (0,\infty) \\ \textit{and} \quad \|a+b\| \le K(\|a\|+\|b\|). \tag{23.86}$$

Then if $\|\cdot\|$ *satisfies properties* (C1)-(C2) *in Definition* 23.5 *together with* (23.85)*, it follows that* $\|\cdot\|$ *is* (η,η')*-compliant.*

In particular, if $\mathbb{Y}$ *is a Generalized Banach Sequence Space on* $(\mathbb{Z},\mu)$*, where* μ *is the counting measure, then the quasinorm* $\|\cdot\|_{\mathbb{Y}}$ *is* (η,η')*-compliant if and only if* $\|\cdot\|_{\mathbb{Y}}$ *satisfies* (23.85) *and* (C1)-(C2) *in Definition 23.5.*

Proof To verify the claim made in item *(a)*, observe that the lattice property of $\|\cdot\|$ along with (C3)-(C4) in Definition 23.5 imply that for every sequence $\{a_\ell\}_{\ell\in\mathbb{Z}} \subseteq \mathbb{R}_+$ we have

$$\Big\|\Big\{\sum_{\ell=k+2}^{\infty} 2_*^{(k-\ell)\eta} a_\ell\Big\}_{k\in\mathbb{Z}}\Big\| \le \Big\|\Big\{\sum_{\ell=k}^{\infty} 2_*^{(k-\ell)\eta} a_\ell\Big\}_{k\in\mathbb{Z}}\Big\| \le C\|\{a_\ell\}_{\ell\in\mathbb{Z}}\| \tag{23.87}$$

and

$$\Big\|\Big\{\sum_{\ell=-\infty}^{k-2} 2_*^{(\ell-k)\eta'} a_\ell\Big\}_{k\in\mathbb{Z}}\Big\| \le \Big\|\Big\{\sum_{\ell=-\infty}^{k} 2_*^{(\ell-k)\eta'} a_\ell\Big\}_{k\in\mathbb{Z}}\Big\| \le C\|\{a_\ell\}_{\ell\in\mathbb{Z}}\|. \tag{23.88}$$

This establishes (23.85).

To justify the claim in item *(b)*, it suffices to show that $\|\cdot\|$ satisfies (C3) and (C4) in Definition 23.5. To this end, fix an arbitrary sequence $\{a_\ell\}_{\ell\in\mathbb{Z}} \subseteq \mathbb{R}_+$. Thanks to (23.85), the quasi-triangle inequality, and the positive homogeneity of $\|\cdot\|$ we may estimate

$$\begin{aligned} \Big\|\Big\{\sum_{\ell=k}^{\infty} 2_*^{(k-\ell)\eta} a_\ell\Big\}_{k\in\mathbb{Z}}\Big\| &= \Big\|\Big\{a_k + 2_*^{-\eta} a_{k+1} + \sum_{\ell=k+2}^{\infty} 2_*^{(k-\ell)\eta} a_\ell\Big\}_{k\in\mathbb{Z}}\Big\| \\ &\le K^2\|\{a_k\}_{k\in\mathbb{Z}}\| + 2_*^{-\eta}K^2\|\{a_{k+1}\}_{k\in\mathbb{Z}}\| \\ &\quad + K\Big\|\Big\{\sum_{\ell=k+2}^{\infty} 2_*^{(k-\ell)\eta} a_\ell\Big\}_{k\in\mathbb{Z}}\Big\| \\ &\le C\|\{a_\ell\}_{\ell\in\mathbb{Z}}\| \end{aligned} \tag{23.89}$$

and

$$
\begin{aligned}
\left\|\left\{\sum_{\ell=-\infty}^{k} 2_*^{(\ell-k)\eta'} a_\ell\right\}_{k\in\mathbb{Z}}\right\| &= \left\|\left\{a_k + 2_*^{-\eta'} a_{k-1} + \sum_{\ell=-\infty}^{k-2} 2_*^{(\ell-k)\eta'} a_\ell\right\}_{k\in\mathbb{Z}}\right\| \\
&\leq K^2\|\{a_k\}_{k\in\mathbb{Z}}\| + 2_*^{-\eta'} K^2\|\{a_{k-1}\}_{k\in\mathbb{Z}}\| \\
&\quad + K\left\|\left\{\sum_{\ell=-\infty}^{k-2} 2_*^{(\ell-k)\eta'} a_\ell\right\}_{k\in\mathbb{Z}}\right\| \\
&\leq C\|\{a_\ell\}_{\ell\in\mathbb{Z}}\|, \qquad (23.90)
\end{aligned}
$$

for some constant $C \in (0,\infty)$ depending on K, η, η', and 2_*. Hence $\|\cdot\|$ satisfies (C3) and (C4) in Definition 23.5, thus concluding the treatment of item *(b)*. This finishes the proof of Proposition 23.7. □

In the context of Generalized Quasi-Banach Sequence Spaces, compliancy of the norm implies compliancy of the associated norm.

Lemma 23.4 *Assume $\mathbb{Y}$ is a Generalized Quasi-Banach Sequence Space on $(\mathbb{Z},\mu)$, where μ denotes the counting measure. Having fixed $\eta,\eta' \in (0,\infty)$, assume that $\|\cdot\|_{\mathbb{Y}}$ is (η,η')-compliant with compliancy constant $C \in (0,\infty)$. Then $\|\cdot\|_{\mathbb{Y}'}$ is (η',η)-compliant with same compliancy constant $C \in (0,\infty)$, where $\mathbb{Y}'$ denotes the associated space of $\mathbb{Y}$.*

Proof Fix an arbitrary sequence $\{a_\ell\}_{\ell\in\mathbb{Z}} \in \mathbb{R}^{\mathbb{Z}}$ along with a sequence $\{b_\ell\}_{\ell\in\mathbb{Z}} \in \mathbb{Y}$ such that $\|\{b_\ell\}_{\ell\in\mathbb{Z}}\|_{\mathbb{Y}} \leq 1$. By the generalized Hölder inequality (23.12), and the compliancy of $\|\cdot\|_{\mathbb{Y}}$ we estimate

$$
\begin{aligned}
\sum_{\ell=-\infty}^{\infty} |a_{\ell+1} b_\ell| &= \sum_{\ell=-\infty}^{\infty} |a_\ell b_{\ell-1}| \leq \|\{a_\ell\}_{\ell\in\mathbb{Z}}\|_{\mathbb{Y}'}\|\{b_{\ell-1}\}_{\ell\in\mathbb{Z}}\|_{\mathbb{Y}} \\
&\leq C\|\{a_\ell\}_{\ell\in\mathbb{Z}}\|_{\mathbb{Y}'}\|\{b_\ell\}_{\ell\in\mathbb{Z}}\|_{\mathbb{Y}} \leq C\|\{a_\ell\}_{\ell\in\mathbb{Z}}\|_{\mathbb{Y}'}. \qquad (23.91)
\end{aligned}
$$

After taking the supremum over all such sequences $\{b_\ell\}_{\ell\in\mathbb{Z}}$, and being mindful of (23.4), we obtain

$$
\|\{a_{\ell+1}\}_{\ell\in\mathbb{Z}}\|_{\mathbb{Y}'} \leq C\|\{a_\ell\}_{\ell\in\mathbb{Z}}\|_{\mathbb{Y}'}. \qquad (23.92)
$$

A similar type of reasoning as in (23.91) also ensures that

$$
\|\{a_{\ell-1}\}_{\ell\in\mathbb{Z}}\|_{\mathbb{Y}'} \leq C\|\{a_\ell\}_{\ell\in\mathbb{Z}}\|_{\mathbb{Y}'}. \qquad (23.93)
$$

This shows that $\|\cdot\|_{\mathbb{Y}'}$ satisfies (C1)-(C2) in Definition 23.5 (with the constant C preserved).

Next, continue to assume $\{b_\ell\}_{\ell\in\mathbb{Z}} \in \mathbb{Y}$ with $\|\{b_\ell\}_{\ell\in\mathbb{Z}}\|_{\mathbb{Y}} \leq 1$. Using the generalized Hölder inequality (23.12), the compliancy of $\|\cdot\|_{\mathbb{Y}}$ and Fubini's Theorem for series, write

$$\sum_{k=-\infty}^{\infty}\Big(\sum_{\ell=-\infty}^{k}2_*^{(\ell-k)\eta}|a_\ell|\Big)|b_k| = \sum_{\ell=-\infty}^{\infty}|a_\ell|\Big(\sum_{k=\ell}^{\infty}2_*^{(\ell-k)\eta}|b_k|\Big)$$
$$\leq \|\{a_\ell\}_{\ell\in\mathbb{Z}}\|_{\mathbb{Y}'}\Big\|\Big\{\sum_{k=\ell}^{\infty}2_*^{(\ell-k)\eta}|b_k|\Big\}_{\ell\in\mathbb{Z}}\Big\|_{\mathbb{Y}}$$
$$\leq C\|\{a_\ell\}_{\ell\in\mathbb{Z}}\|_{\mathbb{Y}'}\|\{b_\ell\}_{\ell\in\mathbb{Z}}\|_{\mathbb{Y}} \leq C\|\{a_\ell\}_{\ell\in\mathbb{Z}}\|_{\mathbb{Y}'} \tag{23.94}$$

and

$$\sum_{k=-\infty}^{\infty}\Big(\sum_{\ell=k}^{\infty}2_*^{(k-\ell)\eta'}|a_\ell|\Big)|b_k| = \sum_{\ell=-\infty}^{\infty}|a_\ell|\Big(\sum_{k=-\infty}^{\ell}2_*^{(k-\ell)\eta'}|b_k|\Big)$$
$$\leq \|\{a_\ell\}_{\ell\in\mathbb{Z}}\|_{\mathbb{Y}'}\Big\|\Big\{\sum_{k=-\infty}^{\ell}2_*^{(k-\ell)\eta'}|b_k|\Big\}_{\ell\in\mathbb{Z}}\Big\|_{\mathbb{Y}}$$
$$\leq C\|\{a_\ell\}_{\ell\in\mathbb{Z}}\|_{\mathbb{Y}'}\|\{b_\ell\}_{\ell\in\mathbb{Z}}\|_{\mathbb{Y}} \leq C\|\{a_\ell\}_{\ell\in\mathbb{Z}}\|_{\mathbb{Y}'}. \tag{23.95}$$

Since these hold for all $\{b_\ell\}_{\ell\in\mathbb{Z}} \in \mathbb{Y}$ with $\|\{b_\ell\}_{\ell\in\mathbb{Z}}\|_{\mathbb{Y}} \leq 1$, after recalling (23.4) we finally arrive at

$$\Big\|\Big\{\sum_{\ell=k}^{\infty}2_*^{(k-\ell)\eta'}|a_\ell|\Big\}_{k\in\mathbb{Z}}\Big\|_{\mathbb{Y}'} \leq C\|\{a_\ell\}_{\ell\in\mathbb{Z}}\|_{\mathbb{Y}'} \tag{23.96}$$

and

$$\Big\|\Big\{\sum_{\ell=-\infty}^{k}2_*^{(\ell-k)\eta}|a_\ell|\Big\}_{k\in\mathbb{Z}}\Big\|_{\mathbb{Y}'} \leq C\|\{a_\ell\}_{\ell\in\mathbb{Z}}\|_{\mathbb{Y}'}. \tag{23.97}$$

This establishes that $\|\cdot\|_{\mathbb{Y}'}$ satisfies (C3)-(C4) in Definition 23.5, which ultimately concludes the proof of the fact that $\|\cdot\|_{\mathbb{Y}'}$ is (η',η)-compliant. □

In the following definition, we adapt the notion of compliancy for functions $\|\cdot\|$ defined over sequences indexed by $\mathbb{N}_0$.

Definition 23.6 Select numbers $\eta,\eta' \in (0,\infty)$, and $2_* \in (1,\infty)$. Consider $\mathbb{R}_+^{\mathbb{N}_0}$ the collection of all sequences in $\mathbb{R}_+$ indexed over $\mathbb{N}_0$. A function $\|\cdot\| : \mathbb{R}_+^{\mathbb{N}_0} \to [0,\infty]$ is called (η,η')-`compliant` on $\mathbb{N}_0$ if there exists some constant $C \in (0,\infty)$, called compliancy constant, with the property that for every sequence $\{a_\ell\}_{\ell\in\mathbb{N}_0} \in \mathbb{R}_+^{\mathbb{N}_0}$ one has (with the convention that $a_{-1} := 0$)

(D1) Boundedness of forward shift:

$$\|\{a_{\ell+1}\}_{\ell\in\mathbb{N}_0}\| \leq C\|\{a_\ell\}_{\ell\in\mathbb{N}_0}\| \tag{23.98}$$

(D2) Boundedness of backward shift:

$$\|\{a_{\ell-1}\}_{\ell\in\mathbb{N}_0}\| \leq C\|\{a_\ell\}_{\ell\in\mathbb{N}_0}\|, \tag{23.99}$$

(D3) Boundedness of upper tail of weighted series:

$$\left\|\left\{\sum_{\ell=k}^{\infty} 2_*^{(k-\ell)\eta} a_\ell\right\}_{k\in\mathbb{N}_0}\right\| \leq C\|\{a_\ell\}_{\ell\in\mathbb{N}_0}\|, \tag{23.100}$$

(D4) Boundedness of finite weighted sums:

$$\left\|\left\{\sum_{\ell=0}^{k} 2_*^{(\ell-k)\eta'} a_\ell\right\}_{k\in\mathbb{N}_0}\right\| \leq C\|\{a_\ell\}_{\ell\in\mathbb{N}_0}\|. \tag{23.101}$$

Remark 23.15 In the same context as in Definition 23.6, it is useful to observe that if $\|\cdot\|$ is (η,η')-compliant on $\mathbb{N}_0$, and satisfies the lattice property, then the fact that

$$\{b_k\}_{k\in\mathbb{N}_0}\cdot \mathbb{1}_{\{k\geq 2\}} \leq \{b_k\}_{k\in\mathbb{N}_0} \quad \text{for any sequence } \{b_k\}_{k\in\mathbb{N}_0}\subseteq \mathbb{R}_+ \tag{23.102}$$

ensures that $\|\cdot\|$ satisfies

$$\begin{aligned}
&\|\{a_{\ell+1}\}_{\ell\in\mathbb{N}_0}\cdot \mathbb{1}_{\{\ell\geq 2\}}\|_{\mathbb{Y}} \leq C\|\{a_\ell\}_{\ell\in\mathbb{N}_0}\|_{\mathbb{Y}},\\
&\|\{a_{\ell-1}\}_{\ell\in\mathbb{N}_0}\cdot \mathbb{1}_{\{\ell\geq 2\}}\|_{\mathbb{Y}} \leq C\|\{a_\ell\}_{\ell\in\mathbb{N}_0}\|_{\mathbb{Y}},\\
&\left\|\left\{\sum_{\ell=k+2}^{\infty} 2_*^{(k-\ell)\eta} a_\ell\right\}_{k\in\mathbb{N}_0}\cdot \mathbb{1}_{\{k\geq 2\}}\right\|_{\mathbb{Y}} \leq C\|\{a_\ell\}_{\ell\in\mathbb{N}_0}\|_{\mathbb{Y}},\\
&\left\|\left\{\sum_{\ell=0}^{k-2} 2_*^{(\ell-k)\eta'} a_\ell\right\}_{k\in\mathbb{N}_0}\cdot \mathbb{1}_{\{k\geq 2\}}\right\|_{\mathbb{Y}} \leq C\|\{a_\ell\}_{\ell\in\mathbb{N}_0}\|_{\mathbb{Y}},
\end{aligned} \tag{23.103}$$

for every $\{a_\ell\}_{\ell\in\mathbb{N}_0}\in\mathbb{R}_+^{\mathbb{N}_0}$.

Remark 23.16 It turns out that whenever $\mathbb{Y}$ is a Generalized Quasi-Banach Sequence Space on $\mathbb{N}_0$ and $\|\cdot\|_{\mathbb{Y}}$ is (η,η')-compliant on $\mathbb{N}_0$ for some $\eta,\eta'\in(0,\infty)$, then $\|\cdot\|_{\mathbb{Y}'}$ is (η',η)-compliant on $\mathbb{N}_0$. This follows by naturally adapting the proof of Lemma 23.4, while bearing in mind Definition 23.6.

23.4 Sequence Spaces

In this section we study certain spaces of sequences that will eventually serve as one of the base spaces for considering Composite Herz spaces. Each example is accompanied by a claim regarding the GQBSS nature of the sequence space under consideration, as well as a claim regarding the compliancy of its quasinorm. Our first example is inspired by the manner in which classical Herz spaces $\dot{\mathrm{K}}^{\alpha}_{p,q}$ are defined (see, e.g., Section 17.3).

Example 23.1 (**ℓ^q_α Spaces**) Suppose $\mathcal{J} \subseteq \mathbb{Z}$ is an infinite set, and denote by μ the counting measure on $\mathcal{J}$. Fix an exponent $q \in (0, \infty]$, an index $\alpha \in \mathbb{R}$, along with some number $2_* \in (1, \infty)$. For each sequence $\{a_j\}_{j\in\mathcal{J}} \in \mathbb{R}^{\mathcal{J}}$ introduce

$$\|\{a_j\}_{j\in\mathcal{J}}\|_{\ell^q_\alpha(\mathcal{J})} := \big\|\{2_*^{j\alpha}a_j\}_{j\in\mathcal{J}}\big\|_{\ell^q(\mathcal{J})} = \begin{cases} \Big[\sum_{j\in\mathcal{J}} 2_*^{j\alpha q}|a_j|^q\Big]^{1/q} & \text{if } 0 < q < \infty, \\ \sup_{j\in\mathcal{J}} \{2_*^{j\alpha}|a_j|\} & \text{if } q = \infty, \end{cases} \tag{23.104}$$

which is well defined in $[0, \infty]$. Define the space of sequences

$$\ell^q_\alpha(\mathcal{J}) := \big\{\{a_j\}_{j\in\mathcal{J}} \in \mathbb{R}^{\mathcal{J}} : \|\{a_j\}_{j\in\mathcal{J}}\|_{\ell^q_\alpha(\mathcal{J})} < \infty\big\}. \tag{23.105}$$

<u>`Claim 1.`</u> The space $\ell^q_\alpha(\mathcal{J})$ is a Generalized Quasi-Banach Sequence Space on $(\mathcal{J}, \mu)$.

Proof of `Claim 1`. It is visible from (23.104) that the function $\rho := \|\cdot\|_{\ell^q_\alpha(\mathcal{J})}$ satisfies (Q1'), (Q2'), (Q4'), and (Q5') in Proposition 23.1, while property (Q3') follows from Lebesgue's Monotone Convergence Theorem for sequences. As such, $\|\cdot\|_{\ell^q_\alpha(\mathcal{J})}$ is a function quasinorm, which in turn implies that $\ell^q_\alpha(\mathcal{J})$ is a Generalized Quasi-Banach Sequence Space on $\mathcal{J}$. □

<u>`Claim 2.`</u> Fix two numbers $\eta, \eta' \in (0, \infty)$. Then

$$\|\cdot\|_{\ell^q_\alpha(\mathbb{Z})} \text{ is } (\eta, \eta')\text{-compliant, whenever } -\eta < \alpha < \eta'. \tag{23.106}$$

Similarly,

$$\|\cdot\|_{\ell^q_\alpha(\mathbb{N}_0)} \text{ is } (\eta, \eta')\text{-compliant on } \mathbb{N}_0, \text{ whenever } -\eta < \alpha < \eta'. \tag{23.107}$$

Proof of `Claim 2`. We only prove (23.106), as the proof (23.107) follows by a similar argument. Assume first that $-\eta < \alpha < \eta'$, and pick a sequence $\{a_\ell\}_{\ell\in\mathbb{Z}} \subseteq \mathbb{R}_+$. Then

$$\begin{aligned} \|\{a_{\ell+1}\}_{\ell\in\mathbb{Z}}\|_{\ell^q_\alpha(\mathbb{Z})} &= \big\|\{2_*^{\ell\alpha q}a_{\ell+1}\}_{\ell\in\mathbb{Z}}\big\|_{\ell^q(\mathbb{Z})} \\ &= 2_*^{-\alpha}\big\|\{2_*^{(\ell+1)\alpha}a_{\ell+1}\}_{\ell\in\mathbb{Z}}\big\|_{\ell^q(\mathbb{Z})} \\ &= 2_*^{-\alpha}\|\{a_\ell\}_{\ell\in\mathbb{Z}}\|_{\ell^q_\alpha(\mathbb{Z})}. \end{aligned} \tag{23.108}$$

Similarly,

$$\|\{a_{\ell-1}\}_{\ell\in\mathbb{Z}}\|_{\ell^q_\alpha(\mathbb{Z})} = 2_*^{\alpha}\|\{a_\ell\}_{\ell\in\mathbb{Z}}\|_{\ell^q_\alpha(\mathbb{Z})}. \tag{23.109}$$

This shows that $\|\cdot\|_{\ell^q_\alpha(\mathbb{Z})}$ satisfies (C1)-(C2) in Definition 23.5.

To verify condition (C3) from Definition 23.5, we write

$$\begin{aligned}\Big\|\Big\{\sum_{\ell=k}^{\infty} 2_*^{(k-\ell)\eta} a_\ell\Big\}_{k\in\mathbb{Z}}\Big\|_{\ell^q_\alpha(\mathbb{Z})} &= \Big\|\Big\{2_*^{k\alpha}\Big(\sum_{\ell=k}^{\infty} 2_*^{(k-\ell)\eta} a_\ell\Big)\Big\}_{k\in\mathbb{Z}}\Big\|_{\ell^q(\mathbb{Z})}\\ &= \Big\|\Big\{\sum_{\ell=k}^{\infty} 2_*^{(k-\ell)(\eta+\alpha)} 2_*^{\ell\alpha} a_\ell\Big\}_{k\in\mathbb{Z}}\Big\|_{\ell^q(\mathbb{Z})} =: \mathrm{S}_1.\end{aligned} \tag{23.110}$$

<u>**Case 1.**</u> Assume $0 < q \le 1$. In this case, use Fubini's Theorem for series along with $\alpha + \eta > 0$ to estimate

$$\begin{aligned}\mathrm{S}_1 &= \Big[\sum_{k=-\infty}^{\infty}\Big(\sum_{\ell=k}^{\infty} 2_*^{(k-\ell)(\eta+\alpha)} 2_*^{\ell\alpha} a_\ell\Big)^q\Big]^{1/q}\\ &\le \Big[\sum_{k=-\infty}^{\infty}\sum_{\ell=k}^{\infty} 2_*^{(k-\ell)(\eta+\alpha)q} 2_*^{\ell\alpha q} a_\ell^q\Big]^{1/q}\\ &= \Big[\sum_{\ell=-\infty}^{\infty} 2_*^{\ell\alpha q} a_\ell^q \sum_{k=-\infty}^{\ell} 2_*^{(k-\ell)(\eta+\alpha)q}\Big]^{1/q}\\ &= \Big[\sum_{\ell=-\infty}^{\infty} 2_*^{\ell\alpha q} a_\ell^q \sum_{j=0}^{\infty} 2_*^{-j(\eta+\alpha)q}\Big]^{1/q}\\ &= C\Big[\sum_{\ell=-\infty}^{\infty} 2_*^{\ell\alpha q} a_\ell^q\Big]^{1/q} = C\|\{a_\ell\}_{\ell\in\mathbb{Z}}\|_{\ell^q_\alpha(\mathbb{Z})}\end{aligned} \tag{23.111}$$

for some constant $C \in (0,\infty)$ depending only on q, α, η, and 2_*.

<u>**Case 2.**</u> Assume $1 < q < \infty$. Denoting by $q' \in (1,\infty)$ the Hölder conjugate exponent of q, use Hölder's inequality, Fubini's Theorem for series, and the fact that $\alpha + \eta > 0$ to obtain

$$\begin{aligned}
\mathrm{S}_1 &= \Bigg[\sum_{k=-\infty}^{\infty}\Big(\sum_{\ell=k}^{\infty}2_*^{(k-\ell)(\eta+\alpha)}2_*^{\ell\alpha}a_\ell\Big)^q\Bigg]^{1/q} \\
&\leq \Bigg[\sum_{k=-\infty}^{\infty}\Big(\sum_{\ell=k}^{\infty}2_*^{\ell\alpha q}2_*^{(k-\ell)(\eta+\alpha)q/2}a_\ell^q\Big)\Big(\sum_{\ell=k}^{\infty}2_*^{(k-\ell)(\eta+\alpha)q'/2}\Big)^{q/q'}\Bigg]^{1/q} \\
&= \Bigg[\sum_{k=-\infty}^{\infty}\Big(\sum_{\ell=k}^{\infty}2_*^{\ell\alpha q}2_*^{(k-\ell)(\eta+\alpha)q/2}a_\ell^q\Big)\Big(\sum_{j=0}^{\infty}2_*^{-j(\eta+\alpha)q'/2}\Big)^{q/q'}\Bigg]^{1/q} \\
&= C\Bigg[\sum_{k=-\infty}^{\infty}\sum_{\ell=k}^{\infty}2_*^{\ell\alpha q}2_*^{(k-\ell)(\eta+\alpha)q/2}a_\ell^q\Bigg]^{1/q} \\
&= C\Bigg[\sum_{\ell=-\infty}^{\infty}2_*^{\ell\alpha q}a_\ell^q\sum_{k=-\infty}^{\ell}2_*^{(k-\ell)(\eta+\alpha)q/2}\Bigg]^{1/q} \\
&= C\Bigg[\sum_{\ell=-\infty}^{\infty}2_*^{\ell\alpha q}a_\ell^q\sum_{j=0}^{\infty}2_*^{-j(\eta+\alpha)q/2}\Bigg]^{1/q} \\
&= C\Bigg[\sum_{\ell=-\infty}^{\infty}2_*^{\ell\alpha q}a_\ell^q\Bigg]^{1/q} = C\|\{a_\ell\}_{\ell\in\mathbb{Z}}\|_{\ell_\alpha^q(\mathbb{Z})}
\end{aligned} \tag{23.112}$$

for some constant $C \in (0,\infty)$ depending only on q, α, η, and 2_*.

<u>**Case 3.**</u> Assume $q = \infty$. In this case, bearing in mind that $\eta + \alpha > 0$, we estimate S_1 from (23.110) as

$$\begin{aligned}
\mathrm{S}_1 &= \sup_{k\in\mathbb{Z}}\Big\{\sum_{\ell=k}^{\infty}2_*^{(k-\ell)(\eta+\alpha)}2_*^{\ell\alpha}a_\ell\Big\} \\
&\leq \Big(\sup_{k\in\mathbb{Z}}\sum_{\ell=k}^{\infty}2_*^{(k-\ell)(\eta+\alpha)}\Big)\|\{2_*^{\ell\alpha}a_\ell\}_{\ell\in\mathbb{Z}}\|_{\ell^\infty(\mathbb{Z})} \\
&= \Big(\sum_{j=0}^{\infty}2_*^{-j(\eta+\alpha)}\Big)\|\{a_\ell\}_{\ell\in\mathbb{Z}}\|_{\ell_\alpha^\infty(\mathbb{Z})} \leq C\|\{a_\ell\}_{\ell\in\mathbb{Z}}\|_{\ell_\alpha^\infty(\mathbb{Z})}
\end{aligned} \tag{23.113}$$

for some constant $C \in (0,\infty)$ depending only on η, α, and 2_*.

All together (23.110), (23.111), (23.112), and (23.113) imply that $\|\cdot\|_{\ell_\alpha^q(\mathbb{Z})}$ satisfies (C3) in Definition 23.5.

We now show that $\|\cdot\|_{\ell_\alpha^q(\mathbb{Z})}$ also satisfies (C4) in Definition 23.5. To this end, notice that

$$\Big\|\Big\{\sum_{\ell=-\infty}^{k} 2_*^{(\ell-k)\eta'} a_\ell\Big\}_{k\in\mathbb{Z}}\Big\|_{\ell^q_\alpha(\mathbb{Z})} = \Big\|\Big\{2_*^{k\alpha}\sum_{\ell=-\infty}^{k} 2_*^{(\ell-k)\eta'} a_\ell\Big\}_{k\in\mathbb{Z}}\Big\|_{\ell^q(\mathbb{Z})}$$

$$= \Big\|\Big\{\sum_{\ell=-\infty}^{k} 2_*^{(\ell-k)(\eta'-\alpha)} 2_*^{\ell\alpha} a_\ell\Big\}_{k\in\mathbb{Z}}\Big\|_{\ell^q(\mathbb{Z})} =: \mathrm{S}_2. \tag{23.114}$$

<u>**Case 1.**</u> Assume $0 < q \leq 1$. Employing Fubini's Theorem for series and the fact that $\eta' - \alpha > 0$ we see that

$$\begin{aligned}
\mathrm{S}_2 &= \Big[\sum_{k=-\infty}^{\infty}\Big(\sum_{\ell=-\infty}^{k} 2_*^{\ell\alpha} 2_*^{(\ell-k)(\eta'-\alpha)} a_\ell\Big)^q\Big]^{1/q} \\
&\leq \Big[\sum_{k=-\infty}^{\infty}\sum_{\ell=-\infty}^{k} 2_*^{\ell\alpha q} 2_*^{(\ell-k)(\eta'-\alpha)q} a_\ell^q\Big]^{1/q} \\
&= \Big[\sum_{\ell=-\infty}^{\infty} 2_*^{\ell\alpha q} a_\ell^q \sum_{k=\ell}^{\infty} 2_*^{(\ell-k)(\eta'-\alpha)q}\Big]^{1/q} \\
&= \Big[\sum_{\ell=-\infty}^{\infty} 2_*^{\ell\alpha q} a_\ell^q \sum_{j=0}^{\infty} 2_*^{-j(\eta'-\alpha)q}\Big]^{1/q} \\
&= C\Big[\sum_{\ell=-\infty}^{\infty} 2_*^{\ell\alpha q} a_\ell^q\Big]^{1/q} = \|\{a_\ell\}_{\ell\in\mathbb{Z}}\|_{\ell^q_\alpha(\mathbb{Z})}
\end{aligned} \tag{23.115}$$

for some constant $C \in (0, \infty)$ depending only on q, η', α, and 2_*.

<u>**Case 2.**</u> Assume $1 < q < \infty$. To deal with this case, use Hölder's inequality, Fubini's Theorem for series, and $\eta' - \alpha > 0$ to estimate

$$
\begin{aligned}
\mathrm{S}_2 &= \left[\sum_{k=-\infty}^{\infty}\Big(\sum_{\ell=-\infty}^{k} 2_*^{\ell\alpha} 2_*^{(\ell-k)(\eta'-\alpha)} a_\ell\Big)^q\right]^{1/q} \\
&= \left[\sum_{k=-\infty}^{\infty}\Big(\sum_{\ell=-\infty}^{k} 2_*^{\ell\alpha q} 2_*^{(\ell-k)(\eta'-\alpha)q/2} a_\ell^q\Big)\Big(\sum_{\ell=-\infty}^{k} 2_*^{(\ell-k)(\eta'-\alpha)q'/2}\Big)^{q/q'}\right]^{1/q} \\
&= \left[\sum_{k=-\infty}^{\infty}\Big(\sum_{\ell=-\infty}^{k} 2_*^{\ell\alpha q} 2_*^{(\ell-k)(\eta'-\alpha)q/2} a_\ell^q\Big)\Big(\sum_{j=0}^{\infty} 2_*^{-j(\eta'-\alpha)q'/2}\Big)^{q/q'}\right]^{1/q} \\
&= C\left[\sum_{k=-\infty}^{\infty}\sum_{\ell=-\infty}^{k} 2_*^{\ell\alpha q} 2_*^{(\ell-k)(\eta'-\alpha)q/2} a_\ell^q\right]^{1/q} \\
&= C\left[\sum_{\ell=-\infty}^{\infty} 2_*^{\ell\alpha q} a_\ell^q \sum_{k=\ell}^{\infty} 2_*^{(\ell-k)(\eta'-\alpha)q/2}\right]^{1/q} \\
&= C\left[\sum_{\ell=-\infty}^{\infty} 2_*^{\ell\alpha q} a_\ell^q \sum_{j=0}^{\infty} 2_*^{-j(\eta'-\alpha)q/2}\right]^{1/q} \\
&= C\left[\sum_{\ell=-\infty}^{\infty} 2_*^{\ell\alpha q} a_\ell^q\right]^{1/q} = C\|\{a_\ell\}_{\ell\in\mathbb{Z}}\|_{\ell_\alpha^q(\mathbb{Z})}.
\end{aligned}
\tag{23.116}
$$

for some constant $C \in (0, \infty)$ depending only on q, η', α, and 2_*.

<u>**Case 3.**</u> Assume $q = \infty$. Since $\eta' - \alpha > 0$, we bound S_2 from (23.114) by

$$
\begin{aligned}
\mathrm{S}_2 &= \sup_{k\in\mathbb{Z}}\Big\{\sum_{\ell=-\infty}^{k} 2_*^{(\ell-k)(\eta'-\alpha)} 2_*^{\ell\alpha} a_\ell\Big\} \\
&\le \Big(\sup_{k\in\mathbb{Z}}\sum_{\ell=-\infty}^{k} 2_*^{(\ell-k)(\eta'-\alpha)}\Big)\|\{2_*^{\ell\alpha} a_\ell\}_{\ell\in\mathbb{Z}}\|_{\ell^\infty(\mathbb{Z})} \\
&= \Big(\sum_{j=0}^{\infty} 2_*^{-j(\eta'-\alpha)}\Big)\|\{a_\ell\}_{\ell\in\mathbb{Z}}\|_{\ell_\alpha^\infty(\mathbb{Z})} \le C\|\{a_\ell\}_{\ell\in\mathbb{Z}}\|_{\ell_\alpha^\infty(\mathbb{Z})}
\end{aligned}
\tag{23.117}
$$

for some constant $C \in (0, \infty)$ depending only on q, η', α, and 2_*.

Collectively, (23.114), (23.115), (23.116), and (23.117) give that $\|\cdot\|_{\ell_\alpha^q(\mathbb{Z})}$ satisfies (C4) in Definition 23.5. This establishes that $\|\cdot\|_{\ell_\alpha^q(\mathbb{Z})}$ is (η, η')-compliant, whenever $-\eta < \alpha < \eta'$. The proof of (23.107) follows via a similar argument, now bearing in mind Definition 23.6. This completes the treatment of Claim 2. □

Claim 3. Fix two numbers $\eta, \eta' \in (0, \infty)$ and suppose that $0 < q < \infty$. If $\|\cdot\|_{\ell^q_\alpha(\mathbb{Z})}$ is (η, η') compliant, then necessarily $-\eta < \alpha < \eta'$. Corresponding to the case $q = \infty$, if $\|\cdot\|_{\ell^\infty_\alpha(\mathbb{Z})}$ is (η, η') compliant then necessarily $-\eta \le \alpha \le \eta'$. Similar type of conclusions hold for $\|\cdot\|_{\ell^q_\alpha(\mathbb{N}_0)}$.

Proof of Claim 3. Assume first $0 < q < \infty$ and that $\|\cdot\|_{\ell^q_\alpha(\mathbb{Z})}$ is (η, η')-compliant. Consider the sequence $\{\delta_{0\ell}\}_{\ell \in \mathbb{Z}}$, where $\delta_{0\ell}$ is the Kronecker delta. Using the assumption that $\|\cdot\|_{\ell^q_\alpha(\mathbb{Z})}$ satisfies (C3) and (C4) in Definition 23.5 with some constant $C \in (0, \infty)$, we have

$$\begin{aligned} C = C\|\{\delta_{0\ell}\}_{\ell\in\mathbb{Z}}\|^q_{\ell^q_\alpha(\mathbb{Z})} &\ge \left\|\left\{\sum_{\ell=k}^{\infty} 2_*^{(k-\ell)\eta}\delta_{0\ell}\right\}_{k\in\mathbb{Z}}\right\|^q_{\ell^q_\alpha(\mathbb{Z})} \\ &= \sum_{k=-\infty}^{\infty} 2_*^{k\alpha q}\left(\sum_{\ell=k}^{\infty} 2_*^{(k-\ell)\eta}\delta_{0\ell}\right)^q \\ &= \sum_{k=-\infty}^{0} 2_*^{k\alpha q}\cdot 2_*^{k\eta q} = \sum_{k=-\infty}^{0} 2_*^{k(\alpha+\eta)q} \end{aligned} \tag{23.118}$$

and

$$\begin{aligned} C = C\|\{\delta_{0\ell}\}_{\ell\in\mathbb{Z}}\|^q_{\ell^q_\alpha(\mathbb{Z})} &\ge \left\|\left\{\sum_{\ell=-\infty}^{k} 2_*^{(\ell-k)\eta'}\delta_{0\ell}\right\}_{k\in\mathbb{Z}}\right\|^q_{\ell^q_\alpha(\mathbb{Z})} \\ &= \sum_{k=-\infty}^{\infty} 2_*^{k\alpha q}\left(\sum_{\ell=-\infty}^{k} 2_*^{(\ell-k)\eta'}\delta_{0\ell}\right)^q \\ &= \sum_{k=0}^{\infty} 2_*^{k\alpha q}\cdot 2_*^{-k\eta' q} = \sum_{k=0}^{\infty} 2_*^{k(\alpha-\eta')q}. \end{aligned} \tag{23.119}$$

The finiteness of the series in (23.119) and (23.118) then forces $-\eta < \alpha < \eta'$. Corresponding to the case when $q = \infty$, we rely on the fact that $\|\cdot\|_{\ell^\infty_\alpha(\mathbb{Z})}$ satisfies (C3) and (C4) in Definition 23.5 to obtain

$$\begin{aligned} C = C\|\{\delta_{0\ell}\}_{\ell\in\mathbb{Z}}\|_{\ell^\infty_\alpha(\mathbb{Z})} &\ge \left\|\left\{\sum_{\ell=k}^{\infty} 2_*^{(k-\ell)\eta}\delta_{0\ell}\right\}_{k\in\mathbb{Z}}\right\|_{\ell^\infty_\alpha(\mathbb{Z})} \\ &= \sup_{k\in\mathbb{Z}} 2_*^{k\alpha}\left(\sum_{\ell=k}^{\infty} 2_*^{(k-\ell)\eta}\delta_{0\ell}\right) = \sup_{k\le 0} 2_*^{k(\alpha+\eta)} \end{aligned} \tag{23.120}$$

and similarly

$$C = C\|\{\delta_{0\ell}\}_{\ell\in\mathbb{Z}}\|_{\ell^\infty_\alpha(\mathbb{Z})} \geq \Big\|\Big\{\sum_{\ell=-\infty}^{k} 2_*^{(\ell-k)\eta'}\delta_{0\ell}\Big\}_{k\in\mathbb{Z}}\Big\|_{\ell^\infty_\alpha(\mathbb{Z})}$$

$$= \sup_{k\in\mathbb{Z}} 2_*^{k\alpha}\Big(\sum_{\ell=-\infty}^{k} 2_*^{(\ell-k)\eta'}\delta_{0\ell}\Big) = \sup_{k\geq 0} 2_*^{k(\alpha-\eta')}. \tag{23.121}$$

Since the suprema in (23.120)-(23.121) are finite we may conclude that $-\eta \leq \alpha \leq \eta'$. This finishes the proof of Claim 3. □

The next example is inspired by the study of Muckenhoupt weighted Herz spaces, a class of Herz-type spaces introduced in [LuYa95].

Example 23.2 (**Muckenhoupt weighted ℓ^q spaces**) Assume $\Sigma \subseteq \mathbb{R}^n$ is a closed unbounded Ahlfors regular set, and abbreviate $\sigma := \mathcal{H}^{n-1}\lfloor\Sigma$. Fix a reference point $x_0 \in \Sigma$ and a number $2_* \in (1,\infty)$. Consider an infinite set $\mathcal{J} \subseteq \mathbb{Z}$, and denote by μ the counting measure on $\mathcal{J}$. Introduce

$$\Delta_j := \Delta(x_0, 2_*^j) \quad \text{for all } j \in \mathbb{Z}. \tag{23.122}$$

Fix exponents $q \in (0,\infty]$ and $\alpha \in \mathbb{R}$. Pick a weight w on Σ such that $w \in L^1_{\rm loc}(\Sigma,\sigma)$. With each sequence $\{a_j\}_{j\in\mathcal{J}} \subseteq \mathbb{R}$ associate a quantity in $[0,\infty]$ by setting

$$\|\{a_j\}_{j\in\mathcal{J}}\|_{\ell^q_\alpha(\mathcal{J};w)} := \begin{cases} \Big[\sum_{j\in\mathcal{J}} \big(w(\Delta_j)\big)^{\alpha q/(n-1)}|a_j|^q\Big]^{1/q} & \text{if } 0<q<\infty, \\ \sup_{j\in\mathcal{J}} \big\{\big(w(\Delta_j)\big)^{\alpha/(n-1)}|a_j|\big\} & \text{if } q=\infty. \end{cases} \tag{23.123}$$

Define the space of sequences

$$\ell^q_\alpha(\mathcal{J};w) := \big\{\{a_j\}_{j\in\mathcal{J}} \in \mathbb{R}^{\mathcal{J}} : \|\{a_j\}_{j\in\mathcal{J}}\|_{\ell^q_\alpha(\mathcal{J};w)} < \infty\big\}. \tag{23.124}$$

Claim 1. The space $\ell^q_\alpha(\mathcal{J};w)$ is a Generalized Quasi-Banach Sequence Space on $(\mathcal{J},\mu)$.

Proof of Claim 1. Bearing in mind that the weight w is a locally integrable function on Σ, and that $0 < w < \infty$ at σ-a.e. point on Σ, it is visible from definitions that $\|\cdot\|_{\ell^q_\alpha(\mathcal{J};w)}$ satisfies (Q1'), (Q2'), (Q4'), and (Q5') in Proposition 23.1. That $\|\cdot\|_{\ell^q_\alpha(\mathcal{J};w)}$ also satisfies (Q3') follows from Lebesgue's Monotone Convergence Theorem for sequences. Hence, $\|\cdot\|_{\ell^q_\alpha(\mathcal{J};w)}$ is a function quasinorm, which further implies that $\ell^q_\alpha(\mathcal{J};w)$ is a Generalized Quasi-Banach Sequence Space on $(\mathcal{J},\mu)$.□

Claim 2. Select two numbers $\eta,\eta' \in (0,\infty)$. Having fixed an integrability exponent $p \in (1,\infty)$ along with a weight $w \in A_p(\Sigma,\sigma)$, it follows that

$$\text{if } -\tfrac{\eta}{p} < \alpha < \tfrac{\eta'}{p}, \text{ then } \|\cdot\|_{\ell^q_\alpha(\mathbb{Z};w)} \text{ is } (\eta,\eta')\text{-compliant} \\ \text{and } \|\cdot\|_{\ell^q_\alpha(\mathbb{N}_0;w)} \text{ is } (\eta,\eta')\text{-compliant on } \mathbb{N}_0. \tag{23.125}$$

Proof of `Claim 2`. Here we only establish compliancy of $\|\cdot\|_{\ell^q_\alpha(\mathbb{Z};w)}$, as the compliancy $\|\cdot\|_{\ell^q_\alpha(\mathbb{N}_0;w)}$ can be shown via a similar argument. Also, assume that $0 < q < \infty$. The proof for the case when $q = \infty$ follows in an analogous fashion. To get started fix a sequence $\{a_\ell\}_{\ell\in\mathbb{Z}} \subseteq \mathbb{R}_+$. Thanks to the doubling character of w (cf. item *(5)* in Proposition 2.3) we see that

$$\begin{aligned}\|\{a_{\ell+1}\}_{\ell\in\mathbb{Z}}\|_{\ell^q_\alpha(\mathbb{Z};w)} &= \Big[\sum_{\ell=-\infty}^{\infty} \big(w(\Delta_\ell)\big)^{\alpha q/(n-1)} a^q_{\ell+1}\Big]^{1/q} \\ &\leq C\Big[\sum_{\ell=-\infty}^{\infty} \big(w(\Delta_{\ell+1})\big)^{\alpha q/(n-1)} a^q_{\ell+1}\Big]^{1/q} \\ &= C\|\{a_\ell\}_{\ell\in\mathbb{Z}}\|_{\ell^q_\alpha(\mathbb{Z};w)}\end{aligned} \tag{23.126}$$

and

$$\begin{aligned}\|\{a_{\ell-1}\}_{\ell\in\mathbb{Z}}\|_{\ell^q_\alpha(\mathbb{Z};w)} &= \Big[\sum_{\ell=-\infty}^{\infty} \big(w(\Delta_\ell)\big)^{\alpha q/(n-1)} a^q_{\ell-1}\Big]^{1/q} \\ &\leq C\Big[\sum_{\ell=-\infty}^{\infty} \big(w(\Delta_{\ell-1})\big)^{\alpha q/(n-1)} a^q_{\ell-1}\Big]^{1/q} \\ &= C\|\{a_\ell\}_{\ell\in\mathbb{Z}}\|_{\ell^q_\alpha(\mathbb{Z};w)}\end{aligned} \tag{23.127}$$

for some constant $C \in (0,\infty)$ depending on p, $[w]_{A_p}$, n, and the Ahlfors regularity constants of Σ. Hence $\|\cdot\|_{\ell^q_\alpha(\mathbb{Z};w)}$ satisfies (C1) and (C2) in Definition 23.5.

To verify that (C3) and (C4) in Definition 23.5 hold for $\|\cdot\|_{\ell^q_\alpha(\mathbb{Z};w)}$ we use the fact that $w \in A_p(\Sigma,\sigma)$ satisfies a Reverse Hölder inequality (cf. item *(6)* in Proposition 2.3), i.e., there exists a constant $C = C(n,p,[w]_{A_p},\Sigma) \in (0,\infty)$ and an exponent $\tau \in (0,\infty)$ with the property that for every surface ball $\Delta \subseteq \Sigma$ and every σ-measurable set $E \subseteq \Delta$ we have

$$\frac{w(E)}{w(\Delta)} \leq C\left(\frac{\sigma(E)}{\sigma(\Delta)}\right)^\tau. \tag{23.128}$$

In particular, upon recalling (23.122), this and the Ahlfors regularity of Σ imply

$$\frac{w(\Delta_\ell)}{w(\Delta_k)} \leq C2_*^{(\ell-k)(n-1)\tau} \quad \text{for each } \ell, k \in \mathbb{Z} \text{ with } \ell \leq k, \tag{23.129}$$

for some constant $C \in (0,\infty)$ depending only on p, $[w]_{A_p}$, n, and the Ahlfors regularity constants of Σ. Granted these, estimate

$$
\begin{aligned}
&\Big\|\Big\{\sum_{\ell=k}^{\infty} 2_*^{(k-\ell)\eta} a_\ell\Big\}_{k\in\mathbb{Z}}\Big\|_{\ell^q_\alpha(\mathbb{Z};w)}\\
&= \Big[\sum_{k=-\infty}^{\infty} \big(w(\Delta_k)\big)^{\alpha q/(n-1)}\Big(\sum_{\ell=k}^{\infty} 2_*^{(k-\ell)\eta} a_\ell\Big)^q\Big]^{1/q}\\
&= \Big[\sum_{k=-\infty}^{\infty} \Big(\sum_{\ell=k}^{\infty} \big(w(\Delta_k)\big)^{\alpha/(n-1)} 2_*^{(k-\ell)\eta} a_\ell\Big)^q\Big]^{1/q}\\
&= \Big[\sum_{k=-\infty}^{\infty} \Big(\sum_{\ell=k}^{\infty} \big(w(\Delta_\ell)\big)^{\alpha/(n-1)} a_\ell \Big(\frac{w(\Delta_k)}{w(\Delta_\ell)}\Big)^{\alpha/(n-1)} 2_*^{(k-\ell)\eta}\Big)^q\Big]^{1/q}\\
&\leq C\Big[\sum_{k=-\infty}^{\infty} \Big(\sum_{\ell=k}^{\infty} \big(w(\Delta_\ell)\big)^{\alpha/(n-1)} a_\ell \Big(\frac{w(\Delta_k)}{w(\Delta_\ell)}\Big)^{\alpha/(n-1)} \Big(\frac{\sigma(\Delta_k)}{\sigma(\Delta_\ell)}\Big)^{\eta/(n-1)}\Big)^q\Big]^{1/q}\\
&\leq C\Big[\sum_{k=-\infty}^{\infty} \Big(\sum_{\ell=k}^{\infty} \big(w(\Delta_\ell)\big)^{\alpha/(n-1)} a_\ell \Big(\frac{w(\Delta_k)}{w(\Delta_\ell)}\Big)^{\alpha/(n-1)} \Big(\frac{w(\Delta_k)}{w(\Delta_\ell)}\Big)^{\eta/[(n-1)p]}\Big)^q\Big]^{1/q}\\
&= C\Big[\sum_{k=-\infty}^{\infty} \Big(\sum_{\ell=k}^{\infty} \big(w(\Delta_\ell)\big)^{\alpha/(n-1)} a_\ell \Big(\frac{w(\Delta_k)}{w(\Delta_\ell)}\Big)^{(\eta/p+\alpha)/(n-1)}\Big)^q\Big]^{1/q}\\
&\leq C\Big[\sum_{k=-\infty}^{\infty} \Big(\sum_{\ell=k}^{\infty} \big(w(\Delta_\ell)\big)^{\alpha/(n-1)} a_\ell \cdot 2_*^{(k-\ell)(\eta/p+\alpha)\tau}\Big)^q\Big]^{1/q} =: \mathrm{S}_1, \qquad (23.130)
\end{aligned}
$$

for some constant $C \in (0, \infty)$ depending only on n, p, α, η, $[w]_{A_p}$, and the Ahlfors regularity of Σ. The first inequality follows Ahlfors regularity of Σ, while in the second inequality is a consequence of the doubling character of w (cf. item *(5)* in Proposition 2.3). The third inequality comes from (23.129) (with the roles of k and ℓ switched).

To continue estimating S_1 in (23.130), we treat the cases $0 < q \leq 1$ and $1 < q < \infty$ separately.

Case 1. Assume $0 < q \leq 1$. Employing Fubini's Theorem for series, and recalling that $\eta/p + \alpha > 0$ we estimate

$$
\begin{aligned}
\mathrm{S}_1 &= C\Big[\sum_{k=-\infty}^{\infty} \Big(\sum_{\ell=k}^{\infty} \big(w(\Delta_\ell)\big)^{\alpha/(n-1)} a_\ell \cdot 2_*^{(k-\ell)(\eta/p+\alpha)\tau}\Big)^q\Big]^{1/q}\\
&\leq C\Big[\sum_{k=-\infty}^{\infty} \sum_{\ell=k}^{\infty} \big(w(\Delta_\ell)\big)^{\alpha q/(n-1)} a_\ell^q \cdot 2_*^{(k-\ell)(\eta/p+\alpha)\tau q}\Big]^{1/q}
\end{aligned}
$$

$$= C\left[\sum_{\ell=-\infty}^{\infty} \big(w(\Delta_\ell)\big)^{\alpha q/(n-1)} a_\ell^q \sum_{k=-\infty}^{\ell} 2_*^{(k-\ell)(\eta/p+\alpha)\tau q}\right]^{1/q}$$

$$= C\left[\sum_{\ell=-\infty}^{\infty} \big(w(\Delta_\ell)\big)^{\alpha q/(n-1)} a_\ell^q \sum_{j=0}^{\infty} 2_*^{-j(\eta/p+\alpha)\tau q}\right]^{1/q}$$

$$= C\left[\sum_{\ell=-\infty}^{\infty} \big(w(\Delta_\ell)\big)^{\alpha q/(n-1)} a_\ell^q\right]^{1/q} = C\|\{a_\ell\}_{\ell\in\mathbb{Z}}\|_{\ell_\alpha^q(\mathbb{Z};w)} \tag{23.131}$$

for some constant $C \in (0,\infty)$ depending on n, p, q, η, α, $[w]_{A_p}$, and the Ahlfors regularity of Σ.

Case 2. Assume $1 < q < \infty$. Denoting by $q' \in (1,\infty)$ the Hölder conjugate exponent of q, using Hölder's inequality, Fubini's Theorem for series, along with $\eta/p + \alpha > 0$ we have

$$\mathrm{S}_1 = C\left[\sum_{k=-\infty}^{\infty}\Big(\sum_{\ell=k}^{\infty} \big(w(\Delta_\ell)\big)^{\alpha/(n-1)} a_\ell \cdot 2_*^{(k-\ell)(\eta/p+\alpha)\tau}\Big)^q\right]^{1/q}$$

$$\leq C\left[\sum_{k=-\infty}^{\infty}\Big(\sum_{\ell=k}^{\infty} \big(w(\Delta_\ell)\big)^{\alpha q/(n-1)} a_\ell^q \cdot 2_*^{(k-\ell)(\eta/p+\alpha)\tau q/2}\Big)\right.$$

$$\left.\times\Big(\sum_{\ell=k}^{\infty} 2_*^{(k-\ell)(\eta/p+\alpha)\tau q'/2}\Big)^{q/q'}\right]^{1/q}$$

$$= C\left[\sum_{k=-\infty}^{\infty}\Big(\sum_{\ell=k}^{\infty} \big(w(\Delta_\ell)\big)^{\alpha q/(n-1)} a_\ell^q \cdot 2_*^{(k-\ell)(\eta/p+\alpha)\tau q/2}\Big)\right.$$

$$\left.\times\Big(\sum_{j=0}^{\infty} 2_*^{-j(\eta/p+\alpha)\tau q'/2}\Big)^{q/q'}\right]^{1/q}$$

$$= C\left[\sum_{k=-\infty}^{\infty}\sum_{\ell=k}^{\infty} \big(w(\Delta_\ell)\big)^{\alpha q/(n-1)} a_\ell^q \cdot 2_*^{(k-\ell)(\eta/p+\alpha)\tau q/2}\right]^{1/q}$$

$$= C\left[\sum_{\ell=-\infty}^{\infty} \big(w(\Delta_\ell)\big)^{\alpha q/(n-1)} a_\ell^q \sum_{k=-\infty}^{\ell} 2_*^{(k-\ell)(\eta/p+\alpha)\tau q/2}\right]^{1/q}$$

$$= C\left[\sum_{\ell=-\infty}^{\infty} \big(w(\Delta_\ell)\big)^{\alpha q/(n-1)} a_\ell^q \sum_{j=0}^{\infty} 2_*^{-j(\eta/p+\alpha)\tau q/2}\right]^{1/q}$$

$$= C\left[\sum_{\ell=-\infty}^{\infty} \big(w(\Delta_\ell)\big)^{\alpha q/(n-1)} a_\ell^q\right]^{1/q} = C\|\{a_\ell\}_{\ell\in\mathbb{Z}}\|_{\ell_\alpha^q(\mathbb{Z};w)} \tag{23.132}$$

for some constant $C \in (0,\infty)$ depending on n, p, q, η, α, $[w]_{A_p}$, and the Ahlfors regularity of Σ. This, (23.130), and (23.131) establish that $\|\cdot\|_{\ell^q_\alpha(\mathbb{Z};w)}$ satisfies property (C3) in Definition 23.5.

Moving on, compute

$$\begin{aligned}
&\Big\|\Big\{\sum_{\ell=-\infty}^{k} 2_*^{(\ell-k)\eta'} a_\ell\Big\}_{k\in\mathbb{Z}}\Big\|_{\ell^q_\alpha(\mathbb{Z};w)}\\
&=\Bigg[\sum_{k=-\infty}^{\infty}\big(w(\Delta_k)\big)^{\alpha q/(n-1)}\Big(\sum_{\ell=-\infty}^{k} 2_*^{(\ell-k)\eta'} a_\ell\Big)^q\Bigg]^{1/q}\\
&=\Bigg[\sum_{k=-\infty}^{\infty}\Bigg(\sum_{\ell=-\infty}^{k}\big(w(\Delta_k)\big)^{\alpha/(n-1)} 2_*^{(\ell-k)\eta'} a_\ell\Bigg)^q\Bigg]^{1/q}\\
&=\Bigg[\sum_{k=-\infty}^{\infty}\Bigg(\sum_{\ell=-\infty}^{k}\big(w(\Delta_\ell)\big)^{\alpha/(n-1)} a_\ell\left(\frac{w(\Delta_k)}{w(\Delta_\ell)}\right)^{\alpha/(n-1)} 2_*^{(\ell-k)\eta'}\Bigg)^q\Bigg]^{1/q}\\
&\leq C\Bigg[\sum_{k=-\infty}^{\infty}\Bigg(\sum_{\ell=-\infty}^{k}\big(w(\Delta_\ell)\big)^{\alpha/(n-1)} a_\ell\left(\frac{w(\Delta_k)}{w(\Delta_\ell)}\right)^{\alpha/(n-1)}\left(\frac{\sigma(\Delta_\ell)}{\sigma(\Delta_k)}\right)^{\eta'/(n-1)}\Bigg)^q\Bigg]^{1/q}\\
&\leq C\Bigg[\sum_{k=-\infty}^{\infty}\Bigg(\sum_{\ell=-\infty}^{k}\big(w(\Delta_\ell)\big)^{\alpha/(n-1)} a_\ell\left(\frac{w(\Delta_k)}{w(\Delta_\ell)}\right)^{\alpha/(n-1)}\left(\frac{w(\Delta_\ell)}{w(\Delta_k)}\right)^{\eta'/[(n-1)p]}\Bigg)^q\Bigg]^{1/q}\\
&= C\Bigg[\sum_{k=-\infty}^{\infty}\Bigg(\sum_{\ell=-\infty}^{k}\big(w(\Delta_\ell)\big)^{\alpha/(n-1)} a_\ell\left(\frac{w(\Delta_\ell)}{w(\Delta_k)}\right)^{(\eta'/p-\alpha)/(n-1)}\Bigg)^q\Bigg]^{1/q}\\
&\leq C\Bigg[\sum_{k=-\infty}^{\infty}\Big(\sum_{\ell=-\infty}^{k}\big(w(\Delta_\ell)\big)^{\alpha/(n-1)} a_\ell\cdot 2_*^{(\ell-k)(\eta'/p-\alpha)\tau}\Big)^q\Bigg]^{1/q} =: S_2 \qquad (23.133)
\end{aligned}$$

constant $C \in (0,\infty)$ appearing in (23.133) depends on n, p, α, η', $[w]_{A_p}$, and the Ahlfors regularity of Σ. The first inequality is a consequence of (23.122) and the Ahlfors regularity of Σ. The second inequality follows from the doubling property of w (cf. item *(5)* in Proposition 2.3), while the last inequality is obtained via (23.129) and the fact that $\eta'/p-\alpha>0$ (see (23.125)).

Case 1. Assume $0<q\leq 1$. Use Fubini's Theorem for series and that $\eta'/p-\alpha>0$ to obtain

$$S_2 = C\Bigg[\sum_{k=-\infty}^{\infty}\Big(\sum_{\ell=-\infty}^{k}\big(w(\Delta_\ell)\big)^{\alpha/(n-1)} a_\ell\cdot 2_*^{(\ell-k)(\eta'/p-\alpha)\tau}\Big)^q\Bigg]^{1/q}$$

$$\leq C\Bigg[\sum_{k=-\infty}^{\infty}\sum_{\ell=-\infty}^{k}\big(w(\Delta_\ell)\big)^{\alpha q/(n-1)}a_\ell^q\cdot 2_*^{(\ell-k)(\eta'/p-\alpha)\tau q}\Bigg]^{1/q}$$

$$= C\Bigg[\sum_{\ell=-\infty}^{\infty}\big(w(\Delta_\ell)\big)^{\alpha q/(n-1)}a_\ell^q\sum_{k=\ell}^{\infty}2_*^{(\ell-k)(\eta'/p-\alpha)\tau q}\Bigg]^{1/q}$$

$$= C\Bigg[\sum_{\ell=-\infty}^{\infty}\big(w(\Delta_\ell)\big)^{\alpha q/(n-1)}a_\ell^q\sum_{j=0}^{\infty}2_*^{-j(\eta'/p-\alpha)\tau q}\Bigg]^{1/q}$$

$$= C\Bigg[\sum_{\ell=-\infty}^{\infty}\big(w(\Delta_\ell)\big)^{\alpha q/(n-1)}a_\ell^q\Bigg]^{1/q} = C\|\{a_\ell\}_{\ell\in\mathbb{Z}}\|_{\ell_\alpha^q(\mathbb{Z};w)} \qquad (23.134)$$

for some constant $C\in(0,\infty)$ depending on n, p, q, η', α, $[w]_{A_p}$, and the Ahlfors regularity of Σ.

Case 2. Assume $1<q<\infty$. Thanks to Fubini's Theorem for series, Hölder's inequality, and $\eta'/p-\alpha>0$ we have

$$S_2 = C\Bigg[\sum_{k=-\infty}^{\infty}\Big(\sum_{\ell=-\infty}^{k}\big(w(\Delta_\ell)\big)^{\alpha/(n-1)}a_\ell\cdot 2_*^{(\ell-k)(\eta'/p-\alpha)\tau}\Big)^q\Bigg]^{1/q}$$

$$\leq C\Bigg[\sum_{k=-\infty}^{\infty}\Big(\sum_{\ell=-\infty}^{k}\big(w(\Delta_\ell)\big)^{\alpha q/(n-1)}a_\ell^q\cdot 2_*^{(\ell-k)(\eta'/p-\alpha)\tau q/2}\Big)\times$$

$$\times\Big(\sum_{\ell=-\infty}^{k}2_*^{(\ell-k)(\eta'/p-\alpha)\tau q'/2}\Big)^{q/q'}\Bigg]^{1/q}$$

$$= C\Bigg[\sum_{k=-\infty}^{\infty}\Big(\sum_{\ell=-\infty}^{k}\big(w(\Delta_\ell)\big)^{\alpha q/(n-1)}a_\ell^q\cdot 2_*^{(\ell-k)(\eta'/p-\alpha)\tau q/2}\Big)\times$$

$$\times\Big(\sum_{j=0}^{\infty}2_*^{-j(\eta'/p-\alpha)\tau q'/2}\Big)^{q/q'}\Bigg]^{1/q}$$

$$= C\Bigg[\sum_{k=-\infty}^{\infty}\sum_{\ell=-\infty}^{k}\big(w(\Delta_\ell)\big)^{\alpha q/(n-1)}a_\ell^q\cdot 2_*^{(\ell-k)(\eta'/p-\alpha)\tau q/2}\Bigg]^{1/q}$$

$$= C\Bigg[\sum_{\ell=-\infty}^{\infty}\big(w(\Delta_\ell)\big)^{\alpha q/(n-1)}a_\ell^q\sum_{k=\ell}^{\infty}2_*^{(\ell-k)(\eta'/p-\alpha)\tau q/2}\Bigg]^{1/q}$$

$$= C\Bigg[\sum_{\ell=-\infty}^{\infty}\big(w(\Delta_\ell)\big)^{\alpha q/(n-1)}a_\ell^q\sum_{j=0}^{\infty}2_*^{-j(\eta'/p-\alpha)\tau q/2}\Bigg]^{1/q}$$

$$= C\left[\sum_{\ell=-\infty}^{\infty} \big(w(\Delta_\ell)\big)^{\alpha q/(n-1)} a_\ell^q\right]^{1/q} = C\|\{a_\ell\}_{\ell\in\mathbb{Z}}\|_{\ell^q_\alpha(\mathbb{Z};w)} \tag{23.135}$$

for some constant $C \in (0,\infty)$ depending on n, p, q, η', α, $[w]_{A_p}$, and the Ahlfors regularity of Σ. All together, (23.133), (23.134), and (23.135), imply that $\|\cdot\|_{\ell^q_\alpha(\mathbb{Z};w)}$ satisfies (C4) in Definition 23.5. Ultimately, this finishes the proof of Claim 2. □

Motivated by the theory of grand variable Herz spaces (see, e.g., [NaRaZa20]) we discuss the following example.

Example 23.3 **(Grand Lebesgue Sequence Spaces with Dyadic Weights)** Assume $\mathcal{J} \subseteq \mathbb{Z}$ is an infinite set, and let μ be the counting measure on $\mathcal{J}$. Fix indices $q \in [1,\infty)$, $\theta \in (0,\infty)$, and $\alpha \in \mathbb{R}$, along with a number $2_* \in (1,\infty)$. For each sequence $\{a_j\}_{j\in\mathcal{J}} \subseteq \mathbb{R}$ define the quantity

$$\|\{a_j\}_{j\in\mathcal{J}}\|_{\ell^{q),\theta}_\alpha(\mathcal{J})} := \sup_{\varepsilon>0}\Big(\varepsilon^\theta \sum_{j\in\mathcal{J}} 2_*^{j\alpha q(1+\varepsilon)}|a_j|^{q(1+\varepsilon)}\Big)^{\frac{1}{q(1+\varepsilon)}} \in [0,\infty]. \tag{23.136}$$

Introduce the space of sequences

$$\ell^{q),\theta}_\alpha(\mathcal{J}) := \Big\{\{a_j\}_{j\in\mathcal{J}} \in \mathbb{R}^{\mathcal{J}} : \|\{a_j\}_{j\in\mathcal{J}}\|_{\ell^{q),\theta}_\alpha(\mathcal{J})} < \infty\Big\}. \tag{23.137}$$

Claim 1. The space $\ell^{q),\theta}_\alpha(\mathbb{Z})$ is a Generalized Banach Sequence Space on $(\mathbb{Z},\mu)$. A similar conclusion holds for $\ell^{q),\theta}_\alpha(\mathbb{N}_0)$.

Proof of Claim 1. It is visible from (23.136) that $\|\cdot\|_{\ell^{q),\theta}_\alpha(\mathbb{Z})}$ is a genuine norm satisfying properties (Q1')-(Q2') in Proposition 23.1. Fix $a = \{a_j\}_{j\in\mathbb{Z}} \subseteq \mathbb{R}_+$. To verify (Q3'), for each $k \in \mathbb{N}$ consider a sequence $a^{(k)} = \{a_j^{(k)}\}_{j\in\mathbb{Z}} \subseteq \mathbb{R}_+$ with the property that for each $j \in \mathbb{Z}$ we have $a_j^{(k)} \nearrow a_j$ as $k \to \infty$. Hence, from (23.136) we see that $\|a^{(k)}\|_{\ell^{q),\theta}_\alpha(\mathbb{Z})} \le \|a\|_{\ell^{q),\theta}_\alpha(\mathbb{Z})}$ for all $k \in \mathbb{N}$, which further implies

$$\limsup_{k\to\infty}\|a^{(k)}\|_{\ell^{q),\theta}_\alpha(\mathbb{Z})} \le \|a\|_{\ell^{q),\theta}_\alpha(\mathbb{Z})}. \tag{23.138}$$

Next, fix a parameter $\rho \in (0,\infty)$ such that

$$0 < \rho < \|a\|_{\ell^{q),\theta}_\alpha(\mathbb{Z})}. \tag{23.139}$$

From (23.136), there exists some $\varepsilon_0 > 0$ satisfying

$$\rho < \left(\varepsilon_0^\theta \sum_{j=-\infty}^{\infty} \big(2_*^{j\alpha} a_j\big)^{q(1+\varepsilon_0)}\right)^{\frac{1}{q(1+\varepsilon_0)}}. \tag{23.140}$$

Thanks to the assumption that $a^{(k)} \nearrow a$ pointwise in $\mathbb{Z}$ as $k \to \infty$, Lebesgue's Monotone Convergence Theorem for series ensures the existence of some $N \in \mathbb{N}$

for which

$$\rho < \left(\varepsilon_0^{\theta} \sum_{j=-\infty}^{\infty} (2_*^{j\alpha} a_j^{(k)})^{q(1+\varepsilon_0)} \right)^{\frac{1}{q(1+\varepsilon_0)}} \leq \|a^{(k)}\|_{\ell_\alpha^{q),\theta}(\mathbb{Z})} \tag{23.141}$$

whenever $k \geq N$. This forces

$$\rho \leq \liminf_{k\to\infty} \|a^{(k)}\|_{\ell_\alpha^{q),\theta}(\mathbb{Z})}. \tag{23.142}$$

Since this holds for every ρ as in (23.139), we can let $\rho \nearrow \|a\|_{\ell_\alpha^{q),\theta}(\mathbb{Z})}$ to obtain

$$\|a\|_{\ell_\alpha^{q),\theta}(\mathbb{Z})} \leq \liminf_{k\to\infty} \|a^{(k)}\|_{\ell_\alpha^{q),\theta}(\mathbb{Z})}. \tag{23.143}$$

In concert with (23.138), this implies $\|a^{(k)}\|_{\ell_\alpha^{q),\theta}(\mathbb{Z})} \nearrow \|a\|_{\ell_\alpha^{q),\theta}(\mathbb{Z})}$ as $k \to \infty$. Hence, $\|\cdot\|_{\ell_\alpha^{q),\theta}(\mathbb{Z})}$ satisfies (Q3') from Proposition 23.1.

Moving on, for each $k \in \mathbb{Z}$, recall that $\mathbb{1}_{\{k\}} = \{\delta_{jk}\}_{j\in\mathbb{Z}}$, and compute

$$\|\mathbb{1}_{\{k\}}\|_{\ell_\alpha^{q),\theta}(\mathbb{Z})} = \sup_{\varepsilon>0} \left(\varepsilon^{\theta} \sum_{j=-\infty}^{\infty} (2_*^{j\alpha} \delta_{jk})^{q(1+\varepsilon)} \right)^{\frac{1}{q(1+\varepsilon)}} \leq 2_*^{k\alpha} \big(\sup_{\varepsilon>0} \varepsilon^{\frac{1}{1+\varepsilon}} \big)^{\theta/q}. \tag{23.144}$$

Elementary calculus shows that the function $(0,\infty) \ni t \mapsto t^{\frac{1}{1+t}}$ is non-negative, continuous, bounded, and has exactly one global maximum. As such, the supremum in (23.144) is attained and finite. This shows that $\|\cdot\|_{\ell_\alpha^{q),\theta}(\mathbb{Z})}$ satisfies (Q4') from Proposition 23.1. Let $\varepsilon_\star > 0$ be the number for which $\sup_{\varepsilon>0} \varepsilon^{\frac{1}{1+\varepsilon}} = \varepsilon_\star^{\frac{1}{1+\varepsilon_\star}} < \infty$. For any sequence $\{a_j\}_{j\in\mathbb{Z}} \subseteq \mathbb{R}_+$, and for each $j \in \mathbb{Z}$ we estimate

$$\begin{aligned} a_j &= \varepsilon_\star^{\frac{-\theta}{q(1+\varepsilon_\star)}} 2_*^{-j\alpha} \left[\varepsilon_\star^{\theta} (2_*^{j\alpha} a_j)^{q(1+\varepsilon_\star)} \right]^{\frac{1}{q(1+\varepsilon_\star)}} \\ &\leq \varepsilon_\star^{\frac{-\theta}{q(1+\varepsilon_\star)}} 2_*^{-j\alpha} \left[\varepsilon_\star^{\theta} \sum_{k=-\infty}^{\infty} (2_*^{k\alpha} a_k)^{q(1+\varepsilon_\star)} \right]^{\frac{1}{q(1+\varepsilon_\star)}} \\ &\leq \varepsilon_\star^{\frac{-\theta}{q(1+\varepsilon_\star)}} 2_*^{-j\alpha} \|\{a_k\}_{k\in\mathbb{Z}}\|_{\ell_\alpha^{q),\theta}(\mathbb{Z})}. \end{aligned} \tag{23.145}$$

Hence, $\|\cdot\|_{\ell_\alpha^{q),\theta}(\mathbb{Z})}$ also satisfies (Q5') from Proposition 23.1. As such, this is a function norm on $\mathbb{Z}$, which in turn guarantees that $\ell_\alpha^{q),\theta}(\mathbb{Z})$ is a Generalized Banach Sequence Space on $\mathbb{Z}$. This completes the proof of Claim 1. The case of $\ell_\alpha^{q),\theta}(\mathbb{N}_0)$ is dealt with in a similar fashion. □

Claim 2. Consider two numbers $\eta, \eta' \in (0,\infty)$, and assume that $-\eta < \alpha < \eta'$. Then $\|\cdot\|_{\ell_\alpha^{q),\theta}(\mathbb{Z})}$ is (η,η')-compliant, and $\|\cdot\|_{\ell_\alpha^{q),\theta}(\mathbb{N}_0)}$ is (η,η')-compliant on $\mathbb{N}_0$.

Proof of `Claim 2.` We only justify the claim made in relation to $\|\cdot\|_{\ell^{q),\theta}_{\alpha}(\mathbb{Z})}$ since for $\|\cdot\|_{\ell^{q),\theta}_{\alpha}(\mathbb{N}_0)}$ follows via a similar reasoning. Start by fixing an arbitrary sequence $\{a_\ell\}_{\ell\in\mathbb{Z}} \subseteq \mathbb{R}_+$. Then, from (23.136) we see that

$$
\begin{aligned}
\|\{a_{\ell+1}\}_{\ell\in\mathbb{Z}}\|_{\ell^{q),\theta}_{\alpha}(\mathbb{Z})} &= \sup_{\varepsilon>0}\Big(\varepsilon^{\theta}\sum_{\ell=-\infty}^{\infty} 2_*^{\ell\alpha q(1+\varepsilon)} a_{\ell+1}^{q(1+\varepsilon)}\Big)^{\frac{1}{q(1+\varepsilon)}}\\
&= 2_*^{-\alpha}\sup_{\varepsilon>0}\Big(\varepsilon^{\theta}\sum_{\ell=-\infty}^{\infty} 2_*^{(\ell+1)\alpha q(1+\varepsilon)} a_{\ell+1}^{q(1+\varepsilon)}\Big)^{\frac{1}{q(1+\varepsilon)}}\\
&= 2_*^{-\alpha}\|\{a_\ell\}_{\ell\in\mathbb{Z}}\|_{\ell^{q),\theta}_{\alpha}(\mathbb{Z})}.
\end{aligned}
\tag{23.146}
$$

Similarly,

$$
\|\{a_{\ell-1}\}_{\ell\in\mathbb{Z}}\|_{\ell^{q),\theta}_{\alpha}(\mathbb{Z})} = 2_*^{\alpha}\|\{a_\ell\}_{\ell\in\mathbb{Z}}\|_{\ell^{q),\theta}_{\alpha}(\mathbb{Z})}.
\tag{23.147}
$$

Hence, $\|\cdot\|_{\ell^{q),\theta}_{\alpha}(\mathbb{Z})}$ satisfies properties (C1)-(C2) in Definition 23.5.

For any $\varepsilon \in (0,\infty)$ we have $1 < q(1+\varepsilon) < \infty$. Let $[q(1+\varepsilon)]' \in (1,\infty)$ denote the Hölder conjugate exponent of $q(1+\varepsilon)$. To verify (C3), use Hölder's inequality, Fubini's Theorem for series, and the fact that $\alpha+\eta>0$ to estimate

$$
\begin{aligned}
&\Big\|\Big\{\sum_{\ell=k}^{\infty} 2_*^{(k-\ell)\eta} a_\ell\Big\}_{k\in\mathbb{Z}}\Big\|_{\ell^{q),\theta}_{\alpha}(\mathbb{Z})}\\
&= \sup_{\varepsilon>0}\Big[\varepsilon^{\theta}\sum_{k=-\infty}^{\infty} 2_*^{k\alpha q(1+\varepsilon)}\Big(\sum_{\ell=k}^{\infty} 2_*^{(k-\ell)\eta} a_\ell\Big)^{q(1+\varepsilon)}\Big]^{\frac{1}{q(1+\varepsilon)}}\\
&= \sup_{\varepsilon>0}\Big[\varepsilon^{\theta}\sum_{k=-\infty}^{\infty}\Big(\sum_{\ell=k}^{\infty} 2_*^{(k-\ell)(\alpha+\eta)} 2_*^{\ell\alpha} a_\ell\Big)^{q(1+\varepsilon)}\Big]^{\frac{1}{q(1+\varepsilon)}}\\
&\le \sup_{\varepsilon>0}\Big[\varepsilon^{\theta}\sum_{k=-\infty}^{\infty}\Big(\sum_{\ell=k}^{\infty} 2_*^{(k-\ell)\frac{(\alpha+\eta)}{2}q(1+\varepsilon)} 2_*^{\ell\alpha q(1+\varepsilon)} a_\ell^{q(1+\varepsilon)}\Big)\times\\
&\qquad\times\Big(\sum_{\ell=k}^{\infty} 2_*^{(k-\ell)\frac{(\alpha+\eta)}{2}[q(1+\varepsilon)]'}\Big)^{\frac{q(1+\varepsilon)}{[q(1+\varepsilon)]'}}\Big]^{\frac{1}{q(1+\varepsilon)}}\\
&\le \sup_{\varepsilon>0}\Big[\varepsilon^{\theta}\sum_{k=-\infty}^{\infty}\Big(\sum_{\ell=k}^{\infty} 2_*^{(k-\ell)\frac{(\alpha+\eta)}{2}q(1+\varepsilon)} 2_*^{\ell\alpha q(1+\varepsilon)} a_\ell^{q(1+\varepsilon)}\Big)\times\\
&\qquad\times\Big(\sum_{j=0}^{\infty} 2_*^{-j\frac{(\alpha+\eta)}{2}[q(1+\varepsilon)]'}\Big)^{\frac{q(1+\varepsilon)}{[q(1+\varepsilon)]'}}\Big]^{\frac{1}{q(1+\varepsilon)}}
\end{aligned}
$$

$$
\begin{aligned}
&\leq \sup_{\varepsilon>0}\left[\varepsilon^{\theta}\sum_{k=-\infty}^{\infty}\sum_{\ell=k}^{\infty}2_*^{(k-\ell)\frac{(\alpha+\eta)}{2}q(1+\varepsilon)}2_*^{\ell\alpha q(1+\varepsilon)}a_\ell^{q(1+\varepsilon)}\right]^{\frac{1}{q(1+\varepsilon)}}\left(\sum_{j=0}^{\infty}2_*^{-j\frac{(\alpha+\eta)}{2}}\right)\\
&= C\cdot\sup_{\varepsilon>0}\left[\varepsilon^{\theta}\sum_{k=-\infty}^{\infty}\sum_{\ell=k}^{\infty}2_*^{(k-\ell)\frac{(\alpha+\eta)}{2}q(1+\varepsilon)}2_*^{\ell\alpha q(1+\varepsilon)}a_\ell^{q(1+\varepsilon)}\right]^{\frac{1}{q(1+\varepsilon)}}\\
&= C\cdot\sup_{\varepsilon>0}\left[\varepsilon^{\theta}\sum_{\ell=-\infty}^{\infty}2_*^{\ell\alpha q(1+\varepsilon)}a_\ell^{q(1+\varepsilon)}\sum_{k=-\infty}^{\ell}2_*^{(k-\ell)\frac{(\alpha+\eta)}{2}q(1+\varepsilon)}\right]^{\frac{1}{q(1+\varepsilon)}}\\
&= C\cdot\sup_{\varepsilon>0}\left[\varepsilon^{\theta}\sum_{\ell=-\infty}^{\infty}2_*^{\ell\alpha q(1+\varepsilon)}a_\ell^{q(1+\varepsilon)}\sum_{j=0}^{\infty}2_*^{-j\frac{(\alpha+\eta)}{2}q(1+\varepsilon)}\right]^{\frac{1}{q(1+\varepsilon)}}\\
&\leq C\cdot\sup_{\varepsilon>0}\left[\varepsilon^{\theta}\sum_{\ell=-\infty}^{\infty}2_*^{\ell\alpha q(1+\varepsilon)}a_\ell^{q(1+\varepsilon)}\right]^{\frac{1}{q(1+\varepsilon)}}\left(\sum_{j=0}^{\infty}2_*^{-j\frac{(\alpha+\eta)}{2}}\right)\\
&\leq C\|\{a_\ell\}_{\ell\in\mathbb{Z}}\|_{\ell_\alpha^{q),\theta}(\mathbb{Z})}
\end{aligned}
\tag{23.148}
$$

for some constant $C = C(2_*, \alpha, \eta) \in (0, \infty)$. This establishes that $\|\cdot\|_{\ell_\alpha^{q),\theta}(\mathbb{Z})}$ satisfies condition (C3) in Definition 23.5.

Lastly, we show that $\|\cdot\|_{\ell_\alpha^{q),\theta}(\mathbb{Z})}$ satisfies (C4). Specifically, we rely on Hölder's inequality, Fubini's Theorem for series, and the fact that $\eta' - \alpha > 0$, to estimate

$$
\begin{aligned}
&\left\|\left\{\sum_{\ell=-\infty}^{k}2_*^{(\ell-k)\eta'}a_\ell\right\}_{k\in\mathbb{Z}}\right\|_{\ell_\alpha^{q),\theta}(\mathbb{Z})}\\
&= \sup_{\varepsilon>0}\left[\varepsilon^{\theta}\sum_{k=-\infty}^{\infty}2_*^{k\alpha q(1+\varepsilon)}\left(\sum_{\ell=-\infty}^{k}2_*^{(\ell-k)\eta'}a_\ell\right)^{q(1+\varepsilon)}\right]^{\frac{1}{q(1+\varepsilon)}}\\
&= \sup_{\varepsilon>0}\left[\varepsilon^{\theta}\sum_{k=-\infty}^{\infty}\left(\sum_{\ell=-\infty}^{k}2_*^{(\ell-k)(\eta'-\alpha)}2_*^{\ell\alpha}a_\ell\right)^{q(1+\varepsilon)}\right]^{\frac{1}{q(1+\varepsilon)}}.\\
&\leq \sup_{\varepsilon>0}\left[\varepsilon^{\theta}\sum_{k=-\infty}^{\infty}\left(\sum_{\ell=-\infty}^{k}2_*^{(\ell-k)\frac{(\eta'-\alpha)}{2}q(1+\varepsilon)}2_*^{\ell\alpha q(1+\varepsilon)}a_\ell^{q(1+\varepsilon)}\right)\right.\\
&\quad\left.\times\left(\sum_{\ell=-\infty}^{k}2_*^{(\ell-k)\frac{(\eta'-\alpha)}{2}[q(1+\varepsilon)]'}\right)^{\frac{q(1+\varepsilon)}{[q(1+\varepsilon)]'}}\right]^{\frac{1}{q(1+\varepsilon)}}\\
&= \sup_{\varepsilon>0}\left[\varepsilon^{\theta}\sum_{k=-\infty}^{\infty}\left(\sum_{\ell=-\infty}^{k}2_*^{(\ell-k)\frac{(\eta'-\alpha)}{2}q(1+\varepsilon)}2_*^{\ell\alpha q(1+\varepsilon)}a_\ell^{q(1+\varepsilon)}\right)\right.
\end{aligned}
$$

$$\times \Big(\sum_{j=0}^{\infty} 2_*^{-j\frac{(\eta'-\alpha)}{2}[q(1+\varepsilon)]'} \Big)^{\frac{q(1+\varepsilon)}{[q(1+\varepsilon)]'}} \Bigg]^{\frac{1}{q(1+\varepsilon)}}$$

$$\leq \sup_{\varepsilon>0} \Bigg[\varepsilon^{\theta} \sum_{k=-\infty}^{\infty} \sum_{\ell=-\infty}^{k} 2_*^{(\ell-k)\frac{(\eta'-\alpha)}{2} q(1+\varepsilon)} 2_*^{\ell \alpha q(1+\varepsilon)} a_\ell^{q(1+\varepsilon)} \Bigg]^{\frac{1}{q(1+\varepsilon)}} \Big(\sum_{j=0}^{\infty} 2_*^{-j\frac{(\eta'-\alpha)}{2}} \Big)$$

$$= C \cdot \sup_{\varepsilon>0} \Bigg[\varepsilon^{\theta} \sum_{\ell=-\infty}^{\infty} 2_*^{\ell \alpha q(1+\varepsilon)} a_\ell^{q(1+\varepsilon)} \sum_{k=\ell}^{\infty} 2_*^{(\ell-k)\frac{(\eta'-\alpha)}{2} q(1+\varepsilon)} \Bigg]^{\frac{1}{q(1+\varepsilon)}}$$

$$= C \cdot \sup_{\varepsilon>0} \Bigg[\varepsilon^{\theta} \sum_{\ell=-\infty}^{\infty} 2_*^{\ell \alpha q(1+\varepsilon)} a_\ell^{q(1+\varepsilon)} \sum_{j=0}^{\infty} 2_*^{-j\frac{(\eta'-\alpha)}{2} q(1+\varepsilon)} \Bigg]^{\frac{1}{q(1+\varepsilon)}}$$

$$\leq C \cdot \sup_{\varepsilon>0} \Bigg[\varepsilon^{\theta} \sum_{\ell=-\infty}^{\infty} 2_*^{\ell \alpha q(1+\varepsilon)} a_\ell^{q(1+\varepsilon)} \Bigg]^{\frac{1}{q(1+\varepsilon)}} \Big(\sum_{j=0}^{\infty} 2_*^{-j\frac{(\eta'-\alpha)}{2}} \Big)$$

$$\leq C \| \{a_\ell\}_{\ell\in\mathbb{Z}} \|_{\ell_\alpha^{q),\theta}(\mathbb{Z})} \tag{23.149}$$

for some constant $C = C(2_*, \eta', \alpha) \in (0, \infty)$. This ensures that $\| \cdot \|_{\ell_\alpha^{q),\theta}(\mathbb{Z})}$ satisfies (C4) in Definition 23.5, thus concluding the proof of Claim 2. □

As a preamble to our next example, we recall several definitions. A non-negative function ω defined in $\mathbb{R}_+$ is said to be `almost increasing` (respectively `almost decreasing`) in $\mathbb{R}_+$ if there exists some constant $C \in [1, \infty)$ such that whenever $s, t \in (0, \infty)$, with $t \leq s$ (respectively $t \geq s$) one has

$$\omega(t) \leq C\omega(s). \tag{23.150}$$

Define the `function class` $\mathrm{M}(\mathbb{R}_+)$ to be the collection of all positive functions ω on $\mathbb{R}_+$ such that whenever $0 < \delta < N < \infty$ there holds

$$0 < \inf_{t\in(\delta,N)} \omega(t) \leq \sup_{t\in(\delta,N)} \omega(t) < \infty, \tag{23.151}$$

and there exist exponents $\alpha_0, \alpha_\infty, \beta_0, \beta_\infty \in \mathbb{R}$ with the property that

(i) $(0, 1] \ni t \mapsto \omega(t)t^{-\alpha_0}$ is almost increasing and $(0, 1] \ni t \mapsto \omega(t)t^{-\beta_0}$ is almost decreasing;

(ii) $[1, \infty) \ni t \mapsto \omega(t)t^{-\alpha_\infty}$ is almost increasing and $[1, \infty) \ni t \mapsto \omega(t)t^{-\beta_\infty}$ is almost decreasing.

Next, the `Matuszewska-Orlicz indices` of a positive function ω on $\mathbb{R}_+$ are defined according to

$$m_0(\omega) := \sup_{0<t<1} \frac{\ln\left(\limsup\limits_{h\to 0^+} \frac{\omega(ht)}{\omega(h)}\right)}{\ln t}, \quad M_0(\omega) := \inf_{0<t<1} \frac{\ln\left(\liminf\limits_{h\to 0^+} \frac{\omega(ht)}{\omega(h)}\right)}{\ln t},$$

$$m_\infty(\omega) := \sup_{1<t<\infty} \frac{\ln\left(\liminf\limits_{h\to \infty} \frac{\omega(ht)}{\omega(h)}\right)}{\ln t}, \quad M_\infty(\omega) := \inf_{1<t<\infty} \frac{\ln\left(\limsup\limits_{h\to \infty} \frac{\omega(ht)}{\omega(h)}\right)}{\ln t}. \tag{23.152}$$

As pointed out in [LiYaHu23, Remark 1.1.5, (ii), p. 7], whenever $\omega \in \mathrm{M}(\mathbb{R}_+)$ one has

$$\begin{aligned} &-\infty < m_0(\omega) \leq M_0(\omega) < \infty \\ \text{and} \quad &-\infty < m_\infty(\omega) \leq M_\infty(\omega) < \infty. \end{aligned} \tag{23.153}$$

The following auxiliary result from [RaSa20, Lemma 3.1, p. 6] gives estimates for functions belonging to $\mathrm{M}(\mathbb{R}_+)$.

Lemma 23.5 *Consider $\omega \in \mathrm{M}(\mathbb{R}_+)$, and let $m_0(\omega)$, $M_0(\omega)$, $m_\infty(\omega)$, and $M_\infty(\omega)$ be its Matuszewska-Orlicz indices (as in (23.152)). Abbreviate*

$$m_{0,\infty} := \min\{m_0(\omega), m_\infty(\omega)\} \quad \textit{and} \quad M_{0,\infty} := \max\{M_0(\omega), M_\infty(\omega)\} \tag{23.154}$$

Then for every $\varepsilon \in (0,\infty)$ there exists a constant $C = C(\varepsilon) \in (0,\infty)$ such that

$$\frac{\omega(t)}{\omega(s)} \leq C\left(\frac{t}{s}\right)^{m_{0,\infty}-\varepsilon} \quad \textit{whenever } 0 < t < s < \infty, \tag{23.155}$$

and

$$\frac{\omega(t)}{\omega(s)} \leq C\left(\frac{t}{s}\right)^{M_{0,\infty}+\varepsilon} \quad \textit{whenever } 0 < s < t < \infty. \tag{23.156}$$

We are now in a position to present our next example.

Example 23.4 **(Function-Class-Weighted Sequence Spaces)** Consider $\mathcal{J} \subseteq \mathbb{Z}$ an infinite set, and denote by μ the counting measure on $\mathcal{J}$. Fix an exponent $q \in (0,\infty]$, a number $2_* \in (1,\infty)$, and select a function $\omega \in \mathrm{M}(\mathbb{R}_+)$. With any $\{a_j\}_{j\in\mathcal{J}} \subseteq \mathbb{R}$ associate the quantity in $[0,\infty]$ defined as

$$\|\{a_j\}_{j\in\mathcal{J}}\|_{\ell^q_{\omega(\cdot)}(\mathcal{J})} := \begin{cases} \left[\sum\limits_{j\in\mathcal{J}} (\omega(2_*^j))^q |a_j|^q\right]^{1/q} & \text{if } 0 < q < \infty, \\ \sup\limits_{j\in\mathcal{J}} \left\{\omega(2_*^j)|a_j|\right\} & \text{if } q = \infty. \end{cases} \tag{23.157}$$

Finally, introduce the space of sequences

$$\ell^q_{\omega(\cdot)}(\mathcal{J}) := \left\{\{a_j\}_{j\in\mathcal{J}} \in \mathbb{R}^{\mathcal{J}} : \|\{a_j\}_{j\in\mathcal{J}}\|_{\ell^q_{\omega(\cdot)}(\mathcal{J})} < \infty\right\}. \tag{23.158}$$

Claim 1. The space $\ell^q_{\omega(\cdot)}(\mathcal{J})$ is a Generalized Quasi-Banach Sequence Space on $(\mathcal{J}, \mu)$.

Proof of Claim 1. Keeping in mind that the function ω is positive and finite, it is visible from definitions that $\|\cdot\|_{\ell^q_{\omega(\cdot)}(\mathcal{J})}$ satisfies (Q1'), (Q2'), (Q4'), and (Q5') in Proposition 23.1. Property (Q3') is a consequence of Lebesgue's Monotone Convergence Theorem for sequences. This ensures that $\|\cdot\|_{\ell^q_{\omega(\cdot)}(\mathcal{J})}$ is a function quasinorm on $\mathcal{J}$, which further implies that $\ell^q_{\omega(\cdot)}(\mathcal{J})$ is a Generalized Quasi-Banach Sequence Space on $\mathcal{J}$. □

Claim 2. Fix parameters $\eta, \eta' \in (0, \infty)$. Bring in the Matuszewska-Orlicz indices $m_0(\omega)$, $M_0(\omega)$, $m_\infty(\omega)$, and $M_\infty(\omega)$ of ω, and assume that

$$-\eta < m_0(\omega) \leq M_0(\omega) < \eta' \tag{23.159}$$

$$\text{and} \quad -\eta < m_\infty(\omega) \leq M_\infty(\omega) < \eta'. \tag{23.160}$$

Then $\|\cdot\|_{\ell^q_{\omega(\cdot)}(\mathbb{Z})}$ is (η, η')-compliant, and $\|\cdot\|_{\ell^q_{\omega(\cdot)}(\mathbb{N}_0)}$ is (η, η')-compliant on $\mathbb{N}_0$.

Proof of Claim 2. Here we only present the proof of the (η, η')-compliancy of $\|\cdot\|_{\ell^q_{\omega(\cdot)}(\mathbb{Z})}$, since the (η, η')-compliancy on $\mathbb{N}_0$ of $\|\cdot\|_{\ell^q_{\omega(\cdot)}(\mathbb{N}_0)}$ can be obtained via a similar argument. Also, work under the assumption that $0 < q < \infty$, as for $q = \infty$ the desired conclusion can be obtained by making natural modifications. To get started, fix $\varepsilon \in (0, \infty)$ such that

$$0 < \varepsilon < \min\left\{\eta' - M_{0,\infty},\, \eta + m_{0,\infty}\right\} \tag{23.161}$$

where $M_{0,\infty}$ and $m_{0,\infty}$ are as in (23.154). Fix an arbitrary sequence $\{a_\ell\}_{\ell\in\mathbb{Z}} \subseteq \mathbb{R}_+$. Thanks to estimate (23.155) we see that

$$\begin{aligned}
\|\{a_{\ell+1}\}_{\ell\in\mathbb{Z}}\|_{\ell^q_{\omega(\cdot)}(\mathbb{Z})} &= \left[\sum_{\ell=-\infty}^{\infty} (\omega(2_*^\ell))^q a_{\ell+1}^q\right]^{1/q} \\
&= \left[\sum_{\ell=-\infty}^{\infty} \left(\frac{\omega(2_*^\ell)}{\omega(2_*^{\ell+1})}\right)^q (\omega(2_*^{\ell+1}))^q a_{\ell+1}^q\right]^{1/q} \\
&\leq C 2_*^{\varepsilon - m_{0,\infty}} \left[\sum_{\ell=-\infty}^{\infty} (\omega(2_*^{\ell+1}))^q a_{\ell+1}^q\right]^{1/q} \\
&= C\|\{a_\ell\}_{\ell\in\mathbb{Z}}\|_{\ell^q_{\omega(\cdot)}(\mathbb{Z})}
\end{aligned} \tag{23.162}$$

for some constant $C = C(2_*, \varepsilon, m_{0,\infty}) \in (0, \infty)$. Hence, $\|\cdot\|_{\ell^q_{\omega(\cdot)}(\mathbb{Z})}$ satisfies (C1) in Definition 23.5. This quasinorm also satisfies (C2) in Definition 23.5, since from (23.156) we have

$$
\begin{aligned}
\|\{a_{\ell-1}\}_{\ell\in\mathbb{Z}}\|_{\ell^q_{\omega(\cdot)}(\mathbb{Z})} &= \left[\sum_{\ell=-\infty}^{\infty} (\omega(2_*^\ell))^q a_{\ell-1}^q\right]^{1/q} \\
&= \left[\sum_{\ell=-\infty}^{\infty} \left(\frac{\omega(2_*^\ell)}{\omega(2_*^{\ell-1})}\right)^q (\omega(2_*^{\ell-1}))^q a_{\ell-1}^q\right]^{1/q} \\
&\le C 2_*^{M_{0,\infty}+\varepsilon}\left[\sum_{\ell=-\infty}^{\infty} (\omega(2_*^{\ell-1}))^q a_{\ell-1}^q\right]^{1/q} \\
&= C\|\{a_\ell\}_{\ell\in\mathbb{Z}}\|_{\ell^q_{\omega(\cdot)}(\mathbb{Z})} \qquad (23.163)
\end{aligned}
$$

for some $C = C(2_*, \varepsilon, M_{0,\infty}) \in (0, \infty)$.

We next verify condition (C3) in Definition 23.5. In light of (23.155) (currently used with $t := 2_*^k$ and $s := 2_*^\ell$), we write

$$
\begin{aligned}
\left\|\left\{\sum_{\ell=k}^{\infty} 2_*^{(k-\ell)\eta} a_\ell\right\}_{k\in\mathbb{Z}}\right\|_{\ell^q_{\omega(\cdot)}(\mathbb{Z})} &= \left[\sum_{k=-\infty}^{\infty} (\omega(2_*^k))^q \left(\sum_{\ell=k}^{\infty} 2_*^{(k-\ell)\eta} a_\ell\right)^q\right]^{1/q} \\
&= \left[\sum_{k=-\infty}^{\infty} \left(\sum_{\ell=k}^{\infty} \omega(2_*^k) 2_*^{(k-\ell)\eta} a_\ell\right)^q\right]^{1/q} \\
&= \left[\sum_{k=-\infty}^{\infty} \left(\sum_{\ell=k}^{\infty} \frac{\omega(2_*^k)}{\omega(2_*^\ell)} 2_*^{(k-\ell)\eta} \omega(2_*^\ell) a_\ell\right)^q\right]^{1/q} \\
&\le \left[\sum_{k=-\infty}^{\infty} \left(\sum_{\ell=k}^{\infty} \left(\frac{2_*^k}{2_*^\ell}\right)^{m_{0,\infty}-\varepsilon} 2_*^{(k-\ell)\eta} \omega(2_*^\ell) a_\ell\right)^q\right]^{1/q} \\
&= C\left[\sum_{k=-\infty}^{\infty} \left(\sum_{\ell=k}^{\infty} 2_*^{(k-\ell)(m_{0,\infty}-\varepsilon+\eta)} \omega(2_*^\ell) a_\ell\right)^q\right]^{1/q} =: \mathrm{S}_1 \qquad (23.164)
\end{aligned}
$$

for some $C = C(\varepsilon) \in (0, \infty)$.

Case 1. Assume $0 < q \le 1$. Using Fubini's Theorem for series along with the fact that $m_{0,\infty} - \varepsilon + \eta > 0$ (see (23.161)) we estimate

$$
\begin{aligned}
\mathrm{S}_1 &= C\Bigg[\sum_{k=-\infty}^{\infty}\Bigg(\sum_{\ell=k}^{\infty} 2_*^{(k-\ell)(m_{0,\infty}-\varepsilon+\eta)}\omega(2_*^{\ell})a_\ell\Bigg)^q\Bigg]^{1/q}\\
&\le C\Bigg[\sum_{k=-\infty}^{\infty}\sum_{\ell=k}^{\infty} 2_*^{(k-\ell)(m_{0,\infty}-\varepsilon+\eta)q}\big(\omega(2_*^{\ell})\big)^q a_\ell^q\Bigg]^{1/q}\\
&= C\Bigg[\sum_{\ell=-\infty}^{\infty}\big(\omega(2_*^{\ell})\big)^q a_\ell^q \sum_{k=-\infty}^{\ell} 2_*^{(k-\ell)(m_{0,\infty}-\varepsilon+\eta)q}\Bigg]^{1/q}\\
&= C\Bigg[\sum_{\ell=-\infty}^{\infty}\big(\omega(2_*^{\ell})\big)^q a_\ell^q \sum_{j=0}^{\infty} 2_*^{-j(m_{0,\infty}-\varepsilon+\eta)q}\Bigg]^{1/q} \le C\|\{a_\ell\}_{\ell\in\mathbb{Z}}\|_{\ell^q_{\omega(\cdot)}(\mathbb{Z})}
\end{aligned}
\tag{23.165}
$$

for some $C = C(2_*, q, \varepsilon, \eta, m_{0,\infty}) \in (0, \infty)$.

<u>**Case 2.**</u> Assume $1 < q < \infty$. Employing Hölder's inequality, Fubini's Theorem for series, together with $m_{0,\infty} - \varepsilon + \eta > 0$ (see (23.161)) we obtain

$$
\begin{aligned}
\mathrm{S}_1 &= C\Bigg[\sum_{k=-\infty}^{\infty}\Bigg(\sum_{\ell=k}^{\infty} 2_*^{(k-\ell)(m_{0,\infty}-\varepsilon+\eta)}\omega(2_*^{\ell})a_\ell\Bigg)^q\Bigg]^{1/q}\\
&\le C\Bigg[\sum_{k=-\infty}^{\infty}\Bigg(\sum_{\ell=k}^{\infty} 2_*^{(k-\ell)(m_{0,\infty}-\varepsilon+\eta)q/2}\big(\omega(2_*^{\ell})\big)^q a_\ell^q\Bigg)\times\\
&\qquad\times\Bigg(\sum_{\ell=k}^{\infty} 2_*^{(k-\ell)(m_{0,\infty}-\varepsilon+\eta)q'/2}\Bigg)^{q/q'}\Bigg]^{1/q}\\
&= C\Bigg[\sum_{k=-\infty}^{\infty}\Bigg(\sum_{\ell=k}^{\infty} 2_*^{(k-\ell)(m_{0,\infty}-\varepsilon+\eta)q/2}\big(\omega(2_*^{\ell})\big)^q a_\ell^q\Bigg)\times\\
&\qquad\times\Bigg(\sum_{j=0}^{\infty} 2_*^{-j(m_{0,\infty}-\varepsilon+\eta)q'/2}\Bigg)^{q/q'}\Bigg]^{1/q}\\
&\le C\Bigg[\sum_{\ell=-\infty}^{\infty}\big(\omega(2_*^{\ell})\big)^q a_\ell^q \sum_{k=-\infty}^{\ell} 2_*^{(k-\ell)(m_{0,\infty}-\varepsilon+\eta)q/2}\Bigg]^{1/q}\\
&\le C\Bigg[\sum_{\ell=-\infty}^{\infty}\big(\omega(2_*^{\ell})\big)^q a_\ell^q \sum_{j=0}^{\infty} 2_*^{-j(m_{0,\infty}-\varepsilon+\eta)q/2}\Bigg]^{1/q} \le C\|\{a_\ell\}_{\ell\in\mathbb{Z}}\|_{\ell^q_{\omega(\cdot)}(\mathbb{Z})}
\end{aligned}
\tag{23.166}
$$

for some constant $C = C(2_*, q, \varepsilon, \eta, m_{0,\infty}) \in (0, \infty)$. Hence, (23.165) and (23.166) imply that $\|\cdot\|_{\ell^q_{\omega(\cdot)}(\mathbb{Z})}$ satisfies property (C3) in Definition 23.5.

Regarding property (C4) in Definition 23.5, we make use of (23.156) to compute

$$\begin{aligned}
\Big\|\Big\{\sum_{\ell=-\infty}^{k} 2_*^{(\ell-k)\eta'} a_\ell\Big\}_{k\in\mathbb{Z}}\Big\|_{\ell^q_{\omega(\cdot)}(\mathbb{Z})} &= \left[\sum_{k=-\infty}^{\infty} (\omega(2_*^k))^q \left(\sum_{\ell=-\infty}^{k} 2_*^{(\ell-k)\eta'} a_\ell\right)^q\right]^{1/q} \\
&= \left[\sum_{k=-\infty}^{\infty} \left(\sum_{\ell=-\infty}^{k} \omega(2_*^k) 2_*^{(\ell-k)\eta'} a_\ell\right)^q\right]^{1/q} \\
&= \left[\sum_{k=-\infty}^{\infty} \left(\sum_{\ell=-\infty}^{k} \frac{\omega(2_*^k)}{\omega(2_*^\ell)} 2_*^{(\ell-k)\eta'} \omega(2_*^\ell) a_\ell\right)^q\right]^{1/q} \\
&\leq \left[\sum_{k=-\infty}^{\infty} \left(\sum_{\ell=-\infty}^{k} \left(\frac{2_*^k}{2_*^\ell}\right)^{M_{0,\infty}+\varepsilon} 2_*^{(\ell-k)\eta'} \omega(2_*^\ell) a_\ell\right)^q\right]^{1/q} \\
&= C\left[\sum_{k=-\infty}^{\infty} \left(\sum_{\ell=-\infty}^{k} 2_*^{(k-\ell)(M_{0,\infty}+\varepsilon-\eta')} \omega(2_*^\ell) a_\ell\right)^q\right]^{1/q} =: \mathrm{S}_2
\end{aligned} \tag{23.167}$$

for some constant $C = C(\varepsilon) \in (0, \infty)$.

Case 1. Assume $0 < q \leq 1$. Thanks to Fubini's Theorem for series, and the inequality $M_{0,\infty} + \varepsilon - \eta' < 0$ (as implied by (23.161)), we bound

$$\begin{aligned}
\mathrm{S}_2 &= C\left[\sum_{k=-\infty}^{\infty} \left(\sum_{\ell=-\infty}^{k} 2_*^{(k-\ell)(M_{0,\infty}+\varepsilon-\eta')} \omega(2_*^\ell) a_\ell\right)^q\right]^{1/q} \\
&\leq C\left[\sum_{k=-\infty}^{\infty} \sum_{\ell=-\infty}^{k} 2_*^{(k-\ell)(M_{0,\infty}+\varepsilon-\eta')q} (\omega(2_*^\ell))^q a_\ell^q\right]^{1/q} \\
&= C\left[\sum_{\ell=-\infty}^{\infty} (\omega(2_*^\ell))^q a_\ell^q \sum_{k=\ell}^{\infty} 2_*^{(k-\ell)(M_{0,\infty}+\varepsilon-\eta')q}\right]^{1/q} \\
&= C\left[\sum_{\ell=-\infty}^{\infty} (\omega(2_*^\ell))^q a_\ell^q \sum_{j=0}^{\infty} 2_*^{j(M_{0,\infty}+\varepsilon-\eta')q}\right]^{1/q} \leq C\|\{a_\ell\}_{\ell\in\mathbb{Z}}\|_{\ell^q_{\omega(\cdot)}(\mathbb{Z})}
\end{aligned} \tag{23.168}$$

for some $C = C(2_*, q, \varepsilon, \eta', M_{0,\infty}) \in (0, \infty)$.

Case 2. Assume $1 < q < \infty$. Using Hölder's inequality, Fubini's Theorem for series, along with $M_{0,\infty} + \varepsilon - \eta' < 0$ (see (23.161)), compute

$$
\begin{aligned}
S_2 &= C\left[\sum_{k=-\infty}^{\infty}\left(\sum_{\ell=-\infty}^{k} 2_*^{(k-\ell)(M_{0,\infty}+\varepsilon-\eta')}\omega(2_*^{\ell})a_\ell\right)^q\right]^{1/q}\\
&\leq C\left[\sum_{k=-\infty}^{\infty}\left(\sum_{\ell=-\infty}^{k} 2_*^{(k-\ell)(M_{0,\infty}+\varepsilon-\eta')q/2}(\omega(2_*^{\ell}))^q a_\ell^q\right)\times\right.\\
&\qquad\left.\times\left(\sum_{\ell=-\infty}^{k} 2_*^{(k-\ell)(M_{0,\infty}+\varepsilon-\eta')q'/2}\right)^{q/q'}\right]^{1/q}\\
&= C\left[\sum_{k=-\infty}^{\infty}\left(\sum_{\ell=-\infty}^{k} 2_*^{(k-\ell)(M_{0,\infty}+\varepsilon-\eta')q/2}(\omega(2_*^{\ell}))^q a_\ell^q\right)\times\right.\\
&\qquad\left.\times\left(\sum_{j=0}^{\infty} 2_*^{j(M_{0,\infty}+\varepsilon-\eta')q'/2}\right)^{q/q'}\right]^{1/q}\\
&\leq C\left[\sum_{\ell=-\infty}^{\infty}(\omega(2_*^{\ell}))^q a_\ell^q\sum_{k=\ell}^{\infty}2_*^{(k-\ell)(M_{0,\infty}+\varepsilon-\eta')q/2}\right]^{1/q}\\
&= C\left[\sum_{\ell=-\infty}^{\infty}(\omega(2_*^{\ell}))^q a_\ell^q\sum_{j=0}^{\infty}2_*^{j(M_{0,\infty}+\varepsilon-\eta')q/2}\right]^{1/q} \leq C\|\{a_\ell\}_{\ell\in\mathbb{Z}}\|_{\ell^q_{\omega(\cdot)}(\mathbb{Z})}
\end{aligned}
\tag{23.169}
$$

for some $C = C(2_*, q, \varepsilon, \eta', M_{0,\infty}) \in (0,\infty)$. Collectively, (23.168) and (23.169) establish that $\|\cdot\|_{\ell^q_{\omega(\cdot)}(\mathbb{Z})}$ satisfies (C4) in Definition 23.5. This concludes the proof of Claim 2. □

The next example is inspired by the theory of Morrey-Herz spaces (see, e.g., [LuXu05]).

Example 23.5 **(Mixed Sequence Spaces with Dyadic Weights)** Consider an infinite set $\mathcal{J} \subseteq \mathbb{Z}$, and denote by μ the counting measure on $\mathcal{J}$. Select an exponent $q \in (0,\infty]$, a number $2_* \in (1,\infty)$, along with indices $\alpha \in \mathbb{R}$ and $\lambda \in (0,\infty)$. For each sequence $\{a_j\}_{j\in\mathcal{J}} \subseteq \mathbb{R}$ define the quantity

$$
\|\{a_j\}_{j\in\mathcal{J}}\|_{\ell^{\infty,q}_{\alpha,\lambda}(\mathcal{J})} := \begin{cases} \sup\limits_{L\in\mathcal{J}}\left\{2_*^{-L\lambda}\left[\sum\limits_{\substack{j\in\mathcal{J}\\ j\leq L}}(2_*^{j\alpha}|a_j|)^q\right]^{1/q}\right\} & \text{if } 1 < q < \infty,\\[2ex] \sup\limits_{L\in\mathcal{J}}\left\{2_*^{-L\lambda}\left[\sup\limits_{\substack{j\in\mathcal{J}\\ j\leq L}} 2_*^{j\alpha}|a_j|\right]\right\} & \text{if } q = \infty, \end{cases}
\tag{23.170}
$$

in $[0,\infty]$. Use this to introduce the following sequence space

$$\ell^{\infty,q}_{\alpha,\lambda}(\mathcal{J}) := \big\{\{a_j\}_{j\in\mathcal{J}} \in \mathbb{R}^{\mathcal{J}} : \|\{a_j\}_{j\in\mathcal{J}}\|_{\ell^{\infty,q}_{\alpha,\lambda}(\mathcal{J})} < \infty\big\}. \tag{23.171}$$

<u>`Claim 1.`</u> The space $\ell^{\infty,q}_{\alpha,\lambda}(\mathbb{Z})$ is a Generalized Quasi-Banach Sequence Space on $(\mathbb{Z}, \mu)$. A similar conclusion holds for $\ell^{\infty,q}_{\alpha,\lambda}(\mathbb{N}_0)$.

Proof of `Claim 1`. It is apparent from (23.170) that $\|\cdot\|_{\ell^{\infty,q}_{\alpha,\lambda}(\mathbb{Z})}$ satisfies properties (Q1') and (Q2') in Proposition 23.1. Fix an arbitrary sequence $a = \{a_j\}_{j\in\mathbb{Z}} \subseteq \mathbb{R}_+$. To verify (Q3') in Proposition 23.1, for each $k \in \mathbb{N}$ assume $a^{(k)} = \{a^{(k)}_j\}_{j\in\mathbb{Z}} \subseteq \mathbb{R}_+$ is such that $a^{(k)}_j \nearrow a_j$ as $k \to \infty$ for all $j \in \mathbb{Z}$. The lattice property then implies $\|a^{(k)}\|_{\ell^{\infty,q}_{\alpha,\lambda}(\mathbb{Z})} \leq \|a\|_{\ell^{\infty,q}_{\alpha,\lambda}(\mathbb{Z})}$. As such, we have

$$\limsup_{k\to\infty} \|a^{(k)}\|_{\ell^{\infty,q}_{\alpha,\lambda}(\mathbb{Z})} \leq \|a\|_{\ell^{\infty,q}_{\alpha,\lambda}(\mathbb{Z})}. \tag{23.172}$$

Select a threshold $\rho \in (0, \infty)$ for which

$$0 < \rho < \|a\|_{\ell^{\infty,q}_{\alpha,\lambda}(\mathbb{Z})}. \tag{23.173}$$

Bearing in mind (23.170), there exists some $L_0 \in \mathbb{Z}$ satisfying

$$\rho < 2_*^{-L_0\lambda}\left[\sum_{j=-\infty}^{L_0} (2_*^{j\alpha} a_j)^q\right]^{1/q}. \tag{23.174}$$

Since $a^{(k)}_j \nearrow a_j$ as $k \to \infty$ for all $j \in \mathbb{Z}$, this and Lebesgue's Monotone Convergence Theorem for series imply that there exists $N \in \mathbb{N}$ such that for every $k \geq N$ we have

$$\rho \leq 2_*^{-L_0\lambda}\left[\sum_{j=-\infty}^{L_0} (2_*^{j\alpha} a^{(k)}_j)^q\right]^{1/q} \leq \|a^{(k)}\|_{\ell^{\infty,q}_{\alpha,\lambda}(\mathbb{Z})}, \tag{23.175}$$

where the last inequality follows from (23.170). This further implies

$$\rho \leq \liminf_{k\to\infty} \|a^{(k)}\|_{\ell^{\infty,q}_{\alpha,\lambda}(\mathbb{Z})}. \tag{23.176}$$

Since this holds for each ρ as in (23.173), we ultimately obtain

$$\|a\|_{\ell^{\infty,q}_{\alpha,\lambda}(\mathbb{Z})} \leq \liminf_{k\to\infty} \|a^{(k)}\|_{\ell^{\infty,q}_{\alpha,\lambda}(\mathbb{Z})}. \tag{23.177}$$

Combining this with (23.172) yields $\|a^{(k)}\|_{\ell^{\infty,q}_{\alpha,\lambda}(\mathbb{Z})} \nearrow \|a\|_{\ell^{\infty,q}_{\alpha,\lambda}(\mathbb{Z})}$ as $k \to \infty$. Therefore, $\|\cdot\|_{\ell^{\infty,q}_{\alpha,\lambda}(\mathbb{Z})}$ satisfies property (Q3') in Proposition 23.1.

Moving on, recall that $\lambda > 0$ and $2_* > 1$. Then for each $k \in \mathbb{Z}$ we compute

$$
\begin{aligned}
\|\mathbb{1}_{\{k\}}\|_{\ell^{\infty,q}_{\alpha,\lambda}(\mathbb{Z})} &= \sup_{L\in\mathbb{Z}}\left\{2_*^{-L\lambda}\Big[\sum_{j=-\infty}^{L}(2_*^{j\alpha}\delta_{kj})^q\Big]^{1/q}\right\}\\
&\leq \sup_{L<k}\left\{2_*^{-L\lambda}\Big[\sum_{j=-\infty}^{L}(2_*^{j\alpha}\delta_{kj})^q\Big]^{1/q}\right\}+\sup_{L\geq k}\left\{2_*^{-L\lambda}\Big[\sum_{j=-\infty}^{L}(2_*^{j\alpha}\delta_{kj})^q\Big]^{1/q}\right\}\\
&= 0+\Big(\sup_{L\geq k}2_*^{-L\lambda}\Big)\cdot 2_*^{k\alpha}=2_*^{-k\alpha}2_*^{k\alpha}=1<\infty.
\end{aligned}
\tag{23.178}
$$

Hence, $\|\cdot\|_{\ell^{\infty,q}_{\alpha,\lambda}(\mathbb{Z})}$ satisfies (Q4') in Proposition 23.1. To see that $\|\cdot\|_{\ell^{\infty,q}_{\alpha,\lambda}(\mathbb{Z})}$ also satisfies (Q5'), observe that for each $j\in\mathbb{Z}$ we have

$$
\begin{aligned}
a_j &= 2_*^{-j\alpha}(2_*^{j\alpha}a_j)\leq 2_*^{-j\alpha}\Big[\sum_{k=-\infty}^{j}(2_*^{j\alpha}a_j)^q\Big]^{1/q}\\
&= 2_*^{-j(\alpha-\lambda)}\cdot 2_*^{-j\lambda}\Big[\sum_{k=-\infty}^{j}(2_*^{k\alpha}a_k)^q\Big]^{1/q}\leq 2_*^{-j(\alpha-\lambda)}\|a\|_{\ell^{\infty,q}_{\alpha,\lambda}(\mathbb{Z})}.
\end{aligned}
\tag{23.179}
$$

As such, this and Proposition 23.1 establish that $\|\cdot\|_{\ell^{\infty,q}_{\alpha,\lambda}(\mathbb{Z})}$ is a function quasinorm on $\mathbb{Z}$. In particular, $\ell^{\infty,q}_{\alpha,\lambda}(\mathbb{Z})$ is a Generalized Quasi-Banach Sequence Space on $(\mathbb{Z},\mu)$. This concludes the proof of Claim 1. The case of $\ell^{\infty,q}_{\alpha,\lambda}(\mathbb{N}_0)$ is dealt with similarly. □

Claim 2. Fix $\eta,\eta'\in(0,\infty)$ and assume that

$$
-\eta+\lambda<\alpha<\eta'+\lambda. \tag{23.180}
$$

Then $\|\cdot\|_{\ell^{\infty,q}_{\alpha,\lambda}(\mathbb{Z})}$ is (η,η')-compliant and $\|\cdot\|_{\ell^{\infty,q}_{\alpha,\lambda}(\mathbb{N}_0)}$ is (η,η')-compliant on $\mathbb{N}_0$.

Proof of Claim 2. We only justify the compliancy of $\|\cdot\|_{\ell^{\infty,q}_{\alpha,\lambda}(\mathbb{Z})}$, as the same type of reasoning can be used to show the compliancy on $\mathbb{N}_0$ of $\|\cdot\|_{\ell^{\infty,q}_{\alpha,\lambda}(\mathbb{N}_0)}$. Also, assume $0<q<\infty$. The proof corresponding to the case $q=\infty$ follows via a similar reasoning. Fix an arbitrary sequence $\{a_\ell\}_{\ell\in\mathbb{Z}}\subseteq\mathbb{R}_+$. To see that $\|\cdot\|_{\ell^{\infty,q}_{\alpha,\lambda}(\mathbb{Z})}$ satisfies (C1) and (C2) in Definition 23.5, compute

$$
\begin{aligned}
\|\{a_{\ell+1}\}_{\ell\in\mathbb{Z}}\|_{\ell^{\infty,q}_{\alpha,\lambda}(\mathbb{Z})} &= \sup_{L\in\mathbb{Z}} \left\{ 2_*^{-L\lambda} \left[\sum_{\ell=-\infty}^{L} (2_*^{\ell\alpha} a_{\ell+1})^q \right]^{1/q} \right\} \\
&= 2_*^{-\alpha+\lambda} \sup_{L\in\mathbb{Z}} \left\{ 2_*^{-(L+1)\lambda} \left[\sum_{\ell=-\infty}^{L} (2_*^{(\ell+1)\alpha} a_{\ell+1})^q \right]^{1/q} \right\} \\
&= 2_*^{-\alpha+\lambda} \sup_{L\in\mathbb{Z}} \left\{ 2_*^{-(L+1)\lambda} \left[\sum_{j=-\infty}^{L+1} (2_*^{j\alpha} a_j)^q \right]^{1/q} \right\} \\
&= 2_*^{-\alpha+\lambda} \|\{a_\ell\}_{\ell\in\mathbb{Z}}\|_{\ell^{\infty,q}_{\alpha,\lambda}(\mathbb{Z})},
\end{aligned}
\tag{23.181}
$$

and similarly

$$
\|\{a_{\ell-1}\}_{\ell\in\mathbb{Z}}\|_{\ell^{\infty,q}_{\alpha,\lambda}(\mathbb{Z})} = 2_*^{\alpha-\lambda} \|\{a_\ell\}_{\ell\in\mathbb{Z}}\|_{\ell^{\infty,q}_{\alpha,\lambda}(\mathbb{Z})}. \tag{23.182}
$$

Next, bearing in mind (23.170), the fact that $\lambda > 0$ and $\eta - \lambda + \alpha > 0$ (see (23.180)), estimate

$$
\begin{aligned}
&\left\| \left\{ \sum_{\ell=k}^{\infty} 2_*^{(k-\ell)\eta} a_\ell \right\}_{k\in\mathbb{Z}} \right\|_{\ell^{\infty,q}_{\alpha,\lambda}(\mathbb{Z})} \\
&= \sup_{L\in\mathbb{Z}} \left\{ 2_*^{-L\lambda} \left[\sum_{k=-\infty}^{L} \left(2_*^{k\alpha} \sum_{\ell=k}^{\infty} 2_*^{(k-\ell)\eta} a_\ell \right)^q \right]^{1/q} \right\} \\
&= \sup_{L\in\mathbb{Z}} \left\{ 2_*^{-L\lambda} \left[\sum_{k=-\infty}^{L} \left(\sum_{\ell=k}^{\infty} 2_*^{(k-\ell)(\eta+\alpha)} 2_*^{\ell\alpha} a_\ell \right)^q \right]^{1/q} \right\} \\
&\le \sup_{L\in\mathbb{Z}} \left\{ 2_*^{-L\lambda} \left[\sum_{k=-\infty}^{L} \left(\sum_{\ell=k}^{\infty} 2_*^{(k-\ell)(\eta+\alpha)} \left(\sum_{j=-\infty}^{\ell} (2_*^{j\alpha} a_j)^q \right)^{1/q} \right)^q \right]^{1/q} \right\} \\
&= \sup_{L\in\mathbb{Z}} \left\{ 2_*^{-L\lambda} \left[\sum_{k=-\infty}^{L} \left(\sum_{\ell=k}^{\infty} 2_*^{(k-\ell)(\eta+\alpha)} 2_*^{\ell\lambda} 2_*^{-\ell\lambda} \left(\sum_{j=-\infty}^{\ell} (2_*^{j\alpha} a_j)^q \right)^{1/q} \right)^q \right]^{1/q} \right\} \\
&\le \|\{a_j\}_{j\in\mathbb{Z}}\|_{\ell^{\infty,q}_{\alpha,\lambda}(\mathbb{Z})} \cdot \sup_{L\in\mathbb{Z}} \left\{ 2_*^{-L\lambda} \left[\sum_{k=-\infty}^{L} \left(\sum_{\ell=k}^{\infty} 2_*^{(k-\ell)(\eta+\alpha)} 2_*^{\ell\lambda} \right)^q \right]^{1/q} \right\} \\
&= \|\{a_j\}_{j\in\mathbb{Z}}\|_{\ell^{\infty,q}_{\alpha,\lambda}(\mathbb{Z})} \cdot \sup_{L\in\mathbb{Z}} \left\{ 2_*^{-L\lambda} \left[\sum_{k=-\infty}^{L} 2_*^{k\lambda q} \left(\sum_{\ell=k}^{\infty} 2_*^{(k-\ell)(\eta-\lambda+\alpha)} \right)^q \right]^{1/q} \right\} \\
&= \|\{a_j\}_{j\in\mathbb{Z}}\|_{\ell^{\infty,q}_{\alpha,\lambda}(\mathbb{Z})} \cdot \sup_{L\in\mathbb{Z}} \left\{ 2_*^{-L\lambda} \left[\sum_{k=-\infty}^{L} 2_*^{k\lambda q} \left(\sum_{j=0}^{\infty} 2_*^{-j(\eta-\lambda+\alpha)} \right)^q \right]^{1/q} \right\}
\end{aligned}
$$

$$\leq C\|\{a_j\}_{j\in\mathbb{Z}}\|_{\ell^{\infty,q}_{\alpha,\lambda}(\mathbb{Z})}\cdot\sup_{L\in\mathbb{Z}}\left\{2_*^{-L\lambda}\left[\sum_{k=-\infty}^{L}2_*^{k\lambda q}\right]^{1/q}\right\}$$

$$\leq C\|\{a_j\}_{j\in\mathbb{Z}}\|_{\ell^{\infty,q}_{\alpha,\lambda}(\mathbb{Z})}\cdot\sup_{L\in\mathbb{Z}}\left\{2_*^{-L\lambda}2_*^{L\lambda}\right\}=C\|\{a_j\}_{j\in\mathbb{Z}}\|_{\ell^{\infty,q}_{\alpha,\lambda}(\mathbb{Z})} \tag{23.183}$$

for some constant $C=C(2_*,q,\eta,\lambda,\alpha)\in(0,\infty)$. This proves that $\|\cdot\|_{\ell^{\infty,q}_{\alpha,\lambda}(\mathbb{Z})}$ satisfy (C3) in Definition 23.5.

Regarding (C4) in Definition 23.5, use the fact that $\lambda>0$ and $\eta'+\lambda-\alpha>0$ (see (23.180)) while keeping in mind (23.170) to bound

$$\begin{aligned}
&\Big\|\Big\{\sum_{\ell=-\infty}^{k}2_*^{(\ell-k)\eta'}a_\ell\Big\}_{k\in\mathbb{Z}}\Big\|_{\ell^{\infty,q}_{\alpha,\lambda}(\mathbb{Z})}\\
&=\sup_{L\in\mathbb{Z}}\left\{2_*^{-L\lambda}\left[\sum_{k=-\infty}^{L}\left(2_*^{k\alpha}\sum_{\ell=-\infty}^{k}2_*^{(\ell-k)\eta'}a_\ell\right)^q\right]^{1/q}\right\}\\
&=\sup_{L\in\mathbb{Z}}\left\{2_*^{-L\lambda}\left[\sum_{k=-\infty}^{L}\left(\sum_{\ell=-\infty}^{k}2_*^{(\ell-k)(\eta'-\alpha)}2_*^{\ell\alpha}a_\ell\right)^q\right]^{1/q}\right\}\\
&\leq\sup_{L\in\mathbb{Z}}\left\{2_*^{-L\lambda}\left[\sum_{k=-\infty}^{L}\left(\sum_{\ell=-\infty}^{k}2_*^{(\ell-k)(\eta'-\alpha)}\Big(\sum_{j=-\infty}^{\ell}(2_*^{j\alpha}a_j)^q\Big)^{1/q}\right)^q\right]^{1/q}\right\}\\
&=\sup_{L\in\mathbb{Z}}\left\{2_*^{-L\lambda}\left[\sum_{k=-\infty}^{L}\left(\sum_{\ell=-\infty}^{k}2_*^{(\ell-k)(\eta'-\alpha)}2_*^{\ell\lambda}2_*^{-\ell\lambda}\Big(\sum_{j=-\infty}^{\ell}(2_*^{j\alpha}a_j)^q\Big)^{1/q}\right)^q\right]^{1/q}\right\}\\
&\leq\|\{a_j\}_{j\in\mathbb{Z}}\|_{\ell^{\infty,q}_{\alpha,\lambda}(\mathbb{Z})}\cdot\sup_{L\in\mathbb{Z}}\left\{2_*^{-L\lambda}\left[\sum_{k=-\infty}^{L}\left(\sum_{\ell=-\infty}^{k}2_*^{(\ell-k)(\eta'-\alpha)}2_*^{\ell\lambda}\right)^q\right]^{1/q}\right\}\\
&=\|\{a_j\}_{j\in\mathbb{Z}}\|_{\ell^{\infty,q}_{\alpha,\lambda}(\mathbb{Z})}\cdot\sup_{L\in\mathbb{Z}}\left\{2_*^{-L\lambda}\left[\sum_{k=-\infty}^{L}2_*^{k\lambda q}\left(\sum_{\ell=-\infty}^{k}2_*^{(\ell-k)(\eta'+\lambda-\alpha)}\right)^q\right]^{1/q}\right\}\\
&=\|\{a_j\}_{j\in\mathbb{Z}}\|_{\ell^{\infty,q}_{\alpha,\lambda}(\mathbb{Z})}\cdot\sup_{L\in\mathbb{Z}}\left\{2_*^{-L\lambda}\left[\sum_{k=-\infty}^{L}2_*^{k\lambda q}\left(\sum_{j=0}^{\infty}2_*^{-j(\eta'+\lambda-\alpha)}\right)^q\right]^{1/q}\right\}\\
&\leq C\|\{a_j\}_{j\in\mathbb{Z}}\|_{\ell^{\infty,q}_{\alpha,\lambda}(\mathbb{Z})}\cdot\sup_{L\in\mathbb{Z}}\left\{2_*^{-L\lambda}\left[\sum_{k=-\infty}^{L}2_*^{k\lambda q}\right]^{1/q}\right\}\\
&\leq C\|\{a_j\}_{j\in\mathbb{Z}}\|_{\ell^{\infty,q}_{\alpha,\lambda}(\mathbb{Z})}\cdot\sup_{L\in\mathbb{Z}}\left\{2_*^{-L\lambda}2_*^{L\lambda}\right\}=C\|\{a_j\}_{j\in\mathbb{Z}}\|_{\ell^{\infty,q}_{\alpha,\lambda}(\mathbb{Z})},
\end{aligned} \tag{23.184}$$

for some constant $C = C(2_*, q, \eta', \lambda, \alpha) \in (0, \infty)$. Putting everything together, (23.181)-(23.184) prove that $\|\cdot\|_{\ell^{\infty,q}_{\alpha,\lambda}(\mathbb{Z})}$ is indeed (η, η')-compliant. This concludes the proof of Claim 2. □

23.5 Function Spaces and their A_∞ Indices

Here we provide a wealth of examples of Generalized Banach Function Spaces and study their A_∞ indices.

Example 23.6 **(Lebesgue Spaces)** Consider a closed unbounded Ahlfors regular set $\Sigma \subseteq \mathbb{R}^n$, and abbreviate $\sigma := \mathcal{H}^{n-1}\lfloor\Sigma$. Given an integrability exponent $p \in (1, \infty)$, the Lebesgue space $L^p(\Sigma, \sigma)$ is a Generalized Banach Function Space on (Σ, σ), and its associated space is $L^{p'}(\Sigma, \sigma)$, where $p' \in (1, \infty)$ denotes the Hölder conjugate exponent of p. Moreover, the Hardy-Littlewood maximal operator $\mathcal{M}_\Sigma$ is bounded on both $L^p(\Sigma, \sigma)$ and $L^{p'}(\Sigma, \sigma)$. It is worth mentioning that at the endpoints, $L^1(\Sigma, \sigma)$ and $L^\infty(\Sigma, \sigma)$ are Generalized Banach Function Spaces, which are associated spaces of one another, and $\mathcal{M}_\Sigma$ is bounded on $L^\infty(\Sigma, \sigma)$, but not on $L^1(\Sigma, \sigma)$.

Next, denoting by $\Upsilon_{L^p(\Sigma,\sigma)}$ and $\Upsilon_{L^{p'}(\Sigma,\sigma)}$ the A_∞ indices (cf. (23.67)) of $L^p(\Sigma, \sigma)$ and $L^{p'}(\Sigma, \sigma)$, respectively, we claim that

$$\Upsilon_{L^p(\Sigma,\sigma)} = \frac{1}{p} \quad \text{and} \quad \Upsilon_{L^{p'}(\Sigma,\sigma)} = \frac{1}{p'}. \tag{23.185}$$

To justify this claim, observe that for any surface ball $\Delta \subseteq \Sigma$ and any σ-measurable set $E \subseteq \Delta$ we have

$$\frac{\|\mathbb{1}_E\|_{L^p(\Sigma,\sigma)}}{\|\mathbb{1}_\Delta\|_{L^p(\Sigma,\sigma)}} = \left(\frac{\sigma(E)}{\sigma(\Delta)}\right)^{1/p}. \tag{23.186}$$

This and (23.67) imply that $\frac{1}{p} \leq \Upsilon_{L^p(\Sigma,\sigma)}$. Assume that $\frac{1}{p} < \Upsilon_{L^p(\Sigma,\sigma)}$. Since $\Upsilon_{L^p(\Sigma,\sigma)}$ is a supremum (cf. (23.67)), we can find an exponent $\tau \in (0, \infty)$ with $\frac{1}{p} < \tau < \Upsilon_{L^p(\Sigma,\sigma)}$, and a constant $C \in (0, \infty)$ with the property that for any surface ball $\Delta \subseteq \Sigma$ and any σ-measurable set $E \subseteq \Delta$ the following inequality holds

$$\left(\frac{\sigma(E)}{\sigma(\Delta)}\right)^{1/p} = \frac{\|\mathbb{1}_E\|_{L^p(\Sigma,\sigma)}}{\|\mathbb{1}_\Delta\|_{L^p(\Sigma,\sigma)}} \leq C\left(\frac{\sigma(E)}{\sigma(\Delta)}\right)^{\tau}. \tag{23.187}$$

In particular, for any $x \in \Sigma$ and $\rho \in (0, 1)$ this holds for the choices $\Delta := \Delta(x, 1)$ and $E := \Delta(x, \rho)$. By Ahlfors regularity of Σ we then find that

$$\rho^{\frac{n-1}{p}} \leq C\rho^{(n-1)\tau} \implies \rho^{(n-1)(\frac{1}{p}-\tau)} \leq C < \infty. \tag{23.188}$$

Since this holds for every $\rho \in (0,1)$ and $\frac{1}{p} - \tau < 0$, if we let $\rho \searrow 0$ then $\rho^{(n-1)(\frac{1}{p}-\tau)} \nearrow \infty$, a contradiction. This establishes that $\Upsilon_{L^p(\Sigma,\sigma)} = \frac{1}{p}$ as desired. The same argument as above shows that $\Upsilon_{L^{p'}(\Sigma,\sigma)} = \frac{1}{p'}$.

Example 23.7 **(Lorentz Spaces)** Assume $\Sigma \subseteq \mathbb{R}^n$ is a closed unbounded Ahlfors regular set, and abbreviate $\sigma := \mathcal{H}^{n-1}\lfloor\Sigma$. Fix two exponents $p, r \in (0,\infty]$. For each σ-measurable function f on Σ, define

$$\|f\|_{L^{p,r}(\Sigma,\sigma)} := \begin{cases} \Big(\displaystyle\int_0^\infty \big[t^{1/p} f_\Sigma^*(t)\big]^r \frac{dt}{t}\Big)^{1/r} & \text{if } 0 < p, r < \infty, \\ \sup\limits_{t>0} \big[t^{1/p} f_\Sigma^*(t)\big] & \text{if } 0 < p \leq \infty,\ r = \infty, \\ \|f\|_{L^\infty(\Sigma,\sigma)} & \text{if } p = \infty,\ 0 < r \leq \infty, \end{cases} \tag{23.189}$$

where

$$\begin{aligned} f_\Sigma^*(t) &:= \sup\Big\{\lambda \geq 0 : \sigma\big(\{x \in \Sigma : |f(x)| > \lambda\}\big) > t\Big\} \\ &= \inf\Big\{\lambda \geq 0 : \sigma\big(\{x \in \Sigma : |f(x)| > \lambda\}\big) \leq t\Big\} \quad \text{for all } t \in [0,\infty). \end{aligned} \tag{23.190}$$

Since $\sup_{t>0}\big[f_\Sigma^*(t)\big] = \|f\|_{L^\infty(\Sigma,\sigma)}$ for every σ-measurable function f on Σ, the overlap in the last two lines in (23.189) causes no problems. Introduce the `Lorentz space` on Σ by setting

$$L^{p,r}(\Sigma,\sigma) := \big\{f \in \mathscr{M}(\Sigma,\sigma) : \|f\|_{L^{p,r}(\Sigma,\sigma)} < \infty\big\}, \tag{23.191}$$

where $\mathscr{M}(\Sigma,\sigma)$ denotes the collection of all σ-measurable functions defined on Σ.

In the regime $p \in (1,\infty)$ and $r \in [1,\infty]$, the Lorentz space $L^{p,r}(\Sigma,\sigma)$ is a classical Banach Function Space, hence a Generalized Banach Function Space on (Σ,σ). Its associated space is the Lorentz space $L^{p',r'}(\Sigma,\sigma)$, where $p' \in (1,\infty)$ and $r' \in [1,\infty]$ are the Hölder conjugate exponents of p and r, respectively (with the understanding that $r' = \infty$ if $r = 1$, and $r' = 1$ if $r = \infty$); cf. [BeSha88, Theorem 4.6, Theorem 4.7, pp. 219–220]. Via real interpolation, the Hardy-Littlewood maximal operator $\mathcal{M}_\Sigma$ is bounded both on $L^{p,r}(\Sigma,\sigma)$ and $L^{p',r'}(\Sigma,\sigma)$ (cf. [GHA-I, Corollary 7.6.3]).

Regarding the A_∞ indices (cf. (23.67)) of $L^{p,r}(\Sigma,\sigma)$ and $L^{p',r'}(\Sigma,\sigma)$, we claim that

$$\Upsilon_{L^{p,r}(\Sigma,\sigma)} = \frac{1}{p} \quad \text{and} \quad \Upsilon_{L^{p',r'}(\Sigma,\sigma)} = \frac{1}{p'}. \tag{23.192}$$

To see that this is the case, we rely on [GHA-I, Lemma 6.3.5 *(b)*, p. 504] which ensures that $\|\mathbb{1}_A\|_{L^{p,r}(\Sigma,\sigma)} \approx \sigma(A)^{1/p}$ uniformly in the class of σ-measurable subsets A of Σ. In particular, for any surface ball $\Delta \subseteq \Sigma$ and any σ-measurable set $E \subseteq \Delta$ we have

$$\frac{\|\mathbb{1}_E\|_{L^{p,r}(\Sigma,\sigma)}}{\|\mathbb{1}_\Delta\|_{L^{p,r}(\Sigma,\sigma)}} \approx \left(\frac{\sigma(E)}{\sigma(\Delta)}\right)^{1/p} \tag{23.193}$$

with proportionality constants independent of Δ and E. In view of (23.67), this implies $\frac{1}{p} \leq \Upsilon_{L^{p,r}(\Sigma,\sigma)}$. The end-game is similar to that in Example 23.6 (as to establish (23.185)). Hence, $\Upsilon_{L^{p,r}(\Sigma,\sigma)} = \frac{1}{p}$ and $\Upsilon_{L^{p',r'}(\Sigma,\sigma)} = \frac{1}{p'}$, as desired.

Example 23.8 **(Morrey Spaces and Block Spaces)** Assume $\Sigma \subseteq \Sigma$ is a closed unbounded Ahlfors regular set, and abbreviate $\sigma := \mathcal{H}^{n-1}\lfloor\Sigma$. Fix an exponent $p \in (1,\infty)$ along with an index $\lambda \in (0, n-1)$. For each σ-measurable function f on Σ, introduce

$$\|f\|_{M^{p,\lambda}(\Sigma,\sigma)} := \sup_{\substack{x\in\Sigma\\ R>0}} \left(\frac{1}{R^\lambda}\int_{\Delta(x,R)} |f|^p\, d\sigma\right)^{1/p} \in [0,\infty]. \tag{23.194}$$

Define the `Morrey space` on Σ as

$$M^{p,\lambda}(\Sigma,\sigma) := \big\{f \in \mathscr{M}(\Sigma,\sigma) : \|f\|_{M^{p,\lambda}(\Sigma,\sigma)} < \infty\big\}. \tag{23.195}$$

Next, fix an exponent $q \in (1,\infty)$ and continue to assume that $\lambda \in (0, n-1)$. A function $b \in L^q(\Sigma,\sigma)$ is said to be a (q,λ)-block on Σ provided there exists a location $x \in \Sigma$ and a radius $R \in (0,\infty)$ with the property that

$$\operatorname{supp} b \subseteq \Delta(x,R) \quad \text{and} \quad \|b\|_{L^q(\Sigma,\sigma)} \leq R^{-\lambda(1-1/q)}. \tag{23.196}$$

Define the `Block space` on Σ as the collection $\mathcal{B}^{q,\lambda}(\Sigma,\sigma)$ consisting of all distributions $f \in \big(\mathrm{Lip}_c(\Sigma)\big)'$ for which there exists a sequence $\{\lambda_j\}_{j\in\mathbb{N}} \in \ell^1(\mathbb{N})$ and a sequence of (q,λ)-blocks $\{b_j\}_{j\in\mathbb{N}}$ such that

$$f = \sum_{j=1}^{\infty} \lambda_j b_j \quad \text{with convergence in } \big(\mathrm{Lip}_c(\Sigma)\big)'. \tag{23.197}$$

For each $f \in \mathcal{B}^{q,\lambda}(\Sigma,\sigma)$ introduce the norm

$$\|f\|_{\mathcal{B}^{q,\lambda}(\Sigma,\sigma)} := \inf\left\{\sum_{j=1}^{\infty} |\lambda_j| : f = \sum_{j=1}^{\infty} \lambda_j b_j \ \text{ in } \big(\mathrm{Lip}_c(\Sigma)\big)'\right\} \tag{23.198}$$

where the infimum is taken over all possible block representations of f as in (23.197).

It has been shown in [GHA-II, Corollary 6.2.11, p. 336] and [GHA-II, Corollary 6.2.13, p. 340] that

$$\text{the Hardy-Littlewood maximal operator } \mathcal{M}_\Sigma \text{ is bounded on both Morrey space } M^{p,\lambda}(\Sigma,\sigma) \text{ and Block space } \mathcal{B}^{q,\lambda}(\Sigma,\sigma). \tag{23.199}$$

Work done in [GHA-II, Proposition 6.2.17, p. 352] ensures that

$$\mathcal{M}^{p,\lambda}(\Sigma,\sigma) \text{ and } \mathcal{B}^{q,\lambda}(\Sigma,\sigma) \text{ are Generalized Banach Function Spaces on } (\Sigma,\sigma) \tag{23.200}$$

and that

$$\text{the spaces } \mathcal{M}^{p,\lambda}(\Sigma,\sigma) \text{ and } \mathcal{B}^{p',\lambda}(\Sigma,\sigma) \text{ are associated spaces to one another,} \tag{23.201}$$

where $p' \in (1,\infty)$ is the Hölder conjugate exponent of p.

Regarding the A_∞ condition (23.57) for these scales, we claim that there exists some constant $C \in (0,\infty)$ such that for any surface ball $\Delta \subseteq \Sigma$, and any σ-measurable set $E \subseteq \Delta$ the following inequalities hold

$$\frac{\|\mathbb{1}_E\|_{\mathcal{M}^{p,\lambda}(\Sigma,\sigma)}}{\|\mathbb{1}_\Delta\|_{\mathcal{M}^{p,\lambda}(\Sigma,\sigma)}} \leq C\left(\frac{\sigma(E)}{\sigma(\Delta)}\right)^{\frac{n-1-\lambda}{p(n-1)}} \quad \text{and} \tag{23.202}$$

$$\frac{\|\mathbb{1}_E\|_{\mathcal{B}^{q,\lambda}(\Sigma,\sigma)}}{\|\mathbb{1}_\Delta\|_{\mathcal{B}^{q,\lambda}(\Sigma,\sigma)}} \leq C\left(\frac{\sigma(E)}{\sigma(\Delta)}\right)^{\frac{1}{q}}. \tag{23.203}$$

To justify (23.202), fix a surface ball $\Delta \subseteq \Sigma$ and a σ-measurable set $E \subseteq \Delta$. Recall from [GHA-II, (6.2.7), p. 315] the following continuous embedding

$$L^s(\Sigma,\sigma) \hookrightarrow \mathcal{M}^{p,\lambda}(\Sigma,\sigma) \quad \text{with} \quad s := \tfrac{p(n-1)}{n-1-\lambda} \in (p,\infty). \tag{23.204}$$

Then there exists a constant $C = C(\Sigma, n, p, \lambda) \in (0,\infty)$ for which

$$\|f\|_{\mathcal{M}^{p,\lambda}(\Sigma,\sigma)} \leq C\|f\|_{L^s(\Sigma,\sigma)} \quad \text{for every } f \in L^s(\Sigma,\sigma). \tag{23.205}$$

In particular,

$$\|\mathbb{1}_E\|_{\mathcal{M}^{p,\lambda}(\Sigma,\sigma)} \leq C\|\mathbb{1}_E\|_{L^s(\Sigma,\sigma)} = C\cdot\sigma(E)^{\frac{n-1-\lambda}{p(n-1)}}. \tag{23.206}$$

Abbreviate $\Delta := \Delta(x_0, R_0)$ for some $x_0 \in \Sigma$ and $R_0 > 0$. From (23.194) and the Ahlfors regularity of Σ we see that

$$\begin{aligned}\|\mathbb{1}_\Delta\|_{\mathcal{M}^{p,\lambda}(\Sigma,\sigma)} &\geq \Big(\frac{1}{R_0^\lambda}\int_\Delta \mathbb{1}_\Delta\,d\sigma\Big)^{1/p} \\ &\geq \frac{C}{\sigma(\Delta)^{\lambda/(n-1)p}}\sigma(\Delta)^{1/p} = C\cdot\sigma(\Delta)^{\frac{n-1-\lambda}{p(n-1)}}\end{aligned} \tag{23.207}$$

for some $C = C(\Sigma, n, p, \lambda) \in (0,\infty)$. This together with (23.206) then implies the validity of (23.202).

Next, to verify (23.203), we rely on [GHA-II, (6.2.71), p. 327] which guarantees that the following continuous embedding holds

$$\mathcal{B}^{q,\lambda}(\Sigma,\sigma) \hookrightarrow L^r(\Sigma,\sigma) \ \text{ with } \ r := \tfrac{q(n-1)}{n-1+\lambda(q-1)} \in (1,q). \tag{23.208}$$

Continue to assume that $\Delta \subseteq \Sigma$ is a surface ball and $E \subseteq \Delta$ is a σ-measurable set. Based on (23.208), there exists some constant $C \in (0,\infty)$ for which

$$\sigma(\Delta)^{\frac{n-1+\lambda(q-1)}{q(n-1)}} = \|\mathbb{1}_\Delta\|_{L^r(\Sigma,\sigma)} \leq C\|\mathbb{1}_\Delta\|_{\mathcal{B}^{q,\lambda}(\Sigma,\sigma)}. \tag{23.209}$$

With $\Delta = \Delta(x_0, R_0)$ as before, observe that $b := R_0^{-\lambda(1-1/q)} \frac{\mathbb{1}_E}{\|\mathbb{1}_E\|_{L^q(\Sigma,\sigma)}}$ satisfies

$$\operatorname{supp} b \subseteq \Delta \quad \text{and} \quad \|b\|_{L^q(\Sigma,\sigma)} = R_0^{-\lambda(1-1/q)}. \tag{23.210}$$

This implies that b is (q,λ)-block on Σ, and in particular $\|b\|_{\mathcal{B}^{q,\lambda}(\Sigma,\sigma)} \leq 1$ (as ensured by (23.198)). This and the definition of b yield

$$\|\mathbb{1}_E\|_{\mathcal{B}^{q,\lambda}(\Sigma,\sigma)} \leq R_0^{\lambda(1-1/q)}\|\mathbb{1}_E\|_{L^q(\Sigma,\sigma)} = R_0^{\frac{\lambda(q-1)}{q}} \cdot \sigma(E)^{1/q}. \tag{23.211}$$

Thanks to this, (23.209), and the Ahlfors regularity character of Σ, we estimate

$$\begin{aligned}\frac{\|\mathbb{1}_E\|_{\mathcal{B}^{q,\lambda}(\Sigma,\sigma)}}{\|\mathbb{1}_\Delta\|_{\mathcal{B}^{q,\lambda}(\Sigma,\sigma)}} &\leq \frac{R_0^{\frac{\lambda(q-1)}{q}}\,\sigma(E)^{1/q}}{\sigma(\Delta)^{\frac{n-1+\lambda(q-1)}{q(n-1)}}} \\ &\leq C\frac{\sigma(\Delta)^{\frac{\lambda(q-1)}{q(n-1)}}\,\sigma(E)^{1/q}}{\sigma(\Delta)^{\frac{n-1+\lambda(q-1)}{q(n-1)}}} = C\left(\frac{\sigma(E)}{\sigma(\Delta)}\right)^{\frac{1}{q}}.\end{aligned} \tag{23.212}$$

This finishes the justification of (23.203).

Having established (23.202)-(23.203), and recalling (23.67), we find that the A_∞ indices $\Upsilon_{\mathcal{M}^{p,\lambda}(\Sigma,\sigma)}$ and $\Upsilon_{\mathcal{B}^{q,\lambda}(\Sigma,\sigma)}$ satisfy

$$\tfrac{n-1-\lambda}{p(n-1)} \leq \Upsilon_{\mathcal{M}^{p,\lambda}(\Sigma,\sigma)} \quad \text{and} \quad \tfrac{1}{q} \leq \Upsilon_{\mathcal{B}^{q,\lambda}(\Sigma,\sigma)}. \tag{23.213}$$

In relation to these, we claim that

$$\Upsilon_{\mathcal{M}^{p,\lambda}(\Sigma,\sigma)} = \frac{n-1-\lambda}{p(n-1)} \quad \text{and} \quad \frac{1}{q} \leq \Upsilon_{\mathcal{B}^{q,\lambda}(\Sigma,\sigma)} \leq \frac{n-1+\lambda(q-1)}{q(n-1)}. \tag{23.214}$$

To verify the equality involving $\Upsilon_{\mathcal{M}^{p,\lambda}(\Sigma,\sigma)}$, we reason by contradiction. As such, assume $\frac{n-1-\lambda}{p(n-1)} < \Upsilon_{\mathcal{M}^{p,\lambda}(\Sigma,\sigma)}$. Then there exists a constant $C \in (0,\infty)$ along with an exponent $\tau \in (0,\infty)$, with $\frac{n-1-\lambda}{p(n-1)} < \tau < \Upsilon_{\mathcal{M}^{p,\lambda}(\Sigma,\sigma)}$, such that for every surface ball $\Delta \subseteq \Sigma$ and every σ-measurable set $E \subseteq \Delta$ we have.

$$\frac{\|\mathbb{1}_E\|_{\mathcal{M}^{p,\lambda}(\Sigma,\sigma)}}{\|\mathbb{1}_\Delta\|_{\mathcal{M}^{p,\lambda}(\Sigma,\sigma)}} \leq C\left(\frac{\sigma(E)}{\sigma(\Delta)}\right)^{\tau}. \tag{23.215}$$

Thanks to this and (23.206)-(23.207), whenever $x \in \Sigma$ and $\rho \in (0,1)$ it follows that

$$\left(\frac{\sigma(\Delta(x,\rho))}{\sigma(\Delta(x,1))}\right)^{\frac{n-1-\lambda}{p(n-1)}} \leq C\frac{\|\mathbb{1}_{\Delta(x,\rho)}\|_{M^{p,\lambda}(\Sigma,\sigma)}}{\|\mathbb{1}_{\Delta(x,1)}\|_{M^{p,\lambda}(\Sigma,\sigma)}} \leq C\left(\frac{\sigma(\Delta(x,\rho))}{\sigma(\Delta(x,1))}\right)^{\tau}. \tag{23.216}$$

This and the Ahlfors regularity of Σ then yield

$$\rho^{\frac{n-1-\lambda}{p}} \leq C\rho^{(n-1)\tau} \implies \rho^{(n-1)[\frac{n-1-\lambda}{p(n-1)}-\lambda]} \leq C \tag{23.217}$$

for some $C = C(n, p, \lambda, \Sigma) \in (0, \infty)$. Since this holds for all $\rho \in (0, 1)$, if we send $\rho \searrow 0$ and recall that $\frac{n-1-\lambda}{p(n-1)} - \lambda < 0$, then $\rho^{(n-1)[\frac{n-1-\lambda}{p(n-1)}-\lambda]} \nearrow \infty$, a contradiction. This forces $\Upsilon_{M^{p,\lambda}(\Sigma,\sigma)} = \frac{n-1-\lambda}{p(n-1)}$, as desired.

It remains to prove the upper bound for $\Upsilon_{\mathcal{B}^{q,\lambda}(\Sigma,\sigma)}$ claimed in (23.214). To verify this, let $\tau \in (0, \infty)$ be such that there exists some $C \in (0, \infty)$ for which whenever $\Delta \subseteq \Sigma$ is a surface ball and $E \subseteq \Delta$ is a σ-measurable set we have

$$\frac{\|\mathbb{1}_E\|_{\mathcal{B}^{q,\lambda}(\Sigma,\sigma)}}{\|\mathbb{1}_\Delta\|_{\mathcal{B}^{q,\lambda}(\Sigma,\sigma)}} \leq C\left(\frac{\sigma(E)}{\sigma(\Delta)}\right)^{\tau}. \tag{23.218}$$

The existence of τ and C is guaranteed by Lemma 23.1, along with (23.199)-(23.201). Notice that the reasoning in (23.209)-(23.211) (used with $E := \Delta$) and the Ahlfors regularity of Σ imply

$$\|\mathbb{1}_\Delta\|_{\mathcal{B}^{q,\lambda}(\Sigma,\sigma)} \approx \sigma(\Delta)^{\frac{n-1+\lambda(q-1)}{q(n-1)}} \tag{23.219}$$

uniformly with respect to the collection of all surface balls Δ on Σ. Fix $x \in \Sigma$, and $\rho \in (0, 1)$. A combination of (23.218) and (23.219) yields

$$\left(\frac{\sigma(\Delta(x,\rho))}{\sigma(\Delta(x,1))}\right)^{\frac{n-1+\lambda(q-1)}{q(n-1)}} \leq C\left(\frac{\sigma(\Delta(x,\rho))}{\sigma(\Delta(x,1))}\right)^{\tau} \tag{23.220}$$

for some constant $C = C(n, q, \lambda, \Sigma) \in (0, \infty)$. This and the Ahlfors regularity of Σ further imply

$$\rho^{(n-1)[\frac{n-1+\lambda(q-1)}{q(n-1)}-\tau]} \leq C \quad \text{for all } \rho \in (0, 1) \tag{23.221}$$

for some constant $C = C(n, q, \lambda, \Sigma) \in (0, \infty)$. In turn, this forces $\tau \leq \frac{n-1+\lambda(q-1)}{q(n-1)}$, otherwise the expression of the left-hand side of the inequality in (23.221) blows up as $\rho \searrow 0$. Since $\tau \in (0, \infty)$ is an arbitrary participant of the set in which the supremum defining $\Upsilon_{\mathcal{B}^{q,\lambda}(\Sigma,\sigma)}$ is taken (cf. (23.67)), we finally obtain $\Upsilon_{\mathcal{B}^{q,\lambda}(\Sigma,\sigma)} \leq \frac{n-1+\lambda(q-1)}{q(n-1)}$, as wanted. This completes the treatment of the claim made in (23.214).

Example 23.9 **(Muckenhoupt Weighted Lebesgue Spaces)** Let $\Sigma \subseteq \mathbb{R}^n$ be a closed unbounded Ahlfors regular set, and abbreviate $\sigma := \mathcal{H}^{n-1}\lfloor\Sigma$. Select an integrability exponent $p \in (1, \infty)$, and fix some weight $w \in A_p(\Sigma, \sigma)$. Then the Muckenhoupt weighted Lebesgue space $L^p(\Sigma, w)$ is a Generalized Banach Function Space on (Σ, σ), and its associated space is $L^{p'}(\Sigma, w')$, where $p' := p/(p-1) \in (1, \infty)$, and

$w' := w^{1-p'} \in A_{p'}(\Sigma, \sigma)$ is the dual weight of w (cf. item *(3)* of Proposition 2.3). Recall from (2.44) that the membership $w \in A_p(\Sigma, \sigma)$ (respectively $w' \in A_{p'}(\Sigma, \sigma)$) is equivalent to the Hardy-Littlewood maximal operator $\mathcal{M}_\Sigma$ being bounded on $L^p(\Sigma, w)$ (respectively on $L^{p'}(\Sigma, w')$).

As a consequence of the Reverse Hölder inequality for Muckenhoupt weights (cf. item *(6)* in Proposition 2.3), there exist exponents $\gamma, \gamma' \in (0,1)$ with the property that

$$\frac{\|\mathbb{1}_E\|_{L^p(\Sigma,w)}}{\|\mathbb{1}_\Delta\|_{L^p(\Sigma,w)}} \leq C\left(\frac{\sigma(E)}{\sigma(\Delta)}\right)^{\gamma/p} \quad \text{and} \quad \frac{\|\mathbb{1}_E\|_{L^{p'}(\Sigma,w')}}{\|\mathbb{1}_\Delta\|_{L^{p'}(\Sigma,w')}} \leq C\left(\frac{\sigma(E)}{\sigma(\Delta)}\right)^{\gamma'/p'} \tag{23.222}$$

for any surface ball $\Delta \subseteq \Sigma$ and any σ-measurable set $E \subseteq \Delta$, for some constant $C = C(\Sigma, n, p, w) \in (0, \infty)$. Bearing in mind (23.67), this ensures that the A_∞ indices $\Upsilon_{L^p(\Sigma,w)}$ of $L^p(\Sigma, w)$ and $\Upsilon_{L^{p'}(\Sigma,w)}$ of $L^{p'}(\Sigma, w')$ satisfy

$$\frac{\gamma}{p} \leq \Upsilon_{L^p(\Sigma,w)} \quad \text{and} \quad \frac{\gamma'}{p'} \leq \Upsilon_{L^{p'}(\Sigma,w')}. \tag{23.223}$$

We claim that $\Upsilon_{L^p(\Sigma,w)}$ and $\Upsilon_{L^{p'}(\Sigma,w')}$ are such that

$$\frac{\gamma}{p} \leq \Upsilon_{L^p(\Sigma,w)} \leq 1 \quad \text{and} \quad \frac{\gamma'}{p'} \leq \Upsilon_{L^{p'}(\Sigma,w')} \leq 1. \tag{23.224}$$

Notice that it suffices to verify the upper bound for the A_∞ index of $L^p(\Sigma, w)$, as for $L^{p'}(\Sigma, w')$ this follows by the same reasoning. To begin, use Lemma 23.1 to obtain the existence of some $\tau \in (0, \infty)$ and some $C \in (0, \infty)$ such that for any surface ball $\Delta \subseteq \Sigma$ and any σ-measurable set $E \subseteq \Delta$ we have

$$\frac{\|\mathbb{1}_E\|_{L^p(\Sigma,w)}}{\|\mathbb{1}_\Delta\|_{L^p(\Sigma,w)}} = \left(\frac{w(E)}{w(\Delta)}\right)^{1/p} \leq C\left(\frac{\sigma(E)}{\sigma(\Delta)}\right)^{\tau}. \tag{23.225}$$

The doubling character of w (cf. item *(5)* in Proposition 2.3) ensures that for any $x \in \Sigma$ and $\rho \in (0, 1)$ we have

$$\frac{\sigma(\Delta(x,\rho))}{\sigma(\Delta(x,1))} \leq [w]_{A_p(\Sigma,\sigma)}^{1/p}\left(\frac{w(\Delta(x,\rho))}{w(\Delta(x,1))}\right)^{1/p}. \tag{23.226}$$

Thanks to this, the Ahlfors regularity of Σ, and using (23.225) (with $\Delta := \Delta(x, 1)$ and $E := \Delta(x, \rho)$) we see that

$$\frac{\sigma(\Delta(x,\rho))}{\sigma(\Delta(x,1))} \leq C\left(\frac{\sigma(\Delta(x,\rho))}{\sigma(\Delta(x,1))}\right)^{\tau} \implies \rho^{n-1} \leq C\rho^{(n-1)\tau} \text{ for all } \rho \in (0,1) \tag{23.227}$$

for some $C = C(\Sigma, n, p, \tau, [w]_{A_p}) \in (0, \infty)$. In turn, this forces

$$\rho^{(n-1)(1-\tau)} \leq C \quad \text{for all } \rho \in (0,1). \tag{23.228}$$

Hence $1-\tau \geq 0$, since otherwise $\rho^{(n-1)(1-\tau)} \nearrow \infty$ as $\rho \searrow 0$. This proves that $\tau \leq 1$. The fact that $\tau \in (0,\infty)$ is arbitrarily selected in such a way that (23.225) holds implies that $\Upsilon_{L^p(\Sigma,w)} \leq 1$, as desired.

Example 23.10 **(Muckenhoupt Weighted Morrey and Block Spaces)** Let $\Sigma \subseteq \mathbb{R}^n$ be a closed unbounded Ahlfors regular set, and abbreviate $\sigma := \mathcal{H}^{n-1}\lfloor\Sigma$. Fix two exponents $p, q \in (1,\infty)$. Consider a weight $w \in A_p(\Sigma,\sigma) \cap RH_q(\Sigma,\sigma)$. The membership $w \in RH_q(\Sigma,\sigma)$ means that there exists some constant $C \in (0,\infty)$ such that w satisfies the following L^q-styled Reverse Hölder inequality:

$$\Big(\fint_\Delta w^q\,d\sigma\Big)^{1/q} \leq C \fint_\Delta w\,d\sigma \tag{23.229}$$

for every surface ball $\Delta \subseteq \Sigma$. With $q' \in (1,\infty)$ denoting the Hölder conjugate exponent of q, select a parameter $\lambda \in \big(0,(n-1)/q'\big)$. The `Muckenhoupt weighted Morrey space` is defined by

$$M^{p,\lambda}(\Sigma,w) := \big\{f : \Sigma \to \mathbb{C} : f \text{ is } \sigma\text{-measurable and } \|f\|_{M^{p,\lambda}(\Sigma,w)} < \infty\big\}, \tag{23.230}$$

where

$$\|f\|_{M^{p,\lambda}(\Sigma,w)} := \sup_{\substack{x\in\Sigma\\ R>0}} \Big(\frac{1}{R^\lambda}\int_{\Delta(x,R)} |f|^p\,dw\Big)^{1/p} \in [0,\infty] \tag{23.231}$$

for each σ-measurable function f on Σ.

Next, fix an exponent $q \in (1,\infty)$ and a Muckenhoupt weight $\omega \in A_q(\Sigma,\sigma)$. Let $\varepsilon \in (0,q-1)$ be such that $\omega \in A_{q-\varepsilon}(\Sigma,\sigma)$ (that this is possible is ensured by the self-improvement property in item *(1)* from Proposition 2.3), and fix a parameter λ belonging to $\big(0,\varepsilon(n-1)/(q-1)\big)$. A σ-measurable function b defined on Σ is called (q,λ,ω)-`block` if $b \in L^q(\Sigma,\omega)$ and there exist $x_0 \in \Sigma$ and $R \in \big(0, 2\,\text{diam}(\Sigma)\big)$ such that

$$\text{supp}\, b \subseteq \Delta(x_0,R) \quad \text{and} \quad \|b\|_{L^q(\Sigma,\omega)} \leq R^{\lambda(1/q-1)}. \tag{23.232}$$

Define the `Muckenhoupt weighted Block space` $\mathcal{B}^{q,\lambda}(\Sigma,\omega)$ as the collection of all functions $g \in L^1\big(\Sigma, \frac{\sigma(x)}{1+|x|^{n-1}}\big)$ that admit a block representation as

$$g = \sum_{j=1}^\infty \lambda_j b_j \ \text{ in } L^1\Big(\Sigma, \frac{\sigma(x)}{1+|x|^{n-1}}\Big) \tag{23.233}$$

for some numerical sequence $\{\lambda_j\}_{j\in\mathbb{N}} \in \ell^1(\mathbb{N})$ and a family $\{b_j\}_{j\in\mathbb{N}}$ of (q,λ,ω)-blocks on Σ. This space is equipped with the norm

$$\|g\|_{\mathcal{B}^{q,\lambda}(\Sigma,\omega)} := \inf\Big\{\sum_{j=1}^{\infty}|\lambda_j| : \sum_{j=1}^{\infty}\lambda_j b_j \text{ is a block-representation of } g\Big\} \quad (23.234)$$

for each $g \in \mathcal{B}^{q,\lambda}(\Sigma,\omega)$.

Having defined Muckenhoupt weighted Morrey and Block spaces, we now recall some results recently established in [LaMi24]. Under the same geometric background as before, select exponents $p, q \in (1,\infty)$, and denote by $p', q' \in (1,\infty)$ their respective Hölder conjugate exponents. Select a weight $w \in A_p(\Sigma,\sigma) \cap RH_q(\Sigma,\sigma)$ and consider its conjugate weight $w' := w^{-p'/p} \in A_{p'}(\Sigma,\sigma)$. Work done in [LaMi24, Proposition 6.35, p. 165] ensures that, for $\varepsilon := (p'-1)/q' \in (0, p'-1)$, one has $w' \in A_{p'-\varepsilon}(\Sigma,\sigma)$. Fix $\lambda \in \big(0, \varepsilon(n-1)/(p'-1)\big) = \big(0,(n-1)/q'\big)$. In the subsequent discussion, assume this setting.

A result in [LaMi24, Theorem 6.28, p. 157] ensures that for every σ-measurable functions f and g defined on Σ, the following generalized Hölder inequality holds:

$$\int_\Sigma |fg|\, d\sigma \leq \|f\|_{M^{p,\lambda}(\Sigma,w)}\, \|g\|_{\mathcal{B}^{p',\lambda}(\Sigma,w')}\,. \quad (23.235)$$

It is also relevant to recall that work done in [LaMi24, Proposition 10.41, pp. 294–295], [LaMi24, Proposition 10.45, p. 299], [LaMi24, Theorem 10.46, p. 302] guarantees that

$$\begin{gathered}\text{the spaces } M^{p,\lambda}(\Sigma,w) \text{ and } \mathcal{B}^{p',\lambda}(\Sigma,w') \text{ are}\\ \text{Generalized Banach Function Spaces on } (\Sigma,\sigma),\end{gathered} \quad (23.236)$$

and they are associated spaces to one another, i.e.,

$$\big(M^{p,\lambda}(\Sigma,w)\big)' = \mathcal{B}^{p',\lambda}(\Sigma,w') \quad \text{and} \quad \big(\mathcal{B}^{p',\lambda}(\Sigma,w')\big)' = M^{p,\lambda}(\Sigma,w). \quad (23.237)$$

It has also been shown in [LaMi24, Proposition 5.1, p. 106] and [LaMi24, Proposition 7.1, p. 167] that

$$\begin{gathered}\mathcal{M}_\Sigma : M^{p,\lambda}(\Sigma,w) \to M^{p,\lambda}(\Sigma,w)\\ \mathcal{M}_\Sigma : \mathcal{B}^{p',\lambda}(\Sigma,w') \to \mathcal{B}^{p',\lambda}(\Sigma,w')\\ \text{are well-defined and bounded sublinear mappings,}\end{gathered} \quad (23.238)$$

where $\mathcal{M}_\Sigma$ denotes Hardy-Littlewood maximal operator on Σ.

In order to compute the A_∞ indices of Muckenhoupt weighted Morrey and Block spaces, we recall a weighted estimate involving Morrey spaces obtained in [LaMi24, Proposition 10.18, pp. 269–270]. Specifically, there exists some constant $C \in (0,\infty)$ such that for every function f defined on Σ, which is σ-measurable and supported in some surface ball $\Delta \subseteq \Sigma$, one has

$$\|f\|_{M^{p,\lambda}(\Sigma,w)} \leq C w(\Delta)^{1/p} \sigma(\Delta)^{-1/(pq')} \|f\|_{L^{s_0}(\Sigma,\sigma)}, \quad (23.239)$$

where $s_0 := \frac{(n-1)pq'}{n-1-\lambda q'} \in (1, \infty)$.

We claim that there exists some $C \in (0, \infty)$ along with some index $\tau \in (0, 1)$ with the property that for any surface ball $\Delta \subseteq \Sigma$, and any σ-measurable set $E \subseteq \Delta$, the following estimates hold:

$$\frac{\|\mathbb{1}_E\|_{M^{p,\lambda}(\Sigma,w)}}{\|\mathbb{1}_\Delta\|_{M^{p,\lambda}(\Sigma,w)}} \leq C\left(\frac{\sigma(E)}{\sigma(\Delta)}\right)^{\frac{n-1-\lambda}{(n-1)pq'}} \quad \text{and} \tag{23.240}$$

$$\frac{\|\mathbb{1}_E\|_{\mathcal{B}^{p',\lambda}(\Sigma,w')}}{\|\mathbb{1}_\Delta\|_{\mathcal{B}^{p',\lambda}(\Sigma,w')}} \leq C\left(\frac{\sigma(E)}{\sigma(\Delta)}\right)^{\tau/p'}. \tag{23.241}$$

To justify these, select an arbitrary surface ball $\Delta = \Delta(x, R)$ for some $x \in \Sigma$, and some $R \in (0, \infty)$, and fix a σ-measurable set $E \subseteq \Delta$. Use (23.239) with $f := \mathbb{1}_E$ to write

$$\begin{aligned}\|\mathbb{1}_E\|_{M^{p,\lambda}(\Sigma,w)} &\leq Cw(\Delta)^{1/p}\sigma(\Delta)^{-1/(pq')}\|\mathbb{1}_E\|_{L^{s_0}(\Sigma,\sigma)} \\ &= Cw(\Delta)^{1/p}\sigma(\Delta)^{-1/(pq')}\sigma(E)^{\frac{n-1-\lambda q'}{(n-1)pq'}}.\end{aligned} \tag{23.242}$$

Also, from (23.231) and the Ahlfors regularity of Σ we see that

$$\|\mathbb{1}_\Delta\|_{M^{p,\lambda}(\Sigma,w)} \geq \left(\frac{1}{R^\lambda}\int_{\Delta(x,R)} \mathbb{1}_{\Delta(x,R)}\, dw\right)^{1/p} \geq C\sigma(\Delta)^{-\lambda/[(n-1)p]}w(\Delta)^{1/p}. \tag{23.243}$$

In concert, (23.242)-(23.243) and the Ahlfors regularity of Σ entail

$$\frac{\|\mathbb{1}_E\|_{M^{p,\lambda}(\Sigma,w)}}{\|\mathbb{1}_\Delta\|_{M^{p,\lambda}(\Sigma,w)}} \leq \frac{Cw(\Delta)^{1/p}\sigma(\Delta)^{-1/(pq')}\sigma(E)^{\frac{n-1-\lambda q'}{(n-1)pq'}}}{\sigma(\Delta)^{-\lambda/[(n-1)p]}w(\Delta)^{1/p}} = C\left(\frac{\sigma(E)}{\sigma(\Delta)}\right)^{\frac{n-1-\lambda}{(n-1)pq'}}, \tag{23.244}$$

for some constant $C \in (0, \infty)$ depending only on n, p, q, λ, w, and the Ahlfors regularity character of Σ. This establishes (23.240).

We now focus on verifying the estimate claimed in (23.241). Employing (23.235) together with (23.239) yields

$$\begin{aligned}\sigma(\Delta) &= \int_\Sigma \mathbb{1}_\Delta\, d\sigma \leq \|\mathbb{1}_\Delta\|_{M^{p,\lambda}(\Sigma,w)}\|\mathbb{1}_\Delta\|_{\mathcal{B}^{p',\lambda}(\Sigma,w')} \\ &\leq Cw(\Delta)^{1/p}\sigma(\Delta)^{-1/(pq')}\|\mathbb{1}_\Delta\|_{L^{s_0}(\Sigma,\sigma)}\|\mathbb{1}_\Delta\|_{\mathcal{B}^{p',\lambda}(\Sigma,w')} \\ &= Cw(\Delta)^{1/p}\sigma(\Delta)^{-1/(pq')}\sigma(\Delta)^{\frac{n-1-\lambda q'}{(n-1)pq'}}\|\mathbb{1}_\Delta\|_{\mathcal{B}^{p',\lambda}(\Sigma,w')}\end{aligned} \tag{23.245}$$

for some constant $C \in (0, \infty)$ depending on only n, p, q, λ, w, and the Ahlfors regularity of Σ. This further gives

$$\|\mathbb{1}_\Delta\|_{\mathcal{B}^{p',\lambda}(\Sigma,w')} \geq C\sigma(\Delta)w(\Delta)^{-1/p}\sigma(\Delta)^{1/(pq')}\sigma(\Delta)^{-\frac{n-1-\lambda q'}{(n-1)pq'}}. \tag{23.246}$$

Since $E \subseteq \Delta = \Delta(x, R)$, we see from definitions (cf. (23.232)) that the function

$$b_E := R^{\lambda(1/p'-1)}w'(E)^{-1/p'} \cdot \mathbb{1}_E \quad \text{is a } (p', \lambda, w')\text{-block on } \Sigma. \tag{23.247}$$

In particular, (23.234) shows that $\|b_E\|_{\mathcal{B}^{p',\lambda}(\Sigma,w')} \leq 1$. This and (23.247) then imply that there exists some $C = C(n, p, \lambda, \Sigma) \in (0, \infty)$ such that

$$\|\mathbb{1}_E\|_{\mathcal{B}^{p',\lambda}(\Sigma,w')} \leq R^{-\lambda(1/p'-1)}w'(E)^{1/p'} \leq C\sigma(\Delta)^{\lambda/[p(n-1)]}w'(E)^{1/p'}, \tag{23.248}$$

where in the last inequality we have used the Ahlfors regularity of Σ and the fact that $1/p = 1 - 1/p'$. A combination of (23.246) and (23.248) then produces

$$\begin{aligned}\frac{\|\mathbb{1}_E\|_{\mathcal{B}^{p',\lambda}(\Sigma,w')}}{\|\mathbb{1}_\Delta\|_{\mathcal{B}^{p',\lambda}(\Sigma,w')}} &\leq C\frac{\sigma(\Delta)^{\lambda/[p(n-1)]}w'(E)^{1/p'}}{\sigma(\Delta)w(\Delta)^{-1/p}\sigma(\Delta)^{1/(pq')}\sigma(\Delta)^{-\frac{n-1-\lambda q'}{(n-1)pq'}}} \\ &= C\frac{w'(E)^{1/p'}w(\Delta)^{1/p}}{\sigma(\Delta)},\end{aligned} \tag{23.249}$$

with $C \in (0, \infty)$ depending on only n, p, q, λ, w, and the Ahlfors regularity of Σ. At this stage, use the fact that w is a Muckenhoupt weight on Σ, along with item *(6)* in Proposition 2.3, to guarantee the existence of some $\tau \in (0, 1)$ satisfying

$$\frac{w'(E)^{1/p'}w(\Delta)^{1/p}}{\sigma(\Delta)} \leq \left(\frac{w'(E)}{w'(\Delta)}\right)^{1/p'} \leq C\left(\frac{\sigma(E)}{\sigma(\Delta)}\right)^{\tau/p'}, \tag{23.250}$$

for some constant $C = C(n, p, w, \Sigma) \in (0, \infty)$. All together, (23.249)-(23.250) imply the validity of (23.241).

The bottom line here is that estimates (23.240)-(23.241) permit us to obtain bounds from below for the A_∞ indices (cf. (23.67)) of Muckenhoupt weighted Morrey and Block spaces. Specifically, we have

$$\frac{n-1-\lambda}{(n-1)pq'} \leq \Upsilon_{M^{p,\lambda}(\Sigma,w)} \quad \text{and} \quad \frac{\tau}{p'} \leq \Upsilon_{\mathcal{B}^{p',\lambda}(\Sigma,w')}. \tag{23.251}$$

This finishes the discussion regarding Muckenhoupt weighted Morrey and Block spaces, and their A_∞ indices.

Example 23.11 **(Variable Exponent Lebesgue Spaces)** Assume $\Sigma \subseteq \mathbb{R}^n$ is a closed unbounded Ahlfors regular set, and abbreviate $\sigma := \mathcal{H}^{n-1}\lfloor\Sigma$. Fix a σ-measurable function $p(\cdot) : \Sigma \to (1, \infty)$. The `variable exponent Lebesgue space` $L^{p(\cdot)}(\Sigma, \sigma)$ is defined as the collection of all σ-measurable functions f on Σ such that

$$\int_{\Sigma} \left(\frac{|f(x)|}{\lambda} \right)^{p(x)} d\sigma(x) < \infty \quad \text{for some } \lambda \in (0, \infty). \tag{23.252}$$

Equip the space $L^{p(\cdot)}(\Sigma, \sigma)$ with the norm

$$\|f\|_{L^{p(\cdot)}(\Sigma,\sigma)} := \inf \left\{ \lambda > 0 : \int_{\Sigma} \left(\frac{|f(x)|}{\lambda} \right)^{p(x)} d\sigma(x) \leq 1 \right\} \tag{23.253}$$

for every σ-measurable function f on Σ. Then $L^{p(\cdot)}(\Sigma, \sigma)$ is a classical Banach Function Spaces, hence a Generalized Banach Function Space on (Σ, σ), and its associated space is $L^{p'(\cdot)}(\Sigma, \sigma)$, where $p'(\cdot) := \frac{p(\cdot)}{p(\cdot)-1}$ on Σ (cf., e.g., [DiHaHaRu17, Theorem 3.2.13, p. 78]).

Next, assume

$$1 < \operatorname*{ess\,inf}_{\Sigma} p(\cdot) \leq \operatorname*{ess\,sup}_{\Sigma} p(\cdot) < \infty, \tag{23.254}$$

and suppose $p(\cdot)$ satisfies a log-Hölder continuity condition on Σ, i.e., there exist constants $C \in (0, \infty)$, $p_\infty \in (0, \infty)$, along with a point $x_o \in \Sigma$ with the property that

$$\begin{aligned} |p(x) - p(y)| &\leq \frac{C}{\log(e + |x - y|^{-1})} \quad \text{and} \\ |p(x) - p_\infty| &\leq \frac{C}{\log(e + |x - x_o|)} \quad \text{for all } \ x, y \in \Sigma. \end{aligned} \tag{23.255}$$

Under these extra assumptions on $p(\cdot)$, the Hardy-Littlewood maximal operator $\mathcal{M}_\Sigma$ is known to be bounded both on $L^{p(\cdot)}(\Sigma, \sigma)$ and $L^{p'(\cdot)}(\Sigma, \sigma)$ (see, e.g., [HaHäPe04] and [AdHaHä15]). As such, from Lemma 23.1 there exist $\tau_{p(\cdot)}, \tau_{p'(\cdot)} \in (0, 1)$ along with a constant $C \in (0, \infty)$ for which

$$\frac{\|\mathbb{1}_E\|_{L^{p(\cdot)}(\Sigma,\sigma)}}{\|\mathbb{1}_\Delta\|_{L^{p(\cdot)}(\Sigma,\sigma)}} \leq C \left(\frac{\sigma(E)}{\sigma(\Delta)} \right)^{\tau_{p(\cdot)}} \quad \text{and} \quad \frac{\|\mathbb{1}_E\|_{L^{p'(\cdot)}(\Sigma,\sigma)}}{\|\mathbb{1}_\Delta\|_{L^{p'(\cdot)}(\Sigma,\sigma)}} \leq C \left(\frac{\sigma(E)}{\sigma(\Delta)} \right)^{\tau_{p'(\cdot)}}, \tag{23.256}$$

for any surface ball $\Delta \subseteq \Sigma$, and any σ-measurable set $E \subseteq \Delta$. Upon recalling (23.67), we have the following estimates involving the A_∞ indices of variable Lebesgue spaces:

$$\tau_{p(\cdot)} \leq \Upsilon_{L^{p(\cdot)}(\Sigma,\sigma)} \quad \text{and} \quad \tau_{p'(\cdot)} \leq \Upsilon_{L^{p'(\cdot)}(\Sigma,\sigma)}. \tag{23.257}$$

This ends the discussion of A_∞ indices on Lebesgue spaces with variable exponents.

Example 23.12 **(Orlicz Spaces)** Assume $\Sigma \subseteq \mathbb{R}^n$ is a closed unbounded Ahlfors regular set, and abbreviate $\sigma := \mathcal{H}^{n-1}\lfloor\Sigma$. Let $\Phi : [0, \infty) \to [0, \infty)$ be a `Young function`, i.e., Φ is convex, does not vanish on $(0, \infty)$, and satisfies

$$\lim_{t \to 0^+} \frac{\Phi(t)}{t} = 0 \quad \text{and} \quad \lim_{t \to \infty} \frac{\Phi(t)}{t} = \infty. \tag{23.258}$$

For each σ-measurable function f on Σ introduce its Luxemburg norm as

$$\|f\|_{L^\Phi(\Sigma,\sigma)} := \inf\Big\{\lambda > 0 : \int_\Sigma \Phi\big(|f(x)|/\lambda\big)\, d\sigma(x) \leq 1\Big\} \in [0,\infty]. \tag{23.259}$$

Define the Orlicz space associated with Φ as

$$L^\Phi(\Sigma,\sigma) := \big\{f \in \mathscr{M}(\Sigma,\sigma) : \|f\|_{L^\Phi(\Sigma,\sigma)} < \infty\big\}. \tag{23.260}$$

As pointed out in [GHA-II, (5.3.62), p. 262], the space $L^\Phi(\Sigma,\sigma)$ is a Generalized Banach Function Space on (Σ,σ). Moreover, its associated space is identified as

$$\big(L^\Phi(\Sigma,\sigma)\big)' = L^{\widetilde{\Phi}}(\Sigma,\sigma), \tag{23.261}$$

that is, the Orlicz space associated with the complementary function of Φ, namely $\widetilde{\Phi} : [0,\infty) \to [0,\infty)$ defined as

$$\widetilde{\Phi}(t) := \sup_{s>0}\big\{st - \Phi(s)\big\} \quad \text{for each } t \in [0,\infty). \tag{23.262}$$

Also, recall that the lower and upper dilation indices of Φ are defined, respectively, as

$$i(\Phi) := \lim_{t\to 0^+} \frac{\ln h_{\Phi(t)}}{\ln t} \quad \text{and} \quad I(\Phi) := \lim_{t\to\infty} \frac{\ln h_{\Phi(t)}}{\ln t}, \tag{23.263}$$

where $h_\Phi(t) := \sup_{s>0} \frac{\Phi(st)}{\Phi(s)}$ for all $t \in (0,\infty)$. Then [GHA-II, Proposition 5.3.15, p. 279] ensures that whenever $1 < i(\Phi) \leq I(\Phi) < \infty$ it follows that

$$\begin{gathered} \mathcal{M}_\Sigma : L^\Phi(\Sigma,\sigma) \to L^\Phi(\Sigma,\sigma) \text{ and } \mathcal{M}_\Sigma : L^{\widetilde{\Phi}}(\Sigma,\sigma) \to L^{\widetilde{\Phi}}(\Sigma,\sigma) \\ \text{are well-defined and bounded operators.} \end{gathered} \tag{23.264}$$

Hence, in light of Lemma 23.1 and the GBFS character of Orlicz spaces, there exist indices $\tau_\Phi, \tau_{\widetilde{\Phi}} \in (0,1)$ along with some constant $C \in (0,\infty)$ with the property that

$$\frac{\|\mathbb{1}_E\|_{L^\Phi(\Sigma,\sigma)}}{\|\mathbb{1}_\Delta\|_{L^\Phi(\Sigma,\sigma)}} \leq C\left(\frac{\sigma(E)}{\sigma(\Delta)}\right)^{\tau_\Phi} \quad \text{and} \quad \frac{\|\mathbb{1}_E\|_{L^{\widetilde{\Phi}}(\Sigma,\sigma)}}{\|\mathbb{1}_\Delta\|_{L^{\widetilde{\Phi}}(\Sigma,\sigma)}} \leq C\left(\frac{\sigma(E)}{\sigma(\Delta)}\right)^{\tau_{\widetilde{\Phi}}}, \tag{23.265}$$

for every surface ball $\Delta \subseteq \Sigma$ and every σ-measurable set $E \subseteq \Delta$. In particular, this and (23.67) imply

$$\tau_\Phi \leq \Upsilon_{L^\Phi(\Sigma,\sigma)} \quad \text{and} \quad \tau_{\widetilde{\Phi}} \leq \Upsilon_{L^{\widetilde{\Phi}}(\Sigma,\sigma)}. \tag{23.266}$$

This concludes the discussion regarding the nature of Orlicz spaces and their A_∞ indices.

23.6 Sublinear Operators on Composite Herz Spaces

The goal here is to study the boundedness of certain classes of sublinear operators on Composite Herz spaces. Paving the way is the following definition which plays a crucial role in the development of the present theory of Composite Herz spaces. The reader is reminded of the notions of compliancy in Definition 23.5 and in Definition 23.6.

Definition 23.7 Assume $\Sigma \subseteq \mathbb{R}^n$ is a closed Ahlfors regular set, and abbreviate $\sigma := \mathcal{H}^{n-1}\lfloor\Sigma$. Consider a Generalized Banach Function Space $\mathbb{X}$ on (Σ, σ), and let $\mathbb{X}'$ be its associated space. Also, let $\mathbb{Y}$ Generalized Quasi-Banach Sequence Space on $\mathbb{Z}$ (respectively $\mathbb{N}_0$). Say that $\mathbb{Y}$ is $\mathbb{X}$-`compliant` if $\|\cdot\|_{\mathbb{Y}}$ is $\big((n-1)\tau, (n-1)\tau'\big)$-compliant (respectively compliant on $\mathbb{N}_0$) for some indices $\tau, \tau' \in \mathbb{R}$ such that

$$0 < \tau \preceq \Upsilon_{\mathbb{X}} \quad \text{and} \quad 0 < \tau' \preceq \Upsilon_{\mathbb{X}'} \tag{23.267}$$

where $\Upsilon_{\mathbb{X}}$ and $\Upsilon_{\mathbb{X}'}$ denote the A_∞ indices (cf. Definition 23.4) of $\mathbb{X}$ and $\mathbb{X}'$, respectively.

Remark 23.17 In the same context as in Definition 23.7, we claim that whenever $\mathbb{Y}$ is $\mathbb{X}$-compliant it follows that the associated space $\mathbb{Y}'$ is $\mathbb{X}'$-compliant. Indeed, bearing in mind that $\mathbb{X}'' = \mathbb{X}$ (cf. [GHA-II, Proposition 5.1.14, pp. 226–227]), Lemma 23.4 ensures that if $\|\cdot\|_{\mathbb{Y}}$ is $\big((n-1)\tau, (n-1)\tau'\big)$-compliant then $\|\cdot\|_{\mathbb{Y}'}$ is $\big((n-1)\tau', (n-1)\tau\big)$-compliant. Hence, $\mathbb{Y}'$ is $\mathbb{X}'$-compliant.

The result recorded below establishes the boundedness of the Hardy-Littlewood maximal operator on Composite Herz spaces.

Theorem 23.1 *Consider $\Sigma \subseteq \mathbb{R}^n$ a closed unbounded Ahlfors regular set, and abbreviate $\sigma := \mathcal{H}^{n-1}\lfloor\Sigma$. Let $\mathbb{X}$ be a Generalized Banach Function Space on (Σ, σ). Assume that*

$$\begin{gathered} \mathcal{M}_\Sigma : \mathbb{X} \to \mathbb{X} \ \textit{and} \ \mathcal{M}_\Sigma : \mathbb{X}' \to \mathbb{X}' \\ \textit{are well defined and bounded,} \end{gathered} \tag{23.268}$$

where $\mathcal{M}_\Sigma$ is the Hardy-Littlewood maximal operator on Σ (cf. (2.26)). Select a Generalized Quasi-Banach Sequence Space $\mathbb{Y}$ on $\mathbb{Z}$ which is $\mathbb{X}$-compliant (in the sense of Definition 23.7). Then

$$\begin{gathered} \mathcal{M}_\Sigma : \dot{\mathbb{K}}(\Sigma; \mathbb{X}; \mathbb{Y}) \to \dot{\mathbb{K}}(\Sigma; \mathbb{X}; \mathbb{Y}) \\ \textit{and} \ \mathcal{M}_\Sigma : \dot{\mathbb{K}}(\Sigma; \mathbb{X}'; \mathbb{Y}') \to \dot{\mathbb{K}}(\Sigma; \mathbb{X}'; \mathbb{Y}') \\ \textit{are well-defined and bounded sublinear operators.} \end{gathered} \tag{23.269}$$

Remark 23.18 Before proving Theorem 23.1, we make the observation to the effect that for every σ-measurable function f defined on Σ we have

$$\mathcal{M}_\Sigma f(x) \le C \int_\Sigma \frac{|f(y)|}{|x-y|^{n-1}}\, d\sigma(y) \quad \text{for all } x \in \Sigma, \tag{23.270}$$

for some $C \in (0, \infty)$ depending only on the (lower) Ahlfors regularity constant of Σ. To see this, fix an arbitrary $x \in \Sigma$. Using the Ahlfors regularity of Σ estimate

$$\begin{aligned}\mathcal{M}_\Sigma f(x) &= \sup_{r>0} ⨍_{\Delta(x,r)} |f(y)|\, d\sigma(y) \le C \cdot \sup_{r>0} \int_{\Delta(x,r)} \frac{|f(y)|}{r^{n-1}}\, d\sigma(y) \\ &\le C \cdot \sup_{r>0} \int_{\Delta(x,r)} \frac{|f(y)|}{|x-y|^{n-1}}\, d\sigma(y) \le C \int_\Sigma \frac{|f(y)|}{|x-y|^{n-1}}\, d\sigma(y), \end{aligned} \tag{23.271}$$

for some $C \in (0, \infty)$ of the desired nature. This justifies (23.270).

Remark 23.19 Assumption (23.268) ensures that $L^\infty_{\rm comp}(\Sigma, \sigma) \subseteq \mathbb{X} \cap \mathbb{X}'$ (see [GHA-II, Lemma 5.2.5, p. 245]). Hence for every surface ball $\Delta \subseteq \Sigma$ we have $\mathbb{1}_\Delta \in \mathbb{X} \cap \mathbb{X}'$, with $\|\mathbb{1}_\Delta\|_{\mathbb{X}} > 0$ and $\|\mathbb{1}_\Delta\|_{\mathbb{X}'} > 0$. In particular, upon recalling the family $\{C_k\}_{k\in\mathbb{Z}}$ from (23.16), we see that for every $k \in \mathbb{Z}$ the function $\mathbb{1}_{C_k}$ belongs to $\mathbb{X} \cap \mathbb{X}'$ and has non-zero norm in both $\mathbb{X}$ and $\mathbb{X}'$.

Proof of Theorem 23.1 Since the space $\mathbb{Y}$ is $\mathbb{X}$-compliant (cf. Definition 23.7), the function $\|\cdot\|_{\mathbb{Y}}$ is $\big((n-1)\tau, (n-1)\tau'\big)$-compliant, for some indices $\tau, \tau' \in \mathbb{R}$ satisfying

$$0 < \tau \preceq \Upsilon_{\mathbb{X}} \ \text{ and } \ 0 < \tau' \preceq \Upsilon_{\mathbb{X}'} \tag{23.272}$$

where $\Upsilon_{\mathbb{X}}$ and $\Upsilon_{\mathbb{X}'}$ denotes the A_∞ indices (cf. Definition 23.4) of $\mathbb{X}$ and $\mathbb{X}'$, respectively. Hence, Remark 23.13 ensures that there exists some constant $C \in (0, \infty)$ with the property that for any surface ball $\Delta \subseteq \Sigma$ and any σ-measurable set $E \subseteq \Delta$ we have

$$\frac{\|\mathbb{1}_E\|_{\mathbb{X}}}{\|\mathbb{1}_\Delta\|_{\mathbb{X}}} \le C\left(\frac{\sigma(E)}{\sigma(\Delta)}\right)^{\tau} \quad \text{and} \quad \frac{\|\mathbb{1}_E\|_{\mathbb{X}'}}{\|\mathbb{1}_\Delta\|_{\mathbb{X}'}} \le C\left(\frac{\sigma(E)}{\sigma(\Delta)}\right)^{\tau'}. \tag{23.273}$$

Fix an arbitrary function $f \in \dot{\mathbb{K}}(\Sigma; \mathbb{X}; \mathbb{Y})$, and decompose it as

$$f = \sum_{\ell=-\infty}^{\infty} f \cdot \mathbb{1}_{C_\ell} \quad \text{pointwise } \sigma\text{-a.e. on } \Sigma. \tag{23.274}$$

Based on this, the fact that the family $\{C_k\}_{k\in\mathbb{Z}}$ (cf. (23.16)) is a disjoint family that covers the entire Σ, and Lebesgue's Monotone Convergence Theorem, we see that

$$\mathcal{M}_\Sigma f \le \sum_{\ell=-\infty}^{\infty} \mathcal{M}_\Sigma(f \cdot \mathbb{1}_{C_\ell}) \quad \text{everywhere on } \Sigma. \tag{23.275}$$

With this in mind, we may now estimate

$$
\begin{aligned}
\|\mathcal{M}_\Sigma f\|_{\dot{\mathbb{K}}(\Sigma;\mathbb{X};\mathbb{Y})} &\leq \left\| \left\{ \left\| \Big[\sum_{\ell=-\infty}^{\infty} \mathcal{M}_\Sigma(f\cdot \mathbb{1}_{C_\ell})\Big] \cdot \mathbb{1}_{C_k} \right\|_{\mathbb{X}} \right\}_{k\in\mathbb{Z}} \right\|_{\mathbb{Y}} \\
&\leq C \left\| \left\{ \left\| \Big[\sum_{\ell=-\infty}^{k-2} \mathcal{M}_\Sigma(f\cdot \mathbb{1}_{C_\ell})\Big] \cdot \mathbb{1}_{C_k} \right\|_{\mathbb{X}} \right\}_{k\in\mathbb{Z}} \right\|_{\mathbb{Y}} \\
&\quad + C \left\| \left\{ \left\| \Big[\sum_{\ell=k-1}^{k+1} \mathcal{M}_\Sigma(f\cdot \mathbb{1}_{C_\ell})\Big] \cdot \mathbb{1}_{C_k} \right\|_{\mathbb{X}} \right\}_{k\in\mathbb{Z}} \right\|_{\mathbb{Y}} \\
&\quad + C \left\| \left\{ \left\| \Big[\sum_{\ell=k+2}^{\infty} \mathcal{M}_\Sigma(f\cdot \mathbb{1}_{C_\ell})\Big] \cdot \mathbb{1}_{C_k} \right\|_{\mathbb{X}} \right\}_{k\in\mathbb{Z}} \right\|_{\mathbb{Y}} \\
&=: \mathrm{I}_1 + \mathrm{I}_2 + \mathrm{I}_3,
\end{aligned}
\tag{23.276}
$$

for some $C \in (0, \infty)$ depending only on the modulus of concavity of $\|\cdot\|_{\mathbb{Y}}$.

To deal with I_2, use the boundedness of $\mathcal{M}_\Sigma$ on $\mathbb{X}$ (cf. (23.268)) and the compliancy of $\|\cdot\|_{\mathbb{Y}}$ to obtain

$$
\begin{aligned}
\mathrm{I}_2 &\leq C \left\| \left\{ \sum_{\ell=k-1}^{k+1} \left\| \big[\mathcal{M}_\Sigma(f\cdot \mathbb{1}_{C_\ell})\big] \cdot \mathbb{1}_{C_k} \right\|_{\mathbb{X}} \right\}_{k\in\mathbb{Z}} \right\|_{\mathbb{Y}} \\
&\leq C \left\| \left\{ \sum_{\ell=k-1}^{k+1} \|\mathcal{M}_\Sigma(f\cdot \mathbb{1}_{C_\ell})\|_{\mathbb{X}} \right\}_{k\in\mathbb{Z}} \right\|_{\mathbb{Y}} \\
&\leq C \left\| \left\{ \sum_{\ell=k-1}^{k+1} \|f\cdot \mathbb{1}_{C_\ell}\|_{\mathbb{X}} \right\}_{k\in\mathbb{Z}} \right\|_{\mathbb{Y}} \\
&\leq C \Big(\left\| \{\|f\cdot \mathbb{1}_{C_{k-1}}\|_{\mathbb{X}}\}_{k\in\mathbb{Z}} \right\|_{\mathbb{Y}} + \left\| \{\|f\cdot \mathbb{1}_{C_k}\|_{\mathbb{X}}\}_{k\in\mathbb{Z}} \right\|_{\mathbb{Y}} \\
&\qquad + \left\| \{\|f\cdot \mathbb{1}_{C_{k+1}}\|_{\mathbb{X}}\}_{k\in\mathbb{Z}} \right\|_{\mathbb{Y}} \Big) \\
&\leq C \left\| \{\|f\cdot \mathbb{1}_{C_k}\|_{\mathbb{X}}\}_{k\in\mathbb{Z}} \right\|_{\mathbb{Y}} = C\|f\|_{\dot{\mathbb{K}}(\Sigma;\mathbb{X};\mathbb{Y})},
\end{aligned}
\tag{23.277}
$$

for some constant $C \in (0, \infty)$ depending only on $\|\mathcal{M}_\Sigma\|_{\mathbb{X}\to\mathbb{X}}$, the modulus of concavity of $\|\cdot\|_{\mathbb{Y}}$ as well as the compliancy constant of $\|\cdot\|_{\mathbb{Y}}$.

Next, to estimate I_1 from (23.276) fix $k, \ell \in \mathbb{Z}$ with $\ell \leq k-2$ along with some $x \in C_k$. Observe that if $y \in C_\ell$ then

$$
|x-y| \geq |x-x_0| - |y-x_0| \geq 2_*^k - 2_*^{\ell+1} \geq 2_*^k - 2_*^{k-1} = c2_*^k,
\tag{23.278}
$$

where $c := 1 - 2_*^{-1} > 0$. Thanks to this, (23.270), and the generalized Hölder inequality (10.149) we have

$$\begin{aligned}\mathcal{M}_\Sigma(f\cdot\mathbb{1}_{C_\ell})(x)&\leq C\int_{C_\ell}\frac{|f(y)|}{|x-y|^{n-1}}\,d\sigma(y)\\&\leq C2_*^{-k(n-1)}\|f\cdot\mathbb{1}_{C_\ell}\|_{L^1(\Sigma,\sigma)}\\&\leq C2_*^{-k(n-1)}\|f\cdot\mathbb{1}_{C_\ell}\|_{\mathbb{X}}\|\mathbb{1}_{C_\ell}\|_{\mathbb{X}'}\end{aligned}\tag{23.279}$$

for some constant $C\in(0,\infty)$ depending on n and Σ. With this in mind, estimate I_1 from (23.276) as

$$\begin{aligned}\mathrm{I}_1&\leq C\left\|\left\{\left\|\left[\sum_{\ell=-\infty}^{k-2}2_*^{-k(n-1)}\|f\cdot\mathbb{1}_{C_\ell}\|_{\mathbb{X}}\|\mathbb{1}_{C_\ell}\|_{\mathbb{X}'}\right]\cdot\mathbb{1}_{C_k}\right\|_{\mathbb{X}}\right\}_{k\in\mathbb{Z}}\right\|_{\mathbb{Y}}\\&=C\left\|\left\{\sum_{\ell=-\infty}^{k-2}2_*^{-k(n-1)}\|f\cdot\mathbb{1}_{C_\ell}\|_{\mathbb{X}}\|\mathbb{1}_{C_\ell}\|_{\mathbb{X}'}\|\mathbb{1}_{C_k}\|_{\mathbb{X}}\right\}_{k\in\mathbb{Z}}\right\|_{\mathbb{Y}}.\end{aligned}\tag{23.280}$$

`Claim 1.` There exists some constant $C=C(n,\Sigma,\|\mathcal{M}_\Sigma\|_{\mathbb{X}\to\mathbb{X}})\in(0,\infty)$ with the property that whenever $k,\ell\in\mathbb{Z}$ with $\ell\leq k-2$ it follows that

$$\|\mathbb{1}_{C_\ell}\|_{\mathbb{X}'}\|\mathbb{1}_{C_k}\|_{\mathbb{X}}\leq C2_*^{(\ell-k)(n-1)\tau'}\cdot 2_*^{k(n-1)}.\tag{23.281}$$

To justify this claim, fix $k,\ell\in\mathbb{Z}$ with $\ell\leq k-2$. Notice that $C_\ell\subseteq\Delta(x_0,2_*^{k+1})$. This, (23.273), Lemma 23.2, the lattice property of $\mathbb{X}$, and the Ahlfors regularity of Σ entail

$$\begin{aligned}\|\mathbb{1}_{C_\ell}\|_{\mathbb{X}'}\|\mathbb{1}_{C_k}\|_{\mathbb{X}}&=\frac{\|\mathbb{1}_{C_\ell}\|_{\mathbb{X}'}}{\|\mathbb{1}_{\Delta(x_0,2_*^{k+1})}\|_{\mathbb{X}'}}\|\mathbb{1}_{C_k}\|_{\mathbb{X}}\|\mathbb{1}_{\Delta(x_0,2_*^{k+1})}\|_{\mathbb{X}'}\\&\leq C\left(\frac{\sigma(C_\ell)}{\sigma\big(\Delta(x_0,2_*^{k+1})\big)}\right)^{\tau'}\|\mathbb{1}_{\Delta(x_0,2_*^{k+1})}\|_{\mathbb{X}}\|\mathbb{1}_{\Delta(x_0,2_*^{k+1})}\|_{\mathbb{X}'}\\&\leq C2_*^{(\ell-k)(n-1)\tau'}\sigma\big(\Delta(x_0,2_*^{k+1})\big)\\&\leq C2_*^{(\ell-k)(n-1)\tau'}\cdot 2_*^{k(n-1)},\end{aligned}\tag{23.282}$$

thus concluding the proof of `Claim 1`.

Going back to (23.280), use (23.281) and the compliancy of $\|\cdot\|_{\mathbb{Y}}$ to finally obtain

$$\begin{aligned}\mathrm{I}_1&\leq C\left\|\left\{\sum_{\ell=-\infty}^{k-2}2_*^{-k(n-1)}\|f\cdot\mathbb{1}_{C_\ell}\|_{\mathbb{X}}2_*^{(\ell-k)(n-1)\tau'}\cdot 2_*^{k(n-1)}\right\}_{k\in\mathbb{Z}}\right\|_{\mathbb{Y}}\\&=C\left\|\left\{\sum_{\ell=-\infty}^{k-2}2_*^{(\ell-k)(n-1)\tau'}\|f\cdot\mathbb{1}_{C_\ell}\|_{\mathbb{X}}\right\}_{k\in\mathbb{Z}}\right\|_{\mathbb{Y}}\\&\leq C\left\|\left\{\|f\cdot\mathbb{1}_{C_\ell}\|_{\mathbb{X}}\right\}_{k\in\mathbb{Z}}\right\|_{\mathbb{Y}}=C\|f\|_{\dot{\mathbb{K}}(\Sigma;\mathbb{X};\mathbb{Y})},\end{aligned}\tag{23.283}$$

for some $C \in (0, \infty)$ depending on n, $\|\mathcal{M}_\Sigma\|_{\mathbb{X}\to\mathbb{X}}$, the compliancy constant of $\|\cdot\|_{\mathbb{Y}}$, the modulus of concavity of $\|\cdot\|_{\mathbb{Y}}$, and the Ahlfors regularity constants of Σ.

Moving on, we focus on estimating I_3 from (23.276). To this end, select $k, \ell \in \mathbb{Z}$ with $\ell \geq k + 2$, and fix an arbitrary $x \in C_k$. Then whenever $y \in C_\ell$ it follows that

$$|x - y| \geq |y - x_0| - |x - x_0| \geq 2_*^\ell - 2_*^{k+1} \geq 2_*^\ell - 2_*^{\ell-1} = c2_*^\ell \tag{23.284}$$

with $c := 1 - 2_*^{-1} > 0$. In view of this, (23.270), and the generalized Hölder inequality (10.149), we estimate

$$\begin{aligned} \mathcal{M}_\Sigma(f \cdot \mathbb{1}_{C_\ell})(x) &\leq C \int_{C_\ell} \frac{|f(y)|}{|x - y|^{n-1}}\, d\sigma(y) \\ &\leq C 2_*^{-\ell(n-1)} \|f \cdot \mathbb{1}_{C_\ell}\|_{L^1(\Sigma,\sigma)} \\ &\leq C 2_*^{-\ell(n-1)} \|f \cdot \mathbb{1}_{C_\ell}\|_{\mathbb{X}} \|\mathbb{1}_{C_\ell}\|_{\mathbb{X}'} \end{aligned} \tag{23.285}$$

for some $C \in (0, \infty)$ depending on n and Σ. Once this has been established, estimate

$$\begin{aligned} \mathrm{I}_3 &\leq C \left\| \left\{ \left\| \left[\sum_{\ell=k+2}^{\infty} 2_*^{-\ell(n-1)} \|f \cdot \mathbb{1}_{C_\ell}\|_{\mathbb{X}} \|\mathbb{1}_{C_\ell}\|_{\mathbb{X}'} \right] \cdot \mathbb{1}_{C_k} \right\|_{\mathbb{X}} \right\}_{k\in\mathbb{Z}} \right\|_{\mathbb{Y}} \\ &= C \left\| \left\{ \sum_{\ell=k+2}^{\infty} 2_*^{-\ell(n-1)} \|f \cdot \mathbb{1}_{C_\ell}\|_{\mathbb{X}} \|\mathbb{1}_{C_\ell}\|_{\mathbb{X}'} \|\mathbb{1}_{C_k}\|_{\mathbb{X}} \right\}_{k\in\mathbb{Z}} \right\|_{\mathbb{Y}}. \end{aligned} \tag{23.286}$$

By switching ℓ with k, as well as τ' with τ, and $\mathbb{X}$ with $\mathbb{X}'$ in the proof of `Claim 1`, we may conclude that there exists some constant $C = C(n, \Sigma, \|\mathcal{M}_\Sigma\|_{\mathbb{X}'\to\mathbb{X}'}) \in (0, \infty)$ such that for every $k, \ell \in \mathbb{Z}$ with $\ell \geq k + 2$ (i.e., $k \leq \ell - 2$) we have

$$\|\mathbb{1}_{C_k}\|_{\mathbb{X}} \|\mathbb{1}_{C_\ell}\|_{\mathbb{X}'} \leq C 2_*^{(k-\ell)(n-1)\tau} \cdot 2_*^{\ell(n-1)}. \tag{23.287}$$

Granted this, and the compliancy of $\|\cdot\|_{\mathbb{Y}}$, we may further augment (23.286) by

$$\begin{aligned} \mathrm{I}_3 &\leq C \left\| \left\{ \sum_{\ell=k+2}^{\infty} 2_*^{-\ell(n-1)} \|f \cdot \mathbb{1}_{C_\ell}\|_{\mathbb{X}} 2_*^{(k-\ell)(n-1)\tau} \cdot 2_*^{\ell(n-1)} \right\}_{k\in\mathbb{Z}} \right\|_{\mathbb{Y}} \\ &= C \left\| \left\{ \sum_{\ell=k+2}^{\infty} 2_*^{(k-\ell)(n-1)\tau} \|f \cdot \mathbb{1}_{C_\ell}\|_{\mathbb{X}} \right\}_{k\in\mathbb{Z}} \right\|_{\mathbb{Y}} \\ &\leq C \left\| \left\{ \|f \cdot \mathbb{1}_{C_\ell}\|_{\mathbb{X}} \right\}_{k\in\mathbb{Z}} \right\|_{\mathbb{Y}} = C \|f\|_{\dot{\mathbb{K}}(\Sigma;\mathbb{X};\mathbb{Y})} \end{aligned} \tag{23.288}$$

for some constant $C \in (0, \infty)$ depending only on n, $\|\mathcal{M}_\Sigma\|_{\mathbb{X}'\to\mathbb{X}'}$, the compliancy constant of $\|\cdot\|_{\mathbb{Y}}$, the modulus of concavity of $\|\cdot\|_{\mathbb{Y}}$, and the Ahlfors regularity of Σ. All together, (23.276), (23.277), (23.283), and (23.288) ensure the existence of some $C \in (0, \infty)$ depending on n, $\|\mathcal{M}_\Sigma\|_{\mathbb{X}\to\mathbb{X}}$, $\|\mathcal{M}_\Sigma\|_{\mathbb{X}'\to\mathbb{X}'}$, the compliancy constant of

$\|\cdot\|_{\mathbb{Y}}$, the modulus of concavity of $\|\cdot\|_{\mathbb{Y}}$, and the Ahlfors regularity character of Σ such that

$$\|\mathcal{M}_\Sigma f\|_{\dot{\mathrm{K}}(\Sigma;\mathbb{X};\mathbb{Y})} \leq C\|f\|_{\dot{\mathrm{K}}(\Sigma;\mathbb{X};\mathbb{Y})}. \tag{23.289}$$

Finally, to obtain the boundedness of $\mathcal{M}_\Sigma$ on $\dot{\mathrm{K}}(\Sigma;\mathbb{X}';\mathbb{Y}')$ run the same argument as before with $\mathbb{X}'$ in place of $\mathbb{X}$ and $\mathbb{X}'' = \mathbb{X}$ in place of $\mathbb{X}'$, now bearing in mind that $\mathbb{Y}'$ is $\mathbb{X}'$-compliant (cf. Remark 23.17). This ultimately proves Theorem 23.1.□

Corresponding to inhomogeneous Composite Herz spaces, we have the following boundedness result:

Theorem 23.2 *Assume $\Sigma \subseteq \mathbb{R}^n$ is a closed unbounded Ahlfors regular set, and abbreviate $\sigma := \mathcal{H}^{n-1}\lfloor\Sigma$. Consider a Generalized Banach Function Space $\mathbb{X}$ on (Σ,σ) such that*

$$\begin{gathered}\mathcal{M}_\Sigma : \mathbb{X} \to \mathbb{X} \ \textit{ and } \ \mathcal{M}_\Sigma : \mathbb{X}' \to \mathbb{X}' \\ \textit{are well defined and bounded,}\end{gathered} \tag{23.290}$$

where $\mathcal{M}_\Sigma$ is the Hardy-Littlewood maximal operator on Σ (cf. (2.26)). Let $\mathbb{Y}$ be a Generalized Quasi-Banach Sequence Space on $\mathbb{N}_0$ which is $\mathbb{X}$-compliant (in the sense of Definition 23.7). Then

$$\begin{aligned}&\mathcal{M}_\Sigma : \mathrm{K}(\Sigma;\mathbb{X};\mathbb{Y}) \to \mathrm{K}(\Sigma;\mathbb{X};\mathbb{Y}) \\ \textit{and } \ &\mathcal{M}_\Sigma : \mathrm{K}(\Sigma;\mathbb{X}';\mathbb{Y}') \to \mathrm{K}(\Sigma;\mathbb{X}';\mathbb{Y}') \\ &\textit{are well-defined and bounded operators.}\end{aligned} \tag{23.291}$$

Proof To begin, select indices $\tau, \tau' \in \mathbb{R}$ for which (23.272) and (23.273) hold. Fix an arbitrary function $f \in \mathrm{K}(\Sigma;\mathbb{X};\mathbb{Y})$ and decompose it as

$$f = \sum_{k=0}^{\infty} f \cdot \mathbb{1}_{\widetilde{C}_\ell} \ \text{ pointwise } \sigma\text{-a.e. on } \Sigma. \tag{23.292}$$

As in the past, this implies

$$\mathcal{M}_\Sigma f \leq \sum_{\ell=0}^{\infty} \mathcal{M}_\Sigma(f\cdot\mathbb{1}_{\widetilde{C}_\ell}) \ \text{ everywhere on } \Sigma. \tag{23.293}$$

In a first stage, write

$$
\begin{aligned}
\|\mathcal{M}_\Sigma f\|_{\mathbb{K}(\Sigma;\mathbb{X};\mathbb{Y})} &= \Big\|\Big\{\|(\mathcal{M}_\Sigma f)\cdot \mathbb{1}_{\widetilde{C}_k}\|_{\mathbb{X}}\Big\}_{k\in\mathbb{N}_0}\Big\|_{\mathbb{Y}} \\
&\le C\Big\|\Big\{\|(\mathcal{M}_\Sigma f)\cdot \mathbb{1}_{\widetilde{C}_k}\|_{\mathbb{X}}\Big\}_{k\in\mathbb{N}_0}\cdot \mathbb{1}_{\{0,1\}}\Big\|_{\mathbb{Y}} \\
&\quad + C\Big\|\Big\{\|(\mathcal{M}_\Sigma f)\cdot \mathbb{1}_{\widetilde{C}_k}\|_{\mathbb{X}}\Big\}_{k\in\mathbb{N}_0}\cdot \mathbb{1}_{\{k\ge 2\}}\Big\|_{\mathbb{Y}} \\
&=: \mathrm{I}_1 + \mathrm{I}_2.
\end{aligned}
\tag{23.294}
$$

To deal with I_1, we use (23.293) to further estimate

$$
\begin{aligned}
\mathrm{I}_1 &\le C\bigg\|\bigg\{\Big\|\Big[\sum_{\ell=0}^{\infty}\mathcal{M}_\Sigma(f\cdot \mathbb{1}_{\widetilde{C}_\ell})\Big]\cdot \mathbb{1}_{\widetilde{C}_k}\Big\|_{\mathbb{X}}\bigg\}_{k\in\mathbb{N}_0}\cdot \mathbb{1}_{\{0,1\}}\bigg\|_{\mathbb{Y}} \\
&\le C\bigg\|\bigg\{\Big\|\Big[\sum_{\ell=0}^{2}\mathcal{M}_\Sigma(f\cdot \mathbb{1}_{\widetilde{C}_\ell})\Big]\cdot \mathbb{1}_{\widetilde{C}_k}\Big\|_{\mathbb{X}}\bigg\}_{k\in\mathbb{N}_0}\cdot \mathbb{1}_{\{0,1\}}\bigg\|_{\mathbb{Y}} \\
&\quad + C\bigg\|\bigg\{\Big\|\Big[\sum_{\ell=3}^{\infty}\mathcal{M}_\Sigma(f\cdot \mathbb{1}_{\widetilde{C}_\ell})\Big]\cdot \mathbb{1}_{\widetilde{C}_k}\Big\|_{\mathbb{X}}\bigg\}_{k\in\mathbb{N}_0}\cdot \mathbb{1}_{\{0,1\}}\bigg\|_{\mathbb{Y}} \\
&=: \mathrm{II}_1 + \mathrm{II}_2.
\end{aligned}
\tag{23.295}
$$

On the one hand, using the boundedness assumption (23.290), the lattice property on $\mathbb{X}$, keeping in mind that $\mathbb{1}_{\{0,1\}} = \mathbb{1}_{\{0\}} + \mathbb{1}_{\{1\}} \in \mathbb{Y}$ (see Remark 23.7), and (Q5') in Proposition 23.1 (currently used with $\rho := \|\cdot\|_{\mathbb{Y}}$), we estimate

$$
\begin{aligned}
\mathrm{II}_1 &\le C\Big\|\Big\{\sum_{\ell=0}^{2}\|\mathcal{M}_\Sigma(f\cdot \mathbb{1}_{\widetilde{C}_\ell})\|_{\mathbb{X}}\Big\}_{k\in\mathbb{N}_0}\cdot \mathbb{1}_{\{0,1\}}\Big\|_{\mathbb{Y}} \\
&= C\sum_{\ell=0}^{2}\|\mathcal{M}_\Sigma(f\cdot \mathbb{1}_{\widetilde{C}_\ell})\|_{\mathbb{X}}\|\mathbb{1}_{\{0,1\}}\|_{\mathbb{Y}} \\
&\le C\big(\|f\cdot \mathbb{1}_{\widetilde{C}_0}\|_{\mathbb{X}} + \|f\cdot \mathbb{1}_{\widetilde{C}_1}\|_{\mathbb{X}} + \|f\cdot \mathbb{1}_{\widetilde{C}_2}\|_{\mathbb{X}}\big) \\
&\le C\Big\|\big\{\|f\cdot \mathbb{1}_{\widetilde{C}_\ell}\|_{\mathbb{X}}\big\}_{\ell\in\mathbb{N}_0}\Big\|_{\mathbb{Y}} = \|f\|_{\mathbb{K}(\Sigma;\mathbb{X};\mathbb{Y})}
\end{aligned}
\tag{23.296}
$$

for some constant $C \in (0,\infty)$ depending only on the operator norm of $\mathcal{M}_\Sigma$ on $\mathbb{X}$, and on the modulus of concavity of $\|\cdot\|_{\mathbb{Y}}$. Next, based on (23.270), the same type of computations as in (23.284)-(23.285) along with (23.287) yield the existence of some $C = C(n, \Sigma, \|\mathcal{M}_\Sigma\|_{\mathbb{X}\to\mathbb{X}}) \in (0,\infty)$ with the property that

$$\begin{gathered}\big[\mathcal{M}_\Sigma(f\cdot\mathbb{1}_{\widetilde{C}_\ell})\big]\cdot\mathbb{1}_{\widetilde{C}_k}\leq C2_*^{-\ell(n-1)}\|f\cdot\mathbb{1}_{\widetilde{C}_\ell}\|_{\mathbb{X}}\|\mathbb{1}_{\widetilde{C}_\ell}\|_{\mathbb{X}'}\ \text{ pointwise on } \Sigma,\\ \text{and } \|\mathbb{1}_{\widetilde{C}_k}\|_{\mathbb{X}}\|\mathbb{1}_{\widetilde{C}_\ell}\|_{\mathbb{X}'}\leq C2_*^{(k-\ell)(n-1)\tau}\cdot 2_*^{\ell(n-1)}\\ \text{whenever } k\in\{0,1\}, \text{ and } \ell\in\mathbb{N}_0 \text{ with } \ell\geq 3.\end{gathered}\tag{23.297}$$

Granted this, and the compliancy on $\mathbb{N}_0$ of $\|\cdot\|_{\mathbb{Y}}$, we may estimate II_2 from (23.295) as

$$\begin{aligned}\mathrm{II}_2&\leq C\Bigg\|\Bigg\{\sum_{\ell=3}^{\infty}2_*^{-\ell(n-1)}\|f\cdot\mathbb{1}_{\widetilde{C}_\ell}\|_{\mathbb{X}}\|\mathbb{1}_{\widetilde{C}_\ell}\|_{\mathbb{X}'}\|\mathbb{1}_{\widetilde{C}_k}\|_{\mathbb{X}}\Bigg\}_{k\in\mathbb{Z}}\cdot\mathbb{1}_{\{0,1\}}\Bigg\|_{\mathbb{Y}}\\ &\leq C\Bigg\|\Bigg\{\sum_{\ell=3}^{\infty}2_*^{-\ell(n-1)}\|f\cdot\mathbb{1}_{\widetilde{C}_\ell}\|_{\mathbb{X}}2_*^{(k-\ell)(n-1)\tau}\cdot 2_*^{\ell(n-1)}\Bigg\}_{k\in\mathbb{Z}}\cdot\mathbb{1}_{\{0,1\}}\Bigg\|_{\mathbb{Y}}\\ &=C\Bigg\|\Bigg\{\sum_{\ell=3}^{\infty}2_*^{(k-\ell)(n-1)\tau}\|f\cdot\mathbb{1}_{\widetilde{C}_\ell}\|_{\mathbb{X}}\Bigg\}_{k\in\mathbb{Z}}\cdot\mathbb{1}_{\{0,1\}}\Bigg\|_{\mathbb{Y}}\\ &\leq C\Bigg\|\Bigg\{\sum_{\ell=k}^{\infty}2_*^{(k-\ell)(n-1)\tau}\|f\cdot\mathbb{1}_{\widetilde{C}_\ell}\|_{\mathbb{X}}\Bigg\}_{k\in\mathbb{Z}}\Bigg\|_{\mathbb{Y}}\\ &\leq C\Big\|\big\{\|f\cdot\mathbb{1}_{\widetilde{C}_\ell}\|_{\mathbb{X}}\big\}_{\ell\in\mathbb{N}_0}\Big\|_{\mathbb{Y}}=\|f\|_{\mathbb{K}(\Sigma;\mathbb{X};\mathbb{Y})}\end{aligned}\tag{23.298}$$

for some constant $C\in(0,\infty)$ depending only on n, the operator norms of $\mathcal{M}_\Sigma$ on $\mathbb{X}$ and $\mathbb{X}'$, the Ahlfors regularity character of Σ, and the compliancy constant of $\|\cdot\|_{\mathbb{Y}}$.

To proceed, we now focus on obtaining a bound for I_2 in (23.294). To this end, recall (23.293), and much as in (23.276), we have

$$\begin{aligned}\mathrm{I}_2\leq\ & C\Bigg\|\Bigg\{\Bigg\|\Big[\sum_{\ell=0}^{k-2}\mathcal{M}_\Sigma(f\cdot\mathbb{1}_{\widetilde{C}_\ell})\Big]\cdot\mathbb{1}_{\widetilde{C}_k}\Bigg\|_{\mathbb{X}}\Bigg\}_{k\in\mathbb{N}_0}\cdot\mathbb{1}_{\{k\geq 2\}}\Bigg\|_{\mathbb{Y}}\\ &+C\Bigg\|\Bigg\{\Bigg\|\Big[\sum_{\ell=k-1}^{k+1}\mathcal{M}_\Sigma(f\cdot\mathbb{1}_{\widetilde{C}_\ell})\Big]\cdot\mathbb{1}_{\widetilde{C}_k}\Bigg\|_{\mathbb{X}}\Bigg\}_{k\in\mathbb{N}_0}\cdot\mathbb{1}_{\{k\geq 2\}}\Bigg\|_{\mathbb{Y}}\\ &+C\Bigg\|\Bigg\{\Bigg\|\Big[\sum_{\ell=k+2}^{\infty}\mathcal{M}_\Sigma(f\cdot\mathbb{1}_{\widetilde{C}_\ell})\Big]\cdot\mathbb{1}_{\widetilde{C}_k}\Bigg\|_{\mathbb{X}}\Bigg\}_{k\in\mathbb{N}_0}\cdot\mathbb{1}_{\{k\geq 2\}}\Bigg\|_{\mathbb{Y}}.\end{aligned}\tag{23.299}$$

An inspection of (23.278)-(23.288) reveals that the same type of computations hold in the current scenario, thanks to (23.273) and Remark 23.15. In view of this, we conclude that

$$\mathrm{I}_2\leq C\|f\|_{\mathbb{K}(\Sigma;\mathbb{X};\mathbb{Y})}\tag{23.300}$$

for some constant $C\in(0,\infty)$ depending on n, the Ahlfors regularity character of Σ, the operator norms of $\mathcal{M}_\Sigma$ on $\mathbb{X}$ and $\mathbb{X}'$, and the compliancy constant of $\|\cdot\|_{\mathbb{Y}}$.

Collectively, (23.294), (23.295), (23.296), (23.298), and (23.300) ensure that

$$\|\mathcal{M}_\Sigma f\|_{\mathrm{K}(\Sigma;\mathbb{X};\mathbb{Y})} \leq C\|f\|_{\mathrm{K}(\Sigma;\mathbb{X};\mathbb{Y})} \tag{23.301}$$

for some constant of same nature as in (23.300), and independent of f.

Lastly, that $\mathcal{M}_\Sigma$ is bounded on $\mathrm{K}(\Sigma;\mathbb{X}';\mathbb{Y}')$ is a consequence of what we have proved so far, and Remark 23.16. This finishes the proof of Theorem 23.2. □

In suitable settings, Generalized Quasi-Banach Function Spaces enjoy useful embeddings. An inspection of the proof of [GHA-II, Lemma 5.2.5, p. 245] reveals that a similar type of embedding, as in the aforementioned reference, holds for Generalized Quasi-Banach Function Spaces, yielding the following result:

Proposition 23.8 *Consider $\Sigma \subseteq \mathbb{R}^n$ an arbitrary Ahlfors regular set, and abbreviate $\sigma := \mathcal{H}^{n-1}\lfloor\Sigma$. Let $\mathbb{X}$ be a Generalized Quasi-Banach Function Space on (Σ, σ). Then, with $\mathcal{M}_\Sigma$ denoting the Hardy-Littlewood maximal operator on Σ (cf. (2.26)), it follows that*

$$\begin{gathered} L^\infty_{\mathrm{comp}}(\Sigma,\sigma) \subseteq \mathbb{X} \subseteq L^1_{\mathrm{loc}}(\Sigma,\sigma) \\ \textit{whenever } \mathcal{M}_\Sigma(\mathbb{X}) \subseteq \mathbb{X}, \end{gathered} \tag{23.302}$$

and

$$\begin{gathered} L^\infty_{\mathrm{comp}}(\Sigma,\sigma) \subseteq \mathbb{X}' \subseteq L^1_{\mathrm{loc}}(\Sigma,\sigma) \\ \textit{whenever } \mathcal{M}_\Sigma(\mathbb{X}') \subseteq \mathbb{X}'. \end{gathered} \tag{23.303}$$

As a direct consequence of Proposition 23.8 and Theorem 23.1, we have the following embedding result for Composite Herz spaces.

Corollary 23.3 *Assume $\Sigma \subseteq \mathbb{R}^n$ is a closed unbounded Ahlfors regular set, and abbreviate $\sigma := \mathcal{H}^{n-1}\lfloor\Sigma$. Consider a Generalized Banach Function Space $\mathbb{X}$ on (Σ, σ) such that*

$$\begin{gathered} \mathcal{M}_\Sigma : \mathbb{X} \to \mathbb{X} \ \textit{ and } \ \mathcal{M}_\Sigma : \mathbb{X}' \to \mathbb{X}' \\ \textit{are well defined and bounded,} \end{gathered} \tag{23.304}$$

where $\mathcal{M}_\Sigma$ is the Hardy-Littlewood maximal operator on Σ (cf. (2.26)). Let $\mathbb{Y}$ be a Generalized Quasi-Banach Sequence Space on $\mathbb{Z}$ which is $\mathbb{X}$-compliant (in the sense of Definition 23.7). Then the following inclusions hold:

$$\begin{gathered} L^\infty_{\mathrm{comp}}(\Sigma,\sigma) \subseteq \dot{\mathrm{K}}(\Sigma;\mathbb{X};\mathbb{Y}) \subseteq L^1_{\mathrm{loc}}(\Sigma,\sigma) \\ \textit{and } \ L^\infty_{\mathrm{comp}}(\Sigma,\sigma) \subseteq \dot{\mathrm{K}}(\Sigma;\mathbb{X}';\mathbb{Y}') \subseteq L^1_{\mathrm{loc}}(\Sigma,\sigma). \end{gathered} \tag{23.305}$$

Proof This follows from Proposition 23.4, Proposition 23.5, Proposition 23.8, and Theorem 23.1. □

The proof of the embedding result recorded in [GHA-II, Proposition 5.2.7, p. 248] in the context of Generalized Banach Function Spaces remains valid, irrespective of the veracity of the triangle inequality property of the norm. With this in mind, we have a similar result for Generalized Quasi-Banach Function Spaces.

Proposition 23.9 *Let Σ be a closed Ahlfors regular set, and denote $\sigma := \mathcal{H}^{n-1}\lfloor\Sigma$. Consider $\mathbb{X}$ a Generalized Quasi-Banach Function Space on (Σ, σ). Then*

$$\begin{gathered}\text{whenever } \mathcal{M}_\Sigma(\mathbb{X}') \subseteq \mathbb{X}' \text{ one has} \\ \mathbb{X} \hookrightarrow L^1\Big(\Sigma, \frac{\sigma(x)}{1+|x|^{n-1}}\Big) \text{ continuously,}\end{gathered} \tag{23.306}$$

and

$$\begin{gathered}\text{whenever } \mathcal{M}_\Sigma(\mathbb{X}) \subseteq \mathbb{X} \text{ one has} \\ \mathbb{X}' \hookrightarrow L^1\Big(\Sigma, \frac{\sigma(x)}{1+|x|^{n-1}}\Big) \text{ continuously,}\end{gathered} \tag{23.307}$$

where $\mathcal{M}_\Sigma$ is the Hardy-Littlewood maximal operator on Σ (cf. (2.26)).

Based on this and the fact that Composite Herz spaces are themselves Generalized Quasi-Banach Function Spaces (cf. Proposition 23.4), we obtain the following useful embedding result.

Corollary 23.4 *Suppose $\Sigma \subseteq \mathbb{R}^n$ is a closed unbounded Ahlfors regular set, and abbreviate $\sigma := \mathcal{H}^{n-1}\lfloor\Sigma$. Consider a Generalized Banach Function Space $\mathbb{X}$ on (Σ, σ) such that*

$$\begin{gathered}\mathcal{M}_\Sigma : \mathbb{X} \to \mathbb{X} \text{ and } \mathcal{M}_\Sigma : \mathbb{X}' \to \mathbb{X}' \\ \text{are well defined and bounded,}\end{gathered} \tag{23.308}$$

where $\mathcal{M}_\Sigma$ is the Hardy-Littlewood maximal operator on Σ (cf. (2.26)). Let $\mathbb{Y}$ be a Generalized Quasi-Banach Sequence Space on $\mathbb{Z}$ that is $\mathbb{X}$-compliant (in the sense of Definition 23.7). Then one has the following continuous embeddings:

$$\begin{gathered}\dot{\mathbb{K}}(\Sigma; \mathbb{X}; \mathbb{Y}) \hookrightarrow L^1\big(\Sigma, \tfrac{\sigma(x)}{1+|x|^{n-1}}\big) \\ \text{and } \dot{\mathbb{K}}(\Sigma; \mathbb{X}'; \mathbb{Y}') \hookrightarrow L^1\big(\Sigma, \tfrac{\sigma(x)}{1+|x|^{n-1}}\big).\end{gathered} \tag{23.309}$$

Proof This follows from Proposition 23.4, Proposition 23.5, Proposition 23.9, and Theorem 23.1. □

The techniques used to prove Theorem 23.1 can be employed to obtain boundedness for a certain class of sublinear operators on $\dot{\mathbb{K}}(\Sigma; \mathbb{X}; \mathbb{Y})$. The following result may then be regarded as a self-improvement of Theorem 23.1.

Theorem 23.3 *Let $\Sigma \subseteq \mathbb{R}^n$ be a closed unbounded Ahlfors regular set, and abbreviate $\sigma := \mathcal{H}^{n-1}\lfloor\Sigma$. Assume $\mathbb{X}$ is a Generalized Banach Function Space on (Σ, σ) such that*

$$\begin{gathered}\mathcal{M}_\Sigma : \mathbb{X} \to \mathbb{X} \text{ and } \mathcal{M}_\Sigma : \mathbb{X}' \to \mathbb{X}' \\ \text{are well defined and bounded,}\end{gathered} \tag{23.310}$$

where $\mathcal{M}_\Sigma$ is the Hardy-Littlewood maximal operator on Σ (cf. (2.26)). Select $\mathbb{Y}$ a Generalized Quasi-Banach Sequence Space on $\mathbb{Z}$ which is $\mathbb{X}$-compliant (in the sense of Definition 23.7). Consider a sublinear operator

$$T : L^1\big(\Sigma, \tfrac{\sigma(x)}{1+|x|^{n-1}}\big) \to \mathscr{M}(\Sigma, \sigma), \tag{23.311}$$

and assume there exists some $\widetilde{C} \in (0, \infty)$ such that for every $f \in L^1\big(\Sigma, \frac{\sigma(x)}{1+|x|^{n-1}}\big)$ one has

$$|Tf(x)| \leq \widetilde{C} \int_\Sigma \frac{|f(y)|}{|x-y|^{n-1}}\, d\sigma(y) \quad \textit{for all } x \in \Sigma \setminus \operatorname{supp} f. \tag{23.312}$$

Then, the Composite Herz spaces $\dot{\mathbb{K}}(\Sigma; \mathbb{X}; \mathbb{Y})$ and $\dot{\mathbb{K}}(\Sigma; \mathbb{X}'; \mathbb{Y}')$ embed continuously into $L^1\big(\Sigma, \frac{\sigma(x)}{1+|x|^{n-1}}\big)$, and the following statements are valid:

(a) Whenever the operator

$$T : \mathbb{X} \to \mathbb{X} \quad \textit{is well defined and bounded,} \tag{23.313}$$

it follows that

$$T : \dot{\mathbb{K}}(\Sigma; \mathbb{X}; \mathbb{Y}) \to \dot{\mathbb{K}}(\Sigma; \mathbb{X}; \mathbb{Y}) \quad \textit{is well defined and bounded,} \tag{23.314}$$

with operator norm satisfying

$$\|T\|_{\dot{\mathbb{K}}(\Sigma;\mathbb{X};\mathbb{Y}) \to \dot{\mathbb{K}}(\Sigma;\mathbb{X};\mathbb{Y})} \leq C\big(\widetilde{C} + \|T\|_{\mathbb{X}\to\mathbb{X}}\big), \tag{23.315}$$

for some $C \in (0, \infty)$ depending only on n, the operator norms of $\mathcal{M}_\Sigma$ on $\mathbb{X}$ and $\mathbb{X}'$, and the Ahlfors regularity constants of Σ.

(b) Whenever the operator

$$T : \mathbb{X}' \to \mathbb{X}' \quad \textit{is well defined and bounded,} \tag{23.316}$$

it follows that

$$T : \dot{\mathbb{K}}(\Sigma; \mathbb{X}'; \mathbb{Y}') \to \dot{\mathbb{K}}(\Sigma; \mathbb{X}'; \mathbb{Y}') \quad \textit{is well defined and bounded,} \tag{23.317}$$

with operator norm satisfying

$$\|T\|_{\dot{\mathbb{K}}(\Sigma;\mathbb{X}';\mathbb{Y}') \to \dot{\mathbb{K}}(\Sigma;\mathbb{X}';\mathbb{Y}')} \leq C\big(\widetilde{C} + \|T\|_{\mathbb{X}'\to\mathbb{X}'}\big), \tag{23.318}$$

for some constant $C \in (0, \infty)$ depending only on n, the operator norm of $\mathcal{M}_\Sigma$ on $\mathbb{X}$ and $\mathbb{X}'$, and the Ahlfors regularity constants of Σ.

Lastly, the same type of conclusion regarding boundedness on inhomogeneous Composite Herz spaces $\mathrm{K}(\Sigma;\mathbb{X};\mathbb{Y})$ *hold whenever* $\mathbb{Y}$ *is a* GQBSS *on* $\mathbb{N}_0$ *which is* $\mathbb{X}$*-compliant.*

Proof The current assumptions allow us to invoke Corollary 23.4 to guarantee that the following embeddings are continuous:

$$\begin{aligned}&\dot{\mathrm{K}}(\Sigma;\mathbb{X};\mathbb{Y}) \hookrightarrow L^1\big(\Sigma, \tfrac{\sigma(x)}{1+|x|^{n-1}}\big)\\ \text{and }\ &\dot{\mathrm{K}}(\Sigma;\mathbb{X}';\mathbb{Y}') \hookrightarrow L^1\big(\Sigma, \tfrac{\sigma(x)}{1+|x|^{n-1}}\big).\end{aligned} \tag{23.319}$$

Hence the action of T is meaningfully defined on $\dot{\mathrm{K}}(\Sigma;\mathbb{X};\mathbb{Y})$.

Consider first the claim made in item *(a)*. Select a function $f \in \dot{\mathrm{K}}(\Sigma;\mathbb{X};\mathbb{Y})$. Recall the family of annuli $\{C_\ell\}_{\ell\in\mathbb{Z}}$ as in (23.16). Having fixed $k \in \mathbb{Z}$, write

$$f = \sum_{\substack{\ell\in\mathbb{Z}\\ |\ell-k|\geq 2}} f\cdot \mathbb{1}_{C_\ell} + \sum_{\ell=k-1}^{k+1} f\cdot \mathbb{1}_{C_\ell} =: f_1 + f_2 \quad \text{pointwise } \sigma\text{-a.e. on } \Sigma. \tag{23.320}$$

Notice that since the function f_1 is supported in $\Sigma\backslash\big(\Delta(x_0, 2_*^{k+2})\backslash\Delta(x_0, 2_*^{k-2})\big)$, we have $\operatorname{supp} f_1 \cap C_k = \varnothing$. Thanks to (23.312), for every $x \in C_k$ we bound

$$\begin{aligned}|Tf_1(x)| &\leq \widetilde{C}\int_\Sigma \frac{\Big|\sum_{\substack{\ell\in\mathbb{Z}\\ |\ell-k|\geq 2}} (f\cdot \mathbb{1}_{C_\ell})(y)\Big|}{|x-y|^{n-1}}\,d\sigma(y)\\
&\leq \widetilde{C}\int_\Sigma \frac{\sum_{\substack{\ell\in\mathbb{Z}\\ |\ell-k|\geq 2}} |f(y)|\mathbb{1}_{C_\ell}(y)}{|x-y|^{n-1}}\,d\sigma(y)\\
&\leq \widetilde{C}\sum_{\substack{\ell\in\mathbb{Z}\\ |\ell-k|\geq 2}} \int_{C_\ell} \frac{|f(y)|}{|x-y|^{n-1}}\,d\sigma(y)\\
&= \widetilde{C}\sum_{\ell=-\infty}^{k-2} \int_{C_\ell} \frac{|f(y)|}{|x-y|^{n-1}}\,d\sigma(y) + \widetilde{C}\sum_{\ell=k+2}^{\infty} \int_{C_\ell} \frac{|f(y)|}{|x-y|^{n-1}}\,d\sigma(y),\end{aligned} \tag{23.321}$$

where the third inequality uses Lebesgue's Monotone Convergence Theorem. To estimate each term of the infinite sums in (23.321), employ the same type of reasoning as in (23.278)-(23.279) and (23.284)-(23.285) to obtain, for every $x \in C_k$,

$$\begin{aligned}|Tf_1(x)| &\leq C\widetilde{C}\sum_{\ell=-\infty}^{k-2} 2_*^{-k(n-1)}\|f\cdot \mathbb{1}_{C_\ell}\|_{\mathbb{X}}\|\mathbb{1}_{C_\ell}\|_{\mathbb{X}'}\\
&\quad + C\widetilde{C}\sum_{\ell=k+2}^{\infty} 2_*^{-\ell(n-1)}\|f\cdot \mathbb{1}_{C_\ell}\|_{\mathbb{X}}\|\mathbb{1}_{C_\ell}\|_{\mathbb{X}'}\end{aligned} \tag{23.322}$$

for some $C \in (0, \infty)$ depending only on n, and the Ahlfors regularity constants of Σ. Based on this, running the same type of argument as in (23.280)-(23.283) and (23.286)-(23.288), and bearing in mind the compliancy of $\|\cdot\|_{\mathbb{Y}}$ we may bound

$$\begin{aligned}
&\Big\|\Big\{\|(Tf_1)\cdot \mathbb{1}_{C_k}\|_{\mathbb{X}}\Big\}_{k\in\mathbb{Z}}\Big\|_{\mathbb{Y}}\\
&\leq C\widetilde{C}\left\|\left\{\sum_{\ell=-\infty}^{k-2} 2_*^{-k(n-1)}\|f\cdot \mathbb{1}_{C_\ell}\|_{\mathbb{X}}\|\mathbb{1}_{C_\ell}\|_{\mathbb{X}'}\|\mathbb{1}_{C_k}\|_{\mathbb{X}}\right\}_{k\in\mathbb{Z}}\right\|_{\mathbb{Y}}\\
&\quad + C\widetilde{C}\left\|\left\{\sum_{\ell=k+2}^{\infty} 2_*^{-\ell(n-1)}\|f\cdot \mathbb{1}_{C_\ell}\|_{\mathbb{X}}\|\mathbb{1}_{C_\ell}\|_{\mathbb{X}'}\|\mathbb{1}_{C_k}\|_{\mathbb{X}}\right\}_{k\in\mathbb{Z}}\right\|_{\mathbb{Y}}\\
&\leq C\widetilde{C}\|f\|_{\dot{\mathrm{K}}(\Sigma;\mathbb{X};\mathbb{Y})},
\end{aligned} \tag{23.323}$$

for some $C \in (0, \infty)$ depending only on n, $\|\mathcal{M}_\Sigma\|_{\mathbb{X}\to\mathbb{X}}$, $\|\mathcal{M}_\Sigma\|_{\mathbb{X}'\to\mathbb{X}'}$, the compliancy constant of of $\|\cdot\|_{\mathbb{Y}}$, the modulus of concavity of $\|\cdot\|_{\mathbb{Y}}$, and the Ahlfors regularity character of Σ.

Next, thanks to the sublinearity of T, the boundedness of T on $\mathbb{X}$, and the compliancy of $\|\cdot\|_{\mathbb{Y}}$, a similar reasoning as in (23.277) presently gives

$$\begin{aligned}
\Big\|\Big\{\|(Tf_2)\cdot \mathbb{1}_{C_k}\|_{\mathbb{X}}\Big\}_{k\in\mathbb{Z}}\Big\|_{\mathbb{Y}} &\leq \left\|\left\{\sum_{\ell=k-1}^{k+1}\|[T(f\cdot \mathbb{1}_{C_\ell})]\cdot \mathbb{1}_{C_k}\|_{\mathbb{X}}\right\}_{k\in\mathbb{Z}}\right\|_{\mathbb{Y}}\\
&\leq \|T\|_{\mathbb{X}\to\mathbb{X}}\left\|\left\{\sum_{\ell=k-1}^{k+1}\|f\cdot \mathbb{1}_{C_\ell}\|_{\mathbb{X}}\right\}_{k\in\mathbb{Z}}\right\|_{\mathbb{Y}}\\
&\leq C\|T\|_{\mathbb{X}\to\mathbb{X}}\Big\|\{\|f\cdot \mathbb{1}_{C_k}\|_{\mathbb{X}}\}_{k\in\mathbb{Z}}\Big\|_{\mathbb{Y}}\\
&= C\|T\|_{\mathbb{X}\to\mathbb{X}}\|f\|_{\dot{\mathrm{K}}(\Sigma;\mathbb{X};\mathbb{Y})}
\end{aligned} \tag{23.324}$$

for some constant $C \in (0, \infty)$ depending only on the compliancy constant of $\|\cdot\|_{\mathbb{Y}}$ and the modulus of concavity of $\|\cdot\|_{\mathbb{Y}}$. Putting it all together, the sublinearity of T, (23.320), (23.323), and (23.324) give

$$\begin{aligned}
\|Tf\|_{\dot{\mathrm{K}}(\Sigma;\mathbb{X};\mathbb{Y})} &= \Big\|\Big\{\|(Tf)\cdot \mathbb{1}_{C_k}\|_{\mathbb{X}}\Big\}_{k\in\mathbb{Z}}\Big\|_{\mathbb{Y}}\\
&\leq C\widetilde{C}\Big\|\Big\{\|(Tf_1)\cdot \mathbb{1}_{C_k}\|_{\mathbb{X}}\Big\}_{k\in\mathbb{Z}}\Big\|_{\mathbb{Y}}\\
&\quad + C\|T\|_{\mathbb{X}\to\mathbb{X}}\Big\|\Big\{\|(Tf_2)\cdot \mathbb{1}_{C_k}\|_{\mathbb{X}}\Big\}_{k\in\mathbb{Z}}\Big\|_{\mathbb{Y}}\\
&\leq C\big(\widetilde{C} + \|T\|_{\mathbb{X}\to\mathbb{X}}\big)\|f\|_{\dot{\mathrm{K}}(\Sigma;\mathbb{X};\mathbb{Y})},
\end{aligned} \tag{23.325}$$

for some $C \in (0, \infty)$ depending only n, the Ahlfors regularity constants of Σ, the operator norms $\|\mathcal{M}_\Sigma\|_{\mathbb{X}\to\mathbb{X}}$, $\|\mathcal{M}_\Sigma\|_{\mathbb{X}'\to\mathbb{X}'}$, the compliancy constant of $\|\cdot\|_{\mathbb{Y}}$, and modulus of concavity of $\|\cdot\|_{\mathbb{Y}}$. This proves (23.314)-(23.315).

The proof of the claim made in *(b)* follows from what we have proved so far, by replacing $\mathbb{X}$ by $\mathbb{X}'$, and $\mathbb{Y}$ by $\mathbb{Y}'$, and recalling Remark 23.17.

The very last claim made in relation to the inhomogeneous Composite Herz spaces, is obtained via an analogous argument as before, and similar computations as in (23.298)-(23.295), while being mindful of Definition 23.6. This concludes the proof of Theorem 23.3. □

23.7 Vanishing Composite Herz Spaces

A natural and relevant environment where one can obtain boundedness results, for the Hardy-Littlewood maximal operator, and large classes of singular integral operators of interest, is the subspace $\mathring{\mathbb{X}}$ of a Generalized Banach Function Space $\mathbb{X}$ (see [GHA-II, Definition 5.2.6, p. 247]), which is defined as being the closure of $L^\infty_{\rm comp}$ in the $\mathbb{X}$ norm (provided this is meaningful). Our goal in this section is to introduce a similar subspace in the context of Composite Herz spaces.

To set the stage, consider a closed unbounded Ahlfors regular set $\Sigma \subseteq \mathbb{R}^n$, and abbreviate $\sigma := \mathcal{H}^{n-1}\lfloor\Sigma$. Denote by $\mathcal{M}_\Sigma$ the Hardy-Littlewood maximal operator on Σ (cf. (2.26)). Assume $\mathbb{X}$ is a Generalized Banach Function Space on (Σ, σ) such that

$$\begin{gathered} \mathcal{M}_\Sigma : \mathbb{X} \to \mathbb{X} \quad \text{and} \quad \mathcal{M}_\Sigma : \mathbb{X}' \to \mathbb{X}' \\ \text{are well defined and bounded,} \end{gathered} \tag{23.326}$$

where $\mathbb{X}'$ denotes the associated space of $\mathbb{X}$ (cf. Definition 10.3). Let $\mathbb{Y}$ be a Generalized Quasi-Banach Sequence Space on $\mathbb{Z}$, and suppose $\mathbb{Y}$ is $\mathbb{X}$-compliant (cf. Definition 23.7). In this context, define the space

$$\mathring{\mathrm{K}}(\Sigma; \mathbb{X}; \mathbb{Y}) := \overline{L^\infty_{\rm comp}(\Sigma, \sigma)}^{\|\cdot\|_{\dot{\mathrm{K}}(\Sigma;\mathbb{X};\mathbb{Y})}}. \tag{23.327}$$

Thanks to Corollary 23.3, this space is meaningfully defined and, by design,

$$\mathring{\mathrm{K}}(\Sigma; \mathbb{X}; \mathbb{Y}) \text{ is a closed subspace of } \dot{\mathrm{K}}(\Sigma; \mathbb{X}; \mathbb{Y}). \tag{23.328}$$

Hence, in light of Corollary 23.1,

$$\begin{gathered} \big(\mathring{\mathrm{K}}(\Sigma; \mathbb{X}; \mathbb{Y}), \|\cdot\|_{\dot{\mathrm{K}}(\Sigma;\mathbb{X};\mathbb{Y})}\big) \text{ is a Quasi-Banach space} \\ \text{in which } L^\infty_{\rm comp}(\Sigma, \sigma) \text{ is a dense subspace.} \end{gathered} \tag{23.329}$$

Furthermore, from (23.327), Remark 23.9, and [GHA-II, Lemma 1.2.20, p. 24], the operator of pointwise multiplication by a given function $b \in L^\infty(\Sigma, \sigma)$ is a bounded mapping from $\mathring{\mathbb{K}}(\Sigma; \mathbb{X}; \mathbb{Y})$ into itself, with operator norm $\leq \|b\|_{L^\infty(\Sigma,\sigma)}$. (23.330)

In a scenario when Composite Herz spaces fall within the framework of Generalized Banach Function Spaces, work done in [GHA-II] permits us to obtain alternative descriptions of vanishing Composite Herz spaces, as well as a boundedness result for the Hardy-Littlewood maximal operator. The details are as follows:

Proposition 23.10 *Consider a closed unbounded Ahlfors regular set $\Sigma \subseteq \mathbb{R}^n$, and abbreviate $\sigma := \mathcal{H}^{n-1}\lfloor\Sigma$. Suppose $\mathbb{X}$ is a Generalized Banach Function Space on (Σ, σ) with the property that*

$$\mathcal{M}_\Sigma : \mathbb{X} \to \mathbb{X} \quad \text{and} \quad \mathcal{M}_\Sigma : \mathbb{X}' \to \mathbb{X}' \quad \text{are well-defined bounded mappings,} \tag{23.331}$$

where $\mathcal{M}_\Sigma$ is the Hardy-Littlewood maximal operator on Σ (cf. (2.26)). Let $\mathbb{Y}$ be a Generalized Banach Sequence Space on $\mathbb{Z}$, and suppose $\mathbb{Y}$ is $\mathbb{X}$-compliant (cf. Definition 23.7).

Then there exists some $p_* = p_*(\Sigma, \mathbb{X}) \in (1, \infty)$ with the following significance. Whenever $q \in (1, p_*)$ and $q' := q/(q-1) \in (1, \infty)$ is its Hölder conjugate exponent, the space $\mathring{\mathbb{K}}(\Sigma; \mathbb{X}; \mathbb{Y})$ originally defined in (23.327) may be described as

$$\mathring{\mathbb{K}}(\Sigma; \mathbb{X}; \mathbb{Y}) = \overline{L^{q'}_{\text{comp}}(\Sigma, \sigma)}^{\|\cdot\|_{\dot{\mathbb{K}}(\Sigma;\mathbb{X};\mathbb{Y})}}, \tag{23.332}$$

as well as

$$\mathring{\mathbb{K}}(\Sigma; \mathbb{X}; \mathbb{Y}) = \overline{\text{Lip}_c(\Sigma)}^{\|\cdot\|_{\dot{\mathbb{K}}(\Sigma;\mathbb{X};\mathbb{Y})}}. \tag{23.333}$$

Furthermore, the Hardy-Littlewood maximal operator

$$\mathcal{M}_\Sigma : \mathring{\mathbb{K}}(\Sigma; \mathbb{X}; \mathbb{Y}) \to \mathring{\mathbb{K}}(\Sigma; \mathbb{X}; \mathbb{Y}) \quad \text{is a well-defined bounded mapping.} \tag{23.334}$$

Proof Observe that the assumption made in (23.331), together with Proposition 23.4, ensure that $\dot{\mathbb{K}}(\Sigma; \mathbb{X}; \mathbb{Y})$ is a Generalized Banach Function Space on (Σ, σ). Moreover, thanks to Proposition 23.5 and Theorem 23.1 we have

$$\begin{aligned} &\mathcal{M}_\Sigma : \dot{\mathbb{K}}(\Sigma; \mathbb{X}; \mathbb{Y}) \to \dot{\mathbb{K}}(\Sigma; \mathbb{X}; \mathbb{Y}) \\ \text{and } &\mathcal{M}_\Sigma : \big(\dot{\mathbb{K}}(\Sigma; \mathbb{X}; \mathbb{Y})\big)' \to \big(\dot{\mathbb{K}}(\Sigma; \mathbb{X}; \mathbb{Y})\big)' \\ &\text{are well-defined and bounded operators.} \end{aligned} \tag{23.335}$$

In light of these, all claims made in (23.332), (23.333), and (23.334) follow from [GHA-II, Lemma 5.2.8, p. 250]. □

23.8 Calderón-Zygmund Theory on Composite Herz Spaces

The relevance of the boundedness result obtained in Theorem 23.3 is that a large family of singular integral operators of Calderón-Zygmund type (in the context of UR sets) fall within the class of (sublinear) operators that satisfy the size condition stipulated in (23.312). We shall therefore use this to obtain the following boundedness result for convolution-type integral operators on Composite Herz spaces.

Theorem 23.4 *Suppose $\Sigma \subseteq \mathbb{R}^n$ is a closed unbounded uniformly rectifiable set, and abbreviate $\sigma := \mathcal{H}^{n-1}\lfloor\Sigma$. For some sufficiently large integer $N = N(n) \in \mathbb{N}$, consider a complex-valued kernel $k \in \mathscr{C}^N(\mathbb{R}^n \setminus \{0\})$ which is odd and positive homogeneous of degree $1-n$. In this setting, for each function $f \in L^1\big(\Sigma, \frac{\sigma(x)}{1+|x|^{n-1}}\big)$ define*

$$T_\varepsilon f(x) := \int_{\substack{y\in\Sigma \\ |x-y|>\varepsilon}} k(x-y)f(y)\,d\sigma(y) \text{ for all } x \in \Sigma \text{ and } \varepsilon > 0, \tag{23.336}$$

$$T_{\max} f(x) := \sup_{\varepsilon>0} |T_\varepsilon f(x)| \text{ for each } x \in \Sigma, \tag{23.337}$$

$$Tf(x) := \lim_{\varepsilon\to 0^+} T_\varepsilon f(x) \text{ for } \sigma\text{-a.e. } x \in \Sigma. \tag{23.338}$$

Assume $\mathbb{X}$ is a Generalized Banach Function Space on (Σ, σ) such that

$$\begin{gathered} \mathcal{M}_\Sigma : \mathbb{X} \to \mathbb{X} \text{ and } \mathcal{M}_\Sigma : \mathbb{X}' \to \mathbb{X}' \\ \text{are well defined and bounded,} \end{gathered} \tag{23.339}$$

where $\mathcal{M}_\Sigma$ is the Hardy-Littlewood maximal operator on Σ (cf. (2.26)). Select $\mathbb{Y}$ a Generalized Quasi-Banach Sequence Space on $\mathbb{Z}$ which is $\mathbb{X}$-compliant (in the sense of Definition 23.7).

Then the maximal operator $T_{\max}$ induces well-defined and bounded mappings

$$\begin{aligned} T_{\max} &: \dot{\mathbb{K}}(\Sigma; \mathbb{X}; \mathbb{Y}) \longrightarrow \dot{\mathbb{K}}(\Sigma; \mathbb{X}; \mathbb{Y}), \\ T_{\max} &: \dot{\mathbb{K}}(\Sigma; \mathbb{X}'; \mathbb{Y}') \longrightarrow \dot{\mathbb{K}}(\Sigma; \mathbb{X}'; \mathbb{Y}') \end{aligned} \tag{23.340}$$

with operator norms satisfying

$$\begin{aligned} \|T_{\max}\|_{\dot{\mathbb{K}}(\Sigma;\mathbb{X};\mathbb{Y})\to\dot{\mathbb{K}}(\Sigma;\mathbb{X};\mathbb{Y})} &\le C\Big(\sum_{|\alpha|\le N} \sup_{S^{n-1}} |\partial^\alpha k|\Big), \\ \|T_{\max}\|_{\dot{\mathbb{K}}(\Sigma;\mathbb{X}';\mathbb{Y}')\to\dot{\mathbb{K}}(\Sigma;\mathbb{X}';\mathbb{Y}')} &\le C\Big(\sum_{|\alpha|\le N} \sup_{S^{n-1}} |\partial^\alpha k|\Big) \end{aligned} \tag{23.341}$$

for some constant $C \in (0, \infty)$ depending exclusively on n, the operator norms of $\mathcal{M}_\Sigma$ on $\mathbb{X}$ and $\mathbb{X}'$, the compliancy constant of $\|\cdot\|_{\mathbb{Y}}$, and the UR *constants of Σ. In* *particular,*

$$\text{the truncated integral operators } T_\varepsilon : \dot{\mathbb{K}}(\Sigma;\mathbb{X};\mathbb{Y}) \longrightarrow \dot{\mathbb{K}}(\Sigma;\mathbb{X};\mathbb{Y}) \text{ and } T_\varepsilon : \dot{\mathbb{K}}(\Sigma;\mathbb{X}';\mathbb{Y}') \longrightarrow \dot{\mathbb{K}}(\Sigma;\mathbb{X}';\mathbb{Y}') \text{ are well defined, linear, and bounded uniformly with respect to the truncation parameter } \varepsilon \in (0,\infty). \tag{23.342}$$

Moreover, the principal-value singular integral operators

$$\begin{aligned} &T : \dot{\mathbb{K}}(\Sigma;\mathbb{X};\mathbb{Y}) \longrightarrow \dot{\mathbb{K}}(\Sigma;\mathbb{X};\mathbb{Y}), \\ &T : \dot{\mathbb{K}}(\Sigma;\mathbb{X}';\mathbb{Y}') \longrightarrow \dot{\mathbb{K}}(\Sigma;\mathbb{X}';\mathbb{Y}') \end{aligned} \tag{23.343}$$

are well defined, linear, and bounded, and have operator norms satisfying

$$\begin{aligned} \|T\|_{\dot{\mathbb{K}}(\Sigma;\mathbb{X};\mathbb{Y})\to\dot{\mathbb{K}}(\Sigma;\mathbb{X};\mathbb{Y})} &\leq C\Big(\sum_{|\alpha|\leq N}\sup_{S^{n-1}}|\partial^\alpha k|\Big), \\ \|T\|_{\dot{\mathbb{K}}(\Sigma;\mathbb{X}';\mathbb{Y}')\to\dot{\mathbb{K}}(\Sigma;\mathbb{X}';\mathbb{Y}')} &\leq C\Big(\sum_{|\alpha|\leq N}\sup_{S^{n-1}}|\partial^\alpha k|\Big) \end{aligned} \tag{23.344}$$

for some constant $C \in (0,\infty)$ depending exclusively on n, the operator norms of $\mathcal{M}_\Sigma$ on $\mathbb{X}$ and $\mathbb{X}'$, the compliancy constant of $\|\cdot\|_{\mathbb{Y}}$, and the UR *constants of* Σ.

Next, if one makes the stronger assumption that $\mathbb{Y}$ is a Generalized Banach Sequence Space on $\mathbb{Z}$ which is $\mathbb{X}$-compliant then, in relation to the vanishing Composite Herz space $\mathring{\mathbb{K}}(\Sigma;\mathbb{X};\mathbb{Y})$ (defined in (23.327)*), one may also conclude that*

$$\begin{aligned} &T_{\max} : \mathring{\mathbb{K}}(\Sigma;\mathbb{X};\mathbb{Y}) \longrightarrow \mathring{\mathbb{K}}(\Sigma;\mathbb{X};\mathbb{Y}) \\ &\text{and}\quad T : \mathring{\mathbb{K}}(\Sigma;\mathbb{X};\mathbb{Y}) \longrightarrow \mathring{\mathbb{K}}(\Sigma;\mathbb{X};\mathbb{Y}) \\ &\text{are well-defined and bounded mappings,} \end{aligned} \tag{23.345}$$

with operator norms

$$\begin{aligned} \|T_{\max}\|_{\mathring{\mathbb{K}}(\Sigma;\mathbb{X};\mathbb{Y})\to\mathring{\mathbb{K}}(\Sigma;\mathbb{X};\mathbb{Y})} &\leq C\Big(\sum_{|\alpha|\leq N}\sup_{S^{n-1}}|\partial^\alpha k|\Big) \\ \|T\|_{\mathring{\mathbb{K}}(\Sigma;\mathbb{X};\mathbb{Y})\to\mathring{\mathbb{K}}(\Sigma;\mathbb{X};\mathbb{Y})} &\leq C\Big(\sum_{|\alpha|\leq N}\sup_{S^{n-1}}|\partial^\alpha k|\Big). \end{aligned} \tag{23.346}$$

Lastly, the same type of conclusions, regarding mapping properties and boundedness, hold in the context of inhomogeneous Composite Herz spaces $\mathbb{K}(\Sigma;\mathbb{X};\mathbb{Y})$.

Proof In the context of (23.339), [GHA-III, item *(11)* in Theorem 2.3.2, pp. 348–357] ensures that $T_{\max}$ induces well-defined and bounded operators

$$\begin{aligned} &T_{\max} : \mathbb{X} \to \mathbb{X} \quad\text{and}\quad T_{\max} : \mathbb{X}' \to \mathbb{X}', \quad \text{with operator norms} \\ &\max\Big\{\|T_{\max}\|_{\mathbb{X}\to\mathbb{X}},\ \|T_{\max}\|_{\mathbb{X}'\to\mathbb{X}'}\Big\} \leq C\Big(\textstyle\sum_{|\alpha|\leq N}\sup_{S^{n-1}}|\partial^\alpha k|\Big), \end{aligned} \tag{23.347}$$

while T induces well-defined and bounded operators

$$\begin{aligned} &T : \mathbb{X} \to \mathbb{X} \quad \text{and} \quad T : \mathbb{X}' \to \mathbb{X}', \quad \text{with operator norms} \\ &\max\Big\{\|T\|_{\mathbb{X}\to\mathbb{X}}, \|T\|_{\mathbb{X}'\to\mathbb{X}'}\Big\} \le C\Big(\textstyle\sum_{|\alpha|\le N} \sup_{S^{n-1}} |\partial^\alpha k|\Big), \end{aligned} \tag{23.348}$$

for some constant $C \in (0, \infty)$ depending only on n, the operator norms of $\mathcal{M}_\Sigma$ on $\mathbb{X}$ and $\mathbb{X}'$, and the UR constants of Σ.

Next, fix a parameter $\varepsilon \in (0, \infty)$. Pick a function $f \in L^1\big(\Sigma, \frac{\sigma(x)}{1+|x|^{n-1}}\big)$. In view of the smoothness and homogeneity of the kernel k, for each $x \in \Sigma \setminus \operatorname{supp} f$ we estimate

$$\begin{aligned} |T_\varepsilon f(x)| &\le \int_{\substack{y\in\Sigma \\ |x-y|>\varepsilon}} |k(x-y)||f(y)|\, d\sigma(y) \\ &= \int_{\substack{y\in\Sigma \\ |x-y|>\varepsilon}} \Big|k\Big(\frac{x-y}{|x-y|}\Big)\Big| \frac{|f(y)|}{|x-y|^{n-1}}\, d\sigma(y) \\ &\le \Big(\sum_{|\alpha|\le N} \sup_{S^{n-1}} |\partial^\alpha k|\Big) \int_\Sigma \frac{|f(y)|}{|x-y|^{n-1}}\, d\sigma(y). \end{aligned} \tag{23.349}$$

Since ε is arbitrary, (23.349), (23.337), and (23.338) ensure that

$$|T_{\max} f(x)| \le \Big(\sum_{|\alpha|\le N} \sup_{S^{n-1}} |\partial^\alpha k|\Big) \int_\Sigma \frac{|f(y)|}{|x-y|^{n-1}}\, d\sigma(y) \tag{23.350}$$

$$\text{and} \quad |Tf(x)| \le \Big(\sum_{|\alpha|\le N} \sup_{S^{n-1}} |\partial^\alpha k|\Big) \int_\Sigma \frac{|f(y)|}{|x-y|^{n-1}}\, d\sigma(y). \tag{23.351}$$

Hence, Theorem 23.3, (23.347), and (23.348) imply all claims made in (23.340)-(23.344).

Next, if we now assume that $\mathbb{Y}$ is actually a Generalized Banach Sequence Space on $\mathbb{Z}$, then from Proposition 23.4 we have that $\dot{\mathrm{K}}(\Sigma; \mathbb{X}; \mathbb{Y})$ is a Generalized Banach Function Space on (Σ, σ). Moreover, thanks to Proposition 23.5 and Theorem 23.1, it follows that

$$\begin{aligned} &\mathcal{M}_\Sigma : \dot{\mathrm{K}}(\Sigma; \mathbb{X}; \mathbb{Y}) \to \dot{\mathrm{K}}(\Sigma; \mathbb{X}; \mathbb{Y}) \\ \text{and } &\mathcal{M}_\Sigma : \big(\dot{\mathrm{K}}(\Sigma; \mathbb{X}; \mathbb{Y})\big)' \to \big(\dot{\mathrm{K}}(\Sigma; \mathbb{X}; \mathbb{Y})\big)' \\ &\text{are well-defined and bounded operators.} \end{aligned} \tag{23.352}$$

Thanks to these, the boundedness claimed in (23.345)-(23.346) becomes a consequence of [GHA-III, item *(11)* in Theorem 2.3.2, pp. 348–257]. This concludes the proof of Theorem 23.4. □

It is also of interest to obtain nontangential maximal function estimates in the context of Composite Herz spaces.

Theorem 23.5 *Assume that $\Omega \subseteq \mathbb{R}^n$ is an open set such that $\partial\Omega$ is an unbounded* UR *set, and abbreviate $\sigma := \mathcal{H}^{n-1}\lfloor\partial\Omega$. For some sufficiently large $N = N(n) \in \mathbb{N}$, consider a complex-valued kernel $k \in \mathscr{C}^N(\mathbb{R}^n \setminus \{0\})$ which is odd and positive homogeneous of degree $1-n$. For each function $f \in L^1\big(\partial\Omega, \frac{\sigma(x)}{1+|x|^{n-1}}\big)$ define*

$$\mathcal{T}f(x) := \int_{\partial\Omega} k(x-y)f(y)\,d\sigma(y) \quad \text{for each } x \in \Omega. \tag{23.353}$$

Suppose $\mathbb{X}$ is a Generalized Banach Function Space on $(\partial\Omega, \sigma)$ such that

$$\begin{gathered} \mathcal{M}_{\partial\Omega} : \mathbb{X} \to \mathbb{X} \ \text{ and } \ \mathcal{M}_{\partial\Omega} : \mathbb{X}' \to \mathbb{X}' \\ \text{are well defined and bounded,} \end{gathered} \tag{23.354}$$

where $\mathcal{M}_{\partial\Omega}$ is the Hardy-Littlewood maximal operator on $\partial\Omega$ (see (2.26) *with $\Sigma := \partial\Omega$). Select $\mathbb{Y}$ a Generalized Quasi-Banach Sequence Space on $\mathbb{Z}$ which is $\mathbb{X}$-compliant (in the sense of Definition 23.7). Finally, fix an aperture parameter $\kappa \in (0,\infty)$.*

Then there exists some $C \in (0,\infty)$ depending only on n, κ, the operator norms of $\mathcal{M}_{\partial\Omega}$ on $\mathbb{X}$ and $\mathbb{X}'$, the compliancy constant of $\|\cdot\|_{\mathbb{Y}}$, and the UR *constants of $\partial\Omega$ with the property that*

$$\begin{gathered} \|\mathcal{N}_\kappa(\mathcal{T}f)\|_{\dot{\mathbb{K}}(\partial\Omega;\mathbb{X};\mathbb{Y})} \le C\Big(\sum_{|\alpha|\le N}\sup_{S^{n-1}}|\partial^\alpha k|\Big)\|f\|_{\dot{\mathbb{K}}(\partial\Omega;\mathbb{X};\mathbb{Y})} \\ \text{for each function } f \in \dot{\mathbb{K}}(\partial\Omega;\mathbb{X};\mathbb{Y}), \end{gathered} \tag{23.355}$$

and

$$\begin{gathered} \|\mathcal{N}_\kappa(\mathcal{T}f)\|_{\dot{\mathbb{K}}(\partial\Omega;\mathbb{X}';\mathbb{Y}')} \le C\Big(\sum_{|\alpha|\le N}\sup_{S^{n-1}}|\partial^\alpha k|\Big)\|f\|_{\dot{\mathbb{K}}(\partial\Omega;\mathbb{X}';\mathbb{Y}')} \\ \text{for each function } f \in \dot{\mathbb{K}}(\partial\Omega;\mathbb{X}';\mathbb{Y}'). \end{gathered} \tag{23.356}$$

Furthermore, for each function $f \in \dot{\mathbb{K}}(\partial\Omega;\mathbb{X};\mathbb{Y})$ the following jump-formula holds

$$\Big(\mathcal{T}f\Big|_{\partial\Omega}^{\kappa-\text{n.t.}}\Big)(x) = \frac{1}{2i}\widehat{k}\big(\nu(x)\big)f(x) + (Tf)(x) \quad \text{at } \sigma\text{-a.e. } x \in \partial_*\Omega, \tag{23.357}$$

where $\widehat{k}$ denotes the Fourier transform (cf. (2.4)*) of k, and T is the principal value singular integral operator associated with k as in* (23.338)*.*

Under the stronger assumption that $\mathbb{Y}$ is a Generalized Banach Sequence Space on $\mathbb{Z}$, and that $\mathbb{Y}$ is $\mathbb{X}$-compliant, it follows that

$$\mathcal{N}_\kappa(\mathcal{T}f) \in \mathring{\mathbb{K}}(\partial\Omega;\mathbb{X};\mathbb{Y}) \quad \text{for each } f \in \mathring{\mathbb{K}}(\partial\Omega;\mathbb{X};\mathbb{Y}). \tag{23.358}$$

Lastly, the same type of conclusions, regarding nontangential maximal function estimates and jump-formulas hold in the context of inhomogeneous Composite Herz spaces $\mathbb{K}(\partial\Omega;\mathbb{X};\mathbb{Y})$.

Proof The assumptions made in (23.354) permit us to invoke [GHA-III, item *(4)* in Theorem 2.4.1, pp. 382–390] to ensure the existence of some $C \in (0,\infty)$, which depends only on n, the operator norms of $\mathcal{M}_{\partial\Omega}$ on $\mathbb{X}$ and $\mathbb{X}'$, and the UR constants of $\partial\Omega$, with the property that

$$\|\mathcal{N}_\kappa(\mathcal{T}f)\|_{\mathbb{X}} \leq C\Big(\sum_{|\alpha|\leq N} \sup_{S^{n-1}} |\partial^\alpha k|\Big)\,\|f\|_{\mathbb{X}} \quad \text{for each } f \in \mathbb{X}, \tag{23.359}$$

and

$$\|\mathcal{N}_\kappa(\mathcal{T}f)\|_{\mathbb{X}'} \leq C\Big(\sum_{|\alpha|\leq N} \sup_{S^{n-1}} |\partial^\alpha k|\Big)\,\|f\|_{\mathbb{X}'} \quad \text{for each } f \in \mathbb{X}'. \tag{23.360}$$

Also, the Cotlar-type inequality from [GHA-III, item *(4)* in Theorem 2.4.1, pp. 382–390] guarantees the existence of some constant $C = C(\partial\Omega,\kappa,n) \in (0,\infty)$ such that for every function $f \in L^1\big(\partial\Omega, \frac{\sigma(x)}{1+|x|^{n-1}}\big)$ and every $x \in \partial\Omega$ we have

$$\mathcal{N}_\kappa(\mathcal{T}f)(x) \leq T_{\max} f(x) + C\Big(\sum_{|\alpha|\leq N} \sup_{S^{n-1}} |\partial^\alpha k|\Big)\mathcal{M}_{\partial\Omega} f(x), \tag{23.361}$$

where $T_{\max}$ is the maximal operator associated with k as in (23.336)-(23.337). Select an arbitrary function $f \in L^1\big(\partial\Omega, \frac{\sigma(x)}{1+|x|^{n-1}}\big)$ and fix a point $x \in \partial\Omega \setminus \operatorname{supp} f$. Based on (23.361), together with the estimates recorded in (23.270) and (23.350), we have

$$\mathcal{N}_\kappa(\mathcal{T}f)(x) \leq C\Big(\sum_{|\alpha|\leq N} \sup_{S^{n-1}} |\partial^\alpha k|\Big) \int_{\partial\Omega} \frac{|f(y)|}{|x-y|^{n-1}}\, d\sigma(y). \tag{23.362}$$

As such, (23.362), (23.355), (23.356), and Theorem 23.3 imply the nontangential maximal function estimates in (23.355)-(23.356).

The jump-formula in (23.357) is a consequence of (6.11) and the embeddings obtained in Corollary 23.4.

Moving on, assume $\mathbb{Y}$ is a Generalized Banach Sequence Space on $\mathbb{Z}$ which is $\mathbb{X}$-compliant. In this case, Proposition 23.4 guarantees that $\dot{\mathbb{K}}(\partial\Omega;\mathbb{X};\mathbb{Y})$ is a Generalized Banach Function Space on $(\partial\Omega,\sigma)$. In addition, Proposition 23.5 and Theorem 23.1 show that

$$\begin{aligned} &\mathcal{M}_{\partial\Omega} : \dot{\mathbb{K}}(\partial\Omega;\mathbb{X};\mathbb{Y}) \to \dot{\mathbb{K}}(\partial\Omega;\mathbb{X};\mathbb{Y}) \\ \text{and } &\mathcal{M}_{\partial\Omega} : \big(\dot{\mathbb{K}}(\partial\Omega;\mathbb{X};\mathbb{Y})\big)' \to \big(\dot{\mathbb{K}}(\partial\Omega;\mathbb{X};\mathbb{Y})\big)' \\ &\text{are well-defined and bounded operators.} \end{aligned} \tag{23.363}$$

As such, the membership in (23.358) follows from [GHA-III, item *(4)* in Theorem 2.4.1, pp. 382–390]. This concludes the proof of Theorem 23.5. □

The next order of business is to obtain operator norm estimates which are sensitive to flatness for singular integral operators of chord-dot-normal type in the context of Composite Herz spaces. The reader is advised to recall the piece of notation defined in (6.113).

Theorem 23.6 *Assume $\Omega \subseteq \mathbb{R}^n$ is a* UR *domain with unbounded boundary. Abbreviate $\sigma := \mathcal{H}^{n-1}\lfloor\partial\Omega$, and denote by ν the geometric measure theoretic outward unit normal to Ω. For some sufficiently large integer $N = N(n) \in \mathbb{N}$, consider a kernel $k \in \mathscr{C}^N(\mathbb{R}^n \setminus \{0\})$ that is even and positive homogeneous of degree $-n$. Recall that the chord-dot-normal singular integral operators T and $T^{\#}$ associated with k act on each function $f \in L^1\big(\partial\Omega, \frac{\sigma(x)}{1+|x|^{n-1}}\big)$ according to*

$$Tf(x) := \lim_{\varepsilon\to 0^+} \int_{\substack{y\in\partial\Omega \\ |x-y|>\varepsilon}} \langle\nu(y), x-y\rangle k(x-y) f(y)\, d\sigma(y) \text{ for } \sigma\text{-a.e. } x \in \partial\Omega, \tag{23.364}$$

and

$$T^{\#} f(x) := \lim_{\varepsilon\to 0^+} \int_{\substack{y\in\partial\Omega \\ |x-y|>\varepsilon}} \langle\nu(x), y-x\rangle k(x-y) f(y)\, d\sigma(y) \text{ for } \sigma\text{-a.e. } x \in \partial\Omega. \tag{23.365}$$

Next, suppose $\mathbb{X}$ is a Generalized Banach Function Space on $(\partial\Omega, \sigma)$ such that

$$\begin{gathered} \mathcal{M}_{\partial\Omega} : \mathbb{X} \to \mathbb{X} \text{ and } \mathcal{M}_{\partial\Omega} : \mathbb{X}' \to \mathbb{X}' \\ \text{are well defined and bounded,} \end{gathered} \tag{23.366}$$

where $\mathcal{M}_{\partial\Omega}$ is the Hardy-Littlewood maximal operator on $\partial\Omega$ (defined as in (2.26) *with $\Sigma := \partial\Omega$). Select $\mathbb{Y}$ a Generalized Quasi-Banach Sequence Space on $\mathbb{Z}$ which is $\mathbb{X}$-compliant (in the sense of Definition 23.7).*

Then the following operators are well defined and bounded:

$$\begin{aligned} &T, T^{\#} : \dot{\mathbb{K}}(\partial\Omega; \mathbb{X}; \mathbb{Y}) \to \dot{\mathbb{K}}(\partial\Omega; \mathbb{X}; \mathbb{Y}), \\ &T, T^{\#} : \dot{\mathbb{K}}(\partial\Omega; \mathbb{X}'; \mathbb{Y}') \to \dot{\mathbb{K}}(\partial\Omega; \mathbb{X}'; \mathbb{Y}'). \end{aligned} \tag{23.367}$$

Moreover, there exists some $C \in (0, \infty)$ depending only on n, the operator norms of $\mathcal{M}_{\partial\Omega}$ on $\mathbb{X}$ and $\mathbb{X}'$, the compliancy constant of $\|\cdot\|_{\mathbb{Y}}$, the modulus of concavity of $\|\cdot\|_{\mathbb{Y}}$, and the UR *constants of $\partial\Omega$, with the property that*

$$
\begin{aligned}
\|T\|_{\dot{\mathbb{K}}(\partial\Omega;\mathbb{X};\mathbb{Y})\to\dot{\mathbb{K}}(\partial\Omega;\mathbb{X};\mathbb{Y})} &\leq C\Big(\sum_{|\alpha|\leq N}\sup_{S^{n-1}}|\partial^\alpha k|\Big),\\
\|T\|_{\dot{\mathbb{K}}(\partial\Omega;\mathbb{X}';\mathbb{Y}')\to\dot{\mathbb{K}}(\partial\Omega;\mathbb{X}';\mathbb{Y}')} &\leq C\Big(\sum_{|\alpha|\leq N}\sup_{S^{n-1}}|\partial^\alpha k|\Big),\\
\|T^\#\|_{\dot{\mathbb{K}}(\partial\Omega;\mathbb{X};\mathbb{Y})\to\dot{\mathbb{K}}(\partial\Omega;\mathbb{X};\mathbb{Y})} &\leq C\Big(\sum_{|\alpha|\leq N}\sup_{S^{n-1}}|\partial^\alpha k|\Big),\\
\|T^\#\|_{\dot{\mathbb{K}}(\partial\Omega;\mathbb{X}';\mathbb{Y}')\to\dot{\mathbb{K}}(\partial\Omega;\mathbb{X}';\mathbb{Y}')} &\leq C\Big(\sum_{|\alpha|\leq N}\sup_{S^{n-1}}|\partial^\alpha k|\Big).
\end{aligned}
\tag{23.368}
$$

Next, if one makes the stronger assumption that $\mathbb{Y}$ *is actually a Generalized Banach Sequence Space on* $\mathbb{Z}$ *which is* $\mathbb{X}$*-compliant, then in addition to the boundedness results in* (23.367)*, one also has*

$$
\begin{gathered}
T,\ T^\# : \mathring{\mathbb{K}}(\partial\Omega;\mathbb{X};\mathbb{Y}) \to \mathring{\mathbb{K}}(\partial\Omega;\mathbb{X};\mathbb{Y})\\
\textit{are well-defined and bounded operators.}
\end{gathered}
\tag{23.369}
$$

Furthermore, for each $m\in\mathbb{N}$*, there exists some* $C_m\in(0,\infty)$ *depending only on* n*,* m*, the operator norms of* $\mathcal{M}_{\partial\Omega}$ *on* $\mathbb{X}$ *and* $\mathbb{X}'$*, the compliancy constant of* $\|\cdot\|_{\mathbb{Y}}$*, the modulus of concavity of* $\|\cdot\|_{\mathbb{Y}}$*, and the Ahlfors regularity constants of* $\partial\Omega$*, with the property that*

$$
\begin{aligned}
\|T\|_{\dot{\mathbb{K}}(\partial\Omega;\mathbb{X};\mathbb{Y})\to\dot{\mathbb{K}}(\partial\Omega;\mathbb{X};\mathbb{Y})} &\leq C_m\Big(\sum_{|\alpha|\leq N}\sup_{S^{n-1}}|\partial^\alpha k|\Big)\|\nu\|^{\langle m\rangle}_{[\mathrm{BMO}(\partial\Omega,\sigma)]^n},\\
\|T\|_{\dot{\mathbb{K}}(\partial\Omega;\mathbb{X}';\mathbb{Y}')\to\dot{\mathbb{K}}(\partial\Omega;\mathbb{X}';\mathbb{Y}')} &\leq C_m\Big(\sum_{|\alpha|\leq N}\sup_{S^{n-1}}|\partial^\alpha k|\Big)\|\nu\|^{\langle m\rangle}_{[\mathrm{BMO}(\partial\Omega,\sigma)]^n},\\
\|T^\#\|_{\dot{\mathbb{K}}(\partial\Omega;\mathbb{X};\mathbb{Y})\to\dot{\mathbb{K}}(\partial\Omega;\mathbb{X};\mathbb{Y})} &\leq C_m\Big(\sum_{|\alpha|\leq N}\sup_{S^{n-1}}|\partial^\alpha k|\Big)\|\nu\|^{\langle m\rangle}_{[\mathrm{BMO}(\partial\Omega,\sigma)]^n},\\
\|T^\#\|_{\dot{\mathbb{K}}(\partial\Omega;\mathbb{X}';\mathbb{Y}')\to\dot{\mathbb{K}}(\partial\Omega;\mathbb{X}';\mathbb{Y}')} &\leq C_m\Big(\sum_{|\alpha|\leq N}\sup_{S^{n-1}}|\partial^\alpha k|\Big)\|\nu\|^{\langle m\rangle}_{[\mathrm{BMO}(\partial\Omega,\sigma)]^n},\\
\|T\|_{\mathring{\mathbb{K}}(\partial\Omega;\mathbb{X};\mathbb{Y})\to\mathring{\mathbb{K}}(\partial\Omega;\mathbb{X};\mathbb{Y})} &\leq C_m\Big(\sum_{|\alpha|\leq N}\sup_{S^{n-1}}|\partial^\alpha k|\Big)\|\nu\|^{\langle m\rangle}_{[\mathrm{BMO}(\partial\Omega,\sigma)]^n},\\
\|T^\#\|_{\mathring{\mathbb{K}}(\partial\Omega;\mathbb{X};\mathbb{Y})\to\mathring{\mathbb{K}}(\partial\Omega;\mathbb{X};\mathbb{Y})} &\leq C_m\Big(\sum_{|\alpha|\leq N}\sup_{S^{n-1}}|\partial^\alpha k|\Big)\|\nu\|^{\langle m\rangle}_{[\mathrm{BMO}(\partial\Omega,\sigma)]^n}.
\end{aligned}
\tag{23.370}
$$

Proof Since we are assuming that (23.366) holds, work done in [GHA-V, Theorem 4.1.3, pp. 184–187] ensures that there exists some constant $C\in(0,\infty)$ depending only on n, the operator norms of $\mathcal{M}_{\partial\Omega}$ on $\mathbb{X}$ and $\mathbb{X}'$, and the UR constants of $\partial\Omega$ with the property that

$$\max\Big\{\|T\|_{\mathbb{X}\to\mathbb{X}},\ \|T\|_{\mathbb{X}'\to\mathbb{X}'},\ \|T^{\#}\|_{\mathbb{X}\to\mathbb{X}},\ \|T^{\#}\|_{\mathbb{X}'\to\mathbb{X}'}\Big\} \leq C\Big(\sum_{|\alpha|\leq N}\sup_{S^{n-1}}|\partial^{\alpha}k|\Big). \tag{23.371}$$

The smoothness and homogeneity of the kernel k imply that for each truncation parameter $\varepsilon \in (0,\infty)$, every function $f \in L^1\big(\partial\Omega, \frac{\sigma(x)}{1+|x|^{n-1}}\big)$, and each $x \in \partial\Omega\setminus\operatorname{supp} f$ we have

$$\begin{aligned}
&\Big|\int_{\substack{y\in\partial\Omega\\|x-y|>\varepsilon}} \langle\nu(y), x-y\rangle k(x-y) f(y)\,d\sigma(y)\Big|\\
&\quad\leq \int_{\substack{y\in\partial\Omega\\|x-y|>\varepsilon}} |x-y||k(x-y)||f(y)|\,d\sigma(y)\\
&\quad= \int_{\substack{y\in\partial\Omega\\|x-y|>\varepsilon}} \Big|k\Big(\frac{x-y}{|x-y|}\Big)\Big|\frac{|f(y)|}{|x-y|^{n-1}}\,d\sigma(y)\\
&\quad\leq \Big(\sum_{|\alpha|\leq N}\sup_{S^{n-1}}|\partial^{\alpha}k|\Big)\int_{\partial\Omega}\frac{|f(y)|}{|x-y|^{n-1}}\,d\sigma(y).
\end{aligned} \tag{23.372}$$

After taking the limit $\varepsilon \to 0^+$, this yields

$$|Tf(x)| \leq \Big(\sum_{|\alpha|\leq N}\sup_{S^{n-1}}|\partial^{\alpha}k|\Big)\int_{\partial\Omega}\frac{|f(y)|}{|x-y|^{n-1}}\,d\sigma(y). \tag{23.373}$$

A similar reasoning shows that $T^{\#}$ also satisfies the same size condition. Granted these, the boundedness claimed in (23.367) and the operator norm estimates in (23.368) become consequences of Theorem 23.3.

Moving on, work under the assumption that $\mathbb{Y}$ is a Generalized Banach Sequence Space on $\mathbb{Z}$ which is $\mathbb{X}$-compliant. In this scenario, Proposition 23.4 yields that $\dot{\mathbb{K}}(\partial\Omega;\mathbb{X};\mathbb{Y})$ is a Generalized Banach Function Space on $(\partial\Omega,\sigma)$. Moreover, Proposition 23.5 and Theorem 23.1 guarantee the boundedness of the Hardy-Littlewood maximal operator both on $\dot{\mathbb{K}}(\partial\Omega;\mathbb{X};\mathbb{Y})$ and $\big(\dot{\mathbb{K}}(\partial\Omega;\mathbb{X};\mathbb{Y})\big)'$. As such, the remaining claims made in (23.369)-(23.370) follow from [GHA-V, Theorem 4.1.3, pp. 184–187]. □

23.9 A Wealth of Examples: Existing Herz-type Spaces and New Scales

The theory of Composite Herz spaces provides a recipe for creating a large variety of new brands of Herz-type spaces. In this section we elaborate on some of these spaces by discussing several specific examples.

Corollary 23.5 *Consider a closed unbounded Ahlfors regular set $\Sigma \subseteq \mathbb{R}^n$, and abbreviate $\sigma := \mathcal{H}^{n-1}\lfloor\Sigma$. Let $\mathbb{X}$ be a Generalized Banach Function Space on (Σ, σ), and $\mathbb{Y}$ a Generalized Quasi-Banach Sequence Space on $\mathbb{Z}$. Assume $\mathbb{X}$ and $\mathbb{Y}$ are as in any of the following cases:*

$$\begin{gathered} \mathbb{X} := L^p(\Sigma,\sigma),\ \mathbb{X}' = L^{p'}(\Sigma,\sigma) \\ \textit{where } p, p' \in (1,\infty) \textit{ satisfy } 1/p + 1/p' = 1, \\ \mathbb{Y} := \ell^q_\alpha(\mathbb{Z}) \textit{ with } q \in (0,\infty] \textit{ and } \tfrac{-(n-1)}{p} < \alpha < \tfrac{n-1}{p'} \end{gathered} \tag{23.374}$$

or

$$\begin{gathered} \mathbb{X} := L^{p,r}(\Sigma,\sigma),\ \mathbb{X}' = L^{p',r'}(\Sigma,\sigma) \textit{ where } p, p' \in (1,\infty) \\ \textit{and } r, r' \in [1,\infty] \textit{ satisfy } 1/p + 1/p' = 1 \textit{ and } 1/r + 1/r' = 1, \\ \mathbb{Y} := \ell^q_\alpha(\mathbb{Z}) \textit{ with } q \in (0,\infty] \textit{ and } \tfrac{-(n-1)}{p} < \alpha < \tfrac{n-1}{p'} \end{gathered} \tag{23.375}$$

or

$$\begin{gathered} \mathbb{X} := L^p(\Sigma,w),\ \mathbb{X}' = L^{p'}(\Sigma,w') \textit{ where } p, p' \in (1,\infty) \\ \textit{satisfy } 1/p + 1/p' = 1,\ w \in A_p(\Sigma,\sigma), \textit{ and } w' := w^{1-p} \in A_{p'}(\Sigma,\sigma), \\ \mathbb{Y} := \ell^q_\alpha(\mathbb{Z}) \textit{ with } q \in (0,\infty],\ \tfrac{-(n-1)\tau}{p} < \alpha < \tfrac{(n-1)\tau'}{p'} \\ \textit{and } \tau, \tau' \in (0,1) \textit{ are as in (23.222)-(23.223)} \end{gathered} \tag{23.376}$$

or

$$\begin{gathered} \mathbb{X} := L^{p(\cdot)}(\Sigma,\sigma),\ \mathbb{X}' = L^{p'(\cdot)}(\Sigma,\sigma) \textit{ where } p(\cdot) : \Sigma \to (1,\infty) \\ \textit{is a } \sigma\textit{-measurable function satisfying (23.254)-(23.255),} \\ \textit{and } p'(\cdot) : \Sigma \to (1,\infty) \textit{ is such that } 1/p(\cdot) + 1/p'(\cdot) = 1, \\ \mathbb{Y} := \ell^q_\alpha(\mathbb{Z}) \textit{ with } q \in (0,\infty],\ -(n-1)\tau_{p(\cdot)} < \alpha < (n-1)\tau_{p'(\cdot)} \\ \textit{where } \tau_{p(\cdot)}, \tau_{p'(\cdot)} \in (0,\infty) \textit{ are as in (23.256)-(23.257)} \end{gathered} \tag{23.377}$$

or

$$\begin{gathered} \mathbb{X} := M^{p,\lambda}(\Sigma,\sigma),\ \mathbb{X}' = \mathcal{B}^{p',\lambda}(\Sigma,\sigma) \textit{ where } \lambda \in (0,n-1), \\ \textit{and } p, p' \in (1,\infty) \textit{ satisfy } 1/p + 1/p' = 1, \\ \mathbb{Y} := \ell^q_\alpha(\mathbb{Z}) \textit{ with } q \in (0,\infty] \textit{ and } \tfrac{-(n-1-\lambda)}{p} < \alpha < \tfrac{n-1}{p'} \end{gathered} \tag{23.378}$$

or

$$\begin{gathered}\mathbb{X} := \mathcal{B}^{p,\lambda}(\Sigma,\sigma),\ \mathbb{X}' = M^{p',\lambda}(\Sigma,\sigma) \textit{ where } \lambda \in (0,n-1),\\ \textit{and } p,p' \in (1,\infty) \textit{ satisfy } 1/p + 1/p' = 1,\\ \mathbb{Y} := \ell^q_\alpha(\mathbb{Z}) \textit{ with } q \in (0,\infty] \textit{ and } \tfrac{-(n-1)}{p} < \alpha < \tfrac{n-1-\lambda}{p'}\end{gathered} \tag{23.379}$$

or

$$\begin{gathered}\mathbb{X} := M^{p,\lambda}(\Sigma,w),\ \mathbb{X}' = \mathcal{B}^{p',\lambda}(\Sigma,w') \textit{ where } p,p',q,q' \in (1,\infty)\\ \textit{satisfy } 1/p + 1/p' = 1,\ 1/q + 1/q' = 1, \textit{ and where}\\ w \in A_p(\Sigma,\sigma) \cap RH_q(\Sigma,\sigma), \textit{ and } w' := w^{1-p} \in A_{p'-\varepsilon}(\Sigma,\sigma),\\ \textit{with } \varepsilon := (p'-1)/q' \in (0,p'-1), \textit{ and } \lambda \in \big(0,(n-1)/q'\big),\\ \mathbb{Y} := \ell^{\widetilde{q}}_\alpha(\mathbb{Z}) \textit{ with } \widetilde{q} \in (0,\infty] \textit{ and } \tfrac{-(n-1-\lambda q')}{pq'} < \alpha < \tfrac{(n-1)\tau}{p'}\\ \textit{where the parameter } \tau \in (0,1) \textit{ is as in } (23.241)\end{gathered} \tag{23.380}$$

or

$$\begin{gathered}\mathbb{X} := \mathcal{B}^{p',\lambda}(\Sigma,w'),\ \mathbb{X}' = M^{p,\lambda}(\Sigma,w) \textit{ where } p,p',q,q' \in (1,\infty)\\ \textit{satisfy } 1/p + 1/p' = 1,\ 1/q + 1/q' = 1, \textit{ and where}\\ w \in A_p(\Sigma,\sigma) \cap RH_q(\Sigma,\sigma), \textit{ and } w' := w^{1-p} \in A_{p'-\varepsilon}(\Sigma,\sigma),\\ \textit{with } \varepsilon := (p'-1)/q' \in (0,p'-1), \textit{ and } \lambda \in \big(0,(n-1)/q'\big),\\ \mathbb{Y} := \ell^{\widetilde{q}}_\alpha(\mathbb{Z}) \textit{ with } \widetilde{q} \in (0,\infty] \textit{ and } \tfrac{-(n-1)\tau}{p'} < \alpha < \tfrac{n-1-\lambda q'}{pq'}\\ \textit{where the parameter } \tau \in (0,1) \textit{ is as in } (23.241)\end{gathered} \tag{23.381}$$

or

$$\begin{gathered}\mathbb{X} := L^\Phi(\Sigma,\sigma),\ \mathbb{X}' = L^{\widetilde{\Phi}}(\Sigma,\sigma) \textit{ where}\\ \Phi \textit{ is a Young function satisfying } 1 < i(\Phi) \le I(\Phi) < \infty,\\ \textit{and } \widetilde{\Phi} \textit{ is the complementary function of } \Phi\ (\textit{cf. } (23.262)),\\ \mathbb{Y} := \ell^q_\alpha(\mathbb{Z}) \textit{ with } q \in (0,\infty] \textit{ and } -(n-1)\tau_\Phi < \alpha < (n-1)\tau_{\widetilde{\Phi}}\\ \textit{where } \tau_\Phi, \tau_{\widetilde{\Phi}} \in (0,1) \textit{ are as in } (23.265)\text{-}(23.266).\end{gathered} \tag{23.382}$$

Moreover, suppose $T : L^1_{\mathrm{loc}}(\Sigma,\sigma) \to \mathcal{M}(\Sigma,\sigma)$ is a sublinear operator satisfying the size condition (23.312)*. Then whenever $T : \mathbb{X} \to \mathbb{X}$ is well defined and bounded one has*

$$\begin{gathered}T : \dot{\mathbb{K}}(\Sigma;\mathbb{X};\mathbb{Y}) \to \dot{\mathbb{K}}(\Sigma;\mathbb{X};\mathbb{Y})\\ \textit{is a well-defined and bounded operator.}\end{gathered} \tag{23.383}$$

Proof In either on the scenarios (23.374)-(23.382), the boundedness claimed in (23.383) is a consequence of Theorem 23.3 combined with Examples 23.6, 23.7, 23.8, 23.9, 23.10, 23.11, 23.12, along with (23.106) in Example 23.1, and Definition 23.7. □

We next make several comments concerning the nature of the resulting Herz space corresponding to each of the cases (23.374)-(23.379) in the Euclidean setting $\Sigma := \mathbb{R}^{n-1} \equiv \mathbb{R}^{n-1} \times \{0\}$, with the choices $x_0 := 0$, and $2_* := 2$ (to define the family of annuli is as in (23.16)).

Comment 1. The choices in (23.374) yield classical Herz spaces, the same scale studied in Chapter 17. Moreover, corresponding to $\Sigma = \mathbb{R}^{n-1}$, we obtain the classical homogeneous Herz space, namely

$$\dot{\mathrm{K}}\big(\mathbb{R}^{n-1}; L^p(\mathbb{R}^{n-1}); \ell^q_\alpha(\mathbb{Z})\big) = \dot{\mathrm{K}}^\alpha_{p,q}(\mathbb{R}^{n-1}), \tag{23.384}$$

which has been extensively studied in several papers (see for instance [LuYa], [LiYa96], [HeWeYa96], [HeYa99], [Li00], [LuYaHu08], and the references therein). In particular, the boundedness (23.383) recovers a result from [LiYa96, Corollary 2.1, p. 488].

Comment 2. The choices made in (23.374) lead to the Lorentz-based Herz space

$$\dot{\mathrm{K}}\big(\mathbb{R}^{n-1}; L^{p,r}(\mathbb{R}^{n-1}); \ell^q_\alpha(\mathbb{Z})\big) = \mathrm{L}\dot{\mathrm{K}}^{\alpha,r}_{p,q}(\mathbb{R}^{n-1}), \tag{23.385}$$

the same scale appearing in [WeYa22] and [BhKoKo23]. The boundedness result (23.383) in this scenario has been proved in [BhKoKo23, Theorem 3.1.2.2], albeit for a more restricted range of q's, namely $1 \leq q \leq \infty$, as opposed to our result that covers the entire range $q \in (0, \infty]$.

Comment 3. In the case (23.376), we obtain the simply-weighted Herz space:

$$\dot{\mathrm{K}}\big(\mathbb{R}^{n-1}; L^p(\mathbb{R}^{n-1}, w); \ell^q_\alpha(\mathbb{Z})\big) = \dot{\mathrm{K}}^\alpha_{p,q}(\mathbb{R}^{n-1}, w), \tag{23.386}$$

which is a particular case of the doubly-weighted Herz spaces introduced in [LuYa95]. Furthermore, our boundedness result (23.383) in this context may be regarded as an improved version of [LuYa95, Theorem 2.1]. More on doubly-weighted Herz spaces will be discussed in Corollary 23.6.

Comment 4. Herz spaces with variable exponents have been first introduced in [Iz09-I], and this is the resulting scale when the choices in (23.377) are made, i.e.,

$$\dot{\mathrm{K}}\big(\mathbb{R}^{n-1}; L^{p(\cdot)}(\mathbb{R}^{n-1}); \ell^q_\alpha(\mathbb{Z})\big) =: \dot{\mathrm{K}}^\alpha_{p(\cdot),q}(\mathbb{R}^{n-1}). \tag{23.387}$$

As such, the boundedness result recorded in (23.383) recovers a result from [Iz10-I, Theorem 3.3].

Comment 5. The scales induced by the choices in (23.378)-(23.379) are the Morrey-based and Block-based Herz spaces, respectively, i.e.,

$$\begin{aligned} &\dot{\mathrm{K}}\big(\mathbb{R}^{n-1}; M^{p,\lambda}(\mathbb{R}^{n-1}); \ell^q_\alpha(\mathbb{Z})\big) =: \widetilde{\mathrm{M}}\dot{\mathrm{K}}^{\alpha,\lambda}_{p,q}(\mathbb{R}^{n-1}), \\ &\dot{\mathrm{K}}\big(\mathbb{R}^{n-1}; \mathcal{B}^{p,\lambda}(\mathbb{R}^{n-1}); \ell^q_\alpha(\mathbb{Z})\big) =: \mathcal{B}\dot{\mathrm{K}}^{\alpha,\lambda}_{p,q}(\mathbb{R}^{n-1}). \end{aligned} \tag{23.388}$$

To the best of our knowledge, these are new scales of Herz-type spaces. The corresponding boundedness result (23.383) is also new in this setting.

Comment 6. The scenario in (23.380)-(23.381) yields a new brand of spaces, namely:

$$\begin{aligned} &\dot{\mathrm{K}}\big(\mathbb{R}^{n-1}; M^{p,\lambda}(\mathbb{R}^{n-1}, w); \ell^{\widetilde{q}}_{\alpha}(\mathbb{Z})\big) =: \mathrm{M}\dot{\mathrm{K}}^{\alpha,\lambda}_{p,\widetilde{q}}(\mathbb{R}^{n-1}, w), \\ &\dot{\mathrm{K}}\big(\mathbb{R}^{n-1}; \mathcal{B}^{p',\lambda}(\mathbb{R}^{n-1}, w'); \ell^{\widetilde{q}}_{\alpha}(\mathbb{Z})\big) =: \mathcal{B}\dot{\mathrm{K}}^{\alpha,\lambda}_{p',\widetilde{q}}(\mathbb{R}^{n-1}, w'). \end{aligned} \tag{23.389}$$

It is natural to refer to these as Muckenhoupt weighted Morrey-based Herz space and Muckenhoupt weighted Block-based Herz space, respectively. Hence, the boundedness (23.383) corresponding to these scales is new. In particular, the case when $w \equiv 1$ and $w' \equiv 1$ recovers (23.388).

Comment 7. The case (23.382) yields Orlicz-based Herz spaces, which is also a new Herz-type scale:

$$\dot{\mathrm{K}}\big(\mathbb{R}^{n-1}; L^{\Phi}(\mathbb{R}^{n-1}); \ell^{q}_{\alpha}(\mathbb{Z})\big) =: \dot{\mathrm{K}}^{\alpha}_{\Phi,q}(\mathbb{R}^{n-1}). \tag{23.390}$$

Hence the boundedness result (23.383) is new in this context.

Pressing on, we augment Corollary 23.6 with additional examples yielding a similar boundedness conclusion.

Corollary 23.6 *Consider a closed unbounded Ahlfors regular set $\Sigma \subseteq \mathbb{R}^n$, and abbreviate $\sigma := \mathcal{H}^{n-1}\lfloor\Sigma$. Let $\mathbb{X}$ be a Generalized Banach Function Space on (Σ, σ), and $\mathbb{Y}$ a Generalized Quasi-Banach Sequence Space on $\mathbb{Z}$. Make the assumption that $\mathbb{X}$ and $\mathbb{Y}$ are as in any of the following cases:*

$$\begin{gathered} \mathbb{X} := L^p(\Sigma, \sigma),\ \mathbb{X}' = L^{p'}(\Sigma, \sigma) \\ \textit{where } p, p' \in (1, \infty) \textit{ satisfy } 1/p + 1/p' = 1, \\ \mathbb{Y} := \ell^q_{\alpha}(\mathbb{Z}; \Sigma; w_1) \textit{ for some } w_1 \in A_{p_1}(\Sigma, \sigma) \textit{ and } p_1 \in (1, \infty), \\ \textit{along with } q \in (0, \infty] \textit{ and } \tfrac{-(n-1)}{p_1 p} < \alpha < \tfrac{n-1}{p_1 p'} \end{gathered} \tag{23.391}$$

or

$$\begin{gathered} \mathbb{X} := L^{p,r}(\Sigma, \sigma),\ \mathbb{X}' = L^{p',r'}(\Sigma, \sigma) \textit{ where } p, p' \in (1, \infty) \\ \textit{and } r, r' \in [1, \infty] \textit{ satisfy } 1/p + 1/p' = 1 \textit{ and } 1/r + 1/r' = 1, \\ \mathbb{Y} := \ell^q_{\alpha}(\mathbb{Z}; \Sigma; w_1) \textit{ for some } w_1 \in A_{p_1}(\Sigma, \sigma) \textit{ and } p_1 \in (1, \infty), \\ \textit{along with } q \in (0, \infty] \textit{ and } \tfrac{-(n-1)}{p_1 p} < \alpha < \tfrac{n-1}{p_1 p'} \end{gathered} \tag{23.392}$$

or

$$
\begin{gathered}
\mathbb{X} := L^p(\Sigma, w),\ \mathbb{X}' = L^{p'}(\Sigma, w') \textit{ where } p, p' \in (1, \infty) \\
\textit{satisfy } 1/p + 1/p' = 1,\ w \in A_p(\Sigma, \sigma), \textit{ and } w' := w^{1-p} \in A_{p'}(\Sigma, \sigma), \\
\mathbb{Y} := \ell^q_\alpha(\mathbb{Z}; \Sigma; w_1) \textit{ for some } w_1 \in A_{p_1}(\Sigma, \sigma) \textit{ and } p_1 \in (1, \infty), \\
\textit{along with } q \in (0, \infty] \textit{ and } \tfrac{-(n-1)\tau}{p_1 p} < \alpha < \tfrac{(n-1)\tau'}{p_1 p'} \\
\textit{and } \tau, \tau' \in (0,1) \textit{ are as in (23.222)-(23.223)}
\end{gathered}
\tag{23.393}
$$

or

$$
\begin{gathered}
\mathbb{X} := L^{p(\cdot)}(\Sigma, \sigma),\ \mathbb{X}' = L^{p'(\cdot)}(\Sigma, \sigma) \textit{ where } p(\cdot) : \Sigma \to (1, \infty) \\
\textit{is a } \sigma\textit{-measurable function satisfying (23.254)-(23.255)}, \\
\textit{and } p'(\cdot) : \Sigma \to (1, \infty) \textit{ is such that } 1/p(\cdot) + 1/p'(\cdot) = 1, \\
\mathbb{Y} := \ell^q_\alpha(\mathbb{Z}; \Sigma; w_1) \textit{ for some } w_1 \in A_{p_1}(\Sigma, \sigma) \textit{ and } p_1 \in (1, \infty), \\
\textit{along with } q \in (0, \infty], \textit{ and } \tfrac{-(n-1)\tau_{p(\cdot)}}{p_1} < \alpha < \tfrac{(n-1)\tau_{p'(\cdot)}}{p_1} \\
\textit{where } \tau_{p(\cdot)}, \tau_{p'(\cdot)} \in (0, \infty) \textit{ are as in (23.256)-(23.257)}
\end{gathered}
\tag{23.394}
$$

or

$$
\begin{gathered}
\mathbb{X} := M^{p,\lambda}(\Sigma, \sigma),\ \mathbb{X}' = \mathcal{B}^{p',\lambda}(\Sigma, \sigma) \textit{ where } \lambda \in (0, n-1), \\
\textit{and } p, p' \in (1, \infty) \textit{ satisfy } 1/p + 1/p' = 1, \\
\mathbb{Y} := \ell^q_\alpha(\mathbb{Z}; \Sigma; w_1) \textit{ for some } w_1 \in A_{p_1}(\Sigma, \sigma) \textit{ and } p_1 \in (1, \infty), \\
\textit{with } q \in (0, \infty] \textit{ and } \tfrac{-(n-1-\lambda)}{p_1 p} < \alpha < \tfrac{n-1}{p_1 p'}
\end{gathered}
\tag{23.395}
$$

or

$$
\begin{gathered}
\mathbb{X} := \mathcal{B}^{p,\lambda}(\Sigma, \sigma),\ \mathbb{X}' = M^{p',\lambda}(\Sigma, \sigma) \textit{ where } \lambda \in (0, n-1), \\
\textit{and } p, p' \in (1, \infty) \textit{ satisfy } 1/p + 1/p' = 1, \\
\mathbb{Y} := \ell^q_\alpha(\mathbb{Z}; \Sigma; w_1) \textit{ for some } w_1 \in A_{p_1}(\Sigma, \sigma) \textit{ and } p_1 \in (1, \infty), \\
\textit{with } q \in (0, \infty] \textit{ and } \tfrac{-(n-1)}{p_1 p} < \alpha < \tfrac{n-1-\lambda}{p_1 p'}
\end{gathered}
\tag{23.396}
$$

or

$$
\begin{gathered}
\mathbb{X} := M^{p,\lambda}(\Sigma, w),\ \mathbb{X}' = \mathcal{B}^{p',\lambda}(\Sigma, w') \textit{ where } p, p', q, q' \in (1, \infty) \\
\textit{satisfy } 1/p + 1/p' = 1,\ 1/q + 1/q' = 1, \textit{ and where} \\
w \in A_p(\Sigma, \sigma) \cap RH_q(\Sigma, \sigma), \textit{ and } w' := w^{1-p} \in A_{p'-\varepsilon}(\Sigma, \sigma), \\
\textit{with } \varepsilon := (p'-1)/q' \in (0, p'-1), \textit{ and } \lambda \in \big(0, (n-1)/q'\big), \\
\mathbb{Y} := \ell^{\widetilde{q}}_\alpha(\mathbb{Z}; \Sigma; w_1) \textit{ for some } w_1 \in A_{p_1}(\Sigma, \sigma) \textit{ and } p_1 \in (1, \infty), \\
\textit{with } \widetilde{q} \in (0, \infty] \textit{ and } \tfrac{-(n-1-\lambda q')}{p_1 p q'} < \alpha < \tfrac{(n-1)\tau}{p_1 p'} \\
\textit{where the parameter } \tau \in (0,1) \textit{ is as in (23.241)}
\end{gathered}
\tag{23.397}
$$

or

$$
\begin{aligned}
&\mathbb{X} := \mathcal{B}^{p',\lambda}(\Sigma, w'),\ \mathbb{X}' = M^{p,\lambda}(\Sigma, w) \textit{ where } p, p', q, q' \in (1,\infty) \\
&\quad \textit{satisfy } 1/p + 1/p' = 1,\ 1/q + 1/q' = 1, \textit{ and where} \\
&w \in A_p(\Sigma,\sigma) \cap RH_q(\Sigma,\sigma), \textit{ and } w' := w^{1-p} \in A_{p'-\varepsilon}(\Sigma,\sigma), \\
&\textit{with } \varepsilon := (p'-1)/q' \in (0, p'-1), \textit{ and } \lambda \in \big(0, (n-1)/q'\big), \\
&\mathbb{Y} := \ell^{\widetilde{q}}_{\alpha}(\mathbb{Z};\Sigma;w_1) \textit{ for some } w_1 \in A_{p_1}(\Sigma,\sigma) \textit{ and } p_1 \in (1,\infty), \\
&\quad \textit{with } \widetilde{q} \in (0,\infty] \textit{ and } \tfrac{-(n-1)\tau}{p_1 p'} < \alpha < \tfrac{(n-1-\lambda q')}{p_1 p q'} \\
&\quad \textit{where the parameter } \tau \in (0,1) \textit{ is as in } (23.241)
\end{aligned}
\tag{23.398}
$$

or

$$
\begin{aligned}
&\mathbb{X} := L^{\Phi}(\Sigma,\sigma),\ \mathbb{X}' = L^{\widetilde{\Phi}}(\Sigma,\sigma) \textit{ where} \\
&\Phi \textit{ is a Young function satisfying } 1 < i(\Phi) \le I(\Phi) < \infty, \\
&\textit{and } \widetilde{\Phi} \textit{ is the complementary function of } \Phi \textit{ (cf. (23.262))}, \\
&\mathbb{Y} := \ell^{q}_{\alpha}(\mathbb{Z};\Sigma;w_1) \textit{ for some } w_1 \in A_{p_1}(\Sigma,\sigma) \textit{ and } p_1 \in (1,\infty), \\
&\quad \textit{with } q \in (0,\infty] \textit{ and } \tfrac{-(n-1)\tau_{\Phi}}{p_1} < \alpha < \tfrac{(n-1)\tau_{\widetilde{\Phi}}}{p_1} \\
&\quad \textit{where } \tau_{\Phi}, \tau_{\widetilde{\Phi}} \in (0,1) \textit{ are as in } (23.265)\text{-}(23.266).
\end{aligned}
\tag{23.399}
$$

Finally, suppose $T : L^1_{\mathrm{loc}}(\Sigma,\sigma) \to \mathcal{M}(\Sigma,\sigma)$ *is a sublinear operator satisfying the size condition in* (23.312). *Then whenever* $T : \mathbb{X} \to \mathbb{X}$ *is well defined and bounded one has*

$$
\begin{aligned}
&T : \dot{\mathbb{K}}(\Sigma;\mathbb{X};\mathbb{Y}) \to \dot{\mathbb{K}}(\Sigma;\mathbb{X};\mathbb{Y}) \\
&\textit{is a well-defined and bounded operator.}
\end{aligned}
\tag{23.400}
$$

Proof The boundedness claimed in (23.400) for any of the choices in (23.391)-(23.399) is a direct consequence of Theorem 23.3, Examples 23.6, 23.7, 23.8, 23.9, 23.10, 23.11, 23.12, (23.125) in Example 23.2, and Definition 23.7. □

Some remarks in relation to the Herz-type scales (in the Euclidean setting, as in the past) contained in Corollary 23.6 are in order:

`Comment 1.` The choices made in (23.393) yield the doubly-weighted Herz space:

$$
\dot{\mathbb{K}}\big(\mathbb{R}^{n-1}; L^p(\mathbb{R}^{n-1}, w); \ell^q_{\alpha}(\mathbb{Z};\mathbb{R}^{n-1};w_1)\big) =: \dot{\mathbb{K}}^{\alpha}_{p,q}(\mathbb{R}^{n-1}; w, w_1). \tag{23.401}
$$

This is a well-known scale of spaces, first introduced in [LuYa95] (see also [LuYa95-01], [LuYa97], [FaYa97], [LuYaYa00], [LuYaHu08], [KoMa09], and the references therein). The boundedness in (23.400) on this scale of spaces is an improvement of [LuYaYa00, Theorem 1] with respect to the range for α and the Muckenhoupt class under consideration. Moreover, the resulting Herz-type scale arising from the choices made in (23.391) is a particular case of (23.393) when $w \equiv 1$.

Comment 2. The resulting Herz-type spaces obtained from the choices in (23.392), (23.394), (23.395), (23.396), (23.397), (23.398), (23.399) are, to the best of our knowledge, new scales. Concretely, we have

(1) Simply-Weighted Lorentz-based Herz spaces:

$$\dot{\mathrm{K}}\big(\mathbb{R}^{n-1}; L^{p,r}(\mathbb{R}^{n-1}); \ell^q_\alpha(\mathbb{Z}; \mathbb{R}^{n-1}; w_1)\big) =: \mathrm{L}\dot{\mathrm{K}}^{\alpha,r}_{p,q}(\mathbb{R}^{n-1}; w_1). \tag{23.402}$$

(2) Variable Exponent Simply-Weighted Herz spaces:

$$\dot{\mathrm{K}}\big(\mathbb{R}^{n-1}; L^{p(\cdot)}(\mathbb{R}^{n-1}); \ell^q_\alpha(\mathbb{Z}; \mathbb{R}^{n-1}; w_1)\big) =: \dot{\mathrm{K}}^{\alpha}_{p(\cdot),q}(\mathbb{R}^{n-1}; w_1). \tag{23.403}$$

(3) Simply-Weighted Morrey-based Herz spaces:

$$\dot{\mathrm{K}}\big(\mathbb{R}^{n-1}; M^{p,\lambda}(\mathbb{R}^{n-1}); \ell^q_\alpha(\mathbb{Z}; \mathbb{R}^{n-1}; w_1)\big) =: \mathrm{M}\dot{\mathrm{K}}^{\alpha,\lambda}_{p,q}(\mathbb{R}^{n-1}; w_1). \tag{23.404}$$

(4) Simply-Weighted Block-based Herz spaces:

$$\dot{\mathrm{K}}\big(\mathbb{R}^{n-1}; \mathcal{B}^{p,\lambda}(\mathbb{R}^{n-1}); \ell^q_\alpha(\mathbb{Z}; \mathbb{R}^{n-1}; w_1)\big) =: \mathcal{B}\dot{\mathrm{K}}^{\alpha,\lambda}_{p,q}(\mathbb{R}^{n-1}; w_1). \tag{23.405}$$

(5) Doubly-Weighted Morrey-based Herz spaces:

$$\dot{\mathrm{K}}\big(\mathbb{R}^{n-1}; M^{p,\lambda}(\mathbb{R}^{n-1}, w); \ell^{\widetilde{q}}_\alpha(\mathbb{Z}; \mathbb{R}^{n-1}; w_1)\big) =: \mathrm{M}\dot{\mathrm{K}}^{\alpha,\lambda}_{p,\widetilde{q}}(\mathbb{R}^{n-1}; w; w_1). \tag{23.406}$$

(6) Doubly-Weighted Block-based Herz spaces:

$$\dot{\mathrm{K}}\big(\mathbb{R}^{n-1}; \mathcal{B}^{p',\lambda}(\mathbb{R}^{n-1}, w'); \ell^{\widetilde{q}}_\alpha(\mathbb{Z}; \mathbb{R}^{n-1}; w_1)\big) =: \mathcal{B}\dot{\mathrm{K}}^{\alpha,\lambda}_{p',\widetilde{q}}(\mathbb{R}^{n-1}; w'; w_1). \tag{23.407}$$

(7) Simply-Weighted Orlicz-based Herz spaces:

$$\dot{\mathrm{K}}\big(\mathbb{R}^{n-1}; L^{\Phi}(\mathbb{R}^{n-1}); \ell^q_\alpha(\mathbb{Z}; \mathbb{R}^{n-1}; w_1)\big) =: \dot{\mathrm{K}}^{\alpha}_{\Phi,q}(\mathbb{R}^{n-1}; w_1). \tag{23.408}$$

Consequently, the boundedness claimed in (23.400) corresponding to each of these spaces is new.

We continue to enhance the body of results recorded so far in Corollary 23.5–23.6 as indicated below.

Corollary 23.7 *Consider a closed unbounded Ahlfors regular set $\Sigma \subseteq \mathbb{R}^n$, and abbreviate $\sigma := \mathcal{H}^{n-1}\lfloor\Sigma$. Let $\mathbb{X}$ be a Generalized Banach Function Space on (Σ, σ), and $\mathbb{Y}$ a Generalized Quasi-Banach Sequence Space on $\mathbb{Z}$. Suppose $\mathbb{X}$ and $\mathbb{Y}$ are as in any of the following cases:*

$$\begin{gathered}\mathbb{X} := L^p(\Sigma,\sigma),\ \mathbb{X}' = L^{p'}(\Sigma,\sigma) \textit{ where}\\ p,p' \in (1,\infty) \textit{ satisfy } 1/p + 1/p' = 1,\\ \mathbb{Y} := \ell^{q),\theta}_{\alpha}(\mathbb{Z}) \textit{ with } q \in (1,\infty),\ \theta \in (0,\infty), \textit{ and } \tfrac{-(n-1)}{p} < \alpha < \tfrac{n-1}{p'}\end{gathered} \tag{23.409}$$

or

$$\begin{gathered}\mathbb{X} := L^{p,r}(\Sigma,\sigma),\ \mathbb{X}' = L^{p',r'}(\Sigma,\sigma) \textit{ where } p,p' \in (1,\infty)\\ \textit{and } r,r' \in [1,\infty] \textit{ satisfy } 1/p + 1/p' = 1 \textit{ and } 1/r + 1/r' = 1,\\ \mathbb{Y} := \ell^{q),\theta}_{\alpha}(\mathbb{Z}) \textit{ with } q \in (1,\infty),\ \theta \in (0,\infty), \textit{ and } \tfrac{-(n-1)}{p} < \alpha < \tfrac{n-1}{p'}\end{gathered} \tag{23.410}$$

or

$$\begin{gathered}\mathbb{X} := L^p(\Sigma,w),\ \mathbb{X}' = L^{p'}(\Sigma,w') \textit{ where } p,p' \in (1,\infty)\\ \textit{satisfy } 1/p + 1/p' = 1,\ w \in A_p(\Sigma,\sigma), \textit{ and } w' := w^{1-p} \in A_{p'}(\Sigma,\sigma),\\ \mathbb{Y} := \ell^{q),\theta}_{\alpha}(\mathbb{Z}) \textit{ with } q \in (1,\infty),\ \theta \in (0,\infty), \textit{ and } \tfrac{-(n-1)\tau}{p} < \alpha < \tfrac{(n-1)\tau'}{p'}\\ \textit{and } \tau,\tau' \in (0,1) \textit{ are as in (23.222)-(23.223)}\end{gathered} \tag{23.411}$$

or

$$\begin{gathered}\mathbb{X} := L^{p(\cdot)}(\Sigma,\sigma),\ \mathbb{X}' = L^{p'(\cdot)}(\Sigma,\sigma) \textit{ where } p(\cdot) : \Sigma \to (1,\infty)\\ \textit{is a } \sigma\textit{-measurable function satisfying (23.254)-(23.255)},\\ \textit{and } p'(\cdot) : \Sigma \to (1,\infty) \textit{ is such that } 1/p(\cdot) + 1/p'(\cdot) = 1,\\ \mathbb{Y} := \ell^{q),\theta}_{\alpha}(\mathbb{Z}) \textit{ with } q \in (1,\infty),\ \theta \in (0,\infty), \textit{ and } -(n-1)\tau_{p(\cdot)} < \alpha < (n-1)\tau_{p'(\cdot)}\\ \textit{where } \tau_{p(\cdot)},\tau_{p'(\cdot)} \in (0,\infty) \textit{ are as in (23.256)-(23.257)}\end{gathered} \tag{23.412}$$

or

$$\begin{gathered}\mathbb{X} := M^{p,\lambda}(\Sigma,\sigma),\ \mathbb{X}' = \mathcal{B}^{p',\lambda}(\Sigma,\sigma) \textit{ where } \lambda \in (0,n-1),\\ \textit{and } p,p' \in (1,\infty) \textit{ satisfy } 1/p + 1/p' = 1,\\ \mathbb{Y} := \ell^{q),\theta}_{\alpha}(\mathbb{Z}) \textit{ with } q \in (1,\infty),\ \theta \in (0,\infty), \textit{ and } \tfrac{-(n-1-\lambda)}{p} < \alpha < \tfrac{n-1}{p'}\end{gathered} \tag{23.413}$$

or

$$\begin{gathered}\mathbb{X} := \mathcal{B}^{p,\lambda}(\Sigma,\sigma),\ \mathbb{X}' = M^{p',\lambda}(\Sigma,\sigma) \textit{ where } \lambda \in (0,n-1),\\ \textit{and } p,p' \in (1,\infty) \textit{ satisfy } 1/p + 1/p' = 1,\\ \mathbb{Y} := \ell^{q),\theta}_{\alpha}(\mathbb{Z}) \textit{with } q \in (1,\infty),\ \theta \in (0,\infty), \textit{ and } \tfrac{-(n-1)}{p} < \alpha < \tfrac{n-1-\lambda}{p'}\end{gathered} \tag{23.414}$$

or

$$
\begin{gathered}
\mathbb{X} := M^{p,\lambda}(\Sigma, w),\ \mathbb{X}' = \mathcal{B}^{p',\lambda}(\Sigma, w') \textit{ where } p, p', q, q' \in (1,\infty) \\
\textit{satisfy } 1/p + 1/p' = 1,\ 1/q + 1/q' = 1, \textit{ and where} \\
w \in A_p(\Sigma,\sigma) \cap RH_q(\Sigma,\sigma), \textit{ and } w' := w^{1-p} \in A_{p'-\varepsilon}(\Sigma,\sigma), \\
\textit{with } \varepsilon := (p'-1)/q' \in (0, p'-1), \textit{ and } \lambda \in \big(0, (n-1)/q'\big), \\
\mathbb{Y} := \ell_\alpha^{\widetilde{q}),\theta}(\mathbb{Z}) \textit{ with } \widetilde{q} \in (1,\infty),\ \theta \in (0,\infty), \textit{ and } \tfrac{-(n-1-\lambda q')}{pq'} < \alpha < \tfrac{(n-1)\tau}{p'} \\
\textit{where the parameter } \tau \in (0,1) \textit{ is as in } (23.241)
\end{gathered}
\tag{23.415}
$$

or

$$
\begin{gathered}
\mathbb{X} := \mathcal{B}^{p',\lambda}(\Sigma, w'),\ \mathbb{X}' = M^{p,\lambda}(\Sigma, w) \textit{ where } p, p', q, q' \in (1,\infty) \\
\textit{satisfy } 1/p + 1/p' = 1,\ 1/q + 1/q' = 1, \textit{ and where} \\
w \in A_p(\Sigma,\sigma) \cap RH_q(\Sigma,\sigma), \textit{ and } w' := w^{1-p} \in A_{p'-\varepsilon}(\Sigma,\sigma), \\
\textit{with } \varepsilon := (p'-1)/q' \in (0, p'-1), \textit{ and } \lambda \in \big(0, (n-1)/q'\big), \\
\mathbb{Y} := \ell_\alpha^{\widetilde{q}),\theta}(\mathbb{Z}) \textit{ with } \widetilde{q} \in (1,\infty),\ \theta \in (0,\infty), \textit{ and } \tfrac{-(n-1)\tau}{p'} < \alpha < \tfrac{n-1-\lambda q'}{pq'} \\
\textit{where the parameter } \tau \in (0,1) \textit{ is as in } (23.241)
\end{gathered}
\tag{23.416}
$$

or

$$
\begin{gathered}
\mathbb{X} := L^{\Phi}(\Sigma,\sigma),\ \mathbb{X}' = L^{\widetilde{\Phi}}(\Sigma,\sigma) \ \textit{ where} \\
\Phi \textit{ is a Young function satisfying } 1 < i(\Phi) \leq I(\Phi) < \infty, \\
\textit{and } \widetilde{\Phi} \textit{ is the complementary function of } \Phi \textit{ (as in } (23.262)), \\
\mathbb{Y} := \ell_\alpha^{q),\theta}(\mathbb{Z}) \textit{ with } q \in (1,\infty),\ \theta \in (0,\infty), \textit{ and } -(n-1)\tau_\Phi < \alpha < (n-1)\tau_{\widetilde{\Phi}} \\
\textit{where } \tau_\Phi, \tau_{\widetilde{\Phi}} \in (0,1) \textit{ are as in } (23.265)\text{-}(23.266).
\end{gathered}
\tag{23.417}
$$

Finally, suppose $T : L^1_{\rm loc}(\Sigma,\sigma) \to \mathcal{M}(\Sigma,\sigma)$ is a sublinear operator satisfying the size condition (23.312). *Then whenever $T : \mathbb{X} \to \mathbb{X}$ is well defined and bounded one has*

$$
\begin{gathered}
T : \dot{\mathrm{K}}(\Sigma; \mathbb{X}; \mathbb{Y}) \to \dot{\mathrm{K}}(\Sigma; \mathbb{X}; \mathbb{Y}) \\
\textit{is a well-defined and bounded operator.}
\end{gathered}
\tag{23.418}
$$

Proof In any of the scenarios (23.409)-(23.417), the boundedness result claimed in (23.418) is a direct consequence of Theorem 23.3, Examples 23.6, 23.7, 23.8, 23.9, 23.10, 23.11, 23.12, Example 23.3, and Definition 23.7. □

<u>`Comment 1.`</u> Grand Variable Herz spaces have first appeared in [NaRaZa20] (see also [NaRaZa21], and the references therein). This scale arises from the choices made in (23.412), namely

$$
\dot{\mathrm{K}}\big(\mathbb{R}^{n-1}; L^{p(\cdot)}(\mathbb{R}^{n-1}); \ell_\alpha^{q),\theta}(\mathbb{Z})\big) = \dot{\mathrm{K}}^{\alpha,\theta}_{p(\cdot),q)}(\mathbb{R}^{n-1}). \tag{23.419}
$$

In particular, a similar boundedness result as in (23.418) was obtained in [NaRaZa20]. If the exponent function $p(\cdot) \equiv p \in (1, \infty)$ is constant, this also recovers the case considered in (23.409).

`Comment 2.` In relation to the other choices made in (23.410), (23.411), (23.413), (23.414), (23.415), (23.416), and (23.417), the resulting Herz-type scales are, to the best of our knowledge, new. Namely, the following function spaces are first introduced in this monograph:

(1) Lorentz-based Grand Herz spaces:

$$\dot{\mathrm{K}}\big(\mathbb{R}^{n-1}; L^{p,r}(\mathbb{R}^{n-1}); \ell^{q),\theta}_{\alpha}(\mathbb{Z})\big) =: \mathrm{L}\dot{\mathrm{K}}^{\alpha,\theta,r}_{p,q)}(\mathbb{R}^{n-1}). \tag{23.420}$$

(2) Simply-weighted Grand Herz spaces:

$$\dot{\mathrm{K}}\big(\mathbb{R}^{n-1}; L^{p}(\mathbb{R}^{n-1}, w); \ell^{q),\theta}_{\alpha}(\mathbb{Z})\big) =: \dot{\mathrm{K}}^{\alpha,\theta}_{p,q)}(\mathbb{R}^{n-1}, w). \tag{23.421}$$

(3) Morrey-based Grand Herz spaces:

$$\dot{\mathrm{K}}\big(\mathbb{R}^{n-1}; M^{p,\lambda}(\mathbb{R}^{n-1}); \ell^{q),\theta}_{\alpha}(\mathbb{Z})\big) =: \mathrm{M}\dot{\mathrm{K}}^{\alpha,\lambda,\theta}_{p,q)}(\mathbb{R}^{n-1}). \tag{23.422}$$

(4) Block-based Grand Herz spaces:

$$\dot{\mathrm{K}}\big(\mathbb{R}^{n-1}; \mathcal{B}^{p,\lambda}(\mathbb{R}^{n-1}); \ell^{q),\theta}_{\alpha}(\mathbb{Z})\big) =: \mathcal{B}\dot{\mathrm{K}}^{\alpha,\lambda,\theta}_{p,q)}(\mathbb{R}^{n-1}). \tag{23.423}$$

(5) Muckenhoupt weighted Morrey-based Grand Herz spaces:

$$\dot{\mathrm{K}}\big(\mathbb{R}^{n-1}; M^{p,\lambda}(\mathbb{R}^{n-1}, w); \ell^{\widetilde{q}),\theta}_{\alpha}(\mathbb{Z})\big) =: \mathrm{M}\dot{\mathrm{K}}^{\alpha,\lambda,\theta}_{p,\widetilde{q})}(\mathbb{R}^{n-1}, w). \tag{23.424}$$

(6) Muckenhoupt weighted Block-based Grand Herz spaces:

$$\dot{\mathrm{K}}\big(\mathbb{R}^{n-1}; \mathcal{B}^{p',\lambda}(\mathbb{R}^{n-1}, w'); \ell^{\widetilde{q}),\theta}_{\alpha}(\mathbb{Z})\big) =: \mathcal{B}\dot{\mathrm{K}}^{\alpha,\lambda,\theta}_{p',\widetilde{q})}(\mathbb{R}^{n-1}, w'). \tag{23.425}$$

(7) Orlicz-based Grand Herz spaces:

$$\dot{\mathrm{K}}\big(\mathbb{R}^{n-1}; L^{\Phi}(\mathbb{R}^{n-1}); \ell^{q),\theta}_{\alpha}(\mathbb{Z})\big) =: \dot{\mathrm{K}}^{\alpha,\theta}_{\Phi,q)}(\mathbb{R}^{n-1}). \tag{23.426}$$

In turn, the version of (23.418) corresponding to each of these scales is also new.

In the next corollary we fix $\mathbb{Y}$ to be the sequence space $\ell^{q}_{\omega(\cdot)}(\mathbb{Z})$ introduced in Example 23.4.

Corollary 23.8 *Consider a closed unbounded Ahlfors regular set $\Sigma \subseteq \mathbb{R}^n$, and abbreviate $\sigma := \mathcal{H}^{n-1}\lfloor\Sigma$. Select a function ω belonging to the function class* $\mathrm{M}(\mathbb{R}_+)$ *(cf.* (23.150)-(23.151)*), and recall its Matuszewska-Orlicz indices $m_0(\omega)$, $M_0(\omega)$, $m_\infty(\omega)$, $M_\infty(\omega)$ (as in* (23.152)*). Let $\mathbb{X}$ be a Generalized Banach Function Space on (Σ,σ), and $\mathbb{Y}$ a Generalized Quasi-Banach Sequence Space on $\mathbb{Z}$. Assume $\mathbb{X}$ and $\mathbb{Y}$ are as in any of the following cases:*

$$\begin{gathered}\mathbb{X} := L^p(\Sigma,\sigma),\ \ \mathbb{X}' = L^{p'}(\Sigma,\sigma)\\ \textit{where } p, p' \in (1,\infty) \textit{ satisfy } 1/p + 1/p' = 1,\\ \mathbb{Y} := \ell^q_{\omega(\cdot)}(\mathbb{Z}) \textit{ with } q \in (0,\infty], \textit{ and Matuszewska-Orlicz indices}\\ \textit{satisfying } \tfrac{-(n-1)}{p} < m_0(\omega) \le M_0(\omega) < \tfrac{-(n-1)}{p'}\\ \textit{and } \tfrac{-(n-1)}{p} < m_\infty(\omega) \le M_\infty(\omega) < \tfrac{-(n-1)}{p'}\end{gathered} \tag{23.427}$$

or

$$\begin{gathered}\mathbb{X} := L^{p,r}(\Sigma,\sigma),\ \ \mathbb{X}' = L^{p',r'}(\Sigma,\sigma) \textit{ where } p, p' \in (1,\infty)\\ \textit{and } r, r' \in [1,\infty] \textit{ satisfy } 1/p + 1/p' = 1 \textit{ and } 1/r + 1/r' = 1,\\ \mathbb{Y} := \ell^q_{\omega(\cdot)}(\mathbb{Z}) \textit{ with } q \in (0,\infty], \textit{ and Matuszewska-Orlicz indices}\\ \textit{satisfying } \tfrac{-(n-1)}{p} < m_0(\omega) \le M_0(\omega) < \tfrac{-(n-1)}{p'}\\ \textit{and } \tfrac{-(n-1)}{p} < m_\infty(\omega) \le M_\infty(\omega) < \tfrac{-(n-1)}{p'}\end{gathered} \tag{23.428}$$

or

$$\begin{gathered}\mathbb{X} := L^p(\Sigma,w),\ \ \mathbb{X}' = L^{p'}(\Sigma,w') \textit{ where } p, p' \in (1,\infty)\\ \textit{satisfy } 1/p + 1/p' = 1,\ w \in A_p(\Sigma,\sigma), \textit{ and } w' := w^{1-p} \in A_{p'}(\Sigma,\sigma),\\ \mathbb{Y} := \ell^q_{\omega(\cdot)}(\mathbb{Z}) \textit{ with } q \in (1,\infty), \textit{ and Matuszewska-Orlicz indices}\\ \textit{satisfying } \tfrac{-(n-1)\tau}{p} < m_0(\omega) \le M_0(\omega) < \tfrac{-(n-1)\tau'}{p'}\\ \textit{and } \tfrac{-(n-1)\tau}{p} < m_\infty(\omega) \le M_\infty(\omega) < \tfrac{-(n-1)\tau'}{p'}\\ \textit{where } \tau, \tau' \in (0,1) \textit{ are as in } (23.222)\text{-}(23.223)\end{gathered} \tag{23.429}$$

or

$$\begin{gathered}\mathbb{X} := L^{p(\cdot)}(\Sigma,\sigma),\ \ \mathbb{X}' = L^{p'(\cdot)}(\Sigma,\sigma) \textit{ where } p(\cdot) : \Sigma \to (1,\infty)\\ \textit{is a } \sigma\textit{-measurable function satisfying } (23.254)\text{-}(23.255),\\ \textit{and } p'(\cdot) : \Sigma \to (1,\infty) \textit{ is such that } 1/p(\cdot) + 1/p'(\cdot) = 1,\\ \mathbb{Y} := \ell^q_{\omega(\cdot)}(\mathbb{Z}) \textit{ with } q \in (0,\infty], \textit{ and Matuszewska-Orlicz indices}\\ \textit{satisfying } -(n-1)\tau_{p(\cdot)} < m_0(\omega) \le M_0(\omega) < (n-1)\tau_{p'(\cdot)}\\ \textit{and } -(n-1)\tau_{p(\cdot)} < m_\infty(\omega) \le M_\infty(\omega) < (n-1)\tau_{p'(\cdot)}\\ \textit{where } \tau_{p(\cdot)}, \tau_{p'(\cdot)} \in (0,\infty) \textit{ are as in } (23.256)\text{-}(23.257)\end{gathered} \tag{23.430}$$

or

$$\begin{aligned}
&\mathbb{X} := M^{p,\lambda}(\Sigma,\sigma),\ \mathbb{X}' = \mathcal{B}^{p',\lambda}(\Sigma,\sigma) \textit{ where } \lambda \in (0,n-1),\\
&\textit{and } p,p' \in (1,\infty) \textit{ satisfy } 1/p + 1/p' = 1,\\
&\mathbb{Y} := \ell^q_{\omega(\cdot)}(\mathbb{Z}) \textit{ with } q \in (0,\infty], \textit{ and Matuszewska-Orlicz indices}\\
&\textit{satisfying } \tfrac{-(n-1-\lambda)}{p} < m_0(\omega) \le M_0(\omega) < \tfrac{n-1}{p'}\\
&\textit{and } \tfrac{-(n-1-\lambda)}{p} < m_\infty(\omega) \le M_\infty(\omega) < \tfrac{n-1}{p'}
\end{aligned} \tag{23.431}$$

or

$$\begin{aligned}
&\mathbb{X} := \mathcal{B}^{p,\lambda}(\Sigma,\sigma),\ \mathbb{X}' = M^{p',\lambda}(\Sigma,\sigma) \textit{ where } \lambda \in (0,n-1),\\
&\textit{and } p,p' \in (1,\infty) \textit{ satisfy } 1/p + 1/p' = 1,\\
&\mathbb{Y} := \ell^q_{\omega(\cdot)}(\mathbb{Z}) \textit{ with } q \in (0,\infty], \textit{ and Matuszewska-Orlicz indices}\\
&\textit{satisfying } \tfrac{-(n-1)}{p} < m_0(\omega) \le M_0(\omega) < \tfrac{n-1-\lambda}{p'}\\
&\textit{and } \tfrac{-(n-1)}{p} < m_\infty(\omega) \le M_\infty(\omega) < \tfrac{n-1-\lambda}{p'}
\end{aligned} \tag{23.432}$$

or

$$\begin{aligned}
&\mathbb{X} := M^{p,\lambda}(\Sigma,w),\ \mathbb{X}' = \mathcal{B}^{p',\lambda}(\Sigma,w') \textit{ where } p,p',q,q' \in (1,\infty)\\
&\textit{satisfy } 1/p + 1/p' = 1,\ 1/q + 1/q' = 1, \textit{ and where}\\
&w \in A_p(\Sigma,\sigma) \cap RH_q(\Sigma,\sigma), \textit{ and } w' := w^{1-p} \in A_{p'-\varepsilon}(\Sigma,\sigma),\\
&\textit{with } \varepsilon := (p'-1)/q' \in (0,p'-1), \textit{ and } \lambda \in \big(0,(n-1)/q'\big),\\
&\mathbb{Y} := \ell^{\widetilde{q}}_{\omega(\cdot)}(\mathbb{Z}) \textit{ with } \widetilde{q} \in (0,\infty], \textit{ and Matuszewska-Orlicz indices}\\
&\textit{satisfying } \tfrac{-(n-1-\lambda q')}{pq'} < m_0(\omega) \le M_0(\omega) < \tfrac{(n-1)\tau}{p'}\\
&\textit{and } \tfrac{-(n-1-\lambda q')}{pq'} < m_\infty(\omega) \le M_\infty(\omega) < \tfrac{(n-1)\tau}{p'}\\
&\textit{where the parameter } \tau \in (0,1) \textit{ is as in } (23.241)
\end{aligned} \tag{23.433}$$

or

$$\begin{aligned}
&\mathbb{X} := \mathcal{B}^{p',\lambda}(\Sigma,w'),\ \mathbb{X}' = M^{p,\lambda}(\Sigma,w) \textit{ where } p,p',q,q' \in (1,\infty)\\
&\textit{satisfy } 1/p + 1/p' = 1,\ 1/q + 1/q' = 1, \textit{ and where}\\
&w \in A_p(\Sigma,\sigma) \cap RH_q(\Sigma,\sigma), \textit{ and } w' := w^{1-p} \in A_{p'-\varepsilon}(\Sigma,\sigma),\\
&\textit{with } \varepsilon := (p'-1)/q' \in (0,p'-1), \textit{ and } \lambda \in \big(0,(n-1)/q'\big),\\
&\mathbb{Y} := \ell^{\widetilde{q}}_{\omega(\cdot)}(\mathbb{Z}) \textit{ with } \widetilde{q} \in (0,\infty], \textit{ and Matuszewska-Orlicz indices}\\
&\textit{satisfying } \tfrac{-(n-1)\tau}{p'} < m_0(\omega) \le M_0(\omega) < \tfrac{(n-1-\lambda q')}{pq'}\\
&\textit{and } \tfrac{-(n-1)\tau}{p'} < m_\infty(\omega) \le M_\infty(\omega) < \tfrac{(n-1-\lambda q')}{pq'}\\
&\textit{where the parameter } \tau \in (0,1) \textit{ is as in } (23.241)
\end{aligned} \tag{23.434}$$

or

$$
\begin{aligned}
&\mathbb{X} := L^{\Phi}(\Sigma,\sigma),\ \mathbb{X}' = L^{\widetilde{\Phi}}(\Sigma,\sigma)\ \textit{where}\\
&\Phi \textit{ is a Young function satisfying } 1 < i(\Phi) \le I(\Phi) < \infty,\\
&\textit{and } \widetilde{\Phi} \textit{ is the complementary function of } \Phi \textit{ (cf. (23.262))},\\
&\mathbb{Y} := \ell^q_{\omega(\cdot)}(\mathbb{Z}) \textit{ with } q \in (0,\infty], \textit{ and Matuszewska-Orlicz indices}\\
&\textit{satisfying } \ -(n-1)\tau_{\Phi} < m_0(\omega) \le M_0(\omega) < (n-1)\tau_{\widetilde{\Phi}}\\
&\textit{and } \ -(n-1)\tau_{\Phi} < m_\infty(\omega) \le M_\infty(\omega) < (n-1)\tau_{\widetilde{\Phi}}\\
&\textit{where } \tau_{\Phi}, \tau_{\widetilde{\Phi}} \in (0,1) \textit{ are as in (23.265)-(23.266).}
\end{aligned}
\tag{23.435}
$$

Finally, suppose $T : L^1_{\text{loc}}(\Sigma,\sigma) \to \mathscr{M}(\Sigma,\sigma)$ *is a sublinear operator satisfying the size condition* (23.312)*. Then whenever* $T : \mathbb{X} \to \mathbb{X}$ *is well defined and bounded one has*

$$
\begin{aligned}
&T : \dot{\mathbb{K}}(\Sigma;\mathbb{X};\mathbb{Y}) \to \dot{\mathbb{K}}(\Sigma;\mathbb{X};\mathbb{Y})\\
&\textit{is a well-defined and bounded operator.}
\end{aligned}
\tag{23.436}
$$

Proof Bearing in mind Examples 23.6, 23.7, 23.8, 23.9, 23.10, 23.11, 23.12, along with Example 23.4 and Definition 23.7, the boundedness claimed in (23.436) in any of the cases (23.427)-(23.435) is a consequence of Theorem 23.3. □

Comment 1. The Herz-type spaces originating from the choices made in (23.427) have been first introduced in [RaSa20] and studied in the monograph [LiYaHu23]. The resulting scale is known as local generalized Herz space:

$$
\dot{\mathbb{K}}(\mathbb{R}^{n-1}; L^p(\mathbb{R}^{n-1}); \ell^q_{\omega(\cdot)}(\mathbb{Z})) = \dot{\mathbb{K}}^{\omega(\cdot)}_{p,q}(\mathbb{R}^{n-1}). \tag{23.437}
$$

The version of (23.436) corresponding to this space has been obtained in [RaSa20, Theorem 4.3] for $p \in (1,\infty)$ and $q \in (1,\infty)$, and subsequently improved in [LiYaHu23, Theorem 1.5.1, p. 74] for $p \in (1,\infty]$, and $q \in (0,\infty]$. As such, our previous corollary recovers both results in the range $p \in (1,\infty)$, and $q \in (0,\infty]$.

Comment 2. Regarding the scales derived from either of the scenarios (23.428)-(23.435), these represent new function spaces of Herz-type along with new boundedness results of sublinear operators. Concretely, to the best of our knowledge, the following Herz-type spaces are new:

(1) Lorentz-based Local Generalized Herz spaces:

$$
\dot{\mathbb{K}}(\mathbb{R}^{n-1}; L^{p,r}(\mathbb{R}^{n-1}); \ell^q_{\omega(\cdot)}(\mathbb{Z})) =: \mathrm{L}\dot{\mathbb{K}}^{\omega(\cdot),r}_{p,q}(\mathbb{R}^{n-1}). \tag{23.438}
$$

(2) Simply-weighted Local Generalized Herz spaces:

$$
\dot{\mathbb{K}}(\mathbb{R}^{n-1}; L^p(\mathbb{R}^{n-1},w); \ell^q_{\omega(\cdot)}(\mathbb{Z})) =: \dot{\mathbb{K}}^{\omega(\cdot)}_{p,q}(\mathbb{R}^{n-1},w). \tag{23.439}
$$

(3) Variable Exponent Local Generalized Herz spaces:

$$\dot{\mathrm{K}}\big(\mathbb{R}^{n-1}; L^{p(\cdot)}(\mathbb{R}^{n-1}); \ell^{q}_{\omega(\cdot)}(\mathbb{Z})\big) =: \dot{\mathrm{K}}^{\omega(\cdot)}_{p(\cdot),q}(\mathbb{R}^{n-1}). \tag{23.440}$$

(4) Morrey-based Local Generalized Herz spaces:

$$\dot{\mathrm{K}}\big(\mathbb{R}^{n-1}; M^{p,\lambda}(\mathbb{R}^{n-1}); \ell^{q}_{\omega(\cdot)}(\mathbb{Z})\big) =: \mathrm{M}\dot{\mathrm{K}}^{\omega(\cdot),\lambda}_{p,q}(\mathbb{R}^{n-1}). \tag{23.441}$$

(5) Block-based Local Generalized Herz spaces:

$$\dot{\mathrm{K}}\big(\mathbb{R}^{n-1}; \mathcal{B}^{p,\lambda}(\mathbb{R}^{n-1}); \ell^{q}_{\omega(\cdot)}(\mathbb{Z})\big) =: \mathcal{B}\dot{\mathrm{K}}^{\omega(\cdot),\lambda}_{p,q}(\mathbb{R}^{n-1}). \tag{23.442}$$

(6) Muckenhoupt weighted Morrey-based Local Generalized Herz spaces:

$$\dot{\mathrm{K}}\big(\mathbb{R}^{n-1}; M^{p,\lambda}(\mathbb{R}^{n-1}, w); \ell^{\widetilde{q}}_{\omega(\cdot)}(\mathbb{Z})\big) =: \mathrm{M}\dot{\mathrm{K}}^{\omega(\cdot),\lambda}_{p,\widetilde{q}}(\mathbb{R}^{n-1}, w). \tag{23.443}$$

(7) Muckenhoupt weighted Block-based Local Generalized Herz spaces:

$$\dot{\mathrm{K}}\big(\mathbb{R}^{n-1}; \mathcal{B}^{p',\lambda}(\mathbb{R}^{n-1}, w'); \ell^{\widetilde{q}}_{\omega(\cdot)}(\mathbb{Z})\big) =: \mathcal{B}\dot{\mathrm{K}}^{\omega(\cdot),\lambda}_{p',\widetilde{q}}(\mathbb{R}^{n-1}, w'). \tag{23.444}$$

(8) Orlicz-based Local Generalized Herz spaces:

$$\dot{\mathrm{K}}\big(\mathbb{R}^{n-1}; L^{\Phi}(\mathbb{R}^{n-1}); \ell^{q}_{\omega(\cdot)}(\mathbb{Z})\big) =: \dot{\mathrm{K}}^{\omega(\cdot)}_{\Phi,q}(\mathbb{R}^{n-1}). \tag{23.445}$$

We record below another list of examples, now in relation to the sequence space $\ell^{\infty,q}_{\alpha,\lambda}$ introduced in Example 23.5.

Corollary 23.9 *Assume $\Sigma \subseteq \mathbb{R}^n$ is a closed unbounded Ahlfors regular set, and abbreviate $\sigma := \mathcal{H}^{n-1}\lfloor\Sigma$. Let $\mathbb{X}$ be a Generalized Banach Function Space on (Σ, σ), and $\mathbb{Y}$ a Generalized Quasi-Banach Sequence Space on $\mathbb{Z}$. Make the assumption that $\mathbb{X}$ and $\mathbb{Y}$ are as in any of the following cases:*

$$\begin{gathered} \mathbb{X} := L^p(\Sigma, \sigma),\ \mathbb{X}' = L^{p'}(\Sigma, \sigma) \text{ where} \\ p, p' \in (1, \infty) \text{ satisfy } 1/p + 1/p' = 1, \\ \mathbb{Y} := \ell^{\infty,q}_{\alpha,\lambda}(\mathbb{Z}) \text{ with } q \in (0, \infty],\ \lambda \in (0, \infty), \\ \text{and } \tfrac{-(n-1)}{p} + \lambda < \alpha < \tfrac{n-1}{p'} + \lambda \end{gathered} \tag{23.446}$$

or

$$\begin{aligned}&\mathbb{X} := L^{p,r}(\Sigma,\sigma),\ \mathbb{X}' = L^{p',r'}(\Sigma,\sigma) \text{ where } p,p' \in (1,\infty)\\&\text{and } r,r' \in [1,\infty] \text{ satisfy } 1/p + 1/p' = 1 \text{ and } 1/r + 1/r' = 1,\\&\mathbb{Y} := \ell^{\infty,q}_{\alpha,\lambda}(\mathbb{Z}) \text{ with } q \in (0,\infty],\ \lambda \in (0,\infty),\\&\text{and } \tfrac{-(n-1)}{p} + \lambda < \alpha < \tfrac{n-1}{p'} + \lambda\end{aligned} \tag{23.447}$$

or

$$\begin{aligned}&\mathbb{X} := L^{p}(\Sigma,w),\ \mathbb{X}' = L^{p'}(\Sigma,w') \text{ where } p,p' \in (1,\infty)\\&\text{satisfy } 1/p + 1/p' = 1,\ w \in A_p(\Sigma,\sigma), \text{ and } w' := w^{1-p} \in A_{p'}(\Sigma,\sigma),\\&\mathbb{Y} := \ell^{\infty,q}_{\alpha,\lambda}(\mathbb{Z}) \text{ with } q \in (1,\infty),\ \lambda \in (0,\infty),\\&\text{and } \tfrac{-(n-1)\tau}{p} + \lambda < \alpha < \tfrac{(n-1)\tau'}{p'} + \lambda\\&\text{where } \tau,\tau' \in (0,1) \text{ are as in (23.222)-(23.223)}\end{aligned} \tag{23.448}$$

or

$$\begin{aligned}&\mathbb{X} := L^{p(\cdot)}(\Sigma,\sigma),\ \mathbb{X}' = L^{p'(\cdot)}(\Sigma,\sigma) \text{ where } p(\cdot) : \Sigma \to (1,\infty)\\&\text{is a } \sigma\text{-measurable function satisfying (23.254)-(23.255)},\\&\text{and } p'(\cdot) : \Sigma \to (1,\infty) \text{ is such that } 1/p(\cdot) + 1/p'(\cdot) = 1,\\&\mathbb{Y} := \ell^{\infty,q}_{\alpha,\lambda}(\mathbb{Z}) \text{ with } q \in (0,\infty],\ \lambda \in (0,\infty), \text{ and}\\&-(n-1)\tau_{p(\cdot)} + \lambda < \alpha < (n-1)\tau_{p'(\cdot)} + \lambda\\&\text{where } \tau_{p(\cdot)}, \tau_{p'(\cdot)} \in (0,\infty) \text{ are as in (23.256)-(23.257)}\end{aligned} \tag{23.449}$$

or

$$\begin{aligned}&\mathbb{X} := M^{p,\lambda}(\Sigma,\sigma),\ \mathbb{X}' = \mathcal{B}^{p',\lambda}(\Sigma,\sigma) \text{ where } \lambda \in (0,n-1),\\&\text{and } p,p' \in (1,\infty) \text{ satisfy } 1/p + 1/p' = 1,\\&\mathbb{Y} := \ell^{\infty,q}_{\alpha,\widetilde{\lambda}}(\mathbb{Z}) \text{ with } q \in (0,\infty],\ \widetilde{\lambda} \in (0,\infty), \text{ and}\\&\tfrac{-(n-1-\lambda)}{p} + \widetilde{\lambda} < \alpha < \tfrac{n-1}{p'} + \widetilde{\lambda}\end{aligned} \tag{23.450}$$

or

$$\begin{aligned}&\mathbb{X} := \mathcal{B}^{p,\lambda}(\Sigma,\sigma),\ \mathbb{X}' = M^{p',\lambda}(\Sigma,\sigma) \text{ where } \lambda \in (0,n-1),\\&\text{and } p,p' \in (1,\infty) \text{ satisfy } 1/p + 1/p' = 1,\\&\mathbb{Y} := \ell^{\infty,q}_{\alpha,\widetilde{\lambda}}(\mathbb{Z}) \text{ with } q \in (0,\infty],\ \widetilde{\lambda} \in (0,\infty), \text{ and}\\&\tfrac{-(n-1)}{p} + \widetilde{\lambda} < \alpha < \tfrac{n-1-\lambda}{p'} + \widetilde{\lambda}\end{aligned} \tag{23.451}$$

or

$$\begin{aligned}
&\mathbb{X} := M^{p,\lambda}(\Sigma, w),\ \mathbb{X}' = \mathcal{B}^{p',\lambda}(\Sigma, w')\ \textit{where}\ p, p', q, q' \in (1,\infty)\\
&\textit{satisfy}\ 1/p + 1/p' = 1,\ 1/q + 1/q' = 1,\ \textit{and where}\\
&w \in A_p(\Sigma,\sigma)\cap RH_q(\Sigma,\sigma),\ \textit{and}\ w' := w^{1-p} \in A_{p'-\varepsilon}(\Sigma,\sigma),\\
&\textit{with}\ \varepsilon := (p'-1)/q' \in (0, p'-1),\ \textit{and}\ \lambda \in \big(0, (n-1)/q'\big),\\
&\mathbb{Y} := \ell^{\infty,\widetilde{q}}_{\alpha,\widetilde{\lambda}}(\mathbb{Z})\ \textit{with}\ \widetilde{q} \in (0,\infty],\ \widetilde{\lambda} \in (0,\infty),\ \textit{and}\\
&\tfrac{-(n-1-\lambda q')}{pq'} + \widetilde{\lambda} < \alpha < \tfrac{(n-1)\tau}{p'} + \widetilde{\lambda}\\
&\textit{where the parameter}\ \tau \in (0,1)\ \textit{is as in}\ (23.241)
\end{aligned} \tag{23.452}$$

or

$$\begin{aligned}
&\mathbb{X} := \mathcal{B}^{p',\lambda}(\Sigma, w'),\ \mathbb{X}' = M^{p,\lambda}(\Sigma, w)\ \textit{where}\ p, p', q, q' \in (1,\infty)\\
&\textit{satisfy}\ 1/p + 1/p' = 1,\ 1/q + 1/q' = 1,\ \textit{and where}\\
&w \in A_p(\Sigma,\sigma)\cap RH_q(\Sigma,\sigma),\ \textit{and}\ w' := w^{1-p} \in A_{p'-\varepsilon}(\Sigma,\sigma),\\
&\textit{with}\ \varepsilon := (p'-1)/q' \in (0, p'-1),\ \textit{and}\ \lambda \in \big(0, (n-1)/q'\big),\\
&\mathbb{Y} := \ell^{\infty,\widetilde{q}}_{\alpha,\widetilde{\lambda}}(\mathbb{Z})\ \textit{with}\ \widetilde{q} \in (0,\infty],\ \widetilde{\lambda} \in (0,\infty),\\
&\textit{and}\ \tfrac{-(n-1)\tau}{p'} + \widetilde{\lambda} < \alpha < \tfrac{(n-1-\lambda q')}{pq'} + \widetilde{\lambda}\\
&\textit{where the parameter}\ \tau \in (0,1)\ \textit{is as in}\ (23.241)
\end{aligned} \tag{23.453}$$

or

$$\begin{aligned}
&\mathbb{X} := L^{\Phi}(\Sigma,\sigma),\ \mathbb{X}' = L^{\widetilde{\Phi}}(\Sigma,\sigma)\ \textit{where}\\
&\Phi\ \textit{is a Young function satisfying}\ 1 < i(\Phi) \le I(\Phi) < \infty,\\
&\textit{and}\ \widetilde{\Phi}\ \textit{is the complementary function of}\ \Phi\ (\textit{cf.}\ (23.262)),\\
&\mathbb{Y} := \ell^{\infty,q}_{\alpha,\lambda}(\mathbb{Z})\ \textit{with}\ q \in (0,\infty],\ \lambda \in (0,\infty),\\
&\textit{and}\ -(n-1)\tau_{\Phi} + \lambda < \alpha < (n-1)\tau_{\widetilde{\Phi}} + \lambda\\
&\textit{where}\ \tau_{\Phi}, \tau_{\widetilde{\Phi}} \in (0,1)\ \textit{are as in}\ (23.265)\text{-}(23.266).
\end{aligned} \tag{23.454}$$

Lastly, let $T : L^1_{\mathrm{loc}}(\Sigma,\sigma) \to \mathcal{M}(\Sigma,\sigma)$ *be a sublinear operator satisfying the size condition* (23.312)*. Then whenever* $T : \mathbb{X} \to \mathbb{X}$ *is well defined and bounded it follows that*

$$\begin{aligned}
&T : \dot{\mathrm{K}}(\Sigma;\mathbb{X};\mathbb{Y}) \to \dot{\mathrm{K}}(\Sigma;\mathbb{X};\mathbb{Y})\\
&\textit{is a well-defined and bounded operator.}
\end{aligned} \tag{23.455}$$

Proof Being mindful of Examples 23.6, 23.7, 23.8, 23.9, 23.10, 23.11, 23.12 as well as Example 23.5 and Definition 23.7, the boundedness result claimed in (23.436) in any of the scenarios (23.446)-(23.454) readily follows from Theorem 23.3. □

`Comment 1.` Morrey-Herz spaces have been introduced first in [LuXu05], and they make up the Herz-type scale corresponding to the choices made in (23.446), i.e.,

$$\dot{\mathrm{K}}\big(\mathbb{R}^{n-1}; L^p(\mathbb{R}^{n-1}); \ell^{\infty,q}_{\alpha,\lambda}(\mathbb{Z})\big) = \mathrm{M}\dot{\mathrm{K}}^{\alpha,\lambda}_{p,q}(\mathbb{R}^{n-1}). \tag{23.456}$$

The boundedness (23.455) on these spaces is a particular case of [LuXu05, Theorem2.2], where a result regarding the boundedness of rough singular integral operators is obtained.

Comment 2. When considering the case (23.449) we recover the variable exponent Morrey-Herz spaces, which have first appeared in [Iz09-II] (see also [Iz10-II]), that is

$$\dot{\mathrm{K}}(\mathbb{R}^{n-1}; L^{p(\cdot)}(\mathbb{R}^{n-1}); \ell^{\infty,q}_{\alpha,\lambda}(\mathbb{Z})) = \mathrm{M}\dot{\mathrm{K}}^{\alpha,\lambda}_{p(\cdot),q}(\mathbb{R}^{n-1}). \tag{23.457}$$

Comment 3. The remaining cases in (23.447), (23.448), (23.450), (23.451), (23.452), (23.453), (23.454) yield the following new Herz-type scales:

(1) Lorentz-based Morrey-Herz spaces:

$$\dot{\mathrm{K}}(\mathbb{R}^{n-1}; L^{p,r}(\mathbb{R}^{n-1}); \ell^{\infty,q}_{\alpha,\lambda}(\mathbb{Z})) =: \mathrm{L}\dot{\mathrm{K}}^{\alpha,\lambda,r}_{p,q}(\mathbb{R}^{n-1}). \tag{23.458}$$

(2) Simply-weighted Morrey-Herz space:

$$\dot{\mathrm{K}}(\mathbb{R}^{n-1}; L^{p}(\mathbb{R}^{n-1}, w); \ell^{\infty,q}_{\alpha,\lambda}(\mathbb{Z})) =: \dot{\mathrm{K}}^{\alpha,\lambda}_{p,q}(\mathbb{R}^{n-1}, w). \tag{23.459}$$

(3) Doubly-Morreyed Herz spaces:

$$\dot{\mathrm{K}}(\mathbb{R}^{n-1}; M^{p,\lambda}(\mathbb{R}^{n-1}); \ell^{\infty,q}_{\alpha,\widetilde{\lambda}}(\mathbb{Z})) =: \mathrm{M}\dot{\mathrm{K}}^{\alpha,\lambda,\widetilde{\lambda}}_{p,q}(\mathbb{R}^{n-1}). \tag{23.460}$$

(4) Block-based Morrey-Herz spaces:

$$\dot{\mathrm{K}}(\mathbb{R}^{n-1}; \mathcal{B}^{p,\lambda}(\mathbb{R}^{n-1}); \ell^{\infty,q}_{\alpha,\widetilde{\lambda}}(\mathbb{Z})) =: \mathcal{B}\dot{\mathrm{K}}^{\alpha,\lambda,\widetilde{\lambda}}_{p,q}(\mathbb{R}^{n-1}). \tag{23.461}$$

(5) Muckenhoupt weighted Doubly-Morreyed Herz spaces:

$$\dot{\mathrm{K}}(\mathbb{R}^{n-1}; M^{p,\lambda}(\mathbb{R}^{n-1}, w); \ell^{\infty,\widetilde{q}}_{\alpha,\widetilde{\lambda}}(\mathbb{Z})) =: \mathrm{M}\dot{\mathrm{K}}^{\alpha,\lambda,\widetilde{\lambda}}_{p,\widetilde{q}}(\mathbb{R}^{n-1}, w). \tag{23.462}$$

(6) Muckenhoupt weighted Block-based Morrey-Herz spaces:

$$\dot{\mathrm{K}}(\mathbb{R}^{n-1}; \mathcal{B}^{p',\lambda}(\mathbb{R}^{n-1}, w'); \ell^{\infty,\widetilde{q}}_{\alpha,\widetilde{\lambda}}(\mathbb{Z})) =: \mathcal{B}\dot{\mathrm{K}}^{\alpha,\lambda,\widetilde{\lambda}}_{p',\widetilde{q}}(\mathbb{R}^{n-1}, w'). \tag{23.463}$$

(7) Orlicz-based Morrey-Herz space:

$$\dot{\mathrm{K}}(\mathbb{R}^{n-1}; L^{\Phi}(\mathbb{R}^{n-1}); \ell^{\infty,q}_{\alpha,\lambda}(\mathbb{Z})) =: \mathrm{L}\dot{\mathrm{K}}^{\alpha,\lambda}_{\Phi,q}(\mathbb{R}^{n-1}). \tag{23.464}$$

The boundedness result recorded in (23.455) corresponding to each of these spaces is also new.

23.10 A New Scale: Vanishing Morrey-based Herz Spaces

Starting with the scale of Morrey-based Herz spaces, i.e., Composite Herz spaces corresponding to the choices in (23.378) (see also (23.388)), our goal in this section is to introduce a new scale of Herz-type spaces. To set the stage, assume $\Sigma \subseteq \mathbb{R}^n$ is a closed unbounded Ahlfors regular set, and abbreviate $\sigma := \mathcal{H}^{n-1}\lfloor\Sigma$. Fix an integrability exponent $p \in (1, \infty)$ along with a parameter $\lambda \in (0, n-1)$. Following [GHA-II, (6.2.14), p. 316], consider the vanishing Morrey space defined as

$$\mathring{M}^{p,\lambda}(\Sigma, \sigma) := \text{the closure of } L^s(\Sigma, \sigma) \text{ with } s := \tfrac{p(n-1)}{n-1-\lambda} \text{ in } M^{p,\lambda}(\Sigma, \sigma). \tag{23.465}$$

As noted in [GHA-II, (6.2.16)–(6.2.17), p. 316],

$$\text{the operator of pointwise multiplication by a function } b \in L^\infty(\Sigma, \sigma) \text{ is a bounded mapping from the space } \mathring{M}^{p,\lambda}(\Sigma, \sigma) \text{ into itself, with operator norm} \leq \|b\|_{L^\infty(\Sigma,\sigma)}, \tag{23.466}$$

and $\mathring{M}^{p,\lambda}(\Sigma, \sigma)$ satisfies the lattice property, i.e.,

$$\text{if } f, g : \Sigma \to \mathbb{C} \text{ are } \sigma\text{-measurable functions with the property that } |g| \leq |f| \text{ at } \sigma\text{-a.e. point on } \Sigma \text{ and } f \in \mathring{M}^{p,\lambda}(\Sigma, \sigma), \text{ then } g \text{ also belongs to } \mathring{M}^{p,\lambda}(\Sigma, \sigma). \tag{23.467}$$

Next, fix an exponent $q \in (0, \infty)$ and an index $\alpha \in \mathbb{R}$. Recall the Morrey-based Herz space $\dot{\mathrm{K}}\big(\Sigma; M^{p,\lambda}(\Sigma, \sigma); \ell^q_\alpha(\mathbb{Z})\big)$ (cf. (23.388)), i.e., the Composite Herz space associated with $M^{p,\lambda}(\Sigma, \sigma)$ and $\ell^q_\alpha(\mathbb{Z})$ (in the sense of Definition 23.2). In relation to this, we claim that

$$\dot{\mathrm{K}}^\alpha_{s,q}(\Sigma, \sigma) \hookrightarrow \dot{\mathrm{K}}\big(\Sigma; M^{p,\lambda}(\Sigma, \sigma); \ell^q_\alpha(\mathbb{Z})\big) \quad \text{continuously}. \tag{23.468}$$

To justify this, recall the continuous embedding mentioned in (23.204), and observe that for every $f \in \mathrm{K}^\alpha_{s,q}(\Sigma, \sigma)$ we have

$$\begin{aligned}\|f\|_{\dot{\mathrm{K}}(\Sigma;M^{p,\lambda}(\Sigma,\sigma);\ell^q_\alpha(\mathbb{Z}))} &= \Big[\sum_{k=-\infty}^{\infty}\big(2_*^{k\alpha}\|f\cdot\mathbb{1}_{C_k}\|_{M^{p,\lambda}(\Sigma,\sigma)}\big)^q\Big]^{1/q} \\ &\leq C\Big[\sum_{k=-\infty}^{\infty}\big(2_*^{k\alpha}\|f\cdot\mathbb{1}_{C_k}\|_{L^s(\Sigma,\sigma)}\big)^q\Big]^{1/q} \\ &= C\|f\|_{\dot{\mathrm{K}}^\alpha_{s,q}(\Sigma,\sigma)} \end{aligned} \tag{23.469}$$

for some constant $C = C(\Sigma, n, p, \lambda) \in (0, \infty)$. This proves (23.468).

Moving on, recall the family of annuli $\{C_k\}_{k\in\mathbb{Z}}$ from (23.16). Introduce the `vanishing Morrey-Based Herz space` as

$$\begin{aligned}&\mathring{\mathrm{K}}\big(\Sigma;\mathring{M}^{p,\lambda}(\Sigma,\sigma);\ell^q_\alpha(\mathbb{Z})\big) \\ &\quad := \Big\{f\in\dot{\mathrm{K}}\big(\Sigma;M^{p,\lambda}(\Sigma,\sigma);\ell^q_\alpha(\mathbb{Z})\big) : f\cdot\mathbb{1}_{C_k}\in\mathring{M}^{p,\lambda}(\Sigma,\sigma),\ \forall\, k\in\mathbb{Z}\Big\},\end{aligned} \tag{23.470}$$

and equip this with the norm inherited from $\dot{\mathrm{K}}\big(\Sigma;M^{p,\lambda}(\Sigma,\sigma);\ell^q_\alpha(\mathbb{Z})\big)$. In relation to this space, we claim that $\mathring{\mathrm{K}}\big(\Sigma;\mathring{M}^{p,\lambda}(\Sigma,\sigma);\ell^q_\alpha(\mathbb{Z})\big)$ is a lattice. Specifically,

$$\begin{aligned}&\text{if } f, g \text{ are two } \sigma\text{-measurable functions defined on } \Sigma \text{ such that} \\ &|g|\leq|f| \text{ at } \sigma\text{-a.e. point on } \Sigma \text{ and } f\in\mathring{\mathrm{K}}\big(\Sigma;\mathring{M}^{p,\lambda}(\Sigma,\sigma);\ell^q_\alpha(\mathbb{Z})\big), \\ &\text{then the function } g \text{ also belongs to } \mathring{\mathrm{K}}\big(\Sigma;\mathring{M}^{p,\lambda}(\Sigma,\sigma);\ell^q_\alpha(\mathbb{Z})\big).\end{aligned} \tag{23.471}$$

Indeed, since $\dot{\mathrm{K}}\big(\Sigma;M^{p,\lambda}(\Sigma,\sigma);\ell^q_\alpha(\mathbb{Z})\big)$ is itself a lattice (cf. Proposition 23.4), it follows that $g\in\dot{\mathrm{K}}\big(\Sigma;M^{p,\lambda}(\Sigma,\sigma);\ell^q_\alpha(\mathbb{Z})\big)$. Also, notice that $|g\cdot\mathbb{1}_{C_k}|\leq|f\cdot\mathbb{1}_{C_k}|$ for every $k\in\mathbb{Z}$, and at σ-a.e. point on Σ. Granted this and (23.467) we have

$$g\cdot\mathbb{1}_{C_k}\in\mathring{M}^{p,\lambda}(\Sigma,\sigma)\ \text{ for all } k\in\mathbb{Z}. \tag{23.472}$$

This shows that $g\in\mathring{\mathrm{K}}\big(\Sigma;\mathring{M}^{p,\lambda}(\Sigma,\sigma);\ell^q_\alpha(\mathbb{Z})\big)$, as desired.

The following lemma establishes that the vanishing Morrey-based Herz space is a closed subspace of the Morrey-based Herz space.

Lemma 23.6 *Suppose $\Sigma\subseteq\mathbb{R}^n$ is a closed unbounded Ahlfors regular set, and abbreviate $\sigma := \mathcal{H}^{n-1}\lfloor\Sigma$. Fix two exponents $p\in(1,\infty)$ and $q\in(0,\infty)$, along with a parameter $\lambda\in(0,n-1)$ and an index $\alpha\in\mathbb{R}$. Then the vanishing Morrey-based Herz space $\mathring{\mathrm{K}}\big(\Sigma;\mathring{M}^{p,\lambda}(\Sigma,\sigma);\ell^q_\alpha(\mathbb{Z})\big)$ (cf. (23.470)) is a closed subspace of the Morrey-based Herz space $\dot{\mathrm{K}}\big(\Sigma;M^{p,\lambda}(\Sigma,\sigma);\ell^q_\alpha(\mathbb{Z})\big)$ (cf. (23.388)).*

Proof Consider a sequence $\{f_j\}_{j\in\mathbb{N}}\subseteq\mathring{\mathrm{K}}\big(\Sigma;\mathring{M}^{p,\lambda}(\Sigma,\sigma);\ell^q_\alpha(\mathbb{Z})\big)$ and assume there exists some function $f\in\dot{\mathrm{K}}\big(\Sigma;M^{p,\lambda}(\Sigma,\sigma);\ell^q_\alpha(\mathbb{Z})\big)$ satisfying

$$f_j\to f\quad\text{in } \dot{\mathrm{K}}\big(\Sigma;M^{p,\lambda}(\Sigma,\sigma);\ell^q_\alpha(\mathbb{Z})\big) \text{ as } j\to\infty. \tag{23.473}$$

In particular, bearing in mind (23.470),

$$f_j \cdot \mathbb{1}_{C_k} \in \mathring{M}^{p,\lambda}(\Sigma, \sigma) \quad \text{for all } j \in \mathbb{N} \text{ and } k \in \mathbb{Z}. \tag{23.474}$$

For each $k \in \mathbb{Z}$ we have

$$\|f \cdot \mathbb{1}_{C_k}\|_{M^{p,\lambda}(\Sigma,\sigma)} \leq 2_*^{-k\alpha} \|f\|_{\dot{\mathrm{K}}(\Sigma; M^{p,\lambda}(\Sigma,\sigma); \ell^q_\alpha(\mathbb{Z}))} < \infty, \tag{23.475}$$

which gives

$$f \cdot \mathbb{1}_{C_k} \in M^{p,\lambda}(\Sigma, \sigma) \quad \text{for all } k \in \mathbb{Z}. \tag{23.476}$$

Moreover, for each $j \in \mathbb{N}$ and $k \in \mathbb{Z}$, estimate

$$\begin{aligned} \|f_j \cdot \mathbb{1}_{C_k} - f \cdot \mathbb{1}_{C_k}\|_{M^{p,\lambda}(\Sigma,\sigma)} &= \|(f_j - f) \cdot \mathbb{1}_{C_k}\|_{M^{p,\lambda}(\Sigma,\sigma)} \\ &\leq 2_*^{-k\alpha} \|f_j - f\|_{\dot{\mathrm{K}}(\Sigma; M^{p,\lambda}(\Sigma,\sigma); \ell^q_\alpha(\mathbb{Z}))}. \end{aligned} \tag{23.477}$$

Based on this and (23.473), we may deduce that

$$f_j \cdot \mathbb{1}_{C_k} \to f \cdot \mathbb{1}_{C_k} \text{ in } M^{p,\lambda}(\Sigma, \sigma) \text{ as } j \to \infty, \text{ and for every } k \in \mathbb{Z}. \tag{23.478}$$

In light of this, (23.474), (23.476), and the fact that $\mathring{M}^{p,\lambda}(\Sigma, \sigma)$ is a closed subspace of the Morrey space $M^{p,\lambda}(\Sigma, \sigma)$, we finally obtain

$$f \cdot \mathbb{1}_{C_k} \in \mathring{M}^{p,\lambda}(\Sigma, \sigma). \tag{23.479}$$

This shows that $f \in \mathring{\mathrm{K}}\big(\Sigma; \mathring{M}^{p,\lambda}(\Sigma, \sigma); \ell^q_\alpha(\mathbb{Z})\big)$, thus establishing Lemma 23.6. □

Remarkably, we are able to describe the vanishing Morrey-based Herz spaces (defined in (23.470)) as being the closure of some (standard) Herz space in the norm of the Morrey-based Herz space. More precisely, we have the following result.

Proposition 23.11 *Assume $\Sigma \subseteq \mathbb{R}^n$ is a closed unbounded Ahlfors regular set, and abbreviate $\sigma := \mathcal{H}^{n-1} \lfloor \Sigma$. Fix two exponents $p \in (1, \infty)$ and $q \in (0, \infty)$, along with a parameter $\lambda \in (0, n-1)$ and an index $\alpha \in \mathbb{R}$. Also, consider $s := \frac{p(n-1)}{n-1-\lambda} \in (p, \infty)$. Then the vanishing Morrey-based Herz space may be identified as*

$$\mathring{\mathrm{K}}\big(\Sigma; \mathring{M}^{p,\lambda}(\Sigma, \sigma); \ell^q_\alpha(\mathbb{Z})\big) = \overline{\dot{\mathrm{K}}^\alpha_{s,q}(\Sigma, \sigma)}^{\|\cdot\|_{\dot{\mathrm{K}}(\Sigma; M^{p,\lambda}(\Sigma,\sigma); \ell^q_\alpha(\mathbb{Z}))}}, \tag{23.480}$$

where $\dot{\mathrm{K}}^\alpha_{s,q}(\Sigma, \sigma)$ is the homogeneous Herz space as in (17.116).

Proof Fix a function $f \in \dot{\mathrm{K}}^\alpha_{s,q}(\Sigma, \sigma)$. We claim that

$$f \in \dot{\mathrm{K}}\big(\Sigma; M^{p,\lambda}(\Sigma, \sigma); \ell^q_\alpha(\mathbb{Z})\big) \quad \text{and} \quad f \cdot \mathbb{1}_{C_k} \in L^s(\Sigma, \sigma) \quad \text{for all } k \in \mathbb{Z}. \tag{23.481}$$

Indeed, the first membership above is a consequence of (23.468), while the second membership is visible from the fact that

$$\|f \cdot \mathbb{1}_{C_k}\|_{L^s(\Sigma,\sigma)} \leq 2_*^{-k\alpha}\|f\|_{\dot{\mathrm{K}}^{\alpha}_{s,q}(\Sigma,\sigma)} \quad \text{for all } k \in \mathbb{Z}. \tag{23.482}$$

Based on (23.481) and (23.465), for every $k \in \mathbb{Z}$ the function $f \cdot \mathbb{1}_{C_k}$ belongs to the vanishing Morrey space $\mathring{M}^{p,\lambda}(\Sigma, \sigma)$. Hence, $f \in \mathring{\mathrm{K}}\big(\Sigma; \mathring{M}^{p,\lambda}(\Sigma,\sigma); \ell^q_\alpha(\mathbb{Z})\big)$, and therefore, the following inclusion holds:

$$\dot{\mathrm{K}}^{\alpha}_{s,q}(\Sigma,\sigma) \subseteq \mathring{\mathrm{K}}\big(\Sigma; \mathring{M}^{p,\lambda}(\Sigma,\sigma); \ell^q_\alpha(\mathbb{Z})\big). \tag{23.483}$$

As such, since $\mathring{\mathrm{K}}\big(\Sigma; \mathring{M}^{p,\lambda}(\Sigma,\sigma); \ell^q_\alpha(\mathbb{Z})\big)$ is a closed subspace of the composite Herz space $\dot{\mathrm{K}}\big(\Sigma; M^{p,\lambda}(\Sigma,\sigma); \ell^q_\alpha(\mathbb{Z})\big)$ (cf. Lemma 23.6), from (23.483) we see that

$$\overline{\dot{\mathrm{K}}^{\alpha}_{s,q}(\Sigma,\sigma)}^{\|\cdot\|_{\dot{\mathrm{K}}(\Sigma;M^{p,\lambda}(\Sigma,\sigma);\ell^q_\alpha(\mathbb{Z}))}} \subseteq \mathring{\mathrm{K}}\big(\Sigma; \mathring{M}^{p,\lambda}(\Sigma,\sigma); \ell^q_\alpha(\mathbb{Z})\big). \tag{23.484}$$

In the opposite direction, select a function $f \in \mathring{\mathrm{K}}\big(\Sigma; \mathring{M}^{p,\lambda}(\Sigma,\sigma); \ell^q_\alpha(\mathbb{Z})\big)$, i.e.,

$$\begin{aligned} &f \in \dot{\mathrm{K}}\big(\Sigma; M^{p,\lambda}(\Sigma,\sigma); \ell^q_\alpha(\mathbb{Z})\big) \text{ and} \\ &f \cdot \mathbb{1}_{C_k} \in \mathring{M}^{p,\lambda}(\Sigma,\sigma) \text{ for all } k \in \mathbb{Z}. \end{aligned} \tag{23.485}$$

In view of (23.465), for each $k \in \mathbb{Z}$ there exists a sequence $\{\varphi_j^{(k)}\}_{j\in\mathbb{N}} \subseteq L^s(\Sigma,\sigma)$ with the property that

$$\|f \cdot \mathbb{1}_{C_k} - \varphi_j^{(k)}\|_{M^{p,\lambda}(\Sigma,\sigma)} \to 0 \quad \text{as } j \to \infty. \tag{23.486}$$

Without loss of generality, assume that

$$\|\varphi_j^{(k)}\|_{M^{p,\lambda}(\Sigma,\sigma)} \leq \|f \cdot \mathbb{1}_{C_k}\|_{M^{p,\lambda}(\Sigma,\sigma)} + \varepsilon_k \quad \text{for all } j \in \mathbb{N} \text{ and } k \in \mathbb{Z}, \tag{23.487}$$

for some sequence $\{\varepsilon_k\}_{k\in\mathbb{Z}}$ to be determined later. Next, for each $j \in \mathbb{N}$ define

$$\varphi_j := \sum_{\ell=-j}^{j} \varphi_j^{(\ell)} \cdot \mathbb{1}_{C_\ell} \text{ on } \Sigma. \tag{23.488}$$

By design, for each $k \in \mathbb{Z}$ and $j \in \mathbb{N}$ we have

$$\varphi_j \cdot \mathbb{1}_{C_k} = \begin{cases} \varphi_j^{(k)} \cdot \mathbb{1}_{C_k} & \text{if } |k| \leq j, \\ 0 & \text{if } |k| > j. \end{cases} \tag{23.489}$$

Thanks to this and the fact that $\{\varphi_j^{(k)}\}_{j\in\mathbb{N}} \subseteq L^s(\Sigma,\sigma)$ for all $k \in \mathbb{Z}$, if $j \in \mathbb{N}$ we see that

$$\|\varphi_j\|_{\dot{\mathrm{K}}^{\alpha}_{s,q}(\Sigma,\sigma)} = \Big[\sum_{k=-\infty}^{\infty}\big(2_*^{k\alpha}\|\varphi_j\cdot\mathbb{1}_{C_k}\|_{L^s(\Sigma,\sigma)}\big)^q\Big]^{1/q}$$
$$= \Big[\sum_{k=-j}^{j}\big(2_*^{k\alpha}\|\varphi_j^{(k)}\cdot\mathbb{1}_{C_k}\|_{L^s(\Sigma,\sigma)}\big)^q\Big]^{1/q}$$
$$\leq \Big[\sum_{k=-j}^{j}\big(2_*^{k\alpha}\|\varphi_j^{(k)}\|_{L^s(\Sigma,\sigma)}\big)^q\Big]^{1/q} < \infty. \tag{23.490}$$

Hence,

$$\varphi_j \in \dot{\mathrm{K}}^{\alpha}_{s,q}(\Sigma,\sigma) \quad \text{for all } j \in \mathbb{N}. \tag{23.491}$$

To proceed, fix an arbitrary $\varepsilon \in (0,\infty)$. Notice that the first membership in (23.485) ensures that there exists some $N \in \mathbb{N}$ sufficiently large so that

$$\Big[\sum_{\substack{k\in\mathbb{Z}\\ |k|>N}}\big(2_*^{k\alpha}\|f\cdot\mathbb{1}_{C_k}\|_{L^s(\Sigma,\sigma)}\big)^q\Big]^{1/q} < \varepsilon. \tag{23.492}$$

Also, in light of the convergence in (23.486), for each $k \in \mathbb{Z}$, with $|k| \leq N$, there exists some $J_k \in \mathbb{N}$ sufficiently large, say $J_k \geq N$, such that

$$\|f\cdot\mathbb{1}_{C_k} - \varphi_j^{(k)}\|_{M^{p,\lambda}(\Sigma,\sigma)} < \frac{\varepsilon}{(2N+1)^{1/q}2_*^{k\alpha}} \quad \text{for every } j \geq J_k. \tag{23.493}$$

From this, (23.489), and (23.487), whenever $j \geq \max\{J_k : |k| \leq N\} \geq N$ we deduce

$$\|f-\varphi_j\|_{\dot{\mathrm{K}}(\Sigma;M^{p,\lambda}(\Sigma,\sigma);\ell^q_\alpha(\mathbb{Z}))}$$
$$= \Big[\sum_{k=-\infty}^{\infty}\big(2_*^{k\alpha}\|(f-\varphi_j)\cdot\mathbb{1}_{C_k}\|_{M^{p,\lambda}(\Sigma,\sigma)}\big)^q\Big]^{1/q}$$
$$\leq C\Big[\sum_{k=-N}^{N}\big(2_*^{k\alpha}\|(f-\varphi_j)\cdot\mathbb{1}_{C_k}\|_{M^{p,\lambda}(\Sigma,\sigma)}\big)^q\Big]^{1/q}$$
$$+ C\Big[\sum_{\substack{k\in\mathbb{Z}\\ |k|>N}}\big(2_*^{k\alpha}\|(f-\varphi_j)\cdot\mathbb{1}_{C_k}\|_{M^{p,\lambda}(\Sigma,\sigma)}\big)^q\Big]^{1/q}$$
$$\leq C\Big[\sum_{k=-N}^{N}\big(2_*^{k\alpha}\|f\cdot\mathbb{1}_{C_k} - \varphi_j^{(k)}\|_{M^{p,\lambda}(\Sigma,\sigma)}\big)^q\Big]^{1/q}$$

$$+ C\Bigg[\sum_{\substack{k\in\mathbb{Z}\\|k|>N}} 2_*^{k\alpha q}\big(\|f\cdot \mathbb{1}_{C_k}\| + \|\varphi_j\cdot \mathbb{1}_{C_k}\|_{M^{p,\lambda}(\Sigma,\sigma)}\big)^q\Bigg]^{1/q}$$

$$\leq C\Bigg[\sum_{k=-N}^{N} \frac{\varepsilon^q}{2N+1}\Bigg]^{1/q}$$

$$+ C\Bigg[\sum_{\substack{k\in\mathbb{Z}\\|k|>N}} 2_*^{k\alpha q}\|f\cdot \mathbb{1}_{C_k}\|^q\Bigg]^{1/q} + C\Bigg[\sum_{\substack{k\in\mathbb{Z}\\|k|>N}} 2_*^{k\alpha q}\varepsilon_k^q\Bigg]^{1/q}$$

$$\leq C\varepsilon + C\Bigg[\sum_{\substack{k\in\mathbb{Z}\\|k|>N}} 2_*^{k\alpha q}\varepsilon_k^q\Bigg]^{1/q}, \tag{23.494}$$

for some constant $C = C(q) \in (0,\infty)$. At this stage, if we choose $\varepsilon_k := \frac{2_*^{-k\alpha}\varepsilon}{2^{|k|/q}}$ for each $k \in \mathbb{Z}$, then (23.494) entails

$$\|f - \varphi_j\|_{\dot{\mathrm{K}}(\Sigma;M^{p,\lambda}(\Sigma,\sigma);\ell^q_\alpha(\mathbb{Z}))} \leq C\varepsilon + C\Bigg(\sum_{\substack{k\in\mathbb{Z}\\|k|>N}} \frac{\varepsilon}{2^{|k|}}\Bigg)^{1/q} \leq C\varepsilon, \tag{23.495}$$

for some $C = C(q) \in (0,\infty)$. After readjusting ε, this shows that

$$\varphi_j \to f \quad \text{in } \dot{\mathrm{K}}\big(\Sigma; M^{p,\lambda}(\Sigma,\sigma); \ell^q_\alpha(\mathbb{Z})\big) \text{ as } j \to \infty. \tag{23.496}$$

Collectively, (23.491) and (23.496) prove that

$$\mathring{\mathrm{K}}\big(\Sigma; \mathring{M}^{p,\lambda}(\Sigma,\sigma); \ell^q_\alpha(\mathbb{Z})\big) \subseteq \overline{\dot{\mathrm{K}}^{\alpha}_{s,q}(\Sigma,\sigma)}^{\|\cdot\|_{\dot{\mathrm{K}}(\Sigma;M^{p,\lambda}(\Sigma,\sigma);\ell^q_\alpha(\mathbb{Z}))}}. \tag{23.497}$$

This concludes the proof of Proposition 23.11. □

In contrast, we next show that the vanishing Morrey-based Herz spaces introduced in (23.470) coincides with the vanishing Composite Herz spaces associated with the Morrey space $M^{p,\lambda}(\Sigma,\sigma)$ and $\ell^q_\alpha(\mathbb{Z})$. Here is a precise statement.

Theorem 23.7 *Assume $\Sigma \subseteq \mathbb{R}^n$ is a closed unbounded Ahlfors regular set, and abbreviate $\sigma := \mathcal{H}^{n-1}\lfloor\Sigma$. Select two exponents $p \in (1,\infty)$, and $q \in (0,\infty)$, along with a parameter $\lambda \in (0, n-1)$. Set $s := \frac{(n-1)p}{n-1-\lambda} \in (p,\infty)$, and pick an index $\alpha \in \mathbb{R}$ satisfying*

$$\tfrac{-(n-1)}{s} < \alpha < \tfrac{n-1}{p'} \tag{23.498}$$

where $p' \in (1,\infty)$ denotes the Hölder conjugate exponent of p. Then one may describe the vanishing Morrey-based Herz space (cf. (23.470)) as

$$\mathring{\mathrm{K}}\big(\Sigma;\mathring{M}^{p,\lambda}(\Sigma,\sigma);\ell^q_\alpha(\mathbb{Z})\big)=\mathring{\mathrm{K}}\big(\Sigma;M^{p,\lambda}(\Sigma,\sigma);\ell^q_\alpha(\mathbb{Z})\big), \tag{23.499}$$

where the space in the right-hand side is the vanishing Composite Herz space associated with $M^{p,\lambda}(\Sigma,\sigma)$ *and* $\ell^q_\alpha(\mathbb{Z})$ *as in* (23.327).

It is of relevance to make the following remark.

Remark 23.20 In the context of Theorem 23.7, observe that Example 23.8 (see (23.200)-(23.201)), together with Example 23.1 (see (23.106)), and Definition 23.7, ensure that the Morrey space $M^{p,\lambda}(\Sigma,\sigma)$ is a Generalized Banach Function Space on (Σ,σ) for which the Hardy-Littlewood maximal operator is bounded both on $M^{p,\lambda}(\Sigma,\sigma)$ and on the associated space $\mathcal{B}^{p',\lambda}(\Sigma,\sigma)$, and that Generalized Quasi-Banach Sequence Space $\ell^q_\alpha(\mathbb{Z})$ is $M^{p,\lambda}(\Sigma,\sigma)$-compliant. As such, the vanishing Composite Herz space $\mathring{\mathrm{K}}\big(\Sigma;M^{p,\lambda}(\Sigma,\sigma);\ell^q_\alpha(\mathbb{Z})\big)$ associated with $M^{p,\lambda}(\Sigma,\sigma)$ and $\ell^q_\alpha(\mathbb{Z})$ (cf. (23.327)) is meaningfully defined.

We now turn to the proof of Theorem 23.7.

Proof of Theorem 23.7 First, thanks to the range of α specified in (23.498), Proposition 17.7 (see in particular (17.134)) ensures that $L^\infty_{\rm comp}(\Sigma,\sigma)\subseteq\dot{\mathrm{K}}^\alpha_{s,q}(\Sigma,\sigma)$. In view of this and Proposition 23.11, we have

$$\begin{aligned}\mathring{\mathrm{K}}\big(\Sigma;M^{p,\lambda}(\Sigma,\sigma);\ell^q_\alpha(\mathbb{Z})\big)&=\overline{L^\infty_{\rm comp}(\Sigma,\sigma)}^{\|\cdot\|_{\dot{\mathrm{K}}(\Sigma;M^{p,\lambda}(\Sigma,\sigma);\ell^q_\alpha(\mathbb{Z}))}}\\&\subseteq\overline{\dot{\mathrm{K}}^\alpha_{s,q}(\Sigma,\sigma)}^{\|\cdot\|_{\dot{\mathrm{K}}(\Sigma;M^{p,\lambda}(\Sigma,\sigma);\ell^q_\alpha(\mathbb{Z}))}}\\&=\mathring{\mathrm{K}}\big(\Sigma;\mathring{M}^{p,\lambda}(\Sigma,\sigma);\ell^q_\alpha(\mathbb{Z})\big).\end{aligned} \tag{23.500}$$

To establish the reverse inclusion, start with an arbitrary function $f\in\dot{\mathrm{K}}^\alpha_{s,q}(\Sigma,\sigma)$, and for each $j\in\mathbb{N}$ define

$$f_j:=\sum_{k=-j}^{j}f\cdot\mathbb{1}_{C_k}\quad\text{on }\Sigma. \tag{23.501}$$

Notice that

$$|f_j|\le|f|\quad\text{at }\sigma\text{-a.e. point on }\Sigma. \tag{23.502}$$

Hence, the lattice property of $\dot{\mathrm{K}}^\alpha_{s,q}(\Sigma,\sigma)$ (cf. (17.5)) implies

$$f_j\in\dot{\mathrm{K}}^\alpha_{s,q}(\Sigma,\sigma)\quad\text{for all }j\in\mathbb{N}. \tag{23.503}$$

In relation to the functions in (23.501), we claim that

$$f_j\to f\quad\text{in }\dot{\mathrm{K}}\big(\Sigma;M^{p,\lambda}(\Sigma,\sigma);\ell^q_\alpha(\mathbb{Z})\big)\text{ as }j\to\infty. \tag{23.504}$$

Indeed, by design (23.501), for each $j\in\mathbb{N}$ and $k\in\mathbb{Z}$ we see that

$$(f - f_j) \cdot \mathbb{1}_{C_k} = \begin{cases} f \cdot \mathbb{1}_{C_k} & \text{if } |k| > j, \\ 0 & \text{if } |k| \leq j. \end{cases} \tag{23.505}$$

Next, fix an arbitrary $\varepsilon' \in (0, \infty)$. Since f belongs to the Herz space $\dot{\mathrm{K}}^{\alpha}_{s,q}(\Sigma, \sigma)$, there exists some $j_0 \in \mathbb{N}$ such that

$$\Bigg(\sum_{\substack{k \in \mathbb{Z} \\ |k| > j}} \big(2_*^{k\alpha} \|f \cdot \mathbb{1}_{C_k}\|_{L^s(\Sigma,\sigma)}\big)^q \Bigg)^{1/q} < \varepsilon' \quad \text{for all } j \in \mathbb{N} \text{ with } j \geq j_0. \tag{23.506}$$

In light of this, (23.505), and the continuous embedding $L^s(\Sigma, \sigma) \hookrightarrow M^{p,\lambda}(\Sigma, \sigma)$ (cf. (23.204)), there exists some $C = C(\Sigma, n, p, \lambda) \in (0, \infty)$ such that whenever $j \in \mathbb{N}$ and $j \geq j_0$, we have

$$\begin{aligned} &\|f - f_j\|_{\dot{\mathrm{K}}(\Sigma; M^{p,\lambda}(\Sigma,\sigma); \ell^q_\alpha(\mathbb{Z}))} \\ &= \Bigg(\sum_{k=-\infty}^{\infty} \big(2_*^{k\alpha} \|(f - f_j) \cdot \mathbb{1}_{C_k}\|_{M^{p,\lambda}(\Sigma,\sigma)}\big)^q \Bigg)^{1/q} \\ &= \Bigg(\sum_{\substack{k \in \mathbb{Z} \\ |k| > j}} \big(2_*^{k\alpha} \|f \cdot \mathbb{1}_{C_k}\|_{M^{p,\lambda}(\Sigma,\sigma)}\big)^q \Bigg)^{1/q} \\ &\leq C \Bigg(\sum_{\substack{k \in \mathbb{Z} \\ |k| > j}} \big(2_*^{k\alpha} \|f \cdot \mathbb{1}_{C_k}\|_{L^s(\Sigma,\sigma)}\big)^q \Bigg)^{1/q} < C\varepsilon'. \end{aligned} \tag{23.507}$$

After readjusting ε', this yields (23.504).

To proceed, fix $j \in \mathbb{N}$. For each $N \in \mathbb{N}$ define

$$f_j^{(N)} := \big(f_j \cdot \mathbb{1}_{\{|f_j| < N\}}\big) \cdot \mathbb{1}_{\bigcup_{|k| \leq N} C_k} \quad \text{on } \Sigma. \tag{23.508}$$

From this, and (23.502) it is visible that the sequence $\{f_j^{(N)}\}_{N \in \mathbb{N}}$ enjoys the following properties

$$\begin{aligned} &\{f_j^{(N)}\}_{j \in \mathbb{N}} \subseteq L^\infty_{\mathrm{comp}}(\Sigma, \sigma), \quad |f_j^{(N)}| \leq |f| \in \dot{\mathrm{K}}^{\alpha}_{s,q}(\Sigma, \sigma) \text{ for all } N \in \mathbb{N}, \\ &\text{and } f_j^{(N)} \to f_j \text{ pointwise } \sigma\text{-a.e. on } \Sigma, \text{ as } N \to \infty. \end{aligned} \tag{23.509}$$

Hence, Proposition 17.9 ensures that

$$f_j^{(N)} \to f_j \quad \text{in } \dot{\mathrm{K}}^{\alpha}_{s,q}(\Sigma, \sigma) \text{ as } N \to \infty. \tag{23.510}$$

Based on this and the continuous embedding recorded in (23.468), we obtain that

$$f_j^{(N)} \to f_j \quad \text{in } \dot{\mathrm{K}}\big(\Sigma; M^{p,\lambda}(\Sigma,\sigma); \ell^q_\alpha(\mathbb{Z})\big) \text{ as } N \to \infty, \text{ for all } j \in \mathbb{N}. \tag{23.511}$$

At this stage, pick $\varepsilon \in (0,\infty)$ arbitrary. Bearing in mind (23.504), we can choose $j_0 \in \mathbb{N}$ such

$$\|f - f_{j_0}\|_{\dot{\mathrm{K}}(\Sigma;M^{p,\lambda}(\Sigma,\sigma);\ell^q_\alpha(\mathbb{Z}))} < \frac{\varepsilon}{2} \tag{23.512}$$

Furthermore, based on (23.511), pick $N_0 \in \mathbb{N}$ sufficiently large so that

$$\|f_{j_0} - f_{j_0}^{(N)}\|_{\dot{\mathrm{K}}(\Sigma;M^{p,\lambda}(\Sigma,\sigma);\ell^q_\alpha(\mathbb{Z}))} < \frac{\varepsilon}{2} \quad \text{for all } N \in \mathbb{N} \text{ with } N \geq N_0. \tag{23.513}$$

Collectively, (23.512)-(23.513) imply that whenever $N \in \mathbb{N}$ and $N \geq N_0$ we have

$$\begin{aligned} &\|f - f_{j_0}^{(N)}\|_{\dot{\mathrm{K}}(\Sigma;M^{p,\lambda}(\Sigma,\sigma);\ell^q_\alpha(\mathbb{Z}))} \\ &\quad \leq \|f - f_{j_0}\|_{\dot{\mathrm{K}}(\Sigma;M^{p,\lambda}(\Sigma,\sigma);\ell^q_\alpha(\mathbb{Z}))} + \|f_{j_0} - f_{j_0}^{(N)}\|_{\dot{\mathrm{K}}(\Sigma;M^{p,\lambda}(\Sigma,\sigma);\ell^q_\alpha(\mathbb{Z}))} < \varepsilon. \end{aligned} \tag{23.514}$$

As such, since $f_{j_0}^{(N)} \in L^\infty_{\mathrm{comp}}(\Sigma,\sigma)$ (cf. (23.509)), this shows that

$$f \in \overline{L^\infty_{\mathrm{comp}}(\Sigma,\sigma)}^{\|\cdot\|_{\dot{\mathrm{K}}(\Sigma;M^{p,\lambda}(\Sigma,\sigma);\ell^q_\alpha(\mathbb{Z}))}} = \mathring{\mathrm{K}}\big(\Sigma; M^{p,\lambda}(\Sigma,\sigma); \ell^q_\alpha(\mathbb{Z})\big), \tag{23.515}$$

and because f has been taken arbitrary in $\dot{\mathrm{K}}^\alpha_{s,q}(\Sigma,\sigma)$ to begin with, this proves the inclusion

$$\dot{\mathrm{K}}^\alpha_{s,q}(\Sigma,\sigma) \subseteq \mathring{\mathrm{K}}\big(\Sigma; M^{p,\lambda}(\Sigma,\sigma); \ell^q_\alpha(\mathbb{Z})\big). \tag{23.516}$$

Upon recalling that $\mathring{\mathrm{K}}\big(\Sigma; M^{p,\lambda}(\Sigma,\sigma); \ell^q_\alpha(\mathbb{Z})\big)$ is a closed subspace of the Composite Herz space $\dot{\mathrm{K}}\big(\Sigma; M^{p,\lambda}(\Sigma,\sigma); \ell^q_\alpha(\mathbb{Z})\big)$ (cf. (23.328)), we finally arrive at

$$\begin{aligned} \mathring{\mathrm{K}}\big(\Sigma; \mathring{M}^{p,\lambda}(\Sigma,\sigma); \ell^q_\alpha(\mathbb{Z})\big) &= \overline{\dot{\mathrm{K}}^\alpha_{s,q}(\Sigma,\sigma)}^{\|\cdot\|_{\dot{\mathrm{K}}(\Sigma;M^{p,\lambda}(\Sigma,\sigma);\ell^q_\alpha(\mathbb{Z}))}} \\ &\subseteq \mathring{\mathrm{K}}\big(\Sigma; M^{p,\lambda}(\Sigma,\sigma); \ell^q_\alpha(\mathbb{Z})\big). \end{aligned} \tag{23.517}$$

In concert with (23.500), this ultimately shows (23.499). The proof of Theorem 23.7 is therefore complete. □

The relevance of obtaining the description of vanishing Morrey-based Herz spaces in Theorem 23.7 is particularly apparent from the wealth of results described in Sections 23.7 and 23.8 regarding vanishing Composite Herz spaces. To further elaborate on this topic, pick two exponents $p \in (1,\infty)$ and $q \in (0,\infty)$, along with a parameter $\lambda \in (0, n-1)$. Set $s := \frac{(n-1)p}{n-1-\lambda} \in (p,\infty)$, and fix an index $\alpha \in \mathbb{R}$ satisfying

$$\tfrac{-(n-1)}{s} < \alpha < \tfrac{n-1}{p'} \tag{23.518}$$

where $p' \in (1,\infty)$ denotes the Hölder conjugate exponent of p. As observed in Remark 23.20 we have that

$$\begin{aligned}&M^{p,\lambda}(\Sigma,\sigma) \text{ is a Generalized Banach Function Space on } (\Sigma,\sigma),\\ &\text{with associated space } \big(M^{p,\lambda}(\Sigma,\sigma)\big)' = \mathcal{B}^{p',\lambda}(\Sigma,\sigma), \text{ and the}\\ &\text{Hardy-Littlewood maximal operator } \mathcal{M}_\Sigma \text{ on } \Sigma \text{ is bounded on}\\ &M^{p,\lambda}(\Sigma,\sigma) \text{ as well as on } \mathcal{B}^{p',\lambda}(\Sigma,\sigma),\end{aligned} \tag{23.519}$$

and also

$$\begin{aligned}&\ell^q_\alpha(\mathbb{Z}) \text{ is a Generalized Quasi-Banach Sequence}\\ &\text{Space on } \mathbb{Z} \text{ which is } M^{p,\lambda}(\Sigma,\sigma)\text{-compliant.}\end{aligned} \tag{23.520}$$

In this context, we wish to make the following comments:

Comment 1. If we restrict the range of q to $(1,\infty)$, a scenario ensuring that $\ell^q_\alpha(\mathbb{Z})$ is actually a Generalized Banach Sequence Space on $\mathbb{Z}$, then Theorem 23.7 and Proposition 23.10 give that the Hardy-Littlewood maximal operator

$$\begin{aligned}&\mathcal{M}_\Sigma : \mathring{\mathrm{K}}\big(\Sigma; \mathring{M}^{p,\lambda}(\Sigma,\sigma); \ell^q_\alpha(\mathbb{Z})\big) \to \mathring{\mathrm{K}}\big(\Sigma; \mathring{M}^{p,\lambda}(\Sigma,\sigma); \ell^q_\alpha(\mathbb{Z})\big)\\ &\qquad\text{is a well-defined bounded sublinear mapping.}\end{aligned} \tag{23.521}$$

Comment 2. Once again, restricting to the range $q \in (1,\infty)$, guarantees that $\ell^q_\alpha(\mathbb{Z})$ is actually a Generalized Banach Sequence Space on $\mathbb{Z}$. In this case, thanks to the identification obtained in Theorem 23.7, the boundedness and mapping properties for singular integral operators described in Theorem 23.5, Theorem 23.4, and Theorem 23.6 hold in the context of vanishing Morrey-Based Herz spaces $\mathring{\mathrm{K}}\big(\Sigma; \mathring{M}^{p,\lambda}(\Sigma,\sigma); \ell^q_\alpha(\mathbb{Z})\big)$.

23.11 Fractional Sublinear Operators

In this section, the principal aim is to investigate the boundedness of certain fractional sublinear operators on Composite Herz spaces. We begin by introducing the fractional Hardy-Littlewood maximal operator. Consider a closed Ahlfors regular set $\Sigma \subseteq \mathbb{R}^n$, and abbreviate $\sigma := \mathcal{H}^{n-1}\lfloor\Sigma$. Having fixed a parameter $\beta \in \mathbb{R}$, define the centered fractional Hardy-Littlewood maximal operator of order β acting on each σ-measurable function $f : \Sigma \to \mathbb{R}$ as

$$\mathcal{M}_{\Sigma,\beta} f(x) := \sup_{r>0} \Big(r^\beta \fint_{\Delta(x,r)} |f|\, d\sigma \Big) \quad \text{for all } x \in \Sigma. \tag{23.522}$$

Also, define the action of the non-centered fractional Hardy-Littlewood maximal operator of order β each σ-measurable function $f : \Sigma \to \mathbb{R}$ according to

$$\widetilde{\mathcal{M}}_{\Sigma,\beta} f(x) := \sup_{\substack{\Delta\subseteq\Sigma\\ \Delta\ni x}} \Big(\sigma(\Delta)^\beta \fint_{\Delta} |f|\, d\sigma \Big) \quad \text{for all } x \in \Sigma. \tag{23.523}$$

Observe that for any given σ-measurable function $f : \Sigma \to \mathbb{R}$ one has

$$\mathcal{M}_{\Sigma,\beta} f(x) \approx \widetilde{\mathcal{M}}_{\Sigma, \frac{\beta}{n-1}} f(x) \quad \text{uniformly for } x \in \Sigma, \tag{23.524}$$

with implicit proportionality constants depending only on n, β, and the Ahlfors regularity of Σ.

The following version of Lemma 23.2 for the fractional Hardy-Littlewood maximal operator proves useful.

Lemma 23.7 *Consider a closed unbounded Ahlfors regular set $\Sigma \subseteq \mathbb{R}^n$, and abbreviate $\sigma := \mathcal{H}^{n-1} \lfloor \Sigma$. Assume $\mathbb{X}$ and $\widetilde{\mathbb{X}}$ are Generalized Banach Function Spaces on (Σ, σ), and let $\mathbb{X}'$ and $\widetilde{\mathbb{X}}'$ be their associated spaces, respectively. Having fixed $\beta \in (0, n-1)$, suppose*

$$\mathcal{M}_{\Sigma,\beta} : \mathbb{X}' \to \widetilde{\mathbb{X}}' \quad \textit{is well defined and bounded.} \tag{23.525}$$

Then there exists some $C \in (0, \infty)$ that depends only on n, β, $\|\mathcal{M}_{\Sigma,\beta}\|_{\mathbb{X}' \to \widetilde{\mathbb{X}}'}$, and the Ahlfors regularity constants of Σ with the property that for every $x_0 \in \Sigma$ and $r \in (0, \infty)$ one has

$$\|\mathbb{1}_{\Delta(x_0,r)}\|_{\mathbb{X}} \cdot \|\mathbb{1}_{\Delta(x_0,r)}\|_{\widetilde{\mathbb{X}}'} \leq C \cdot r^{-\beta} \cdot \sigma\big(\Delta(x_0, r)\big). \tag{23.526}$$

Proof Begin by fixing $x_0 \in \Sigma$ and $r \in (0, \infty)$, and abbreviate $\Delta := \Delta(x_0, r)$. Select a function $f \in \mathbb{X}'$ and define

$$P_\Delta f := \sigma(\Delta)^{\beta/(n-1)} \Big(\fint_\Delta |f| \, d\sigma \Big) \cdot \mathbb{1}_\Delta \quad \text{on } \Sigma. \tag{23.527}$$

From (23.523) and (23.524) we see that

$$P_\Delta f \leq \widetilde{\mathcal{M}}_{\Sigma, \frac{\beta}{n-1}} f \leq C \cdot \mathcal{M}_{\Sigma,\beta} f \quad \text{everywhere on } \Sigma, \tag{23.528}$$

and for some constant $C = C(n, \beta, \Sigma) \in (0, \infty)$. This and (23.525) then gives

$$\|P_\Delta f\|_{\widetilde{\mathbb{X}}'} \leq C \|\mathcal{M}_{\Sigma,\beta} f\|_{\widetilde{\mathbb{X}}'} \leq C \|f\|_{\mathbb{X}'} \tag{23.529}$$

for some constant $C \in (0, \infty)$ depending on n, β, $\|\mathcal{M}_{\Sigma,\beta}\|_{\mathbb{X}' \to \widetilde{\mathbb{X}}'}$, and the Ahlfors regularity of Σ. Bearing in mind (23.527), and employing the Ahlfors regularity of Σ we estimate

$$\begin{aligned} \|P_\Delta f\|_{\widetilde{\mathbb{X}}'} &= \sigma(\Delta)^{\beta/(n-1)} \Big(\fint_\Delta |f| \, d\sigma \Big) \cdot \|\mathbb{1}_\Delta\|_{\widetilde{\mathbb{X}}'} \\ &\geq C \frac{r^\beta}{\sigma(\Delta)} \Big(\int_\Sigma |f \cdot \mathbb{1}_\Delta| \, d\sigma \Big) \cdot \|\mathbb{1}_\Delta\|_{\widetilde{\mathbb{X}}'} \end{aligned} \tag{23.530}$$

which in concert with (23.529) entails

$$\Big(\int_{\Sigma} |f \cdot \mathbb{1}_{\Delta}| \, d\sigma\Big) \cdot \|\mathbb{1}_{\Delta}\|_{\widetilde{\mathbb{X}}'} \leq C \cdot r^{-\beta} \cdot \sigma(\Delta) \|f\|_{\mathbb{X}'} \tag{23.531}$$

for some constant $C \in (0, \infty)$ of same nature as in (23.529). After taking the supremum over all functions $f \in \mathbb{X}'$ with $\|f\|_{\mathbb{X}'} \leq 1$, and using the Köthe reflexivity of $\mathbb{X}$ (see [GHA-II, Proposition 5.1.14, pp. 226–227]), we find that

$$\|\mathbb{1}_{\Delta}\|_{\mathbb{X}} \cdot \|\mathbb{1}_{\Delta}\|_{\widetilde{\mathbb{X}}'} \leq C \cdot r^{-\beta} \cdot \sigma(\Delta). \tag{23.532}$$

This finishes the proof of (23.526). □

It is of interest to introduce a notion of compliancy for pairs of quasinorms in sequence spaces that aligns well with Definition 23.5.

Definition 23.8 Fix $\eta, \eta' \in (0, \infty)$ along with a number $2_* \in (1, \infty)$, and consider two functions $\|\cdot\|_1, \|\cdot\|_2 : \mathbb{R}_+^{\mathbb{Z}} \to [0, \infty]$. Then $\|\cdot\|_2$ is said to be (η, η')-compliant with respect to $\|\cdot\|_1$ if there exists some constant $C \in (0, \infty)$, with the property that for every sequence $\{a_\ell\}_{\ell \in \mathbb{Z}} \in \mathbb{R}_+^{\mathbb{Z}}$ one has

$$\begin{aligned}
\|\{a_\ell\}_{\ell\in\mathbb{Z}}\|_2 &\leq C\|\{a_\ell\}_{\ell\in\mathbb{Z}}\|_1 \\
\|\{a_{\ell+1}\}_{\ell\in\mathbb{Z}}\|_2 &\leq C\|\{a_\ell\}_{\ell\in\mathbb{Z}}\|_1, \\
\|\{a_{\ell-1}\}_{\ell\in\mathbb{Z}}\|_2 &\leq C\|\{a_\ell\}_{\ell\in\mathbb{Z}}\|_1, \\
\Big\|\Big\{\sum_{\ell=k+2}^{\infty} 2_*^{(k-\ell)\eta} a_\ell\Big\}_{k\in\mathbb{Z}}\Big\|_2 &\leq C\|\{a_\ell\}_{\ell\in\mathbb{Z}}\|_1, \\
\Big\|\Big\{\sum_{\ell=-\infty}^{k-2} 2_*^{(\ell-k)\eta'} a_\ell\Big\}_{k\in\mathbb{Z}}\Big\|_2 &\leq C\|\{a_\ell\}_{\ell\in\mathbb{Z}}\|_1.
\end{aligned} \tag{23.533}$$

Call the smallest number C doing the hob above the compliancy constant of $\|\cdot\|_2$.

Our next result concerns the boundedness of the fractional Hardy-Littlewood maximal operator on Composite Herz spaces.

Theorem 23.8 *Consider a closed unbounded Ahlfors regular set $\Sigma \subseteq \mathbb{R}^n$, and abbreviate $\sigma := \mathcal{H}^{n-1}\lfloor\Sigma$. Let $\mathbb{X}$ and $\widetilde{\mathbb{X}}$ be Generalized Banach Function Spaces on (Σ, σ), and $\mathbb{Y}, \widetilde{\mathbb{Y}}$ Generalized Quasi-Banach Sequence Spaces on $(\mathbb{Z}, \mu)$, where μ denotes the counting measure. Assume that*

$$\begin{gathered}
\mathcal{M}_\Sigma : \mathbb{X} \to \mathbb{X} \ \textit{ and } \ \mathcal{M}_\Sigma : \widetilde{\mathbb{X}}' \to \widetilde{\mathbb{X}}' \\
\textit{are well defined and bounded.}
\end{gathered} \tag{23.534}$$

Moreover, suppose $\|\cdot\|_{\widetilde{\mathbb{Y}}}$ is $\big((n-1)\tau, (n-1)\tau'\big)$-compliant with respect to $\|\cdot\|_{\mathbb{Y}}$, for some indices $\tau, \tau' \in \mathbb{R}$ satisfying

$$0 < \tau \preceq \Upsilon_{\widetilde{\mathbb{X}}} \quad and \quad 0 < \tau' \preceq \Upsilon_{\mathbb{X}'}. \tag{23.535}$$

Lastly, having fixed $\beta \in (0, n-1)$ *make the assumption that*

$$\mathcal{M}_{\Sigma,\beta} : \mathbb{X} \to \widetilde{\mathbb{X}} \text{ is well defined and bounded.} \tag{23.536}$$

Then

$$\mathcal{M}_{\Sigma,\beta} : \dot{\mathbb{K}}(\Sigma; \mathbb{X}; \mathbb{Y}) \to \dot{\mathbb{K}}(\Sigma; \widetilde{\mathbb{X}}; \widetilde{\mathbb{Y}}) \text{ is well defined and bounded.} \tag{23.537}$$

Remark 23.21 Notice that whenever $\beta \in (0, n-1)$, for every σ-measurable function f on Σ the operator $\mathcal{M}_{\Sigma,\beta}$ has the property that

$$\mathcal{M}_{\Sigma,\beta} f(x) \leq C \int_\Sigma \frac{|f(y)|}{|x-y|^{n-1-\beta}} \, d\sigma(y) \quad \text{for all } x \in \Sigma, \tag{23.538}$$

for some constant $C \in (0, \infty)$ depending only on the Ahlfors regularity character of Σ. This can be seen by using the Ahlfors regularity of Σ, bearing in mind that $n - 1 - \beta > 0$, to bound

$$\begin{aligned} \mathcal{M}_{\Sigma,\beta} f(x) &= \sup_{r>0} \Big(r^\beta \fint_{\Delta(x,r)} |f(y)| \, d\sigma \Big) \leq C \cdot \sup_{r>0} \int_{\Delta(x,r)} \frac{|f(y)|}{r^{n-1-\beta}} \, d\sigma \\ &\leq C \cdot \sup_{r>0} \int_{\Delta(x,r)} \frac{|f(y)|}{|x-y|^{n-1-\beta}} \, d\sigma \leq C \int_\Sigma \frac{|f(y)|}{|x-y|^{n-1-\beta}} \, d\sigma(y). \end{aligned} \tag{23.539}$$

We next turn to the task of giving the proof of Theorem 23.8.

Proof of Theorem 23.8 In view of (23.535) and Remark 23.13, there exists some constant $C \in (0, \infty)$ with the property that for any surface ball $\Delta \subseteq \Sigma$ and any σ-measurable set $E \subseteq \Delta$ we have

$$\frac{\|\mathbb{1}_E\|_{\widetilde{\mathbb{X}}}}{\|\mathbb{1}_\Delta\|_{\widetilde{\mathbb{X}}}} \leq C \left(\frac{\sigma(E)}{\sigma(\Delta)} \right)^\tau \quad \text{and} \quad \frac{\|\mathbb{1}_E\|_{\mathbb{X}'}}{\|\mathbb{1}_\Delta\|_{\mathbb{X}'}} \leq C \left(\frac{\sigma(E)}{\sigma(\Delta)} \right)^{\tau'}. \tag{23.540}$$

Fix an arbitrary function $f \in \dot{\mathbb{K}}(\Sigma; \mathbb{X}; \mathbb{Y})$ and decompose it as

$$f = \sum_{\ell=-\infty}^{\infty} f \cdot \mathbb{1}_{C_\ell} \quad \text{pointwise } \sigma\text{-a.e. on } \Sigma. \tag{23.541}$$

Based on this, (23.522), Lebesgue's Monotone Convergence Theorem, and the fact that $\{C_\ell\}_{\ell \in \mathbb{Z}}$ (cf. (23.16)) is a disjoint family of σ-measurable sets on Σ, we bound

$$(\mathcal{M}_{\Sigma,\beta} f)(x) \leq \sum_{\ell=-\infty}^{\infty} \big[\mathcal{M}_{\Sigma,\beta}(f \cdot \mathbb{1}_{C_\ell})\big](x) \quad \text{for every } x \in \Sigma. \tag{23.542}$$

Then, much as in (23.276), estimate

$$\begin{aligned}\|\mathcal{M}_{\Sigma,\beta}f\|_{\dot{\mathbb{K}}(\Sigma;\widetilde{\mathbb{X}};\widetilde{\mathbb{Y}})} &\leq C\Big\|\Big\{\Big\|\Big[\sum_{\ell=-\infty}^{k-2}\mathcal{M}_{\Sigma,\beta}(f\cdot\mathbb{1}_{C_\ell})\Big]\cdot\mathbb{1}_{C_k}\Big\|_{\widetilde{\mathbb{X}}}\Big\}_{k\in\mathbb{Z}}\Big\|_{\widetilde{\mathbb{Y}}}\\ &\quad+C\Big\|\Big\{\Big\|\Big[\sum_{\ell=k-1}^{k+1}\mathcal{M}_{\Sigma,\beta}(f\cdot\mathbb{1}_{C_\ell})\Big]\cdot\mathbb{1}_{C_k}\Big\|_{\widetilde{\mathbb{X}}}\Big\}_{k\in\mathbb{Z}}\Big\|_{\widetilde{\mathbb{Y}}}\\ &\quad+C\Big\|\Big\{\Big\|\Big[\sum_{\ell=k+2}^{\infty}\mathcal{M}_{\Sigma,\beta}(f\cdot\mathbb{1}_{C_\ell})\Big]\cdot\mathbb{1}_{C_k}\Big\|_{\widetilde{\mathbb{X}}}\Big\}_{k\in\mathbb{Z}}\Big\|_{\widetilde{\mathbb{Y}}}\\ &=:\mathrm{I}_1+\mathrm{I}_2+\mathrm{I}_3,\end{aligned}\tag{23.543}$$

for some constant $C\in(0,\infty)$ depending on the modulus of concavity of $\|\cdot\|_{\widetilde{\mathbb{Y}}}$.

To deal with I_2, use the boundedness (23.536), the lattice property of $\widetilde{\mathbb{X}}$, the triangle inequality for $\|\cdot\|_{\widetilde{\mathbb{X}}}$, the quasi-triangle inequality for $\|\cdot\|_{\widetilde{\mathbb{Y}}}$, and the compliancy of $\|\cdot\|_{\widetilde{\mathbb{Y}}}$ with respect to $\|\cdot\|_{\mathbb{Y}}$ to obtain

$$\begin{aligned}\mathrm{I}_2 &\leq C\Big\|\Big\{\sum_{\ell=k-1}^{k+1}\Big\|[\mathcal{M}_{\Sigma,\beta}(f\cdot\mathbb{1}_{C_\ell})]\cdot\mathbb{1}_{C_k}\Big\|_{\widetilde{\mathbb{X}}}\Big\}_{k\in\mathbb{Z}}\Big\|_{\widetilde{\mathbb{Y}}}\\ &\leq C\Big\|\Big\{\sum_{\ell=k-1}^{k+1}\|\mathcal{M}_{\Sigma,\beta}(f\cdot\mathbb{1}_{C_\ell})\|_{\widetilde{\mathbb{X}}}\Big\}_{k\in\mathbb{Z}}\Big\|_{\widetilde{\mathbb{Y}}}\\ &\leq C\Big\|\Big\{\sum_{\ell=k-1}^{k+1}\|f\cdot\mathbb{1}_{C_\ell}\|_{\mathbb{X}}\Big\}_{k\in\mathbb{Z}}\Big\|_{\widetilde{\mathbb{Y}}}\\ &\leq C\Big\|\big\{\|f\cdot\mathbb{1}_{C_{k-1}}\|_{\mathbb{X}}\big\}_{k\in\mathbb{Z}}\Big\|_{\widetilde{\mathbb{Y}}}+C\Big\|\big\{\|f\cdot\mathbb{1}_{C_k}\|_{\mathbb{X}}\big\}_{k\in\mathbb{Z}}\Big\|_{\widetilde{\mathbb{Y}}}\\ &\quad+C\Big\|\big\{\|f\cdot\mathbb{1}_{C_{k+1}}\|_{\mathbb{X}}\big\}_{k\in\mathbb{Z}}\Big\|_{\widetilde{\mathbb{Y}}}\\ &\leq C\Big\|\big\{\|f\cdot\mathbb{1}_{C_k}\|_{\mathbb{X}}\big\}_{k\in\mathbb{Z}}\Big\|_{\mathbb{Y}}=C\|f\|_{\dot{\mathbb{K}}(\Sigma;\mathbb{X};\mathbb{Y})}\end{aligned}\tag{23.544}$$

for some constant $C\in(0,\infty)$ depending only on $\|\mathcal{M}_{\Sigma,\beta}\|_{\mathbb{X}\to\widetilde{\mathbb{X}}}$, the modulus of concavity of $\|\cdot\|_{\widetilde{\mathbb{Y}}}$, and the compliancy constant of $\|\cdot\|_{\widetilde{\mathbb{Y}}}$.

Regarding I_1 in (23.543), estimates (23.278), (23.538), and the generalized Hölder inequality (10.149), yield the existence of some $C=C(n,\beta)\in(0,\infty)$ such that whenever $k,\ell\in\mathbb{Z}$ with $\ell\leq k-2$ and $x\in C_k$ we have

$$\begin{aligned}\mathcal{M}_{\Sigma,\beta}(f\cdot\mathbb{1}_{C_\ell})(x)&\leq C\int_{C_\ell}\frac{|f(y)|}{|x-y|^{n-1-\beta}}\,d\sigma(y)\\&\leq C\cdot 2_*^{-k(n-1-\beta)}\|f\cdot\mathbb{1}_{C_\ell}\|_{L^1(\Sigma,\sigma)}\\&\leq C\cdot 2_*^{-k(n-1-\beta)}\|f\cdot\mathbb{1}_{C_\ell}\|_{\mathbb{X}}\|\mathbb{1}_{C_\ell}\|_{\mathbb{X}'}.\end{aligned}\tag{23.545}$$

Next, in view of the boundedness of $\mathcal{M}_\Sigma$ on $\mathbb{X}$ (see (23.534)), Lemma 23.7 (whose applicability is ensured by (23.536)), and the Ahlfors regularity of Σ, run the same type of argument as the one in (23.282) with $\mathbb{X}$ replaced by $\widetilde{\mathbb{X}}$, while keeping in mind the second estimate in (23.540), to guarantee the existence of some constant $C = C(n,\Sigma,\|\mathcal{M}_\Sigma\|_{\mathbb{X}\to\mathbb{X}},\|\mathcal{M}_{\Sigma,\beta}\|_{\mathbb{X}\to\widetilde{\mathbb{X}}})\in(0,\infty)$ with the property that for all $k,\ell\in\mathbb{Z}$ with $\ell\leq k-2$ we have

$$\|\mathbb{1}_{C_\ell}\|_{\mathbb{X}'}\cdot\|\mathbb{1}_{C_k}\|_{\widetilde{\mathbb{X}}}\leq C2_*^{(\ell-k)(n-1)\tau'}\cdot 2_*^{k(n-1-\beta)}.\tag{23.546}$$

Combining (23.545), (23.546), the lattice property of $\widetilde{\mathbb{X}}$, and the compliancy of $\|\cdot\|_{\widetilde{\mathbb{Y}}}$ with respect to $\|\cdot\|_{\mathbb{Y}}$, we estimate I_1 from (23.543) as

$$\begin{aligned}\mathrm{I}_1&=C\Big\|\Big\{\Big\|\Big[\sum_{\ell=-\infty}^{k-2}\mathcal{M}_{\Sigma,\beta}(f\cdot\mathbb{1}_{C_\ell})\Big]\cdot\mathbb{1}_{C_k}\Big\|_{\widetilde{\mathbb{X}}}\Big\}_{k\in\mathbb{Z}}\Big\|_{\widetilde{\mathbb{Y}}}\\&\leq C\Big\|\Big\{\Big\|\Big[\sum_{\ell=-\infty}^{k-2}2_*^{-k(n-1-\beta)}\|f\cdot\mathbb{1}_{C_\ell}\|_{\mathbb{X}}\|\mathbb{1}_{C_\ell}\|_{\mathbb{X}'}\Big]\cdot\mathbb{1}_{C_k}\Big\|_{\widetilde{\mathbb{X}}}\Big\}_{k\in\mathbb{Z}}\Big\|_{\widetilde{\mathbb{Y}}}\\&=C\Big\|\Big\{\sum_{\ell=-\infty}^{k-2}2_*^{-k(n-1-\beta)}\|f\cdot\mathbb{1}_{C_\ell}\|_{\mathbb{X}}\|\mathbb{1}_{C_\ell}\|_{\mathbb{X}'}\|\mathbb{1}_{C_k}\|_{\widetilde{\mathbb{X}}}\Big\}_{k\in\mathbb{Z}}\Big\|_{\widetilde{\mathbb{Y}}}\\&\leq C\Big\|\Big\{\sum_{\ell=-\infty}^{k-2}2_*^{-k(n-1-\beta)}\|f\cdot\mathbb{1}_{C_\ell}\|_{\mathbb{X}}2_*^{(\ell-k)(n-1)\tau'}2_*^{k(n-1-\beta)}\Big\}_{k\in\mathbb{Z}}\Big\|_{\widetilde{\mathbb{Y}}}\\&=C\Big\|\Big\{\sum_{\ell=-\infty}^{k-2}2_*^{(\ell-k)(n-1)\tau'}\|f\cdot\mathbb{1}_{C_\ell}\|_{\mathbb{X}}\Big\}_{k\in\mathbb{Z}}\Big\|_{\widetilde{\mathbb{Y}}}\\&\leq C\Big\|\Big\{\|f\cdot\mathbb{1}_{C_\ell}\|_{\mathbb{X}}\Big\}_{k\in\mathbb{Z}}\Big\|_{\mathbb{Y}}=C\|f\|_{\dot{\mathbb{K}}(\Sigma;\mathbb{X};\mathbb{Y})}\end{aligned}\tag{23.547}$$

for some constant $C\in(0,\infty)$ depending exclusively on n, $\|\mathcal{M}_\Sigma\|_{\mathbb{X}\to\mathbb{X}}$, $\|\mathcal{M}_{\Sigma,\beta}\|_{\mathbb{X}\to\widetilde{\mathbb{X}}}$, the modulus of concavity of $\|\cdot\|_{\widetilde{\mathbb{Y}}}$, the compliancy constant of $\|\cdot\|_{\widetilde{\mathbb{Y}}}$, and the Ahlfors regularity of Σ.

We now turn to the task of estimating I_3 in (23.543). To this end, notice that whenever $k,\ell\in\mathbb{Z}$ and $\ell\geq k+2$, and $x\in C_k$, estimates (23.284), (23.538), and the generalized Hölder inequality (10.149), ensure that there exists some constant $C=C(n,\beta)\in(0,\infty)$ for which

$$\begin{aligned}\mathcal{M}_{\Sigma,\beta}(f\cdot\mathbb{1}_{C_\ell})(x) &\le C\int_{C_\ell}\frac{|f(y)|}{|x-y|^{n-1-\beta}}\,d\sigma(y)\\ &\le C\cdot 2_*^{-\ell(n-1-\beta)}\|f\cdot\mathbb{1}_{C_\ell}\|_{L^1(\Sigma,\sigma)}\\ &\le C\cdot 2_*^{-\ell(n-1-\beta)}\|f\cdot\mathbb{1}_{C_\ell}\|_{\mathbb{X}}\|\mathbb{1}_{C_\ell}\|_{\mathbb{X}'}. \end{aligned}\tag{23.548}$$

By switching k with ℓ, as well as $\mathbb{X}'$ with $\widetilde{\mathbb{X}}$, and τ' with τ in (23.546), while bearing in mind the boundedness of $\mathcal{M}_\Sigma$ on $\widetilde{\mathbb{X}}'$ (see (23.534)), we conclude that there exists some constant $C = C(n,\Sigma,\|\mathcal{M}_\Sigma\|_{\widetilde{\mathbb{X}}\to\widetilde{\mathbb{X}}},\|\mathcal{M}_{\Sigma,\beta}\|_{\mathbb{X}\to\widetilde{\mathbb{X}}})\in(0,\infty)$ so that for all $k,\ell\in\mathbb{Z}$ with $\ell\ge k+2$ there holds

$$\|\mathbb{1}_{C_\ell}\|_{\mathbb{X}'}\cdot\|\mathbb{1}_{C_k}\|_{\widetilde{\mathbb{X}}}\le C2_*^{(k-\ell)(n-1)\tau}\cdot 2_*^{\ell(n-1-\beta)}.\tag{23.549}$$

Making use of (23.548), (23.549), the lattice property of $\widetilde{\mathbb{X}}$, and the compliancy of $\|\cdot\|_{\widetilde{\mathbb{Y}}}$ with respect to $\|\cdot\|_{\mathbb{Y}}$, estimate I_3 in (23.543) by

$$\begin{aligned}\mathrm{I}_3 &= C\Big\|\Big\{\Big\|\Big[\sum_{\ell=k+2}^{\infty}\mathcal{M}_{\Sigma,\beta}(f\cdot\mathbb{1}_{C_\ell})\Big]\cdot\mathbb{1}_{C_k}\Big\|_{\widetilde{\mathbb{X}}}\Big\}_{k\in\mathbb{Z}}\Big\|_{\widetilde{\mathbb{Y}}}\\ &\le C\Big\|\Big\{\Big\|\Big[\sum_{\ell=k+2}^{\infty}2_*^{-\ell(n-1-\beta)}\|f\cdot\mathbb{1}_{C_\ell}\|_{\mathbb{X}}\|\mathbb{1}_{C_\ell}\|_{\mathbb{X}'}\Big]\cdot\mathbb{1}_{C_k}\Big\|_{\widetilde{\mathbb{X}}}\Big\}_{k\in\mathbb{Z}}\Big\|_{\widetilde{\mathbb{Y}}}\\ &= C\Big\|\Big\{\sum_{\ell=k+2}^{\infty}2_*^{-\ell(n-1-\beta)}\|f\cdot\mathbb{1}_{C_\ell}\|_{\mathbb{X}}\|\mathbb{1}_{C_\ell}\|_{\mathbb{X}'}\|\mathbb{1}_{C_k}\|_{\widetilde{\mathbb{X}}}\Big\}_{k\in\mathbb{Z}}\Big\|_{\widetilde{\mathbb{Y}}}\\ &\le C\Big\|\Big\{\sum_{\ell=k+2}^{\infty}2_*^{-\ell(n-1-\beta)}\|f\cdot\mathbb{1}_{C_\ell}\|_{\mathbb{X}}2^{(k-\ell)(n-1)\tau}2^{\ell(n-1-\beta)}\Big\}_{k\in\mathbb{Z}}\Big\|_{\widetilde{\mathbb{Y}}}\\ &= C\Big\|\Big\{\sum_{\ell=k+2}^{\infty}2^{(k-\ell)(n-1)\tau}\|f\cdot\mathbb{1}_{C_\ell}\|_{\mathbb{X}}\Big\}_{k\in\mathbb{Z}}\Big\|_{\widetilde{\mathbb{Y}}}\\ &\le C\Big\|\Big\{\|f\cdot\mathbb{1}_{C_\ell}\|_{\mathbb{X}}\Big\}_{k\in\mathbb{Z}}\Big\|_{\mathbb{Y}} = C\|f\|_{\dot{\mathrm{K}}(\Sigma;\mathbb{X};\mathbb{Y})}\end{aligned}\tag{23.550}$$

for some constant $C\in(0,\infty)$ depending only on n, the compliancy constant of $\|\cdot\|_{\widetilde{\mathbb{Y}}}$, the operator norms $\|\mathcal{M}_\Sigma\|_{\widetilde{\mathbb{X}}'\to\widetilde{\mathbb{X}}'}$, $\|\mathcal{M}_{\Sigma,\beta}\|_{\mathbb{X}\to\widetilde{\mathbb{X}}}$ and the Ahlfors regularity of Σ. All together, (23.544), (23.547), and (23.550) we finally conclude that

$$\|\mathcal{M}_{\Sigma,\beta}f\|_{\dot{\mathrm{K}}(\Sigma;\widetilde{\mathbb{X}};\widetilde{\mathbb{Y}})}\le C\|f\|_{\dot{\mathrm{K}}(\Sigma;\mathbb{X};\mathbb{Y})},\tag{23.551}$$

for some constant $C\in(0,\infty)$ depending only on n, the modulus of concavity of $\|\cdot\|_{\widetilde{\mathbb{Y}}}$, the compliancy constant of $\|\cdot\|_{\widetilde{\mathbb{Y}}}$, $\|\mathcal{M}_\Sigma\|_{\mathbb{X}\to\mathbb{X}}$, $\|\mathcal{M}_\Sigma\|_{\widetilde{\mathbb{X}}'\to\widetilde{\mathbb{X}}'}$, $\|\mathcal{M}_{\Sigma,\beta}\|_{\mathbb{X}\to\widetilde{\mathbb{X}}}$, and the Ahlfors regularity of Σ. This finishes the proof of Theorem 23.8. □

The above result self-improves to the boundedness of a class of "fractional" sublinear operators satisfying a similar size condition as in (23.538). A formal statement is as follows.

Theorem 23.9 *Consider an arbitrary unbounded Ahlfors regular set $\Sigma \subseteq \mathbb{R}^n$, and abbreviate $\sigma := \mathcal{H}^{n-1}\lfloor\Sigma$. Assume $\mathbb{X}$, $\widetilde{\mathbb{X}}$ are Generalized Banach Function Spaces on (Σ, σ) such that*

$$\mathcal{M}_\Sigma : \mathbb{X} \to \mathbb{X} \ \textit{and} \ \mathcal{M}_\Sigma : \widetilde{\mathbb{X}}' \to \widetilde{\mathbb{X}}' \quad \textit{are well defined and bounded.} \tag{23.552}$$

Select $\mathbb{Y}$, $\widetilde{\mathbb{Y}}$ Generalized Quasi-Banach Sequence Spaces on $\mathbb{Z}$, equipped with the counting measure, and suppose $\|\cdot\|_{\widetilde{\mathbb{Y}}}$ is $\big((n-1)\tau, (n-1)\tau'\big)$-compliant with respect to $\|\cdot\|_{\mathbb{Y}}$, for some indices $\tau, \tau' \in \mathbb{R}$ satisfying

$$0 < \tau \preceq \Upsilon_{\widetilde{\mathbb{X}}} \quad \textit{and} \quad 0 < \tau' \preceq \Upsilon_{\mathbb{X}'} \tag{23.553}$$

where $\Upsilon_{\widetilde{\mathbb{X}}}$ and $\Upsilon_{\mathbb{X}'}$ are the A_∞ indices of $\widetilde{\mathbb{X}}$ and $\mathbb{X}'$, respectively (cf. Definition 23.4).

Having fixed $\beta \in (0, n-1)$, consider a sublinear operator

$$T_\beta : \dot{\mathrm{K}}(\Sigma; \mathbb{X}; \mathbb{Y}) \to \mathscr{M}(\Sigma, \sigma) \tag{23.554}$$

with the property that

$$|T_\beta f(x)| \leq C \int_\Sigma \frac{|f(y)|}{|x-y|^{n-1-\beta}}\, d\sigma(y) \quad \textit{for all } f \in \dot{\mathrm{K}}(\Sigma; \mathbb{X}; \mathbb{Y}) \textit{ and } x \in \Sigma \setminus \operatorname{supp} f. \tag{23.555}$$

Make the assumption that

$$\mathcal{M}_{\Sigma,\beta} : \mathbb{X} \to \widetilde{\mathbb{X}} \ \textit{is well defined and bounded,} \tag{23.556}$$

where $\mathcal{M}_{\Sigma,\beta}$ is the fractional Hardy-Littlewood maximal operator of order β on Σ (cf. (23.522)). Also, suppose

$$T_\beta : \mathbb{X} \to \widetilde{\mathbb{X}} \ \textit{is well defined and bounded.} \tag{23.557}$$

Then T_β is a well-defined operator from $\dot{\mathrm{K}}(\Sigma; \mathbb{X}; \mathbb{Y})$ into itself, and there exists some $C \in (0, \infty)$, depending only on n, the operator norms of $\mathcal{M}_\Sigma$ on $\mathbb{X}$ and $\widetilde{\mathbb{X}}'$, $\|\mathcal{M}_{\Sigma,\beta}\|_{\mathbb{X}\to\widetilde{\mathbb{X}}}$, the compliancy constant of $\|\cdot\|_{\widetilde{\mathbb{Y}}}$, the modulus of concavity of $\widetilde{\mathbb{Y}}$, and the Ahlfors regularity of Σ, with the property that

$$\|T_\beta f\|_{\dot{\mathrm{K}}(\Sigma;\widetilde{\mathbb{X}};\widetilde{\mathbb{Y}})} \leq C\|f\|_{\dot{\mathrm{K}}(\Sigma;\mathbb{X};\mathbb{Y})} \quad \textit{for all } f \in \dot{\mathrm{K}}(\Sigma; \mathbb{X}; \mathbb{Y}). \tag{23.558}$$

Proof Recall the family of annuli $\{C_\ell\}_{\ell\in\mathbb{Z}}$ defined in (23.16). Fix an arbitrary function $f \in \dot{\mathbb{K}}(\Sigma;\mathbb{X};\mathbb{Y})$ along with some $k \in \mathbb{Z}$, and write

$$f = \sum_{\substack{\ell\in\mathbb{Z}\\ |\ell-k|\geq 2}} f\cdot \mathbb{1}_{C_\ell} + \sum_{\ell=k-1}^{k+1} f\cdot \mathbb{1}_{C_\ell} =: f_1 + f_2 \quad \text{pointwise } \sigma\text{-a.e. on } \Sigma. \tag{23.559}$$

Since $\{C_\ell\}_{\ell\in\mathbb{Z}}$ is a disjoint family of σ-measurable sets that covers the entire Σ, we see that $|f_1| \leq |f|$ and $|f_2| \leq |f|$. Then the lattice property of $\dot{\mathbb{K}}(\Sigma;\mathbb{X};\mathbb{Y})$ implies that both f_1 and f_2 belong to $\dot{\mathbb{K}}(\Sigma;\mathbb{X};\mathbb{Y})$. As such, the action of T_β on f_1 and f_2 is meaningfully defined. Notice that f_1 is supported in $\Sigma\setminus\big(\Delta(x_0, 2_*^{k+2})\setminus\Delta(x_0, 2_*^{k-2})\big)$, which then implies that $(\operatorname{supp} f_1)\cap C_k = \varnothing$. Hence, bearing in mind the size condition (23.555), similar computations as in (23.321), combined with (23.545) and (23.548) ensure that for every $x \in C_k$ we have

$$\begin{aligned}
|T_\beta f_1(x)| &\leq C \sum_{\ell=-\infty}^{k-2} \int_{C_\ell} \frac{|f(y)|}{|x-y|^{n-1-\beta}}\, d\sigma(y) + C \sum_{\ell=k+2}^{\infty} \int_{C_\ell} \frac{|f(y)|}{|x-y|^{n-1-\beta}}\, d\sigma(y). \\
&\leq C \sum_{\ell=-\infty}^{k-2} 2_*^{-k(n-1-\beta)} \|f\cdot \mathbb{1}_{C_\ell}\|_{\mathbb{X}} \|\mathbb{1}_{C_\ell}\|_{\mathbb{X}'} \\
&\quad + C \sum_{\ell=k+2}^{\infty} 2_*^{-\ell(n-1-\beta)} \|f\cdot \mathbb{1}_{C_\ell}\|_{\mathbb{X}} \|\mathbb{1}_{C_\ell}\|_{\mathbb{X}'}
\end{aligned} \tag{23.560}$$

for some constant $C \in (0,\infty)$. In view of this, the compliancy of $\|\cdot\|_{\widetilde{\mathbb{Y}}}$ with respect to $\|\cdot\|_{\mathbb{Y}}$, (23.552), and (23.556), we can employ the same type of argument as in (23.547) and (23.550) to obtain

$$\begin{aligned}
&\Big\|\Big\{\|(T_\beta f_1)\cdot \mathbb{1}_{C_k}\|_{\widetilde{\mathbb{X}}}\Big\}_{k\in\mathbb{Z}}\Big\|_{\widetilde{\mathbb{Y}}} \\
&\quad\leq C\left\|\left\{\sum_{\ell=-\infty}^{k-2} 2_*^{-k(n-1-\beta)} \|f\cdot \mathbb{1}_{C_\ell}\|_{\mathbb{X}} \|\mathbb{1}_{C_\ell}\|_{\mathbb{X}'} \|\mathbb{1}_{C_k}\|_{\widetilde{\mathbb{X}}}\right\}_{k\in\mathbb{Z}}\right\|_{\widetilde{\mathbb{Y}}} \\
&\quad\quad + C\left\|\left\{\sum_{\ell=k+2}^{\infty} 2_*^{-\ell(n-1-\beta)} \|f\cdot \mathbb{1}_{C_\ell}\|_{\mathbb{X}} \|\mathbb{1}_{C_\ell}\|_{\mathbb{X}'} \|\mathbb{1}_{C_k}\|_{\widetilde{\mathbb{X}}}\right\}_{k\in\mathbb{Z}}\right\|_{\widetilde{\mathbb{Y}}} \\
&\quad\leq C\|f\|_{\dot{\mathbb{K}}(\Sigma;\mathbb{X};\mathbb{Y})}
\end{aligned} \tag{23.561}$$

for some $C \in (0,\infty)$ depending only on n, β, the operator norms of $\mathcal{M}_\Sigma$ on $\mathbb{X}$ and $\widetilde{\mathbb{X}}$, $\|\mathcal{M}_{\Sigma,\beta}\|_{\mathbb{X}\to\widetilde{\mathbb{X}}}$, the compliancy constant of $\|\cdot\|_{\widetilde{\mathbb{Y}}}$, the modulus of concavity of $\widetilde{\mathbb{Y}}$, and the Ahlfors regularity constants of Σ.

To proceed, recall f_2 from (23.559) and notice that the membership of f to $\dot{\mathbb{K}}(\Sigma;\mathbb{X};\mathbb{Y})$ implies that $f_2 \in \mathbb{X}$. Keeping this in mind, use the boundedness assumption made in (23.557), the compliancy of $\|\cdot\|_{\widetilde{\mathbb{Y}}}$ with respect to $\|\cdot\|_{\mathbb{Y}}$, and the same type of argument as in (23.544) to estimate

$$\begin{aligned}\Big\|\Big\{\|(T_\beta f_2)\cdot \mathbb{1}_{C_k}\|_{\widetilde{\mathbb{X}}}\Big\}_{k\in\mathbb{Z}}\Big\|_{\widetilde{\mathbb{Y}}} &\leq \Bigg\|\Bigg\{\sum_{\ell=k-1}^{k+1}\|T_\beta(f\cdot \mathbb{1}_{C_\ell})\|_{\widetilde{\mathbb{X}}}\Bigg\}_{k\in\mathbb{Z}}\Bigg\|_{\widetilde{\mathbb{Y}}} \\ &\leq C\Bigg\|\Bigg\{\sum_{\ell=k-1}^{k+1}\|f\cdot \mathbb{1}_{C_\ell}\|_{\mathbb{X}}\Bigg\}_{k\in\mathbb{Z}}\Bigg\|_{\widetilde{\mathbb{Y}}} \\ &\leq C\|f\|_{\dot{\mathbb{K}}(\Sigma;\mathbb{X};\mathbb{Y})}\end{aligned} \tag{23.562}$$

for some $C\in(0,\infty)$ depending on the compliancy constant of $\|\cdot\|_{\widetilde{\mathbb{Y}}}$, the modulus of concavity of of $\|\cdot\|_{\widetilde{\mathbb{Y}}}$, and the operator norm of T_β from $\mathbb{X}$ to $\widetilde{\mathbb{X}}$. The sublinearity of T_β, combined with (23.559), (23.561), and (23.562) imply (23.558). This ultimately proves Theorem 23.9. □

We conclude by recording some friendlier (albeit weaker) versions of Theorem 23.9. To state them, denote by $x_0\in\Sigma$ the reference point used to define $\dot{\mathbb{K}}(\Sigma;\mathbb{X};\mathbb{Y})$ (cf. Definition 23.2). Since the present assumptions imply that the space $\dot{\mathbb{K}}(\Sigma;\mathbb{X};\mathbb{Y})$ is included in $L^1_{\rm loc}(\Sigma\backslash\{x_0\},\sigma)$ (thanks to [GHA-II, Proposition 5.2.4, p. 245], (23.18), and item (Q5') in Proposition 23.1 used with $\rho:=\|\cdot\|_{\mathbb{Y}}$), the conclusion in (23.558) holds if we replace $\dot{\mathbb{K}}(\Sigma;\mathbb{X};\mathbb{Y})$ by $L^1_{\rm loc}(\Sigma\backslash\{x_0\},\sigma)$ in (23.554) and (23.555). Moreover, under the additional assumptions that $\mathcal{M}_\Sigma$ is bounded on $\mathbb{X}'$, and $\mathbb{Y}$ is $\mathbb{X}$-compliant (in the sense of Definition 23.5), then in light of Corollary 23.4 one may actually replace $\dot{\mathbb{K}}(\Sigma;\mathbb{X};\mathbb{Y})$ by $L^1\big(\Sigma,\frac{\sigma(x)}{1+|x|^{n-1}}\big)$ in (23.554) and (23.555) and still have the same conclusion in (23.558).

23.12 Extrapolation on Composite Herz Spaces

The primary goal of this section is to produce an extrapolation result converting norm estimates for Muckenhoupt weighted Lebesgue spaces into norm estimates for Composite Herz spaces, in the case when the base spaces are both Generalized Banach Function Spaces. This result, which is of basic importance to the present work reads as follows:

Theorem 23.10 *Assume $\Sigma\subseteq\mathbb{R}^n$ is a closed unbounded Ahlfors regular set, and abbreviate $\sigma:=\mathcal{H}^{n-1}\lfloor\Sigma$. Let $\mathbb{X}$ be a Generalized Banach Function Space on (Σ,σ) and make the assumption that*

$$\begin{gathered}\mathcal{M}_\Sigma:\mathbb{X}\to\mathbb{X}\ \textit{and}\ \mathcal{M}_\Sigma:\mathbb{X}'\to\mathbb{X}'\\ \textit{are well-defined and bounded operators,}\end{gathered} \tag{23.563}$$

where $\mathcal{M}_\Sigma$ is the Hardy-Littlewood maximal operator on Σ (cf. (2.26)*). Consider $\mathbb{Y}$ a Generalized Banach Sequence Space on $(\mathbb{Z}, \mu)$, where μ is the counting measure, and suppose $\mathbb{Y}$ is $\mathbb{X}$-compliant (in the sense of Definition 23.7).*

Then there exists some constant $C_\Sigma \in (0, \infty)$ depending only on the ambient Σ with the following significance. Having fixed an integrability exponent $p_0 \in (1, \infty)$, introduce

$$
\begin{aligned}
W_{\mathbb{X},\mathbb{Y},p_0} &:= C_\Sigma^{p_0} \|\mathcal{M}_\Sigma\|^{p_0-1}_{\dot{\mathrm{K}}(\Sigma;\mathbb{X};\mathbb{Y})\to\dot{\mathrm{K}}(\Sigma;\mathbb{X};\mathbb{Y})} \|\mathcal{M}_\Sigma\|_{\dot{\mathrm{K}}(\Sigma;\mathbb{X}';\mathbb{Y}')\to\dot{\mathrm{K}}(\Sigma;\mathbb{X}';\mathbb{Y}')} \\
\textit{and} \quad W_{\mathbb{X}',\mathbb{Y}',p_0} &:= C_\Sigma^{p_0} \|\mathcal{M}_\Sigma\|^{p_0-1}_{\dot{\mathrm{K}}(\Sigma;\mathbb{X}';\mathbb{Y}')\to\dot{\mathrm{K}}(\Sigma;\mathbb{X}';\mathbb{Y}')} \|\mathcal{M}_\Sigma\|_{\dot{\mathrm{K}}(\Sigma;\mathbb{X};\mathbb{Y})\to\dot{\mathrm{K}}(\Sigma;\mathbb{X};\mathbb{Y})}.
\end{aligned}
\tag{23.564}
$$

Consider two σ-measurable functions f and g defined on Σ such that for each weight $w \in A_{p_0}(\Sigma, \sigma)$ with $[w]_{A_{p_0}} \leq W_{\mathbb{X},\mathbb{Y},p_0}$ one has

$$
\|f\|_{L^{p_0}(\Sigma,w)} \leq C_w \|g\|_{L^{p_0}(\Sigma,w)} \tag{23.565}
$$

for some constant $C_w \in (0, \infty)$ that depends only on n, p_0, w, and the Ahlfors regularity of Σ. Under these considerations, it follows that

$$
\|f\|_{\dot{\mathrm{K}}(\Sigma;\mathbb{X};\mathbb{Y})} \leq 2^{2-1/p_0} \Bigg(\sup_{\substack{w \in A_{p_0}(\Sigma,\sigma) \\ [w]_{A_{p_0}} \leq W_{\mathbb{X},\mathbb{Y},p_0}}} C_w \Bigg) \|g\|_{\dot{\mathrm{K}}(\Sigma;\mathbb{X};\mathbb{Y})}. \tag{23.566}
$$

Furthermore, if (23.565) *holds for all $w \in A_{p_0}(\Sigma, \sigma)$ with $[w]_{A_{p_0}} \leq W_{\mathbb{X}',\mathbb{Y}',p_0}$ then*

$$
\|f\|_{\dot{\mathrm{K}}(\Sigma;\mathbb{X}';\mathbb{Y}')} \leq 2^{2-1/p_0} \Bigg(\sup_{\substack{w \in A_{p_0}(\Sigma,\sigma) \\ [w]_{A_{p_0}} \leq W_{\mathbb{X}',\mathbb{Y}',p_0}}} C_w \Bigg) \|g\|_{\dot{\mathrm{K}}(\Sigma;\mathbb{X}';\mathbb{Y}')}. \tag{23.567}
$$

Lastly, the same type of conclusions are valid in the case when $\mathbb{Y}$ is a Generalized Banach Sequence Space on $\mathbb{N}_0$ now with the inhomogeneous Composite Herz space $\mathrm{K}(\Sigma; \mathbb{X}; \mathbb{Y})$ playing the role of $\dot{\mathrm{K}}(\Sigma; \mathbb{X}; \mathbb{Y})$.

Proof To get started, observe that the assumptions made in the first paragraph of the statement of Theorem 23.10 allow us to invoke Theorem 23.1, Proposition 23.4, and Remark 23.19 to ensure that $\dot{\mathrm{K}}(\Sigma; \mathbb{X}; \mathbb{Y})$ is a Generalized Banach Function Space on (Σ, σ) and

$$
\begin{aligned}
&\mathcal{M}_\Sigma : \dot{\mathrm{K}}(\Sigma; \mathbb{X}; \mathbb{Y}) \to \dot{\mathrm{K}}(\Sigma; \mathbb{X}; \mathbb{Y}) \\
\text{and } &\mathcal{M}_\Sigma : \big(\dot{\mathrm{K}}(\Sigma; \mathbb{X}; \mathbb{Y})\big)' \to \big(\dot{\mathrm{K}}(\Sigma; \mathbb{X}; \mathbb{Y})\big)'
\end{aligned}
\tag{23.568}
$$

are well-defined and bounded operators.

As such, the constants appearing in (23.564) are meaningfully defined and finite. Granted these, all conclusions follow from the extrapolation result for Generalized Banach Function Spaces from [GHA-II, Theorem 5.2.1, p. 233]. □

23.13 Boundary Value Problems on Composite Herz Spaces

As a culmination of the work up to this stage, we are now positioned to treat boundary value problems in Ahlfors regular domains for weakly elliptic systems when boundary data are prescribed from Composite Herz spaces. We start by dealing with the Dirichlet Problem with boundary data from Composite Herz spaces. As a preamble, we introduce the scale of boundary Sobolev spaces of order one styled after Composite Herz spaces.

Definition 23.9 Assume Ω is an Ahlfors regular domain with unbounded boundary, and abbreviate $\sigma := \mathcal{H}^{n-1}\lfloor\partial\Omega$. Consider $\mathbb{X}$ a Generalized Banach Function Space on $(\partial\Omega, \sigma)$ and, with $\mathbb{X}'$ denoting its associated space, suppose

$$\begin{gathered} \mathcal{M}_{\partial\Omega} : \mathbb{X} \to \mathbb{X} \quad \text{and} \quad \mathcal{M}_{\partial\Omega} : \mathbb{X}' \to \mathbb{X}' \\ \text{are well defined and bounded,} \end{gathered} \tag{23.569}$$

where $\mathcal{M}_{\partial\Omega}$ is the Hardy-Littlewood maximal operator on $\partial\Omega$. Let $\mathbb{Y}$ be a Generalized Banach Sequence Space on $(\mathbb{Z}, \mu)$, where μ is counting measure, and suppose $\mathbb{Y}$ is $\mathbb{X}$-compliant (in the sense of Definition 23.7). Introduce the `Composite Herz-based Sobolev space of order one` on $\partial\Omega$ as

$$\begin{aligned} &\dot{\mathrm{K}}_1(\partial\Omega; \mathbb{X}; \mathbb{Y}) \\ &\quad := \Big\{ f \in \dot{\mathrm{K}}(\partial\Omega; \mathbb{X}; \mathbb{Y}) : \partial_{\tau_{jk}} f \in \dot{\mathrm{K}}(\partial\Omega; \mathbb{X}; \mathbb{Y}), \ \text{ if } \ 1 \le j, k \le n \Big\} \end{aligned} \tag{23.570}$$

and equip it with the norm

$$\begin{gathered} \|f\|_{\dot{\mathrm{K}}_1(\partial\Omega;\mathbb{X};\mathbb{Y})} := \|f\|_{\dot{\mathrm{K}}(\partial\Omega;\mathbb{X};\mathbb{Y})} + \sum_{j,k=1}^{n} \|\partial_{\tau_{jk}} f\|_{\dot{\mathrm{K}}(\partial\Omega;\mathbb{X};\mathbb{Y})} \\ \text{for every function } f \in \dot{\mathrm{K}}_1(\partial\Omega; \mathbb{X}; \mathbb{Y}). \end{gathered} \tag{23.571}$$

Finally, in the case when $\mathbb{Y}$ is a Generalized Banach Sequence Space on $\mathbb{N}_0$, define $\mathrm{K}_1(\partial\Omega; \mathbb{X}; \mathbb{Y})$ in a similar fashion with now the inhomogeneous Composite Herz space $\mathrm{K}(\partial\Omega; \mathbb{X}; \mathbb{Y})$ playing the role of $\dot{\mathrm{K}}(\partial\Omega; \mathbb{X}; \mathbb{Y})$.

In the setting of the above definition, the Composite Herz space $\dot{\mathrm{K}}(\partial\Omega; \mathbb{X}; \mathbb{Y})$ is included in $L^1_{\mathrm{loc}}(\Sigma, \sigma)$ (cf. Corollary 23.3). As such, the weak tangential derivatives (cf. Definition 14.1) appearing in the definition of the space in (23.570) are meaningful.

Here is the result advertised earlier.

Theorem 23.11 *Suppose $\Omega \subseteq \mathbb{R}^n$ is an Ahlfors regular domain, and abbreviate $\sigma := \mathcal{H}^{n-1}\lfloor\partial\Omega$. Denote by ν the geometric measure theoretic outward unit normal to Ω. Let $\mathbb{X}$ be a Generalized Banach Function Space on $(\partial\Omega, \sigma)$ and let $\mathbb{X}'$ be its associated space. Assume*

$$\begin{gathered} \mathcal{M}_{\partial\Omega} : \mathbb{X} \to \mathbb{X} \quad \text{and} \quad \mathcal{M}_{\partial\Omega} : \mathbb{X}' \to \mathbb{X}' \\ \text{are well defined and bounded,} \end{gathered} \tag{23.572}$$

where $\mathcal{M}_{\partial\Omega}$ is the Hardy-Littlewood maximal operator on $\partial\Omega$. Let $\mathbb{Y}$ be a Generalized Banach Sequence Space on $(\mathbb{Z}, \mu)$, where μ is counting measure, and suppose $\mathbb{Y}$ is $\mathbb{X}$-compliant (in the sense of Definition 23.7). Given L a second-order, homogeneous, constant complex coefficient, weakly elliptic, $M \times M$ system in $\mathbb{R}^n$, and having fixed an aperture parameter $\kappa \in (0, \infty)$, formulate the Dirichlet Problem for L in Ω with boundary data in the Composite Herz space $[\dot{\mathrm{K}}(\partial\Omega; \mathbb{X}; \mathbb{Y})]^M$ as:

$$\begin{cases} u \in \big[\mathscr{C}^\infty(\Omega)\big]^M, \\ Lu = 0 \ \text{ in } \Omega, \\ \mathcal{N}_\kappa u \in \dot{\mathrm{K}}(\partial\Omega; \mathbb{X}; \mathbb{Y}), \\ u\big|_{\partial\Omega}^{\kappa-\mathrm{n.t.}} = f \in \big[\dot{\mathrm{K}}(\partial\Omega; \mathbb{X}; \mathbb{Y})\big]^M. \end{cases} \tag{23.573}$$

In relation to this, the following statements are true:

(a) [Existence, Integral Representation, and Estimates] *If $\mathfrak{A}_L^{\mathrm{dis}} \neq \varnothing$ then there exists some threshold $\delta \in (0, 1)$ depending on n, M, L, the operator norms of $\mathcal{M}_{\partial\Omega}$ on $\mathbb{X}$ and $\mathbb{X}'$, the compliancy constant of $\|\cdot\|_{\mathbb{Y}}$, and the Ahlfors regularity constants of $\partial\Omega$ with the property that if*

$$\|\nu\|_{[\mathrm{BMO}(\partial\Omega,\sigma)]^n} < \delta \tag{23.574}$$

it follows that the Dirichlet Problem (23.573) *is solvable. More specifically, for any $A \in \mathfrak{A}_L^{\mathrm{dis}}$ the boundary-to-boundary double layer potential K_A associated with A (cf.* (7.23)*) is well defined, linear, and bounded from $\big[\dot{\mathrm{K}}(\partial\Omega; \mathbb{X}; \mathbb{Y})\big]^M$ into itself, the operator*

$$\tfrac{1}{2}I + K_A \quad \text{is invertible on } \big[\dot{\mathrm{K}}(\partial\Omega; \mathbb{X}; \mathbb{Y})\big]^M, \tag{23.575}$$

and the function $u : \Omega \to \mathbb{C}^M$ defined as

$$u := \mathcal{D}_A\Big[\big(\tfrac{1}{2}I + K_A\big)^{-1} f\Big] \ \text{ in } \Omega \tag{23.576}$$

solves the Dirichlet Problem (23.573) *and satisfies*

$$\|\mathcal{N}_\kappa u\|_{\dot{\mathrm{K}}(\partial\Omega;\mathbb{X};\mathbb{Y})} \approx \|f\|_{[\dot{\mathrm{K}}(\partial\Omega;\mathbb{X};\mathbb{Y})]^M} \tag{23.577}$$

with implicit proportionality constants independent of f.

(b) [Additional Regularity] *The function u defined in* (23.576) *satisfies*

$$\mathcal{N}_\kappa(\nabla u) \in \dot{\mathrm{K}}(\partial\Omega; \mathbb{X}; \mathbb{Y}) \iff f \in \big[\dot{\mathrm{K}}_1(\partial\Omega; \mathbb{X}; \mathbb{Y})\big]^M, \tag{23.578}$$

and if either of these conditions hold it follows that

$$\begin{gathered}(\nabla u)\Big|_{\partial\Omega}^{\kappa-\mathrm{n.t.}} \text{ exists at } \sigma\text{-a.e. point on } \partial\Omega \text{ and} \\ \|\mathcal{N}_\kappa u\|_{\dot{\mathrm{K}}(\partial\Omega;\mathbb{X};\mathbb{Y})} + \|\mathcal{N}_\kappa(\nabla u)\|_{\dot{\mathrm{K}}(\partial\Omega;\mathbb{X};\mathbb{Y})} \approx \|f\|_{[\dot{\mathrm{K}}_1(\partial\Omega;\mathbb{X};\mathbb{Y})]^M}.\end{gathered} \tag{23.579}$$

(c) [Uniqueness] *If $\mathfrak{A}^{\mathrm{dis}}_{L^\top} \neq \varnothing$, there exists a threshold $\delta \in (0,1)$ depending exclusively on n, L, the operator norms of $\mathcal{M}_{\partial\Omega}$ on $\mathbb{X}$ and $\mathbb{X}'$, the compliancy constant of $\|\cdot\|_{\mathbb{Y}}$, and the Ahlfors regularity constants of $\partial\Omega$, with the property that whenever $\|\nu\|_{[\mathrm{BMO}(\partial\Omega,\sigma)]^n} < \delta$ then the Dirichlet Problem* (23.573) *has at most one solution.*

(d) [Well-Posedness] *If $\mathfrak{A}^{\mathrm{dis}}_{L} \neq \varnothing$ and $\mathfrak{A}^{\mathrm{dis}}_{L^\top} \neq \varnothing$ then there exists some threshold $\delta \in (0,1)$ depending solely on n, L, the operator norms of $\mathcal{M}_{\partial\Omega}$ on $\mathbb{X}$ and $\mathbb{X}'$, the compliancy constant of $\|\cdot\|_{\mathbb{Y}}$, and the Ahlfors regularity constants of $\partial\Omega$ such that whenever $\|\nu\|_{[\mathrm{BMO}(\partial\Omega,\sigma)]^n} < \delta$ it follows that the Dirichlet Problem* (23.573) *is well posed.*

(e) [Sharpness] *If $\mathfrak{A}^{\mathrm{dis}}_{L} = \varnothing$ then the Dirichlet Problem* (23.573) *may fail to be solvable (actually for boundary data belonging to an infinite dimensional subspace of the corresponding Sobolev space styled after the given Composite Herz space). Also, if $\mathfrak{A}^{\mathrm{dis}}_{L^\top} = \varnothing$ then the Dirichlet Problem* (23.573) *may have more than one solution (in fact, the linear space of null-solutions may actually be infinite dimensional).*

Lastly, the same type of conclusions regarding mapping properties, existence, uniqueness, integral representation, estimates, regularity, well-posedness, and sharpness also hold for the inhomogeneous Composite Herz space $\mathrm{K}(\partial\Omega; \mathbb{X}; \mathbb{Y})$, *now assuming that $\mathbb{Y}$ is a Generalized Banach Sequence Space on* $(\mathbb{N}_0, \mu)$.

The above result is new even for the basic case when the differential operator in question is the Laplacian, and the underlying domain is the upper half-space $\mathbb{R}^n_+$. In addition, every choice of the base spaces $\mathbb{X}$ and $\mathbb{Y}$ leads to a new result.

Proof of Theorem 23.11 The assumptions made in the statement of the present theorem ensure that $\dot{\mathrm{K}}(\partial\Omega; \mathbb{X}; \mathbb{Y})$ is a Generalized Banach Function Space on $(\partial\Omega, \sigma)$ (cf. Proposition 23.4) such that the Hardy-Littlewood maximal operator $\mathcal{M}_{\partial\Omega}$ is bounded both on $\dot{\mathrm{K}}(\partial\Omega; \mathbb{X}; \mathbb{Y})$ and on its associated space $\dot{\mathrm{K}}(\partial\Omega; \mathbb{X}'; \mathbb{Y}')$ (cf. Theorem 23.1 and Proposition 23.5). As such, the validity of all claims follows from [GHA-V, Theorem 8.4.40, pp. 869–871]. □

Next we discuss the Inhomogeneous Regularity Problem with boundary data in Composite Herz spaces.

Theorem 23.12 *Suppose $\Omega \subseteq \mathbb{R}^n$ is an Ahlfors regular domain, set $\sigma := \mathcal{H}^{n-1}\lfloor\partial\Omega$, and denote by ν the geometric measure theoretic outward unit normal to Ω. Let $\mathbb{X}$ be a Generalized Banach Function Space on $(\partial\Omega, \sigma)$ and make the assumption that*

$$\begin{gathered} \mathcal{M}_{\partial\Omega} : \mathbb{X} \to \mathbb{X} \quad \textit{and} \quad \mathcal{M}_{\partial\Omega} : \mathbb{X}' \to \mathbb{X}' \\ \textit{are well defined and bounded,} \end{gathered} \tag{23.580}$$

where $\mathcal{M}_{\partial\Omega}$ is the Hardy-Littlewood maximal operator on $\partial\Omega$. Select $\mathbb{Y}$ a Generalized Banach Sequence Space on $(\mathbb{Z}, \mu)$, where μ is the counting measure, and assume that $\mathbb{Y}$ is $\mathbb{X}$-compliant (in the sense of Definition 23.7). Also, recall the Composite Herz-based Sobolev space of order one on $\partial\Omega$ (cf. Definition 23.9). Given L a second-order, homogeneous, constant complex coefficient, weakly elliptic, $M \times M$ system in $\mathbb{R}^n$, and an aperture parameter $\kappa \in (0, \infty)$, consider the Inhomogeneous Regularity Problem for L in Ω with boundary data in the Composite Herz-based Sobolev space $[\dot{\mathrm{K}}_1(\partial\Omega; \mathbb{X}; \mathbb{Y})]^M$, namely:

$$\begin{cases} u \in [\mathscr{C}^\infty(\Omega)]^M, \\ Lu = 0 \;\; \textit{in } \Omega, \\ \mathcal{N}_\kappa u, \; \mathcal{N}_\kappa(\nabla u) \in \dot{\mathrm{K}}(\partial\Omega; \mathbb{X}; \mathbb{Y}), \\ u\big|_{\partial\Omega}^{\kappa-\mathrm{n.t.}} = f \in \big[\dot{\mathrm{K}}_1(\partial\Omega; \mathbb{X}; \mathbb{Y})\big]^M. \end{cases} \tag{23.581}$$

Then the following conclusions are valid:

(a) [Existence, Integral Representation, and Estimates] *Whenever $\mathfrak{A}_L^{\mathrm{dis}} \neq \emptyset$ there exists some $\delta \in (0, 1)$ depending on n, M, L, the operator norms of $\mathcal{M}_{\partial\Omega}$ on $\mathbb{X}$ and $\mathbb{X}'$, the compliancy constant of $\|\cdot\|_{\mathbb{Y}}$, and the Ahlfors regularity constants of $\partial\Omega$ with the property that if*

$$\|\nu\|_{[\mathrm{BMO}(\partial\Omega, \sigma)]^n} < \delta \tag{23.582}$$

then the Inhomogeneous Regularity Problem (23.581) *is solvable. Concretely, if $A \in \mathfrak{A}_L^{\mathrm{dis}}$ then the boundary-to-boundary double layer potential K_A associated with A (cf.* (7.23)) *is well defined, linear, and bounded from $\big[\dot{\mathrm{K}}_1(\partial\Omega; \mathbb{X}; \mathbb{Y})\big]^M$ into itself, and*

$$\tfrac{1}{2}I + K_A \quad \textit{is invertible on } \big[\dot{\mathrm{K}}_1(\partial\Omega; \mathbb{X}; \mathbb{Y})\big]^M. \tag{23.583}$$

Also, the function $u : \Omega \to \mathbb{C}^M$ defined as

$$u := \mathcal{D}_A\big[\big(\tfrac{1}{2}I + K_A\big)^{-1} f\big] \;\; \textit{in } \Omega \tag{23.584}$$

solves the Inhomogeneous Regularity Problem (23.581) *and satisfies*

$$\begin{aligned}\|\mathcal{N}_\kappa u\|_{\dot{\mathbb{K}}(\partial\Omega;\mathbb{X};\mathbb{Y})} &\approx \|f\|_{[\dot{\mathbb{K}}(\partial\Omega;\mathbb{X};\mathbb{Y})]^M} \quad \text{and}\\ \|\mathcal{N}_\kappa u\|_{\dot{\mathbb{K}}(\partial\Omega;\mathbb{X};\mathbb{Y})} + \|\mathcal{N}_\kappa(\nabla u)\|_{\dot{\mathbb{K}}(\partial\Omega;\mathbb{X};\mathbb{Y})} &\approx \|f\|_{[\dot{\mathbb{K}}_1(\partial\Omega;\mathbb{X};\mathbb{Y})]^M}.\end{aligned} \tag{23.585}$$

(b) [Uniqueness] *Whenever* $\mathfrak{A}^{\rm dis}_{L^\top} \neq \varnothing$ *there exists some threshold* $\delta \in (0,1)$ *depending only on n, L, the operator norms of* $\mathcal{M}_{\partial\Omega}$ *on* $\mathbb{X}$ *and* $\mathbb{X}'$*, the compliancy constant of* $\|\cdot\|_{\mathbb{Y}}$*, and the Ahlfors regularity constants of* $\partial\Omega$ *with the property that whenever* $\|\nu\|_{[\mathrm{BMO}(\partial\Omega,\sigma)]^n} < \delta$ *then the Inhomogeneous Regularity Problem* (23.581) *admits at most one solution.*

(c) [Well-Posedness] *If* $\mathfrak{A}^{\rm dis}_{L} \neq \varnothing$ *and* $\mathfrak{A}^{\rm dis}_{L^\top} \neq \varnothing$*, then there exists some* $\delta \in (0,1)$ *depending exclusively on n, L, the operator norms of* $\mathcal{M}_{\partial\Omega}$ *on* $\mathbb{X}$ *and* $\mathbb{X}'$*, the compliancy constant of* $\|\cdot\|_{\mathbb{Y}}$*, and the Ahlfors regularity constants of* $\partial\Omega$ *for which whenever* $\|\nu\|_{[\mathrm{BMO}(\partial\Omega,\sigma)]^n} < \delta$*, it follows that the Inhomogeneous Regularity Problem* (23.581) *is well posed.*

(d) [Sharpness] *If* $\mathfrak{A}^{\rm dis}_{L} = \varnothing$ *the Inhomogeneous Regularity Problem* (23.581) *may fail to be solvable* (*actually for boundary data belonging to an infinite dimensional subspace of* $[\dot{\mathbb{K}}_1(\partial\Omega;\mathbb{X};\mathbb{Y})]^M$) *even when* Ω *is a half-space, and if* $\mathfrak{A}^{\rm dis}_{L^\top} = \varnothing$ *the Inhomogeneous Regularity Problem* (23.581) *may possess more than one solution* (*in fact, the linear subspace of null-solutions may actually be infinite dimensional*) *even when* Ω *is a half-space. In particular, if either* $\mathfrak{A}^{\rm dis}_{L} = \varnothing$ *or* $\mathfrak{A}^{\rm dis}_{L^\top} = \varnothing$ *then the Inhomogeneous Regularity Problem* (23.581) *may fail to be well posed even when* Ω *is a half-space.*

Lastly, the same type of conclusions regarding mapping properties, existence, uniqueness, integral representation, estimates, and well-posedness are valid when $\mathbb{Y}$ *is a Generalized Banach Sequence Space on* $(\mathbb{N}_0,\mu)$*, and with the inhomogeneous Composite Herz space* $\mathbb{K}(\partial\Omega;\mathbb{X};\mathbb{Y})$ *now replacing* $\dot{\mathbb{K}}(\partial\Omega;\mathbb{X};\mathbb{Y})$.

Proof All claims can be obtained by invoking [GHA-V, Theorem 8.4.41, pp. 871–872], bearing in mind that $\dot{\mathbb{K}}(\partial\Omega;\mathbb{X};\mathbb{Y})$ is a Generalized Banach Function Space (cf. Proposition 23.4), and that the Hardy-Littlewood maximal operator $\mathcal{M}_{\partial\Omega}$ is bounded on $\dot{\mathbb{K}}(\partial\Omega;\mathbb{X};\mathbb{Y})$ and on its associated space $\big(\dot{\mathbb{K}}(\partial\Omega;\mathbb{X};\mathbb{Y})\big)' = \dot{\mathbb{K}}(\partial\Omega;\mathbb{X}';\mathbb{Y}')$ (cf. Theorem 23.1 and Proposition 23.5). □

We next consider the Homogeneous Regularity Problem with boundary data in Composite Herz-based homogeneous Sobolev spaces, defined below:

Definition 23.10 Consider an Ahlfors regular domain $\Omega \subseteq \mathbb{R}^n$ with unbounded boundary, and set $\sigma := \mathcal{H}^{n-1}\lfloor\partial\Omega$. Let $\mathbb{X}$ be a Generalized Banach Function Space on $(\partial\Omega,\sigma)$ and $\mathbb{Y}$ a Generalized Banach Sequence Space on $(\mathbb{Z},\mu)$, where μ is counting measure. Introduce the `Composite Herz-based homogeneous Sobolev space` on $\partial\Omega$ as

$$\dot{\mathbb{K}}_{(1)}(\partial\Omega;\mathbb{X};\mathbb{Y}) := \Big\{ f \in L^1\big(\partial\Omega, \tfrac{\sigma(x)}{1+|x|^n}\big) \cap \dot{\mathbb{K}}_{\rm loc}(\partial\Omega;\mathbb{X};\mathbb{Y}) : \\ \partial_{\tau_{jk}} f \in \dot{\mathbb{K}}(\partial\Omega;\mathbb{X};\mathbb{Y}) \;\text{ for }\; j,k \in \{1,\dots,n\}\Big\}, \tag{23.586}$$

where the membership $f \in \dot{\mathbb{K}}_{\rm loc}(\partial\Omega;\mathbb{X};\mathbb{Y})$ means $f \cdot \mathbb{1}_\Delta \in \dot{\mathbb{K}}(\partial\Omega;\mathbb{X};\mathbb{Y})$ for every surface ball $\Delta \subseteq \partial\Omega$. Equip this space with the seminorm

$$\|f\|_{\dot{\mathbb{K}}_{(1)}(\partial\Omega;\mathbb{X};\mathbb{Y})} := \sum_{j,k=1}^{n} \|\partial_{\tau_{jk}} f\|_{\dot{\mathbb{K}}(\partial\Omega;\mathbb{X};\mathbb{Y})} \quad \text{for each } f \in \dot{\mathbb{K}}_{(1)}(\partial\Omega;\mathbb{X};\mathbb{Y}). \tag{23.587}$$

We are now in a position to state and prove the following well-posedness result.

Theorem 23.13 *Consider $\Omega \subseteq \mathbb{R}^n$ an Ahlfors regular domain, set $\sigma := \mathcal{H}^{n-1}\lfloor\partial\Omega$, and denote by ν the geometric measure theoretic outward unit normal to Ω. Let $\mathbb{X}$ be a Generalized Banach Function Space on $(\partial\Omega,\sigma)$ with the property that*

$$\mathcal{M}_{\partial\Omega} : \mathbb{X} \to \mathbb{X} \;\; \textit{and} \;\; \mathcal{M}_{\partial\Omega} : \mathbb{X}' \to \mathbb{X}' \\ \textit{are well defined and bounded,} \tag{23.588}$$

where $\mathcal{M}_{\partial\Omega}$ is the Hardy-Littlewood maximal operator on $\partial\Omega$. Select $\mathbb{Y}$ a Generalized Banach Sequence Space on $(\mathbb{Z},\mu)$, where μ is the counting measure, and suppose that $\mathbb{Y}$ is $\mathbb{X}$-compliant (in the sense of Definition 23.7).

Given L a second-order, homogeneous, constant complex coefficient, weakly elliptic, $M \times M$ system in $\mathbb{R}^n$, and an aperture parameter $\kappa \in (0,\infty)$, formulate the Homogeneous Regularity Problem for L in Ω with boundary data in the Composite Herz-based homogeneous Sobolev space $[\dot{\mathbb{K}}_{(1)}(\partial\Omega;\mathbb{X};\mathbb{Y})]^M$ as follows:

$$\begin{cases} u \in [\mathscr{C}^\infty(\Omega)]^M, \\ Lu = 0 \;\; \textit{in } \Omega, \\ \mathcal{N}_\kappa(\nabla u) \in \dot{\mathbb{K}}(\partial\Omega;\mathbb{X};\mathbb{Y}), \\ u\big|_{\partial\Omega}^{\kappa-\mathrm{n.t.}} = f \in \big[\dot{\mathbb{K}}_{(1)}(\partial\Omega;\mathbb{X};\mathbb{Y})\big]^M. \end{cases} \tag{23.589}$$

In relation to this, the following properties are true:

(a) [Existence, Estimate, and Integral Representation] *If $\mathfrak{A}_L^{\rm dis} \neq \varnothing$ there exists some $\delta \in (0,1)$ depending on n, M, L, the operator norms of $\mathcal{M}_{\partial\Omega}$ on $\mathbb{X}$ and $\mathbb{X}'$, the compliancy constant of $\|\cdot\|_{\mathbb{Y}}$, and the Ahlfors regularity constants of $\partial\Omega$ with the property that whenever*

$$\|\nu\|_{[\mathrm{BMO}(\partial\Omega,\sigma)]^n} < \delta \tag{23.590}$$

then the Homogeneous Regularity Problem (23.589) *is solvable. More transparently, if* $A \in \mathfrak{A}_L^{\text{dis}}$ *then the following properties are valid: First, upon recalling the modified boundary-to-boundary single layer potential* S_{mod} *in* (7.57) *it follows that*

$$\begin{gathered}[S_{\text{mod}}] : \big[\dot{\mathbb{K}}(\partial\Omega; \mathbb{X}; \mathbb{Y})\big]^M \longrightarrow \big[\dot{\mathbb{K}}_{(1)}(\partial\Omega; \mathbb{X}; \mathbb{Y})/\!\sim\big]^M \\ \textit{defined as } \ [S_{\text{mod}}]g := [S_{\text{mod}}g] \ \textit{ for all } \ g \in \big[\dot{\mathbb{K}}(\partial\Omega; \mathbb{X}; \mathbb{Y})\big]^M\end{gathered} \tag{23.591}$$

is a well-defined linear isomorphism. Second, denoting the equivalence class (modulo constants) of the boundary datum f *by* $[f] \in \big[\dot{\mathbb{K}}_{(1)}(\partial\Omega; \mathbb{X}; \mathbb{Y})/\!\sim\big]^M$, *and selecting* $g \in \big[\dot{\mathbb{K}}(\partial\Omega; \mathbb{X}; \mathbb{Y})\big]^M$ *such that* $[S_{\text{mod}}]g = [f]$, *then there exists some constant* $c \in \mathbb{C}^M$ *making the function* $u : \Omega \to \mathbb{C}^M$ *given by*

$$u := \mathscr{S}_{\text{mod}}g + c \ \text{ in } \Omega \tag{23.592}$$

a solution of the boundary value problem (23.589)*. Third, such a solution satisfies*

$$\|\mathcal{N}_\kappa(\nabla u)\|_{\dot{\mathbb{K}}(\partial\Omega;\mathbb{X};\mathbb{Y})} \approx \|\nabla_{\tan} f\|_{[\dot{\mathbb{K}}(\partial\Omega;\mathbb{X};\mathbb{Y})]^{n\times M}} \tag{23.593}$$

with implicit proportionality constants independent of f*.*

Alternatively, for each coefficient tensor $A \in \mathfrak{A}_L^{\text{dis}}$ *the operator* (*recall* (7.23))

$$\tfrac{1}{2}I + [K_{A,\text{mod}}] : \big[\dot{\mathbb{K}}_{(1)}(\partial\Omega; \mathbb{X}; \mathbb{Y})/\!\sim\big]^M \longrightarrow \big[\dot{\mathbb{K}}_{(1)}(\partial\Omega; \mathbb{X}; \mathbb{Y})/\!\sim\big]^M \tag{23.594}$$

is an isomorphism, and the Homogeneous Regularity Problem (23.589) *may also be solved by*

$$u := \mathcal{D}_{A,\text{mod}}h + c \ \text{ in } \Omega, \tag{23.595}$$

for some constant $c \in \mathbb{C}^M$ *and some function*

$$h \in \big[\dot{\mathbb{K}}_{(1)}(\partial\Omega; \mathbb{X}; \mathbb{Y})\big]^M \ \textit{ satisfying } \ [h] = \Big(\tfrac{1}{2}I + [K_{A,\text{mod}}]\Big)^{-1}[f]. \tag{23.596}$$

Furthermore, the solution u *from* (23.595) *continues to satisfy* (23.593)*.*

(b) [Uniqueness] *Whenever* $\mathfrak{A}_{L^\top}^{\text{dis}} \neq \varnothing$, *there exists* $\delta \in (0,1)$ *depending only on* n, L, *the operator norms of* $\mathcal{M}_{\partial\Omega}$ *on* $\mathbb{X}$ *and* $\mathbb{X}'$, *the compliancy constant of* $\|\cdot\|_{\mathbb{Y}}$, *and the Ahlfors regularity constants of* $\partial\Omega$ *such that whenever* $\|\nu\|_{[\text{BMO}(\partial\Omega,\sigma)]^n} < \delta$ *then the Homogeneous Regularity Problem* (23.589) *has at most one solution.*

(c) [Well-Posedness and Additional Regularity] *If* $\mathfrak{A}_L^{\text{dis}} \neq \varnothing$ *and* $\mathfrak{A}_{L^\top}^{\text{dis}} \neq \varnothing$, *then there exists a threshold* $\delta \in (0,1)$ *depending exclusively on* n, L, *the operator norms of* $\mathcal{M}_{\partial\Omega}$ *on* $\mathbb{X}$ *and* $\mathbb{X}'$, *the compliancy constant of* $\|\cdot\|_{\mathbb{Y}}$, *and the Ahlfors regularity constants of* $\partial\Omega$ *for which whenever* $\|\nu\|_{[\text{BMO}(\partial\Omega,\sigma)]^n} < \delta$, *it follows that the Homogeneous Regularity Problem* (23.589) *is well posed. In addition,*

the unique solution u of (23.589) *satisfies*

$$\mathcal{N}_\kappa u \in \dot{\mathrm{K}}(\partial\Omega;\mathbb{X};\mathbb{Y}) \iff f \in \big[\dot{\mathrm{K}}_1(\partial\Omega;\mathbb{X};\mathbb{Y})\big]^M \tag{23.597}$$

where $\dot{\mathrm{K}}_1(\partial\Omega;\mathbb{X};\mathbb{Y})$ *denotes the Composite Herz-based Sobolev space of order one on* $\partial\Omega$ *defined in* (23.570).

(*d*) [Sharpness] *If* $\mathfrak{A}_L^{\rm dis} = \varnothing$ *the Homogeneous Regularity Problem* (23.589) *may fail to be solvable (actually for boundary data belonging to an infinite dimensional subspace of the corresponding homogeneous Sobolev space styled after the Composite Herz space* $\dot{\mathrm{K}}(\partial\Omega;\mathbb{X};\mathbb{Y})$*), and if* $\mathfrak{A}_{L^\top}^{\rm dis} = \varnothing$ *the Homogeneous Regularity Problem* (23.589) *may possess more than one solution (in fact, the linear subspace of null-solutions may actually be infinite dimensional) even in the case when* $\Omega = \mathbb{R}^n_+$. *In particular, if either* $\mathfrak{A}_L^{\rm dis} = \varnothing$ *or* $\mathfrak{A}_{L^\top}^{\rm dis} = \varnothing$ *then the Homogeneous Regularity Problem* (23.589) *may fail to be well posed, again, even in the case when* $\Omega = \mathbb{R}^n_+$.

Lastly, similar conclusions regarding existence, estimates, integral representations, mapping properties, uniqueness, regularity, and well-posedness also hold when $\mathbb{Y}$ *is a Generalized Banach Sequence Space on* $(\mathbb{N}_0,\mu)$ *now with the inhomogeneous Composite Herz space* $\mathrm{K}(\partial\Omega;\mathbb{X};\mathbb{Y})$ *playing the role of* $\dot{\mathrm{K}}(\partial\Omega;\mathbb{X};\mathbb{Y})$.

Proof Bearing in mind that the current assumptions ensure that $\dot{\mathrm{K}}(\partial\Omega;\mathbb{X};\mathbb{Y})$ is a Generalized Banach Function Space (cf. Proposition 23.4), and that $\mathcal{M}_{\partial\Omega}$ is bounded on $\dot{\mathrm{K}}(\partial\Omega;\mathbb{X};\mathbb{Y})$ and on $\big(\dot{\mathrm{K}}(\partial\Omega;\mathbb{X};\mathbb{Y})\big)' = \dot{\mathrm{K}}(\partial\Omega;\mathbb{X}';\mathbb{Y}')$ (cf. Theorem 23.1 and Proposition 23.5), the veracity of all claims is a consequence of [GHA-V, Theorem 8.4.42, pp. 869–871]. □

We now treat the Neumann Problem on Composite Herz spaces.

Theorem 23.14 *Assume* $\Omega \subseteq \mathbb{R}^n$ *is an Ahlfors regular domain, set* $\sigma := \mathcal{H}^{n-1}\lfloor\partial\Omega$, *and denote by* ν *the geometric measure theoretic outward unit normal to* Ω. *Select a Generalized Banach Function Space* $\mathbb{X}$ *on* $(\partial\Omega,\sigma)$ *and suppose that*

$$\begin{gathered}\mathcal{M}_{\partial\Omega} : \mathbb{X} \to \mathbb{X} \quad \text{and} \quad \mathcal{M}_{\partial\Omega} : \mathbb{X}' \to \mathbb{X}' \\ \text{are well defined and bounded,}\end{gathered} \tag{23.598}$$

where $\mathcal{M}_{\partial\Omega}$ *is the Hardy-Littlewood maximal operator on* $\partial\Omega$. *Also, let* $\mathbb{Y}$ *be a Generalized Banach Sequence Space on* $(\mathbb{Z},\mu)$*, where* μ *is the counting measure, and assume* $\mathbb{Y}$ $\mathbb{X}$*-compliant (in the sense of Definition 23.7). Next, select a second-order, homogeneous, constant complex coefficient, weakly elliptic,* $M \times M$ *system* L *in* $\mathbb{R}^n$. *Finally, fix some coefficient tensor* $A \in \mathfrak{A}_L$ *and an aperture parameter* $\kappa \in (0,\infty)$. *Formulate the Neumann Problem for* L *in* Ω *with boundary data in the Composite Herz space* $[\dot{\mathrm{K}}(\partial\Omega;\mathbb{X};\mathbb{Y})]^M$ *as:*

$$\begin{cases} u \in [\mathscr{C}^\infty(\Omega)]^M, \\ Lu = 0 \ \ in\ \Omega, \\ \mathcal{N}_\kappa(\nabla u) \in \dot{\mathbb{K}}(\partial\Omega; \mathbb{X}; \mathbb{Y}), \\ \partial_\nu^A u = f \in [\dot{\mathbb{K}}(\partial\Omega; \mathbb{X}; \mathbb{Y})]^M. \end{cases} \tag{23.599}$$

Then the following statements are valid:

(a) [Existence, Integral Representation, and Estimates] *If* $A^\top \in \mathfrak{A}_{L^\top}^{\text{dis}}$ *then there exists some* $\delta \in (0,1)$ *depending only on* n, M, L, *the operator norms of* $\mathcal{M}_{\partial\Omega}$ *on* $\mathbb{X}$ *and* $\mathbb{X}'$, *the compliancy constant of* $\|\cdot\|_{\mathbb{Y}}$, *and the Ahlfors regularity constants of* $\partial\Omega$ *with the property that whenever*

$$\|\nu\|_{[\text{BMO}(\partial\Omega,\sigma)]^n} < \delta \tag{23.600}$$

then the Neumann Problem (23.599) *is solvable. Specifically, the boundary-to boundary double layer potential* $K^\#_{A^\top}$ *(cf.* (7.24)*) is well defined and bounded from* $[\dot{\mathbb{K}}(\partial\Omega; \mathbb{X}; \mathbb{Y})]^M$ *into itself, and the operator* $-\frac{1}{2}I + K^\#_{A^\top}$ *is invertible on* $[\dot{\mathbb{K}}(\partial\Omega; \mathbb{X}; \mathbb{Y})]^M$. *Furthermore, the function* $u : \Omega \to \mathbb{C}^M$ *defined as*

$$u := \mathscr{S}_{\text{mod}}\big[\big(-\tfrac{1}{2}I + K^\#_{A^\top}\big)^{-1} f\big] \ \ in\ \Omega \tag{23.601}$$

is a solution of the Neumann Problem (23.599) *and satisfies*

$$\big\|\mathcal{N}_\kappa(\nabla u)\big\|_{\dot{\mathbb{K}}(\partial\Omega;\mathbb{X};\mathbb{Y})} \leq C\|f\|_{[\dot{\mathbb{K}}(\partial\Omega;\mathbb{X};\mathbb{Y})]^M} \tag{23.602}$$

for some constant $C \in (0,\infty)$ *independent of* f.
Alternatively, if $A^\top \in \mathfrak{A}_{L^\top}^{\text{dis}}$ *then the operator*

$$\begin{gathered} \partial_\nu^A \mathcal{D}_{A,\text{mod}} : \big[\dot{\mathbb{K}}_{(1)}(\partial\Omega; \mathbb{X}; \mathbb{Y})/\sim\big]^M \to \big[\dot{\mathbb{K}}(\partial\Omega; \mathbb{X}; \mathbb{Y})\big]^M \\ \textit{defined as}\ \ (\partial_\nu^A \mathcal{D}_{A,\text{mod}})[g] := \partial_\nu^A(\mathcal{D}_{A,\text{mod}} g) \\ \textit{for each function}\ g \in \big[\dot{\mathbb{K}}_{(1)}(\partial\Omega; \mathbb{X}; \mathbb{Y})\big]^M \end{gathered} \tag{23.603}$$

is an isomorphism, hence

$$\begin{gathered} \textit{there exists}\ g \in \big[\dot{\mathbb{K}}_{(1)}(\partial\Omega; \mathbb{X}; \mathbb{Y})\big]^M\ \textit{such that}\ \partial_\nu^A(\mathcal{D}_{A,\text{mod}} g) = f \\ \textit{and}\ \|g\|_{[\dot{\mathbb{K}}_{(1)}(\partial\Omega;\mathbb{X};\mathbb{Y})]^M} \leq C\|f\|_{[\dot{\mathbb{K}}(\partial\Omega;\mathbb{X};\mathbb{Y})]^M}\ \textit{for some}\ C \in (0,\infty), \end{gathered} \tag{23.604}$$

and the Neumann Problem (23.599) *may also be solved as* $u := \mathcal{D}_{A,\text{mod}} g$ *in* Ω, *and this solution continues to satisfy* (23.602).

(b) [Uniqueness (modulo constants)] *If* $A \in \mathfrak{A}_L^{\text{dis}}$, *there exists* $\delta \in (0,1)$ *depending on* n, M, A, *the operator norms of* $\mathcal{M}_{\partial\Omega}$ *on* $\mathbb{X}$ *and* $\mathbb{X}'$, *the compliancy constant*

of $\|\cdot\|_{\mathbb{Y}}$, *and the Ahlfors regularity constants of* $\partial\Omega$ *with the property that whenever* $\|\nu\|_{[\mathrm{BMO}(\partial\Omega,\sigma)]^n} < \delta$ *then any two solutions of the Neumann Problem* (23.599) *differ by a* $\mathbb{C}^M$*-constant.*

(c) [Well-Posedness] *If* $A \in \mathfrak{A}_L^{\mathrm{dis}}$ *and* $A^\top \in \mathfrak{A}_{L^\top}^{\mathrm{dis}} \neq \varnothing$, *then there exists* $\delta \in (0,1)$ *depending on* n, M, L, *the operator norms of* $\mathcal{M}_{\partial\Omega}$ *on* $\mathbb{X}$ *and* $\mathbb{X}'$, *the compliancy constant of* $\|\cdot\|_{\mathbb{Y}}$, *and the Ahlfors regularity constants of* $\partial\Omega$ *with the property that whenever* $\|\nu\|_{[\mathrm{BMO}(\partial\Omega,\sigma)]^n} < \delta$ *the Neumann Problem* (23.599) *has a solution that is unique modulo constants.*

Finally, the same type of conclusions regarding mapping properties, existence, uniqueness, integral representation, estimates, regularity, and well-posedness also hold in the context of the inhomogeneous Composite Herz space $\mathrm{K}(\partial\Omega;\mathbb{X};\mathbb{Y})$, *if one now assumes that* $\mathbb{Y}$ *is a Generalized Banach Sequence Space on* $(\mathbb{N}_0,\mu)$.

Proof Thanks to Proposition 23.4, the Composite Herz space $\dot{\mathrm{K}}(\partial\Omega;\mathbb{X};\mathbb{Y})$ is a Generalized Banach Function Space on $(\partial\Omega,\sigma)$. Also, from Theorem 23.1 and Proposition 23.5 we have that the Hardy-Littlewood maximal operator $\mathcal{M}_{\partial\Omega}$ is bounded both on $\dot{\mathrm{K}}(\partial\Omega;\mathbb{X};\mathbb{Y})$ and on its associated space $\dot{\mathrm{K}}(\partial\Omega;\mathbb{X}';\mathbb{Y}')$. Granted these, the veracity of all claims in *(a)-(c)* follows from [GHA-V, Theorem 8.4.43, pp. 875–876]. □

We conclude by presenting a well-posedness result for the Transmission Problem with boundary data in Composite Herz spaces.

Theorem 23.15 *Suppose* $\Omega \subseteq \mathbb{R}^n$ *is a two-sided* NTA *domain with an unbounded Ahlfors regular boundary. Abbreviate* $\sigma := \mathcal{H}\lfloor\partial\Omega$, *denote by* ν *the geometric measure theoretic outward unit normal to* Ω, *and set*

$$\Omega_+ := \Omega \text{ and } \Omega_- := \mathbb{R}^n\setminus\overline{\Omega}. \tag{23.605}$$

Let $\mathbb{X}$ *be a Generalized Banach Function Space on* $(\partial\Omega,\sigma)$ *with the property that*

$$\begin{gathered}\mathcal{M}_{\partial\Omega} : \mathbb{X} \to \mathbb{X} \quad \text{and} \quad \mathcal{M}_{\partial\Omega} : \mathbb{X}' \to \mathbb{X}' \\ \text{are well defined and bounded,}\end{gathered} \tag{23.606}$$

where $\mathcal{M}_{\partial\Omega}$ *is the Hardy-Littlewood maximal operator on* $\partial\Omega$. *Let* $\mathbb{Y}$ *be a Generalized Banach Function Space on* $(\mathbb{Z},\mu)$, *where* μ *is the counting measure, and assume* $\mathbb{Y}$ *is* $\mathbb{X}$*-compliant (in the sense of Definition 23.7). Next, pick a second-order, homogeneous, constant complex coefficient, weakly elliptic,* $M \times M$ *system* L *in* $\mathbb{R}^n$. *Having selected some coefficient tensor* $A \in \mathfrak{A}_L$ *along with an aperture parameter* $\kappa \in (0,\infty)$ *and a transmission parameter* $\eta \in \mathbb{C}$, *consider the following Transmission Problem:*

$$\begin{cases} u^{\pm} \in [\mathscr{C}^{\infty}(\Omega)]^M, \ \ Lu^{\pm} = 0 \ \ in\ \Omega_{\pm}, \\ \mathcal{N}_{\kappa}(\nabla u^{\pm}) \in \dot{\mathbb{K}}(\partial\Omega; \mathbb{X}; \mathbb{Y}), \\ u^{+}\Big|_{\partial\Omega}^{\kappa-\mathrm{n.t.}} - u^{-}\Big|_{\partial\Omega}^{\kappa-\mathrm{n.t.}} = f \in [\dot{\mathbb{K}}_{(1)}(\partial\Omega; \mathbb{X}; \mathbb{Y})]^M, \\ \partial_{\nu}^{A} u^{+} - \eta \cdot \partial_{\nu}^{A} u^{-} = g \in [\dot{\mathbb{K}}(\partial\Omega; \mathbb{X}; \mathbb{Y})]^M, \end{cases} \tag{23.607}$$

where $\dot{\mathbb{K}}_{(1)}(\partial\Omega; \mathbb{X}; \mathbb{Y})$ is the Composite Herz-based homogeneous Sobolev space in Definition 23.10. Lastly, suppose that either

$$A^{\top} \in \mathfrak{A}_{L^{\top}}^{\mathrm{dis}} \ \ and\ \eta \in \mathbb{C}\setminus\{-1\} \tag{23.608}$$

or

$$A \in \mathfrak{A}_{L}^{\mathrm{dis}} \ \ and\ \eta \in \mathbb{C}\setminus\{0, -1\}. \tag{23.609}$$

Then for each $m \in \mathbb{N}$ there exists some threshold $\delta \in (0, 1)$, depending solely on n, M, A, the operator norms of $\mathcal{M}_{\partial\Omega}$ on $\mathbb{X}$ and $\mathbb{X}'$, the compliancy constant of $\|\cdot\|_{\mathbb{Y}}$, the two-sided NTA *constants of Ω, and the Ahlfors regularity constants of $\partial\Omega$ with the property that whenever*

$$\Big|\tfrac{1-\eta}{1+\eta}\Big| \cdot \|\nu\|_{[\mathrm{BMO}(\partial\Omega,\sigma)]^n}^{\langle m\rangle} < \delta \tag{23.610}$$

it follows that the Transmission Problem (23.607) *is well posed. Specifically, there exists a pair of functions u^{+} and u^{-} solving* (23.607)*, which are unique modulo additive constants (in $\mathbb{C}^M$), and such that*

$$\|\mathcal{N}_{\kappa}(\nabla u^{\pm})\|_{\dot{\mathbb{K}}(\partial\Omega;\mathbb{X};\mathbb{Y})} \leq C\big(\|f\|_{[\dot{\mathbb{K}}_{(1)}(\partial\Omega;\mathbb{X};\mathbb{Y})]^M} + \|g\|_{[\dot{\mathbb{K}}(\partial\Omega;\mathbb{X};\mathbb{Y})]^M}\big), \tag{23.611}$$

for some constant $C \in (0, \infty)$ independent of the boundary data f and g. Finally, the same type of conclusions hold in the case of inhomogeneous Composite Herz spaces.

Proof Since $\dot{\mathbb{K}}(\partial\Omega; \mathbb{X}; \mathbb{Y})$ is a Generalized Banach Function Space (see Proposition 23.4), and the Hardy-Littlewood maximal operator $\mathcal{M}_{\partial\Omega}$ is bounded both on $\dot{\mathbb{K}}(\partial\Omega; \mathbb{X}; \mathbb{Y})$ and on its associated space $\big(\dot{\mathbb{K}}(\partial\Omega; \mathbb{X}; \mathbb{Y})\big)' = \dot{\mathbb{K}}(\partial\Omega; \mathbb{X}'; \mathbb{Y}')$ (see Theorem 23.1 and Proposition 23.5), the well-posedness of the Transmission Problem (23.607) follows from [GHA-V, Theorem 8.4.44, p. 877]. □

The well-posedness results presented in Theorem 23.11, Theorem 23.13, Theorem 23.12, Theorem 23.14, and Theorem 23.15 are applicable whenever the function space $\mathbb{X}$ and the sequence space $\mathbb{Y}$ are as described in any of the scenarios (23.374)-(23.382) (for $q \in (1, \infty)$), or (23.391)-(23.399) (for $q \in (1, \infty)$), or (23.409)-(23.417), or (23.427)-(23.435) (for $q \in (1, \infty)$), or (23.446)-(23.454) (for $q \in (1, \infty)$).

Composite Herz spaces provide a recipe for creating a multitude of Herz-type spaces, thus unifying and generalizing a large body of work in the literature. Most significantly, the theory developed here for this brand of Herz spaces constitutes an appropriate functional analytic environment where a great variety of basic elliptic boundary value problems for complex second-order systems are shown to be well-posed in domains of a very general geometric nature, which goes beyond the class of Lipschitz domains. (23.612)

Chapter 24
Herz Spaces on Bounded Ahlfors Regular Sets

So far in this monograph we have only considered Herz-type spaces over unbounded Ahlfors regular sets. To accommodate the case when the Ahlfors regular set under consideration is bounded, we introduce a new brand of Herz spaces.

Consider $\Sigma \subseteq \mathbb{R}^n$ an arbitrary Ahlfors regular set, and abbreviate $\sigma := \mathcal{H}^{n-1}\lfloor\Sigma$. Fix a reference point $x_0 \in \Sigma$, along with a dilation factor $2_* > (C_\Sigma/c_\Sigma)^{1/(n-1)} \geq 1$, where $C_\Sigma, c_\Sigma \in (0,\infty)$ denotes the upper and lower Ahlfors regularity constants of Σ. Define the `truncation index of` Σ, denoted by $k_* = k_*(\Sigma)$, as

$$k_* = \infty \quad \text{if } \Sigma \text{ is unbounded,} \tag{24.1}$$

and as the greatest integer satisfying

$$k_* < \min\left\{\frac{\ln(\operatorname{diam}\Sigma)}{\ln(2_*)}, \frac{\ln\left(\sigma(\Sigma)/C_\Sigma\right)}{\ln(2_*^{n-1})}\right\} \quad \text{if } \Sigma \text{ is bounded.} \tag{24.2}$$

In particular, whenever Σ is a bounded set,

$$\begin{gathered} k_* \in \mathbb{Z} \ \text{ is a well-defined and finite number which satisfies} \\ k_* < \frac{\ln(\operatorname{diam}\Sigma)}{\ln(2_*)} \quad \text{and} \quad k_* < \frac{\ln\left(\sigma(\Sigma)/C_\Sigma\right)}{\ln(2_*^{n-1})}. \end{gathered} \tag{24.3}$$

For each $k \in \mathbb{Z}$, introduce the sets

$$\begin{aligned} C_{k_*} &:= \Sigma\setminus\Delta(x_0, 2_*^{k_*}) \quad \text{and} \\ C_k &:= \Delta(x_0, 2_*^{k+1})\setminus\Delta(x_0, 2_*^k) \quad \text{if } k < k_*. \end{aligned} \tag{24.4}$$

Corresponding to the case when Σ is unbounded, we have $C_{k_*} = \varnothing$, and in such a case the family of annuli $\{C_k\}_{k\in\mathbb{Z}}$ is as in the past. Then it is visible that

© The Author(s), under exclusive license to Springer Nature Switzerland AG 2026

M. Mitrea and P. Takemura, *Singular Integrals, Herz-Type Function Spaces, and Boundary Problems*, Progress in Mathematics 362,
https://doi.org/10.1007/978-3-032-12516-3_24

$$\bigcup_{k\le k_*} C_k = \Sigma \setminus \{x_0\} \quad \Longrightarrow \quad \sigma\Big(\Sigma \setminus \bigcup_{k\le k_*} C_k\Big) = 0. \tag{24.5}$$

Also, regardless of whether Σ is bounded or unbounded, we always have

$$\sigma(C_k) < \infty \quad \text{for all } k \in \mathbb{Z}, \text{ with } k < k_*, \tag{24.6}$$

while

$$\begin{aligned} \sigma(C_{k_*}) &= \infty \quad \text{if } \Sigma \text{ is unbounded, and} \\ \sigma(C_{k_*}) &< \infty \quad \text{if } \Sigma \text{ is bounded.} \end{aligned} \tag{24.7}$$

It is also worth noticing that

$$\sigma(C_k) > 0 \quad \text{for all } k \in \mathbb{Z} \text{ with } k \le k_*. \tag{24.8}$$

Indeed, in the case when $k = k_*$, and Σ is unbounded, this is clear from the first line in (24.7). So assume Σ is bounded. Observe that the second inequality in (24.3) ensures that $2_*^{k_*(n-1)} < \sigma(\Sigma)/C_\Sigma$. As such, by the Ahlfors regularity of Σ we have

$$\sigma\big(\Sigma \setminus \Delta(x_0, 2_*^{k_*})\big) \ge \sigma(\Sigma) - \sigma\big(\Delta(x_0, 2_*^{k_*})\big) \ge \sigma(\Sigma) - C_\Sigma \cdot 2_*^{k_*(n-1)} > 0, \tag{24.9}$$

as desired. Next, assume $k < k_*$. This and the first inequality in (24.3) imply

$$2_*^{k+1} \le 2_*^{k_*} \le \operatorname{diam}\Sigma < 2\operatorname{diam}\Sigma \quad \Longrightarrow \quad 2_*^{k+1} \le 2\operatorname{diam}\Sigma. \tag{24.10}$$

Hence, using the Ahlfors regularity of Σ, and recalling that $2_* > (C_\Sigma/c_\Sigma)^{1/(n-1)}$, we find that

$$\begin{aligned} \sigma(C_k) &\ge \sigma\big(\Delta(x_0, 2_*^{k+1})\big) - \sigma\big(\Delta(x_0, 2_*^{k})\big) \\ &\ge c_\Sigma \cdot 2_*^{(k+1)(n-1)} - C_\Sigma \cdot 2_*^{k(n-1)} \\ &= 2_*^{k(n-1)}\big(c_\Sigma \cdot 2_*^{n-1} - C_\Sigma\big) > 0. \end{aligned} \tag{24.11}$$

This justifies (24.8).

Moving on, retain the same assumptions as before, select two exponents $p, q \in (0, \infty]$, along with an index $\alpha \in \mathbb{R}$. For each σ-measurable function f on Σ, define

$$\|f\|_{\dot{\mathrm{K}}^{\alpha}_{p,q}(\Sigma,\sigma;k_*)} := \begin{cases} \left[\displaystyle\sum_{k=-\infty}^{k_*} \big(2^{k\alpha} \|f \cdot \mathbb{1}_{C_k}\|_{L^p(\Sigma,\sigma)}\big)^q \right]^{1/q} & \text{if } q < \infty, \\ \displaystyle\sup_{\substack{k\in\mathbb{Z} \\ k\le k_*}} \big\{2^{k\alpha} \|f \cdot \mathbb{1}_{C_k}\|_{L^p(\Sigma,\sigma)}\big\} & \text{if } q = \infty. \end{cases} \tag{24.12}$$

Introduce the `homogeneous Herz space of` k_*`-index` on Σ as

$$\dot{\mathrm{K}}^{\alpha}_{p,q}(\Sigma,\sigma;k_*) := \big\{f \in \mathscr{M}(\Sigma,\sigma) : \|f\|_{\dot{\mathrm{K}}^{\alpha}_{p,q}(\Sigma,\sigma;k_*)} < \infty\big\}, \tag{24.13}$$

where $\mathscr{M}(\Sigma, \sigma)$ denotes the collection of all σ-measurable functions on Σ. It is worth to emphasize that the above considerations (24.1)-(24.8) yield

$$\dot{\mathrm{K}}^{\alpha}_{p,q}(\Sigma, \sigma; k_*) = \dot{\mathrm{K}}^{\alpha}_{p,q}(\Sigma, \sigma; \infty) \equiv \dot{\mathrm{K}}^{\alpha}_{p,q}(\Sigma, \sigma) \quad \text{whenever } \Sigma \text{ is an unbounded set,} \tag{24.14}$$

i.e., the homogeneous Herz space of ∞-index coincides with the (standard) homogeneous Herz space (studied in Section 17). Furthermore, it follows from definitions that

$$\dot{\mathrm{K}}^{0}_{p,p}(\Sigma, \sigma; k_*) = L^p(\Sigma, \sigma) \quad \text{whenever } p \in (0, \infty]. \tag{24.15}$$

The first observation that we wish to make in relation to the scale of homogeneous Herz spaces of k_*-index is that

$$\dot{\mathrm{K}}^{\alpha}_{p,q}(\Sigma, \sigma; k_*) \text{ is a Generalized Banach Function Space on } (\Sigma, \sigma). \quad \text{for any exponents } p, q \in (1, \infty) \text{ and any index } \alpha \in \mathbb{R}. \tag{24.16}$$

Taking into consideration the family of annuli $\{C_k\}_{\substack{k\in\mathbb{Z},\\ k\leq k_*}}$ this claim follows via a similar argument as in the proof of Proposition 17.5. In addition, the same type of reasoning used to prove *(a)* in Proposition 17.1 ensures that if $p, q \in (1, \infty)$, and $\alpha \in \mathbb{R}$, then for every two functions f and g that are σ-measurable one has

$$\int_{\Sigma} |fg| \, d\sigma \leq \|f\|_{\dot{\mathrm{K}}^{\alpha}_{p,q}(\Sigma,\sigma;k_*)} \|g\|_{\dot{\mathrm{K}}^{-\alpha}_{p',q'}(\Sigma,\sigma;k_*)} \tag{24.17}$$

where $p', q' \in (1, \infty)$ are the Hölder conjugate exponents of p and q, respectively. As such, for the same range of p and q, this suggests that the associated space of k_*-homogeneous Herz space $\dot{\mathrm{K}}^{\alpha}_{p,q}(\Sigma, \sigma; k_*)$ can be identified as

$$\big(\dot{\mathrm{K}}^{\alpha}_{p,q}(\Sigma, \sigma; k_*)\big)' = \dot{\mathrm{K}}^{-\alpha}_{p',q'}(\Sigma, \sigma; k_*). \tag{24.18}$$

This is indeed the case since we can naturally adapt the reasoning used in the proof of Proposition 10.12 to obtain

$$\|f\|_{\dot{\mathrm{K}}^{-\alpha}_{p',q'}(\Sigma,\sigma;k_*)} = \sup\Big\{ \int_{\Sigma} |fg| \, d\sigma : g \in \dot{\mathrm{K}}^{\alpha}_{p,q}(\Sigma, \sigma; k_*), \text{ with } \|g\|_{\dot{\mathrm{K}}^{\alpha}_{p,q}(\Sigma,\sigma;k_*)} \leq 1 \Big\}. \tag{24.19}$$

Our next goal is to obtain the boundedness of the Hardy-Littlewood maximal operator on k_*-homogeneous Herz spaces.

Theorem 24.1 *Assume $\Sigma \subseteq \mathbb{R}^n$ is an arbitrary Ahlfors regular set. Select exponents $p \in (1, \infty)$, and $q \in (0, \infty]$, along with an index $\alpha \in \mathbb{R}$ satisfying*

$$\frac{-(n-1)}{p} < \alpha < \frac{n-1}{p'}, \tag{24.20}$$

where $p' \in (1, \infty)$ denotes the Hölder conjugate exponent of p. Also, pick k_* as in (24.2). *Then, in this context, it follows that the operator*

$$\begin{gathered}\mathcal{M}_\Sigma : \dot{\mathrm{K}}^{\alpha}_{p,q}(\Sigma, \sigma; k_*) \to \dot{\mathrm{K}}^{\alpha}_{p,q}(\Sigma, \sigma; k_*) \\ \textit{is well defined and bounded.}\end{gathered} \tag{24.21}$$

Proof In the case when Σ is unbounded, the considerations made in (24.14) along with Theorem 17.4 readily imply (24.21). As such, assume that Σ is a bounded set, a case which forces k_* to be a finite integer. Henceforth, select an arbitrary function f belonging to $\dot{\mathrm{K}}^{\alpha}_{p,q}(\Sigma, \sigma; k_*)$, and write it as

$$f = \sum_{\ell=-\infty}^{k_*} f \cdot \mathbb{1}_{C_\ell} \quad \text{pointwise } \sigma\text{-a.e. on } \Sigma. \tag{24.22}$$

Then, from the definition of the Hardy-Littlewood maximal operator, (24.22), and Lebesgue's Monotone Convergence Theorem

$$\mathcal{M}_\Sigma f \le \sum_{\ell=-\infty}^{k_*} \mathcal{M}_\Sigma(f \cdot \mathbb{1}_{C_\ell}) \text{ everywhere on } \Sigma. \tag{24.23}$$

Hence, in view of the lattice property of $\dot{\mathrm{K}}^{\alpha}_{p,q}(\Sigma, \sigma; k_*)$ we have:

$$\begin{aligned}
&\|\mathcal{M}_\Sigma f\|_{\dot{\mathrm{K}}^{\alpha}_{p,q}(\Sigma,\sigma;k_*)} \\
&\le \Bigg[\sum_{k=-\infty}^{k_*}\Big(2_*^{k\alpha}\sum_{\ell=-\infty}^{k_*}\big\|[\mathcal{M}_\Sigma(f\cdot\mathbb{1}_{C_\ell})]\cdot\mathbb{1}_{C_k}\big\|_{L^p(\Sigma,\sigma)}\Big)^q\Bigg]^{1/q} \\
&\le \Bigg[\sum_{k=k_*-1}^{k_*}\Big(2_*^{k\alpha}\sum_{\ell=-\infty}^{k_*}\big\|[\mathcal{M}_\Sigma(f\cdot\mathbb{1}_{C_\ell})]\cdot\mathbb{1}_{C_k}\big\|_{L^p(\Sigma,\sigma)}\Big)^q\Bigg]^{1/q} \\
&\quad + \Bigg[\sum_{k=-\infty}^{k_*-2}\Big(2_*^{k\alpha}\sum_{\ell=-\infty}^{k_*}\big\|[\mathcal{M}_\Sigma(f\cdot\mathbb{1}_{C_\ell})]\cdot\mathbb{1}_{C_k}\big\|_{L^p(\Sigma,\sigma)}\Big)^q\Bigg]^{1/q} \\
&=: \mathrm{I}_1 + \mathrm{I}_2.
\end{aligned} \tag{24.24}$$

By adapting the computations in (23.278), (23.279), and (23.281), for the choice $\mathbb{X} := L^p(\Sigma, \sigma)$ we obtain

$$\big\|[\mathcal{M}_\Sigma(f\cdot \mathbb{1}_{C_\ell})]\cdot \mathbb{1}_{C_k}\big\|_{L^p(\Sigma,\sigma)} \leq C 2_*^{-k(n-1)/p'} 2_*^{\ell(n-1)/p'} \|f\cdot \mathbb{1}_{C_\ell}\|_{L^p(\Sigma,\sigma)}$$
$$\text{for every } k,\ell\in\mathbb{Z} \text{ satisfying } k\leq k_* \text{ and } \ell\leq k-2. \tag{24.25}$$

The reasoning in (23.284), (23.285), and (23.287) entails

$$\big\|[\mathcal{M}_\Sigma(f\cdot \mathbb{1}_{C_\ell})]\cdot \mathbb{1}_{C_k}\big\|_{L^p(\Sigma,\sigma)} \leq C 2_*^{k(n-1)/p} 2_*^{-\ell(n-1)/p} \|f\cdot \mathbb{1}_{C_\ell}\|_{L^p(\Sigma,\sigma)}$$
$$\text{for every } k,\ell\in\mathbb{Z} \text{ satisfying } k\leq k_*-2 \text{ and } k+2\leq \ell\leq k_*. \tag{24.26}$$

Granted these, we can estimate I_1 and I_2 in (24.24). First, notice that

$$\begin{aligned}\mathrm{I}_1 &\leq C\sum_{\ell=-\infty}^{k_*}\big\|[\mathcal{M}_\Sigma(f\cdot \mathbb{1}_{C_\ell})]\cdot \mathbb{1}_{C_{k_*-1}}\big\|_{L^p(\Sigma,\sigma)}\\ &\quad + C\sum_{\ell=-\infty}^{k_*}\big\|[\mathcal{M}_\Sigma(f\cdot \mathbb{1}_{C_\ell})]\cdot \mathbb{1}_{C_{k_*}}\big\|_{L^p(\Sigma,\sigma)},\end{aligned} \tag{24.27}$$

for some constant $C\in(0,\infty)$ depending only on q, α, and k_*. Also, from the boundedness of $\mathcal{M}_\Sigma$ on $L^p(\Sigma,\sigma)$, (24.12), (24.20), and (24.25),

$$\begin{aligned}&\sum_{\ell=-\infty}^{k_*}\big\|[\mathcal{M}_\Sigma(f\cdot \mathbb{1}_{C_\ell})]\cdot \mathbb{1}_{C_{k_*-1}}\big\|_{L^p(\Sigma,\sigma)}\\
&\quad\leq \big\|[\mathcal{M}_\Sigma(f\cdot \mathbb{1}_{C_{k_*}})]\cdot \mathbb{1}_{C_{k_*-1}}\big\|_{L^p(\Sigma,\sigma)} + \big\|[\mathcal{M}_\Sigma(f\cdot \mathbb{1}_{C_{k_*-1}})]\cdot \mathbb{1}_{C_{k_*-1}}\big\|_{L^p(\Sigma,\sigma)}\\
&\qquad + \big\|[\mathcal{M}_\Sigma(f\cdot \mathbb{1}_{C_{k_*-2}})]\cdot \mathbb{1}_{C_{k_*-1}}\big\|_{L^p(\Sigma,\sigma)}\\
&\qquad + \sum_{\ell=-\infty}^{k_*-3}\big\|[\mathcal{M}_\Sigma(f\cdot \mathbb{1}_{C_\ell})]\cdot \mathbb{1}_{C_{k_*-1}}\big\|_{L^p(\Sigma,\sigma)}\\
&\quad\leq C\big\|f\cdot \mathbb{1}_{C_{k_*}}\big\|_{L^p(\Sigma,\sigma)} + C\big\|f\cdot \mathbb{1}_{C_{k_*-1}}\big\|_{L^p(\Sigma,\sigma)} + C\big\|f\cdot \mathbb{1}_{C_{k_*-2}}\big\|_{L^p(\Sigma,\sigma)}\\
&\qquad + C\sum_{\ell=-\infty}^{k_*-3} 2_*^{-(k_*-1)(n-1)/p'} 2_*^{\ell(n-1)/p'}\|f\cdot \mathbb{1}_{C_\ell}\|_{L^p(\Sigma,\sigma)}\\
&\quad\leq C\Big[\sum_{j=-\infty}^{k_*}\big(2_*^{\alpha j}\|f\cdot \mathbb{1}_{C_j}\|_{L^p(\Sigma,\sigma)}\big)^q\Big]^{1/q}\\
&\qquad + C\sum_{\ell=-\infty}^{k_*-3} 2_*^{\ell[(n-1)/p'-\alpha]} 2_*^{\alpha\ell}\|f\cdot \mathbb{1}_{C_\ell}\|_{L^p(\Sigma,\sigma)}\\
&\quad\leq C\|f\|_{\dot{\mathrm{K}}^{\alpha}_{p,q}(\Sigma,\sigma;k_*)} + C\Big(\sum_{\ell=-\infty}^{k_*-3} 2_*^{\ell[(n-1)/p'-\alpha]}\Big)\|f\|_{\dot{\mathrm{K}}^{\alpha}_{p,q}(\Sigma,\sigma;k_*)}\end{aligned}$$

$$\leq C\|f\|_{\dot{\mathrm{K}}^{\alpha}_{p,q}(\Sigma,\sigma;k_*)} \tag{24.28}$$

for some constant $C \in (0,\infty)$ depending only on n, p, α, k_*, and the Ahlfors regularity constants of Σ. A similar reasoning as in (24.28) shows that

$$\begin{aligned}
&\sum_{\ell=-\infty}^{k_*} \big\|[\mathcal{M}_\Sigma(f\cdot \mathbb{1}_{C_\ell})]\cdot \mathbb{1}_{C_{k_*}}\big\|_{L^p(\Sigma,\sigma)}\\
&\quad\leq \big\|[\mathcal{M}_\Sigma(f\cdot \mathbb{1}_{C_{k_*}})]\cdot \mathbb{1}_{C_{k_*}}\big\|_{L^p(\Sigma,\sigma)} + \big\|[\mathcal{M}_\Sigma(f\cdot \mathbb{1}_{C_{k_*-1}})]\cdot \mathbb{1}_{C_{k_*}}\big\|_{L^p(\Sigma,\sigma)}\\
&\qquad + \sum_{\ell=-\infty}^{k_*-2} \big\|[\mathcal{M}_\Sigma(f\cdot \mathbb{1}_{C_\ell})]\cdot \mathbb{1}_{C_{k_*}}\big\|_{L^p(\Sigma,\sigma)}\\
&\quad\leq C\|f\|_{\dot{\mathrm{K}}^{\alpha}_{p,q}(\Sigma,\sigma;k_*)}
\end{aligned} \tag{24.29}$$

for some $C \in (0,\infty)$ of the same nature as in (24.28). At this stage, combine (24.27)-(24.29) to show that there exists some $C \in (0,\infty)$ depending only on n, p, q, α, k_*, and the Ahlfors regularity character of Σ for which

$$\mathrm{I}_1 \leq C\|f\|_{\dot{\mathrm{K}}^{\alpha}_{p,q}(\Sigma,\sigma;k_*)}. \tag{24.30}$$

Pressing on, we deal with I_2 in (24.24), by estimating

$$\begin{aligned}
\mathrm{I}_2 &\leq C\Bigg[\sum_{k=-\infty}^{k_*-2}\Big(2_*^{k\alpha}\sum_{\ell=-\infty}^{k-2}\big\|[\mathcal{M}_\Sigma(f\cdot \mathbb{1}_{C_\ell})]\cdot \mathbb{1}_{C_k}\big\|_{L^p(\Sigma,\sigma)}\Big)^q\Bigg]^{1/q}\\
&\quad + C\Bigg[\sum_{k=-\infty}^{k_*-2}\Big(2_*^{k\alpha}\sum_{\ell=k-1}^{k+1}\big\|[\mathcal{M}_\Sigma(f\cdot \mathbb{1}_{C_\ell})]\cdot \mathbb{1}_{C_k}\big\|_{L^p(\Sigma,\sigma)}\Big)^q\Bigg]^{1/q}\\
&\quad + C\Bigg[\sum_{k=-\infty}^{k_*-2}\Big(2_*^{k\alpha}\sum_{\ell=k+2}^{k_*}\big\|[\mathcal{M}_\Sigma(f\cdot \mathbb{1}_{C_\ell})]\cdot \mathbb{1}_{C_k}\big\|_{L^p(\Sigma,\sigma)}\Big)^q\Bigg]^{1/q}\\
&=: \mathrm{II}_1 + \mathrm{II}_2 + \mathrm{II}_3.
\end{aligned} \tag{24.31}$$

For II_2, use the boundedness of $\mathcal{M}_\Sigma$ on $L^p(\Sigma,\sigma)$ and (24.12) to obtain

$$
\begin{aligned}
\mathrm{II}_2 &\leq C\left[\sum_{k=-\infty}^{k_*-2}\Big(2_*^{k\alpha}\sum_{\ell=k-1}^{k+1}\|f\cdot\mathbb{1}_{C_\ell}\|_{L^p(\Sigma,\sigma)}\Big)^q\right]^{1/q} \\
&\leq C2_*^{\alpha}\left[\sum_{k=-\infty}^{k_*-2}\Big(2_*^{(k-1)\alpha}\|f\cdot\mathbb{1}_{C_{k-1}}\|_{L^p(\Sigma,\sigma)}\Big)^q\right]^{1/q} \\
&\quad + C\left[\sum_{k=-\infty}^{k_*-2}\Big(2_*^{k\alpha}\|f\cdot\mathbb{1}_{C_k}\|_{L^p(\Sigma,\sigma)}\Big)^q\right]^{1/q} \\
&\quad + C2_*^{-\alpha}\left[\sum_{k=-\infty}^{k_*-2}\Big(2_*^{(k+1)\alpha}\|f\cdot\mathbb{1}_{C_{k+1}}\|_{L^p(\Sigma,\sigma)}\Big)^q\right]^{1/q} \\
&\leq C\|f\|_{\dot{\mathrm{K}}^{\alpha}_{p,q}(\Sigma,\sigma;k_*)}
\end{aligned}
\tag{24.32}
$$

for some $C = C(n, p, q, \alpha, \Sigma) \in (0, \infty)$.

Turning our attention to II_1 in (24.31), use (24.25) to bound

$$
\begin{aligned}
\mathrm{II}_1 &\leq C\left[\sum_{k=-\infty}^{k_*-2}\Big(2_*^{k\alpha}\sum_{\ell=-\infty}^{k-2}2_*^{-k(n-1)/p'}2_*^{\ell(n-1)/p'}\|f\cdot\mathbb{1}_{C_\ell}\|_{L^p(\Sigma,\sigma)}\Big)^q\right]^{1/q} \\
&= C\left[\sum_{k=-\infty}^{k_*-2}\Big(\sum_{\ell=-\infty}^{k-2}2_*^{(k-\ell)[\alpha-(n-1)/p']}\big(2_*^{\ell\alpha}\|f\cdot\mathbb{1}_{C_\ell}\|_{L^p(\Sigma,\sigma)}\big)\Big)^q\right]^{1/q}.
\end{aligned}
\tag{24.33}
$$

At this point the analysis branches out, depending on the size of q. Consider the case when $q > 1$, and denote by $q' \in (1, \infty)$ its Hölder conjugate exponent. By utilizing $\alpha - (n-1)/p' < 0$ (cf. (24.20)), (24.33), Hölder's inequality, and Fubini's Theorem for series, we have

$$
\begin{aligned}
\mathrm{II}_1 &\leq C\left[\sum_{k=-\infty}^{k_*-2}\Big(\sum_{\ell=-\infty}^{k-2}2_*^{(k-\ell)[\alpha-(n-1)/p']q/2}\big(2_*^{\ell\alpha}\|f\cdot\mathbb{1}_{C_\ell}\|_{L^p(\Sigma,\sigma)}\big)^q\Big)\right. \\
&\qquad\left.\times\Big(\sum_{\ell=-\infty}^{k-2}2_*^{(k-\ell)[\alpha-(n-1)/p']q'/2}\Big)^{q/q'}\right]^{1/q} \\
&= C\left[\sum_{k=-\infty}^{k_*-2}\sum_{\ell=-\infty}^{k-2}2_*^{(k-\ell)[\alpha-(n-1)/p']q/2}\big(2_*^{\ell\alpha}\|f\cdot\mathbb{1}_{C_\ell}\|_{L^p(\Sigma,\sigma)}\big)^q\right]^{1/q} \\
&= C\left[\sum_{\ell=-\infty}^{k_*-4}\big(2_*^{\ell\alpha}\|f\cdot\mathbb{1}_{C_\ell}\|_{L^p(\Sigma,\sigma)}\big)^q\sum_{k=\ell+2}^{k_*-2}2_*^{(k-\ell)[\alpha-(n-1)/p']q/2}\right]^{1/q}
\end{aligned}
$$

$$= C\Bigg[\sum_{\ell=-\infty}^{k_*-4}\big(2_*^{\ell\alpha}\|f\cdot\mathbb{1}_{C_\ell}\|_{L^p(\Sigma,\sigma)}\big)^q\Bigg]^{1/q} \leq C\|f\|_{\dot{\mathrm{K}}^\alpha_{p,q}(\Sigma,\sigma;k_*)} \tag{24.34}$$

for some $C = C(n, p, q, \alpha, k_*, \Sigma) \in (0, \infty)$. In the scenario when $0 < q \leq 1$, employ Fubini's Theorem for series, and recall that $\alpha - (n-1)/p' < 0$ (cf. (24.20)), to bound II_1 in (24.33) as

$$\begin{aligned}
\mathrm{II}_1 &\leq C\Bigg[\sum_{k=-\infty}^{k_*-2}\sum_{\ell=-\infty}^{k-2} 2_*^{(k-\ell)[\alpha-(n-1)/p']q}\big(2_*^{\ell\alpha}\|f\cdot\mathbb{1}_{C_\ell}\|_{L^p(\Sigma,\sigma)}\big)^q\Bigg]^{1/q}\\
&\leq C\Bigg[\sum_{\ell=-\infty}^{k_*-4}\big(2_*^{\ell\alpha}\|f\cdot\mathbb{1}_{C_\ell}\|_{L^p(\Sigma,\sigma)}\big)^q\sum_{k=\ell+2}^{k_*-2} 2_*^{(k-\ell)[\alpha-(n-1)/p']q}\Bigg]^{1/q}\\
&= C\Bigg[\sum_{\ell=-\infty}^{k_*-4}\big(2_*^{\ell\alpha}\|f\cdot\mathbb{1}_{C_\ell}\|_{L^p(\Sigma,\sigma)}\big)^q\Bigg]^{1/q} \leq C\|f\|_{\dot{\mathrm{K}}^\alpha_{p,q}(\Sigma,\sigma;k_*)}
\end{aligned} \tag{24.35}$$

for some constant $C = C(n, p, q, \alpha, k_*, \Sigma) \in (0, \infty)$.

We now deal with II_3 from (24.31). In light of (24.26), it follows that

$$\begin{aligned}
\mathrm{II}_3 &\leq C\Bigg[\sum_{k=-\infty}^{k_*-2}\Big(2_*^{k\alpha}\sum_{\ell=k+2}^{k_*} 2_*^{k(n-1)/p}2_*^{-\ell(n-1)/p}\|f\cdot\mathbb{1}_{C_\ell}\|_{L^p(\Sigma,\sigma)}\Big)^q\Bigg]^{1/q}\\
&= C\Bigg[\sum_{k=-\infty}^{k_*-2}\Big(\sum_{\ell=k+2}^{k_*} 2_*^{(k-\ell)[(n-1)/p+\alpha]}\big(2_*^{\ell\alpha}\|f\cdot\mathbb{1}_{C_\ell}\|_{L^p(\Sigma,\sigma)}\big)\Big)^q\Bigg]^{1/q}.
\end{aligned} \tag{24.36}$$

When $q > 1$, use Hölder's inequality, Fubini's Theorem for series, as well as the fact that $\alpha > -(n-1)/p$ (cf. (24.20)) to estimate

$$\begin{aligned}
\mathrm{II}_3 &\leq C\Bigg[\sum_{k=-\infty}^{k_*-2}\Big(\sum_{\ell=k+2}^{k_*} 2_*^{(k-\ell)[(n-1)/p+\alpha]q/2}\big(2_*^{\ell\alpha}\|f\cdot\mathbb{1}_{C_\ell}\|_{L^p(\Sigma,\sigma)}\big)^q\Big)\\
&\qquad\times\Big(\sum_{\ell=k+2}^{k_*} 2_*^{(k-\ell)[(n-1)/p+\alpha]q'/2}\Big)^{q/q'}\Bigg]^{1/q}\\
&= C\Bigg[\sum_{k=-\infty}^{k_*-2}\sum_{\ell=k+2}^{k_*} 2_*^{(k-\ell)[(n-1)/p+\alpha]q/2}\big(2_*^{\ell\alpha}\|f\cdot\mathbb{1}_{C_\ell}\|_{L^p(\Sigma,\sigma)}\big)^q\Bigg]^{1/q}\\
&= C\Bigg[\sum_{\ell=-\infty}^{k_*}\big(2_*^{\ell\alpha}\|f\cdot\mathbb{1}_{C_\ell}\|_{L^p(\Sigma,\sigma)}\big)^q\sum_{k=-\infty}^{\ell-2} 2_*^{(k-\ell)[(n-1)/p+\alpha]q/2}\Bigg]^{1/q}
\end{aligned}$$

$$= C\left[\sum_{\ell=-\infty}^{k_*}\left(2_*^{\ell\alpha}\|f\cdot\mathbb{1}_{C_\ell}\|_{L^p(\Sigma,\sigma)}\right)^q\right]^{1/q} = C\|f\|_{\dot{\mathrm{K}}^\alpha_{p,q}(\Sigma,\sigma;k_*)} \tag{24.37}$$

for some constant $C = C(n, p, q, \alpha, k_*, \Sigma) \in (0, \infty)$. In the scenario when $0 < q \leq 1$, we once again employ Fubini's Theorem for series, along with $\alpha > -(n-1)/p$ (cf. (24.20)), to continue estimating II_3 in (24.36) as

$$\begin{aligned}
\mathrm{II}_3 &\leq \left[\sum_{k=-\infty}^{k_*-2}\sum_{\ell=k+2}^{k_*} 2_*^{(k-\ell)[(n-1)/p+\alpha]q}\left(2_*^{\ell\alpha}\|f\cdot\mathbb{1}_{C_\ell}\|_{L^p(\Sigma,\sigma)}\right)^q\right]^{1/q} \\
&= C\left[\sum_{\ell=-\infty}^{k_*}\left(2_*^{\ell\alpha}\|f\cdot\mathbb{1}_{C_\ell}\|_{L^p(\Sigma,\sigma)}\right)^q \sum_{k=-\infty}^{\ell-2} 2_*^{(k-\ell)[(n-1)/p+\alpha]q}\right]^{1/q} \\
&= C\left[\sum_{\ell=-\infty}^{k_*}\left(2_*^{\ell\alpha}\|f\cdot\mathbb{1}_{C_\ell}\|_{L^p(\Sigma,\sigma)}\right)^q\right]^{1/q} = C\|f\|_{\dot{\mathrm{K}}^\alpha_{p,q}(\Sigma,\sigma;k_*)},
\end{aligned} \tag{24.38}$$

for some $C = C(n, p, q, \alpha, k_*, \Sigma) \in (0, \infty)$.

Collectively, (24.24), (24.30), (24.31), (24.32), (24.34), (24.35), (24.37), and (24.38) finally imply that there exists some constant $C \in (0, \infty)$ depending only on n, p, q, α, k_*, and the Ahlfors regularity character of Σ with the property that

$$\|\mathcal{M}_\Sigma f\|_{\dot{\mathrm{K}}^\alpha_{p,q}(\Sigma,\sigma;k_*)} \leq C\|f\|_{\dot{\mathrm{K}}^\alpha_{p,q}(\Sigma,\sigma;k_*)}. \tag{24.39}$$

This ultimately proves Theorem 24.1. □

Chapter 25
Boundary Value Problems in Domains with Compact Boundary

In [GHA-V] various boundary value problems in domains with compact boundary have been treated, in particular when the boundary data is taken from Generalized Banach Functions Spaces. With this in mind, our goal in this section is to study the Dirichlet, Regularity, and Neumann Problems for weakly elliptic systems in domains with compact boundaries with boundary data prescribed in k_*-homogeneous Herz spaces. To set the stage, we recall some definitions. Consider a closed Ahlfors regular set $\Sigma \subseteq \mathbb{R}^n$, and abbreviate $\sigma := \mathcal{H}^{n-1}\lfloor\Sigma$. Define the Sarason space $\mathrm{VMO}(\Sigma,\sigma)$ of functions of vanishing mean oscillations on Σ as

$$\mathrm{VMO}(\Sigma,\sigma) := \text{closure of } \mathrm{UC}(\Sigma) \cap \mathrm{BMO}(\Sigma,\sigma) \text{ in } \mathrm{BMO}(\Sigma,\sigma), \tag{25.1}$$

where $\mathrm{UC}(\Sigma)$ stands for the space of uniformly continuous functions on Σ. Then, according to [GHA-II, Corollary 3.1.6, p. 92] for every function $f \in \mathrm{BMO}(\Sigma,\sigma)$, and $p \in [1,\infty)$, it follows that

$$\operatorname{dist}\Big(f, \mathrm{VMO}(\Sigma,\sigma)\Big) \approx \lim_{R\to 0^+} \sup_{\substack{x\in\Sigma \\ r\in(0,R)}} \left(⨍_{\Delta(x,r)} \left| f - ⨍_{\Delta(x,r)} f\,d\sigma \right|^p d\sigma \right)^{1/p}, \tag{25.2}$$

with implicit proportionality constants independent of f. In particular, this entails

$$\begin{aligned} & f \in \mathrm{VMO}(\Sigma,\sigma) \\ & \Longleftrightarrow \lim_{R\to 0^+} \sup_{\substack{x\in\Sigma \\ r\in(0,R)}} \left(⨍_{\Delta(x,r)} \left| f - ⨍_{\Delta(x,r)} f\,d\sigma \right|^p d\sigma \right)^{1/p} = 0, \end{aligned} \tag{25.3}$$

for some (or all) $p \in [1,\infty)$.

We next recall a definition from [GHA-V, Definition 3.4.1, p. 158] which will play a significant role in the treatment of boundary value problems later on in this section.

© The Author(s), under exclusive license to Springer Nature Switzerland AG 2026

M. Mitrea and P. Takemura, *Singular Integrals, Herz-Type Function Spaces, and Boundary Problems*, Progress in Mathematics 362,
https://doi.org/10.1007/978-3-032-12516-3_25

Definition 25.1 Fix $n \in \mathbb{N}$, with $n \geq 2$, and consider an Ahlfors regular domain $\Omega \subseteq \mathbb{R}^n$. Abbreviate $\sigma := \mathcal{H}^{n-1}\lfloor\partial\Omega$, and denote by ν the geometric measure theoretic outward unit normal to Ω.

(1) Given $\delta \in (0,1)$, call Ω a `δ-oscillating AR domain` provided

$$\operatorname{dist}\Big(\nu, [\mathrm{VMO}(\partial\Omega,\sigma)]^n\Big) < \delta, \tag{25.4}$$

where the distance is measured in the John-Nirenberg space $[\mathrm{BMO}(\Sigma,\sigma)]^n$.

(2) Call Ω an `infinitesimally flat AR domain` provided

$$\nu \in [\mathrm{VMO}(\partial\Omega,\sigma)]^n. \tag{25.5}$$

Heuristically, the class of infinitesimally flat AR domains is a sharp version, from a geometric measure theoretic point of view, of the class of $\mathscr{C}^1$ domains with compact boundaries.

Our first result in this section establishes the Fredholmness of the Regularity Problem. To state it, we recall that the notion of relative flatness radius of an Ahlfors regular domain $\Omega \subseteq \mathbb{R}^n$, denoted by $\wp(\Omega)$, has been introduced in [GHA-V, Definition 3.1.9, p. 123]

Theorem 25.1 *Fix two integers $n, M \in \mathbb{N}$ with $n \geq 2$ and suppose $\Omega \subseteq \mathbb{R}^n$ is an Ahlfors regular domain with compact boundary. Abbreviate $\sigma := \mathcal{H}^{n-1}\lfloor\partial\Omega$ and denote by ν the geometric measure theoretic outward unit normal to Ω. Fix exponents $p, q \in (1,\infty)$ along with an index $\alpha \in \mathbb{R}$ satisfying*

$$\frac{-(n-1)}{p} < \alpha < \frac{n-1}{p'} \tag{25.6}$$

where $p' \in (1,\infty)$ denotes the Hölder conjugate exponent of p. Also, pick k_* as in (24.2). Introduce the space

$$\big(\dot{\mathrm{K}}^{\alpha}_{p,q}(\partial\Omega,\sigma;k_*)\big)^{\circ} := \overline{L^{\infty}_{\mathrm{comp}}(\partial\Omega,\sigma)}^{\|\cdot\|_{\dot{\mathrm{K}}^{\alpha}_{p,q}(\partial\Omega,\sigma;k_*)}}, \tag{25.7}$$

and the k_*-homogeneous Herz-styled Sobolev spaces:

$$\dot{\mathrm{K}}^{\alpha,1}_{p,q}(\partial\Omega,\sigma;k_*) := \Big\{f \in \dot{\mathrm{K}}^{\alpha}_{p,q}(\partial\Omega,\sigma;k_*) : \partial_{\tau_{jk}} f \in \dot{\mathrm{K}}^{\alpha}_{p,q}(\partial\Omega,\sigma;k_*) \text{ whenever } 1 \leq j,k \leq n\Big\}, \tag{25.8}$$

and

$$\big(\dot{\mathrm{K}}^{\alpha,1}_{p,q}(\partial\Omega,\sigma;k_*)\big)^{\circ} := \Big\{f \in \big(\dot{\mathrm{K}}^{\alpha}_{p,q}(\partial\Omega,\sigma;k_*)\big)^{\circ} : \partial_{\tau_{jk}} f \in \big(\dot{\mathrm{K}}^{\alpha}_{p,q}(\partial\Omega,\sigma;k_*)\big)^{\circ} \text{ whenever } 1 \le j,k \le n\Big\}, \tag{25.9}$$

(these are known to be Banach spaces when equipped with natural norms; cf. [GHA-V, (5.4.84), p. 333]). Finally, consider a homogeneous, second-order, constant complex coefficient, $M \times M$ system L in $\mathbb{R}^n$, and fix an arbitrary aperture parameter $\kappa > 0$.

Then there exists some threshold $\delta \in (0,1)$, which depends only on n, p, q, α, k_*, the system L, the AR *constants of $\partial\Omega$, and the relative flatness ratio $\wp(\Omega)/\operatorname{diam}\partial\Omega$, with the property that when*

$$\operatorname{dist}\Big(\nu\,,\ [\mathrm{VMO}\,(\partial\Omega,\sigma)]^n\Big) < \delta \tag{25.10}$$

where the distance is considered in the space $\big[\mathrm{BMO}\,(\partial\Omega,\sigma)\big]^n$, hence if Ω is a δ-oscillating AR *domain*[1] *(cf. Definition 25.1), the following results are true:*

(1) If one assumes that

$$L \text{ is weakly elliptic with } \mathfrak{A}^{\mathrm{dis}}_L \neq \varnothing \text{ and } \mathfrak{A}^{\mathrm{dis}}_{L^\top} \neq \varnothing, \tag{25.11}$$

then the Regularity Problem

$$\begin{cases} u \in \big[\mathscr{C}^\infty(\Omega)\big]^M, \quad Lu = 0 \text{ in } \Omega, \\ \mathcal{N}_\kappa u,\ \mathcal{N}_\kappa(\nabla u) \in \dot{\mathrm{K}}^{\alpha}_{p,q}(\partial\Omega,\sigma;k_*), \\ u\big|^{\kappa-\mathrm{n.t.}}_{\partial\Omega} = f \in \big[\dot{\mathrm{K}}^{\alpha,1}_{p,q}(\partial\Omega,\sigma;k_*)\big]^M, \\ u(x) = O(|x|^{2-n}) \text{ as } |x| \to \infty \text{ if } \Omega \text{ is unbounded}, \end{cases} \tag{25.12}$$

is Fredholm solvable, in the following precise sense: First, the space of admissible boundary data for the boundary problem (25.12)*, i.e.,*

$$\Big\{u\big|^{\kappa-\mathrm{n.t.}}_{\partial\Omega} : u \in \big[\mathscr{C}^\infty(\Omega)\big]^M,\ Lu = 0 \text{ in } \Omega,\ \mathcal{N}_\kappa u,\ \mathcal{N}_\kappa(\nabla u) \in \dot{\mathrm{K}}^{\alpha}_{p,q}(\partial\Omega,\sigma;k_*), \text{ and } u(x) = O(|x|^{2-n}) \text{ as } |x| \to \infty \text{ if } \Omega \text{ is unbounded}\Big\}, \tag{25.13}$$

is a closed linear subspace of finite codimension in $\big[\dot{\mathrm{K}}^{\alpha,1}_{p,q}(\partial\Omega,\sigma;k_*)\big]^M$. Second, a solution u of (25.12) *exists if and only if the boundary datum f belongs to* (25.13)*, in which scenario a solution u may be found such that*

$$\|\mathcal{N}_\kappa u\|_{\dot{\mathrm{K}}^{\alpha}_{p,q}(\partial\Omega,\sigma;k_*)} + \|\mathcal{N}_\kappa(\nabla u)\|_{\dot{\mathrm{K}}^{\alpha}_{p,q}(\partial\Omega,\sigma;k_*)} \approx \|f\|_{[\dot{\mathrm{K}}^{\alpha,1}_{p,q}(\partial\Omega,\sigma;k_*)]^M} \tag{25.14}$$

[1] in particular, if Ω is an infinitesimally flat AR domain.

with implicit proportionality constants independent of the boundary datum f. Third, the space of null-solutions of the boundary problem (25.12), *i.e.,*

$$\Big\{u \in \big[\mathscr{C}^\infty(\Omega)\big]^M : Lu = 0 \ \text{in}\ \Omega,\ \mathcal{N}_\kappa u,\ \mathcal{N}_\kappa(\nabla u) \in \dot{\mathrm{K}}^{\alpha}_{p,q}(\partial\Omega, \sigma; k_*), \quad u\Big|_{\partial\Omega}^{\kappa-\mathrm{n.t.}} = 0 \ \text{at}\ \sigma\text{-a.e. point on}\ \partial\Omega, \text{and} \quad u(x) = O(|x|^{2-n}) \ \text{as}\ |x| \to \infty \ \text{if}\ \Omega \ \text{is unbounded}\Big\}, \tag{25.15}$$

is finite dimensional.

(2) If (25.11) *is strengthened to*

> *L satisfies the Legendre-Hadamard ellipticity condition* (*cf.* (7.10)) *and one also has* $\mathfrak{A}_L^{\mathrm{dis}} \neq \varnothing$, (25.16)

the Regularity Problem (25.12) *is actually well posed. More specifically, the boundary value problem* (25.12) *has a unique solution, and this solution enjoys the additional property that there exists* $C = C(\Omega, L, n, p, q, \alpha, k_*) \in (0, \infty)$ *such that* (25.14) *holds.*

(3) Similar results to those described in items (1) and (2) above also hold for the Regularity Problem with boundary data in $\big(\dot{\mathrm{K}}^{\alpha,1}_{p,q}(\partial\Omega, \sigma; k_*)\big)^\circ$, *i.e.,*

$$\begin{cases} u \in \big[\mathscr{C}^\infty(\Omega)\big]^M, \quad Lu = 0 \ \text{in}\ \Omega, \\ \mathcal{N}_\kappa u,\ \mathcal{N}_\kappa(\nabla u) \in \big(\dot{\mathrm{K}}^{\alpha}_{p,q}(\partial\Omega, \sigma; k_*)\big)^\circ, \\ u\Big|_{\partial\Omega}^{\kappa-\mathrm{n.t.}} = f \in \big[\big(\dot{\mathrm{K}}^{\alpha,1}_{p,q}(\partial\Omega, \sigma; k_*)\big)^\circ\big]^M, \\ u(x) = O(|x|^{2-n}) \ \text{as}\ |x| \to \infty \ \text{if}\ \Omega \ \text{is unbounded}. \end{cases} \tag{25.17}$$

(4) There exist symmetric weakly elliptic systems L with $\mathfrak{A}_L^{\mathrm{dis}} = \varnothing$, *as well as indices p, q, and α, for which the Regularity Problems* (25.12) *and* (25.17) *fail to be Fredholm solvable even when the underlying domain Ω is the unit disk in the plane* (*or the complement of its closure*)*. In fact, the aforementioned failure of Fredholm solvability may be total, in the sense that for each of these problems, both the space of admissible boundary data may have infinite codimension in the full space of boundary data, and the space of null-solutions may be infinite dimensional.*

Proof A combination of (24.15), (24.16), (24.18), Theorem 24.1, and [GHA-V, Theorem 8.1.5, pp. 582–585] ensures that all claims *(1)-(4)* hold. □

Moving on to the Dirichlet problem, we have the following result.

Theorem 25.2 *Fix two integers $n, M \in \mathbb{N}$ with $n \geq 2$ and suppose $\Omega \subseteq \mathbb{R}^n$ is an Ahlfors regular domain with compact boundary. Abbreviate* $\sigma := \mathcal{H}^{n-1}\lfloor\partial\Omega$ *and* *denote by ν the geometric measure theoretic outward unit normal to Ω. Fix exponents $p, q \in (1, \infty)$ along with an index $\alpha \in \mathbb{R}$ satisfying*

$$\frac{-(n-1)}{p} < \alpha < \frac{n-1}{p'} \tag{25.18}$$

where $p' \in (1, \infty)$ denotes the Hölder conjugate exponent of p. Also, pick k_* as in (24.2).

Furthermore, suppose L is a homogeneous, second-order, constant complex coefficient, $M \times M$ system in $\mathbb{R}^n$ and, for some fixed arbitrary aperture parameter $\kappa > 0$, consider the boundary value problem

$$\begin{cases} u \in \big[\mathscr{C}^\infty(\Omega)\big]^M, \quad Lu = 0 \ \ in \ \ \Omega, \\ \mathcal{N}_\kappa u \in \dot{\mathrm{K}}^\alpha_{p,q}(\partial\Omega, \sigma; k_*), \\ u\Big|^{\kappa-\mathrm{n.t.}}_{\partial\Omega} = f \in \big[\dot{\mathrm{K}}^\alpha_{p,q}(\partial\Omega, \sigma; k_*)\big]^M, \\ u(x) = O(|x|^{2-n}) \ \ as \ \ |x| \to \infty \ \ if\, \Omega \ is \ unbounded. \end{cases} \tag{25.19}$$

Then there exists some threshold $\delta \in (0, 1)$, which depends only on n, p, q, α, k_*, the system L, the AR *constants of $\partial\Omega$, and the relative flatness ratio $\wp(\Omega)/\mathrm{diam}\,\partial\Omega$, with the property that when*

$$\mathrm{dist}\Big(\nu\,,\ [\mathrm{VMO}\,(\partial\Omega, \sigma)]^n\Big) < \delta \tag{25.20}$$

where the distance is considered in the space $\big[\mathrm{BMO}\,(\partial\Omega, \sigma)\big]^n$, hence if Ω is a δ-oscillating AR *domain*[2] *(cf. Definition 25.1), the following statements are true:*

(1) If one assumes that

$$L \text{ is weakly elliptic with } \mathfrak{A}^{\mathrm{dis}}_L \neq \varnothing \text{ and } \mathfrak{A}^{\mathrm{dis}}_{L^\top} \neq \varnothing, \tag{25.21}$$

then the Dirichlet Problem (25.19) *is Fredholm solvable, in the following precise sense: First, if the system L is weakly elliptic with $\mathfrak{A}^{\mathrm{dis}}_L \neq \varnothing$, the space of admissible boundary data for the Dirichlet Problem* (25.19), *i.e.,*

$$\Big\{u\Big|^{\kappa-\mathrm{n.t.}}_{\partial\Omega} : u \in \big[\mathscr{C}^\infty(\Omega)\big]^M,\ Lu = 0 \,in\, \Omega,\ \mathcal{N}_\kappa u \in \dot{\mathrm{K}}^\alpha_{p,q}(\partial\Omega, \sigma; k_*),\ u\Big|^{\kappa-\mathrm{n.t.}}_{\partial\Omega} \text{ exists } \sigma\text{-a.e., and } u(x) = O(|x|^{2-n}) \text{ as } |x| \to \infty \text{ if } \Omega \text{ is unbounded}\Big\}, \tag{25.22}$$

is a closed linear subspace of finite codimension in $\big[\dot{\mathrm{K}}^\alpha_{p,q}(\partial\Omega, \sigma; k_*)\big]^M$, and a solution u of (25.19) *exists if and only if the boundary datum f belongs to*

[2] in particular, if Ω is an infinitesimally flat AR domain

(25.22). *Second, if L is weakly elliptic with* $\mathfrak{A}^{\rm dis}_{L^\top} \neq \varnothing$, *the space of null-solutions of the boundary problem* (25.19), *i.e.*,

$$\Big\{u \in \big[\mathscr{C}^\infty(\Omega)\big]^M : Lu = 0 \;\text{ in }\; \Omega,\; \mathcal{N}_\kappa u \in \dot{\rm K}^\alpha_{p,q}(\partial\Omega, \sigma; k_*),\quad u\big|^{\kappa-{\rm n.t.}}_{\partial\Omega} = 0 \text{ at } \sigma\text{-a.e. point on } \partial\Omega, \text{ and } u(x) = O(|x|^{2-n}) \;\text{ as }\; |x| \to \infty \;\text{ if } \Omega \text{ is unbounded}\Big\}, \tag{25.23}$$

is finite dimensional; in addition,

for any null-solution u of the boundary problem (25.19) (*i.e., for any function u belonging to the space described in* (25.23)), $\mathcal{N}_\kappa(\nabla u)$ *belongs to* $\dot{\rm K}^\alpha_{p,q}(\partial\Omega, \sigma; k_*)$ *and the non-tangential trace* $(\nabla u)\big|^{\kappa-{\rm n.t.}}_{\partial\Omega}$ *exists at* σ*-a.e. point on* $\partial\Omega$, (25.24)

so, in particular,

the space of null-solutions for the Dirichlet Problem (25.19) (*cf.* (25.23)) *coincides with the space of null-solution for the Regularity Problem* (25.12) (*cf.* (25.15)). (25.25)

Lastly, the Fredholm solvability result stated above is optimal, since in the class of operators L for which $\mathfrak{A}^{\rm dis}_{L} = \varnothing$ *or* $\mathfrak{A}^{\rm dis}_{L^\top} = \varnothing$ *it may happen that the Dirichlet Problem* (25.19) *fails to be Fredholm solvable even when the underlying domain* Ω *is the unit disk in the plane (or the complement of its closure). In fact, the aforementioned failure of Fredholm solvability may be total, in the sense that for the problem* (25.19), *both the space of admissible boundary data may have infinite codimension in the full space of boundary data, and the space of null-solutions may be infinite dimensional.*

(2) *If* (25.21) *is strengthened to assuming that*

L satisfies the Legendre-Hadamard strong ellipticity condition (*cf.* (7.10)) *and one has* $\mathfrak{A}^{\rm dis}_{L} \neq \varnothing$, (25.26)

then the Dirichlet Problem (25.19) *is actually well posed. More specifically, the boundary problem* (25.19) *has a unique solution, and this solution enjoys the additional property that*

$$\big\|\mathcal{N}_\kappa u\big\|_{\dot{\rm K}^\alpha_{p,q}(\partial\Omega,\sigma;k_*)} \approx \|f\|_{[\dot{\rm K}^\alpha_{p,q}(\partial\Omega,\sigma;k_*)]^M}, \tag{25.27}$$

with implicit proportionality constants independent of f.

(3) *Similar properties like the ones discussed in items (1)-(2) above are valid for the Dirichlet Problem with boundary data in the space* $\big(\dot{\rm K}^\alpha_{p,q}(\partial\Omega, \sigma; k_*)\big)^\circ$ (*defined in* (25.7)), *i.e.*,

$$\begin{cases} u \in \big[\mathscr{C}^{\infty}(\Omega)\big]^M, \quad Lu = 0 \ \ in \ \ \Omega, \\ \mathcal{N}_{\kappa} u \in \big(\dot{\mathrm{K}}^{\alpha}_{p,q}(\partial\Omega, \sigma; k_*)\big)^{\circ}, \\ u\big|_{\partial\Omega}^{\kappa-\mathrm{n.t.}} = f \in \big[\big(\dot{\mathrm{K}}^{\alpha}_{p,q}(\partial\Omega, \sigma; k_*)\big)^{\circ}\big]^M, \\ u(x) = O(|x|^{2-n}) \ \ as \ \ |x| \to \infty \ \ if\, \Omega \ is \ unbounded, \end{cases} \tag{25.28}$$

Proof In light of (24.16), (24.18), and Theorem 24.1, the veracity of all claims may be obtained from [GHA-V, Theorem 8.1.9, pp. 614–615]. □

Our next goal is to treat the Neumann Problem with boundary data in k_*-homogeneous Herz spaces. Stating this requires the following definitions. Given an arbitrary open set $\Omega \subseteq \mathbb{R}^n$, define

$$\Omega_+ := \Omega \ \text{ and } \ \Omega_- := \mathbb{R}^n \setminus \overline{\Omega}. \tag{25.29}$$

In relation to these sets, the following function spaces arise naturally:

$$\begin{aligned} \mathbb{C}_{\Omega_\pm} &:= \mathrm{span}_{\mathbb{C}} \big\{ \mathbf{1}_{O} : O \text{ bounded connected component of } \Omega_\pm \big\}, \\ \mathbb{C}_{\partial\Omega_\pm} &:= \mathrm{span}_{\mathbb{C}} \big\{ \mathbf{1}_{\partial O} : O \text{ bounded connected component of } \Omega_\pm \big\}, \\ \mathbb{C}_{\partial\Omega} &:= \mathrm{span}_{\mathbb{C}} \big\{ \mathbf{1}_{\Sigma} : \Sigma \text{ connected component of } \partial\Omega \big\}. \end{aligned} \tag{25.30}$$

Also, for any space X of distributions on $\partial\Omega$ we agree to abbreviate

$$X \cap \big(\mathbb{C}_{\partial\Omega_\pm}\big)^{\perp} := \Big\{ f \in X : \langle f, \chi \rangle = 0 \ \text{ for each } \ \chi \in \mathbb{C}_{\partial\Omega_\pm} \Big\}, \tag{25.31}$$

$$X \cap \big(\mathbb{C}_{\partial\Omega}\big)^{\perp} := \Big\{ f \in X : \langle f, \chi \rangle = 0 \text{ for each } \chi \in \mathbb{C}_{\partial\Omega} \Big\}. \tag{25.32}$$

In addition, we shall consider

$$\begin{aligned} \mathbb{C}^{\infty}_{\Omega_\pm} &:= \mathrm{span}_{\mathbb{C}} \big\{ \mathbf{1}_{O} : O \text{ connected component of } \Omega_\pm \big\}, \\ \mathbb{C}^{\infty}_{\partial\Omega_\pm} &:= \mathrm{span}_{\mathbb{C}} \big\{ \mathbf{1}_{\partial O} : O \text{ connected component of } \Omega_\pm \big\}, \end{aligned} \tag{25.33}$$

and, for any given space X of distributions on $\partial\Omega$, set

$$X \cap \big(\mathbb{C}^{\infty}_{\partial\Omega_\pm}\big)^{\perp} := \Big\{ f \in X : \langle f, \chi \rangle = 0 \ \text{ for each } \ \chi \in \mathbb{C}^{\infty}_{\partial\Omega_\pm} \Big\}, \tag{25.34}$$

as well as

$$X \cap \mathbb{C}^{\perp} := \Big\{ f \in X : \langle f, c \rangle = 0 \ \text{ for each } \ c \in \mathbb{C} \Big\}. \tag{25.35}$$

We are now in a position to state and prove the aforementioned solvability theorem for the Neumann Problem. The reader is reminded that the conormal derivative operator has been defined in (7.55).

Theorem 25.3 *Select $n, M \in \mathbb{N}$ with $n \geq 2$ and suppose $\Omega \subseteq \mathbb{R}^n$ is an Ahlfors regular domain with compact boundary. Abbreviate $\sigma := \mathcal{H}^{n-1}\lfloor\partial\Omega$ and denote by ν the geometric measure theoretic outward unit normal to Ω. Select exponents $p, q \in (1, \infty)$, along with an index $\alpha \in \mathbb{R}$ satisfying*

$$\frac{-(n-1)}{p} < \alpha < \frac{n-1}{p'} \tag{25.36}$$

where $p' \in (1, \infty)$ denotes the Hölder conjugate exponent of p. Also, pick k_* as in (24.2).

Finally, let L be a homogeneous, second-order, constant complex coefficient, $M \times M$ system in $\mathbb{R}^n$, and pick a coefficient tensor $A \in \mathfrak{A}_L$.

In this setting, having fixed an arbitrary aperture parameter $\kappa \in (0, \infty)$, consider the Neumann Problem

$$\begin{cases} u \in \left[\mathscr{C}^\infty(\Omega)\right]^M, \quad Lu = 0 \ \text{ in } \ \Omega, \\ \mathcal{N}_\kappa(\nabla u) \in \dot{\mathrm{K}}^\alpha_{p,q}(\partial\Omega, \sigma; k_*), \\ \partial^A_\nu u = f \in \left[\dot{\mathrm{K}}^\alpha_{p,q}(\partial\Omega, \sigma; k_*)\right]^M, \\ u(x) = O(|x|^{2-n}) \ \text{ as } \ |x| \to \infty \ \text{ if } \Omega \text{ is unbounded.} \end{cases} \tag{25.37}$$

Then there exists some threshold $\delta \in (0, 1)$, which depends only on n, A, p, q, α, k_*, the AR *constants of $\partial\Omega$, and the relative flatness ratio $\wp(\Omega)/\mathrm{diam}\,\partial\Omega$, with the property that whenever*

$$\mathrm{dist}\Big(\nu\,,\ [\mathrm{VMO}\,(\partial\Omega, \sigma)]^n\Big) < \delta, \tag{25.38}$$

where the distance is considered in the space $\left[\mathrm{BMO}\,(\partial\Omega, \sigma)\right]^n$, hence whenever Ω is a δ-oscillating AR *domain*[3] *(cf. Definition 25.1), the set Ω becomes a* UR *domain and the following statements are true:*

(1) If one assumes that

$$L \text{ is weakly elliptic with } \mathfrak{A}^{\mathrm{dis}}_L \neq \varnothing \text{ and } \mathfrak{A}^{\mathrm{dis}}_{L^\top} \neq \varnothing, \tag{25.39}$$

then whenever $A \in \mathfrak{A}^{\mathrm{dis}}_L$ it follows that the boundary value problem (25.37) *is Fredholm solvable, in the following precise sense: First, the space of admissible boundary data for the boundary problem* (25.37)*, i.e.,*

$$\Big\{\partial^A_\nu u : u \in \left[\mathscr{C}^\infty(\Omega)\right]^M, \ Lu = 0 \ \text{ in } \ \Omega, \ \mathcal{N}_\kappa(\nabla u) \in \dot{\mathrm{K}}^\alpha_{p,q}(\partial\Omega, \sigma; k_*), \\ \text{and } \ u(x) = O(|x|^{2-n}) \ \text{ as } \ |x| \to \infty \ \text{ if } \Omega \text{ is unbounded}\Big\} \tag{25.40}$$

[3] in particular, if Ω is an infinitesimally flat AR domain

is a closed linear subspace of finite codimension in $\big[\dot{\mathrm{K}}^{\alpha}_{p,q}(\partial\Omega,\sigma;k_*)\big]^M$. *Second, a solution u of* (25.37) *exists if and only if the boundary datum f belongs to* (25.40), *in which scenario a solution u may be found such that*

$$\|\mathcal{N}_\kappa(\nabla u)\|_{\dot{\mathrm{K}}^{\alpha}_{p,q}(\partial\Omega,\sigma;k_*)} \leq C\|f\|_{[\dot{\mathrm{K}}^{\alpha}_{p,q}(\partial\Omega,\sigma;k_*)]^M}, \tag{25.41}$$

for some constant $C \in (0,\infty)$ *independent of f. Third, the space of null-solutions of the boundary problem* (25.37), *i.e.,*

$$\Big\{u \in \big[\mathscr{C}^\infty(\Omega)\big]^M : Lu = 0 \ \text{ in } \ \Omega,\ \mathcal{N}_\kappa(\nabla u) \in \dot{\mathrm{K}}^{\alpha}_{p,q}(\partial\Omega,\sigma;k_*), \tag{25.42}$$
$$\partial^A_\nu u = 0 \text{ at } \sigma\text{-a.e. point on } \partial\Omega, \text{ and}$$
$$u(x) = O(|x|^{2-n}) \ \text{ as } \ |x| \to \infty \ \text{ if } \Omega \text{ is unbounded}\Big\}$$

is finite dimensional.

(2) *Suppose* (25.39) *is strengthened* (*cf. [GHA-V, Proposition 1.4.1, p. 25]*) *to the demand that*

$$L \text{ satisfies the Legendre-Hadamard strong ellipticity condition (cf.(7.10)) and one has } \mathfrak{A}^{\mathrm{dis}}_L \neq \varnothing. \tag{25.43}$$

Then for each $A \in \mathfrak{A}^{\mathrm{dis}}_L$ *the index of the Neumann Problem* (25.37), *i.e., the dimension of* (25.42) *minus the codimension of* (25.40) *in* $\big[\dot{\mathrm{K}}^{\alpha}_{p,q}(\partial\Omega,\sigma;k_*)\big]^M$, *is equal to zero. Moreover, the space of admissible boundary data for the Neumann Problem* (25.37), *defined in* (25.40),

$$\text{is equal to } \mathrm{Im}\Big(-\tfrac{1}{2}I + K^{\#}_{A^\top};\ \big[\dot{\mathrm{K}}^{\alpha}_{p,q}(\partial\Omega,\sigma;k_*)\big]^M\Big),$$
$$\text{if either } n \geq 3, \text{ or } n = 2 \text{ and the set } \Omega \text{ is bounded,}$$
$$\text{while if } n = 2 \text{ and } \Omega \text{ is an exterior domain it may be described as} \tag{25.44}$$
$$\Big\{\big(-\tfrac{1}{2}I + K^{\#}_{A^\top}\big)g \,:\, g \in \big[\dot{\mathrm{K}}^{\alpha}_{p,q}(\partial\Omega,\sigma;k_*)\big]^M \ \text{ with } \int_{\partial\Omega} g\,d\sigma = 0\Big\}.$$

Furthermore, there exists a constant $C \in (0,\infty)$ *with the property that whenever the boundary datum f belongs to the space of admissible boundary data, defined in* (25.40), *then a solution u of the boundary problem* (25.37) *may be found such that* (25.41) *holds.*

Also, with $\mathcal{D}_A$ *denoting the boundary-to-domain double layer potential operator associated with A and* Ω *as in* (7.21), *when either* $n \geq 3$, *or* $n = 2$ *and the set* Ω *is bounded, it follows that the space of null-solutions of the Neumann Problem* (25.37) (*described in* (25.42)) *is equal to the space*

$$\Big\{\mathcal{D}_A\phi \,:\, \phi \in \mathrm{Ker}\Big(-\tfrac{1}{2}I + K_A;\ \big[\dot{\mathrm{K}}^{\alpha,1}_{p,q}(\partial\Omega,\sigma;k_*)\big]^M\Big)\Big\}, \tag{25.45}$$

the mapping associating to each function u belonging to the space of null-solutions of the Neumann Problem (25.37) (*described in* (25.42)) *its nontangential boundary trace* $u\Big|_{\partial\Omega}^{\kappa-\mathrm{n.t.}}$ *is an isomorphism onto the finite dimensional space*

$$\mathrm{Ker}\Big(-\tfrac{1}{2}I + K_A;\ \big[\dot{\mathrm{K}}^{\alpha,1}_{p,q}(\partial\Omega,\sigma;k_*)\big]^M\Big), \tag{25.46}$$

and the inverse of said isomorphism $u \mapsto u\Big|_{\partial\Omega}^{\kappa-\mathrm{n.t.}}$ *is the boundary-to-domain double layer potential operator* $\mathcal{D}_A$ *mapping* $\mathrm{Ker}\Big(-\tfrac{1}{2}I+K_A;\ \big[\dot{\mathrm{K}}^{\alpha,1}_{p,q}(\partial\Omega,\sigma;k_*)\big]^M\Big)$ *isomorphically onto the space of null-solutions of the Neumann Problem* (25.37) (*described in* (25.42)).

Finally, in the case when $n = 2$ *and* Ω *is an exterior domain it follows that the space of null-solutions of the Neumann Problem* (25.37) *is equal to*

$$\Big\{\mathcal{D}_A\phi + c : \phi \in \big[\dot{\mathrm{K}}^{\alpha,1}_{p,q}(\partial\Omega,\sigma;k_*)\big]^M,\ c \in \mathbb{C}^M,\ \big(-\tfrac{1}{2}I+K_A\big)\phi = -c\Big\} \tag{25.47}$$

the mapping associating to each function u belonging to the space of null-solutions of the Neumann Problem (25.37) (*described in* (25.42)) *the pair* $\Big(u\Big|_{\partial\Omega}^{\kappa-\mathrm{n.t.}},\ \lim_{|x|\to\infty} u(x)\Big)$ *is a well-defined isomorphism onto the space*

$$\Big\{(\phi,c) \in \big[\dot{\mathrm{K}}^{\alpha,1}_{p,q}(\partial\Omega,\sigma;k_*)\big]^M \oplus \mathbb{C}^M : \big(-\tfrac{1}{2}I + K_A\big)\phi = -c\Big\} \tag{25.48}$$

and the inverse of said linear isomorphism $u \mapsto \Big(u\Big|_{\partial\Omega}^{\kappa-\mathrm{n.t.}},\ \lim_{|x|\to\infty} u(x)\Big)$ *in the above context is the assignment* $(\phi, c) \longmapsto \mathcal{D}_A\phi + c$ *mapping the last space in* (25.48) *isomorphically onto the space of null-solutions of the Neumann Problem* (25.37) (*described in* (25.42)).

(3) *If* (25.43) *is further strengthened to*

$$A \in \mathfrak{A}^{\mathrm{dis}}_L \text{ is positive definite,} \tag{25.49}$$

which in particular implies that L satisfies the Legendre-Hadamard strong ellipticity condition (*cf.* (7.10)), *then the Neumann Problem* (25.37) *has the following additional properties* (*with the spaces* $\mathcal{C}_{\Omega_\pm}$, $\mathcal{C}_{\partial\Omega_\pm}$, $\mathcal{C}^\infty_{\Omega_\pm}$, $\mathcal{C}^\infty_{\partial\Omega_\pm}$ *defined as in* (25.30), (25.33), *and with the conventions* (25.31), (25.34) *in effect*). *First, if either* $n \geq 3$, *or* $n = 2$ *and the set* Ω *is bounded, then*

$$\begin{aligned} &\text{the space of admissible boundary data for the Neumann Problem} \\ &\text{(25.37) (see (25.40)) is equal to } \big[\dot{\mathrm{K}}^{\alpha}_{p,q}(\partial\Omega,\sigma;k_*) \cap (\mathcal{C}_{\partial\Omega_+})^{\perp}\big]^M, \\ &\text{and the space of null-solutions of the Neumann Problem (25.37)} \\ &\text{(described in (25.42)) is equal to } \big(\mathcal{C}_{\Omega_+}\big)^M. \end{aligned} \tag{25.50}$$

Second, if $n = 2$ *and* Ω *is an exterior domain, then* *the space of admissible boundary data for the Neumann Problem (25.37) (see (25.40)) is equal to* $\big[\dot{\mathrm{K}}^{\alpha}_{p,q}(\partial\Omega,\sigma;k_*)\cap\big(\mathbb{C}^{\infty}_{\partial\Omega_+}\big)^{\perp}\big]^M$, *and the space of null-solutions of the Neumann Problem (25.37) (described in (25.42)) is equal to* $\big(\mathbb{C}^{\infty}_{\Omega_+}\big)^M$. (25.51)

(4) Similar results as in items (1)-(3) above are true for the Neumann Problem with boundary data in $\big(\dot{\mathrm{K}}^{\alpha}_{p,q}(\partial\Omega,\sigma;k_*)\big)^{\circ}$ *(defined in (25.7)), namely*

$$\begin{cases} u\in\big[\mathscr{C}^{\infty}(\Omega)\big]^M, \quad Lu=0 \ \ in\ \ \Omega,\\ \mathcal{N}_{\kappa}(\nabla u)\in\big(\dot{\mathrm{K}}^{\alpha}_{p,q}(\partial\Omega,\sigma;k_*)\big)^{\circ},\\ \partial^A_{\nu}u=f\in\big[\big(\dot{\mathrm{K}}^{\alpha}_{p,q}(\partial\Omega,\sigma;k_*)\big)^{\circ}\big]^M,\\ u(x)=O(|x|^{2-n}) \ \ as \ \ |x|\to\infty \ \ if\ \Omega \ is\ unbounded, \end{cases} \tag{25.52}$$

(5) Even when $L:=\Delta$, *the Laplace operator (which obviously satisfies (25.39)), if* $A\notin\mathfrak{A}^{\rm dis}_L$ *then it may happen that the Neumann Problems (25.37), (25.52) fail to be Fredholm solvable even when the underlying domain* Ω *is the unit disk in the plane (or the complement of its closure). In fact, the aforementioned failure of Fredholm solvability may be total, in the sense that for each of these problems, both the space of admissible boundary data may have infinite codimension in the full space of boundary data, and the space of null-solutions may be infinite dimensional. Finally, a similar lack of Fredholm solvability occurs in higher dimensions.*

Proof All claims may be obtained via [GHA-V, Theorem 8.1.17, pp. 680–684], whose applicability in the current context is ensured by (24.16), (24.18), and Theorem 24.1. □

We close by noting that the boundary condition in (25.37) is naturally formulated. Indeed, choosing the threshold $\delta\in(0,1)$ small permits us to invoke [GHA-V, Theorem 3.4.2, pp. 160–161] which ensures that Ω is a two-sided NTA domain. Granted this, the Fatou-type result proved in [GHA-III, *(g)* in Theorem 3.3.4, pp. 702–705] (currently used with $\mathbb{X}:=\dot{\mathrm{K}}^{\alpha}_{p,q}(\partial\Omega,\sigma;k_*)$) gives that any function satisfying

$$u\in\big[\mathscr{C}^{\infty}(\Omega)\big]^M \ \text{ with } \ Lu=0 \ \text{ in } \ \Omega \ \text{ and } \ \mathcal{N}_{\kappa}(\nabla u)\in\dot{\mathrm{K}}^{\alpha}_{p,q}(\partial\Omega,\sigma;k_*) \tag{25.53}$$

has the property that the nontangential boundary trace

$$\begin{gathered} (\nabla u)\big|^{\kappa-\mathrm{n.t.}}_{\partial\Omega} \ \text{exists (in } \mathbb{C}^{n\times M}\text{) at } \sigma\text{-a.e. point on } \partial\Omega \ \text{ and}\\ \text{the function } (\nabla u)\big|^{\kappa-\mathrm{n.t.}}_{\partial\Omega} \ \text{ belongs to } \ \big[\dot{\mathrm{K}}^{\alpha}_{p,q}(\partial\Omega,\sigma;k_*)\big]^{n\times M}. \end{gathered} \tag{25.54}$$

In particular, the conormal derivative $\partial^A_{\nu}u$, considered in a pointwise sense, is meaningful in this case and satisfies

$$\partial_\nu^A u \in \big[\dot{\mathrm{K}}^\alpha_{p,q}(\partial\Omega, \sigma; k_*)\big]^M. \tag{25.55}$$

In summary, any function satisfying the properties listed in the first two lines of (25.37) has a well-defined conormal derivative, considered in a pointwise sense, and as a function this belongs to $\big[\dot{\mathrm{K}}^\alpha_{p,q}(\partial\Omega, \sigma; k_*)\big]^M$ (the space from which the boundary data for the problem (25.37) are selected).

References

[AdHaHä15] T. Adamowicz, P. Harjulehto, P. Hästö, *Maximal operator in variable exponent Lebesgue spaces on unbounded quasimetric measure spaces*, Math. Scand. **116**(1), 5–22 (2015)

[AlDr12] A. Almeida, D. Drihem, *Maximal, potential and singular type operators on Herz spaces with variable exponents*, J. Math. Anal. Appl. **394**, 781–795 (2012)

[AlMi15] R. Alvarado, M. Mitrea, *Hardy Spaces on Ahlfors-Regular Quasi Metric Spaces: A Sharp Theory*, Lecture Notes in Mathematics, Vol. 2142, Springer, Cham, 2015

[AlLaGu00] J. Alvarez, J. Lakey, M. Guzmán-Partida, *Spaces of Bounded λ-Central Mean Oscillation, Morrey Spaces and λ-Central Carleson Measures*, Collect. Math **51**(3), 1-47 (2000)

[AzHoEtAl16] J. Azzam, S. Hofmann, J.M. Martell, S. Mayboroda, M. Mourgoglou, X. Tolsa, A. Volberg, *Rectifiability of harmonic measure*, Geom. Funct. Anal. **26**(3), 703–728 (2016)

[BaMa13] A.E. Barton and S. Mayboroda, *The Dirichlet problem for higher order equations in composition form*, J. Funct. Anal. **265**(1), 49–107 (2013)

[BaMa14] A.E. Barton and S. Mayboroda, *Boundary-value problems for higher-order elliptic equations in non-smooth domains*, in Concrete operators, spectral theory, operators in harmonic analysis and approximation, 53–93, Oper. Theory Adv. Appl., 236, Birkhäuser/Springer, Basel, (2014)

[BaMa16-I] A.E. Barton and S. Mayboroda, *Layer potentials and boundary-value problems for second order elliptic operators with data in Besov spaces*, Mem. Amer. Math. Soc. **243**(1149), v+110pp. (2016)

[BaMa16-II] A.E. Barton and S. Mayboroda, *Higher-order elliptic equations in non-smooth domains: a partial survey*, in Harmonic analysis, partial differential equations, complex analysis, Banach spaces, and operator theory. Vol. 1, 55–121, Assoc. Women Math. Ser., 4, Springer Cham, 2016

[BaSa85] A. Baernstein II, E.T. Sawyer, *Embedding and Multiplier Theorems for $H^p(\mathbb{R})$*, Mem. Amer. Math. Soc. **53**(318), iv+82pp. (1985)

[BeSha88] C. Bennett, R. Sharpley, *Interpolation of Operators*, Pure and Applied Mathematics, Vol. 129, Academic Press, Boston, 1988

[BeLo76] J. Bergh and J. Löfström, *Interpolation Spaces. An Introduction*, Springer, Berlin/New York, 1976

[Be64] A. Beurling, *Construction and analysis of some convolution algebras*, Ann. Inst. Fourier (Grenoble) **14**, 1–32 (1964)

[BhKoKo23] M.A. Bhat, P. Kolwicz, G.S.R. Kosuru, *Köthe-Herz Spaces: The Amalgam-Type Spaces of Infinite Direct Sums*, arXiv:2209.05897 (2023)

[Bre11] H. Brezis, *Functional Analysis, Sobolev Spaces and Partial Differential Equations*, Univesitext, Springer, New York, 2011

[Bro89] R.M. Brown, *The method of layer potentials for the heat equation in Lipschitz cylinders*, Amer. J. Math., **111**(2), 339–379 (1989)

© The Editor(s) (if applicable) and The Author(s), under exclusive license to Springer Nature Switzerland AG 2026

M. Mitrea and P. Takemura, *Singular Integrals, Herz-Type Function Spaces, and Boundary Problems*, Progress in Mathematics 362,
https://doi.org/10.1007/978-3-032-12516-3

[Bro94] R. Brown, *The mixed problem for Laplace's equation in a class of Lipschitz domains*, Comm. Partial Differential Equations **19**, 1217—1233 (1994)

[Bro95] R.M. Brown, *The Neumann Problem on Lipschitz domains in Hardy spaces of order less than one*, Pacific J. Math. **171**(2), 389–408 (1995)

[BrHu01] R.M. Brown and W. Hu, *Boundary value problems for higher order parabolic equations*, Trans. Amer. Math. Soc. **353**(2), 809–838 (2001)

[BrMi09] R.M. Brown, I. Mitrea, *The mixed problem for the Lamé system in a class of Lipschitz domains*, J. Differential Equations **246**(7), 2577—2589 (2009)

[BrMiMiWr10] R. Brown, I. Mitrea, M. Mitrea, M. Wright, *Mixed boundary value problems for the Stokes system*, Trans. Amer. Math. Soc. **362**(3), 1211–1230 (2010)

[BrOtTa13] R.M. Brown, K.A. Ott, J.L. Taylor *The mixed problem in Lipschitz domains with general decompositions of the boundary*, Trans. Amer. Math. Soc. **365**(6), 2895–2930 (2013)

[BrSy01] R.M. Brown, J.D. Sykes, *The mixed boundary problem in L^p and Hardy spaces for Laplace's equation on a Lipschitz domain*, in Harmonic analysis and boundary value problems (Fayetteville, AR, 2000), 1–18, Contemp. Math., **277**, Amer. Math. Soc., (Providence, RI, 2001)

[BuMa69] Yu.D. Burago, V.G. Maz'ya. *Potential Theory and Function Theory for Irregular Regions* Seminars in mathematics, New York, Consultants Bureau (1969)

[BMS62] Yu.D. Burago, V.G. Maz'ya, and V.D. Sapozhnikova, *On the double layer potential for nonregular regions*, Sov. Math. Dokl., 3 (1962), 1640–1642 (translation from Dokl. Akad. Nauk SSSR 147 (1962), 523–525).

[BMS66] Yu.D. Burago, V.G. Maz'ya, and V.D. Sapozhnikova, *On the theory of simple and double layer potentials for domains with irregular boundaries*, Probl. Mat. Anal., 1 (1968), 1–30 (translation from Probl. Mat. Analiza, Kraevye Zadachi, Integral. Uravn. Leningrad (1966), 3–34).

[Ca54] A.P. Calderón, *The multipole expansion of radiation fields*, J. Rational Mech. Anal.**3**, 523–537 (1954)

[Ca77] A.P. Calderón, *Cauchy integrals on Lipschitz curves and related operators*, Proc. Nat. Acad. Sci. U.S.A. **74**(4), 1324–1327 (1977)

[Ca78] A.P. Calderón, *Commutators, singular integrals on Lipschitz curves and applications*, Proceedings of the International Congress of Mathematicians (Helsinki, 1978). Helsinki: Acad. Sci. Fennica, 1978

[CaZy52] A.P. Calderón, A. Zygmund, *On the existence of certain singular integrals,* Acta Math. **88**, 85–139 (1952)

[CaZy56] A.P. Calderón, A. Zygmund, *On singular integrals*, Amer. J. Math. **78**, 289–309 (1956)

[Ca75] C.P. Calderón, *On commutators of singular integrals*, Studia Math. **53**(2), 139–174 (1975)

[ChLa89] Y.Z. Chen, K.S. Lau, *Some New Classes of Hardy Spaces*, J. Funct. Anal. **84**, 255-278 (1989)

[CoGo83] J. Cohen and J.A. Gosselin, *The Dirichlet problem for the biharmonic equation in a C^1 domain in the plane*, Indiana Univ. Math. J. **32**(5), 635–685 (1983)

[CoMcMe82] R. Coifman, A. McIntosh and Y. Meyer, *L'intégrale de Cauchy definit un opérateur borné sur L^2 pour les courbes lipschitziennes*, Ann. Math., **116** 361–388 (1982)

[CoWe71] R.R. Coifman, G. Weiss, *Analyse harmonique non-commutative sur certains espaces homogènes. Étude de certaines intégrales singulières*, Lecture Notes in Mathematics, Vol. 242 Springer-Verlag, Berlin-New York, 1971

[CoWe77] R. R. Coifman, G. Weiss,*Extensions of Hardy spaces and their use in analysis*, Bull. Amer. Math. Soc. **83**(4), 569—645 (1977)

[CrHeMa16] D. Cruz-Uribe, E. Hernández, and J.M. Martell, *Greedy bases in variable Lebesgue spaces*, Monatsh Math **179**, 355—378 (2016)

[Da77] B. Dahlberg, *Estimates of harmonic measure*. Arch. Rat. Mech. Anal., **65**, 278–288 (1977)

[Da79] B. Dahlberg, *On the Poisson integral for Lipschitz and C^1-domains*, Stud. Math., **66**, 13–24 (1979)

[DaJeKe84] B.E.J. Dahlberg, D.S. Jerison, C.E. Kenig, *Area integral estimates for elliptic differential operators with nonsmooth coefficients*, Ark. Mat. **22**(1), 97–108 (1984)

[DaKeVePi97] B.E.J. Dahlberg, C. Kenig, G. Verchota, J. Pipher *Area integral estimates for higher order elliptic equations and systems*, Ann. Inst. Fourier (Grenoble) **47**, 1425–1461 (1997)

[DaKeVe86] B. Dahlberg, C. Kenig, G. Verchota, *The Dirichlet problem for the biharmonic equation in a Lipschitz domain.* Ann. Inst. Fourier **36**(3), 109–135 (1986)

[DaKeVe88] B. Dahlberg, C. Kenig, G. Verchota, Boundary value problems for the systems of elastostatics in Lipschitz domains, Duke Math. J. **57**(3), 795–818 (1988)

[DaKe87] B.E.J. Dahlberg, C.E. Kenig, *Hardy Spaces and the Neumann Problem in L^p for Laplace's Equation in Lipschitz Domains,* Annals of Mathematics, **125**(3), 437–465 (1987)

[Da84] G. David, *Opérateurs intégraux singuliers sur certaines coubes du plan complexe*, Ann. Sci. École Norm. Sup., **17**, 157–189 (1984)

[Da88] G. David, *Opérateurs d'intégrale singulière sur les surfaces régulières*, Ann. Sci. École Norm. Sup., **21**, 225–258 (1988)

[DaJe90] G. David and D. Jerison, *Lipschitz approximation to hypersurfaces, harmonic measure, and singular integrals*, Indiana Univ. Math. J., **39**(3), 831–845 (1990)

[DaSe91] G. David, S. Semmes, *Singular Integral and Rectifiable Sets in $\mathbb{R}^n$. Audelà des graphes lipschitziens.* Astérisque, **193** (1991)

[DaSe93] G. David and S. Semmes, *Analysis of and on Uniformly Rectifiable Sets*, Mathematical Surveys and Monographs, **38**, American Mathematical Society, Providence, RI, 1993

[DiHaHaRu17] L. Diening, P. Harjulehto, P. Hästö, M. Ruzicka *Lebesgue and Sobolev spaces with variable exponents*, Lecture Notes in Mathematics, Springer, Heidelberg, 2017

[Du01] J. Duoandikoetxea, *Fourier Analysis*, Vol. 29, American Mathematical Society, 2001

[DyIhLeTuVa19] B. Dyda, L. Ihnatsyeva, J. Lehrbäck, H. Tuominen, and A.V. Vähäkangas, *Muckenhoupt A_p-properties of distance functions and applications to Hardy-Sobolev–type inequalities*, Potential Analysis, **50**, 83–105 (2019)

[EsMi04] L. Escauriaza and M. Mitrea, *Transmission problems and spectral theory for singular integral operators on Lipschitz domains*, J. Funct. Anal., **216**(1), 141–171 (2004)

[EvGa92] L.C. Evans, R.F Gariepy, *Measure Theory and Fine Properties of Functions*, Studies in Advanced Mathematics (CRC Press, Boca Raton, 1992)

[FaJoRi78] E.B. Fabes, M. Jodeit, N.M. Rivière, *Potential techniques for boundary value problems on C1-domains.* Acta Mathematica, **141**, 165–186, (1978)

[FaJoLe77] E.B. Fabes, M. Jodeit, J.E. Lewis *Double layer potentials for domains with corners and edges*, Indiana Univ. Math. J. **26**(1), 95—114 (1977)

[FaJoNe76] E.B. Fabes, R.L. Johnson, U. Neri, *Spaces of Harmonic Functions Representable by Poisson Integrals of Functions in* BMO *and $\mathcal{L}^{p,\lambda}$,* Indiana Univ. Math. J. **25**(2), 159–170 (1976)

[FaKe81] E.B. Fabes, C.K. Kenig, *On the Hardy Space H^1 of a C^1 Domain*, Arkiv för matematik **19**, 1–22 (1981)

[FaKeVe88] E. Fabes, C. Kenig, G. Verchota, *The Dirichlet problem for the Stokes system on Lipschitz domains*, Duke Math. J., **57**(3), 769–793 (1988)

[FaMeMi98] E. Fabes, O. Mendez, M. Mitrea, *Boundary layers on Sobolev-Besov spaces and Poisson's equation for the Laplacian in Lipschitz domains*, J. Funct. Anal., **159**(2), 323–368 (1998)

[FaNe80] E.B. Fabes, U. Neri, *Dirichlet Problem in Lipschitz Domains with BMO Data*, Proc. Amer. Math. Soc., **78**(1), 33—93 (1980)

[FaYa97] D. Fan, D. Yang, *The Weighted Herz-type Hardy Spaces $h\dot{K}_q^{\alpha,p}(\omega_1, \omega_2)$*, Approx. Theory & Appl. **13**, 19–41 (1997)

[FeSt71] C. Fefferman, E.M. Stein, *Some Maximal Inequalities*, American Journal of Mathematics, **93**(1), 107—115 (1971)

[FeSt72] C. Fefferman, E.M. Stein, *H^p spaces of several variables*, Acta Math. **129**, 137—193 (1972)

[Fe87] H.G. Feichtinger, *An elementary approach to Wiener's third Tauberian theorem for the Euclidean n-space*, In Symposia Math., Vol. XXIX of Analisa Armonica, pages 267–301, Cortona, 1988.

[FeMaZh21] J. Feneuil, S. Mayboroda, Z. Zhao, *The Dirichlet problem in domains with lower dimensional boundaries*, Rev. Mat. Iberoam. **37**(3) 821–910 (2021)

[Fl74] T.M. Flett, *Some elementary inequalities for integrals with applications to Fourier transforms*, Proc. London Math. Soc. **29**(3), 538–556 (1974)

[Fo99] G. Folland, *Real Analysis: Modern Techniques and Their Applications*, John Wiley & Sons, New York, 1999

[Ga89] J. García-Cuerva, *Hardy Spaces and Beurling Algebras*, J. London Math. Soc. **39**(2), 499–513 (1989)

[GaHe94] J. García-Cuerva, M.J.L. Herrero, *A theory of Hardy spaces associated to the Herz spaces*, Proc. London Math. Soc. **69**(3), 605—628 (1994)

[GaRu85] J. García-Cuerva, J. Rubio de Francia. *Weighted Norm Inequalities and Related Topics*, North-Holland **116**, Amsterdam, 1985

[GeGoKoKr97] I. Genebashvili, A. Gogatishvili, V. Kokilashvili, and M. Krbec, *Weight Theory for Integral Transforms on Spaces of Homogeneous Type*, In: Pitman Monographs and Surveys. Pure and Appl. Math. **92**, CRC Press, 1997

[Go79] D. Goldberg, *A local version of real Hardy spaces*, Duke Mathematical Journal **46**(1), 27–42 (1979)

[Gr14-I] L. Grafakos, *Classical Fourier Analysis*, Third Edition, Graduate Texts in Math. **249**, Springer, New York, 2014

[Gr14-II] L. Grafakos, *Modern Fourier Analysis*, Third Edition, Graduate Texts in Math. **250**, Springer, New York, 2014

[GrLiYa98] L. Grafakos, X. Li, D. Yang, *Bilinear Operators on Herz-type Hardy Spaces*, Transactions of the AMS, **350**(3), 1249–1275 (1998)

[Gu13] M. Guzmán-Partida, *Harmonic functions with boundary values on Herz spaces*, Front. Math. China, **8**, 1265—1274 (2013)

[HaHäPe04] P. Harjulehto, P. Hästö, and M. Pere *Variable exponent Lebesgue spaces on metric spaces: the Hardy-Littlewood maximal operator*, Real Anal. Exchange, **30**(1), 87–104, (2004)

[HeWeYa96] E. Hernández, G. Weiss, D. Yang, *The ϕ-Transform and Wavelet Characterization of Herz-type Spaces*, Collect. Math., **47**(3) 285–320 (1996),

[HoLiMiMiMi21] G. Hoepfner, P. Liboni, D. Mitrea, I. Mitrea, and M. Mitrea *Multilayer potentials for higher-order systems in rough domains*, Anal. PDE **14**(4), 1233–1308 (2021)

[HeYa98] E. Hernández, D. Yang, *Interpolation of Herz-type Hardy spaces*, Illinois J. Math. **42**(4) 564–581 (1998)

[HeYa99] E. Hernández, D. Yang, *Interpolation of Herz Spaces and Applications*, Math. Nachr. **205**(1), 69–87 (1999)

[He68] C.S. Herz, *Lipschitz Spaces and Bernstein's Theorem on Absolutely Convergent Fourier Transforms*, J. Math. Mech., **18**(4), 283–323 (1968)

[HoMiTa10] S. Hofmann, M. Mitrea, M. Taylor, *Singular integrals and elliptic boundary problems on regular Semmes-Kenig-Toro domains*, Int. Math. Res. Not. IMRN **2010**(14), 2567–2865 (2010)

[HoMiMo15] S. Hofmann, M. Mitrea, A. Morris, *The method of layer potentials in L^p and endpoint spaces for elliptic operators with L^∞ coefficients*, Proceedings of the London Mathematical Society **111**(3), 681–716 (2015)

[HyKa12] T. Hytönen, A. Kairema, *Systems of dyadic cubes in a doubling metric space*, Colloquium Mathematicae, **126**(1), 1–33 (2012)

[Iz09-I] M. Izuki, *Herz and amalgam spaces with variable exponent, the Haar wavelets and greediness of the wavelet system*, East J. Approx., **15**(1), 87—110 (2009)

[Iz09-II] M. Izuki, *Boundedness of vector-valued sublinear operators on Herz–Morrey spaces with variable exponent*, Math. Sci. Res. J. **13**(10), 243—253 (2009)

[Iz10-I] M. Izuki, *Boundedness of sublinear operators on Herz spaces with variable exponent and application to wavelet characterization*, Anal. Math., **36**(1), 33–50 (2010)

[Iz10-II] M. Izuki, *Fractional Integrals on Herz-Morrey spaces with variable exponent*, Hiroshima Math. J. **40**(3), 343–355 (2010)

[IzNo16] M. Izuki, T. Noi, *Boundedness of Fractional Integrals on Weighted Herz Spaces with Variable Exponent*, Journal of Inequalities and Applications, **2016**(1), 1–15 (2016)

[IzNo17] M. Izuki, T. Noi *An intrinsic square function on weighted Herz spaces with variable exponent*, Journal of Mathematical Inequalities, **11**(3), 799–816 (2017)

[JaMi95] B. Jawerth, M. Mitrea, *Higher dimensional scattering electromagnetic scattering theory on $C1$ and Lipschitz domains*, Amer. J. Math. **117**(4), 929–963 (1995)

[JeKe80] D. Jerison, C. Kenig, *An identity with applications to harmonic measure*, Bull. Amer. Math. **2**(3), 447–451 (1980)

[JeKe81] D. Jerison, C. Kenig, *The Neumann Problem on Lipschitz domains*, Bull. Amer. Math. Soc.(NS), **4**(2), 203–207 (1981)

[JeKe82] D. Jerison and C. Kenig, *Boundary behavior of harmonic functions in nontangentially accessible domains*, Adv. in Math., **46**(1), 80–147 (1982)

[JeKe95] D. Jerison, C. Kenig, *The inhomogeneous Dirichlet problem in Lipschitz domains*, J. Funct. Anal. **130**(1), 161–219 (1995)

[Jo74] R. Johnson, *Lipschitz Spaces, Littlewood-Paley Spaces, and Convoluteurs*, Proc. London Math. Soc. **3**(1), 127–141 (1974)

[Jo78] R. Johnson, *Multipliers of H^p Spaces*, Ark. Mat. **16**(1), 235-–249 (1978)

[Jo83] J.L. Journé, *Calderón-Zygmund Operators, Pseudo-Differential Operators and the Cauchy Integral of Calderón*, Lecture Notes in Mathematics Vol. 994, Springer-Verlag, 1983

[KaMi98] N. Kalton, M. Mitrea, *Stability Results on Interpolation Scales of Quasi-Banach Spaces and Applications*, Transactions of the American Mathematical Society, **350**(10), 3903–3922 (1998)

[Ke87] C. Kenig, *Elliptic Boundary Value Problems on Lipschitz Domains*. In: E. Stein (Ed.), Beijing Lectures in Harmonic Analysis **112**, Princeton: Princeton University Press, 131–184, 1987

[Ke94] C. E. Kenig, *Harmonic Analysis Techniques for Second Order Elliptic Boundary Value Problems*, CBMS Regional Conference Series in Mathematics, No. 83, American Mathematical Soc., Providence, RI, 1994

[KiSh11] J. Kilty and Z. Shen, *A bilinear estimate for biharmonic functions in Lipschitz domains*, Math. Ann. **349**, 367–394 (2011)

[KoMa09] Y. Komori, K. Matsuoka, *Boundedness of Several Operators on Weighted Herz Spaces*, Journal of Function Spaces and Applications, **7**(1), 1–12 (2009)

[Kr64] J. Král, *On the logarithmic potential of the double distribution*, Czechoslovak Math. J., 14 (1964), 306–321.

[Kr65] J. Král, *The Fredholm radius of an operator in potential theory*, Czechoslovak Math. J., 15 (1965), 454–473 and 565–588.

[Kr80] J. Král, *Integral Operators in Potential Theory*, Lecture Notes in Mathematics, Vol. 823, Springer-Verlag, 1980.

[KoMaRo97] V.A. Kozlov, V.G. Maz'ya and J. Rossmann, *Elliptic boundary value problems in domains with point singularities*, Mathematical Surveys and Monographs, Vol. 52, Amer. Math. Soc., Providence, RI, 1997

[LaSh05] L. Lanzani, Z. Shen, *On the Robin boundary condition for Laplace's equation in Lipschitz domains*, Comm. Partial Differential Equations **29**(1-2), 91-–109, (2005)

[LaMi24] M. Laurel, M. Mitrea, *Weighted Morrey Spaces: Calderón-Zygmund Theory and Boundary Problems*, Berlin, Boston: De Gruyter, 2024

[Le13] H. Lebesgue. *Sur des cas d'impossibilitié du probléme du Dirichlet ordinaire*. C.R. Séances Soc. Math. France, **17**(2), (1913)

[Le90] J.E. Lewis, *Layer potentials for elastostatics and hydrostatics in curvilinear polygonal domains*, Trans. Amer. Math. Soc. **320**(1), 53-–76 (1990)

[LiYa96] X. Li, D. Yang, *Boundedness of Some Sublinear Operators on Herz Spaces*, Illinois Journal of Mathematics, **40**(3), 484–501 (1996)

[LiYaHu23] Y. Li, D. Yang, L. Huang, *Real-variable Theory of Hardy Spaces Associated with Generalized Herz Spaces of Rafeiro and Samko*. Lecture Notes in Math., Vol. 2320 Springer, Singapore, 2023

[Li00] Z. Liu, *Boundedness of Commutators of Fractional Integration on Herz-type Spaces*, Acta Mathematica Scientia, **20**(4), 461–470 (2000)

[LoNi24] E. Lorist, Z. Nieraeth, *Banach Function Spaces Done Right*, Indagationes Mathematicae **35**(2), 247–268 (2024)

[LuYa92] S. Lu, D. Yang, *The Littlewood-Paley function and φ-transform characterization of a new Hardy space HK_2 associated with the Herz spaces*, Studia Mathematica **101**(3), 284–298 (1992)

[LuYa] S. Lu, D. Yang, *The Local Versions of $H^p(\mathbb{R}^n)$ spaces at the origin*, Studia Mathematica **116**(2), 103–131 (1995)

[LuYa95] S. Lu, D. Yang, *The Decomposition of Weighted Herz Spaces on $\mathbb{R}^n$ and its Applications*, Science in China (Series A), **38**(2), (1995)

[LuYa95-01] S. Lu, D. Yang, *The weighted Herz-type Hardy space and its applications*, Science in China (Series A), **38**(6), 662–673 (1995)

[LuYa95-02] S. Lu, D. Yang, *The central BMO spaces and Littlewood-Paley operators*. Approx. Theory Appl.(NS) **11**(3), 72-–94 (1995)

[LuYa96] S. Lu, D. Yang, *The continuity of commutators on Herz-type spaces*, Michigan Mathematical Journal **44**(2), 255–281 (1997)

[LuYa97] S. Lu, D. Yang, *Some Characterizations of Weighted Herz-type Hardy Spaces and Their Applications*, Acta Mathematica Sinica, **13**(1), 45–58 (1997)

[LuYaYa00] S. Lu, K. Yabuta, D. Yang *Boundedness of some sublinear operators in weighted Herz-type spaces*, Kodai Mathematical Journal **23**(3), 391–410 (2000)

[LuYaHu08] S. Lu, D. Yang, G. Hu, *Herz Type Spaces and Their Applications*, Mathematics Monograph Series 10, Science Press Beijing, 2008

[LuXu05] S. Lu, L. Xu, *Boundedness of Rough Singular Integral Operators on Homogeneous Morrey-Herz Spaces*, Hokkaido Math. J., **34**(2), 299–314 (2005)

[MaMi^3] J.M. Martell, D. Mitrea, I. Mitrea, M. Mitrea, (2016). *The Dirichlet problem for elliptic systems with data in Köthe function spaces*. Rev. Mat. Iberoam., **32**(3), 913–970 (2016)

[MMMi^3] J.J. Marín, J.M. Martell, D. Mitrea, I. Mitrea, M.Mitrea, *Singular Integral Operators, Quantitative Flatness, and Boundary Problems*, Progress in Mathematics, **344**, Birkhäuser, 2022

[MaRo10] V.G. Maz'ya, J. Rossmann, *Elliptic equations in polyhedral domains*, Mathematical Surveys and Monographs, **162**, Amer. Math. Soc., 2010

[MeMi00] O. Méndez, M. Mitrea, *The Banach envelopes of Besov and Triebel-Lizorkin spaces and applications to partial differential equations*, J. Fourier Anal. Appl. **6**(5) 503–531 (2000)

[Me92] Y. Meyer, *Wavelets and operators*, Cambridge Studies in Advanced Math., Vol. 1, N. 37, Cambridge Univ. Press, Cambridge, 1992

[Mik48] S. G. Mikhlin, *Singular Integral Equations*, Uspekhi Mat. Nauk, **3**(3), 29—112 (1948)

[Mik86] S. G. Mikhlin, S. Prössdorf *Singular Integral Operators*, Berlin, Boston: De Gruyter, 1986.

[Mi02] D. Mitrea, *Layer potentials and Hodge decompositions in two dimensional Lipschitz domains*, Math. Ann. **322**, 75-–101 (2002)

[Mi18] D. Mitrea, *Distributions, Partial Differential Equations, and Harmonic Analysis*, Second Edition, Universitext, Springer 2018

[MiMi03] D. Mitrea, I. Mitrea, *On the Besov regularity of conformal maps and layer potentials on nonsmooth domains*, J. Funct. Anal. **201**(2), 380–429 (2003)

[MiMiMi20] D. Mitrea, I. Mitrea, M. Mitrea, *A Sharp Divergence Theorem with Nontangential Traces*, Notices of the American Mathematical Society, **67**(9), 1295–1305 (2020)

[GHA-I] D. Mitrea, I. Mitrea, M. Mitrea, *Geometric Harmonic Analysis - Volume I: A Sharp Divergence Theorem with Nontangential Pointwise Traces*, Developments in Mathematics Series, 72 Springer, Cham, 2022

[GHA-II] D. Mitrea, I. Mitrea, M. Mitrea, *Geometric Harmonic Analysis - Volume II: Function Spaces Measuring Size and Smoothness on Rough Sets*, Developments in Mathematics Series, 73 Springer, Cham, 2022

[GHA-III] D. Mitrea, I. Mitrea, M. Mitrea, *Geometric Harmonic Analysis - Volume III: Integral Representations, Calderón-Zygmund Theory, Fatou Theorems, and Applications to Scattering*, Developments in Mathematics Series, 74 Springer, Cham, 2022

[GHA-IV] D. Mitrea, I. Mitrea, M. Mitrea, *Geometric Harmonic Analysis - Volume IV: Boundary Layer Potentials in Uniformly Rectifiable Domains, and Applications to Complex Analysis*, Developments in Mathematics Series, 75 Springer, Cham, 2022

[GHA-V] D. Mitrea, I. Mitrea, M. Mitrea, *Geometric Harmonic Analysis - Volume V: Fredholm Theory and Finer Estimates for Integral Operators, with Applications to Boundary Problems*, Developments in Mathematics Series, 76 Springer, Cham, 2022

[MiMiMiMo13] D. Mitrea, I. Mitrea, M. Mitrea, S. Monnoiaux, *Groupoid Metrization Theory: With Applications to Analysis on Quasi-Metric Spaces and Functional Analysis*, Birkhäuser, 2013

[MiMiMiTa16] D. Mitrea, I. Mitrea, M. Mitrea, M. Taylor, *The Hodge-Laplacian: Boundary Value Problems on Riemannian Manifolds*, Berlin, Boston: De Gruyter, 2016

[MiMi02] D. Mitrea and M. Mitrea, *Finite energy solutions of Maxwell's equations and constructive Hodge decompositions on nonsmooth Riemannian manifolds*, J. Funct. Anal. **190**(2), 339–417 (2002)

[MiMiPi97] D. Mitrea, M. Mitrea, J. Pipher, *Vector potential theory on Lipschitz domains in $\mathbb{R}^3$ and applications to electromagnetic scattering*, J. Four. Anal. Appl., **3**, 131-192 (1997)

[MiMiTa01] D. Mitrea, M. Mitrea, M.E. Taylor, *Layer potentials, the Hodge Laplacian, and global boundary problems in nonsmooth Riemannian manifolds*, Mem. Amer. Math. Soc. **150**(713), 1–120 (2001)

[Mi99] I. Mitrea, *Spectral radius properties for layer potentials associated with the elastostatics and hydrostatics equations in nonsmooth domains*, J. Fourier Anal. Appl. **5**(4), 385—408 (1999)

[Mi00] I. Mitrea, *On the spectra of elastostatic and hydrostatic layer potentials on curvilinear polygons*, J. Fourier Anal. Appl.**8**(5), 442—488 (2002)

[MiMi07] I. Mitrea, M. Mitrea, *The Poisson problem with mixed boundary conditions in Sobolev and Besov spaces in non-smooth domains*, Trans. Amer. Math. Soc., **359**(9), 4143–4182 (2007)

[MiMi13] I. Mitrea, M. Mitrea. *Multi-Layer Potentials and Boundary Problems for Higher-Order Elliptic Systems in Lipschitz domains*, Lecture Notes in Mathematics, Springer, Berlin, 2013

[MiTu03] I. Mitrea, W. Tucker, *Some counterexamples for the spectral-radius conjecture*, Differential and Integral Equations **16**(12), 1409-–1439 (2003)

[Mi95] M. Mitrea, *The method of layer potentials in electromagnetic scattering theory on nonsmooth domains*, Duke Math. J. **77**(1), 111–133 (1995)

[Mi04] M. Mitrea, *Sharp Hodge decompositions, Maxwell's equations, and vector Poisson problems on nonsmooth, three-dimensional Riemannian manifolds*, Duke Math. J. **125**(3) 467–547 (2004)

[MiMoWr11] M. Mitrea, S. Monniaux, M. Wright, *The Stokes operator with Neumann boundary conditions in Lipschitz domains*, Journal of Mathematical Sciences **176**(3), 409–457 (2011)

[MiTa1] M. Mitrea, M. Taylor, *Boundary layer methods for Lipschitz domains in Riemannian manifolds*, J. Funct. Anal., **163**(2), 181–251 (1999)

[MiTa2] M. Mitrea, M. Taylor, *Potential theory on Lipschitz domains in Riemannian manifolds: Hölder continuous metric tensors*, Commun. Partial. Differ. Equ., **25**(7-8), 1487–1536 (2000)

[MiTa3] M. Mitrea, M. Taylor, *Potential theory on Lipschitz domains in Riemannian manifolds: Sobolev-Besov space results and the Poisson problem*, J. Funct. Anal., **176**(1), 1–79 (2000)

[MiTa4] M. Mitrea, M. Taylor, *Potential theory on Lipschitz domains in Riemannian manifolds: L^p, Hardy, and Hölder space results*, Commun. Partial. Differ. Equ., **9**(2), 369–421 (2001)

[MiTa5] M. Mitrea, M. Taylor, *Potential theory for Lipschitz subdomains of Riemannian manifolds equipped with Dini continuous metric tensors*, Trans. Amer. Math. Soc., **355**(5), 1961–1985 (2003)

[MiTa6] M. Mitrea, M. Taylor, *The Poisson problem in weighted Sobolev spaces on Lipschitz domains*, Indiana Univ. Math. J., **55**(3), 1063–1089 (2006)

[MiTa7] M. Mitrea, M. Taylor, *Boundary problems for the Navier-Stokes system on non-smooth Riemannian manifolds*, Math. Annalen., **321**(4), 955–987 (2001)

[MiWr12] M. Mitrea, M. Wright, *Boundary value problems for the Stokes system in arbitrary Lipschitz domains*, Astérisque Soc. Math. de France, **344**, 2012

[MoTo24-I] M. Mourgoglou, X. Tolsa, *The regularity problem for the Laplace equation in rough domains*, Duke Math. J. **173**(9), 1731–1837 (2024)

[MoTo24-II] M. Mourgoglou, X. Tolsa, *Solvability of the Neumann problem for elliptic equations in chord-arc domains with very big pieces of good superdomains*, arXiv:2407.20385 (2024)

[Ne67] J. Nečas, *Les Méthodes Directes en Théorie des Équations Elliptiques*, Academia, Académie Tchécoslovaque des Sciences, 1967

[NaToVo14] F.L. Nazarov, X. Tolsa and A.L. Volberg, *On the uniform rectifiability of AD-regular measures with bounded Riesz transform operator: the case of codimension 1*, Acta Math. **213**, 237–321 (2014)

[NaRaZa20] H. Nafis, H. Rafeiro, M.A. Zaighum, *A Note on the Boundedness of Sublinear Operators on Grand Variable Herz Spaces*. J. Inequal. Appl. **2020**(1), 1–13 (2020)

[NaRaZa21] H. Nafis, H. Rafeiro, M.A. Zaighum, *Boundedness of the Marcinkiewicz Integral on Grand Variable Herz Spaces*. J. Math. Ineq., **15**(2), 739–753 (2021)

[NaRaZa22] H. Nafis, H. Rafeiro, M.A. Zaighum, *Boundedness of Multilinear Calderón-Zygmund Operators on Grand Variable Herz Spaces*, Journal of Function Spaces, **2022**(1), 4845507, 11 pages (2022)

[PaWe54] L.E. Payne, H.F. Weinberger, *New Bounds in Harmonic and Biharmonic Problems*, J. of Math. and Phy. **33**(1-4), 291–307 (1955)

[PiVe91] J. Pipher, G. Verchota, *Area integral estimates for the biharmonic operator in Lipschitz domains*, Trans. Amer. Math. Soc., **327**(2), 903–917 (1991)

[PiVe92] J. Pipher, G. Verchota, *The Dirichlet Problem in L^p for the biharmonic equation on Lipschitz domains*, Amer. J. Math., **114**(5), 923–972 (1992)

[PiVe93] J. Pipher, G. Verchota, *A maximum principle for biharmonic functions in Lipschitz and C^1 domains*, Comment. Math. Helv. **68**(1), 385—414 (1993)

[PiVe95-I] J. Pipher, G. Verchota, *Dilation invariant estimates and the boundary Gårding inequality for higher order elliptic operators*, Ann. of Math., **142**(1), 1–38 (1995)

[PiVe95-II] J. Pipher, G. Verchota, *Maximum principles for the polyharmonic equation on Lipschitz domains*, Potential Anal. **4**, 615—636 (1995)

[RaSa20] H. Rafeiro, S. Samko, *Herz spaces meet Morrey type spaces and complementary Morrey type spaces*, J. Fourier Anal. Appl., **26**(5), 74 (2020)

[RaSa22] H. Rafeiro, S. Samko, *Coincidence of Variable Exponent Herz Spaces with Variable Exponent Morrey Type Spaces and Boundedness of Sublinear Operators in these Spaces*, Potential Anal. **56**, 437–457 (2022)

[Ra09] M.A. Ragusa, *Homogeneous Herz spaces and regularity results*, Nonlinear Analysis: Theory, Methods & Applications, **71**(12), e1909-e1914 (2009)

[RiNa52] F. Riesz and B. Sz.-Nagy, *Functional Analysis*, Ungar Publishing Co., New York, 1955, French original: Akad. Kiado, Budapest, 1952.

[Ru91] W. Rudin, *Function Analysis*, Second Edition, McGraw-Hill Science, 1991

[Sa13] S. Samko, *Variable Exponent Herz Spaces*, Mediterr. J. Math. **10**, 2007—2025 (2013)

[Sc19] A. Scapellato, *Regularity of solutions to elliptic equations on Herz spaces with variable exponents*, Bound. Value Probl. **2019**, 2 (2019)

[Sh91] Z. Shen, *Boundary value problems for parabolic Lamé systems and a nonstationary linearized system of Navier-Stokes equations in Lipschitz cylinders*, Amer. J. Math., **113**(2), 293–373 (1991)

[Sh06-I] Z. Shen, *Necessary and sufficient conditions for the solvability of the L^p Dirichlet problem on Lipschitz domains*, Math. Ann. **336**(3) 697—725 (2006)

[Sh06-II] Z. Shen, *The L^p Dirichlet problem for elliptic systems on Lipschitz domains*, Math. Res. Lett. **13**(1), 143–159 (2006)

[ShZh22] M. Shi, J. Zhou *Homogeneous mixed Herz-Morrey spaces and its Applications*, arXiv:2207.01087 (2022)

[SeSi81] R. Selvaggi, I. Sisto, *An existence theorem for the Dirichlet problem with respect to the operator Δ^2 in certain domains of class C^1*, Boll. Un. Mat. Ital. B (5) **18**(2), 473–483 (1981)

[St57] E.M. Stein, *Note on Singular Integrals*, Proceedings of the American Mathematical Society, **8**(2), 250–254 (1957)

[St93] E.M. Stein, *Harmonic Analysis. Real-Variable Methods, Orthogonality, and Oscillatory Integrals*, Princeton University Press **43**, 1993

[SuAzSuMeMl22] B. Sultan, F. Azmi, M. Sultan, M. Mehmood, N. Mlaiki *Boundedness of Riesz Potential Operator on Grand Herz-Morrey Spaces*, Axioms. **11**(11), 583 (2022)

[SuSuMeAzAlMl22] B. Sultan, M. Sultan, M. Mehmood, F. Azmi, M.A. Alghafli, and N. Mlaiki, *Boundedness of fractional integrals on grand weighted Herz spaces with variable exponent*, AIMS Mathematics 8(1), 752–764 (2022)

[SuAzSuMaMlSo22] B. Sultan, F.M. Azmi, M. Sultan, T. Mahmood, N. Mlaiki, N. Souayah, *Boundedness of Fractional Integrals on Grand Weighted Herz–Morrey Spaces with Variable Exponent*, Fractal Fract. **6**(11), 660, (2022)

[TaShZh09] X. Tao, Y.L. Shi, S.Y. Zhang, *Boundedness of multilinear Riesz potential on the product of Morrey and Herz-Morrey spaces*, Acta Math. Sinica (Chinese Ser.), **52**(3), 535–548 (2009)

[TaWe80] M.H. Taibleson, G. Weiss, *The molecular characterization of certain Hardy spaces*, Astérique **77**, 67–149 (1980)

[TaYa00] L. Tang, D. Yang *Boundedness of vector-valued operators on weighted Herz spaces*. Approx. Theory & its Appl. **16**(2), 58—70 (2000)

[Tr95] H. Triebel, *Interpolation Theory, Function Spaces, Differential Operators*, Huethig & Wepf, 1995

[Ts11] Y. Tsutsui, *The Navier-Stokes equations and weak Herz spaces*, Advances in Differential Equations, Adv. Differential Equations **16**(11-12), 1049–1085 (2011)

[Ve84] G. Verchota, *Layer potentials and regularity for the Dirichlet problem for Laplace's equation in Lipschitz domains*, J. Funct. Anal., **59**(3), 572–611 (1984)

[Ve87] G.C. Verchota, *The Dirichlet problem for the biharmonic equation in* C^1 *domains*, Indiana Univ. Math. J. **36**(4), 867–895 (1987)

[Ve90] G.C. Verchota, *The Dirichlet problem for the polyharmonic equation in Lipschitz domains*, Indiana Univ. Math. J. **39**(3), 671–702 (1990)

[WaLi12] H. Wang, Z. Liu, *The Herz-type Hardy Spaces with Variable Exponent and their Applications*, Taiwanese J. Math. **16**(4), 1363–1389 (2012)

[WaSh18] L. Wang, L. Shu, *Boundedness of some sublinear operators on weighted variable Herz-Morrey spaces*, Journal of Mathematical Inequalities, **12**(1), 31–42 (2018)

[WaXu21] S. Wang, J. Xu, *Boundedness of vector-valued sublinear operators on weighted Herz-Morrey spaces with variable exponents*, Open Mathematics, **19**(1), 412–426 (2021)

[We22] M. Wei, *Mapping properties for operators on Herz-Morrey and Herz-Morrey-Hardy spaces*, J. Pseudo-Differ. Oper. Appl., **14**, 4 (2023)

[WeYa22] M. Wei, D. Yan, *Operators on Herz-type Spaces Associated with Ball Quasi-Banach Function Spaces*, arXiv:2209.04323 (2022)

[Wu09] J. Wu, *Boundedness of multilinear singular integral operators on the homogeneous Morrey-Herz spaces*, Ukrainian Math. J., **61**, 1684–1693 (2009)

[YeHo20] T.L. Yee, K.P. Ho, *Hardy's inequalities and integral operators on Herz-Morrey spaces*, Open Mathematics, **18**(1), 106–121 (2020)

[YuLi22] X. Yu, Z. Liu, *Weighted Grand Herz-Type Spaces and Its Applications*, Journal of Function Spaces, **2022**(1), 1369159 (2022)

Subject Index

© The Editor(s) (if applicable) and The Author(s), under exclusive license to Springer Nature Switzerland AG 2026

M. Mitrea and P. Takemura, *Singular Integrals, Herz-Type Function Spaces, and Boundary Problems*, Progress in Mathematics 362,
https://doi.org/10.1007/978-3-032-12516-3

Symbol Index

© The Editor(s) (if applicable) and The Author(s), under exclusive license to Springer Nature Switzerland AG 2026
M. Mitrea and P. Takemura, *Singular Integrals, Herz-Type Function Spaces, and Boundary Problems*, Progress in Mathematics 362,
https://doi.org/10.1007/978-3-032-12516-3

GPSR Compliance

The European Union's (EU) General Product Safety Regulation (GPSR) is a set of rules that requires consumer products to be safe and our obligations to ensure this.

If you have any concerns about our products, you can contact us on ProductSafety@springernature.com

In case Publisher is established outside the EU, the EU authorized representative is:

Springer Nature Customer Service Center GmbH
Europaplatz 3
69115 Heidelberg, Germany

Batch number: 10755308

Printed by Printforce, the Netherlands